Werner Heisenberg

GESAMMELTE WERKE / COLLECTED WORKS

Abteilung / Series A

Original Scientific Papers
Wissenschaftliche Originalarbeiten

Abteilung / Series B

Scientific Review Papers, Talks, and Books
Wissenschaftliche Übersichtsartikel, Vorträge und Bücher

Abteilung / Series C

Philosophical and Popular Writings
Allgemeinverständliche Schriften

The *Gesammelte Werke / Collected Works* of Werner Heisenberg
are published jointly by R. Piper GmbH & Co KG Verlag, München, Zürich
and Springer-Verlag, Berlin, Heidelberg, New York, Tokyo.
The volumes of Series A and B are issued by Springer-Verlag,
those of Series C by R. Piper Verlag.

Werner Heisenberg

GESAMMELTE WERKE
COLLECTED WORKS

Edited by
W. Blum, H.-P. Dürr, and H. Rechenberg

Series B
Scientific Review Papers, Talks, and Books
Wissenschaftliche Übersichtsartikel, Vorträge und Bücher

Volume 1

Springer-Verlag Berlin Heidelberg GmbH 1984

Dr. Walter Blum
Professor Dr. Hans-Peter Dürr
Dr. Helmut Rechenberg

Max-Planck-Institut für Physik und Astrophysik, Werner-Heisenberg-Institut für Physik
Föhringer Ring 6, D-8000 München 40, Fed. Rep. of Germany

ISBN 978-3-642-61743-0   ISBN 978-3-642-61742-3 (eBook)
DOI 10.1007/978-3-642-61742-3

Library of Congress Cataloging in Publication Data. Heisenberg, Werner, 1901–1976. Gesammelte Werke =
Collected works. Contents: – ser. B. Scientific review papers, talks, and books. 1. Physics–Collected works.
I. Blum, W. (Walter), 1937-. II. Dürr, H.-P. (Hans-Peter), 1929-. III. Rechenberg, H. (Helmut), 1937-.
IV. Title. V. Title: Collected works. QC3.H33 1984   530   84-10633

Reproduction and typesetting: Druckservice K. Teichmann, 6901 Mauer

2153/3130-5 4 3 2 1 0

Editors' Note

As the *Editors' Preface* in the first volume of *Series A* of the *Collected Works* explains, the publications of Werner Heisenberg have been divided into three general categories or series, *Series A* and *B* containing the scientific papers and books and *Series C* the writings addressed to a wider public. In *Series B* in particular we reprint the published lectures delivered to scientific conferences and meetings of scientific societies, together with review articles summarizing the status of particular problems in physical research. These reviews originally appeared in several *Festschriften* and review journals or were produced as special books and monographs. Quite a number of Heisenberg's scientific talks have not appeared fully in print; only their abstracts can be found in the literature. In order to give an account, as complete as possible, of Heisenberg's role as conference speaker in the *Collected Works*, we also reprint in *Series B* the available abstracts of his scientific lectures, except those signed by other authors.

The volume starts with a biographical sketch of Heisenberg's life and work by D. Cassidy and H. Rechenberg, and a short outline of Heisenberg's role as lecturer and author of scientific review papers and books by H. Rechenberg, in order to acquaint the reader with the background of the reprinted publications of Heisenberg. These then follow in strict chronological order, determined by the submission or presentation date. Besides the published papers we have included a report on the general properties of elementary particles that Heisenberg completed in June 1939 for the Eighth Solvay Conference in Physics, planned for October 1939 but cancelled because of the war in Europe.

We would like to thank the Max-Planck-Gesellschaft zur Förderung der Wissenschaften e. V. for generous financial support.

Munchen, July 1984 *W. Blum, H.-P. Dürr, and H. Rechenberg*

Contents*

Scientific Review Papers, Talks, and Books

---

* The papers are reprinted so as to follow the original publication as closely as possible, being photo-graphically reproduced. For most papers we have kept strict pagewise correspondence, but in the case of three column originals we have broken up the pages; thus the beginning of a new page in the original print is indicated by giving the page number at the margin of the corresponding line in the reprint. In reproducing of the two books, Nos. 8 and 33, four pages of the original have been placed on one page of the reprint.

VIII

Biographical Data

Werner Heisenberg (1901–1976)

1. Youth (1901–1920)

Werner Karl Heisenberg was born in Würzburg, Germany, on 5 December 1901. His father, August Heisenberg (1869–1930), stemmed from a family of master craftsmen in Osnabrück, and had studied classical philology in Munich; at the time of Werner's birth he held the dual positions of teacher at the *Altes Gymnasium* in Würzburg and *Privatdozent* for Greek philology at the University of Würzburg. His mother was Anna Wecklein, the daughter of Nikolaus Wecklein (1843–1926), a classical philologist and rector of the *Maximilians-Gymnasium* in Munich. They married in 1899 and their first son, Erwin, was born in Munich in 1900; he later became a chemist and worked in industry (he died in 1965).

The family returned to Munich in April 1910 when August Heisenberg was called (in January) to succeed his teacher Karl Krumbacher in the chair for medieval and modern Greek philology at the university. On 18 September 1911 Werner entered the *Maximilians-Gymnasium,* still under his grandfather's rectorship and at that time the best humanistic gymnasium in the Bavarian capital. Werner studied diligently and rapidly and was regarded as an outstanding, ambitious, self-confident pupil. He excelled particularly in mathematics; far exceeding the teaching program, he taught himself differential and integral calculus, worked with elliptic functions, and studied abstract number theory. Besides his academic achievements Heisenberg learned to play the piano and by the age of 13 he was playing master compositions. He remained an excellent player throughout his life.

The First World War and its consequences intruded into Heisenberg's boyhood life. His father, a reserve infantry officer, was immediately called to duty and remained away from the family for nearly the entire war. An increasing shortage of food and fuel prevailed throughout Germany and forced the occasional closing of schools. Because of his family's lack of food, Werner signed up for the war assistance service in the summer of 1918 and helped to bring in the harvest with schoolmates on a farm near Miesbach in Upper Bavaria.

The loss of the war and the abdication of the monarchy generated revolutionary unrest throughout Germany. In Bavaria a socialist republic came to power on 7 November 1918; it developed into a Soviet Republic *(Räterepublik)* on 7 April 1919 that was crushed in early May 1919 by troops dispatched by the Reich Government. During the fighting in Munich and the ensuing restoration of moderate socialist rule, Heisenberg, along with many of his schoolmates, voluntarily served in support of one of the republican units, Cavalry Rifle Command No. 11 *(Kavallerie-Schützen-Kommando No. 11)*. He carried messages, guarded the street from the roof of the Catholic seminary across from the university, and read Plato's *Timaeus* in Greek. Shortly afterwards Heisenberg became associated, as the elected leader of a small group of younger boys, with the New Boy Scouts *(Bund deutscher Neupfadfinder),* a representative of the postwar German youth movement that strove for renewal of personal and social life in Germany. Until the prohibition of independent youth groups in 1933 Heisenberg spent nearly all of his leisure time with his group, hiking and camping within Germany and on trips to neighboring countries (e. g. to Finland in summer 1923). He loved hiking, skiing, mountain climbing, and enjoyed the beauty of nature all his life.

2. Student and Postdoctoral Years (1920–1925)

After a brilliant performance on his final gymnasium examinations *(Abitur),* for which he was accepted as a scholar of the prestigious Maximilianeum Foundation, Heisenberg entered the University of Munich during the winter semester of 1920/21. At first he planned to study pure mathematics, but a discomforting meeting with the famous professor of that subject, Ferdinand von Lindemann, resulted in his turning to theoretical physics. Arnold Sommerfeld, professor of that field, immediately recognized his talents and admitted him to his *Seminar,* composed of advanced students and postdoctoral researchers. Among the *Seminar* members at that time were Gregor Wentzel and Wolfgang Pauli. Otto Laporte and Karl Bechert arrived a little later.

During his three years of university study Heisenberg attended Sommerfeld's lectures covering all fields of theoretical physics, including the qantum and relativity theories, as well as the special lectures offered by Karl Herzfeld. He showed less interest in the lectures and laboratory exercises in experimental physics provided by Wilhelm Wien. For one of his two minors he attended mathematics courses offered by Lindemann, Alfred Pringsheim, Artur Rosenthal and Aurel Voss. For his second minor he studied astronomy with Hugo von Seeliger.

In Sommerfeld's *Seminar* Heisenberg studied the latest literature in physics and carried on intensive discussions of the problems arising from it with his teacher and a group of excellent fellow students. Under Sommerfeld's direction he delved into the problems of atomic theory, a difficult and abstract discipline that tended to attract the best and most dedicated young theorists of the day. As early as his first semester Heisenberg presented a quantum-theoretical analysis of the anomalous Zeeman effect, the results of which he utilized a year later in his first publication (submitted in November 1921). At the same time he examined problems in hydrodynamics, a subject on which he published his second paper (on Kármán vortices, 1922) held his first public lecture in September 1922 (at a topical conference in Innsbruck), and wrote his doctoral dissertation (finished in July 1923).

In June 1922 Sommerfeld brought his star pupil to Göttingen where Niels Bohr presented a series of lectures on quantum theory and atomic structure – the "Bohr Festival" (12–22 June 1922). There Heisenberg met the leading representatives of atomic physics in Germany and Europe, among them Max Born, Paul Ehrenfest, James Franck, Hendrik Anthony Kramers and Alfred Landé. Bohr and Born began to take notice of the young physicist. When Sommerfeld left for a guest professorship at the University of Wisconsin, Madison, during the winter of 1922/23, Born invited Heisenberg to Göttingen as his personal assistant (November 1922– March 1923). In Göttingen Heisenberg learned the rigorous mathematical methods of the Hilbert school and worked with Born on the many-body problem in atomic theory, especially on the energy states of the helium atom.

During the summer semester of 1923 Heisenberg completed his dissertation in Munich. In it he successfully treated the problem of the onset of turbulence, which had resisted all previous efforts made by mathematicians and physicists. Sommerfeld was very pleased with this result and the mathematical methods Heisenberg employed. However, Wilhelm Wien, the examiner in experimental physics, came to quite a different conclusion in his field during the final orals (on 23 July 1923) and wanted to fail the candidate. Heisenberg thus received his doctorate "cum laude" rather than "magna" or "summa cum laude", the two highest marks.

In October 1923 Heisenberg became Born's assistant at the University of Göttingen. He continued to work with Born on atomic and molecular models. By means of a modification of the Bohr-Sommerfeld quantum rules he obtained certain advances in the explanation of the anomalous Zeeman effect. He got his *Habilitation* at the University of Göttingen on 28 July 1924. Although only 22 years old, he thus became a *Privatdozent* and fully qualified as a university lecturer.

Heisenberg made his first visit to Bohr's institute in Copenhagen during the spring of 1924 (15 March to early April). With the help of a Rockefeller fellowship (International Education Board) he resided in Copenhagen during the seven months from middle of September 1924 to early April 1925, and during the fall of 1925 (September–October), and otherwise returned to Born's institute in Göttingen. At Bohr's institute Heisenberg met a number of talented young physicists from a variety of nations, among them Christian Møller (Denmark), Svein Rosseland (Norway), G. H. Dieke (Holland), Ralph Fowler (England), Ralph de Laer Kronig and David Dennison (both from U. S. A.). During his visits he learned to speak Danish and English and worked intensively with Bohr and his closest collaborator Kramers on the most difficult problems of atomic theory. This work led to Heisenberg's papers on resonance fluorescence (submitted in November 1924), the dispersion of light by atoms (with Kramers, December 1924) and the structure of complex spectra and their Zeeman effects (April 1925).

During his years of study and postdoctoral research Heisenberg had received in Munich and Göttingen with Sommerfeld and Born a thorough education in all aspects of theoretical physics, while simultaneously familiarizing himself with the main problems of atomic and quantum theory. With Bohr in Copenhagen he deepened his understanding of the foundations of quantum physics. As his friend Pauli wrote (to Kramers on 27 July 1925), Heisenberg "learned a little of philosophical thinking" and moved "noticeably away from the purely formal". Heisenberg himself later summarized the influence of his teachers thus: "From Sommerfeld I learned optimism, from the Göttingen people mathematics, and from Bohr physics".

3. The Development of Quantum Mechanics (1925–1927)

The early 1920s witnessed fundamental difficulties in atomic physics. The quantum theory of atomic structure, founded by Bohr and largely developed by Bohr and Sommerfeld, did not describe the properties of complicated atoms and molecules. Moreover, the discovery of the Compton effect at the end of 1922 focussed attention on the problem of the nature of radiation. Its interpretation in the light-quantum hypothesis contradicted classical radiation theory, and the radical attempt by Bohr, Kramers and Slater in early 1924 to resolve the difficulty by assuming only statistical conservation of energy and momentum was refuted by the experiment of Walther Bothe and Hans Geiger in April 1925.

Heisenberg was growing ever more concerned with these and other difficulties in atomic theory. His works on the anomalous Zeeman effect, only successful in part, and his unsuccessful calculation of the helium states with Born had sensitized him by early 1925 to the "crisis" of current theory. Nevertheless, his latest calculations in Copenhagen on dispersion theory and on complex spectra, especially the principle of "sharpened" correspondence applied in these works, seemed to point toward a future satisfactory theory.

With characteristic optimism the Göttingen *Privatdozent* took on a new and difficult problem at the beginning of May 1925, the calculation of the line intensities in the hydrogen spectrum. Heisenberg began with a Fourier analysis of the classical hydrogen orbits, intending to translate them into a quantum theoretical scheme – just as he had done with Kramers for the dispersion of light by atoms. But the hydrogen problem proved much too difficult, and he replaced it with the simpler one of an anharmonic oscillator. With the help of a new multiplication rule for a quantum-theoretical Fourier series he succeeded in writing down a solution for the equations of motion for this system. On 7 June 1925 he went to the island of Helgoland to recover from a severe attack of hay fever. There he completed the calculation of the anharmonic oscillator, determining all the constants of the motion. He made use, in particular, of a modified quantum condition that was later called by Born, Pascual Jordan and himself a "commutation relation", and he proved that the new theory yielded stationary states (conservation of energy). Returning to Göttingen on 19 June 1925 Heisenberg composed his fundamental paper "Über die quantentheoretische Umdeutung kinematischer und mechanischer Beziehungen" ("On a Quantum Theoretical Reinterpretation of Kinematic and Mechanical Relations"), which was completed on 9 July 1925. In this paper, the starting point for a new quantum mechanics, Heisenberg announced as the leading philosophical principle of quantum mechanics that only observable quantities are allowed in the theoretical description of atoms. Heisenberg reported his new results during visits shortly thereafter with Paul Ehrenfest in Leiden and with Ralph Fowler in Cambridge.

After Born and Jordan managed in August and September 1925 to develop the mathematical content of Heisenberg's work into a consistent theory with the help of infinite Hermitian matrices (Z. Phys. *34*, 858, 1925), Heisenberg participated, starting in September 1925, in the completion and application of the new "matrix mechanics", culminating in the long "three-man-paper", by Born, Heisenberg and Jordan, submitted on 16 November 1925. Further developments followed rapidly: Pauli calculated the stationary states of the hydrogen atom in October 1925; Cornelius Lanczos in Frankfurt and Born and Norbert Wiener in the USA extended the method of operator mechanics to describe continuous motions (December 1925); and Paul Adrien Maurice Dirac in Cambridge developed independently of the Göttingen school a different scheme based upon Heisenberg's July paper, the method of q-numbers (November 1925), in which many-electron atoms and the relativistic Compton effect could be handled successfully (spring 1926). In addition Heisenberg and Jordan utilized electron spin and matrix mechanics to solve the old problems of hydrogen fine structure and the anomalous Zeeman effect (April 1926); and finally Heisenberg discovered the phenomenon of quantum-mechanical resonance (June 1926), which played a decisive role in his subsequent calculation of the term system of the helium atom (July 1926).

In May 1926 Niels Bohr offered Heisenberg a position at his institute in Copenhagen as Lector and successor to his assistant Kramers. There Heisenberg delivered lectures at the university (in Danish) on contemporary physical theories, directed beginning students, helped guest researchers with their problems, and discussed with Bohr the most important results of quantum mechanics. In the summer and fall of 1926 the main topic of discussion was wave mechanics, the quantum atomic theory that Erwin Schrödinger began introducing in January 1926. The complete mathematical equivalence between Göttingen's matrix mechanics or Dirac's q-number scheme and Schrödinger's wave mechanics was proved by Jordan and

Dirac in December 1926, after preparatory work by Schrödinger (March 1926), Pauli (April 1926), and Carl Eckart (June 1926). However, Schrödinger's physical interpretation of the square of the wave amplitude as the continuously distributed charge density of the electron was rejected by Born, Bohr and Heisenberg and replaced on Born's proposal by the interpretation that it is the probability for finding the electron at each location (June 1926). In close contact with Pauli, and in intense discussion with Bohr, Heisenberg analyzed what he termed the "perceptual content of the quantum-theoretical kinematics and mechanics". As a result of the analysis he presented in March 1927 the so-called "indeterminacy" or "uncertainty relations", which limit the simultaneous measurement of canonically conjugate variables, such as the position and momentum of a particle. Bohr, on the other hand, pondered the simultaneous use of the physical pictures of particles and waves, which resulted in his general principle of "complementarity" announced in fall of 1927.

Born's statistical interpretation of Schrödinger's wave function, Heisenberg's uncertainty relations, and Bohr's complementary principle formed the basis of the physical interpretation of the new quantum mechanics, as explicated by Bohr in his lectures at the Volta Conference in Como (September 1927) and at the Solvay Congress in Brussels (24–29 October 1927). This "Copenhagen Interpretation" of quantum mechanics, as it was later called, found acceptance by most physicists, but not by all: Albert Einstein in particular raised serious objections to it at the 1927 and 1930 Solvay conferences and later, for example, in his paper with Boris Podolsky and Nathan Rosen (Phys. Rev. *47,* 777, 1935).

4. Professor in Leipzig (1927–1942)

Through his fundamental contributions to quantum mechanics Heisenberg soon became known far beyond the circle of the Munich, Göttingen and Copenhagen physicists as a leading representative of the new, successful atomic theory. At the Meeting of the German Scientists and Physicians in Düsseldorf (19–25 September 1926) he delivered the main address on quantum mechanics. In October 1927 Heisenberg, then only 25 years old, was called from Copenhagen to the chair for theoretical physics at the University of Leipzig, where Peter Debye had just accepted the chair for experimental physics. In spring 1929 Friedrich Hund joined Heisenberg's theoretical physics institute, and in 1931 the mathematician Bartel Leendert van der Waerden, who developed group theoretical methods in quantum mechanics, came to Leipzig as professor of mathematics.

Together Heisenberg and his colleagues created in Leipzig a leading center for research on atomic and quantum physics that attracted numerous talented students and collaborators. Felix Bloch, Rudolf Peierls and Edward Teller were among Heisenberg's first students. Hans Euler, Léon Rosenfeld, Victor Weisskopf, Carl Friedrich von Weizsäcker and Gian Carlo Wick also studied under Heisenberg. As part of their research program Debye and Heisenberg set up an annual symposium (until 1933) on current research in physics, the famous Leipzig University Week *(Universitätswoche).* The first of these took place on 18–23 June 1928 with, among others, Paul Dirac speaking on his relativistic electron equation and Fritz London on the theory of the chemical bond.

Heisenberg's research in Leipzig concentrated first on applications and extensions of quantum mechanics. In May 1928 he provided an explanation of ferromagnetism from first principles in which he showed that the quantum mechanical exchange integral that had played a decisive role in his solution of the helium problem could account for the strong molecular magnetic field in the interior of ferromagnetic materials. A little later Felix Bloch supplemented Heisenberg's theory of ferromagnetism with the invention of spin waves. During the period 1928–1931 Heisenberg and his pupils presented a series of basic contributions to the theories of molecules and metals. At the same time Heisenberg collaborated with Wolfgang Pauli, who held the chair in theoretical physics at the E. T. H. (Swiss Polytechnic) in Zurich since spring 1928, on the formulation of relativistic quantum field theory. In March and September 1929 they completed two basic papers that allowed a systematic treatment of interacting quantum fields, but that also indicated the existence of severe difficulties and divergences in the theory, such as infinite self energies of electrons. The discovery of the neutron in early 1932 occasioned Heisenberg's development of the neutron-proton model of the nucleus (independently of Dimitri Iwanenko), in which he introduced the concept of nuclear exchange force and the formalism of isotopic spin (June and July 1932). In recognition of his work on quantum mechanics and its applications Heisenberg received the Nobel Prize for physics in 1933 (for the year 1932).

Besides his intensive research during the early years in Leipzig, Heisenberg undertook a number of trips abroad. In spring of 1929 he delivered a series of lectures at the University of Chicago on "The Physical Principles of the Quantum Theory" which were published as his first book. He made the return trip with Dirac, who had also come to the United States, through the American West (August 1929) to Japan and India (September–November 1929). In 1932 he lectured at the Ann Arbor summer school. Heisenberg participated as well in numerous conferences, such as the Solvay congresses of 1927 (24–29 October), 1930 (20–25 October) and 1933 (22–29 October) in Brussels, the centenary meeting of the British Association in London (29 September 1931) and the nuclear physics conference in Rome (October 1931).

Throughout the 1930s Heisenberg very frequently visited Niels Bohr in Copenhagen and occasionally Wolfgang Pauli in Zurich; his close relations with both men and their colleagues were for him irreplaceable in the difficult period.

Quantum mechanics led to entirely new conceptions in the description of the microscopic world and to a revision of some basic physical and philosophical notions, such as the principle of causality. After the completion of the theory by Born, Bohr and Heisenberg, the latter two, as did others, devoted considerable attention to obtaining many of the intellectual consequences of the new theory. These consequences often went far beyond the realm of physics, extending into chemistry, biology, and even into social and ethical phenomena. To many laymen, however, the new ideas seemed as strange and disturbing as the revisions brought about earlier by the relativity theory – if not more so. In order to explain the new conceptions to nonspecialists and to relieve the public of any bewilderment, which – as with relativity theory – grew dangerously infected with ideology, Heisenberg wrote for and spoke often before wide audiences of laymen interested in science during the 1930s. His widely read first collection of essays, *Wandlungen in den Grundlagen der Naturwissenschaft* (1935), contained two such addresses.

Despite the political situation in Germany, which also affected his collaborators, Heisenberg's physics continued to flourish. The main focus of his research concentrated during the 1930s on the analysis of problems in relativistic quantum field theory. A major component of this work involved the analysis of elementary processes in high energy cosmic-ray interactions, where the new elementary particles positron and meson were discovered in 1932 and 1937, respectively. For the solution of problems in field theory and the explanation of cosmic-ray phenomena, fundamentally new ideas appeared to Heisenberg unavoidable. In cosmic-ray physics he argued in favor of the existence of particular multiple processes, which he called "explosion showers", that could be comprehended only by means on a nonlinear quantum field theory (June 1936, May 1939). He further claimed that the divergence difficulties in quantum field theory had to be resolved by assuming a fundamental length that limited the validity of field theory at very small distances (June 1938).

In the years following 1933 Heisenberg became the main spokesman in Germany for modern theoretical physics. Although the number of his students decreased, Heisenberg presented systematic lectures on all fields of theoretical physics, including the politically disfavored relativity – both the special and the general theories. He defended relativity theory even in an article published in the Nazi Party newspaper *Völkischer Beobachter* on 26 February 1936. Together with two colleagues in experimental physics, Hans Geiger in Berlin and Max Wien in Jena, Heisenberg authored a memorandum to the Reich Minister for Education and Science *(Reichsminister für Erziehung und Wissenschaft)* opposing the official discrediting of theoretical physics and emphasizing its great importance for the training of needed physicists. His opponents, who gathered around Philipp Lenard and Johannes Stark, sought to dismiss relativity and quantum theory as "Jewish" and "non-German".

After Arnold Sommerfeld reached retirement age in 1934, attacks by representatives of the so-called "German physics" greatly increased when Heisenberg was named by the faculty of the University of Munich as the first candidate to succeed him. On 25 July 1937 an article entitled "Weiße Juden in der Wissenschaft" ("White Jews in Science") and signed by Stark appeared in *Das schwarze Korps,* the journal of the SS. It contained wild accusations against modern theoretical physics and theoreticians, such as Heisenberg, and closed with the demand that Heisenberg and his colleagues be made to "disappear" like the Jews. Heisenberg responded with a letter to the head of the SS, Heinrich Himmler, urgently requesting a termination of the campaign against him. After a long series of investigations and interrogations by the SS, Heisenberg attained his goal in a letter from Himmler on 21 July 1938. The campaign ceased, but the Munich chair went in December 1939 to the aerodynamicist Wilhelm Müller. Heisenberg managed, however, to have an article on the significance of modern theoretical physics published in the organ of the Reich student's organization. The article, entitled "Die Bedeutung der 'modernen' theoretischen Physik" ("The meaning of 'modern' theoretical physics"), finally appeared in 1942 after considerable delay, and was eagerly read by many young people.

In spite of the official discrediting of theoretical physics in the Third Reich, Heisenberg and Hund managed to keep an unusually stimulating atmosphere in their institute, made even more lively by many foreign guests. Besides the German members of the institute (E. Bagge, S. Flügge, B. Kockel and H. Voltz), there were a number of visitors from America (R. S. Mulliken, J. C. Slater, J. H. Van Vleck, C. Zener), Italy (U. Fano, G. Gentile), Japan (K. Ariyama, Y. Fujioka, S. Kikuchi, S. Tomonaga, K. Umeda, S. Watanabe), Scandinavia (B. O. Grönblom, H. Wergeland), Hungary (L. Tisza) and Yugoslavia (I. Supek), to name only a few. Heisenberg also continued his relations with foreign physicists through lectures and travels to scientific conferences. In 1937 he attended the Galvani Conference in Bologna (18–21 October), in April 1934 and March 1938 he traveled to England (to Cambridge and Manchester, respectively), and in summer of 1939 he went to America where he presented his latest results on the theory of cosmic-ray showers at the Symposium on Cosmic Rays at the University of Chicago (27–30 June). Although he received several tempting offers of positions in America, he decided to return to Germany.

In April 1937 Heisenberg married Elisabeth Schumacher, daughter of Hermann Schumacher, a professor of economics at the University of Berlin. In early 1938 the twins Maria and Wolfgang were born; the family was later increased by Jochen (1939), Martin (1940), Barbara (1942), Christine (1944) and Verena (1950). For a secure place of refuge during the imminent war the Heisenbergs bought in 1939 a summer house in Urfeld on Lake Walchen, south of Munich, that had previously belonged to the painter Lovis Corinth. When Heisenberg transferred to Berlin in 1942 to assume directorship of the Kaiser Wilhelm Institut für Physik, his family did not accompany him but moved to Urfeld rather than to Berlin, for safety against bombing raids.

5. The War Years (1939–1945)

The outbreak of war on 1 September 1939 directly affected Heisenberg's scientific career. He was called to military service, and to his surprise was ordered to report to the Army Weapons Bureau (*Heereswaffenamt*) in Berlin. There he and other leading German atomic scientists, the so-called Uranium Club *(Uranverein)*, were asked to investigate whether the fission of uranium, discovered in December 1938 by Otto Hahn and Fritz Strassmann in Berlin, could be used for gaining energy on a large scale. Within two months Heisenberg completed a comprehensive report in which he developed the theory of a chain reaction with uranium fission by neutrons (6 December 1939); he followed it soon by a second report refining the theory and taking into account recently improved data (29 February 1940).

The Army Weapons Bureau designated the *Kaiser-Wilhelm-Institut (KWI) für Physik* in Berlin-Dahlem as the center of its uranium research. Peter Debye, director of the institute and a Dutch citizen, was prevented as a foreigner from working on the secret research. He placed himself on leave and accepted a guest professorship at Cornell University in January 1940. Kurt Diebner assumed Debye's office as head of the Berlin uranium project. At the suggestion of C. F. von Weizsäcker and especially Karl Wirtz, two of Debye's former collaborators now working on the project, Heisenberg acted as scientific advisor. He thus traveled weekly to Berlin from Leipzig and wrote several reviews on the progress of research at the Berlin institute (December 1940–February 1942). Other, non-secret research begun under Debye (for example on X-ray analysis and on nuclear and low temperature physics) continued at the *KWI für Physik* under the formal supervision of Max von Laue, vice-director. During 1941 and 1942 Heisenberg also directed a seminar at the institute in which he and other members and guests of the institute discussed the recent progress and problems of cosmic-ray physics. These talks were published in 1943 in the collection *Kosmische Strahlung* (Cosmic Radiation) dedicated to Sommerfeld on his 75th birthday.

In spring of 1940 Heisenberg began experiments with Robert and Klara Döpel at the University of Leipzig in which they studied possible arrangements of uranium and neutron moderator substances, since for uranium fission slow neutrons turned out to be most effective. These experiments substantiated the usefulness of heavy water as moderator (28 October 1941) and indicated in early 1942 that a spherical arrangement of natural uranium and heavy water led to a small multiplication of the irradiating neutrons (confirmed in June 1942). Thus the prerequisites for the construction of a functioning reactor were fulfilled. On 4 June 1942 a meeting was held between Albert Speer, Minister for Armament and War Production *(Reichsminister für Rüstung und Kriegsproduktion)*, and other officials and military leaders and the scientists, including O. Hahn, W. Heisenberg, K. Diebner, P. Harteck and K. Wirtz. There it was decided that the uranium project should continue with the goal of constructing a nuclear reactor, but not an atomic bomb. As a result the Army Weapons Bureau returned the *KWI für Physik* to the Kaiser Wilhelm Society *(Kaiser-Wilhelm-Gesellschaft)*, but the reactor project remained secret. Heisenberg, the chief scientific advisor to the institute since 1940 and theoretical director of the successful Leipzig experiments, was named head of the leading reactor research group in Germany and appointed on July 1942 director *at the KWI für Physik* (to substitute for the absent director Debye). He was assisted by many of Debye's assistants, including Horst Korsching and Karl Wirtz, as well as by Erich Bagge and Fritz Bopp. Kurt Diebner, previous head of the Berlin project, went with several collaborators to the army laboratory at Gottow where he set up his own uranium experiments. C. F. von Weizsäcker accepted a professorship for theoretical physics at the University of Strasbourg.

The work on the uranium project did not exhaust Heisenberg's activities during the war years. He occupied himself particularly with problems in cosmic-ray physics and developed in several papers a new approach to the theory of elementary particles based on the concept of the scattering matrix. Three papers, received in September and October 1942 and May 1944, respectively, were published in the *Zeitschrift für Physik.* He regularly gave lecture courses at the universities of Leipzig and Berlin (after summer 1942) on topics of theoretical physics. And he traveled, as far as the war restrictions allowed, to foreign countries, e. g. to Budapest (April 1941), to Zurich and Bern (November 1942) and to Leiden and Utrecht (October 1943), to meet with colleagues and friends and to deliver talks on his (non-secret) scientific research or on more general questions of physics and science. A particular visit was paid in October 1941 to Niels Bohr in Copenhagen. Without violating the secrecy of his work on the uranium project, Heisenberg tried to convey the message that in Germany the construction of the atomic bomb was not being considered. The failure of this visit cast a shadow on the future relations between Heisenberg and his teacher and friend.

Despite the early successes, which at first gave the German reactor project a slight edge over Fermi's American project, progress was very slow. Although the basic scientific and technological problems had been overcome and only the task remained of assembling enough uranium and heavy water to create an energy producing reactor, the project was increasingly hindered by the war conditions. Germany could barely provide enough uranium and not nearly enough heavy water, deficiencies made more acute by the fact that at least three groups performed competing experiments: besides the largest group in Berlin, a group in Hamburg under Paul Harteck and the group in Gottow under Diebner. The increasing air raids on Berlin forced the transfer of all apparatus to a bunker in the Dahlem institute, and finally, in late 1944, the entire project was shipped out of Berlin and housed in a rock bunker under the castle chapel in the southern German town of Haigerloch. The *KWI für Physik* had already been transferred earlier (since 1943) to the nearby town of Hechingen. In the first months of 1945 the model reactor B_8 nearly went critical (the point at which the reaction is self-sustaining). But the lack of a few hundred kilograms of uranium (out of 1.5 tons) and of several hundred liters of heavy water prevented success before war's end.

At the end of April 1945 members of an American science intelligence unit, the Alsos Mission, reached Tailfingen (where Otto Hahn's *KWI für Chemie* had moved), Hechingen and Haigerloch just ahead of the advancing French troops. The German atomic scientists gathered there – Bagge, Hahn, Korsching, von Laue, von Weizsäcker and Wirtz – were taken prisoner. Heisenberg, however, had gone a few days earlier by bicycle to Urfeld to join his family. He was taken prisoner there on 3 May 1945 by a small unit of the Alsos Mission led by Colonel

Boris T. Pash. Heisenberg and Walther Gerlach, Reich Commissioner *(Reichsbeauftragter)* for the reactor project, along with Diebner and Harteck, soon joined the Tailfingen-Hechingen group in France. All ten scientists were transferred to Castle Facquerel in Belgium and finally to England (on 23 July 1945), while the Americans dismantled the Haigerloch reactor and transported all of the apparatus, uranium, heavy water, and numerous documents to the United States.

6. The Period of Reconstruction and Renewal (1946–1958)

In April 1945 Soviet troops reached Berlin, the Reich collapsed and the Russians began sending to the Soviet Union the remaining apparatus in the Dahlem institute, the institute library, and the several German scientists they found there (among them Ludwig Bewilogua who had built the low temperature laboratory and had run the high voltage generator during the war). The Berlin *KWI für Physik* ceased to exist, while the Hechingen division barely survived with a handful of scientists – including Fritz Bopp, the crystallographer Georg Menzer and the spectroscopist Hermann Schüler – and a handful of experiments unrelated to nuclear physics.

Heisenberg was held along with the other prominent members of the Uranium Club at the country estate Farm Hall near Cambridge, England, well treated and equally well watched. During their stay at Farm Hall the atomic bomb was dropped on the Japanese city of Hiroshima on 6 August 1945. From news reports the German scientists learned that the Allies had, in their efforts to gain energy from atomic fission, obviously progressed much further than the Germans and had achieved military uses, an application that the Germans had believed to lie only far in the future. Their many discussions about the consequences of military applications of nuclear energy contributed to their later stance against nuclear weapons for the German Army.

At Farm Hall Heisenberg also set the stage for his future research. After years of difficulty, political pressure and reduced exchange of ideas he took the opportunity to devote himself in leisure to a wide range of physical problems. Included were the still unresolved problems of explaining superconductivity, which he discussed with von Laue, and the application of statistical methods to the description of turbulence, discussed with von Weizsäcker. He also decided to remain at the *KWI für Physik* if possible, despite lucrative offers to come to the United States.

On 3 January 1946, the German scientists were released from Farm Hall and returned to Western Germany. Heisenberg went with Hahn, von Laue, von Weizsäcker and Wirtz to Göttingen in the British Zone, where Max Planck and Ernst Telschow were seeking to continue the Kaiser Wilhelm Society and its institutes. Heisenberg devoted himself upon his return to two large tasks: the reconstruction of the *KWI für Physik* as a center for experimental and theoretical research in physics and the renewal of research in Germany. In particular he pondered a suitable vehicle for mediating between scientific research and the future German government in order to insure technical prosperity and the protection of science as the source of technical progress.

During the immediate post-war years Germans had to rebuild practically from the ground up. With the collapse of economic and political organization and the division of Germany into four occupation zones, the reconstruction of research in Germany was extremely difficult.

The conditions were perhaps most favorable in the British Zone, where the authorities, especially the liaison officer Colonel Bertie Blount, appeared quite open to the needs of the German scientists. As early as 1 January 1946 scientists and officials formed the German Scientific Council *(Deutscher Wissenschaftlicher Rat)* under the chairmanship of the Nobel Prize chemist Adolf Windaus. It mediated between the British military government and German scientific institutes and enabled, among other activities, the rebuilding of the Physico-Technical Institute *(Physikalisch-Technische Reichsanstalt)* in Braunschweig and the refounding of the German Physical Society *(Deutsche Physikalische Gesellschaft)* within the British Zone. The old British Hanoverian university city of Göttingen, largely untouched by the war, became once again a scientific center. During a meeting with Göttingen scientists in October 1946 the British authorities permitted the installation in Göttingen of several Kaiser Wilhelm institutes on the grounds of the former Aerodynamical Research Institute *(Aerodynamische Versuchsanstalt)* which had been completely dismantled as war-related. The institutes for physics, physical chemistry and medical research, previously all in Berlin, were thus revived. Finally the Kaiser Wilhelm Society was renamed and reinstituted in the three western zones as the Max Planck Society with seat in Göttingen; after 16 February 1948 all its institutes in western Germany were renamed Max Planck institutes (MPI).

In July 1946 Heisenberg began reequipping the institute for physics, whose direction he assumed after the official director Debye declined an offer to return. Two departments were set up initially: one for theoretical physics under von Weizsäcker, the other for experimental physics under Wirtz. Research was limited by the directives of the Allied Control Commission in that Germans were forbidden to work in a variety of fields considered war related, especially areas of nuclear physics involving slow neutrons and nuclear fission. Despite the limitations Heisenberg and his colleagues, many of whom had worked with him earlier in Berlin, did not lack research topics. Foremost remained the study of cosmic-ray physics, begun in Leipzig, continued in Berlin, and now picked up once again in Göttingen. Most of the institute resources were directed at first into theoretical research, since financing did not suffice for experimental studies. Only gradually did the institute catch up with the rapid developments occurring elsewhere in cosmic-ray and elementary particle physics at that time. The second edition of *Kosmische Strahlung,* published in 1953, indicated a near approximation to international standards in this subject. During the early post-war years Heisenberg also presented a theory of superconductivity (1946–1948), a statistical theory of turbulence (1946–1948), and examined the theoretical properties of elementary particles (after 1946).

Besides the direction of his institute and his own research Heisenberg devoted himself with great energy to the revival of scientific research in Western Germany and especially to his conception of governmental science policy. Learning from his experiences in the Third Reich, his observations of the close cooperation between scientists and government in Great Britain, and believing that a modern industrial state required a central science policy, Heisenberg sought a direct role for the government of the new Federal Republic of Germany in forming a national policy of support for science and technology and a role for science advisors to the chancellor. Such conceptions found expression in the German Research Council *(Deutschen Forschungsrat),* founded on 9 March 1949 by the Max Planck Society and the surviving academies of science in West Germany. It was composed of 15 leading scientists with Heisenberg as president. The new council, supported by Chancellor Konrad Adenauer, represented German science in international affairs and directly in the chancellor's office. Among its successes were the acquisition of Marshall Plan money for the support of German research, the admission of the Federal Republic to the International Union of Scientific Councils and to

UNESCO, the provision of federal responsibility for science in the West German constitution, and the assurance of mutual support between German industry and science. Yet the institution of the German Research Council contradicted the long tradition in Germany that science fell under the authority of the cultural ministers of each state *(Land)*. Hence the council was challenged increasingly by the Emergency Association of German Science *(Notgemeinschaft der Deutschen Wissenschaft),* revived in January 1949 by the ministers of culture and the university rectors. To avoid rivalry and the damaging overlap of interests, Heisenberg, with much regret, gave up his idea of a federal science body. The two bodies, i. e. the Research Council and the Emergency Association, were joined into the present-day German Research Association *(Deutsche Forschungsgemeinschaft)* in August 1951. The task of representing German science was assumed by the senate of the new body, composed mainly of the old Research Council. Heisenberg was elected to the presidential committee of the new Research Association; he also directed after February 1952 its Commission for Atomic Physics *(Kommission für Atomphysik)* which coordinated nuclear research (with the exception of research in nuclear fission) nationwide.

Heisenberg's constant involvement in German science policy found its counterpart in his involvement in the organization of research in his own institute. On 1 July 1947 a new astrophysical department was added under the direction of Ludwig Biermann, who had earlier worked at the Babelsberg observatory and the University of Hamburg. Biermann and his collaborators concentrated in Göttingen on cosmic plasma physics and on quantum theoretical problems in astrophysics. They soon attained significant results, in particular the prediction of the so-called "solar wind" from an analysis of the behavior of comet tails (Biermann and Rhea Lüst) and a new formation of the basic equations of plasma physics (Arnulf Schlüter). Their work required, as did other topics treated at Heisenberg's institute, the increased application of electronic data handling. Since such tools were not available at that time in Germany, the Max Planck Society began its own development program. Heinrich Billing, since 1 June 1950 head of the computer group at the Institute for Instrument Studies *(Institut für Instrumentenkunde)* in Göttingen and after 1957 a member of the *MPI für Physik,* developed and constructed the programmed calculating machines G1 and G2; the former was used after October 1952 at the MPI in Göttingen. Billing's breakthroughs in computer technology were unfortunately not realized by German industry.

The work produced by the Göttingen MPI during the years of reconstruction won a growing international reputation, and the institute's director Heisenberg attained increasing recognition as the leading representative of German science in the international arena. The reestablishment of international relations, also of great importance to Heisenberg, began officially with his invitation to lecture at the British universities in Cambridge, Edinburgh and Bristol in December 1947. During the following years he repeatedly visited Niels Bohr in Copenhagen and in summer 1950 he participated in the International Congress of Mathematicians in Cambridge, Massachusetts (30 August–6 September). In 1954 he served as the West German delegate to the conference on Atoms for Peace in Geneva. When in 1952 the European Council for Nuclear Research came into being, Heisenberg headed the German delegation and participated in the decision to locate the large European research center for high energy physics, CERN, in Geneva, Switzerland. The Scientific Policy Committee of CERN, responsible for planning the research program, elected Heisenberg as its chairman.

Immediately after his return to post-war Germany Heisenberg emphasized in a programmatic speech before Göttingen students the role of "science as a tool of mutual understanding between peoples" (13 July 1946). He plunged into the task of getting German scientists reaccepted as members of the international family of scientists and of renewing personal relations after the isolation and alienation in the previous period of the Third Reich and World War II. In this endeavor he assigned a particularly important role to the reestablished Alexander von Humboldt Foundation, whose first president he was appointed on 10 December 1953 by Chancellor Adenauer. The purpose of the foundation is to enable young scholars and scientists from around the world to collaborate with German colleagues while guests at German research institutes. Numerous Humboldt fellows were invited to Heisenberg's institute in Göttingen and later in Munich. He held the president's office until a severe illness forced him to resign in October 1975.

On 5 May 1955 the Paris Accords came into effect, whereby the Western Allies granted the Federal Republic full sovereignty and full membership in the NATO alliance. All restrictions on West German research were thus removed and, after ten years of restraint, the development of German nuclear energy resumed in full. In October 1955 Adenauer created a Federal Ministry for Atomic Questions *(Bundesministerium für Atomfragen),* forerunner of the wider ranged, present-day Ministry for Research and Technology *(Bundesministerium für Forschung und Technologie).* Heisenberg served as a leading member of the German Atomic Commission *(Deutsche Atomkommission)* set up by the new ministry and composed of scientists, industrialists, and politicians with the purpose of advising the ministry on nuclear energy policy. Heisenberg also directed the committee on nuclear physics within the Atomic Commission, served as a member of the Bavarian Atomic Commission, and acted as the main impetus toward the construction of Germany's first nuclear reactor, a research model set up at Garching near Munich in October 1957.

While Heisenberg supported the development of nuclear energy for peaceful uses, he and other scientists equally energetically opposed the plans of Chancellor Adenauer to equip the West German army with tactical nuclear weapons. Adenauer met with stiff resistance from Hahn, Heisenberg, von Weizsäcker and other atomic scientists, eighteen of whom issued a public declaration (basically formulated by von Weizsäcker and Heisenberg) from Göttingen on 12 April 1957, opposing research on or possession of nuclear weapons by West Germany. The West German army has since remained non-nuclear.

Heisenberg's main scientific interest in the early fifties increasingly focused on the search for a consistent quantum field theory of elementary particles. After an unsuccessful attempt at obtaining a nonlocal theory, he became concerned after 1952 with the investigation of nonlinear field equations, in which the mathematical space of states was extended beyond that used since the early days of quantum mechanics. In this theory, which Heisenberg and his collaborators developed in a series of papers between November 1953 and December 1956, the conditions of relativistic invariance could be immediately introduced and finite results could always be obtained without the use of supplementary subtraction or normalization procedures. After Heisenberg had demonstrated the consistency of his ideas and methods in the case of a model field theory, the so-called Lee model (October 1957), he entered into a close collaboration with Pauli that yielded in early 1958 the proposal of a nonlinear spinor equation designed to describe the properties and the behavior of all known elementary particles (dubbed the "world formula" by eager journalists). While Heisenberg expounded the equation in April 1958 and on several later occasions, Pauli withdrew his support.

Heisenberg's efforts to obtain a consistent quantum field theory for all elementary particles harmonized with the philosophical views that he presented at that time in many public lectures and lecture series (e. g., in the Gifford Lectures delivered in the winter term 1955/56 at St. Andrew's University in Scotland).

In September 1958 Heisenberg's *MPI für Physik* moved from Göttingen into a large new building in Munich near the English Garden, which solved the problem of overcrowding at the old institute. Differing from original plans, Karl Wirtz and his reactor group did not move to Munich, but left the MPI in March 1957 and settled at the nuclear research center of the *Kernreaktor Bau- und Betriebsgesellchaft mbH.* close to Karlsruhe. C. F. von Weizsäcker did not go to Munich either, as he accepted in June 1957 a call to a philosophy chair at the University of Hamburg. Yet he remained a regular guest in the Munich MPI during his semester vacations.

7. The Munich Years (1958–1976)

When Heisenberg moved to Munich in 1958, he characteristically broadened the program of his institute, which ultimately led to the establishment of a series of new institutes. First, the former astrophysics department under Ludwig Biermann was raised to an *Institut für Astrophysik* which, together with the *Institut für Physik,* formed Heisenberg's *Max-Planck-Institut für Physik und Astrophysik.* The program of the Göttingen institute was divided between the two institutes. The groups for experimental and theoretical elementary particle physics (and related fields) constituted, together with a group on experimental plasma physics (under Gerhard von Gierke), the physics institute. The astrophysics institute housed a department for electronic computers (Billing), one for theoretical astrophysics (Reimar Lüst) and one for theoretical plasma physics (Schlüter). With the addition of experimental plasma physics the institute compensated for the loss of Wirtz' reactor group and kept an open option on contributing to fundamental research in nuclear fusion. Yet it soon overflowed the available laboratory space in Munich and in June 1960 began moving to Garching, to form the present *MPI für Plasmaphysik* (move completed in 1968). Heisenberg remained, however, in the scientific administration of the new institute. The *MPI für extraterrestrische Physik* under Reimar Lüst also emerged (in 1964) from Heisenberg's institute and settled on the grounds of the Garching unit, but it remained an official sub-division of the mother institute.

While Heisenberg played an active role in these extensions of his institute, he remained convinced that the large experimental and theoretical tasks in elementary particle physics could only be solved by international efforts. He thus continued to aid and advise the European laboratory CERN near Geneva, which began operations in January 1959. Many members of his institute performed experimental work at CERN. During the later expansion of CERN Heisenberg actively supported the construction of the storage rings, which he dedicated on 16 October 1971. He also provided essential support during the early planning stages for the electron synchrotron (DESY) near Hamburg, whose first high-energy accelerator went into operation with a 6 GeV electron beam in February 1962.

Among Heisenberg's theoretical researches in Munich the development and evaluation of the nonlinear spinor theory occupied center place. In March 1959 he completed with his collaborators Hans-Peter Dürr, Heinrich Mitter, Siegfried Schlieder and Kazuo Yamazaki a long paper enunciating the principles of the theory together with several consequences; they obtained, in particular, its resonance states, one of which, the η-meson, was found more than a year later. Further investigations yielded an organizing of nucleons and hyperons with the help of the spurion concept (December 1964) and a calculation of the electromagnetic fine structure coupling constant (January 1965). Heisenberg wrote an introductory textbook on nonlinear spinor theory, which was published in 1966, and he retained his belief in it until the end of his life. He argued that all experimental researches on the structure of matter during the past decades had yielded either the already known elementary particles, or new ones having the same qualities. Moreover, the assumption of successive layers of further, ever more elementary objects, such as quarks, would not allow the achievement of the main goal of a fundamental theory, i. e., to explain the dynamical behavior of matter; it would rather relegate the problem to the next deeper level. It appeared to Heisenberg that his nonlinear spinor theory agreed in spirit with Plato's representation of the structure of matter by simple geometrical forms. He proposed to replace Plato's form by a highly symmetrical field equation. All properties of matter should follow from this field equation and the imposed conditions, although the detailed computations involved nonlinear and often very complicated approximation procedures.

In Heisenberg's later work the foundations of physics, which guarantee the unity of the description of nature, tended to merge with the conceptions of Plato's world view. He tirelessly presented his ideas about the intimate connection between physics and philosophy to wider audiences. He saw this connection verified in the historical development of quantum physics, describing parts of his own role in that development in his recollections *Der Teil und das Ganze* (1969, in English: *Physics and Beyond,* 1971). But the connection was also reflected in Heisenberg's standpoint on questions of art, and even on questions of religion and society.

On 31 December 1970 Heisenberg resigned the directorship of the *MPI für Physik und Astrophysik,* the institute that he had directed for almost 30 years. Yet he still came regularly to his office and continued to work on scientific papers and on more general articles, lectures and the second edition of his book on the nonlinear spinor theory. He participated at selected conferences, such as the symposium in honor of Dirac's 70th birthday in Miramare near Trieste (18–25 September 1972) and the colloquium celebrating the 200th anniversary of the Brussels Academy (16–17 May 1973). He traveled to the United States for the last time in April 1973. In the middle of 1973 he fell seriously ill. He slowly improved and a year later he appeared to have fully recovered. But in July 1975 he suffered a severe relapse. He died at his home in Munich on 1 February 1976.

Heisenberg received numerous national and international awards in recognition of his work and his influence on science and society. He received, besides the Nobel Prize for physics, the Barnard Medal of Columbia University (New York, 1929), the Matteucci medal of the *Accademia Nationale dei Lincei* (Rome, 1929) the Planck medal of the German Physical Society (1933), the Copernicus prize of the University of Königsberg (1943), the Hugo Grotius medal (1956), the order *Pour le mérite für Wissenschaften und Künste* (1957), the *Kulturpreis* of the City of Munich (1958), the Niels Bohr medal (Copenhagen, 1970) and the Guardini prize (Munich, 1973). In addition he received high national and international honors. He was a member of more than 30 scientific societies, including the Saxonian Academy (Leipzig, 1930), the *Kaiserlich Leopoldinisch-Carolinische Akademie der Naturforscher* (Halle, 1933), Norwegian Academy of Sciences (Oslo, 1936), Göttingen Academy of Sciences (1938), the Royal Swedish Academy (Uppsala, 1938), the *Société Philomatique* (Paris, 1938), the Royal Dutch Academy of Sciences (Amsterdam, 1939), the Prussian Academy of Science (Berlin, 1943), the *Accademia Nazionale dei Lincei* (Rome, 1947), the Bavarian Academy of Sciences (Munich, 1949), the Royal Danish Academy of Sciences (Copenhagen, 1951), the Pontificial Academy of Sciences (1955), the Royal Society of London (1955), the American Academy of Arts and Sciences (Boston, 1958) and the US National Academy of Sciences (Washington, 1961). At the very beginning of Heisenberg's studies Sommerfeld reminded him of the words of Schiller: „Wenn Könige bauen, haben Kärrner zu tun." ("When kings build, wagoners have to work.") By this he meant that his pupil had first to work along side the wagoners. Little did he suspect how quickly Heisenberg would begin to build and to contribute, as did few others in our century, to the construction of modern theoretical physics.

David C. Cassidy and Helmut Rechenberg

Werner Heisenberg

as Lecturer and Author of Scientific Review Articles and Books

The first talks which Heisenberg gave on physical problems were to Arnold Sommerfeld's *Seminar* at the University of Munich. Probably the earliest seminar he ever presented occurred towards the end of his first semester in winter 1920/21; in it, he discussed a paper of Hendrik Kramers on the Stark effect of hydrogen (Z. Phys. *3*, 199, 1920). In summer 1922 he appeared before the experts assembled at the Innsbruck Conference on Hydro- and Aerodynamics, where he spoke on preliminary results obtained for his doctoral thesis on the problem of the stability of laminar flow (item No. 1 in the Table of Contents).

In the years after World War I the German Physical Society *(Deutsche Physikalische Gesellschaft)* was regionally organized into Sections *(Gauvereine),* each of which organized several meetings per year at varying places in the region. At these meetings young scientists and even students were especially invited to discuss the results of their research. The titles of the talks, often together with short abstracts were published in the *Verhandlungen der Deutschen Physikalischen Gesellschaft,* third series, which first appeared in 1919. In most cases, a detailed paper would subsequently appear in a physics journal, preferably the *Zeitschrift für Physik* (founded in 1919). Heisenberg gave several contributions to the Section *Niedersachsen* in his early Göttingen years. The first entry in the *Verhandlungen* is the title of a paper presented at the Hannover meeting on 10–11 February 1923: 'Normalzustand des Heliums' (see *Verh. d. Deutsch. Phys. Ges.* (3) *4*, 7, 1923). Later that year he spoke in Göttingen on 'Eine neue Vorschrift zur Quantelung und ihre Anwendung auf den Zeemaneffekt' (ibid., p. 40). No abstract is provided in either case. Below we reproduce all existing abstracts of Heisenberg's later contributions to various section meetings of the German Physical Society, beginning with the one on 'Quantitatives über die Deformierbarkeit edelgasartiger Ionen' (No. 2). The last abstract is one of a lecture contributed to the first official German physicists' meeting after World War II in Göttingen, entitled 'Zur Elektronentheorie der Supraleitung' (No. 30). We might add that in most cases Heisenberg spoke on results of his recent research, on which he also published an original paper, i. e., one of those reprinted in *Series A;* the abstracts from the *Verhandlungen* reproduced here in *Series B* thus provide to some extent summaries of these papers.

The pioneering work on quantum mechanics made Heisenberg's name known far beyond the local circles of Munich, Göttingen and Copenhagen. Already in 1926 he was invited to give the main report on quantum mechanics at the 89th *Naturforscherversammlung* (Meeting of the German Scientists and Physicians) in Düsseldorf; like many other main reports of the *Naturforscherversammlungen* in those days it was printed in full in the then weekly scientific journal *Naturwissenschaften* (No. 4). Later he spoke several times at large national science meetings, like the *Naturforscherversammlungen* or the *Physikertagungen* (see Nos. 43, 47, 49), but even more frequently at international conferences. Very often he could report important progress on various problems of quantum, nuclear and elementary particle physics: thus, for example, on the structure of atomic nuclei to the Solvay Conference in fall 1933 (No. 12), on the theory of showers in cosmic radiation to the Copenhagen Conference in 1936 (No. 15), and on problems of quantum field theory and elementary particle physics to several conferences from the late thirties to the middle sixties (Nos. 26, 38, 45, 46, 48, 50, 51, 55, 56).

As a member of the physics establishment, especially as a pioneer in quantum mechanics – which was confirmed by the award of the 1932 Nobel prize for physics in 1933 – and later as the director of a prestigious German physics institute, Heisenberg was asked regularly to contribute to the major national and international physics conferences. However, he did not always accept the invitations. He disliked large meetings, having grown up during a time when only a handful of researchers were interested in the difficult problems of atomic and quantum theory. Conferences with hundreds of participants, whose program had to be split into numerous parallel sessions, would rather disturb him. He did not feel comfortable in crowds of people not personally known to him; and it could happen that, at such an occasion, he discussed a certain physical point with a person from his own institute, although he might have obtained much better information from the expert in this question who was also present at the conference, but had not been introduced to him. In any case, he withdraw increasingly in later years from the annual elementary particle physics conferences (the Rochester Conferences and their successors), although he supported their purposes. He still showed up at smaller international meetings where he could converse with all participants; he even took an active part in the creation and organization of such meetings, e. g. the Feldafing Seminar on Unified Theories of Elementary Particles (see his talks, Nos. 58 and 59).

The prototype of the particular type of meeting that Heisenberg preferred to attend, had come into existence already in 1911, the so-called Solvay Conferences in Physics. Their purpose was to assemble about a dozen physicists deeply interested in solving specific, fundamental problems of unusual difficulty. The participants were supposed to especially discuss at the meeting reports on the status of the chosen problems which had been prepared by selected rapporteurs and distributed in advance. The first Solvay Conference in which Heisenberg participated was the 1927 conference on 'Electrons and Photons', where Niels Bohr, Max Born and Erwin Schrödinger presented reports on the new quantum mechnanics and its interpretation. Although Heisenberg did not give a report at this conference – Born presented a common contribution (No. 5) – or at the following one in 1930, he participated eagerly in the serious debate about the consistency and interpretation of the theory he had helped to create. Especially the arguments of Albert Einstein and their refutation by Bohr, with the help of his assistants Heisenberg and Wolfgang Pauli, left a lasting impression on Heisenberg. For the subsequent Solvay Conference, dealing with the structure of the nucleus, Heisenberg was asked to give a main report (No. 12), and he was supposed to present, together with Pauli, a review of the properties of elementary particles at the following conference planned for fall 1939 (No. 27). Of the post World War II Solvay Conferences in Physics he did not attend those of 1948, 1951, 1954 and 1958 (the last three dealing with topics in which he was not interested at that time) and he could not come, due to illness, to the 1973 conference. Heisenberg prepared and presented a report only in 1967 (No. 62), but took a very active part in the debates at other conferences (e. g. in 1961, where the idea and meaning of Regge poles was discussed).

While he left out many opportunities to address big audiences on specific problems of physics, Heisenberg felt obliged and even enjoyed talking about the general aspects and philosophical implications of the results obtained in quantum and elementary particle physics before scientists and laymen alike. So he spoke at the 1930 *Naturforscherversammlung* in the philosophical section on 'Kausalität und Quantenmechanik' (Erkenntnis *2*, 172, 1931), at the

1950 *Naturforscherversammlung* on '50 Jahre Quantentheorie' (Naturwissenschaften *38*, 49, 1951), and at the 1968 Trieste symposium on 'Contemporary physics', where he delivered an evening lecture, in which he outlined his views and experiences as a scientist (reprinted in *From a Life of Physics,* Vienna: International Atomic Energy Agency, 1968). On other occasions he took the opportunity to acquaint wider audiences with the outstanding results of twentieth century physics and its cultural and social consequences. These talks and essays will be included in *Series C* of the *Collected Papers.*

Among Heisenberg's publications summarizing certain topics or fields of physics, we find, besides contributions to *Festschriften* – for example, to those for Sommerfeld in 1928 (No. 6), Peter Zeeman in 1935 (No. 14) and Hideki Yukawa in 1965 (No. 35) – and several smaller pieces written for various occasions (e. g., Nos. 9, 11) only a very few extended summaries of mainly his own work. Among the latter is the FIAT Review of the German reactor project, coauthored with Karl Wirtz (No. 32); and there are several reviews on problems of cosmic ray physics published between 1938 and 1953 (Nos. 21, 29, 37). Some reviews arose from seminar lectures (Nos. 29, 37). In addition Heisenberg wrote several books on physical topics, all of which emerged from lectures: his book *The Physical Principles of the Quantum Theory* (No. 8) was based on the lecture course he presented in summer 1929 at the University of Chicago; the book on *Kernphysik* (Braunschweig: Fr. Vieweg & Sohn, 1943; reprinted in *Series C)* originated from lectures in 1942 before German electrical engineers, and the *Introduction to the Unified Theory of Elementary Particles* (No. 60) from a lecture course in the summer semester 1965 at the University of Munich. Even most books for the general public owe their existence to lectures, especially that on *Physics and Philosophy* (New York, Harper and Row, 1958; reprinted in *Series C)* which contains his Gifford Lectures of 1955 at the University of St. Andrews.

The published lectures mentioned above constitute only a small part of Heisenberg's activity as an academic teacher of theoretical physics, whose career extended over half a century: in summer 1924 he became *Privatdozent* in Göttingen, in spring 1926 he succeeded Hendrik Kramers as lecturer at the University of Copenhagen, in fall 1927 he became ordinary professor at the University of Leipzig; when he moved to Berlin in summer 1942 to direct the *Kaiser-Wilhelm-Institut für Physik,* he was appointed simultaneously professor at the University of Berlin; after World War II in Göttingen and since 1958 in Munich he continued to teach courses on various fields of theoretical physics as a honorary professor. Heisenberg left detailed notes and manuscripts on nearly all theoretical topics that were usually taught at universities in his time, as well as on special topics of his own research, e. g., on the problems of cosmic radiation; they have been preserved and are kept now in the *Werner-Heisenberg-Archiv* at the *Werner-Heisenberg-Institut für Physik,* Föhringer Ring 6, D-8000 Munich 40*. The notes of some of his lecture courses were made available to a wider circle as mimeographed scripts or even printed books. As a rule those lecture notes were worked out by an advanced student participating in the lectures, a tradition which might be traced back in Germany at least to the mathematicians in Göttingen. Especially Felix Klein had arranged that elaborated notes from the main mathematical lecture courses were placed in the *Mathematisches Lesezimmer* at the University of Göttingen. Heisenberg's teacher Born, who had been a student there, had continued this habit for the lecture courses in theoretical physics and had passed it on to Heisenberg. Examples of such lecture notes, available as mimeographed or printed books, are: *Quan-*

---

* The *Archiv* will be transferred in near future to the *Bibliothek und Archiv zur Geschichte der Max-Planck-Gesellschaft,* Boltzmannstraße 4, D-1000 Berlin 33

tentheorie der Wellenfelder, lectures given in the winter semester 1946/47, elaborated by
K. Wildermuth; *Theorie des Atomkerns,* lectures given in the summer semester 1950, elaborated by W. Macke, Göttingen 1951; *Theorie der Neutronen,* lectures given in the winter semester 1950/51, elaborated by K. Wildermuth, Göttingen 1952; *Einführung in die Theorie der Elementarteilchen,* lectures given in the summer semester 1961, elaborated by H. Rechenberg and K. Lagally. We do not reprint these lectures on theoretical physics in the *Collected Works.* The reader is, however, compensated for the loss by finding in *Series B* the reprints of two lecture courses by Heisenberg which the author himself published as books (Nos. 8 and 60).

In comparison to his colleague and friend Wolfgang Pauli, Heisenberg composed only few survey and review papers on physical topics. The main explanation for this fact may be sought in Heisenberg's particular approach to research problems. Being always completely involved, even immersed, in his own ideas, he spent less time in following the paths of others; hence he was less prepared for summarizing the work of his contemporaries, especially when their methods or aims deviated from the ones he had in mind. In later years he certainly did not read or study in detail the increasing physical literature on the topics in which he was interested, but rather asked his associates and students to do so and to report to him the main contents and results. As a consequence Heisenberg never acted as a rapporteur at one of the large high energy conferences, where his task would have consisted in reviewing and summarizing some hundred papers submitted to the conference on a given topic. Also he did not exhibit the persuasive and brilliant performance of some of his contemporaries and of younger physicists when presenting and advertising their own and other's ideas before big crowds. His talks like his writings never sounded pretentious, but rested on solid ground and basically on simple physical or mathematical arguments. So he transmitted, even on the illuminated stage, the impression of a reliable craftsman. This does not mean, however, that these presentations lacked the intellectual and sometimes the artist's touch; but what held his audience frequently from the first to the last moment when he spoke was the sequence and order of arguments given and their logical, at times highly artistic, combination to often unexpected, even magically appearing, results. It is perhaps this aspect of the speaker Heisenberg that is most difficult to recognize in the printed text of his lectures.

H. Rechenberg

Scientific Review Papers, Talks, and Books

In *Vorträge aus dem Gebiete der Hydro- und Aerodynamik (Innsbruck 1922)*, ed. by Th. v. Kármán, T. Levi-Civita (Springer, Berlin 1924) pp. 139–142

Nichtlaminare Lösungen der Differentialgleichungen für reibende Flüssigkeiten.

Von W. Heisenberg in München.

Die bisherigen Untersuchungen über die Stabilität oder Labilität der laminaren Strömung haben, in Übereinstimmung mit experimentellen Untersuchungen von Ekman, zum Ergebnis geführt, daß die laminare Strömung gegenüber kleinen Störungen — kleinen Schwingungen im Sinn der gewöhnlichen Mechanik — im allgemeinen stabil[1] sei. Es bleiben also nur noch zwei Möglichkeiten, dem Turbulenzproblem näherzukommen: Erstens kann man die Stabilität der laminaren Strömung gegenüber Störungen untersuchen, die nicht unter den Begriff der „kleinen Schwingungen" fallen. Dies Problem hat Noether[2] in Angriff genommen. Zweitens aber kann man die Stabilitätsfrage ganz offen lassen und die Frage stellen: Gibt es noch eine andere Lösung der hydrodynamischen Differentialgleichungen, als die laminare? Diese Frage wollen wir im folgenden zu beantworten versuchen. Wir fragen also nach der turbulenten Bewegung selbst, nach ihrem Aussehen und nach dem Wertbereich der Reynoldsschen Zahl R, für den sie möglich ist.

Wir schreiben die hydrodynamischen Differentialgleichungen an

$$\left. \begin{aligned} \frac{\partial u}{\partial t} + u\frac{\partial u}{\partial x} + v\frac{\partial u}{\partial y} &= -\frac{1}{\varrho}\left(\frac{\partial p}{\partial x} - \mu\,\varDelta u\right) \\ \frac{\partial v}{\partial t} + u\frac{\partial v}{\partial x} + v\frac{\partial v}{\partial y} &= -\frac{1}{\varrho}\left(\frac{\partial p}{\partial y} - \mu\,\varDelta v\right) \end{aligned} \right\} \quad \ldots \ldots \quad (1)$$

und wollen den Couetteschen Fall behandeln — Strömung des Wassers zwischen zwei parallelen, relativ zueinander mit der Geschwindigkeit $\overline{u}$ bewegten Platten. Die X-Achse legen wir in die Mitte zwischen die beiden Platten, parallel zu ihnen so, daß die eine

---

[1] Vgl. jedoch Prandtl, L.: Bemerkung über die Entstehung der Turbulenz, Phys. Z. 23, 1922, S. 19.

[2] Noether, F.: Über die Entstehung einer turbulenten Flüssigkeitsbewegung. Sitzungsber. d. bayr. Ak. d. W. 1913, II, S. 309 (dazu vgl. Blumenthal, O.: Zum Turbulenzproblem, ebd. 1913, III, S. 563) u. Gött. Nachr. 1917: „Zur Theorie der Turbulenz."

                  W. Heisenberg:

sich mit $+\dfrac{\overline{u}}{2}$, die andere mit $-\dfrac{\overline{u}}{2}$ relativ zur X-Achse bewegt. Der Plattenabstand sei h.

Wir führen nun in bekannter Weise dimensionslose Variable ein durch den Ansatz:

$$\mathfrak{u} = \frac{u}{\overline{u}}, \qquad \mathfrak{v} = \frac{v}{\overline{u}}, \qquad \mathfrak{t} = t\,\frac{\overline{u}}{h}, \qquad \xi = \frac{x}{h}, \qquad \eta = \frac{y}{h}$$

und setzen wegen der Divergenzbedingung:

$$\mathfrak{u} = +\frac{\partial \varphi}{\partial \eta}; \qquad \mathfrak{v} = -\frac{\partial \varphi}{\partial \xi}.$$

Dann folgt aus (1) ($R = $ Reynoldssche Zahl):

$$\frac{\partial}{\partial t}\,\varDelta\varphi + \frac{\partial\varphi}{\partial\eta}\,\frac{\partial}{\partial\xi}\,\varDelta\varphi - \frac{\partial\varphi}{\partial\xi}\,\frac{\partial}{\partial\eta}\,\varDelta\varphi = \frac{1}{R}\,\varDelta\varDelta\varphi . \quad \ldots \quad (2)$$

Bis hierher gilt alles ganz allgemein.

Zur turbulenten Strömung speziell wollen wir nun folgendermaßen gelangen:

Wir zerlegen die Geschwindigkeit entsprechend den experimentellen Ergebnissen über die Turbulenz in zwei wesentlich verschiedene Teile: der eine gibt die mittlere Geschwindigkeitsverteilung — hat also nur eine X-Komponente — und soll mit $\mathfrak{u}_1 = w\ (\mathfrak{v}_1 = 0)$ bezeichnet werden. Der andere stellt die dieser Geschwindigkeitsverteilung überlagerten Schwingungen dar und soll als rein periodisch und harmonisch in ξ und t betrachtet werden. Mathematisch bedeutet dies: Wir nehmen die turbulente Bewegung als periodisch in ξ und t an, entwickeln φ, $\mathfrak{u}$ und $\mathfrak{v}$ dementsprechend in eine Fourierreihe und behalten in erster Näherung von dieser Fourierreihe nur die ersten beiden Glieder — das konstante und das rein harmonische — bei. Eine exakte Behandlung des Turbulenzproblems hat natürlich auch die höheren Fourierglieder mit zu berücksichtigen. Doch läßt gerade das experimentell bekannte Bild der turbulenten Bewegung vermuten, daß die Fourierreihen gut konvergent sind[1]).

Wir setzen also:

$$\varphi = \varphi_0(\eta) + e^{i(\alpha\xi - \beta t)}\,\varphi_1(\eta) + e^{-i(\alpha\xi - \beta t)}\,\overline{\varphi}_1(\eta) + \cdots \quad \ldots \quad (3)$$

Das konjugierte Glied mußten wir hinzufügen, um ein reelles φ zu erhalten; für φ_0 gilt:

$$\frac{\partial\varphi_0}{\partial\eta} = w .$$

Setzt man diesen Ansatz für φ in (2) ein und vergleicht gleiche

---

[1]) Dieser Ansatz findet sich schon bei N o e t h e r (Jahresber. der deutschen Math.-Ver. 23, 1914, S. 138), jedoch ohne ins Einzelne gehende physikalische Folgerungen.

Nichtlaminare Lösungen der Differentialgleich. für reibende Flüssigkeiten. 141

Potenzen von $e^{i(\alpha\xi-\beta t)}$ auf beiden Seiten, so ergeben sich die beiden grundlegenden Differentialgleichungen:

$$\varphi_1'''' - \alpha^2\,\varphi_1'' + \alpha^4\,\varphi_1 = i\cdot\alpha\cdot R\left[\left(w-\frac{\beta}{\alpha}\right)(\varphi_1''-\alpha^2\,\varphi_1)-w''\,\varphi_1\right] \quad . \quad (4)$$

$$w''' = i\,\alpha R\,(\overline{\varphi}_1\,\varphi_1''' - \overline{\varphi}_1'''\,\varphi_1 + \overline{\varphi}_1'\,\varphi_1'' - \varphi_1'\,\overline{\varphi}_1'') \;.\;.\;.\;(5)$$

Letztere Gleichung (5) läßt sich noch zweimal integrieren und liefert:

$$w' = i\cdot\alpha\cdot R\,(\overline{\varphi}_1\,\varphi_1' - \overline{\varphi}_1'\,\varphi_1) + \delta + \delta_1\cdot\eta\,;\;.\;.\;.\;.\;(6)$$

δ und δ_1 sind die Integrationskonstanten, δ_1 ist jedoch $=0$ zu setzen; aus (1) folgt nämlich, daß w'' am Rand verschwindet (dies ist nur bei der Couetteschen Anordnung der Fall, wo am Rand $\frac{\partial p}{\partial x}=0$ ist; dagegen ist z. B. bei der Strömung des Wassers in Röhren ein Druckgefälle vorhanden. Da nun am Rande auch φ_1 und φ_1' verschwindet (wegen $\mathfrak{u}=0$ und $\mathfrak{v}=0$), so folgt, daß in

$$w'' = i\,\alpha\,R\,(\overline{\varphi}_1\,\varphi_1'' - \overline{\varphi}_1''\,\varphi_1) + \delta_1$$

tatsächlich $\delta_1=0$ zu setzen ist.

Zur Integration der beiden simultanen Differentialgleichungen (4) und (6) kann ein Näherungsverfahren etwa in folgender Weise versucht werden: Wir wissen, daß w um den Punkt $\eta=0$ schiefsymmetrisch ist und wir kennen für $\eta=\pm\frac{1}{2}$ die Grenzbedingungen. Also entwickeln wir etwa φ_1 und w in der Umgebung von $\eta=0$ und $\eta=\pm\frac{1}{2}$ nach Potenzen von η und suchen so Näherungen für die Lösungen zu bekommen, deren Konvergenz allerdings zunächst ganz ungewiß ist.

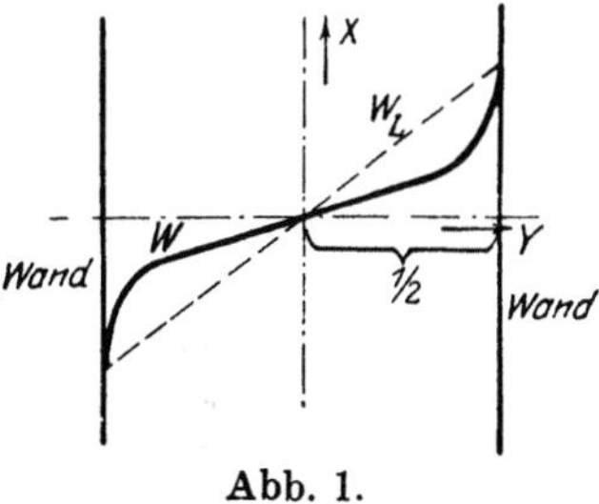

Abb. 1.

Das Aneinanderfügen der beiden von $\eta=0$ und $\eta=\frac{1}{2}$ kommenden Näherungen geschieht durch die Bedingung, daß an der betreffenden Stelle, wo die Näherungen ineinander übergehen sollen, $\varphi_1, \varphi_1', \varphi_1''$, und φ_1''' und w stetig sein müssen. Die Stetigkeit der übrigen Differentialquotienten von φ_1 und w folgt dann aus (4) und (6). Durch die Grenzbedingungen $w=_{(-)}^{+}\frac{1}{2}$ für $\eta=_{(-)}^{+}\frac{1}{2}$ und $\varphi_1=\varphi_1'=0$ für $\eta=\pm\frac{1}{2}$ werden jetzt φ_1 und w eindeutig festgelegt, wenn α und R gegeben sind.

Aus den Schlußresultaten, die diese Näherungsrechnung liefert, sei folgendes Wichtigste hervorgehoben:

1. Das Geschwindigkeitsprofil w weicht vom laminaren Profil $w_\mathfrak{L}$ wesentlich ab und schmiegt sich mit wachsendem R immer mehr den Wänden der Flüssigkeit an. Die Gesamtenergie der Flüssigkeit ist bei der turbulenten Bewegung geringer, als bei der laminaren.

2. Im mittleren Teil des Flüssigkeitsstromes, in dem w merklich linear in η geht, vollziehen sich die Bewegungen — in Über-

142 W. Heisenberg: Nichtlam. Lös. d. Differentialgleich. f. reib. Flüssigkeiten.

einstimmung mit der Prandtlschen Grenzschichttheorie — ganz wie in einer reibungslosen Flüssigkeit. Der Einfluß der Reibung macht sich erst in der unmittelbaren Nähe der Wände geltend. Hier sind auch die der Grundströmung überlagerten Wirbelgeschwindigkeiten am größten.

3. Nicht für alle Werte der Reynoldsschen Zahl ist die turbulente Strömung möglich. Unterhalb eines gewissen Wertes von R können sich über das Grundprofil w keine ungedämpften Schwingungen mehr überlagern, alle vorkommenden Störungen würden abklingen. Die angedeutete Näherungsrechnung liefert für diesen kritischen Wert

$$R \sim 1560.$$

Alle diese Ergebnisse stehen in Übereinstimmung mit der Erfahrung.

Am Schluß sei aber noch hervorgehoben, daß wir allen Grund haben, die Resultate mit großer Vorsicht aufzunehmen, solange die Konvergenzbeweise für die angewandten Näherungsverfahren ausstehen. Zunächst scheint mir der Wert der Resultate im wesentlichen nur darin zu liegen, daß sie die Hoffnung bestärken, daß der zur Behandlung des Turbulenzproblems eingeschlagene Weg richtig ist.

Anm. bei der Korrektur. Einen sehr viel allgemeineren Überblick über die Lösungen von (4) und (6) gewinnt man durch die Anwendung der asymptotischen Formeln für die Integrale von (4), wie sie z. B. im Spezialfall $w = y^3$, $\beta = 0$ Blumenthal l. c. verwendet hat. Man kann dann allgemeine Kriterien dafür angeben, wann ein Profil ungedämpfte und labile Schwingungen zuläßt und wann nicht, und kann für die labilen Profile die kritische Reynoldssche Zahl berechnen. An den obigen Resultaten ändert sich qualitativ bei Anwendung dieser Methoden nichts; es zeigt sich aber, daß man, um zu einigermaßen sicheren quantitativen Ergebnissen zu gelangen, wenigstens noch ein weiteres Glied der Fourierreihe (3) zu berücksichtigen hat; hierdurch wird die Möglichkeit, durch Aneinanderfügen der von der Mitte und von den Rändern kommenden Lösungen in der oben beschriebenen Weise eine gute Näherung für die Lösung von (4) und (6) und die Reynoldssche Zahl zu erhalten, in Frage gestellt. Die Größenordnung der kritischen Reynoldsschen Zahl bleibt aber jedenfalls erhalten. (Die ausführliche Veröffentlichung erscheint demnächst in den Ann. d. Physik.)

Wanderung suspendierter fester Körper in Flüssigkeiten unter dem Einflusse eines Stromes erwähnen. Zweifellos werden noch Akademien und wissenschaftliche Zeitschriften seine Lebensarbeit ausführlicher würdigen.

Obwohl der Verstorbene in Frankfurt a. O. geboren war, steht wohl jedem, der QUINCKE persönlich näher kannte und besonders auch sich seiner vornehmen Gastfreundschaft erfreuen durfte, seine Persönlichkeit als die eines humorvollen und liebenswürdigen Berliners lebhaft vor Augen und so wird er auch unserer Gesellschaft als Forscher wie als Mensch unvergessen bleiben.

Vorträge: 1. Hr. **E. Lau**: Über die Spektren des Wasserstoffs und das Spektrum des ionisierten Heliums. Zum Teil nach gemeinsamen Versuchen mit Hrn. E. GEHRCKE.

2. Hr. **Wohl**: Über die chemischen Konstanten der Halogene.

Gemeinschaftliche Tagung des Gauvereins Niedersachsen der Deutschen Physikalischen Gesellschaft in Braunschweig am 9. und 10. Februar 1924.

In der geschäftlichen Sitzung wird Hr. E. HOPPE-Göttingen zum ersten Vorsitzenden und Hr. P. P. KOCH-Hamburg zum zweiten Vorsitzenden gewählt; dritter Vorsitzender bleibt Hr. M. REICH.

Die gemeinsame Tagung des Sommersemesters soll in Hamburg gehalten werden.

Vorträge: 1. Hr. **W. Grotrian** (Potsdam): Über Absorptionsspektren von Chrom- und Eisendampf (nach gemeinsamen Versuchen mit Frl. GIESELER (erscheint in ZS. f. Phys.)

2. Hr. **F. Hund** (Göttingen): Rydbergkorrektion und Größe des Atomrumpfes.

Die Terme der geordneten Spektren lassen eine deutliche Scheidung der Bahnen des Leuchtelektrons zu in solche, die in den Rumpf eintauchen und solche, die ganz im Äußeren verlaufen. Aus der Stelle des Übergangs der einen in die anderen kann man die Größe des Rumpfes abschätzen. Dabei kommt man zu Widersprüchen mit den aus anderen Erscheinungsgebieten bekannten Ionengrößen, wenn man die azimutale Quantenzahl k des Leuchtelektrons ganzzahlig annimmt. Setzt man jedoch, wie ja in mehreren neueren Arbeiten vorgeschlagen wird, k halbzahlig an (gleich $1/_2$ bei den s-, $3/_2$ bei den p-Bahnen usw.), so fallen die Widersprüche fort. Eine ausführlichere Darstellung soll in der ZS. f. Phys. erscheinen.

3. Hr. W. Heisenberg (Göttingen): **Quantitatives über die Deformierbarkeit edelgasähnlicher Ionen.** (Nach einer gemeinsam mit Hrn. M. BORN ausgeführten Arbeit). Es wird gezeigt, daß die RYDBERG-RITZsche Korrektion in den Spektren alkaliähnlicher Atome und Ionen wesentlich von der Deformierbarkeit des edelgasähnlichen Rumpfes (Ion) herrührt und ein Maß für diese ist. Aus der Reihe Na, Mg^+, Al^{++}, Si^{+++} kann man extrapolatorisch auf die Deformierbarkeit des Ne-Atoms schließen, und analog in den anderen Reihen; die so gefundenen Werte stehen in guter Übereinstimmung mit den gemessenen Atomrefraktionen der Edelgase, wenn man die azimutale Quantenzahl halbzahlig annimmt. Nunmehr kann man die Extrapolation auf die Halogenionen ausdehnen und hat damit das Material zur Berechnung der Molekularrefraktionen zahlreicher Salze. Als Ergebnis der Ausgleichung der verschiedenen Methoden erhalten wir für das Verhältnis α vom elektrischen Moment zu erzeugender Feldstärke eines Ions (Dimension cm^3):

$$\alpha \cdot 10^{25} =
\begin{array}{cc|cccccccc}
\text{He} & \text{Li}^+ & \text{O}^{--} & \text{Fl}^- & \text{Ne} & \text{Na}^+ & \text{Mg}^{++} & \text{Al}^{+++} & \text{Si}^{++++} \\
2{,}2 & 0{,}75 & (31) & 10 & 4 & 2{,}1 & 1{,}2 & 0{,}64 & 0{,}43
\end{array}$$

$$
\begin{array}{cccccc|cccc}
\text{S}^{--} & \text{Cl}^- & \text{A} & \text{K}^+ & \text{Ca}^{++} & \text{Br}^- & \text{Kr} & \text{Rb}^+ & \text{Sr}^{++} & \text{J}^- & \text{X} & \text{Cs}^+ & \text{Ba}^{++} \\
(72) & 31 & 17 & 8{,}7 & 6 & 45 & 26 & 18 & 14 & 67 & 40 & 28 & (21)
\end{array}$$

Die hieraus berechneten α-Werte der Salze werden in der folgenden Tabelle mit den aus der Molekularrefraktion berechneten verglichen:

$10^{25} \cdot \alpha$	F		Cl		Br		J	
	ber.	Refr.	ber.	Refr.	ber.	Refr.	ber.	Refr.
Li	10,7	9,7	31,7	29,5	45,7	41,4	68	64,5
Na	12,1	12	33,1	32,4	47,1	46,5	69,1	64,5
K	19	20,3	40	40,5	54	53,5	76	76
Rb	28	—	49	49,5	63	61,5	85	85

4. Hr. **M. Born** (Göttingen): Der Einfluß der Ionendeformation auf physikalische und chemische Konstanten. (Nach einer gemeinsam mit Hrn. W. HEISENBERG ausgeführten Arbeit.)

Im Gitter eines Alkalihaloids ist jedes Ion so regelmäßig von den Nachbarn umgeben, daß es keine einseitige Deformation (höchstens eine geringfügige Volumänderung) erleiden kann. In der Molekel des Salzdampfes aber üben beide Ionen aufeinander gerichtete Kräfte aus, durch die sie deformiert werden. Die Energie der Molekel (Trennungsarbeit in die Ionen) läßt sich mit Hilfe der im vorigen Vortrage (HEISENBERG) mitgeteilten α-Werte der Ionen berechnen, wobei außer den elektrostatischen Anziehungen der Ladungen und der induzierten Dipole die aus dem Gitter ableitbare Abstoßungskraft $b\,r^{-10}$ berücksichtigt wird. Die so gefundenen Energiewerte Φ müssen gleich der Differenz von Gitterenergie U und Sublimationswärme S (gemessen von V. WARTENBERG und ALBRECHT) sein. Folgende Tabelle zeigt die Übereinstimmung:

	Φ ber.	$U-S$	Φ ber.	$U-S$	Φ ber.	$U-S$	Φ ber.	$U-S$
Na	168	163	142	139	137	130	130,5	122
K	144	148	126	124,5	121	116	116	107
Rb	126	127	123,5	123	115	114	112,5	104

Über quantentheoretische Kinematik und Mechanik.

Von

W. Heisenberg in Göttingen.

———

Die vorliegende Arbeit soll eine zusammenfassende Darstellung der Grundzüge einer neuen physikalischen Theorie geben, die den Versuch unternimmt, durch eine genaue Diskussion der Frage nach den prinzipiell beobachtbaren Größen die Gesetze der für Atomsysteme gültigen Kinematik und Mechanik aus der Erfahrung herzuleiten; es sollen also in dieser Theorie die Grundpostulate der Quantentheorie, die in der Bohrschen Atomtheorie zu so außerordentlichen Fortschritten in unserem Wissen über den Bau der Atome geführt haben, in einfache mathematische Verbindung gebracht werden mit jenen hinsichtlich der Kinematik und der Mechanik abgeänderten Raum-Zeitvorstellungen der Quantentheorie. Da die Theorie auch zu einer interessanten Anwendung mancher in der Physik bisher wenig gebrauchten mathematischen Methoden geführt hat, so mag eine Darstellung der Theorie in dieser Zeitschrift berechtigt erscheinen.

Die physikalischen Grundlagen der im folgenden darzustellenden Theorie wurden, zusammen mit der mathematischen Formulierung der kinematischen und mechanischen Grundgleichungen gegeben in einer Arbeit des Verfassers[1]; die Durchführung dieser Ansätze und der mathematische Ausbau der Theorie gelang Born und Jordan[2], ein Teil der von Born und Jordan abgeleiteten Gesetzmäßigkeiten und andere neue Folgerungen der Theorie wurden unabhängig angegeben von Dirac[3], der weitere mathematische Ausbau der Theorie wurde veröffentlicht in einer gemeinsamen Arbeit von Born, Jordan und dem Verfasser[4]; die physikalische Bedeutung der von Born, Jordan und Dirac angegebenen Vertauschungsrelationen

---

[1]) W. Heisenberg, Ztschr f. Phys. **33** (1925), S. 879.
[2]) M. Born und P. Jordan, Ztschr. f. Phys. **34** (1925), S. 858.
[3]) P. Dirac, Proceedings Roy. Soc. London **109** (1925), S. 642.
[4]) M. Born, W. Heisenberg und P. Jordan, Ztschr. f. Phys. **35** (1926), S. 557.

W. Heisenberg.

(vgl. § 2) behandelte Kramers[5]); schließlich gelang es Pauli[6]), die Gesetze des Wasserstoffatoms aus der neuen Theorie herzuleiten.

Unsere Erfahrungen über die Gesetzmäßigkeiten des Atombaues schöpfen wir zunächst aus zwei Quellen: den Strahlungsphänomenen und den Untersuchungen über Stoßvorgänge: die Gesetze der Wärmestrahlung haben Planck zur Einführung der Quantenhypothese veranlaßt, Einstein zeigte die Anwendbarkeit dieser Hypothese auf den lichtelektrischen Effekt; durch Versuche über den Zusammenstoß von α- und β-Partikeln mit Atomen gelangten Lenard und Rutherford zur Vorstellung vom Kernatom. Durch Zusammenfassung aller Erfahrungen über die Gesetzmäßigkeiten des Atombaues gelang es Bohr, die optischen Äußerungen der Atomgesetze in Verbindung zu bringen mit der Vorstellung vom Kernatom und dadurch die wichtigsten Fortschritte in unserem Wissen über die Bedeutung des periodischen Systems der Elemente und über alle Gesetze der Atomstruktur zu erzielen. Trotz dieser Erfolge stieß die quantitative Deutung der optischen Erfahrungen, also insbesondere die Berechnung der Serienspektra der Elemente auf unüberwindliche Schwierigkeiten; überdies forderten die Strahlungsphänomene an sich schon eine tiefgehende Revision unserer Raum-Zeitvorstellungen bei ihrer Anwendung auf die Vorgänge in den Atomen[7]). Diese Sachlage legte den Schluß nahe, daß zwar zweifellos die Grundpostulate der Quantentheorie und das Bohrsche Korrespondenzprinzip, denen wir ja fast allen Aufschluß über die Struktur der Atome verdanken, schon eine sinngemäße Beschreibung der Erfahrung darstellen, daß aber die speziellen Rechenregeln der bisherigen Theorie, die zur Deutung der Balmerserie geführt hatten, noch nicht die richtige Darstellung der Gesetzmäßigkeiten des Atombaues enthalten könnten.

Wenn man mit Bohr die Annahme macht, daß die Balmerserie den einer periodischen Bewegung eines Elektrons entsprechenden Strahlungsvorgang repräsentiert, so stößt man auf die bekannte grundsätzliche Schwierigkeit, daß das Spektrum jeder beliebigen periodischen Bewegung ein Spektrum äquidistanter Linien sein muß (alle Frequenzen sind als harmonische Obertöne ganzzahlige Vielfache einer Grundfrequenz), während die Balmerserie eine Linienfolge mit Häufung der Linien an einer endlichen Grenze darstellt. Anscheinend kann man dieser Schwierig-

---

[5]) H. Kramers, Physika 5 (1925), S. 369.

[6]) W. Pauli jr., Ztschr. f. Phys. (Im Erscheinen.)

(Die hier von [1]) bis [6]) zitierten Arbeiten werden wir im folgenden als (l. c. [1])), (l. c. [2])) usw. anführen.)

[7]) Für die allgemeine Diskussion dieser Probleme vgl. insbesondere N. Bohr, Vortrag vor dem Mathematikerkongreß in Kopenhagen, Sept. 1925. Nature 116 (1925), S. 845. Naturwiss. 14 (1926), S. 1.

keit· nur auf zweierlei Weise entgehen: entweder nimmt man an, daß das Strahlungsspektrum nichts mit dem mathematischen Spektrum der Bewegung des Elektrons im Atom zu tun habe, oder aber, man verläßt die Gesetze der klassischen Kinematik bei der Deutung der Atomprobleme. Die erstere Alternative scheint kaum möglich im Hinblick auf die großen Erfolge, welche die klassische Strahlungstheorie bei der Beschreibung optischer Phänomene erzielt hat. Die zweite Alternative führt zum Problem, die wirklichen Gesetze der quantentheoretischen Kinematik aus den Erfahrungen abzuleiten; dieses Problem ist der Gegenstand der hier darzustellenden Theorie.

§ 1.

Quantentheoretische Kinematik (l. c. [1]).

Die grundlegende Erfahrung, die im folgenden zur Ableitung der kinematischen Gesetze der Quantentheorie notwendig ist, bietet das Rydberg-Ritzsche Kombinationsprinzip. Es besagt, daß jede Frequenz $\nu(nm)$ des beobachteten Spektrums darstellbar sei als Differenz zweier „Terme" T_n und T_m und daß umgekehrt jede Differenz zweier Terme aus der Gesamtheit der Terme als Frequenz einer beobachtbaren Spektrallinie auftreten kann, daß also gilt

$$(1) \qquad \begin{aligned} \nu(nm) &= T_n - T_m \\ \nu(nk) + \nu(km) &= \nu(nm), \end{aligned}$$

wo die $\nu(nk)$, $\nu(km)$ und $\nu(nm)$ beobachtbare Frequenzen des Spektrums bedeuten.

Aus den optischen Phänomenen kann ganz allgemein geschlossen werden, daß zu den prinzipiell beobachtbaren Größen gehört: die Frequenz und die zu dieser Frequenz gehörige Amplitude und Phase. Es soll sich zunächst handeln um ein Problem von einem Freiheitsgrad, also um ein Elektron, das längs einer Geraden innerhalb gewisser Grenzen periodisch schwingen kann. Die Elongation des Elektrons heiße x. Diese Größe x würde in der klassischen Theorie in eine Fourierreihe nach t zerlegt werden können:

$$(2) \qquad x = x(n, t) = \sum_\tau x(n)_\tau \cdot e^{2\pi i \nu(n) \cdot \tau \cdot t},$$

wo n eine Konstante und τ die Nummer der Oberschwingung bedeutet. Die einzelnen Glieder dieser Reihe

$$(3) \qquad x(n)_\tau \, e^{2\pi i \nu(n) \tau t}$$

würden direkt die als beobachtbar hervorgehobenen Größen Frequenz,

Amplitude und Phase zusammenfassen. — Aus dem Erfahrungsgesetz (1) folgt als sinngemäße Erweiterung dieser klassischen Auffassung, daß bei einer wirklichen Beschreibung der atomtheoretischen Phänomene an Stelle der Größen (3) Ausdrücke der Form

$$(4) \qquad x(nm)\, e^{2\pi i \nu(nm)t}$$

treten müssen. Eine Zusammenfassung solcher Größen $x(nm)$ zu einer Summe analog der Gleichung (2) scheint hier zunächst nicht sinnvoll. Man wird vielmehr die Frage nach der Zusammenfassung der Größen (4) zunächst offenlassen müssen und den Sinn der Größen (4) so ausdrücken: *Die Elongation x wird repräsentiert durch die Gesamtheit*

$$(5) \qquad \boldsymbol{x} : \left(x(nm)\, e^{2\pi i \nu(nm)t} \right).$$

Mit dieser Aufstellung wäre aber offenbar physikalisch wie mathematisch nichts gewonnen, wenn man nicht zugleich angäbe, wie die Beziehungen solcher Gesamtheiten lauten, die den Beziehungen zwischen Fourierschen Reihen in der klassischen Theorie analog sind.

Denken wir uns also eine zweite Größe y (z. B. etwa die Entfernung des Elektrons von einem gegebenen Punkt im Raum), die ebenfalls klassisch in eine Fourierreihe

$$(6) \qquad y = y(n,\, t) = \sum_\tau y(n)_\tau\, e^{2\pi i \nu(n)\cdot \tau \cdot t}$$

entwickelbar wäre. Quantentheoretisch würde man auch hier wieder aus (1) und aus der Tatsache, daß Frequenz, Amplitude und Phase für y prinzipiell beobachtbar sein dürften, schließen, daß y repräsentiert wird durch eine Gesamtheit:

$$(7) \qquad \boldsymbol{y} : \left(y(nm)\, e^{2\pi i \nu(nm)t} \right).$$

Allgemein wird man — und dies soll auch für Systeme von beliebig vielen Freiheitsgraden gelten — jede zu dem betreffenden Atomsystem gehörige quantentheoretische Größe z, sofern nur ihr klassisches Analogon in eine Fourierreihe entsprechend (2) entwickelbar ist, darstellen durch die Gesamtheit

$$(8) \qquad \boldsymbol{z} : \left(z(nm)\, e^{2\pi i \nu(nm)t} \right).$$

Die denkbar einfachste kinematische Fragestellung ist offenbar die: Gegeben seien die x und y repräsentierenden Gesamtheiten; welche Gesamtheit stellt $x + y$ und welche xy dar?

Die naturgemäße Anwort lautet offenbar (l. c. [1]):

$$\boldsymbol{x} + \boldsymbol{y} : \left((x+y)(nm)\, e^{2\pi i \nu(nm)t} \right); \qquad \boldsymbol{x} \cdot \boldsymbol{y} : \left((xy)(nm)\, e^{2\pi i \nu(nm)t} \right), \text{ wobei}$$

$$(9) \qquad (x+y)(nm)\, e^{2\pi i \nu(nm)t} = \left[x(nm) + y(nm) \right] e^{2\pi i \nu(nm)t},$$

$$(10) \qquad (x\,y)(n\,m)\,e^{2\pi i\nu(nm)t} = \sum_k x(n\,k)\,y(k\,m)\,e^{2\pi i[\nu(nk)+\nu(km)]t}$$
$$= \sum_k x(n\,k)\,y(k\,m)\,e^{2\pi i\nu(nm)t}.$$

Die erste (9) dieser Relationen leuchtet ohne weiteres ein, die zweite scheint eine sinngemäße Folge von (1); in Analogie zur entsprechenden klassischen Formel wird man nämlich erwarten, daß auf der rechten Seite von (10) eine Summe steht von Produkten je zweier Glieder der Reihen (5) und (7), wobei diese Produkte der Summe sämtlich die gleiche e-Potenz enthalten sollen. Dies ist nach (1) naturgemäß durch Gleichung (10) zu erreichen. Da für alle zum selben Atomsystem gehörigen Gesamtheiten der exponentielle Faktor $e^{2\pi i\nu(nm)t}$ derselbe ist, so kann man ihn der Einfachheit halber weglassen und die bisherigen Resultate in den Formeln zusammenfassen:

$$(11) \qquad \begin{cases} \boldsymbol{x}:(x(n\,m)) \\ \boldsymbol{y}:(y(n\,m)). \end{cases}$$

$$(12) \qquad (\boldsymbol{x}+\boldsymbol{y})(n\,m) = x(n\,m)+y(n\,m).$$

$$(13) \qquad (\boldsymbol{x}\boldsymbol{y})(n\,m) = \sum_k x(n\,k)\,y(k\,m).$$

Diese in (11), (12), (13) enthaltenen Rechenregeln entsprechen vollständig den aus der Theorie der linearen Gleichungen wohlbekannten für Matrizen gültigen Rechenregeln (l. c. [2])). Man wird also eine quantentheoretische Größe mit einer (übrigens unendlichen) quadratischen Matrix vergleichen können. Für ihre Multiplikation gilt bekanntlich das distributive und das assoziative Gesetz:

$$(14) \qquad \boldsymbol{z}(\boldsymbol{x}+\boldsymbol{y}) = \boldsymbol{z}\boldsymbol{x}+\boldsymbol{z}\boldsymbol{y},$$

$$(15) \qquad (\boldsymbol{x}\boldsymbol{y})\boldsymbol{z} = \boldsymbol{x}(\boldsymbol{y}\boldsymbol{z}),$$

aber im allgemeinen nicht das kommutative Gesetz; also im allgemeinen: $\boldsymbol{x}\boldsymbol{y}+\boldsymbol{y}\boldsymbol{x}$; die Forderung $\boldsymbol{x}\boldsymbol{y}-\boldsymbol{y}\boldsymbol{x}=0$ dürfte zum Spezialfall der klassischen Theorie zurückführen.

Irgendeine Funktion $f(\boldsymbol{x},\boldsymbol{y},\boldsymbol{z})$ wird nach (12), (13) quantentheoretisch dann definiert werden können, wenn ihr klassisches Analogon in eine Potenzreihe nach $\boldsymbol{x},\boldsymbol{y},\boldsymbol{z}\ldots$ entwickelbar ist.

Zur Vervollständigung der kinematischen Gesetzmäßigkeiten sei noch hervorgehoben, daß die Relationen (5, 12, 13) auch (in etwas modifizierter Form) noch Gültigkeit behalten, wenn die zugrunde gelegte Bewegung nicht als periodisch betrachtet werden kann (l. c. [1])). Es tritt dann klassisch an Stelle der Fourierreihe das Fourierintegral; dem entspricht es, daß in der Matrix $(x(n\,m))$ die Indizes n und m auch kontinuierliche Wertebereiche durchlaufen können; die Summe der rechten Seite von (13) wird in diesen kontinuierlichen Wertebereichen durch ein Integral zu ersetzen sein.

Außer der für quantentheoretische Größen gültigen Algebra haben noch die Operationen der Differentiation für die Formulierung der mechanischen Gesetzmäßigkeiten eine entscheidende Bedeutung. Allerdings wird sich später herausstellen, daß der Prozeß der Differentiation nur künstlich in die Theorie eingeführt werden kann und daß eine andere Operation (vgl. § 3) das naturgemäße Analogon zur Differentiation der klassischen Theorie darstellt. Wir werden im folgenden für die Differentiation nach der Zeit in einfacher Analogie zur klassischen Theorie annehmen:

Wird eine quantentheoretische Größe z dargestellt durch die Gesamtheit

$$z : (z(n\,m)\,e^{2\pi i\nu(nm)t}) \quad \text{oder einfacher} \quad (z(n\,m)),$$

so wird die Größe $\dot{z} = \dfrac{dz}{dt}$ repräsentiert durch:

$$\dot{z} : (z(n\,m)\,2\pi i\,\nu(n\,m)\,e^{2\pi i\nu(nm)t})$$

oder einfacher:

$$(16) \qquad\qquad \dot{z} : (2\pi i\,\nu(n\,m)\,z(n\,m)).$$

Der Differentialquotient einer Funktion $f(x, y, z \ldots)$ nach einer der durch Matrizen repräsentierten quantentheoretischen Größen $x, y, z \ldots$ wird am einfachsten definiert werden können durch:

$$(17) \qquad \frac{\partial f(x, y, z \ldots)}{\partial y} = \lim_{\alpha \to 0} \frac{f(x + \alpha \cdot 1, y, z \ldots) - f(x, y, z \ldots)}{\alpha},$$

wobei α eine Konstante, 1 die „Einheitsmatrix" bedeutet $1 : (1(n\,m))$, bei der alle Glieder außerhalb der Diagonale Null sind, während die Diagonalglieder sämtlich den Wert 1 haben:

$$1(n\,m) = \delta_{nm} = \begin{cases} 1 & \text{für} \quad n = m \\ 0 & \text{für} \quad n \neq m. \end{cases}$$

Die Differentiation genügt offenbar den Beziehungen:

$$\frac{\partial}{\partial x}(f + g) = \frac{\partial f}{\partial x} + \frac{\partial g}{\partial x},$$

$$(17\,a) \qquad \frac{\partial}{\partial x}(f\,g) = \frac{\partial f}{\partial x}\,g + f\,\frac{\partial g}{\partial x}.$$

Durch die Annahmen (5), (12), (13), (16), (17) scheinen sämtliche Rechenoperationen für quantentheoretische Größen definiert. Es entsteht die weitere Aufgabe, auf der Basis dieser kinematischen Relationen eine Quantenmechanik aufzubauen. Der leitende Gesichtspunkt dabei wird wieder entsprechend Bohrs Korrespondenzprinzip der möglichst enge Anschluß an die klassische Theorie sein müssen.

§ 2.
Quantenmechanische Vertauschungsrelationen.

Die Rechenregeln, die in der bisherigen Theorie bedingt-periodischer quantentheoretischer Probleme üblich waren[8]), lassen sich folgendermaßen kurz skizzieren: Entsprechend den Gesetzen der Hamilton-Jacobischen Mechanik suche man (unter Vernachlässigung der Strahlungskräfte) als kanonische Variable die Winkelvariablen $w_k(w_1 \ldots w_f)$ bei einem Problem von f (Freiheitsgraden) und die kanonisch-konjugierten Wirkungsvariablen $J_k(J_1 \ldots J_f)$ auf. Die w_k sind lineare Funktionen der Zeit $w_k = \nu_k t + \varphi_k$, die J_k sind Konstante der Bewegung und genügen der Gleichung

$$\nu_k = \frac{\partial H}{\partial J_k}$$

($H =$ Hamiltonsche Funktion). Dem Korrespondenzprinzip wurde Genüge geleistet durch die Annahme, daß ein stationärer Zustand charakterisiert sei durch $J_k = n_k \cdot h$ (h bedeutet die Plancksche Konstante, n_k eine ganze Zahl), dann stimmte gerade die quantentheoretische Bohrsche Frequenzbedingung

$$\nu(n\,m) = \frac{H_n - H_m}{h}$$

(H die Energie) im Limes $h = 0$ und für hohe Quantenzahlen mit der klassischen Beziehung

$$\nu_k = \frac{\partial H}{\partial J_k}$$

überein.

Unter Benutzung dieser durch die letzten Beziehungen angedeuteten Korrespondenz soll jetzt das Verhalten des quantentheoretischen Ausdrucks $xy - yx$ für hohe Quantenzahlen untersucht werden (l. c. [3])). Im Grenzfall hoher Quantenzahlen wird man entsprechend dem Übergang zur klassischen Theorie (vgl. (2)) als Näherung setzen dürfen $x(n, n - \alpha) = x(J)_\alpha$, wo $J_k = n_k \cdot h$ oder, was praktisch hier dasselbe bedeuten soll, $J_k = (n_k + \alpha)h$.

Wir befinden uns also in einem Gebiet, wo die Unterschiede zwischen Anfangs- und Endbahn relativ gering sind; in diesem Gebiet werden die Resultate der klassischen Theorie, wie die der bisherigen Quantentheorie näherungsweise richtig sein. Daher kann man die n_k als Quantenzahlen, die J_k als Wirkungsvariablen der klassischen Theorie, α_k als die Nummer der Oberschwingung deuten. Wir denken uns also, wie schon durch die Bezeichnungsweise angedeutet, die Indizes n entsprechend den bisherigen quantentheoretischen Rechenregeln als f-dimensionale Mannigfaltigkeit von „Quantenzahlen" $n_k(n_1 \ldots n_f)$ geordnet.

---

[8]) Vgl. z. B. A. Sommerfeld, Atombau und Spektrallinien. 4. Aufl. Braunschweig, Vieweg, 1924. M. Born, Vorlesungen über Atommechanik. Berlin, Springer, 1924.

Dann ergibt sich:

$$(19) \quad x(n, n-\alpha)\,y(n-\alpha, n-\alpha-\beta) - y(n, n-\beta)\,x(n-\beta, n-\alpha-\beta)$$
$$= \{x(n, n-\alpha) - x(n-\beta, n-\alpha-\beta)\}\,y(n-\alpha, n-\alpha-\beta)$$
$$- \{y(n, n-\beta) - y(n-\alpha, n-\alpha-\beta)\}\,x(n-\beta, n-\alpha-\beta)$$
$$\approx h \cdot \sum_k \left\{ \beta_k \frac{\partial x\,(J)_{\alpha_k}}{\partial J_k}\, y(J)_{\beta_k} - \alpha_k \frac{\partial y\,(J)_{\beta_k}}{\partial J_k} \cdot x\,(J)_{\alpha_k} \right\}.$$

Nun ist aber

$$(20) \qquad 2\pi i\,\beta_k \cdot y(J)_{\beta_k} \cdot e^{2\pi i w_k \beta_k} = \frac{\partial}{\partial w_k}\{y(J)_{\beta_k} e^{2\pi i w_k \beta_k}\},$$

wo die w_k wieder die zu den J_k kanonisch konjugierten Winkelvariablen bedeuten.

Daraus folgt mit (19):

$$(21) \qquad \lim (\boldsymbol{x}\boldsymbol{y} - \boldsymbol{y}\boldsymbol{x}) = \frac{h}{2\pi i} \sum_k \left(\frac{\partial x}{\partial J_k}\frac{\partial y}{\partial w_k} - \frac{\partial y}{\partial J_k}\frac{\partial x}{\partial w_k} \right)$$

im Grenzfall „hoher Quantenzahlen" oder $\lim h = 0$. In den x, y der rechten Seite ist durch die Schreibweise ausgedrückt, daß man hier von den Matrizen der Quantentheorie zu den Fouriergliedern der klassischen Mechanik übergehen kann. Denn im eben betrachteten Grenzfall entspricht je ein Glied der Matrix einem Fourierglied der klassischen Mechanik und kann näherungsweise diesem gleich gesetzt werden.

Die Summe der rechten Seite von (21) stellt das Jacobische Klammersymbol dar:

$$(22) \quad [x, y] = \sum_k \left(\frac{\partial x}{\partial J_k}\frac{\partial y}{\partial w_k} - \frac{\partial y}{\partial J_k}\frac{\partial x}{\partial w_k} \right) = \sum_k \left(\frac{\partial x}{\partial p_k}\frac{\partial y}{\partial q_k} - \frac{\partial y}{\partial p_k}\frac{\partial x}{\partial q_k} \right),$$

wo p_k, q_k irgendwelche kanonische Variable bedeuten. Führen wir nun als quantentheoretische Klammersymbole die Ausdrücke

$$(23) \qquad (\boldsymbol{x}, \boldsymbol{y}) = \frac{2\pi i}{h}(\boldsymbol{x}\boldsymbol{y} - \boldsymbol{y}\boldsymbol{x})$$

ein, so läßt sich Gleichung (21) schreiben:

$$(21\,\mathrm{a}) \qquad \lim (\boldsymbol{x}, \boldsymbol{y}) = [x, y] \text{ für hohe Quantenzahlen.}$$

In der klassischen Theorie gelten für die $[x, y]$ die Beziehungen:

$$
\begin{aligned}
&\text{a)} \quad [x+y, z] = [x, z] + [y, z],\\
(24) \quad &\text{b)} \quad [x, y] \quad\;\; = -[y, x],\\
&\text{c)} \quad [xy, z] \quad = [x, z]\,y + x\,[y, z].
\end{aligned}
$$

Ferner gilt

$$(25) \quad [q_r, q_s] = 0;\quad [p_r, p_s] = 0;\quad [p_r, q_s] = \delta_{rs} = \begin{cases} 1 \text{ für } r = s. \\ 0 \text{ für } r \neq s. \end{cases}$$

Wegen der Gleichungen (24) läßt sich klassisch ·das Symbol $[x, y]$ ausdrücken durch $[q_r, q_s]$, $[p_r, p_s]$, $[p_r, q_s]$, wenn x und y als Potenzreihen in p und q darstellbar sind.

Die quantentheoretischen Klammersymbole zeigen nun eine weitgehende Analogie zu den Jacobischen Symbolen der klassischen Theorie.

Es gilt nämlich wegen (14) und (15):

$$\begin{aligned} &\text{a)} \quad (x+y, z) = (x, z) + (y, z), \\ (26) \quad &\text{b)} \quad (x, y) \quad\;\; = -(y, x), \\ &\text{c)} \quad (xy, z) \quad = (x, z)y + x(y, z). \end{aligned}$$

Wieder kann hier, wenn x und y nach Potenzen der p und q entwickelt werden können, (x, y) ausgedrückt werden durch die elementaren Symbole

$$(p_r, p_s), \quad (q_r, q_s), \quad (p_r, q_s).$$

Diese enge Verwandtschaft der klassischen Symbole $[x, y]$ und der quantentheoretischen (x, y), zusammen mit Gleichung (21 a) legt *die für alles folgende fundamentale Annahme* nahe (l. c. [2]), [3]), [4])):

$$(27) \qquad (p_r, p_s) = 0; \quad (q_r, q_s) = 0; \quad (p_r, q_s) = \delta_{rs}.$$

Durch diese Hypothese ist die Korrespondenzbeziehung (21 a) gesichert und eine vollkommene Analogie der (x, y) zu den $[x, y]$ hergestellt.

Die Gleichungen (27) entsprechen korrespondenzmäßig den „Quantenbedingungen" der bisherigen Quantentheorie. Daß die Hypothesen (27) wirklich als neue unabhängige Annahmen erlaubt sind, daß sie also keine Widersprüche enthalten und die Lösungen des gestellten Problems eindeutig bestimmt machen, soll in § 5 gezeigt werden.

Man könnte auch quantentheoretisch die Symbole $[x, y]$ gemäß Gleichung (22) einführen. Diese Symbole genügen dann den Gleichungen (24 a, b) und (25); Gleichung (24 c) ist wegen der Nichtgültigkeit des kommutativen Gesetzes in der Quantentheorie nur dann erfüllt, wenn z eine lineare Funktion der p und q ist.

Es folgt aus dieser Überlegung ohne weiteres der Satz: Ist z eine *lineare* Funktion der p und q, so gilt:

$$(28) \qquad\qquad [x, z] = (x, z).$$

Es stellen aber, wie schon oben betont, die quantentheoretischen $[x, y]$ nur künstliche Bildungen dar (was z. B. aus der allgemeinen Nichtgültigkeit von (24 c) hervorgeht); das sinngemäße Analogon zu den klassischen $[x, y]$ sind die Symbole (x, y).

Den Gleichungen (27) (l. c. [2]), [3]), [4])), die man als quantenmechanische Vertauschungsrelationen bezeichnen kann, muß in der Quantenmechanik eine viel allgemeinere Bedeutung zukommen, als den „Quantenbedingungen" der bisherigen Theorie. Die Quantenbedingungen der bisherigen Theorie waren nur auf mechanische Probleme von bedingt-periodischem Charakter anwendbar und dienten dort dazu, aus der kontinuierlichen Mannigfaltigkeit möglicher Lösungen eine diskrete Anzahl quantentheoretischer Lösungen auszusondern. In der hier darzustellenden Theorie sind die Relationen (27) notwendig, um dem Problem der Integration der Bewegungsgleichungen einen eindeutigen Sinn zu geben, sie sollen für alle denkbaren Lösungen erfüllt sein, unabhängig davon, ob die Lösungen selbst periodischen oder nichtperiodischen Charakter tragen. Zugleich erscheinen die Gleichungen (27) als die exakte Formulierung des Bohrschen Korrespondenzprinzips. Übrigens lassen die Relationen (27) noch vom Standpunkt der Dispersionstheorie aus eine einfache physikalische Interpretation zu (l. c. [1]) und [5])).

§ 3.

Die kanonischen Bewegungsgleichungen; Energiesatz und Frequenzbedingung.

Zur Definition eines mechanischen Problems genügt in der klassischen Theorie die Angabe der Hamiltonschen Funktion $H(p, q)$. Die Bewegungsgleichungen lauten dann:

$$(29) \qquad \dot{p}_k = -\frac{\partial H}{\partial q_k}; \quad \dot{q}_h = \frac{\partial H}{\partial p_k} \quad \text{oder} \quad \dot{p}_k = [H, p_k]; \quad \dot{q}_k = [H, q_k].$$

Es scheint die einfachste Annahme, diese Gleichungen direkt in die Quantenmechanik zu übernehmen (vgl. l. c. [1-4])):

$$(30) \qquad \dot{\boldsymbol{p}}_k = -\frac{\partial H}{\partial \boldsymbol{q}_k} = [H, \boldsymbol{p}_k]; \quad \dot{\boldsymbol{q}}_k = \frac{\partial H}{\partial \boldsymbol{p}_k} = [H, \boldsymbol{q}_k]$$

oder vgl. (28):

$$(30\,a) \qquad \dot{\boldsymbol{p}}_k = (H, \boldsymbol{p}_k); \quad \dot{\boldsymbol{q}}_k = (H, \boldsymbol{q}_k).$$

Auch hier ist zu betonen, daß der Gleichung (30 a) vor (30) der Vorzug gegeben werden muß, da (30 a) keine Differentialquotienten nach $\boldsymbol{p}$ und $\boldsymbol{q}$ enthält. In (30) und (30 a) wurde außerdem bisher noch die Zusatzannahme gemacht (l. c. [1])), daß für ein gegebenes Problem die Energiefunktion $\boldsymbol{H}$ in kartesischen Koordinaten quantenmechanisch dieselbe Form haben solle, wie in der klassischen Theorie. Der physikalische Sinn dieser Zusatzannahme ist folgender: Die Form der Hamiltonschen Funktion ist im Prinzip durch Versuche mit freien Elektronen, durch Stoßversuche usw. unabhängig vom Spektrum experimentell prüfbar. Handelt es sich dabei z. B. um Versuche mit sehr schnellen Elektronen, so kann man, wie bisher. näherungsweise die Gesetze der klassischen Mechanik der Bestimmung von $\boldsymbol{H}$ zugrunde legen. Bei der Extrapolation der gefundenen Funktion $\boldsymbol{H}$

auf die Fälle, in denen die klassische Theorie nicht mehr näherungsweise gültig ist, wird die Form der Energiefunktion dann erhalten bleiben müssen (und daher gleich der Form der „klassischen" Energiefunkticn sein), wenn nur Produkte von vertauschbaren Größen in H vorkommen. Dies ist im allgemeinen in einem kartesischen Koordinatensystem der Fall, da in ihm die Energie additiv in zwei nur von den Variablen einer Sorte abhängige Teile $H_1(p) + H_2(q)$ zerfällt.

Wir werden im folgenden die Gestalt der Hamiltonschen Funktion H ganz offen lassen und annehmen, daß sie durch Experimente oder quantenmechanische Überlegungen irgendwie gefunden sei.

Dann kann aus den quantenmechanischen Gleichungen (30) die zeitliche Konstanz von H sowie die Bohrsche Frequenzbedingung abgeleitet werden, wie folgt (l. c. [2]), [3]), [4])):

Führt man eine quantentheoretische Größe W ein, die mit den Termen T_n der Gleichung (1) zusammenhängt nach der Formel

$$(31) \qquad W(nm) = \begin{cases} T_n \cdot h & \text{für} \quad n = m, \\ 0 & \text{für} \quad n \neq m, \end{cases}$$

so folgt aus (1), (16) allgemein:

$$(32) \qquad \dot{x} = \frac{2\pi i}{h}(Wx - xW) = (W, x).$$

Die Bewegungsgleichungen (30), (30 a) gehen also direkt über in

$$(33) \qquad \begin{aligned} (W, q_k) &= (H, q_k); & (W - H, q_k) &= 0; \\ (W, p_k) &= (H, p_k); & (W - H, p_k) &= 0. \end{aligned}$$

Da H eine Funktion der p und q ist, so folgt

$$(W - H, H) = 0; \qquad (W, H) = 0$$

oder wegen (32):

$$(34) \qquad \begin{aligned} \dot{H} &= 0, \\ H &= \text{konst.} \end{aligned}$$

und wegen (33):

$$(H(nn) - H(mm))\, q(nm) = (W(nn) - W(mm))\, q(nm),$$

d. h. (vgl. (31))

$$(35) \qquad \frac{H(nn) - H(mm)}{h} = \nu(nm).$$

Die Gleichungen (34) und (35) enthalten den Energiesatz und die Frequenzbedingung der Quantenmechanik (l. c. [2]), [3]), [4]). Diese Resultate bedeuten, daß die in der hier vorzutragenden Theorie enthaltenen quantenmechanischen Gesetze bereits eine sinngemäße mathematische Formulierung der Grundpostulate der Quantentheorie ermöglichten. Es muß noch der Beweis nachgeholt werden, daß die Werte $H(nn)$ allgemein sowohl eine diskrete wie eine kontinuierliche Mannigfaltigkeit bilden können; dann sind

694 W. Heisenberg.

alle Aussagen der quantentheoretischen Grundpostulate in der hier zu beschreibenden Theorie enthalten.

§ 4.
Kanonische Transformationen.

In der klassischen Mechanik ist ein System neu einzuführender Variabler P, Q dann kanonisch, wenn für dieses System gilt:

$$[P_r, P_s] = 0; \quad [Q_r, Q_s] = 0; \quad [P_r, Q_s] = \delta_{rs}.$$

Ein ganz analoger Satz gilt in der Quantenmechanik (l. c. [2]) und [4])):

Wenn ein System neuer Variabler P, Q den Bedingungen genügt:

$$(36) \qquad (P_r, P_s) = 0; \quad (Q_r, Q_s) = 0; \quad (P_r, Q_s) = \delta_{rs},$$

so gelten für dieses System die kanonischen Gleichungen (30), vorausgesetzt, daß die Energie H nach Potenzen der P, Q entwickelbar ist.

Zum Beweise bemerken wir, daß sich bei Gültigkeit von (36) das Symbol (H, P_k) bzw. (H, Q_k) einerseits gemäß (32), (34), (35) interpretieren läßt als $\dot{P}_k$ bzw. $\dot{Q}_k$, andererseits nach (28) als

$$[H, P_k] = - \frac{\partial H}{\partial Q_k}; \quad [H, Q_k] = + \frac{\partial H}{\partial P_k}.$$

Es entsteht jetzt das Problem, die allgemeinste kanonische Transformation der Quantenmechanik anzugeben, welche die Variablen p_k, q_k in neue Variablen P_k, Q_k überführt.

Eine sehr allgemeine derartige Transformation, welche die Gleichungen (27) invariant läßt, lautet (l. c. [4])):

$$(37) \qquad \begin{cases} P_k = S p_k S^{-1}, \\ Q_k = S q_k S^{-1}, \end{cases}$$

wo S eine ganz beliebige quantentheoretische Größe bedeutet. Aus der Gültigkeit von (27) und (37) folgt in der Tat (36). Daß (37) auch die allgemeinste kanonische Transformation darstellt, konnte bisher nicht bewiesen werden.

Aus (37) folgt noch, wenn f irgendeine nach Potenzen von p, q entwickelbare Funktion bedeutet:

$$(38) \qquad f(PQ) = S f(pq) S^{-1},$$

wobei $f(PQ)$ aus $f(pq)$ dadurch hervorgeht, daß p, q unter Beibehaltung der Funktionsform f durch P, Q ersetzt werden.

Die Wichtigkeit der kanonischen Transformationen beruht auf folgendem Satze: Sei p_k^0, q_k^0 irgendein System von Variablen, das den Gleichungen (27) genügt, so kann das Problem der Integration der Bewegungsgleichungen (30) zurückgeführt werden auf das andere Problem (l. c. [4])):

Es ist eine Funktion S so zu bestimmen, daß

$$(39) \qquad S H(p^0, q^0) S^{-1} = W = \text{konst.}$$

zu einer Diagonalmatrix wird; hierin soll $H(p^0, q^0)$ aus $H(p, q)$ hervorgehen, indem p, q durch p^0, q^0 unter Beibehaltung der Funktionsform H ersetzt werden.

Ist eine solche Funktion S gefunden, so erhält man durch die Gleichungen:

$$(40) \qquad \begin{cases} p_k = S\, p_k^0\, S^{-1}, \\ q_k = S\, q_k^0\, S^{-1} \end{cases}$$

direkt die Lösungen der Bewegungsgleichungen (30). (39) stellt das quantenmechanische Analogon dar zur Hamilton-Jacobischen partiellen Differentialgleichung.

S entspricht in gewisser Weise der Wirkungsfunktion der klassischen Mechanik. Wir werden im folgenden (39) als die für das mechanische Problem „charakteristische Gleichung" bezeichnen.

§ 5.

Lösung der charakteristischen Gleichung (l. c. [4]).

Wir nehmen im fogenden irgendwelche Matrizen p^0, q^0, die den Vertauschungsrelationen (27) genügen und sonst völlig willkürlich gewählt werden können, als gegeben an.

Es wird z. B. zunächst am einfachsten sein, für die p_k^0, q_k^0 $(k=1, \ldots, f)$ die Koordinaten eines Systems von f ungekoppelten harmonischen Oszillatoren zu nehmen. Dann sind die Gleichungen (27) sämtlich für die p^0, q^0 erfüllt und wenn eine Transformation (39) angegeben werden kann, so ist dadurch auch der Beweis der Widerspruchsfreiheit von (27) nachträglich erbracht. Diese Wahl der Koordinaten von f ungekoppelten Oszillatoren als p^0, q^0 entspricht in gewisser Weise der Einführung von Normalkoordinaten J, w in der klassischen Theorie; doch muß hier gleich hervorgehoben werden, daß ein direktes Analogon zu diesen Normalkoordinaten, also quantentheoretische Variable, bei denen die Variablen der einen Art konstant, die der anderen Art lineare Funktionen der Zeit sind, in der Quantenmechanik noch nicht gefunden ist.

Der eigentlichen Lösung von (39) soll eine kurze Diskussion der Eigenschaften der in der Quantentheorie gebrauchten Matrizen vorausgeschickt werden. Alle physikalischen Größen müssen reell sein. Dem entspricht es, daß die quantentheoretischen Größen stets durch Hermitesche Matrizen dargestellt werden. Eine Matrix q ist dann vom Hermiteschen Typus, wenn $q(nm) = \overset{*}{q}(mn)$. (* bedeutet Übergang zur konjugiert komplexen Größe). Trotzdem wird sich gelegentlich die Einführung auch allgemeinerer Matrizen als nützlich erweisen. Jedenfalls aber müssen Größen wie p, q, H stets durch Hermitesche Matrizen dargestellt sein (l. c. [2]) u. [4])).

Die quantentheoretischen Matrizen sind stets unendlich; d. h. die Anzahl der durch Indizes n, m charakterisierten Zustände ist stets unendlich. Dies folgt z. B. aus der Gleichung

$$(41) \qquad p_r q_r - q_r p_r = \sum_k (p_r(nk) q_r(kn) - q_r(nk) p_r(kn)) = \frac{h}{2\pi i}.$$

Wäre die Anzahl der Zustände endlich, so würde die Summation über alle Zustände n der Gleichung (41) auf der linken Seite offenbar Null geben, da es bei einer endlichen Summe auf die Reihenfolge der Glieder nicht ankommt und sich daher die Glieder paarweise fortheben.

$$(42) \qquad \sum_n \sum_k [p_r(nk) q_r(kn) - q_r(nk) p_r(kn)] = 0,$$

während die rechte Seite $\sum_k \frac{h}{2\pi i}$ sicher einen endlichen Wert hat; also enthält die Annahme einer endlichen Anzahl „stationärer Zustände" einen Widerspruch.

Zu jeder Matrix $a : (a(nm))$ gehört eine bilineare Form (l. c. [4])):

$$(42) \qquad A(st) = \sum_{nm} a(nm) s_n t_m$$

zweier Reihen von Variablen s_n, t_n.

Ist die Matrix von Hermiteschem Typus, d. h.

$$(44) \qquad a(nm) = a^*(mn) \quad \text{oder} \quad a = \tilde{a}^*$$

($*$ bedeutet Übergang zur konjugiert komplexen Größe; mit $\tilde{\ }$ bezeichnen wir diejenige Matrix, die aus der ursprünglichen durch Vertauschung der Indizes hervorgeht[9])), so nimmt die Form $A(st)$ reelle Werte an, wenn man für die t_n die zu s_n konjugiert komplexen Werte setzt.

$$(45) \qquad A(ss^*) = \sum_{nm} a(nm) s_n s_m^*.$$

Nach diesen Vorbemerkungen behaupten wir: Das Problem der Auflösung der charakteristischen Gleichung (39) ist identisch mit dem Problem, die zu der Hermiteschen Matrix $H(p^0 q^0) : (H(nm))$ gehörige Form

$$\sum_{nm} H(nm) s_n s_m^*$$

auf Hauptachsen zu transformieren (l. c. [4])).

In der Tat, wenn eine orthogonale Transformation v der s_n in neue Variable t_n gefunden ist:

$$(46) \qquad s_n = \sum_l v(ln) t_l,$$

---

[9]) Diese von der üblichen abweichende Bezeichnungsweise ist hier eingeführt, um mit den zugrunde liegenden Arbeiten (l. c. [2]) u. [4])) hinsichtlich der Bezeichnung in Übereinstimmung zu bleiben.

$$(47) \quad \begin{cases} \sum_l v(n\,l)\,v^*(n\,l) = 1, \\[2mm] \sum_l v(n\,l)\,v^*(m\,l) = 0 \quad \text{für} \quad n \neq m, \end{cases}$$

d. h.
$$v \cdot \tilde{v}^* = 1,$$

derart, daß gilt:

$$(48) \qquad \sum_{nm} H(n\,m)\,s_n\,s_m^* = \sum_n W(n\,n)\,t_n\,t_n^*,$$

so gibt die Annahme $v = S$ direkt die Lösung von (39). Denn es gilt dann (vgl. (46), (47)):

$$(49) \qquad v = S, \quad \tilde{v}^* = S^{-1},$$

$$(50) \quad \begin{cases} W(nn) = \sum_{l\,k} H(l\,k)\,v(n\,l)\,v^*(n\,k), \\[2mm] \text{d. h.} \\[1mm] W = v \cdot H \cdot \tilde{v}^* = S \cdot H \cdot S^{-1}. \end{cases}$$

Die Energiewerte $W_n = W(n\,n)$ der stationären Zustände erscheinen also direkt als die *Eigenwerte* der Hermiteschen Form $H(n\,m)$. Aus der Theorie dieser Formen kann man daher das Resultat entnehmen, daß zu einer unendlichen Hermiteschen Form — und mit einer solchen hat man es hier stets zu tun — im allgemeinen diskrete und kontinuierliche Folgen von Eigenwerten gehören. Das Termspektrum der Physik ist daher identisch mit dem mathematischen Spektrum der Hermiteschen Form, bestehend aus Punkt- und Streckenspektrum.

Was die Frage nach der Existenz der Lösungen (46), (47), (48) betrifft, so ist der Beweis hierfür geliefert nur für den Fall beschränkter Formen[10]): Ist eine beschränkte unendliche Hermitesche Form:

$$\sum_{nm} H(n\,m)\,s_n\,s_m^*$$

gegeben, so gibt es stets eine orthogonale Transformation, welche diese Form überführt in:

$$(51) \qquad \sum_{nm} H(n\,m)\,s_n\,s_m^* = \sum_n W\,t_n\,t_n^* + \int W(\varphi)\,t(\varphi)\,t^*(\varphi)\,d\varphi.$$

Für die genauere Diskussion der Frage, was zu verstehen sei unter einer orthogonalen Transformation für den Fall, daß ein Streckenspektrum auftritt[11]), und für die Frage des Verhaltens der Lösungen p, q in solchen Fällen verweisen wir auf die betreffenden Arbeiten (l. c. [4])).

Die in der Quantentheorie vorkommenden Formen $H(n\,m)$ werden wohl im allgemeinen nicht zur Klasse der „beschränkten Formen" ge-

---

[10]) Siehe z. B. D. Hilbert, Grundzüge einer allgemeinen Theorie der linearen Integralgleichungen.

[11]) Vgl. z. B. E. Hellinger, Crelles Journal **136** (1910), S. 1.

hören. Wir nehmen im folgenden an, daß sich die Resultate bei diesen allgemeinen Formen nicht wesentlich von den hier aus der Theorie der beschränkten Formen übernommenen Ergebnissen unterscheiden.

Die Hauptachsentransformation (51) kann prinzipiell durch folgende Methode aufgefunden werden:

Man sucht Lösungen der linearen Gleichungen mit unendlich vielen Unbekannten:

$$(52) \qquad W v_k - \sum_l H(k\,l)\, v_l = 0.$$

Diejenigen Werte von W, für die solche Lösungen existieren, sind die Eigenwerte der Hermiteschen Form $H(n\,m)$. Seien W_n und W_m irgend zwei Eigenwerte, so gilt

$$(53) \qquad \begin{cases} \text{a)} \quad W_n v_{kn} - \sum_l H(k\,l)\, v_{ln} = 0, \\[2mm] \text{b)} \quad W_m v_{km}^* - \sum_l H(k\,l)\, v_{lm}^* = 0. \end{cases}$$

Multipliziert man (53 a) mit v_{km}^*, (53 b) mit v_{kn} und summiert über k, so folgt wegen des Hermiteschen Charakters von H:

$$(54) \qquad \begin{cases} (W_n - W_m) \sum_k v_{kn} v_{km}^* = 0. \\[2mm] \text{Normiert man dann die } v \text{ noch durch} \\[1mm] \qquad \sum_k v_{kn} v_{kn}^* = 1, \end{cases}$$

so gilt $v\,\tilde{v}^* = 1$ und v stellt die Transformationsfunktion S dar (vgl. (49)).

Die Wichtigkeit der Gleichung (51) beruht wesentlich darauf, daß es Methoden gibt, die Eigenwerte der Form $H(n\,m)$ zu bestimmen, ohne die Transformation v wirklich aufsuchen zu müssen. Im Falle Hermitescher Formen von endlich vielen Variablen sind die Eigenwerte Wurzeln einer algebraischen Gleichung. Bei Formen unendlich vieler Variablen kann man das Verfahren von Graeffe und Bernoulli[12]) benutzen. Die bisher nach den hier geschilderten Methoden behandelten Probleme, insbesondere die Paulische Theorie des Wasserstoffatoms (1. c. [6])) scheinen allerdings zu zeigen, daß andere speziellere Methoden zum Aufsuchen der Eigenwerte erheblich dem eben angeführten allgemeinen Verfahren überlegen sind.

Bei der Transformation (51) kann der Fall mehrfacher Eigenwerte W_n auftreten; d. h. zu einem bestimmten W_n gibt es mehrere linear unabhängige Lösungssysteme v_{kn} der Gleichung (53). Dieser Fall ist analog zum Problem der Entartung in der klassischen Theorie; in der klassischen Theorie tritt Entartung ein, wenn eins oder mehrere der ν_k verschwinden;

---

[12]) Vgl. z. B. Courant-Hilbert, Methoden der math. Physik. Berlin, Springer, 1925, S. 15.

dem entspricht hier nach (33), daß zwei oder mehrere W_n einander gleich werden.

Sei bei einem r-fachen Eigenwert ein System $v_{kn}^{(1)} \ldots v_{kn}^{(r)}$ von Lösungen der Gleichung (53) gegeben, so kann man durch lineare Kombinationen der $v_{kn}^{(s)}$ beliebige neue Lösungssysteme $v_{kn}^{(s)'}$ erzeugen, die nur der Bedingung (54) $\left(\text{Invarianz der Summe} \sum_{s=1}^{r} v_{kn}^{(s)} v_{kn}^{(s)*} \text{ gegen die oben ge-}\right.$ schilderten Transformationen$\Big)$ unterworfen sind. Dem entspricht es, daß die Lösungen p, q (vgl. (40))

$$(55) \quad \begin{cases} \boldsymbol{p}_k = \boldsymbol{v}\, p_k^0\, \tilde{\boldsymbol{v}}^*, & p_k(ln)^{(s)} = \sum_{h,i} v_{lh}\, p_k^0(hi)\, v_{ni}^{*(s)} \\[2mm] \boldsymbol{q}_k = \boldsymbol{v}\, q_k^0\, \tilde{\boldsymbol{v}}'^*, & q_k(ln)^{(s)} = \sum_{h,i} v_{lh}\, q_k^0(hi)\, v_{ni}^{*(s)} \end{cases}$$

unbestimmt werden. Durch Wahl eines anderen Systems linear unabhängiger $v_{ni}^{*(s)}$ kann man von (55) wesentlich verschiedene Lösungen erhalten.

Dagegen sind die Quadratsummen

$$(56) \quad \begin{cases} \sum_{s=1}^{r} p_k(ln)^{(s)}\, p_k^*(ln)^{(s)} \\[2mm] \sum_{s=1}^{r} q_k(ln)^{(s)}\, q_k^*(ln)^{(s)} \end{cases}$$

wieder invariant gegen diese genannten Transformationen der $v_{kn}^{(s)}$ unter sich (l. c.[4])).

Die eben geschilderte Unbestimmtheit der Lösungen (55) beim Auftreten mehrfacher Eigenwerte entspricht ganz der Unbestimmtheit, die in der klassischen Theorie bei Entartung stets auftritt. Die Invarianz der Quadratsummen (56) hat empirisch bei der Diskussion der sogenannten spektroskopischen Stabilität eine wichtige Rolle gespielt.

Die Vielfachheit eines Eigenwertes W_n bestimmt das statistische Gewicht des zu dem betreffenden Energiewerte gehörigen stationären Zustandes.

Durch die Gleichungen (53), (54) werden auch bei nicht entarteten Systemen (Nichtauftreten mehrfacher Eigenwerte) die Lösungen v_{kn} nur bis auf einen Faktor vom Absolutbetrage 1 der Form $e^{i(\varphi_k - \varphi_n)}$ bestimmt sein. Ebenso sind auch die Lösungen $p(nm)$, $q(nm)$ nur bis auf Faktoren der Form $e^{i(\varphi_n - \varphi_m)}$ bestimmt. Es bleibt also für jeden stationären Zustand eine Phasenkonstante willkürlich. Aus Analogie zur klassischen Theorie kann man vermuten, daß, abgesehen von dieser Unbestimmtheit, die Lösungen p, q bei nichtentarteten Systemen eindeutig durch die Bewegungsgleichungen und die Bedingungen (27) bestimmt sind. Ein Beweis für diese Vermutung ist bisher nicht erbracht worden.

§ 6.
Störungstheorie (l. c.[4]).

Die Möglichkeit der Lösung der Bewegungsgleichungen durch Lösung der charakteristischen Gleichung (39) führt auch zu einer Störungstheorie, die der Störungstheorie der Hamilton-Jacobischen Mechanik weitgehend analog ist.

Nehmen wir an, die Energiefunktion H sei gegeben als Potenzreihe nach einem Parameter λ:

$$(57) \qquad H(pq) = H_0(pq) + \lambda H_1(pq) + \lambda^2 H_2(pq) + \ldots$$

Sei ferner eine Lösung p^0, q^0 des durch H_0 charakteristischen ungestörten Problems gegeben derart, daß $H_0(p^0 q^0) = W_0$ eine Diagonalmatrix wird und für p^0, q^0 die Gleichungen (27) erfüllt sind.

Dann suchen wir eine Funktion S

$$(58) \qquad \begin{cases} p_k = S p_k^0 S^{-1} \\ q_k = S q_k^0 S^{-1} \end{cases}$$

so zu bestimmen, daß

$$(59) \qquad S H(p^0, q^0) S^{-1} = W$$

eine Diagonalmatrix wird. Wir versuchen den Ansatz

$$(60) \qquad \begin{cases} S = 1 + \lambda S_1 + \lambda^2 S_2 + \ldots, \qquad \text{also} \\ S^{-1} = 1 - \lambda S_1 + \lambda^2 (S_1^2 - S_2) + \ldots \end{cases}$$

und

$$(61) \qquad W = W_0 + \lambda W_1 + \lambda^2 W_2 + \ldots$$

Dann ergibt (59) durch Vergleichung der einzelnen Koeffizienten von λ, λ^2, … auf beiden Seiten die Näherungsgleichungen zur Bestimmung von S:

$$(62) \quad \begin{cases} H_0(p^0 q^0) = W_0 \\ S_1 H_0 - H_0 S_1 + H_1 = W_1 \\ S_2 H_0 - H_0 S_2 + H_0 S_1^2 - S_1 H_0 S_1 + S_1 H_1 - H_1 S_1 + H_2 = W_2 \\ \cdots \cdots \cdots \cdots \cdots \cdots \cdots \cdots \cdots \cdots \cdots \\ S_r H_0 - H_0 S_r + F_r(H_0 \ldots H_r, S_0 \ldots S_{r-1}) = W_r. \end{cases}$$

Für den Fall, daß das Ausgangssystem nicht entartet ist, lassen sich diese Gleichungen ohne weiteres auflösen: Die erste Gleichung (62) ist von selbst erfüllt. Für die anderen bildet man den zeitlichen Mittelwert

$$(63) \qquad \overline{F_r} = W_r \quad \text{oder} \quad F_r(nn) = W_r(nn),$$

und findet dann ohne weiteres die Lösung:

$$(64) \qquad S_r(mn) = \frac{F_r(mn)}{h \nu_0(mn)} (1 - \delta_{mn}).$$

Ist das Ausgangssystem entartet, sei z. B. ein r-facher Eigenwert vorhanden:

$$W_0(n+1, n+1) = W_0(n+2, n+2) = \ldots = W_0(n+r, n+r),$$

so ist der Ansatz (60) zweckmäßig zu ersetzen durch den folgenden:

$$(65) \quad \begin{cases} S = S_0(1 + \lambda S_1 + \lambda^2 S_2 + \ldots), \\ S^{-1} = (1 - \lambda S_1 + \lambda^2(S_1^2 - S_2) + \ldots)S_0^{-1}, \end{cases}$$

wo S_0 eine Matrix bedeutet, die der Gleichung genügt

$$(66) \qquad\qquad S_0\,\tilde{S}_0^* = 1.$$

Dann liefern die ersten Näherungsgleichungen:

$$(67) \quad \begin{cases} S_0\,H_0\,S_0^{-1} = W_0, \\ S_0(S_1 H_0 - H_0 S_1)\,S_0^{-1} + S_0 H_1 S_0^{-1} = W_1 \\ \cdot \ \cdot \ \cdot \ \cdot \ \cdot \ \cdot \ \cdot \ \cdot \ \cdot \ \cdot \ \cdot \ \cdot \end{cases}$$

Die erste Gleichung (67) sagt aus, daß S_0 zeitlich konstant sein soll ($H_0 = W_0$, also $S_0 W_0 - W_0 S_0 = 0$), d. h. daß es nur Glieder der Form

$$(68) \quad \begin{cases} S_0(n+1, n+1),\ S_0(n+1, n+2),\ldots, S_0(n+1, n+r) \\ S_0(n+2, n+1) \ \cdot \ \cdot \ \cdot \ \cdot \ \cdot \ \cdot \ \cdot \ \cdot \ \cdot \ \cdot \ \cdot \\ \cdot \ \cdot \ \cdot \ \cdot \ \cdot \ \cdot \ \cdot \ \cdot \ \cdot \ \cdot \ \cdot \ \cdot \ \cdot \ \cdot \ \cdot \\ S_0(n+r, n+1) \ \cdot \ \cdot \ \cdot \ \cdot \ \cdot \ \cdot \ \cdot \ \cdot \ \cdot \ \cdot \end{cases}$$

und Diagonalglieder (wegen (66))

$$|S_0(m, m)| = 1 \qquad (m \neq n+1, \ldots, n+r)$$

enthält, aber keine Glieder der Form $S_0(kl)$, wo k, l $(k \neq l)$ Indizes sind, von denen einer nicht in der Reihe $n+1, n+2, \ldots, n+r$ enthalten ist.

Aus der zweiten Gleichung (67) ergibt sich durch zeitliche Mittelbildung über die ungestörte Bewegung

$$(69) \qquad\qquad S_0\,\bar{H}_1\,S_0^{-1} = W_1,$$

wobei $\bar{H}_1$ offenbar wie S_0 Glieder der Form

$$(70) \quad \begin{cases} \bar{H}_1(n+1, n+1),\ \bar{H}_1(n+1, n+2)\ldots \bar{H}_1(n+1, n+r) \\ \cdot \ \cdot \ \cdot \ \cdot \ \cdot \ \cdot \ \cdot \ \cdot \ \cdot \ \cdot \ \cdot \ \cdot \ \cdot \ \cdot \ \cdot \ \cdot \ \cdot \\ \bar{H}_1(n+r, n+1) \ \cdot \ \cdot \ \cdot \ \cdot \ \cdot \ \cdot \ \cdot \ \cdot \ \cdot \ \cdot \end{cases}$$

und solche $\bar{H}_1(m, m)$ enthält.

Es folgt also nach (67)

$$\bar{H}_1(m, m) = W_1(m, m),$$

wenn m keine Zahl der Reihe $n+1, \ldots, n+r$ ist.

                      W. Heisenberg.

Zur Bestimmung der übrigen $W_1(n+1, n+1)$ bis $W_1(n+r, n+r)$ kann man wieder das Verfahren der Hauptachsentransformation (§ 5) anwenden.

Man suche die Lösungen der linearen Gleichungen mit r Unbekannten S_l:

$$(71) \qquad W_1 S_k - \sum_l \overline{H}_1(kl) S_l = 0 \quad \begin{cases} k = n+1, \ldots, n+r \\ l = n+1, \ldots, n+r \end{cases}$$

auf. Dieses System wird lösbar sein für r Werte von W, die sich aus der Gleichung r-ten Grades

$$(72) \quad \begin{vmatrix} [W_1 - \overline{H}_1(n+1, n+1)] & -\overline{H}_1(n+1, n+2) \ldots -\overline{H}_1(n+1, n+r) \\ -\overline{H}_1(n+2, n+1) [W - H_1(n+2, n+2)] \ldots -\overline{H}_1(n+2, n+r) \\ \cdot\cdot\cdot\cdot\cdot\cdot\cdot\cdot\cdot\cdot\cdot\cdot\cdot\cdot\cdot\cdot\cdot\cdot\cdot\cdot\cdot\cdot\cdot\cdot\cdot\cdot \\ \cdot\cdot\cdot\cdot\cdot\cdot\cdot\cdot\cdot\cdot\cdot\cdot\cdot\cdot\cdot\cdot\cdot\cdot\cdot\cdot\cdot\cdot\cdot\cdot\cdot\cdot \\ -\overline{H}_1(n+r, n+1) \ldots\ldots\ldots [W_1 - \overline{H}_1(n+r, n+r)] \end{vmatrix} = 0$$

bestimmen.

Die zu diesen r Werten W_1 gehörigen Lösungen S_l geben dann analog zu (50) die noch fehlenden Komponenten von S_0 und die Lösung von (69) an.

Ist die Bestimmung von W_1 und S_0 durchgeführt, so geschieht die Berechnung von S_1 ganz analog zu Gleichung (64); verschwindende Nenner treten in der rechten Seite von (64) jetzt nicht auf, da wir in (69) nur über die Frequenzen der ungestörten Bewegung gemittelt und die zu den verschwindenden Frequenzen $\nu(n+1, n+2)$ usw. gehörigen Glieder in (69) getrennt behandelt haben.

Das wesentlichste Ergebnis der quantentheoretischen Störungstheorie scheint uns, daß die bekannten Konvergenzeigenschaften der klassischen störungstheoretischen Reihen, die zu den berüchtigten Schwierigkeiten des Dreikörperproblems Anlaß geben, sich in den entsprechenden quantentheoretischen Reihen (64) wahrscheinlich nicht wiederfinden. Zwar tritt auch in (64) die Frequenz $\nu(nm)$ im Nenner auf. Aber diese Frequenzen können im allgemeinen nicht durch Wahl hinreichend hoher geeigneter n, m beliebig klein gemacht werden.

Das Termspektrum wird ja im allgemeinen, wie beim Wasserstoffproblem, eine Häufung der Terme und Grenze im Endlichen haben. Daher wird es zwar möglich sein, durch Wahl zweier Terme in der Nähe der Grenze die $\nu(n, m)$ klein zu machen, doch dies entspricht nur der bekannten Tatsache, daß die Perioden der weit außen liegenden Bahnen sehr lange werden und hat mit jener Konvergenzschwierigkeit nichts zu tun. Um den Unterschied zwischen klassischer und Quantentheorie deutlich zu machen, zeichnen wir jeweils zwei die Terme bzw. die Frequenzen repräsentierende Punktreihen untereinander.

Der klassischen Theorie entspricht es, wenn die Punkte der Reihe äquidistant gewählt werden, da die Frequenzen der Oberschwingungen ganzzahlige Vielfache der Grundfrequenz sind.

1. Klassisch:

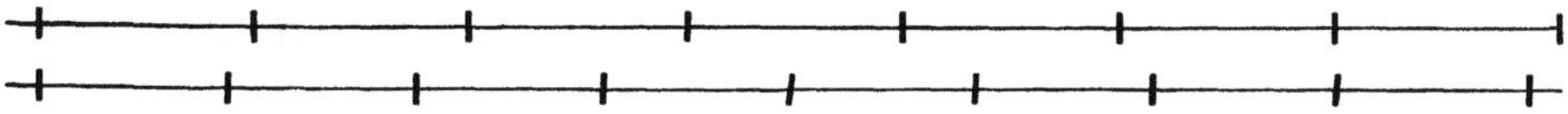

2. Quantentheoretisch:

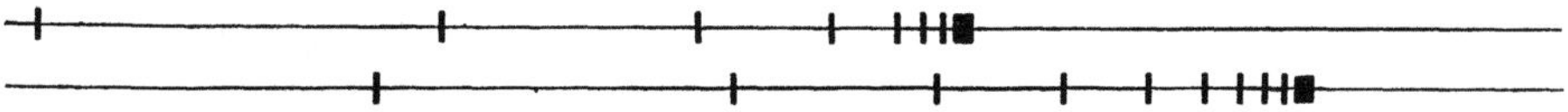

Quantentheoretisch tragen wir die Punkte in einer etwa den Wasserstofftermen entsprechenden Reihe auf. Das Auftreten sehr kleiner Nenner, d. h. kleiner Kombinationsfrequenzen wird dann in der Figur dadurch veranschaulicht, daß zwei Punkte entsprechender Reihen sehr genau übereinander stehen. Nun leuchtet es anschaulich ein, daß man in der Reihe 1, wenn man die Reihe beliebig weit nach rechts verfolgt, immer wieder Punktpaare finden wird, deren Punkte sehr genau übereinander liegen, und zwar bei Nichtkommensurabilität der Grundfrequenzen beliebig genau. In der Reihe 2 aber ist es ebenso anschaulich evident, daß nirgends zwei Punkte sehr genau übereinander liegen. Es gibt hier eine *kleinste* Kombinationsfrequenz.

Allerdings sind auch quantentheoretisch die Verhältnisse in Wirklichkeit komplizierter, als die Figur sie darstellt, da wegen des kontinuierlichen Termspektrums Kommensurabilitäten auftreten können; auch werden neue Komplikationen entstehen, wenn die Häufungsstellen der beiden Punktreihen 2. zusammenfallen. Obwohl wir also von einer strengen Theorie der Konvergenzeigenschaften der quantentheoretischen Reihen (59), (60), (61) weit entfernt sind, scheint es doch wohl nicht zu kühn, wenn wir die Vermutung aussprechen, daß alle im Endlichen verlaufenden Bahnen der Quantentheorie auch periodischen Charakter tragen und daß die Schwierigkeiten des Dreikörperproblems in der quantentheoretischen Mechanik unbekannt sind.

§ 7.

Die Impulssätze der Quantenmechanik; Kritik der Theorie.

Sei ein aus f Massenpunkten bestehendes System gegeben, wobei die kartesischen Koordinaten der Punkte mit $x_k, y_k, z_k; p_{x_k}, p_{y_k}, p_{z_k},$ bezeichnet werden sollen $(k = 1, \ldots, f)$, so folgt aus den Bewegungsgleichungen (30), daß für ein abgeschlossenes System gilt:

$$(73) \qquad \dot{p}_x = \dot{p}_y = \dot{p}_z = 0,$$

wobei

$$p_x = \sum_{k=1}^{f} p_{x_k}; \qquad p_y = \sum_{k=1}^{f} p_{y_k}; \qquad p_z = \sum_{k=1}^{f} p_{z_k}.$$

Denn auch in der Qantentheorie kann die potentielle Energie bei abgeschlossenen Systemen nur von den relativen Koordinaten $x_i - x_k \ldots$ abhängen.

Ein analoger Satz läßt sich für die Komponenten des Drehimpulses $\mathfrak{M}$:

$$(75) \qquad \begin{cases} M_x = \sum_k (y_k p_{z_k} - z_k p_{y_k}) \\[2mm] M_y = \sum_k (z_k p_{x_k} - x_k p_{z_k}) \\[2mm] M_z = \sum_k (x_k p_{y_k} - y_k p_{x_k}) \end{cases}$$

aufstellen. Da nämlich die Energiefunktion in kartesischen Koordinaten additiv in zwei Teile $H = H_1(p) + H_2(q)$ zerfällt, die jeweils nur von den Variablen der einen Sorte abhängen, so wird nach (30) jedes Glied $\dot{p}$ eine Funktion der q allein, jedes Glied $\dot{q}$ eine Funktion der p allein. Also zerfällt der Wert von $\dot{M}_x, \dot{M}_y, \dot{M}_z$, wenn wir die $\dot{p}, \dot{q}$ aus (30) einsetzen, nach (75) additiv in zwei Teile

$$(76) \qquad \dot{M}_x = f_1(p) + f_2(q) \quad \text{usw.}$$

Da aber alle p untereinander, sowie alle q untereinander nach (27) vertauschbar sind, so wird der Ausdruck (76) Null unter denselben Bedingungen, wie der entsprechende Ausdruck in der klassischen Theorie. Daher sind bei abgeschlossenen Systemen die M_x, M_y, M_z zeitlich konstant.

    Es kann wohl als ein wichtiger Erfolg der hier kurz dargestellten Theorie angesehen werden, daß sie einerseits durch alleinige Benutzung der beobachtbaren Frequenzen, Amplituden und Phasen eine enge Analogie ermöglicht zur klassischen wellentheoretischen Beschreibung der Strahlungsvorgänge (vgl. die Bohr-Kramers-Slatersche Theorie der Strahlung), daß sie andererseits wegen der Gültigkeit der Erhaltungssätze (Energie und Impulssatz vgl. § 3 und § 7) auch mit der Einsteinschen Lichtquantentheorie nicht in Widerspruch zu stehen scheint; es ist ferner befriedigend, daß die Grundpostulate der Quantentheorie im formalen Schema dieser Theorie einen sinngemäßen mathematischen Ausdruck finden[13].

---

[13] Es mag hier darauf hingewiesen werden, daß die Theorie auch eine Reihe von Erfahrungstatsachen zu beschreiben gestattet, die nach der bisherigen Theorie nicht gedeutet werden konnten. Vgl l. c. [4]; [6].

Doch muß die hier beschriebene Theorie noch als unvollständig angesehen werden. Der eigentliche geometrische oder kinematische Sinn der Grundannahme (5) ist noch nicht restlos geklärt. Eine erhebliche Schwierigkeit liegt besonders darin, daß die Zeit in der Theorie scheinbar eine andere Rolle spielt und formal anders behandelt wird als die räumlichen Koordinaten. Der formale Charakter der Zeitkoordinate in dem mathematischen Gebäude der Theorie geht besonders deutlich aus der Tatsache hervor, daß in der Theorie bis jetzt die Frage nach dem zeitlichen Ablauf eines Ereignisses keinen unmittelbaren Sinn hat und daß der Begriff des früher oder später kaum exakt definiert werden kann. Trotzdem wird man diese Schwierigkeiten nicht als Einwand gegen die Theorie aufzufassen brauchen, da das Auftreten eben solcher Schwierigkeiten nach dem Wesen der für Atomsysteme gültigen Raum-Zeitverhältnisse durchaus zu erwarten war.

Göttingen, den 21. Dez. 1925, Institut für theor. Physik.

(Eingegangen am 21. 12. 1925.)

Die Naturwissenschaften *14*, 989–994 (1926)

Quantenmechanik [1]).

Von Werner Heisenberg, Kopenhagen.

(Aus dem Institut für theoretische Physik der Universität.)

Nach unserer gewöhnlichen „Anschauung", d. h. bei Anwendung der üblichen Raum-Zeitbegriffe, werden Raum und Materie als in letzter Linie kontinuierlich und *im Prinzip* in beliebig kleine Teile zerlegbar vorgestellt — von der etwaigen technischen Undurchführbarkeit einer solchen Zerlegung abgesehen. Die physikalischen und chemischen Erfahrungen haben aber, entgegen dieser Folgerung aus unserer einfachen Anschauung, ergeben, daß bei den Vorgängen in ganz kleinen Räumen und Zeiten ein typisch diskontinuierliches Element eine hervorragende Rolle spielt. Die „ganzzahligen Proportionen" in der Chemie schon legten den Gedanken an eine atomistische Struktur der Materie nahe, die sog. „Schwankungserscheinungen" (Brownsche Bewegung, Streuung des Lichtes usw.) führten zur Vorstellung vom Aufbau der Materie aus Korpuskeln von *wohldefinierter, endlicher* Größe, in den Versuchen über Korpuskularstrahlen (Kathodenstrahlen, α-, β-Strahlen) gelangten diese kleinsten Bausteine der Materie *direkt* zur Beobachtung. Wegen dieser unmittelbaren experimentellen Evidenz der atomistischen Vorstellungen lag es daher nahe, den Grundbausteinen der Materie — also in letzter Linie dem positiven und dem negativen Elektron — die gleiche Art von Realität zuzusprechen, wie etwa den Gegenständen der uns umgebenden täglichen Welt; man stellte sich also diese Grundbausteine als außerordentlich kleine Körperchen bekannter (und zwar immer ein und derselben) Ladung und Masse, doch noch unbekannter innerer Struktur vor, die sich nach näher zu ergründenden Gesetzen in Raum und Zeit, und zwar in der unserer Anschauung entsprechenden bekannten kontinuierlichen Raum-Zeitwelt — bewegten. Diese Vorstellung hat sich wohl im Laufe der Zeit als falsch erwiesen, was ja auch im Hinblick auf die eigentliche prinzipielle Unanschaulichkeit jenes diskontinuierlichen Elements keineswegs zu verwundern war; die Elektronen bzw. die Atome besitzen *nicht* jenen Grad von unmittelbarer Realität, wie die Gegenstände der täglichen Erfahrung. Die Untersuchung der Art von physikalischer Realität, die den Elektronen und Atomen zukommt, ist eben der Gegenstand der Atomphysik und damit auch der „Quantenmechanik" (Qu.M.). Jenes typisch diskontinuierliche Element, über das wir oben gesprochen haben, findet nicht nur in der Tatsache der atomistischen Struktur der Materie ihren Ausdruck, sondern auch in den Gesetzmäßigkeiten des Atombaues. Aus der Bohrschen Theorie, sowie experimentell aus den Franck-Hertzschen Stoßversuchen und dem Stern-Gerlachschen Molekularstrahlversuch schließen wir die Existenz diskreter stationärer Zustände der Atome, wobei die Übergangsprozesse von einem solchen Zustand zu einem andern als typisch diskontinuierliche Akte zu betrachten sind. Schließlich treffen wir dieses diskontinuierliche Element bei den Strahlungsphänomenen. Dies hat zuerst Planck aus dem von ihm gefundenen Gesetz der schwarzen Strahlung erschlossen, Einstein hat gezeigt, daß eben wieder die Schwankungserscheinungen auf die Vorstellung von „Lichtkorpuskeln" ganz bestimmter wohldefinierter Energie und wohldefinierten Impulses führen und die Experimente über den lichtelektrischen Effekt, der Compton-Effekt und insbesondere das Geiger-Bothesche Experiment über den Compton-Effekt veranschaulichen die Fruchtbarkeit der Lichtquantenhypothese aufs Deutlichste. Trotzdem hat man den Lichtquanten — im Gegensatz zu den Materiepartikeln — von vornherein nie die Art von Realität zugesprochen, die den Gegenständen der täglichen Welt zukommt, da man sich durch eine solche Vorstellung in allzu große Widersprüche mit den bewährten Gesetzen der klassischen Optik verwickelt hätte. Wohl aber bestehen Anzeichen dafür — was besonders von Einstein betont worden ist —, daß umgekehrt den *Elektronen* ein ähnlicher Grad von Realität zukommt, wie den *Lichtquanten;* doch soll auf diese Frage erst später eingegangen werden. Hier kam es uns darauf an, zu betonen, daß die Untersuchung jenes typisch diskontinuierlichen Elementes und jener „Art von Realität" das eigentliche Problem der Atomphysik und daher auch der Inhalt aller quantenmechanischen Überlegungen ist.

I. Durch die Experimente von Lenard und Rutherford und durch die großen Erfolge der Bohrschen Theorie konnte es als bewiesen gelten, daß die Atome aus positiven und negativen Elektronen aufgebaut seien, so wie diese Theorie es annimmt. Gleichzeitig bedeuteten die Bohrschen Grundpostulate der Quantentheorie schon einen endgültigen Bruch mit den Begriffen der klassischen Theorie. Es lag trotzdem nahe, die Benutzung klassischer Begriffe und Bilder so weit zu versuchen, als es logisch möglich schien. Die so entstehende Form der Bohrschen Theorie, die durch das Korrespondenzprinzip eine vollständige *qualitative* Beschreibung fast aller Züge des Atom-

---

[1]) Vortrag, gehalten auf der 89. Versammlung Deutscher Naturforscher und Ärzte, Düsseldorf.

baues bis in die Einzelheiten ermöglichte, genügte jedoch nicht zu einer quantitativen Beschreibung der Atomvorgänge; auch stellten sich bei der Anwendung dieser Form der Theorie auf gewisse Probleme (Dispersion, Wasserstoffatom in gekreuzten Feldern) große gedankliche Schwierigkeiten heraus. Die eigentliche Ursache dieser Schwierigkeiten war eben die dem Wesen der quantentheoretischen Grundpostulate fremde Übertragung klassischer Begriffe und Vorstellungen auf die Probleme des Atombaus, die Benützung einfach anschaulicher Modelle und Bilder zur Deutung physikalischer Gesetzmäßigkeiten, deren anschaulicher Inhalt in Wirklichkeit durchaus nicht zu übersehen war.

Das Programm der Qu.M. mußte daher sein, sich zunächst von diesen anschaulichen Bildern freizumachen und an Stelle der bisher benutzten Gesetze der klassischen Kinematik und Mechanik einfache Beziehungen zu setzen zwischen experimentell gegebenen Größen. Während also die frühere Theorie den Vorteil der unmittelbaren Anschaulichkeit und des Gebrauchs bewährter physikalischer Prinzipien mit dem Nachteil verband, im allgemeinen mit Beziehungen zu rechnen, die prinzipiell nicht prüfbar waren und daher zu inneren Widersprüchen führen konnten, sollte die neue Theorie auf die Anschaulichkeit zunächst ganz verzichten, dafür aber nur ganz konkrete Beziehungen enthalten, die einer unmittelbaren experimentellen Prüfung zugänglich sind, und deswegen kaum in die Gefahr innerer Widersprüche kommen. Um dieses Ziel zu erreichen, mußte man sich von der klassischen Anschauung offenbar sehr weit entfernen: Denkt man etwa an das Spektrum des Wasserstoffatoms, so bestand ja der bekannte Widerspruch, daß das Spektrum *jeder* periodischen Bewegung einer Partikel nach der klassischen Kinematik ein Spektrum äquidistanter Linien sein muß, daß aber in Wirklichkeit ein Linienspektrum mit Häufung der Linien im endlichen beobachtet wird und daß wir trotzdem schon um des Korrespondenzprinzips willen von einer periodischen Bewegung des Elektrons sprechen möchten. Wenn überhaupt die Korpuskularvorstellung beibehalten werden sollte, konnte man dieser Schwierigkeit offenbar nur entgehen, wenn man überhaupt darauf verzichtete, dem Elektron oder dem Atom einen bestimmten Punkt im Raum als Funktion der Zeit zuzuordnen; zur Rechtfertigung muß angenommen werden, daß ein solcher Punkt auch nicht direkt beobachtet werden kann. Dieser Verzicht bedeutet die *erste entscheidende Einschränkung* bei der Diskussion der Realität der Korpuskeln.

An Stelle des aufgegebenen Begriffs vom „*Ort des Elektrons*" versuchte die Qu.M. eine Gesamtheit physikalisch wohldefinierter Größen zu setzen, die in der *klassischen* Theorie dem „Ort des Elektrons" mathematisch äquivalent wäre. Die gesamte Ausstrahlung des Elektrons würde in der klassischen Theorie durch die Fourierentwick-

lung der Bewegung des Elektrons gegeben sein und insofern als Repräsentant der Bewegung des Elektrons angesehen werden können. Frequenz, Amplitude und Polarisation einer Spektrallinie sind jedenfalls wohldefinierte beobachtbare Größen. Daher wurde in der Qu.M. die Gesamtheit der beobachtbaren Strahlungsgrößen, die der klassischen Fourierreihe entsprechen, als Repräsentant des „Ortes des Elektrons" angesehen. Da nach den Grundpostulaten der Quantentheorie die Ausstrahlung einer Linie mit dem Übergang von einem stationären Zustand des Atoms zu einem anderen verknüft ist, so ist jede Strahlungsgröße zwei Termen oder Zuständen zugeordnet. An Stelle der klassischen „Koordinate des Elektrons" trat also in der Qu.M. eine zweidimensionale „*Tabelle*" von Strahlungsgrößen, eine sog. „Matrix".

Es lag der weitere Schritt nahe, an Stelle aller aus der klassischen Theorie übernommenen und vielleicht nicht direkt beobachtbaren Begriffe wie *Impuls, Energie* usw. solche Tabellen von konkreten, beobachtbaren Größen einzuführen. Der Gesamtenergie des Atoms entsprach dann z. B. die Tabelle aller Energiewerte der stationären Zustände des Atoms. Diese eben genannte Annahme bedeutete eine weitere Einschränkung der oben diskutierten Realitätsverhältnisse der Atome. Gleichzeitig ermöglichte diese Annahme eine merkwürdig enge Verbindung der Qu.-M. mit den in den Grundpostulaten der BOHRschen Theorie enthaltenen Zügen von Diskontinuität. Zunächst stellte sich heraus, daß die Existenz diskreter Energiewerte für die Qu.M. ebenso natürlich war, wie etwa die Existenz diskreter Eigenschwingungen einer Membran für die klassischen Theorie. Weiter zeigte sich, daß einfache Beziehungen, die aus Überlegungen etwa über die Häufigkeit von Übergängen oder über Zeitmittelwerte von diskontinuierlich veränderlichen Größen gewonnen waren, auch als mathematisches Resultat des Rechnens mit solchen Tabellen von Strahlungsgrößen gewonnen werden konnten. Dies scheint mir eine der allerwichtigsten Eigenschaften der Qu.M.; leider ist eben diese Eigenschaft der Qu.M. bisher wenig untersucht worden.

Um zu einer geschlossenen Theorie zu kommen, war es noch notwendig, die mathematischen *Beziehungen* zwischen jenen Tabellen von Strahlungsgrößen zu finden, die den Beziehungen der klassischen Mechanik korrespondenzmäßig entsprachen; es stellte sich heraus, daß diese Beziehungen formal sehr einfach waren: Aus physikalischen Analogieschlüssen fand man, daß Addition und Multiplikation dieser Tabellen nach den wohlbekannten Rechenregeln der *Matrizenalgebra* erfolgen mußten, rein *formal* bestand also der Unterschied der neuen von der alten Theorie zunächst nur in der Nichtgültigkeit der Kommutativität der Multiplikation. Die HAMILTONschen Gleichungen der Mechanik konnten der Form nach vollständig in die neue Theorie übernommen werden. Wegen der Nichtkommutativität der

Die Naturwissenschaften *14*, 989–994 (1926)

Multiplikation mußten noch gewisse Vertauschungsrelationen angegeben werden, um das mathematische Schema der Theorie vollständig zu machen. Diese Relationen entsprachen in gewissem Sinne den Quantenbedingungen der früheren Theorie und enthielten als einzige Relationen der neuen Theorie die PLANCKsche Konstante.

Damit ist dann das mathematische Schema der Qu.M. schon vollständig gegeben. Bei seiner näheren Durchführung, auf die ich gleich zu sprechen kommen werde, stellt sich heraus, daß die Qu.M. in vieler Beziehung der klassischen Theorie so ähnlich ist, wie man überhaupt verlangen konnte. Es gelten Energiesatz und Impulssätze, wie in der klassischen Theorie. Es läßt sich eine Theorie der kanonischen Transformationen und damit eine vollständige Störungstheorie entwickeln, die ganz den Methoden der Astronomie entspricht. Die Qu.M. ist sogar in dieser Beziehung viel einfacher als die klassische Mechanik. Die Störungsreihen beim Mehrkörperproblem zeigen in der klassischen Theorie die berüchtigten Konvergenzschwierigkeiten, in der Qu.M. *konvergieren* diese Reihen im allgemeinen und das Mehrkörperproblem hat hier keine prinzipielle Schwierigkeiten.

So mag die Qu.M. wenigstens prinzipiell dem Stand unserer tatsächlichen physikalischen Erfahrungen über Atome ganz weitgehend entsprechen. Für makroskopische Vorgänge geht die Qu.M. formal in die klassische Mechanik über — so, daß sich auch die Realitätsverhältnisse denen unserer gewöhnlichen Anschauung außerordentlich weit annähern können. Für mikroskopische Vorgänge bleiben nur noch Relationen zwischen beobachtbaren experimentell gegebenen Größen übrig, eine unmittelbare anschauliche Deutung kann den zugrunde liegenden physikalischen Vorgängen einstweilen nicht gegeben werden.

Zur experimentellen Prüfung der Theorie liegt ein außerordentlich umfangreiches Material vor: die Spektra aller Elemente, Energiemessungen usw. Zum Vergleich der Theorie mit dem Experiment war aber zunächst ein mathematischer Ausbau der Theorie erforderlich. Er ist auf drei verschiedene Weisen — unabhängig — durchgeführt worden.

1. BORN-JORDAN: Da von vornherein die quantenmechanischen Größen als Matrizen gegeben sind, werden die bekannten Methoden der höheren linearen Algebra angewendet. Die Lösung eines quantenmechanischen Problems wird zurückgeführt auf ein Eigenwertproblem, nämlich eine Hauptachsentransformation einer HERMITEschen Form. Behandelt: Wasserstoffatom, Zeemaneffekt, Starkeffekt (PAULI), Intensitäten beim Zeemaneffekt, Multipletts und anomaler Zeemaneffekt, Feinstruktur, Bandenspektra: Oszillator, Rotator; Dispersionstheorie.

2. DIRAC. Es wird Algebra und Analysis von Größen, für die das kommutative Gesetz nicht gilt („q-Zahlen") — unabhängig von ihrer Deutung als „Matrizen" — soweit ausgebaut, daß einfache Rechenverfahren, die denen der klassischen Theorie weitgehend analog sind, zur Behandlung mechanischer Probleme gefunden werden können (Einführung der Wirkungs- und Winkelvariabeln, Fourierreihen; vgl. auch Arbeiten von LONDON). Behandelt: Wasserstoffatom, Intensitätsformeln bei den Multipletts, g-Werte, relativistische Qu.M. Comptoneffekt, $\frac{h\nu}{c}$ Rückstoß, Dispersion.

3. SCHRÖDINGER. Auf die physikalischen Grundlagen der SCHRÖDINGERschen Theorie werde ich nachher zu sprechen kommen; einstweilen sei nur auf ihre Bedeutung für den mathematischen Ausbau der Qu.M. hingewiesen. Nach bekannten Prinzipien der höheren linearen Algebra und der Analysis ist ein Eigenwertproblem einer unendlichen quadratischen Form, wie das BORN-JORDANsche, im allgemeinen äquivalent mit einem durch eine lineare Differentialgleichung und Randbedingungen charakterisierten analytischen Eigenwertproblem. Wenn es gelang, diese lineare Differentialgleichung und dieses Eigenwertproblem zu finden, so war die mathematische Behandlung quantenmechanischer Probleme zurückgeführt auf außerordentlich weitentwickelte mathematische Disziplinen. Dies wurde durchgeführt von SCHRÖDINGER, der aber zu diesem mathematischen Schema auf einem *anderen*, als dem hier angegebenen Wege und *unabhängig* von den früheren quantenmechanischen Arbeiten gekommen ist. Behandelt: Wasserstoffatom (ohne magnetische Feinstruktur), Starkeffekt, Zeemaneffekt, Bandenspektra, Dispersion. Die Stärke des SCHRÖDINGERschen Verfahrens besteht hauptsächlich darin, daß es eine einfache Bestimmung der Übergangswahrscheinlichkeiten gestattet, was bei den anderen mathematischen Methoden im allgemeinen sehr schwierig sein dürfte. Behandelt: Intensitäten im Starkeffekt des Wasserstoffatoms, Intensitätsformeln bei den Zeemaneffekten. Intensitäten in der LYMAN- und der BALMERserie (PAULI).

II. Kehren wir von dem mathematischen Ausbau der Theorie wieder zur physikalischen Bedeutung dieses Formalismus, also zur Diskussion der Aussagen, die sich über die Realität und die Gesetze der Korpuskeln machen lassen, zurück. Von einer ganz anderen Seite her ist ein Angriff auf dieses Problem unternommen worden durch die Theorie DE BROGLIES, aus der ja SCHRÖDINGERS Überlegungen hervorgegangen sind, und die EINSTEINsche Anwendung der BOSEschen Statistik. Die Einschränkungen, die wir in Teil I über die Realität der Korpuskeln gemacht haben, insbesondere die Aussage, daß es unmöglich sei, einer Korpuskel einen bestimmten Ort als Funktion der Zeit, eine bestimmte Energie usw. zuzuordnen, lassen es schon als möglich erscheinen, daß die Realität der Materiekorpuskeln eine große Ähnlichkeit aufweist mit der Realität der Lichtquanten, denen ja auch wegen der Interferenz- und Beugungsphänomene der Lichtwellen nicht eine bestimmte Bahn und ein bestimmter

Ort zugeschrieben werden kann. Eine Analogie im Verhalten von Materiekorpuskeln und Lichtquanten wird besonders deutlich, wenn man die Reflexion eines Materieteilchens etwa an einem Beugungsgitter quantenmechanisch untersucht. Das Materieteilchen erleidet dann, wie vor kurzem auf Grund einer Überlegung DUANES von JORDAN gezeigt worden ist, nach der Qu.M. Reflexion an dem Gitter nur in ganz bestimmten, diskreten Richtungen, ebenso wie ein Lichtstrahl in ganz bestimmten diskreten Richtungen abgebeugt wird. Ähnliche Analogien hatten DE BROGLIE zu folgender Annahme geführt, und zwar lange *vor* dem Entstehen der Qu.M.: Ebenso, wie bei der Theorie des Lichtes einstweilen jener merkwürdige Dualismus besteht, nach dem viele Erscheinungen durch die Wellentheorie des Lichtes, andere durch die Lichtquantentheorie, manche durch beide Theorien beschrieben werden können, so mag ein solcher Dualismus auch bei den Materiepartikeln einstweilen berechtigt sein; DE BROGLIE ordnete also jedem Materieteilchen eine Welle bestimmter Frequenz zu; diese Frequenz bestimmte sich, wie bei den Lichtquanten, aus der Energie der Partikeln nach der $h\nu$-Relation. Nach EINSTEIN sollten diese Wellen, ebenso wie Lichtwellen, zu Interferenzen Anlaß geben; eine Schar von Elektronen sollte also bei der Reflexion an einem Gitter eben in bestimmten diskreten Richtungen abgebeugt werden. Ein solches Ergebnis wurde später — wie oben erwähnt — auch auf Grund der Qu.M. gefunden und dieser Sachverhalt schon legt den Gedanken nahe, daß die Qu.M. und die DE BROGLIEschen Wellen in engstem Zusammenhang stehen. DE BROGLIE wies auch schon darauf hin, daß man die BOHRschen Bahnen im Wasserstoffatom erhalten könnte, wenn man forderte, daß der dem Elektron entsprechende Wellenzug um den Kern herum eine eindeutige Funktion des Raumes sei. Der eigentliche Zusammenhang der DE BROGLIEschen Theorie und der Qu.M. wurde aufgedeckt durch SCHRÖDINGER. Dieser Forscher hat die von DE BROGLIE und EINSTEIN herrührenden Gedanken weiter ausgebaut, indem er die Differentialgleichung der DE BROGLIEschen Wellen aufstellte und zeigte, daß das ihr entsprechende Eigenwertproblem eben das Eigenwertproblem der Qu.M. sei. Dabei hat es sich allerdings herausgestellt, daß eine dreidimensionale Wellengleichung, wie in der Theorie des Lichtes, in der Theorie der Materie nicht angegeben werden konnte, da ja die Wellengeschwindigkeit immer noch von der Anwesenheit anderer Partikel beeinflußt werden kann. Dagegen war es möglich, bei einem Problem über die Bewegung von f-Korpuskeln eine Wellengleichung im Koordinatenraum von 3·f-Dimensionen aufzustellen, die dann das quantenmechanische Problem mathematisch völlig ersetzte. Es ist bisher nicht allgemein gelungen, einen direkten Zusammenhang der SCHRÖDINGERschen Wellen im Phasenraum mit den DE BROGLIEschen Wellen im gewöhnlichen Raum, die den Lichtwellen analog sein sollten, herzustellen. Die

Wellen im q-Raum haben also bis jetzt nur eine formale Bedeutung. Ebenso, wie seinerzeit eine große formale Ähnlichkeit der klassischen Mechanik mit einer geometrischen Optik in mehrdimensionalen Räumen von HAMILTON aufgedeckt und zur Grundlage für die wirksamste mathematische Behandlungsweise klassischer Probleme ausgebaut wurde, so besteht nach SCHRÖDINGER eine große formale Ähnlichkeit der Qu.M. mit einer Wellenoptik in mehrdimensionalen Räumen, die auch hier zur wirksamsten mathematischen Behandlungsweise quantenmechanischer Probleme führt. Es ist in letzter Zeit manchmal (SCHRÖDINGER, FLAMM) die Vermutung ausgesprochen worden, daß auf Grund der SCHRÖDINGERschen Differentialgleichung eine rein kontinuierliche Beschreibung der quantentheoretischen Erfahrungen im Sinne etwa der klassischen Theorie möglich sei, daß also die Quantentheorie in ihrer bisherigen Form illusorisch sei. Bei einer konsequenten Durchführung dieses Gesichtspunktes verläßt man aber eben die Grundlagen der DE BROGLIEschen Theorie, damit der Qu.M. und aller Quantentheorie überhaupt und gerät meines Erachtens in vollständigen Widerspruch mit der Erfahrung (Gesetz der schwarzen Strahlung; Dispersionstheorie). Dieser Weg ist also nicht gangbar. Die eigentliche Realität der DE BROGLIEschen Wellen liegt vielmehr in den oben genannten Interferenzphänomenen, die jeder Deutung auf Grund klassischer Begriffe spotten. Die außerordentliche physikalische Bedeutung der SCHRÖDINGERschen Ergebnisse liegt in der Feststellung, daß eine anschauliche Interpretation der quantenmechanischen Formeln sowohl typische Züge einer Korpuskulartheorie, wie typische Züge einer Wellentheorie enthält.

III. EINSTEIN hat, ausgehend von der BOSEschen Statistik der Lichtquanten, eine Statistik der Materiepartikel vorgeschlagen, die die Realität der DE BROGLIEschen Wellen noch von einer anderen Seite her beleuchtet. BOSE hatte gezeigt, daß man zu einer mit der Erfahrung übereinstimmenden Statistik der Lichtkorpuskeln kommen könnte, wenn man darauf verzichtete, den Ort eines Partikels im Phasenraum zur Bestimmung des „Zustandes" zu verwenden, und wenn man statt dessen einen Zustand als durch die Angabe, *wieviele* gleichartige Partikel sich in einer Zelle des Phasenraums befinden, bestimmt ansieht. Diese Grundannahme der BOSEschen Statistik, die in einer Korpuskulartheorie zunächst keinen Platz zu haben scheint und jedenfalls eine sehr merkwürdige weitere Einschränkung der Realität der Korpuskeln bedeutet, wird einigermaßen verständlich, wenn·man von den Lichtkorpuskeln zu den in irgendeiner bisher nicht bekannten Weise „entsprechenden" Lichtwellen übergeht; an Stelle der „Zahl der Korpuskeln" tritt dann die „Energie einer Eigenschwingung" als Bestimmungsstück eines Zustandes, was durchaus der gewöhnlichen Statistik entspricht. Diese Grundannahme der BOSEschen Statistik hat EINSTEIN direkt auf die Statistik gleicher Materieteilchen übertragen; vom

Die Naturwissenschaften *14*, 989–994 (1926)

Standpunkt der DE BROGLIESchen Wellentheorie aus ist eine solche Statistik wieder verständlich; in einer Korpuskulartheorie bedeutet sie, daß es im allgemeinen nicht möglich sei, eine Korpuskel auf ihrem Wege zu verfolgen und wiederzuerkennen d. h. die Individualität einer Korpuskel kann verloren gehen. Eine solche Annahme liegt durchaus im Sinn der oben gelegentlich der Ableitung der Grundlagen der Quantenmechanik gemachten Einschränkungen bei Begriffen, wie Ort des Elektrons usw. Trotzdem stand dieser EINSTEINschen Statistik die Qu.M. zunächst fremd gegenüber. Denn da die Qu.M. mit Korpuskeln rechnet bzw. bei SCHRÖDINGER mit Wellen in 3-f-dimensionalen Räumen, so ergibt eine Abzählung der Zustände z. B. eines Atoms immer zunächst das der *klassischen* Statistik entsprechende Resultat.

Um diesen Widerspruch aufzuklären, wurden die bei den Atomsystemen auftretenden Mehrkörperprobleme genauer untersucht. Zwar handelt es sich hier um ein mechanisches Problem, das von dem der EINSTEINschen Statistik zugrunde liegenden verschieden ist. Man konnte aber doch erwarten, die wesentlichen Züge, die die EINSTEINsche Statistik von der klassischen unterscheiden, auch hier wiederzufinden. Zunächst zeigte sich — unter diesem Vorbehalt — daß die experimentell gefundene Anzahl von stationären Zuständen bei Atomen mit mehreren Elektronen eine eindeutige Entscheidung *zugunsten* einer Reduktion der statistischen Gewichte im EINSTEINschen Sinne und *zuungunsten* der klassischen Statistik lieferte. Wäre die klassische Statistik richtig, so wäre die Anzahl der stationären Zustände um ein vielfaches größer, als die beobachtete. Weiter zeigte sich, daß die quantenmechanische Lösung eines Mehrkörperproblems eine charakteristische Unbestimmtheit aufweist: Das gesamte Termsystem oder Termspektrum des Problems zerfällt in verschiedene *Teil*systeme. Unter diesen Teilsystemen ist eines, das keine „äquivalenten Bahnen" enthält und von den anderen in der Weise getrennt ist, daß auf keine Weise Übergänge von Termen dieses Teilsystems zu Termen anderer Teilsysteme vorkommen können. Nun kann sowohl das gesamte Termsystem, als auch dieses Teilsystem als vollständige quantentheoretische Lösung angesehen werden. Denn von einer quantenmechanischen Lösung eines Problems wird ja nur verlangt, daß das ihr entsprechende Termsystem „geschlossen" sei, d. h. daß es nur Übergänge innerhalb des Systems geben dürfte. Wählt man nun, zunächst ohne eigentliche Begründung, dieses eine Teilsystem als endgültige quantenmechanische Lösung aus, so reduziert man damit einerseits die statistischen Gewichte eben in dem von EINSTEIN vorgeschlagenen Sinne und erfüllt andererseits von selbst PAULIS Verbot äquivalenter Bahnen. Wie weit das PAULIsche Verbot umgekehrt eine Modifikation der EINSTEINschen Statistik notwendig macht, ließ sich aus dieser Untersuchung nicht entscheiden. Ein enger Zusammenhang mit der

EINSTEINschen Statistik ist daran zu erkennen, daß diese Reduktion der statistischen Gewichte nur möglich ist bei völliger Gleichheit der Partikel des Mehrkörperproblems (bei Elektronen ist diese Gleichheit natürlich gegeben). Bei auch sehr kleinen Verschiedenheiten der Partikel treten *Übergänge* zwischen den genannten Teilsystemen auf, als quantenmechanische Lösung kommt dann nur das *gesamte* Termsystem in Betracht — was der klassischen Abzählung entspricht. Der mechanische Grund für die oben diskutierte Einteilung des Termspektrums in Teilsysteme, die nicht miteinander kombinieren, ist ein charakteristisches Resonanzphänomen, das darin besteht, daß in allen Mehrkörperproblemen die das System bildenden Partikel kontinuierlich die Plätze tauschen; z. B. ist eine Unterscheidung zwischen inneren und äußeren Elektronen bei Atomen sinnlos. Wenn, wie oben angegeben, das eine Teilsystem als quantentheoretische Lösung ausgewählt wird, so bedeutet dies nach dem Gang der Rechnung, daß nur symmetrische Funktionen der Elektronen des Atoms physikalischen Sinn haben, daß es nicht möglich ist, über die Bewegung eines einzelnen Elektrons, auch nicht über die diese Bewegung repräsentierende Matrix zu sprechen. Dies bedeutet eine neue Einschränkung in der Frage nach der Realität der Korpuskeln. Als Beispiel für Mehrkörperprobleme sind bisher untersucht worden die Spektra der Atome mit zwei Elektronen, also He und $\overset{+}{\text{Li}}$ mit befriedigender Übereinstimmung mit der Erfahrung.

Die verschiedenen Aussagen, die sich nach den hier besprochenen Überlegungen über das typisch diskontinuierliche Element bei Vorgängen in kleinen Dimensionen, über die Korpuskeln und ihre Realitätsverhältnisse machen lassen, sollen noch einmal kurz zusammengefaßt werden:

1. Aus allen Versuchen über α- und β-Strahlen, WILSONaufnahmen, Molekularstrahlen usw. folgt die unmittelbare experimentelle Evidenz der Materiekorpuskeln. Ähnlich lassen die Experimente über den lichtelektrischen Effekt, der GEIGER-BOTHEsche Versuch (vgl. auch den neueren Versuch von BOTHE und die Untersuchungen von KIRCHNER) direkt die Realität der Lichtquanten in Erscheinung treten. Die Existenz *diskreter* stationärer Zustände bei den Atomen wird durch die FRANCK-HERTZschen Stoßversuche und die STERN-GERLACHschen Molekularstrahlexperimente dargetan.

2. Es ist nicht möglich, einer Korpuskel einen bestimmten Ort als Funktion der Zeit zuzuordnen, doch kann ihr eine Gesamtheit von Strahlungsgrößen, die die Fourrierreihe der klassischen Theorie ersetzen, zugeordnet werden. Es ist ferner unter einer Reihe gleichartiger Korpuskeln prinzipiell nicht möglich, eine bestimmte Korpuskel immer wieder zu identifizieren.

3. Es besteht in unserer anschaulichen Interpretation des physikalischen Geschehens und der

mathematischen Formeln ein Dualismus zwischen Wellentheorie und Korpuskulartheorie derart, daß viele Phänomene am natürlichsten durch eine Wellentheorie sowohl des Lichtes wie der Materie beschrieben werden, insbesondere Interferenz- und Beugungsphänomene, während andere Phänomene wieder nur auf Grund der Korpuskulartheorie gedeutet werden können.

Diese Feststellungen sollten den jetzigen Stand unseres Wissens über das bei den Vorgängen in sehr kleinen Räumen auftretende typisch diskontinuierliche Element in groben Umrissen wiedergeben. Die in dem bisherigen Schema enthaltenen Widersprüche der anschaulichen Deutungen verschiedener Phänomene sind ganz unbefriedigend. Zu einer widerspruchsfreien anschaulichen Interpretation der ja an sich widerspruchsfreien Experimente fehlt bis jetzt noch irgendein wesentlicher Zug in unserem Bilde vom Bau der Materie.

Über den Vagusstoff.

Von O. Loewi, Graz.

(Aus dem pharmakologischen Institut der Universität.)

Im Jahrgang 1922 (Heft 3) dieser Wochenschrift wurde über den damaligen Stand meiner Untersuchungen „über die humorale Übertragbarkeit der Herznervenwirkungen" berichtet. Über Wunsch der Schriftleitung seien im folgenden die Ergebnisse derjenigen seitdem durchgeführten Untersuchungen kurz mitgeteilt, die der Erforschung der Natur des „Vagusstoffs" galten.

Wir fanden zunächst[1]), daß der Vagusstoff, wie wir der Kürze halber den noch nicht isolierten chemischen Träger der Vaguswirkung in den Herzextrakten und -diffusaten bezeichnen, durch Collodiummembranen dialysiert, alkaliempfindlich aber säureunempfindlich ist und bei saurer Reaktion des Mediums bei $40°$ ohne, bei $100°$ mit sehr geringem Wirkungsverlust zur Trockne gebracht werden kann. Der trockne Stoff ist auch nicht spurenweise in Äther, wohl aber völlig in absolutem Alkohol löslich. Was die Art des Stoffes anbetrifft, so konnten wir in Bestätigung unserer früheren Versuche[2]) erneut feststellen, daß der Vagusstoff nicht Cholin ist; denn eine Cholinmenge von gleicher Wirkungsstärke wie der Stoff ist nach der Acetylierung viel stärker wirksam als der acetylierte Stoff. Seine Natur aufzudecken konnten wir mit Rücksicht auf die nach unseren Erfahrungen aus welchem Material immer zu gewinnende minimale Ausbeute nicht daran denken, ihn zu isolieren und rein darzustellen. Wir mußten also versuchen auf anderem Weg zum Ziel zu gelangen. Den Ausgangspunkt für diesen Weg bildete die Tatsache, daß die Wirkung ins Herz eingefüllten Vagusstoffs sehr rasch schwindet[3]), und daß nach dem Wirkungsschwund die Herzfüllung völlig unwirksam ist, also keinen Vagusstoff mehr enthält. Am nächsten lag es als Ursache des Schwundes Zerstörung des Vagusstoffs durch das Herz anzunehmen.

Zu prüfen, ob das Herz Vagusstoff zerstört, gingen wir derart vor[4]), daß wir ein wässeriges Herzextrakt darstellten, dies mit Vagusstoff beschickten und bei $18°$ bzw. $38°$ stehen ließen: die zu verschiedenen Zeiten vom Moment der Mischung an angestellte Prüfung ergab in der Tat eine mit der Zeit zunehmende Abnahme der Wirksamkeit; das Herz ist also fähig Vagusstoff zu zerstören. Wodurch bewirkt das Herz bzw. sein wässeriger Extrakt diese Zerstörung? Es stellte sich heraus, daß alle Eingriffe, die Fermente unwirksam machen, als Erhitzen auf $56°$, Bestrahlen mit Fluorescenz- und Ultraviolettlicht auch die Fähigkeit des Herzextraktes Vagusstoff zu zerstören aufheben: die Zerstörung ist also durch ein Ferment bedingt.

Es kam nun alles darauf an, festzustellen, welcher Art die Wirkung dieses Fermentes ist, weil die Klarlegung derselben auch zur Klarlegung des Charakters des Vagusstoffs beitragen mußte.

Schon von Anfang an hatte ich an die Möglichkeit gedacht, daß der Vagusstoff ein Cholinester sei[1]). Diese Anschauung wurde später[2]) gestützt durch die Beobachtung, daß Wirkungsbild und Wirkungsablauf einer bestimmten Vagusstoffmenge und einer damit gleich wirksamen Acetylcholinmenge identisch sind.

Wir fanden nun ferner[3]), daß ebenso wie Vagusstoff auch Acetylcholin durch Stehenlassen mit Herzextrakt zerstört wird. Nach Reacetylierung fanden wir das gesamte im ursprünglichen Acetylcholin gebunden gewesene Cholin wieder: daraus geht mit Sicherheit hervor, daß das Acetylcholin und Vagusstoff zerstörende Ferment eine Esterase ist, wonach mindestens der Rückschluß gestattet ist, daß der Vagusstoff ein Ester ist.

In den nunmehr zu besprechenden Versuchen[3]) haben wir noch eine weitere Analogie im Verhalten dieser beiden Stoffe aufgedeckt: bekanntlich wird der Erfolg der Vagusreizung durch vorgängige Verabfolgung von Physostigmin wesentlich verstärkt, vor allem verlängert. Nachdem die Vagusreizung durch den Vagusstoff bedingt ist,

[1]) VIII. Mitteilung: Pflügers Arch. f. d. ges. Physiol. 208, 694. 1925.

[2]) II. Mitteilung: Pflügers Arch. f. d. ges. Physiol. 193, 201. 1921.

[3]) VII. Mitteilung: Pflügers Arch. f. d. ges. Physiol. 206, 135. 1924.

[4]) X. Mitteilung: Pflügers Arch. f. d. ges. Physiol. im Druck.

[1]) II. Mitteilung: Pflügers Arch. f. d. ges. Physiol. 193, 201. 1921.

[2]) X. Mitteilung: Pflügers Arch. f. d. ges. Physiol. im Druck.

[3]) XI. Mitteilung: Pflügers Arch. f. d. ges. Physiol. im Druck.

In *Électrons et Photons: Rapports et Discussions du Cinquième Conseil de Physique Tenu a Bruxelles du 24 au 29 Octobre 1927 sous les Auspices de l'Institut International de Physique Solvay,* ed. by Institut International de Physique Solvay (Gauthier-Villars, Paris 1928) pp. 143–184

LA

MÉCANIQUE DES QUANTA

Par MM. Max BORN et Werner HEISENBERG

INTRODUCTION.

La mécanique des quanta est fondée sur cette idée que la physique atomique se distingue essentiellement de la physique classique par l'existence de discontinuités (*voir* spéc. [1, 4, 58-63])[1]. La mécanique des quanta doit être considérée comme une extension directe de la théorie des quanta, établie par Planck, Einstein et Bohr. Déjà avant la naissance de la mécanique des quanta, Bohr surtout avait insisté souvent sur ce point que ces discontinuités devaient conduire à l'introduction de nouvelles notions en cinématique et en mécanique, de sorte que de toutes façons la mécanique classique et le système de concepts qui y correspondait devaient être abandonnés [1, 4]. La mécanique des quanta essaie d'introduire les nouvelles notions par une analyse précise de ce qui est « essentiellement observable ». Cela ne revient pas à établir le principe qu'il est possible et même nécessaire de faire une séparation nette entre ce qui est « observable » et ce qui est « inobservable ». Dès qu'un système de concepts est donné, on peut conclure des observations à d'autres faits qui à proprement parler ne sont pas directement observables, et la limite entre ce qui est observable et ce qui ne l'est pas devient tout à fait indéterminée. Mais lorsque le système des concepts est lui-même

---

[1] Les nombres entre crochets se rapportent à la bibliographie donnée à la fin.

encore inconnu, on ne s'intéresse tout naturellement qu'aux observations elles-mêmes, sans en tirer des conclusions, parce qu'autrement des idées fausses et d'anciens préjugés s'opposent à l'intelligence des relations physiques. Le nouveau système de concepts donne en même temps le contenu intuitif de la nouvelle théorie. D'une théorie intuitive en ce sens on doit donc demander uniquement qu'elle soit en elle-même sans contradiction et qu'elle permette de prédire sans ambiguïté les résultats de toutes les expériences imaginables dans son domaine. La mécanique des quanta sera dans ce sens une théorie intuitive et complète des processus micromécaniques [47].

Deux espèces de discontinuités sont caractéristiques pour la physique de l'atome : l'existence de corpuscules (électrons, quanta de lumière) d'une part, l'existence d'états stationnaires séparés (valeurs déterminées de l'énergie, valeurs de l'impulsion etc.) d'autre part. Les deux espèces de discontinuités ne peuvent être introduites dans la théorie classique que par des hypothèses auxiliaires fort artificielles. Pour la mécanique des quanta l'existence d'états stationnaires et de valeurs d'énergie déterminés est tout aussi naturelle que l'est, par exemple, l'existence de vibrations caractéristiques déterminées dans un problème de vibrations classique [4]. Il est possible que plus tard l'existence des corpuscules pourra être ramenée d'une façon tout aussi peu forcée à des états stationnaires séparés des processus ondulatoires (quantification des ondes électromagnétiques d'une part, des ondes de de Broglie d'autre part) [4], [54].

Ainsi que le montre déjà la notion de « probabilités de transformation », les discontinuités introduisent dans la physique de l'atome un élément statistique. Cet élément statistique constitue une partie *essentielle* des bases de la mécanique des quanta (*voir* spécialement [4, 30, 38, 39, 46, 61, 62]); d'après celle-ci l'allure d'une expérience ne peut, dans beaucoup de cas, être déduite des conditions initiales que d'une manière statistique; du moins, si l'on considère, pour établir les conditions initiales, uniquement les expériences imaginables en principe jusqu'ici. Cette conséquence de la mécanique des quanta peut être contrôlée empiriquement. Malgré son caractère statistique la théorie rend compte de la détermination en apparence tout à fait causale des processus microscopiques. En particulier les principes de conservation de l'énergie et de la quantité de mouvement sont exacts même dans la mécanique·des quanta. Il semble donc qu'il n'y ait aucun argument d'ordre empirique qui s'oppose à ce qu'on admette en principe l'indétermination du microcosme.

I. — LES MÉTHODES MATHÉMATIQUES DE LA MÉCANIQUE DES QUANTA.

Le phénomène, à l'étude duquel le formalisme mathématique de la mécanique des quanta doit son premier développement, est le rayonnement libre d'un atome excité. Lorsque d'innombrables tentatives faites en vue d'expliquer par des modèles empruntés à la mécanique classique la structure des spectres de raies se furent montrées infructueuses, on en revint à la description immédiate du phénomène basée sur ses lois empiriques les plus simples (Heisenberg [1]). En tête de ces lois se trouve le principe de combinaison de Ritz, d'après lequel la fréquence de chaque raie spectrale d'un atome se présente comme différence de deux termes : $\nu_{ik} = T_i - T_k$; l'ensemble de toutes les raies de l'atome se décrit donc le mieux au moyen d'un schéma en forme de carré, et comme chaque raie possède en dehors de sa fréquence encore une intensité et une phase, on doit écrire à chaque place du schéma une fonction vibratoire élémentaire à amplitude complexe :

$$(1) \qquad \begin{vmatrix} q_{11}\, e^{2\pi i \nu_{11} t} & q_{12}\, e^{2\pi i \nu_{12} t} & \dots \\ q_{21}\, e^{2\pi i \nu_{21} t} & q_{22}\, e^{2\pi i \nu_{22} t} & \dots \\ \dots\dots\dots & \dots\dots\dots & \dots \end{vmatrix}.$$

Ce schéma définit une coordonnée q comme fonction du temps de la même manière, à peu près, que dans la théorie classique l'ensemble des termes de la série de Fourier

$$q(t) = \sum_n q_n\, e^{2\pi i \nu_n t} \qquad \text{avec } \nu_n = n\nu_0;$$

seulement, à cause des deux indices, la sommation n'a maintenant plus aucun sens.

On peut se demander quelles configurations correspondent à des fonctions des coordonnées classiques, par exemple au carré q^2. Or, en mathématiques de pareils schémas ordonnés suivant deux indices se présentent dans la théorie des formes quadratiques et des transformations linéaires comme *matrices*; à la composition de deux transformations linéaires,

$$x_k = \sum_l a_{kl} y_l, \qquad y_l = \sum_j b_{lj} z_j$$

en une nouvelle

$$x_k = \sum_j c_{kj} z_j$$

correspond alors la composition ou la multiplication des matrices

$$(2) \qquad ab = c, \qquad \text{c'est-à-dire} \qquad \sum_l a_{kl} b_{lj} = c_{kj}.$$

En général, cette multiplication *n'est pas* commutative.

Il est tout indiqué d'appliquer cette recette au schéma des vibrations atomiques (Born et Jordan [2], Dirac [3]); on trouve alors immédiatement qu'en vertu de la formule $\nu_{ik} = T_i - T_k$ il n'apparaît pas de nouvelles fréquences, pas plus que dans la théorie classique, lorsqu'on multiplie entre elles deux séries de Fourier, et c'est là une première justification du procédé. Par une application répétée d'additions et de multiplications, on peut définir des fonctions de matrices quelconques.

L'analogie avec la théorie classique conduit ensuite à considérer comme représentants de grandeurs réelles uniquement ces matrices là qui sont « hermitiques », c'est-à-dire dont les éléments passent, par permutation des indices, dans les nombres complexes conjugués. Ici la nature discontinue des processus atomiques est introduite dès le début dans la théorie comme un fait établi par l'expérience.

Mais, par là n'est pas encore établie la relation avec la théorie des quanta et sa constante caractéristique h. Or, cela aussi réussit par une extension logique des conditions de quanta de Bohr et Sommerfeld sous une forme donnée par Kuhn et Thomas, dans laquelle ces conditions sont écrites sous forme de relation entre les coefficients de Fourier des coordonnées q et des quantités

LA MÉCANIQUE DES QUANTA.   147

de mouvement p. On obtient de cette manière l'équation de matrices :

$$(3) \qquad pq - qp = \frac{h}{2\pi i}\,\mathbf{1},$$

où $\mathbf{1}$ représente la matrice unité. La matrice p n'est donc pas permutable avec q. Dans le cas d'un nombre plus grand de degrés de liberté la relation de commutation (3) est valable pour chaque paire de grandeurs conjuguées, tandis que les q_k sont permutables entre eux ainsi que les p_k entre eux et aussi les p_k avec les q_k non correspondants.

Pour édifier sur ces bases la nouvelle mécanique (Born, Heisenberg et Jordan [4]), on étend le plus possible les notions de la théorie classique. On parvient à définir la différentiation d'une matrice par rapport au temps et celle d'une fonction de matrices par rapport à une matrice argument. On peut donc étendre à la théorie des matrices les équations canoniques

$$\frac{dq}{dt} = \frac{\partial \mathcal{H}}{\partial p}, \qquad \frac{dp}{dt} = -\frac{\partial \mathcal{H}}{\partial q};$$

on doit alors entendre par $\mathcal{H}(p, q)$ la même fonction des matrices p et q qui se présente dans la théorie classique comme fonction des nombres p et q ([1]). Ce procédé fut vérifié en l'appliquant à des exemples simples (oscillateurs harmonique et anharmonique). Ensuite, on peut démontrer le principe de l'énergie, qui prend ici, pour des systèmes non dégénérés (tous les termes T_k sont différents les uns des autres, ou bien toutes les fréquences ν_{ik} sont différentes de zéro), la forme suivante. Pour les solutions p, q des équations canoniques la fonction hamiltonienne $\mathcal{H}(p, q)$ devient une matrice diagonale W. Il en résulte alors tout naturellement que les éléments de cette matrice diagonale représentent les termes T_n, multipliés par h, de la formule de Ritz (condition de fréquence de Bohr). Particulièrement importante est la constatation qu'inversement la condition

$$\mathcal{H}(p, q) = \mathrm{W} \qquad \text{(matrice diagonale)}$$

([1]) Il peut évidemment se présenter une ambiguïté à cause de la non-commutativité de la multiplication, c'est ainsi que $p^2 q$ n'est pas la même chose que pqp.

est un remplaçant parfait des équations canoniques du mouvement et conduit à des solutions déterminées sans ambiguïté même dans le cas où l'on admet des dégénérescences (égalité de termes, disparition de fréquences).

Au moyen d'une matrice dont les éléments sont des fonctions harmoniques du temps, on ne peut évidemment représenter que des grandeurs (coordonnées) qui correspondent à des grandeurs périodiques dans le temps de la théorie classique. Des coordonnées cycliques (angles), qui croissent proportionnellement au temps, ne peuvent donc pas encore être traitées de cette façon pour le moment. On réussit néanmoins aisément à soumettre à la méthode des matrices des systèmes tournants en représentant par des matrices les composantes rectangulaires de la quantité de mouvement de rotation. On obtient alors pour l'énergie des expressions qui s'écartent d'une manière caractéristique des expressions classiques correspondantes; ainsi, par exemple, l'impulsion de rotation totale n'est pas égale à

$$\frac{h}{2\pi} j \qquad (j = 0,\ 1,\ 2,\ \ldots),$$

mais à

$$\frac{h}{2\pi} \sqrt{j(j+1)},$$

en concordance avec les règles empiriques que Landé et d'autres déduisirent des décompositions de termes dans l'effet Zeeman. Ensuite, on obtient pour les changements des nombres quantiques de rotation, les bonnes règles de sélection et les formules d'intensité, telles qu'elles furent déjà établies antérieurement par des considérations de correspondance, et vérifiées par les observations faites à Utrecht.

Pauli [6] réussit même, en évitant les variables angulaires, à faire au moyen de la mécanique des matrices le calcul complet pour l'atome d'hydrogène, du moins pour ce qui concerne les valeurs de l'énergie et quelques traits des intensités.

La question de savoir quelles sont les coordonnées les plus générales pour lesquelles les lois de la mécanique quantique sont valables conduit à la généralisation des notions de variables canoniques et de transformations canoniques de la théorie classique. Dirac [3] a observé que les expressions comme

$$\frac{2\pi i}{h}(p_k q_l - q_l p_k) - \delta_{kl},$$

qui figurent dans les relations de commutation du type (3), sont des extensions logiques des symboles à parenthèses de Poisson, dont la disparition caractérise dans la mécanique classique un système de variable comme canonique. Voilà pourquoi, on doit aussi qualifier de canoniques dans la mécanique des quanta tout système de paires de matrices q et p qui satisfait aux relations de commutation, et toute transformation qui laisse cette relation invariante est une transformation canonique. Au moyen d'une matrice quelconque S de pareilles transformations peuvent être mises sous la forme

$$(4) \qquad \mathcal{P} = S^{-1} p\, S, \qquad \mathcal{Q} = S^{-1} q\, S$$

et dans un certain sens on a ainsi la transformation canonique la plus générale. On a, d'ailleurs, pour une fonction quelconque

$$f(\mathcal{P}, \mathcal{Q}) = S^{-1} f(p, q)\, S.$$

Maintenant, on peut passer aussi à l'extension de l'idée fondamentale de la théorie de Hamilton-Jacobi [4]. En effet, si la fonction hamiltonienne $\mathcal{H}$ est donnée comme fonction de certaines matrices canoniques connues p_0 et q_0, la solution du problème mécanique défini par $\mathcal{H}$ revient à trouver une matrice S qui satisfait à l'équation

$$(5) \qquad S^{-1} \mathcal{H}(p_0, q_0)\, S = W.$$

C'est l'analogue de l'équation différentielle de Hamilton-Jacobi de la mécanique classique.

Tout comme dans cette mécanique classique la théorie des perturbations peut être traitée ici de la manière la plus intuitive à l'aide de l'équation (5). Si $\mathcal{H}$ est développé en une série de puissances par rapport à un petit paramètre

$$\mathcal{H} = \mathcal{H}_0 + \lambda \mathcal{H}_1 + \lambda^2 \mathcal{H}_2 + \ldots$$

et si le problème mécanique est résolu pour $\lambda = 0$, c'est-à-dire si $\mathcal{H}_0 = W_0$ est connu comme matrice diagonale, la solution

 ÉLECTRONS ET PHOTONS.

de (5) peut s'obtenir aisément sous forme d'une série de puissances

$$\mathscr{S} = 1 + \lambda \mathscr{S}_1 + \lambda^2 \mathscr{S}_2 + \ldots$$

par approximations successives. Parmi les nombreuses applications de ce procédé, je ne rappellerai ici que la déduction de la formule de dispersion de Kramers, que l'on obtient en supposant que le système émetteur de lumière et le système diffusant sont faiblement couplés et en calculant la perturbation du dernier système sans tenir compte de sa réaction sur le premier [4].

La théorie des transformations canoniques conduit à une conception plus profonde, qui plus tard devint essentielle pour l'intelligence de la signification physique du formalisme.

A chaque matrice $a = (a_{nm})$ on peut associer une forme quadratique (plus exactement, hermitique) [1]

$$\sum_{nm} a_{nm} \varphi_n \overline{\varphi_m}$$

d'une série de variables φ_1, φ_2, ..., ou aussi une transformation linéaire de la série de variables φ_1, φ_2, ..., en une autre ψ_1, ψ_2, ...

$$(6) \qquad \psi_n = \sum_n a_{nm} \varphi_m.$$

La signification de ces variables φ_n et ψ_n reste provisoirement indéterminée; nous y reviendrons.

Une transformation (6) est dite « orthogonale » lorsqu'elle transforme en elle-même la forme

$$(7) \qquad \sum_n \varphi_n \overline{\varphi_n} = \sum_n \psi_n \overline{\psi_n}.$$

Or, on reconnaît immédiatement que ces transformations orthogonales des variables auxiliaires φ_n sont en réalité identiques aux transformations canoniques des matrices q, p; le caractère hermétique et les relations de commutation sont conservées. Ensuite, on peut remplacer l'équation de matrices (5) par la

---

(1) $\overline{\varphi}$ signifie le nombre complexe conjugué de φ.

condition équivalente [4]. La forme

$$\sum_{nm} \mathcal{H}_{nm}(q_0,\, p_0)\varphi_n \overline{\varphi_m}$$

doit être transformée orthogonalement en une somme de carrés

$$(8) \qquad \sum_n \mathrm{W}_n \psi_n \overline{\psi_n}.$$

Le problème mécanique fondamental n'est donc autre chose que le problème des axes principaux des surfaces de second degré dans l'espace à infinité de dimensions, qui se rencontre partout dans les mathématiques pures et appliquées et qui fut souvent étudié. On sait que ce problème équivaut à la recherche des valeurs du paramètre W pour lequel les équations linéaires

$$(9) \qquad \mathrm{W}\varphi_n = \sum_m \mathcal{H}_{nm}\varphi_m$$

ont une solution qui n'est pas identiquement nulle. Ces valeurs $\mathrm{W} = \mathrm{W}_1,\ \mathrm{W}_2,\ \ldots$ sont ce qu'on appelle les valeurs caractéristiques de la forme $\mathcal{H}$; ce sont les valeurs de l'énergie (termes) du système mécanique. A chaque valeur caractéristique W_n correspond une solution caractéristique $\varphi_k = \varphi_{kn}$. L'ensemble de ces' solutions caractéristiques constitue évidemment une nouvelle matrice, et l'on voit aisément que celle-ci est identique à la matrice de transformation $\mathcal{S}$ qui figure dans (5).

On sait que les valeurs caractéristiques sont invariantes vis-à-vis des transformations orthogonales φ_k, et comme celles-ci correspondent aux substitutions canoniques des matrices p, q, on comprend immédiatement l'invariance canonique des valeurs d'énergie W_n.

Il est vrai que les matrices de la théorie des quanta n'appartiennent pas à la classe des matrices (finies et partiellement infinies), étudiées par les mathématiciens (spécialement par Hilbert et ses élèves), mais on peut néanmoins étendre au cas plus général les traits principaux de la théorie connue. La formulation précise de ces principes fut donnée récemment par J. v. Neumann [42], dans un travail sur lequel nous n'avons pas à revenir.

Le résultat le plus important auquel on arriva de cette manière est ce théorème, qu'une forme ne peut pas du tout être décomposée toujours en une somme de carrés (8), mais qu'il se présente en outre des intégrales invariantes

$$(10) \qquad \int W \, \psi(W) \overline{\psi(W)} \, dW,$$

où la série de variables ψ_1, ψ_2, ... doit être complétée par la distribution continue ψ (W).

De cette façon les spectres continus entrent dans la théorie de la manière la plus naturelle. Mais cela ne signifie pas du tout que dans ce domaine la théorie classique soit de nouveau rendue exactement. Ici encore subsistent les discontinuités caractéristiques de la théorie des quanta; un changement d'état (spontané) consiste, dans le spectre continu aussi, en un « saut » du système d'un point W' à un autre W'', avec émission d'une onde $q\,(W, W')\, 2\pi i \nu t$ de fréquence $\nu = \frac{1}{h}\,(W' - W'')$.

Le défaut principal de la mécanique des matrices consiste dans sa difficulté, voire même son impuissance à traiter des grandeurs non périodiques, comme des variables angulaires ou des coordonnées qui atteignent des valeurs infiniment grandes (trajectoires hyperboliques, par exemple). Pour surmonter cette difficulté on a suivi deux voies essentiellement différentes, le calcul des opérateurs de Born et Wiener [21], et la théorie du nombre q de Dirac [7].

Cette dernière part de l'idée qu'une grande partie des relations de matrices peuvent s'obtenir sans qu'il soit nécessaire de représenter explicitement les matrices, simplement en se basant sur les règles du calcul au moyen des symboles des matrices. Ces règles ne s'écartent de celles du calcul des nombres qu'en ceci qu'en général la multiplication n'est pas commutative. Dirac considère pour cette raison des grandeurs abstraites, qu'il appelle des nombres q (par opposition aux nombres ordinaires c), et avec lesquels il opère suivant les règles de l'algèbre non commutative. Il s'agit donc d'une espèce de système hypercomplexe. Les relations de commutation sont évidemment conservées. Cette théorie devient d'une ressemblance extraordinaire à la théorie classique; c'est ainsi qu'on peut introduire des variables d'angle et d'action w, $\mathscr{F}$, et développer un nombre quelconque q en une série de Fourier par rapport à w; les coefficients sont des fonctions de $\mathscr{F}$ et se montrent identiques aux éléments des matrices, lorsqu'on remplace les $\mathscr{F}$ par des multiples entiers de h. Par sa méthode, Dirac est arrivé à des résultats importants; il a, par exemple, indépendamment de Pauli, soumis l'atome d'hydrogène au calcul [7], et déterminé les rapports d'intensité dans l'effet Compton [12]. Un inconvénient de ce formalisme — abstraction faite de l'application fort laborieuse de l'algèbre non commutative — est la nécessité de remplacer à un certain endroit du calcul les nombres q par des nombres ordinaires (par exemple $\mathscr{F} = hn$), afin d'obtenir des résultats comparables aux résultats expérimentaux. Il faut donc de nouveau des « conditions de quanta » spéciales, qui avaient disparu de la théorie des matrices.

Le calcul par opérateurs se distingue de la méthode du nombre q en ceci qu'il introduit des nombres hypercomplexes non point abstraits, mais concrets, des configurations mathématiques constructibles, qui obéissent aux mêmes lois, savoir des opérateurs ou des fonctions dans l'espace à une infinité de variables. La méthode est d'Eckart [22] et, comme suite à la mécanique ondulatoire de Schrödinger, elle fut développée par plusieurs autres auteurs, spécialement par Dirac [38] et Jordan [39], et mise sous une forme mathématique irréprochable par J. v. Neumann [42]; elle repose sur l'idée suivante, à peu près.

Une série de variables peut être représentée par un point dans un espace à ∞ dimensions. Si la somme de carrés $\sum_{n} |\varphi_n|^2$ est convergente, elle représente une mesure de distance, une mesure euclidienne dans cet espace; cet espace métrique à ∞ dimensions est appelé brièvement un espace hilbertien. Les transformations canoniques de la mécanique des matrices correspondent donc aux rotations de l'espace hilbertien. Or, un point de cet espace peut être fixé d'une autre manière encore que par l'indication de coordonnées séparées φ_1, φ_2, ... Prenons, par exemple, un système orthogonal complet, normé, de fonctions $f_1(q)$, $f_2(q)$, ..., c'est-à-dire un système pour lequel

$$(11) \qquad \int f_n(q) \overline{f_m(q)}\, dq = \delta_{nm} = \begin{cases} 1 \text{ pour } n = m, \\ 0 \text{ pour } n \text{ différent de } m\,; \end{cases}$$

la variable q est supposée pouvoir parcourir un domaine quelconque, également à plusieurs dimensions.

Si l'on pose alors (Lanczos [23]

$$(12) \qquad \begin{cases} \varphi(q) = \sum_n \varphi_n f_n(q), \\[2mm] \mathcal{H}(q', q'') = \sum_{nm} \mathcal{H}_{nm}\, f_n(q')\, \overline{f_m(q'')}. \end{cases}$$

les équations linéaires (9) se transforment en l'équation intégrale

$$W\,\varphi(q') = \int \mathcal{H}(q', q'')\, \varphi(q'')\, dq''.$$

Cette relation, obtenue au moyen de (12), ne signifie donc autre chose qu'un changement du système de coordonnées dans l'espace de Hilbert, déterminé par la matrice de transformation orthogonale $f_n(q)$ avec un indice discontinu et un indice continu.

On voit donc que la favorisation de systèmes de coordonnées « séparés » dans la conception originale de la théorie des matrices n'a absolument rien d'essentiel. On peut se servir tout aussi bien de « matrices continues », comme $\mathcal{H}(q', q'')$. Il ne s'agit donc pas du tout de représenter d'une façon spéciale un point dans l'espace hilbertien par sa projection sur des axes de coordonnées orthogonaux déterminés; il s'agit bien plutôt de pouvoir réunir les équations (9) et (13) en une seule, plus générale,

$$(14) \qquad\qquad\qquad W\,\varphi = \mathcal{H}\,\varphi,$$

où $\mathcal{H}$ représente un opérateur linéaire, qui transforme le point φ de l'espace hilbertien en un autre. L'équation pose l'exigence de trouver les points φ qui par l'opération $\mathcal{H}$ ne subissent qu'un déplacement de leur droite de jonction avec l'origine. Les points qui satisfont à cette condition déterminent un système d'axes orthogonal, le système d'axes principaux de l'opérateur $\mathcal{H}$; le nombre de ces axes est fini ou infini; dans le dernier cas ces axes sont distribués d'une façon séparée ou d'une façon continue et les valeurs propres W sont les longueurs des axes principaux. Les opérateurs linéaires dans l'espace de Hilbert sont donc le concept général qui peut servir à représenter mathématiquement une grandeur physique. Le calcul au moyen des opérateurs s'effectue évidemment suivant les mêmes règles que pour les nombres q de Dirac; ces règles représentent une réalisation de cette notion abstraite.

Nous avons fait jusqu'ici l'exposé en prenant comme exemple la fonction d'Hamilton, mais nos considérations s'appliquent à toute grandeur de la mécanique quantique. Toute coordonnée q peut, au lieu d'être mise sous la forme d'une matrice à indices déterminés q_{nm}, être écrite, par projection sur un système orthogonal de fonctions, comme fonction de deux variables continues $q\,(q', q'')$ ou, d'une manière générale, être considérée comme un opérateur linéaire dans l'espace hilbertien; elle possède alors des valeurs caractéristiques invariantes et des solutions caractéristiques par rapport à chaque système de coordonnées orthogonales. Il en est de même pour une impulsion p et pour toute fonction de q et p, en somme pour toute « grandeur » de la mécanique des quanta. Alors que dans la théorie classique des grandeurs physiques sont représentées par des variables qui peuvent prendre des valeurs numériques dans n'importe quel domaine de valeurs, une grandeur physique de la théorie des quanta est représentée par un opérateur linéaire et la collection de valeurs qu'elle peut prendre est constituée par les valeurs caractéristiques du problème des axes principaux correspondant dans l'espace hilbertien.

Dans cette manière de voir la mécanique de Schrödinger [24] se montre formellement comme cas particulier. L'opérateur le plus simple, dont les valeurs caractéristiques sont tous les nombres réels, est, notamment, la multiplication d'une fonction $\mathscr{F}(q)$ par le nombre réel q : on l'écrit simplement q. Les fonctions caractéristiques sont donc à vrai dire des fonctions « non véritables », car elles doivent, d'après (14), avoir la propriété d'être toujours nulles, sauf quand $W = q$. Pour représenter de pareilles fonctions impropres, Dirac [38] a introduit la « fonction unité », $\delta(s)$, qui doit être toujours nulle lorsque s est différent de zéro, mais pour laquelle il faut néanmoins

$$\int_{-\infty}^{+\infty} \delta(s)\, ds = 1.$$

Les fonctions caractéristiques (normées) appartenant à l'opérateur q peuvent alors s'écrire

$$(15) \qquad \varphi(q,\, \mathrm{W}) = q\, \delta(\mathrm{W} - q).$$

A cet opérateur q est conjugué l'opérateur différentiel

$$(16) \qquad p = \frac{h}{2\pi i} \frac{\partial}{\partial q};$$

car, on a la relation de commutation (3), qui ne signifie pas autre chose que la simple identité

$$(pq - qp)\,\mathcal{F}(q) = \frac{h}{2\pi i}\left[\frac{d}{dq}(q\,\mathcal{F}) - q\frac{d\mathcal{F}}{dq}\right] = \frac{h}{2\pi i}\mathcal{F}(q).$$

Si maintenant on construit au moyen de p, q (ou de plusieurs paires conjuguées de ce genre) une fonction hamiltonienne, l'équation (14) devient une équation différentielle pour la grandeur $\varphi(q)$,

$$(17) \qquad \mathcal{H}\left(q,\, \frac{h}{2\pi i}\frac{\partial}{\partial q}\right)\varphi(q) = \mathrm{W}\,\varphi(q).$$

C'est l'équation des ondes de Schrödinger, qui apparaît ici comme cas particulier de la théorie des opérateurs. Le caractère le plus important de cette manière de formuler les lois des quanta (sans parler du grand avantage du raccordement à des méthodes mathématiques connues) est le remplacement de toutes les « conditions des quanta », qui étaient encore nécessaires dans la théorie des nombres q de Dirac, par la simple exigence que la fonction caractéristique $\varphi(q) = \varphi(q, \mathrm{W})$ soit finie en tout point du domaine de définition de la variable q : par là il se produit tout naturellement, dans certaines circonstances, un spectre discontinu de valeurs caractéristiques W_n (à côté d'un spectre continu).

Mais au fond, la fonction earactéristique de Schrödinger $\varphi(q, \mathrm{W})$ n'est autre chose que la matrice de transformation $\mathcal{S}$ de l'équation (5), que l'on peut mettre encore sous la forme

$$\mathcal{H}\mathcal{S} = \mathcal{S}\,W,$$

analogue à (17).

Dirac [38] a rendu ce rapport plus clair encore en écrivant les opérateurs q et p, et par suite aussi $\mathcal{H}$, sous forme d'opérateurs intégraux comme dans (13); on a alors

$$(18) \quad \begin{cases} q\,\mathcal{F}(q') = \int q''\,\delta(q'-q'')\mathcal{F}(q'')\,dq'' = q'\,\mathcal{F}(q'), \\[2ex] p\,\mathcal{F}(q') = \int \dfrac{h}{2\pi i}\,\delta'(q'-q'')\mathcal{F}(q'')\,dq'' = \dfrac{h}{2\pi i}\dfrac{d\mathcal{F}}{dq'}, \end{cases}$$

où la présence de la dérivée de la fonction singulière δ doit évidemment être prise par-dessus le marché. Alors l'équation (17) de Schrödinger prend la forme (13).

Le passage direct à la représentation par matrices dans le sens restreint se fait par renversement des formules (12), dans lesquelles le système orthogonal $f_n(q)$ est identifié aux fonctions caractéristiques $\varphi(q, W_n)$ appartenant au spectre discontinu. Si T est un opérateur quelconque $\left(\text{composé de } q \text{ et } p = \dfrac{h}{2\pi i}\dfrac{\partial}{\partial q}\right)$, on définit la matrice T_{nm} correspondante au moyen des coefficients du développement

$$(19) \qquad T\,\varphi_n(q) = \sum_m T_{nm}\,\varphi_m(q)$$

ou

$$(19^a) \qquad T_{nm} = \int \overline{\varphi_m(q)}\,T\,\varphi_n(q)\,dq;$$

on reconnaît alors aisément que l'équation (17) est équivalente à (9).

La suite du développement de la théorie formelle s'est faite en relation étroite avec son interprétation physique; c'est pourquoi nous allons maintenant nous occuper de celle-ci.

II. — L'INTERPRÉTATION PHYSIQUE DE LA THÉORIE.

Le défaut le plus apparent de la mécanique des matrices originale consiste en ceci qu'elle ne semble pas donner directement des renseignements sur des phénomènes réels, mais seulement sur des états et des processus possibles. Elle permet de calculer les états stationnaires possibles d'un système; ensuite, elle se prononce au süjet de la vibration harmonique qui peut se manifester sous forme d'onde lumineuse, lors d'un saut quantique. Mais elle ne nous apprend pas quand un état déterminé existe, ni quand un changement peut être attendu. La raison en est claire : la mécanique des matrices ne s'occupe que de systèmes fermés, périodiques, et dans ceux-ci il ne se produit pas en réalité de changements. Pour avoir de véritables phénomènes on doit, aussi longtemps qu'on reste dans le domaine de la mécanique des matrices, porter son attention sur une *partie* du système; alors celle-ci n'est plus un système fermé, mais entre en interaction avec le reste du système. La question est de savoir ce que la mécanique des matrices peut apprendre à ce sujet.

Considérons par exemple deux systèmes 1 et 2 faiblement couplés l'un à l'autre (Heisenberg [35], Jordan [36]). Le principe de l'énergie vaut alors pour le système tout entier, c'est-à-dire que $\mathcal{H}$ est une matrice diagonale. Mais pour un système partiel, 1 par exemple, $\mathcal{H}_1$ n'est pas constant et la matrice a des membres en dehors de la diagonale. Or, l'échange d'énergie peut être interprété de deux manières. Dans l'une des manières de voir les membres périodiques de la matrice de $\mathcal{H}_1$ (ou de $\mathcal{H}_2$) représentent un long battement, un va-et-vient continu de l'énergie; mais on peut aussi décrire le processus au moyen des notions de la théorie du discontinu et dire que le système 1 effectue des sauts quantiques et transporte l'énergie libérée dans ces sauts sous forme de quanta sur le système 2, et inversement. Or, on peut démontrer que ces deux manières de voir, en apparence si différentes, ne sont aucunement contradictoires. Cela repose sur un théorème de mathématiques, qui dit ce qui suit :

Soit $f(W_n^{(1)})$ une fonction des valeurs d'énergie $W_n^{(1)}$ du système isolé 1 ; si l'on forme la même fonction de la matrice $\mathcal{H}^{(1)}$, qui représente l'énergie du système 1 dans le cas d'existence d'un couplage au système 2, $f(\mathcal{H}^{(1)})$ est une matrice qui ne se compose pas uniquement de membres diagonaux $f[\mathcal{H}^{(1)}]_{nn}$. Mais ceux-ci représentent la valeur moyenne dans le temps de la grandeur $f(\mathcal{H}^{(1)})$. L'action du couplage est donc mesurée par la différence

$$\overline{\delta f_n} = f[\mathcal{H}^{(1)}{}_{nn}] - f[W_n^{(1)}].$$

Or, la première partie du théorème en question dit que $\overline{\delta f_n}$ peut être mise sous la forme :

$$(20) \qquad \overline{\delta f_n} = \sum_m [f(W_n) - f(W_m)] \Phi_{nm}.$$

Cela peut s'interpréter ainsi : La moyenne dans le temps du changement de f par suite du couplage est la moyenne arithmétique de tous les sauts possibles de f dans le système isolé, ces sauts étant affectés de certains poids Φ_{nm}. Ces Φ_{nm} devront être appelés des « probabilités de transition ».

La seconde partie du théorème détermine ces Φ_{nm} par les propriétés du couplage. Si, notamment, p_1^0, q_1^0, p_2^0, q_2^0 sont des coordonnées qui satisfont aux équations de mouvement des systèmes non couplés, pour lesquels $\mathcal{H}_1$ et $\mathcal{H}_2$ sont donc eux-mêmes des matrices diagonales, on peut se représenter l'énergie, y compris l'action mutuelle, comme exprimée par une fonction de ces grandeurs. Alors la solution du problème mécanique aboutit d'après (2) à la formation d'une matrice $\mathcal{S}$, qui satisfait à l'équation

$$\mathcal{S}^{-1} \mathcal{H}(p_1^0, q_1^0, p_2^0, q_2^0) \mathcal{S} = W.$$

Si l'on représente les états du système 1 par n_1, ceux du système 2 par n_2, un état du système total est donné par n_1, n_2 et à tout passage $n_1 n_2 \to m_1 m_2$ correspond un élément de $\mathcal{S}$, $\mathcal{S}_{n_1 n_2,\, m_1 m_2}$. Le résultat est alors

$$(21) \qquad \Phi_{nm} = \sum_{n_2 m_2} |\mathcal{S}_{n_1 n_2,\, m_1 m_2}|^2.$$

Les carrés des éléments de la matrice $\mathcal{S}$ déterminent donc les probabilités de transition. Chaque terme $|\mathcal{S}_{n_1 n_2,\, m_1 m_2}|^2$ de la somme (21), pris individuellement, représente évidemment la partie de la probabilité de transition pour le saut $n_1 \to m_1$ du système 1, qui est induit par le saut $n_2 \to m_2$ du système 2.

Ces résultats lèvent la contradiction entre les deux points de vue d'où nous sommes partis. Car pour les valeurs moyennes seulement observables la représentation du battement continu conduit toujours au même résultat que la représentation des sauts quantiques.

　　　　　　　ÉLECTRONS ET PHOTONS.

Si l'on demande *quand* se produit un saut quantique, la théorie reste sans réponse à cette question. Il semblait d'abord qu'il y avait là une lacune qui se comblerait lorsqu'on approfondirait la théorie. Mais bientôt on dut reconnaître qu'il n'en était pas ainsi et qu'il s'agit ici plutôt d'une impuissance essentielle, profondément ancrée dans la nature de notre pouvoir de comprendre les phénomènes physiques.

On voit que la mécanique des quanta fournit des valeurs moyennes avec exactitude, mais ne peut prédire la venue d'un événement isolé. Le déterminisme qui jusqu'ici a été admis comme base des sciences exactes de la nature semble ne plus pouvoir être admis sans conteste. Chaque nouveau progrès dans l'interprétation des formules a montré que le système des formules de la mécanique des quanta ne peut être interprété sans contradiction que du point de vue d'un indéterminisme fondamental, mais qu'en même temps aussi l'ensemble des faits qui peuvent être établis par l'expérience peut être rendu par le système de la théorie.

En fait, presque toutes les observations dans le domaine de la physique de l'atome ont un caractère statistique; ce sont des numérations, par exemple d'atomes dans un état détermi é. La détermination du processus individuel est bien admise par la physique classique, mais pratiquement elle ne joue aucun rôle, parce que les microcoordonnées qui fixent exactement un processus atomique ne peuvent jamais être données toutes; c'est pourquoi on les élimine des formules par le calcul d'une moyenne, et par là ces formules deviennent des déclarations statistiques. On a reconnu que la mécanique des quanta représente un fusionnement de la mécanique et de la statistique, d'où les microcoordonnées non observables sont chassées.

La lourdeur de la théorie des matrices dans sa description de processus qui se déroulent dans le temps peut être évitée en recourant au formalisme plus général décrit ci-dessus. Dans l'équation générale (14) on peut aisément introduire le temps explicitement en invoquant ce principe de la mécanique classique, que l'énergie W et le temps t se comportent l'un vis-à-vis de l'autre comme des grandeurs canoniquement conjuguées; à cela correspond qu'on doit avoir dans la mécanique des quanta une relation de commutation

$$\mathrm{W}\,t - t\,\mathrm{W} = \frac{h}{2\pi i}.$$

On peut donc affecter W de l'opérateur $\dfrac{h}{2\pi i}\dfrac{\partial}{\partial t}$. L'équation (14) devient alors

$$(22)\qquad \frac{h}{2\pi i}\frac{\partial\varphi}{\partial t} = \mathcal{H}\varphi,$$

où l'on peut considérer $\mathcal{H}$ comme dépendant explicitement du temps. Un cas particulier de cette équation est l'équation suivante, donnée par Schrödinger :

$$(22^a)\qquad \left[\mathcal{H}\left(q,\frac{h}{2\pi i}\frac{\partial}{\partial q}\right) - \frac{h}{2\pi i}\frac{\partial}{\partial t}\right]\varphi(q) = 0,$$

qui est à (17) ce que (22) est à (14), et l'équation

$$(22^b)\qquad \frac{h}{2\pi i}\frac{\partial\varphi(q')}{\partial t} = \int \mathcal{H}(q',q'')\varphi(q'')\,dq'',$$

qui est souvent employée par Dirac et qui se rattache à la formule intégrale (13).

Au fond l'introduction du temps comme variable numérique a pour conséquence qu'on se figure le système considéré comme couplé à un autre et qu'on néglige la réaction sur ce dernier. Mais ce formalisme est très commode et conduit à un nouveau développement de la conception statistique, notamment par considération du cas ou s'ajoute à une fonction d'énergie $\mathcal{H}^0$, qui est indépendante du temps, une perturbation $V(t)$ qui dépend explicitement du temps, de sorte qu'on a l'équation(Dirac, Born)

$$(23)\qquad \frac{h}{2\pi i}\frac{\partial\varphi}{\partial t} = [\mathcal{H}^0 + V(t)]\varphi.$$

Si maintenant φ_n^0 sont les fonctions caractéristiques de l'opérateur $\mathcal{H}^0$, que pour plus de simplicité nous supposerons discontinues, la grandeur cherchée φ peut être développée en série comme suit :

$$(24)\qquad \varphi(t) = \sum_n c_n(t)\varphi_n^0.$$

Les $c(t)$ sont alors les coordonnées de φ dans l'espace hilbertien rapportées au système orthogonal φ_n^0; elles peuvent être déduites par calcul de l'équation différentielle (23), lorsque leurs valeurs initiales $c_n(o)$ sont données. Le résultat peut encore se traduire ainsi :

$$(25) \qquad c_n(t) = \sum_m \mathcal{S}_{nm}(t)\, c_m(o).$$

où $\mathcal{S}_{nm}(t)$ est une matrice orthogonale indépendante de t, déterminée par $V(t)$.

Le processus temporaire est donc représenté par une rotation de l'espace de Hilbert ou par une transformation canonique (4) avec la matrice $\mathcal{S}$ qui dépend du temps.

Comment cela doit-il être interprété ?

Du point de vue de la théorie de Bohr un système ne peut jamais être que dans un seul état quantique. A chacun de ces états appartient une solution caractéristique du système φ_n^0 non perturbé. Or, si l'on veut calculer ce que devient un système qui se trouve au début dans un certain état, disons-le $k^{-\text{ième}}$, on doit choisir comme condition initiale pour l'équation (23) $\varphi = \varphi_k^0$, c'est-à-dire $c_n(o) = o$ pour n différent de k et $c_k(o) = 1$. Mais alors on a à la fin du trouble $c_n(t) = \mathcal{S}_{nk}(t)$ et la solution se compose d'une superposition de solutions caractéristiques. D'après les principes de Bohr cela n'a aucun sens que de dire qu'un système se trouve à la fois dans plusieurs états. La seule interprétation possible paraît être l'interprétation statistique :

La superposition de plusieurs solutions caractéristiques exprime que l'état initial peut passer par la perturbation à un autre état quantique et il est clair que l'on doit prendre comme mesure de la probabilité de transition la grandeur

$$\Phi_{nk} = |\,\mathcal{S}_{nk}(t)\,|^2;$$

car on retrouve ainsi pour le changement moyen d'une fonction d'état quelconque l'équation (20).

Cette interprétation est confirmée par le fait qu'on démontre l'exactitude du principe adiabatique d'Ehrenfest (Born [34]); on peut dire que dans le cas d'une action infiniment lente

$$\Phi_{nn} \to 1, \qquad \Phi_{nk} \to o \qquad (n \text{ différent de } k).$$

c'est-à-dire que la probabilité d'un saut tend vers zéro.

Cette thèse conduit d'ailleurs immédiatement à une interprétation du $c_n(t)$ même : les $|c_n(t)|^2$ doivent être les probabilités d'état.

On se heurte cependant ici à une difficulté, qui a une grande importance en principe, dès qu'on part d'un état initial pour lequel tous les $c_n(0)$ ne disparaissent pas à l'exception d'un seul. En physique ce cas se présenterait si l'on donnait un système dont on ne sait pas exactement dans quel état quantique il se trouve, mais dont on connaît uniquement la probabilité $|c_n(0)|^2$ de chaque état quantique. Alors évidemment, les arguments des grandeurs complexes $c_n(0)$ restent encore indéterminés; si l'on pose

$$c_n(0) = |c_n(0)|\, e^{i\gamma_n},$$

les γ_n représentent des phases, dont le sens doit être établi. La distribution des probabilités à la fin de la perturbation est alors, d'après (25),

$$(26) \qquad |c_n(t)|^2 = \left| \sum_m \mathcal{S}_{nm}(t)\, c_m(0) \right|^2$$

et non

$$(24) \qquad \sum_m |\mathcal{S}_{nm}(t)|^2\, |c_m(0)|^2,$$

ainsi qu'on pourrait le supposer d'après le calcul des probabilités ordinaire.

La formule (26) peut être appelée, d'après Pauli, le théorème de *l'interférence des probabilités;* sa signification a été rendue claire par la mécanique ondulatoire de de Broglie et de Schrödinger; nous y reviendrons tantôt. Mais avant d'en parler je voudrais faire remarquer que cette « interférence » ne signifie pas une contradiction avec les règles du calcul des probabilités, c'est-à-dire avec l'idée que les $|\mathcal{S}_{nk}|^2$ sont de simples probabilités. La règle de combinaison (27) se déduit notamment, pour le problème traité ici, de la notion de probabilité lorsque — et alors seulement — le nombre relatif [c'est-à-dire la probabilité $(c_{nk})^2$] des atomes dans l'état nk a été préalablement *établi par l'expérience.* Dans ce cas les phases p_{nk} sont essentiellement inconnues, de sorte (26) se transforme alors tout naturellement en (27) [46].

Je ferai remarquer ensuite que la formule (26) se transforme en (27) lorsque la fonction de perturbation est une fonction du temps tout à fait irrégulière. Tel est, par exemple, le cas lorsque le trouble est produit par de la « lumière blanche ». Alors les termes excédents de (26) disparaissent en faisant la moyenne et l'on obtient (27). De cette façon il est aisé de trouver le coefficient B_{nm} d'Einstein pour la probabilité par unité de rayonnement des sauts quantiques déterminés par l'absorption de lumière (Dirac [37], Born [30]). Mais en général il ne suffit pas du tout, d'après (26), de connaître les probabilités $|c_n(o)|^2$ pour calculer la marche de la perturbation; on doit connaître encore les phases γ_n.

Cette circonstance rappelle vivement ce qui arrive dans les phénomènes d'interférence. L'intensité d'éclairement d'un écran n'est pas du tout toujours égale à la somme des intensités lumineuses des divers faisceaux de rayons qui frappent l'écran, ou bien, comme on peut le dire à bon droit, elle n'est pas égale à la somme des quanta de lumière qui se meuvent dans les divers faisceaux; elle dépend essentiellement des phases des ondes. A cet endroit se manifeste donc une analogie entre la mécanique quantique des corpuscules et la théorie ondulatoire de la lumière.

Mais en réalité cette relation a été trouvée par de Broglie d'une toute autre manière. Il ne m'appartient pas d'entrer dans des détails à ce sujet. Il suffira que j'aie formulé le résultat des considérations de de Broglie et de leur développement par Schrödinger et que j'aie mis ce résultat en rapport avec la mécanique des quanta.

A la nature dualiste de la lumière — ondes, quanta de lumière — correspond la nature dualiste analogue des particules matérielles. Celles-ci aussi se comportent à un certain point de vue comme des ondes. Schrödinger a établi les lois de la propagation de ces ondes [24] et est arrivé à l'équation (11), qui a été déduite ici d'autre façon. Son idée, que ces ondes constituent l'essence même de la matière, que les particules ne sont que des paquets d'ondes, n'est pas seulement en contradiction avec les principes de la théorie de Bohr, si bien fondée, mais elle conduit aussi à des conséquences impossibles; c'est pourquoi nous n'en parlerons pas ici. En revanche, nous attribuons à la matière aussi une double nature : leur description exige que nous parlions de corpuscules (discontinuités) aussi bien que d'ondes (processus continus). Du point de vue de la conception statistique de la mécanique des quanta on voit clairement comment cela est conciliable : Les ondes sont des ondes de probabilité. Ce ne sont pas, à vrai dire, les probabilités elles-mêmes, mais certaines « amplitudes de probabilité » qui se propagent continûment et satisfont à des équations différentielles ou intégrales, comme dans la physique du continu classique. Mais à côté de cela il y a des discontinuités, des corpuscules, dont la fréquence est réglée par le carré de ces amplitudes.

La confirmation la meilleure de cette manière de voir se trouve dans les phénomènes de choc entre les particules matérielles (Born [30]). Einstein [16], en déduisant de la théorie hardie de de Broglie la possibilité d'une « diffraction » de particules matérielles, avait déjà admis tacitement que c'est le nombre des particules qui est déterminé par l'intensité des ondes. On retrouve cette manière de voir dans l'interprétation qu'Elsasser [17] donne des expériences de Davisson et Kunsman [18, 19] sur la réflexion d'électrons par des cristaux; ici encore on admet, sans plus, que dans les maxima de diffraction le *nombre* des électrons est maximum. Il en est de même des expériences de Dymond [20] sur la diffraction d'électrons par des atomes d'hélium.

L'application de la mécanique ondulatoire au calcul de phénomènes de choc prend une forme analogue à la théorie de la diffraction de la lumière par de petites particules. On doit chercher la solution de l'équation des ondes de Schrödinger (17) qui passe à l'infini à une onde incidente plane donnée; cette solution se comporte partout à l'infini comme une onde sphérique. L'intensité de cette onde sphérique dans une direction quelconque, comparée à l'intensité de l'onde incidente, détermine le nombre relatif des particules d'un faisceau de rayons parallèles, déviées dans cette direction. Comme mesure de l'intensité on doit prendre un « vecteur radiant » constructible au moyen de la solution $\varphi(q, W)$, qui est tout à fait analogue au vecteur de Poynting de la théorie électromagnétique de la lumière et qui donne la mesure du nombre de particules qui traversent l'unité de surface dans l'unité du temps.

De cette manière Wentzel [31] et Oppenheiner [32] ont déduit par la voie de la mécanique ondulatoire la loi de Rutherford, de la diffusion de particules α par des noyaux lourds.

  ÉLECTRONS ET PHOTONS.

Si l'on veut calculer les probabilités d'excitation et d'ionisation d'atomes [30], on doit introduire les coordonnées des électrons atomiques à côté de celles de l'électron choquant comme variables ayant les mêmes droits. Les ondes ne se propagent donc plus dans l'espace à trois dimensions, mais dans un espace de configuration polydimensionnel. On voit par là que les ondes de la mécanique quantique sont tout à fait autre chose que les ondes lumineuses de la théorie classique, par exemple.

Si l'on construit le vecteur radiant, défini ci-dessus, pour une solution de l'équation de Schrödinger généralisée (22), qui représente la variation avec le temps, on voit que la dérivée par rapport au temps de l'intégrale

$$\int |\varphi|^2 \, dq',$$

étendue à un certain domaine de la variable numérique indépendante q', se laisse transformer dans l'intégrale de surface du vecteur radiant étendue aux limites de ce domaine. Il s'ensuit que $|\varphi|^2$ doit être considéré comme densité des particules ou plutôt comme densité de probabilité. La solution φ elle-même est appelée « amplitude de probabilité ».

L'amplitude $\varphi(q', W')$ qui appartient à un état stationnaire fournit donc par l'expression $|\varphi(q', W')|^2$ la probabilité qu'à énergie donnée W' la coordonnée q' est située dans un élément dq' donné d'avance.

Mais cela est susceptible d'une généralisation immédiate. En effet, $\varphi(q', W')$ est la projection de l'axe principal W' de l'opérateur $\mathcal{H}$ sur l'axe principal q' de l'opérateur q. On peut donc dire en toute généralité (Jordan [39]) : Si l'on donne deux grandeurs physiques par les opérateurs q, Q et si l'on connaît les axes principaux de l'un, par exemple en grandeur et en direction, on peut déterminer, au moyen de l'équation

$$Q \, \varphi(q', Q') = Q' \, \varphi(q', Q'),$$

les axes principaux Q' et Q et leurs projections $\varphi(q', Q')$ sur les axes de q. Alors

$$|\varphi(q', Q')|^2 \, dQ'$$

est la probabilité que pour une valeur donnée de Q' la valeur de q' se trouve dans un intervalle donné dq'.

Si, inversement, on suppose que les axes principaux de Q soient donnés, ceux de q' s'obtiennent par la rotation inverse; par là on reconnaît aisément que $\overline{\varphi(Q', q')}$ est l'amplitude correspondante, de sorte que

$$| \varphi(Q', q') |^2 dQ'$$

représente la probabilité que, q' étant donné, la valeur de Q' se trouve dans dQ'. Prend-on par exemple pour Q l'opérateur

$$p = \frac{h}{2\pi i} \frac{\partial}{\partial q},$$

on a alors l'équation

$$\frac{h}{2\pi i} \frac{\partial \varphi}{\partial q'} = p' \varphi,$$

c'est-à-dire

$$(28) \qquad \varphi = C \, e^{\frac{2\pi i}{h} q' p'}.$$

Telle est donc l'amplitude de probabilité pour une paire de grandeurs conjuguées. Pour la densité de probabilité on obtient

$$| \varphi |^2 = C,$$

c'est-à-dire que pour une valeur donnée de q' toutes les valeurs de p' sont également probables.

C'est là un résultat important, car il permet de conserver la notion de « grandeurs conjuguées », même lorsque la définition différentielle n'existe plus, c'est-à-dire quand la grandeur q n'a qu'un spectre discontinu ou n'admet qu'un nombre fini de valeurs. Ce dernier cas se présente, par exemple, lorsqu'on a affaire à des angles quantifiés quant à la direction, pour l'électron magnétique par exemple, ou dans les expériences de Stern et Gerlach. On peut alors, avec Jordan, dire, par définition, qu'une grandeur p est conjuguée à q, lorsque l'amplitude de probabilité correspondante est donnée par l'expression (28).

Comme les amplitudes sont les éléments de la matrice de rotation d'un système orthogonal dans un autre, elles se composent suivant la règle des matrices :

$$(29) \qquad \varphi(q', Q') = \int \psi(q', \beta') \chi(\beta', Q') \, d\beta';$$

dans le cas de spectres discontinus on a, à la place de l'intégrale,

 ÉLECTRONS ET PHOTONS.

des sommes finies ou infinies. C'est là la façon générale de formuler le théorème de l'interférence des probabilités.

Comme application nous considérons de nouveau la formule (24). Dans cette formule $c_n(t)$ était l'amplitude de la probabilité que le système a à l'instant l'énergie W_n; $\varphi_n^0(q')$ est l'amplitude de la probabilité qu'à énergie donnée W_n la coordonnée q' ait une valeur donnée d'avance.

$$\varphi(q',\, t) = \sum_n c_n(t)\, \varphi_n^0(q')$$

représente donc l'amplitude de la probabilité que q' ait à l'instant t une valeur donnée.

A côté de la notion de probabilité d'état relative $|\varphi(q',\, Q')|^2$ vient encore se placer celle de probabilité de transformation, et cela notamment chaque fois que l'on considère un système comme fonction d'un paramètre extérieur, que ce paramètre soit le temps ou une propriété quelconque d'un système extérieur faiblement couplé au premier. Alors le système d'axes principaux d'une grandeur quelconque devient dépendant de ce paramètre; il subit une rotation, représentée par une transformation orthogonale $\mathcal{S}(q',\, q'')$, dans laquelle entre le paramètre [comme dans la formule (25)]. Les grandeurs $|(q',\, q'')|^2$ sont les « probabilités de transformation »; en général, elles ne sont pas indépendantes, car les « amplitudes de transformation » se composent d'après la règle des interférences.

III. — Formulation des principes et délimitation de leur portée.

Maintenant que par l'analyse des faits établis par l'expérience ont été développées les notions générales de la théorie, nous sommes placés devant la double tâche, d'abord d'indiquer un système de principes les plus simples possibles et directement rattachés aux observations, d'où la théorie tout entière puisse être déduite comme d'un système d'axiomes mathématiques, ensuite d'examiner d'un œil critique toute notre connaissance empirique, pour nous assurer qu'aucune expérience faite par les moyens actuels n'est en contradiction avec ces principes.

Jordan [39] a formulé un pareil « système d'axiomes », qui est à la base des théorèmes suivants :

1º Il faut que pour chaque paire de grandeurs q, Q de la mécanique quantique il existe une amplitude de probabilité $\varphi\,(q', Q')$ telle que φ^2 indique la densité de la probabilité que pour une valeur donnée de Q' la valeur de q' tombe dans un intervalle infinitésimal donné.

2º En permutant q et Q on doit trouver l'amplitude correspondante $\overline{\varphi\,(Q', q')}$.

3º Le théorème (29) de la combinaison des amplitudes des probabilités.

4º A chaque grandeur q doit appartenir une grandeur canonique conjuguée p, définie par l'amplitude (28). C'est là le seul endroit où apparaît la constante quantique h.

Enfin, on admet encore comme évident que, si les grandeurs q et Q sont identiques, l'amplitude $\varphi\,(q',\ q'')$ devient égale à la « matrice unité » $\partial\,(q' - q'')$, c'est-à-dire est toujours nulle, sauf quand $q' = q''$. Cette hypothèse caractérise avec le théorème de multiplication (3) les amplitudes ainsi définies comme les coefficients d'une transformation orthogonale; on obtient les conditions d'orthogonalité simplement en exprimant que la composition de l'amplitude appartenant à q, Q avec celle appartenant à Q, q doit fournir l'identité.

On peut alors ramener à des amplitudes toutes les grandeurs qui se présentent, même les opérateurs, en les écrivant sous forme d'opérateurs intégraux, comme dans la formule (13). La multiplication d'opérateurs non commutative est alors une conséquence des axiomes et perd toute la singularité qu'elle présentait dans la théorie des matrices originale.

La méthode de Dirac [38] est tout à fait équivalente à la façon de formuler de Jordan, sauf que Dirac ne met pas les principes sous forme axiomatique.

Cette théorie résume en effet toute la mécanique des quanta en un système dans lequel la simple notion de la probabilité calculable joue le rôle principal pour un événement déterminé. Mais elle présente aussi quelques inconvénients. Un inconvénient formel est la présence de fonctions impropres, comme la fonction

δ de Dirac, dont on a besoin pour représenter la matrice unité pour des domaines de variables continus. Plus grave est cette circonstance que les amplitudes ne sont pas des grandeurs directement mesurables, mais leurs carrés; les facteurs de phase sont bien essentiels pour la relation entre les divers phénomènes, mais ils ne peuvent être établis qu'indirectement, tout à fait comme en optique les phases sont déduites par combinaison des mesures d'intensité. Or c'est précisément un principe auquel on tient beaucoup en mécanique quantique, d'introduire autant que possible comme notions fondamentales d'une théorie uniquement des grandeurs directement observables. Ce défaut tient mathématiquement à ce fait que la définition de la probabilité au moyen des amplitudes ne traduit pas l'invariance pour des transformations orthogonales de l'espace hilbertien (transformations canoniques).

Ces lacunes de la théorie ont été comblées par von Neumann [41, 42]. Cet auteur donne une définition invariante du spectre des valeurs caractéristiques pour un opérateur quelconque et des probabilités relatives, sans présupposer l'existence de fonctions caractéristiques ni même employer des fonctions impropres. Bien que cette théorie n'ait pas encore été élaborée dans toutes les directions, on peut cependant dire avec certitude qu'il est possible d'établir une mécanique des quanta parfaite au point de vue mathématique.

Et maintenant la réponse à la seconde question : Cette théorie est-elle d'accord avec l'ensemble de nos faits d'expérience ? En particulier comment, vu la détermination uniquement statistique des processus individuels dans les phénomènes macroscopiques compliqués, l'ordre déterministe auquel nous sommes accoutumés peut-il être conservé ?

Le pas le plus important dans la voie de la vérification du nouveau système de concepts dans ce sens consiste dans la fixation des limites dans lesquelles est permise l'application des termes et concepts anciens (classiques), tels que « lieu, vitesse, impulsion, énergie d'une particule (électrons) » (Heisenberg [46]). Or, on trouve que toutes ces grandeurs, prises *séparément*, peuvent être mesurées et définies avec précision, comme dans la théorie classique, mais que quand on veut mesurer simultanément des grandeurs

LA MÉCANIQUE DES QUANTA.

171

canoniquement conjuguées (plus généralement : des grandeurs dont les opérateurs ne sont pas permutables) on ne peut pas descendre au-dessous d'une limite d'indétermination caractéristique. Pour déterminer cette limite on peut, avec Bohr [47], partir d'une façon très générale du dualisme empiriquement établi entre ondes et corpuscules. Au fond, on a déjà le même phénomène à chaque diffraction de la lumière à travers une fente. Lorsqu'une onde vient frapper normalement une fente (infiniment étendue) de largeur q_1, la distribution de lumière comme fonction de l'angle de déviation φ est donnée, d'après Kirchhoff, par le carré de la valeur de la grandeur

$$a \int_{-\frac{q_1}{2}}^{+\frac{q_1}{2}} e^{\frac{2\pi i}{\lambda}\sin\varphi.q}\, dq = 2a \frac{\sin\left(\frac{\pi q_1}{\lambda}\sin\varphi\right)}{\frac{\pi q_1}{\lambda}\sin\varphi};$$

elle s'étend donc sur un domaine dont l'ordre de grandeur est donné par $\sin\varphi_1 = \frac{q_1}{\lambda}$ et qui devient de plus en plus grand à mesure que la largeur q_1 de la fente diminue. Si l'on considère ce phénomène du point de vue de la théorie corpusculaire et si les relations établies par de Broglie entre la fréquence et la longueur des ondes et l'énergie et l'impulsion du quantum de lumière

$$h\nu = \mathrm{W}, \qquad \frac{h}{\lambda} = \mathfrak{X},$$

sont exactes, la composante de l'impulsion perpendiculaire à la direction de la fente est

$$p = \mathfrak{X}\sin\varphi = \frac{l}{k}\sin\varphi.$$

On voit donc qu'après le passage à travers la fente les quanta de lumière ont une distribution dont l'amplitude est donnée par

$$e^{\frac{2\pi i}{\lambda}\sin\varphi.q} = e^{\frac{2\pi i}{h}pq},$$

tout comme la mécanique des quanta l'exige pour deux variables canoniquement conjuguées, et la largeur du domaine de la variable p

qui contient le nombre principal des quanta de lumière est

$$p_1 = \mathcal{R} \sin \varphi_1 = \frac{\mathcal{R}\lambda}{q_1} = \frac{h}{q_1}.$$

Par des raisonnements généraux de ce genre on arrive à cette idée que les inexactitudes (erreur moyenne) de deux variables canoniquement conjuguées p et q sont toujours liées entre elles par la relation

$$(30) \qquad p_1 q_1 \geqq h.$$

En rétrécissant le domaine d'une des variables qui constitue l'essence d'une mesure, on élargit inévitablement le domaine de l'autre. Cela résulte immédiatement du formalisme mathématique de la mécanique des quanta d'après la formule (28). Le véritable sens de la constante h de Planck est donc celui-ci, qu'elle constitue la mesure universelle de l'indétermination qui est introduite dans les lois naturelles par le dualisme des ondes et corpuscules.

Que la théorie des quanta est un mélange de principes strictement mécaniques et de principes statistiques, on peut le considérer comme une conséquence de cette indétermination. Dans la théorie classique on fixerait notamment l'état d'un système mécanique par exemple ainsi, qu'on mesure les valeurs initiales de p et q à un instant déterminé. Dans la mécanique des quanta une pareille mesure de l'état initial n'est possible qu'avec la précision (30). La valeurs de p et q ne sont donc également connues que statistiquement aux instants suivants.

Le rapport entre l'ancienne théorie et la nouvelle peut donc être décrit comme suit :

Dans la mécanique classique on admet la déterminabilité précise de l'état initial; ce qui se passe dans la suite est alors déterminé par les lois elles-mêmes.

Dans la mécanique des quanta on peut, en se basant sur la relation d'imprécision, exprimer le résultat de chaque mesure par le choix des valeurs initiales convenables de fonctions de probabilité; les lois de la mécanique quantique déterminent la variation (propagation ondulatoire) de ces fonctions de probabilité. Mais le résultat d'expériences futures reste en général indéterminé et seule l'attente de ce résultat est statistiquement restreinte.

Chaque nouvelle expérience met à la place des fonctions de probabilité valables jusqu'alors de nouvelles fonctions, qui correspondent au résultat de l'observation; elle partage les grandeurs physiques, d'une façon qui est caractéristique pour l'expérience, en grandeurs connues et en grandeurs inconnues (connues avec grande ou faible précision).

Le fait que dans cette manière de voir certaines lois, comme les principes de conservation de l'énergie et de la quantité de mouvement, sont encore strictement valables, provient de ce qu'il s'agit de relations entre grandeurs permutables (uniquement d'espèce q ou d'espèce p).

Le passage de la micromécanique à la macromécanique résulte tout naturellement de la relation d'imprécision à cause de la petitesse de la constante h de Planck. Le fait de l'extension, de la « fusion » d'un « paquet d'ondes ou de probabilités » est essentiel pour cela. Dans quelques systèmes mécaniques simples (électron libre dans un champ magnétique ou dans un champ électrique) (Kennard [50]), oscillateur harmonique (Schrödinger [25]), l'extension du paquet d'ondes en mécanique quantique est d'accord avec l'extension des trajectoires du système, qui se présenterait dans la théorie classique si les conditions initiales n'étaient connues qu'avec la restriction de précision (3o). Par là le traitement purement classique des particules α et β, comme dans la discussion des épreuves de Wilson, par exemple, trouve immédiatement sa justification. Mais en général les lois statistiques de l'extension d'un « paquet » sont différentes pour la théorie classique et la théorie des quanta; on en a des exemples particulièrement extrêmes dans les cas de « diffraction » ou d' « interférence » de rayons matériels, comme dans les expériences déjà mentionnées de Davisson, Kunsman et Germer [18, 19] sur la réflexion des électrons par des cristaux métalliques.

Que l'ensemble des faits d'expérience cadre dans le système de cette théorie, on ne peut évidemment le prouver qu'en calculant et discutant tout les cas accessibles à l'expérience. Des essais isolés, dans lesquels on pourrait soupçonner une contradiction avec la limite de précision (3o), ont été discutés [46, 47]; chaque fois on a pu montrer nettement la cause de l'impossibilité d'établir exactement toutes les parties déterminantes.

 ÉLECTRONS ET PHOTONS.

Il ne reste plus qu'à jeter un coup d'œil sur les dédùctions les plus importantes de la théorie et leur vérification expérimentale.

IV. — Applications de la mécanique des quanta.

Je traiterai rapidement, dans ce chapitre, ces applications-là de la mécanique des quanta qui sont en relation étroite avec les questions de principe. Je mentionnerai d'abord la théorie de l'électron magnétique de Uhlenbeck et Goudsmit.

Il n'y a aucune difficulté à formuler cette théorie ni à traiter l'effet Zeeman anormal par le calcul des matrices [11]; le calcul par la méthode des vibrations propres ne réussit qu'à l'aide de la théorie générale de Dirac-Jordan (Pauli [45]). On associe alors à chaque électron deux fonctions ondulatoires tridimensionnelles. Il est tout indiqué, de cette façon, de chercher dans les ondes matérielles une analogie avec les ondes lumineuses polarisées, ce qu'on peut faire, d'ailleurs, jusqu'à un certain point (Darwin [49], Jordan [53]). Ce qu'il y a de commun aux deux phénomènes, c'est que le nombre des termes est fini, de sorte que la matrice représentative est également finie (deux positions de l'électron, deux directions de polarisation de la lumière). Ici est donc en défaut la définition des grandeurs conjuguées par la différentiation; on doit recourir à la définition de Jordan à l'aide des amplitudes de probabilité [formule (28)].

Parmi les autres applications je citerai la mécanique quantique du problème de plusieurs corps [28, 29, 40]. Dans un système qui contient un grand nombre de particules toutes égales, il se produit une espèce de « résonance » entre ces particules et cela donne lieu à une décomposition du système des termes en systèmes partiels qui ne se combinent pas (Heisenberg, Dirac [28, 37]). Wigner a étudié ce phénomène systématiquement en se servant de méthodes basées sur la théorie des groupes et a établi l'ensemble des systèmes de termes qui ne se combinent pas [40]; Hund a réussi à déduire la plupart de ces résultats par des moyens relativement élémentaires [48]. Un rôle particulier est joué par les systèmes partiels « symétrique » et « antisymétrique » des termes; dans le premier une fonction caractéristique ne change pas lorsqu'on permute deux particules semblables, tandis que dans le dernier le signe change par une pareille permutation. En appli-quant cette théorie aux spectres d'atomes à plusieurs électrons on trouve que la règle d'équivalence ne permet pas le système antisymétrique. En se basant sur cette constatation on peut établir, par la mécanique des quanta, la systématique des spectres de raies et le groupement des électrons dans tout le système périodique des éléments.

Quand on a un grand nombre de particules toutes pareilles, qui doivent être soumises à un traitement statistique (théorie des gaz), on obtient des statistiques différentes suivant qu'on choisit la fonction ondulatoire correspondante conformément à l'un ou à l'autre de ces systèmes partiels. Le système symétrique est caractérisé par ceci que par permutation des particules dans un état représenté par une fonction caractéristique symétrique on n'obtient pas un nouvel état; toutes les permutations qui appartiennent au même système de nombres quantiques (qui se trouvent dans la même « cellule ») ont donc toujours ensemble le poids total 1. Cela correspond à la statistique de Bose-Einstein [56, 16]. Dans le système de termes antisymétriques deux nombres quantiques ne peuvent jamais devenir égaux, parce qu'autrement la fonction caractéristique disparaît; à un système de nombres quantiques correspond donc ou bien pas de fonction caractéristique ou bien une tout au plus; le poids d'un état est donc 0 ou 1. C'est la statistique de Fermi-Dirac [57, 37].

La statistique de Bose-Einstein est valable pour les quanta de lumière, comme il résulte de la validité de la formule du rayonnement de Planck. La statistique de Fermi-Dirac est certainement applicable aux électrons (négatifs), comme il résulte de la systématique des spectres, ci-dessus mentionnée, basée sur la règle d'équivalence de Pauli, et selon toute probabilité aussi à des particules élémentaires positives (protons); on peut le conclure des observations sur les spectres de bandes [28, 43], et en particulier de la chaleur spécifique de l'hydrogène à basse température [56]. L'application de la statistique de Fermi-Dirac aux particules élémentaires positives et négatives de la matière a pour conséquence qu'à toutes les configurations neutres (molécules, par exemple) s'applique la statistique de Bose-Einstein (symétrie des fonctions caractéristiques dans le cas de permutation d'un nombre entier de particules matérielles. Dans la mécanique des quanta, dans laquelle on traite un problème de plusieurs corps dans l'espace de configuration, la nouvelle statistique de Bose-Einstein et celle de Fermi-Dirac ont une place tout indiquée, contrairement à la théorie classique, où un changement arbitraire de la statistique normale est impossible; cependant, les restrictions faites quant à la forme des fonctions caractéristiques se présentent comme des hypothèses additionnelles arbitraires. En particulier, l'exemple des quanta de lumière indique que la nouvelle statistique est essentiellement liée aux propriétés ondulatoires de la matière et de la lumière. Lorsqu'on décompose les vibrations électromagnétiques d'un corps noir en composantes harmoniques spatiales, chacune d'elles se comporte vis-à-vis du temps comme un oscillateur harmonique; or, on trouve qu'en quantifiant ce système d'oscillateurs il vient une solution qui se comporte exactement comme un système de quanta de lumière obéissant à la statistique de Bose-Einstein [4]. Dirac s'est servi de cette circonstance pour traiter d'une manière conséquente des problèmes d'électrodynamique [51, 52]; nous y reviendrons un moment.

La structure corpusculaire de la lumière se présente donc comme une quantification des ondes lumineuses, tout comme inversement la nature ondulatoire de la matière se manifeste dans la « quantification » du mouvement corpusculaire. Jordan a montré [54] qu'on peut faire une chose analogue pour les électrons; on doit alors décomposer la fonction de Schrödinger du corps noir en vibrations fondamentales et harmoniques, et chacune de ces vibrations doit être quantifiée comme un oscillateur harmonique et cela notamment de telle manière qu'on trouve la statistique de Fermi-Dirac. Les nouveaux nombres quantiques, qui représentent les « poids » dans la théorie ordinaire du problème de plusieurs corps, n'ont donc que les valeurs o et 1. On a donc ici de nouveau un cas de matrices finies, qui ne peut être traité que par le théorie générale de Jordan. L'existence des électrons joue donc dans l'édification formelle de la théorie le même rôle que celle des quanta de lumière; tous deux sont des discontinuités de même espèce que les états stationnaires d'un

LA MÉCANIQUE DES QUANTA.   177

système quantifié. Lorsque les particules matérielles exercent des actions les unes sur les autres, des difficultés plus profondes pourraient évidemment s'opposer au développement de cette idée.

Les résultats des recherches de Dirac [51, 52] sur l'électro-dynamique des quanta consistent avant tout en une déduction rigoureuse des probabilités de transformation einsteinienne pour le rayonnement libre. On y considère le champ électromagnétique (décomposé en vibrations harmoniques quantifiées) et l'atome comme système couplé et l'on applique la mécanique des quanta sous la forme de l'équation intégrale (13). On trouve l'énergie d'action mutuelle qui s'y présente par l'application de formules classiques. Conformément à cela on peut expliquer la nature de l'absorption et de la diffusion de la lumière par des atomes. Finalement, Dirac [52] réussit à déduire une formule de dispersion avec terme d'amortissement; il donna en même temps l'inter-prétation par la mécanique des quanta des expériences de Wien sur la cessation de la luminescence des rayons canaux. Sa méthode consiste en ceci qu'il considère le processus de la diffusion de la lumière par des atomes comme un phénomène de choc de quanta de lumière. Comme on peut bien attribuer à un quantum de lumière une énergie et une quantité de mouvement, mais qu'il est difficile de lui attribuer une situation géométrique, on ne réussit pas à baser la théorie des chocs sur la mécanique ondula-toire (Born [30]), dans laquelle on suppose la connaissance des actions réciproques entre les corps qui s'entre-choquent comme fonction de leur situation relative. Il est donc nécessaire de considérer l'impulsion comme une variable indépendante et d'employer à la place de l'équation des ondes de Schrödinger une équation d'opérateurs ayant le caractère de matrices. On a ici un cas où l'application des points de vue généraux, que nous avons relevés dans ce rapport, ne saurait être évitée. En même temps cette théorie de Dirac prouve de nouveau la profonde analogie entre électrons et quanta de lumière.

Conclusion.

En résumé, nous voudrions faire observer que nous ne consi-dérons pas les recherches mentionnées en dernier lieu, qui sont

 ÉLECTRONS ET PHOTONS.

en rapport avec le traitement du champ électromagnétique par la mécanique des quanta, comme terminées, il est vrai, mais que nous tenons la *mécanique des quanta* pour une théorie complète, dont les hypothèses fondamentales physiques et mathématiques ne sont plus susceptibles de modification. Des suppositions au sujet de la signification physique des grandeurs de la mécanique quantique, qui seraient en contradiction avec les postulats de Jordan ou d'autres équivalents, seraient aussi, à notre avis, en contradiction avec les faits expérimentaux. (De pareilles contradictions peuvent se présenter, par exemple, lorsqu'on interprète le carré de la valeur de la fonction caractéristique comme densité électrique.) Dans la question de la « validité de la loi de causalité » notre opinion est celle-ci : Aussi longtemps qu'on ne considère que des expériences qui tombent dans le domaine de nos connaissances physiques et mécaniques acquises jusqu'ici, notre hypothèse fondamentale de l'indéterminisme essentiel est d'accord avec l'expérience. Le développement ultérieur de la théorie du rayonnement ne changera rien à cet état de choses. Car le dualisme entre corpuscules et ondes, qui dans la mécanique des quanta apparaît comme une partie d'une théorie complète, sans contradictions, existe d'une façon tout à fait semblable pour le rayonnement. La relation entre les quanta de lumière et les ondes électromagnériques doit être tout aussi statistique que celle entre les ondes de de Broglie et les électrons. Les difficultés, qui s'opposent encore aujourd'hui au développement complet de la théorie du rayonnement, ne résident donc pas dans le dualisme entre quanta de lumière et ondes — dualisme parfaitement compréhensible, d'ailleurs — mais elles ne se présentent que lorsqu'on essaie d'arriver à une façon parfaite, invariante en relativité, de formuler les lois électromagnétiques; toutes les questions, dans lesquelles une pareille formulation est inutile, peuvent être traitées par la méthode de Dirac [51, 52]. Mais déjà les premiers pas ont été faits dans la voie de la répression de ces difficultés.

Bibliographie.

1. W. Heisenberg, *Ueber quantenmechanische Umdeutung kinematischer and mechanischer Beziehungen* (*Z. f. Phys.*, 33, 1925, p. 879).

2. M. Born et P. Jordan, *Zur Quantenmechanik*, I (*Z. f. Phys.*, 34, 1925, p. 858).

3. P. Dirac, *The fundamental equations of quantum mechanics* (*Proc. Roy. Soc.*, A, 109, 1925, p. 642).

4. M. Born, W. Heisenberg et P. Jordan, *Zur Quantenmechanik*, II (*Z. f. Phys.*, 35, 1926, p. 557).

5. N. Bohr, *Atomtheorie und Mechanik* (*Naturwiss.*, 14, 1926, p. 1).

6. W. Pauli, *Ueber das Wasserstoffspektrum vom Standpunkt der neuen Quantenmechanik* (*Z. f. Phys.*, 36, 1926, p. 336).

7. P. Dirac, *Quantum mechanics and a preliminary investigation of the hydrogen atom* (*Proc. Roy. Soc.*, A, 110, 1926, p. 561).

8. G. E. Uhlenbeck et S. Goudsmit, *Ersetzung der Hypothese vom unmechanischen Zwang durch eine Forderung bezüglich des inneren Verhaltens jedes einzelnen Elektrons (Magnetelektron)* (*Nature*, 13, 1925, p. 953).

9. L. H. Thomas, *The motion of the spinning electron* (*Nature*, 117, 1926, p. 514).

10. J. Frenkel, *Die Elektrodynamik des rotierenden Elektrons* (*Z. f. Phys.*, 37, 1926, p. 243).

11. W. Heisenberg et P. Jordan, *Anwendung der Quantenmechanik auf das Problem der anomalen Zeemaneffekte* (*Z. f. Phys.*, 37, 1926, p. 263).

12. P. Dirac, *Relativity quantum mechanics with an application to Compton scattering* (*Proc. Roy. Soc.*, 111, 1926, p. 405).

13. P. Dirac, *The elimination of the nodes in quantum mechanics* (*Proc. Roy. Soc.*, 111, 1926, p. 281).

14. P. Dirac, *On quantum algebra* (*Proc. Cambridge Phil. Soc.*, 23, 1926, p. 412).

15. J. Dirac, *The Compton effect in wave mechanics* (*Proc. Cambridge Phil. Soc.*, 23, 1926, p. 500).

16. A. Einstein, *Quantentheorie des einatomigen idealen Gases*, II (*Berl. Ber.*, 1925, p. 5).

17. W. Elsasser, *Bemerkungen zur Quantentheorie freier Elektronen* (*Naturw.*, 13, 1925, p. 711).

18. C. Davisson et Kunsman, *Scattering of low speed electrons by* Pt *and* Mg (*Phys. Rev.*, 22, 1923, p. 243).

19. C. Davisson et L. Germer, *The scattering of electrons by a single crystal of Nickel* (*Nature*, 119, 1927, p. 558).

20. E. G. Dymond, *On electron scattering in Helium* (*Phys. Rev.*, 29, 1927, p. 433).

21. M. Born et N. Wiener, *Eine neue Formulierung der Quantengesetze für periodische und nicht periodische Vorgänge* (Z. f. Phys., 36, 1926, p. 174).

22. C. Eckart, *Operator calculus and the solution of the equation of quantum dynamics* (Phys. Rev., 28, 1926, p. 711).

23. K. Lauczos, *Ueber eine feldmässige Darstellung der neuen Quantenmechanik* (Z. f. Phys., 35, 1926, p. 812).

24. E. Schrödinger, *Quantisierung als Eigenwertproblem*, I à IV (I. Ann. d. Phys., 79, 1926, p. 361; II. Ibid., p. 489; III. Ibid., 80, 1926, p. 437; IV, Ibid., 81, 1926, p. 109).

25. E. Schrödinger, *Der stetige Uebergang von der Mikro zur Makromechanik* (Naturw., 14, 1926, p. 664).

26. E. Schrödinger, *Ueber das Verhältnis der Heisenberg-Born-Jordanschen Quantenmechanik zu der meinen* (Ann. d. Phys., 79, 1926, p. 734).

27. P. Jordan, *Bemerkung über einem Zusammenhang zwischen Duanes Quantentheorie der Interferenz und den de Broglieschen Wellen* (Z. f. Phys., 37, 1926, p. 376).

28. W. Heisenberg, *Mehrkörperproblem und Resonanz in der Quantenmechanik*, I et II (Z. f. Phys., 38, 1926, p. 411; II, Ibid., 41, 1927, p. 239).

29. W. Heisenberg, *Ueber die Spektren von Atomsystemen mit zwei Elektronen* (Z. f. Phys., 39, 1926, p. 499).

30. M. Born, *Zur Quantenmechanik der Stossvorgänge* (Z. f. Phys., 37, 1926, p. 863; 38, 1926, p. 803).

31. G. Wentzel, *Zwei Bemerkungen über die Zerstreuung korpuskularer Strahlen als Beugungserscheinung* (Z. f. Phys., 40, 1926, p. 590).

32. J. R. Oppenheimer, *Bemerkung zur Zerstreuung der α-Teilchen* (Z. f. Phys., 43, 1927, p. 413).

33. W. Elsasser, *Diss. Göttingen, Z. f. Phys.*

34. M. Born, *Das Adiabatenprinzip in den Quanten* (Z. f. Phys., 40, 1927, p. 167).

35. W. Heisenberg, *Schwankungserscheinungen und Quantenmechanik* (Z. f. Phys., 40, 1926, p. 501).

36. P. Jordan, *Ueber quantenmechanische Darstellung von Quantensprüngen* (Z. f. Phys., 40, 1926, p. 661).

37. P. Dirac, *On the theory of quantummechanics* (Proc. Roy. Soc., A, 112, 1926, p. 661).

38. P. Dirac, *The physical interpretation of quantum dynamics* (Proc. Roy. Soc., A, 113, 1926, p. 621).

39. P. Jordan, *Ueber eine neue Begründung der Quantenmechanik* (Z. f. Phys., 40, 1926, p. 899; Zweiter Teil : Z. f. Phys., 44, 1927, p. 1).

40. E. Wigner, *Ueber nichtkombinierende Terme in der neueren Quantentheorie* (Erster Teil : Z. f. Phys., 40, 1926, p. 492; Zweiter Teil : 40, 1927, p. 883).

41. D. Hilbert, I. v. Neumann et L. Nordheim, *Mathem. Ann.*, 98, 1927, p. 1.

42. I. v. Neumann, *Gött. Nachr.*, 20 mai 1927.

43. F. Hund, *Zur Deutung der Molekelspektra* (*Z. f. Phys.*, 40, 1926, p. 742; 42, 1927, p. 93; 43, 1927, p. 805).

44. W. Pauli, *Ueber Gasentartung u. Paramagnetismus* (*Z. f. Phys.*, 41, 1927, p. 81).

45. W. Pauli, *Zur Quantenmechanik des magnetischen Elektrons* (*Z. f. Phys.*, 43, 1927, p. 601).

46. W. Heisenberg, *Ueber den anschaulichen Inhalt der Quantenmechanik* (*Z. f. Phys.*, 43, 1927, p. 172).

47. N. Bohr, *Ueber den begrifflichen Aufbau der Quantentheorie* (Im Erscheinen).

48. F. Hund, *Symmetriecharaktere von Termen bei Systemen mit gleichen Partikeln in der Quantenmechanik* (*Z. f. Phys.*, 43, 1927, p. 788).

49. Darwin, *The electron as a vector wave* (*Nature*, 119, 1927, p. 282).

50. E. Kennard, *Zur Quantenmechanik einfacher Bewegungstypen* (*Z. f. Phys.*, 44, 1927, p. 326).

51. P. Dirac, *The quantum theory of emission and absorption of radiation* (*Proc. Roy. Soc.*, A, 114, 1927, p. 243).

52. P. Dirac, *The quantum theory of dispersion* (*Proc. Roy. Soc.*, A, 114, 1927, p. 710).

53. P. Jordan, *Ueber die Polärisation der Lichtquanten* (*Z. f. Phys.*, 44, 1927, p. 292).

54. P. Jordan, *Zur Quantenmechanik der Gasentartung* (*Z. f. Phys.*, Im Erscheinen).

55. D. Dennison, *A note on the specific heat of hydrogen* (*Proc. Roy. Soc.*, A, 114, 1927, p. 483).

   Zur Statistik noch :

56. N. S. Bose, *Plancks Gesetz und Lichtquantenhypothese* (*Z. f. Phys.*, 26, 1924, p. 178).

57. E. Fermi, *Lincei Rend.*, 3, 1926, p. 145.

   Allgemeine zusammenfassende Darstellungen :

58. M. Born (Theorie des Atombaus ?), *Vorlesungen gehalten am Massachusetts Institute of Technology* (Springer, 1927).

59. W. Heisenberg, *Quantentheoretische Kinematik u. Mechanik* (*Math. Ann.*, 95, 1926, p. 683).

60. W. Heisenberg, *Quantenmechanik* (*Naturw.*, 14, 1926, p. 990).

61. M. Born, *Quantenmechanik und Statistik* (*Naturw.*, 15, 1927, p. 238).

62. P. Jordan, *Kausalität und Statistik in der modernen Physik* (*Naturw.*, 15, 1927, p. 105).

63. P. Jordan, *Die Entwicklung der neuen Quantenmechanik* (*Naturw.*, 15, 1927, p. 636).

DISCUSSION DU RAPPORT DE MM. BORN ET HEISENBERG.

M. Dirac. — Je voudrais faire ressortir nettement la nature de la correspondance entre la mécanique des matrices et la mécanique classique. En mécanique classique on peut traiter un problème par deux méthodes : 1° en considérant toutes les variables comme des nombres et en calculant le mouvement, par exemple par les lois de Newton; 2° en considérant les variables comme des fonctions des J (les variables d'action) et employant la théorie générale de transformation de la dynamique. Par la première méthode on calcule le mouvement qui résulte d'un système particulier de valeurs numériques pour les coordonnées et quantités de mouvement initiales, tandis que par la seconde on détermine en même temps le mouvement qui résulte de toutes les conditions initiales possibles.

La théorie des matrices correspond à cette seconde méthode classique. Elle donne des renseignements à la fois sur tous les états du système. Il y a cependant une différence entre la méthode des matrices et la seconde méthode classique en ce sens que dans cette dernière on demande de traiter en même temps uniquement tous les états qui ont à peu près le même J $\left(\text{comme cela arrive,}\right.$ par exemple, quand on fait usage de l'opérateur $\left.\frac{\partial}{\partial J}\right)$, tandis que dans la théorie des matrices on doit traiter simultanément des états dont les J diffèrent de quantités finies.

Afin d'obtenir des résultats comparables à ceux fournis par l'expérience, on doit remplacer par des valeurs numériques les J dans les fonctions des J obtenues par le traitement général. On doit faire la même chose dans la théorie des matrices. Cela donne lieu à une difficulté, parce que les résultats du traitement général sont maintenant des éléments de matrices, dont chacun se rapporte en général à deux systèmes différents de valeurs de J. Ce sont uniquement les éléments diagonaux, pour lesquels ces deux systèmes de valeurs coïncident, qui sont susceptibles d'une interprétation physique directe.

M. Lorentz. — J'ai été fort surpris de voir que les matrices satisfont à des équations de mouvement. En théorie cela est très beau, mais c'est pour moi un grand mystère, qui, je l'espère, s'éclaircira. On me dit que par toutes ces considérations on est parvenu à former des matrices qui représentent ce que l'on peut observer dans l'atome, par exemple les fréquences des radiations émises. Toutefois, la circonstance que les coordonnées, l'énergie potentielle, etc., sont maintenant représentées par des matrices, indique que ces grandeurs on perdu leur sens initial et qu'on a fait un pas énorme dans la voie de l'abstraction.

Permettez-moi d'attirer l'attention sur un autre point qui m'a frappé. Considérons les éléments des matrices qui représentent les coordonnées d'une particule dans un atome, un atome d'hydrogène par exemple, et qui satisfont aux équations du mouvement. On peut alors changer la phase dans chaque élément des matrices sans que celles-ci cessent de satisfaire aux équations du mouvement ; on peut, par exemple, changer le temps. Mais on peut aller plus loin encore et changer les phases, non pas d'une façon arbitraire, mais en multipliant chaque élément par un facteur de la forme $e^{i(\delta_m - \delta_n)}$, et c'est encore autre chose qu'un changement de l'origine du temps.

Maintenant ces éléments des matrices devraient signifier des radiations émises. Si ces radiations émises étaient ce qu'il y a au fond de tout cela, on pourrait s'attendre à ce que toutes les phases puissent être changées d'une manière arbitraire. Le fait mentionné nous conduit alors tout naturellement à l'idée que ce ne sont pas ces radiations qui sont la chose fondamentale ; il nous conduit à penser que derrière ces vibrations émises se cachent des vibrations véritables, dont les vibrations émises sont des vibrations de différence.

Ainsi donc, au fond il y aurait des vibrations dont les vibrations émises sont les différences, comme dans la théorie de Schrödinger, et il me semble que cela est contenu dans les matrices. Cette circonstance indique l'existence d'un substratum ondulatoire plus simple.

M. Born. — M. Lorentz s'étonne de ce que les matrices satisfont aux équations de mouvement ; je voudrais signaler à ce propos l'analogie avec les nombres complexes. Ici aussi nous avons un cas où dans une extension du système des nombre les lois formelles se conservent presque complètement. Les matrices sont une espèce de nombres hypercomplexes, qui se distinguent des nombres ordinaires par le fait que la loi de commutativité n'est plus valable.

M. Dirac. — Les phases arbitraires qui se présentent dans la méthode des matrices trouvent une analogie dans la seconde méthode classique, où les variables sont des fonctions des J et des w (variables d'action et d'angle). Il y a dans les w des phases arbitraires, qui peuvent avoir des valeurs différentes pour chaque système différent de valeurs des J. Ceci est tout à fait analogue à la théorie des matrices, dans laquelle chaque phase arbitraire est associée à une rangée et à une colonne et par conséquent à une série de valeurs pour les J.

M. Born. — Les phases α_n dont M. Lorentz vient de parler sont associées aux divers niveaux d'énergie, tout à fait comme dans la mécanique classique. Je ne pense pas qu'il se cache derrière tout cela quelque chose de mystérieux.

M. Bohr. — La question de la signification des phases arbitraires, soulevée par M. Lorentz, est d'une très grande importance, je pense, dans la discussion de la cohérence des méthodes de la théorie des quanta. Bien que la notion de phase soit indispensable dans le calcul, c'est à peine si elle intervient dans l'interprétation des observations.

In *Probleme der modernen Physik: Arnold Sommerfeld zum 60. Geburtstage gewidmet von seinen Schülern,* ed. by P. Debye (S. Hirzel, Leipzig 1928) pp. 114–122

Zur Quantentheorie des Ferromagnetismus

Von

W. Heisenberg

§ 1. Vergleich der Theorie mit der Pauli-Sommerfeldschen Behandlung der Elektronen im Metall

Das molekulare Feld, welches die Grundlage der Weissschen Theorie des Ferromagnetismus[1]) bildet, kann zurückgeführt werden auf Austauschwechselwirkungen zwischen den Leitungselektronen des ferromagnetischen Metalles[2]). Da das Modell des Metalles, das zu diesem Resultat führte, verschieden ist von dem Modell, welches von Pauli[3]) oder in der Sommerfeldschen Theorie[4]) der Leitfähigkeit zugrunde gelegt wird, so sei zunächst der Übergang vom Pauli-Sommerfeldschen Modell zu dem hier benutzten in einzelnen Schritten durchgeführt.

In erster Näherung können die Elektronen im Metall als völlig frei behandelt werden. Ihre Energieniveaus bestimmen sich dann aus der Lage der Wände des Elementarbereichs und aus der Elektronenmasse. In jeder „Quantenzelle" können zwei Elektronen entgegengesetzter Spinrichtung Platz finden; solange die Wechselwirkung der Elektronen unberücksichtigt bleibt, muß also ganz unabhängig von der Verteilung der Energieniveaus nach Pauli bei tiefen Temperaturen Paramagnetismus eintreten. Berücksichtigt man die Wechselwirkung der Elektronen mit dem Gitter, wie dies in der Sommerfeldschen Theorie Houston[5]) und etwas eingehender mit gruppentheoretischen Hilfsmitteln Bloch[6]) getan hat, so ordnen sich die Termwerte der Elektronen in Gruppen, die jeweils zu einem bestimmten Anregungszustand der Metallatome gehören. Befinden sich im Elementarbereich ursprünglich n Atome mit n Valenzelektronen (Spezialfall der Alkalien!), so besteht jede Gruppe aus n Termen, deren relativer Abstand nun wesentlich durch die Bindungsstärke der Elektronen im Atom gegeben ist. Solange von der Wechselwirkung der Elektronen abgesehen wird, ändert sich an dem Paulischen Resultat über den Paramagnetismus nichts. Allerdings rücken die Termwerte einer Gruppe mit wachsender Bindungsstärke immer mehr zusammen. Von einem gewissen Grenzwert ab werden die Abstände benachbarter Termwerte gegen die Wechselwirkungsenergien der Elektronen vernachlässigt werden können. Als ungestörtes System für die Berücksichtigung der Wechselwirkung wird man also, wenn die Bindungsstärke groß genug ist, die Verteilung von n Elektronen auf n energetisch gleich-

---

[1]) P. Weiss, Journ. de phys. **6**, 661, 1907.
[2]) W. Heisenberg, Ztschr. f. Phys. **49**, 619, 1928, im folgenden als „l. c. I" zitiert.
[3]) W. Pauli, Ztschr. f. Phys. **41**, 81, 1928.
[4]) A. Sommerfeld, Ztschr. f. Phys. **47**, 1, 1928.
[5]) W. Houston, Ztschr. f. Phys. **48**, 449, 1928.
[6]) F. Bloch, Leipziger Dissertation. (Im Erscheinen.)

wertige Quantenzellen wählen müssen. Man braucht dann die Quantenzellen nicht mehr in Beziehung zu den Wänden des Elementarbereichs zu setzen, sondern kann direkt die „Elektronenbahnen" um die einzelnen n Atome als die n energetisch gleichwertigen (aber örtlich verschiedenen) Quantenzellen benützen. Der Unterschied unserer Theorie von der Paulischen Betrachtungsweise wird nun von dem Moment an wesentlich, wo wir die Wechselwirkung der Elektronen mit berücksichtigen; diese Wechselwirkung wird nämlich in erster Linie zur Folge haben, daß nur mehr ein Elektron in einer Quantenzelle Platz hat, d. h. die Zustände, bei denen zwei Elektronen in einer Zelle zu finden sind, haben erheblich höhere Energiewerte als die, bei denen die Zellen nur einfach besetzt sind; bei einer statistischen Betrachtung spielen daher nur die letzteren (bei einigermaßen tiefen Temperaturen) eine Rolle.

Für den Unterschied der geschilderten verschiedenen Modellvorstellungen ist auch das Verhalten der spezifischen Wärme des „Elektronengases" lehrreich. In der Sommerfeldschen Theorie (l. c.), wo die Elektronen als Fermi-Gas behandelt werden, steigt ihre spezifische Wärme linear mit wachsender Temperatur an. Berücksichtigt man die Wechselwirkung mit dem Gitter, so steigt nach Bloch (l. c.) die spezifische Wärme zunächst linear an, nimmt aber dann nach Erreichung eines Maximalwertes wieder zu Null ab; bei unserem Modell schließlich ist die spezifische Wärme des Elektronengases identisch mit der sog. Magnetisierungswärme, auch sie steigt bis zum Curiepunkt an und sinkt dann unstetig zu Null[1]).

§ 2. Zusammenhang der Theorie mit der quantenmechanischen Deutung der Valenzkräfte

Die Weissschen Molekularkräfte sind auf Austauschwirkungen ähnlicher Art zurückzuführen, wie die, durch welche Heitler und London[2]) einen Teil der homöopolaren Valenzkräfte gedeutet haben. Beim Überschreiten des Curiepunktes entkoppelt man also sozusagen diese Valenzkräfte, und man könnte sich füglich darüber wundern, weshalb der Krystall dann überhaupt noch zusammenhält. Die Metallatome sind aber im Gitter noch durch andere Kräfte aneinander gebunden als durch jene Austauschglieder. Zunächst spielen die Polarisationswirkungen (die „van der Waalschen Kräfte") eine große Rolle, sie sind keineswegs klein gegen die Glieder „erster Ordnung". Dann ist zu beachten, daß die Atomreste im allgemeinen nicht kugelsymmetrisch sind, ihr resultierender „Bahnimpuls l" wird nicht verschwinden. Diese Abweichung von der Kugelsymmetrie führt zu erheblichen Quadrupolmomenten; diesen Quadrupolmomenten ist es schließlich auch zu verdanken, daß die „Bahnimpulse l" im Krystall nicht frei drehbar sind, sich im ganzen kompensieren und zum Magnetismus des Krystalls nichts beitragen. Wenn man noch bedenkt, daß von den Austauschgliedern der verschiedenen Elektronen eines Atoms mit den Elektronen eines andern manche positiv, manche negativ sein werden, so versteht man, daß die Austauschglieder nur einen kleinen Beitrag zur Krystallbindung liefern. Man kann seine relative Größe schätzen aus dem Verhältnis der gesamten Magnetisierungswärme zur Verdampfungswärme des Krystalls und erhält etwa einige Prozent. Um einen ähnlichen relativen Betrag müssen sich auch die elastischen Konstanten des

---

[1]) Vgl. z. B. P. Debye, Zusammenfassender Bericht über elektrische und magnetische Molekulareigenschaften im Handbuch der Radiologie, Leipzig 1925, Bd. VI, S. 597; insbes. S. 668ff.

[2]) W. Heitler und F. London, Ztschr. f. Phys. **44**, 455, 1927.

Krystalls beim Durchgang durch den Curiepunkt ändern, worauf mich Herr London in einer Unterredung freundlicherweise hinwies. Experimentell ist dieser Effekt leider niemals untersucht worden; wenn er später einmal gemessen sein wird, so hätte man eine ganz augenfällige mechanische Folge der Entkoppelung von Valenzkräften gefunden. Wie schon erwähnt (vgl. auch l. c. I) dürften bei den bekannten ferromagnetischen Substanzen die Austauschglieder teils positiv, teils negativ sein, und man muß es als eine Art Zufall ansehen, daß sie sich bei Fe, Co, Ni gerade zu einem positiven Gesamtbetrag addieren; natürlich hängt dieser Gesamtbetrag auch von der relativen Lage der Atome im Gitter ab; es ist sehr charakteristisch, daß bei γ-Eisen im Gegensatz zu α-, β- und δ-Eisen der Gesamtbetrag negativ wird. Bei manchen Stoffen wird der bei tiefen Temperaturen beobachtete Paramagnetismus auch auf ein solches negatives Molekularfeld zurückzuführen sein. Der Gesamteindruck, den man von einer Diskussion des experimentellen Materials bekommt, ist allerdings der, daß es noch eine gute Weile dauern wird, bis man diese vielen und komplizierten Einzelheiten wirklich theoretisch wird herleiten können. Bei Eisen z. B. hat man mindestens mit vier Valenzelektronen, die zum Magnetismus beitragen, zu rechnen, also gehen in die Theorie zunächst 16 Austauschglieder als unbekannte Konstanten ein, und man kann damit wohl jedes gewünschte Resultat erreichen, also nichts Positives aussagen. Es erschien mir daher richtig, die Theorie zunächst nur für den Fall eines Valenzelektrons pro Atom durchzuführen. Ich will jedoch noch auf einen bestimmten Punkt in der komplizierteren Theorie hinweisen. Nach den bisherigen Rechnungen schien es, als ob die Langevinsche Funktion $\mathfrak{Cotg}\,x - 1/x$ allgemein wegen der zwei möglichen Einstellungen des Elektronenspins durch $\mathfrak{Tg}\,x$ zu ersetzen sei. Empirisch trifft aber sicher die Kurve $\mathrm{Tg}\,x$ nicht das Richtige, und es lohnt sich daher, zu zeigen, daß im Fall mehrerer Valenzelektronen pro Atom auch andere Langevinfunktionen resultieren.

§ 3. Gruppentheorie für den Fall mehrerer Valenzelektronen pro Atom

Die gruppentheoretische Methode können wir wieder von Heitler[1]) oder London[2]) übernehmen. Im ungestörten System möge jedes Atom y Valenzelektronen besitzen, deren Spinrichtungen alle (der Einfachheit halber) parallel sein sollen, das Atom befindet sich also in einem Term, der zum antisymmetrischen System (partitio $1 + 1 + 1 + \cdots + 1$) gehört. Die Gesamtzahl der Atome im Gitter sei wieder $2n$, die Gesamtzahl der Valenzelektronen $2n \cdot y = 2N$. Die Permutationsgruppe der $2N$ Elektronen enthält als Untergruppen die Permutationsgruppe der y Elektronen im Atom 1, die der y Elektronen im Atom 2 usw. Diejenigen der irreduziblen Darstellungsmatrizen (zu einer gegebenen partitio), welche zu einer solchen Untergruppe gehören, lassen sich für sich wieder ausreduzieren, bis sie den irreduziblen Darstellungen dieser Untergruppe entsprechen. Wir denken uns diese Reduktion in sämtlichen Untergruppen ausgeführt. Macht man dann die Annahme, daß die Wechselwirkung der y Elektronen eines Atoms sehr viel größer sei, als die Wechselwirkung der Elektronen ver-

---

[1]) W. Heitler, Ztschr. f. Phys. **46**, 835, 1927.

[2]) F. London, Ztschr. f. Phys., **50**, 24, 1928; die Londonschen Untersuchungen würden wohl unter geringfügigen Erweiterungen auch die Behandlung des Falles zulassen, bei dem das einzelne Atom im ungestörten Zustand sich nicht in einem Term des antisymmetrischen Systems befindet. Die hier durchgeführten Rechnungen behandeln dagegen den eben genannten Spezialfall und schließen sich eng an die Arbeit von Heitler l. c. an.

Gruppentheorie für den Fall mehrerer Valenzelektronen pro Atom   117

schiedener Atome, so darf man für die Lösung des Säkularproblems die gesamte Säkulardeterminante

$$\left| \sum_P b_{ik}^P J_P - \delta_{ik} W \right|$$

zerspalten in ein Produkt von Teildeterminanten, wobei jede Teildeterminante aus den Zeilen und Kolonnen sich zusammensetzt, die zu bestimmten irreduziblen Darstellungen jeder einzelnen Untergruppe gehören. In unserem speziellen Fall interessieren nur die irreduziblen Darstellungen der Untergruppen mit der partitio $1 + 1 + \cdots + 1$, also einreihige Matrizen mit dem Element $+ 1$ bzw. $- 1$.

Es muß zunächst festgestellt werden, wie oft diese speziellen Darstellungen in der Darstellung mit einer bestimmten partitio der $2N$ Elektronen vorkommen. Es gilt hier nach Heitler (l. c.) die einfache Regel, daß nur diejenigen partitiones der einzelnen Untergruppen zusammen in einer partitio der $2N$ Elektronen vorkommen können, die sich additiv zur Gesamtpartitio zusammensetzen lassen. Für drei Atome (A, B, C) mit je drei Elektronen ($y = 3$) können z. B. die partitiones $1 + 1 + 1$ in vier verschiedenen Weisen zur Gesamtpartitio $1 + 1 + 1 + 2 + 2 + 2$ zusammentreten, wie das folgende Schema zeigt:

Atom A	$1+1+1$		$1+1+1$		$1+1+1$		$1+1+1$	
„ B		$1+1+1$		$1+1+1$		$1+1+1$	$1+1+1$	
„ C		$1+1+1$		$1+1+1$		$1+1+1$		$1+1+1$
	$1+1+1+2+2+2$		$1+1+2+2+2+1$		$1+2+2+2+1+1$		$2+2+2+1+1+1$

Nehmen wir also an, daß allgemein die partitiones $1 + 1 + 1 + \cdots + 1$ in g_σ verschiedenen Weisen zur gegebenen Gesamtpartitio σ zusammentreten können. Dann interessiert uns von der Säkulardeterminante nur eine Teildeterminante mit g_σ Zeilen und Kolonnen (durch eine einfache Umordnung läßt sich erreichen, daß es gerade die ersten g_σ Zeilen und Kolonnen sind); eine Darstellungsmatrix b_{ik}^P, welche zu einer Permutation der Elektronen eines einzelnen Atoms gehört, hat also dann in den ersten g_σ Zeilen und Kolonnen nur Diagonalelemente, und zwar nur $+ 1$ oder $- 1$ stehen, je nachdem die Permutation gerade oder ungerade ist.

Wir berechnen im folgenden wieder die Summe der Wurzeln sowie die Summe der Wurzelquadrate dieser reduzierten Säkulardeterminante, aus diesen beiden Größen dann den Termschwerpunkt und das mittlere Schwankungsquadrat. Von den Austauschintegralen J_P berücksichtigen wir nur die, welche einfachen Transpositionen von Elektronen verschiedener Atome entsprechen. Denn die Austauschglieder, welche den Permutationen der Elektronen eines einzelnen Atoms entsprechen, können für unsere Rechnung als additive Konstanten betrachtet werden. Die Summe der Wurzeln ist in bekannter Weise gegeben durch

$$\sum_1^{g_\sigma} E_k = \sum_{i=1}^{g_\sigma} \sum_P b_{ii}^P J_P . \tag{1}$$

Wir benutzen dann die Gleichung [Heitler, l. c. Gl. (18)]

$$\sum_R b_{ii}^R = \chi_\sigma^R \frac{h_R}{f_\sigma}, \tag{2}$$

links ist die Summe über alle Permutationen einer Klasse R zu erstrecken, h_R bedeutet die Anzahl der Permutationen in R, die übrigen Größen haben die übliche Bedeutung.

Bezeichnet man Transpositionen der Elektronen am Atom 1 mit T_1, solche am Atom 2 mit T_2 usw., alle übrigen Transpositionen mit T, so gilt also

$$\sum_{T_1} b_{ii}^{T_1} + \sum_{T_2} b_{ii}^{T_2} + \cdots + \sum_{T} b_{ii}^{T} = \chi_{(12)}^{\sigma} \frac{\binom{2N}{2}}{f_\sigma}.$$

Nun ist
$$b_{ii}^{T_1} = b_{ii}^{T_2} = \cdots = -1,$$

also
$$\sum_{T_1} b_{ii}^{T_1} = \sum_{T_2} b_{ii}^{T_2} = \cdots = -\binom{y}{2}$$

und daher
$$\sum_{T} b_{ii}^{T} = \chi_\sigma^{(12)} \frac{\binom{2N}{2}}{f_\sigma} + 2n\binom{y}{2}. \tag{3}$$

Die Zahl der verschiedenen T ist

$$\binom{2N}{2} - 2n\binom{y}{2} = \frac{2N(2N-1) - 2ny(y-1)}{2} = N\cdot(2N-y).$$

Aus Symmetriegründen werden die verschiedenen T in gleicher Weise ins Endresultat eingehen, also ist

$$\sum_{h=1}^{g_\sigma} E_k = \frac{g_\sigma}{f_\sigma} \sum_{T} J_T \cdot \frac{\chi_\sigma^{(12)}\binom{2N}{2} + 2n f_\sigma \binom{y}{2}}{N(2N-y)}. \tag{4}$$

Zur Berechnung von $\sum_{h=1}^{g_\sigma} E_k^2$ ist zu beachten, daß allgemein aus dem Bestehen der Gleichung
$$\left| \sum b_{ik}^{P} J_P - W\cdot\delta_{ik} \right| = 0$$

durch Multiplikation mit der Determinante $\left| \sum b_{ik}^{P'} J_{P'} + W\delta_{ik} \right|$ geschlossen werden kann

$$\left| \sum_{PP'}\sum_{l} b_{il}^{P} b_{lk}^{P'} J_P J_{P'} - W^2 \delta_{ik} \right| = 0.$$

Es wird in ähnlicher Weise ganz allgemein $\left(\sum_{l} b_{il}^{P} b_{lk}^{P'} = b_{ik}^{P\cdot P'} \right)$

$$\sum_{l=1}^{f_\sigma} W_l^k = \sum_{P,\,P'\ldots P^{(k)}} \chi^{P\cdot P'\ldots P^{(k)}} J_P \cdot J_{P'} \cdots J_{P^{(k)}}. \tag{5}$$

In unserem Spezialfall berücksichtigen wir wieder nur die g_σ ersten Zeilen und Kolonnen und erhalten

$$\sum_{k=1}^{g_\sigma} E_k^2 = \sum_{P,\,P'} \sum_{i=1}^{g_\sigma} b_{ii}^{P\cdot P'} J_P J_{P'}. \tag{6}$$

Bei der folgenden Berechnung der $\sum_{i=1}^{g_\sigma} b_{ii}^{P\,P'}$ muß geschieden werden zwischen den drei möglichen Fällen 1) $P = P'$; 2) P und P' haben ein Element gemeinsam; 3) P und P' haben kein Element gemeinsam. Da als P nur Transpositionen in Betracht kommen, so ist im Falle 1)

$$b_{ii}^{P\,P'} = b_{ii}^{E} = 1.$$

Gruppentheorie für den Fall mehrerer Valenzelektronen pro Atom 119

Im Fälle 2) gibt es drei Gruppen von b-Größen: Typus $b_{ii}^{T_1 \cdot T_1'}$; Typus $b_{ii}^{T_1 \cdot T}$; Typus $b_{ii}^{T \cdot T'}$.

Es gilt $b_{ii}^{T_1 \cdot T_1'} = 1$; $b_{ii}^{T_1 \cdot T} = -b_{ii}^{T}$, also

$$b_{ii}^{T_1 \cdot T} = -\frac{1}{N(2N-y)} \cdot \left\{ \chi_\sigma^{(12)} \frac{\binom{2N}{2}}{f_\sigma} + 2n\binom{y}{2} \right\}.$$

Die Größe $b_{ii}^{TT'}$ berechnen wir nach Gl. (2).

Es kommt der Typus $b_{ii}^{T_1 T_1'}$ im ganzen $2\binom{y-1}{2} y \cdot 2n$ mal, der Typus $b_{ii}^{T_1 T}$ kommt $2(y-1)(2N-y) \cdot 2N$ mal, der Typus $b_{ii}^{TT'}$ schließlich $2 \cdot \binom{2N-y}{2}$ $\cdot 2N$ mal vor. Unter den letzteren gibt es aber noch $2 \cdot \binom{y}{2}(2n-1) \cdot 2N$ Größen $b_{ii}^{TT'}$, bei denen die beiden nichtgemeinsamen Elemente von T und T' zum gleichen Atom gehören; diese $b_{ii}^{TT'}$ sind gleich denen des Typus $b_{ii}^{T_1 T}$. Also gilt schließlich

$$\left. \begin{aligned} &2\binom{y-1}{2} y\, 2n - \frac{2}{N(2N-y)} \left\{ \frac{\chi_\sigma^{(12)} \binom{2N}{2}}{f_\sigma} + 2n\binom{y}{2} \right\} \\ &\quad \cdot \left\{ (y-1)(2N-y)2N + \binom{y}{2}(2n-1)2N \right\} \\ &\quad + 2\left\{ \binom{2N-y}{2} 2N - \binom{y}{2}(2n-1)2N \right\} b_{ii}^{TT'} = \chi_\sigma^{(123)} \frac{3 \cdot h^{(123)}}{f_\sigma} \cdot \end{aligned} \right\} \quad (7)$$

Auf der rechten Seite muß hier $3\,h^{(123)}$ statt $h^{(123)}$ stehen, weil jede Permutation (123) auf drei verschiedene Weisen als Produkt von zwei Transpositionen geschrieben werden kann; ferner ist $h^{(123)} = 2\binom{2N}{3}$.

Also gilt

$$b_{ii}^{TT'} = \frac{\dfrac{\chi_\sigma^{(123)}}{f_\sigma} \cdot N(2N-1)(2N-2) + 3(y-1)\left\{ \dfrac{\chi_\sigma^{(12)}}{f_\sigma}\binom{2N}{2} + 2n\binom{y}{2} \right\} - \binom{y-1}{2} 2N}{2N(2N-y)(N-y)}. \quad (8)$$

Eine ähnliche Rechnung muß nun noch für den 3. Fall (P und P' haben kein Element gemeinsam) angestellt werden; wieder sind die oben genannten drei Typen zu unterscheiden. Hier kommt nun der Typus $b^{T_1 T_1'}$

$$\left\{ \left[2n\binom{y}{2}\right]^2 - 2n\binom{y}{2} - 2\binom{y-1}{2}2N \right\} \text{mal,}$$

der Typus $b^{T_1 T}$

$$2\left\{ 2n\binom{y}{2} \cdot N(2N-y) - (y-1)(2N-y)2N \right\} \text{mal,}$$

schließlich $b^{T\,T'}$

$$\left\{ N^2(2N-y)^2 - N(2N-y) - 2\binom{2N-y}{2}2N \right\} \text{mal}$$

                    Zur Quantentheorie des Ferromagnetismus

vor. Es folgt daher, wegen

$$h^{(12)(34)} = \frac{N(2N-1)\{N(2N-5)+3\}}{2},$$

$$b_{ii}^{TT'} = \frac{\frac{\chi_\sigma^{(12)(34)}}{f_\sigma}-N(2N-1)[N(2N-5)+3]+2(N-2)(y-1)\left\{\frac{\chi_\sigma^{(12)}}{f_\sigma}\binom{2N}{2}+2n\binom{y}{2}\right\}-N(y-1)[N(y-1)-1-2y]}{N(2N-y)[N(2N-y-4)+1+2y]}.$$
$$\tag{9}$$

Glücklicherweise können diese Formeln für die weiteren Rechnungen erheblich vereinfacht werden. Zunächst folgt aus Gl. (4)

$$\overline{E} = \sum_T J_T b_{ii}^T,\tag{10}$$

ebenso aus Gl. (6)

$$\overline{E^2} = \sum_T \sum_{T'} b_{ii}^{TT'} J_T J_{T'}$$

also

$$\overline{\Delta E^2} = \sideset{}{'}\sum_T \sideset{}{'}\sum_{T'}(b_{ii}^{TT'}-b_{ii}^T\cdot b_{ii}^{T'})J_T J_{T'} = A_{TT'}J_T J_{T'}.\tag{11}$$

Wir berücksichtigen in $\overline{E}$ und $\overline{\Delta E^2}$ nur die höchsten Potenzen von N. Also folgt nach Einsetzen der χ_σ-Werte [Gesamtpartitio $(N+s)+(N-s)$] aus l. c. I:

$$\overline{E} = -\frac{s^2+N^2}{2N^2}\sum_T J_T.\tag{12}$$

Ähnlich ergibt sich für die drei Arten der $A_{TT'}$-Werte:

$$A_{(12)(12)} = \frac{(N^2-s^2)(3N^2+s^2)}{4N^4},\tag{13}$$

$$A_{(12)(13)} = \frac{(N^2-s^2)s^2}{4N^4}.\tag{14}$$

Daß die Gl. (12), (13), (14) mit den entsprechenden Gleichungen für $y=1$ (l. c. I) übereinstimmen, ist selbstverständlich, weil die y-Werte in allen Formeln erst in den Gliedern mit der zweithöchsten Potenz in N vorkommen. Nur für den Wert von $A_{(12)(34)}$, bei dem die höchsten Glieder in N sich kompensieren, ist eine Abweichung zu erwarten.

Es ergibt die Potenzentwicklung von b_{ii} bis zum zweiten Gliede:

$$b_{ii}^{(12)} = \frac{1}{2}\left(1+\frac{y}{2N}\right)\left(1+\left(\frac{s}{N}\right)^2+\frac{y-3+\frac{s}{N}}{N}\right)$$

$$b_{ii}^{(12)(34)} = \frac{1}{4}\left(1+\frac{y+2}{N}\right)\left(\left[1+\left(\frac{s}{N}\right)^2\right]^2+\frac{2y-8+2\frac{s}{N}+(2y-12)\left(\frac{s}{N}\right)^2+2\left(\frac{s}{N}\right)^3}{N}\right)$$

und

$$A_{(12)(34)} = \frac{-(N^2-s^2)s^2}{2N^5}.\tag{15}$$

Obwohl in den b_{ii} Abweichungen auftreten, kompensieren sie sich in der Formel für $A_{(12)(34)}$, die daher mit der entsprechenden Formel in l. c. I übereinstimmt.

Die Berechnung der Werte von $\overline{E}$ und $\overline{\Delta E^2}$ für eine bestimmte partitio $(N-s)+(N+s)$ ist mit Hilfe der Formeln (12) bis (15) möglich, wenn die

Statistische Ableitung der Langerin-Weissschen Formeln. **121**

Austauschintegrale J_T bekannt sind. Nimmt man z. B. an, daß sämtliche J_T zweier Nachbaratome gleich seien ($= J_0$), so multipliziert sich die „Anzahl der Nachbarn z" (vgl. l. c.) einfach mit y, und die Endformeln würden lauten

$$\overline{E} = -z \cdot y \frac{s^2 + N^2}{2N} J_0 (+ \text{const}),$$

$$\overline{\Delta E^2} = + J_0^2 z \cdot y \frac{(N^2 - s^2)(3N^2 - s^2)}{4N^3}.$$

Die Annahme gleicher Austauschintegrale für verschiedene Elektronen scheint aber physikalisch ganz ungerechtfertigt, wir müssen hier die Austauschintegrale als unbestimmte Konstanten stehen lassen.

§ 4. Statistische Ableitung der Langevin-Weissschen Formeln

Die statistische Durchführung der Theorie ist mit den bisherigen Daten nur möglich unter der willkürlichen Annahme der Gaußverteilung (vgl. l. c. I) für die Termwerte, die zu einer partitio gehören. Diese Annahme ist zweifellos nur ein Notbehelf und eine recht bedenkliche Lücke der bisherigen Theorie, deren Ausfüllung mir bisher nicht gelungen ist. Genauere Rechnungen von Herrn Peierls über die höheren Schwankungsmittelwerte $\overline{\Delta E^l}$ zeigten, daß die spezielle Gaußsche Form der Verteilungskurve eine wesentliche Rolle im Endresultat der Rechnung spielt und daß die Verteilungskurve durch die nächsten Schwankungsgrößen $\overline{\Delta E^3}$ usw. nicht verbessert werden kann. Eine solche Verbesserung wäre erst möglich auf Grund der Momente $\overline{\Delta E^l}$, wo l von der Größenordnung N ist, und deren Berechnung würde außerordentliche mathematische Schwierigkeiten bereiten. Es spielt also die Art des Abfalls der Verteilungskurve in weitem Abstand vom Maximum eine entscheidende Rolle, und gerade für diesen Abfall geben die Werte $\overline{E}$ und $\overline{\Delta E^2}$ kaum einen Anhaltspunkt. Mathematisch läuft die gestellte Frage hinaus auf die Berechnung der Summe $\sum_l e^{-\frac{E_l}{kT}}$ genommen über alle zu einer bestimmten irreduzibeln Darstellung gehörigen Termwerte E_l. Es läßt sich leicht zeigen, daß diese Summe als Potenzreihe von $\frac{1}{kT}$ geschrieben werden kann, deren Koeffizienten durch die Gruppencharaktere zur betreffenden partitio und die Austauschintegrale dargestellt werden [vgl. Gl. (5)]. Hierdurch wird aber die numerische Berechnung dieser Funktion von kT nicht erleichtert. Methoden zur Lösung dieses rein mathematischen Problems habe ich in der Literatur nicht gefunden. Einstweilen wird man also auf eine befriedigende Herleitung der Weissschen Formeln verzichten und sich mit dem Notbehelf der Gaußverteilung begnügen müssen. Streng läßt sich nur beweisen, daß für hinreichend tiefe Temperaturen sicher alle Spinvektoren parallel stehen, d. h. daß völlige Sättigung eintritt ($s = N$ liefert den tiefsten Energiewert). Dieses Resultat ist allerdings nach Lenz und Ising[1] nicht identisch mit Ferromagnetismus, denn es könnte ja die „kritische Temperatur Θ" noch von N abhängen (beim Isingschen Modell [lineare Kette] ist $\Theta \sim \dfrac{\text{const}}{\lg N}$); das hier zugrunde gelegte Modell weist tatsächlich

---

[1] E. Ising, Z. S. f. Phys. **31**, 253, 1925.

                     Zur Qantentheorie des Ferromagnetismus.

große Ähnlichkeit mit dem Isingschen Modell (Wechselwirkung nur zwischen Nachbaratomen) auf und unterscheidet sich von jenem im wesentlichen nur durch den Wert von z, d. h. durch die Anzahl der Nachbarn, welche ein Atom umgeben. Daß der Wert von z wesentlich ist, kann man aus der Formel für $\overline{\varDelta E^2}$ schließen; ob aber etwa für $z \cdot 8$ die Weissschen Formeln schon zu Recht bestehen, vermag die bisherige Theorie nicht definitiv zu entscheiden.

Kehren wir nach dieser kritischen Betrachtung wieder zum vorläufigen Programm zurück: Ableitung der Weissschen Formeln unter Annahme der Gaußverteilung für $y \cdot 1$; der einzige Unterschied von der entsprechenden Rechnung für $y = 1$ (l. c. I) besteht wegen Gl. (12) bis (15) darin, daß an Stelle von f_n die Gesamtanzahl der Termwerte g_n tritt. Wie dort kann aus einer einfachen Modellbetrachtung geschlossen werden, daß $g_{(x+s)+(x+s)} = f(s) - f(s+1)$, wo $f(s)$ die Gesamtzahl der möglichen Terme bedeutet, die bei gegebenem y zu einem bestimmten Werte der magnetischen Quantenzahl m, nämlich $m = s$ gehören. In genau der gleichen Weise, wie in l. c. I, läßt sich auch die Zustandssumme zurückführen auf die folgende Form:

$$S = F \sum_{-N}^{+N} e^{2x \cdot m} \cdot f(m), \tag{16}$$

wo x eine Funktion des Mittelwertes m_0 von m bedeutet [vgl. l. c. Gl. (23)], die in erster Näherung linear in

$$\alpha = \frac{1}{kT} \frac{e}{\mu c} \cdot \frac{h}{2\pi} H \qquad \text{und} \qquad \beta_T = \frac{J_T}{kT}$$

ist ($H =$ äußeres Magnetfeld); aus der modellmäßigen Bedeutung von $f(m)$ folgt dann sofort, daß sich Gl. (16) umschreiben läßt in

$$S = F \cdot (e^{xy} + e^{x(y-1)} + \cdots + e^{-xy})^{2n}; \tag{17}$$

hieraus folgt

$$m_0 = \frac{\partial}{\partial x} \log S \approx n \cdot \frac{y e^{xy} + (y-1) e^{x(y-1)} + \cdots - y e^{-xy}}{e^{xy} + e^{x(y-1)} + \cdots + e^{-xy}}. \tag{18}$$

An Stelle der von Lenz[1]) zum erstenmal verwendeten Funktion $\mathfrak{Tg}\,x$, die auch l. c. I abgeleitet wurde, tritt also hier eine kompliziertere Langevinsche Funktion, die sich schließlich für hohe Werte von y der Kurve $\mathfrak{Cotg}\,x - 1/x$ immer mehr annähert. Für den empirischen Vergleich der verschiedenen Langevinschen Kurven mit der Erfahrung sei auf den zusammenfassenden Bericht von Debye (l. c.) hingewiesen.

Es ist hierdurch gezeigt, daß, wie zu erwarten war, kompliziertere Langevinsche Funktionen in die Formeln des Ferromagnetismus eingehen, wenn mehrere Valenzelektronen eines Atoms magnetisch wirksam sind. Die Deutung der Weissschen Molekularkräfte als Austauschwirkungen dürfte also den experimentellen Tatsachen auch in diesem Punkte gerecht werden.

Zusammenfassend möchte ich schließen, daß man die Analogie der Weissschen Kräfte mit den homöopolaren Valenzkräften nach den vorliegenden Resultaten wohl als gesichert ansehen darf, daß aber die quantitativen Endformeln einerseits wegen der geschilderten mathematischen Schwierigkeiten einer Nachprüfung bedürfen, andererseits wegen unserer Unkenntnis der Austauschintegrale noch keiner quantitativen Anwendung auf empirische Ergebnisse fähig sind.

---

[1]) W. Lenz, Phys. Z. S. **21**, 613, 1920.

Die Naturwissenschaften *17*, 490–496 (1929)

Die Entwicklung der Quantentheorie 1918—1928.

Von W. HEISENBERG, Leipzig.

Als MAX PLANCK im Jahre 1900 den Satz schrieb: *„Es ist notwendig, die Energie eines Resonators nicht als eine stetige, unbeschränkt teilbare, sondern als eine diskrete, aus einer ganzen Zahl von gleichen Teilen zusammengesetzte Größe aufzufassen"*, da ahnte er wohl kaum, daß sich im Lauf von nicht ganz dreißig Jahren aus diesem, den damals bekannten physikalischen Prinzipien widersprechenden Satz eine Theorie der atomaren Vorgänge entwickeln würde, die an physikalischer Geschlossenheit und mathematischer Einfachheit den klassischen Disziplinen der theoretischen Physik um nichts nachsteht. Vor elf Jahren feierten die Physiker den sechzigsten Geburtstag des Schöpfers der Quantentheorie. Damals wußte man, daß seine Theorie in neue, unerschlossene Gebiete der Physik führte, daß die PLANCKsche Hypothese der Leitfaden war, mit dessen Hilfe man nach BOHRS Entdeckungen den Weg durch das Labyrinth der Atomphysik und der Serienspektren finden würde. Aber selbst damals mußte noch jedem Erfolge der jungen Theorie ein „Aber" beigefügt werden. Denn noch trennte eine unüberbrückbare Kluft die Quantenphysik von der klassischen Physik; noch fehlte der Quantentheorie die Klarheit und die harmonische Geschlossenheit der klassischen Disziplinen. Diese Vollendung des von PLANCK begonnenen Werkes ist nun in den letzten zehn Jahren gelungen. Die endgültige Klärung der Quantenphysik geschah durch die zum Teil ganz überraschende Kombination scheinbar widersprechender Forschungsrichtungen und durch die intensive Zusammenarbeit der Physiker der verschiedensten Nationen. Es war eine ganz abenteuerliche Zeit, voll von Überraschungen und Enttäuschungen, von Erfolgen und von tiefliegenden Schwierigkeiten, deren Diskussion uns bis an die Grundlagen aller physikalischen Erkenntnis geführt hat.

I. Die Erfolge der älteren Form der Quantentheorie.
1918—1923.

In den ersten Jahren der Epoche, die hier geschildert werden soll, waren die Physiker damit beschäftigt, dasNeuland zu durchforschen und fruchtbar zu machen, das von BOHR durch Anwendung der Quantenhypothese auf das RUTHERFORD-BOHRsche Atommodell entdeckt worden war; diese Tätigkeit mußte naturgemäß die Physiker auch an die Grenzen dieses Neulandes führen, an denen begriffliche Schwierigkeiten zunächst dem weiteren Vordringen Halt geboten. BOHR selbst hatte schon in seiner ersten Arbeit über das Wasserstoffatom diese Grenzen aufgezeigt.

Im Anfang der zu schildernden Epoche der Quantentheorie gelang es zunächst SOMMERFELD, seine Theorie der Feinstruktur des Wasserstoffspektrums auch auf die Röntgenspektren anzuwenden und damit die Röntgenspektren auf ein einfaches theoretisches Schema zurückzuführen.

Zwar hat sich später, durch die Entdeckung des Elektronenspins, die physikalische Deutung der SOMMERFELDschen Formel etwas geändert, die Formel selbst ist ohne Änderung in die neuere Quantentheorie übergegangen. Gleichzeitig wurden auch in der Deutung der optischen Spektren wesentliche Fortschritte erzielt. SCHRÖDINGER führte die Idee der „Tauchbahnen" zur Erklärung der charakteristischen Eigenschaften der Alkalispektren ein, SOMMERFELD erzielte durch Annahme einer „inneren Quantenzahl" die formale Deutung der Triplet- und Dublettfeinstrukturen.

Der am meisten überzeugende Beweis für die Richtigkeit des von der Theorie eingeschlagenen Weges war jedoch die von BOHR entdeckte Deutung des periodischen Systems durch die Quantenhypothese. Es war charakteristisch für die BOHRschen Untersuchungen, daß BOHR die Begriffe der klassischen Mechanik nur qualitativ anwandte, nur soweit, wie ihre Anwendbarkeit in der Quantenphysik durch das Korrespondenzprinzip gerechtfertigt werden konnte. Eben diese Freiheit gegenüber den klassischen Begriffen — oder besser: gegenüber denjenigen Konsequenzen der klassischen Begriffe, die den Prinzipien der Quantentheorie widersprachen — ermöglichte eine qualitative Verbindung der chemischen und der spektralen Eigenschaften der Atome. Die Basis der Untersuchungen bildeten also stets die Grundpostulate der *Quanten*theorie, d. h. die der klassischen Physik *fremden* Züge des atomaren Geschehens. Freilich hatte die Quantentheorie vor dieser Zeit auch eine Reihe von Erfolgen in der quantitativen Erklärung der Spektra zu verzeichnen, und man mußte deshalb der klassischen Mechanik einen gewissen Anwendungsbereich zugestehen, wollte man diese Erfolge nicht aufgeben. So hatte man versucht, mit Hilfe des EHRENFESTschen Adiabatensatzes den Gültigkeitsbereich der klassischen Mechanik abzugrenzen. Man hoffte, daß eine auf die klassische Mechanik gegründete Berechnung die Energiewerte der stationären Zustände auch für kompliziertere Atome quantitativ richtig wiedergeben würde. Andererseits war man sich klar darüber, daß eben die Wechselwirkung der Elektronen im Atom eigentlich nicht den Bedingungen des EHRENFESTschen Adiabatensatzes entsprachen.

Die qualitativen Folgerungen des Korrespondenzprinzips blieben jedoch von diesem Dilemma unberührt. Die Analyse der Serienspektren auf Grund der BOHRschen Theorie machte schnelle Fortschritte. Eine besonders wichtige Rolle spielten damals die von PASCHEN und BACK aufs genaueste untersuchten Zeemaneffekte der komplizierteren Atome. Sie stellten nicht nur experimentell das wichtigste Hilfsmittel dar zur Einordnung zweifelhafter Linien, sie waren auch theoretisch wegen des offenbaren Widerspruches

Die Naturwissenschaften *17*, 490–496 (1929)

sogar gegen die qualitativen Folgerungen des Korrespondenzprinzips von größtem Interesse. Die formale Zusammenfassung der experimentellen Ergebnisse gelang LANDÉ; die LANDÉschen Formeln erwiesen sich als außerordentlich fruchtbar für die Ordnung der verwickelten Spektra, obwohl ihre modellmäßige Deutung noch unbekannt war. Charakteristischerweise konnte man schon aus der mathematischen Gestalt der LANDÉschen Formeln erkennen, daß die klassische Mechanik, selbst auf Grund neuer Modellvorstellungen, niemals diese Formeln liefern würde. Dieser Umstand wies wieder die Theoretiker von der klassischen Mechanik weg in Richtung auf eine noch unbekannte Quantenphysik, deren Konturen im Korrespondenzprinzip undeutlich zu erkennen waren. Eben im Hinblick auf diese noch unbekannte Theorie waren alle rein formalen Zusammenhänge von großem Wert. Ein wichtiger Fortschritt wurde hier in Zusammenhang mit einer Untersuchung STONERS erzielt durch PAULI. PAULI schrieb zur Erklärung der anomalen Zeemaneffekte dem einzelnen Elektron formal vier Freiheitsgrade zu; diese Annahme gab hinsichtlich der Anzahl der Terme und der zugeordneten Quantenzahlen den richtigen Zusammenhang zwischen dem Spektrum eines Ions und dem des zugehörigen Atoms. Ferner konnte PAULI mit Hilfe der Hypothese, daß 2 Elektronen niemals im gleichen Quantenzustand vorkommen können (PAULIS Ausschließungsprinzip), das periodische System direkt durch sukzessive Bindung von Elektronen an einen geladenen Kern deduzieren. Obwohl die Bedeutung des 4. Elektronenfreiheitsgrades merkwürdigerweise auch damals noch nicht erkannt wurde, führte die PAULIsche Systematik in kurzer Zeit durch die Arbeiten von RUSSEL und SAUNDERS, GOUDSMIT, HUND und die Untersuchungen der SOMMERFELDschen Schule zu einer vollständigen theoretischen Ordnung der komplizierten Spektren.

Inzwischen war aber auch der mathematische Apparat zur Behandlung der Mehrkörperprobleme der Atomphysik weiter ausgebildet worden. Die theoretischen Physiker studierten aus CHARLIERS und POINCARÉS Werken die Methoden der Himmelsmechanik und fanden dort eben die mathematischen Hilfsmittel vor, die für die Atomphysik angemessen schienen. Allerdings legten die mathematischen Schwierigkeiten der klassischen Mehrkörperprobleme (mangelnde Konvergenz der Reihen der Störungstheorie!) wieder den Verdacht nahe, daß der wirklichen Quantenphysik ein einfacheres mathematisches Schema zugrunde liegen müsse. Dieses Problem wurde entschieden mit der Durchrechnung des Heliumatommodells nach der klassischen Mechanik. Die Anwendung von Quantenbedingungen auf das klassische Modell ergab *nicht* die richtigen Energiewerte. Damit war erwiesen, daß zum mindesten die formalen Methoden der Quantentheorie einer grundlegenden Revision bedurften und daß den klassischen Begriffen in der

späteren Theorie nur ein Recht als qualitativen Analogien zukommen würde.

II. *Die Krisis der Quantentheorie. 1923—1927.*

Mit diesem definitiven Beweis des Versagens klassischer Methoden zur Berechnung der stationären Zustände beginnt das eigentlich interessanteste Stadium der Quantentheorie. Nach den außerordentlichen Erfolgen der BOHRschen Theorie in der qualitativen Deutung der Atomeigenschaften konnte kaum ein Zweifel daran aufkommen, daß der eingeschlagene Weg im großen und ganzen richtig war. Und doch konnten als gesicherte Grundpfeiler der Theorie nur die Postulate der Quantentheorie und das Korrespondenzprinzip angesehen werden. Die quantitative Deutung des Wasserstoffspektrums erschien mehr als ein zufälliger und unverständlicher Erfolg der Theorie. Man erinnerte sich auch wieder daran, daß eine quantitative Berechnung der Intensitäten der Spektrallinien auf Grund der damaligen Methoden nicht möglich war. Offenbar war die Lösung dieses Problems eng verknüpft mit der korrekten Behandlung der Wechselwirkung der Elektronen im Atom. Denn die periodisch veränderlichen Kräfte, die die Elektronen aufeinander ausübten, waren offenbar in Frequenz und Amplitude eng verknüpft mit der von den Elektronen ausgesandten Strahlung. Die Idee einer von der Strahlung verschiedenen Umlaufsfrequenz des Elektrons wurde schon damals wegen des Versagens der klassischen Mechanik aufgegeben.

Zu dieser Zeit kam das Experiment der Theorie zu Hilfe mit einer Entdeckung, die später von großer Bedeutung für die Entwicklung der Theorie werden sollte. COMPTON fand, daß bei der Streuung von Röntgenlicht an freien Elektronen das Streulicht um einen meßbaren Betrag langwelliger war als das einfallende Licht. Dieser Effekt konnte nach COMPTON und DEBYE auf Grund der EINSTEINschen Lichtquantenhypothese zwanglos gedeutet werden; die Wellentheorie des Lichtes versagte diesem Experiment gegenüber. Damit wurden die Probleme der Strahlungstheorie aufgerollt, die seit den EINSTEINschen Arbeiten aus den Jahren 1906, 1909 und 1917 kaum gefördert worden waren. Da die Physik inzwischen auch in der Atomtheorie auf grundsätzliche Schwierigkeiten gestoßen war, wandte man sich mit erneutem Interesse den ungelösten Fragen der Strahlungstheorie zu, um vielleicht durch den Vergleich der Schwierigkeiten in den verschiedenen Gebieten etwas zu lernen.

In der Theorie der Strahlung fand man einen merkwürdigen Dualismus vor zwischen zwei anschaulichen Bildern, dem Wellenbild und dem Corpuscularbild. Die Beugungs- und Interferenzversuche ließen keinen Zweifel an der weitgehenden Anwendbarkeit der klassischen Lichttheorie aufkommen. Trotzdem zeigte der photoelektrische Effekt ein vollkommenes Versagen dieser Theorie. Wollte man nicht den Energiesatz aufgeben, so folgte aus dem Photoeffekt mit Notwendigkeit die Einsteinsche Corpusculartheorie des Lichtes. Die beiden Bilder waren zweifellos unvereinbar, wenn man sie kritiklos hinnahm. Die Physiker wurden also durch die ganz augenscheinlichen Schwierigkeiten in der anschaulichen Deutung der Experimente zu einer Revision ihrer klassischen Begriffswelt gezwungen. Verfolgt man die theoretisch-physikalische Literatur der damaligen Zeit, so erkennt man deutlich eine allmähliche Auflockerung der klassischen Begriffe, ein zunehmendes Sichfreimachen von denjenigen Vorurteilen aus der früheren Physik, die offenbar die Schuld an den Widersprüchen trugen. Hand in Hand mit dieser Auflockerung überkommener Begriffe geht ein mehr gefühlsmäßiges Eindringen in die Begriffswelt der Quantenphysik. Trotz ganz offenbarer innerer Widersprüche erwarben die Physiker ein physikalisches Taktgefühl, das ihnen ermöglichte, die Quantentheorie stets in korrekter, wenn auch qualitativer Weise auf die Atomphysik anzuwenden. Insbesondere hat Bohr als Führer des Kopenhagener Kreises damals eine „Atmosphäre der Quantentheorie" geschaffen, die geradezu die Vorbedingung für alle folgende Entwicklung der Quantenphysik war. Den prägnantesten Ausdruck fanden die außerordentlichen begrifflichen Schwierigkeiten, mit denen die Theorie in der damaligen Zeit zu kämpfen hatte, in der bekannten Arbeit von Bohr, Kramers und Slater über die Strahlungstheorie. Diese Untersuchung stellt den eigentlichen Höhepunkt in der Krisis der Quantentheorie dar und hat, obwohl sie nicht den richtigen Ausweg aus den Schwierigkeiten geben konnte, mehr als irgendeine andere Arbeit jener Zeit zur Klärung der Situation in der Quantentheorie beigetragen. Zunächst betonten Bohr, Kramers und Slater in aller Schärfe die Unmöglichkeit, von den Resultaten der klassischen Theorie in Interferenz- und Beugungserscheinungen abzuweichen. Zur Erklärung des photoelektrischen Effektes mußten sie also den Energiesatz aufgeben; die Wellen wurden als Wahrscheinlichkeitswellen für die Absorption und Emission von Energiebeträgen angesehen. Es hat äußerlich den Anschein, als ob man den Energiesatz und die Interferenzerscheinungen vereinigen könnte, indem man einfach fordert: „daß die Wellenfelder die Wahrscheinlichkeit für die Bahnen der Lichtquanten angeben." Diese Auffassung ist aber oberflächlich und kann leicht ad absurdum geführt werden, wenn man nicht vorher eine eingehende erkenntnistheoretische Kritik der im eben ausgesprochenen Satze verwendeten Begriffe treibt. Trotzdem war die Bohr-Kramers-Slatersche Theorie nur der eine von zwei offenbar gleichberechtigten Auswegen aus den Schwierigkeiten der Strahlungstheorie. Man konnte den Energiesatz beibehalten und dafür die klassische Theorie der Interferenzerscheinungen an gewissen Punkten aufgeben. Diese Möglichkeit wurde von Einstein eine Zeitlang vertreten und später abgelehnt. In der Tat waren beide hier

diskutierten Auswege nicht mit der Erfahrung vereinbar. Den eigentlichen Schlüssel zur Lösung dieser Schwierigkeiten bildete vielmehr später der Umstand, daß genau die gleichen Paradoxien auch in der Theorie der Materie auftraten. Bevor wir jedoch zu dieser Weiterentwicklung der Theorie übergehen, sollen die Fortschritte geschildert werden, die inzwischen in der Atomtheorie gemacht worden waren.

Die Grundlagen der Quantentheorie, die Existenz diskreter stationärer Zustände und die Frequenzbedingung, wurden um diese Zeit durch wichtige Experimente aufs neue bestätigt. Stern und Gerlach konnten einen Silberatomstrahl in einem inhomogenen Magnetfeld aufspalten in zwei getrennte Strahlen, die den beiden möglichen Richtungen des magnetischen Moments des Silberatoms relativ zum äußeren Felde entsprachen. Aus Experimenten der Franckschen Schule sowie aus Versuchen Woods ging eindeutig hervor, daß die Intensität von Spektrallinien, die mit einem Übergang von einen stationären Zustand zu einem anderen verknüpft wird, nur von der Anzahl der Atome im energiereicheren Zustand abhängt, nicht von der Anzahl im energieärmeren Zustand. Beide Experimente waren in bester Übereinstimmung mit den Postulaten der Quantentheorie und standen in Widerspruch zur klassischen Theorie. Auch die Theoretiker verzichteten um diese Zeit bereits auf jeden quantitativen Gebrauch der klassischen Theorie. Vielmehr ging die allgemeine Tendenz theoretischer Untersuchungen jetzt dahin, auf Grund der Quantenpostulate das Bohrsche Korrespondenzprinzip soweit zu „verschärfen", daß es quantitative Resultate lieferte. In Weiterführung der Arbeiten Ladenburgs über die Dispersionskurven der Atome gelang Kramers die Aufstellung einer Dispersionsformel, die nur die Einsteinschen Koeffizienten der Übergangswahrscheinlichkeit als Konstanten enthielt und die doch im Grenzfall hoher Quantenzahlen in die klassische Formel überging. Ferner wurde auf Grund der quantitativen Intensitätsmessungen Ornsteins und seiner Mitarbeiter eine weitgehende Verschärfung des Korrespondenzprinzips für das Intensitätsproblem möglich. Es gelang auf diese Weise Burger und Dorgelo, Kronig, Sommerfeld, Hönl und Russel eine formelmäßige Darstellung der Multiplettintensitäten. Damit war ein wichtiger Schritt in Richtung auf den eigentlich der Quantentheorie angemessenen Formalismus getan. Inzwischen waren auch die physikalischen Grundlagen klarer geworden. Insbesondere zeigten die Versuche von Geiger und Bothe, sowie die von Compton und Simon, daß der Dualismus von Wellenbild und Corpuscularbild in der Strahlung nicht ohne weitgehenden Verzicht auf die Begriffe unserer Raum-Zeitwelt des täglichen Lebens würden verstanden werden können. — Wie weit übrigens das „gefühlsmäßige Verständnis" der Quantentheorie damals schon fortgeschritten war, geht am deutlichsten aus dem PAULISchen Buch über Quantentheorie hervor, an dessen physikalischen Inhalt auch heute kaum ein Wort zu ändern ist. Trotzdem konnte natürlich die vollständige physikalische Klarheit erst nach Vollendung des mathematischen Formalismus gewonnen werden.

III. Klärung der formalen Zusammenhänge 1925—1927.

Die eigentlichen formalen Zusammenhänge der Quantentheorie sind auf zwei ganz unabhängigen Wegen ungefähr gleichzeitig gefunden worden. In konsequenter Weiterentwicklung der oben erwähnten Ansätze zur ,,Verschärfung" des Korrespondenzprinzips gelang es dem *Göttinger* Kreis und DIRAC, ein in sich konsequentes mathematisches Schema aufzufinden, das als die quantitative Formulierung des Korrespondenzprinzips aufgefaßt werden kann. Der Grundgedanke dieser Theorie war, daß zwischen den wirklich beobachtbaren Amplituden und Frequenzen der Atome ein ähnlicher Zusammenhang bestehen müsse, wie zwischen den korrespondierenden Größen der klassischen Modelle. Die mathematischen Hilfsmittel der Theorie waren die elementare nichtkommutative Algebra und ihre Darstellungen durch Matrizen. — Gleichzeitig mit den Arbeiten des Göttinger Kreises hatte DE BROGLIE mit dem größten Erfolg versucht, den Dualismus Welle-Corpuscel von der Strahlung auch auf die Materie zu übertragen. Er ordnete, wie in der Lichtquantentheorie, den Begriffen ,,Energie und Impuls" im Partikelbild die Begriffe ,,Frequenz und Wellenlänge" im Wellenbild zu. Nach der Theorie DE BROGLIES sollten Elektronenstrahlen homogener Geschwindigkeit von einem Beugungsgitter reflektiert werden, wie einfarbige Lichtbündel — was in gewisser Weise übrigens schon vorher aus den Untersuchungen DUANES hervorging. Eine vorläufige experimentelle Bestätigung der DE BROGLIEschen Hypothesen ließ sich aus den früheren Experimenten von DAVISSON und KUNSMAN einerseits, aus dem gleichfalls wesentlich früher entdeckten RAMSAUER-Effekt andererseits herleiten. Der Durchbruch zum vollständigen mathematischen System der Quantentheorie gelang wenige Monate später SCHRÖDINGER, indem er die Differentialgleichung der DE BROGLIEschen Wellen aufstellte. SCHRÖDINGER zeigte, daß die Bestimmung der stationären Zustände eines Atoms gleichbedeutend war mit der Lösung eines durch eine Differentialgleichung charakterisierten Eigenwertproblems. Allerdings war die SCHRÖDINGERsche Theorie in einem wichtigen Punkte keine konsequente Weiterbildung der DE BROGLIEschen Theorie. Während DE BROGLIE offenbar eine Wellentheorie der Materie anstrebte von ähnlicher Einfachheit und Anschaulichkeit, wie die MAXWELLsche Theorie der Strahlung, stellte die SCHRÖDINGERsche Theorie eine Wellentheorie im vieldimensionalen Konfigurationsraum des Partikelbildes dar; die Differentialgleichung selbst konnte nur mit Hilfe der HAMILTONschen Funktion des Partikelbildes gewonnen werden. Aus diesem Grunde kann die SCHRÖDINGERsche Theorie — im Gegensatz zu DE BROGLIES Ansätzen und zu später zu diskutierenden Weiterbildungen der Theorie — *nicht* mit einer klassischen raumzeitlichen Wellentheorie verglichen werden. Eine solche raumzeitliche Theorie könnte ja auch nie den Postulaten der PLANCKschen Quantentheorie gerecht werden. Die SCHRÖDINGERsche Theorie dagegen gab mit einem Schlage die Möglichkeit, die Linienspektra der Atome auszurechnen und sie stellt heute noch das wirksamste mathematische Instrument für alle Anwendungen der Quantentheorie dar. SCHRÖDINGER und ECKART bewiesen kurze Zeit nach Entdeckung der ,,Wellenmechanik" deren mathematische Äquivalenz mit der aus dem Korrespondenzprinzip entstandenen ,,Quantenmechanik". Damit konnten auch nach der Methode der Eigenschwingungen die Resultate des Korrespondenzprinzips für die physikalische Interpretation des mathematischen Schemas ausgenützt werden; z. B. zur Berechnung der Linienintensitäten. Andererseits fanden die vorher scheinbar unüberwindlichen mathematischen Schwierigkeiten der Matrizentheorie ihre überraschend einfache Lösung. Die Wellenmechanik und die Quantenmechanik wurden, sozusagen in einer höheren Einheit, zusammengefaßt in der Transformationstheorie von DIRAC, JORDAN und LONDON. In dieser Theorie konnten insbesondere die verschiedenen aus dem Korrespondenzprinzip folgenden physikalischen Interpretationen des Formalismus auf eine einzige Annahme zurückgeführt werden. Diese Annahme ist statistischer Art, enthält also Aussagen über die Wahrscheinlichkeit dafür, daß ein bestimmter Vorgang eintritt oder nicht eintritt. In einem Spezialfall war diese Annahme schon vorher von BORN in seiner Theorie der Stoßvorgänge angewandt worden. Trotz dieser schnellen Fortschritte der Quantentheorie war aber doch der Gedanke an eine raumzeitliche Kontinuumstheorie der Materie nicht aufgegeben worden. Es zeigte sich vielmehr, daß diese Kontinuumstheorie *formal* erheblich weiter ausgebaut werden konnte. So ließen sich gewisse einfache quadratische Ausdrücke der Wellenamplituden als Strom- und Ladungsdichte der Materie interpretieren; in der relativistisch invarianten Formulierung von SCHRÖDINGER, GORDON, KLEIN, FOCK, KUDAR konnte auch ein Energie-Impulstensor der Materiewellen aufgestellt werden, der zusammen mit dem MAXWELLschen Energie- und Impulstensor der Strahlung die Erhaltungssätze erfüllte. Trotz der formalen Schönheit dieser Theorie war es aber klar, daß sie nicht mit der Erfahrung in Einklang sein konnte, denn sie enthielt ja überhaupt noch kein quantentheoretisches Element. Die Existenz der Elektronen war in dieser Theorie ebenso unverständlich, wie die Existenz der Lichtquanten in der MAXWELLschen Theorie. Man kann also nicht genug den Unterschied betonen zwischen dieser konsequenten Feldtheorie und der vorher geschilderten SCHRÖDINGERschen Wellenmechanik, in der die Existenz der Elektronen geradezu der notwendige Ausgangspunkt der Betrachtungen ist. Wenn aber, wie dies aus physikalischen Gründen äußerst wahrscheinlich war, volle Symmetrie zwischen Wellen- und Partikelbild bestand, so mußte man die konsequente Feldtheorie genau so fruchtbar verwenden können wie früher die klassische Mechanik. Man fügte also zu den Feldgleichungen noch Quantenbedingungen (die Gesamtladung wurde als ganzzahliges Vielfaches der Elektronenladung angesetzt) und erhielt dann durch Anwendung des Korrespondenzprinzips(KLEIN) befriedigende qualitative Resultate (siehe z. B. die Untersuchungen HARTREES).

Der entscheidende Schritt von der klassischen zur Quantentheorie der Wellenfelder geschah in den Arbeiten von DIRAC, PAULI, JORDAN, KLEIN und WIGNER. Genau analog zur Quantentheorie des Partikelbildes treten in dieser Quantendynamik der Wellenfelder an Stelle der Wellenamplituden nichtkommutative Größen, die durch Matrizen oder Operatoren dargestellt werden; damit geht naturgemäß wieder die raumzeitliche Anschaulichkeit der Theorie verloren; die Lösung des entstehenden mathematischen Problems geschieht wieder nach Methoden, die denen der SCHRÖDINGERschen Wellenmechanik analog sind. Nach diesen Untersuchungen steht also die Quantentheorie des Wellenbildes der Quantentheorie des Partikelbildes *völlig gleichberechtigt* gegenüber. Es ist nun für die Widerspruchsfreiheit der Quantentheorie im ganzen von grundlegender Bedeutung, daß nach JORDAN, KLEIN und WIGNER die beiden genannten Theorien (unter gewissen, später zu erwähnenden Einschränkungen) mathematisch äquivalent, d. h. ineinander überführbar sind. Damit ist gezeigt, daß im Formalismus der Quantentheorie Partikel- und Wellenbild nur als zwei verschiedene Erscheinungsformen ein und derselben physikalischen Realität auftreten.

Gleichzeitig mit dieser Entwicklung des mathematischen Schemas der Quantentheorie, aber zunächst unabhängig davon, wurde auch das Problem der anomalen Zeemaneffekte endlich gelöst, deren physikalische Bedeutung trotz der formalen Ordnung der Serien- und Multiplettspektra bis dahin noch unverstanden war. UHLENBECK und GOUDSMIT erklärten durch die Hypothese, daß das Elektron ein magnetisches Moment der Größe $\frac{e}{2mc}\frac{h}{2\pi}$ und ein Drehmoment der Größe $\frac{1}{2}\frac{h}{2\pi}$ besitzt, qualitativ alle Einzelheiten der Theorie der Multipletts. Übrigens war die Hypothese des Elektronenspins, allerdings ohne Verbindung mit der Theorie der Spektra, schon früher von COMPTON und anderen Forschern diskutiert worden. Die Behandlung des UHLENBECK-GOUDSMITschen Modells nach der Quantenmechanik

gab auch quantitativ die richtigen Resultate. Eine schöne experimentelle Bestätigung fand die ganze Theorie der Spektra um diese Zeit in den Messungen von MILLIKAN und BOWEN über die Ultraviolettspektren hochionisierter Atome.

IV. Klärung der physikalischen Grundlagen 1927.

Trotz der geschilderten außerordentlichen Fortschritte auf dem Gebiete der quantentheoretischen Mathematik war man bis zum Ende des Jahres 1926 kaum über die intuitiv richtige Anwendung des Formalismus hinausgekommen. Über die physikalischen Grundlagen herrschte noch keine Klarheit. Obwohl der Formalismus der Quantentheorie keine raumzeitliche kausale Verknüpfung von physikalischen Phänomenen zuläßt, sich vielmehr in hochdimensionalen mathematischen Räumen abspielt, verwandte man zur Beschreibung der Fakta kritiklos die aus unserer anschaulichen Raum-Zeitwelt übernommenen Begriffe und verwickelte sich dabei naturgemäß in Widersprüche. Andererseits schien es hoffnungslos, die Konstruktion einer Sprache, d. h. einer Begriffswelt zu versuchen, die diesen geschilderten mathematischen Zusammenhängen adäquat wäre; denn unser ganzes Denken ist untrennbar mit der anschaulichen Raum-Zeitvorstellung verknüpft; wir beschreiben das Ergebnis unserer Experimente stets in Worten, die dieser anschaulichen Raum-Zeitwelt entlehnt sind. Die einzig mögliche Lösung dieser Schwierigkeiten bestand also darin, zwar die bisherigen anschaulichen Begriffe beizubehalten, aber ihren Anwendbarkeitsbereich, soweit es notwendig war, einzuschränken. Dieses Programm wurde durchgeführt in den Arbeiten des Kopenhagener Kreises. Grundlage der Untersuchungen waren wieder die ausführlichen Diskussionen über die fundamentalen Begriffe der Quantentheorie, die BOHR im Anschluß an die Arbeiten über die Strahlungstheorie durchgeführt hatte. Es stellte sich heraus, daß die Begriffe der Partikeltheorie durch die sog. „Unbestimmtheitsrelationen" in ihrer Anwendbarkeit beschränkt werden müssen. Die Unbestimmtheitsrelationen sind der physikalische Ausdruck für die ursprüngliche PLANCKsche Einteilung des Phasenraumes in Zellen der Größe h. Diese Relationen genügten aber noch nicht zu einer Durchführung des genannten Programms. Vielmehr zeigte BOHR, daß eben die *gleichzeitige* Benützung des Partikelbildes und des Wellenbildes (unter Wellen*bild* ist in diesem Satz die anschauliche raum-zeitliche Wellenvorstellung von MAXWELL und DE BROGLIE gemeint, nicht die Phasenraumtheorie) notwendig und hinreichend ist, um in allen Fällen die Grenzen abzustecken, bis zu denen die klassischen Begriffe anwendbar sind. Aus dieser Symmetrie zwischen Wellenbild und Corpuscularbild folgt übrigens, daß auch für das Wellenbild Unbestimmtheitsrelationen existieren, die mathematisch aus den Vertauschungsrelationen der Wellenamplituden hergeleitet werden können. Es handelte sich aber nicht nur darum,

Die Naturwissenschaften 17, 490–496 (1929)

die klassischen Begriffe einzuschränken, sondern es war insbesondere die Aufgabe, zu zeigen, daß mit Hilfe der so präzisierten Begriffe eine widerspruchsfreie physikalische Interpretation des quantentheoretischen Formalismus möglich war. Dieser Beweis wurde von BOHR in allen Einzelheiten erbracht. Der wichtigste Unterschied der Quantentheorie von den klassischen Theorien besteht darin, daß bei der Beobachtung irgendeiner physikalischen Größe die Störung wesentlich in Betracht gezogen werden muß, die das zur Beobachtung ausgeführte Experiment am zu messenden System hervorruft. Eine Beobachtung verändert im allgemeinen das physikalische Verhalten des Systems. Die vorher scheinbar unlösbaren Paradoxien der Quantentheorie beruhten alle darauf, daß man diese mit jeder Beobachtung notwendig verbundene Störung vernachlässigt hatte; die Paradoxien fanden also nach BOHR eine einfache Erklärung durch sorgfältige Diskussion der Wechselwirkung zwischen den Meßapparaten und dem zu messenden System. Eine weitere bedeutungsvolle Konsequenz der Quantentheorie war die Unmöglichkeit, das Kausalgesetz so präzis zu formulieren, wie dies in der klassischen Theorie üblich war. Vielmehr erwiesen sich die physikalischen Zusammenhänge der Quantentheorie als wesentlich statistischer Art.

Das physikalische Verständnis der Quantentheorie war also, ähnlich wie das der Relativitätstheorie, nur möglich auf Grund einer Revision und einer Erweiterung der klassischen Begriffswelt, d. h. auf Grund einer sorgfältigen erkenntnistheoretischen Untersuchung der in die Theorie einzuführenden Begriffe. Wegen dieser notwendigen Erweiterung der klassischen Begriffswelt bedurfte es einer Reihe von Diskussionen, um über die physikalische Deutung der Quantentheorie völlige Klarheit zu erlangen. Wenn man über die Entwicklung der Quantentheorie erzählt, darf man insbesondere nicht die Diskussionen des von LORENTZ geleiteten SOLVAY-Kongresses in Brüssel 1927 vergessen. Dieser Kongreß hat durch die Möglichkeit der Aussprache der Vertreter verschiedener Forschungsrichtungen außerordentlich zur Klärung der physikalischen Grundlagen der Quantentheorie beigetragen; er bildet gewissermaßen äußerlich den Abschluß der Quantentheorie, die nunmehr als in sich geschlossene Theorie auf alle Probleme der Atomphysik ohne Bedenken angewendet werden kann und die eine Basis für weitere Untersuchungen bildet.

V. Anwendungen und experimentelle Bestätigungen. 1926—1928.

Über die Anwendungen der Theorie soll hier im folgenden nur ein ganz summarischer Überblick gegeben werden; das Gebiet der Anwendungen ist schon zu umfangreich geworden; außerdem wird es zum Teil in anderen Artikeln in diesem Heft beschrieben.

Auch für die neue Form der Quantentheorie waren die Atomspektra der erste Prüfstein. Der erste Erfolg war die Ableitung der KRAMERSschen Dispersionsformel und der Multiplettintensitäten. PAULI, DIRAC und SCHRÖDINGER gaben dann die Theorie des Wasserstoffspektrums nach dem neuen Formalismus. Insbesondere erwies sich die SCHRÖDINGERsche Wellenmechanik in der folgenden Zeit als das für alle Anwendungen zweckmäßigste und erfolgreichste mathematische Schema. Die Intensitäten des Wasserstoffspektrums (mit oder ohne äußeres elektrisches Feld) wurden von SCHRÖDINGER berechnet.

Für das Mehrkörperproblem gab die neuere Form der Quantentheorie ein in der klassischen Theorie nicht enthaltenes, qualitativ neues Phänomen. Dieses sog. Resonanz- oder Austauschphänomen beruht auf der Gleichheit der Elektronen und besteht darin, daß — wenn man überhaupt den Gebrauch des Partikelbildes so weit treiben will — die Elektronen im Atom periodisch die Plätze wechseln. Dieser Austausch hat dann weiter zur Folge, daß das ganze Eigenwertspektrum des Atoms eingeteilt werden kann in nichtkombinierende Termsysteme. Zu jedem Termsystem gehören SCHRÖDINGERsche Eigenfunktionen einer bestimmten Symmetrieklasse. Die Auswahl desjenigen Termsystems, das (bei Einbeziehung des Elektronenspins) zu SCHRÖDINGERfunktionen gehört, die in den Elektronenkoordinaten symmetrisch sind, entspricht der BOSE-EINSTEINschen Statistik der Elektronen, die Auswahl der antisymmetrischen Eigenfunktionen entspricht dem PAULIschen Ausschließungsprinzip. (FERMI-DIRAC-Statistik.) Die BOSE-EINSTEINsche Statistik war zunächst von ·BOSE für die Lichtquanten postuliert worden, um das PLANCKsche Strahlungsgesetz zu erhalten, EINSTEIN hat sie dann in Zusammenhang mit DE BROGLIES Wellen auf die Materie übertragen. Das PAULIsche Ausschließungsprinzip wurde von FERMI und DIRAC auf die Statistik des idealen Gases angewandt. Geht man vom Partikelbild zum Wellenbild über, so entsprechen den beiden Möglichkeiten der Statistik nach JORDAN, KLEIN und WIGNER zwei verschiedene Möglichkeiten der Vertauschungsrelationen der Wellenamplituden. Das Resonanzphänomen wurde zunächst für die angeregten Zustände des Heliumatoms quantitativ durchgerechnet. KELLNER behandelte später den Normalzustand desselben Atoms. Der weitere Ausbau der Theorie des Austausches geschah unter Anwendung gruppentheoretischer Hilfsmittel durch WIGNER, NEUMANN, HUND, LONDON, HEITLER und WEYL. Insbesondere wendeten HUND, LONDON und HEITLER die Ergebnisse der neueren Quantentheorie auf die Chemie mit großem Erfolge an.

Von den SCHRÖDINGERschen Methoden ausgehend gab BORN eine Theorie der Stöße von Elektronen auf Atome und gab damit ein Verfahren, die FRANCK-HERTZschen Experimente quantitativ zu behandeln. SOMMERFELD, ECKART,

HOUSTON, FOWLER und NORDHEIM übertrugen die Ergebnisse der Quantentheorie auf die Elektronentheorie der Metalle. Der *Ferromagnetismus* wurde auf das diskutierte Austauschphänomen zurückgeführt. Es ist unmöglich, im Rahmen dieses Artikels auf die vielen Erfolge der Quantentheorie im einzelnen einzugehen.

Die Grundlagen der Quantentheorie wurden um diese Zeit noch einmal durch wichtige Experimente bestätigt. DAVISSON und GERMER wiesen die Beugung von Elektronenstrahlen an Krystallgittern nach, THOMSON gab das Analogon der DEBYE-SCHERRER-Methode für Elektronenstrahlen, RUPP erhielt Beugung am geritzten Gitter, KIKUCHI schließlich zeigte durch Variation der Schichtdicken der durchstrahlten Materie alle Übergänge von der Beugung am Flächengitter bis zur LAUE-Aufnahme. Auch die Entdeckung des RAMAN-Effektes entsprach den Voraussagen der Quantentheorie (SMEKAL) und bestätigte insbesondere auch wesentliche Züge in der physikalischen Interpretation des Formalismus.

Im ganzen gewährt also die Quantentheorie den Eindruck einer in sich abgeschlossenen, konsequenten Theorie; trotzdem gibt es — natürlicherweise — Gebiete der Atomphysik, in denen die Quantentheorie noch nicht ausreicht zum Verständnis der physikalischen Zusammenhänge. Es ist bisher nicht möglich gewesen, den Forderungen der Relativitätstheorie und denen der Quantentheorie gleichzeitig in befriedigender Weise Rechnung zu tragen. Wesentliche Fortschritte in dieser Richtung sind von DIRAC in seiner Theorie des Drehelektrons erzielt worden. Aber es bleiben noch ungelöste Schwierigkeiten übrig. Vielleicht hängen die Probleme der Kernstruktur unmittelbar mit diesen Schwierigkeiten zusammen. Jedenfalls kann man schließen, daß für eine Theorie der Kernstruktur die strenge relativistische Behandlung der Elektronen eine notwendige Bedingung ist. Ein wichtiger Erfolg der Quantentheorie im Gebiet der Radioaktivität ist zwar durch die GAMOW-CONDON-GURNEYsche Theorie des GEIGER-NUTALLschen Gesetzes erzielt worden; aber dieser Erfolg war nur möglich, weil in die genannte Theorie nur ganz allgemeine Annahmen über den Aufbau der Kerne eingehen.

An das Gebiet der durch PLANCKS berühmte Konstante h charakterisierten Quantentheorie grenzen also andere noch unerforschte Gebiete der Atomphysik, in denen neben dem PLANCKschen h andere grundlegende Konstanten (Elektronenladung e, Elektronenmasse m, Lichtgeschwindigkeit c, Protonenmasse M) eine entscheidende Rolle spielen werden. Die Physiker werden also auch in den nächsten Jahrzehnten noch nicht gezwungen sein, sich auf die Nutzbarmachung schon durchforschten Gebietes zu beschränken. Vielmehr werden sie auch in Zukunft auf Abenteuer in unbekanntem Neuland ausgehen können, wie in den ersten dreißig Jahren der PLANCKschen Theorie.

Literatur: Eine auch nur annähernd vollständige Literaturangabe wäre im Rahmen dieses Artikels unmöglich, da sie mehrere Seiten füllen würde. Ich verweise daher auf A. SOMMERFELDS Buch über Atombau und Spektrallinien, insbesondere auch „Wellenmechanischer Ergänzungsband" (VIEWEG), ferner M. BORN, Atommechanik (Springer 1926), F. HUND, Linienspektra und periodisches System der Elemente (Springer 1927), W. PAULI, Quantentheorie (Geiger-Scheelsches Handbuch der Physik **23**, Springer). In diesen Werken findet man wohl alle wichtigen Arbeiten zitiert.

Die Bedeutung der Planckschen Quantentheorie für die Experimentalphysik.

Von G. HERTZ, Berlin.

Das Verfahren der statistischen Behandlung physikalischer Vorgänge besteht darin, daß man von den als bekannt vorausgesetzten Gesetzen der Einzelvorgänge ausgehend die Gesetzmäßigkeiten für solche Vorgänge ableitet, die sich durch das regellose Zusammenwirken einer sehr großen Zahl von Einzelvorgängen ergeben. Durch Vergleich der so gefundenen Gesetzmäßigkeiten mit den Ergebnissen des Experiments ist es dann möglich, wieder rückwärts auf die für die Einzelvorgänge geltenden Gesetze zu schließen. Auf diese Weise gelingt es, in diesen Gesetzen noch vorhandene Unbestimmtheiten zu beseitigen, etwa durch zahlenmäßige Festlegung bisher unbestimmter Konstanten, und zu entscheiden, ob an der Form der Gesetze selbst noch Änderungen anzubringen sind. Die auf solche Weise gewonnenen Erkenntnisse haben aber nicht nur für diejenigen Erscheinungen Bedeutung, aus welchen sie abgeleitet worden sind, sondern für die Gesamtheit aller Vorgänge, bei denen der betreffende Einzelvorgang eine Rolle spielt, und als gesichert können sie erst angesehen werden, wenn sie für alle diese Vorgänge zu Folgerungen führen, die mit den experimentellen Befunden im Einklange sind. So sind z. B. bei den Gasen die Diffusion, die innere Reibung und die Wärmeleitung alle durch denselben Einzelvorgang bestimmt, nämlich den Zusammenstoß zwischen Molekülen. Folgerungen, welche aus Beobachtungen an einer dieser Erscheinungen mit Bezug auf die zwischen den Molekülen wirkenden Kräfte gezogen werden, können nur richtig sein, wenn sich die durch sie ermöglichten Voraussagen bezüglich der beiden anderen als experimentell richtig erweisen.

Je tiefgreifender eine auf solche Weise gewonnene neue Erkenntnis ist, um so größer ist der Kreis der Erscheinungen, für welche sie von Bedeutung ist. Als PLANCK auf Grund der statistischen Behandlung des Problems der Wärmestrahlung zu der Überzeugung kam, daß für die Wechselwirkung zwischen Strahlung und Atomen die Gesetze der klassischen Physik nicht gültig sein könnten, und seine Quantenhypothese aufstellte, hatte er eine Erkenntnis gewonnen, die

The Physical Principles of the Quantum Theory (The University of Chicago Press, Chicago 1930)

THE
PHYSICAL PRINCIPLES
OF THE QUANTUM
THEORY

By

WERNER HEISENBERG

Professor of Physics, University of Leipzig

Translated into English by

CARL ECKART AND FRANK C. HOYT

Department of Physics, University of Chicago

THE UNIVERSITY OF CHICAGO PRESS
CHICAGO, ILLINOIS

vii

FOREWORD TO THE ENGLISH EDITION

It is an unusual pleasure to present Professor Heisenberg's Chicago lectures on "The Physical Principles of the Quantum Theory" to a wider audience than could attend them when they were originally delivered. Professor Heisenberg's leading place in the development of the new quantum mechanics is well recognized by those who have been following its growth. It was in fact he who first saw clearly that in the older forms of quantum theory we were describing our spectra in terms of atomic mechanisms regarding which we could gain no definite knowledge, and who first found a way to interpret (or at least describe) spectroscopic phenomena without assuming the existence of such atomic mechanisms. Likewise, "the uncertainty principle" has become a household phrase throughout our universities, and it is especially fortunate to have this opportunity of learning its significance from one who is responsible for its formulation.

The power of the new quantum mechanics in giving us a better understanding of events on an atomic scale is becoming increasingly evident. The structure of the helium atom, the existence of half-quantum numbers in band spectra, the continuous spatial distribution of photo-electrons, and the phenomenon of radioactive disintegration, to mention only a few examples, are achievements of the new theory which had baffled the old. While the writing of this chapter of the history of physics is

viii

doubtless not yet complete, it has progressed to such a stage that we may profitably pause and consider the significance of what has been written. As we make this survey, we are indeed fortunate to have Professor Heisenberg to guide our thoughts.

ARTHUR H. COMPTON

PREFACE

The lectures which I gave at the University of Chicago in the spring of 1929 afforded me the opportunity of reviewing the fundamental principles of quantum theory. Since the conclusive studies of Bohr in 1927 there have been no essential changes in these principles, and many new experiments have confirmed important consequences of the theory (for example, the Raman effect). But even today the physicist more often has a kind of faith in the correctness of the new principles than a clear understanding of them. For this reason the publication of these Chicago lectures in the form of a small book seems justified.

Since the formal mathematical apparatus of the quantum theory is already available in several excellent texts and is more familiar to many than the physical principles, I have placed it at the end of the book, in what is little more than a collection of formulas.[1] In the text itself I have been at pains to use only elementary formulas and calculations, so far as this is possible.

In the body of the text particular emphasis has been

ix

[1] TRANSLATORS' NOTE.—In the English edition, Professor Heisenberg's lectures on the mathematical part of the theory have been reproduced in more detail. This seemed advisable since a treatment of the general transformation theory and the quantum theory of wave fields was not available in English at the time the manuscript was prepared. The former has since been treated in several texts (E. U. Condon and P. M. Morse, *Quantum Mechanics;* A E. Ruark and H. C. Urey, *Atoms, Molecules and Quanta;* both published by McGraw-Hill).

The English text also deviates in several other points from the German, but these are felt to be unessential changes.

placed on the complete equivalence of the corpuscular and wave concepts, which is clearly reflected in the newer formulations of the mathematical theory. This symmetry of the book with respect to the words "particle" and "wave" shows that nothing is gained by discussing fundamental problems (such as causality) in terms of one rather than the other. I have also attempted to make the distinction between waves in space-time and the Schrödinger waves in configuration space as clear as possible.

On the whole the book contains nothing that is not to be found in previous publications, particularly in the investigations of Bohr. The purpose of the book seems to me to be fulfilled if it contributes somewhat to the diffusion of that *"Kopenhagener Geist der Quantentheorie,"* if I may so express myself, which has directed the entire development of modern atomic physics.

My thanks are due in the first place to Drs. C. Eckart and F. Hoyt, of the University of Chicago, who have taken on themselves not only the labor of preparing the English translation, but have also contributed essentially to the improvement of the book by working over several sections and giving me the benefit of their advice. I am also indebted to Dr. G. Beck for reading proof of the German edition and for valuable assistance in the preparation of the manuscript.

W. HEISENBERG

LEIPZIG
March 3, 1930

x

CONTENTS

CHAPTER I

INTRODUCTORY

§ 1. THEORY AND EXPERIMENT

The experiments of physics and their results can be described in the language of daily life. Thus if the physicist did not demand a theory to explain his results and could be content, say, with a description of the lines appearing on photographic plates, everything would be simple and there would be no need of an epistemological discussion. Difficulties arise only in the attempt to classify and synthesize the results, to establish the relation of cause and effect between them—in short, to construct a theory. This synthetic process has been applied not only to the results of scientific experiment, but, in the course of ages, also to the simplest experiences of daily life, and in this way all concepts have been formed. In the process, the solid ground of experimental proof has often been forsaken, and generalizations have been accepted uncritically, until finally contradictions between theory and experiment have become apparent. In order to avoid these contradictions, it seems necessary to demand that no concept enter a theory which has not been experimentally verified at least to the same degree of accuracy as the experiments to be explained by the theory. Unfortunately it is quite impossible to fulfil this requirement, since the commonest ideas and words would often be excluded. To avoid these insurmountable difficulties it is found ad-

2 visable to introduce a great wealth of concepts into a physical theory, without attempting to justify them rigorously, and then to allow experiment to decide at what points a revision is necessary.

Thus it was characteristic of the special theory of relativity that the concepts "measuring rod" and "clock" were subject to searching criticism in the light of experiment; it appeared that these ordinary concepts involved the tacit assumption that there exist (in principle, at least) signals that are propagated with an infinite velocity. When it became evident that such signals were not to be found in nature, the task of eliminating this tacit assumption from all logical deductions was undertaken, with the result that a consistent interpretation was found for facts which had seemed irreconcilable. A much more radical departure from the classical conception of the world was brought about by the general theory of relativity, in which only the concept of coincidence in space-time was accepted uncritically. According to this theory, ordinary language (i.e., classical concepts) is applicable only to the description of experiments in which both the gravitational constant and the reciprocal of the velocity of light may be regarded as negligibly small.

Although the theory of relativity makes the greatest of demands on the ability for abstract thought, still it fulfils the traditional requirements of science in so far as it permits a division of the world into subject and object (observer and observed) and hence a clear formulation of the law of causality. This is the very point at which the difficulties of the quantum theory begin. In atomic physics, the concepts "clock" and "measuring rod" need no

immediate consideration, for there is a large field of phe- 3 nomena in which $1/c$ is negligible. The concepts "space-time coincidence" and "observation," on the other hand, do require a thorough revision. Particularly characteristic of the discussions to follow is the interaction between observer and object; in classical physical theories it has always been assumed either that this interaction is negligibly small, or else that its effect can be eliminated from the result by calculations based on "control" experiments. This assumption is not permissible in atomic physics; the interaction between observer and object causes uncontrollable and large changes in the system being observed, because of the discontinuous changes characteristic of atomic processes. The immediate consequence of this circumstance is that in general every experiment performed to determine some numerical quantity renders the knowledge of others illusory, since the uncontrollable perturbation of the observed system alters the values of previously determined quantities. If this perturbation be followed in its quantitative details, it appears that in many cases it is impossible to obtain an exact determination of the simultaneous values of two variables, but rather that there is a lower limit to the accuracy with which they can be known.[1]

The starting-point of the critique of the relativity theory was the postulate that there is no signal velocity greater than that of light. In a similar manner, this lower limit to the accuracy with which certain variables can be known simultaneously may be postulated as a law of nature (in the form of the so-called uncertainty relations)

[1] W. Heisenberg, *Zeitschrift für Physik*, **43**, 172, 1927.

and made the starting-point of the critique which forms 4 the subject matter of the following pages. These uncertainty relations give us that measure of freedom from the limitations of classical concepts which is necessary for a consistent description of atomic processes. The program of the following considerations will therefore be: first, to obtain a general survey of all concepts whose introduction is suggested by the atomic experiments; second, to limit the range of application of these concepts; and third, to show that the concepts thus limited, together with the mathematical formulation of quantum theory, form a self-consistent scheme.

§ 2. THE FUNDAMENTAL CONCEPTS OF QUANTUM THEORY

The most important concepts of atomic physics can be induced from the following experiments:

a) Wilson[1] *photographs.*—The α- and β-rays emitted by radioactive elements cause the condensation of minute droplets when allowed to pass through supersaturated water vapor. These drops are not distributed at random, but are arranged along definite tracks which, in the case of α-rays (Fig. 1), are nearly straight lines, in the case of β-rays, are irregularly curved. The existence of the tracks and their continuity show that the rays may appropriately be regarded as streams of minute particles moving at high speeds. As is well known, the mass and charge of these particles may be determined from the deflection of the rays by electric and magnetic fields.

[1] *Proceedings of the Royal Society*, A, **85**, 285, 1911; see also *Jahrbuch der Radioaktivität*, **10**, 34, 1913.

5   *b) Diffraction of matter waves (Davisson and Germer,[1] Thomson,[2] Rupp[3]).*—After the conception of β-rays as streams of particles had remained unchallenged for more than fifteen years, another series of experiments was per-

Fig. 1.—Tracks of α-particles in Wilson Chamber

formed which indicated that they could be diffracted and were capable of interference as if they were waves. Typical of these experiments is that of G. P. Thomson, in which a narrow beam of artificial β-rays of moderate

[1] *Physical Review*, 30, 705, 1927; *Proceedings of the National Academy*, 14, 317, 1928.

[2] *Proceedings of the Royal Society*, A, 117, 600, 1928; A, 119, 651, 1928.

[3] *Annalen der Physik*, 85, 981, 1928.

6   energy is passed through a thin foil of matter. The foil is composed of minute crystals oriented at random, but the atoms in each crystal are regularly arranged. A photographic plate receiving the emergent rays exhibits rings of blackening (Fig. 2), as though the rays were waves and were diffracted by the minute crystals. From the diame-

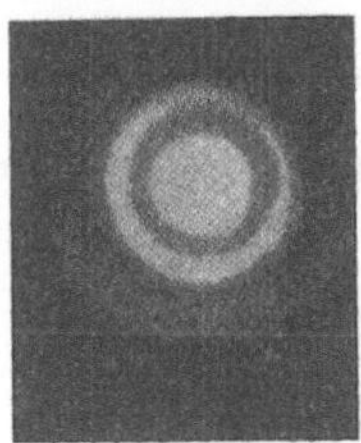

Fig. 2.—Diffraction of electrons on passing through a thin foil of matter.

ters of the rings and the structure of the crystals, the length of these waves may be determined and is found to be $\lambda = h/mv$, where m is the mass and v the velocity of the particles as determined by the above-mentioned experiments. Similar experiments were performed by Davisson and Germer, Kikuchi,[1] and Rupp.

   c) The diffraction of X-rays.—The same dual interpretation is necessary in the case of light and electromagnetic radiation in general. After Newton's objections to

[1] *Japanese Journal of Physics*, 5, 83, 1928.

7   the wave theory of light had been refuted and the phenomena of interference explained by Fresnel, this theory dominated all others for many years, until Einstein[1] pointed out that the experiments of Lenard on the photoelectric effect could only be explained by a corpuscular theory. He postulated that the momentum of the hypothetical particles was related to the wave-length of the radiation by the formula $p = h/\lambda$ (cf. § 2b). The necessity for both interpretations is particularly clear in the case of X-rays: If a homogeneous beam of X-rays is passed through a crystalline mass, and the emergent rays received on a photographic plate (Fig. 3), the result is much like the result of G. P. Thomson's experiment, and it may be concluded that X-rays are a form of wave motion, with a determinable wave-length.

   d) The Compton-Simon[2] experiment.—When a beam of X-rays passes through supersaturated water vapor, it is scattered by the molecules. Secondary products of the scattering are the "recoil" electrons, which are apparently particles of considerable energy, since they form tracks of condensed droplets as do the β-rays. These tracks are not very long, however, and occur with random direction. They apparently originate within the region traversed by the primary X-ray beam. Other secondary products of the scattering are the photoelectrons, which again make themselves evident by longer tracks of condensed water droplets. Under suitable conditions these tracks originate at points outside the primary X-ray beam, but the two secondary products are not unrelated.

[1] *Annalen der Physik*, 17, 145, 1905.   [2] *Physical Review*, 25, 306, 1925.

8   If it be assumed that the X-ray beam consists of a stream of light-particles (photons) and that the scattering process is the collision of a photon with one of the electrons of the molecule, as a result of which the electron recoils in the observed direction, Einstein's postulate regarding the

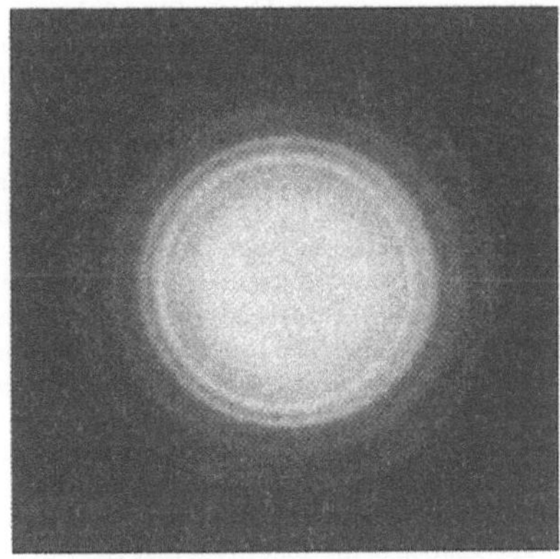

Fig. 3.—Diffraction of X-rays by MgO powder

energy and momentum of the photons enables the direction of the photon after the collision to be calculated. This photon then collides with a second molecule, and gives up its remaining energy to an electron (the photoelectron). This assumption has been quantitatively verified (Fig. 4).

9   *e) The collision experiments of Franck and Hertz.*[1]— When a beam of slow electrons with homogeneous velocity passes through a gas, the electronic current as function of the velocity changes discontinuously at certain values of the velocity (energy). The analysis of these experiments leads to the conclusion that the atoms in the

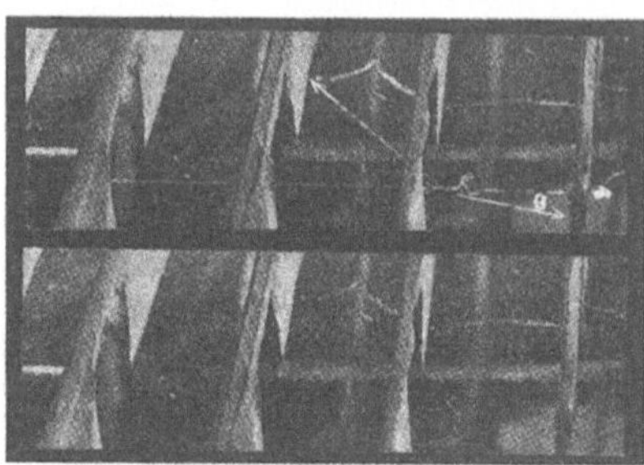

Fig. 4.—Photograph showing recoil electron and associated photo electron liberated by X-rays. The upper photograph is retouched.

gas can only assume discrete energy values (Bohr's postulate). When the energy of the atom is known, one speaks of a "stationary state of the atom." When the kinetic energy of the electron is too small to change the atom from its stationary state to a higher one, the electron makes only elastic collisions with the atoms, but when the kinetic energy suffices for excitation some electrons will transfer their energy to the atom, so the elec-

[1] *Verhandlungen der Deutschen Physikalische Gesellschaft*, **15**, 613, 1913.

10   tronic current as a function of the velocity changes rapidly in the critical region. The concept of stationary states, which is suggested by these experiments, is the most direct expression of the discontinuity in all atomic processes.

From these experiments it is seen that both matter and radiation possess a remarkable duality of character, as they sometimes exhibit the properties of waves, at other times those of particles. Now it is obvious that a thing cannot be a form of wave motion and composed of particles at the same time—the two concepts are too different. It is true that it might be postulated that two separate entities, one having all the properties of a particle, and the other all the properties of wave motion, were combined in some way to form "light." But such theories are unable to bring about the intimate relation between the two entities which seems required by the experimental evidence. As a matter of fact, it is experimentally certain only that light sometimes behaves as if it possessed some of the attributes of a particle, but there is no experiment which proves that it possesses all the properties of a particle; similar statements hold for matter and wave motion. The solution of the difficulty is that the two mental pictures which experiments lead us to form—the one of particles, the other of waves—are both incomplete and have only the validity of analogies which are accurate only in limiting cases. It is a trite saying that "analogies cannot be pushed too far," yet they may be justifiably used to describe things for which our language has no words. Light and matter are both single entities, and the apparent duality arises in the limitations of our language.

It is not surprising that our language should be incapable of describing the processes occurring within the atoms, for, as has been remarked, it was invented to describe the experiences of daily life, and these consist only of processes involving exceedingly large numbers of atoms. Furthermore, it is very difficult to modify our language so that it will be able to describe these atomic processes, for words can only describe things of which we can form mental pictures, and this ability, too, is a result of daily experience. Fortunately, mathematics is not subject to this limitation, and it has been possible to invent a mathematical scheme—the quantum theory—which seems entirely adequate for the treatment of atomic processes; for visualization, however, we must content ourselves with two incomplete analogies—the wave picture and the corpuscular picture. The simultaneous applicability of both pictures is thus a natural criterion to determine how far each analogy may be "pushed" and forms an obvious starting-point for the critique of the concepts which have entered atomic theories in the course of their development, for, obviously, uncritical deduction of consequences from both will lead to contradictions. In this way one obtains the limitations of the concept of a particle by considering the concept of a wave. As N. Bohr[1] has shown, this is the basis of a very simple derivation of the uncertainty relations between co-ordinate and momentum of a particle. In the same manner one may derive the limitations of the concept of a wave by comparison with the concept of a particle.

It must be emphasized that this critique cannot be car-

[1] *Nature*, **121**, 580, 1928; *Naturwissenschaften*, **16**, 245, 1928.

ried through entirely without using the mathematical apparatus of the quantum theory, for the development of the latter preceded the clarification of the physical principles in the historic sequence. In order to avoid obscuring the essential relationships by too much mathematics, however, it has seemed advisable to relegate this formalism to the Appendix. The exposition of mathematical principles given there does not pretend to be complete, but only to furnish the reader with those formulas which are essential for the argument of the text. References to this Appendix are given as A (16), etc.

13

CHAPTER II

CRITIQUE OF THE PHYSICAL CONCEPTS OF THE CORPUSCULAR THEORY OF MATTER

§ 1. THE UNCERTAINTY RELATIONS

The concepts of velocity, energy, etc., have been developed from simple experiments with common objects, in which the mechanical behavior of macroscopic bodies can be described by the use of such words. These same concepts have then been carried over to the electron, since in certain fundamental experiments electrons show a mechanical behavior like that of the objects of common experience. Since it is known, however, that this similarity exists only in a certain limited region of phenomena, the applicability of the corpuscular theory must be limited in a corresponding way. According to Bohr,[1] this restriction may be deduced from the principle that the processes of atomic physics can be visualized equally well in terms of waves or particles. Thus the statement that the position[2] of an electron is known to within a certain accuracy Δx at the time t can be visualized by the picture of a wave packet in the proper position with an approximate extension Δx. By "wave packet" is meant a wavelike disturbance whose amplitude is appreciably different from

[1] N. Bohr, *Nature*, **121**, 580, 1928.

[2] The following considerations apply equally to any of the three space co-ordinates of the electron, therefore only one is treated explicitly.

14   zero only in a bounded region. This region is, in general, in motion, and also changes its size and shape, i.e., the disturbance spreads. The velocity of the electron corresponds to that of the wave packet, but this latter cannot be exactly defined, because of the diffusion which takes place. This indeterminateness is to be considered as an essential characteristic of the electron, and not as evidence of the inapplicability of the wave picture. Defining momentum as $p_x = \mu v_x$ (where μ = mass of electron, v_x = x-component of velocity), this uncertainty in the velocity causes an uncertainty in p_x of amount Δp_x; from the simplest laws of optics, together with the empirically established law $\lambda = h/p$, it can readily be shown that

$$\Delta x \Delta p_x \gtrsim h . \tag{1}$$

Suppose the wave packet made up by superposition of plane sinusoidal waves, all with wave-lengths near λ_0. Then, roughly speaking, $n = \Delta x / \lambda_0$ crests or troughs fall within the boundary of the packet. Outside the boundary the component plane waves must cancel by interference; this is possible if, and only if, the set of component waves contains some for which at least $n+1$ waves fall in the critical range. This gives

$$\frac{\Delta x}{\lambda_0 - \Delta \lambda} \gtrsim n+1 ,$$

where $\Delta \lambda$ is the approximate range of wave-lengths necessary to represent the packet. Consequently

$$\frac{\Delta x \Delta \lambda}{\lambda_0^2} \gtrsim 1 . \tag{2}$$

On the other hand, the group velocity of the waves (i.e., the velocity of the packet) is by A (85)   15

$$v_g = \frac{h}{\mu \lambda_0} , \tag{3}$$

so that the spreading of the packet is characterized by the range of velocities

$$\Delta v_g = \frac{h}{\mu \lambda_0^2} \Delta \lambda .$$

By definition $\Delta p_x = \mu \Delta v_g$ and therefore by equation (2),

$$\Delta x \Delta p_x \gtrsim h .$$

This uncertainty relation specifies the limits within which the particle picture can be applied. Any use of the words "position" and "velocity" with an accuracy exceeding that given by equation (1) is just as meaningless as the use of words whose sense is not defined.[1]

The uncertainty relations can also be deduced without explicit use of the wave picture, for they are readily obtained from the mathematical scheme of quantum theory

[1] In this connection one should particularly remember that the human language permits the construction of sentences which do not involve any consequences and which therefore have no content at all—in spite of the fact that these sentences produce some kind of picture in our imagination; e.g., the statement that besides our world there exists another world, with which any connection is impossible in principle, does not lead to any experimental consequence, but does produce a kind of picture in the mind. Obviously such a statement can neither be proved nor disproved. One should be especially careful in using the words "reality," "actually," etc., since these words very often lead to statements of the type just mentioned.

and its physical interpretation.[1] Any knowledge of the   16 co-ordinate q of the electron can be expressed by a probability amplitude $S(q')$, $|S(q')|^2 dq'$ being the probability of finding the numerical value of the co-ordinate of the electron between q' and $q' + dq'$. Let

$$\bar{q} = \int q' |S(q')|^2 dq' \tag{4}$$

be the average value of q. Then Δq defined by

$$(\Delta q)^2 = 2 \int (q' - \bar{q})^2 |S(q')|^2 dq' \tag{5}$$

can be called the uncertainty in the knowledge of the electron's position. In an exactly analogous way $|T(p')|^2 dp'$ gives the probability of finding the momentum of the electron between p' and $p' + dp'$; again $\bar{p}$ and Δp may be defined as

$$\bar{p} = \int p' |T(p')|^2 dp' , \tag{6}$$

$$(\Delta p)^2 = 2 \int (p' - \bar{p})^2 |T(p')|^2 dp' . \tag{7}$$

By equation A(169), the probability amplitudes are related by the equations

$$\left.\begin{array}{l} T(p') = \int S(q') R(q'p') dq' , \\ S(q') = \int T(p') R^*(q'p') dp' , \end{array}\right\} \tag{8}$$

where $R(q'p')$ is the matrix of the transformation from a Hilbert space in which q is a diagonal matrix to one in which p is diagonal. From equation A(41) we have

$$\int p(q'q'') R(q''p') dq'' = \int R(q'p'') p(p''p') dp'' ,$$

[1] Kennard, *Zeitschrift für Physik*, **44**, 326, 1927.

17 and by equation A(42) this is equivalent to

$$\frac{h}{2\pi i}\frac{\partial}{\partial q'}R(q'p')=p'R(q'p') ,\qquad (9)$$

whose solution is

$$R=ce^{\frac{2\pi i}{h}p'q'} .\qquad (10)$$

Normalizing gives c the value $1/\sqrt{h}$. The values of Δp, Δq are thus not independent. To simplify further calculations, we introduce the following abbreviations:

$$\left.\begin{aligned}
x&=q'-\bar{q} , \qquad y=p'-\bar{p} , \\
s(x)&=S(q')e^{\frac{2\pi i}{h}\bar{p}q'} , \\
t(y)&=T(p')e^{-\frac{2\pi i}{h}\bar{q}(p'-\bar{p})} .
\end{aligned}\right\}\qquad (11)$$

Then equations (5) and (7) become

$$(\Delta q)^2=2\int x^2|s(x)|^2 dx ,\qquad (5a)$$

$$(\Delta p)^2=2\int y^2|t(y)|^2 dy ,\qquad (7a)$$

while equations (8) become

$$t(y)=\frac{1}{\sqrt{h}}\int s(x)e^{\frac{2\pi i}{h}xy}dx ,$$

$$s(x)=\frac{1}{\sqrt{h}}\int t(y)e^{-\frac{2\pi i}{h}xy}dy .\qquad (8a)$$

18 Combining (5a), (7a), and (8a), the expression for $(\Delta p)^2$ may be transformed, giving

$$\begin{aligned}
\tfrac{1}{2}(\Delta p)^2&=\frac{1}{\sqrt{h}}\int y^2 t^*(y)dy\int s(x)e^{\frac{2\pi i}{h}xy}dx , \\
&=\frac{1}{\sqrt{h}}\int t^*(y)dy\int s(x)\left(\frac{h}{2\pi i}\frac{d}{dx}\right)^2 e^{\frac{2\pi i}{h}xy}dx , \\
&=\frac{1}{\sqrt{h}}\left(\frac{h}{2\pi i}\right)^2\int t^*(y)dy\int\frac{d^2s}{dx^2}e^{\frac{2\pi i}{h}xy}dx , \\
&=\left(\frac{h}{2\pi i}\right)^2\int s^*(x)\frac{d^2s}{dx^2}dx ,
\end{aligned}$$

or

$$\tfrac{1}{2}(\Delta p)^2=\frac{h^2}{4\pi^2}\int\left|\frac{ds}{dx}\right|^2 dx .\qquad (12)$$

Now

$$\left|\frac{ds}{dx}\right|^2\geq\frac{1}{(\Delta q)^2}|s(x)|^2-\frac{d}{dx}\left(\frac{x}{(\Delta q)^2}|s(x)|^2\right)$$
$$-\frac{x^2}{(\Delta q)^4}|s(x)|^2 ,\qquad (13)$$

as may be proved by rearranging the obvious relation

$$\left|\frac{x}{(\Delta q)^2}s(x)+\frac{ds}{dx}\right|^2\geq 0 .\qquad (13a)$$

Hence it follows from equation (12) that

$$\left.\begin{aligned}
\tfrac{1}{2}(\Delta p)^2&\geq\tfrac{1}{2}\frac{h^2}{4\pi^2}\frac{1}{(\Delta q)^2} , \\
\Delta p\Delta q&\geq\frac{h}{2\pi} .
\end{aligned}\right\}\qquad (14)$$

19 which was to be proved. The equality can be true in (14) only when the left side of (13a) vanishes, i.e., when

$$\left.\begin{aligned}
s(x)&=ce^{-\frac{x^2}{2(\Delta q)^2}} , \\
\text{or}\quad S(q')&=ce^{-\frac{(q'-\bar{q})^2}{2(\Delta q)^2}-\frac{2\pi i}{h}\bar{p}q'} ,
\end{aligned}\right\}\qquad (15)$$

where c is an arbitrary constant. Thus the Gaussian probability distribution causes the product $\Delta p\Delta q$ to assume its minimum value.

It must be emphasized again that this proof does not differ at all in mathematical content from that given at the beginning of this section on the basis of the duality between the wave and corpuscular pictures of atomic phenomena. The first proof, if carried through precisely, would also involve all the equations (4)–(14). Physically, the last proof appears to be more general than the former, which was proved on the assumption that x was a cartesian co-ordinate and applies specifically only to free electrons because of the relation $\lambda=h/\mu v_g$ which enters into the proof. Equation (14), on the other hand, applies to any pair of canonic conjugates p and q. This greater generality of (14) is rather specious, however. As Bohr[1] has emphasized, if a measurement of its co-ordinate is to be possible at all, the electron must be practically free.

[1] Loc. cit.

§ 2. ILLUSTRATIONS OF THE UNCERTAINTY RELATIONS 20

The uncertainty principle refers to the degree of indeterminateness in the possible present knowledge of the simultaneous values of various quantities with which the quantum theory deals; it does not restrict, for example, the exactness of a position measurement alone or a velocity measurement alone. Thus suppose that the velocity of a free electron is precisely known, while the position is completely unknown. Then the principle states that every subsequent observation of the position will alter the momentum by an unknown and undeterminable amount such that after carrying out the experiment our knowledge of the electronic motion is restricted by the uncertainty relation. This may be expressed in concise and general terms by saying that every experiment destroys some of the knowledge of the system which was obtained by previous experiments. This formulation makes it clear that the uncertainty relation does not refer to the past; if the velocity of the electron is at first known and the position then exactly measured, the position for times previous to the measurement may be calculated. Then for these past times $\Delta p\Delta q$ is smaller than the usual limiting value, but this knowledge of the past is of a purely speculative character, since it can never (because of the unknown change in momentum caused by the position measurement) be used as an initial condition in any calculation of the future progress of the electron and thus cannot be subjected to experimental verification. It is a matter of personal belief whether such a calculation concerning the past history of the electron can be ascribed any physical reality or not.

21 *a) Determination of the position of a free particle.*—As a first example of the destruction of the knowledge of a particle's momentum by an apparatus determining its position, we consider the use of a microscope.[1] Let the particle be moving at such a distance from the microscope that the cone of rays scattered from it through the objective has an angular opening ϵ. If λ is the wave-length of the light illuminating it, then the uncertainty in the measurement of the x-co-ordinate (see Fig. 5) according to the laws of optics governing the resolving power of any instrument is:

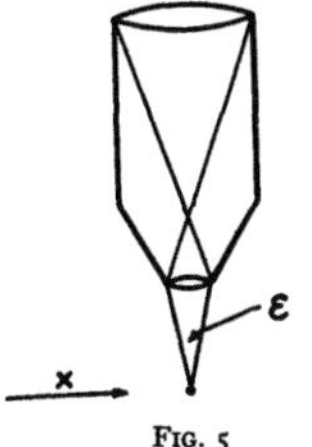

FIG. 5

$$\Delta x = \frac{\lambda}{\sin \epsilon} . \tag{16}$$

But, for any measurement to be possible at least one photon must be scattered from the electron and pass through the microscope to the eye of the observer. From this photon the electron receives a Compton recoil of order of magnitude h/λ. The recoil cannot be exactly known, since the direction of the scattered photon is undetermined within the bundle of rays entering the microscope. Thus there is an uncertainty of the recoil in the x-direction of amount

$$\Delta p_x \sim \frac{h}{\lambda} \sin \epsilon , \tag{17}$$

and it follows that for the motion after the experiment

$$\Delta p_x \Delta x \sim h . \tag{18}$$

[1] N. Bohr, *loc. cit.*

22 Objections may be raised to this consideration; the indeterminateness of the recoil is due to the uncertain path of the light quantum within the bundle of rays, and we might seek to determine the path by making the microscope movable and measuring the recoil it receives from the light quantum. But this does not circumvent the uncertainty relation, for it immediately raises the question of the position of the microscope, and its position and momentum will also be found to be subject to equation (18). The position of the microscope need not be considered if the electron and a fixed scale be simultaneously observed through the moving microscope, and this seems to afford an escape from the uncertainty principle. But an observation then requires the simultaneous passage of at least two light quanta through the microscope to the observer—one from the electron and one from the scale—and a measurement of the recoil of the microscope is no longer sufficient to determine the direction of the light scattered by the electron. And so on *ad infinitum.*

One might also try to improve the accuracy by measuring the maximum of the diffraction pattern produced by the microscope. This is only possible when many photons co-operate, and a calculation shows that the error in measurement of x is reduced to $\Delta x = \lambda / \sqrt{m} \sin \epsilon$ when m photons produce the pattern. On the other hand, each photon contributes to the unknown change in the electron's momentum, the result being $\Delta p_x = \sqrt{m}\, h \sin \epsilon / \lambda$ (addition of independent errors). The relation (18) is thus not avoided.

It is characteristic of the foregoing discussion that simultaneous use is made of deductions from the corpuscular and wave theories of light, for, on the one hand, we speak of resolving power, and, on the other hand, of

23 photons and the recoils resulting from their collision with the particle under consideration. This is avoided, in so far as the theory of light is concerned, in the following considerations.

If electrons are made to pass through a slit of width d (Fig. 6), then their co-ordinates in the direction of this width are known at the moment after having passed it with the accuracy $\Delta x = d$. If we assume the momentum in this direction to have been zero before passing through the slit (normal incidence), it would appear that the uncertainty relation is not fulfilled. But the electron may also be considered to be a plane de Broglie wave, and it is at once apparent that diffraction phenomena are necessarily produced by the slit. The emergent beam has a finite angle of divergence a, which is, by the simplest laws of optics,

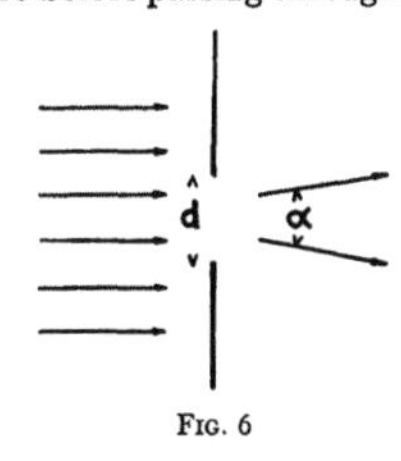

FIG. 6

$$\sin a \sim \frac{\lambda}{d} , \tag{19}$$

where λ is the wave-length of the de Broglie waves. Thus the momentum of the electron parallel to the screen is uncertain, after passing through the slit, by an amount

$$\Delta p = \frac{h}{\lambda} \sin a \tag{20}$$

since h/λ is the momentum of the electron in the direction of the beam. Then, since $\Delta x = d$,

$$\Delta x \Delta p \sim h .$$

24 In this discussion we have avoided the dual character of light, but have made extensive use of the two theories of the electron.

As a last method of determining position we discuss the well-known method of observing scintillations produced by a-rays when they are received on a fluorescent screen or of observing their tracks in a Wilson chamber. The essential point of these methods is that the position of the particle is indicated by the ionization of an atom; it is obvious that the lower limit to the accuracy of such a measurement is given by the linear dimension Δq_s of the atom, and also that the momentum of the impinging particle is changed during the act of ionization. Since the momentum of the electron ejected from the atom is measurable, the uncertainty in the change of momentum of the impinging particle is equal to the range Δp_s within which the momentum of this electron varies while moving in its un-ionized orbit. This variation in momentum is again related to the size of the atom by the inequality

$$\Delta p_s \Delta q_s \gtrsim h .$$

Later discussion will show, in fact, that quite generally[1]

$$\Delta p_s \Delta q_s \sim nh ,$$

where n is the quantum number of the stationary state concerned (cf. § 2c below). Thus the uncertainty relation also governs this type of position measurement; here the dualism of treatment is relegated to the background, and

[1] N Bohr, *loc. cit.*

25 the uncertainty relation appears rather to be the result of the Bohr quantum conditions determining the stationary state, but naturally the quantum conditions are themselves manifestations of the duality.

b) Measurement of the velocity or momentum of a free particle.—The simplest and most fundamental method of measuring velocity depends on the determination of posi-

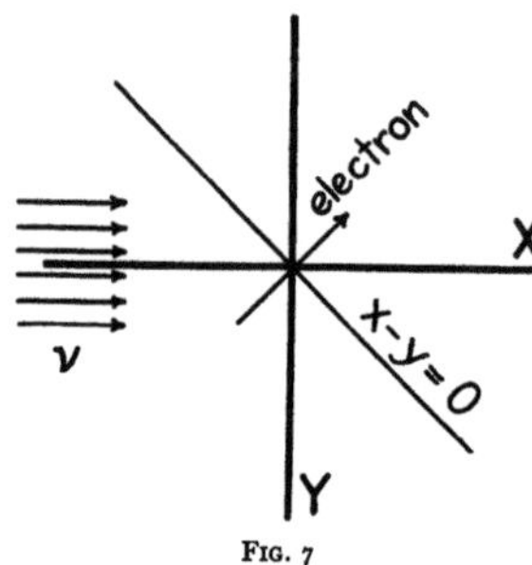

Fig. 7

tion at two different times. If the time interval elapsing between the position measurements is sufficiently large, it is possible to determine the velocity before the second was made with any desired accuracy, but it is the velocity after this measurement which alone is of importance to the physicist, and this cannot be determined with exactness. The change in momentum which is necessarily produced by the last observation is subject to such an indeterminateness that the uncertainty relation is again fulfilled, as has been shown in the last section.

26 Another common method of determining the velocity of charged particles makes use of the Doppler effect. Figure 7 shows the experimental arrangement in its essentials. The component, p_x, of the electron's momentum may be supposed to be known with ideal exactness, its x-co-ordinate therefore completely unknown. On the other hand, the y-co-ordinate of the electron will be assumed to have been accurately determined, and p_y correspondingly unknown. The problem is therefore to determine the velocity in the y-direction, and it is to be shown that the knowledge of the y-co-ordinate is destroyed by this measurement to the extent demanded by the uncertainty relation. The light may be supposed incident along the x-axis, and the scattered light observed in the y-direction. (It is to be noted that the Doppler effect vanishes, under these conditions, if the electron moves along the straight line $x - y = 0$.) The theory of the Doppler effect is in this case identical with that of the Compton effect, and it is only necessary to use the laws of conservation of energy and momentum of the electron and light quantum. Letting E denote the energy of the electron, ν the frequency of the incident light, and using primes to distinguish the same quantity before and after the collision, we have

$$\left.\begin{aligned}
h\nu + E &= h\nu' + E' , \\
\frac{h\nu}{c} + p_x &= p_x' , \\
p_y &= \frac{h\nu'}{c} + p_y' ,
\end{aligned}\right\} \tag{21}$$

whence

$$\left.\begin{aligned}
h(\nu - \nu') &= E' - E , \\
&= \frac{1}{2\mu}[p_x'^2 + p_y'^2 - p_x^2 - p_y^2] , \\
&\sim \frac{1}{\mu}[(p_x' - p_x)p_x + (p_y' - p_y)p_y] , \\
&= \frac{1}{\mu}\left[\frac{h\nu}{c}p_x - \frac{h\nu'}{c}p_y\right] , \\
&\sim \frac{h\nu}{\mu c}(p_x - p_y) .
\end{aligned}\right\} \tag{22}$$

Since it is assumed that p_x and ν are known, the accuracy of the determination of p_y is conditioned only by the accuracy with which the frequency ν' of the scattered light is measured:

$$\Delta p_y' = \frac{\mu c}{\nu}\Delta\nu' . \tag{23}$$

To determine ν' with this accuracy, it is necessary to observe a train of waves of finite length, which in turn demands a finite time:

$$T = \frac{1}{\Delta\nu'} .$$

As it is unknown whether the photon collided with the electron at the beginning or at the end of this time interval, it is also unknown whether the electron moved with the velocity $(1/\mu)p_y$ or $(1/\mu)p_y'$ during this time. The uncertainty in the position of the electron which is produced by this cause is thus

$$\Delta y = \frac{1}{\mu}(p_y - p_y')T = \frac{h\nu}{c\mu}T ,$$

whence

$$\Delta p_y \Delta y \sim h .$$

A third method of velocity measurement depends on the deflection of charged particles by a magnetic field. For this purpose a beam must be defined by a slit, whose width will be designated by d. This ray then enters a homogeneous magnetic field, whose direction is to be taken perpendicular to the plane of Figure 8. The length of that part of the ray which lies in the region of the field may be a; after leaving this region, the ray traverses a field-free region of length l and then passes through a second slit also of width d, whose position determines the angle of deflection a. The velocity of the particles in the direction of the beam is to be determined from the equation

$$a = \frac{\dfrac{a}{v}He\dfrac{v}{c}}{\mu v} = \frac{aHe}{\mu vc} . \tag{24}$$

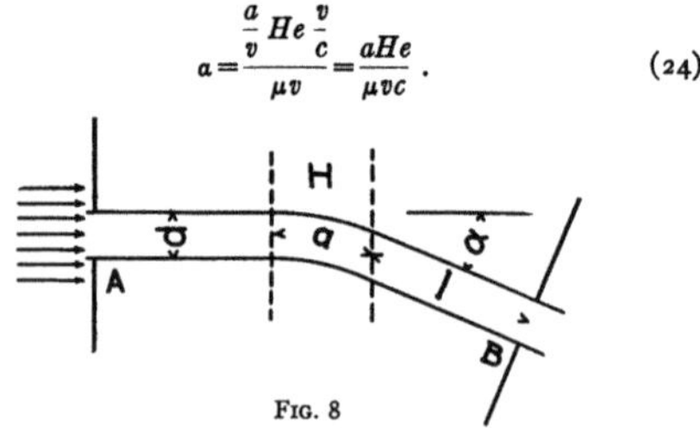

Fig. 8

The corresponding errors in measurement are related by

$$\Delta a = \frac{aHe}{\mu c}\frac{\Delta v}{v^2} .$$

It may be supposed that the position of the particle in the direction of the ray was initially known with great accuracy. This may be achieved, for example, by opening the first slit only during a very brief interval. It will again be shown that this knowledge is lost during the experiment in such a manner that the relation $\Delta p \Delta q \sim h$ is fulfilled after the experiment. To begin with, the accuracy with which the angle a can be determined is obviously $d/(l+a)$, but even this accuracy can only be attained if the natural de Broglie scattering of the ray is less than this. Therefore

$$\Delta a \geq \frac{d}{l+a} , \qquad \Delta a \geq \frac{\lambda}{d} ,$$

whence

$$(\Delta a)^2 \geq \frac{\lambda}{l+a} .$$

The uncertainty in the position of the particle in the ray after the experiment is equal to the product of the time required to pass through the field and reach the second slit and the uncertainty in the velocity. Thus

$$\Delta q \sim \frac{l+a}{v} \Delta v ,$$

whence

$$\Delta q \Delta v \sim \frac{l+a}{v} (\Delta v)^2 ,$$

$$\sim \frac{l+a}{v} \left(\frac{\mu c v}{aHe}\right)^2 (\Delta a)^2 ,$$

$$\geq \frac{\lambda}{v} \left(\frac{\mu c v^2}{aHe}\right)^2 .$$

The terms in the parentheses are equal to v/a and $\lambda = h/\mu v$, whence

$$\mu \Delta q \Delta v \geq \frac{h}{a^2} \geq h ,$$

since equation (24) is valid only for small values of a. For large angles of deflection, this derivation requires radical modification. One must remember, among other things, that the experiment as described here would not distinguish between $a = 0$ and $a = 2\pi$.

c) *Bound electrons.*—If it be required to deduce the uncertainty relations for the position, q, and momentum, p, of bound electrons, two problems must be clearly distinguished. The first assumes that the energy of the system, i.e., its stationary state, is known, and then inquires what accuracy of knowledge of p and q is implied in, or is compatible with, this knowledge of the energy. The second, distinct problem disregards the possibility of determining the energy of the system and merely inquires what the greatest accuracy is with which p and q may simultaneously be known. In this second case, the experiments necessary for the measurement of p and q may produce transitions from one stationary state to another; in the first case, the methods of measurement must be so chosen that transitions are not induced.

We consider the first problem in some detail, and assume an atom in a given stationary state. As Bohr has shown,[1] the corpuscular theory then forces one to conclude that $\Delta p \Delta q$ is in general greater than h. For it is obvious that we are concerned with the variation of p and q as the electron moves in its orbit, and it follows from

$$\int p\,dq = nh \qquad (25)$$

that

$$\Delta q_s \Delta p_s \sim nh . \qquad (26)$$

This may most readily be comprehended from a diagram of the orbit in phase space as given by classical mechanics

[1] *Ibid.*

(Fig. 9). The integral is nothing else than the area inclosed by the orbit, and $\Delta p_s \Delta q_s$ is obviously of the same order of magnitude. The index s which accompanies these uncertainties is to indicate that they are not the absolute minima of these quantities, but are the special values which are assumed by them when the stationary state of the atom is known simultaneously and exactly. This uncertainty is of practical importance, for example, in the discus-

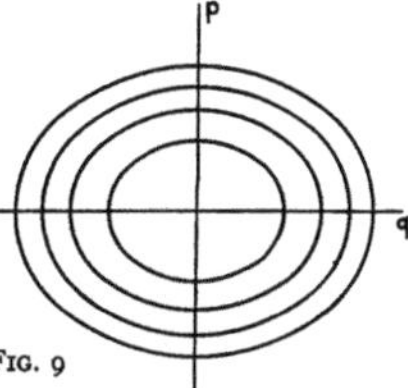

FIG. 9

sion of the scintillation method of counting a-particles (chap. ii, § 2a). In the classical theory, it would seem strange to consider this as an essential uncertainty, for further experiments could be made without disturbing the orbit. The quantum theory, however, shows that a knowledge of the energy is a "determinate case" (*reiner Fall*),[1] i.e., a case which is represented in the mathematical scheme by a definite wave packet (in configuration space) which does not involve any undetermined constants. This wave packet is the Schrödinger function of the stationary state. If the calculation of pages 16–19 is carried through for this packet, the value of $\Delta p_s \Delta q_s$ is found to be greater in proportion to the number of nodes possessed by the characteristic function. If we consider a function s in equation (12) which possesses n nodes, the calculation would show that

$$\Delta p_s \Delta q_s \sim nh .$$

[1] The translators believe that the literal rendering of the German phrase ("pure case") does not at all convey the concept involved.

To pass on to the second problem: The maximum accuracy is obviously given by $\Delta p \Delta q \sim h$ if all knowledge of the stationary states be disregarded. Then the measurements can be carried out by such violent agents that the electron can be regarded as free (acted on only by negligible forces). The momentum of the electron can most readily be measured by suddenly rendering the interaction of the electron with the nucleus and neighboring electrons negligible. It will then execute a straight-line motion and its momentum can be measured in the manner already explained. The disturbance necessary for such a measurement is therefore obviously of the same order of magnitude as the binding energy of the electron.

The relation [eq. (6)] is of importance, as Bohr points out, for the equivalence of classical and quantum mechanics in the limit of large quantum numbers. This is seen when the validity of the concept of an "orbit" is examined. As the highest accuracy attainable is $\Delta p \Delta q \sim h$, the orbit must be the path of a probability packet whose cross-section ($|S(p')|^2 |S(q')|^2$) is approximately h. Such a packet can describe a well-defined, approximately closed path only if the area inclosed by this path is much greater than the cross-section of the wave packet. This, according to equation (26), is possible only in the limit of large quantum numbers; for small n, on the other hand, the concept of an orbit loses all significance, in phase space as well as in configuration space. It is thus seen to be essential for this limiting equivalence of the two theories that the factor n occurs on the right side of equation (26).

The inapplicability of the concept of an orbit in the region of small quantum numbers can be made clear from direct physical considerations in the following manner: The orbit is the temporal sequence of the points in space at which the electron is observed. As the dimensions of the atom in its lowest state are of the order 10^{-8} cm, it will be necessary to use light of wave-length not greater than 10^{-9} cm in order to carry out a position measurement of sufficient accuracy for the purpose. A single photon of such light is, however, sufficient to remove the electron from the atom, because of the Compton recoil. Only a single point of the hypothetical orbit is thus observable. One can, however, repeat this single observation on a large number of atoms, and thus obtain a probability distribution of the electron in the atom. According to Born, this is given mathematically by $\psi\psi^*$ (or, in the case of several electrons, by the average of this expression taken over the co-ordinates of the other electrons in the atom). This is the physical significance of the statement that $\psi\psi^*$ is the probability of observing the electron at a given point. This result is stranger than it seems at first glance. As is well known, ψ diminishes exponentially with increasing distance from the nucleus; there is thus always a small but finite probability of finding the electron at a great distance from the center of the atom. The potential energy of the electrons is negative at such a point, but very small. The kinetic energy is always positive; so that the total energy is therefore certainly greater than the energy of the stationary state under consideration. This paradox finds its resolution when the energy imparted to the electron by the photon used in making the position measurement is taken into account. This energy is considerably greater than the ionization energy of the electron, and

thus suffices to prevent any violation of the law of conservation of energy, as is readily calculated explicitly from the theory of the Compton effect.

This paradox also serves as a warning against carrying out the "statistical interpretation" of quantum mechanics too schematically. Because of the exponential behavior of the Schrödinger function at infinity, the electron will sometimes be found as much as, say, 1 cm from the nucleus. One might suppose that it would be possible to verify the presence of the electron at such a point by the use of red light. This red light would not produce any appreciable Compton recoil and the foregoing paradox would arise once more. As a matter of fact, the red light will not permit such a measurement to be made; the atom as a whole will react with the light according to the formulas of dispersion theory, and the result will not yield any information regarding the position of a given electron in the atom. This may be made plausible if one remembers that (according to the corpuscular theory) the electron will execute a number of rotations about the nucleus during one period of the red light. The statistical predictions of quantum theory are thus significant only when combined with experiments which are actually capable of observing the phenomena treated by the statistics. In many cases it seems better not to speak of the probable position of the electron, but to say that its size depends upon the experiment being performed.

The orbital concept has a significance when applied to highly excited states of the atom; therefore it must be possible to carry out the determination of the position of the electron with an uncertainty less than the dimension of the atom. It does not follow any longer that the elec-

tron will be removed from the atom by the Compton recoil, as may be seen from the following equations. It is necessary that the wave-length of the light, λ, be much less than Δq_s, or by equation (26),

$$\frac{h}{\lambda} \gg \frac{\Delta p_s}{n} .$$

The energy imparted to the electron by its recoil is approximately

$$\frac{h}{\lambda} \frac{\Delta p_s}{\mu} \gg \frac{(\Delta p_s)^2}{n\mu} \sim \frac{|E|}{n} \qquad (26a)$$

(E is the energy of the atom, μ, the mass of the electron); for large values of n, this recoil energy is much less than $|E|$, the ionization energy of the electron. On the other hand, this energy will always be great compared to the energy differences between neighboring stationary states in this region of the spectrum, which is also, in general, of the order $|E|/n$. As a matter of fact, from equation $(26a)$ it follows at once that

$$h\nu \gg \frac{|E|}{n} ,$$

so that the frequency of the light used in making the measurement is great compared to the frequency of the electron in its orbit.

The Compton effect has as its consequence that the electron is caused to jump from a state, say $n = 1000$, to some other state for which n is, say, greater than 950 and less than 1050. The particular orbit to which the electron jumps remains essentially indeterminate because of the considerations of chapter ii, § 1b. The result of the position measurement is therefore to be represented in the mathe-

matical scheme by a probability packet in configuration space, which is built up of characteristic functions of the states between $n = 950$ and 1050. Its size is determined by the exactitude of the position measurement. This packet describes an orbit analogous to that of a corpuscle of classical mechanics, but, in general, spreads and increases in size with the time. The result of a future measurement of position can therefore only be predicted statistically. The mathematical representation of the physical process changes discontinuously with each new measurement; the observation singles out of a large number of possibilities one of which is the one which has happened. The wave packet which has spread out is replaced by a smaller one which represents the result of this observation. As our knowledge of the system does change discontinuously at each observation its mathematical representation must also change discontinuously; this is to be found in classical statistical theories as well as in the present theory.

The motion and spreading of probability packets has been studied by various authors,[1] and therefore no mathematical discussion of it need be given here. A simple consideration of Ehrenfest's[2] may be mentioned, however. Consider the motion of a single electron moving in a field of force whose potential is $V(q)$. The wave function satisfies [cf. eq. A (80)]

$$-\frac{h^2}{8\pi^2\mu} \nabla^2\psi + eV\psi = -\frac{h}{2\pi i} \frac{\partial\psi}{\partial t} , \qquad (27)$$

[1] Kennard, *loc. cit.*; C. G. Darwin, *Proceedings of the Royal Society,* A, **117**, 258, 1927.

[2] P. Ehrenfest, *Zeitschrift für Physik*, **45**, 455, 1927.

37

and the probable value of q is given by equation (4) with $\psi = S$; q is one of the rectangular co-ordinates x, y, z. Then differentiating by t:

$$\mu \dot{q} = \mu \int q \left(\frac{\partial \psi}{\partial t} \psi^* + \psi \frac{\partial \psi^*}{\partial t} \right) d\tau \, ;$$

on substituting the value of $\partial \psi / \partial t$ and $\partial \psi^* / \partial t$ from (27):

$$\mu \dot{q} = \frac{h}{4\pi} \int q (-\psi^* \nabla^2 \psi + \psi \nabla^2 \psi^*) d\tau \, ;$$

integrating by parts:

$$\mu \dot{q} = \frac{h}{4\pi} \int \left(\psi^* \frac{\partial \psi}{\partial q} - \psi \frac{\partial \psi^*}{\partial q} \right) d\tau \, .$$

This process may be repeated a second time to obtain $\mu \ddot{q}$. As the calculation is lengthy, but simple, we give only the result:

$$\mu \ddot{q} = -e \int \frac{\partial V}{\partial q} \psi \psi^* d\tau \, . \tag{28}$$

If ψ represents a wave packet whose spatial dimension is small compared to the distance within which $\partial V / \partial q$ changes appreciably, this may be written

$$\mu \ddot{q} = -e \frac{\partial V(\bar{q})}{\partial \bar{q}} \, . \tag{29}$$

This proves that, so long as the wave packet remains small, its center will move according to the classical equations of motion of the electron.

38

A remark concerning the rate of spreading of the wave packet may not be out of place at this point. If the classical motion of the system is periodic, it may happen that the size of the wave packet at first undergoes only periodic changes. The number of revolutions which the packet may perform before it spreads completely over the whole region of the atom can be calculated qualitatively as follows: If there were no spreading at all, it would be possible to make a Fourier analysis of the probability density into which only integral multiples of the fundamental frequency of the orbit enter. As a matter of fact, however, the "overtones" of quantum theory are not exactly integral multiples of this fundamental frequency. The time in which the phase of the quantum theoretical overtones is completely shifted from that of the classical overtones will be qualitatively the same as the time required for the spreading of the wave packet. Let J be the action variable of classical theory, then this time will be

$$t \sim \frac{1}{h \frac{\partial \nu}{\partial J}} \, ,$$

and the number of revolutions performed in this time is

$$N \sim \frac{\nu}{h \frac{\partial \nu}{\partial J}} \, . \tag{30}$$

In the special case of the harmonic oscillator, N becomes infinite—the wave packet remains small for all time. In general, however, N will be of the order of magnitude of the quantum number n.

In relation to these considerations, one other idealized experiment (due to Einstein) may be considered. We imagine a photon which is represented by a wave packet built up out of Maxwell waves.[1] It will thus have a certain spatial extension and also a certain range of frequency. By reflection at a semi-transparent mirror, it is possible to decompose it into two parts, a reflected and a transmitted packet. There is then a definite probability for finding the photon either in one part or in the other part of the divided wave packet. After a sufficient time the two parts will be separated by any distance desired; now if an experiment yields the result that the photon is, say, in the reflected part of the packet, then the probability of finding the photon in the other part of the packet immediately becomes zero. The experiment at the position of the reflected packet thus exerts a kind of action (reduction of the wave packet) at the distant point occupied by the transmitted packet, and one sees that this action is propagated with a velocity greater than that of light. However, it is also obvious that this kind of action can never be utilized for the transmission of signals so that it is not in conflict with the postulates of the theory of relativity.

d) Energy measurements.—The measurement of the energy of a free electron is identical with the measurement of its velocity, so that most of the possible methods have already been treated. A method not yet discussed for measuring the energy of free electrons is that in which

39

[1] For a single photon the configuration space has only three dimensions; the Schrödinger equation of a photon can thus be regarded as formally identical with the Maxwell equations.

they are caused to move against a retarding field. If the electron passes through the field it is customary to assume the result of classical theory, that its energy E is certainly greater than the energy V corresponding to the highest potential of the field, and if it is reflected, that its energy is smaller than this critical value. Such a conclusion is certainly incorrect in the quantum theory, and a brief discussion of the method will therefore be given here. If the width of the potential barrier is comparable to the de Broglie wave-length, λ, of the electron, a certain number of electrons will penetrate it even though their energies E are less than the critical value necessary on the classical theory. This number decreases exponentially as the width of the barrier and $V - E$ increase. Conversely, when $E > V$, a certain number will be reflected if the potential changes appreciably in a distance λ. In any practicable experiment, these conditions are not realizable, and the conclusions of the classical theory can be used without

40

appreciable error. The mathematical treatment of the situation just sketched is important, however, and will therefore be illustrated in the case of an abrupt discontinuity in the potential distribution. The Schrödinger equation for a single electron will be used; this is not identical with the wave theory of matter, for this latter would take the reaction of the wave on itself into account. The potential distribution is shown in Figure 10. For the

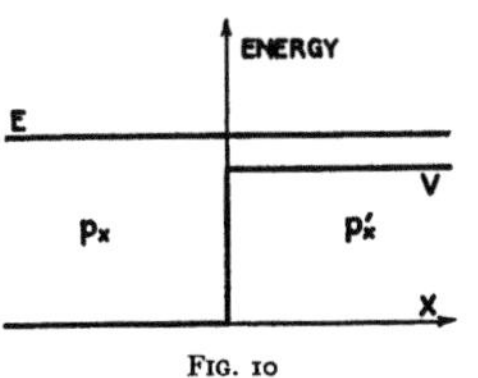

FIG. 10

41 incident ψ-wave in the region I ($x<0$), we then readily obtain the expression

$$\psi_i = a e^{\frac{2\pi i}{h}(px - Et)}, \qquad \frac{1}{2\mu} p^2 = E, \qquad p > 0 ; \quad (31a)$$

for the wave penetrating into the region II ($x>0$),

$$\psi_t = a' e^{\frac{2\pi i}{h}(p'x - Et)}, \qquad \frac{1}{2\mu} p'^2 = E - V ; \quad (31b)$$

and for the reflected wave in I,

$$\psi_r = a'' e^{\frac{2\pi i}{h}(-px - Et)} . \quad (31c)$$

If p' is real, it is to be taken greater than zero; if it is imaginary, total reflection occurs and it is to be taken as positive imaginary, since ψ_t must remain finite as $x \to \infty$. At the discontinuity ($x=0$), ψ must be continuous and possess a continuous first derivative; hence

$$\left.\begin{array}{c} \psi_i + \psi_r = \psi_t \\[2mm] \dfrac{\partial \psi_i}{\partial x} + \dfrac{\partial \psi_r}{\partial x} = \dfrac{\partial \psi_t}{\partial x} , \end{array}\right\} \text{ when } x = 0 ;$$

or

$$a + a'' = a'$$
$$p(a - a'') = a' p' .$$

Solving these equations for a' and a'':

$$\left.\begin{array}{c} a'' = a \dfrac{p - p'}{p + p'} , \\[3mm] a' = a \dfrac{2p}{p + p'} . \end{array}\right\} \quad (32)$$

42 The number of electrons that pass through a given cross-section per unit time is given by the square of the absolute magnitude of the wave amplitude multiplied by the momentum provided it is real. Thus, when $E > V$, the intensities of the incident, transmitted and reflected waves are respectively proportional to

$$\left.\begin{array}{l} I_i = |a|^2 p ; \\[2mm] I_t = |a|^2 \left(\dfrac{2p}{p + p'}\right)^2 ; \\[3mm] I_r = -|a|^2 \left(\dfrac{p - p'}{p + p'}\right)^2 . \end{array}\right\} \quad (33)$$

For imaginary values of p', the wave ψ_t does not represent a current of electrons, but a stationary charge distribution, and $I_t = 0$. As $|a''| = |a|$ in this case, $I_r = -I_i$. In both cases

$$I_i = I_t - I_r .$$

The relative probabilities for reflection and penetration of the electron are, by (33) and (31),

$$\left.\begin{array}{l} P'' = \dfrac{I_r}{I_i} = \left|\dfrac{\sqrt{E} - \sqrt{E - V}}{\sqrt{E} + \sqrt{E - V}}\right|^2 , \\[4mm] P' = \dfrac{I_t}{I_i} = \sqrt{\dfrac{E - V}{E}} \left|\dfrac{2\sqrt{E}}{\sqrt{E} + \sqrt{E - V}}\right|^2 . \end{array}\right\} \quad (34)$$

These expressions are plotted as solid lines in Figure 11; the curves expected from the classical theory are the dotted lines.

For the elucidation of the physical principles of the quantum theory a consideration of the mesaurement of the energy of atoms is more important than that of free electrons, and this will be given in greater detail than the preceding. As the phase of the electronic motion is the

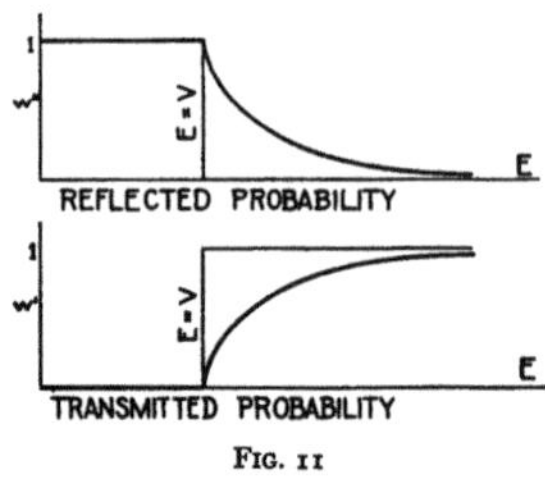

Fig. 11

variable which is canonically conjugate to the energy, it follows from the uncertainty principle that this must be completely unknown if the energy is precisely determined. Since the phase of the electronic motion determines the phase of the radiation emitted, it is this latter which is to enter the physical discussion. It will be shown that any experiment which separates atoms that are in the stationary state n from those in m necessarily destroys any pre-existing knowledge of the phase of the radiation corresponding to the transition $n \rightleftarrows m$.

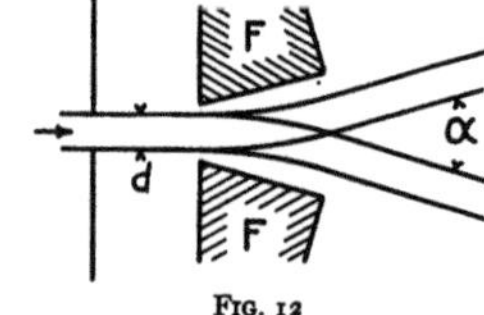

Fig. 12

Let S be a beam of atoms (Fig. 12), of width d in the x-direction, which is sent through an inhomogeneous field F (which is not necessarily a magnetic field, as in the experiment of Stern-Gerlach, but may be electric or gravitational). The energy of the atoms in state m will be designated by E_m; it will depend on the magnitude of the field F at the center of gravity of the atom, so that the deflecting force of the field in the x-direction is $\partial(E_m(F))/\partial x = (dE_m/dF)(dF/dx)$, and is different for atoms in different states. If T be the time required by the atoms to pass through the field, and p the momentum of the atoms in the direction of the beam, the angular deflection of the atoms will be

$$\frac{\partial E_m}{\partial x} \frac{T}{p} .$$

The original beam will thus be divided into several, each containing only atoms in one state; the angular separation α of the two beams containing atoms in states n and m, respectively, will then be

$$\alpha = \left(\frac{\partial E_m}{\partial x} - \frac{\partial E_n}{\partial x}\right)\frac{T}{p} .$$

This angle must be greater than the natural scattering of the atomic beams if the two kinds of atoms are to be separated; hence

$$\alpha \geq \frac{\lambda}{d} = \frac{h}{pd} . \quad (35)$$

The Schrödinger function ψ_n contains the periodic factor $e^{\frac{2\pi i}{h} E_n t}$. As E_n is a function of F, the frequency and phase of the wave are changed while passing through the field. This change is indeterminate, to a certain extent, since it is impossible to tell in what part of the beam the atom is moving and F varies from point to point. The uncertainty, $\Delta\varphi$, of the phase change of the radiation of frequency $(E_m - E_n)/h$ during the time T is therefore

$$\Delta\varphi \sim 2\pi \left(\frac{\partial E_m}{\partial x} - \frac{\partial E_n}{\partial x} \right) \frac{Td}{h} = \frac{pd}{h}\, 2\pi a \; .$$

From equation (35) it follows at once that

$$\Delta\varphi \geq 1 \; . \tag{36}$$

This means complete indeterminateness in the phases.

The calculation can be carried through more concretely if it is restricted to apply only to magnetic fields. Neglecting the electron spin, it is known that the atom precesses like a rigid body when under the influence of a magnetic field H; the velocity of this precession is

$$\omega = \frac{e}{2\mu c}\, H \; ,$$

and its axis coincides with the direction of the field. This velocity is different for various atoms because of the width of the beam and the inhomogeneity of the field. This difference in the precession of different atoms tends to destroy any phase relation which may initially be present. For the uncertainty in ω, we readily obtain

$$\Delta\omega = \frac{ed}{2\mu c}\, \frac{\partial H}{\partial x} \; ,$$

and the angular separation of the two beams is

$$a = \frac{e}{2\mu c}\, \frac{\partial H}{\partial x}\, \frac{hT}{2\pi p} \; ;$$

as a must be greater than h/pd,

$$T\Delta\omega \geq 2\pi \; .$$

All trace of the original phase has thus been destroyed by the experiment. Some atoms will have executed one rotation more than others, and all intermediate angles are possible. This does not follow if the apparatus is incapable of resolving the two beams, as then a may be less than h/pd.

Bohr[1] has shown that the foregoing consideration resolves one of the paradoxes introduced by the assumption of stationary states. If a beam of atoms, all initially in the normal state, be excited to fluorescence by illumination with light of a resonance frequency, we are compelled to assume that they will radiate coherently. That is, each atom will scatter a spherical wave, whose phase is determined by that of the incident plane wave at the atom. The elementary spherical waves will then be so related that their superposition results in a refracted plane wave. From the observation of this wave it is impossible to determine the quantum state of the emitter—or even its atomic character. But if the beam leaves the illuminated region and is analyzed by means of an inhomogeneous field, only the beam of atoms in the excited state will be luminous. This beam will contain relatively few atoms, widely spaced compared to the probable length of the train of waves emitted. Their radiation must therefore be practically identical with that from independent point sources. This action of the magnetic field was quite incomprehensible as long as the assumption was retained that the resolving power of the apparatus could be increased indefinitely by decreasing the width of the beam of atoms.

[1] *Loc. cit.*

CHAPTER III

CRITIQUE OF THE PHYSICAL CONCEPTS OF THE WAVE THEORY

In the foregoing chapter the simplest concepts of the wave theory, which are well established by experiment, were assumed without question to be "correct." They were taken as the basis of a critique of the corpuscular picture, and it appeared that this picture is only applicable within certain limits, which were determined. The wave theory, as well, is only applicable with certain limitations, which will now be determined. Just as in the case of particles the limitations of a wave representation were not originally taken into account, so that historically we first encounter attempts to develop *three-dimensional* wave theories that could be readily visualized (Maxwell and de Broglie waves). For these theories the term "classical wave theories" will be used; they are related to the quantum theory of waves in the same way as classical mechanics to quantum mechanics. The mathematical scheme of the classical and quantum theories of waves will be found in the Appendix. (The reader must be warned against an unwarrantable confusion of classical wave theory with the Schrödinger theory of waves in a phase space.) After a critique of the wave concept has been added to that of the particle concept all contradictions between the two disappear—provided only that due regard is paid to the limits of applicability of the two pictures.

§ 1. THE UNCERTAINTY RELATIONS FOR WAVES

The concepts of wave amplitude, electric and magnetic field strengths, energy density, etc., were originally derived from primitive experiences of daily life, such as the observation of water waves or the vibrations of elastic bodies. These concepts are also widely applicable to light and even, as we now know, to matter waves. But since we also know that the concepts of the corpuscular theory are applicable to radiation and matter, it follows that the wave picture also has its limitations, which may be derived from the particle representation. These will now be considered, first for the case of radiation.

Before proceeding to the subject proper, however, we must first discuss briefly what is meant by an exact knowledge of a wave amplitude—for instance, that of an electric or magnetic field strength. Such an exact knowledge of the amplitude at every point of a region of space (in the strict mathematical sense) is obviously an abstraction that can never be realized. For every measurement can yield only an average value of the amplitude in a very small region of space and during a very short interval of time. Although it is perhaps possible in principle to diminish these space and time intervals without limit by refinement of the measuring instruments, nevertheless for the physical discussion of the concepts of the wave theory it is advantageous to introduce finite values for the space and time intervals involved in the measurements and only pass to the limit zero for these intervals at the end of the calculations. This is, in fact, exactly the procedure adopted in treating the mathematical theory of wave fields (cf. A, § 9). It is possible that future developments of the quantum theory will show that the limit zero for such intervals is an abstraction without physical meaning; for the present, however, there seems no reason for imposing any limitations.

For precision of thought we therefore assume that our measurements always give average values over a very small space region of volume $\delta v = (\delta l)^3$, which depends on the method of measurement. Since it is a question of the measurement of the field strengths, light of wave-length λ much less than δl will not be detected by the experiment. The measurement gives, say, the values E and H for the field strengths (averaged over δv). If these values E and H were exactly known there would be a contradiction to the particle theory, since the energy and momentum of the small volume δv are

$$E = \delta v \,\frac{\mathrm{I}}{8\pi}\,(E^2 + H^2)\;, \qquad G = \delta v\,\frac{\mathrm{I}}{4\pi c}\,E \times H\;, \qquad (37)$$

and the right-hand members could be made as small as desired by taking δv sufficiently small. This is inconsistent with the particle theory, according to which the energy and momentum content of the small volume is made up of discrete and finite amounts $h\nu$ and $h\nu/c$, respectively. For the highest frequency detectable $h\nu \leq (hc/\delta l)$ so that it is clear that the right-hand members of equation (37) must be uncertain by just the magnitudes of these quanta ($h\nu$ and $h\nu/c$) in order that there be no contradiction to the particle theory. Accordingly there must be uncertainty relations between the components of E and H which give rise to an uncertainty in the value of E of the order of magnitude $hc/\delta l$ and in G

of the order of magnitude $h/\delta l$ when E and G are calculated by equations (37). Let ΔE and ΔH be the uncertainties in E and H; then the uncertainties in E and G are

$$\Delta E = \frac{\delta v}{8\pi}\,\{2\,|E\cdot\Delta E| + 2\,|H\cdot\Delta H| + (\Delta E)^2 + (\Delta H)^2\}\;,$$

$$\Delta G_x = \frac{\delta v}{4\pi c}\,\{\,|(E\times\Delta H)_x| + |(\Delta E\times H)_x| + |(\Delta E\times\Delta H)_x|\,\}\;,$$

with cyclic permutation for the y- and z-directions.

Since the most probable values of E and H may possibly be zero the terms on the right which contain only ΔE and ΔH must alone be sufficient to give the necessary uncertainty to E and G. This is attained if

$$\Delta E_x \Delta H_y \geq \frac{hc}{\delta v\,\delta l} = \frac{hc}{(\delta l)^4}\;, \qquad (38)$$

with cyclic permutation for the other components. These uncertainty relations refer to a simultaneous knowledge of E_x and H_y in the same volume element; in different volume elements E_x and H_y can be known to any degree of accuracy.

The relations (38), as in the case of the particle theory, can also be derived directly from the exchange relations for E and H (cf. A, §§ 9, 12). If a division of space into finite cells of magnitude δv is used, the integration with respect to dv in the Lagrangian of A (97) becomes a sum over all the cells δv. The momentum conjugate to $\psi_a(r)$ in the rth cell is then [cf. A(104)]

$$\delta v\,\frac{\partial L}{\partial \dot\psi_a(r)} = \delta v\,\Pi_a(r)\;, \qquad (39)$$

and in place of A(111),

$$\Pi_a(r)\psi_\beta(s) - \psi_\beta(s)\,\Pi_a(r) = \delta_{\alpha\beta}\delta_{rs}\,\frac{h}{2\pi i}\,\frac{\mathrm{I}}{\delta v}\;, \qquad (40)$$

where δ_{rs} is now the usual δ-function,

$$\delta_{rs} = \begin{cases} \mathrm{I} & \text{for } r = s\;, \\ \mathrm{o} & \text{for } r \neq s\;. \end{cases}$$

In the limit $\delta v \to \mathrm{o}$ (40) becomes A(111).

From (40) and A(134) applied to the case of electric and magnetic fields it follows that

$$E_i(r)\Phi_a(s) - \Phi_a(s)E_i(r) = -\,2hci\delta_{rs}\delta_{ai}\,\frac{\mathrm{I}}{\delta v}\;. \qquad (41)$$

When it is remembered that an uncertainty $\Delta\Phi_k$ gives an uncertainty of order of magnitude $\Delta\Phi_k/\delta l$ for the field strengths resulting from Φ_k, it will be seen that (41) leads immediately to the uncertainty relations (38).

Matter waves may be treated in an entirely similar way. It must be noted, however, that no experiment can ever measure the amplitude directly, as is evident from the fact that the de Broglie waves are complex. If exchange relations for the wave amplitudes are derived formally from those for ψ and ψ^*, the result is, to be sure, a physically reasonable one in the case of the Bose-Einstein statistics. However, use of the experimentally correct Fermi-Dirac statistics gives the meaningless result that ψ and ψ^* cannot be exactly measured simultaneously at different points of space. It is thus highly satisfactory that there is no experiment which will measure ψ at a given point at a given time. The mathematical reason for this is that even for the interaction of

radiation and matter the part of the Lagrangian referring to matter contains only terms of the form $\psi\psi^*$. From the considerations just given it can also be seen that the Bose-Einstein statistics is a physical necessity for light-quanta if one makes the apparently very natural assumption that measurements of the electric and magnetic fields at different points of space must be independent of each other.

§ 2.   DISCUSSION OF AN ACTUAL MEASUREMENT
OF THE ELECTROMAGNETIC FIELD

As in the case of the corpuscular picture, it must be possible to trace the origin of the uncertainty in a measurement of the electromagnetic field to its experimental

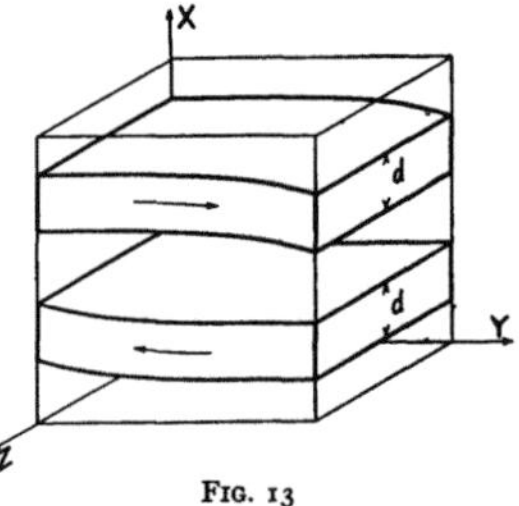

Fig. 13

source. We therefore discuss an experiment which is capable of simultaneously measuring E_x and H_z in the same element of volume δv. This can be accomplished by the observation of the deflection in the direction of x of two beams of cathode rays which traverse the volume in opposite directions along the y-axis (cf. Fig. 13). It may be assumed that the width of both beams in the z-direction is δl, i.e., the whole width of the volume element, but their widths in the perpendicular direction must be less than this, say d, so that they may traverse δv without mutual disturbance. If the distance between the two rays is of order of magnitude δl, the small inhomogeneities of the field in this direction are also averaged out; it would also be possible to vary the distance between them for this purpose. This experimental arrangement will enable the measurement of E_x and H_z in δl provided only that the fields are not too inhomogeneous; should this condition not be fulfilled, the method is incapable of giving a definite result, for the field must not vary appreciably across the width of the rays, or else these will become diffuse and no simple method of determining the deflections is then available.

The angular deflection, a, of the rays in the distance δl is to be observed, and the field can be calculated from the formulas

$$a_{\pm} = \frac{e}{p_v}\left(E_x \pm \frac{p_v}{\mu c} H_z\right)\frac{\mu \delta l}{p_v}\,.$$

Because of the natural spreading of the matter rays, the accuracy of the measurements is given by

$$\Delta E_x \geq \frac{h}{ed}\frac{p_v}{\mu \delta l}\,, \qquad \Delta H_z \geq \frac{h}{ed}\frac{p_v}{\mu \delta l}\frac{\mu c}{p_v}\,. \tag{42}$$

One essential factor remains to be considered, however. Each of the two electrons which pass through δv simultaneously modifies the field, and hence the path of the other electron. The amount of this modification is uncertain to some extent, since it is not known at which point

in the cathode ray the electron is to be found. The uncertainty as to the actual fields which arises from this fact is thus

$$\Delta E_x \geq \frac{ed}{(\delta l)^3}\,, \qquad \Delta H_z \geq \frac{ed}{(\delta l)^3}\frac{p_v}{\mu c}\,, \tag{43}$$

whence

$$\Delta E_x \Delta H_z \geq \frac{hc}{(\delta l)^4}\,,$$

which was to be shown. It is to be noted that the simultaneous consideration of both the corpuscular and wave picture of the process taking place is again fundamental. If the corpuscular picture of the cathode rays had not been invoked, and a continuous distribution of charge assumed as the picture of the rays, then the uncertainty (43) would have disappeared.

55

CHAPTER IV

THE STATISTICAL INTERPRETATION OF QUANTUM THEORY

§ 1. MATHEMATICAL CONSIDERATIONS

It is instructive to compare the mathematical apparatus of quantum theory with that of the theory of relativity. In both cases there is an application of the theory of linear algebras. One can therefore compare the matrices of quantum theory with the symmetric tensors of the special theory of relativity. The greatest difference is the fact that the tensors of quantum theory are in a space of infinitely many dimensions, and that this space is not real but imaginary. The orthogonal transformations are replaced by the so-called "unitary" transformations. In order to obtain a picture of this space, we abstract from such differences, fundamental

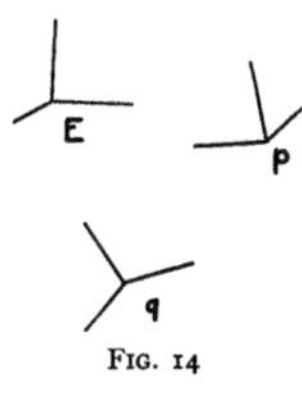

FIG. 14

though they be. Then every quantum theoretical "quantity" is characterized by a tensor whose principal directions may be drawn in this space (cf. Fig. 14). In order to obtain a clear picture, one may recall the tensor of the moments of inertia of a rigid body. The principal directions are, in general, different for each quantity; only matrices which commute with one another have coincident principal directions. The exact knowledge of

the numerical value of any dynamical variable corresponds to the determination of a definite direction in this space, in the same manner as the exact knowledge of the moment of inertia of a solid body determines the principal direction to which this moment belongs (it is assumed that there is no degeneracy). This direction is thus parallel to the kth principal axis of the tensor T, along which the component T_{kk} has the value measured. The exact knowledge of the direction (except for a factor of absolute magnitude unity) in unitary space is the maximum information regarding the quantum dynamical variable which can be obtained. Weyl[1] has called this degree of knowledge a determinate case (*reiner Fall*). An atom in a (non-degenerate) stationary state presents such a determinate case: The direction characterizing it is that of the kth principal axis of the tensor E, which belongs to the energy value E_{kk}. There is obviously no significance to be attached to the terms "value of the coordinate q," etc., in this direction, just as the specification of the moment of inertia about an axis not coinciding with one of the principal directions is insufficient to determine any type of motion of the rigid body, no matter how simple. Only tensors whose principal axes coincide with those of E have a value in this direction. The total angular momentum of the atom, for example, can be determined simultaneously with its energy. If a measurement of the value of q is to be made, then the exact knowledge of the direction must be replaced by inexact information, which can be considered as a "mixture" of the original directions E_{kk}, each with a certain probability coefficient.

[1] H. Weyl, *Zeitschrift für Physik*, 46, 1, 1927.

57 For example, the indeterminate recoil of the electron when its position is measured by a microscope converts the determinate case E_{kk} into such a mixture (cf. chap. ii, § 2a). This mixture must be of such a kind that it may also be considered as a mixture of the principal directions of q, though with other probability coefficients. The measurement singles a particular value q' out of this as being the actual result. It follows from this discussion that the value of q' cannot be uniquely predicted from the result of the experiment determining E, for a disturbance of the system, which is necessarily indeterminate to a certain degree, must occur between the two experiments involved.

This disturbance is qualitatively determined, however, as soon as one knows that the result is to be an exact value of q. In this case, the probability of finding a value q' after E has been measured is given by the square of the cosine of the angle between the original direction E_k and the direction q'. More exactly one should say by the analogue to the cosine in the unitary space, which is $|S(E_k, q')|$. This assumption is one of the formal postulates of quantum theory and cannot be derived from any other considerations. It follows from this axiom that the values of two dynamical quantities are causally related if, and only if, the tensors corresponding to them have parallel principal axes. In all other cases there is no causal relationship. The statistical relation by means of probability coefficients is determined by the disturbance of the system produced by the measuring apparatus. Unless this disturbance is produced, there is no significance to be given the terms "value" or "probable value" of a variable in a

58 given direction of unitary space which is not parallel to a principal axis of the corresponding tensor. Thus one becomes entangled in contradictions if one speaks of the probable position of the electron without considering the experiment used to determine it (cf. the paradox of negative kinetic energy, chap. ii, § 2d). It must also be emphasized that the statistical character of the relation depends on the fact that the influence of the measuring device is treated in a different manner than the interaction of the various parts of the system on one another. This last interaction also causes changes in the direction of the vector representing the system in the Hilbert space, but these are completely determined. If one were to treat the measuring device as a part of the system—which would necessitate an extension of the Hilbert space—then the changes considered above as indeterminate would appear determinate. But no use could be made of this determinateness unless our observation of the measuring device were free of indeterminateness. For these observations, however, the same considerations are valid as those given above, and we should be forced, for example, to include our own eyes as part of the system, and so on. The chain of cause and effect could be quantitatively verified only if the whole universe were considered as a single system—but then physics has vanished, and only a mathematical scheme remains. The partition of the world into observing and observed system prevents a sharp formulation of the law of cause and effect. (The observing system need not always be a human being; it may also be an inanimate apparatus, such as a photographic plate.)

As examples of cases in which causal relations do exist

the following may be mentioned: The conservation theorems for energy and momentum are contained in the quantum theory, for the energies and momenta of different parts of the same system are commutative quantities. Furthermore, the principal axes of q at time t are only infinitesimally different from the principal axes of q at time $t+dt$. Hence, if two position measurements are carried out in rapid succession, it is practically certain that the electron will be in almost the same place both times. 59

§ 2. INTERFERENCE OF PROBABILITIES

Many paradoxical conclusions may be deduced from the foregoing principles if the perturbation introduced by measuring instruments is not adequately considered. The following idealized experiment furnishes a typical example of such a paradox.

A beam of atoms, all of which are initially in the state n, is directed through a field F_1 (Fig. 15). This field will

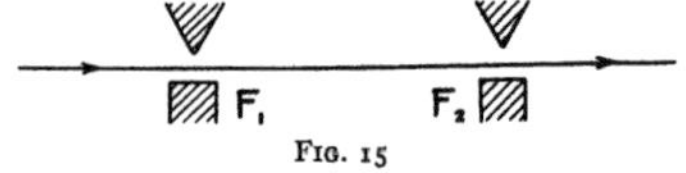

FIG. 15

cause transitions to other states if it is inhomogeneous in the direction of the beam, but will not separate atoms of one state from those in another. Let S'_{nm} be the transformation function for the transitions in the field F_1 so that $|S'_{nm}|^2$ is the probability of finding an atom in the state m after it has emerged from the field F_1. Farther on the atoms encounter a second field F_2, similar in properties to F_1 for which the corresponding transformation func-

tion is S''_{ml}. This field is again incapable of separating the atoms in different states, but beyond F_2 a determination of the stationary state is made by means of a third field of force. Now, for those atoms that are in the state m after passing through F_1 the probability of a transition to state l on passing F_2 is given by $|S_{ml}|^2$. Hence the probable fraction of the atoms in the state l beyond F_2 should be given by 60

$$\sum_m |S'_{nm}|^2 |S''_{ml}|^2 . \qquad (44)$$

On the other hand, according to equation A(69), the transformation function for the combined fields F_1 and F_2 is $S'''_{nl} = \sum_m S'_{nm} S''_{ml}$, which results in the value

$$|S'''_{nl}|^2 = \left| \sum_m S'_{nm} S''_{ml} \right|^2 \qquad (45)$$

for the same probability as represented by equation (44).

The contradiction disappears when it is remarked that the formulas (44) and (45) really refer to two different experiments. The reasoning leading to (44) is correct only when an experiment permitting the determination of the stationary state of the atom is performed between F_1 and F_2. The performance of such an experiment will necessarily alter the phase of the de Broglie wave of the atom in state m by an unknown amount of order of magnitude one, as has been shown in chapter ii, § 2d. In applying (45) to this experiment each member $S'_{nm} S''_{ml}$ in the summation must thus be multiplied by the arbitrary factor $exp(i\varphi_m)$ and then averaged over all values of φ_m. This

61 phase average agrees with (44), which thus applies to this experiment. The rules of the calculus of probabilities can be applied to $|S_{nm}|^2$ only when the causal chain has actually been broken by an observation in the manner explained in the foregoing section. If no break of this sort has occurred it is not reasonable to speak of the atom as having been in a stationary state between F_1 and F_2, and the rules of quantum mechanics apply.

Three general cases may be illustrated by this experiment, and they must be carefully distinguished in any application of the general principles. They are:

CASE I: The atoms remain undisturbed between F_1 and F_2. The probability of observing the state l beyond F_2 is then

$$\left| \sum_m S'_{nm} S''_{ml} \right|^2 .$$

CASE II: The atoms are disturbed between F_1 and F_2 by the performance of an experiment which would have made possible the determination of the stationary state. The result of the experiment is not observed, however. The probability of the state l is then

$$\sum_m |S'_{nm}|^2 |S''_{ml}|^2 .$$

CASE III: The additional experiment of Case II is performed and its result is observed. The atom is known to have been in state m while passing from F_1 to F_2. The probability of the state l is then given by

$$|S''_{ml}|^2 .$$

62 The difference between Cases II and III is recognized in all treatments of the theory of probability, but the difference between I and II does not exist in classical theories which assume the possibility of observation without perturbation. When stated in a sufficiently generalized form, this distinction is the center of the whole quantum theory.

§ 3. BOHR'S CONCEPT OF COMPLEMENTARITY[1]

With the advent of Einstein's relativity theory it was necessary for the first time to recognize that the physical world differed from the ideal world conceived in terms of everyday experience. It became apparent that ordinary concepts could only be applied to processes in which the velocity of light could be regarded as practically infinite. The experimental material resulting from modern refinements in experimental technique necessitated the revision of old ideas and the acquirement of new ones, but as the mind is always slow to adjust itself to an extended range of experience and concepts, the relativity theory seemed at first repellantly abstract. None the less, the simplicity of its solution for a vexatious problem has gained it universal acceptance. As is clear from what has been said, the resolution of the paradoxes of atomic physics can be accomplished only by further renunciation of old and cherished ideas. Most important of these is the idea that natural phenomena obey exact laws—the principle of causality. In fact, our ordinary description of nature, and the idea of exact laws, rests on the assumption that it is

[1] *Nature*, 121, 580, 1928.

63 possible to observe the phenomena without appreciably influencing them. To co-ordinate a definite cause to a definite effect has sense only when both can be observed without introducing a foreign element disturbing their interrelation. The law of causality, because of its very nature, can only be defined for isolated systems, and in atomic physics even approximately isolated systems cannot be observed. This might have been foreseen, for in atomic physics we are dealing with entities that are (so far as we know) ultimate and indivisible. There exist no infinitesimals by the aid of which an observation might be made without appreciable perturbation.

Second among the requirements traditionally imposed on a physical theory is that it must explain all phenomena as relations between objects existing in space and time. This requirement has suffered gradual relaxation in the course of the development of physics. Thus Faraday and Maxwell explained electromagnetic phenomena as the stresses and strains of an ether, but with the advent of the relativity theory, this ether was dematerialized; the electromagnetic field could still be represented as a set of vectors in space-time, however. Thermodynamics is an even better example of a theory whose variables cannot be given a simple geometric interpretation. Now, as a geometric or kinematic description of a process implies observation, it follows that such a description of atomic processes necessarily precludes the exact validity of the law of causality—and conversely. Bohr[1] has pointed out that it is therefore impossible to demand that both require-

[1] *Ibid.*

64 ments be fulfilled by the quantum theory. They represent complementary and mutually exclusive aspects of atomic phenomena. This situation is clearly reflected in the theory which has been developed. There exists a body of exact mathematical laws, but these cannot be interpreted as expressing simple relationships between objects existing in space and time. The observable predictions of this theory can be approximately described in such terms, but not uniquely—the wave and the corpuscular pictures both possess the same approximate validity. This indeterminateness of the picture of the process is a direct result of the interdeterminateness of the concept "observation"—it is not possible to decide, other than arbitrarily, what objects are to be considered as part of the observed system and what as part of the observer's apparatus. In the formulas of the theory this arbitrariness often makes it possible to use quite different analytical methods for the treatment of a single physical experiment. Some examples of this will be given later. Even when this arbitrariness is taken into account the concept "observation" belongs, strictly speaking, to the class of ideas borrowed from the experiences of everyday life.[1] It can only be carried over to atomic phenomena when due regard is paid to the limitations placed on all space-time descriptions by the uncertainty principle.

The general relationships discussed here may be summarized in the following[2] diagrammatic form:

[1] It need scarcely be remarked that the term "observation" as here used does not refer to the observation of lines on photographic plates, etc., but rather to the observation of "the electrons in a single atom," etc. Cf. p. 1.

[2] N. Bohr, *loc. cit.*

65

CLASSICAL THEORY
CAUSAL RELATIONSHIPS OF PHENOMENA DESCRIBED IN TERMS OF SPACE AND TIME

QUANTUM THEORY

Either ... *Or*

Phenomena described in terms of space and time *But* Uncertainty principle	Alternatives related statistically	Causal relationship expressed by mathematical laws *But* Physical description of phenomena in space-time impossible

It is only after attempting to fit this fundamental complementarity of space-time description and causality into one's conceptual scheme that one is in a position to judge the degree of consistency of the methods of quantum theory (particularly of the transformation theory). To mold our thoughts and language to agree with the observed facts of atomic physics is a very difficult task, as it was in the case of the relativity theory. In the case of the latter, it proved advantageous to return to the older philosophical discussions of the problems of space and time. In the same way it is now profitable to review the fundamental discussions, so important for epistemology, of the difficulty of separating the subjective and objective aspects of the world. Many of the abstractions that are characteristic of modern theoretical physics are to be found discussed in the philosophy of past centuries. At that time these abstractions could be disregarded as mere mental exercises by those scientists whose only concern was with reality, but today we are compelled by the refinements of experimental art to consider them seriously.

66

CHAPTER V

DISCUSSION OF IMPORTANT EXPERIMENTS

In the preceding chapters the principles of the quantum theory have all been discussed, but a real understanding of them is obtainable only through their relation to the body of experimental facts which the theory must explain. This is particularly true of the general principle of complementarity. A discussion of further experiments of a less idealized type than those previously used to illustrate the separate principles is therefore necessary at this point.

§ 1. THE C. T. R. WILSON EXPERIMENTS

The essential features of the C. T. R. Wilson photographs may be most easily explained with the help of the classical corpuscular picture. This explanation is also completely justified from the standpoint of the quantum theory. The uncertainty relations are not essential to the explanation of the primary fact of the rectilinearity of the tracks of α-particles. It is always correct to apply the classical theory to such semi-macroscopic phenomena, and the quantum theory is necessary only for the explanation of the finer features.

Nevertheless it will be profitable to discuss the quantum theory of the Wilson photograph. We encounter at once the arbitrariness in the concept of observation already mentioned, and it appears purely as a matter of expediency whether the molecules to be ionized are re-

garded as belonging to the observed system or to the observing apparatus. Consider first the latter alternative. The system to be observed then consists of one α-particle only, and the position measurement resulting from the ionization will be represented in the mathematical scheme of the theory by a probability packet $|\psi(q')|^2$ in the co-ordinate space $q = x, y, z$, of the α-particle. The calculation will be carried out only for one of the three degrees of freedom.

If the time of this determination be taken as $t = 0$, and if a previous determination at a known time is also available, the momentum of the particle at time $t = 0$ may be determined: let $\bar{p}$ and $\bar{q}$ denote the most probable values of the momentum and co-ordinate at this time, and Δp, Δq the probable errors. The value of the uncertainty product will be considerably greater than h in any actual case, but we may assume that $\Delta p \Delta q = h/2\pi$ (cf. the remarks concerning scintillation measurements, chap. ii, § 2a). This is a determinate case; it is then known [eq. (15)] that

$$\psi(q_0') = e^{-(q_0'-\bar{q})^2/2(\Delta q)^2 - \frac{2\pi i}{h} \bar{p}(q'-\bar{q})} .$$

(The index o indicates that q_0' is the value of the co-ordinate at $t = 0$.) The quantum theoretical equations of motion are then

$$p = p_0 = \text{Const.},$$

$$\dot{q} = \frac{1}{\mu} p .$$

68

Although p and q do not commute, the latter equation may nevertheless be integrated[1] to

$$q = \frac{1}{\mu} pt + q_0 .$$

To obtain the probability amplitude $\psi(q')$ at time t the transformation function must be calculated from A(41) and A(42):

$$\left(\frac{t}{\mu} \frac{h}{2\pi i} \frac{\partial}{\partial q_0'} + q_0' \right) S(q_0' q') = q' S(q_0' q') .$$

The solution of this equation is

$$S(q_0' q') = a e^{\frac{2\pi i \mu}{ht}(q' q_0' - q_0'^2/2)} ; \tag{46}$$

by A(69) the distribution at time t is then to be found from

$$\psi(q') = \int_{-\infty}^{+\infty} \psi(q_0') S(q_0' q') dq_0' ,$$

which becomes, on evaluation of the integral,

$$\psi(q') = b e^{[\bar{q} + i(q' - \bar{p} t/\mu)]^2/[2(\Delta q)^2(1 + i/\beta)]} , \tag{47}$$

where

$$\beta = \frac{h}{2\pi} \frac{t}{\mu} \frac{1}{(\Delta q)^2} = \Delta p \frac{t}{\mu \Delta q} .$$

It follows that

$$|\psi(q')|^2 = b' e^{-(q' - p' t/\mu - \bar{q})^2/[(\Delta q)^2 + (t \Delta p/\mu)^2]} . \tag{48}$$

[1] Kennard, *Zeitschrift für Physik*, **44**, 326, 1927.

69 The most probable value for q' is thus $(t/\mu)\bar{p}+\bar{q}$, which is the result to be expected on the classical theory. The mean square error $(\Delta q)^2+(t\Delta p/\mu)^2$ for q' is made up of two terms corresponding to the uncertainties in q'_0 and p'_0; its value again agrees with that which would be calculated classically.

If these methods are applied to all three degrees of freedom, x, y, z, it is seen at once that the path of the center of the probability packet is a straight line. It is to be noted, however, that this result applies only while the α-particle is undisturbed in its motion. Each successive ionization of a water molecule transforms the packet (48) into an aggregate of such packets (Case II, p. 61). If the ionization is accompanied by an observation of the position, a smaller probability packet of the same form as (48) but with new parameters is separated out of the aggregate (Case III, p. 61). This forms the starting-point of a new orbit—and so on. The angular deviations between successive orbital segments are determined by the relative momenta of the particle and the atomic electron with which it interacts, which accounts for the differences between the paths of α- and β-particles.

As regards the formal aspect of the foregoing calculations, it may be noted that the transformation from q'_0 to q' can also be carried out by way of the energy. By equation A(70):

$$S(q'_0 q') = \int S(q'_0 E) S(E q') dE ,$$

and therefore

$$\psi(q') = \int S(E q') dE \int \psi(q'_0) S(q'_0 E) dq'_0 .$$

70 The functions $S(q'E)$, $S(Eq'_0)$ are the normalized Schrödinger wave functions for the free electron; the function $\psi(q')$ can thus be built up by superposition of such Schrödinger functions. This method has been used by Darwin in an investigation of the motion of probability packets.

To complete this discussion we shall finally carry through a mathematical treatment of the Wilson photographs under the assumption that the molecules to be ionized are regarded as part of the system. This procedure is more complicated than the preceding method, but has the advantage that the discontinuous change of the probability function recedes one step and seems less in conflict with intuitive ideas. In order to avoid complication we consider only two molecules and one α-particle, and suppose the centers of mass of the former to be fixed at the points x_1, y_1, z_1; x_2, y_2, z_2. The α-particle is in motion with the momenta p_x, p_y, p_z, and its co-ordinates are x, y, z. The co-ordinates of the electrons in the molecules may be denoted by the single symbols q_1 and q_2, respectively; the configuration space will thus involve only x, y, z, q_1, and q_2. We inquire for the probability that both molecules will be ionized and show that it is negligibly small unless the line joining them has nearly the same direction as the vector $(p_x p_y p_z)$. All interaction between the two molecules will be neglected, and their interaction with the α-particle will be treated as a perturbation;[1] the energy of this interaction may be written

$$\left. \begin{aligned} H^{(1)}(1)+H^{(1)}(2) = &\, H^{(1)}(x-x_1,\ y-y_1,\ z-z_1,\ q_1) \\ &+ H^{(1)}(x-x_2,\ y-y_2,\ z-z_2,\ q_2) , \end{aligned} \right\} \quad (49)$$

[1] M. Born, *Zeitschrift für Physik*, **38**, 803, 1926.

regarded as operators acting on the Schrödinger function. The wave equation is then

$$\left. \begin{aligned} &-\frac{h^2}{8\pi^2\mu}\nabla^2\psi + \underbrace{H^\circ(q_1)\psi}_{} + \underbrace{H^\circ(q_2)\psi}_{} + \underbrace{\epsilon[H^{(1)}(1)+H^{(1)}(2)]\psi}_{} \\ &\qquad\qquad + \frac{h}{2\pi i}\frac{\partial\psi}{\partial t}=0 , \end{aligned} \right\} \quad (50)$$

$$\underbrace{}_{\alpha\text{-Particle}} \quad \underbrace{}_{\text{Molecules}} \quad \underbrace{}_{\text{Interaction}}$$

in which $\nabla^2 = \partial^2/\partial x^2 + \partial^2/\partial y^2 + \partial^2/\partial z^2$, $H^\circ(q_i)$ is the energy operator of the molecule i, and ϵ is the perturbation parameter in powers of which the wave function is to be expanded: $\psi = \psi^{(0)} + \epsilon\psi^{(1)} + \epsilon^2\psi^{(2)} \ldots \ldots$ Substituting this series into the wave equation and equating each power of ϵ to zero, we obtain

$$\left. \begin{aligned} &-\frac{h^2}{8\pi^2\mu}\nabla^2\psi^{(0)} + H^{(0)}(1)\psi^{(0)} + H^{(0)}(2)\psi^{(0)} + \frac{h}{2\pi i}\frac{\partial\psi^{(0)}}{\partial t} \\ &\qquad\qquad = 0 , \\ &-\frac{h^2}{8\pi^2\mu}\nabla^2\psi^{(1)} + H^{(0)}(1)\psi^{(1)} + H^{(0)}(2)\psi^{(1)} + \frac{h}{2\pi i}\frac{\partial\psi^{(1)}}{\partial t} \\ &\qquad\qquad = -[H^{(1)}(1)+H^{(1)}(2)]\psi^{(0)} , \\ &-\frac{h^2}{8\pi^2\mu}\nabla^2\psi^{(2)} + H^{(0)}(1)\psi^{(2)} + H^{(0)}(2)\psi^{(2)} + \frac{h}{2\pi i}\frac{\partial\psi^{(2)}}{\partial t} \\ &\qquad\qquad = -[H^{(1)}(1)+H^{(1)}(2)]\psi^{(1)} , \\ &\qquad \cdot\ \cdot\ \cdot\ \cdot\ \cdot\ \cdot\ \cdot\ \cdot\ \cdot\ \cdot\ \cdot\ \cdot\ \cdot\ \cdot \end{aligned} \right\} \quad (51)$$

The characteristic solutions of the first equation are

$$\psi^{(0)} = e^{\frac{2\pi i}{h}\,p\cdot x}\,\varphi_{n_1}(q_1)\varphi_{n_2}(q_2)\,e^{-\frac{2\pi i}{h}E^{(0)}t} , \quad (52)$$

where

$$H^{(0)}(q)\varphi_n(q) = E_n\varphi_n(q) , \quad (53)$$

and

$$E^0 = \frac{1}{2\mu}\,p^2 + E_{n_1} + E_{n_2} . \quad (54)$$

These solutions correspond to the case in which the momentum of the α-particle is known to be exactly p, its position therefore entirely unknown, while the molecules are known to be in the states n_1, n_2, respectively. All interaction is neglected, and the problem is to determine how the interaction modifies this state of affairs.

This may be solved by determining $\psi^{(1)}$, $\psi^{(2)}$ according to the method of Born. These quantities are first expanded in terms of the orthogonal functions $\varphi_{m_1}(q_1)$ $\varphi_{m_2}(q_2)$,

$$\psi^{(i)} = \sum_{m_1}\sum_{m_2} v^{(i)}_{m_1 m_2}\,\varphi_{m_1}(q_1)\,\varphi_{m_2}(q_2) , \quad (55)$$

in which the $v^{(i)}_{m_1 m_2}$ are of course functions of x, y, z, and t. The significance of these quantities is that

$$\left| \sum_i \epsilon^i v^{(i)}_{m_1 m_2} \right|^2 \quad (56)$$

is the probability of observing the molecule 1 in the state m_1, molecule 2 in the state m_2, and the electron at x, y, z.

Substituting equation (55) for $i=1$ into the first of equations (51), we obtain

$$\left(-\frac{h^2}{8\pi^2\mu}\nabla^2 + E_{n_1} + E_{n_2} + \frac{h}{2\pi i}\frac{\partial}{\partial t}\right) v^{(1)}_{m_1 m_2}$$
$$= -[h_{n_1 m_1}(1)\delta_{n_2 m_2} + h_{n_2 m_2}(2)\delta_{n_1 m_1}]e^{\frac{2\pi i}{h}[p\cdot x - E^0 t]} ,$$

73 in which the abbreviations

$$h_{n_1 m_1}(1) = \int \varphi_{m_1}^*(q_1) H^{(1)}(1) \varphi_{n_1}(q_1) dq_1 \\ h_{n_2 m_2}(2) = \int \varphi_{m_2}^*(q_2) H^{(1)}(2) \varphi_{n_2}(q_2) dq_2 \quad (57)$$

have been used. The co-ordinates q_1 and q_2 have thus been eliminated from further consideration; the functions $h(1)$, $h(2)$ are functions of x, y, z, and of x_1, y_1, z_1 or x_2, y_2, z_2, respectively. These equations may be further simplified by writing

$$v_{m_1 m_2}^{(1)}(xyzt) = w_{m_1 m_2}^{(1)}(xyz) e^{-\frac{2\pi i}{h} E^0 t} ,$$

whence

$$(\nabla^2 + k_{m_1 m_2}^2) w_{m_1 m_2}^{(1)} = \frac{8\pi^2 \mu}{h^2} (h_{n_1 m_1}(1)\delta_{n_2 m_2} + h_{n_2 m_2}(2)\delta_{n_1 m_1}) e^{\frac{2\pi i}{h} p \cdot x} \quad (58)$$

where

$$\frac{h^2}{8\pi^2 \mu} k_{m_1 m_2}^2 = \left[E_{n_1} + E_{n_2} + \frac{1}{2\mu} p^2 - E_{m_1} - E_{m_2} \right] . \quad (59)$$

In this expression the kinetic energy of the α-particle is so much greater than the other terms that, to a sufficient approximation, we may take

$$k_{m_1 m_2}^2 = k^2 = \frac{4\pi^2 p^2}{h^2} = \frac{4\pi^2}{\lambda_0^2} . \quad (60)$$

Equations (58) are then all of the form

$$(\nabla^2 + k^2) w_{m_1 m_2}^{(1)} = \rho_{m_1 m_2}(xyz) , \quad (61)$$

which is the ordinary equation of wave-motion; $\rho_{m_1 m_2}(xyz)$ is the density of the oscillators producing the wave, and,

74 as it is complex, also determines their phase. The solution of equation (61) is given by Huyghen's principle:

$$w_{m_1 m_2}^{(1)} = \iiint \rho_{m_1 m_2}(x'y'z') \frac{e^{-ikR}}{R} dx' dy' dz' ,$$

where R is the distance from x', y', z' to x, y, z.

Since, according to (58), $\rho_{m_1 m_2}$ is zero unless $m_1 = n_1$ or $m_2 = n_2$, all the $w_{m_1 m_2}^{(1)}$ will be zero except $w_{n_1 m_2}^{(1)}$ and $w_{m_1 n_2}^{(1)}$; to the first approximation, only one of the two

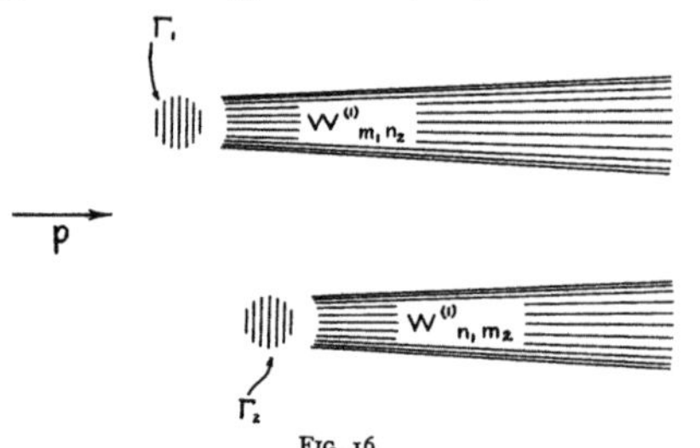

FIG. 16

molecules will be excited. This is in agreement with the classical theory, which says that the probability of two collisions is of second order. The character of the functions $w_{n_1 m_2}^{(1)}$ and $w_{m_1 n_2}^{(1)}$ is readily determined qualitatively; by equation (57)

$$\rho_{m_1 n_2} = \frac{8\pi^2 \mu}{h^2} h_{n_1 m_1}(x - x_1, y - y_1, z - z_1) e^{\frac{2\pi i}{h} p \cdot x} .$$

The (fictitious) oscillators producing the wave are thus all located in the region Γ_1 about x_1, y_1, z_1 (cf. Fig. 16) in

75 which $h_{n_1 m_1}$ is appreciably different from zero. They vibrate coherently, their phase being determined essentially by the factor $e^{\frac{2\pi i}{h} p \cdot x}$; in the figure the lines of equal phase are drawn perpendicular to p. They are spaced at distances λ_0. According to equation (61) the wave-length emitted by the oscillators is also λ_0, and a simple application of Huyghen's principle shows that the wave disturbance will have an appreciable amplitude only in the conical region which is shaded and whose axis is in the direction of p. The cross-section of this region near x_1, y_1, z_1 is determined by the cross-section of the molecule: Γ_1. Its angular opening also depends on Γ_1, being greater when Γ_1 is small—i.e., the uncertainty relation $\Delta p_x \Delta x \sim h/2\pi$ is fulfilled. Similar considerations apply to $w_{n_1 m_2}^{(1)}$; it is different from zero only in a beam originating in Γ_2 and also having the direction p.

We now pass to the second approximation: $v_{m_1 m_2}^{(2)}$ may also be written $w_{m_1 m_2}^{(2)} exp(-2\pi i/h) E^0 t$ and equation (51) reduces to

$$(\nabla^2 + k^2) w_{m_1 m_2}^{(2)} = -\frac{8\pi^2 \mu}{h^2} \left\{ \sum_l w_{l m_2}^{(1)} h_{l m_1}(1) + \sum_i w_{m_1 i}^{(1)} h_{m_2 i}(2) \right\} , \\ = \frac{8\pi^2 \mu}{h^2} \{ w_{n_1 m_2}^{(1)} h_{n_1 m_1}(1) + w_{m_1 n_2}^{(1)} h_{m_2 n_2}(2) \} . \quad (62)$$

The right-hand side of this equation will always be practically zero unless one of the two molecules lies in the beam originating at the other, for $w_{n_1 m_2}^{(1)}$ is different from zero only in the beam originating in Γ_2 and $h_{n_1 m_1}(1)$ only in Γ_1. Unless these two regions intersect, the first term will be zero; similarly the second term. Thus the prob-

76 ability of simultaneous ionization or excitation of the two atoms will vanish even in the second approximation unless the line joining their centers of gravity is practically parallel to the direction of motion of the α-particle. These considerations may be extended to the case of any number of molecules without essential modification. For each additional molecule the approximation must be carried one step farther, but the principles and results will be the same. It has thus been proved that the ionized molecules will lie practically on straight lines, and that the deviations from rectilinearity satisfy the uncertainty relations. In thus including the molecules in the observed system, it has not been necessary to introduce the discontinuously changing probability packet, but if we wish to consider the methods by which the excitation of the molecule can actually be observed, these discontinuous changes (now of a probability packet in the configuration space x, y, z, q_1, q_2) will again play a rôle.

§ 2. DIFFRACTION EXPERIMENTS

The diffraction of light or matter (Davisson-Germer, Thomson, Rupp, Kikuchi) by gratings may be explained most simply by the aid of the classical wave theories. The application of space-time wave theories to these experiments is justified from the point of view of the quantum theory, since the uncertainty relations do not in any way affect the purely geometrical aspects of the waves, but only their amplitude (cf. chap. iii, § 1). The quantum theory need only be invoked when discussing the dynamical relations involving the energy and momentum content of the waves.

77   The quantum theory of the waves being thus certainly in agreement with the classical theory in so far as the geometric diffraction pattern is concerned, it seems useless to prove it by detailed calculation. On the other hand, Duane has given an interesting treatment of diffraction phenomena from the quantum theory of the corpuscular picture. We imagine for simplicity that the corpuscle is reflected from a plane ruled grating, whose constant is d.

Let the grating itself be movable. Its translation in the x-direction may be looked upon as a periodic motion, in so far as only the interaction of the incident particles with the grating is considered; for the displacement of the whole grating by an amount d will not change this interaction. Thus we may conclude that the motion of the grating in this direction is quantized and that its momentum p_z may assume only the values nh/d (as follows at once from the earlier form of the theory: $\int p\,dq = nh$). Since the total momentum of grating and particle must remain unchanged, the momentum of the particle can be changed only by an amount mh/d (m an integer):

$$p_z' = p_z + \frac{mh}{d} \,.$$

Furthermore, because of its large mass, the grating cannot take up any appreciable amount of energy, so that

$$p_z'^2 + p_y'^2 = p_z^2 + p_y^2 = p^2 \,.$$

If θ is the angle of incidence, θ' that of reflection, we have

$$\cos\theta = \frac{p_y}{p}, \qquad \cos\theta' = \frac{p_y'}{p},$$

78   whence

$$\sin\theta' - \sin\theta = \frac{mh}{pd} \,.$$

From equation A(83) for the wave-length of the wave associated with a particle it then follows that

$$d(\sin\theta' - \sin\theta) = m\lambda \,,$$

in agreement with the ordinary wave theory.

The dual characters of both matter and light gave rise to many difficulties before the physical principles involved were clearly comprehended, and the following paradox was often discussed. The forces between a part of the grating and the particle certainly diminish very rapidly with the distance between the two. The direction of reflection should therefore be determined only by those parts of the grating which are in the immediate neighborhood of the incident particle, but none the less it is found that the most widely separated portions of the grating are the important factors in determining the sharpness of the diffraction maxima. The source of this contradiction is the confusion of two different experiments (Cases I and II, p. 61). If no experiment is performed which would permit the determination of the position of the particle before its reflection, there is no contradiction with observation if the whole of the grating does act on it. If, on the other hand, an experiment is performed which determines that the particle will strike on a section of length Δx of the grating, it must render the knowledge of the particle's momentum essentially uncertain by an amount $\Delta p \sim h/\Delta x$. The direction of its reflection will therefore

become correspondingly uncertain. The numerical value of this uncertainty in direction is precisely that which would be calculated from the resolving power of a grating of $\Delta x/d$ lines. If $\Delta x \ll d$ the interference maxima disappear entirely; not until this case is reached can the path of the particle properly be compared with that expected on the classical particle theory, for not until then can it be determined whether the particle will impinge on a ruling or on one of the plane parts of the surface, etc.

§ 3. THE EXPERIMENT OF EINSTEIN AND RUPP[1]

Another paradox was thought to be presented by the following experiment: An atom (canal ray) is made to pass a slit S of width d with the velocity v, and emits light while doing so. This light is analyzed by a spectroscope behind S. Since the light can reach the spectroscope only during the time $t = d/v$, the train of waves to be analyzed has a finite length, and the spectroscope will show it as a line whose width corresponds to a frequency range

$$\Delta\nu = \frac{1}{t} = \frac{v}{d} \,.$$

On the other hand, the corpuscular theory seems to prohibit such a broadening. The atom emits monochromatic radiation, the energy of each particle of which is $h\nu$, and the diaphragm (because of its great mass) will not be able to change the energy of the particles.

The fallacy lies in neglecting the Doppler effect and the diffraction of the light at the slit. Those photons which reach P from the atom are not all emitted perpendicularly

[1] A. Einstein, *Berliner Berichte*, p. 334, 1926; A. Rupp, *ibid.*, p. 341, 1926.

to the canal ray; the angular aperture of the beam of photons is $\sin a \sim \lambda/d$ because of the diffraction. The Doppler change of frequency due to this is

$$\Delta\nu = \sin a \frac{v}{c} \nu \,,$$

or

$$\Delta\nu = \frac{\lambda v}{cd} \nu = \frac{v}{d} \,,$$

in agreement with the previous result. In this experiment the exact validity of the energy law for corpuscles is thus in conformity with the requirements of classical optics.

§ 4. EMISSION, ABSORPTION, AND DISPERSION OF RADIATION

a) Application of the conservation laws.—The postulate of the existence of stationary states, combined with the theory of photons, is sufficient to give a qualitative explanation of the interaction of atoms and radiation. This was the first decisive success of the Bohr theory. The most important results of this theory may be briefly summarized here. Let the stationary states of the atom be numbered $1, 2, 3 \ldots n \ldots$ (Fig. 17), counting from the normal state.

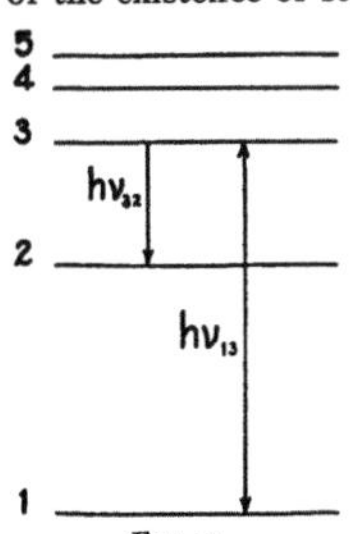

FIG. 17

An atom in state 3, for example, can spontaneously perform a transition to state 2, and emit a photon of energy $h\nu_{32} = E_3 - E_2$. In the

81 same way, an atom in state 1 may absorb a photon of energy $h\nu_{31} = E_3 - E_1$ and thus be excited to the state 3. It must be emphasized that these statements are to be taken quite literally, and not as having only a symbolic significance, for it is possible (e.g., by a Stern-Gerlach experiment) to determine the stationary state of the atoms both before and after the emission. It therefore follows that the intensity of an emission line is proportional to the number of atoms in the upper of the two states associated with it, while the intensity of an absorption line is proportional to the number of atoms in the lower state. These results, which have certainly been amply confirmed by experiment, are entirely characteristic of the quantum theory and can be deduced from no classical theory, either of the wave or particle representation, since even the existence of discrete energy values can never be explained by the classical theory.

An exactly similar situation is met with in the case of scattering. If an atom in state 1 is excited by a photon $h\nu$ it can re-emit the same light quantum without change of state (the mass of the nucleus being assumed infinite), or it can send out the light quantum of energy $h\nu' = h\nu - E_2 + E_1$ by transition to state 2 (Smekal[1] transition; see Fig. 18). The intensity of both kinds of scattered light is proportional to the number of atoms in state 1. If an atom in state 2 is irradiated with light of frequency ν it can emit a photon of energy $h\nu' = h\nu + E_2 - E_1$ of shorter wave-length by transition to state 1, and again the intensity of this "anti-Stokes" scattered light is propor-

[1] *Naturwissenschaften*, 11, 873, 1923.

82 tional to the number of atoms in state 2. This has been confirmed by Raman's[1] experiments.

b) Correspondence principle and the method of virtual charges.—The postulate of stationary states and the theory of photons, because of their very nature, cannot yield any information either regarding the interference of the emitted light or even regarding the a priori probability of the transitions involved. The interference properties can be completely accounted for by the classical wave theory, but it is in turn unable to account for the transitions. To treat these successfully a self-consistent quantum theory of radiation is necessary. It is true that an

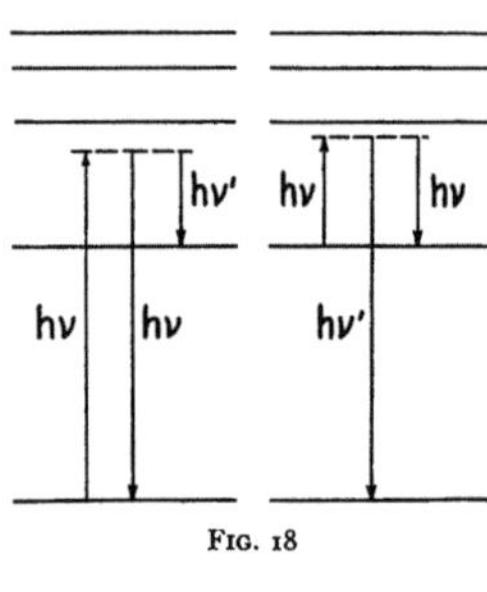

FIG. 18

ingenious combination of arguments based on the correspondence principle can make the quantum theory of matter together with a classical theory of radiation furnish quantitative values for the transition probabilities, i.e., either by the use of Schrödinger's virtual charge density or its equivalent, the element of the matrix representing the electric dipole moment of the atom. Such a formulation of the radiation problem is far from satisfactory, however, and easily leads to false conclusions. These

[1] *Nature*, 121, 501; 122, 12, 1928.

methods may only be applied with the greatest caution, 83 as the following examples may illustrate.

Consider first the case of an atom containing a single electron, and whose nucleus has an infinite mass. If $x \equiv (x, y, z)$ be the co-ordinate of the electron, and $\psi_0(x)$ the Schrödinger function, then

$$-e x_{nm} = -e \int x \psi_n \psi_m^* d\tau \qquad (63)$$

is the element of the matrix representing the dipole moment of the atom. This matrix can enter, strictly speaking, only into calculations based on the principles of the quantum theory of the electron, which in no way involve radiation. It may none the less be interpreted as the dipole moment of the virtual oscillator producing the radiation which is emitted during the transition $n \rightarrow m$. This may be deduced from the correspondence principle by remembering that it has been shown that $x_{nm} \rightarrow x_n(n-m)$ in the limit of large quantum numbers, where $x_n(n-m)$ is a Fourier coefficient of the classical motion. It may thus be presumed that x_{nm} will enter into the formulas determining the intensity of the radiation in the same way as $x_n(n-m)$, i.e., that $|x_{nm}|^2$ will be the a priori probability of the transition $n \rightarrow m$. It must be emphasized that this is a purely formal result; it does not follow from any of the physical principles of quantum theory.

It may be made plausible by another consideration which brings out its unsatisfactory character more clearly. It has been pointed out that the solutions ψ_n of the Schrödinger equation are first approximations to the solutions of the classical matter-wave equations [cf. A(8)]. Denoting by ψ^c a true solution of the latter, the radiation

from the charge distribution thus represented will be de- 84 termined by its dipole moment

$$-e \int \psi^c \psi^{c*} x d\tau$$

provided the extension of this distribution is small compared to the wave-length of the radiation emitted. Now

$$\psi^c \sim \sum_n a_n \psi_n e^{-\frac{2\pi i}{h} E_n t},$$

whence the radiation, calculated by means of this classical distribution, should be determined by

$$-e \sum_{nm} a_n a_m^* x_{nm} e^{\frac{2\pi i}{h}(E_n - E_m)t} . \qquad (64)$$

This formula is certainly wrong since it is derived from a purely classical theory; the intensity of the radiation of frequency $(E_n - E_m)/n$ depends on the coefficient a_m of the final state, as well as on a_n of the initial state. This is in direct contradiction to Bohr's fundamental postulate. The contradiction may be eliminated by arbitrarily dissecting the sum into its separate terms, omitting the offending factors and relating each term to the upper level. The formula (63) for the moment of the virtual dipole associated with the transition then appears once more.

c) The complete treatment of radiation and matter.—The consistent treatment of radiation phenomena requires the simultaneous application of the quantum theory to radiation and matter, in which case it is naturally immaterial whether the particle or wave representation is used.

85 Dirac,[1] in his radiation theory, employs the language of the particle representation, but makes use of conclusions drawn from the wave theory of radiation in his derivation of the Hamiltonian function. The fundamental ideas of this theory are briefly outlined here.

The atom will be represented by a single electron moving in an electrostatic force field ϕ_0. The relativistically invariant equation of the one electron problem is, according to Dirac[2] (ϕ_0 scalar potential, ϕ_i [$i = 1, 2, 3$], electromagnetic potentials),

$$p_0 + \frac{e}{c}\phi_0 + a_i\left(p_i + \frac{e}{c}\phi_i\right) + a_4 mc = 0 ,\qquad (65)$$

or

$$H = -e\phi_0 - a_i c\left(p_i + \frac{e}{c}\phi_i\right) - a_4 mc^2 .\qquad (66)$$

(The usual summation convention is adopted.) Here, as before, the p_i's are the momenta canonically conjugate to the q_i, and the a's are operators which satisfy the equations

$$a_i a_k + a_k a_i = 2\delta_{ik} ;\quad a_i a_4 + a_4 a_i = 0 ;\quad a_4^2 = 1 .\qquad (67)$$

From the equations of motion it follows that

$$\dot{p}_i = -\frac{\partial H}{\partial q_i} = a_k c\,\frac{\partial\phi_k}{\partial x_i} ;\quad \dot{q}_i = \frac{\partial H}{\partial p_i} = -a_i c .\qquad (68)$$

Except for a factor $(-c)$ the a_i's are thus identical with the velocity matrices. From (66) it follows that the inter-

[1] *Proceedings of the Royal Society*, A, **114**, 243, 710, 1927.
[2] *Ibid.*, **117**, 610, 1928.

86 action energy of atoms and radiation field can be written in the simple form

$$-a_i e\phi_i = \frac{e}{c}\,\dot{q}_i\phi_i .\qquad (69)$$

The Hamiltonian function of the complete system atom plus radiation field is thus

$$H_{total\ system} = H_{atom} + \frac{e}{c}\,\dot{q}_i\phi_i + H_{radiation\ field} .\qquad (70)$$

The problem is brought into a simple mathematical form by assuming the radiation field to be in an inclosure, thus providing an orthogonal system of functions on solution of the Maxwell equations subject to the appropriate boundary conditions. The ϕ_i may be developed in this system, and the coefficients [cf. A(123) and (124)] may be written in the form

$$a_r = e^{-\frac{2\pi i}{h}\Theta_r}N_r^{1/2} ,$$

where N_r is the number of light quanta belonging to the rth characteristic vibration. The total energy of the radiation field before considering its interaction with the atom is simply

$$H_{radiation\ field} = \sum_r N_r h\nu_r .\qquad (71)$$

In the development of the ϕ_i in the orthogonal system the individual terms still depend on the position of the atom in the inclosure. Since the dependence averages out in the final result when the inclosure is sufficiently large, it is convenient to introduce a mean-square amplitude ob-

tained by averaging the square of the true amplitude over 87 all possible positions of the atom. This yields the following expression for ϕ_i:

$$\phi_i = \left(\frac{h}{2\pi c}\right)^{1/2}\sum_r \cos a_{ir}\left(\frac{\nu_r}{\sigma_r}\right)^{1/2}\left[N_r^{1/2}e^{\frac{2\pi i}{h}\Theta_r} + e^{-\frac{2\pi i}{h}\Theta_r}N_r^{1/2}\right] .\qquad (72)$$

Here a_{ir} is the angle between the electric vector of the rth characteristic vibration and the q_i-axis, and σ_r is the number of characteristic vibrations in the frequency interval $\Delta\nu_r$ and solid angle $\Delta\omega_r$ divided by $\Delta\nu_r\Delta\omega_r$. Thus the Hamiltonian function for the complete system is

$$\left.\begin{aligned}H = H_{atom} &+ \sum_r N_r h\nu_r \\ &+ \frac{e}{c}\left(\frac{h}{2\pi c}\right)^{1/2}\sum_r \dot{q}_r\left(\frac{\nu_r}{\sigma_r}\right)^{1/2}\left[N_r^{1/2}e^{\frac{2\pi i}{h}\Theta_r} + e^{-\frac{2\pi i}{h}\Theta_r}N_r^{1/2}\right] ,\end{aligned}\right\}\ (73)$$

where $\dot{q}_r$ is the component of the vector $\dot{q}$ in the direction of the electric vector of the rth characteristic vibration.

From equation (73) all the results obtained above by the use of the conservation laws may immediately be deduced. Thus the constancy of H may be proved as in the Appendix (§ 1, p. 121), and it further follows that for the emission or absorption of a light quantum $h\nu_r$, the essential factor is the matrix element of $\dot{q}_r$ corresponding to the transition concerned. Except for certain numerical factors which will not be calculated here the transition probability is given directly by the square of this matrix element. If the calculation is carried out (the interaction terms being regarded as perturbations), emission and absorption processes appear as first-order effects and dis-

persion phenomena as second order. For the details of the 88 calculation the reader is referred to the papers of Dirac.[1]

The formulation of the Hamiltonian of the radiation problem in equation (73) has the disadvantage that it does not appear to involve the interference and coherence properties of the radiation. This is only the case, however, when mean amplitudes are used, as in the foregoing. If the correct amplitudes resulting from the development of the Φ_i in the orthogonal functions are retained, then the fact that these functions are solutions of the Maxwell equations assures interference and coherence properties for the radiation that correspond to the Maxwell equations. For example, solutions of the Maxwell equations appear as factors of the quantities a_r in A(113) and these factors disappear at the position occupied by the atom when the vector potential vanishes there because of interference. Thus there will be no absorption of light in regions where there would be none according to the classical interference theory. From these considerations it follows at once that the classical wave theory is sufficient for the discussion of all questions of coherence and interference.

§ 5. INTERFERENCE AND THE CONSERVATION LAWS

It is very difficult for us to conceive the fact that the theory of photons does not conflict with the requirements of the Maxwell equations. There have been attempts to avoid the contradiction by finding solutions of the latter which represent "needle" radiation (unidirectional

[1] Dirac (*loc. cit.*) uses the original Schrödinger form in place of the Hamiltonian function (73). With the use of (73) the calculation is somewhat simpler, since the quadratic terms in ϕ_i drop out of the interaction energy. The results are the same as those of Dirac.

beams), but the results could not be satisfactorily interpreted until the principles of the quantum theory had been elucidated. These show us that whenever an experiment is capable of furnishing information regarding the direction of emission of a photon, its results are precisely those which would be predicted from a solution of the Maxwell equations of the needle type (cf. the reduction of wave-packets, II, § 2c).

As an example, the recoil produced by the emission of a photon will be discussed. Let an atom go from stationary state n to m with the emission of a photon, and an appropriate change of its total momentum. As we are only concerned with the coherence properties of the emitted radiation, we use the correspondence-principle method, in which the radiation is calculated classically. As source of the radiation we take a charge distribution which is modeled after the expression which would be given by the classical theory of matter waves. The atom will be supposed to consist of one electron (of mass μ, charge $-e$, co-ordinates r_e) and a nucleus (of mass M, charge $+e$, co-ordinates r_n). The Schrödinger function of the nth state, in which the atom has the total momentum P, is

$$e^{\frac{2\pi i}{h} P \cdot r_e} \psi_n(r_e - r_n) e^{\frac{2\pi i}{h} Et} ,$$

where $r_c = (\mu r_e + M r_n)/(\mu + M)$ is the vector to the center of gravity of the atom. If the matrix element of the probability density associated to the transition $n \to m$, $P \to P'$, $E \to E'$, be calculated, one obtains

$$e^{\frac{2\pi i}{h}(P-P') \cdot r_c} \psi_n(r_e - r_n) \psi_m^*(r_e - r_n) e^{-\frac{2\pi i}{h}(E-E')t} .$$

By averaging over the co-ordinates of the nucleus, one obtains the charge density due to the electron, by averaging over the co-ordinates of the electron, that due to the nucleus; the total charge density is their sum. This density is to be considered as the virtual source of the emitted radiation, at least in so far as its coherence properties are concerned. The two component densities are [the common factor e is omitted, $r = r_e - r_n$ is the variable of integration, dv the volume element, and $\gamma = M/(\mu + M)$]

$$\rho_e = e^{\frac{2\pi i}{h}(P-P') \cdot r_e} \int e^{\frac{2\pi i}{h} \gamma(P-P') \cdot r} \psi_n \psi_m^* dv \cdot e^{\frac{2\pi i}{h}(E-E')t} ,$$

$$\rho_n = e^{\frac{2\pi i}{h}(P-P') \cdot r_n} \int e^{\frac{2\pi i}{h} \gamma(P-P') \cdot r} \psi_n \psi_m^* dv \cdot e^{\frac{2\pi i}{h}(E-E')t} .$$

The total density is thus

$$\rho = \text{Const.} \; e^{\frac{2\pi i}{h}[(P-P') \cdot r - (E-E')t]} ,$$

in which the value of the constant does not interest us. The current densities are given by analogous expressions. The radiation emitted by these charges is to be calculated from the retarded potentials:

$$\Phi_0 = \int \rho(t - R'/c)/R' \cdot dv$$

is the scalar potential and analogous expressions may be obtained for the vector potentials Φ_i (R' is the distance from the point of integration, r, to the point of observation R). The result is therefore

$$\Phi_0 = \text{Const.} \int \frac{\exp \frac{2\pi i}{h}[(P-P') \cdot r - (E-E')(t - R'/c)]}{R'} \, dv .$$

If one supposes that an experiment has determined the position of the atom with a given accuracy (the value of the momentum P must then be correspondingly uncertain), then this means that the density ρ is given by the foregoing expression only in a finite volume Δv, and is zero elsewhere. If the radiation at a great distance from Δv is required, R' may be expanded in terms of R (the co-ordinates of the point of observation) and r (the co-ordinates of the point of integration):

$$R' = R - R_1 \cdot r ,$$

where $R_1 = R/R$. The scalar potential is then given by

$$\Phi_0 = \text{Const.} \; e^{\frac{2\pi i}{h}(t - R/c)} \int (1/R) e^{\frac{2\pi i}{h}(P - P' - h\nu R_1/c) \cdot r} dv ,$$

in which $h\nu = E - E'$.

The integral is appreciably different from zero only in that regions for which the factor of r in the exponential is less in absolute magnitude than the reciprocal of Δl, the linear dimension of Δv. In all other regions, the radiation from different portions of Δv is destroyed by interference. Hence

$$P - P' = h\nu R_1/c \pm h/\Delta l ,$$

and the atom recoils with the momentum $h\nu R_1/c$ (except for the natural uncertainty $h/\Delta l$). If the direction of recoil is determined by some experimental procedure, the emitted radiation thus behaves like a unidirectional beam. This is only a special case, however, which is realized only when P and P' are determined with sufficient accuracy, and the co-ordinates of the center of gravity are correspondingly unknown. The other extreme is realized when the experiment fixes the position of the atom more precisely than $\Delta l = h/|P - P'| = c/\nu$, i.e., more precisely than one wave-length of the emitted radiation. The expression for Φ_0 then represents a regular spherical wave and no conclusions can be drawn concerning the recoil, since its uncertainty is greater than its probable value.

This example illustrates very clearly how the quantum theory strips even the light waves of the primitive reality which is ascribed to them by the classical theory. The particular solution of the Maxwell equation which represents the emitted radiation depends on the accuracy with which the co-ordinates of the center of mass of the atom are known.

§ 6. THE COMPTON EFFECT AND THE EXPERIMENT OF COMPTON AND SIMON

There are analogous relations in the theory of the Compton effect, but even though the calculations are the same as those of the preceding paragraph, a summary of the essential results will be given here. It is more interesting to consider bound electrons than free electrons, for then (if one assumes the position of the stationary atomic nucleus as given) there is a certain a priori knowledge concerning the position of the scattering electron. The laws of conservation result in the equations

$$\left. \begin{array}{l} h\nu + E = h\nu' + E' , \\[4pt] \dfrac{h\nu}{c} e \pm \sim \Delta p = \dfrac{h\nu'}{c} e' + p' , \end{array} \right\} \tag{74}$$

The unprimed letters refer to variables before the collision, and the primed ones to variables after the collision;

93 p is the linear momentum of the electron, and e and e' signify unit vectors in the direction of motion of the light quantum; Δp gives the range of momentum of the electron in the atom. If $\sim\Delta p$ is small compared with p and $h\nu/c$, then (74) enables correspondingly exact conclusions regarding the relation between the directions e' and p' to be drawn. If, for example, p' be measured in a Wilson chamber, then the radiation will have all the properties of needle radiation, since the direction of emission of the light quantum is determined. If $p' \gg \Delta p$, then the translational wave function may be regarded as that of a plane wave, namely, $exp\ 2\pi i/h \cdot (p' \cdot r - E't)$, where r is the vector specifying the position of the electron. Let the wave function of the unperturbed state E, which will be assumed to be the normal state, be $\psi_E(r)\ exp\ 2\pi i/h \cdot Et$, where ψ_E is different from zero in an interval $\Delta l [\Delta l \cdot \Delta p \sim h]$.

These wave functions are perturbed by the incident wave of frequency ν, and the perturbation function is a periodic space function of wave-length $\lambda = c/\nu$. Therefore, as the final result for the perturbed charge distribution, one obtains an expression of the form

$$\left.\begin{aligned}
\rho &= c f_E(r) e^{-\frac{2\pi i}{h} Et}\, e^{\frac{2\pi i}{h}\left(\frac{r \cdot e}{\lambda} - \nu t\right)}\, e^{-\frac{2\pi i}{h}(p \cdot r - E't)} \\
&= c f_E(r) e^{\frac{2\pi i}{h}\left[\left(\frac{h\nu}{c} e - p'\right) \cdot r - (E - E' + h\nu)t\right]} ,
\end{aligned}\right\} \quad (75)$$

Where f_E is different from zero only in the interval Δl. If one writes the retarded potentials for points at a great distance from the atom, then[1]

$$\Phi_0(R) = c\, e^{-2\pi i \nu'\left(t - \frac{R}{c}\right)} \int_{\text{atom}} \frac{dv'}{R'} f_E(r') e^{\frac{2\pi i}{h}\left(\frac{h\nu}{c} e - p' - \frac{h\nu'}{c} e'\right) \cdot r} . \quad (76)$$

[1] G. Breit, *Journal of the Optical Society of America*, 14, 324, 1927.

94 In this equation $h\nu' = E - E' + h\nu$, r' is the vector to the point of integration, R to the point of observation, and $R' = R - r'$. The time factor in equation (76) shows that the frequency of the scattered radiation is ν' and corresponds to that of equation (74). Furthermore, the integral on the right-hand side of equation (76) vanishes because of interference, if the factor of r' is materially greater than the reciprocal atomic diameter. Accordingly, since $\Delta l \Delta p \sim h$,

$$\frac{h\nu}{c} e = \frac{h\nu'}{c} e' + p' \pm \sim\Delta p , \quad (77)$$

in agreement with the second equation of (74). The scattered radiation behaves, therefore, in so far as its coherence properties are concerned, like needle radiation. However, the direction of the light quantum is not exactly prescribed, which may be regarded as a consequence of the indeterminateness of the momentum in the original stationary state. This indeterminateness can be diminished if one experiments with more loosely bound electrons, but then the atomic cross-section will be correspondingly greater. If one applies the considerations to an excited state, then $\Delta l \Delta p \sim nh$ appears in place of $\Delta l \Delta p \sim h$ and in the evaluation of the retarded potentials one must take the number of nodes of $\psi(r')$ into account. Since this involves only nonessential complications, we have confined ourselves to the normal state.

If one wishes to explain the Geiger-Bothe experiment on the simultaneity of emission of recoil electron and scattered photon, then if the correspondence principle methods sketched here are used, one must deal with

95 charge distributions which radiate only during a definite time interval. The initial state of the electron will be given, by a wave-packet at rest, whose size depends on the experimental arrangement. The final state will be represented by a morning wave-packet, and the charge density, given by the product of the two wave functions, will then be different from zero only during the time the two packets overlap. The radiation produced will then be a finite wave train moving in a definite direction. A more consequent explanation of the Geiger-Bothe experiment, even though it is equivalent in all its essential points, can only be obtained from the quantum theory of radiation. Moreover, as already shown, in this theory the laws of conservation applied to light quanta and electrons hold, so that one can, without any misgivings, use the customary corpuscular theory of this experiment.

§ 7. RADIATION FLUCTUATION PHENOMENA

The large mean-square fluctuations, which belong to a corpuscular theory, are contained in the mathematical framework of the quantum theory, as shown in the Appendix. It is especially instructive, however, to study the relations between the various physical pictures with which the quantum theory operates by calculating the fluctuation of a radiation field. Let there be given a black cavity, of volume V, containing radiation in temperature equilibrium. The mean energy $\overline{\mathbb{E}}$ contained in a small volume element ΔV in the frequency range between ν and $\nu + \Delta \nu$ is, according to Planck's formula,

$$\overline{\mathbb{E}} = \frac{8\pi^2 h\nu}{c^3} \frac{\Delta\nu\Delta V}{e^{h\nu/kT} - 1} ; \quad (78)$$

96 k is the Boltzmann constant and T the temperature. According to general thermodynamic laws,[1] the following relation holds for the mean-square fluctuation of $\mathbb{E}$:

$$\overline{\Delta\mathbb{E}^2} = kT^2 \frac{d\overline{\mathbb{E}}}{dT} .$$

Substituting into equation (78), it was shown by Einstein that

$$\overline{\Delta\mathbb{E}^2} = \underbrace{h\nu\overline{\mathbb{E}}}_{\text{corpuscle}} + \underbrace{\frac{c^3}{8\pi^2\nu^2\Delta\nu\Delta V}\overline{\mathbb{E}}^2}_{\text{wave}} . \quad (79)$$

This value for the mean-square fluctuation can only be derived partially with the help of the classical theory. The corpuscular viewpoint yields

$$\overline{\mathbb{E}} = h\nu\bar{n} . \quad (80)$$

The classical particle theory thus results only in the first part of formula (79). The classical wave theory of radiation, on the other hand, leads exactly to the second part of (79). The calculations for this will be given later in connection with the quantum theory. Thus, the quantum theory proper is necessary for the derivation of formula (79), in which it is naturally immaterial whether one uses the wave or the corpuscular picture.

If, in particular, one treats the problem by means of the configuration space of the particles (although it is true that this has not been done in a detailed manner for

[1] J. W. Gibbs, *Elementary Principles in Statistical Mechanics*, pp. 70–72, 1902.

97 light quanta), then one must note that the whole term system of the problem can be subdivided into non-combining partial systems, from which a definite one can be chosen as a solution. Because of the exchange relations (84), which become apparent from the corresponding uncertainty relations, that term system must be taken whose characteristic functions are symmetric in the co-ordinates of the light quanta. This choice leads to the Bose statistics for the light quanta and also, as Bose[1] has shown, to equation (78).

If the wave picture be used, then one obtains the number of light quanta corresponding to the vibration concerned from the amplitudes of the characteristic vibrations, and therefore the same mathematical scheme. In order to avoid unnecessary complications in the calculations, let us treat a vibrating string of length l instead of the black radiation cavity. Let $\varphi(x, t)$ be its lateral displacement, and c the velocity of sound in the string. The Lagrangian function becomes

$$L = \tfrac{1}{2}\left[\frac{1}{c^2}\left(\frac{\partial \varphi}{\partial t}\right)^2 - \left(\frac{\partial \varphi}{\partial x}\right)^2\right] , \tag{81}$$

whence (A § 9)

$$\Pi = \frac{1}{c^2}\frac{\partial \varphi}{\partial t} , \tag{82}$$

and

$$\bar{H} = \tfrac{1}{2}\int_0^l \left\{c^2\Pi^2 + \left(\frac{\partial \varphi}{\partial x}\right)^2\right\} = \tfrac{1}{2}\int_0^l \left\{\frac{1}{c^2}\left(\frac{\partial \varphi}{\partial t}\right)^2 + \left(\frac{\partial \varphi}{\partial x}\right)^2\right\} dx . \tag{83}$$

The following exchange relations are to be used:

$$\Pi(x)\varphi(x') - \varphi(x')\Pi(x) = \delta(x-x')\frac{h}{2\pi i} . \tag{84}$$

[1] *Zeitschrift für Physik*, 26, 178, 1924.

98 With the introduction of

$$\varphi(x,t) = \sqrt{\frac{2}{l}}\sum_k q_k(t)\sin\frac{k\pi x}{l} ,$$

$\bar{H}$ goes over into

$$\bar{H} = \tfrac{1}{2}\sum_k \left\{\frac{1}{c^2}\dot{q}_k^2 + \left(\frac{k\pi}{l}\right)^2 q_k^2\right\} . \tag{85}$$

On introducing the momenta associated to q_k,

$$p_k = \frac{1}{c^2}\dot{q}_k , \tag{86}$$

equation (84) becomes

$$p_k q_l - q_l p_k = \delta_{kl}\frac{h}{2\pi i} \tag{87}$$

or

$$\left. \begin{aligned} p_k &= \sqrt{\frac{k\pi}{l}}\sqrt{\frac{h}{2\pi}}\left\{N_k^{\frac{1}{2}}e^{\frac{2\pi i}{h}\Theta_k} + e^{-\frac{2\pi i}{h}\Theta_k}N_k^{\frac{1}{2}}\right\} \\ q_k &= \sqrt{\frac{k\pi}{l}}\sqrt{\frac{h}{2\pi}}\left\{N_k^{\frac{1}{2}}e^{\frac{2\pi i}{h}\Theta_k} - e^{-\frac{2\pi i}{h}\Theta_k}N_k^{\frac{1}{2}}\right\}\frac{1}{i} . \end{aligned} \right\} \tag{88}$$

The characteristic frequencies of the string are $\nu_k = k(c/2l)$, and therefore

$$\bar{H} = \sum_k h\nu_k(N_k + \tfrac{1}{2}) . \tag{89}$$

For the energy in a small section $(0, a)$ of the string, one obtains, however,

$$\mathbb{E} = \frac{1}{l}\int_0^a \sum_{j,k}\left\{\frac{1}{c^2}\dot{q}_j\dot{q}_k\sin\frac{j\pi x}{l}\sin\frac{k\pi x}{l}\right.$$
$$\left. + q_j q_k jk\left(\frac{\pi}{l}\right)^2\cos\frac{j\pi x}{l}\cos\frac{k\pi x}{l}\right\}dx . \tag{90}$$

99 If the terms of this sum with $j=k$ be singled out, then under the explicit hypothesis that the wave-lengths to be considered are all small with respect to a, one obtains the value

$$\bar{\mathbb{E}} = \frac{a}{l}\bar{H} .$$

One thus finds the fluctuation $\Delta\mathbb{E} = \mathbb{E} - \bar{\mathbb{E}}$ by neglecting the terms with $j=k$ in (90). The integration results in

$$\Delta\mathbb{E} = \frac{1}{2l}\sum_{j\neq k}\left\{\frac{1}{c^2}\dot{q}_j\dot{q}_k K_{jk} + jk\left(\frac{\pi}{l}\right)^2 q_j q_k K'_{jk}\right\} , \tag{91}$$

where

$$\left. \begin{aligned} K_{jk} &= c\,\frac{\sin(\nu_j-\nu_k)a/c}{\nu_j-\nu_k} - c\,\frac{\sin(\nu_j+\nu_k)a/c}{\nu_j+\nu_k} , \\ K'_{jk} &= c\,\frac{\sin(\nu_j-\nu_k)a/c}{\nu_j-\nu_k} + c\,\frac{\sin(\nu_j+\nu_k)a/c}{\nu_j+\nu_k} . \end{aligned} \right\} \tag{92}$$

Accordingly, the mean-square fluctuation is given by

$$\overline{\Delta\mathbb{E}^2} = \frac{1}{2l^2}\sum_{j\neq k}\left\{\frac{1}{c^4}\overline{\dot{q}_j^2}\,\overline{\dot{q}_k^2}K_{jk}^2 + j^2 k^2\left(\frac{\pi}{l}\right)^4\overline{q_j^2}\,\overline{q_k^2}K'_{jk}\right.$$
$$\left. + \left(\frac{\pi}{l}\right)^2\frac{jk}{c^2}\left(\overline{q_j\dot{q}_j}\,\overline{q_k\dot{q}_k} + \overline{\dot{q}_j q_j}\,\overline{\dot{q}_k q_k}\right)K_{jk}K'_{jk}\right\} .$$

The sums over j and k may be replaced by an integral over the frequencies ν_j and ν_k, respectively, if it be assumed that the string l is very long, so that its characteristic frequencies are close together. In addition, one finally assumes that a is large and uses the relation

$$\lim_{a\to\infty}\frac{1}{a}\int_{-n}^{n}\frac{\sin^2\nu a}{\nu^2}f(\nu)d\nu = \pi f(0) \tag{93}$$

100 if $\nu_1 > 0$, $\nu_2 > 0$. The double integral then becomes a simple integral and one finds that

$$\overline{\Delta\mathbb{E}^2} = \frac{a}{c}\int d\nu\left\{\frac{1}{c^4}(\overline{\dot{q}_\nu^2})^2 + \left[\left(\frac{2\pi\nu}{c}\right)^2\overline{q_\nu^2}\right]^2\right.$$
$$\left. + \frac{1}{c^2}\left(\frac{2\pi\nu}{c}\right)^2\left[(\overline{q_\nu\dot{q}_\nu})^2 + (\overline{\dot{q}_\nu q_\nu})^2\right]\right\} . \tag{94}$$

Because of the exchange relations (84),

$$\overline{q_\nu\dot{q}_\nu} = -\overline{\dot{q}_\nu q_\nu} = c^2\frac{h}{4\pi i} , \tag{95}$$

so that

$$\bar{\mathbb{E}} = \frac{a}{l}\int d\nu\, Z_\nu\, h\nu(N_\nu + \tfrac{1}{2}) , \tag{96}$$

where $Z_\nu d\nu$ denotes the number of characteristic frequencies in the interval $d\nu$, or, in this case, $Z_\nu = 2l/c$. If the integral be taken over the frequency interval $\Delta\nu$, one obtains

$$\bar{\mathbb{E}} = \frac{a}{l}Z_\nu\,\Delta\nu\, h\nu(N_\nu + \tfrac{1}{2}) , \tag{97}$$

$$\overline{\Delta\mathbb{E}^2} = \frac{a}{c}\Delta\nu\left[\tfrac{1}{2}\left(\frac{\bar{\mathbb{E}}c}{a\Delta\nu}\right)^2 - \tfrac{1}{2}(h\nu)^2\right] . \tag{98}$$

One then subdivides $\bar{\mathbb{E}}$ into the thermal energy $\bar{\mathbb{E}}^*$ and the zero point energy:

$$\bar{\mathbb{E}} = \bar{\mathbb{E}}^* + \frac{a}{l}Z_\nu\,\Delta\nu\,\frac{h\nu}{2} = \bar{\mathbb{E}}^* + a\,\Delta\nu\, h\nu ,$$

and finds

$$\overline{\Delta\mathbb{E}^2} = \frac{a}{2c}\Delta\nu\left[\left(\frac{\bar{\mathbb{E}}^*c}{a\Delta\nu}\right)^2 + 2\frac{\bar{\mathbb{E}}^*c}{a\Delta\nu}h\nu\right]$$
$$= h\nu\bar{\mathbb{E}}^* + \frac{\overline{\mathbb{E}}^{*2}}{\Delta\nu}Z_\nu\frac{a}{l} . \tag{99}$$

This value corresponds exactly to formula (79). The corresponding relation in the classical wave theory may be obtained by passing to the limit $h = 0$ in (99). The classical wave theory thus leads only to the second term of equation (99). The quantum theory, which one can interpret as a particle theory or as a wave theory as one sees fit, leads to the complete fluctuation formula.

§ 8. RELATIVISTIC FORMULATION OF THE QUANTUM THEORY

The conditions imposed on all physical theories by the principle of relativity have been neglected in most of the foregoing discussions, and consequently the results obtained are applicable only under those conditions in which the velocity of light may be regarded as infinite. The reason for this neglect is that all relativistic effects belong to the *terra incognita* of quantum theory; the physical principles which have been elucidated in this book must be valid in this region also and thus it seemed proper not to obscure them with questions that cannot be aswered definitely at the present time. None the less, this book would be incomplete without a brief discussion of the attempts to construct theories which shall embody both sets of principles, and the difficulties which have arisen in these attempts.

Dirac[1] has set up a wave equation which is valid for one electron and is invariant under the Lorentz transformation. It fulfils all requirements of the quantum theory, and is able to give a good account of the phenomena of the "spinning" electron, which could previously only be

[1] P. A. M. Dirac, *Proceedings of the Royal Society*, A, 117, 610, 1928.

treated by *ad hoc* assumptions. The essential difficulty which arises with all relativistic quantum theories is not eliminated however. This arises from the relation

$$\frac{1}{c^2} E^2 = \mu^2 c^2 + p_x^2 + p_y^2 + p_z^2 \qquad (100)$$

between the energy and momentum of a free electron. According to this equation there are two values of E which differ in sign associated with each set of values of p_x, p_y, p_z. The classical theory could eliminate this by arbitrarily excluding the one sign, but this is not possible according to the principles of quantum theory. Here spontaneous transitions may occur to the states of negative energy; as these have never been observed, the theory is certainly wrong. Under these conditions it is very remarkable that the positive energy-levels (at least in the case of one electron) coincide with those actually observed.

The difficulty inherent in formula (100) is also shown by a calculation of O. Klein,[1] who proves that if the electron is governed by any equation based on this relation it will be able to pass unhindered through regions in which its potential energy is greater than $2mc^2$. If only motion in the x-direction be considered the formulas (31a) (31c) become

$$\frac{E^2}{c^2} = \mu^2 c^2 + p_x^2 ,$$

$$\frac{(E-V)^2}{c^2} = \mu^2 c^2 + p_x'^2 ,$$

[1] *Zeitschrift für Physik*, 53, 157, 1929.

whence

$$p_x'^2 = p_x^2 + \frac{(E-v)^2 - E^2}{c^2} ,$$

while the wave function has the form

$$e^{\frac{2\pi i}{h}(p_x' x - Et)} .$$

For very small values of V, p_x' is real and there are transmitted waves, just as in chapter ii, §2f. For larger values, p_x' becomes a pure imaginary, so that the wave is totally reflected at the discontinuity and decreases exponentially in region II. But for very large values of V, p_x' again becomes real, i.e., the electron wave again penetrates into the region II with constant amplitude. A more exact calculation verifies this result.

A difficulty of a somewhat different character arises in the calculation of the energy of the field of the electron according to the relativistic theory. For a point electron (one of zero radius) even the classical theory yields an infinite value of the energy, as is well known, so that it becomes necessary to introduce a universal constant of the dimension of a length—the "radius of the electron." It is remarkable that in the non-relativistic theory this difficulty can be avoided in another way—by a suitable choice of the order of non-commutative factors in the Hamiltonian function. This has hitherto not been possible in the relativistic quantum theory.

The hope is often expressed that after these problems have been solved the quantum theory will be seen to be based, in a large measure at least, on classical concepts. But even a superficial survey of the trend of the evolution of physics in the past thirty years shows that it is far more likely that the solution will result in further limitations on the applicability of classical concepts than that it will result in a removal of those already discovered. The list of modifications and limitations of our ideal world—which now contains those required by the relativity theory (for which c is characteristic) and the uncertainty relations (symbolized by Planck's constant h)—will be extended by others which correspond to e, μ, M. But the character of these is as yet not to be anticipated.

APPENDIX[1]
THE MATHEMATICAL APPARATUS OF THE QUANTUM THEORY[2]

For the derivation of the mathematical scheme of the quantum theory, whether based on the wave or the particle picture, two sources are available: empirical facts and the correspondence principle. The correspondence principle, which is due to Bohr,[3] postulates a detailed analogy between the quantum theory and the classical theory appropriate to the mental picture employed. This analogy does not merely serve as a guide to the discovery of formal laws; its special value is that it furnishes the interpretation of the laws that are found in terms of the mental picture used.

We commence with a derivation of the mathematical structure of quantum mechanics from the corpuscular analogy.[4]

§ 1. THE CORPUSCULAR CONCEPT OF MATTER

The fundamental equations of classical mechanics for a system of f-degrees of freedom may be written in the so-called "canonical" form,

$$\dot{p}_k = -\frac{\partial H}{\partial q_k}, \quad \dot{q}_k = \frac{\partial H}{\partial p_k}, \qquad (k=1, 2, \ldots, f), \qquad (1)$$

[1] Unless otherwise indicated equation numbers and section numbers refer to the Appendix.

[2] Cf. Translators' note in Preface.

[3] Cf. N. Bohr, *Zeitschrift für Physik*, **13**, 117, 1923.

[4] W. Heisenberg, *ibid.*, **33**, 879, 1925; M. Born and P. Jordan, *ibid.*, **34**, 858, 1925; M. Born, W. Heisenberg, and P. Jordan, *ibid.*, **35**, 557, 1926. Cf. also W. Heisenberg, *Mathematische Annalen*, **95**, 683, 1926.

where $q_1, q_2, \ldots, q_f$ are the generalized co-ordinates, $p_1, p_2, \ldots, p_f$ their conjugate momenta, and H the Hamiltonian function. When H does not depend explicitly on the time the energy equation

$$H(p, q) = W, \qquad (2)$$

where W, the total energy, is a constant, follows at once. For simplicity it may be assumed that the system is multiply periodic, in which case any co-ordinate q_k as a function of the time may be written as a Fourier series, that is, as a sum of harmonic terms in the form

$$q_k = \sum_{\tau_1=-\infty}^{+\infty} \sum_{\tau_2=-\infty}^{+\infty} \cdots \sum_{\tau_f=-\infty}^{+\infty} q_{\tau_1, \tau_2, \ldots, \tau_f}^{(k)} e^{2\pi i(\tau_1\nu_1 + \tau_2\nu_2 + \cdots + \tau_f\nu_f)t}. \qquad (3)$$

The $q_{\tau_1, \tau_2, \ldots, \tau_f}^{(k)}$ are amplitudes independent of the time and $\nu_1, \nu_2, \ldots, \nu_f$ are the fundamental frequencies of the motion. Similar expressions involving the same frequencies may be written for the p_k and in general for any function of the p_k and q_k.

By a canonical transformation—that is, one which leaves invariant the form of equations (1)—it is possible to introduce a new set of canonical conjugates J_k, w_k, known as "action-angle variables." These are essentially defined by the following properties: The Hamiltonian H depends on the J_k only and the w_k are related to the fundamental frequencies of the motion by equations of the form

$$w_k = \nu_k t + \beta_k$$

where the β_k are constants. In these variables the equations of motion therefore become

$$\dot{J}_k = -\frac{\partial H}{\partial w_k}, \quad \dot{w}_k = \nu_k = \frac{\partial H}{\partial J_k}, \qquad (k=1, 2, \ldots, f). \qquad (4)$$

According to classical electrodynamics the frequencies of the spectral lines emitted by an atom will be the frequencies of the harmonic terms in equation (3) and the amplitudes will determine the corresponding intensities.

According to the correspondence principle there must exist a close relationship between the mechanics of classical particles as outlined above and the mechanics of the quantum theory. For the latter we must therefore seek a set of equations analogous in form to the equations of classical theory, but which also take account of certain well-established empirical facts of atomic physics. Primary among these are the following:

1. *The Rydberg-Ritz combination principle.*—The observed spectral frequencies of an atom possess a characteristic term structure. That is, all the spectral lines of an element may be represented as the differences of a relatively small number of terms. If these terms are arranged in a one-dimensional array $T_1, T_2, \ldots$, the atomic frequencies form a two-dimensional array

$$\nu(nm) = T_n - T_m, \qquad (5)$$

from which follows at once the combination principle

$$\nu(nk) + \nu(km) = \nu(nm). \qquad (6)$$

2. *The existence of discrete energy values.*—The fundamental experiments of Franck and Hertz on electronic impacts show that the energy of an atom can take on only certain definite discrete[1] values, $W_1, W_2, \ldots$

3. *The Bohr frequency relation.*—The characteristic frequencies of an atom are related to its characteristic energies by the equation

$$\nu(nm) = \frac{1}{h}(W_n - W_m). \qquad (7)$$

We shall now sketch the deduction of the fundamental equations of the new quantum mechanics, following the program outlined above. It should be distinctly understood, however, that this cannot be a deduction in the mathematical sense of the word, since the equations to be obtained form themselves the postulates of the theory. Although made highly plausible by the following considerations, their ultimate justification lies in the agreement of their predictions with experiment.

A profound modification, not only of classical dynamics, but of classical kinematics, is evidently necessary if the simple experimental facts mentioned above are to be incorporated in the foundations of a new theory. In the classical theory all possible motions of the co-ordinates may be built up by addition from Fourier terms of the kind contained in equation (3), and these may be termed the "kinematic elements," since the quantities with which the theory deals, and in particular the energy,

[1] In general, the atomic energy can also take on continuous values in a certain range. For the time being this "continuous spectrum" may be disregarded, corresponding to the assumption that the system is multiply periodic.

can be expressed in terms of them. Their amplitudes and frequencies are functions of continuously variable constants of integration as well as of the integers $\tau_1 \ldots \tau_f$, which determine the order of the harmonics. This is in direct contradiction to the existence of only discrete values of the atomic energies and frequencies and, in fact, to the very existence of sharply defined spectral lines.

Similar elements must be assumed in quantum mechanics if a correspondence is to be preserved between the two theories. To assure the existence of discrete energy values at the outset, the elements will be taken to be functions of integers. Corresponding to the Rydberg-Ritz combination principle, a dependence on two sets of integers is required, while the f-fold character of the classical harmonics suggests that each set contain f integers. We therefore postulate elements of the form

$$q(n_1 \ldots n_f \, ; \; m_1 \ldots m_f)e^{2\pi i\nu(n_1 \ldots n_f; \, m_1 \ldots m_f)t} \, , \qquad (8)$$

in which the complexes $n_1 \ldots n_f$ and $m_1 \ldots m_f$ replace the single integers n and m in an easily understandable way. Furthermore, the amplitudes and frequencies are assumed to be directly those which are given by a spectral analysis of the emitted radiation, so that the new theory may be described as a calculus of observable quantities. The frequencies $\nu(n_1 \ldots n_f; \, m_1 \ldots m_f)$ are therefore assumed to have the term structure (5); they accordingly obey the combination principle (6).

There can clearly be no question of the addition of such elements to form a Fourier series as in the classical theory; there must, however, be an analogue to the representation

of a co-ordinate by such a series. A sufficiently general and flexible method is afforded by taking simply the ensemble of all elements of the form (8) as the entity which, in the quantum theory of the particle picture, replaces mathematically the classical representation of a co-ordinate given in equation (3). The ensemble may be written as a matrix,

$$\left\| q(n_1 \ldots n_f \, ; \; m_1 \ldots m_f)e^{2\pi i\nu(n_1 \ldots n_f; \, m_1 \ldots m_f)t} \right\| \, ,$$

that is, as an infinite quadratic array, ordered according to the integers n_i, m_i, which take on all real values. The new kinematics is accordingly based on a matrix representation of the co-ordinates, with

$$q_k = \left\| q_k(nm)e^{2\pi i\nu(nm)t} \right\| \qquad (9)$$

corresponding to q_k. As here, the complexes $n_1 \ldots n_f$ and $m_1 \ldots m_f$ will, in general, be replaced by single letters n and m. For the momenta p_k a similar matrix representation is assumed, with the same frequencies, as is the case in classical Fourier series.[1]

Such a representation is, however, meaningless both mathematically and physically until properties and rules of operation for the matrices have been defined. The correspondence principle must be our guide here. In the first place, the classical expression (3) must have a real value; since the terms are complex this can be the case only if for each term there occurs the conjugate imaginary. This

<hr>

[1] For a system which is not multiply periodic, matrices with continuously variable indices must be used, corresponding to a classical representation by Fourier integrals.

will also be true of the elements of the matrix (9) if we assume

$$q_k(mn) = q_k^*(nm) \, ,$$

since by (6) $\nu(mn) = -\nu(nm)$. The asterisk denotes the conjugate imaginary. Matrices with this type of symmetry are called Hermitian and in the quantum theory all co-ordinate matrices are assumed to be of this kind.

The time derivative $\dot{q}_k$ of any co-ordinate is represented classically by the Fourier series whose terms are the time derivatives of those of the series representing q_k. Hence for the quantum-theory matrices

$$\dot{q} \equiv \left\| 2\pi i\nu(nm)q(nm)e^{2\pi i\nu(nm)t} \right\| \, , \qquad (10)$$

which is again a Hermitian matrix of the form (9).

It must be possible in the quantum theory to answer such elementary kinematical questions as the following. Given the matrices representing, say, a momentum p and a co-ordinate q, what matrices represent $p+q$, pq, and in general any function of p and q? In the case of addition the answer is obvious from the classical analogue. Since the sum of two Fourier series of the form (3) is again a series of the same kind and with the same frequencies, but with amplitudes which are the sums of the component amplitudes, we must expect for the elements of the quantum-theory matrices

$$(p+q)(nm) \equiv \left\| [p(nm)+q(nm)]e^{2\pi i\nu(nm)t} \right\| \, .$$

The rule for multiplication is defined from similar considerations with, however, a characteristic difference

from classical multiplication, due to the fact that the quantum frequencies obey the Rydberg-Ritz combination principle. The product of two Fourier series in the classical theory may be written as the double sum

$$pq = \sum_\sigma \sum_{\sigma'} p_\sigma q_{\sigma'} e^{2\pi i[(\sigma+\sigma')\nu]t} \, ,$$

where σ replaces the complex $\sigma_1 \ldots \sigma_f$ and $[(\sigma+\sigma')\nu]$ stands for $(\sigma_1+\sigma_1')\nu_1 + \ldots + (\sigma_f+\sigma_f')\nu_f$. To write this again in the form of equation (3) terms of the same frequency must be collected, i.e., those for which $\sigma+\sigma' = \tau$, giving

$$pq = \sum_\tau (pq)_\tau e^{2\pi i[\tau\nu]t} \, ,$$

where

$$(pq)_\tau = \sum_\sigma p_\sigma q_{\tau-\sigma} \, . \qquad (11)$$

In the quantum theory the matrix representing pq must be an ensemble made up of terms $p(nm)e^{2\pi i\nu(nm)t}$ and $q(nm)e^{2\pi i\nu(nm)t}$. A matrix of the type (9) is again obtained if all elements with the same frequency are added together, i.e., those for which $\nu(nk)+\nu(km) = \nu(nm)$ by the combination principle (6). The new amplitudes are therefore taken to be

$$pq(nm) = \sum_k p(nk)q(km) \, , \qquad (12)$$

and the elements are then $pq(nm)e^{2\pi i\nu(nm)t}$.

113   This is the well-known mathematical rule for the multiplication of matrices or tensors, and justifies the use of these terms here. As is obvious from equation (12), $pq(nm) \neq qp(nm)$, so that multiplication in the quantum theory is non-commutative—a result of great importance for the further development.

By means of the rules for addition and multiplication a meaning is given to any function $x(p, q)$ of the co-ordinate and momentum matrices, at least in so far as the function may be expressed as a power series. The elements of the function x will always be of the form $x(nm)e^{2\pi i\nu(nm)t}$ and the array of frequencies $\nu(nm)$ will always be the same for a given atomic system. Hence a matrix is sufficiently well represented by its amplitudes $x(nm)$ alone, the exponential terms being understood.

The customary definitions and conventions of the theory of matrices are adopted in the quantum theory. Equality of two matrices means equality of corresponding elements. The unit matrix is defined as the matrix whose diagonal elements are all unity and whose non-diagonal elements are zero. It is conveniently written

$$1 \equiv \| \delta_{nm} \| ,$$

where

$$\delta_{nm} = \begin{cases} 1 \text{ when } n = m , \\ 0 \text{ when } n \neq m . \end{cases}$$

The reciprocal x^{-1} of a matrix x is the matrix satisfying the equations

$$x^{-1}x = xx^{-1} = 1 .$$

114   The transpose $\bar{x}$ of x is the matrix $\|x(mn)\|$ obtained by interchanging the rows and columns of x.

We are now in possession of the elements of a quantum algebra, in which it is readily seen that all the rules of ordinary algebra remain valid with the exception of the commutative law. Thus if x, y, and z represent any functions of the dynamical variables they obey, in the quantum theory, the rules of matrix algebra:

$$x+y=y+x ,$$
$$x(y+z)=xy+xz ,$$
$$x(yz)=(xy)z ,$$
$$(x+y)+z=x+(y+z) ,$$

but, in general,

$$xy \neq yx .$$

So far the Planck constant h, which must play a fundamental rôle, has not been introduced into the theory. Its appearance proves to be closely related to the non-commutativity of the variables which forms so striking a contrast to the classical theory. In fact, it has been found by Dirac[1] that in the quantum theory the expression $(2\pi i/h)(xy-yx)$ is the analogue of the Poisson bracket

$$[xy] = \sum_{k=1}^{f} \left(\frac{\partial x}{\partial q_k} \frac{\partial y}{\partial p_k} - \frac{\partial y}{\partial q_k} \frac{\partial x}{\partial p_k} \right)$$

in classical mechanics. The invariance of this expression with respect to canonical transformations of the p_k and

[1] P. A. M. Dirac, *Proceedings of the Royal Society*, A, **109**, 642, 1925.

q_k is well known. In order to make plausible this significant connection it will be shown that in the limiting region where the integers n and m are large compared to their differences there is asymptotic agreement between the matrix elements of $(2\pi i/h)(xy-yx)$ and the harmonic elements of the classical bracket expression $[xy]$. It is first necessary, however, to state more exactly the connection between the matrix elements and the Fourier amplitudes. 115

It will be recalled that in the theory of stationary states, which formed a preliminary stage in the development of the present quantum mechanics, the existence of only discrete energy values is attained through the fixation of "stationary" classical motions. If these are defined from among the continuum of possible motions by the equations[1]

$$J_k = n_k h \qquad (k=1, 2, \ldots, f) , \qquad (13)$$

where the J_k are the action variables and the n_k integers, the Bohr frequency condition (7) then appears as the analogue of the classical relation

$$\nu_k = \frac{\partial H}{\partial J_k} .$$

For since H is a function of the n_k only by equations (4), $\partial H/\partial J_k$ may be written

$$\frac{\partial H}{\partial J_k} = \lim_{a_k \to 0} \frac{H(n_1 \ldots n_f) - H(n_1 \ldots n_k - a_k, \ldots, n_f)}{a_k h} ,$$

[1] A possible degeneracy is here neglected.

and in the limiting region where the n_k are very large compared to the a_k, 116

$$\nu(n_1 \ldots n_f; m_1 \ldots m_f) = \frac{1}{h}[H(n_1 \ldots n_f) - H(n_1 - a_1, \ldots, n_f - a_f)]$$
$$\sim a_1 \frac{\partial H}{\partial n_1} + \ldots + a_f \frac{\partial H}{\partial n_f}$$
$$= a_1 \nu_1 + \ldots + a_f \nu_f .$$

There is therefore asymptotic agreement in this region, which may be briefly referred to as that of large quantum integers, between the spectral frequency $\nu(n_1 \ldots n_f; m_1 \ldots m_f)$ and the harmonic $(n_1-m_1)\nu_1 + \ldots + (n_f-m_f)\nu_f$ in the $(n_1 \ldots n_f)$ or $(m_1 \ldots m_f)$ stationary state. Since the harmonic elements of the matrices of quantum mechanics represent the spectral lines this suggests a general co-ordination between the matrix element $q(n_1 \ldots n_f; n_1-a_1, \ldots, n_f-a_f)e^{2\pi i\nu(n_1 \ldots n_f; n_1-a_1 \ldots n_f-a_f)t}$ and the harmonic $(a_1 \ldots a_f)$ in the $(n_1 \ldots n_f)$ stationary state. More briefly,

$$q(n, n-a)e^{2\pi i\nu(n, n-a)t} \text{ corresponds to } q_a(n)e^{2\pi i[a\nu]t} \qquad (14)$$

in the region of large quantum numbers. This co-ordination is further justified by the approximate agreement found empirically in this region between the intensities calculated classically from the Fourier amplitudes $q_a(n)$ in the stationary states and the intensity of the spectral line $\nu(n, n-a)$. The indices n and m of the matrix elements thus correspond to the quantum numbers of two stationary states, while the diagonal elements $(n=m)$ correspond to the stationary states themselves.

With the aid of the co-ordination (14) the above-mentioned correspondence with the Poisson brackets is readily shown. The (nm) element of $(2\pi i/h)(xy-yx)$ may be written as a sum over α and β of terms of the form $(2\pi i/h)\{x(n, n-\alpha)y(n-\alpha, n-\alpha-\beta)-y(n, n-\beta)x(n-\beta, n-\alpha-\beta)\}$, where $\alpha+\beta=n-m$. On adding and subtracting $x(n-\beta, n-\alpha-\beta)y(n-\alpha, n-\alpha-\beta)$ this becomes

$$\left(\frac{2\pi i}{h}\right)\{[x(n, n-\alpha)-x(n-\beta, n-\alpha-\beta)]y(n-\alpha, n-\alpha-\beta)$$
$$-[y(n, n-\beta)-y(n-\alpha, n-\alpha-\beta)]x(n-\beta, n-\alpha-\beta)\} .$$

Now in the region of "large quantum numbers" where $\alpha, \beta \ll n$,

$$x(n, n-\alpha)-x(n-\beta, n-\alpha-\beta)\sim h\beta\,\frac{\partial x_\alpha(n)}{\partial J} ,$$

and

$$y(n-\alpha, n-\alpha-\beta)\sim \frac{1}{2\pi i\beta}\frac{\partial y_\beta(n-\alpha)}{\partial w}\sim\frac{1}{2\pi i\beta}\frac{\partial y_\beta(n)}{\partial w}$$

since the harmonics of y are of the form $y_\beta(n)e^{2\pi i\beta w}$ by equations (4). Hence the foregoing matrix element is approximately[1]

$$\sum_{\alpha+\beta=n-m}\sum_{k=1}^{f}\left[\frac{\partial x_\alpha(n)}{\partial J_k}\frac{\partial y_\beta(n)}{\partial w_k}-\frac{\partial y_\beta(n)}{\partial J_k}\frac{\partial x_\alpha(n)}{\partial w_k}\right] ,$$

[1] The summation necessarily extends into the region where the quantum numbers are not large compared to their difference; hence for numerical agreement the matrix elements far removed from the diagonal must be assumed negligible, since they correspond to high harmonics in the classical theory. The formal agreement, which is of most importance here, is, of course, unaffected.

which by the rule (12) for the multiplication of Fourier amplitudes is the $(n-m)$ harmonic of $[xy]$, expressed in terms of the action-angle variables.

In the classical theory the Poisson brackets of canonically conjugate variables p_k and q_k satisfy the relations

$$[p_k\,q_l]=\begin{cases}1 \text{ when } k=l\\ 0 \text{ when } k\neq l\end{cases}, \quad [p_k,\,p_l]=0 , \quad [q_k,\,q_l]=0 .$$

The analogous relations will therefore be assumed for conjugate variables in the quantum theory, that is,

$$\begin{aligned}p_k q_l - q_l p_k &= \begin{cases}\dfrac{h}{2\pi i}\,1 \text{ when } k=l\\ 0 \quad\text{ when } k\neq l\end{cases}\\ p_k p_l - p_l p_k &= 0 ,\\ q_k q_l - q_l q_k &= 0 .\end{aligned} \qquad (15)$$

These "exchange relations," by means of which h is introduced into the equations, are of fundamental importance for quantum mechanics. They correspond to the quantum conditions of the theory of stationary classical motions, but whereas these conditions could be applied only to a multiply periodic system, the present exchange relations must be regarded as generally valid for any motion. In fact, as will appear later, they are necessary in order to give meaning to the problem of integration of the equations of motion, which will now be established.

The canonical equations (1) of the classical theory, if expressed in terms of the Poisson brackets, become

$$\dot{p}_k=[Hp_k] , \qquad \dot{q}_k=[Hq_k] .$$

The simplest assumption is to take over these equations formally into the quantum theory, replacing the Poisson brackets by their quantum analogues. We therefore assume the equations of motion in the quantum theory to be[1]

$$\begin{aligned}\dot{p}_k&=\frac{2\pi i}{h}\,(Hp_k-p_kH) ,\\ \dot{q}_k&=\frac{2\pi i}{h}\,(Hq_k-q_kH) .\end{aligned}\Bigg\} \qquad (16)$$

Clearly the equations (15) and (16) are not independent of each other. Strictly speaking, it is only permissible to assume equation (15) to be true at a single instant of time. The exchange relations at any other time must then be determined by the solution of equations (16); however, a calculation shows that equations (15) are really independent of the time.

The formal basis of the new mechanics is now completed; for any physical application, however, the form of the Hamiltonian corresponding to the special dynamical problem must be known. It is in general sufficient, in the spirit of the correspondence principle, to assume the same form as in the classical theory. The ambiguity as to the

[1] The equations of motion may be written directly in the classical form (1) without the use of the Poisson brackets if partial differentiation is defined in a rational way for matrices. The relations

$$\frac{h}{2\pi i}\frac{\partial f}{\partial q}=pf-fp , \qquad \frac{h}{2\pi i}\frac{\partial f}{\partial p}=fq-qf$$

for any function f are then easily established from the exchange relations (15). The more useful form (16) then follows at once.

order of factors in a product which may occur here seldom arises; when it does special considerations suffice to determine the correct form.

The law of the conservation of energy and the Bohr frequency condition are not contained explicitly in the postulates of the theory; it is therefore necessary to show that they may be derived from them. We commence by forming a diagonal matrix W with elements

$$W(nm)=\begin{cases}T_n h & \text{when } n=m\\ 0 & \text{when } n\neq m\end{cases} \qquad (17)$$

where the T_n are the term values of equation (5). The time derivative of any quantity x may be expressed in terms of this matrix by the equation

$$\dot{x}=\frac{2\pi i}{h}\,(Wx-xW) , \qquad (18)$$

since the (nm) element of $(2\pi i/h)(wx-xw)$ is

$$\frac{2\pi i}{h}\sum_k[W(nk)x(km)-x(nk)W(km)]=2\pi i(T_n-T_m)x(nm)$$
$$=2\pi i\nu(nm)x(nm)=\dot{x}(nm)$$

by equation (10). From equation (18) and the equations of motion (16) it follows that $Wp-pW=Hp-pH$ and $Wq-qW=Hq-qH$, or

$$(W-H)p=p(W-H) , \quad (W-H)q=q(W-H) . \quad (18')$$

That is, the matrix $W-H$ "commutes" with both p and q, and it is readily shown that it therefore commutes with

121   any function of p and q that can be represented as a power series. In particular it commutes with H, so that

$$(W-H)H-H(W-H)=WH-HW=0 , \qquad (19)$$

which, by equation (18), means

$$\dot{H}=0 , \qquad (20)$$

expressing the conservation of energy.

Equation (20) gives for the elements of H the infinite set of equations $\nu(nm)H(nm)=0$. If $\nu(nm)=0$ only when $n=m$, all the non-diagonal elements of H are zero and H is necessarily a diagonal matrix. In this case, the system is said to be "non-degenerate." It may happen, however, that $\nu(nm)=0$ for $n\neq m$; the corresponding elements of H are then undetermined and H is not necessarily diagonal. The system is then said to be "degenerate."

It follows further from equation (18′) that

$$(W_n-H_n)p(nm)=p(nm)(W_m-H_m) ,$$
$$(W_n-H_n)q(nm)=q(nm)(W_m-H_m) ,$$

i.e., $W_n-H_n=W_m-H_m$ for any value of n and m. Therefore

$$H=W+C ,$$

where C is the unity matrix, multiplied by an arbitrary constant. It is most convenient to put

$$H=W . \qquad (21)$$

The mathematical apparatus belonging to the particle picture has been outlined above. Its physical interpretation is discussed in detail elsewhere, but the two most im-

122   portant rules follow naturally at this point from the correspondence principle.

1. The time average of a quantity represented as a Fourier series is given by the terms independent of t. Hence, for a non-degenerate system, the diagonal elements of the matrix representing any variable give the time averages corresponding to the various stationary states.

2. The radiation process, when the particle picture is used, may be regarded as the emission of photons with the spectral frequencies $\nu(nm)$ accompanied by a simultaneous transition of the atom from the initial state with energy W_n to the final state with energy W_m, $(W_n>W_m)$. The intensity (rate of emission of energy) may then be represented statistically as $A(nm)h\nu(nm)$ where $A(nm)$ is the probability of spontaneous transition from state n to state m with emission of a photon. On the other hand, the classical theory gives for the average intensity corresponding to the τth harmonic $2/3(e^2/c^3)(2\pi)^4[\tau\nu]^4|r_\tau|^2\cdot 2$ where er is the vector dipole moment of the electrons (r is the vector with components $x=\sum_k q_k^{(x)}, y=\sum_k q_k^{(y)}, z=\sum_k q_k^{(z)}$, $q_k^{(x)}, q_k^{(y)}, q_k^{(z)}$ being the rectangular co-ordinates of the electrons). On equating the expressions of the two theories and replacing Fourier terms by matrix elements we obtain for the transition probability

$$A(nm)=\frac{1}{h\nu(nm)}\frac{2}{3}\frac{e^2}{c^3}[2\pi\nu(nm)]^4|r(nm)|^2\cdot 2 . \qquad (22)$$

The justification of this second rule is not obvious since the Maxwell theory also requires reconsideration. How-

123   ever, equation (22) determines only the time average of the emitted radiation, and it has been shown in chapter v, § 4, that the Maxwell theory is competent to furnish this information exactly.

§ 2. THE TRANSFORMATION THEORY

The mathematical scheme of quantum mechanics has been derived in § 1 in a way which displays its analogy to classical mechanics; it is not, however, as yet in an easily usable form. In this section it will be shown that the solution of a dynamical problem in the quantum theory is equivalent to the principal axis transformation of a Hermitian form or tensor. This provides the basis for a practicable method of solution and shows the consistency of the conditions imposed.

Suppose a set of Hermitian matrices p_k, q_k can be found which are independent of the time, satisfy the exchange relations, and make $H(p, q)$ a diagonal matrix. The dynamical problem is then solved, for if the matrices are provided with the time factors $e^{\frac{2\pi i}{h}(H_n-H_m)t}$, where H_n and H_m are the diagonal elements of H, it is readily seen that the equations of motion (16) are satisfied. If $p_k^{(0)}$, $q_k^{(0)}$ is any set of matrices satisfying the exchange relations, the transformations

$$p_k=S^{-1}p_k^{(0)}S , \qquad q_k=S^{-1}q_k^{(0)}S , \qquad (23)$$

where S is any matrix, give a new set likewise satisfying the exchange relations. This is seen algebraically on substituting equations (23) in the exchange relations for the new variables; in a similar way it is easily proved that if f

124   is any function of the $p_k^{(0)}$ and $q_k^{(0)}$ that can be written as a power series, then

$$f(p_k,q_k)=f(S^{-1}p_k^{(0)}S , S^{-1}q_k^{(0)}S)=S^{-1}f(p_k^{(0)}, q_k^{(0)})S . \qquad (24)$$

Since special Hermitian matrices satisfying the exchange relations can be found, the problem reduces to that of finding a transformation function S such that

$$S^{-1}H(p_k^{(0)},q_k^{(0)})S=W , \qquad (25)$$

where W is a diagonal matrix.

The transformations (23) are analogous to the canonical transformations of classical mechanics; but they have also a geometrical interpretation of great importance if the matrices of the quantum theory are interpreted as tensors in a unitary space of infinitely many dimensions (Hilbert space). This not only furnishes an analytical method of representing the transformations (23) and equation (25) but also provides a convenient language for the physical interpretation of the theory, as shown in chapter iv, § 1. For present purposes a purely abstract formulation will suffice.

Let $u_1^{(0)}$, $u_2^{(0)}$, ... , be an infinite set of unit orthogonal vectors. The space used is that of all vectors

$$t=\sum_n t_n^{(0)}u_n^{(0)} ,$$

where the components $t_n^{(0)}$ are complex numbers. A tensor q then expresses a linear relation between two vectors according to the equations

$$t=qs, \text{ or } t_n^{(0)}=\sum_m q^{(0)}(nm)s_m^{(0)} .$$

125 Consider now a transformation from the foregoing co-ordinate system $U_0(u_1^{(0)}, u_2^{(0)}, \ldots)$ to a new co-ordinate system $U(u_1, u_2, \ldots)$, the new vectors being given in terms of the old ones by the linear equations

$$u_n = \sum_m S(mn) u_m^{(0)} . \qquad (26)$$

The components t_n of any vector t and $q(nm)$ of any matrix q in the new system are then given by the equations

$$t_n^{(0)} = \sum_m S(nm) t_m , \qquad (27)$$

$$q(nm) = \sum_{k,l} S^{-1}(nk) q^{(0)}(kl) S(lm) , \qquad (28)$$

where S^{-1} is the matrix of the transformation $t_n = \sum_m S^{-1}(nm) t_m^{(0)}$ inverse to equation (27). [S is assumed to be non-singular.] Of special importance are the so-called "unitary" transformations, i.e., those which leave invariant the quadratic form $\sum_n t_n t_n^*$ which is the analogue of distance in unitary space. It is readily verified that for such unitary transformations

$$\sum_k S(nk) S^*(mk) = \sum_k S(kn) S^*(km) = \delta_{nm} ,$$

which means that $S^{-1} = \tilde{S}^*$, or

$$S\tilde{S}^* = \tilde{S}^* S = \mathbf{1} . \qquad (29)$$

They are the analogue in unitary space of rotations of rectangular co-ordinate systems in real, three-dimensional space.

126 It is now seen that equations (23) are of precisely the form of equations (28), by virtue of the rule (12) for quantum multiplication; p_k, q_k may therefore be regarded as the same matrices or tensors as $p_k^{(0)}, q_k^{(0)}$ expressed in a new co-ordinate system U', the new co-ordinates being related to the co-ordinates in the original system U_0 by equations (27). Equation (25) then expresses the condition on the transformation matrix S that in the new system the tensor H is in the diagonal form—i.e., the co-ordinate vectors of the new system are the principal axes of H. It is sufficient to consider only unitary transformations [S satisfying eq. (28)] since under these conditions it is well known that the principal axis transformation problem, at least for finite matrices, always has a solution.

A word is necessary as to the notation. In general it is not expedient to distinguish matrices in different co-ordinate systems by new symbols; they are more conveniently characterized by using a distinguishing letter for the indices of the components in each co-ordinate system. Different numerical values of the indices will be indicated by primes; thus $p(l'l'')$, say, represents the components of p in the "l" system and $p(a'a'')$ the components in another "a" system of co-ordinates. The first of equations (23), for example, is to be written

$$p(a'a'') = \sum_{l'} \sum_{l''} S^{-1}(a'l') p(l'l'') S(l''a'') .$$

The indices of the transformation matrix S then refer naturally to different co-ordinate systems.

The solution of a quantum-mechanical problem given by the equations of motion (16) and the exchange rela-

tions (15) thus reduces to the problem of the principal 127 axis transformation of the Hermitian matrix H. It remains to state briefly the method of solution, which is a well-known one. The equation (25) may be written

$$HS - SW = 0 , \qquad (30)$$

which gives for the elements of S the equations

$$\sum_{l'} H(l'l'') S(l''a') - \sum_{a''} S(l'a'') W(a''a') = 0$$
$$\begin{pmatrix} l' = 1, 2, \ldots \\ a' = 1, 2, \ldots \end{pmatrix} ,$$

or, since W is diagonal, an infinite set of homogeneous linear equations

$$\sum_{l''} H(l'l'') S(l''a') - S(l'a') W_{a'} = 0 \quad (l' = 1, 2, \ldots) , \quad (31)$$

for the determination of the elements of any column of the matrix $S(l'a')$. The $W_{a'}$'s, which appear as parameters, are also determined, and, in fact, independently of the $S(l'a')$, since the equations (31) will have a solution when and only when the determinant of the left-hand member is zero, that is, when the $W_{a'}$'s are solutions of the algebraic equation

$$\begin{vmatrix} H(11)-W & H(12) & H(13) & \cdot & \cdot \\ H(21) & H(22)-W & H(23) & \cdot & \cdot \\ H(31) & H(32) & H(33)-W & \cdot & \cdot \\ \cdot & \cdot & \cdot & \cdot & \cdot \\ \cdot & \cdot & \cdot & \cdot & \cdot \\ \cdot & \cdot & \cdot & \cdot & \cdot \end{vmatrix} = 0 . \qquad (32)$$

The roots $W_{a'}$ of this equation are thus characteristic 128 values of equation (30) or equations (31) and are always real. They are the diagonal elements of W and therefore give the energy levels of the system; when the roots of equation (32) are multiple the system is degenerate, for there is then coincidence of frequencies by equation (7).

To each $W_{a'}$ corresponds a characteristic solution $C_{a'} S(1a'), C_{a'} S(2a'), \ldots$, of equations (31) and hence a column of the matrix S, the arbitrary constant $C_{a'}$ occurring because of the homogeneity of the equations (31). In case the system is not degenerate it is readily seen that any two characteristic solutions are orthogonal to each other, i.e.,

$$\sum_{l'} S(l'a') S^*(l'a'') = 0 \qquad \text{when } a' \neq a'' .$$

The relation (29) is thus satisfied for the non-diagonal elements. It may also be satisfied for the diagonal elements by proper choice of the $C_{a'}$, although this "normalization" obviously determines only the absolute magnitude of the $C_{a'}$. There is therefore always an undetermined factor of absolute magnitude one common to the elements of each column of S. In case of degeneracy there is a further indeterminateness, but equation (29) may always be satisfied.

From the transformation function S the co-ordinates and momenta which form the solution are given by equations (23). The extended discussion of the physical interpretation of S is, however, reserved for § 5.

In the preceding it has been tacitly assumed that theorems for finite matrices and sets of equations are true for the infinite ones of quantum mechanics. This may be directly justified only under certain conditions, but the more rigorous treatment shows that the results of the formal treatment above are essentially correct.[1] There is one important distinction, however, in the case of infinite matrices: The characteristic value "spectrum" may contain a continuous sequence of values as well as the discontinuous one hitherto exclusively considered. In the case of the energy this accounts for the existence of continuous optical spectra. The occurrence of continuous characteristic values also means that in certain co-ordinate systems the elements of the matrices will have continuously variable indices, or indices discontinuous in a certain range and continuous in another. Our matrix relations must accordingly be extended to include this case. The methods of Dirac[2] will be used for this purpose; though somewhat formal in character they have the advantage of great clarity and may be rigorously justified in all cases which occur practically.

In the first place sums must be replaced by integrals in a range where the indices are continuously variable, the elements becoming functions of two sets of variables. Thus when the range is wholly a continuous one the product rule, for example, becomes

$$pq(nm) = \int dk\, p(nk)q(km) ,$$

while in the case of mixed ranges there will occur a sum and an integral. To represent the unit matrix in the con-

[1] In many practical problems, however, a principal axis transformation with a finite number of variables suffices, as in the perturbation method (§ 4).

[2] *Proceedings of the Royal Society*, A, 113, 621, 1927.

tinuous case Dirac has introduced a function $\delta(\xi)$, corresponding to δ_{nm} defined by the following properties:

$$\xi\delta(\xi) = 0 ,$$

so that $\delta(\xi) = 0$ for $\xi \neq 0$,

$$\delta(-\xi) = \delta(\xi) , \qquad (33)$$

and

$$\int_{\xi_1}^{\xi_2} \delta(\xi)d\xi = 1 , \qquad (34)$$

when the value zero lies between ξ_1 and ξ_2. It is thus a function with a singularity at $\xi = 0$ and is only possible as the limit of a sequence of functions. From the foregoing properties it follows readily that

$$\int_{-\infty}^{+\infty} f(\xi)\delta(a-\xi)d\xi = f(a) , \qquad (35)$$

$$\int_{-\infty}^{+\infty} f(\xi)\delta'(a-\xi)d\xi = f'(a) , \qquad (36)$$

where $f(\xi)$ is any regular function and $\delta'(\xi) = (d/d\xi)\delta(\xi)$. Equation (35) results from an integration by parts. Furthermore, since

$$\int_{-\infty}^{+\infty} \delta(a-\xi)\delta(\xi-b)d\xi = 0$$

when $a \neq b$ and

$$\int db \int \delta(a-\xi)\delta(\xi-b)d\xi = \int \delta(a-\xi)d\xi \int \delta(\xi-b)db = 1 ,$$

$$\int_{-\infty}^{+\infty} \delta(a-\xi)\delta(\xi-b)d\xi = \delta(a-b) , \qquad (37)$$

since the integral has all the properties of the δ-function of $a-b$.

The elements of the unit matrix in the continuous case may be expressed in terms of the δ-function, for $\delta(a'-a'')$ has, by equation (37), the property that

$$\int \delta(a'-a''')x(a'''a'')da''' = x(a'a'') . \qquad (38)$$

Hence

$$1(a'a'') = \delta(a'-a'') .$$

A diagonal matrix with continuous indices is one of the form $q(a'a'')\delta(a'-a'')$. The extension to multiple indices causes no difficulty; the unit matrix, for example, becomes

$$1(a'a'') = \delta(a_1'-a_1'')\delta(a_2'-a_2'') \dots \delta(a_f'-a_f'')$$

and may again be written simply $\delta(a'-a'')$.

For the quantum theory those co-ordinate systems in which quantities other than the energy take the diagonal form are also of importance. In such a system it often proves convenient to replace the indices of all matrices by corresponding diagonal elements of matrices which are diagonal in that system. Rows and columns are thus designated by characteristic values of the matrices which define the co-ordinate system. This is equivalent to replacing quantum numbers by the energies of the corresponding stationary states in a system of one degree of freedom; by the energy and, for example, the angular momentum in a system of two degrees of freedom, etc. In general, if the matrices $x_1, x_2, \dots, x_f$ have the diagonal form, the matrix elements of q will be written

$$q(x'x'') = q(x_1'x_2' \dots x_f' ;\ x_1''x_2'' \dots x_f'') ,$$

the primed letters denoting characteristic values of the corresponding matrices; in particular, the diagonal matrices x, when the indices are continuous, have the form

$$x(x'x'') = x'\delta(x_1'-x_1'')\delta(x_2'-x_2'') \dots \delta(x_f'-x_f'') . \qquad (39)$$

The question naturally arises as to what matrices can simultaneously have the diagonal form in a given co-ordinate system. The answer is well known from the theory of Hermitian forms, and is highly significant for the quantum theory: Any set of matrices all of which commute with any other of the set can be simultaneously brought to the diagonal form by a unitary transformation. Thus it will always be possible to find a co-ordinate system in which the position co-ordinates $q_1 \dots q_f$ are diagonal, but if the exchange relations are satisfied the momenta $p_1 \dots p_f$ cannot also have the diagonal form.

§ 3. THE SCHRÖDINGER EQUATION

The admission of continuous matrices into the mathematical scheme permits a new formulation of the principal axis transformation problem. If, namely, the original co-ordinate system in which the exchange relations are satisfied is taken to be one in which the q_k are continuous diagonal matrices the equation determining the transformation function S to a system in which any function F is diagonal becomes a partial differential equation, which is the analogue of equations (31). While a rigorous justification of the method used here (that of Dirac[1]) is difficult, the results may be confirmed by more exact, though also more cumbersome, methods.

[1] *Ibid.*

133 Since the original co-ordinate system need only be one which the co-ordinate matrices are diagonal and bears no necessary relation to any special dynamical problem, we may assume for the q_k the general diagonal form

$$q_k = q_k' \delta(q_1' - q_1'') \ldots \delta(q_f' - q_f'') , \qquad (40a)$$

the indices being designated by the characteristic values q_k' of q_k. To represent the conjugate momenta a set of matrices must be found which satisfies the exchange relations (15) with the foregoing q_k. A possible set is obtained by taking

$$\left. \begin{aligned} p_k(q'q'') = \frac{h}{2\pi i}\,\delta'(q_k - q_k')\delta(q_1' - q_1'') \ldots \\ \delta(q_{k-1}' - q_{k-1}'')\delta(q_{k+1}' - q_{k+1}'') \ldots \delta(q_f' - q_f'') , \end{aligned} \right\} (40b)$$

for it may be shown by calculating $p_k q_l - q_l p_k$ that the exchange relations are then satisfied. The proof for one degree of freedom is as follows: The $(q'q'')$ element of $pq - qp$ is

$$\frac{h}{2\pi i}\int dq'''[\delta'(q' - q''')q'''\delta(q''' - q'') - q'\delta(q' - q''')\delta'(q''' - q'')] .$$

The first term, on integration by parts, becomes

$$\int dq'''\delta(q' - q''') \frac{\partial}{\partial q'''} [q'''\delta(q''' - q'')]$$
$$= \int dq'''[q'''\delta'(q''' - q'')\delta(q' - q''') + \delta(q' - q''')\delta(q''' - q'')] .$$

Therefore,

$$(pq - qp)(q'q'') = \frac{h}{2\pi i}\int dq'''[(q''' - q')\delta(q' - q''')\delta'(q''' - q'')]$$
$$+ \frac{h}{2\pi i}\int \delta(q' - q''')\delta(q''' - q'')dq''' .$$

134 The first integral vanishes by equation (33), while the second is $(h/2\pi i)\delta(q' - q'')$ by equation (37). Hence

$$(pq - qp)(q'q'') = \left(\frac{h}{2\pi i}\right)\delta(q' - q'') = \left(\frac{h}{2\pi i}\right)1(q'q'')$$

and the exchange relations are satisfied. The extension to several degrees of freedom follows without difficulty.

Consider now the general problem of transforming any function $F(p, q)$ to the diagonal form by a unitary transformation S. As in the discontinuous case S is essentially determined by equation (25), which now becomes

$$S^{-1}FS = F'\delta(F' - F'') ,$$

the indices in the new system where F is diagonal being denoted by F' and F''. Again this may be written in the form of equation (30):

$$FS = S[F'\delta(F' - F'')]$$

or

$$\int F(q'q'')S(q''F')dq'' = S(q'F')F' , \qquad (41)$$

which is an integral equation corresponding to the infinite set of linear equations (31). This, however, becomes a partial differential equation when the particular values of p_k, q_k given by equations (40) are substituted in the left-hand member. Carrying out the integration, using the properties of the δ-functions, gives

$$\int F(p_k, q_k)(q'q'')S(q''F')dq'' = F\left(\frac{h}{2\pi i}\,\frac{\partial}{\partial q_k'}, q_k'\right)S(q'F') , \quad (42)$$

where $F([h/2\pi i][\partial/\partial q_k'], q_k')$ is the operator obtained from 135
F by the substitution

$$p_k \to \frac{h}{2\pi i}\,\frac{\partial}{\partial q_k'} , \qquad q_k \to q_k' . \qquad (43)$$

Only the proof for one degree of freedom need be given. For the special cases $F = q$ and $F = p$ the result follows at once, since by equations (36) and (35)

$$\int \frac{h}{2\pi i}\delta'(q' - q'')S(q''F')dq'' = \frac{h}{2\pi i}\,\frac{\partial S(q'F')}{\partial q'} ,$$
$$\int q'\delta(q' - q'')S(q''F')dq'' = q'S(q'F') .$$

Since all functions which need be considered can be built up by multiplication and addition from p and q, it only remains to show that if equation (42) holds for F_1 and F_2 it holds for $F_1 + F_2$ and $F_1 F_2$. That it holds for $F_1 + F_2$ is trivial. For $F_1 F_2 = \int F_1(q'q'') F_2(q'''q'')dq'''$ substitution in equation (42) gives

$$\iint F_1(q'q''')dq''' F_2(q'''q'')dq'' S(q''F')$$
$$= \int F_1(q'q''')dq''' \int F_2(q'''q'')S(q''F')dq'' ,$$
$$= \int F_1(q'q''')dq''' F_2\left(\frac{h}{2\pi i}\,\frac{\partial}{\partial q'''}, q'''\right)S(q'''F') ,$$
$$= F_1\left(\frac{h}{2\pi i}\,\frac{\partial}{\partial q'}, q'\right)F_2\left(\frac{h}{2\pi i}\,\frac{\partial}{\partial q'}, q'\right)S(q'F') ,$$
$$= F_1 F_2\left(\frac{h}{2\pi i}\,\frac{\partial}{\partial q'}, q'\right)S(q'F') ,$$

and the theorem is therefore proved.

The required transformation function $S(q'F')$ must 136
therefore be a solution of the partial differential equation

$$F\left(\frac{h}{2\pi i}\,\frac{\partial}{\partial q_k'}, q_k'\right)S(q'F') - F'S(q'F') = 0 , \qquad (44)$$

in which F' is a parameter, corresponding to $W_{\alpha'}$ in equations (31) of which equation (44) is the analogue. Here also there will be only certain discrete values or continuous ranges of F for which a solution is possible; these characteristic values give the diagonal elements of F. The conditions that the transformation be unitary $(\widetilde{S}^* = S^{-1})$ are of importance in determining the character of the solutions of equation (44). When S is continuous in both indices they may be written

$$\int S^*(q'F')S(q'F'')dq' = \delta(F' - F'') , \qquad (45)$$
$$\int S^*(q'F')S(q''F')dF' = \delta(q' - q'') , \qquad (46)$$

analogously to equations (28). There are corresponding summations when the characteristic value spectrum contains a discrete part.

The mathematical problem just treated is a very general one. That there are corresponding physical ones will appear after the extended physical interpretation of the transformation function has been given in § 5. For the present we only note that the foregoing method, when applied to the Hamiltonian H, yields a solution of the equations of motion.

When H is substituted for F in equation (44) the resulting differential equation is the Schrödinger[1] equation,

[1] E. Schrödinger, *Annalen der Physik*, **79**, 361, 489, 1926.

137 originally discovered in an entirely different manner. The corresponding transformation function $S(q'H')$ is in this case customarily written $\psi_W(q)$. The Schrödinger equation is then

$$H\left(\frac{h}{2\pi i}\frac{\partial}{\partial q_k},\ q_k\right)\psi_W(q)-W\psi_W(q)=0 \qquad (47)$$

and its characteristic values given the energy levels of the system.

The solutions $\psi_W(q)$ form the columns of the transformation matrix, which should be compared with the S of § 2. Both represent transformations to a system in which the energy is diagonal—in the present case, however, the initial system is a particular one in which the co-ordinates are diagonal, corresponding to a particular choice of $p_k^{(0)}$, $q_k^{(0)}$ in § 2.

In the typical case of a discrete characteristic value spectrum the orthogonality conditions (45) become

$$\int \psi^*_{W'}(q)\psi_{W''}(q)dq=0 \qquad (48)$$

when $W'\neq W''$,

$$\int |\psi_W(q)|^2 dq=1 . \qquad (49)$$

Equation (49) is in general equivalent to boundary conditions, and the orthogonality of the characteristic solutions $\psi_W(q)$, which usually follows, then assures the validity of equations (48). As in the case of the transformation matrix S of § 2 there remains in each "column" $\psi_W(q)$ an undetermined phase factor $e^{i\varphi_W}$ not fixed by the normalization (49).

138 The co-ordinate and momentum matrices in the system in which the energy is diagonal are, by equations (23),

$$p(W'W'')=\int \psi^*_{W'}\frac{h}{2\pi i}\frac{\partial \psi_{W''}}{\partial q}\,dq , \qquad (50)$$

$$q(W'W'')=\int \psi^*_{W'}(q)q\psi_{W''}(q)dq . \qquad (51)$$

Equations (47), (50), and (51) constitute the most effective mathematical method for treatment of the dynamical problems of quantum mechanics, but they contribute nothing new to the physical interpretation. Special considerations are necessary to make clear the physical meaning of the transformation matrix (cf. § 5).

§ 4. THE PERTURBATION METHOD

A description of the principal features of the perturbation theory in quantum mechanics is necessary at this point. This method may be used when the Hamiltonian H can be developed in terms of a small parameter λ in the form

$$H=H_0+\lambda H_1+\lambda^2 H_2+\ldots, \qquad (52)$$

and the solution of the problem corresponding to the Hamiltonian H_0 is known, i.e., when the matrices p and q, and any function of p and q, are known in that system in which H_0 is diagonal (H_0-system). In the following the letter H will be used for the energy matrix in this co-ordinate system, while W will stand for the energy matrix in the system in which the complete Hamiltonian is diagonal (H-system). Corresponding to equation (52) W may be written in the form

$$W=W_0+\lambda W_1+\lambda^2 W_2+\ldots \qquad (53)$$

where $W_0=H_0$. The required transformation function 139 which leads from the H_0-system to the H-system may also be written

$$S=S_0+\lambda S_1+\lambda^2 S_2+\ldots, \qquad (54)$$

and S will be unitary to zeroth approximation if

$$S_0\tilde{S}_0^*=1 . \qquad (55)$$

A set of equations will now be found from which S may be determined. As in § 2, S must satisfy the equation $HS=SW$, W being diagonal; substituting the developments (52), (53), and (54) in this equation and equating coefficients of equal powers of λ gives the equations

$$\left.\begin{array}{l}
H_0S_0=S_0W_0 , \\
H_0S_1-S_1H_0=S_0W_1 , \\
H_0S_2-S_2H_0+H_2S_0-S_1W_1=S_0W_2 , \\
\qquad\vdots \\
H_0S_r-S_rH_0+F_r(S_1\ldots S_{r-1},\ H_1\ldots H_r)=S_0W_r , \\
\qquad\vdots
\end{array}\right\} \qquad (56)$$

which may be solved in sequence for $S_0,\ S_1,\ \ldots$, and $W_0,\ W_1,\ \ldots$

The first equation gives, for the elements of S_0,

$$S_0(nn)[H_0(nn)-H_0(mm)]=S_0(nm)h\nu_0(nm)=0 , \qquad (57)$$

where the $\nu_0(nm)$ are the frequencies of the unperturbed system.[1] A distinction must be made at this point be-

[1] For simplicity it is assumed that all matrices are discontinuous in their indices. The method is equally applicable for continuous indices and hence for the Schrödinger equation.

140 tween non-degenerate and degenerate unperturbed systems. In the former case [$\nu_0(nm)\neq 0$ when $n\neq m$] it follows at once from equation (57) that S_0 is a diagonal matrix; in the latter the non-diagonal terms of S_0 do not necessarily vanish. Since the treatment of the two cases differs from here on it will be assumed at first that the unperturbed system is non-degenerate.

When S_0 is diagonal, equation (55) requires $|S_0(nm)|=1$; hence, disregarding the undetermined phases always present in S, we may take $S_0=1$. The second of equations (56) then becomes

$$H_0S_1-S_1H_0+H_1=W_1 ,$$

or, for the elements

$$h\nu_0(nm)S_1(nm)+H_1(mm)=W_1(nm)\delta_{nm} . \qquad (58)$$

For the diagonal elements this gives the determination of the perturbation energy to first approximation:

$$W_1(nn)=H_1(nn) . \qquad (59)$$

When $n\neq m$ equation (58) determines the non-diagonal elements of S_1; the diagonal elements of S_1 are undetermined by equation (58) but the condition $S\tilde{S}^*=1$ is satisfied to first approximation if they are taken to be zero. Hence

$$S_1(nm)=-\frac{H_1(nm)}{h\nu_0(nm)}\ (1-\delta_{nm}) .$$

The similarity of these results to those of the perturbation theory in classical mechanics will be noted. In par-

141   ticular equation (59) corresponds to the well-known classical theorem that the perturbation function is to first order the average of the perturbation energy, since the diagonal elements of H_1 are its time average. The equation may accordingly be written

$$W_1 = \overline{H}_1 .$$

The remaining equations in (56), when treated in the same way, give

$$W_r(nn) = F_r(nn) ,$$
$$S_r(nm) = -\frac{F_r(nm)}{h\nu_0(nm)} (1 - \delta_{nm}) ,$$

each F_r being determined by the equations preceding the rth one.

If the unperturbed solution is degenerate it no longer follows from $W_0 S_0 = S_0 W_0$ that S_0 is diagonal. When, for example, $W_0(n+1) = W_0(n+2) = \ldots = W_0(n+k)$, equation (57) shows that S_0 can still contain elements that correspond to transitions between the states $n+1$, $n+2$, $\ldots$, $n+k$. The second of equations (56), however, provides a system of homogeneous linear equations giving these non-vanishing elements of S_0 and at the same time W_1. Again forming the time mean over the unperturbed motion (i.e., picking out the rows n and columns m for which the corresponding $\nu(nm)$ vanish) gives the equation

$$\overline{H}_1 S_0 = S_0 W_1 , \qquad (60)$$

which provides a system of homogeneous linear equations precisely analogous to equations (31). As there W_1 may

142   be found independently of S_0 from the so-called "secular equation,"

$$\begin{vmatrix} H_1(n+1, n+1) - W_1 & .. & H_1(n+1, n+k) \\ H_1(n+2, n+1) & .. & H_1(n+2, n+k) \\ \cdot & & \\ \cdot & & \\ \cdot & & \\ H_1(n+k, n+1) & .. & H_1(n+k, n+k) - W_1 \end{vmatrix} = 0 . \quad (61)$$

The roots give the elements of W_1 and the corresponding linear equations determine S_0 except for a phase factor in each column. From here on the calculation may be carried out as for a non-degenerate system.

§ 5. RESONANCE BETWEEN TWO ATOMS: THE PHYSICAL INTERPRETATION OF THE TRANSFORMATION MATRICES

The completed scheme for the interpretation of the mathematics of the quantum theory depends on certain assumptions as to the physical meaning of the transformation functions. To illustrate the nature of these assumptions and to make them plausible a simple problem will first be discussed—that of the interaction of two atoms in resonance.[1]

Consider two atoms, I and II, with the characteristic value spectra $W_I(n)$ and $W_{II}(i)$ which have a common characteristic frequency, so that, for instance, $\nu_I(nm) = \nu_{II}(ik)$ or $W_I(n) - W_I(m) = W_{II}(i) - W_{II}(k)$; they are thus in resonance. An energy interchange can then occur between the two atoms, even if the coupling between them

[1] W. Heisenberg, *Zeitschrift für Physik*, 40, 501, 1926.

is very weak, the interaction taking place as follows: 143   Atom I goes from the state n to the state m, giving up energy $h\nu(nm)$, while atom II takes up the same energy $h\nu(nm) = h\nu(ik)$ in going from state k to state i, the process being reversible.

If the uncoupled atoms are considered as the "unperturbed" system the interaction energy H_1 may be treated as a perturbation by the method of § 4. A state of the combined atoms, in the system in which $W_I + W_{II}$ is diagonal, may be specified by two integers (nk), the first giving the state of atom I, the second the state of atom II. The states (nk) and (mi) of the unperturbed system then have equal energies by virtue of the relation

$$W_0(nk) = W_I(n) + W_{II}(k) = W_I(m) + W_{II}(i) = W_0(mi) \quad (62)$$

resulting from the equality of frequencies; the resonance thus introduces a characteristic degeneracy. The secular equation for the determination of the perturbation W_1 in the energy may be set up as in § 2 by picking out the elements of the interaction energy $H_1(nk; mi)$ for which the frequencies $\nu(nk; mi) = (1/h)[W_0(nk) + W_0(mi)]$ vanish by equation (62). This gives, corresponding to equation (61),

$$\begin{vmatrix} H_1(nk; nk) - W_1 & H_1(nk; mi) \\ H_1(mi; nk) & H_1(mi; mi) - W_1 \end{vmatrix} = 0 . \quad (63)$$

The two solutions of this equation are the perturbation energies $W_1(a)$ and $W_1(b)$ of the two states of the coupled system which replace the states (nk) and (mi) of equal energy for the uncoupled system. (The more symmetric notation $W(nk; mi)$, etc., is likely to lead to confusion,

since there is not one-to-one correspondence with the unperturbed states.) To each root of equation (63) corresponds a column of the transformation matrix S (obtained 144   by solution of the linear equations) which will be of the form

$$\left.\begin{aligned} S(nk; a) &= s(nk; a)e^{i\phi_a} \\ S(mi; a) &= s(mi; a)e^{i\phi_a} \end{aligned}\right\} \text{ for } W_1(a) ,$$

$$\left.\begin{aligned} S(nk; b) &= s(nk; b)e^{i\phi_b} \\ S(mi; b) &= s(mi; b)e^{i\phi_b} \end{aligned}\right\} \text{ for } W_1(b) .$$

The ϕ's are real quantities undetermined by the "normalization" $S\tilde{S}^* = 1$. The orthogonal matrix

$$S = \left\| \begin{matrix} s(nk; a)e^{i\phi_a} & s(nk; b)e^{i\phi_b} \\ s(mi; a)e^{i\phi_a} & s(mi; b)e^{i\phi_b} \end{matrix} \right\| \quad (64)$$

is thus the zeroth approximation to the transformation function leading from the system in which the energies W_I and W_{II} are diagonal to the system in which the total energy $W_I + W_{II} = W$ is diagonal.

It may be noted parenthetically that in the case of two equivalent atoms resonance will always occur. This special case is obtained from the foregoing by setting $i = n$ and $k = m$; it is then readily shown that

$$H_1(nm; nm) = H_1(mn; mn) ,$$
$$H_1(nm; mn) = H_1(mn; nm) ,$$

when the interaction is symmetric in the two systems. Since H_1 is Hermitian the non-diagonal terms in the de-

145 terminant of equation (63) are real, and the solutions are seen to be

$$W_1(a) = H_1(nm;\ nm) + H_1(mn;\ nm)\ , \\ W_1(b) = H_1(nm;\ nm) - H_1(mn;\ nm)\ . \qquad (65)$$

For the corresponding matrix of the s's the calculation gives, after normalization,

$$\begin{array}{c c c} & (a) & (b) \\ nm & \dfrac{1}{\sqrt{2}} & -\dfrac{1}{\sqrt{2}} \\ mn & \dfrac{1}{\sqrt{2}} & \dfrac{1}{\sqrt{2}} \end{array} \qquad (66)$$

We return now to the general case.

We shall next discuss the further physical information that may be obtained from these results. Consider, for instance, the question of what may be said in the quantum theory as to the energy of atom I alone as a function of the time. Classically there would occur between two coupled oscillators of equal frequency a periodic and harmonic energy interchange with a frequency proportional to the coupling force; the energy of one of the oscillators would be given by a curve like that of Figure 19a. In the quantum theory, on the other hand, it is to be expected that the energy of atom I has either the value $W_I(n)$ or $W_I(m)$, with a probability of transition between these values depending again on the strength of coupling; $H_I(t)$ should therefore be represented by a curve like that of Figure 19b. To be sure, this curve cannot be calculated in the quantum theory, nor can it be experimentally determined; nevertheless the rules so far obtained for the

146 physical interpretation of quantum mechanics are sufficient to permit a calculation of the time mean and the mean-square fluctuations of $H_I(t)$ or any function of $H_I(t)$.

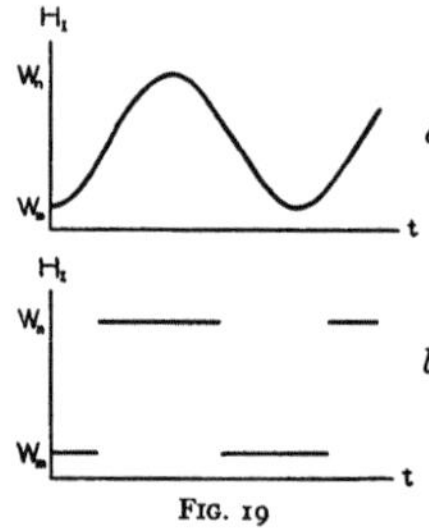

FIG. 19

The calculation of the time mean of any function of $H_I(t)$ may be made as follows. By rule 1 of § 1 the diagonal elements of the matrix representing any quantity give directly the time averages in the corresponding states. The average $\overline{f(H_I)}_a$ in the state a may therefore be calculated in terms of the diagonal elements $f(W_I(n))$ and $f(W_I(m))$ of $f(H_I)$ in the system in which H_I is itself diagonal (the unperturbed system) by making use of the transformation function S of equation (64):

$$\overline{f(H_I)}_a = [f(H_I)](aa) = S^*(nk;\ a)f(nk;\ nk)S(nk;\ a) \\ + S^*(mi;\ a)f(mi;\ mi)S(mi;\ a) \\ = |S(nk;\ a)|^2 f(W_I(n)(n)) + |S(mi;\ a)|^2 f(W_I(m))\ . \qquad (67)$$

This is precisely the expression for the time average which would result from the assumption that $f(H_I)$ can have only the values $f(W_I(n))$ and $f(W_I(m))$ and that these values occur with relative frequencies $|S(nk;\ a)|^2$ and $|S(mi;a)|^2$, respectively. Since $f(W_I(n))$ and $f(W_I(m))$ are the elements of $f(H_I)$ in the system in which $f(H_I)$ is diagonal, the first part of the foregoing assumption is equivalent to the hypothesis that the possible values of f are the diagonal elements of its matrix in the system in which it is itself diagonal. The second part, on the other hand, is a consequence of supposing that $|S(nk;\ a)|^2$ is the relative probability of finding the value $f(W_I(n))$ for $f(H_I)$ when the total system is in the state a. (The index (nk) corresponds to the value $f(W_I(n))$ since it is the label of a stationary state in the system in which f is diagonal.) The interpretation as relative probabilities is consistent because by the normalization $|S(nk;\ a)|^2 + |S(mi;\ a)|^2 = 1$. 147

While a special problem has been treated here the formal relations are the same in the general transformation problem. Thus if $S(a'\beta')$ is the transformation matrix from a system in which any quantity a is diagonal to a system in which β is diagonal[1] the time average of $f(a)$ will always appear in the form (67); i.e.,

$$f(a)_{\beta'} = [f(a)](\beta'\beta') = \sum_{a'} S^*(a'\beta')[f(a)](a'a')S(a'\beta')$$

$$= \sum_{a'} |S(a'\beta')|^2 f_a(a'a')$$

is the time average of $f(a)$ corresponding to the state β'. It is therefore reasonable to generalize the assumptions made above in a special case and to make the following hypotheses as regards the physical interpretation of the transformation scheme:[2]

The values which a quantity a can take on are given by

[1] The practice of labeling rows and columns by the elements of the diagonal matrices is used here again.

[2] P. Jordan, *Zeitschrift für Physik*, 40, 809, 1927; 44, 1, 1927; P. A. M. Dirac, *Proceedings of the Royal Society*, A, 113, 621, 1927.

its characteristic value spectrum, i.e., by the elements of its matrix in the system in which it is itself diagonal. 148

If $S(a'\beta')$ is the unitary transformation matrix from a system in which a is diagonal to a system in which β is diagonal then

$$|S(a'\beta')|^2 \qquad (68)$$

is the relative probability of finding the value a' of a when it is known that the value β' must be ascribed to β.

The foregoing assumptions of course apply equally well to the case of continually varying indices and hence to the case in which S is found by solution of a Schrödinger equation.

The detailed discussion of the physical interpretation of the statistical elements thus introduced into the theory will be found in the body of the text and especially in chapter iv. Here it will only be noted that we must add the express condition that the experiment under consideration actually affords a determination of a'. At first sight this condition appears trivial; it is, however, essential, for an application of the foregoing interpretation of the quantities (68) without consideration of the experiment leading to the measurement of a' gives rise at once to logical inconsistencies.

Having established the basis for its physical interpretation, we proceed to the further development of the general transformation theory.

The elements of the transformation matrix S give probabilities only on forming the squares of their absolute magnitudes; they may themselves be called "probability amplitudes." Carrying out successively a transformation from the system α (the system in which a is diagonal) to a system β and then a transformation from the system β to the system γ gives, since transformations combine by the rule for matrix multiplication,

$$S(a'\gamma') = \sum_{\beta'} S(a'\beta')S(\beta'\gamma') . \tag{69}$$

Thus quite independently of γ the probability amplitude $S(a'\gamma')$ can always be represented as a linear function of the set of probability amplitudes $S(a'\beta')$. The probability amplitude for finding a' regardless of the predetermined quantity γ', which may be written simply $S(a')$, is therefore, even in the most general case, a linear function of the elements of the transformation matrix $S(a'\beta')$, and the system β may be chosen arbitrarily. In particular β may be taken to be the energy, and $S(a')$ can then always be expressed in the form

$$S(a') = \sum_{W'} c_{W'} S_{W'}(a') , \tag{70}$$

where the $c_{W'}$'s are constants and $S_{W'}(a')$ is the transformation matrix to the system in which W is diagonal.

While the probabilities $S_{W'}(a')$ are always constant in time, referring to a *stationary* state W', this is not true in general for $|S(a')|^2$ (i.e., when something other than the energy is specified). The proper time dependence of $S(a')$ may be deduced from the following considerations:

According to (9) each matrix element $x(nm)$ has a time

factor $e^{\frac{2\pi i}{h}(W_n - W_m)t}$ in the system in which the energy is diagonal. Since on transforming to this system from any other system

$$x(nm) = \sum_{a'a''} S^*(a'n)x(a'a'')S(a''m) , \tag{71}$$

the correct time dependence will be obtained by providing each element $S(a'n)$ with the time factor $e^{-\frac{2\pi i}{h}W_n t}$. This is possible since hitherto $S(a'n)$ has contained an arbitrary phase factor of absolute magnitude 1; from now on it will be understood that $S(a'n) = S_{W'}(a')$ contains this time factor.

The most general probability amplitude $S(a')$, since it can be expressed in the form (70), must satisfy the equation $HS - SW = 0$ determining the $S_{W'}(a')$. Since $SW = -(h/2\pi i)(\partial S/\partial t)$ when S has the time factor introduced above, the equation for $S(a')$ becomes

$$\sum_{a''} H(a'a'')S(a'') + \frac{h}{2\pi i}\frac{\partial S(a')}{\partial t} = 0 . \tag{72}$$

In particular taking a to be a co-ordinate q, this becomes the wave equation of Schrödinger,

$$H\left(\frac{h}{2\pi i}\frac{\partial}{\partial q} , q\right)\psi(q) + \frac{h}{2\pi i}\frac{\partial\psi(q)}{\partial t} = 0 . \tag{73}$$

Characteristic solutions of the form $\psi_{W'}(q) = u_{W'}(q)e^{-\frac{2\pi i}{h}W't}$ correspond to the elements $S_{W'}(a')$ with the time factor, and by (70) the most general probability amplitude is

$$\psi = \sum_{W'} c_{W'} u_{W'} e^{-\frac{2\pi i}{h}W't} . \tag{74}$$

As an example of the application of equation (72) consider again the example of coupled atoms. Suppose a measurement at time $t = 0$ gives the result that atom I is in state n and atom II in state k. Equation (72) then gives the variation with time of the matrix S given by equation (64), in which the time is contained only in the phases ϕ_a and ϕ_b. Substitution in equation (72), since the matrix s of the constant amplitudes satisfies the equation $Hs + sW = 0$, gives

$$\frac{h}{2\pi i}\frac{\partial\phi_a}{\partial t} = -W_a , \qquad \frac{h}{2\pi i}\frac{\partial\phi_b}{\partial t} = -W_b .$$

Hence $\phi_a = -2\pi i/h \cdot W_a t + \text{Const.}$ and $\phi_b = -2\pi i/h \cdot W_b t + \text{Const.}$ and the characteristic solutions of equation (70) are $S(nk; a) = \text{Const.} \times s(nk; a)e^{-\frac{2\pi i}{h}W_a t}$, etc. The general probability amplitudes are then by equation (70),

$$S(nk) = c_a s(nk; a)e^{-\frac{2\pi i}{h}W_a t} + c_b s(nk; b)e^{-\frac{2\pi i}{h}W_b t} ,$$

$$S(mi) = c_a s(mi; a)e^{-\frac{2\pi i}{h}W_a t} + c_b s(mi; b)e^{-\frac{2\pi i}{h}W_b t} ,$$

where the c's are constants which may be determined by the initial conditions. Since in this case the initial conditions are $S(nk) = 1$, $S(mi) = 0$, and the determinant of the s's is 1, we readily find

$$S(nk) = s(mi; b)s(nk; a)e^{-\frac{2\pi i}{h}W_a t} - s(mi; a)s(nk; b)e^{-\frac{2\pi i}{h}W_b t} ,$$

$$S(mi) = s(mi; b)s(mi; b)\left[e^{-\frac{2\pi i}{h}W_a t} - e^{-\frac{2\pi i}{h}W_b t}\right] .$$

For the special case of equivalent atoms, where s has the form (66),

$$S(nm) = \tfrac{1}{2}\left(e^{-\frac{2\pi i}{h}W_a t} + e^{-\frac{2\pi i}{h}W_b t}\right)$$

$$S(mn) = \tfrac{1}{2}\left(e^{-\frac{2\pi i}{h}W_a t} - e^{-\frac{2\pi i}{h}W_b t}\right) .$$

From this follow the probabilities

$$|S(nm)|^2 = \tfrac{1}{2}\left[1 + \cos\frac{2\pi}{h}(W_a - W_b)t\right]$$

$$|S(mn)|^2 = \tfrac{1}{2}\left[1 - \cos\frac{2\pi}{h}(W_a - W_b)t\right] .$$

These formulas give the probabilities of finding (nm) or (mn) as functions of the time. As $W_a - W_b$ is small to the order of magnitude of the interaction energy of the atoms, the probabilities vary only slowly. Shortly after the first measurement (i.e., for small values of t) it is extremely probable that we find again the configuration (nm). If, however, the second measurement is made exactly at time $t = \tfrac{1}{2}h(W_a - W_b)$, the result will certainly be the configuration (mn). All of these considerations are valid only when the system actually remains unperturbed in the interval between the two measurements; that is, actually remains governed by equation (72). This condition is, of course, quite trivial. It is specially mentioned here, however, as it is of decisive importance for the consistency of the theory.

The interpretation of the transformation matrices as probability functions just sketched gives a complete scheme for the application of the mathematics of the quantum mechanics to all physical problems

§ 6. THE CORPUSCULAR CONCEPT FOR RADIATION

The corpuscular theory of radiation is too well known in its general outlines to require extended discussion at this point. It is essentially Einstein's theory of light quanta according to which radiation can be regarded as the action of rapidly moving particles (quanta) whose velocity is always c. Energy E and momentum p are related by the fundamental equation

$$E = cp , \qquad (75)$$

and the color is given by the quantum relation

$$\nu = \frac{E}{h} .$$

Light quanta can appear and disappear, so that in contradistinction to the particle picture of matter their number is variable. No interaction takes place between different light quanta (when gravitation is disregarded), but the interaction between light quanta and matter is responsible for the phenomena of absorption, emission, and dispersion.

§ 7. QUANTUM STATISTICS

Consider a system of n identical particles that are entirely indistinguishable from each other (e.g., electrons or photons). For simplicity it will be assumed that the system has only a discrete characteristic value spectrum, and the interaction between the particles will at first be neglected. The problem may be treated by first determining the possible states and corresponding characteristic functions $\psi_a(r)$ for the individual particles and then considering the distribution of the n particles among these

states. In order to treat such a statistical distribution it is necessary to define what constitutes a distinct state of the system.

In classical statistics (Boltzmann statistics) a distribution of n particles among n different states has a relative probability $n!$, since obviously every permutation of the n particles represents an independent realization of the given distribution. In the quantum theory this means that every distribution of n particles among n different states corresponds to an $n!$-fold degenerate term of the total system. The corresponding $n!$ linearly independent characteristic functions are obtained by performing the $n!$ permutations of the r_{β_k} with the a_i fixed, in the expression

$$\psi_{a_1}(r_{\beta_1})\psi_{a_2}(r_{\beta_2}) \, . \, . \, . \, . \, \psi_{a_n}(r_{\beta_n}) . \qquad (76)$$

Instead of the functions (76) any other system of $n!$ linearly independent linear aggregates may of course be used to describe the n-body problem. One is led to such a system of functions, for example, on attempting to treat the interaction of the particles as a perturbation. Among the $n!$ linear aggregates thus obtained two are singled out by a particularly simple structure:

$$\sum_{\substack{\text{All} \\ \text{permutations}}} \psi_{a_1}(r_1)\psi_{a_2}(r_2) \, . \, . \, . \, . \, \psi_{a_n}(r_n) , \qquad (77)$$

and the determinant

$$|\psi_{a_i}(r_k)| \qquad (i, k = 1, 2, \ldots , n) . \qquad (78)$$

The first is unaltered by any interchange of two particles and is called the "symmetric characteristic function" of the system; the second only changes its sign on such an interchange and is called the "antisymmetric characteristic function." If it is assumed that the ψ_a's are normalized, then it is readily shown that the characteristic functions (77) and (78) of the total system are also normalized if multiplied by $\sqrt{1/n!}$.

These relations are clearly illustrated in the simplest case of $n = 2$. Corresponding to one particle in state a_1 and the other in state a_2, there is then a doubly degenerate term with the two characteristic functions

$$\psi_s(1, 2) = \frac{1}{\sqrt{2}} \left[\psi_{a_1}(r_1)\psi_{a_2}(r_2) + \psi_{a_1}(r_2)\psi_{a_2}(r_1) \right] ,$$

$$\psi_a(1, 2) = \frac{1}{\sqrt{2}} \left[\psi_{a_1}(r_1)\psi_{a_2}(r_2) - \psi_{a_1}(r_2)\psi_{a_2}(r_1) \right] .$$

In the first place it is readily seen that no intercombinations can take place between terms with symmetric and terms with antisymmetric characteristic functions. The probability of such a transition is always given by an integral of the form

$$\int\int f(1, 2)\psi_s(1, 2)\psi_a(1, 2)d\tau_1 d\tau_2 \qquad (79)$$

in which $f(1, 2)$ is a function which is not altered when the particles are interchanged, since the two particles are indistinguishable. If now the two electrons are interchanged in (79) the value of the integral is clearly unaltered, since it is only the designation of the variables of integration that is changed. On the other hand, the sign of $\psi_a(1, 2)$

is reversed while all other quantities in the integrand remain the same. Accordingly (79) must vanish.

A more thorough mathematical investigation based on the theory of the representation of groups shows that this special result must be generalized to the following:[1]

The terms of a system of n equal particles may always be divided into partial systems in such a way that only the terms belonging to a given partial system can combine with each other. In particular, there will always occur two partial systems in one of which the characteristic functions are symmetric, while in the other they are antisymmetric.

This result remains valid for any interaction between the particles provided only that the interaction of the particles is a symmetric function of their co-ordinates.

The fact that intercombinations cannot occur between two different term systems leaves open the possibility of introducing further hypotheses which exclude all but one of these systems from physical significance.

Consider, for example, the symmetric term system alone. A definite distribution of the particles among the individual states of the single particles (again neglecting the interaction) corresponds, in this term system, to only a single characteristic function. The possibilities that are represented in the symmetric term system therefore correspond to those states which are distinguished in the Bose-Einstein[2] statistics.

In the term system made up of antisymmetric char-

---

[1] E. Wigner, *Zeitschrift für Physik*, **40**, 883, 1927.

[2] S. N. Bose, *ibid.*, **26**, 178, 1924; A. Einstein, *Berliner Berichte*, p. 261, 1924

¹⁵⁷ acteristic functions, on the other hand, any function which corresponds to two particles in the same state necessarily vanishes. This is the expression in the quantum theory of the Pauli[1] exclusion of equivalent orbits, which applies to electrons and protons. The choice of an antisymmetric term system corresponds to the use of the Fermi[2]-Dirac[3] statistics.

Quantum statistics thus singles out one term system from the possible term manifolds of an n-body problem, of either symmetric or antisymmetric characteristic functions, as the only physically significant one; each term of the manifold thus singled out represents a distinct state of the physical system of n-bodies. The first case corresponds to the Bose-Einstein statistics, which applies to light quanta; the second to the Pauli-Fermi-Dirac statistics. It is important to remember that this formulation remains valid for any interaction of the particles.

In applying the Pauli exclusion principle to electrons or protons it must not be forgotten that r_k, in $\psi_a(r_k)$, represents not only the three space co-ordinates of the kth particle, but also the fourth variable describing the spin which can only have the values $+\frac{1}{2}$ and $-\frac{1}{2}$.

The formulation of quantum statistics in the wave picture will be treated in § 10.

§ 8. THE WAVE CONCEPT FOR MATTER AND RADIATION: CLASSICAL THEORY

The classical wave theory is that of the de Broglie waves for matter and of electromagnetic waves for radiation. This section will treat primarily those waves which

[1] W. Pauli, *Zeitschrift für Physik*, **31**, 765, 1925.
[2] E. Fermi, *ibid.*, **36**, 902, 1926.
[3] P. A. M. Dirac, *Proceedings of the Royal Society*, A, **112**, 661, 1926.

¹⁵⁸ are associated with the electron (the proton waves can be treated in an entirely similar manner), though light waves will also be considered briefly. No attempt will be made to include relativistic effects, and it is then logical to treat only electrostatic forces and to neglect magnetic and retardational phenomena.

The proper wave equation for matter waves was first discovered by Schrödinger,[1] and is most simply obtained from the transformation equation (73) of § 5. This general Schrödinger equation (73) cannot itself be properly regarded as a true wave equation, since it is an equation in $3N$-dimensional co-ordinate space for N particles; however, for $N = 1$ this space reduces to ordinary 3-space, and it is therefore reasonable to try to regard the equation in this special case as the space-time (i.e., the classical) equation for matter waves. The transformation function $\psi(xyz)$ is then to be considered as a "field scalar."

For one (corpuscular) electron the total Hamiltonian is made up of the kinetic energy $E_{kin} = (1/2\mu)(p_x^2 + p_y^2 + p_z^2)$ and the potential energy $E_{pot} = -eV$, where e and μ are the charge and mass of the electron respectively and V is the electrostatic potential. Hence equation (73) in this case reduces to

$$\frac{h^2}{8\pi^2\mu} \nabla^2\psi + eV\psi - \frac{h}{2\pi i} \frac{\partial\psi}{\partial t} = 0 , \qquad (80)$$

where ∇^2 is the Laplacian operator $(\partial^2/\partial x^2) + (\partial^2/\partial y^2) + (\partial^2/\partial z^2)$. The conjugate complex equation

$$\frac{h^2}{8\pi^2\mu} \nabla^2\psi^* + eV\psi^* + \frac{h}{2\pi i} \frac{\partial\psi^*}{\partial t} = 0 \qquad (81)$$

is implicitly contained in equation (80).

[1] *Annalen der Physik*, **79**, 361 (1926).

The mathematical theory of these equations can be regarded as a "classical" theory of matter waves, though of course in this case the interpretation of the mathematics is essentially different from that of the foregoing sections. The quantities entering into these equations can all be visualized in terms of space and time just as can the quantities in the Maxwell equations, since they are all functions only of the four variables x, y, z, t. ¹⁵⁹

The wave theory does not consider electrons, and e and μ are merely universal constants of the wave equation. Although equations (80) and (81) were obtained from the one-electron problem of the corpuscular theory, they are now in no manner restricted "to apply to one electron only," for the phrase is meaningless in the wave theory. On the contrary they have complete generality in so far as "waves of negative electricity" are concerned. From this remark it follows at once that, in contrast to the quantum theory of the one-electron problem, V no longer simply represents the potential of the external forces but also includes the potential of the matter waves themselves, that is, it takes account of the reaction of one part of the charge distribution upon another part. This theory will be as unable to represent the phenomena of atomic physics as the Maxwell theory. Its value is exclusively heuristic in that it is related to the quantum theory of waves in the same way that classical mechanics is related to the quantum theory of particles.

As a first example the case of very small wave amplitude, i.e., very low density of matter, will be treated. It will assume that the external potential is also zero, so that

V vanishes to the requisite approximation. Then equation (80) becomes ¹⁶⁰

$$\frac{h^2}{8\pi^2\mu} \nabla^2\psi - \frac{h}{2\pi i} \frac{\partial\psi}{\partial t} = 0 , \qquad (82)$$

which possesses the solution

$$\psi = e^{\frac{2\pi i}{h}(p_x x + p_y y + p_z z - Et)} ,$$

where

$$E = \frac{1}{2\mu}(p_x^2 + p_y^2 + p_z^2) = \frac{1}{2\mu}p^2 .$$

These have the form of plane waves, the direction of the wave normal being given by p_x, p_y, p_z and the wavelength and frequency being

$$\lambda = \frac{h}{p} , \qquad \nu = \frac{E}{h} . \qquad (83)$$

The phase velocity v_ϕ of the waves is

$$v_\phi = \frac{E}{p} = \frac{p}{2\mu} , \qquad (84)$$

while the group velocity v_g can be calculated from elementary optical principles to be

$$v_g = \frac{dE}{dp} = \frac{p}{\mu} = \frac{h}{\lambda\mu} . \qquad (85)$$

According to de Broglie,[1] these are the equations which govern the interference of matter waves for very low

[1] L. de Broglie, *Annales de Physique*, 10 Série, **2**, 22, 1925; *Ondes et Mouvement*, Paris, 1926.

density. The relationship between group velocity and wave-length permits an association of wave-length to moving complexes of negative electricity without in any way appealing to the particle picture. This theory of de Broglie therefore gives a simple qualitative account of the experiments of Davisson and Germer, Thomson, Rupp, and others. This is precisely analogous to the success of the classical mechanics in explaining the Wilson photographs, the deflection of cathode rays by electric fields, etc. Nevertheless one can regard these achievements of classical theories only as proof of the similarity of the classical and quantum theories, in the sense of the correspondence principle; for the answer to all quantitative questions an appeal must be made to the exact quantum theory.

Before passing on to the quantum theory of waves it will be necessary to elaborate this classical wave theory somewhat further. For this purpose we return to the wave equation (80) which is not restricted to low density of matter, and make the following assumptions for the interpretation of the wave function ψ:

$$\left.\begin{array}{l} \text{Charge density: } \rho = -e\psi^*\psi\,, \\[2mm] \text{Current density: } \sigma = -\dfrac{eh}{4\pi i\mu}\,(\psi^*\nabla\psi - \psi\nabla\psi^*)\,, \\[2mm] \text{Energy density: } u = \dfrac{h^2}{8\pi^2\mu}\,\nabla\psi^*\cdot\nabla\psi\,. \end{array}\right\} \quad (86)$$

The strict justification of these assumptions can be found only in the later developments of the quantum theory of waves. None the less they are plausible at this point because the quantities ρ, σ, and u thus introduced obey by virtue of equations (80) and (81) the following conservation laws of the kind which must be demanded of any classical theory:

Conservation of charge: $\dfrac{d}{dt}\int\rho\,dv = 0\,,$ $\qquad$ (87a)

Conservation of momentum: $\dfrac{d}{dt}\int\sigma\,dv = -e\int\nabla V\psi^*\psi\,dv\,,$ (87b)

Conservation of energy: $\dfrac{d}{dt}\int u\,dv = \int eV\dfrac{\partial}{\partial t}(\psi^*\psi)\,dv\,.$ (87c)

In these equations dv is the volume element and the integrals are over all space. It is assumed that ψ vanishes over the infinite sphere so that whenever Green's theorem is applied the surface integral vanishes. To deduce (87a) multiply equation (80) by ψ^* and equation (81) by ψ, subtract the two equations thus obtained, integrate over all space and apply Green's theorem. To deduce (87b) multiply equation (80) by $\partial\psi^*/\partial x$, differentiate equation (81) with respect to x, multiply by ψ, and then subtract and integrate as before. Finally, (87c) is deduced in the same manner as (87a) except that the equations are added instead of subtracted.

Besides the waves of negative electricity other charges may be present in space, such as atomic nuclei, charged condensers, etc. The density of these charges will be designated by ρ_0. The total electric potential must then be determined by Poisson's equation $\nabla\cdot\mathbf{E} = 4\pi(\rho+\rho_0)$, or

$$\nabla^2 V = -4\pi(\rho+\rho_0)\,. \qquad (88)$$

For the purpose of the quantum theory of wave fields to be developed in the next sections it is necessary to note that equations (80), (81), and (88) can all three be deduced from a single variation principle. The proper Lagrangian function is seen to be

$$\left.\begin{array}{l} L = -\dfrac{h^2}{8\pi^2\mu}\,\nabla\psi^*\cdot\nabla\psi - \dfrac{h}{4\pi i}\Big(\dfrac{\partial\psi}{\partial t}\,\psi^* - \dfrac{\partial\psi^*}{\partial t}\,\psi\Big) \\[3mm] \qquad\qquad + eV\psi\psi^* - \rho_0 V + \dfrac{1}{8\pi}\,\nabla V\cdot\nabla V\,, \end{array}\right\} \quad (89)$$

since on varying ψ and ψ^* the condition

$$\iint L\,dv\,dt = \text{Extremum}$$

gives the equations (80) and (81), respectively, and on varying V gives equation (88).

The total energy of the system is composed of the energy of the matter waves and that of the electromagnetic field. Hence the total energy density $\mathscr{H}$ is given by the equation

$$\mathscr{H} = \dfrac{h^2}{8\pi^2\mu}\,\nabla\psi^*\cdot\nabla\psi + \dfrac{1}{8\pi}\,\nabla V\cdot\nabla V\,, \qquad (90)$$

and the conservation law

$$H = \int\mathscr{H}\,dv = \text{Const.} \qquad (91)$$

is readily proved, provided ρ_0 is independent of the time. The proof is as follows: From equations (90), (88), and (87c)

$$\dfrac{dH}{dt} = \int dv\Big[\dfrac{\partial u}{\partial t} - \dfrac{1}{4\pi}\,V\dfrac{\partial}{\partial t}\,(\nabla V\cdot\nabla V)\Big]\,,$$

$$= \int dv\Big[\dfrac{\partial u}{\partial t} - V\dfrac{\partial}{\partial t}\,(e\psi\psi^*)\Big]\,,$$

$$= 0\,.$$

This self-consistent space-time theory, built according to the model of a classical field theory, does not as yet contain a single corpuscular element. This is evident above all from the fact that the total charge of the system

$$\int\rho\,dv = -e\int\psi^*\psi\,dv \qquad (92)$$

can take on any desired value, and not merely the values $-e$, $-2e$, $-3e$, $\ldots\ldots$, as must be required of any true theory of atomic (or quantized) systems. Furthermore, the total energy and the characteristic frequencies can also have any value, since the differential equations are non-linear and the characteristic frequencies therefore depend on the amplitudes of ψ. In spite of these defects (which are those of any classical theory), the present theory can be used to account for atomic phenomena in a manner precisely analogous to that used by Bohr and Sommerfeld in applying the classical corpuscular theory. Just as these authors introduced the conditions $\int p_k dq_k = n_k h$ into classical mechanics, so Hartree[1] has been able to give an approximate account of atomic spectra by imposing the "quantum conditions"

$$\int\psi_k^*\psi_k\,dv = n_k \qquad (93)$$

in the present field theory.[2] The quantity n_k is an integer, and the suffix k refers to a characteristic vibration of the system. Hartree is able to obtain satisfactory results only

[1] D. R. Hartree, *Proceedings of the Cambridge Philosophical Society*, **24**, 89, 1928.

[2] Hartree has shown that satisfactory results are obtained only if the energy of the interaction of the electron with its own field is subtracted from the total energy.

upon neglecting the periodic time-variations in V, which are produced by the periodic character of ψ. This is analogous to the difficulties encountered by the Bohr-Sommerfeld theory. It is characteristic that this field theory is quite as difficult to treat mathematically as the classical mechanics; at any rate it is far more difficult than the quantum theory of either particles or waves.

It is probably unnecessary to enter into a detailed account of the classical theory of radiation, since this is the well-known Maxwell theory. It contains no quantum element whatsoever, as witnessed by the fact that the energy $\int (E^2 + H^2)dv$ is continuously variable. Again the difficulty may be avoided by quantum conditions like those of Hartree, which make possible only discontinuous energy changes of amount $h\nu$; this does not, however, lead to a quantum theory of the field.

§ 9. QUANTUM THEORY OF WAVE FIELDS[1]

The mathematical apparatus necessary for the quantum theory of wave fields may be put in a form completely analogous to that of the quantum mechanics of particles provided the classical wave theory is first brought into a form analogous to the Hamiltonian form of classical mechanics. The present section treats the general problem of a classical wave theory that can be derived from a variation principle. The Lagrangian function of this variation principle may contain a number of wave functions $\psi_a = \psi_a(x, y, z, t)$, $(a = 1, 2, 3, \ldots)$ [e.g., ψ, ψ^*, and V of § 8], their first order space derivatives $(\partial \psi_a / \partial x_i)$

[1] P. Jordan and W. Pauli, *Zeitschrift für Physik*, **47**, 151, 1928; W. Heisenberg and W. Pauli, *ibid.*, **56**, 1, 1929; **59**, 168, 1930.

$(i = 1, 2, 3$ for $x, y, z)$, and their first-order time derivatives $(\partial \psi_a / \partial t) = \dot\psi_a$. The variation principle will then be

$$\int \int L\left(\psi_a, \frac{\partial \psi_a}{\partial x_i}, \dot\psi_a\right) dv\, dt = \text{Extremum}, \qquad (94)$$

and the wave equations are the corresponding Eulerian equations

$$\frac{\partial L}{\partial \psi_a} - \sum_i \frac{\partial}{\partial x_i} \frac{\partial L}{\partial \left(\frac{\partial \psi_a}{\partial x_i}\right)} - \frac{\partial}{\partial t} \frac{\partial L}{\partial \dot\psi_a} = 0 \quad (a = 1, 2, \ldots). \qquad (95)$$

The classical mechanics of a system of particles may be derived entirely from Hamilton's variation principle

$$\int L(q_k, \dot q_k) dt = \text{Extremum}. \qquad (96)$$

The variation principle (94) for a continuous field may be made formally similar to the variation principle (96) for a discrete set of particles by introducing the quantity

$$\bar L = \int L\left(\psi_a, \frac{\partial \psi_a}{\partial x_i}, \dot\psi_a\right) dv, \qquad (97)$$

and then writing (94) in the form

$$\int \bar L\left(\psi_a, \frac{\partial \psi_a}{\partial x_i}, \dot\psi_a\right) dt = \text{Extremum}. \qquad (98)$$

Now while $L(q_k, \dot q_k)$ depends on the q_k for all values of the index k, $\bar L[\psi_a, (\partial \psi_a / \partial x_i), \psi_a]$ is determined by the values of ψ_a and $\dot\psi_a$ at all points of space. Hence the analogy between the two quantities is complete *if the points P of*

the space be regarded as the indices of the wave function. The complete wave function may then be regarded as the complex of quantities $\psi_a(P)$ dependent on two kinds of indices: a discrete set a and a continuously variable set P. (P, of course, takes the place of the *three* indices x, y, z.)

The Eulerian equations (95) may now be expressed in terms of the Lagrangian $\bar L$, which is the analogue of the Lagrangian for a system of particles. As the analogue of the ordinary derivative $(\partial / \partial q_k)L(q_i, \dot q_i)$, which may be written

$$\frac{\partial L}{\partial q_k} = \lim_{\Delta q \doteq 0} \frac{L(q_i + \delta_{ik}\Delta q, \dot q_i) - L(q_i, \dot q_i)}{\Delta q},$$

we may define the derivative

$$\frac{\delta \bar L\left[\psi_\beta(P'), \dfrac{\partial \psi_\beta(P')}{\partial x_i}, \dot\psi_\beta(P')\right]}{\delta \psi_a(P)} =$$
$$\left. \begin{aligned} &\lim_{\Delta\psi \doteq 0}\frac{1}{\Delta\psi}\left\{\bar L\left[\psi_\beta(P') + \delta_{a\beta}\delta(P - P')\Delta\psi(P'),\right.\right. \\ &\qquad \frac{\partial}{\partial x_i}[\psi_\beta(P') + \delta_{a\beta}\delta(P - P')\Delta\psi(P'), \dot\psi_\beta(P')]\Big] \\ &\qquad \left.\left. - \bar L\left[\psi_\beta(P'), \frac{\partial \psi_\beta(P')}{\partial x_i}, \dot\psi_\beta(P')\right]\right\}\right.\end{aligned} \right\} \quad (99)$$

The symbol $\delta(P - P')$ stands for a function analogous to Dirac's δ-function (cf. § 2) having the properties

$$\left.\begin{aligned} \delta(P - P') &= 0 \quad \text{when } P \neq P', \\ \int \delta(P - P')dv &= 1 \text{ or } 0, \end{aligned}\right\} \qquad (100)$$

and

according to whether the volume of integration contains or does not contain the point P'. From the definition (97) of $\bar L$ it is readily seen that

$$\frac{\delta \bar L}{\delta \psi_a} = \frac{\partial L}{\partial \psi_a} - \sum_i \frac{\partial}{\partial x_i} \frac{\partial L}{\partial \left(\frac{\partial \psi_a}{\partial x_i}\right)}. \qquad (101)$$

Since it is obvious that

$$\frac{\delta \bar L}{\delta \dot\psi_a} = \frac{\partial L}{\partial \dot\psi_a},$$

the Eulerian equations become

$$\frac{\delta \bar L}{\delta \psi_a} - \frac{\partial}{\partial t} \frac{\delta \bar L}{\delta \dot\psi_a} = 0, \qquad (102)$$

in complete analogy to the Lagrangian equations of classical mechanics.

The transition from the Lagrangian to the Hamiltonian form in classical particle mechanics is brought about by introducing the Hamiltonian

$$H = \sum_k p_k \dot q_k - L, \qquad (103)$$

where $p_k = \partial L/\partial \dot q_k$; the equations then take the Hamiltonian form (1). The same procedure will now be used for the wave equations (95). A conjugate Π_a to the wave function ψ_a may be introduced by the relation

$$\Pi_a = \frac{\delta \bar L}{\delta \dot\psi_a} = \frac{\partial L}{\partial \dot\psi_a}, \qquad (104)$$

169 and the Hamiltonian will then be, by analogy to (103),

$$\bar{H} = \int \sum_a \Pi_a \dot{\psi}_a \, dv - L \ .$$ (105)

Analogously to the relations between L and $\bar{L}$,

$$\bar{H} = \int H \, dv$$ (106)

if

$$H = \sum_a \Pi_a \dot{\psi}_a - L \ .$$ (107)

The wave equations (95) now take the Hamiltonian form

$$\dot{\psi}_a = \frac{\delta \bar{H}}{\delta \Pi_a} \ , \qquad \dot{\Pi}_a = -\frac{\delta \bar{H}}{\delta \psi_a} \ .$$ (108)

Conservation laws may be deduced as in particle mechanics. Directly from (108) follows the conservation of energy,

$$\frac{d\bar{H}}{dt} = 0 \ ,$$ (109)

while the equations

$$\frac{d}{dt} \int \sum_a \Pi_a \frac{\partial \psi_a}{\partial x_i} \, dv = 0 \quad (i = 1, 2, 3) \ ,$$ (110)

expressing the conservation of momentum follow from (108) and (101), since

$$\frac{d}{dt} \int dv \sum_a \Pi_a \frac{\partial \psi_a}{\partial x_i} = \int dv \left[\Pi_a \frac{\partial}{\partial x_i} \frac{\delta \bar{H}}{\delta \Pi_a} - \frac{\partial \psi_a}{\partial x_i} \frac{\delta \bar{H}}{\delta \psi_a} \right] ,$$

$$= -\int dv \sum_a \left[\frac{\partial \Pi_a}{\partial x_i} \frac{\delta \bar{H}}{\delta \Pi_a} + \frac{\partial \psi_a}{\partial x_i} \frac{\delta \bar{H}}{\delta \psi_a} \right] ,$$

$$= -\int dv \frac{\partial H}{\partial x_i} = 0 \ .$$

170 In both cases it is assumed that H contains no function of space and time other than Π_a, ψ_a, and their derivatives.

The transition from classical theory to quantum theory can now be accomplished without difficulty by analogy to the procedure of § 1. Just as the co-ordinates were there replaced by matrices, so here the wave functions may be replaced by non-commutative variables, which can be represented as matrices in a suitably chosen Hilbert space. (Such quantities have been called "q-numbers" by Dirac.) To the differential equations (108) must then be added the exchange relations analogous to (15):

$$\left.\begin{aligned} \Pi_a(P)\psi_\beta(P') - \psi_\beta(P')\Pi_a(P) &= \delta_{a\beta}\delta(P-P')\,\frac{h}{2\pi i} , \\ \Pi_a(P)\Pi_\beta(P') - \Pi_\beta(P')\Pi_a(P) &= 0 \ , \\ \psi_a(P)\psi_\beta(P') - \psi_\beta(P')\psi_a(P) &= 0 \ . \end{aligned}\right\}$$ (111)

In this quantum theory of wave fields the space-time co-ordinates x, y, z, t are thus parameters (like the time in the particle theory); they are therefore numbers in the ordinary sense (called "c-members" by Dirac), and of course commute with each other and all other quantities.

The conservation laws

$$\bar{H} = \text{Const.}, \qquad \int \sum_a \Pi_a \frac{\partial \psi_a}{\partial x_i} \, dv = \text{Const.}$$ (112)

remain valid, as is readily proved with the help of relations (111).

The simplest method for the mathematical treatment of a wave problem defined by the equations (108) and (111) is to develop the wave functions in a suitably 171 chosen set of orthogonal function $u_a^r(P)$:

$$\psi_a = \sum_r a_r(t) u_a^r(P) \ , \qquad \Pi_a = \sum_r b_r(t) u_a^r(P) \ .$$ (113)

The $u_a^r(P)$ are ordinary c-numbers and the coefficients a_r, b_r must then be regarded as q-numbers dependent on the time.

In order that ψ_a and Π_a when written in this form shall obey the exchange relations (111), the a_r and b_r must satisfy the exchange relations

$$\left.\begin{aligned} b_s a_r - a_r b_s &= \frac{h}{2\pi i}\,\delta_{rs} \ , \\ a_s a_r - a_r a_s &= 0 \ , \\ b_s b_r - b_r b_s &= 0 \ , \end{aligned}\right\}$$ (114)

which are formally analogous to equations (15). This is readily proved by substituting the developments (113) in equation (111), multiplying both sides by $u_a^s(P)u_\beta^r(P')$, integrating over P and P' and summing over a and β. In the integration use must be made of the orthogonality relations for the u_a^r:

$$\int dv_P \sum_a u_a^r(P) u_a^s(P) = \delta_{rs} \ .$$

The Hamiltonian H and the equations of motion (108) may now be expressed in terms of the a_r and b_r. The methods previously described for solution of a quantum dynamical problem are then available here—in fact, the only difference between the quantum theory of wave 172 fields and of particles is that in the former the number of variables is infinite while in the latter it is finite.

§ 10. APPLICATION TO WAVES OF NEGATIVE CHARGE

The method of the last section will now be applied to the waves of negative charge treated in § 8. The classical Lagrangian is then

$$L = -\frac{h^2}{8\pi^2\mu}\,\nabla\psi^* \cdot \nabla\psi + \frac{1}{8\pi}\,\nabla V \cdot \nabla V + eV\psi^*\psi$$

$$- \rho_0 V - \frac{h}{4\pi i}\left(\frac{\partial \psi}{\partial t}\,\psi^* - \frac{\partial \psi^*}{\partial t}\,\psi\right) \ .$$

Corresponding to the division of the charge density into that of the given external charges (ρ_0) and that of the internal charges (ρ) the potential V may be written $V = V_0 + V_1$, where

$$\nabla^2 V_0 = -4\pi\rho_0 \ , \qquad \nabla^2 V_1 = 4\pi e\psi^*\psi \ .$$ (115)

The foregoing Lagrangian may then be modified to a more convenient form by adding the total derivatives $(h/4\pi i)(\partial/\partial t)(\psi^*\psi)$ and $-(1/4\pi)\nabla \cdot (V_1\nabla V_0)$ and discarding terms involving only the known function ρ_0. This does not alter the variation problem, and in the Lagrangian

$$L = -\frac{h^2}{8\pi^2\mu}\,\nabla\psi^* \cdot \nabla\psi - \frac{h}{2\pi i}\,\frac{\partial \psi}{\partial t}\,\psi^* + \frac{1}{8\pi}\,\nabla V_1 \cdot \nabla V_1$$

$$+ e(V_0 + V_1)\psi^*\psi$$ (116)

thus obtained only ψ, ψ^*, and V_1 are to be varied.

173 A slight difficulty arises because of the fact that the time derivative of V_1 does not occur in (116), thus making it impossible to introduce the exchange relations (111), since the conjugate to V_1 defined by equation (104) would vanish. The dilemma is easily avoided, however, by not regarding V_1 as an independent wave function but rather treating the equation resulting from the variation of V_1 as a secondary condition. With its help V_1 may be expressed as a function of ψ and ψ^*. Since the equation obtained by varying V_1 is $\nabla^2 V_1 = 4\pi e \psi^* \psi$, V_1 is given in terms of ψ and ψ^* by the well-known solution of this equation:

$$V(P) = -e \int G(PP') \psi^*(P') \psi(P') dv_{P'} , \tag{117}$$

where $G(PP')$ is the Green's function (in general, simply $1/r_{PP'}$) of the region in which the waves occur. On substituting this in the Lagrangian (116) the result is, after a slight modification involving again the addition of total derivatives,

$$\left. \begin{aligned} L = -\frac{h^2}{8\pi^2\mu} \nabla\psi^* \cdot \nabla\psi - \frac{h}{2\pi i} \frac{\partial\psi}{\partial t} \psi^* - eV_0 \psi^* \psi \\ -\frac{e^2}{2} \int dv_{P'} \psi^*(P)\psi(P)\psi^*(P')\psi(P')G(PP') . \end{aligned} \right\} \tag{118}$$

The momentum conjugate to ψ is [cf. eq. (104)]

$$\Pi = \frac{\partial L}{\partial \dot\psi} = -\frac{h}{2\pi i} \psi^* ,$$

and consequently the Hamiltonian is

$$H = -\frac{h}{2\pi i} \psi^* \frac{\partial\psi}{\partial t} - L ,$$

174 giving

$$\left. \begin{aligned} \overline{H} = \int dv \left[\frac{h^2}{8\pi^2\mu} \nabla\psi^* \cdot \nabla\psi - eV_0 \, \psi^*\psi \right] \\ + \frac{e^2}{2} \iint dv_P dv_{P'} G(PP') \psi^*(P)\psi(P)\psi^*(P')\psi(P') . \end{aligned} \right\} \tag{119}$$

From this classical Hamiltonian form the transition to quantum theory may be made as in § 9, by introducing the exchange relations

$$\left. \begin{aligned} \psi(P)\psi^*(P') - \psi^*(P')\psi(P) = \delta(P-P') , \\ \psi(P)\psi(P') - \psi(P')\psi(P) = 0 , \\ \psi^*(P)\psi^*(P') - \psi^*(P')\psi^*(P) = 0 . \end{aligned} \right\} \tag{120}$$

The Hamiltonian may again be taken over from the expression (119) of the classical theory. However, the order of factors, which is now of importance, is not determined in this way; in fact, the correct form, in so far as it involves the order of factors, can only be determined empirically. It has been found by Jordan and Klein[1] that the proper Hamiltonian for matter waves is

$$\left. \begin{aligned} \overline{H} = \int dv \left[\frac{h^2}{8\pi^2\mu} \nabla\psi^* \cdot \nabla\psi - eV_0 \, \psi^*\psi \right] \\ + \frac{e^2}{2} \iint dv_P dv_{P'} G(PP') \psi^*(P)\psi^*(P')\psi(P)\psi(P') . \end{aligned} \right\} \tag{121}$$

It should be remarked that the definition of ψ^* as the conjugate of ψ requires some modification when ψ is a q-number. If ψ is given as a function of Hermitian ma-

---

[1] P. Jordan and O. Klein, *Zeitschrift für Physik*, **45**, 751, 1927.

trices, then ψ^* is obtained from it by replacing i by $-i$ and also interchanging the order of factors, e.g., 175

$$(pq)^* = q^* p^* .$$

In this quantum theory of matter waves the total charge

$$-e \int dv \psi^* \psi$$

is again a constant in time, as is most readily proved by showing that it commutes with $\overline{H}$. As must also be the case, its characteristic values are integral multiples of $-e$. This may be shown in the following manner. As in § 9, if we put

$$\left. \begin{aligned} \psi = \sum_r a_r u_r(P) , \qquad \psi^* = \sum_r a_r^* u_r(P) , \\ \int u_r u_s \, dv = \delta_{rs} , \end{aligned} \right\} \tag{122}$$

the a_r and a_r^* satisfy the exchange relations

$$\left. \begin{aligned} a_r a_s^* - a_s^* a_r = \delta_{rs} , \\ a_r a_s - a_s a_r = 0 , \\ a_r^* a_s^* - a_s^* a_r^* = 0 , \end{aligned} \right\} \tag{123}$$

analogous to equations (114). The foregoing exchange relations may be satisfied by setting

$$a_r = e^{-\frac{2\pi i}{h}\Theta_r} N_r^{\frac{1}{2}} , \qquad a_r^* = N_r^{\frac{1}{2}} e^{\frac{2\pi i}{h}\Theta_r} , \tag{124}$$

provided N_r and Θ_r are Hermitian operators satisfying the exchange relations

$$\Theta_r N_s - N_s \Theta_r = \delta_{rs} .$$

It is then possible to prove that 176

$$e^{-\frac{2\pi i}{h}\Theta_r} f(N_r) = f(N_r+1) e^{-\frac{2\pi i}{h}\Theta_r} , \tag{125}$$

and that the characteristic values of the N_r are positive integers. It then follows from equation (122) that

$$e \int dv \psi^* \psi = e \int dv \sum_{r,s} a_r^* a_s u_r u_s$$

$$= e \sum_{r,s} a_r^* a_s = e \sum_r N_r .$$

The quantum theory of matter waves thus accounts for the existence of the electron. At the same time it is evident that the Hartree "quantum conditions" 93) are the analogue, in the sense of the correspondence principle, of the exchange relations (123). Since ΣN_r is a constant of integration of the equations of motion it is possible to consider separately those stationary states for which this quantity has the numerical value N. (It may be remarked that ΣN_r is a constant even when V_0 depends on the time.) It has been shown by Jordan and Klein (cf. § 11)[1] that the solutions of the wave problem with Hamiltonian (119) for which this condition is fulfilled are mathematically and physically equivalent to the solutions of the N-electron problem of the corpuscular theory, i.e., to the solutions of the Schrödinger equation (47). However, they do not correspond to all the solutions of this equation but only to those of the possible solutions in which the transformation function ψ is symmetric in the

---

[1] *Ibid.*

co-ordinates of the electrons. These solutions themselves form a closed term system, namely, that one for which the Bose-Einstein statistics is valid. The quantum theory of matter waves [especially the exchange relations (111)] thus requires the Bose-Einstein statistics for the corresponding particle picture.

The exchange relations (111) are, however, only one possibility out of many. Another equally justifiable set is obtained by changing the minus sign into a plus sign, so that the wave functions satisfy the equations

$$\left.\begin{aligned}
&\psi(P)\psi^*(P') + \psi^*(P')\psi(P) = \delta(P - P') ,\\
&\psi(P)\psi(P') + \psi(P')\psi(P) = 0 ,\\
&\psi^*(P)\psi^*(P') + \psi^*(P')\psi^*(P) = 0 .
\end{aligned}\right\} \quad (126)$$

According to Jordan and Wigner,[1] the quantum theory of waves based on these exchange relations is equivalent to the antisymmetric solutions of the Schrödinger equation; that is, these relations lead to the Pauli exclusion principle and the corresponding Fermi-Dirac statistics.

§ 11. PROOF OF THE MATHEMATICAL EQUIVALENCE OF THE QUANTUM THEORY OF PARTICLES AND OF WAVES

The problem of quantum theory centers on the fact that the particle picture and the wave picture are merely two different aspects of one and the same physical reality. Although this is a problem of purely physical nature it is satisfying to find a counterpart to this duality in the

[1] P. Jordan and E. Wigner, *Zeitschrift für Physik*, 47, 631, 1928.

mathematical apparatus of the theory. The analogy consists in the fact that one and the same set of mathematical equations can be interpreted at will in terms of either picture.

The proof of this assertion may be made perfectly general without regard to the particular form of Hamiltonian considered. The Schrödinger equation of the particle picture for N equivalent particles may be written

$$\left\{\sum_{n=1}^{N} O^n + \sum_{n>m}^{N} O^{nm} + \cdot\cdot + \frac{h}{2\pi i}\frac{\partial}{\partial t}\right\}\varphi(x_1, \ldots, x_N) = 0 \quad (127)$$

where O^n is an operator acting only on the space co-ordinates x_n of the nth particle, and O^{nm} one acting on the co-ordinates of both the nth and mth. Furthermore, it may be assumed that a certain system of orthogonal functions $u_r(x)$ has been found, in terms of which all functions in 3-space satisfying the boundary conditions can be expanded; it will then be possible to expand $\varphi(x_1, \ldots, x_N)$ in terms of products of these functions:

$$\varphi(x_1, \ldots, x_N) = \sum_{r_1 \ldots r_N} b(r_1, \ldots, r_N, t) u_{r_1}(x_1) \ldots u_{r_N}(x_N) . \quad (128)$$

The quantities $|b(r_1 \ldots r_N, t)|^2$ may be regarded as determining the probability that the particle 1 is in the r_1-state, particle 2 in the r_2-state, etc. If this expression for φ be substituted in equation (127), the result multiplied by $u_{s_1}(x_1) u_{s_2}(x_2) \ldots u_{s_N}(x_N)$ and then integrated over

$x_1, x_2, \ldots, x_N$ there results the following differential equation for the b's:

$$\left.\begin{aligned}
0 =\ & \frac{h}{2\pi i}\frac{\partial}{\partial t} b(s_1, s_2, \ldots, s_N, t)\\
&+ \sum_{n}\sum_{r_m} O^n_{s_n r_m} b(s_1 \ldots r_m \ldots s_N)\\
&+ \sum_{n>m}\sum_{r_n r_m} O^{nm}_{s_n s_m;\, r_n r_m} b(s_1 \ldots r_n \ldots r_m \ldots s_N) + \cdot\cdot\cdot
\end{aligned}\right\} (129)$$

Use has here been made of the orthogonality relations for the $u_r(x)$ and the quantities

$$O^n_{s_n r_n} = \int u_{s_n} O^n u_{r_n} dv_n ,$$

$$O^{nm}_{s_n s_m;\, r_n r_m} = \int\int u_{s_n} u_{s_m} O^{nm} u_{r_n} u_{r_m} dv_n dv_m ,$$

are the elements of the matrices representing the corresponding operators in the co-ordinate system characterized by the functions $u_r(x)$. Because of the symmetry of the Hamiltonian in the co-ordinates of the particles, the numerical values of the matrix elements depend only on the indices r and s, and not explicitly on n and m. In the case of the Bose-Einstein statistics the $b(s_1 \ldots s_N)$ are symmetric in the quantum numbers of the particles, so that they can also be expressed as functions of the number N_r of particles in the rth state. Since the a priori probability of finding N_1 particles in the first state, N_2 in the second, etc., is then given by $Z^2 = N!/(N_1! N_2! \ldots)$, it is convenient to define the quantity

$$b(N_1, N_2 \ldots) = Z b(r_1, r_2, \ldots, r_N). \quad (130)$$

The operators $e^{-\frac{2\pi i}{h}\Theta_r}$ of equation (125) which change N_r to $N_r + 1$ are useful here; with their aid equation (129) may be written

$$0 = \left\{\frac{h}{2\pi i}\frac{\partial}{\partial t} + \sum_{s,r} N_s O_{sr} e^{\frac{2\pi i}{h}(\Theta_s - \Theta_r)}\right.$$

$$+ \tfrac{1}{2}\sum_{ss';\, rr'} N_s(N_{s'} - \delta_{ss'}) O_{ss';\, rr'} e^{\frac{2\pi i}{h}(\Theta_s + \Theta_{s'} - \Theta_r - \Theta_{r'})}$$

$$\left. + \ldots \right\}\frac{1}{Z} b(N_1, N_2, \ldots) .$$

On multiplying this equation from the left by Z and commuting $1/Z$ to the left equation (125) yields

$$\left.\begin{aligned}
0 =\ & \left\{\frac{h}{2\pi i}\frac{\partial}{\partial t} + \sum_{s,r} N_s^{\frac{1}{2}}(N_s - \delta_{rs} + 1)^{\frac{1}{2}} O_{sr} e^{\frac{2\pi i}{h}(\Theta_s - \Theta_r)}\right.\\
&+ \tfrac{1}{2}\sum_{ss';\, rr'} N_s^{\frac{1}{2}}(N_{s'} - \delta_{ss'})^{\frac{1}{2}}(N_r + 1 - \delta_{rs} - \delta_{rs'})^{\frac{1}{2}}\\
&(N_{r'} + 1 + \delta_{rr'} - \delta_{r's} - \delta_{r's'})^{\frac{1}{2}} \cdot e^{\frac{2\pi i}{h}(\Theta_s + \Theta_{s'} - \Theta_r - \Theta_{r'})}\right\}\\
&\cdot b(N_1, N_2, \ldots) .
\end{aligned}\right\} (131)$$

We turn now to the corresponding problem expressed in the wave theory; the Hamiltonian corresponding to (127) is then

$$\bar{H} = \int dv_P \psi_P^* O^P \psi_P + \tfrac{1}{2}\int\int dv_P dv_{P'} \psi_P^* \psi_{P'}^* O^{P'P} \psi_{P'} \psi_P + \cdots\cdots$$

By (122) this may also be written

$$\bar{H} = \sum_{s,r} a_s^* a_r O_{sr} + \tfrac{1}{2}\sum_{ss';\, rr'} a_s^* a_{s'}^* a_r a_{r'} O_{ss';\, rr'} + \cdots\cdots .$$

181 Then on substituting equations (124) in the equation

$$\bar{H}S + \frac{h}{2\pi i}\frac{\partial S}{\partial t} = 0 \, ,$$

we obtain

$$0 = \left\{ \frac{h}{2\pi i}\frac{\partial}{\partial t} + \sum_{s,r} N_s^{\frac{1}{2}} O_{sr} e^{\frac{2\pi i}{h}(\Theta_s - \Theta_r)} N_r^{\frac{1}{2}} \right.$$
$$\left. + \tfrac{1}{2}\sum_{ss';\, rr'} N_s^{\frac{1}{2}} e^{\frac{2\pi i}{h}\Theta_{s'}} O_{ss';\, rr'} e^{\frac{2\pi i}{h}\Theta_r} N_r^{\frac{1}{2}} e^{-\frac{2\pi i}{h}\Theta_{r'}} N_r^{\frac{1}{2}} \right.$$
$$\left. + \cdots \right\} S(N_1, N_2, \ldots .) \, .$$

Commutation of the operators $e^{\frac{2\pi i}{h}\Theta}$ to the right gives

$$\left.\begin{aligned}
0 = \Big\{ & \frac{h}{2\pi i}\frac{\partial}{\partial t} + N_s^{\frac{1}{2}}(N_r - \delta_{sr} + 1)^{\frac{1}{2}} O_{sr} e^{\frac{2\pi i}{h}(\Theta_s - \Theta_r)} \\
& + \tfrac{1}{2}\sum_{ss';\, rr'} N_s^{\frac{1}{2}}(N_{s'} - \delta_{ss'})^{\frac{1}{2}}(N_r + 1 - \delta_{rs'} - \delta_{rs})^{\frac{1}{2}} \\
& (N_{r'} + 1 + \delta_{rr'} - \delta_{r's} - \delta_{r's'})^{\frac{1}{2}} e^{\frac{2\pi i}{h}(\Theta_s + \Theta_{s'} - \Theta_r - \Theta_{r'})} \Big\} S \, .
\end{aligned}\right\} \quad (132)$$

This equation is identical with equation (131), and the mathematical equivalence of the particle and wave pictures has therefore been proved. A similar proof may be given in the case of the Pauli exclusion principle and the exchange relations (126).

Although the classical theories of the corpuscular and wave pictures are so entirely different, both physically and mathematically, the quantum theories of the two are identical.

182   § 12. APPLICATION TO THE THEORY OF RADIATION[1]

It will be recalled that the Maxwell equations, which govern the classical wave theory of radiation, can be derived by variation of the potentials in the Lagrangian

$$L = \frac{1}{8\pi}(E^2 - H^2) + \sum_{a=1}^{4} \Phi_a s_a \, .$$

The $s_a (a = 1, 2, 3, 4)$ are the components of the 4-current density, the Φ_a the 4-potentials $(\Phi_4 = i\Phi_0,\ x_4 = ict)$; hence the Lagrangian becomes, when written explicitly in terms of the potentials,

$$\left.\begin{aligned}
L = \frac{1}{8\pi}\Bigg[& \sum_i \left(\frac{1}{c}\frac{\partial \Phi_i}{\partial t} + \frac{\partial \Phi_0}{\partial x_i} \right)^2 - \sum_{i>k} \left(\frac{\partial \Phi_i}{\partial x_k} - \frac{\partial \Phi_k}{\partial x_i} \right)^2 \Bigg] \\
& + \sum_a \Phi_a s_a \, .
\end{aligned}\right\} \quad (133)$$

(In this and the following equations Latin indices run from 1 to 3, Greek indices from 1 to 4.)

The momentum conjugate to Φ_i is, by (104),

$$\Pi_i = \frac{\partial L}{\partial \dot{\Phi}_i} = \frac{1}{4\pi c}\left(\frac{1}{c}\frac{\partial \Phi_i}{\partial t} + \frac{\partial \Phi_0}{\partial x_i} \right) = \frac{1}{4\pi c} E_i \, . \qquad (134)$$

Since the Bose-Einstein statistics applies to light quanta, the proper exchange relations are

$$E_i(P)\Phi_a(P') - \Phi_a(P')E_i(P) = -2hci\,\delta(P - P')\delta_{ia} \, ,$$

which give on differentiating

$$\left.\begin{aligned}
& E_i(P)E_k(P') - E_k(P')E_i(P) = 0 \, , \\
& H_i(P)H_k(P') - H_k(P')H_i(P) = 0 \, , \\
& E_i(P)H_k(P') - H_k(P')E_i(P) = -2hci\,\frac{\partial \delta(P - P')}{\partial x_j} \, ,
\end{aligned}\right\} \quad (135)$$

where i, j, k is any cyclic permutation of 1, 2, 3.

A difficulty arises from the circumstance that $\dot{\Phi}_0$ does not occur in the Lagrangian; this affects, however, only the exchange relations between potentials and field components, and not the exchange relations (135).

If the Φ_a be developed in a set of suitably chosen orthogonal functions (e.g., standing waves in an inclosure), then the energy content of a vibration of frequency ν becomes an integral multiple of $h\nu$. Dirac[1] has shown that this makes it possible to consider the number of light quanta in each state as the variables of the system; this constitutes the link with the particle picture.

[1] *Proceedings of the Royal Society*, A, 114, 710, 1927.

---

[1] W. Heisenberg and W. Pauli, *Zeitschrift für Physik*, 56, 1, 1929.

INDEX

W. Heisenberg, Leipzig

Fortschritte in der Theorie des Ferromagnetismus

Der Zustand eines ferromagnetischen Stoffes, z. B. eines Stückes Eisen, läßt sich etwa in folgender Weise kurz beschreiben: Das Material zerfällt hinsichtlich seines magnetischen Verhaltens in kleine Elementarbezirke, die nahezu bis zur Sättigung magnetisiert sind. Die Richtungen der magnetischen Sättigungsmomente der einzelnen Bezirke sind jedoch beim Fehlen äußerer Einwirkungen statistisch regellos verteilt, so daß im Mittel kein magnetisches Moment des Materials entsteht. Gerät das Ferromagnetikum unter den Einfluß eines äußeren Magnetfeldes, so suchen sich die Sättigungsmomente der Elementarbezirke nach dem äußeren Felde zu orientieren, teils durch stetige Richtungsänderung, teils durch diskontinuierliches Umklappen (B a r k h a u s e n effekt). Die Größe der Elementarbezirke hängt von der Regelmäßigkeit des Gitteraufbaues des Materials ab und variiert von makroskopischen zu mikroskopischen Dimensionen.

Eine Theorie des Ferromagnetismus hat daher zwei wesentlich verschiedene Fragen zu klären:

Zunächst muß die Magnetisierung des Elementarbezirkes theoretisch zurückgeführt werden auf berechenbare Eigenschaften der atomaren Struktur des ferromagnetischen Materials; dieser Teil der Theorie wird insbesondere die Größe der magnetischen Momente in den Elementarbereichen als Funktion der Temperatur liefern.

In zweiter Linie hat die Theorie die Phänomene zu klären, die bei der Einstellung der Einzelmomente in äußeren Feldern auftreten. Erst dieser zweite Teil der Theorie gibt näheren Aufschluß über den technischen Magnetisierungsvorgang (Hysteresisschleife, Magnetostriktion, B a r k h a u s e n effekt).

Die magnetischen Vorgänge im Elementarbezirk

Die Tatsache, daß in einem Elementarbezirk die magnetischen Momente nahezu aller Atome parallel zueinander stehen, hat W e i ß[1]) durch die Annahme erklärt, daß die atomaren Magnete Kräfte aufeinander ausüben, die sie parallel zu stellen suchen. Z. B. haben bereits die magnetischen Wechselwirkungen der atomaren Magnete die Tendenz, die Magnete parallel zueinander zu orientieren. Diesen Kräften wirkt jedoch die Wärmebewegung der Atome entgegen, so daß oberhalb einer kritischen Temperatur, des „C u r i e punktes", keine Parallelstellung der Magnete mehr eintritt. Unterhalb der kritischen Temperatur stellt sich auch bei Abwesenheit äußerer Felder ein mittleres Moment ein, das von der Temperatur und von den Eigenschaften des Körpers abhängt. Die Lage des Curiepunktes wird durch die Größe der Wechselwirkungen zwischen den atomaren Magneten bedingt. Hier zeigten sich nun erhebliche Schwierigkeiten in der Weiß= schen Theorie. Die magnetischen Wechselwirkungen erwiesen sich als viel zu schwach, um Curietemperaturen von der Ord= nung $\sim 500°$ zu erklären. Ferner schien es schwer vorstellbar, daß sich in einem Kristallgefüge die Atome so leicht drehen lassen wie nach den Experimenten das magnetische Moment der Einzelbezirke.

Diese Schwierigkeiten sind von der neueren Quantentheorie des Atombaus beseitigt worden. Das magnetische Moment des einzelnen Atoms setzt sich zusammen aus dem „Bahn"= moment der Elektronen, das von den Elektronen bei ihrem Umlauf um den Kern nach dem Biot=Savartschen Gesetz er= zeugt wird, und den Eigenmomenten der Elektronen, dem „Elektronenspin". Dieses Eigenmoment eines Elektrons ist seiner Richtung nach nur durch schwache magnetische Wech= selwirkungen an die Atombahn und an das Kristallgefüge ge= bunden, es ist also in erster Näherung frei drehbar. Die Ver= suche, die Barnett[2]) und später Einstein und de Haas[3]) über das Verhältnis von magnetischem Moment und Drehimpulsmoment in einem Ferromagnetikum anstell= ten, führen daher auch zu dem Resultat, daß die atomaren Magnete der Weiß schen Theorie mit den Eigenmomenten der Elektronen identisch sein müssen. Auch Compton und Rognley[4]) hatten schon früher wegen der freien Dreh= barkeit der atomaren Magnete die Existenz des Elektronen= spins vermutet.

Obwohl nur zwar die Richtung des Elektronendralles relativ zum Atom nur durch schwache magnetische Wirkungen fest= gelegt wird, so wird doch nach der neueren Quantentheorie die relative Richtung der Eigenmomente verschiedener Elektronen durch sehr viel stärkere Kräfte[5]) bestimmt, die elektrosta= tischen Ursprungs sind und die mit der Möglichkeit des Aus= tausches zweier Elektronen zusammenhängen. Auf die ge= nauere Diskussion dieser Kräfte kann im Rahmen dieses Artikels nicht eingegangen werden. Wesentlich ist, daß diese Kräfte, die die Eigenmomente in ferromagnetischen Sub= stanzen parallel zu stellen suchen, ungefähr die aus der Tem= peratur des Curiepunkts gefolgerte Größe haben und daß da= her die Weiß sche Theorie in allen wesentlichen Punkten

beibehalten werden kann. Die mathematische Durchführung der Theorie stößt allerdings noch auf einige Schwierigkeiten: sichergestellt scheint bisher nur eine Formel von Bloch[6]),

welche das mittlere Moment J per Volumeneinheit als Funk=
tion der Temperatur T für tiefe Temperaturen darstellt:

$$J = J_0 - C \cdot T^{3/2}$$

(C bedeutet eine Material=Konstante).

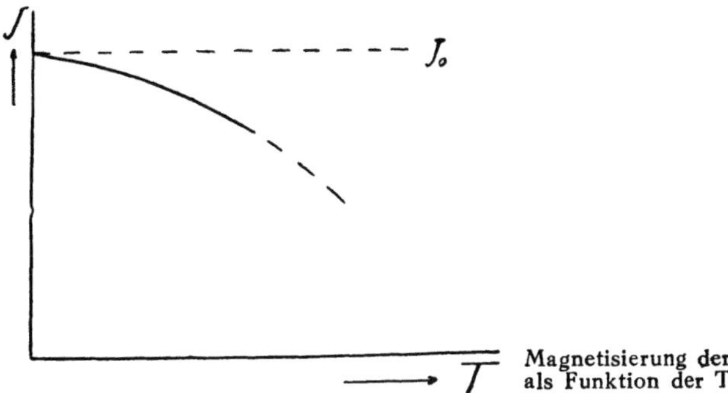

Die thermischen Effekte in der Nähe des Curiepunktes
können nach F o w l e r und K a p i t z a[7]) ähnlich, wie in der
W e i ß schen Theorie[1]) ermittelt werden. Der entscheidende
Unterschied der neuen Formulierung von der W e i ß schen
besteht nur darin, daß die Kräfte, die eine Parallelstellung der
atomaren Magnete bewirken, doch die Richtung des Gesamt=
moments im Kristall völlig unbestimmt lassen. Die Richtung
des Gesamtmoments hängt vielmehr erst mit den magne=
tischen Wechselwirkungen zusammen und kann durch Dis=
kussion der magnetischen Kräfte ziemlich ohne Anwendung
quantentheoretischer Vorstellungen berechnet werden.

Das makroskopische Verhalten der Elementarbezirke

In einem kubischen Gitter, wie wir es bei Eisen und Nickel
antreffen, hängt (sofern das Gitter wirklich exakt kubische
Symmetrie besitzt) die magnetische Wechselwirkung der zu=
einander parallelen atomaren Magnete nicht von der Richtung
des Gesamtmoments relativ zu den Kristallachsen ab. Da=
gegen entsteht eine Kraft, die die Richtung des Gesamt=
moments beeinflußt, durch die Wechselwirkung des Elektro=
nenspins mit der Elektronenbahn; die Elektronenbahn wird
durch die Felder der Nachbaratome verzerrt, so daß sie höch=
stens kubische Symmetrie behält. Ferner führt nach
B e c k e r[8]) auch die magnetische Wechselwirkung der ato=
maren Magnete zu richtenden Kräften, wenn die Abweichun=
gen von der kubischen Symmetrie berücksichtigt werden, die
durch Störungen und Verzerrungen des Gitters beim Be=
arbeiten entstehen. Die beiden eben genannten Arten richten=
der Kräfte dürften, wie eine Überschlagsrechnung zeigt, unge=
fähr von der gleichen Größenordnung sein. Quantitative
Rechnungen liegen bisher aber nur für den letztgenannten
Effekt vor. Die magnetische Wechselwirkung der atomaren

Magnete führt dabei nach A k u l o v[9]) und F o w l e r und
K a p i t z a[7]) zu einer qualitativ befriedigenden Beschreibung
der „Magnetostriktion", d. h. der elastischen Verzerrungen
des ganzen Gitters, die auftreten, wenn man die Richtung des
gesamten magnetischen Moments relativ zu den Kristallachsen
verändert. Bei B e c k e r[8]) werden dagegen die Verzerrungen
als durch chemische Verunreinigungen oder durch die Vor=
gänge bei der Bearbeitung hervorgerufene Störungen ange=
sehen, die für das Material charakteristisch sind und die im
Kristall von Ort zu Ort variieren. Diese Störungen bestimmen
in kleinen Bereichen die Richtung des magnetischen Moments.
Die Durchrechnung der mit den Störungen verknüpften ma=
gnetischen Wechselwirkungskräfte gibt nach B e c k e r be=
friedigende Werte für die Koerzitivkraft.

Einige ferromagnetische Kristalle (z. B. Pyrrhotin) zeigen
eine sehr viel stärkere Richtungsabhängigkeit des magne=
tischen Gesamtmoments, als nach den ebengenannten Unter=
suchungen theoretisch zu erwarten wäre. Eine befriedigende
Erklärung dieses experimentellen Befundes liegt bisher nicht
vor. Abgesehen von diesen speziellen Fällen sprechen aber
die Experimente allgemein dafür, daß die Parallelstellung der
atomaren Magnete einerseits und die Einstellung der Gesamt=
momente relativ zu den Kristallachsen andererseits aus ver=
schiedenen Kraftwirkungen entspringt, wie es die neuere
Quantentheorie auch fordert.

Institut für theoretische Physik der Universität Leipzig.

Literatur:

[1]) P. W e i s s, Journ. de phys. **6**, 661, 1907 u. Phys. Ztschr. **9**, 358, 1908.
[2]) S. J. B a r n e t t, Phys. Rev. **6**, 239, 1915.
[3]) A. E i n s t e i n u. d e H a a s, Verh. d. Deutsch. Phys. Ges. **17**, 152, 1915.
[4]) A. H. C o m p t o n u. O. R o g n l e y, Phys. Rev. **16**, 464, 1920.
[5]) W. H e i s e n b e r g, Ztschr. f. Phys. **49**, 619, 1928.
[6]) F. B l o c h, Ztschr. f. Phys. **61**, 206, 1930.
[7]) R. F o w l e r u. P. K a p i t z a, Proc. Roy. Soc. **A 124**, 1, 1929.
[8]) R. B e c k e r, Ztschr. f. Phys. **62**, 253, 1930.
[9]) N. S. A k u l o v, Ztschr. f. Phys. **52**, 389, 1928; **54**, 582, 1929; **57**, 249, 1929; **59**, 254, 1930.

In *Chemistry at the Centenary (1931) Meeting of the British Association for the Advancement of Science* (W. Heffer & Sons, Cambridge 1932) pp. 247–248

CONTRIBUTION TO DISCUSSION ON THE STRUCTURE OF SIMPLE MOLECULES.

Professor W. HEISENBERG

As a contribution to the lecture which Mr. Fowler has given on the quantum theory of valency forces, I would like to emphasise a few points which Mr. Fowler has already mentioned. The theory of valency, which was given by London and Heitler and later has been extended by Born, Weyl, Slater, Pauling and others, has the great advantage of leading exactly to the concept of valency which is used by the chemist. But it seems questionable to me, whether the quantum theory would have found or would have been able to derive the chemical results about valency, if it had not known them before. The method of approximation in the calculations about a molecule, used by London and Heitler, is not the usual one, nor the one which looks most natural to the physicist. In the theory of the atom we get the best approximation, in calculating the energy-levels, when we first neglect the interaction energy between the electrons and later consider it as a small perturbation compared with the action of the central field. In the London–Heitler method, however, we consider the interaction of two electrons in the same atom as very large, and take only the interaction between electrons in different atoms as a small perturbation. When we apply to the molecule the other method (neglect of interaction energy), which has proved so successful for the atom (Hartree, Thomas and Fermi), we get the scheme to account for the valency-forces put forward by Hund, Lennard-Jones and others. This theory of Hund does not lead generally to the chemical concept of valency, but it leads in many cases to the right experimental result. Now the question arises which of the two aspects of the valency forces, that of Lennard-Jones or that of London and Heitler is more correct. I think that in all cases in which the two methods do not agree, we have to state that the quantum theory does not yet allow a conclusive explanation of the experimental facts. Especially the consequences which have been drawn from the London–Heitler model by Slater and Pauling on the question of "directed valencies" seem to me somewhat doubtful. Obviously I do not criticise in this case the mathematical calculations, but the underlying assumptions, which might be correct, but which are difficult to prove. For example,

248   CHEMISTRY AT THE CENTENARY (1931)–

in the case of the water molecule, the method of Hund does in the first approximation not lead to the same result, as the method of Pauling and Slater.

In order to start the discussion I have tried to make the present quantum theory of valency still more suspicious to the chemist than it already is. But we may hope that after some time the theoretical physicists will be able to give a more accurate explanation of what corresponds to the chemical valency.

Ergebnisse der technischen Röntgenkunde *3*, 26–31 (1933)

Über die Streuung von Röntgenstrahlen an Molekülen und Kristallen.

Von W. HEISENBERG, Leipzig.

Der vorliegende Aufsatz soll eine kurze zusammenfassende Übersicht geben über die Prinzipien, die einer theoretischen Behandlung der Streuung von Röntgenstrahlen an komplizierteren atomaren Systemen zugrunde liegen. Es soll erläutert werden, wie die verschiedenen Schritte, in denen die Berechnung der Streustrahlung gewöhnlich ausgeführt wird (Berechnung der kohärenten Strahlung, der inkohärenten Strahlung, Mittelung über die Lagen der Kerne), aus den Grundformeln der Quantenmechanik folgen. Dabei wird sich Gelegenheit ergeben, die übliche Einteilung der Streustrahlung in kohärente und inkohärente Strahlung kritisch zu diskutieren. Wir beginnen damit, die von WALLER (1) abgeleiteten quantenmechanischen Grundformeln für die Streuung von Röntgenstrahlen zu wiederholen:

Wenn das streuende atomare System, Molekül oder Kristall, aus verschiedenen Partikeln der Ladungen e_1, e_2 .. und der Massen m_1, m_2 ... besteht, so wird die Streustrahlung, die zu einem Übergang des Systems vom Zustand n in den Zustand n' gehört, von einem schwingenden Dipol der Stärke:

$$d_{nn'} = -\frac{1}{8\,\pi^2\,\nu\,\nu'}\,E_0\,\Gamma_{nn'}\,e^{2\,\pi\,i\nu't} \tag{1}$$

emittiert, wo ν bzw. ν' die Frequenz des einfallenden bzw. des gestreuten Lichtes, E_0 die Amplitude des elektrischen Vektors des einfallenden Strahles bedeutet. Ferner ist

$$\Gamma_{nn'} = \int \Phi_n^* \Phi_{n'} \sum_{\varkappa=1}^{N} \frac{e_\varkappa^2}{m_\varkappa} e^{i\mathfrak{s}\mathfrak{r}_\varkappa} \mathrm{d}V_1 \cdots \mathrm{d}V_N; \tag{2}$$

N ist die Anzahl der Partikeln, Φ_n die SCHRÖDINGERfunktion des Systems im Zustand n, der Vektor $\mathfrak{s}$ ist durch

$$\mathfrak{s} = \frac{2\,\pi}{c}\,(\nu\,\mathfrak{n} - \nu'\,\mathfrak{n}')$$

gegeben, wobei $\mathfrak{n}$ einen Einheitsvektor in Richtung des einfallenden Lichtes, $\mathfrak{n}'$ einen in Richtung des gestreuten Lichtquants bedeutet.

$\mathrm{d}\,V_1 \cdots \mathrm{d}\,V_N$ ist das Volumelement im Koordinatenraum der Partikeln.

Da die Masse der Kerne groß ist gegen die der Elektronen, kann man in (2) die Streuung der Kerne vernachlässigen und nur die Streuung der Elektronen beibehalten. Es genügt also, an Stelle des $\Gamma_{nn'}$ die Größe

$$\gamma_{nn'} = \int \Phi_n^* \Phi_{n'} \sum_{\varkappa=1}^{M} e^{i\mathfrak{s}\mathfrak{r}_\varkappa}\, \mathrm{d}\,V_1 \cdots \mathrm{d}\,V_N \tag{3}$$

zu betrachten, wobei die Summe nur über die Elektronen zu erstrecken ist, deren Anzahl zu M angenommen wird. Um nun die Intensität der gesamten, kohärenten und inkohärenten, Streustrahlung zu berechnen, genügt es, unter der Bedingung, daß die Frequenz des einfallenden Lichtes groß ist gegen die Eigenfrequenzen des streuenden Systems, die Summe

$$S = \sum_{n'=0}^{\infty} \gamma_{nn'}^2 \tag{4}$$

auszuwerten. Durch Anwendung der Orthogonalitätsrelationen erhält man aus (*3*) und (*4*):

$$S = \int \Phi_n^* \Phi_n \sum_{\varkappa,\,\lambda}^{M} e^{i\mathfrak{s}(\mathfrak{r}_\varkappa - \mathfrak{r}_\lambda)}\, \mathrm{d}\,V_1 \cdots \mathrm{d}\,V_N. \tag{5}$$

Man wird nun annehmen dürfen, daß die Eigenfunktion im Zustand n näherungsweise als Produkt einer Eigenfunktion X_i der Kernkoordinaten allein und einer Eigenfunktion Ψ_n der Elektronen- und Kernkoordinaten geschrieben werden kann, wobei Ψ_n aus der SCHRÖDINGERgleichung für die Elektronen bei festgehaltenen Kernen zu berechnen ist. Dann folgt:

$$S = \int \mathrm{d}\,V_{M+1} \cdots \mathrm{d}\,V_N\, X_n^* X_n \int \Psi_n^* \Psi_n \sum_{\varkappa,\,\lambda}^{M} e^{i\mathfrak{s}(\mathfrak{r}_\varkappa - \mathfrak{r}_\lambda)}\, \mathrm{d}\,V_1 \cdots \mathrm{d}\,V_M \tag{6}$$

(Die Elektronen sind von 1 bis M, die Kerne von $M+1$ bis N numeriert.)

Nach Gl. (*6*) kann man die gesamte Streustrahlung ermitteln, indem man zunächst bei festgehaltener Lage der Kerne die Intensitätsverteilung berechnet und dann über alle Lagen der Kerne (mit dem zugehörigen Gewicht $X_n^* X_n$) mittelt. Dieses Resultat ist keineswegs trivial und gilt zunächst nur für die gesamte Streustrahlung S, bevor eine Einteilung in den kohärenten und inkohärenten Teil vorgenommen wurde. Für das Verhalten der einzelnen Teile ist eine genaue Diskussion dieser Einteilung der Strahlung in „kohärente" und „inkohärente" Strahlung erforderlich. Die gestreute Strahlung wird sich hinsichtlich ihrer Frequenz stets etwas von der einfallenden Strahlung unterscheiden, entweder um Elektroneneigenschwingungen, oder um Schwingungs- oder Rotationsfrequenzen der Kerne, oder nur um 'einen minimalen Betrag, der vom Compton-Rückstoß des ganzen streuenden Systems herrührt. Gewöhnlich bezeichnet man die Streustrahlung, deren Frequenz sich um eine Elektronenfrequenz von der Frequenz der einfallenden Strahlung unterscheidet, als inkohärent, die übrige Strahlung als kohärent. Diese Einteilung ist aber nur richtig, wenn man die Lage der Kerne als fest annimmt. Tut man das nicht, so kann man als kohärente Streustrahlung streng genommen nur die bezeichnen, bei der außer dem Compton-Rückstoß keine weitere Energie auf das streuende System übertragen wird. Diese Streustrahlung macht aber nur einen unwesentlichen Bruchteil der Gesamtstrahlung aus. Es scheint daher nicht zweckmäßig, prinzipiell eine Einteilung der Streustrahlung in kohärente und inkohärente Strahlung vorzunehmen; trotzdem kann man die übliche Einteilung mit Vorteil bei der Berechnung der Gesamtstrahlung benützen; denn in der Rechnung ist es gerechtfertigt, wie oben gezeigt wurde, zunächst die Lage der Kerne als fest zu betrachten — dann hat die diskutierte Einteilung einen einfachen Sinn — und später über die Lagen der Kerne zu mitteln.

Es handelt sich also jetzt zunächst um die Berechnung des Integrals

$$I = \int \Psi_n^* \, \Psi_n \sum_{\lambda, \varkappa=1}^{M} e^{i\mathfrak{s}(\mathfrak{r}_\lambda - \mathfrak{r}_\varkappa)} \, \mathrm{d}\, V_1 \cdots \mathrm{d}\, V_M \qquad (7)$$

für eine vorgegebene Lage der Kerne. Die Integration über $\mathrm{d}\, V_1 \cdots \mathrm{d}\, V_M$ schließt die Summation über die Spinvariabeln ein.

Vernachlässigt man die Wechselwirkung der Elektronen, so läßt sich Ψ_n als antisymmetrische Kombination von Produkten von Eigenfunktionen darstellen, wobei jede Eigenfunktion abhängt von Orts- und Spinvariable des betreffenden Elektrons. Die Zustände seien mit s und t numeriert, die Eigenfunktion werde $\varphi_s\,(\mathfrak{r},\,\sigma)$ genannt, σ ist die Spinvariable. Aus (7) folgt dann:

$$I = \sum_{s,\,t} \int\!\!\int \mathrm{d}\, V \, \mathrm{d}\, V' \sum_\sigma \sum_{\sigma'} \varphi_s^*\,(\mathfrak{r}\sigma)\, \varphi_s(\mathfrak{r}\sigma)\, \varphi_t^*\,(\mathfrak{r}'\sigma')\, \varphi_t(\mathfrak{r}'\sigma')\, e^{i\mathfrak{s}(\mathfrak{r}-\mathfrak{r}')}$$

$$+\, M - \sum_{s,\,t} \int\!\!\int \mathrm{d}\, V \, \mathrm{d}\, V' \sum_\sigma \sum_{\sigma'} \varphi_s^*\,(\mathfrak{r}\sigma)\, \varphi_t(\mathfrak{r}\sigma)\, \varphi_t^*\,(\mathfrak{r}'\sigma')\, \varphi_s(\mathfrak{r}'\sigma')\, e^{i\mathfrak{s}(\mathfrak{r}-\mathfrak{r}')}. \qquad (8)$$

Die Summation ist über alle besetzten Zustände auszuführen. Das erste Doppelintegral in (8) gibt die sogenannte „kohärente" Streustrahlung, die zweite Zeile die „inkohärente" Strahlung. Wenn man annimmt, daß näherungsweise jeder Elektronenzustand zu einem bestimmten Kern gehört, und daß die Eigenfunktionen, die zu verschiedenen Kernen gehören, sich nicht merklich durchdringen, so fallen die Glieder des zweiten Doppelintegrals, bei denen s und t zu verschiedenen Kernen gehören, weg; in dem ersten Doppelintegral geben die entsprechenden Glieder die Effekte wieder, die durch Interferenz der von verschiedenen Atomen gestreuten Strahlung entstehen. In dieser Näherung reduziert sich dann

die Berechnung der kohärenten Streustrahlung in bekannter Weise
auf die Auswertung des Integrals

$$F = \int \mathrm{d}\,V \sum_t \sum_\sigma \varphi_t^* (\mathfrak{r}, \sigma)\, \varphi_t(\mathfrak{r}, \sigma)\, e^{i\mathfrak{s}\mathfrak{r}}, \tag{9}$$

wobei die Summe über t nur über die zu einem Kern gehörigen Zustände
zu erstrecken ist. Statt (*9*) kann man auch schreiben:

$$F = \int \mathrm{d}\,V \varrho\,(\mathfrak{r})\, e^{i\mathfrak{s}\mathfrak{r}}. \tag{10}$$

Hierin bedeutet $\varrho\,(\mathfrak{r})$ die Ladungsdichte des betreffenden Atoms, die z. B.
nach der Methode von THOMAS und FERMI ermittelt werden kann (2).

Ähnlich genügt für die Berechnung des zweiten Doppelintegrals, also
der inkohärenten Strahlung, die Ermittlung von

$$G = \iint \mathrm{d}\,V \mathrm{d}\,V' \sum_\sigma \sum_{\sigma'} \left| \sum_t \varphi_t^*(\mathfrak{r}\,\sigma)\, \varphi_t(\mathfrak{r}'\,\sigma') \right|^2 e^{i\mathfrak{s}\,(\mathfrak{r}-\mathfrak{r}')}, \tag{11}$$

wobei wieder die Summe über t nur über die zu einem Kern gehörigen
Terme zu erstrecken ist. Auch das Integral (*11*) läßt sich nach der
THOMAS-FERMI-Methode (3) behandeln.

Die Resultate der Anwendung der THOMAS-FERMI-Methode auf die
Integrale (*10*) und (*11*) sind tabuliert in den Arbeiten von DEBYE (2)
und BEWILOGUA (4); eine ausführliche Diskussion dieser Resultate soll
hier nicht mehr gegeben werden.

Dagegen soll noch die in der Endformel (*6*) auftretende Mittelung
über die Lagen der Kerne besprochen werden. Bei der Ableitung der
Gl. (*6*) wurde angenommen, daß der stationäre Zustand des streuenden
Systems genau bekannt sei; daher ließ sich die Mittelung über die Kerne
in der Form schreiben:

$$S = \int \mathrm{d}\,V_{M+1} \cdots \mathrm{d}\,V_N X_n^*\, X_n \cdot I, \tag{6'}$$

wobei X_n die Eigenfunktion des stationären Zustandes der Kerne bezeichnet. Im allgemeinen ist nun nicht der stationäre Zustand des Systems bekannt, sondern seine Temperatur; d. h. man weiß von dem
System, daß es irgendein Exemplar ist aus einer kanonischen Gesamtheit
von gleichartigen Systemen. An Stelle der Wahrscheinlichkeitsdichte
$X_n^* X_n$ tritt in diesem Fall eine Funktion $W\,(T)$ der Temperatur, die aus
$X_n^* X_n$ durch Mittelung über die stationären Zustände des Systems unter
Berücksichtigung des BOLTZMANN-Faktors entsteht:

$$W\,(T) = \frac{\sum\limits_n X_n^* X_n\, e^{-\frac{E_n}{k\,T}}}{\sum e^{-\frac{E_n}{k\,T}}}. \tag{12}$$

Die Summe ist hier über alle stationären Zustände der Kerne zu erstrecken, E_n bedeutet die Energie des Systems im Zustand n. Näherungs-

          W. HEISENBERG.

weise kann man die Bewegung der Kerne zusammensetzen aus Translation und Rotation des Gesamtsystems und aus harmonischen Schwingungen um die Ruhelage, die sich nach den „Hauptschwingungen" des Systems entwickeln lassen. Dementsprechend zerfällt X_n in ein Produkt von Eigenfunktionen, deren jede zu einer bestimmten Hauptschwingung des Systems gehört. Bezeichnet man die Amplitude der τ-ten Hauptschwingung mit Q_τ, so lassen sich die Koordinaten des n-ten Kerns linear durch die Q_τ darstellen:

$$X_n = \sum_\tau a_{n\,\tau} Q_\tau$$

$$Y_n = \sum_\tau \beta_{n\,\tau} Q_\tau \tag{13}$$

$$Z_n = \sum_\tau \gamma_{n\,\tau} Q_\tau \,,$$

ferner ist

$$X_n(X_1 \cdots Z_n) = \Pi_\tau \, \varphi_{n_\tau}(Q_\tau) \,, \tag{14}$$

wobei $\varphi_{n_\tau}(Q_\tau)$ die Eigenfunktion eines harmonischen Oszillators im Zustand n_τ bedeutet. Die Berechnung von (12) kann also zurückgeführt werden auf die Ermittelung von

$$W_\tau(T, Q_\tau) = \frac{\displaystyle\sum_{n_\tau=0}^{\infty} \varphi_{n_\tau}^* \, \varphi_{n_\tau} \, e^{-\frac{h\nu_\tau n_\tau}{kT}}}{\displaystyle\sum_{n_\tau=0}^{\infty} e^{-\frac{h\nu_\tau n_\tau}{kT}}} \,. \tag{15}$$

Für diese Funktion $W_\tau(T, Q_\tau)$ hat BLOCH (5) den folgenden Ausdruck angegeben:

$$W_\tau(T, Q_\tau) = \sqrt{\frac{4\,\pi\,\nu_\tau\,m_\tau}{h} \operatorname{tg} h \frac{h\,\nu_\tau}{2\,k\,T}} \; e^{-\frac{4\,\pi^2\nu_\tau\,m_\tau}{h} Q_\tau^2 \operatorname{tg} h \frac{h\nu_\tau}{2\,k\,T}} \,. \tag{16}$$

Hierin bedeutet m_τ die effektive Masse, ν_τ die Frequenz des Oszillators τ.

Die Funktion $W(T)$ in Gl. (12) kann also als Produkt von Ausdrücken der Form (16) geschrieben werden, die dann außerdem noch mit einer entsprechenden Wahrscheinlichkeitsfunktion für Translation und Rotation des ganzen streuenden Systems zu multiplizieren sind. Die Art der Mittelung über Translation und Rotation hängt ganz vom speziellen System ab und kann nicht allgemein weiter vereinfacht werden. Der Anteil der Funktion $W(T)$, der von den Schwingungen herrührt, läßt sich dagegen erheblich einfacher darstellen, wenn $k\,T$ für alle Eigenschwingungen groß gegen $h\,\nu_\tau$ ist. In diesem Grenzfall geht der Exponentialteil von $W(T, Q_\tau)$ über in

$$e^{-\frac{2\pi^2\nu_\tau^2\,m_\tau\,Q_\tau^2}{kT}}$$

und das Produkt aller dieser Exponentialfunktionen hat die einfache Form

$$e^{-\dfrac{E_{\text{pot}}(Q_1 \cdots Q_\tau)}{kT}} = e^{-\dfrac{E_{\text{pot}}(X_1 \cdots Z_n)}{kT}}.$$

E_{pot} bedeutet die potentielle Energie der Kerne, die eine quadratische Funktion der Q_τ oder der X_n, Y_n, Z_n ist. Im Grenzfall hoher Temperatur ist also die praktische Auswertung von (*12*) besonders einfach.

Der Einfluß der einzelnen Faktoren: inkohärente Streuung, Temperaturbewegung, Nullpunktsschwingung auf die Gesamtstreuung ist in der Literatur experimentell und theoretisch mehrfach ausführlich studiert worden (BRAGG, DEBYE, WALLER). Der vorliegende Aufsatz beabsichtigte nur eine systematische Übersicht über die an sich bekannten Resultate.

Zusammenfassung.

Es wird eine systematische Übersicht gegeben über die Prinzipien, die einer theoretischen Behandlung der Streuung von Röntgenstrahlen an komplizierteren atomaren Systemen zugrunde liegen. Die verschiedenen Schritte zur Berechnung (kohärente Strahlung, inkohärente Strahlung, Mittelung über die Lagen der Kerne, Berücksichtigung von Temperaturbewegungen) folgen, wie gezeigt werden kann, aus den Grundformeln der Quantenmechanik.

Literaturverzeichnis.

(1) J. WALLER, Z. Physik **51**, 213. 1928.
(2) Vgl. P. DEBYE, Physikal. Z **31**, 419. 1930.
(3) Vgl. W. HEISENBERG, Physikal. Z **32**, 737. 1931.
(4) BEWILOGUA, Physikal. Z **32**, 740. 1931.
(5) F. BLOCH, Z. Physik **74**, 309. 1932.

In *Structure et Propriétés des Noyaux Atomiques: Rapports et Discussions du Septième Conseil de Physique Tenu à Bruxelles du 22 au 29 Octobre 1933,* ed. by Institut International de Physique Solvay (Gauthier-Villars, Paris 1934) pp. 289–335 .

CONSIDÉRATIONS THÉORIQUES GÉNÉRALES

SUR

LA STRUCTURE DU NOYAU

Par M. W. HEISENBERG.

§ 1. Rappel de principes.

§ 2. Hypothèses sur la structure des noyaux :

a. Le modèle de la « goutte » de Gamow.
b. Introduction des neutrons comme constituants nucléaires.
c. Les lois d'action d'échange.
d. Conséquences générales concernant les isotopes.

§ 3. Applications :

a. Défaut de masse et stabilité des noyaux.
b. Diffusion et désintégration.

§ 1. — Rappel de principes.

Puisque les données expérimentales concernant la structure du noyau atomique ne nous ont pas apporté jusqu'ici de notions physiques nouvelles allant au delà de la mécanique quantique, il est nécessaire d'examiner au début d'un exposé théorique sur le noyau dans quelle mesure la mécanique quantique ou ondulatoire peut être utilisée dans ce nouveau domaine. Une limitation aussi précise que possible des possibilités d'application de la mécanique quantique est une des premières tâches de la théorie du noyau.

Gamow, Condon et Gurney ([1]) ont montré par leur théorie de la désintégration α que les constituants lourds du noyau (particules α et protons) sont soumis à l'intérieur de celui-ci à des conditions énergétiques analogues à celles qui concernent les électrons dans l'atome autour du noyau. Les énergies de désintégration des éléments radioactifs et celles qui sont nécessaires pour la désintégration des noyaux légers sont petites par rapport à l'énergie propre des corpuscules lourds (si M est la masse du proton, c la vitesse de la lumière, on a $Mc^2 = 1,5 . 10^{-3}$ erg; les énergies de désintégration sont comprises entre 10^{-6} et 10^{-5} erg). Corrélativement, les rayons nucléaires mesurés sont notablement plus grands que la longueur caractéristique des effets de relativité sur les protons $\dfrac{\hbar}{Mc} = 2 . 10^{-14}$ cm. On peut par conséquent s'attendre à ce que la mécanique quantique, sous sa forme actuelle, soit applicable au mouvement des particules lourdes dans le noyau atomique et que les notions fondamentales introduites par Bohr dans la théorie des quanta (états stationnaires, relation de fréquence, probabilité de transition), puissent s'étendre à ces particules, et enfin que les développements de la mécanique quantique dans le sens de la théorie de la relativité ne jouent ici qu'un rôle secondaire. En fait, l'existence de séries discontinues d'énergies de désintégration, leur liaison avec les raies des spectres γ et le succès de la théorie de la désintégration α (*voir* le rapport de Gamow) montrent que les constituants lourds du noyau se comportent bien de manière conforme à la mécanique quantique.

Bohr ([2]) a indiqué une relation qualitative simple entre la grandeur des noyaux et leur défaut de masse, comme conséquence de l'applicabilité de la mécanique quantique au mouvement des particules lourdes à l'intérieur du noyau. Si par exemple r_0 représente le rayon du noyau d'hélium, il résulte de la relation d'indétermination l'expression suivante pour le domaine Δp de variation de la quantité de mouvement d'un proton à l'intérieur d'un noyau

---

([1]) G. Gamow, *Der Bau des Atomkerns und die Radioaktivität* (Leipzig, 1932).
([2]) N. Bohr, *Atomic stability and conservation laws, Convegno di Fisica nucleare* (Rome, 1932).

d'hélium :

$$(1) \qquad \Delta p \sim \frac{\hbar}{r_0},$$

et par suite pour son énergie cinétique moyenne

$$(2) \qquad \overline{E}_{\text{cin.}} \sim \frac{1}{2\,M} \left(\frac{\hbar}{r_0} \right)^2.$$

Comme en général l'énergie cinétique moyenne d'une particule est du même ordre de grandeur que l'énergie potentielle moyenne et que l'énergie totale, on obtient pour le défaut de masse du noyau d'hélium mesuré en énergie une valeur de l'ordre de $4 \cdot \frac{1}{2\,M} \left(\frac{\hbar}{r_0} \right)^2$. Si l'on admet d'après les expériences de Chadwick :

$$r_0 \sim \frac{e^2}{mc^2} = 2,81 \cdot 10^{-13} \text{ cm}$$

(m masse de l'électron), il en résulte

$$\text{défaut de masse} \sim 4 \times \frac{1}{2\,M} \, (mc)^2 \left(\frac{\hbar c}{e^2} \right)^2 \sim 0,01 \, M c^2,$$

du même ordre de grandeur que la valeur expérimentale $0,029 \, M c^2$.

Ces considérations appellent deux remarques qui seront utiles pour la discussion ultérieure : en premier lieu, nous avons complètement négligé la contribution au défaut de masse des charges négatives qui peuvent être contenues dans le noyau, et ceci ne peut pas être justifié théoriquement du point de vue de la mécanique quantique ; en second lieu, il faut souligner que les forces assurant la cohésion du noyau sont certainement d'autre nature que celles de Coulomb, qui s'introduisent dans l'application de la mécanique des quanta aux électrons extérieurs. En effet les forces de Coulomb donneraient lieu dans le noyau d'hélium à un défaut de masse de l'ordre de grandeur

$$\frac{(2e)^2}{r_0} \sim 4 mc^2 = 0,002 \, M c^2,$$

c'est-à-dire seulement la quinzième partie environ du défaut de masse expérimental. Nous déduisons de là que, dans les atomes légers, les forces de Coulomb n'ont qu'une importance secondaire par rapport aux autres actions nucléaires inconnues, et qu'elles ne peuvent devenir importantes que dans les noyaux lourds, puisqu'elles augmentent comme le carré de la charge du noyau. Nous ne possédons encore aucun moyen théorique pour étudier les forces qui agissent dans le noyau entre les divers constituants lourds; cependant les données expérimentales actuelles permettent d'atteindre quelques conclusions générales sur la nature de ces forces. Nous examinerons ultérieurement ce point de manière plus détaillée.

Quand on s'efforce d'interpréter le fait que certains noyaux atomiques se désintègrent avec émission de rayons β, en admettant que les électrons négatifs figurent au même titre que les particules α et les protons comme constituants indépendants du noyau, on se heurte immédiatement à tout un ensemble de difficultés de principe dont la solution ne paraît pas possible dans l'état actuel de la théorie. On sait que la mécanique quantique traite les électrons comme des charges ponctuelles qui agissent les unes sur les autres conformément aux lois de la théorie de Maxwell. Cette manière de procéder n'est cependant admissible que si les distances entre les charges ponctuelles sont grandes par rapport à $\frac{e^2}{mc^2}$. En effet, la théorie de l'inertie de l'énergie permet d'affirmer que la loi de Coulomb cesse d'être exacte à des distances du centre de l'électron d'un ordre inférieur à $\frac{e^2}{mc^2}$, m étant la masse de l'électron. Ce résultat peut encore s'exprimer en disant que le rayon de l'électron est de l'ordre de $\frac{e^2}{mc^2}$. Comme d'autre part les dimensions linéaires des noyaux atomiques ne sont guère plus grandes que $\frac{e^2}{mc^2}$, il en résulte qu'il ne saurait être question d'appliquer la mécanique quantique au mouvement des électrons à l'intérieur des noyaux. Ainsi les circonstances dans lesquelles se trouvent les électrons nucléaires sont si éloignées du domaine d'application des lois connues jusqu'ici qu'aucune conclusion concernant le mouvement des constituants légers dans les noyaux atomiques ne peut être obtenue ni par la théorie électronique de Lorentz, ni par la mécanique quantique, ni même par application du principe de correspondance. Il en résulte aussi, comme dernière conséquence, que l'affirmation : les électrons figurent comme constituants nucléaires, ne possède aucune signification définie en dépit du fait rappelé plus haut que beaucoup de noyaux émettent des rayons β.

A cette situation correspond le fait expérimental que, partout où intervient le comportement des charges négatives à l'intérieur du noyau, l'expérience conduit à des résultats sans analogie avec les lois déjà connues. Par exemple, alors que la mécanique quantique prévoit la validité de la statistique de Bose pour des systèmes constitués de protons et d'électrons de *charge* totale paire et de moment de quantité de mouvement entier, et celle de la statistique de Fermi lorsque la *charge* totale est impaire et le moment de quantité de mouvement multiple impair de $\frac{1}{2}$, il semble que la statistique et le moment de quantité de mouvement d'un noyau déterminé dépendent en réalité de la parité de sa *masse* [1].

De plus, alors que l'énergie d'une particule projetée au cours d'une désintégration devrait être entièrement définie par la différence des masses du noyau avant et après cette désintégration, il semble qu'aucune relation aussi simple ne soit satisfaite dans les cas d'émission de rayons β primaires [2].

A cette difficulté résultant de l'expérience immédiate s'en ajoutent encore d'autres, lorsqu'on s'efforce d · développer un examen théorique, à l'aide de la mécanique des quanta, du comportement des électrons nucléaires. Il apparaît ainsi tout d'abord impossible au point de vue de la théorie de l'électron de Dirac qu'un électron soit lié à un proton dans un domaine de l'ordre de grandeur $\frac{e^2}{mc^2}$ (*cf.* paradoxe de Klein). En outre, une application des relations d'indétermination au mouvement de l'électron dans le noyau conduit, par analogie avec les équations (1) et (2), à des quantités de mouvement de l'ordre de

$$(3) \qquad \Delta p \sim \frac{\hbar}{r_0} = \frac{\hbar c}{e^2}\, mc.$$

---

[1] *Cf.* S. Goudsmit, *Phys. Rev.*, t. 43, 1933, p. 636, et E. Fermi et E. Segré, *Zeits. f. Phys.*, t. 82, 1933, p. 729.

[2] *Cf.* Bohr, *Convegno di Fisica nucleare* (Rome, 1932).

et à une énergie cinétique moyenne

$$(4) \qquad \overline{E}_{\text{cin.}} \sim \Delta p \times c \sim \frac{\hbar c}{e^2}\, mc^2 \sim 137\, mc^2.$$

Si l'on veut conclure, comme pour le proton, que le défaut de masse doit être du même ordre de grandeur que l'énergie cinétique moyenne, on obtient des valeurs beaucoup trop grandes pour le défaut de masse résultant de la présence d'un électron dans le noyau. D'ailleurs, Bechert [1] a montré que même dans le champ de Coulomb il n'est pas légitime d'évaluer l'énergie totale $\mathscr{E}$ à partir de l'énergie cinétique. De l'équation

$$(5) \qquad \overline{\sum_k (\dot{p}_k q_k + p_k \dot{q}_k)} = \overline{\sum_\kappa \frac{d}{dt}(p_k q_k)} = 0$$

résulte dans le cas du champ de Coulomb

$$+ \overline{E}_{\text{pot.}} + \overline{\sum_k \frac{m_k \dot{q}_k^2}{\sqrt{1-\beta_k^2}}} = \overline{E}_{\text{pot.}} + \overline{E}_{\text{cin.}} + \overline{\sum_s \left(m_s c^2 - m_s c^2 \sqrt{1-\beta_s^2}\right)} = 0$$

et par suite

$$(6) \qquad \mathscr{E} = - \overline{\Sigma\, m_s c^2 \left(1 - \sqrt{1-\beta_s^2}\right)}.$$

Le défaut de masse par électron reste ainsi toujours inférieur à mc^2 dans le cas du champ de Coulomb. Il est naturellement impossible de dire si ce résultat est applicable à la théorie du noyau, puisque, comme on l'a vu plus haut, les actions intérieures au noyau ne suivent certainement pas la loi de Coulomb.

L'affirmation que les lois de la mécanique quantique peuvent être appliquées aux seuls constituants lourds du noyau, à l'exclusion des électrons, comporte quelque réserve du fait que, d'après la théorie quantique elle-même, les protons ne peuvent pas à eux seuls constituer un noyau, puisqu'ils se repoussent suivant la loi de Coulomb, et que c'est grâce à la présence de charges négatives que l'association des protons dans le noyau est rendue possible. En tout cas, une séparation nette des domaines dans lesquels la mécanique quantique est ou non applicable ne peut

---

[1] Je suis très obligé à M. Bechert pour sa communication par lettre du raisonnement qui suit.

s'obtenir que grâce à des hypothèses nettement définies concernant la structure des noyaux atomiques.

§ 2. — Hypothèses sur la structure des noyaux.

Si l'on cherche à se représenter plus en détail la structure des noyaux atomiques, on doit tenir compte tout d'abord de ce fait essentiel que les masses de ces noyaux sont très sensiblement des multiples entiers du quart de la masse atomique de l'hélium, et non des multiples entiers de la masse du proton. Il semble en résulter que les noyaux atomiques sont formés pour la plus grande partie de particules α, ce qui est confirmé par le fait que les noyaux radioactifs peuvent se désintégrer avec émission de rayons α. A la question de savoir quels sont les autres constituants du noyau, on peut répondre de diverses manières entre lesquelles l'expérience ne permet pas encore de trancher de manière décisive.

a. — Le modèle de la « goutte » de Gamow.

Gamow a admis que les particules α, qui agissent les unes sur les autres avec des forces très rapidement décroissantes en fonction de la distance, peuvent s'associer encore des charges négatives (électrons nucléaires); en dehors de ces électrons, des protons libres peuvent encore exister dans le noyau. Les actions entre particules α sont envisagées par Gamow comme analogues aux forces de Van der Waals entre les molécules; il admet ainsi tout d'abord un rayon fini pour la particule α, avec une action attractive rapidement décroissante remplacée, quand la distance augmente, par la répulsion de Coulomb. Le noyau atomique apparaît ainsi comme un système, qui peut être comparé à une gouttelette liquide, dont la cohésion se traduit par l'action de la tension superficielle. Cette conception de Gamow rend bien compte du fait expérimental que la densité moyenne du noyau semble à peu près indépendante des dimensions de celui-ci (les rayons nucléaires augmentent comme la racine cubique des masses atomiques). En outre, elle permet de prévoir, au moins qualitativement, la diminution observée du défaut de masse par particule α lorsque la masse atomique augmente et elle l'attribue à l'influence des forces de Coulomb. Au contraire, l'hypothèse des électrons nucléaires conduit à des difficultés : les forces de Coulomb sur ces électrons ne peuvent pas rendre compte de la grandeur du défaut de masse par électron nucléaire, la valeur expérimentale apparaissant comme trop grande. En outre, le résultat expérimental indiqué plus haut concernant la statistique et le spin des noyaux oblige à admettre que les électrons nucléaires possèdent un moment de quantité de mouvement entier et suivent la statistique de Bose, en contradiction avec les propriétés habituelles des électrons. Enfin, il est difficile de comprendre pourquoi, malgré les actions mutuelles énergiques entre les électrons et les particules α, les énergies de désintégration avec émission α ont des valeurs bien définies, tandis que le spectre des rayons β primaires est nettement continu.

Le modèle de « goutte » de Gamow conduit à attribuer à un noyau composé seulement de particules α une énergie de la forme

$$(7) \qquad \mathscr{E} = - \mathrm{C}.\mathrm{N}_\alpha + \frac{(2e\,\mathrm{N}_\alpha)^2}{r},$$

où N_α représente le nombre des particules α et r le rayon du noyau. Le premier terme à gauche correspond aux forces d'attraction rapidement décroissantes entre les particules α, et le second aux forces de Coulomb. Si l'on écrit :

$$r = \mathrm{R}\sqrt[3]{\mathrm{N}_\alpha},$$

il en résulte

$$(8) \qquad \mathscr{E} = - \mathrm{C}\mathrm{N}_\alpha + 4\frac{e^2}{\mathrm{R}}(\mathrm{N}_\alpha)^{\frac{5}{3}}.$$

La courbe du défaut de masse représentée par l'équation (8) présente un minimum et fait prévoir que les noyaux contenant un grand nombre de particules α doivent se désintégrer spontanément avec émission α. Pour les noyaux qui contiennent des électrons à côté des particules α, Gamow admet des relations analogues à (8), mais qui ne peuvent pas se déduire directement du modèle. De l'hypothèse que la stabilité d'un noyau par rapport à une désintégration β peut être déterminée par le bilan d'énergie du processus de désintégration considérée, et en utilisant les défauts de masse observés par Aston, Gamow obtient le schéma connu de la figure 1 pour le défaut de masse en fonction du nombre des électrons et des particules α. Nous reviendrons

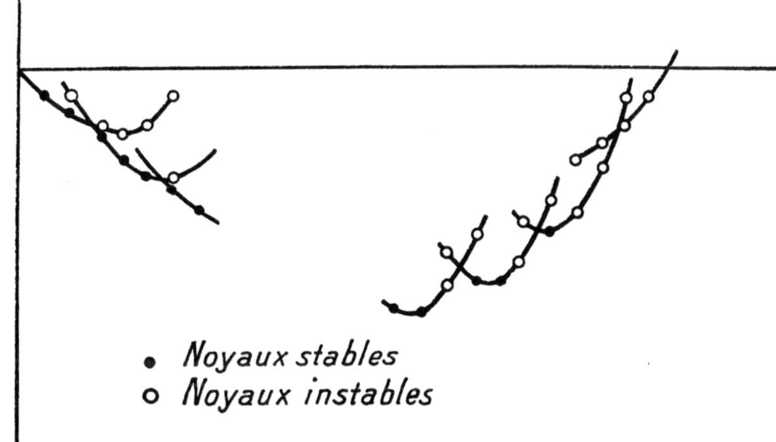

Fig. 1.

ultérieurement sur la légitimité de l'hypothèse concernant la stabilité par rapport à la désintégration β.

Les nombreuses difficultés auxquelles conduit, comme on vient de la voir, l'introduction d'électrons nucléaires rendent nécessaire que nous examinions aussi les autres hypothèses possibles sur la structure des noyaux.

b. — Introduction des neutrons comme constituants nucléaires.

En premier lieu, de nouvelles possibilités résultent de la découverte du neutron par Curie et Joliot [1] et par Chadwick [2]. Cette découverte ne concerne pas seulement l'existence d'un corpuscule de masse 1 et de charge 0, mais montre de manière certaine que ces neutrons peuvent figurer comme constituants indépendants du noyau à côté des protons et des particules α. Les lois expérimentales relatives au spin et à la statistique des

---

[1] I. CURIE et F. JOLIOT, *C. R. Acad. Sc.,* t. **194**, 1932, p. 273, 876.
[2] J. CHADWICK, *Nature*, **129**, 1932, p. 312; *Proc. Roy. Soc.*, **17136**, 1932, p. 692.

noyaux conduisent à admettre que le neutron suit la statistique de Fermi et possède un spin multiple impair de $\frac{1}{2}$. Cela conduit à admettre pour le spin du neutron la valeur $\frac{1}{2}$.

Plusieurs schémas de la structure du noyau sont compatibles avec cette hypothèse et s'obtiennent en introduisant comme constituants nucléaires à côté des particules α des neutrons, des protons et des électrons, ou seulement des neutrons et des protons. Ces schémas ont été discutés très complètement par Perrin [1], Iwanenko [2], Gapon [3], Bartlett [4] et Landé [5]; il nous suffira dans ce rapport de montrer sur quelques noyaux atomiques choisis comme exemples les différences caractéristiques de ces divers schémas. (Nous employons ici les symboles noyau He $= {}^{4}_{2}$He, proton $= {}^{1}_{1}$H, neutron ${}^{1}_{0}$n, électron $= \varepsilon^{-}$.

TABLEAU I.

	1. Gamow.			2. F. Perrin.				3. IwanenkoG-apon.			4.	
	${}^{4}_{2}$He.	${}^{1}_{1}$H,	ε^{-}.	${}^{4}_{2}$He.	${}^{1}_{0}$n.	${}^{1}_{1}$H.	ε^{-}.	${}^{4}_{2}$He.	${}^{1}_{0}$n.	${}^{1}_{1}$H.	${}^{1}_{0}$n.	${}^{1}_{1}$H.
${}^{9}_{4}$Be....	2	1	1	2	1	–	–	2	1	–	5	4
${}^{10}_{5}$B....	2	2	1	2	1	1	–	2	1	1	5	5
${}^{41}_{19}$K...	10	1	2	10	1	–	1	9	4	1	22	19
${}^{208}_{82}$Pb .	52	–	22	52	–	–	22	41	44	–	126	82

La dernière colonne contient une expression différente du contenu de la précédente; on y considère aussi les particules α comme composées de deux neutrons et de deux protons. Le schéma de la colonne 2 considère les neutrons comme des constituants élémentaires non dissociables et, pour rendre compte de la désintégration β des éléments radioactifs, introduit explicitement, à côté des neutrons, des électrons comme constituants

[1] F. PERRIN, *Soc. Franc. d. Physique*, t. 324, 1932, p. 96; *C. R. Acad. Sc.*, t. 194, 1932, p. 343; t. 194, 1932, p. 2211; t. 195, 1932, p. 236.

[2] D. IWANENKO, *Nature*, t. 129, 1932, p. 312.

[3] E. GAPON, *Zeis. f. Phys.*, t. 79, 1932, p. 676; t. 81, 1933, p. 419; t. 82, 1933, p. 404; *E. N. G.* et D. IWANENKO, *Naturw.*, t. 20, 1932, p. 792.

[4] J. BARTLETT, *Phys. Rev.*, t. 41, 1932, p. 370, *cf.* les travaux antérieurs de BARTON, *Phys. Rev.*, t. 35, 1930, p. 408; UREY, *Journ. Am. Chem. Soc.*, t. 53, 1931, p. 2872.

[5] A. LANDÉ, *Phys. Rev.*, t. 43, 1933, p. 620 et 624.

des noyaux. Bien que cette conception conduise à des points de vue intéressants sur la radioactivité β des éléments, — F. Perrin relie par exemple l'activité β de K_{41} à la première apparition d'un électron nucléaire (*cf.* Tableau I) — elle soulève les mêmes objections que le schéma 1 de Gamow en raison de l'introduction d'électron nucléaires libres. Au contraire, les conceptions 3 et 4 interprètent les lois empiriques concernant le spin et la statistique des noyaux en faisant appel aux propriétés simples du neutron, mais se heurtent à des difficultés en ce qui concerne l'activité β; en effet, il faut admettre dans ces conceptions 3 et 4 que le neutron peut, dans des circonstances favorables, se décomposer en un proton et un électron. Il est vrai que, même dans cette hypothèse, il est difficile de donner un sens bien précis à l'affirmation qu'un neutron est composé d'un électron et d'un proton, puisqu'on serait conduit, en l'interprétant au sens littéral, à des conclusions inexactes en ce qui concerne le spin et la statistique du neutron; en outre, le neutron manifeste expérimentalement une stabilité beaucoup plus grande qu'il ne semblerait résulter de son défaut de masse par rapport à la somme d'un proton et d'un électron (ce défaut de masse du neutron est, d'après Chadwick, d'environ 1 à 3 millions d'électron-volts, alors que, dans la désintégration de $^{9}_{4}$Be des neutrons sont chassés du noyau avec des énergies allant jusqu'à 8 millions d'électron-volts. F. Perrin, dans son rapport pour le Congrès de Leningrad, a émis l'hypothèse vraisemblable que l'apparition d'une particule β dans la désintégration β doit être rapprochée de la production d'une paire d'électrons positif et négatif à partir d'une quantum γ (*cf.* le rapport de Joliot) et que, par conséquent, dans des conditions énergétiques favorables, on peut avoir aussi bien décomposition d'un neutron en proton et électron négatif que celle d'un proton en neutron et électron positif. Bien que cette hypothèse reste jusqu'ici sans base théorique précise, elle semble permettre de concilier la stabilité du neutron et du proton avec le fait expérimental de la désintégration β de certains éléments. Si l'on considère ces dernières difficultés comme des conséquences nécessaires de l'impossibilité d'appliquer la mécanique des quanta aux électrons dans le noyau, les schémas 3 et 4 semblent présenter sur 1 et 2 l'avantage de faire apparaître clairement les limites d'applicabilité de cette mécanique. Les schémas 3 et 4 mettent en évidence le fait que les théories actuelles ne permettent pas d'aborder la question des actions mutuelles entre neutrons et protons, ainsi que le problème de l'activité β. D'autre part, si l'on introduit une loi d'action déterminée entre neutrons et protons, la question de la structure du noyau peut être étudiée complètement par application des lois de la mécanique quantique. Bien que cette conception 3-4 soit à peine mieux justifiée expérimentalement que les deux premières, il nous paraît utile d'en développer les conséquences par application de la mécanique quantique.

c. — *Les lois d'action mutuelle.*

L'action mutuelle entre neutrons et protons peut, soit être envisagée comme une force ordinaire, soit, par analogie avec le cas des molécules, être considérée comme une action d'échange; la première hypothèse correspond à l'idée du neutron particule élémentaire indissociable, tandis que la deuxième s'applique de manière naturelle aux schémas 3 et 4. Diverses hypothèses sont d'ailleurs possibles quant à la nature de cette action d'échange. On peut tout d'abord admettre une analogie aussi étroite que possible entre l'action mutuelle proton-neutron et celle qui intervient dans les molécules $H — H^+$. Il s'agit dans ce cas d'une action d'échange où la charge négative passe d'une particule à l'autre sans modification du spin de chacune d'elles. Au contraire, on peut partir des lois expérimentales les plus importantes concernant le noyau et chercher quelle action d'échange permet d'en rendre compte le plus exactement possible. Majorana a montré qu'on aboutit ainsi à un type d'action dans lequel la charge négative et le spin sont échangés simultanément entre les particules. Nous examinerons plus loin l'expression mathématique de cette conception.

Diverses hypothèses sont également possibles en ce qui concerne l'interaction mutuelle des neutrons. Il semble d'ailleurs que cette action entre deux neutrons dans le noyau soit beaucoup plus petite qu'entre un neutron et un proton, de sorte qu'on obtient une approximation raisonnable de la réalité en la négligeant tout d'abord complètement.

En introduisant cette hypothèse et en admettant d'autre part que, dans les noyaux légers, les forces de Coulomb entre protons peuvent également être négligées en première approximation, on obtient les résultats suivants : pour un noyau de masse donnée, la condition la plus favorable à la stabilité énergétique est l'égalité des nombres de protons et de neutrons (ceci résulte de la symétrie du problème par rapport aux neutrons et aux protons). Pour les noyaux lourds, la répulsion électrique des protons déplace la configuration de moindre énergie vers un plus petit nombre de protons et un plus grand nombre de neutrons. Le fait que ces résultats sont en bon accord avec les données expérimentales relatives au noyau apporte, inversement, un argument en faveur de l'hypothèse que l'action mutuelle neutron-neutron est beaucoup plus faible que celle d'un neutron sur un proton.

La représentation mathématique de l'action d'échange peut se développer de deux manières différentes.

1° On peut introduire pour chaque particule nucléaire cinq coordonnées : trois coordonnées de position $\vec{r}_k$, une variable de spin σ_k, et une nouvelle variable ρ_k qui prend la valeur $+1$ ou -1 suivant que la particule est un neutron ou un proton;

2° Chaque particule est caractérisée par quatre variables : $\vec{r}_k$ et σ_k, mais les coordonnées seront désignées de manière différente pour les neutrons et les protons (par exemple $\vec{r}_K, \sigma_K$, et $\vec{r}_k, \sigma_k$).

La fonction de Schrödinger s'écrira ainsi : dans le premier cas

$$\varphi\left(\vec{r}_1, \sigma_1, \rho_1; \vec{r}_2, \sigma_2, \rho_2 \ldots\right)$$

et dans le second

$$\varphi\left(\vec{r}_I, \sigma_I; \vec{r}_{II}, \sigma_{II}; \ldots \vec{r}_1, \sigma_1; \vec{r}_2, \sigma_2 \ldots\right).$$

Les relations entre les deux schémas sont exprimées par les équations

$$(9) \quad \begin{cases} \varphi\left(\vec{r}_1, \sigma_1, +1; \vec{r}_2, \sigma_2, +1; \ldots\right) = \varphi\left(\vec{r}_1, \sigma_1; \vec{r}_{II}, \sigma_{II}; \ldots\right), \\ \varphi\left(\vec{r}_1, \sigma_1, +1; \vec{r}_2, \sigma_2, -1; \ldots\right) = \varphi\left(\vec{r}_1, \sigma_1; \vec{r}_1, \sigma_1; \ldots\right). \\ \varphi\left(\vec{r}_1, \sigma_1, -1; \vec{r}_2, \sigma_2, -1; \ldots\right) = \varphi\left(\vec{r}_1, \sigma_1; \vec{r}_2, \sigma_2; \ldots\right). \\ \cdots\cdots\cdots\cdots\cdots\cdots\cdots\cdots\cdots\cdots\cdots\cdots\cdots\cdots \end{cases}$$

     STRUCTURE ET PROPRIÉTÉS DES NOYAUX ATOMIQUES.

Si l'on introduit les matrices :

$$\rho^{\xi}=\begin{vmatrix} 0 & 1 \\ 1 & 0 \end{vmatrix}; \qquad \rho^{\eta}=\begin{vmatrix} 0 & -i \\ i & 0 \end{vmatrix},$$

le terme d'action mutuelle dans la fonction d'Hamilton sous l'hypothèse d'un simple échange de la charge négative (*fig.* 2)

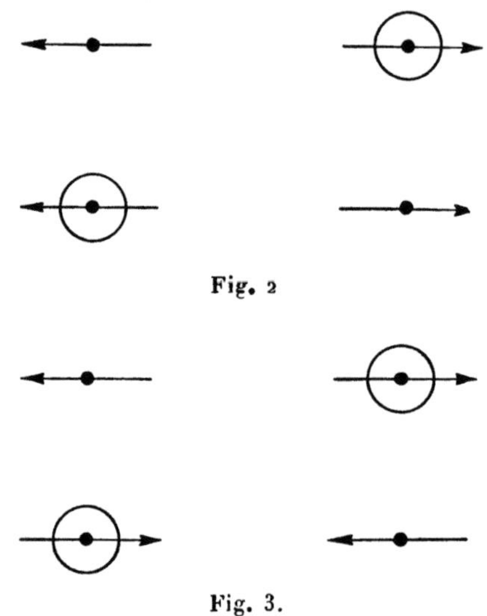

Fig. 2

Fig. 3.

et avec la représentation **1°** devient :

$$(10) \qquad \mathcal{J}(r_{kl})\frac{1}{2}\left[\rho_k^{\xi}\rho_l^{\xi}+\rho_k^{\eta}\rho_l^{\eta}\right],$$

ou, avec la représentation **2°** :

$$(11) \qquad -\mathcal{J}(r_{Kl}).P'_{Kl},$$

où P'_{Kl} représente l'opérateur de permutation des variables r_K, σ_K avec r_l, σ_l.

L'action d'échange introduite par Majorana et dont il sera question plus loin (elle est représentée schématiquement figure 3) conduit au contraire dans la fonction d'Hamilton à des termes qui s'écrivent, dans la représentation **1°** :

$$(12) \qquad \mathcal{J}(r_{kl})\frac{1}{4}\left[\rho_k^{\xi}\rho_l^{\xi}+\rho_k^{\eta}\rho_l^{\eta}\right]\left[1+(\sigma_k\sigma_l)\right],$$

CONSIDÉRATIONS THÉORIQUES GÉNÉRALES SUR LA STRUCTURE DU NOYAU.   3o3

et dans la représentation 2^0

$$(13) \qquad\qquad - \mathfrak{J}(r_{\mathrm{K}l})\mathrm{P}_{\mathrm{K}l},$$

où $\mathrm{P}_{\mathrm{K}l}$ est l'opérateur correspondant à une permutation des coordonnées d'espace r_{K} et r_l.

Pour réduire le nombre des hypothèses possible sur les actions d'échange, Majorana [1] s'appuie sur les faits expérimentaux les plus simples. Un des traits les plus caractéristiques de la structure nucléaire consiste en ce que le rayon du noyau varie sensiblement comme la racine cubique de la masse (*voir* § 2, *a*), c'est-à-dire que la densité de la matière du noyau paraît être sensiblement indépendante de la grandeur de celui-ci. Ce résultat suggère que le noyau n'est pas, comme l'atome lui-même, un système central à l'intérieur duquel un point particulier joue le rôle de centre de force, mais au contraire, présente une analogie avec l'état liquide, ainsi que nous l'avons indiqué à propos du modèle de la goutte de Gamow, où la grosseur de la goutte n'a pas d'influence sur les conditions dans lesquelles se présente l'adjonction d'une nouvelle molécule. La théorie des liquides montre que l'existence d'un rayon moléculaire fini et la présence d'actions mutuelles du type de Van der Waals constituent les bases essentielles d'une représentation de l'état liquide. De la même manière, les propriétés expérimentales du noyau ne peuvent être représentées dans l'hypothèse de forces ordinaires entre protons et neutrons que si l'on introduit un rayon fini pour le neutron, c'est-à-dire une distance minimum au-dessous de laquelle apparaît une force répulsive importante entre le proton et le neutron. Cette conséquence, assez difficilement acceptable, peut être évitée par l'introduction d'une action d'échange entre protons et neutrons. En effet, comme l'a montré la théorie des molécules de London et Heitler, ces actions d'échange donnent lieu à une saturation des liaisons, qui conduit à des résultats analogues à l'hypothèse d'un rayon fini du neutron. Une telle saturation intervient lorsque le signe de $J(r_{\mathrm{K},l})$ dans les équations (10) à (13) est positif. On trouvera plus loin le développement mathématique.

Majorana en tire légitimement la conclusion, — en opposition

---

[1] E. MAJORANA, *Zeits. f. Phys.*, t. 82, 1933, p. 137.

avec une hypothèse antérieure du rapporteur — que, pour
des raisons d'ordre expérimental, le signe positif est le plus vraisemblable pour J $(r_{\mathrm{K},l})$. Un argument de plus en faveur de la conception (12), (13) contre (10), (11) résulte du fait que dans cette
dernière conception le noyau $^2_1\mathrm{H}$ représente déjà un système
fermé, tandis que sous l'action des forces représentées par (12), (13)
cette saturation n'apparaît qu'avec le noyau d'hélium.

Indiquons maintenant la justification mathématique de ces
résultats. En suivant l'exemple de Majorana, nous appliquerons
à un noyau composé d'un grand nombre de particules la méthode
de Thomas-Fermi sous la forme que lui a donnée Dirac. Nous
choisirons avec Majorana la deuxième des représentations mathématiques indiquées à la page 3o1. Dans ces conditions, la fonction
de Schrödinger, pour un noyau composé de n_1 neutrons et de n_2 protons, peut se mettre en première approximation sous la forme

$$
\Phi =
\begin{vmatrix}
\varphi_1\!\left(\vec{r}_{\mathrm{I}}, \sigma_{\mathrm{I}}\right) & \ldots & \varphi_{n_1}\!\left(\vec{r}_{\mathrm{I}}, \sigma_{\mathrm{I}}\right) \\
\varphi_1\!\left(\vec{r}_{\mathrm{II}}, \sigma_{\mathrm{II}}\right) & \ldots & \varphi_{n_1}\!\left(\vec{r}_{\mathrm{II}}, \sigma_{\mathrm{II}}\right) \\
\cdots & \ldots & \cdots \\
\varphi_1\!\left(\vec{r}_{n_1}, \sigma_{n_1}\right) & \ldots & \varphi_{n_1}\!\left(\vec{r}_{n_1}, \sigma_{n_1}\right)
\end{vmatrix}
\begin{vmatrix}
\varphi_1\!\left(\vec{r}_{1}, \sigma_{1}\right) & \ldots & \varphi_{n_2}\!\left(\vec{r}_{1}, \sigma_{1}\right) \\
\cdots & \ldots & \cdots \\
\cdots & \ldots & \cdots \\
\varphi_1\!\left(\vec{r}_{n_2}, \sigma_{n_2}\right) & \ldots & \varphi_{n}\!\left(\vec{r}_{n_2}, \sigma_{n_2}\right)
\end{vmatrix}.
$$

L'énergie potentielle totale correspondant à Φ pour les actions
d'échange (13) devient :

$$
\begin{aligned}
(15)\qquad E_{\mathrm{pot.}} &= -\int \Phi^{\star} \sum_{\mathrm{K},k} \mathcal{J}(r_{\mathrm{K},k})\, \mathrm{P}_{\mathrm{K},k}\,\Phi\, d\omega \\
&= \sum_{\sigma\sigma'} \iint d\vec{r}\, d\vec{r'} \sum_{\mathrm{K}=1}^{n_1} \varphi_{\mathrm{K}}^{\star}\!\left(\vec{r}, \sigma\right) \varphi_{\mathrm{K}}\!\left(\vec{r'}, \sigma\right) \\
&\quad \times \sum_{k=1}^{n_2} \varphi_{k}^{\star}\!\left(\vec{r'}, \sigma'\right) \varphi_{k}\!\left(\vec{r}, \sigma'\right) \mathcal{J}\!\left(\vec{r}-\vec{r'}\right).
\end{aligned}
$$

Si l'on néglige les actions exercées sur le spin des particules et
si l'on écrit pour $\varphi_{\mathrm{K}}\left(\vec{r}, \sigma\right)$ par conséquent $\chi_{\mathrm{K}}(\vec{r})\, f_{\mathrm{K}}(\sigma)$ l'équation (15)
prend la forme

$$
(16)\quad E_{\mathrm{pot.}} = -\iint d\vec{r}\, d\vec{r'} \sum_{\mathrm{K}=1}^{n_1} \chi_{\mathrm{K}}^{\star}\!\left(\vec{r}\right) \chi_{\mathrm{K}}\!\left(\vec{r'}\right) \mathcal{J}\!\left(\left|\vec{r}-\vec{r'}\right|\right) \sum_{k=1}^{n_2} \chi_{k}^{\star}\!\left(\vec{r'}\right) \chi_{k}\!\left(\vec{r}\right).
$$

Les expressions

$$(17) \qquad \begin{cases} \rho_N\!\left(\vec{r}\right) = \sum_{K=1}^{n_1} \chi_K^{*}\chi_K, \\[2ex] \rho_P\!\left(\vec{r}\right) = \sum_{k=1}^{n_2} \chi_k^{*}\chi_k, \end{cases}$$

qui représentent respectivement la densité des neutrons et celle des protons sont remplacées, comme on le sait, dans la méthode de Thomas-Fermi par

$$(18) \qquad \rho_N\!\left(\vec{r}\right) = \frac{2}{h^3}\int_0^{p_N(\vec{r})} d\vec{p}, \qquad \rho_P\!\left(\vec{r}\right) = \frac{2}{h^3}\int_0^{p_P(\vec{r})} d\vec{p},$$

où les limites p_N et p_P sont déterminées par l'énergie potentielle au point r. De manière analogue on écrira avec Dirac [1] :

$$(19) \quad \begin{cases} \displaystyle\sum_{K=1}^{n_1} \chi_K^{*}\!\left(\vec{r}\right)\chi_K\!\left(\vec{r'}\right) = \rho_N\!\left(\vec{r},\vec{r'}\right) = \frac{2}{h^3}\int_0^{p_N\left(\frac{r+r'}{2}\right)} d\vec{p}\, e^{\frac{i}{\hbar}\vec{p}\left(\vec{r'}-\vec{r}\right)}, \\[3ex] \displaystyle\rho_P\!\left(\vec{r},\vec{r'}\right) = \frac{2}{h^3}\int_0^{p_P\left(\frac{r+r'}{2}\right)} d\vec{p}\, e^{\frac{i}{\hbar}\vec{p}\left(\vec{r'}-\vec{r}\right)}, \end{cases}$$

d'où il résulte

$$(20) \quad E_{pot.} = -\iint d\vec{r}\, d\vec{r'}\, \frac{4}{h^6}\int_0^{p_N} d\vec{p}\int_0^{p_P} d\vec{p'}\, e^{\frac{i}{\hbar}\left(\vec{p}-\vec{p'}\right)\left(\vec{r'}-\vec{r}\right)}\, \mathcal{J}(\,|\,\vec{r'}-\vec{r}\,|).$$

Comme d'ailleurs $\rho_N\!\left(\vec{r},\vec{r'}\right)$ et $\rho_P\!\left(\vec{r},\vec{r'}\right)$ ne sont différents de zéro que dans un petit intervalle autour de $\left(\vec{r'}-\vec{r}\right) = 0$, ou lorsque la densité des particules est grande, et comme, en outre, pour la plupart des noyaux $p_N > p_P$, c'est-à-dire $\rho_N > \rho_P$, on peut donner à (20) la forme suivante si $J\left(\left|\vec{r'}-\vec{r}\right|\right)$ varie avec $\left|\vec{r'}-\vec{r}\right|$ régulièrement et plus lentement que $\rho_N\!\left(\vec{r},\vec{r'}\right)$ autour

---

[1] P. A. M. DIRAC, *Proc. Cambr. Phil. Soc.*, t. 26, 1930, p. 376.

INSTITUT SOLVAY (PHYSIQUE).

du point $\left(\vec{r} - \vec{r'}\right) = 0$:

$$(21) \qquad \mathrm{E_{pot.}} \approx - 2 \int d\vec{r}\, \rho\!\left(\vec{r}\right) \mathcal{J}(0) = -2\,n_2\, \mathcal{J}(0).$$

Cette approximation grossière n'est cependant pas suffisante pour la détermination de ρ_N et ρ_P dans un noyau donné. Pour obtenir une meilleure approximation on peut, grâce à la variation rapide de $\rho_N\left(\vec{r},\,\vec{r'}\right)$ avec $\left(\vec{r} - \vec{r'}\right)$, remplacer l'équation (20) par

$$(21^a) \quad \mathrm{E_{pot.}} \approx - \int d\vec{r} \int d\vec{s}\, \frac{4}{h^6} \int_0^{(p_N r)} d\vec{p} \int_0^{p_P(r)} d\vec{p'}\, e^{\frac{i}{\hbar}\left(\vec{p} - \vec{p'}\right)\vec{s}}\, \mathcal{J}(|\,\boldsymbol{s}\,|)$$

$$= - \int d\vec{r}\, f\!\left[\rho_N\!\left(\vec{r}\right),\, \rho_P\!\left(\vec{r}\right)\right].$$

où f est une fonction symétrique en ρ_N et ρ_P, qui s'annule avec ρ_N ou ρ_P et qui tend, pour les grandes densités, vers les limites $2\,\rho_N\, \mathrm{J}(0)$ ou $2\,\rho_P\, \mathrm{J}(0)$ suivant que l'on a $\rho_N < \rho_P$ ou $\rho_N > \rho_P$.

Comme d'autre part l'énergie cinétique, selon Thomas et Fermi, est donnée par

$$\mathrm{E_{cin.}} = \frac{h^2}{\mathrm{M}}\, \frac{4\,\pi}{5}\left(\frac{3}{8\,\pi}\right)^{\frac{5}{3}} \int d\vec{r}\,\Big(\rho_N^{\frac{5}{3}} + \rho_P^{\frac{5}{3}}\Big),$$

il en résulte

$$(22) \qquad \mathrm{E} = \int d\vec{r}\left[\frac{h^2}{\mathrm{M}}\, \frac{4\,\pi}{5}\left(\frac{3}{8\,\pi}\right)^{\frac{5}{3}}\Big(\rho_N^{\frac{5}{3}} + \rho_P^{\frac{5}{3}}\Big) - f(\rho_N,\,\rho_P)\right].$$

On obtient les fonctions $\rho_N\!\left(\vec{r}\right)$ et $\rho_P\!\left(\vec{r}\right)$ en faisant varier ρ_N ou ρ_P dans E sous les conditions

$$(23) \qquad n_1 = \int \rho_N\, d\vec{r}, \qquad n_2 = \int \rho_P\, d\vec{r}.$$

Il en résulte

$$\rho_N\!\left(\vec{r}\right) = \text{const.} = c_1, \qquad \rho_P\!\left(\vec{r}\right) = \text{const.} = c_2.$$

Si V est le volume du noyau on a

$$c_1 = \frac{n_1}{\mathrm{V}}, \qquad c_2 = \frac{n_2}{\mathrm{V}}$$

et par suite

$$(24) \qquad \mathrm{E} = \frac{h^2}{\mathrm{M}}\, \frac{4\,\pi}{5}\left(\frac{3}{8\,\pi}\right)^{\frac{5}{3}}\Big(n_1^{\frac{5}{3}} + n_2^{\frac{5}{3}}\Big)\mathrm{V}^{-\frac{2}{3}} - \mathrm{V} f\!\left(\frac{n_1}{\mathrm{V}},\, \frac{n_2}{\mathrm{V}}\right)$$

et il résulte de $\dfrac{d\mathrm{E}}{d\mathrm{V}} = \mathrm{o}$:

$$(25) \qquad -\frac{h^2}{\mathrm{M}}\frac{8\,\pi}{15}\left(\frac{3}{8\,\pi}\right)^{\frac{5}{3}}\left[\left(\frac{n_1}{\mathrm{V}}\right)^{\frac{5}{3}}+\left(\frac{n_2}{\mathrm{V}}\right)^{\frac{5}{3}}\right]$$

$$-f\left(\frac{n_1}{\mathrm{V}},\frac{n_2}{\mathrm{V}}\right)+\frac{n_1}{\mathrm{V}}\frac{\partial f}{\partial \varrho_{\mathrm{N}}}+\frac{n_2}{\mathrm{V}}\frac{\partial f}{\partial \varrho_{\mathrm{P}}}=\mathrm{o},$$

condition qui détermine le volume du noyau. Si le rapport $\dfrac{n_1}{n_2}$ est maintenu constant, il résulte de (25) que le volume du noyau varie proportionnellement à la masse. Ceci démontre que les actions d'échange introduites par Majorana conduisent, pour la matière du noyau, à des caractères analogues à ceux d'un liquide. Du reste la saturation indiquée plus haut pour les forces de liaison résulte déjà de l'équation (21) d'après laquelle l'action d'échange d'un proton ne peut s'exercer en moyenne que sur deux neutrons.

On voit d'après ce résultat pourquoi un noyau d'hélium présente l'aspect d'un système fermé. En vertu du principe de Pauli, il n'y a place dans un même état que pour deux neutrons ou pour deux protons. Les neutrons dans un état bien défini peuvent se lier à un proton, l'énergie totale de liaison étant simplement proportionnelle à leur nombre. Les neutrons dans d'autres états n'apportent en moyenne aucune contribution à l'énergie de liaison. Si l'on imagine les noyaux construits par apports successifs de neutrons et de protons, un nouveau système fermé doit être constitué chaque fois que deux protons et deux neutrons nouveaux ont été apportés, ce qu'on peut considérer comme correspondant à la formation d'une nouvelle particule d'hélium.

Si l'on avait utilisé les actions d'échange représentées par (10), (11) au lieu de (12), (13), on aurait obtenu d'après Majorana, au lieu de (15), l'expression :

$$\mathrm{E}=-\sum_{\sigma\sigma'}\iint d\vec{r}\,d\vec{r}'\sum_{\mathrm{K}=1}^{n_1}\varphi_{\mathrm{K}}^{\star}(\vec{r},\sigma)\varphi_{\mathrm{K}}(\vec{r}',\sigma')\mathcal{J}\left(\left|\vec{r}-\vec{r}'\right|\right)\sum_{k=1}^{n_2}\varphi_k^{\star}(\vec{r}',\sigma')\varphi_k(\vec{r},\sigma)$$

$$=-\frac{1}{2}\iint d\vec{r}\,d\vec{r}'\sum_{\mathrm{K}=1}^{n_1}\chi_{\mathrm{K}}^{\star}(\vec{r})\;\chi_{\mathrm{K}}(\vec{r}')\;\mathcal{J}\left(\left|\vec{r}-\vec{r}'\right|\right)\sum_{k=1}^{n_2}\chi_k^{\star}(\vec{r}')\;\chi_k(\vec{r}).$$

$$(26)$$

L'énergie moyenne de liaison par proton serait par conséquent moitié moindre qu'avec les actions (12), (13). Pour les forces de liaison (10), (11) l'isotope $^2_1$H représente déjà un système fermé, puisque des neutrons doivent ici coïncider au point de vue de leurs positions et de leurs spins pour que leurs énergies de liaison avec un proton soient additives.

Pour le calcul du défaut de masse correspondant à une loi d'action donnée J (r), la méthode statistique utilisée par Majorana n'est pas la mieux appropriée, puisque la densité moyenne du noyau conserve la même valeur, même pour des valeurs élevées de n_1 et n_2; les erreurs de la méthode statistique restent par conséquent sensiblement les mêmes pour de petites ou pour de grandes valeurs de n_1 et n_2. On pourrait cependant obtenir une meilleure approximation en calculant tout d'abord de manière approximative la fonction propre de la particule α à partir de l'équation de Schrödinger, et en déduisant de là l'action mutuelle moyenne entre les particules α ainsi que l'action entre ces particules et les neutrons ou les protons. On pourrait ensuite appliquer la méthode statistique à un système composé de neutrons et de particules α et réaliser ainsi un progrès sur la méthode décrite jusqu'ici.

Malgré l'insuffisance du procédé qui vient d'être indiqué, on peut essayer d'en déduire, de manière qualitative, la variation du défaut de masse avec J (r), n_1 et n_2. Il peut être utile, en vue d'application ultérieure, de poursuivre les conséquences d'une hypothèse particulière simple sur la forme de J (r) et de calculer la forme correspondante pour la fonction $f(\rho_l, \rho_P)$.

A la forme proposée par Majorana

$$(27) \qquad \mathcal{J}(r) = a\, e^{-br}$$

correspond d'après (21) :

$$(28) \quad f(\rho_N, \rho_P) = a\,\frac{b^3}{6\pi^3}\left\{ 4NP + [1 + 3(N^2 + P^2)]\log\frac{1 + (N - P)^2}{1 + (N + P)^2} \right.$$
$$+ 4(N^3 + P^3)\,\text{arc tang}(N + P)$$
$$\left. - 4(N^3 - P^3)\,\text{arc tang}(N - P) \right\},$$

où l'on a posé

$$N = \frac{1}{b}\sqrt[3]{3\rho_N\pi^2} \qquad \text{et} \qquad P = \frac{1}{b}\sqrt[3]{3\rho_P\pi^2}.$$

Suivant que a est

$$\ll \frac{\hbar^2 b^2}{M} \qquad \text{ou} \qquad \gg \frac{\hbar^2 b^2}{M},$$

on aura, d'après (25),

$$N \quad \text{et} \quad P \ll 1 \quad \text{ou} \quad \gg 1.$$

Pour que la méthode statistique donne des résultats utilisables, il est par conséquent nécessaire de supposer

$$a \gg \frac{\hbar^2 b^2}{M}.$$

Si cette condition est remplie, on peut remplacer approximativement l'expression (28) par

$$(29) \qquad f = a \frac{b^3}{3\pi^3} \Big\{ \pi(N^3 + P^3) - 2(N - P)^2$$
$$- \frac{3}{2}(N^2 + P^2) \log \frac{(N + P)^2}{1 + (N - P)^2}$$
$$- 2(N^3 - P^3) \arctan(N - P) \Big\}.$$

Ces calculs doivent être modifiés dans le cas des noyaux lourds pour tenir compte de la force de Coulomb. A la place de (22) on obtient l'équation

$$(30) \qquad E = \int \vec{dr} \left\{ \frac{h^2}{M} \frac{4\pi}{5} \left(\frac{3}{8\pi} \right)^{\frac{5}{3}} \left(\rho_N^{\frac{5}{3}} + \rho_P^{\frac{5}{3}} \right) - f(\rho_N, \rho_P) \right\}$$
$$+ \frac{1}{2} \int\int \vec{dr}\, \vec{dr'}\, \frac{e^2}{|r - r'|} \rho_P(\vec{r}) \rho_P(\vec{r'}).$$

En faisant varier ρ_N et ρ_P dans E on obtient des équations qui se transforment par la méthode ordinaire de Thomas-Fermi en équations différentielles pour ρ_N et ρ_P. Sous l'influence des forces de Coulomb, les densités ρ_N et ρ_P varient à l'intérieur du noyau, ces forces de Coulomb tendant à accumuler la charge positive sur la surface extérieure. Tant que l'influence de ces forces reste faible, c'est-à-dire pour des noyaux pas trop lourds, nous pouvons traiter leur action comme de petites perturbations et pour le calcul des défauts de masse, introduire les densités non perturbées dans le terme complémentaire de l'équation (30).

Le terme de Coulomb dans l'énergie devient ainsi :

$$(31) \qquad \iint \vec{dr}\, \vec{dr}' \; \frac{e^2}{\left|\vec{r}-\vec{r}'\right|}\, \rho_P\!\left(\vec{r}\right)\rho_P\!\left(\vec{r}'\right) = \frac{3}{5}\,(n_2 e)^2 \left(\frac{3\,V}{4\,\pi}\right)^{-\frac{1}{3}}.$$

Dans le chapitre suivant nous comparerons les formules ainsi obtenues avec les données expérimentales.

d. — *Conséquences générales pour le tableau des isotopes.*

Les hypothèses sur la structure du noyau, qui ont été développées dans les parties *a* à *c* du paragraphe 2, permettent de déduire certaines conclusions générales sur les particularités du tableau des isotopes et sur les lois de la désintégration radioactive.

Les diverses conceptions conduisent toutes à ce résultat que les propriétés du noyau, telles que stabilité, abondance dans la nature, spin, etc., présentent une certaine périodicité pour une variation 4 de la masse et 2 de la charge, ce que l'expérience confirme bien de manière générale. Le modèle de Gamow et le schéma 2, conduisent en particulier à une influence très marquée de la masse sur les propriétés du noyau, suivant que cette masse est de la forme $4\,n$, $4\,n + 1$, $4\,n + 2$ ou $4\,n + 3$. Au contraire, la charge du noyau pour une masse donnée et en particulier le fait que cette charge soit paire ou impaire, ne semble pas jouer un rôle aussi important. Dans la mesure où l'on peut parler des propriétés des électrons nucléaires, l'expérience semble indiquer qu'ils suivent la statistique de Bose; on ne doit par conséquent pas prévoir une périodicité pour la variation 2 de la charge. Si d'autre part on considère le noyau comme composé de neutrons et de protons (schémas 3 et 4), on doit prévoir en premier lieu une périodicité des propriétés nucléaires pour une variation 2 de la charge, puisqu'on a toujours en présence au moins autant de neutrons que de protons et il suffit toujours de deux charges élémentaires nouvelles pour la formation d'une particule α; en second lieu pour une charge donnée une périodicité plutôt moins marquée pour la variation 2 de la masse par application du principe de Pauli au proton et au neutron. Dans le sens de la périodicité prévue par les schémas 3 et 4, on doit prévoir par exemple des analogies entre les noyaux $^{124}_{54}Xe$, $^{126}_{54}Xe$, $^{128}_{54}Xe$, $^{130}_{54}Xe$ et $^{132}_{54}Xe$

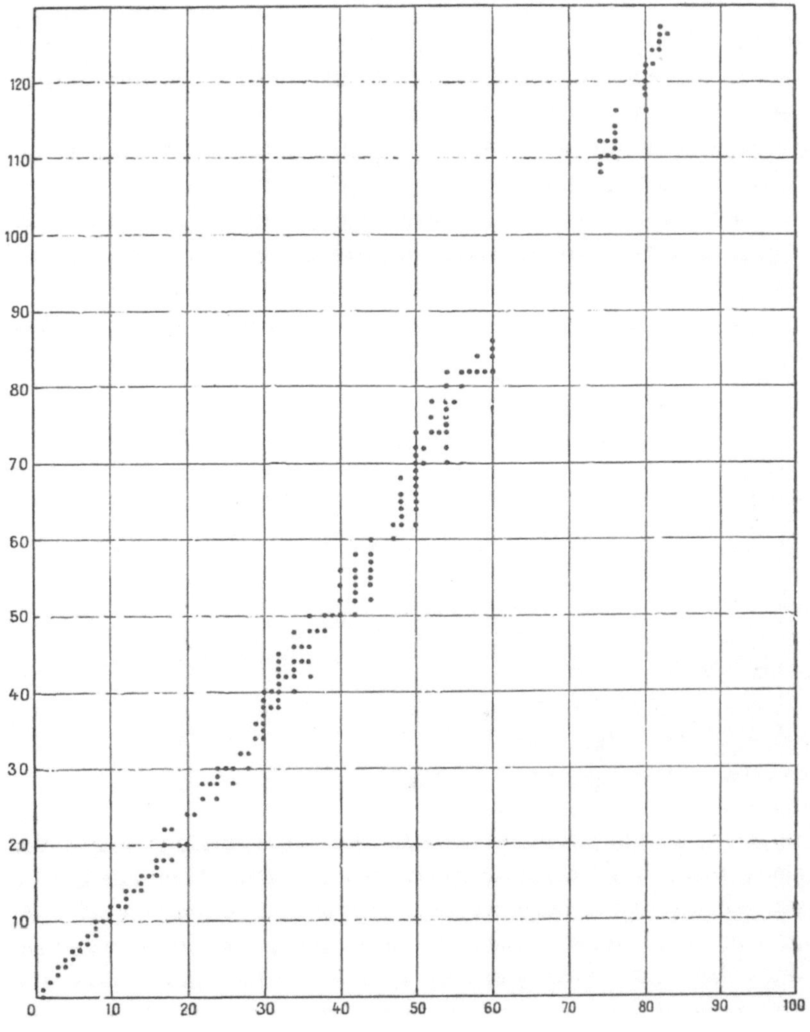

Fig. 4.

tandis que la périodicité qui résulte du schéma de Gamow doit prévoir pour les noyaux $^{124}_{54}Xe$, $^{128}_{54}Xe$ et $^{132}_{54}Xe$ des caractères communs n'appartenant pas aux noyaux $^{126}_{54}Xe$ et $^{130}_{54}Xe$. Dans

      STRUCTURE ET PROPRIÉTÉS DES NOYAUX ATOMIQUES.

la figure 4 les noyaux connus sont représentés en portant en abscisses le nombre de protons n_2 et en ordonnées le nombre des neutrons n_1, de sorte qu'un point correspond à chaque noyau. Le tableau ainsi obtenu indique nettement une prédominance des charges paires et, de manière moins marquée, une prédominance des nombres pairs de neutrons. Cette constatation, bien qu'elle ne soit pas en contradiction avec le schéma de Gamow, apporte un argument en faveur des hypothèses 3 ou 4 puisqu'elle peut être prévue par ces hypothèses.

Dans cette conception, l'on se représentera de la manière suivante la construction du noyau à partir de neutrons et de protons : chaque fois que deux protons et deux neutrons ont été ajoutés, une nouvelle couche se trouve complétée par la formation d'un

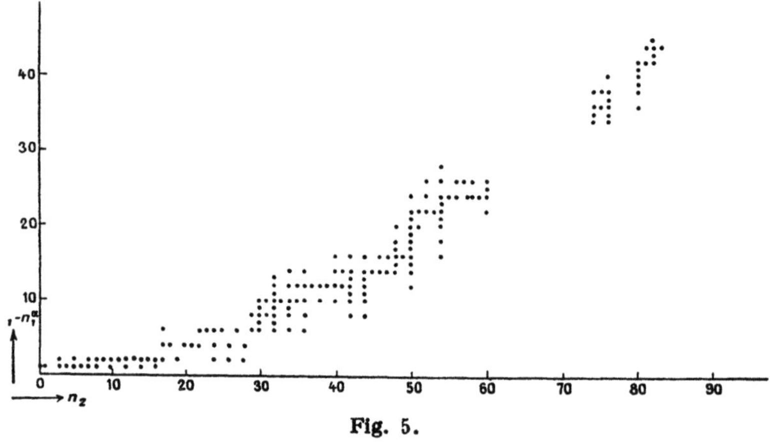

Fig. 5.

noyau d'hélium. Pour les noyaux qui ne contiennent que des particules α, il n'y a lieu d'admettre la formation d'aucune couche au delà de cette construction de noyaux d'hélium; en effet la grande énergie de formation des particules α permet de les considérer dans une certaine mesure comme des éléments constitutifs du noyau, et le fait qu'elles suivent la statistique de Bose a pour conséquence qu'elles se placent toutes au niveau d'énergie le plus bas. Il en est autrement, fait sur lequel Landé [1] a justement insisté, pour les neutrons qui n'entrent pas dans la constitution

---

[1] A. LANDÉ, *Phys. Rev.*, t. 43, 1933, p. 620 et 624.

de particules α. Si l'on admet que la construction du noyau à partir de particules α, de neutrons et éventuellement d'un proton, permet d'obtenir une bonne approximation pour la solution du problème déjà compliqué de mécanique quantique qui concerne les noyaux lourds, on peut comparer le mouvement des neutrons dans le champ d'action des particules α avec celui des électrons périphériques dans le champ qui entoure le noyau de l'atome et par conséquent s'attendre à une structure en couches successives de ce système de neutrons. Landé a représenté dans la figure 5 le nombre expérimental des neutrons indépendants des particules α pour les divers nombres atomiques, et il considère que cette figure conduit immédiatement à l'interprétation suivante : jusqu'à la charge $n_2 = 16$, le niveau le plus bas du système des neutrons est seul occupé et peut constituer une couche de deux neutrons; entre $n_2 = 17$ et $n_2 = 28$ cette couche doit être toujours complète et une couche ultérieure de quatre neutrons peut se compléter; de $n_2 = 32$ jusqu'à $n_2 = 36$ une couche ultérieure de huit se compléterait, ainsi de suite. Si l'on cherche d'autre part à prévoir théoriquement cette formation de couches successives, on doit tout d'abord examiner les états stationnaires possibles dans le champ intérieur au noyau. La répartition du potentiel qui détermine le mouvement d'un neutron à l'intérieur du noyau est connue qualitativement. La densité des particules α doit être à peu près constante à l'intérieur du

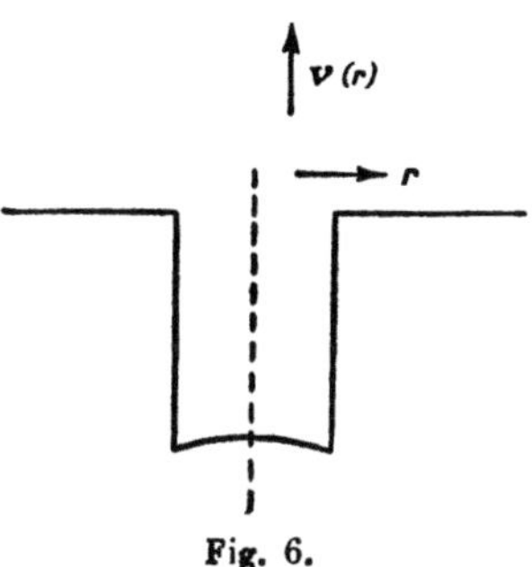

Fig. 6.

noyau avec un léger accroissement vers la surface à cause des actions de Coulomb. Il en résulte pour le neutron une variation de potentiel de la forme représentée dans la figure 6. Les états

     STRUCTURE ET PROPRIÉTÉS DES NOYAUX ATOMIQUES.

de plus faible énergie qui correspondent à ce potentiel représentent, si l'on néglige la réaction du spin sur le moment de circulation, un terme $S(l = 0)$, un terme $P(l = 1)$ et un terme $D(l = 2)$ (*cf.* rapport de Gamow, *fig.* 10). Si l'on fait intervenir l'hypothèse d'une réaction de spin, on obtiendra pour les termes les plus

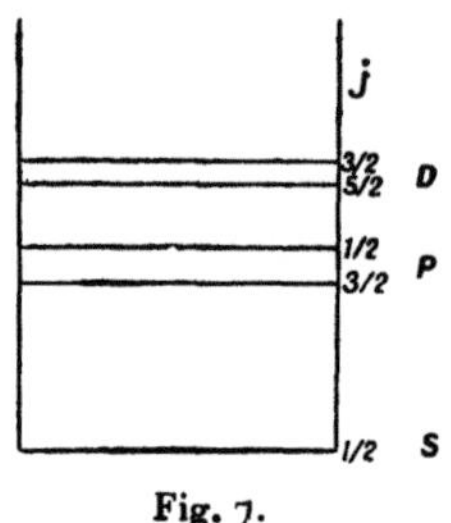

Fig. 7.

bas, un schéma analogue à celui de la figure 7. D'après ce schéma, on obtiendrait tout d'abord une couche de deux, et ensuite une de quatre neutrons $\left(\text{apparition du terme } j = \dfrac{3}{2}\right)$, puis de nouveau une couche de deux, une de six, une de quatre, etc. Il n'est pas encore possible de saisir dans quelle mesure un semblable schéma construit à partir des orbites de neutrons individuels permet de rendre compte des faits expérimentaux concernant le système des isotopes.

La question de la stabilité du noyau au point de vue désintégration α est complètement élucidée par la théorie de Gamow, Condon et Gurney (*cf.* rapport de Gamow); la théorie de Gamow s'applique d'ailleurs de manière analogue aux divers modèles de noyaux. Comme cette théorie fait dépendre la stabilité d'un noyau, au point de vue désintégration α, uniquement du bilan énergétique de ce processus, on peut admettre que, pour une masse donnée du noyau, les forces de Coulomb rendent la désintégration d'autant plus probable que la charge du noyau est plus grande. Le fait qu'il existe expérimentalement, pour une masse donnée, une limite supérieure de la charge placée nettement au-dessous de la moitié de la masse atomique, conduit naturellement à l'hypothèse que des noyaux de forte charge se dissocieraient avec émission de particules α. On constate d'ailleurs que, parmi les noyaux radioactifs de même masse, ce sont en général ceux de plus forte charge qui manifestent la plus grande énergie de désintégration.

Il est difficile actuellement de traiter de manière satisfaisante la question de la stabilité d'un noyau par rapport à la désintégration β. Dans la théorie primitive de Gamow, le bilan énergétique devait déterminer la stabilité au point de vue de l'émission d'un rayon β aussi bien que d'un rayon α. Cette conception ne se justifie pas simplement, puisque c'est un fait expérimental que les particules β émises par le noyau se distribuent suivant un spectre continu d'énergie et qu'il ne semble exister pour les désintégrations β aucune relation entre l'énergie moyenne de désintégration et la durée moyenne de vie, qui soit analogue à celle de Geiger-Nuttall pour les désintégrations α. Cependant, cette question reste ouverte, puisque l'existence d'une limite supérieure nettement définie du spectre β continu représente une difficulté pour toute théorie reposant sur le principe que les particules β sont émises par le noyau avec une énergie indéterminée. Pauli a examiné l'hypothèse que l'émission des rayons β soit toujours accompagnée par celle d'un rayonnement très pénétrant consistant par exemple en un « neutrino » de masse égale à celle de l'électron, ce qui permettrait de maintenir dans le cas des désintégrations β les principes de conservation de l'énergie et de la quantité de mouvement. Bohr ([1]) considère au contraire comme plus vraisemblable une exception au principe de conservation de l'énergie lors de l'émission β par le noyau.

L'aspect théorique de cette question de la désintégration β n'est pas modifié quand on considère le noyau comme constitué par des protons et des neutrons. Dans ce cas, la question de la stabilité par rapport à une désintégration β se ramène à celle de savoir si un neutron peut se dissocier en un proton et un électron lorsqu'il se trouve dans un champ de force énergétiquement favorable à cette décomposition. Pour la solution de ce problème, nous devons, pour les raisons indiquées plus haut, nous contenter des constatations expérimentales. Si le bilan énergétique peut déterminer la stabilité β, il doit exister pour chaque charge nucléaire donnée une limite supérieure de la masse, au delà de laquelle le

---

([1]) N. Bohr, *loc. cit.*

noyau se dissocie en émettant un électron. Il est par conséquent utile d'établir en partant des défauts de masse, dans la mesure où ils sont connus empiriquement ou théoriquement, les bilans énergétiques pour les désintégrations β, et d'en comparer le résultat avec le tableau des noyaux connus comme stables. Nous donnerons le résultat de cette comparaison à propos des applications de la formule de Majorana.

Si l'on considère la désintégration β comme déterminée par le bilan énergétique, la conception 3-4 donne une interprétation satisfaisante du fait que les émissions β à partir du noyau de charge initiale paire vont toujours par couples consécutifs. Si, en effet, les conditions énergétiques sont favorables à l'émission d'une première particule β, une seconde émission, après laquelle une nouvelle particule α peut se former dans le noyau, trouvera des conditions énergétiques au moins aussi favorables ([1]). C'est seulement pour une charge initiale impaire du noyau qu'une seule particule β peut éventuellement être émise (Ac). Le fait qu'il n'existe au-dessus de $^{14}_{7}$N aucun noyau stable avec des nombres impairs de protons et de neutrons peut de la même manière recevoir une interprétation énergétique simple.

§ 3. — APPLICATIONS.

a. — Défaut de masse et stabilité des noyaux.

On obtient des indications très précises au sujet de la variation du défaut de masse en partant des hypothèses sur la structure du noyau qui correspondent aux schémas 3-4, en particulier quand on se place au point de vue de Majorana, qu'on peut considérer comme correspondant à une forme du modèle de goutte de Gamow précisée par l'hypothèse du neutron. On peut essayer de déterminer les deux constantes de la loi de force introduite comme exemple : $J(r) = ae^{-br}$ de manière que les défauts de masse de deux noyaux convenablement choisis soient exactement

---

([1]) HEISENBERG, *Zeits. f. Phys.*, t. 77, 1932, p. 1; t. 78, 1932, p. 156; t. 80, 1933, p. 587.

représentés par l'équation (3o), et voir de quelle manière les défauts de masse calculés pour les autres noyaux par cette même formule se placent par rapport à la courbe expérimentale d'Aston. Dans cette comparaison on doit se souvenir que la méthode de Thomas-Fermi utilisée pour obtenir l'équation (3o) ne peut pas donner de résultats exacts pour les noyaux légers. Cette équation doit cependant représenter de manière qualitative les défauts de masse pour les divers noyaux au-dessus de $n = 20$, par exemple, bien qu'elle doive donner une valeur absolue trop élevée de l'énergie de liaison, puisque cette énergie, par particule, est certainement beaucoup plus faible pour les noyaux très légers que pour les noyaux lourds. En ajoutant un terme constant au second membre de (3o), on doit pouvoir se rapprocher de la réalité.

Si l'on pose par exemple dans $J = ae^{-br}$:

$$a = Mc^2 \times 0,0273 = 4,05 . 10^{-5} \text{ erg},$$

$$b = \frac{Mc}{h} \times 0,165 = 1,25 . 10^{12} \text{ cm}^{-1},$$

on déduit après quelques calculs, à partir des équations (3o) et (31), l'expression approchée :

$$(33) \qquad \frac{E}{Mc^2} = 0,00347\, n_2 - 0,0364\, n_1 + 0,01211\, \frac{n_1^2}{n_2}$$
$$+ n_2^{\frac{3}{2}}\left(3,19 - 0,715\, \frac{n_1}{n_2}\right) . 10^{-4}(+ 0,049).$$

La première ligne représente l'influence des actions d'échange, la deuxième comprend l'action des répulsions de Coulomb entre les protons et le terme additif dont il vient d'être question et qui n'est pas contenu dans l'équation (3o); ce terme est choisi de manière à retrouver le mieux possible les valeurs d'Aston. La figure 8 donne le résultat de la comparaison des valeurs ainsi calculées avec les mesures d'Aston, la masse du neutron ayant été, pour cette comparaison, supposée égale à celle du proton.

On ne doit pas attribuer une trop grande signification à l'accord relativement bon qui résulte de cette comparaison puisque, pour l'établissement de la formule (33), on a disposé de trois constantes arbitraires et que, d'autre part, comme y a insisté Majorana et comme nous l'avons indiqué plus haut, l'application de la méthode de Thomas-Fermi peut comporter des erreurs importantes, même pour des nombres de particules assez élevés. On peut cependant conclure que la formule (3o) donne, de manière qualitative au

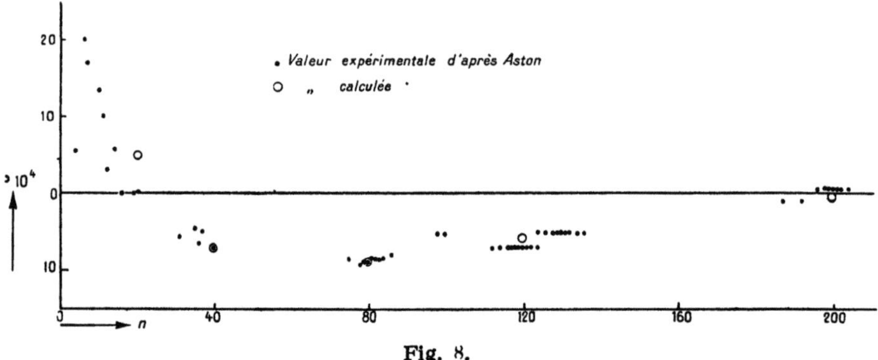

Fig. 8.

moins, la variation correcte de l'énergie avec n_1 et n_2. Il faut attacher une importance plus grande aux conséquences qui peuvent être déduites des équations (3o) et (33) concernant la stabilité des noyaux atomiques. L'énergie nécessaire pour extraire du noyau deux protons et deux neutrons est donnée d'après (33), en fractions de l'énergie intrinsèque du proton, par

$$(34) \quad -2\left(\frac{\partial E}{\partial n_1} + \frac{\partial E}{\partial n_2}\right) = 0,0658 - 0,04844\,\frac{n_1}{n_2} + 0,02422\left(\frac{n_1}{n_2}\right)^2$$
$$+ \left(9,2 - 0,954\,\frac{n_1}{n_2}\right) n_2^{\frac{2}{3}}.10^{-4}.$$

Si cette énergie est plus grande que l'énergie de liaison du noyau d'hélium, le noyau considéré est stable par rapport à une désintégration; si au contraire elle est plus petite, il peut émettre spontanément une particule α.

Dans la figure 9 on a tracé les courbes d'énergie de désintégration constante en partant de la valeur expérimentale du défaut de masse de la particule α et en admettant pour le neutron un défaut de masse négligeable à partir du proton et de l'électron.

Une de ces courbes, celle qui correspond à $\Delta E = 0$, doit représenter la limite entre les noyaux atomiques stables et instables. Si l'on porte sur la figure les points correspondant aux noyaux atomiques expérimentalement stables, on constate que la limite inférieure de ces points concorde bien de manière générale avec la courbe d'énergie de désintégration nulle; la courbe $\Delta E = -0,003$ conviendrait mieux que $\Delta E = 0$, mais étant données les approximations arbitrairement faites (masse du neutron égale à celle du proton, etc.), ce résultat ne peut pas être invoqué contre la validité de la formule (30).

On peut calculer de la même manière la différence d'énergie entre les états initial et final d'un processus β de désintégration :

$$(35) \qquad \frac{\partial E}{\partial n_1} - \frac{\partial E}{\partial n_2} = -0,0399 + 0,02422\,\frac{n_1}{n_2} + 0,01211\left(\frac{n_1}{n_2}\right)^2$$
$$-\left(6,04 - 0,477\,\frac{n_1}{n_2}\right)n_2^{\frac{2}{3}}.10^{-4}.$$

Les courbes d'énergie de désintégration constante (*fig.* 9) se trouvent ici à peu près parallèles à la limite supérieure des

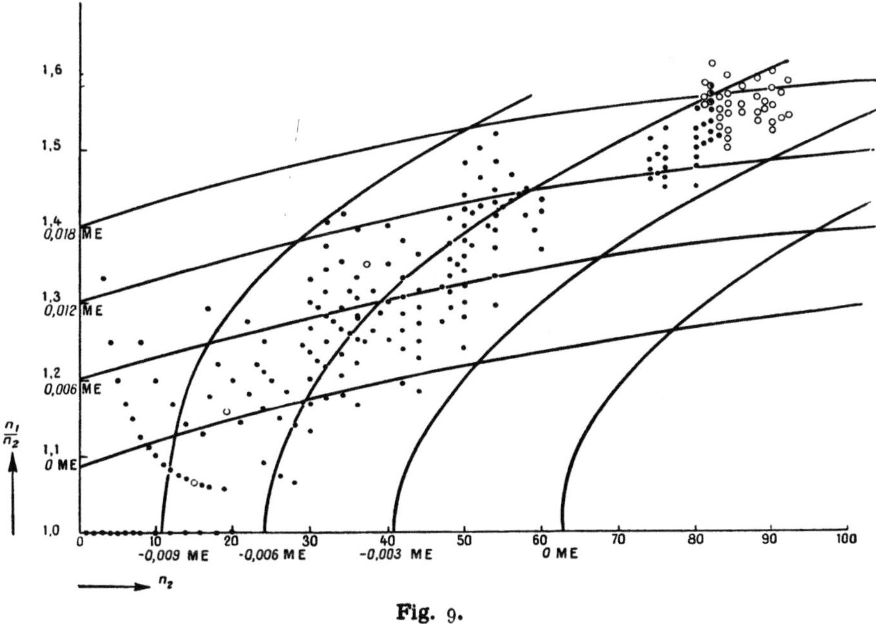

Fig. 9.

points qui correspondent aux noyaux atomiques stables, et ceci représente un argument en faveur d'un criterium énergétique de stabilité au point de vue d'une émission β. La courbe $\Delta E = 0$ se trouve encore ici sensiblement trop bas. On doit d'ailleurs remarquer que, aussi bien pour les désintégrations α que pour les β, la courbe $\Delta E = 0$ ne doit pas représenter exactement la limite des noyaux atomiques expérimentalement instables. En effet, il résulte de la théorie de Gamow pour la désintégration, que les noyaux atomiques lourds, même lorsqu'ils peuvent émettre

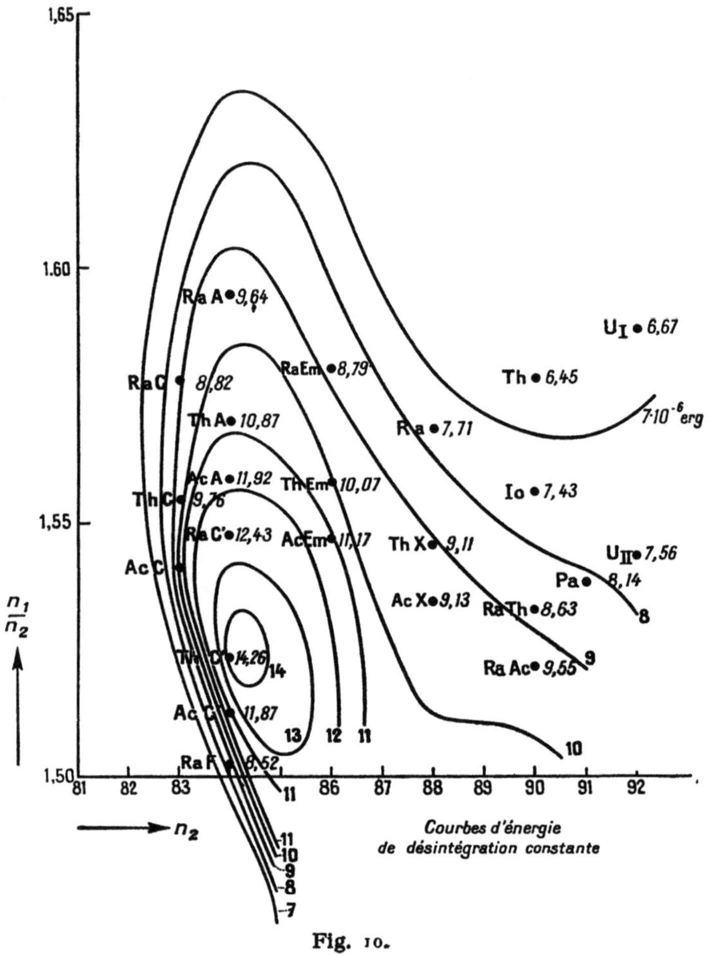

Fig. 10.

une particule α avec une énergie cinétique de 5.10^{-6} erg se comporteront comme pratiquement stables en raison de leur longue vie moyenne; des circonstances analogues se présentent probablement dans le cas des émissions β ([1]).

([1]) *Voir* la Note à la fin de ce rapport.

CONSIDÉRATIONS THÉORIQUES GÉNÉRALES SUR LA STRUCTURE DU NOYAU. 321

La figure 9 montre clairement pourquoi il n'existe plus de noyau stable au-dessus d'une certaine charge nucléaire déterminée par l'intersection des deux courbes de stabilité. On doit d'ailleurs rappeler ici qu'une méthode de calcul aussi approximative que celle de Thomas-Fermi ne peut pas rendre compte des particularités présentées par les familles radioactives. Pour mettre ce fait en évidence, on a tracé dans la figure 10 les courbes d'énergie de désintégration α constante dans le domaine de ces séries radioactives, à partir des données expérimentales. On constate immédiatement certaines particularités (maximum de l'énergie de désintégration pour ThC′) dont ne peut pas rendre compte une théorie sommaire de la structure du noyau. C'est seulement l'allure générale de la variation d'énergie que peut donner la méthode de Thomas-Fermi.

Nous ne disposons jusqu'ici que d'un très petit nombre de données théoriques sur le défaut de masse des noyaux légers. Wigner ([1]) a examiné dans quelle mesure l'hypothèse d'une action ordinaire (pas une action d'échange) entre le neutron et le proton peut rendre compte du fait que le défaut de masse de l'isotope $^2_1$H est quinze fois plus petit que celui du noyau d'hélium. Il trouve que, pour une certaine forme de la loi d'action, l'énergie du noyau $^2_1$H est notablement plus faible en valeur absolue que l'énergie potentielle moyenne ou que l'énergie cinétique moyenne des particules dans ce noyau. S'il en est ainsi, le rapport des défauts de masse de $^4_2$He et de $^2_1$H devient très grand. Les résultats de Wigner peuvent s'étendre immédiatement au modèle de Majorana.

b. — Diffusion et désintégration.

L'hypothèse que les noyaux sont composés de neutrons et de protons et qu'une loi simple d'action existe entre ces deux sortes de particules permet aussi des applications faciles à la théorie des chocs entre noyaux et particules α, protons et neutrons. En particulier, on peut déduire facilement de la loi d'action entre protons et neutrons celle de la diffusion des neutrons par des protons; inversement les résultats expérimentaux concernant cette diffusion permettent d'atteindre directement la loi d'action

---

([1]) E. WIGNER, *Phys. Rev.*, t. 43, 1933, p. 252.

supposée inconnue. Malheureusement le matériel expérimental actuellement disponible ne permet pas une détermination précise de cette loi d'action. Comme la théorie de la diffusion et de la désintégration est traitée dans le rapport de M. Gamow, il ne sera examiné ici que certains points concernant plus particulièrement l'hypothèse du neutron.

Iwanenko [1] a interprété de la manière suivante, en s'appuyant sur l'hypothèse du neutron, le fait que la désintégration des noyaux du béryllium sous l'action des rayons α n'émet que des neutrons et des rayons γ sans aucun proton : le noyau de béryllium de masse 9 se compose, dans l'hypothèse admise ici, de deux particules α et d'un neutron; tant que l'énergie de la particule α incidente ne suffit pas pour désintégrer les particules α elles-mêmes, le noyau ne peut émettre que des neutrons. Si l'on poursuit les conséquences de cette conception, l'on prévoit également une émission de neutrons dans le bombardement du carbone par les particules α ($^{12}_{6}C$ ne peut pas être désintégré, mais les mêmes considérations que pour $^{9}_{4}Be$ s'appliquent à l'isotope rare $^{13}_{6}C$); on prévoit au contraire l'émission d'un mélange de neutrons et de protons avec Li, B, N et O. Si l'on utilise avec Cockroft et Walton [2] des protons pour produire les désintégrations, le rayonnement émis peut se composer de particules α et de neutrons. On doit en particulier prévoir l'apparition de neutrons avec des noyaux plus lourds ($^{37}_{17}Cl$ par exemple) qui contiennent plusieurs neutrons à côté des particules α.

Un résultat particulièrement significatif des expériences de Cockroft et Walton [3] montre que les protons peuvent également produire la désintégration des noyaux lourds (Co, Pb, U) quoique avec un très faible rendement. Ce résultat paraît incompréhensible si l'on considère une pénétration du proton dans le noyau comme condition préalable nécessaire à la désintégration. En fait,

---

[1] IWANENKO, *loc. cit.*

[2] COCKROFT et WALTON, *Proc. Roy. Soc.*, A, t. 137, 1932, p. 229.

[3] Les recherches récentes de Rutherford et Oliphant semblent montrer que les résultats de Cockroft et Walton sur la désintégration d'éléments très lourds sont dus en réalité à des effets secondaires (*cf.* rapport de Cockroft). Les considérations de Teller auxquelles il vient d'être fait allusion peuvent néanmoins conserver leur importance pour la discussion des désintégrations d'éléments plus légers.

la probabilité pour qu'un proton d'énergie 500 000 électron-volts puisse pénétrer dans un noyau d'uranium est infiniment faible. Teller ([1]) a fait remarquer à ce sujet que l'hypothèse d'action d'échange entre neutrons et protons rend possibles des processus dans lesquels un échange a lieu entre le proton incident et un neutron du noyau rencontré. Le neutron ainsi produit a ensuite la possibilité de pénétrer dans le noyau sans avoir à franchir la barrière de Gamow due à l'action des forces de Coulomb. Il en résulte que la diminution de la probabilité de désintégration, quand le nombre atomique augmente, n'est pas représenté de manière générale par le facteur exponentiel de Gamow, et que la partie correspondant à l'action d'échange diminue à peu près comme $|\,\mathrm{J}\,(r_{\mathrm{min.}})\,|^2$, où $r_{\mathrm{min.}}$ représente la distance jusqu'à laquelle le proton peut s'approcher du noyau selon la théorie classique :

$$\frac{\mathrm{Z}\,e^2}{r_{\mathrm{min.}}} = \mathrm{E}_{\mathrm{cin.}}$$

Si donc $\mathrm{J}\,(r)$ ne diminue pas trop rapidement quand le rayon augmente, la diminution de probabilité peut être beaucoup plus faible que ne l'indique le facteur de Gamow. Les données expérimentales correspondant à $\mathrm{J} = ae^{-br}$, $b = 1,25\cdot10^{12}\,\mathrm{cm}^{-1}$, conduisent à une diminution relativement lente de J lorsque r augmente. Il reste d'ailleurs douteux que ces considérations suffisent pour interpréter les résultats de Cockroft et Walton sur les noyaux lourds.

NOTE AJOUTÉE SUR LES ÉPREUVES (le 30-6-1934). — La découverte de la radioactivité avec émission de positrons (Joliot-Curie) oblige à modifier un peu les considérations précédentes (p. 320). A l'approximation grossière, qui est à la base des équations (30) à (35), les seuls noyaux stables devraient être en fait ceux pour lesquels $\left|\dfrac{\partial\mathrm{E}}{\partial n_1} - \dfrac{\partial\mathrm{E}}{\partial n_2}\right| \leqq mc^2$. Les noyaux stables devraient se distribuer dans la figure 9 symétriquement de part et d'autre de la courbe $\dfrac{\partial\mathrm{E}}{\partial n_1} - \dfrac{\partial\mathrm{E}}{\partial n_2} = 0$. La largeur de la bande des éléments stables de part et d'autre de cette courbe ne sera cependant pas donnée par la condition $\dfrac{\partial\mathrm{E}}{\partial n_1} - \dfrac{\partial\mathrm{E}}{\partial n_2} \leqq mc^2$; au contraire, la différence énergétique entre noyaux de charge paire et impaire a pour effet de donner à cette bande, dans le cas des noyaux de masse paire, une largeur notablement plus grande que ne le ferait prévoir la condition précédente. On trouvera des détails sur ce sujet dans W. HEISENBERG, *loc. cit.*, II.

([1]) Je suis très obligé à M. Teller pour une intéressante discussion sur ces questions.

DISCUSSION DU RAPPORT DE M. HEISENBERG.

M. Pauli. — La difficulté provenant de l'existence du spectre continu des rayons β consiste, comme on sait, en ce que la durée moyenne de vie des noyaux qui émettent ces rayons, comme celle des noyaux des corps radioactifs qui en résultent, possède des valeurs bien déterminées. On en conclut nécessairement que l'état, ainsi que l'énergie et la masse du noyau qui reste après l'expulsion de la particule β sont aussi bien déterminés. Je n'insiste pas sur les efforts qu'on pourrait tenter pour échapper à cette conclusion, mais je crois, conformément à l'opinion de M. Bohr, qu'on se heurtera toujours à des difficultés insurmontables dans l'explication des faits expérimentaux.

Dans cet ordre d'idées, deux interprétations des expériences se présentent. Celle que défend M. Bohr admet que les lois de la conservation de l'énergie et de l'impulsion sont en défaut quand il s'agit d'un processus nucléaire où des particules légères jouent un rôle essentiel. Cette hypothèse ne me paraît pas satisfaisante, ni même plausible. D'abord, la charge électrique est conservée dans le processus, et je ne vois pas pourquoi la conservation de la charge serait plus fondamentale que celle de l'énergie et de l'impulsion. Ensuite, ce sont précisément des relations énergétiques qui règlent plusieurs propriétés caractéristiques des spectres β (existence d'une limite supérieure et rapport avec les spectres γ, critère de stabilité de Heisenberg). Si les lois de conservation n'étaient pas valables, il faudrait bien conclure de ces relations qu'une désintégration β est toujours accompagnée d'une perte d'énergie et jamais d'un gain; cette conclusion implique une irréversibilité des processus à l'égard du temps, qui ne me paraît guère acceptable.

En juin 1931, à l'occasion d'une conférence à Pasadena, j'ai proposé l'interprétation suivante : les lois de conservation restent valables, l'expulsion des particules β étant accompagnée d'une radiation très pénétrante de particules neutres, qui n'a pas été observée jusqu'ici. La somme des énergies de la particule β et de la particule neutre (ou des particules neutres, puisqu'on ne sait pas s'il n'y en a qu'une ou s'il y en a plusieurs) émises par le noyau dans un seul processus, sera égale à l'énergie qui correspond à la limite supérieure du spectre β. Il va sans dire que nous n'admettons pas seulement la conservation de l'énergie, mais aussi celle de l'impulsion, celle de l'impulsion angulaire et celle du caractère de la statistique dans tous les processus élémentaires.

Quant aux propriétés de ces particules neutres, les poids atomiques des éléments radioactifs nous apprennent tout d'abord que leur masse ne peut pas dépasser beaucoup celle de l'électron. Pour les distinguer des neutrons lourds, M. Fermi a proposé le nom « neutrino ». Il est possible que la masse propre des neutrinos soit égale à zéro, de sorte qu'ils devraient se propager avec la vitesse de la lumière, comme les photons. Toutefois leur pouvoir pénétrant dépasserait de beaucoup celui des photons de même énergie. Il me paraît admissible que les neutrinos possèdent un spin $\frac{1}{2}$ et qu'ils obéissent à la statistique de Fermi, bien que les expériences ne nous fournissent aucune preuve directe de cette hypothèse. Nous ne savons rien de l'interaction des neutrinos avec les autres particules matérielles et avec les photons : l'hypothèse qu'ils possèdent un moment magnétique, comme je l'avais proposé autrefois (la théorie de Dirac conduit à prévoir la possibilité de l'existence de particules neutres magnétiques), ne me paraît du tout fondée.

Dans cet ordre d'idées, l'étude expérimentale du bilan de l'impulsion dans les désintégrations β constitue un problème de la plus haute importance; on peut prévoir que les difficultés seront très grandes à cause de la petitesse de l'énergie du noyau de recul.

M. CHADWICK. — Je voudrais faire mention d'une expérience que M. Lea et moi nous avons faite dans le but de voir s'il y a production de neutrinos dans les désintégrations radioactives. Nous avons examiné la désintégration du radium E, qui émet des particules β avec des énergies variant depuis une valeur très faible jusqu'à un million d'électron-volts environ. Si cette variation doit être compensée par l'émission de neutrinos, l'énergie moyenne du neutrino doit être d'environ 600 000 électron-volts (en admettant l'émission d'un neutrino par désintégration). Ce rayonnement de neutrinos, s'il existe, doit être très pénétrant, puisque l'énergie moyenne par désintégration, mesurée calorimétriquement, concorde avec la valeur moyenne déduite de la distribution de l'énergie des rayons β. M. Lea et moi nous avons employé une source de 100 millicuries de radium $(D + E)$ et nous avons cherché un rayonnement plus pénétrant que les rayons γ du radium E, en employant une chambre à haute pression pour mesurer l'ionisation. Nous n'avons pu trouver aucune preuve de l'existence d'un tel rayonnement. De nos résultats nous déduisons que le neutrino ne peut produire plus d'une paire d'ions sur un trajet de 150^{km} dans l'air, en admettant l'émission d'un neutrino pour chaque atome de radium E désintégré. On peut, en s'appuyant sur les calculs de Carlson et Oppenheimer, évaluer le moment magnétique du neutrino. Donnant au neutrino une masse égale à celle d'un électron, on trouve que le moment magnétique ne peut pas être plus grand qu'un millième à peu près du magnéton de Bohr.

Il est certain que le neutrino, s'il existe, sera excessivement difficile à déceler.

Mlle MEITNER. — Pour décider si le rayonnement β est accompagné d'une émission de neutrinos, M. Frisch a proposé une expérience basée sur l'observation d'un rayon β de recul. On produit par la méthode de Stern un faisceau de rayons atomiques de Ra E qui passe à côté d'un compteur à pointe. Lorsqu'un rayon β pénètre dans le compteur, le rayon de recul doit pénétrer, en vertu de la conservation de l'impulsion, dans un second compteur situé en face du premier, pour autant qu'il y ait émission de rayons β seulement et non de neutrinos. Pour donner au rayon β de recul l'énergie nécessaire pour traverser la fenêtre du compteur, on l'accélère par un champ électrique de 30 000 volts. Donc, s'il n'y a pas de neutrinos, on doit observer des coïncidences; dans le cas contraire, où il y a émission simultanée de neutrinos, les coïncidences doivent, en général, faire défaut, parce que le rayon β de recul n'est plus lancé dans la direction opposée au rayon β primitif.

M. F. PERRIN. — La forme de la distribution d'intensité dans les spectres continus β permet de faire sur la masse du neutrino une hypothèse raisonnable. On peut en effet, au lieu d'expliquer la position du maximum par l'émission simultanée de l'électron et de plusieurs neutrinos de même masse que l'électron comme l'a proposé M. Mott, l'expliquer en admettant qu'un seul neutrino est émis à chaque désintégration, mais en supposant que la masse du neutrino est inférieure à celle de l'électron, et que ces deux particules partent en moyenne avec des *impulsions* égales. Cette égalité d'impulsion serait toujours réalisée si l'électron et le neutrino se séparaient sans interaction avec d'autres corps, et l'on peut penser qu'elle se retrouvera en moyenne à peu près en présence du reste du noyau. Dans ces conditions, l'électron, particule la plus massive, partira en moyenne avec une énergie moindre que le neutrino. En dynamique classique le partage de l'énergie, pour des impulsions égales, se ferait proportionnellement à l'inverse des masses et l'on obtiendrait un maximum d'intensité dans le spectre β pour le tiers de l'énergie limite si le neutrino avait une masse moitié de celle de l'électron. Mais il faut évidemment appliquer à ces particules la dynamique relativiste. On trouve alors ([1]) que, pour rendre approximativement compte de la position du maximum ou plutôt de la valeur moyenne de l'énergie de l'électron, dans le cas le mieux connu (radium E), il faut admettre que le neutrino a une *masse intrinsèque nulle*, comme le photon.

M. BOHR. — En discutant le problème des rayons β, MM. Heisenberg et Pauli s'efforcent d'appliquer jusqu'au bout les lois connues de la théorie des quanta, et je suis parfaitement d'accord avec cette tendance.

Dans le même ordre d'idées, il n'est pas inutile, peut-être, de remarquer que tout effort pour établir une relation entre le

---

([1]) Remarque ajoutée sur les épreuves (*cf.* F. PERRIN, *Comptes rendus*, **197**, 1933, p. 1625).

      STRUCTURE ET PROPRIÉTÉS DES NOYAUX ATOMIQUES.

spectre continu des rayons β et le principe d'incertitude de Heisenberg est basé sur un malentendu. En effet, ce principe exige un défaut de précision en ce qui concerne la valeur de l'énergie d'une particule, si l'expérience nous permet de constater la présence de cette particule à un instant bien déterminé; ce manque de précision est la conséquence d'une interaction incontrôlable de la particule et de l'appareil qui nous permet de déterminer cet instant. Dans ces conditions, la loi de conservation de l'énergie n'est pas violée; mais elle échappe au contrôle de l'expérience. Or, dans le cas de l'expulsion d'une particule β, la situation est toute différente; on mesure son énergie d'une façon bien définie, et la question du bilan de l'énergie s'impose nécessairement.

Étant données les difficultés théoriques que soulève la solution de ce problème, c'est peut-être une question de goût que de savoir quel point de vue l'on préfère. Aussi longtemps que nous n'aurons pas de nouvelles données expérimentales, il est sage de ne pas abandonner les lois de conservation, mais, d'autre part, personne ne sait quelles surprises nous attendent encore.

M. DIRAC. — Je pense qu'il n'y a aucune preuve décisive contraire à l'hypothèse que des électrons possédant un spin et auxquels la statistique de Fermi serait applicable, pourraient entrer comme particules élémentaires dans la constitution de certains noyaux. Si nous considérons les protons et les neutrons comme des particules élémentaires, nous aurions ainsi trois espèces de particules élémentaires à partir desquelles les noyaux seraient formés. Ce nombre peut paraître grand, mais, de ce point de vue, deux est déjà un grand nombre.

Dans la plupart des noyaux, il y aurait un nombre pair d'électrons, de sorte que leur influence sur le spin et la statistique, passerait inaperçue. Mais quelques noyaux auraient un nombre impair d'électrons. Si, par exemple, nous admettons que des noyaux comme $^{37}_{17}\mathrm{Cl}$ et $^{41}_{19}\mathrm{K}$ contiennent un électron, nous pouvons former une particule α de plus que si nous les constituons uniquement de protons et neutrons et, par conséquent, nous augmentons leur stabilité. Ces noyaux obéiraient alors à la statistique de Bose et auraient un spin entier, alors que d'après M. Heisenberg ce serait exactement le contraire.

M. Heisenberg. — M. Dirac préfère considérer les noyaux de $^{37}_{17}$Cl et $^{41}_{19}$K comme composés, en dehors d'un certain nombre de particules α, d'un neutron et d'un électron; en conséquence, il prévoit pour ces noyaux un nombre quantique de spin entier et une statistique du type Bose-Einstein. Je me demande si ses arguments ne s'appliqueraient pas tout aussi bien au cas du $^{87}_{37}$Rb, et n'exigeraient pas la présence dans ce noyau, à côté des particules α, de onze neutrons et d'un seul électron. Or, ce serait en contradiction avec la valeur du spin de ce noyau, pour laquelle on a trouvé $\frac{3}{2}$. D'ailleurs je ne vois pas pourquoi M. Dirac admet la présence d'un seul électron dans un noyau comme $^{41}_{19}$K; pourquoi ne pourrait-on pas se figurer que $^{41}_{19}$K contient dix particules α, un proton et deux électrons ?

M. Bohr. — La question de la radioactivité du K et du Rb est loin d'être éclaircie. A cause de la longue durée de vie de ces éléments, les idées de M. Beck ne sont guère applicables ici. M. Jacobsen, à Copenhague, est en train de chercher à l'aide de la méthode des coïncidences, si dans ces corps deux particules ne sont pas parfois émises en même temps. Reste aussi la question de savoir si le K et le Rb n'émettent pas de particules positives.

Mlle Meitner. — M. Stefan Meyer a constaté, par la méthode de la déviation dans un champ magnétique, la présence d'un rayonnement négatif. Mais la présence de particules positives n'est pas exclue pour cela.

M. Chadwick. — Il y a quelques mois M. Occhialini a examiné s'il y avait une émission d'électrons positifs par le potassium, mais il n'en trouva pas la moindre indication.

M. Bohr. — M. Hevesy, qui a effectué une séparation partielle des isotopes du potassium et a pu conclure que la radioactivité est due à l'isotope lourd, m'a dit que les expériences sont tellement difficiles, qu'on ne pourrait affirmer avec certitude que c'est l'isotope ^{41}K, et non pas un isotope ^{42}K ou ^{43}K qui est radioactif.

M. HEISENBERG. — Que ce soit un pareil isotope ou non qui soit la cause de l'activité du potassium, ma théorie prévoit en tout cas l'expulsion d'électrons négatifs et non pas de positrons.

M. CHADWICK. — Je ne pense pas que le spin de l'isotope rare du potassium puisse être déduit de l'observation du spectre de bande du potassium ordinaire. M. Oliphant a projeté au laboratoire Cavendish une expérience par laquelle il serait capable de séparer les isotopes du potassium et d'obtenir à peu près 1^{mg} de $^{41}_{19}$K par jour. Cela lui permettrait non seulement de décider quel est celui des deux isotopes qui est radioactif, mais encore d'obtenir le spectre d'absorption de ^{41}K et de déterminer son spin.

M. PAULI. — Je m'intéresse beaucoup au moment magnétique du noyau du lithium. La structure fine du spectre atomique de $^{7}_{3}$Li a été étudiée par M. Güttinger et moi. Quant au $^{6}_{3}$Li, une structure fine n'a pas été observée pour cet isotope, mais cela ne prouve pas que son spin soit nul. En effet, la situation pourrait être assez analogue à celle du $^{14}_{7}$N, où le spin est 1, tandis que le moment magnétique semble être négligeable (absence de structure fine dans le spectre atomique). Il importe donc d'étudier expérimentalement le spectre de bandes dû à la molécule $^{6}_{3}$Li — $^{6}_{3}$Li, afin de connaître la grandeur du spin et le type de la statistique.

M. DIRAC. — Je ne pense pas que les résultats de M. Heisenberg fussent modifiés considérablement s'il y avait des électrons dans les noyaux. En effet, ces électrons seraient toujours fort peu nombreux, et nous pourrions considérer les noyaux comme constitués essentiellement par des protons et des neutrons.

Je voudrais présenter maintenant quelques remarques sur le type d'interaction entre proton et neutron que M. Heisenberg a discuté dans son rapport. On peut dire que cette interaction est du type de l'échange, mais je ne pense pas qu'on puisse l'interpréter comme due à une action d'échange. Les interactions du type de l'échange sont caractérisées par le fait qu'elles entrent en jeu seulement quand les fonctions d'onde des particules agissantes se superposent.

Comparons une pareille interaction avec l'interaction entre électrons et photons. Cette dernière aussi ne se produit que lorsqu'il y a superposition de l'onde électromagnétique et de l'onde électronique de de Broglie. A première vue l'analogie ne paraît pas très étroite à cause du caractère différent des deux ondes considérées; mais il se peut que cette différence ne provienne que des imperfections de l'électrodynamique actuelle. L'analogie pourrait donc être réelle, et dans ce cas elle serait un argument en faveur du type d'interaction de Majorana, plutôt que du type de Heisenberg.

Les interactions par échange représentent un type d'action tout à fait fondamental qu'il n'est pas possible de décrire dans la théorie classique. Si les protons et les neutrons sont tous deux des particules élémentaires, il est logique que leur interaction soit de ce type fondamental.

M. PEIERLS. — Pour quelle raison M. Heisenberg pense-t-il devoir conclure que le neutron est une particule vraiment élémentaire; pourquoi ne pas constater simplement que le neutron ne se désintègre jamais en un proton et un électron ?

M. HEISENBERG. — C'est une impression; le défaut de masse du neutron semble être tellement petit, qu'on devrait pouvoir observer expérimentalement de pareilles désintégrations si le neutron était complexe.

M. BOHR. — A mon avis, le sens qu'on doit attacher à la distinction entre particules élémentaires et particules complexes ne peut pas être indiqué sans ambiguïté.

M. F. PERRIN. — Il me semble difficile de dire que l'une des particules neutron et proton est simple et l'autre complexe, car si par exemple le neutron peut être décomposé en un proton et un électron négatif ($^{0}_{1}\mathbf{n} \rightarrow {}^{1}_{1}\mathrm{H} + \varepsilon^{-}$) et reconstitué inversement par l'union de ces particules (complexité du neutron), on pourra, après avoir matérialisé une paire d'électrons près d'un proton (avec l'apport énergétique d'un photon par exemple), unir au proton l'électron négatif pour former un neutron en laissant libre l'électron positif. On aura ainsi « décomposé » un proton en un neutron et un électron positif ($^{1}_{1}\mathrm{H} + h\nu \rightarrow {}^{1}_{1}\mathrm{H} + \varepsilon^{-} + \varepsilon^{+} \rightarrow {}^{1}_{0}\mathbf{n} + \varepsilon^{+}$) et inversement on pourra unir un neutron et un électron positif pour former un proton, en décomposant d'abord le neutron en un proton et un électron négatif, puis en dématérialisant ensemble cet électron négatif et l'électron positif.

Si l'on peut dire que le neutron est composé d'un proton et d'un électron négatif, on pourra donc aussi bien dire que le proton est constitué d'un neutron et d'un électron positif, les deux réactions

$$^{1}_{0}\mathbf{n} \rightleftharpoons {}^{1}_{1}\mathrm{H} + \varepsilon^{-}. \qquad {}^{1}_{1}\mathrm{H} \rightleftharpoons {}^{1}_{0}\mathbf{n} + \varepsilon^{+}.$$

étant nécessairement toutes deux possibles ou toutes deux impossibles.

Il y a donc sans doute une symétrie complète au point de vue de la complexité entre le neutron et le proton, ces deux particules étant ou bien toutes deux élémentaires et indépendantes, ou bien toutes deux complexes.

M. Heisenberg. — Suivant ma théorie, l'énergie disponible pour l'expulsion d'une seconde particule β par un noyau dépasse de beaucoup celle qui était disponible avant l'expulsion de la première. Or, ceci semble être d'accord avec le fait que les spectres γ qui correspondent à de grandes différences d'énergie, sont émis précisément par les noyaux qui restent après l'expulsion du second électron.

Nous avons discuté mercredi la question de savoir si le noyau peut émettre des radiations dipolaires. Or, si les forces entre protons et neutrons sont du type ordinaire, il n'y a aucune raison d'exclure ces transitions, quoique leur probabilité puisse être assez petite à cause du grand nombre de particules. Si, d'autre part, ces forces sont de pures actions d'échange de l'un des deux types envisagés dans mon rapport, les niveaux seront caractérisés par certaines propriétés de symétrie; une étude approfondie nous montre que dans ce cas les transitions dipolaires entre les divers niveaux du noyau peuvent être exclues à la rigueur (le cas du noyau $^2_1$H fournit un exemple simple).

Du moment que l'on ne néglige plus les forces exercées par les protons entre eux (et les forces entre les neutrons), la symétrie dont je viens de parler n'est plus réalisée que d'une façon approximative, et l'exclusion théorique des transitions dipolaires ne se justifie plus.

M. FERMI. — Il va sans dire, qu'avec une loi du type ae^{-br}, et avec l'hypothèse $m_{\text{proton}} = m_{\text{neutron}}$, on n'a pas le droit d'espérer que l'ensemble des faits expérimentaux trouve son explication théorique. Toutefois, la valeur de $b = 1{,}25.10^{12}$, calculée par Heisenberg à l'aide des données sur le défaut de masse, me paraît trop petite, parce qu'elle conduit à une densité moyenne trop faible des noyaux. En m'appuyant sur la densité expérimentale et en m'arrangeant de façon à être d'accord avec la première partie de la courbe du défaut de masse, je trouve $a = 1{,}46.10^{-4}$, $b = 6{,}8.10^{12}$. Ce résultat correspond à un rayon d'action beaucoup plus petit $\left(\dfrac{1}{b} = 1{,}5.10^{-13} \text{ cm environ}\right)$ que chez Heisenberg; Au point de vue des autres faits (chocs entre proton et neutron, par exemple), j'espère que cette petite valeur ne soulève pas de difficultés réelles.

M. COCKCROFT. — Le nombre de désintégrations produites par un nombre donné de protons de 500 kilovolts, bombardant du fluor, est d'environ un dixième du nombre correspondant pour le lithium, alors que la formule de Gamow fait prévoir que la probabilité relative de pénétration est à peu près un millième. Est-il possible que les forces d'interaction entre un proton qui s'approche du noyau et les neutrons de l'intérieur soient suffisamment grandes pour affecter de façon appréciable la probabilité de passage du proton à travers la barrière de potentiel et augmenter ainsi le nombre de désintégrations prévues pour les éléments lourds ?

M. FERMI. — J'éprouve de grandes difficultés, quand je m'efforce de mettre d'accord les diverses considérations théoriques à l'aide desquelles on essaie de comprendre les propriétés des noyaux. D'une part, on compare le noyau à une agglomération de particules (modèle de la goutte), d'autre part, on introduit l'idée d'états individuels d'une particule qui se meut dans un champ intra-nucléaire.

Si je veux considérer le noyau principalement comme un agrégat de particules α, la densité constante des noyaux nous conduit à admettre que ces particules se comportent plus ou moins comme des globules impénétrables. Même s'il était permis de penser qu'il reste encore assez d'espace libre entre ces particules, il me semble que cette manière de voir nous défend d'introduire les états individuels et d'appliquer la statistique de Bose.

M. BOHR. — La comparaison d'un noyau avec une gouttelette liquide, faite par Gamow, est très schématique à cause du nombre relativement petit de particules α entrant dans la constitution des noyaux. Même dans les noyaux les plus lourds, ce nombre ne dépasse pas la cinquantaine, et si l'on imagine réalisé l'entassement le plus dense possible, on trouve qu'il ne peut y avoir qu'une dizaine de particules à l'intérieur du noyau, les autres formant la surface. Pour cette même raison un modèle tel que celui qu'a proposé M. Delbrück, où le noyau est considéré comme un cristal, n'a pas un caractère bien défini.

Mais toutes les représentations de ce genre ne jouent d'ailleurs qu'un rôle secondaire pour l'interprétation donnée par la mécanique quantique de la relation de Geiger-Nuttall; car cette interprétation est déterminée essentiellement par le comportement énergétique de tout le système et sa charge électrique totale.

M. FERMI. — Si je préfère considérer avec MM. Heisenberg et Majorana le noyau comme composé de neutrons et de protons, la possibilité d'introduire des états individuels dans un champ efficace du noyau entier me paraît encore douteuse. Pour cela il faudrait que l'énergie cinétique moyenne d'une particule dépassât de beaucoup l'énergie avec laquelle une particule est liée à ses voisines. Or cette énergie de liaison doit être de l'ordre de grandeur de l'énergie qui lie les particules dans le noyau de He (quelques millions de volts); elle n'est pas petite par rapport aux différences d'énergie qui jouent un rôle dans les spectres γ.

Peut-être les difficultés que j'ai soulevées ne sont-elles pas aussi graves que je le crois; toutefois je n'entrevois pas leur solution.

Dans la discussion sur ce sujet, MM. HEISENBERG, MOTT et GAMOW prennent la parole.

M. HEISENBERG suggère que, si l'on ne regarde un noyau comme formé principalement de particules α, il est peut-être permis d'admettre que les neutrons en excès se trouvent dans des états individuels correspondant à un même champ efficace, sans qu'on se heurte aux objections de M. Fermi.

M. DEBYE. — J'ai l'impression que si nous avions à l'intérieur du noyau une structure comparable à celle d'un cristal ou même à celle d'un liquide, des expériences de diffusion effectuées avec des rayons α d'énergie suffisamment élevée pour pénétrer dans le noyau, devraient donner des interférences comparables à celles que l'on obtient avec des rayons X dans les cristaux ou les liquides par suite de leur structure moléculaire. Jusqu'à présent les expériences de diffusion faites au Cavendish Laboratory n'ont apporté aucun appui à une pareille idée; mais peut-être n'ont-elles pas été faites avec une précision suffisante.

M. DE DONDER. — Sur la mécanique statistique générale.

Extension en phase et quanta h_l. — Considérons un modèle électromécanique atomique et moléculaire, relativiste ou non, ayant l degrés de liberté. Désignons par $q_1, ..., q_l$ les variables de configuration et par $p_1, ..., p_l$ les moments (ou quantités de mouvement généralisées) associés respectivement à $q_1, ..., q_l$. Nous dirons, d'après W. Gibbs, que les variables q_λ, p_λ $(\lambda = 1, ..., l)$ définissent *la phase* du système (moléculaire) considérée à l'instant t, si la dynamique du modèle moléculaire est régie par les *équations canoniques* ou *hamiltoniennes*

$$(1) \qquad \frac{dq_\lambda}{dt} = \frac{\partial H}{\partial p_\lambda}, \quad \frac{dp_\lambda}{dt} = -\frac{\partial H}{\partial q_\lambda} \qquad (\lambda = 1, ..., l).$$

In *Nobel Lectures Including Presentation Speeches and Laureates' Biographies: Physics 1922-1941,* ed. by The Nobel Foundation (Elsevier, Amsterdam–London–New York 1965) pp. 290-301

WERNER HEISENBERG

The development of quantum mechanics

Nobel Lecture, December 11, 1933

Quantum mechanics, on which I am to speak here, arose, in its formal content, from the endeavour to expand Bohr's principle of correspondence to a complete mathematical scheme by refining his assertions. The physically new viewpoints that distinguish quantum mechanics from classical physics were prepared by the researches of various investigators engaged in analysing the difficulties posed in Bohr's theory of atomic structure and in the radiation theory of light.

In 1900, through studying the law of black-body radiation which he had discovered, Planck had detected in optical phenomena a discontinuous phenomenon totally unknown to classical physics which, a few years later, was most precisely expressed in Einstein's hypothesis of light quanta. The impossibility of harmonizing the Maxwellian theory with the pronouncedly visual concepts expressed in the hypothesis of light quanta subsequently compelled research workers to the conclusion that radiation phenomena can only be understood by largely renouncing their immediate visualization. The fact, already found by Planck and used by Einstein, Debye, and others, that the element of discontinuity detected in radiation phenomena also plays an important part in material processes, was expressed systematically in Bohr's basic postulates of the quantum theory which, together with the Bohr-Sommerfeld quantum conditions of atomic structure, led to a qualitative interpretation of the chemical and optical properties of atoms. The acceptance of these basic postulates of the quantum theory contrasted uncompromisingly with the application of classical mechanics to atomic systems, which, however, at least in its qualitative affirmations, appeared indispensable for understanding the properties of atoms. This circumstance was a fresh argument in support of the assumption that the natural phenomena in which Planck's constant plays an important part can be understood only by largely foregoing a visual description of them. Classical physics seemed the limiting case of visualization of a fundamentally unvisualizable microphysics, the more accurately realizable the more Planck's constant vanishes relative to the parameters of the system. This view of classical mechanics as a limiting case of quantum mechanics also gave rise to Bohr's principle of correspondence which, at least in qualitative terms, transferred a number of conclusions formulated in classical mechanics to quantum mechanics. In connection with the principle of correspondence there was also discussion whether the quantum-mechanical laws could in principle be of a statistical nature; the possibility became particularly apparent in Einstein's derivation of Planck's law of radiation. Finally, the analysis of the relation between radiation theory and atomic theory by Bohr, Kramers, and Slater resulted in the following scientific situation:

According to the basic postulates of the quantum theory, an atomic system is capable of assuming discrete, stationary states, and therefore discrete energy values; in terms of the energy of the atom the emission and absorption of light by such a system occurs abruptly, in the form of impulses. On the other hand, the visualizable properties of the emitted radiation are described by a wave field, the frequency of which is associated with the difference in energy between the initial and final states of the atom by the relation

$$E_1 - E_2 = h\nu$$

To each stationary state of an atom corresponds a whole complex of parameters which specify the probability of transition from this state to another. There is no direct relation between the radiation classically emitted by an orbiting electron and those parameters defining the probability of emission; nevertheless Bohr's principle of correspondence enables a specific term of the Fourier expansion of the classical path to be assigned to each transition of the atom, and the probability for the particular transition follows qualitatively similar laws as the intensity of those Fourier components. Although therefore in the researches carried out by Rutherford, Bohr, Sommerfeld and others, the comparison of the atom with a planetary system of electrons leads to a qualitative interpretation of the optical and chemical properties of atoms, nevertheless the fundamental dissimilarity between the atomic spectrum and the classical spectrum of an electron system imposes the need to relinquish the concept of an electron path and to forego a visual description of the atom.

The experiments necessary to define the electron-path concept also furnish an important aid in revising it. The most obvious answer to the question how the orbit of an electron in its path within the atom could be observed namely, will perhaps be to use a microscope of extreme resolving power. But since the specimen in this microscope would have to be illuminated with light having an extremely short wavelength, the first light quantum from the light source to reach the electron and pass into the observer's eye would eject the electron completely from its path in accordance with the laws of the Compton effect. Consequently only one point of the path would be observable experimentally at any one time.

In this situation, therefore, the obvious policy was to relinquish at first the concept of electron paths altogether, despite its substantiation by Wilson's experiments, and, as it were, to attempt subsequently how much of the electron-path concept can be carried over into quantum mechanics.

In the classical theory the specification of frequency, amplitude, and phase of all the light waves emitted by the atom would be fully equivalent to specifying its electron path. Since from the amplitude and phase of an emitted wave the coefficients of the appropriate term in the Fourier expansion of the electron path can be derived without ambiguity, the complete electron path therefore can be derived from a knowledge of all amplitudes and phases. Similarly, in quantum mechanics, too, the whole complex of amplitudes and phases of the radiation emitted by the atom can be regarded as a complete description of the atomic system, although its interpretation in the sense of an electron path inducing the radiation is impossible. In quantum mechanics, therefore, the place of the electron coordinates is taken by a complex of parameters corresponding to the Fourier coefficients of classical motion along a path. These, however, are no longer classified by the energy of state and the number of the corresponding harmonic vibration, but are in each case associated with two stationary states of the atom, and are a measure for the transition probability of the atom from one stationary state to another. A complex of coefficients of this type is comparable with a matrix such as occurs in linear algebra. In exactly the same way each parameter of classical mechanics, e.g. the momentum or the energy of the electrons, can then be assigned a corresponding matrix in quantum mechanics. To proceed from here beyond a mere description of the empirical state of affairs it was necessary to associate systematically the matrices assigned to the various parameters in the same way as the corresponding parameters in classical mechanics are associated by equations of motions. When, in the interest of achieving the closest possible correspondence between classical and quantum mechanics, the addition and multiplication of Fourier series were tentatively taken as the example for the addition and multiplication of the quantum-theory complexes, the product of two parameters represented by matrices appeared to be most naturally represented by the product matrix in the sense of linear algebra – an assumption already suggested by the formalism of the Kramers-Ladenburg dispersion theory.

It thus seemed consistent simply to adopt in quantum mechanics the equations of motion of classical physics, regarding them as a relation between the matrices representing the classical variables. The Bohr-Sommerfeld quantum conditions could also be re-interpreted in a relation between the matrices, and together with the equations of motion they were sufficient to define all matrices and hence the experimentally observable properties of the atom.

Born, Jordan, and Dirac deserve the credit for expanding the mathematical scheme outlined above into a consistent and practically usable theory. These investigators observed in the first place that the quantum conditions can be written as commutation relations between the matrices representing the momenta and the coordinates of the electrons, to yield the equations (p_r, momentum matrices; q_r, coordinate matrices):

$$p_r q_s - q_s p_r = \frac{h}{2\pi i}\, \delta_{rs} \qquad q_r q_s - q_s q_r = 0 \qquad p_r p_s - p_s p_r = 0$$

$$\delta_{rs} = \begin{cases} 1 \text{ for } r = s \\ 0 \text{ for } r \neq s \end{cases}$$

By means of these commutation relations they were able to detect in quantum mechanics as well the laws which were fundamental to classical mechanics: the invariability in time of energy, momentum, and angular momentum.

The mathematical scheme so derived thus ultimately bears an extensive formal similarity to that of the classical theory, from which it differs outwardly by the commutation relations which, moreover, enabled the equations of motion to be derived from the Hamiltonian function.

In the physical consequences, however, there are very profound differences between quantum mechanics and classical mechanics which impose the need for a thorough discussion of the physical interpretation of quantum mechanics. As hitherto defined, quantum mechanics enables the radiation emitted by the atom, the energy values of the stationary states, and other parameters characteristic for the stationary states to be treated. The theory hence complies with the experimental data contained in atomic spectra. In all those cases, however, where a visual description is required of a transient event, e.g. when interpreting Wilson photographs, the formalism of the theory does not seem to allow an adequate representation of the experimental state of affairs. At this point Schrödinger's wave mechanics, meanwhile developed on the basis of de Broglie's theses, came to the assistance of quantum mechanics.

In the course of the studies which Mr. Schrödinger will report here himself he converted the determination of the energy values of an atom into an eigenvalue problem defined by a boundary-value problem in the coordinate space of the particular atomic system. After Schrödinger had shown the mathematical equivalence of wave mechanics, which he had discovered, with quantum mechanics, the fruitful combination of these two different areas of physical ideas resulted in an extraordinary broadening and enrichment of the formalism of the quantum theory. Firstly it was only wave mechanics which made possible the mathematical treatment of complex atomic systems, secondly analysis of the connection between the two theories led to what is known as the transformation theory developed by Dirac and Jordan. As it is impossible within the limits of the present lecture to give a detailed discussion of the mathematical structure of this theory, I should just like to point out its fundamental physical significance. Through the adoption of the physical principles of quantum mechanics into its expanded formalism, the transformation theory made it possible in completely general terms to calculate for atomic systems the probability for the occurrence of a particular, experimentally ascertainable, phenomenon under given experimental conditions. The hypothesis conjectured in the studies on the radiation theory and enunciated in precise terms in Born's collision theory, namely that the wave function governs the probability for the presence of a corpuscle, appeared to be a special case of a more general pattern of laws and to be a natural consequence of the fundamental assumptions of quantum mechanics. Schrödinger, and in later studies Jordan, Klein, and Wigner as well, had succeeded in developing as far as permitted by the principles of the quantum theory de Broglie's original concept of visualizable matter waves occurring in space and time, a concept formulated even before the development of quantum mechanics. But for that the connection between Schrödinger's concepts and de Broglie's original thesis would certainly have seemed a looser one by this statistical interpretation of wave mechanics and by the greater emphasis on the fact that Schrödinger's theory is concerned with waves in multidimensional space. Before proceeding to discuss the explicit significance of quantum mechanics it is perhaps right for me to deal briefly with this question as to the existence of matter waves in three-dimensional space, since the solution to this problem was only achieved by combining wave and quantum mechanics.

A long time before quantum mechanics was developed Pauli had inferred from the laws in the Periodic System of the elements the well-known principle that a particular quantum state can at all times be occupied by only a single electron. It proved possible to transfer this principle to quantum mechanics on the basis of what at first sight seemed a surprising result: the entire complex of stationary states which an atomic system is capable of adopting breaks down into definite classes such that an atom in a state belonging to one class can never change into a state belonging to another class under the action of whatever perturbations. As finally clarified beyond question by the studies of Wigner and Hund, such a class of states is characterized by a definite symmetry characteristic of the Schrödinger eigenfunction with respect to the transposition of the coordinates of two electrons. Owing to the fundamental identity of electrons, any external perturbation of the atom remains unchanged when two electrons are exchanged and hence causes no transitions between states of various classes. The Pauli principle and the Fermi-Dirac statistics derived from it are equivalent with the assumption that only that class of stationary states is achieved in nature in which the eigenfunction changes its sign when two electrons are exchanged. According to Dirac, selecting the symmetrical system of terms would lead not to the Pauli principle, but to Bose-Einstein electron statistics.

Between the classes of stationary states belonging to the Pauli principle or to Bose-Einstein statistics, and de Broglie's concept of matter waves there is a peculiar relation. A spatial wave phenomenon can be treated according to the principles of the quantum theory by analysing it using the Fourier theorem and then applying to the individual Fourier component of the wave motion, as a system having one degree of freedom, the normal laws of quantum mechanics. Applying this procedure for treating wave phenomena by the quantum theory, a procedure that has also proved fruitful in Dirac's studies of the theory of radiation, to de Broglie's matter waves, exactly the same results are obtained as in treating a whole complex of material particles according to quantum mechanics and selecting the symmetrical system of terms. Jordan and Klein hold that the two methods are mathematically equivalent even if allowance is also made for the interaction of the electrons, i.e. if the field energy originating from the continuous space charge is included in the calculation in de Broglie's wave theory. Schrödinger's considerations of the energy-momentum tensor assigned to the matter waves can then also be adopted in this theory as consistent components of the formalism. The studies of Jordan and Wigner show that modifying the commutation relations underlying this quantum theory of waves results in a formalism equivalent to that of quantum mechanics based on the assumption of Pauli's exclusion principle.

These studies have established that the comparison of an atom with a planetary system composed of nucleus and electrons is not the only visual picture of how we can imagine the atom. On the contrary, it is apparently no less correct to compare the atom with a charge cloud and use the correspondence to the formalism of the quantum theory borne by this concept to derive qualitative conclusions about the behaviour of the atom. However, it is the concern of wave mechanics to follow these consequences.

Reverting therefore to the formalism of quantum mechanics; its application to physical problems is justified partly by the original basic assumptions of the theory, partly by its expansion in the transformation theory on the basis of wave mechanics, and the question is now to expose the explicit significance of the theory by comparing it with classical physics.

In classical physics the aim of research was to investigate objective processes occurring in space and time, and to discover the laws governing their progress from the initial conditions. In classical physics a problem was considered solved when a particular phenomenon had been proved to occur objectively in space and time, and it had been shown to obey the general rules of classical physics as formulated by differential equations. The manner in which the knowledge of each process had been acquired, what observations may possibly have led to its experimental determination, was completely immaterial, and it was also immaterial for the consequences of the classical theory, which possible observations were to verify the predictions of the theory. In the quantum theory, however, the situation is completely different. The very fact that the formalism of quantum mechanics cannot be interpreted as visual description of a phenomenon occurring in space and time shows that quantum mechanics is in no way concerned with the objective determination of space-time phenomena. On the contrary, the formalism of quantum mechanics should be used in such a way that the probability for the outcome of a further experiment may be concluded from the determination of an experimental situation in an atomic system, providing that the system is subject to no perturbations other than those necessitated by performing the two experiments. The fact that the only definite known result to be ascertained after the fullest possible experimental investigation of the system is the probability for a certain outcome of a second experiment shows, however, that each observation must entail a discontinuous change in the formalism describing the atomic proces sand therefore also a discontinuous change in the physical phenomenon itself. Whereas in the classical theory the kind of observation has no bearing on the event, in the quantum theory the disturbance associated with each observation of the atomic phenomenon has a decisive role. Since, furthermore, the result of an observation as a rule leads only to assertions about the probability of certain results of subsequent observations, the fundamentally unverifiable part of each perturbation must, as shown by Bohr, be decisive for the non-contradictory operation of quantum mechanics. This difference between classical and atomic physics is understandable, of course, since for heavy bodies such as the planets moving around the sun the pressure of the sunlight which is reflected at their surface and which is necessary for them to be observed is negligible; for the smallest building units of matter, however, owing to their low mass, every observation has a decisive effect on their physical behaviour.

The perturbation of the system to be observed caused by the observation is also an important factor in determining the limits within which a visual description of atomic phenomena is possible. If there were experiments which permitted accurate measurement of all the characteristics of an atomic system necessary to calculate classical motion, and which, for example, supplied accurate values for the location and velocity of each electron in the system at a particular time, the result of these experiments could not be utilized at all in the formalism, but rather it would directly contradict the formalism. Again, therefore, it is clearly that fundamentally unverifiable part of the perturbation of the system caused by the measurement itself which hampers accurate ascertainment of the classical characteristics and thus permits quantum mechanics to be applied. Closer examination of the formalism shows that between the accuracy with which the location of a particle can be ascertained and the accuracy with which its momentum can simultaneously be known, there is a relation according to which the product of the probable errors in the measurement of the location and momentum is invariably at least as large as Planck's constant divided by 4π. In a very general form, therefore, we should have

$$\Delta p \, \Delta q \geqslant \frac{h}{4\pi}$$

where p and q are canonically conjugated variables. These uncertainty relations for the results of the measurement of classical variables form the necessary conditions for enabling the result of a measurement to be expressed in the formalism of the quantum theory. Bohr has shown in a series of examples how the perturbation necessarily associated with each observation indeed ensures that one cannot go below the limit set by the uncertainty relations. He contends that in the final analysis an uncertainty introduced by the concept of measurement itself is responsible for part of that perturbation remaining fundamentally unknown. The experimental determination of whatever space-time events invariably necessitates a fixed frame – say the system of coordinates in which the observer is at rest – to which all measurements are referred. The assumption that this frame is «fixed» implies neglecting its momentum from the outset, since «fixed» implies nothing other, of course, than that any transfer of momentum to it will evoke no perceptible effect. The fundamentally necessary uncertainty at this point is then transmitted via the measuring apparatus into the atomic event.

Since in connection with this situation it is tempting to consider the possibility of eliminating all uncertainties by amalgamating the object, the measuring apparatuses, and the observer into one quantum-mechanical system, it is important to emphasize that the act of measurement is necessarily visualizable, since, of course, physics is ultimately only concerned with the systematic description of space-time processes. The behaviour of the observer as well as his measuring apparatus must therefore be discussed according to the laws of classical physics, as otherwise there is no further physical problem whatsoever. Within the measuring apparatus, as emphasized by Bohr, all events in the sense of the classical theory will therefore be regarded as determined, this also being a necessary condition before one can, from a result of measurements, unequivocally conclude what has happened. In quantum theory, too, the scheme of classical physics which objictifies the results of observation by assuming in space and time processes obeying laws is thus carried through up to the point where the fundamental limits are imposed by the unvisualizable character of the atomic events symbolized by Planck's constant. A visual description for the atomic events is possible only within certain limits of accuracy – but within these limits the laws of classical physics also still apply. Owing to these limits of accuracy as defined by the uncertainty relations, moreover, a visual picture of the atom free from ambiguity has not been determined. On the contrary the corpuscular and the wave concepts are equally serviceable as a basis for visual interpretation.

The laws of quantum mechanics are basically statistical. Although the parameters of an atomic system are determined in their entirety by an experiment, the result of a future observation of the system is not generally accurately predictable. But at any later point of time there are observations which yield accurately predictable results. For the other observations only the probability for a particular outcome of the experiment can be given. The degree of certainty which still attaches to the laws of quantum mechanics is, for example, responsible for the fact that the principles of conservation for energy and momentum still hold as strictly as ever. They can be checked with any desired accuracy and will then be valid according to the accuracy with which they are checked. The statistical character of the laws of quantum mechanics, however, becomes apparent in that an accurate study of the energetic conditions renders it impossible to pursue at the same time a particular event in space and time.

For the clearest analysis of the conceptual principles of quantum mechanics we are indebted to Bohr who, in particular, applied the concept of complementarity to interpret the validity of the quantum-mechanical laws. The uncertainty relations alone afford an instance of how in quantum mechanics the exact knowledge of one variable can exclude the exact knowledge of another. This complementary relationship between different aspects of one and the same physical process is indeed characteristic for the whole structure of quantum mechanics. I had just mentioned that, for example, the determination of energetic relations excludes the detailed description of space-time processes. Similarly, the study of the chemical properties of a molecule is complementary to the study of the motions of the individual electrons in the molecule, or the observation of interference phenomena complementary to the observation of individual light quanta. Finally, the areas of validity of classical and quantum mechanics can be marked off one from the other as follows: Classical physics represents that striving to learn about Nature in which essentially we seek to draw conclusions about objective processes from observations and so ignore the consideration of the influences which every observation has on the object to be observed; classical physics, therefore, has its limits at the point from which the influence of the observation on the event can no longer be ignored. Conversely, quantum mechanics makes possible the treatment of atomic processes by partially foregoing their space-time description and objectification.

So as not to dwell on assertions in excessively abstract terms about the interpretation of quantum mechanics, I would like briefly to explain with a well-known example how far it is possible through the atomic theory to achieve an understanding of the visual processes with which we are concerned in daily life. The interest of research workers has frequently been focused on the phenomenon of regularly shaped crystals suddenly forming from a liquid, e.g. a supersaturated salt solution. According to the atomic theory the forming force in this process is to a certain extent the symmetry characteristic of the solution to Schrödinger's wave equation, and to that extent crystallization is explained by the atomic theory. Nevertheless this process retains a statistical and – one might almost say – historical element which cannot be further reduced: even when the state of the liquid is completely known before crystallization, the shape of the crystal is not determined by the laws of quantum mechanics. The formation of regular shapes is just far more probable than that of a shapeless lump. But the ultimate shape owes its genesis partly to an element of chance which in principle cannot be analysed further.

Before closing this report on quantum mechanics, I may perhaps be allowed to discuss very briefly the hopes that may be attached to the further development of this branch of research. It would be superfluous to mention that the development must be continued, based equally on the studies of de Broglie, Schrödinger, Born, Jordan, and Dirac. Here the attention of the research workers is primarily directed to the problem of reconciling the claims of the special relativity theory with those of the quantum theory. The extraordinary advances made in this field by Dirac about which Mr. Dirac will speak here, meanwhile leave open the question whether it will be possible to satisfy the claims of the two theories without at the same time determining the Sommerfeld fine-structure constant. The attempts made hitherto to achieve a relativistic formulation of the quantum theory are all based on visual concepts so close to those of classical physics that it seems impossible to determine the fine-structure constant within this system of concepts. The expansion of the conceptual system under discussion here should, furthermore, be closely associated with the further development of the quantum theory of wave fields, and it appears to me as if this formalism, notwithstanding its thorough study by a number of workers (Dirac, Pauli, Jordan, Klein, Wigner, Fermi) has still not been completely exhausted. Important pointers for the further development of quantum mechanics also emerge from the experiments involving the structure of the atomic nuclei. From their analysis by means of the Gamow theory, it would appear that between the elementary particles of the atomic nucleus forces are at work which differ somewhat in type from the forces determining the structure of the atomic shell; Stern's experiments seem, furthermore, to indicate that the behaviour of the heavy elementary particles cannot be represented by the formalism of Dirac's theory of the electron. Future research will thus have to be prepared for surprises which may otherwise come both from the field of experience of nuclear physics as well as from that of cosmic radiation. But however the development proceeds in detail, the path so far traced by the quantum theory indicates that an understanding of those still unclarified features of atomic physics can only be acquired by foregoing visualization and objectification to an extent greater than that customary hitherto. We have probably no reason to regret this, because the thought of the great epistemological difficulties with which the visual atom concept of earlier physics had to contend gives us the hope that the abstracter atomic physics developing at present will one day fit more harmoniously into the great edifice of Science.

In *Pieter Zeeman, 1865 – 25 Mei – 1935, Verhandelingen op 25 Mei 1935 Aangeboden aan Prof. Dr. P. Zeeman* (Martinus Nijhoff, The Hague 1935) pp. 108–116

BEMERKUNGEN ZUR THEORIE DES ATOMKERNS

von W. HEISENBERG, Leipzig

Da die Fülle des experimentellen Materials, das über die Atomkerne gesammelt wird, ein immer deutlicheres Bild der Naturgesetze entstehen lässt, die den Bau der Atomkerne beherrschen, so soll im Folgenden versucht werden, eine kurze Zusammenfassung der im Einzelnen längst bekannten theoretischen Gesichtspunkte zu geben, nach denen heute der Bau des Atomkerns betrachtet wird.

Zu diesem Zweck soll ein Vergleich der einfachsten empirischen Daten über die Atomhülle mit den entsprechenden Daten über die Kerne durchgeführt und eine Art von korrespondenzmässiger Analogie zwischen den schon bekannten Gesetzen der Atomhülle und den noch unbekannten der Kerne gesucht werden.

Die R u t h e r f o r d-B o h r sche Theorie behauptet, das Atom bestehe aus Kern und Elektronen. Diese Aussage wird gerechtfertigt durch die experimentelle Tatsache, dass ein Atom durch Stossversuche in diese Teile: Kern und Elektronen zerlegt werden kann. Dagegen werden die Lichtquanten nicht als Bausteine des Atoms betrachtet, obwohl sie vom Atom emittiert werden können und obwohl sie in der Form des M a x w e l l schen Feldes, das die Elektronen an den Kern bindet, einen erheblichen Teil zur Gesamtenergie des Atoms beitragen. Zur Rechtfertigung dieses Sprachgebrauchs wird man anführen können, dass bei Stossversuchen *im allgemeinen* keine Lichtquanten das Atom verlassen. Vielmehr werden die Lichtquanten, die etwa in Stossversuchen eine Rolle spielen, im allgemeinen erst lange nach dem Stossprozess, nach dem Ablauf der natürlichen Lebensdauer der angeregten Zustände, vom Atom emittiert, ihre Entstehung hat mit dem Stossprozess unmittelbar nichts zu tun. Man sagt daher zweckmässiger, die Lichtquanten werden vom Atom beim Übergang von einem stationären Zustand zu einem anderen erschaffen; im Atom selbst existieren sie nicht als Individuen; es existiert nur als ihnen zugeordnetes Feld das M a x w e l l sche Feld von

238

Kern und Elektronen. Diese ziemlich strenge Scheidung zwischen den Partikeln, aus denen ein Atom besteht und denen, die von ihm erschaffen werden, wird ermöglicht durch die von B o h r als wesentlich hervorgehobene Kleinheit der Strahlungskräfte im Verhältnis zu den C o u l o m b schen Kräften; diese wiederum beruht auf der Kleinheit der Elektronengeschwindigkeiten im Atom gegenüber der Lichtgeschwindigkeit, also letzten Endes auf der Kleinheit von $e^2/\hbar c$.

Sucht man nun in ähnlicher Weise die Struktur der Atomkerne zu beschreiben, so wird man auf Grund der Stossversuche, also hier der Versuche über Atomumwandlung, nur die schweren Teilchen: Protonen, Neutronen und eventuell α-Teilchen als Bestandteile des Atomkerns betrachten. Denn nur diese Partikeln treten unmittelbar als Kerntrümmer nach Zusammenstössen in Erscheinung. Die leichten Teilchen: Elektronen, Positronen und Lichtquanten werden nach den Versuchen über künstliche Radioaktivität und über die Lebensdauern angeregter Kernzustände im allgemeinen erst lange *nach* der Kernumwandlung emittiert, ihre Entstehung hat also mit dem Zertrümmerungsprozess unmittelbar nichts zu tun. Es wird daher zweckmässig sein, von einer Entstehung von Elektronen beim Übergang eines Kerns von einem stationären Zustand zu einem anderen zu sprechen. Dass diese Beschreibungsweise vernünftig ist, wird durch die Versuche von J o l i o t und C u r i e , nach denen auch Positronen radioaktiv emittiert werden können, überzeugend dargetan, da es wohl nicht angeht, positive und negative Elektronen gleichzeitig als Kernbausteine zuzulassen.

Ebenso wie ferner die Entstehung der Lichtquanten zusammenhängt mit dem M a x w e l l schen Feld, das für den Zusammenhalt der Atomhülle sorgt, so muss die Entstehungsmöglichkeit leichter geladener Teilchen aus dem Kern mit der Existenz eines Wellenfeldes verknüpft sein, das dann wieder für den Zusammenhalt des Atomkerns wesentlich sein dürfte. Dies ist auch das Resultat der F e r m i schen [1]) Untersuchungen über die Theorie des β-Zerfalls und ihrer Erweiterung durch T a m m [2]) und I w a n e n k o [3]), über die weiter unten noch zu sprechen sein wird. Auch bei den Kernen wird die

---

1) E. F e r m i , Zs. f. Phys. **88**, 161, 1934.
2) I. T a m m , Nature **133**, 981, 1934.
3) D. I w a n e n k o , Nature **133**, 981, 1934.

reinliche Scheidung zwischen den Teilchen, aus denen der Kern besteht, und denen, die von ihm bei Übergängen erschaffen werden, ermöglicht durch die Kleinheit der „β-Strahlungskräfte" gegen die Wechselwirkungskräfte im Kern, die sich in der Länge der natürlichen Lebensdauern β-aktiver Kerne äussert. Dieser Sachverhalt hängt, wie man z.B. aus der F e r m i schen Theorie des β-Zerfalls nachrechnen kann, wieder damit zusammen, dass die als kinetische Energie der emittierten Teilchen in Erscheinung tretenden Übergangsenergien der Kerne klein gegen die Ruhenergie der Kernbausteine, d.h. dass die Geschwindigkeiten der Protonen und Neutronen im Kern klein gegen die Lichtgeschwindigkeit sind.

Die Analogie zwischen dem Verhalten der Atomhülle und dem des Atomkerns, die hier betrachtet werden soll, kann also summarisch in dem folgenden Schema dargestellt werden:

	Atomhülle	Atomkern	
Elementare Bausteine	Kern, Elektronen	Protonen, Neutronen	
Bei Übergängen emittierte Teilchen	Lichtquanten	Elektronen Positronen Neutrino's	Lichtquanten
Das ihnen zugeordnete Feld	Maxwellsches Feld	Fermisches Feld	Maxwellsches Feld
Erste Näherung der Wechselwirkungskräfte	Coulombkräfte	Austauschkräfte	Coulombkräfte

Neben den Elektronen und Positronen sind hier auch die P a u l ischen Neutrino's erwähnt. Ihre Existenz wird bekanntlich durch die kontinuierlichen β-Spektren und durch die empirischen Regeln über die Statistik und den Spin der Kerne sehr wahrscheinlich gemacht; auch die Länge der Lebensdauern β-aktiver Kerne im Verhältnis zu der Lebensdauer angeregter Zustände bei γ-Strahl-Emission deutet darauf hin, dass bei der β-Emission zwei Teilchen gleichzeitig den Kern verlassen.

Wenn man sich einen Ansatz für die Art der Wechselwirkungen verschaffen will, die zur Emission der Elektronen und Neutrinos führt, so muss man im Sinne der hier besprochenen Analogie an eine Wechselwirkungsenergie denken, die vom Produkt der Wellenfunktionen der schweren Teilchen und der entsprechenden leichten Teilchen *an der gleichen Raumstelle* abhängt. Denn ähnlich wie Lichtquanten und Elektronen nur an der gleichen Raumstelle in Wechselwirkung treten können — die M a x w e l l sche Theorie ist eine „Nahewirkungs"-theorie — so werden auch die leichten geladenen Elementarteilchen und die schweren Kernbausteine nur an der gleichen Stelle — von den Wirkungen ihrer M a x w e l l schen Felder abgesehen — aufeinander wirken können. Diese Annahme hat in der Tat F e r m i seiner Theorie des β-Zerfalls zu Grunde gelegt. Dabei zeigt sich wieder, dass die Annahme der gleichzeitigen Emission eines Elektrons und eines Neutrinos die notwendige Voraussetzung für die mathematische Formulierung des β-Zerfalls ist. Die Wechselwirkungsenergie wird nämlich dargestellt durch das Raumintegral des Produkts von vier Wellenfunktionen, die jeweils zu Proton, Neutron, Elektron und Neutrino gehören. Die Wellenfunktion des Neutrinos ist hierbei notwendig, um zu einem Ausdruck zu gelangen, der sich relativistisch wie eine Energiedichte verhält.

F e r m i hat des weiteren noch angenommen, dass in der Wechselwirkungsenergie nur die Wellenfunktionen, nicht aber deren Ableitungen vorkommen sollen. Diese Annahme ist jedoch, wie B e t h e und P e i e r l s [1]) hervorgehoben haben, wahrscheinlich schon zu speziell.

Wenn die Wechselwirkungsenergie bekannt ist, so sind damit nicht nur die Gesetze des β-Zerfalls bestimmt. Vielmehr werden, ähnlich wie durch die M a x w e l l schen Gleichungen alle elektrischen Kraftwirkungen zwischen Materieteilchen festgelegt sind, durch die Wechselwirkungsenergie zwischen den schweren Kernbestandteilen und den leichten Elementarteilchen alle Kraftwirkungen zwischen den schweren Partikeln bestimmt sein, die mit dem Elektron- und Neutrinofeld zu tun haben. In der Näherung, in der man von den Strahlungsrückwirkungen absehen kann — also dann, wenn die Elektronen sich langsam im Verhältnis zur Lichtgeschwindigkeit bewegen — können die Wirkungen des M a x w e l l schen Feldes durch eine Fernkraft, nämlich die C o u l o m b sche Kraft ersetzt werden. Ganz in der gleichen Weise führt die F e r m i sche Wechselwirkung, wenn die Bewegung der schweren Partikeln langsam gegen die Lichtgeschwindigkeit erfolgt, zu einer Fernkraft zwischen

---

1) Die Ergebnisse von B e t h e und P e i e r ls wurden im September 1934 auf der Konferenz in Kopenhagen von Herrn B e t h e vorgetragen. Vgl. auch die Berichte vom Kongress in London, Herbst 1934.

Neutronen und Protonen, wie unabhängig von F e r m i [1]), T a m m und I w a n e n k o (l. c.) bemerkt worden ist. Die so entstehenden Kräfte sind ihrer Art nach Austauschkräfte; die F e r m i sche Theorie gestattet also im Prinzip die mathematische Durchführung des Gedankens, dass aus der Möglichkeit des β-Zerfalls die Existenz von Austauschkräften folge [2]). Die Abhängigkeit dieser Kräfte vom Abstand der schweren Teilchen und von deren Spinkoordinaten ist allerdings durch die genaue Form der F e r m i schen Wechselwirkungsenergie bedingt und kann daher einstweilen nicht definitiv angegeben werden. T a m m und I w a n e n k o (l. c.) haben gezeigt, dass bei der Annahme der speziell von F e r m i gewählten Wechselwirkungsenergie die Austauschkräfte viel zu klein werden, um den Aufbau der Kerne zu erklären. Diese Schwierigkeit kann jedoch verschwinden, wenn, wie B e t h e und P e i e r l s (l. c.) vorgeschlagen haben, in der Wechselwirkungsenergie die Ableitungen der Wellenfunktionen von Elektron und Neutrino vorkommen. Wegen dieser Unsicherheit in der Form der Wechselwirkungsenergie ist auch noch keine Entscheidung darüber möglich, ob die Austauschkräfte die von M a j o r a n a [3]) oder die vom Verfasser früher vorgeschlagene Form haben; die empirischen Daten über die Massendefekte sprechen wohl eher für die M a j o r a n a'sche Wechselwirkung.

Die wichtigste Aufgabe der Theorie des Kernaufbaus wird also in der nächsten Zeit sein, durch den Vergleich aller empirischen Daten — über die Gestalt der kontinuierlichen β-Strahl-Spektren, über die Massendefekte u.s.w. — die genaue Form der F e r m i schen Wechselwirkung zwischen schweren und leichten Teilchen zu ermitteln.

Freilich bleibt, selbst wenn diese Wechselwirkung bekannt ist, in unserem Verständnis des Kernaufbaus noch eine Lücke, die in der Theorie der Atomhülle *kein* Analogon hat. Wegen der C o u l o m b-schen Form der Wechselwirkung hängt nämlich der Aufbau der Atomhülle nur ganz unwesentlich von den kleinen Abweichungen ab, die bei sehr kleinen Partikelabständen vom C o u l o m b schen Gesetz auftreten. Die Austauschkräfte der Kerntheorie dagegen variieren mit einer viel höheren Potenz der Partikelabstände (bei Annahme der F e r m i schen Form der Wechselwirkung nach T a m m

---

1) Für die freundliche briefliche Mitteilung seiner Überlegungen bin ich Herrn F e r m i zu Dank verpflichtet. Vgl. auch C. W i c k, Rendic. R. Nat. Acad. Lincei **19**, 319, 1934.

2) W. H e i s e n b e r g, Zs. f. Phys. **77**, 1; **78**, 156; **80**, 587, 1932.

3) E. M a j o r a n a, Zs. f. Phys. **82**, 137, 1933.

únd I w a n e n k o mit r^{-5}; wenn die Ableitungen der Wellenfunktionen in der Wechselwirkungsenergie vorkommen, eventuell mit r^{-7} oder r^{-9}), sodass gerade die Abweichungen von diesem einfachen Kraftgesetz bei kleinen Partikelabständen von ausschlaggebender Bedeutung für den Kernaufbau sind. Die Bestimmung dieser Abweichungen ist jedoch nahe verwandt dem Problem der unendlichen Selbstenergie, die für die schweren Teilchen aus dem F e r m i schen Feld — genau so wie für ein geladenes Teilchen aus dem M a x w e l l feld — resultiert. Für die Lösung dieser Frage sind jedoch bisher keine Ansätze vorhanden.

Überbrückt man diese Schwierigkeit vorläufig rein formal durch die Einführung eines geeignet gewählten Protonen- oder Neutronenradius, — die Berechtigung eines solchen Verfahrens bleibt freilich zweifelhaft — so hat die Wechselwirkung, die zur Entstehung von Elektronen und Neutrinos führt, nach der Untersuchung von W i c k [1]) noch eine andere wichtige Konsequenz: sie führt zu einem Zusatz zum magnetischen Moment des Protons und zu einem magnetischen Moment des Neutrons.

Es entsteht die Frage, wieweit sich für diese magnetischen Momente der schweren Teilchen die Analogie zu den magnetischen Momenten der Elektronen in der Atomhülle durchführen lässt; insbesondere muss, um die Analogie zwischen Atomhülle und Kern vollständig zu beschreiben, festgestellt werden, inwieweit auch die magnetischen Momente der schweren Teilchen im Kern in einem Vektormodell zum gesamten magnetischen Moment des Kerns zusammengesetzt werden können.

Die W i c k schen Resultate beruhen auf Überlegungen, die denen von T a m m und I w a n e n k o sehr ähnlich sind; sie sollen hier kurz wiederholt werden:

Führt man die Störungsrechnung mit der F e r m i schen Wechselwirkungsenergie zwischen leichten und schweren Partikeln als Störung durch und berechnet das magnetische Moment des Protons bezw. Neutrons, so entstehen Zusatzglieder zweiter Näherung, die in der Form ganz den Selbstenergien dieser Teilchen gleichen und die dem virtuellen Entstehen und Wiederverschwinden eines Paares Positron-Neutrino beim Proton, bzw. Elektron-Neutrino beim Neutron entsprechen. Diese Zusatzmomente würden unendlich gross werden, wenn man nicht wie bei den Selbstenergien einen endlichen Ra-

---

1) C. W i c k, Rend. R. Nat. Acad. Lincei. Im Erscheinen.

dius der schweren Teilchen einführte. Wird dieser Radius so eingerichtet, dass die Selbstenergie die Grössenordnung der relativistischen Ruhenergie der schweren Teilchen erhält, so kommt der Hauptbeitrag zu den magnetischen Momenten von Elektron-Neutrinopaaren bzw. Positron-Neutrinopaaren, deren Energie etwa das hundert- bis tausendfache der Ruhenergie des Elektrons beträgt. Der genaue Wert hängt wieder von der genauen Form der F e r m ischen Wechselwirkung ab. Wählt man eine Form für diese Wechselwirkung, die die richtige Grössenordnung für die Austauschkräfte liefert, so werden die magnetischen Zusatzmomente von der Grössenordnung eines Kernmagnetons.

Für das Neutron könnte allerdings dieses magnetische Moment wieder zu Null kompensiert werden, wenn das Neutron nicht nur durch die Emission eines Elektrons und Neutrinos in ein Proton, sondern auch durch die Emission eines Positrons und Neutrinos in ein hypothetisches negatives Proton übergehen könnte. Die empirischen Resultate [1]) über die magnetischen Momente von ^1_1H und ^2_1D scheinen jedoch gegen diese Möglichkeit zu sprechen, das Neutron scheint ein magnetisches Moment zu besitzen.

Nimmt man also an, es gebe nur den erstgenannten Emissionsprozess, so werden wegen des Ladungsvorzeichens der virtuell entstehenden Teilchen die Zusatzmomente von Proton und Neutron in erster Näherung entgegengesetzt gleich; eine geringe Abweichung von dieser Gleichheit erhält man erst, wenn man die Verschiedenheit der Massen von Proton und Neutron berücksichtigt; diese Abweichung ist daher ihrer relativen Grösse nach etwa durch das Verhältnis des Massenunterschiedes (in energetischem Mass) zum hundert- bis tausendfachen der Ruhenergie des Elektrons gegeben; sie sollte also nur wenige Prozent betragen und zwar sollte das magnetische Moment des Neutrons absolut genommen etwas grösser sein, wenn die Masse des Neutrons, wie nach C h a d w i c k anzunehmen ist, grösser ist als die des Protons. Dieses Zusatzmoment muss beim Proton dem gewöhnlichen magnetischen Moment, das nach der D i r a cschen Theorie dem Spin $\frac{1}{2}\hbar$ entspricht, hinzugefügt werden.

Die magnetischen Momente von Neutron und Proton sollten nun

1)   J. E s t e r m a n n, R. F r i s c h u. O. S t e r n, Nature **132**, 169, 1933;
J. E s t e r m a n n u. O. S t e r n, Nature **133**, 911, 1934;
J. R a b i, J. K e l l o g, J. Z a c h a r i a s, Phys. Rev. **46**, 157 u. 163, 1934;
F. K a l c k a r u. E. T e l l e r, Nature **134**, 183, 1934.

in guter Näherung genau wie in der Atomhülle vektoriell zum magnetischen Moment des gesamten Kerns zusammengesetzt werden können. Denn der Hauptbeitrag zu den magnetischen Momenten stammt, wie oben erwähnt, von *den* Bahnen des virtuellen Elektron-Neutrinopaares, deren Energie das hundert- bis tausendfache der Ruhenergie des Elektrons beträgt. Diese Bahnen werden durch die Anwesenheit anderer schwerer Teilchen in der Nachbarschaft nur wenig gestört. Die relative Grösse der Abweichung von der strengen Additivität der magnetischen Momente wird also etwa durch das Verhältnis der C o u l o m b schen Wechselwirkungen und der Austauschenergien zum hundert- bis tausendfachen der Elektronenruhenergie gegeben; es sollte also bis auf Fehler von einigen Prozent möglich sein, die magnetischen Momente der Kerne durch Vektorzusammensetzung der Bahn- und Spinmomente zu berechnen, so wie es etwa in den Arbeiten von L a n d é [1]), S c h ü l e r und K a l l m a n n [2]), T a m m und A l t s c h u l e r [3]), I n g l i s [4]) versucht wird. Allerdings ist hierbei zu beachten, dass das Vektormodell für die Atomkerne wegen der fehlenden Zentralkraft wesentlich komplizierter aussehen kann als für die Atomhülle und dass daher die in den eben genannten Arbeiten gezogenen Schlüsse nicht den Grad von Sicherheit beanspruchen können, wie etwa die Bestimmung einer Elektronenkonfiguration aus dem Z e e m a n effekt.

Kehrt man nun wieder zu der oben besprochenen Analogie zwischen Atomkern und Atomhülle zurück, so vervollständigen die Überlegungen über die magnetischen Momente das Bild von einem Aufbau des Atomkerns aus Protonen und Neutronen, der weitgehend nach den Gesetzen der Quantenmechanik beschrieben werden kann. Eine solche Beschreibung kann zunächst nur in der Näherung gelten, in der die Geschwindigkeit der schweren Teilchen als sehr klein gegen die Lichtgeschwindigkeit betrachtet werden kann. Diese Näherung ist nach den empirischen Daten über die Grösse des Atomkerns und die Massendefekte erheblich schlechter als ihr Analogon in der Atomhülle. Man kann aber, wenn jene Überlegungen über die

---

1) A. L a n d é, Phys. Rev. **44**, 1028, 1933; D. I n g l i s u. A. L a n d é, Phys. Rev. **45**, 842; **46**, 76, 1934.

2) H. K a l l m a n n u. H. S c h ü l e r, Zs. f. Phys. **88**, 210, 1934 u. H. S c h ü l e r, Zs. f. Phys. **88**, 323, 1934.

3) J. T a m m u. A l t s c h ü l e r, Acad. U.S.S.R. **1**, 455, 1934.

4) D. I n g l i s, Phys. Rev. **47**, 84, 1935. Vgl. auch G. B r e i t, Phys. Rev. **46**, 230, 1934.

magnetischen Momente schon richtig sind, ebenso wie in der
Atomhülle noch einen Schritt über diese Näherung hinausgehen
und die relativistischen Effekte, insbesondere den Spin, in erster
Näherung mitberücksichtigen. Erst dieser Umstand hat durch das
reiche empirische Material, das der Z e e m a n effekt zur Kenntnis
der Energieniveaus und ihrer Quantenzahlen in der Atomhülle bei-
steuerte, zu einer genauen theoretischen Kenntnis der Struktur der
Atomhülle geführt. Man darf vielleicht hoffen, dass in ähnlicher
Weise die magnetischen Momente der Kerne später die genauesten
Aussagen über die Struktur des Atomkerns liefern werden.

Eingegangen: 15 Feb. 1935

n Protonen an Protonen.''
:tad[9]) der Streuung von
gen zur Folgerung, daß
ibschen Wechselwirkung
m Typus der Neutronen-
Nach Rechnungen von
intial, das zwar geringer,
st, wie die der Theorie
virkungsenergie zwischen

rmischen Theorie des β-

:erfalls auf Grund der
dene Ausdrücke für die
mit dem Neutrinofeld
ürlicher und künstlicher
nd, können gut wieder-
insatz von *Konopinski-*
Verlauf nach dem Ansatz
e theoretischen Kurven
torische Berechnung der
mntnis der Grenzenergie
ge zu, ob beim β-Zerfall
iesatz erfüllt.

ektrum der β-Elektronen

onen der β-Strahler war
ig wegen experimenteller
Vortr. gelungen, eine gas-
en und den radioaktiven

ysic. Rev. **49**, 432 [1936].
mäß brieflicher Mitteilung.
Konopinski u. *Uhlenbeck*,

Tagung der Faraday-Society.

Leeds, 20. bis 22. April 1936.

Vorsitzender: Präsident Prof. Dr. Whytlaw-Gray.

Zerfall in der *Wilson*-Kammer zu studieren. ThC geht unter β-Zerfall in ThC' über, dieses sendet danach praktisch spontan ein α-Teilchen aus. Die Spuren der Elektronen und α-Teilchen haben deshalb paarweise einen gemeinsamen Ursprung, dies gestattet eine sichere Zählung der Zerfallselektronen. Die Zahl der Elektronen mit kleiner Energie ist verhältnismäßig groß, im Gegensatz zu früher von anderen Experimentatoren mit dem Zählrohr erhaltenen vorläufigen Daten. Das β-Spektrum liegt gut auf der *Uhlenbeck-Konopinski*schen Kurve und weicht stark von der *Fermi*schen ab.

Heisenberg, Leipzig: „*Theorie der Schauer*[12]).''

Die erste Möglichkeit der Partikelerzeugung ist in der Theorie des Positrons gegeben. Dieser Prozeß ist jedoch nicht zur Erklärung des Auftretens der Schauer in der Höhenstrahlung geeignet, da hierfür ein Prozeß höherer Ordnung notwendig ist, bei dem eine große Anzahl Teilchen zugleich erzeugt werden müßten. Die Quantenelektrodynamik ergibt aber, daß die Wahrscheinlichkeit dafür, daß in einem Einzelakt n Pärchen erzeugt werden, bei großen Energien proportional der n-ten Potenz der Feinstrukturkonstanten $e^2/hc = 1/_{137}$ ist, also für große n verschwindend klein. Die *Fermi*sche Theorie des β-Zerfalls ergibt jedoch eine neue Möglichkeit. Ein Elementarprozeß wäre beispielsweise ein solcher, bei dem unter Verbrauch der Primärenergie ein Proton im Felde eines Kerns sich in ein Neutron plus Positron plus Neutrino verwandelt; die Iteration des Prozesses ist: Neutron → Proton + Elektron + Neutrino; Proton → Neutron u.s.f. Wichtig ist, daß nach der *Fermi*schen Theorie bei genügend hoher Energie ein Prozeß n-ter Ordnung, bei dem n Pärchen + 2n Neutrinos entstehen, ebenso wahrscheinlich wird wie ein Elementarprozeß. Die mittlere Zahl der erzeugten Teilchen und die mittlere Energie eines Schauerpartikels lassen sich abschätzen aus den Konstanten, die in die *Fermi*sche Theorie eingehen, und mit dem Experiment in Einklang bringen.

————

[12]) Erscheint demnächst in der Z. Physik.

laufende Koagulation. Der Koagulationsverlauf einer großen Reihe von Aerosolen folgt dem einfachen Gesetz:

$$1/n_1 - 1/n_2 = K (t_2 - t_1).$$

n_1 und n_2 sind die im Kubikzentimeter zu den Zeiten t_1 und t_2 vorhandenen Teilchen, K ist die Koagulationskonstante. Ihr Wert hängt ab: 1. von der mittleren Teilchengröße; feinteilige Aerosole koagulieren rascher als grobteilige. 2. Von dem Teilchengrößenbereich und der Verteilung innerhalb desselben; anisodisperse Systeme koagulieren schneller als isodisperse. 3. Von der Form der Teilchen und der von ihnen gebildeten Aggregate.

Über das erste Stadium der Aerosolbildung ist fast nichts

, Rauch und Nebel[1]).

vie technische Bedeutung
l englischer und nicht-
ge zu der Tagung über-

uppen unterteilt:

das Verhalten disperser

ichtigen Teilchen, wie

angehörige Absorptionsbande bei 765 mμ von stärkerer anomaler Dispersion begleitet ist als die intensiveren Banden bei 630 und 578 mμ, die dem $(O_2)_2$-Molekül zugeschrieben werden.

6. Hr. W. Heisenberg (Leipzig): **Theoretische Untersuchungen zur Ultrastrahlung.**

Die theoretischen Ergebnisse über Bremsstrahlung und Paarerzeugung einerseits (Bhabha-Heitler; Oppenheimer und Carlson), über die durch die Kernkräfte hervorgerufenen Prozesse andererseits führen zu der Vermutung, daß energiereiche Elektronen, Positronen und Lichtquanten ihre Energie auf relativ kurzen Strecken (wenige cm Blei) durch „Kaskaden"-prozesse verlieren, daß aber energiereiche schwere Teilchen und Neutrinos häufig ihre Energie durch eine explosionsartige Schauerbildung in einem einzigen Akt abgeben. Das vorliegende experimentelle Material über Koinzidenzen, Hoffmannsche Stöße usw. wird nach diesen Gesichtspunkten gedeutet. Dabei wird auf die Möglichkeit hingewiesen, bei Ultrastrahlungsuntersuchungen in großer Tiefe die Wirkung der Neutrinos nachzuweisen.

7. Hr. E. Wilhelmy (Heidelberg): **Resonanzaustritt von Kerntrümmern.**

Bei der Untersuchung von Kernreaktionen mit schnellen Neutronen wurden Resultate erhalten, die als Umkehrung des Resonanzeintritts geladener Teilchen in einen Kern gedeutet werden: Es wurden die bei den Umwandlungen $_7N^{14}$ (n, α) $_5B^{11}$, $_8O^{16}$ (n, α) $_6C^{13}$ und $_9F^{19}$ (n, p) $_8O^{19}$ auftretenden Energiesummen von α-Teilchen plus Rückstoßkern bzw. Proton plus Rückstoßkern mit Hilfe von Ionisationskammer, Proportionalverstärker und Oszillograph gemessen, und zwar mit Neutronen aus (Rn + Be) und (Po + Be). Kurven, in denen die Anzahl der Oszillographenausschläge bestimmter Größe über den entsprechenden Energiesummen aufgetragen sind, zeigen für jede der Reaktionen charakteristische Maxima, für die Umwandlung von N bei 1,4; 2,0; 2,5; 3,1 e-MV, von O bei 0,85 e-MV und von F bei 1,75 und 2,3 e-MV. Die Energieverteilung der benutzten Neutronen ist völlig verwaschen und kann demnach nicht zum Auftreten solcher Gruppen führen. Die Maxima erschienen auch unabhängig davon, ob (Rn + Be)- oder (Po + Be)-Neutronen verwendet wurden, an den gleichen Stellen. Die beobachteten Energien sind durchweg kleiner, als der Höhe des betreffenden Potentialwalls entspricht. Dieser Befund legt die Auffassung nahe, daß die Gruppen angeregten Zuständen der entstehenden Zwischenkerne N^{15}, O^{17} und F^{20} entspringen. Aus dem kontinuierlichen Neutronenspektrum werden gewissermaßen die Teilchen ausgesucht, die gerade eine zu einer „Resonanzemission" eines Protons oder α-Teilchens passende Energie besitzen. Zur Entscheidung, ob es sich bei O um eine (n, α)- oder (n, p)-Reaktion handelte, wurde eine Prüfung auf Aktivität des

III

WAHRSCHEINLICHKEITSAUSSAGEN IN DER QUANTEN-THEORIE DER WELLENFELDER

par Von W. HEISENBERG
(Leipzig).

Die Gesamtheit der Probleme, die der Mathematiker mit dem Begriff « Wahrscheinlichkeit », verbindet und von denen in diesem Kongress in erster Linie die Rede ist, unterscheidet sich· erheblich von den Fragen, deren Untersuchung die Physiker veranlasst hat, in ihrer Wissenschaft statistische Gesetzmässigkeiten zu betrachten. Während es sich in der Mathematik darum handelt, entweder den Begriff Wahrscheinlichkeit selbst zu untersuchen, oder komplizierte statistische Probleme einer mathematischen Analyse zugänglich zu machen, steht für den Physiker die Frage im Vordergrund, ob die fundamentalen Naturgesetze statistischer Art sind, und in welchen Gebieten der Naturbeschreibung der statistische Charakter der Gesetze die entscheidende Rolle spielt. Ich möchte im Folgenden zunächst von einem abgeschlossenen Gebiete der modernen Physik, nämlich der unrelativistischen Quantenmechanik ausgehen und darlegen, warum wir in diesem Gebiet den statistischen Charakter der Naturgesetze für endgültig halten. Im zweiten Teil will ich über die einstweilen noch tastenden Versuche sprechen, in denen die Physik unternimmt, die Naturgesetze, deren Wirken in Experimenten über die Atomkerne und die Kosmische Strahlung in Erscheinung tritt, mathematisch zu formulieren.

Die prinzipielle Situation der Quantenmechanik soll an einem einfachen Beispiel auseinandergesetzt werden. Von einem Punkte P (fig. 1) sollen nach allen Richtungen Elektronen einer genau bekannten Geschwindigkeit ausgesandt werden. Nach den Erfahrungen über die Beugung der Materiewellen kann dieser Sachverhalt auch durch die Feststellung beschrieben werden, von

Actualités Scientifiques et Industrielles *734/1*, 45–51 (1938)

dem Punkte P werde eine monochromatische Materiewelle ausgesandt. Diese Materiewelle soll nun auf einen Schirm fallen, der
an 2 Stellen S_1 und S_2 durchlöchert ist. Wenn der Durchmesser
der Löcher vergleichbar mit der Wellenlänge der Materiewelle
ist, so wird die Welle auf der anderen Seite des Loches wieder
über einen grösseren Winkelbereich verteilt. Die von den beiden
Spalten herkommenden Wellen fallen dann auf einen zweiten
Schirm. Ein gewisses Gebiet dieses Schirms kann sowohl von
Wellen getroffen werden, die von S_1 ausgehen, wie von solchen,

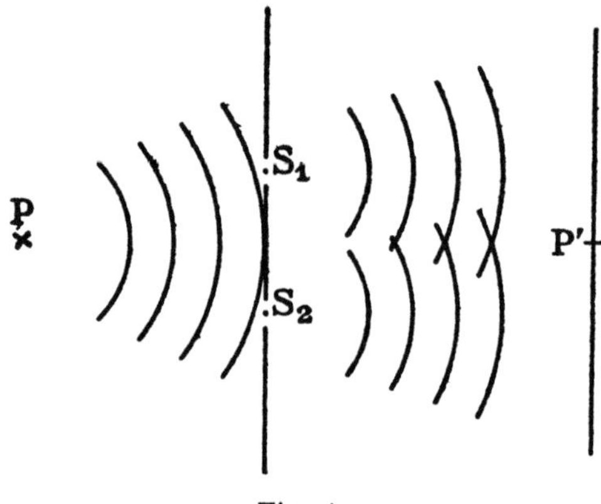

Fig. 1.

die von S_2 herkommen. In diesem Gebiet des Schirms muss eine
Interferenzerscheinung eintreten, da der Gangunterschied der
beiden auftretenden Wellenbewegungen die Intensitität entweder
verstärken oder schwächen kann. Insbesondere wird es auf dém
zweiten Schirm Punkte geben, an denen die in der Phase entgegengesetzten Wellen sich gerade aufheben. Die Erfahrung zeigt,
nun, dass die Materiewellen in einem solchen Sinne als Wahrscheinlichkeitswellen oder Führungswellen für die Elektronen
betrachtet werden können, dass das Quadrat der Wellenamplitude
ein Mass für die Häufigkeit gibt, mit der an der betreffenden Stelle
Elektronen angetroffen werden ; dass also s. B. die Stellen des
zweiten Schirms, an denen die beiden Wellenzüge durch Interferenz sich eben auslöschen, auch nicht von Elektronen getroffen
werden können.

Der eben geschilderte Vorgang soll nun als Vorgang, der sich
an einer grossen Anzahl von Elektronen abspielen kann, nach
den einfachsten statistischen Gesichtspunkten interpretiert wer-

den. Der Bruchteil der Fälle, in denen ein Elektron, das von P ausgesandt wird, den Spalt S_1 bzw. S_2 trifft, soll w_1 bzw. w_2 heissen. Ferner nennen wir den Bruchteil der Fälle, in denen ein Elektron, das durch S_1 gelangt ist, einen bestimmten Punkt P' des zweiten Schirmes trifft, v_1, die entsprechende Zahl für den zweiten Spalt v_2. Die relative Anzahl der Elektronen, die nach P' gelangt, wenn der Spalt 2 geschlossen ist, wird also $w_1 v_1$ betragen, bei geschlossenem Spalt 1 jedoch beträgt sie $w_2 v_2$. Nach den einfachsten Regeln der Abzählung würde folgen, dass bei Oeffnung beider Spalte die relative Gesamtzahl der Elektronen $w_1 v_1 + w_2 v_2$ betrüge. Dies ist jedoch wegen der Interferenz der Materiewellen nicht der Fall. Eben dieser Widerspruch gegen die gewöhnlichen Regeln der Wahrscheinlichkeit ist der eigentliche Grund für die heutige Physik, das Auftreten statistischer Zusammenhänge in den Naturgesetzen für endgültig zu halten. In der Tat könnte, wenn die relative Anzahl der Elektronen im Punkte P' $w_1 v_1 + w_2 v_2$ betrüge, kein Einwand gegen die Hoffnung geäussert werden, es werde in späterer Zeit doch möglich sein, das Verhalten der Elektronen mit bisher vielleicht noch unbekannten Hilfsmitteln exakt vorherzusagen, also z. B. aus den Vorgängen, die sur Aussendung der Elektronen führen, genau vorherzusagen, ob ein bestimmtes Elektron den Spalt S_1 oder S_2 treffen wird. Daraus, dass die Anzahl der Elektronen in P' jedoch von der Summe $w_1 v_1 + w_2 v_2$ verschieden ist, folgt, dass es sich bei diesem Interferenzversuch um einen anderen Beobachtungszusammenhang handelt, bei dem keine Festellung darüber getroffen werden kann, durch welchen Spalt hindurch das Elektron den Punkt P' erreicht hat. Es lässt sich auch leicht zeigen, dass jede Abänderung der Versuchsanordnung, die darauf abzielt, den Elektronenweg festzustellen, gleichzeitig das Interferenzbild verwischen müsste. Die Interferenzanordnung stellt also einen Beobachtungszusammenhang her, bei dem die Unkenntnis des Elektronenweges einen wesentlichen Bestandteil in der physikalischen Beschreibung der beobachteten Vorgänge bildet. Dieser Folgerung kann man sich auch nicht durch den Einwand entziehen, die Verwendung eines anschaulichen Bildes, bei dem die Elektronen als kleine beobachtbare Korpuskeln vorgestellt werden, sei unzulässig, und dieses Bild müsste durch ein anderes ersetzt werden. Das, was schliesslich beobachtet wird, ist ja doch die Anwesenheit eines Elektrons an einem bestimmten

Punkte. So liefert die Quantenmechanik, die ja in ihrem mathematischen und physikalischen Inhalt schon als abgeschlossen betrachtet werden darf, ein erstes Beispiel dafür, dass die Naturgesetze dort, wo wir über den Bereich der einfachen, anschaulich beschreibbaren Vorgänge hinausgehen, statistische Elemente bereits in ihrer Grundlage enthalten können.

Es entsteht nun die Frage, an welcher Stelle der zukünftigen Theorie, mit der wir die Erscheinungen bei den Atomkernen und der Höhenstrahlung zu beschreiben hoffen, statistische Zusammenhänge wesentlich werden. Der Versuch, die Quantenmechanik auf die Strahlungserscheinungen und auf die Atomkerne anzuwenden, führt stets auf eine charakteristische formale Schwierigkeit. Eine relativistisch invariante Wellengleichung, die der Tatsache Rechnung trägt, dass Partikel neu entstehen und verschwinden und auf einander Kräfte ausüben können, muss notwendig nichtlineare Glieder enthalten. Wendet man auf eine solche Wellengleichung die Vorschriften der Quantentheorie, an, so führt die mathematische Analyse stets zu divergenten Resultaten, die zeigen, dass es sich hier überhaupt nicht um ein widerspruchsfreies mathematisches System von Gleichungen handelt. In der klassischen Elektronentheorie war bereits eine ähnliche Schwierigkeit bekannt. Wenn man die Energie des elektromagnetischen Feldes einer Punktladung berechnet, so wird diese Energie unendlich, und man ist daher genötigt, dem Elektron einen endlichen Radius zuzuschreiben. In der Quantentheorie kann zwar die Einführung eines Elektronenradius nicht in dieser Weise begründet werden, denn die Kleinheit der Elektronenladung im Verhältnis zu der aus Wirkungsquantum und Lichtgeschwindigkeit gebildeten Grösse gleicher Dimension $\left(\sqrt{h \cdot c}\right)$ hat zur Folge, dass der Feldbegriff in der Umgebung des Elektrons nicht mehr einfach angewendet werden kann. Trotzdem könnte aber auch in der Quantentheorie nicht erreicht werden, dass die Energie des elektromagnetischen Feldes, das ein Elektron umgibt, zu der richtigen beobachteten Ruhmasse des Elektrons führt, es sei denn, dass man in irgendeiner Form einen Radius oder eine Masse in die Grundannahmen der Theorie hineinsteckt. Denn aus den Konstanten h und c allein lässt sich dimensionsmässig keine Masse bilden. Die genannte Divergenzschwierigkeit kann man ungenau in die Behauptung zusammenfassen, dass die bisher aus der an-

schaulichen Physik übernommenen Naturgesetze, wie wir sie etwa
aus der klassischen Elektronentheorie kennen, zusammen mit den
quantentheoretischen Gesetzen ihre Gültigkeit nur bis zu sehr
kleinen räumlichen Entfernungen, etwa bis zum klassischen Elek-
trone nradius behalten; dass aber bei noch kleineren Entfernungen
an die Stelle dieser Gesetzmässigkeiten neue und uns einstweilen
noch unbekannte treten. Obwohl uns also die neuen Züge der Naturbeschreibung, die erst bei kleinsten räumlichen Abständen in
Erscheinung treten, einstweilen noch unbekannt sind, entnimmt
man aus ihrer vermuteten Existenz doch das Recht, die in den
bisherigen Theorien auftretenden Divergenzen dadurch zu beseitigen, dass man etwa auftretende divergente Integrale nur bis
zu Abständen der kritischen Grössenordnung hinab erstreckt
und dadurch konvergente und in einem gewissen Erfahrungsbe-
reich anwendbare Ergebnisse erhält.

Die Situation der heutigen Physik kann in diesem Punkte
verglichen werden mit dem Zustand der Strahlungstheorie kurz
vor der Jahrhundertwende. Damals hatte die Maxwell'sche
Theorie der Strahlung sowie die statistische Interpretation der
Wärmelehre ihre endgültige Form erreicht. Die konsequente
Anwendung der statistischen Mechanik auf die Strahlungstheorie
führte jedoch für das Problem der Strahlung im thermodynamischen Gleichgewicht zu einem divergenten Resultat, nämlich
zum sog. Rayleigh-Jeans'schen Strahlungsgesetz. Der hier zutage
tretende Widerspruch mit der Erfahrung konnte nur durch die
Hypothese behoben werden, dass in der Theorie der Strahlung
auf eine neue universelle Konstante von der Dimension einer
Wirkung Rücksicht genommen werden müsse. Die mit dieser
Konstanten verknüpften, damals noch unbekannten Erscheinungen
konnten zum Anlass genommen werden, das Rayleigh-Jeans'sche
Gesetz an einer durch diese Konstante bestimmten Stelle abzu-
schneiden und damit den Anschluss an die Erfahrung herzustellen.
In dieser Weise entstand schliesslich das Planck'sche Strahlungsgesetz. Die geschilderte Situation der heutigen Atomphysik deutet in ähnlicher Weise darauf hin, dass in den Erscheinungen der
Kernphysik und der Höhenstrahlung auf eine neue universelle
Konstante von der Dimension einer Länge Rücksicht genommen
werden muss. Das Wirken einer solchen Konstanten tritt uns ja
in der Kernphysik an verschiedenen Stellen entgegen. Ausser

den oben genannten Divergenzen, die zur Einführung einer universellen Länge zwingen, ist in erster Linie die Tatsache zu erwähnen, dass zwischen den Elementarteilchen bei Abständen der Grössenordnung 10^{-13} cm neue Kräfte auftreten, die für den Bau der Atomkerne entscheidend sind und bei grösseren Abständen verschwinden. Zweitens aber kann die Existenz von Elementarteilchen einer bestimmten Ruhmasse als Auswirkung einer universellen Konstanten von der Dimension einer Länge und der Grössenordnung 10^{-13} cm aufgefasst werden. Denn aus den Konstanten : Wirkungsquantum, Lichtgeschwindigkeit und universelle Länge lässt sich eine Grösse von der Dimension einer Masse bilden, die in der Grössenordnung mit den Ruhmassen der Elementarteilchen übereinstimmt. Gerade die Existenz verschiedener Ruhmassen für die verschiedenen Sorten der Elementarteilchen deutet ja darauf hin, dass die Ruhmassen nicht eigentlich fundamentale Konstanten sind, sondern aus einer anderen, in die Grundlage der Naturgesetze eingehenden Konstante hergeleitet werden müssen.

Wenn man das Auftreten einer Konstanten von der Dimension einer Länge in dieser Weise interpretiert, so wird man schliessen müssen, dass in allen physikalischen Messungen der Begriff der Länge im anschaulichen Sinne nicht genauer als bis auf Grössen der Ordnung 10^{-13} cm festgelegt werden kann. Und dies müsste wohl bedeuten, dass die zukünftige Theorie bei Längen von der Grössenordnung 10^{-13} cm neue statistische Elemente enthalten wird, die der bisherigen Theorie unbekannt waren. Vielleicht muss man aber auch an dieser Stelle damit rechnen, dass die Natur uns bei Längen der Grössenordnung 10^{-13} cm zu einer neuen Abstraktion in der Naturbeschreibung zwingt, zu der es weder in der Quantentheorie noch in der Relativitätstheorie ein Analogon gibt.

Da der Rahmen dieses Vortrages nicht erlaubt, über diese kurzen und nur unscharf formulierten Andeutungen hinauszugehen, so mögen sie nur als Hinweis darauf betrachtet werden, dass die Entwicklung der Kernphysik nirgends Veranlassung gibt zu der Hoffnung, es könnten die statistischen Gesetzmässigkeiten der Quantentheorie in der zukünftigen Theorie der Elementarteilchen durch streng deterministiche ersetzt werden. Vielmehr wird die neue, in der Kernphysik zutagetretende universelle Konstante entweder zu neuen statistisschen Zusammenhängen

· führen, oder sie wird, was wohl noch wahrscheinlicher ist, in Verbindung mit den statistischen Gesetzmässigkeiten der Quantentheorie zu neuen Abstraktionen der Naturbeschreibung zwingen, die in der bisherigen Physik kein Analogon haben.

IMPRIMERIE R. BUSSIÈRE. — SAINT-AMAND (CHER), FRANCE. — 13-9-1938

Der Durchgang sehr energiereicher Korpuskeln durch den Atomkern (*).

Prof. W. HEISENBERG (Leipzig).

Auf Nebelkammeraufnahmen der Höhenstrahlung werden häufig Prozesse beobachtet, bei denen gleichzeitig mehrere schwere Teilchen verhältnismassig geringer Energie auftreten und die man deshalb als Kernzertrümmerungen bezeichnet hat ([1]). Um diese Prozesse näher zu studieren, wurde die Frage untersucht, welche Vorgänge nach der üblichen Theorie der Kernkräfte theoretisch zu erwarten sind, wenn ein Neutron oder Proton sehr hoher Energie den Atomkern durchdringt. Als Grundlage für diese Berechnungen wurde vorausgesetzt, dass zwischen je zwei Kernbausteinen ein Potential der Form

$$V_{12} = 0{,}038\, Mc^2\left[\frac{5}{24}(\rho_1\rho_2)(\sigma_1\sigma_2) + \frac{1}{8}(\rho_1\rho_2)\right] e^{-\frac{r^2}{a^2}}$$

wirkt, wobei $a = 0{,}8\,\dfrac{e^2}{mc^2} = 0{,}8\,r_0$ gesetzt wurde ([2]). (*M*, *m* Pro-

(*) Dieser Bericht des Vortrages in Bologna ist auch am 12-11-37 in den « Naturwissenschaften », **25**, 749, 1937 erschienen. Die Voraussetzung über die Kernkräfte und damit die numerischen Resultate sind gegenüber dem Vortrage etwas verändert.

([1]) Z. B. C. P. ANDERSON u. S. H. NEDDERMEYER, « Physic. Rev. », **50**, 263, 1936; ferner die in der Gelatineschicht einer photographischen Platte beobachteten Kernzertrümmerungen von M. BLAU u. H. WAMBACHER, « Nature », (Lond.), **140**, 585, 1937. — Vgl. auch G. HERZOG u. P. SCHERRER, « J. Physique et Radium », **6**, 489, 1935. — G. H. RUMBOUGH u. G. L. LOCHER, « Physic. Rev. », **49**, 855, 1936. — E. FÜNFER, « Naturwiss. », **25**, 235, 1937. — E. SCHOPPER, « Naturwiss. », **25**, 557, 1937.

([2]) Der Ansatz für die Kernkräfte unterliegt einstweilen noch einer gewissen Willkür. Der obige Ansatz berücksichtigt die von G. BREIT u. E. FEENBERG,, « Physic. Rev. », **50**, 850, 1937, H. VOLZ, « Z. Physik », **105**, 551, 1937 tonen- bzw. Elektronenmasse, σ Spinkoordinate, ρ unterscheidet Proton und Neutron).

Ferner wurde angenommen, dass bei sehr grossen gegenseitigen Geschwindigkeiten der Stosspartner die Wirkungsquerschnitte für den Stoss aus diesem Potential nach der Bornschen Methode berechnet werden können.

Als Ergebnis dieser Berechnung zeigte sich, dass der Durchgang eines sehr schnellen Protons oder Neutrons durch den Atomkern grosse Ähnlichkeit aufweist mit dem Durchgang schneller Elektronen durch gewöhnliche Materie. Eine sehr energiereiche Korpuskel gibt längs ihrer ganzen Bahn durch den Atomkern Energie an die Teilchen ab, an denen es vorbeifliegt, wobei die Energie, die auf ein getroffenes Teilchen im Mittel übertragen wird, ungefähr 22 MEV beträgt. Die Wahrscheinlichkeit für die Übertragung sehr hoher Energien wird nach der Rechnung gering. Sei $\bar{n}(E)$ die mittlere Anzahl der Sekundärteilchen, die beim Durchgang eines sehr schnellen Protons durch einen Atomkern einen so grossen Stoss erhalten, dass ihre Energie nach dem Verlassen des Atomkerns den Wert E übersteigt, so ergibt sich für Pb-Kerne und höchste Geschwindigkeit ($v \backsim c$) des eindringenden Teilchens:

$$\varepsilon = \frac{1000E}{Mc^2} = \quad 20 \quad\quad 40 \quad\quad 60 \quad\quad 80 \quad\quad 100$$

$$\bar{n}(E) = \quad 0{,}26 \quad 0{,}052 \quad 0{,}012 \quad 0{,}0019 \quad 0{,}00031.$$

Die Wahrscheinlichkeit w_n dafür, dass bei einem Durchgang gleichzeitig n Sekundärteilchen des betreffenden Energiebereiches ausgelöst werden, wird unter der Voraussetzung $\bar{n} << 1$:

$$n = \quad 1 \quad\quad 2 \quad\quad 3 \quad\quad 4$$

$$w_n = \quad \bar{n} \quad \frac{9}{16}\bar{n}^2 \quad \frac{9}{40}\bar{n}^3 \quad \frac{9}{128}\bar{n}^4.$$

N. KEMMER, « Nature », (Lond.), **146**, 192, 1937, und E. FEENBERG, « Physic. Rev. », **50**, 667, 1937, aufgestellten Bedingungen. (In der VOLZschen Bezeichnung wurde $a_2 = -\,^1/_{16}$, $a_4 = \,^1/_8$ gesetzt, was etwa der Grenze des nach VOLZ zulässigen Bereichs entspricht, während bei VOLZ noch $a_2 = a_4 = 0$ möglich schien). Die Einwände E. FEENBERGS, « Physic. Rev. », **52**, 759, 1937, gegen die VOLZschen Bedingungen scheinen mir nicht genügend begründet, dagegen wurde in dem obigen Ansatz von der Darstellung der Sättigungsbedingungen bei E. FEENBERG, « Physic. Rev. », **51**, 777, 1937, Gebrauch gemacht. Die Reichweite der Kernkräfte wurde so festgesetzt, dass sie zu den älteren GAMOWschen Kernradien passt (vgl. H. EULER, « Z. Physik », **105**, 553, 1937, da BOHR (Besprechung auf der Konferenz in Kopenhagen, September 1937) gezeigt hat, dass die neueren BETHEschen Radien einer eingehenden Kritik nicht standhalten.

Bei dem Vergleich dieser Zahlen mit der Erfahrung ist allerdings zu beachten, dass die ausgelösten Sekundärteilchen in vielen Fällen den Kern nicht verlassen, da sie vorher selbst wieder durch die anderen Kernteilchen gebremst werden und ihre Energie schliesslich im Sinne der Bohrschen Deutung der Kernprozesse zu einer Temperaturerhöhung des ganzen Kernes verwenden. Nur die Teilchen, die an der Kernoberfläche sitzen, werden im allgemeinen aus dem Kern unmittelbar herausgeschlagen werden können. Da beim Pb-Kern etwa zwei Drittel aller Teichen an der Oberfläche sitzen und jedes getroffene Oberflächenteilchen nur in der Hälfte der Fälle den Stoss in der Richtung vom Kern weg empfangen wird, darf man die Wahrscheinlichkeit dafür, dass ein Sekundärteilchen der Energie grösser als E während des Stossaktes den Kern wirklich verlässt, zu etwa $w_1/3$ ansetzen. Eine entsprechende Reduktion tritt für w_2 usw. ein. Nach dem Durchgang des schnellen Teilchens wird allerdings ausserdem infolge der Temperaturerhöhung eine Kernverdampfung einsetzen, bei der Kernbausteine geringerer Energie den Kern verlassen können.

Der mittlere Energieverlust pro Zentimeter des durchdringenden Teilchens wird durch den Ausdruck

$$\frac{\partial \varepsilon}{\partial x} = 1660 \rho r_0{}^2 \frac{\left(1 + \dfrac{\varepsilon}{1000}\right)^2}{\varepsilon\left(1 + \dfrac{\varepsilon}{2000}\right)}$$

gegeben. Hierin ist $\varepsilon = \dfrac{1000E}{Mc^2}$, r_0 der klassische Elektronenradius und ρ die Anzahl der Kernbausteine pro Kubikzentimeter. Diesen Ausdruck kann man in zwei Weisen ausnützen: Man kann erstens die Reichweite R eines Teilchens in der dichten Kernmaterie ausrechnen und erhält hierfür:

$$R = r_0 \cdot 1{,}8 \cdot 10^{-4} \frac{\varepsilon^2}{1 + \dfrac{\varepsilon}{1000}}.$$

Das stossende Teilchen muss also mindestens eine Energie von etwa 150-200 MEV haben, um einen Bleikern zu durchdringen. Man kann zweitens den mittleren Energieverlust pro Zentimeter Bleischicht ausrechnen und mit dem Energieverlust durch Ionisation vergleichen. Der durch Kernprozesse hervorgerufene Energieverlust beträgt in Blei etwa das 0,21 fache, in Aluminium nur das etwa 0,135 fache des Energieverlustes durch Ionisation.

W. HEISENBERG

Beim Durchgang eines Teilchens höchster Geschwindigkeit nimmt ein Bleikern im Mittel eine Energie von etwa 23,7 MEV auf. Langsamere Teilchen werden noch stärker gebremst, so dass der Bleikern maximal bis zu etwa 200 MEV Energie übernehmen kann. Bei der auf eine solche Energieaufnahme folgenden Verdampfung werden naturgemäss viele Sekundärteilchen emittiert.

Die hier aufgezählten Ergebnisse hängen zum Teil stark von den speziellen Voraussetzungen über die Kernkräfte ab, die der Rechnung zugrunde gelegt wurden, sie können also nur als grobe Abschätzungen gelten. Ferner bedarf die Frage, inwieweit die angenommenen Kräfte bei relativistischen Geschwindigkeiten der beteiligten Teilchen Gültigkeit behalten, einer besonderen Diskussion, die zusammen mit den Einzelheiten der Rechnung in einer späteren Arbeit durchgeführt werden soll.

DISCUSSIONSBEMERKUNGEN.

Prof. NIELS BOHR. — Wir verstehen alle, dass die schönen Berechnungen von Prof. HEISENBERG über die Zusammenstösse zwischen Kernen und schweren Elementarteilchen hoher Energie einen sehr wertvollen Beitrag zu der Analyse der bezüglichen Phänomene darstellen. Insbesondere ist es ja sehr interessant, dass es, wie er zeigt, möglich sein dürfte, von dem fortgesetzten Studium der Spuren von schweren Partikeln in Höhenstrahlaufnahmen Schlüsse auf die Kernkräfte zu ziehen. Weiter liefern Heisenbergs Berechnungen eine sehr willkommene Stütze für die neueren Anschauungen uber Kernzusammenstösse, die mehrmals auf diesem Kongress diskutiert worden sind. In der Tat zeigen seine Berechnungen in überzeugender Weise, dass schwere Teilchen nur Kerne durchdringen können, wenn ihre Energie bedeutend grösser ist als diejenige, welche in den gewöhnlichen Versuchen über Kernzertrümmerungen vorkommt.

In *Jahresbericht des Physikalischen Vereins zu Frankfurt am Main für das Rechnungsjahr 1937/38* (Frankfurt 1938) p. 27

— 27 —

IV. Vortrag des Herrn Professor Dr. W. H e i s e n b e r g (Leipzig):

F o r t s c h r i t t e   i n   d e r   T h e o r i e   d e s   A t o m k e r n s. (15. XI. 37.)

Der Vortrag geht aus von den einfachen anschaulichen Vorstellungen vom Atomkern, in denen der Kern mit einem Flüssigkeitstropfen aus sehr dichter Materie verglichen wird. Die Eigenschaften dieser Kernmaterie, die als Lösung mit zwei Komponenten: Protonen und Neutronen betrachtet wird, werden diskutiert. Es wird auseinandergesetzt, wie die Begriffe Temperatur, Dichte, Energie und Entropie mit Vorteil zur Ordnung der experimentellen Erfahrungen über den Atomkern verwendet werden können. Insbesondere werden die neueren Bohrschen Vorstellungen über die Atomkernumwandlung besprochen. Den Schluß des Vortrages bildet eine kurze Schilderung der Bedeutung der Kernphysik für die Probleme der kosmischen Strahlung.

V. Vortrag des Herrn Professor Dr. F. B e r g i u s (Heidelberg):

D i e   A u f s c h l i e ß u n g   v o n   H o l z   n a c h   d e m   B e r g i u s - R h e i n a u - V e r f a h r e n. (18. XI. 37.)

Im Hinblick auf die zur Verzuckerung des Holzes verwendete Säure nimmt das Bergius-Rheinau-Verfahren eine Mittelstellung ein: nicht sehr stark verdünnte und nicht wasserfreie Säuren werden benutzt, sondern eine konzentrierte, aber noch reichlich Wasser enthaltende Salzsäure.

Für die technische Durchführbarkeit sind die neueren Entwicklungen auf dem Gebiete des Steinzeugs und der säurefesten Metalle sehr wichtig gewesen, so daß man jetzt für alle in Betracht kommenden Behandlungen: bei gewöhnlicher oder bei erhöhter Temperatur, bei Zu- oder Abfuhr von Wärme, bei Adsorption konzentriertester und verdünntester Salzsäuregase, leistungsfähige Apparaturen betreiben kann.

Die einzelnen Stufen des Verfahrens: Aufschließung in der Diffusionsbatterie, Verdampfung zum Syrup, Trocknung im Heißluftstrom, sind vom Vortragenden schon ausführlich beschrieben worden[1]).

Das Ergebnis der Heißlufttrocknung ist ein hochkonzentriertes Kohlenhydratgemisch von sehr geringem Salzsäuregehalt, welches man nach Neutralisation verfüttern kann. Sein Wert ist in großen wissenschaftlichen Versuchsreihen erprobt worden.

Durch eine kurze Druckkochung in verdünnter Lösung wird der Zucker vergärbar. Besonderes Interesse wird der Herstellung von Hefe aus den Zuckerlösungen zugewandt, weil man dadurch eiweißreiche Futtermittel mit hohem Vitamingehalt erzeugen kann. Hier ergab das neue Ausgangsmaterial mit seinen besonderen Eigenschaften Anlaß zu neuen grundsätzlichen F o r t s c h r i t t e n in der Herstellungsweise eines schon lange bekannten Materials. Auch dabei ergab

---

[1]) Z. f. angew. Chem. 26, 1928; Zellstoff-Faser 4, 1935; Ind. u. Eng. Chemistry 29, 1937.

2   **Verhandlungen der Deutschen Physikalischen Gesellschaft.**   [Heft 1

Tagung des Gauvereins Thüringen=Sachsen=Schlesien der Deutschen Physikalischen Gesellschaft am 8. und 9. Januar 1938 in Dresden.

Vorträge:

1. Hr. W. Heisenberg (Leipzig): Durchgang energiereicher Korpuskeln durch den Atomkern.

Der Vortrag behandelt die Theorie der Kernzertrümmerungen, die Blau und Wambacher in der Gelatineschicht der photographischen Platte beobachtet haben. Insbesondere wird darauf hingewiesen, daß die Versuche von Blau und Wambacher in kurzer Zeit wohl zu der bisher genauesten Bestimmung der Reichweite der Kernkräfte führen dürften; denn die Energieverteilung der durch sehr energiereiche Korpuskeln ausgelösten Sekundärteilchen hängt nicht von der Größe der Kraft, sondern nur von ihrer Reichweite ab. (Ausführliche Veröffentlichung in Ber. Sächs. Akad., math.-phys. Kl. **89**, 369, 1937.)

2. Hr. F. Hund (Leipzig): Über den Diamagnetismus von Metallstücken.

Im Anschluß an London und Slater werden zwei Fragen gestellt: Wieweit kann man stark diamagnetisches Verhalten von Metallen (etwa von der Stärke des Diamagnetismus der Supraleiter) modellmäßig verstehen? Kann man quantentheoretisch begründen die übliche begriffliche Zerlegung des Stromes in einen „Leitungsstrom" (der durch einen vollen Querschnitt des Leiters Ladung trägt, aber Reibung erfährt) und in einen „Ampère-Weberschen Strom" (der die Magnetisierung bestimmt, keine Reibung hat, aber durch einen vollen Querschnitt des magnetisierten Körpers keine Ladung trägt)? Dazu wird das magnetische Verhalten kleiner Metallstücke bei tiefen Temperaturen untersucht. Es ergeben sich Zustandsgebiete mit starkem Diamagnetismus, die zum Teil durch Unstetigkeitslinien der Magnetisierung begrenzt sind. (Eine Mitteilung erscheint in den Ann. d. Phys.)

3. Hr. Adolf Smekal (Halle): Über die Molekularvorgänge an der Elastizitätsgrenze.

Zur Ermittlung der ersten irreversiblen Molekularvorgänge in mechanisch beanspruchten Einkristallen wurden mit Prof. Yôîchi Kidani aus Port Arthur Messungen der Spannungsdoppelbrechung von Steinsalzkristallen ausgeführt, die nur sehr geringen Drucken (bis 27 g/mm²) senkrecht zu einem Würfelflächenpaar unterworfen waren. Die Kristalle zeigten geringe Selbstspannungen sowie einen nichtlinearen Zusammenhang zwischen Doppelbrechung und Druckbeanspruchung, der auf das Vorhandensein elastischer Spannungsspitzen innerhalb optisch nicht auflösbarer Kristall-

In *Ergebnisse der exakten Naturwissenschaften, 17,* ed. by F. Hund, F. Trendelenburg (Springer, Berlin 1938)
pp. 1-69

Theoretische Gesichtspunkte zur Deutung der kosmischen Strahlung.

Von **H. Euler** und **W. Heisenberg,** Leipzig.

Mit 27 Abbildungen.

Inhaltsverzeichnis.

Die verwickelten Erscheinungen, die der kosmischen Strahlung ihren Ursprung verdanken, sind in den letzten Jahren soweit geordnet und geklärt worden, daß es auf Grund der bisher bekannten Theorie möglich erscheint, ein in sich zusammenhängendes, wenn auch noch nicht in allen Einzelheiten korrektes Bild dieser Erscheinungen zu zeichnen. Die wichtigsten Fortschritte, die diese Klärung möglich gemacht haben, waren die Erkenntnis, daß die Quantentheorie das Verhalten energiereicher Elektronen und Lichtquanten bis zu den höchsten vorkommenden Energien weitgehend richtig beschreibt und, im Zusammenhang mit dieser Erkenntnis, die Entdeckung eines neuen Elementarteilchens, das für die durchdringende Komponente der Höhenstrahlung verantwortlich ist und dessen Masse zwischen der des Elektrons und des Protons liegt. In der letzten Zeit ist dieses Elementarteilchen auch in Einzelaufnahmen in der Wilson-Kammer direkt nachgewiesen worden. Wir beginnen den folgenden Bericht mit einer kurzen Übersicht über die Ergebnisse, die sich aus der Wilson-Aufnahme eines einzelnen ionisierenden Teilchens unter der Wirkung eines Magnetfeldes ermitteln lassen (I). Es folgen die Ergebnisse der vorliegenden Theorie über das Verhalten und die sekundären Wirkungen der leichten Teilchen (II) und der schweren Teilchen (III). Die theoretischen Ergebnisse werden dann ausführlich mit den experimentellen verglichen, und zwar werden zunächst die Spektren der einzelnen Teilchensorten und ihre Verwandlung in der Atmosphäre (IV), dann ihre Sekundärwirkungen (Schauer, Stöße, Kernverwandlungen) (V) behandelt. Eine einigermaßen vollständige Diskussion des umfangreichen experimentellen Materials ist im Rahmen dieses Berichtes nicht angestrebt worden; auf sie konnte um so leichter verzichtet werden, als vor kurzem ein ausführlicher Bericht von Miehl-Nickel (*M 3*) über die Höhenstrahlung erschienen ist, der eine gründliche und umfassende Darstellung der bisher vorliegenden Experimente enthält.

I. Übersicht über das Verhalten einzelner Teilchen.

1. Magnetische Ablenkung und Impulsmessung. Abb. 1 zeigt eine Wilson-Aufnahme der Bahn eines Teilchens im Magnetfeld. Eine solche Aufnahme ermöglicht zunächst eine Messung der Bahnkrümmung. Dies bedeutet eine Messung des Teilchenimpulses p, welcher mit dem Krümmungsradius ϱ im Magnetfeld H durch die bekannte Gleichung verknüpft ist:

$$p c = e H \varrho. \tag{1}$$

(e = Ladung des Teilchens, c = Lichtgeschwindigkeit.)

Es ist üblich, die „magnetische Steifigkeit" $p c / e$ (und damit den Impuls p) in Gauß · cm oder in Volt zu messen. Zwischen beiden Einheiten besteht die Beziehung

$$\frac{p c}{e H \varrho} = 1 \frac{\text{erg}}{\text{e.-s.-Lad. Gauß} \cdot \text{cm}} = 300 \frac{\text{Volt}}{\text{Gauß} \cdot \text{cm}}. \tag{1'}$$

Bei Teilchen, deren kinetische Energie $E = \sqrt{(m c^2)^2 + (p c)^2}$ groß ist gegen ihre Ruhenergie $m c^2$, ist der Ausdruck $p c$ auch ungefähr der

Energie gleich und wird daher meist in Elektronen-Volt (eV) angegeben. Insbesondere kann man für Elektronen in der Höhenstrahlung pc
praktisch stets mit der Energie gleichsetzen; für stärker ionisierende
Teilchen ist jedoch pc von der Energie zu unterscheiden.

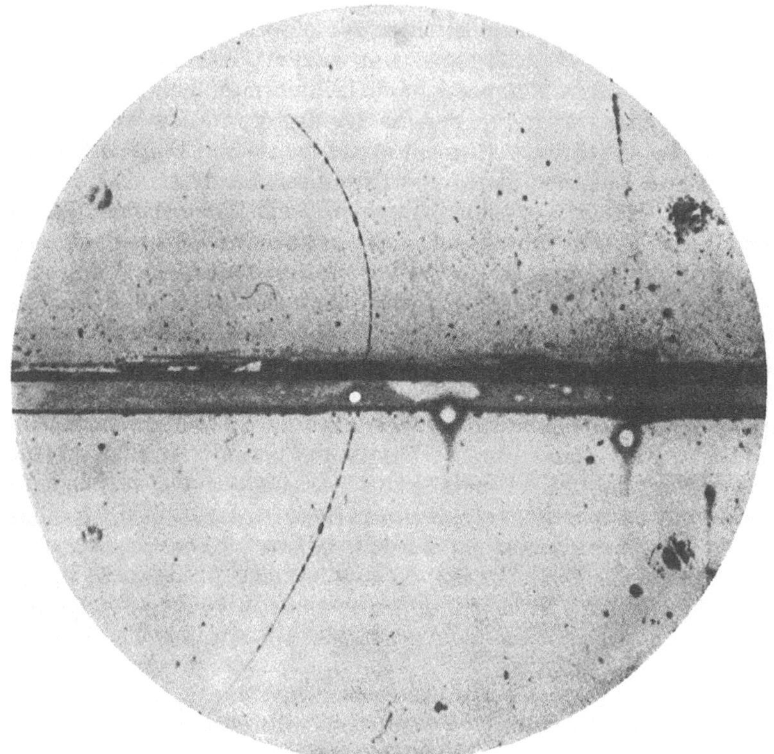

Abb. 1. WILSON-Aufnahmen von ANDERSON. Ein Positron von $6,3 \cdot 10^7$ eV geht von unten durch eine
6 mm dicke Bleiplatte und tritt als $2,3 \cdot 10^7$ eV-Positron aus.

2. Ionisation und Geschwindigkeitsmessung (vgl. *H 9*). Die Aufnahme Abb. 1 ermöglicht ferner eine Zählung der Tröpfchendichte in
der Spur, also eine Messung des Energieverlustes, den das beobachtete
Teilchen durch Ionisation der Luftmoleküle erleidet. Der Ionisations-
verlust pro cm $\left.\dfrac{\partial E}{\partial x}\right)_J$ ist im wesentlichen ein Maß für die Geschwindigkeit
$v = \beta c$ des Teilchens. Er ist durch

$$-\left.\frac{\partial E}{\partial x}\right)_J = \frac{a}{\beta^2} \tag{2}$$

gegeben, worin die Größe a nach BETHE und BLOCH (*B 4*, *B 15*) nur
wenig von der Masse abhängt und nur logarithmisch mit der Energie
des Teilchens ansteigt. Die Größenordnung von a ist 10^7 eV/cm Pb

und $2 \cdot 10^6$ eV/cm Wasser. Die genauere Beschreibung des Ionisationsverlustes erfolgt in Abb. 2, welche den Energieverlust pro cm Wasser und Blei als Funktion des Impulses für Teilchen der Elementarladung mit verschiedenen Massen nach BLOCH und BHABHA (B 6) angibt. Die in der Abbildung behandelten Massen (m = Elektronenmasse, $100\,m$, $1840\,m = M$ = Protonenmasse) zeichnen die Energien $m c^2 = 0{,}51 \cdot 10^6$ eV, $100\,m c^2 = 0{,}51 \cdot 10^8$ eV, $M c^2 = 0{,}91 \cdot 10^9$ eV als Ruhenergie aus.

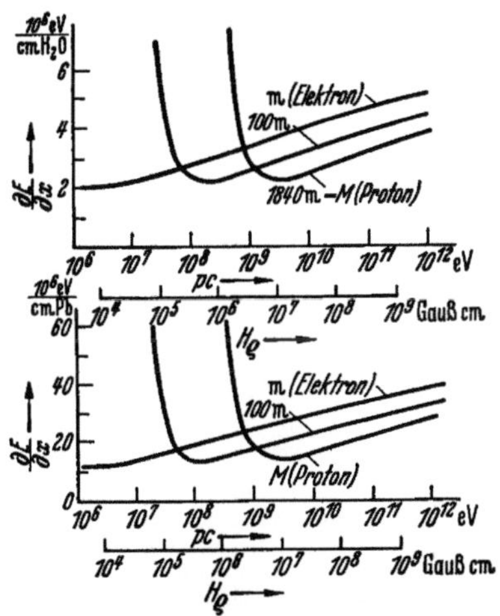

Abb. 2. Ionisationsverlust in Wasser und Blei nach der BLOCHschen Formel. Abszisse: Impuls des Teilchens in eV oder Gauß-cm. Ordinate: Energieverlust in 10^6 eV pro cm. Die drei Kurven gehören zu Teilchen verschiedener Masse: m = Masse des Elektrons, $100\,m$, $1840\,m$ = Masse des Protons.

Nach Formel (2) und Abb. 2 können Elektronen und Protonen bei einem Impuls von $p c = 2 \cdot 10^8$ eV deutlich durch ihre Ionisation unterschieden werden: Das Elektron ($m c^2 = \frac{1}{2} \cdot 10^6$ eV) hat dann nahezu Lichtgeschwindigkeit und erzeugt daher eine dünne Spur, das Proton ($M c^2 = 10^9$ eV) aber bewegt sich bei demselben Impuls langsamer als das Licht und erzeugt daher eine viel stärkere Spur. Die deutliche Unterscheidungsmöglichkeit zwischen Elektronen und Protonen durch die Ionisation hört aber auf bei Impulsen von mehr als $M c^2 = 10^9$ eV, da sich dann beide Teilchen nahezu mit Lichtgeschwindigkeit bewegen, und daher beide eine dünne Spur erzeugen, deren Tröpfchendichte im wesentlichen durch die Ladung bestimmt ist. Die Diskussion der feineren Einzelheiten erfolgt in § 9.

3. Strahlungsbremsung und Ruhmasse (vgl. H 9). Schließlich kann aus einer WILSON-Aufnahme Abb. 1 die Veränderung der Bahnkrümmung in der Bleiplatte entnommen werden. Dieser Impulsverlust in der Bleiplatte setzt sich in der Hauptsache aus zwei Teilen zusammen, dem Ionisationsverlust, der im vorigen Abschnitt angegeben wurde, und dem Strahlungsverlust. Während die Ionisation einem Teilchen mit Lichtgeschwindigkeit immer denselben *absoluten* Betrag von etwa 10^7 eV/cm Pb entzieht, entnimmt die Bremstrahlung einem solchen Teilchen immer denselben *Bruchteil* seiner Energie, welcher dem Quadrat seiner Ruhmasse umgekehrt proportional ist (B 3). Ein Elektron von mehr als 10^7 eV strahlt z. B. auf 4 mm Pb immer etwa die Hälfte seiner Energie aus, während ein Proton auf derselben Strecke nur den unmerklichen Bruchteil von 0,01% durch Strahlung verliert. Der Strahlungsverlust eines Elektrons in 1 cm Pb überwiegt also seinen Ionisationsverlust bei weitem, sobald nur die Energie des Elektrons mehr als 10^7 eV beträgt. Eine genauere Behandlung der Bremsstrahlung erfolgt in § 7.

4. Leichte und schwere Elektronen. Wir kommen nun zu den Resultaten der bisherigen WILSON-Aufnahmen einzelner Bahnen:

1. Die Statistik der Impulsmessungen [KUNZE ($K\,4$), BLACKETT ($B\,12$), HERZOG und SCHERRER ($H\,10$, $H\,11$), ANDERSON ($A\,2$)], welche bis zu $2 \cdot 10^{10}$ eV hinauf erstreckt worden sind, ergab ein *Spektrum* ionisierender Teilchen, welches kontinuierlich zu den hohen Impulsen hin abfällt.

2. Die Spuren der meisten Teilchen zeigen eine *Ionisation*, welche sich nur wenig von der eines Elektrons unterscheidet. Nur etwa $1\,^0/_{00}$ ($A\,3$) aller Bahnen erzeugen eine stärkere Spur. Dies ist ein Argument dafür, daß die Ladung der beobachteten Teilchen nicht wesentlich kleiner als die eines Elektrons sein kann; es wird darüber hinaus angenommen, daß die Ladung aller Teilchen gleich der elektrischen Elementarladung ist.

Aus der Schwäche der Spuren im Impulsgebiet $pc < \tfrac{1}{2} \cdot 10^9$ eV folgt ferner, daß nahezu alle Teilchen unterhalb $\tfrac{1}{2} \cdot 10^9$ eV leichter sind als Protonen. Denn ein Proton mit einem Impuls $pc < \tfrac{1}{2}\,10^9$ eV müßte bereits eine merklich dickere Spur erzeugen als die Mehrzahl der Teilchen, die in diesem Gebiet gefunden werden (Abb. 2).

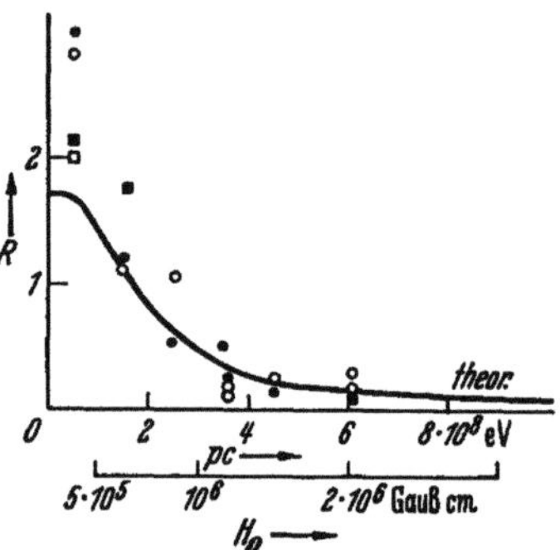

Abb. 3. Bremsverlust als Funktion des Impulses. Abszisse: Impuls. Ordinate: Relative Impulsänderung pro cm Blei. o ● Messung von BLACKETT in $^1/_3$ cm Blei. □ ■ Messung von BLACKETT in 1 cm Blei. ——— Theoretischer Verlauf für eine unendlich dünne Bleiplatte (§ 21).

3. Der Impulsverlust in einer Bleiplatte, welcher bei hohen Impulsen von BLACKETT und WILSON ($B\,10$, $B\,11$), NEDDERMEYER und ANDERSON ($N\,1$), CRUSSARD und LEPRINCE-RINGUET ($C\,5$) gemessen wurde, ist nach den Messungen von BLACKETT in Abb. 3 angegeben. Er ist bei niedrigen Impulsen $p\,c < 2 \cdot 10^8$ eV so stark wie für strahlungsfähige Elektronen (§ 3), nimmt aber bei hohen Impulsen derart ab, daß er bei $\tfrac{1}{2} \cdot 10^9$ eV nur noch etwa den 10. Teil des anfänglichen Wertes hat. Hieraus folgt ($N\,1$), wenn man die Richtigkeit der Strahlungstheorie voraussetzt, daß die meisten Teilchen oberhalb $pc = 2 \cdot 10^8$ eV schwerer sind als Elektronen.

Aus der Tatsache ($B\,10$, $B\,11$, $N\,1$, $C\,5$), daß die Mehrzahl der Teilchen, welche im Impulsgebiet $2 \cdot 10^8$ eV $< pc < \tfrac{1}{2} \cdot 10^9$ eV beobachtet wurden, schwächer ionisieren als Protonen und schwächer strahlen als Elektronen, haben NEDDERMEYER und ANDERSON den Schluß gezogen, daß es sich hier um eine bisher unbekannte Art „schwerer Elektronen" handelt, deren Masse zwischen der des Elektrons und des Protons liegt.

Im Gegensatz zu dieser Interpretation der Messungen durch die Annahme schwerer Elektronen hatten BLACKETT und WILSON zur Deutung ihrer Messungen zuerst angenommen, daß die beobachteten

Teilchen Elektronen sind, welche oberhalb eines Impulses von einigen 10^8 eV ihre Strahlungsfähigkeit verlieren. Da aber die Gültigkeit der Strahlungsformeln, wie später auf Grund der Hoffmannschen Stöße (§ 24) und der Stratosphärenmessungen (§ 19) gezeigt wird, bis zu sehr hohen Energien von 10^{11} eV hinauf bestätigt ist, und da ferner ein Versagen der Strahlungstheorie bei der Herleitung der Bremsformel (§ 3) theoretisch nicht verständlich gemacht werden kann ($W\,2$), wird man jetzt wohl die erste Interpretation für richtig halten müssen.

5. Die Masse der schweren Elektronen. Die Masse der schweren Elektronen kann aus der Kombination einer Impuls- und einer Geschwindigkeitsmessung entnommen werden. Dazu ist es allerdings notwendig, ein schweres Elektron am Ende seiner Bahn zu beobachten, damit es eine Geschwindigkeit hat, welche wesentlich kleiner ist als die des Lichts und eine Spur erzeugt, welche wesentlich stärker ist als die eines Elektrons.

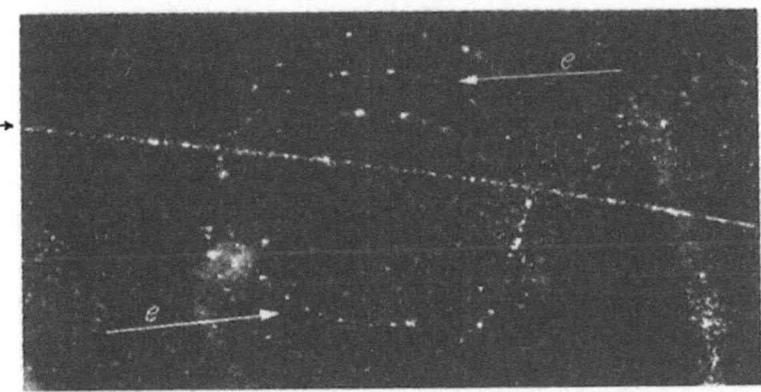

Abb. 4. Wilson-Aufnahme eines schweren Elektrons von Williams und Pickup. [Nature (Lond.) 141, 684 (1938).] d schweres Elektron mit starker Spur. e leichtes Elektron zum Vergleich.

Trotz der Seltenheit dieses Ereignisses sind bereits mehreren Autoren Aufnahmen der Spuren langsamer Elektronen gelungen, aus denen wir in Abb. 4 ein Bild von Williams und Pickup reproduzieren ($W\,4$).

d ist die Spur eines schweren Elektrons, e die zum Vergleich in das Bild kopierte Spur eines Elektrons von $\frac{1}{2} \cdot 10^6$ eV. Man erkennt deutlich, daß die Ionisation des Teilchens d etwa 3mal so stark ist wie die des Elektrons e. Die Krümmung der Spur d zeigt andererseits, daß das Teilchen d kein Proton ist. Krümmung und Ionisation erlauben dann eine Berechnung der Masse schwerer Elektronen, welche in Tabelle 1 von Williams und Pickup durchgeführt ist:

Tabelle 1.

Größe: Einheiten:	Krümmung 10^6 Gauß cm	Ionisation Ion. d. 10^6 eV Elektr.	Impuls p mc	Geschwin- digkeit c	Masse m	Ladung
Spur a	1,10	5	65	0,30	220 ± 50	—
Spur b	1,83	> 7	107	0,25	430 (< 800)	+
Spur c	1,47	3	85	0,45	190 ± 60	+
Spur d	1,15	3,3	67	0,41	160 ± 30	+

Andere Autoren geben die folgenden Ruhmassen μ (in Einheiten der Elektronenmasse m) an:

$$\mu/m = 250 \quad \text{Corson und Brode } (C\ 4),$$
$$125 \quad \text{Brode und Starr } (B\ 19),$$
$$160 \quad \text{Street und Stevenson } (S\ 10),$$
$$\sim 350 \quad \text{Anderson und Neddermeyer } (A\ 3),$$
$$120 \quad \text{Ruhlig und Crane } (R\ 15),$$
$$200 \quad \text{Ehrenfest } (E\ 4),$$
$$100 \quad \text{Auger } (A\ 8),$$
$$180-250 \quad \text{Nishina, Takeuchi, Ichimiya.}$$

Wir setzen nun voraus, daß in der kosmischen Strahlung leichte und schwere Elektronen vorkommen und diskutieren zunächst (Kap. II, III) theoretisch die Wirkungen, die wir erwarten, wenn solche Teilchen auf Materie fallen. Später (Kap. IV, V) vergleichen wir dann die erwarteten Wirkungen als Ganzes mit der Erfahrung, welche uns erlaubt, einige theoretisch offengelassene Parameter nachträglich einzufügen.

II. Theorie der leichten Teilchen (vgl. *H* 9).

6. Übersicht über die elektromagnetischen Prozesse. Wenn ein Elektron hoher Energie ein Stück Materie durchquert, so erfährt es, wie schon besprochen wurde, vor allem zweierlei Wirkungen. Erstens stößt es die Elektronen der Atomhüllen an und verliert Energie durch *Ionisation*. Zweitens wird es von den Coulomb-Feldern der Atomkerne abgelenkt, und diese Ablenkung führt zur *Ausstrahlung* von Licht, wenn die Geschwindigkeit des Teilchens der Lichtgeschwindigkeit nahekommt.

Ebenso wie die elektromagnetischen Kräfte einen Teil der Energie eines Elektrons in Strahlung verwandeln, können sie auch den umgekehrten Prozeß hervorrufen, d. h. sie können dazu führen, daß ein Lichtquant, welches auf einen Atomkern trifft, ein Positron-Elektron-Paar erzeugt.

Die beiden Umwandlungen, Bremsstrahlung und Paarbildung, finden mit vergleichbarer Häufigkeit statt, sobald die Energie des Quantums oder Elektrons die Ruhenergie $\frac{1}{2} \cdot 10^6$ eV des Elektrons genügend übersteigt.

Um nun diese Prozesse der Bremsstrahlung, Paarbildung und Ionisation übersichtlich zu beschreiben, legen wir zunächst mit Bhabha und Heitler $(B\ 5,\ C\ 1)$ in jedem Material diejenige Strecke X_0 als Längeneinheit zugrunde, auf der ein Elektron durch Bremsstrahlung im Mittel seine halbe Energie verliert. Diese Strecke ist in der ersten Zeile der Tabelle 2 angegeben $(B\ 3)$; wir nennen sie „Strahlungseinheit" und bezeichnen die Schichtdicke, gemessen in diesen Einheiten, mit l, $\left(l = \dfrac{X}{X_0} \right)$. Wir bemerken dann ferner mit Bhabha und Heitler $(B\ 5, C\ 1)$, daß in jedem Material eine bestimmte Energie ausgezeichnet ist, nämlich diejenige Energie E_j, bei welcher der Ionisationsverlust eines Elektrons gleich seinem Strahlungsverlust wird. Oberhalb der Energie E_j überwiegt dann der Strahlungsverlust und unterhalb E_j überwiegt der Ionisationsverlust, da der Energieverlust durch Strahlung proportional zur Energie anwächst, während der Energieverlust durch Ionisation nur noch logarithmisch mit der Energie ansteigt, also praktisch konstant ist. Die Energie E_j, welche in der Größenordnung mit dem Ionisationsverlust pro Strahlungseinheit übereinstimmt, ist in Zeile 2 der Tabelle 2 eingetragen. Wir werden im folgenden die Energie E_j auch als „Ionisierungsgrenze" bezeichnen.

Tabelle 2.

	Pb	Fe	Al	H_2O	Luft	
$X_0 =$	0,4	1,4	7,8	34	27500	cm
$E_j =$	1	3	6	15	15	10^7 eV

$$\left(l = \frac{X}{X_0}\right).$$

Der reziproke Wert der Strahlungseinheit ist annähernd proportional zum Quadrat der Atomnummer Z und zur Zahl ϱ/A der Atome pro cm³:

$$X_0 \approx \text{const}\, \frac{A}{\varrho} \cdot \frac{1}{Z^2}\,. \tag{3}$$

Der reziproke Wert der Ionisierungsgrenze ist annähernd proportional zur Atomnummer:

$$E_j \approx \frac{\text{const}}{Z}\,, \tag{4}$$

weil der Ionisierungsverlust in erster Näherung linear mit der Zahl Z der Elektronen im Atom zunimmt, während der Strahlungsquerschnitt mit dem Quadrat der Atomnummer wächst. Eine Abweichung vom Gesetz (4), welche sich in Tabelle 2 bemerkbar macht, liegt an der Abhängigkeit der Ionisierungsenergie eines Atoms von der Atomnummer (§ 9; Abb. 2).

Außer den genannten Prozessen spielt noch der Compton-Effekt der Lichtquanten und die Zerstrahlung der Positronen eine Rolle (vgl. H 9). Da diese Prozesse aber erst bei kleineren Energien der Größenordnung $< 10^6$ eV merklich einsetzen, können sie bei der Diskussion der kosmischen Strahlung, deren Phänomene sich im allgemeinen oberhalb 10^7 eV abspielen, vernachlässigt werden.

7. Die Bremsstrahlung. Wir nehmen nun an, daß ein Elektron der Energie E auf eine dünne Schicht dl fällt und fragen nach der Zahl der Lichtquanten im Energieintervall k, $k + dk$, welche es emittiert. Diese Zahl ist nach Bethe und Heitler ($B\,3$, $B\,5$) annähernd

$$J(k)\,dl\,dk = dl\,\frac{dk}{k}\ln 2 \tag{5}$$

in dem für die kosmische Strahlung wichtigen Energiegebiet oberhalb der Ruhenergie des Elektrons. Hieraus folgt, daß die Gesamtzahl der auf der Strecke dl abgegebenen Lichtquanten oberhalb einer Energie E_1:

$$dn = dl \int_{E_1}^{E} J(k)\,dk = dl\ln\frac{E}{E_1}\ln 2 \tag{6}$$

Theoretische Gesichtspunkte zur Deutung der kosmischen Strahlung. 9

und die auf der Strecke dl im Mittel abgegebene Energie

$$dE = dl \int_0^E J(k)\, k\, dk = dl\, E \ln 2 \tag{7}$$

ist.

Der relative Strahlungsverlust dE/E eines Elektrons hoher Energie ist also von der Energie unabhängig, was wir in § 3 bereits voraussetzten. Hieraus folgt, daß die mittlere Energie des strahlenden Elektrons mit der durchlaufenen Schicht exponentiell abnimmt:

$$\overline{E}_l = E_0\, 2^{-l} \tag{8}$$

($E_0 = $ Anfangsenergie; $E_l = $ Energie nach der Schicht l; $l = X/X_0$).

Bei der Beurteilung der Experimente ist allerdings auch die Schwankung um diesen mittleren Energieverlust zu berücksichtigen. Diese Schwankung wird durch Angabe der Wahrscheinlichkeit beschrieben, mit der für ein Elektron der Anfangsenergie E_0 die Energie im Intervall E_l, $E_l + dl$, liegt, nachdem es eine Strecke l durchlaufen hat.

Diese Wahrscheinlichkeit ist nach BETHE und HEITLER ($B\ 3$)

$$W(E_l)\, dE_l = \frac{dE_l}{E_0} \left(\ln \frac{E_0}{E_l} \right)^{l-1} \frac{1}{(l-1)!}\ . \tag{9}$$

Insbesondere ist also in einer Schicht $l = 1$ jede Energieabgabe zwischen der Abgabe der vollen Energie und der Energieabgabe 0 gleich wahrscheinlich.

Die aus der quantisierten Strahlungstheorie berechneten Formeln (8), (9) für den mittleren Energieverlust eines Elektrons und für die Schwankung des Strahlungsverlustes um seinen Mittelwert konnten durch direkte Messungen in der WILSON-Kammer bis zu $2 \cdot 10^8$ eV hinauf bestätigt werden:

Tabelle 3 gibt den mittleren Energieverlust in einer 0,35-cm-Pb-Platte nach Messungen von ANDERSON und NEDDERMEYER ($A\ 3$) wieder.

Tabelle 3. Mittlerer Energieverlust von Elektronen in Pb nach C. D. ANDERSON und S. H. NEDDERMEYER [Physic. Rev. **50**, 267 (1936)].

7900 Gauß, 0,35-cm-Pb-Platte (Pike's Peak)				
Energieintervall	< 50	50—100	100—150	150—200	$\cdot 10^6$ eV
Zahl der Spuren . . .	29	65	18	13	
Mittlere Anfangsenergie	31	75	123	177	$\cdot 10^6$ eV
Mittlerer Energieverlust					
experimentell . . .	42	82	178	191	$\cdot 10^6$ eV/cm Pb
theoretisch.	50	110	175	248	$\cdot 10^6$ eV/cm Pb

4500 Gauß, 0,35-cm-Pb-Platte (Pasadena)				
Energieintervall	< 50	50—100	100—150	150—200	$\cdot 10^6$ eV
Zahl der Spuren . . .	22	28	15	16	
Mittlere Anfangsenergie	26	71	117	170	$\cdot 10^6$ eV
Mittlerer Energieverlust					
experimentell . . .	37	84	124	207	$\cdot 10^6$ eV/cm Pb
theoretisch.	43	105	167	240	$\cdot 10^6$ eV/cm Pb

Abb. 5 zeigt die Schwankung des Energieverlustes im Energiegebiet $E < 2 \cdot 10^8$ eV nach Messungen von BLACKETT ($B\ 11$). In dieser Abbildung wurde auf der Abszisse der relative Strahlungsverlust R pro cm Pb aufgetragen, welcher sich nach Abzug des geringen Ionisationsverlustes aX für Elektronen aus dem gesamten Energieverlust ergibt:

$$R = \frac{E_0 - E_l - aX}{\dfrac{(E_0 + E_l)}{2} \cdot X} \cdot$$

Der linke Teil der Abb. 5 zeigt die Häufigkeit des Energieverlustes R für eine dünne Platte von $^1/_3$ cm Pb, welche etwas weniger als eine Strahlungseinheit beträgt, und daher einen großen Energieverlust besonders selten zeigt. In der rechten Teilfigur wurde die Verteilung für eine Platte von mehreren Strahlungseinheiten ($X = 1$ cm Pb, $l = 2,5$) aufgetragen, in welcher ein großer Energieverlust besonders häufig ist. Die ausgezogenen Kurven bezeichnen die theoretische Verteilung (9), deren Mittelwert (8) durch die gestrichelte Vertikale angedeutet ist. Diejenigen Meßpunkte, welche einem negativen Impulsverlust entsprechen, beruhen auf den Fehlern bei den Krümmungsmessungen und geben ein ungefähres Maß in der Unsicherheit der Impulsmessung. Rechnet man diese Bahnen als Beispiele für einen kleinen positiven Energieverlust mit, so ergibt sich ein deutliches Maximum bei kleinen Energieverlusten, das mit der Bremsstrahlung nicht erklärt werden kann und offenbar den durchdringenden Teilchen zugeschrieben werden muß.

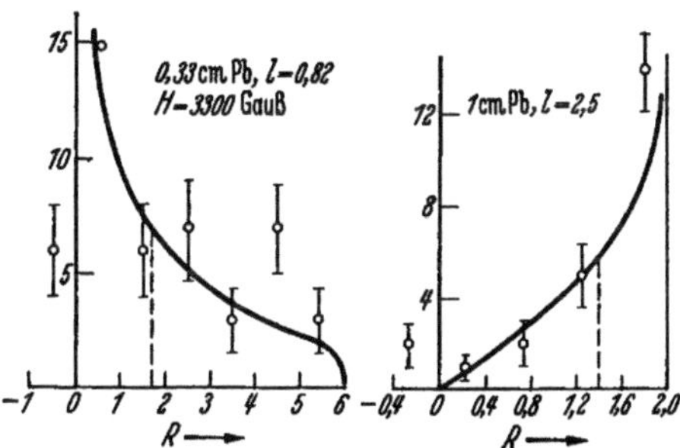

Abb. 5. Schwankung des Strahlungsverlustes für Energien $E < 2 \cdot 10^8$ eV. [Aus BLACKETT: Proc. Roy. Soc. Lond. **165**, 11 (1938).] Abszisse: Relativer Strahlungsverlust R pro cm Blei. Ordinate: Häufigkeit des Strahlungsverlustes pro dR in einer Platte von $^1/_3$ cm Pb (links), bzw. 1 cm Pb (rechts). ○ Messung von BLACKETT. ——— Theorie von BETHE-HEITLER.

8. Die Paarbildung. Die Wahrscheinlichkeit, mit der ein Lichtquant auf der Strecke dl ein Paar erzeugt, ist bei hohen Energien $h\nu \gg mc^2$ unabhängig von der Energie des Lichtquants (s. $H\ 9$)

$$\omega\,dl = 0,6\,dl. \tag{10}$$

Bei niedrigeren Energien nimmt die Wahrscheinlichkeit der Paarerzeugung ab, bis sie bei der Ruhenergie des Paars $h\nu = 2\,mc^2$ den Wert 0 erreicht. Dieses aus der DIRACschen Theorie berechnete Gesetz für die Paarerzeugung ist durch direkte Messungen nur im Gebiet der γ-Strahlung von CURIE-JOLIOT, CHADWICK, BLACKETT und OCCHIALINI (s. $H\ 9$) geprüft. Im Gebiet hoher Energie werden wir seine Extrapolation (10) zunächst voraussetzen, um sie an ihren Folgerungen zu kontrollieren.

9. Ionisation. Die Ionisation der Atomhüllen durch Elektronen kann an den folgenden Erscheinungen studiert werden:

A. Der *Energieverlust* kann aus der Krümmungsänderung einer Bahn im Magnetfeld entnommen werden, wenn die Geschwindigkeit des Elektrons so klein ist, daß die Strahlungsbremsung noch keine Rolle spielt. Die Theorie des Energieverlustes von BETHE (*B 4*) und BLOCH (*B 15*), deren Ergebnis in Abb. 2 angegeben wurde, konnte von WILLIAMS unterhalb $^1/_4$ Leichtgeschwindigkeit aufs genaueste bestätigt werden (s. *B 4*).

Für die *Reichweite R* eines Teilchens der Masse μ vom Impuls p erhält man, wenn man in Formel (2) die Größe a näherungsweise durch einen konstanten (für $p \sim 3{,}5\,\mu c$ genau gültigen) Wert ersetzt:

$$
\left.
\begin{array}{l}
\begin{array}{c|ccc}
a = & \text{Blei} & \text{Wasser} & \text{Luft} \\
\hline
 & 1{,}2\cdot 10^7 & 2\cdot 10^6 & 2{,}5\cdot 10^3 \ \ \text{e V/cm}
\end{array} \\[2em]
R\,(p) = \dfrac{\mu c^2}{a}\left(\dfrac{2+\left(\dfrac{p}{\mu c}\right)^2}{\sqrt{1+\left(\dfrac{p}{\mu c}\right)^2}} - 2\right) \\[2em]
\begin{array}{c|ccccc}
\dfrac{p}{\mu c} = & 0 & 1 & 2 & 3 & \gg 1 \\
\hline
\dfrac{aR}{\mu c^2} = & 0 & 0{,}1 & 0{,}7 & 1{,}5 & \dfrac{p}{\mu c}-2
\end{array}
\end{array}
\right\} \qquad (11)
$$

B. Die *Energieverteilung* der Sekundärelektronen ist von ISHINO (*J 2*) im Gebiet niedriger Energien $E < 300$ eV gemessen worden. Die Messungen sind in guter Übereinstimmung mit der Theorie, welche bei höheren Primär- und Sekundärenergien eine Verteilung von annähernd const$/E$ Sekundärelektronen der Energie $> E$ erwarten läßt (vgl. *B 4, B 1*).

C. Die *sekundäre Ionisation* (*s*), d. h. die Zahl der Ionen, welche von dem primären Elektron direkt gebildet werden, kann durch Zählung der Ionennester in scharfen, noch nicht diffundierten Spuren in der WILSON-Kammer experimentell bestimmt werden.

Die theoretische Zahl der sekundären Ionen, welche ein Elektron der Geschwindigkeit βc pro cm eines Gases unter Normalbedingungen erzeugt, ist (*B 4, B 1*):

$$
\left.
\begin{array}{l}
s = \dfrac{J}{\beta^2}\left(14{,}24 + \ln\dfrac{\beta^2}{1-\beta^2} - \beta^2\right) \\[1.5em]
\begin{array}{c|ccc l}
 & \text{N}_2 & \text{O}_2 & \text{H}_2 & \\
\hline
J = & 1 & 1{,}15 & 0{,}29 & \text{Ionen pro cm bei Atmosphären-} \\
 & & & & \text{druck und Zimmertemperatur.}
\end{array}
\end{array}
\right\} \qquad (12)
$$

Die sekundäre Ionisation soll ebenso wie der Energieverlust bei Impulsen $p < mc$ wie das reziproke Quadrat der Geschwindigkeit zunehmen, in der Nähe des Impulses $3\,mc$ ein Minimum haben und oberhalb dieses Minimums logarithmisch mit dem Impuls $p = mc\dfrac{\beta}{\sqrt{1-\beta^2}}$ ansteigen. Die Konstante J kann theoretisch nur für Wasserstoff mit

Sicherheit angegeben werden $(B\,4)$. Benutzt man nach Bagge $(B\,1)$ die Konstanten (12), so erhält man für die sekundäre Ionisation eines Elektrons in Luft ein Minimum bei der Geschwindigkeit

$$\beta = 0{,}97, \qquad \frac{p}{mc} = \frac{\beta}{\sqrt{1-\beta^2}} = 3{,}5, \qquad H\varrho = 5{,}8 \cdot 10^3 \text{ Gauß-cm.}$$

Die Zahl s der sekundären Ionen pro cm ist bei diesem Minimum:

	theoretisch (12)	experimentell
N_2	17	14—18 Corson und Brode $(C\,4)$
O_2	19,5	20 Williams und Terroux $(W\,5)$
H_2	4,8	5 Williams und Terroux $(W\,5)$

D. Die sekundäre Ionisation ist zu unterscheiden von der *gesamten Ionisation i*, d. h. der Zahl der Ionenpaare, welche pro cm vom primären Elektron und seinen Sekundären, Tertiären usw. gebildet werden.

Die gesamte Ionisation ist von Bagge $(B\,1)$ im unrelativistischen Gebiet theoretisch berechnet worden, wobei sich ungefähr Übereinstimmung mit den Messungen von Gerbes $(G\,6)$ u. a. ergab. Ein Elektron von $2 \cdot 10^4$ eV verbraucht danach auf seiner gesamten Bahn zur Bildung eines Ionenpaares im Mittel

	N_2	H_2
Theoretisch [Bagge $(B\,1)$] . . .	28,6 eV	34 eV
Experimentell [Gerbes $(G\,6)$] .	34 eV	37 eV

Corson und Brode $(C\,4)$ finden beim Minimum der Ionisation $\beta \sim 0{,}96$ in diffusen Spuren unter Fortlassung unauflösbarer Ionennester 25 Ionenpaare pro cm Luft, also ein Verhältnis $> 1{,}4$ bis $1{,}8$ der gesamten zur sekundären Ionisation.

In den Messungen von Corson und Brode $(C\,4)$ konnte zum ersten Male der von der Theorie behauptete *logarithmische Anstieg* der Ionisation mit dem Impuls p festgestellt werden. Die Meßpunkte in Abb. 6 zeigen die Ergebnisse von Corson und Brode, während die Kurve in derselben Abbildung den theoretischen Verlauf darstellt, welchen man durch Hinzufügen eines geeigneten als konstant angenommenen Faktors zur Formel (12) erhält.

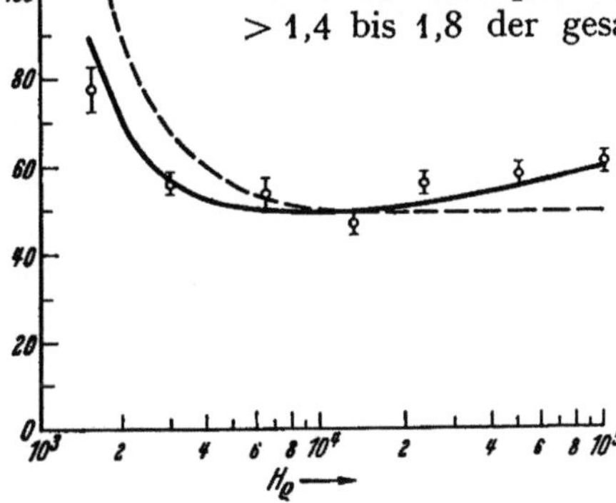

Abb. 6. Spezifische Ionisation der Elektronen nach Corson und Brode. [Physic. Rev. **53**, 215 (1938).] Abszisse: Impuls $H\varrho$ der Elektronen in Gauß·cm. Ordinate: Zahl der Ionen pro cm Luft unter Normalbedingungen. ⚲ Messung von Corson und Brode. ———— Theoretische Kurve (12) mit willkürlichem konstantem Faktor. — — — Näherung der theoretischen Kurve (2) mit $a = $ const.

Die experimentelle Verfolgung des logarithmischen Anstiegs der Ionisation kann nur unterhalb $H\varrho = 2 \cdot 10^5$ Gauß-cm durchgeführt werden, da die Teilchen, welche oberhalb dieses Impulses einfallen, zum Teil schon schwere Elektronen sind, deren Ionisationsminimum sich dem Anstieg der Elektronenionisation überlagert (Abb. 2, 3). Wir können daher seit der Entdeckung der schweren Elektronen verstehen, daß frühere Versuche, den logarithmischen Anstieg der Elektronenionisation oberhalb $2 \cdot 10^5$ Gauß-cm festzustellen, fehlschlugen.

E. Neben den Nebelkammerzählungen, welche die Ionisation als Funktion des Impulses angeben, gibt es noch andere Methoden der Ionisationsmessung, welche allerdings nur eine über die Impulse der kosmischen Strahlung *gemittelte gesamte Ionisation* $\bar{i}$ pro cm ergeben.

STUHLINGER ($S\,14$) vergleicht die Zahl der Strahlen in einem Proportionalzählrohr mit der Ionisation und erhält $\bar{i} = 30 - 35$ Ionenpaare pro cm Luft für durchdringende Teilchen, und $\bar{i} = 50$ Ionenpaare pro cm Luft für die Strahlen in einem Schauer. Da die Strahlen im Schauer meist gewöhnliche Elektronen sind (§ 23), die durchdringenden Teilchen aber schwere Elektronen (§ 17), kann man die höhere Ionenmenge als Beweis für den logarithmischen Anstieg der Ionenzahl mit dem Verhältnis $\dfrac{\text{Energie}}{\text{Ruhmasse}}$ ansehen ($S\,14$) (Abb. 2).

Über die mittlere gesamte Ionisation $\bar{i}$ der kosmischen Strahlen können ferner Anhaltspunkte entnommen werden aus dem Vergleich von Ionisationsstrom und Koinzidenzzahl. Diese Messungen ergeben höhere Werte, welche zwischen 70 und 135 Ionenpaare-cm Luft schwanken (vgl. *M 3, G 1*). Die letztere Messungsart ist aber schwieriger zu beurteilen als die von STUHLINGER, weil man nicht sicher ausschließen kann, daß mehrere Strahlen bei einer Koinzidenz registriert wurden. Vermutlich ist hierdurch der hohe Wert der Resultate zu verstehen.

Bei der theoretischen Behandlung des Ionisationsverlustes in *großen Schichten* ist neben der bisher besprochenen Ionisation der Elektronenhüllen noch die Kernionisation (§ 16) und die Diffusion durch elastische Streuung zu berücksichtigen. Die Kernionisation wird in einem späteren Kapitel (III) besprochen; die elastische Streuung ist durch WILLIAMS (W 6) theoretisch und durch BLACKETT und WILSON (*B 13*) experimentell untersucht worden; sie kann die effektive Absorption beim Durchgang durch dickere Schichten bei kleineren Impulsen der stoßenden Teilchen erheblich vergrößern (*H 2*). Bei hohen Energien spielt sie jedoch nur noch eine geringe Rolle.

10. Die Multiplikationsschauer. Die Prozesse der Bremsstrahlung und Paarbildung beruhen korrespondenzmäßig auf den elektromagnetischen Kräften. Nach der Quantentheorie der elektromagnetischen Felder sind sie im allgemeinen Einfachprozesse, d. h. ein Lichtquant erzeugt meist an einem Kern nur ein Paar, und ein Elektron erzeugt an einem einzelnen Kern nur ein Lichtquant. Höhere Prozesse, in denen an einem einzelnen Kern mehrere Paare oder mehrere Lichtquanten entstehen, sind nach der Strahlungstheorie unwahrscheinlicher um höhere

Potenzen der Sommerfeldschen Feinstrukturkonstanten $\frac{e^2}{\hbar c} = \frac{1}{137}$ und wachsen mit der Energie E des stoßenden Teilchens nur logarithmisch an ($K\,2$).

Trotz der Einfachheit der elektromagnetischen Prozesse können die Vorgänge der Bremsstrahlung und Paarbildung doch in einer endlichen Materieschicht wegen ihrer großen Häufigkeit zur Entwicklung mehrerer Sekundärstrahlen aus einem Primärstrahl führen, wie Carlson-Oppenheimer ($C\,1$) und Bhabha-Heitler ($B\,5$) ungefähr gleichzeitig gezeigt haben (s. auch $L\,2$).

Denn wenn ein Elektron hoher Energie auf eine Schicht Blei fällt, so wird es in den ersten 4 mm ein Lichtquant gleicher Größenordnung abspalten. Dieses Lichtquant wird nach einigen weiteren mm ein Paar erzeugen. Elektron und Positron werden wieder weitere Lichtquanten abspalten, und es wird aus der Bleischicht eine ganze Garbe von Elektronen, Positronen und Lichtquanten

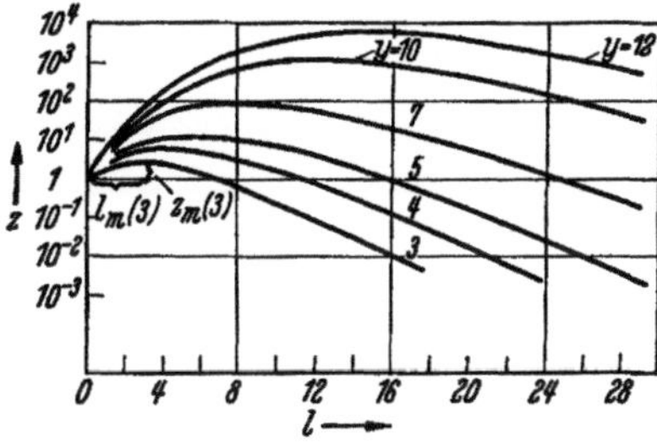

Abb. 7. Die Multiplikation der Elektronenzahl. Abszisse: Schichtdicke l des Materials in Einheiten der Tabelle 2, S. 8. Ordinate: Zahl $z\,(l, y)$ der Elektronen und Positronen, welche aus einem Elektron der Energie $E_j\,e^y$ im Energiegebiet $> E_j$ hinter einer Schicht l entstehen.

herauskommen. Die Prozesse der Bremsstrahlung und Paarbildung führen also durch „Multiplikation" zur Ausbildung „kaskaden"artiger Schauer. Als Vorläufer der „Kaskadentheorie" kann das von Geiger und Fünfer 1935 angegebene Strahlenschema ($G\,3$, $G\,1$) betrachtet werden, welches zur Ordnung der Schauerphänomene erdacht wurde. Carlson und Oppenheimer, Bhabha und Heitler berechneten statistisch aus den Formeln (§ 7, 8) die Größe und Energieverteilung der Kaskadenschauer, d. h. die Zahl z der Elektronen und Positronen oberhalb einer Energie E_1, welche ein Elektron der Anfangsenergie E hinter einer Schicht l erzeugt.

$$z = z\,(l, E, E_1).$$

(Unter „Elektronen" werden wir im folgenden immer Elektronen positiver *und* negativer Ladung verstehen.)

Das Ergebnis ist in Abb. 7 aufgetragen. Die Abszisse enthält die Schichtdicke l, die Ordinate die Teilchenzahl z. Der Parameter y, welcher zu den einzelnen Kurven gehört, bezeichnet den Logarithmus des Verhältnisses der Anfangsenergie zur Endenergie:

$$y = \ln \frac{E}{E_1}\,.$$

Es ist leicht einzusehen, daß die Teilchenzahl nur von diesem Verhältnis abhängt, wenn für $E_1 \gg E_j$ die Wirkung der Ionisation vernachlässigt wird:

$$z = z\,(l, y)\,, \tag{13}$$

da dann die relative Energieteilung nach (7) und (10) bei hohen Energien von der Energie unabhängig ist.

Die in Abb. 7 gegebene „Multiplikationsfunktion" $z = z(l, y)$ enthält nun vor allem zwei Angaben: Erstens gibt sie die mittlere *Größe des Schauers* an, welchen ein Teilchen der Energie E hinter einer Schicht l erzeugt; diese Schauergröße geht aus der Funktion $z(l, y)$ hervor, wenn wir die Endenergie E_1 gleich der Ionisierungsgrenze E_j (Tabelle 2, S. 8) des betreffenden Materials setzen.

$$y = \ln \frac{E}{E_j}:$$

$y =$	3	4	5	7	10	12	
Pb . . $E =$	$2 \cdot 10^8$	$5,5 \cdot 10^8$	$1,5 \cdot 10^9$	$1,1 \cdot 10^{10}$	$2,2 \cdot 10^{11}$	$1,6 \cdot 10^{12}\,\mathrm{eV}$	(14)
Fe . . $E =$	$6 \cdot 10^8$	$1,6 \cdot 10^9$	$4,5 \cdot 10^9$	$3,4 \cdot 10^{10}$	$6,7 \cdot 10^{11}$	$4,8 \cdot 10^{12}\,\mathrm{eV}$	
$\left.\begin{array}{l}\mathrm{H_2O} \\ \mathrm{Luft}\end{array}\right\}\ E =$	$3 \cdot 10^9$	$8,2 \cdot 10^9$	$2,2 \cdot 10^{10}$	$1,7 \cdot 10^{11}$	$3,3 \cdot 10^{12}$	$2,4 \cdot 10^{13}\,\mathrm{eV}$	

Dies bedeutet, daß wir die Wirkung der Ionisation näherungsweise berücksichtigen, indem wir annehmen, daß ein Elektron unterhalb der Ionisierungsgrenze E_j sofort infolge der Ionisation steckenbleibt, daß aber ein Elektron oberhalb der Ionisierungsgrenze E_j von der Wirkung der Ionisation unbeeinflußt ist. Zur genaueren Berechnung der Schauergröße muß zu dieser Zahl der Elektronen $> E_j$ noch die Zahl der Elektronen $< E_j$ hinzugefügt werden, welche in einer Arbeit von ARLEY ($A\ 4$) tabuliert ist (vgl. auch $C\ 1$); für $y = \ln \dfrac{E}{E_j} = 4$ ist zum Beispiel ($A\ 4$):

Bei $l =$	1	2	3	5	10	
$z\,(> E_j) = z\,(l, 4) =$	1,84	3,35	4,66	5,17	1,11	(14a)
$z\,(< E_j)\quad\quad =$	0,09	0,66	1,75	4,39	3,00	

Der Verlauf der mittleren Schauergröße $z(l, y)$ als Funktion der Schichtdicke, welcher mit (14) aus Abb. 7 hervorgeht, ist in seinen qualitativen Zügen verständlich: Bei dünnen Schichten steigt die mittlere Schauergröße mit wachsender Schichtdicke an, da die Anfangsenergie um so mehr zerteilt werden kann, je mehr Atomkerne in Wirkung treten. Bei größeren Schichten aber macht sich der Energieverlust durch Ionisation bemerkbar, welcher alle die Teilchen abbremst, deren Energie durch Strahlungsteilung auf die Ionisierungsgrenze E_j herabgesunken ist. Dieser Energieverlust tritt zunächst ins Gleichgewicht mit der Teilchenvermehrung bei einem Maximum l_m, um dann später zu einem Absinken der Kurve $z(l)$ zu führen. Das Gleichgewicht tritt z. B. für ein Elektron von 10^{11} eV bei einer Schicht von 5 cm Pb ($l_m = 12$) ein, hinter der es einen Schauer von 1000 Teilchen erzeugt. Hinter größeren Schichten überwiegt dann die Absorption der Anfangsenergie 10^{11} eV derart, daß bei 10 cm Pb die Schauergröße auf den 10. Teil reduziert ist.

Abb. 7 gibt ferner Auskunft über das *Spektrum*, welches die verschiedenen Strahlen in einem Multiplikationsschauer bilden. Die Gestalt dieses Sekundärspektrums, in welches sich ein einzelnes Elektron zerteilt, geht aus der Funktion $z(l, y)$ hervor, wenn man die Schichtdicke l und die Anfangsenergie konstant hält und die Endenergie E_1 im Ausdruck $y = \ln \dfrac{E}{E_1}$ variiert. Die nähere Diskussion im nächsten Paragraphen zeigt dann, daß die Zahl der Teilchen oberhalb einer Energie E_1 in einem Schauer ungefähr wie $\dfrac{\text{const}}{E_1^a}$ $[1 < a < 2, \ (19), \ (19\,a), \ (15)]$ abfällt, wenn der Schauer bei großen Schichten oder bei der „Gleichgewichtsschicht" l_m erzeugt wird.

Wenn an die Stelle des schauererzeugenden Elektrons ein Lichtquant tritt, so entstehen im wesentlichen dieselben Schauer, da ja das Lichtquant in den ersten Millimetern Blei ein Paar bildet, das dann seinerseits die oben angegebenen Wirkungen ausübt. Aus dem gleichen Grunde ist die Zahl der Lichtquanten, welche einem Schauer beigemengt sind, von derselben Größenordnung wie die der Elektronen und Positronen.

11. Mathematische Ergänzungen zu 10. α) *Die mittlere Größe der Multiplikationsschauer.* Es sollen nun die mathematischen Eigenschaften der Multiplikationsfunktion Abb. 7

$$z = z(l, y), \qquad y = \ln \frac{E}{E_l},$$

welche die mittlere Größe z eines Schauers der Anfangsenergie E hinter einer Schicht l angibt, noch etwas genauer diskutiert werden.

a) Die *maximale Schauergröße* z_m (d. h. das Maximum der Kurven in Abb. 7) ist annähernd proportional zur Anfangsenergie E des schauererzeugenden Elektrons $(B\ 5)$:

$$z_m(y) = z(l_m, y) \approx \frac{1}{8} \left(\frac{E}{E_l} \right)^{0,93}, \qquad \left(\text{für } \frac{\partial z}{\partial l} = 0 \right). \tag{15}$$

Diese Formel besagt, daß beim Maximum die Energie des Elektrons in lauter Teile von einer mittleren Energie der Größenordnung E_l zerstückelt wird.

b) Die Lage der *Gleichgewichtsschicht* l_m, bei der das Maximum in Abb. 7 eintritt, wächst logarithmisch mit der Anfangsenergie E

$$l_m \approx 1{,}2\, y \approx 3{,}10 \log z_m + 2{,}7 , \qquad \left(\text{für } \frac{\partial z}{\partial l} = 0 \right). \tag{16}$$

Dies ist verständlich, wenn wir uns nach Carlson und Oppenheimer $(C\ 1)$ den wirklichen Prozeß der Schauererzeugung bei der Gleichgewichtsschicht l_m durch einen Modellprozeß veranschaulichen, in welchem jedesmal auf der Strecke $l = 1$ eine Verdoppelung der Zahl ionisierender Strahlen durch Teilung stattfindet. Wir erhalten dann hinter der Schicht l_m eine Schauergröße $z_m \sim 2^{l_m}$, also einen logarithmischen Gang der Gleichgewichtsschichtdicke l_m mit der Teilchenzahl z_m oder der Anfangsenergie E.

Theoretische Gesichtspunkte zur Deutung der kosmischen Strahlung.   17

c) Die *Breite* der Maxima in den Kurven der Abb. 7 ist nur wenig von der Energie abhängig. Dies wird am deutlichsten durch die Beziehung

$$\int_0^\infty dl\, z(l, y) \approx \frac{3}{4}\frac{E}{E_i} \tag{17}$$

zum Ausdruck gebracht, von welcher wir später öfters Gebrauch machen werden.

d) Bei *dünnen Schichten* und hoher Anfangsenergie kann die Wirkung des Energieverlustes bei der Berechnung der Schauergröße vernachlässigt werden. Man erhält dann

$$\left. \begin{aligned} z(l, y) &\approx \sum_{n=0}^\infty \frac{x^n}{n!\,(2n)!}\,, \qquad (\text{für } l \lesssim 2,\ y > 1) \\ x &= 0{,}83\, y\, l^2. \end{aligned} \right\} \tag{18}$$

Die einzelnen Glieder dieser Reihe geben den Beitrag der einzelnen „Generationen" an, durch welche die Vermehrung der Teilchenzahl nacheinander erfolgt. Die Summe (18) kann für große Werte des Produkts x annähernd durch

$$z(x) \approx 0{,}20\, e^{1{,}89\sqrt[3]{x}} \cdot x^{-\frac{1}{6}}, \qquad (\text{für } x \gtrsim 1) \tag{18a}$$

ersetzt werden. Die Schauergröße hinter dünnen Schichten steigt also bei niedrigen Energien nur logarithmisch mit der Anfangsenergie an, wächst aber bei höheren Energien oder dickeren Schichten immer stärker mit der Energie, um schließlich den nahezu linearen Anstieg (14) mit der Energie bei dickeren Schichten zu erreichen.

e) Bei *großen Schichten,* hinter welchen die Absorption überwiegt, fällt die Schauergröße z exponentiell mit der Schichtdicke l ab und steigt nach einem Potenzgesetz mit der Anfangsenergie E an:

$$z(l, y) \approx e^{a\,y - b\,l - 3{,}2}. \tag{19}$$

Dabei sind a und b nur langsam mit y und l veränderliche Größen ($E\,6$):

$$\text{in der Nähe von} \left\{ \begin{array}{llll} l = 9 & 15 & 21 & 30 \\ y = 3 & 5 & 7 & 10 \end{array} \right.$$
$$\text{ist} \left\{ \begin{array}{llll} a = 2{,}14 & 2{,}00 & 1{,}79 & 1{,}44 \\ b = 0{,}48 & 0{,}44 & 0{,}37 & 0{,}25. \end{array} \right\} \tag{19a}$$

Genauere Tabellen der Funktion (13) befinden sich in den Arbeiten von CARLSON-OPPENHEIMER ($C\,1$), BHABHA-HEITLER ($B\,5$) und ARLEY ($A\,4$).

β) *Die Schwankung der Schauergröße.* Die bisher abgeleiteten Gesetze über die mittlere Schauergröße können im allgemeinen noch nicht unmittelbar mit der Erfahrung verglichen werden aus zweierlei Gründen. Erstens ist nur in den seltensten Fällen die Energie des schauererzeugenden Elektrons bekannt, und es können nur Mittelwerte beobachtet werden, welche ein kontinuierliches Spektrum schauererzeugender Elektronen hervorruft. Wir werden daher (in § 11 γ) die bisher abgeleiteten Wirkungen noch über ein Spektrum von Elektronen zu mitteln haben. Zweitens kann über die mittlere Größe der Schauer hinaus die Häufigkeit bestimmter Schauergrößen beobachtet werden, zu deren Berechnung wir die Schwankung in der Schauergröße kennen müssen.

Wir setzen also jetzt voraus, daß die mittlere Größe $\overline{N}$ des Schauers, welchen ein Elektron der Energie $E = E_j\, e^y$ hinter einer Schicht l erzeugt, bekannt sei:

$$\overline{N} = z\,(l,\, y), \qquad (\S\,11\,\alpha)$$

und fragen nach der Schwankung ΔN der Schauergröße um ihren Mittelwert

$$\Delta N = \left(\overline{N^2} - \overline{N}^2\right)^{\frac{1}{2}}.$$

Diese Schwankung wurde zunächst von Bhabha und Heitler $(B\,5)$ abgeschätzt unter der Annahme, daß die einzelnen Schauerstrahlen wie unabhängige Ereignisse behandelt werden können. Sie erhielten daher eine Poissonsche Formel:

$$\frac{\Delta N}{\overline{N}} = \frac{1}{\sqrt{\overline{N}}}. \qquad (20)$$

Furry $(F\,7)$ verknüpfte sodann die Vorgänge bei der Schauerbildung durch eine Modellvorstellung, in der jedes ionisierende Teilchen sich mit einer bestimmten Wahrscheinlichkeit pro Schichtdicke in zwei verwandelt und erhielt die viel größere Schwankung

$$\frac{\Delta N}{\overline{N}} = 1. \qquad (21)$$

Wie eine nähere Untersuchung zeigt $(E\,6)$, kann die wirkliche Schwankung durch eine Formel beschrieben werden, welche im allgemeinen zwischen den beiden Extremfällen (20), (21) liegt, und welche bei ganz dünnen Schichten, beim Maximum der Kurve (Abb. 7) und bei ganz dicken Schichten $(l \gtrsim 2\,y)$ durch die Poisson-Schwankung (20) zu ersetzen ist:

$$\left.\begin{aligned}
\frac{\Delta N}{\overline{N}} &\approx \frac{1}{\sqrt{\overline{N}}} && \text{für} && l \lesssim 1 \\[2mm]
\frac{\Delta N}{\overline{N}} &\approx \frac{1}{\sqrt{l\,y\ln 2}} && \text{für} && 1 \lesssim l \lesssim \tfrac{1}{2}\,y
\end{aligned}\right\} \quad (22a)$$

$$\left.\begin{aligned}
\Delta N &\approx \left|\frac{\partial \overline{N}}{\partial l}\right| \cdot 0{,}92 + \sqrt{\overline{N}} && \text{für} && \tfrac{1}{2}\,y \lesssim l \lesssim 2\,y \\[2mm]
\frac{\Delta N}{\overline{N}} &\approx \frac{1}{\sqrt{\overline{N}}} && \text{für} && l \gtrsim 2\,y.
\end{aligned}\right\} \quad (22b)$$

Die Formel (22a) erhält man, wenn man berücksichtigt, daß in dem Gebiet, in welchem die Absorption noch nicht wesentlich ist $(l < y)$ die Schwankung in der Schauergröße hauptsächlich durch die Schwankung in der Zahl n der Lichtquanten „erster Generation" bedingt ist $\left(\dfrac{\Delta N}{\overline{N}} \approx \dfrac{\Delta n}{\overline{n}} \approx \dfrac{1}{\sqrt{\overline{n}}} = \dfrac{1}{\sqrt{l\,y\,\ln 2}}\right.$ nach (6)$\Big)$. In der Formel (22b), welche im Gebiet stärkerer Absorption gelten soll, ist als wesentlich vorausgesetzt, daß die Abgabe der y energiereichsten Lichtquanten aus dem ursprünglichen Elektron einmal etwas früher und einmal etwas später erfolgen kann, so daß dieselbe Wirkung entsteht wie durch eine Schwankung der Schichtdicke um $\Delta l \approx 0{,}9$ $(C\,1)$.

Die *Verteilung* der Schauergrößen kann bei großen Schauern durch ein GAUSSsches Gesetz angenähert werden: Die Wahrscheinlichkeit dafür, daß ein Elektron der Energie $E = E_j\,e^y$ hinter einer Schicht der Dicke l einen Schauer von N Teilchen erzeugt, ist dann:

$$W(N, l, y)\,dN = \frac{e^{-\frac{(N-\overline{N})^2}{2\,(\Delta N)^2}}}{\sqrt{2\,\pi}\,\Delta N}\,dN, \qquad (\overline{N} = z(l, y)). \tag{23}$$

γ) *Die Häufigkeit der Kaskadenschauer.* Wir nehmen nun an, daß ein Spektrum von

$$F(E) = J_0\left(\frac{E_0}{E}\right)^\gamma = J_0\left(\frac{E_0}{E_j}\right)^\gamma e^{-\gamma y} \tag{24}$$

Elektronen oberhalb der Energie $E = E_j\,e^y$ einfällt[1], und fragen, mit welcher Häufigkeit

$$H(N, l)$$

es Schauer von mehr als N Teilchen hinter der Schicht l erzeugt. Die Funktion $H(N, l)$ wird auch als *Schauerauslösekurve* $H(l)$ bezeichnet, wenn bei konstanter Schauergröße N nur die Variation mit der Schichtdicke l betrachtet wird und als „*Schauerverteilungskurve*" $H(N)$, wenn bei konstanter Schichtdicke nur die Schauergröße N variiert.

Die Schauerhäufigkeit ist allgemein durch

$$\frac{\partial}{\partial N}H(N\,l) = \int\limits_0^\infty dE\,\frac{\partial F(E)}{\partial E}\,W\left(N, l, \frac{E}{E_j}\right) \tag{25}$$

gegeben, worin der erste Faktor die Zahl der einfallenden Elektronen im Energieintervall E, $E + dE$ und der zweite Faktor die durch die Schwankungsformel (22), (23) gegebene Wahrscheinlichkeit bedeutet, mit der das Elektron der Energie E in der Schicht l einen Schauer von N Teilchen erzeugt.

Wie die nähere Diskussion des Ausdrucks (25) zeigt, ist die Schwankung nur wesentlich bei kleinen Schauern hinter dünneren Schichten. Die Schwankung kann aber (für normale Spektren $\gamma < 4$) vernachlässigt werden bei der Behandlung größerer Schauer $N > 100$ oder größerer

---

[1] Die Konstanten J_0 und γ, welche die Intensität und die Abfallspotenz des Spektrums bezeichnen, werden später empirisch bestimmt. E_0 kann willkürlich festgesetzt werden, z. B. E_0: 10^8 eV.

Schichten $l > 3_{10} \log N$ ($E\,6$). (25) geht dann in die einfachere Formel über:

$$H(N\,l) \approx F(E) = J_0 \left(\frac{E_0}{E_j}\right)^{\gamma} e^{-\gamma y}; \qquad N = z(l, y), \qquad (25\,\mathrm{a})$$

welche besagt, daß die Häufigkeit der Schauer von mehr als N Teilchen hinter der Schicht l gleich der Häufigkeit der einfallenden Elektronen oberhalb derjenigen Energie E ist, die im Mittel den Schauer N hinter dieser Schicht l erzeugen würde.

Die Schauerhäufigkeit, welche man auf diese Weise erhält, ist in Abb. 8 und in den folgenden Formeln angegeben. Abb. 8 zeigt die „Schauerauslösekurve" (25 a) für $N = 200$ Teilchen, d. h. die Häufigkeit der Schauer von mehr als 200 Teilchen als Funktion der Schichtdicke l (Tabelle 2) für ein Spektrum

$$F(E) = \left(\frac{E_0}{E}\right)^{1,5} \quad (24)\ (E\,7).$$

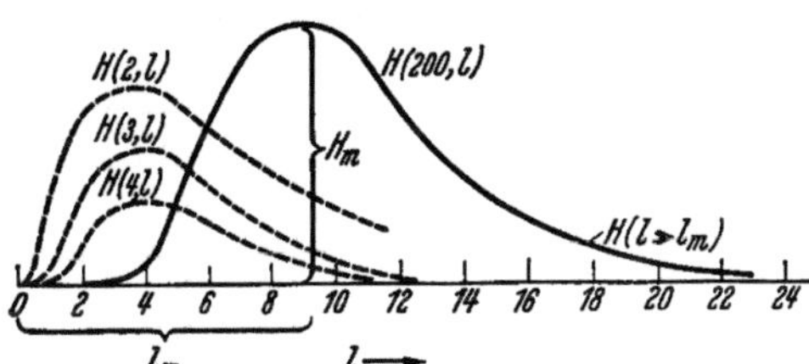

Abb. 8. Schauerauslösekurven. Abszisse: Schichtdicke l in Einheiten der Tabelle 2, S. 8. Ordinate: Theoretische Häufigkeit $H(N, l)$ der Schauer von mehr als N Teilchen. Ausgezogene Kurve: für große Schauer $N = 200$; gestrichelte Kurve: für kleine Schauer $N = 2, 3, 4$ nach ARLEY. Der Ordinatenmaßstab der gestrichelten Kurven ist nicht mit dem der ausgezogenen Kurve vergleichbar.

Abb. 8 zeigt ferner gestrichelt die entsprechenden Auslösekurven der kleineren Schauer von mehr als $N = 2, 3, 4$ Teilchen, welche von ARLEY ($A\,4$) unter Benutzung einer für kleine Schauer gültigen Schwankungsformel (20) und einer Korrektur (14a) für die energiearmen Elektronen aus einem Spektrum

$$F(E) = \left(\frac{E_0}{E}\right)^{1,5} \text{für } E > E_0 = 2 \cdot 10^8\,\mathrm{eV}, \quad F(E) = 15\,\frac{E}{E_0} \text{ für } E < E_0 \ (24)\ (26)$$

berechnet wurde, welche aber nur wenig vom Spektrum abhängen. Die allgemeinen Eigenschaften der Schauerhäufigkeit, welche zum Teil aus der Abbildung abgelesen werden können, sind für ein Spektrum (24) die folgenden ($E\,6$):

a) Bei *dünnen Schichten* steigt die Häufigkeit der großen Schauer in sehr starker Schmiegung mit der Schichtdicke an. Die Schauerverteilung $H(N)$ fällt hier um so schneller zu großen Schauern hin ab, je dünner die Schicht ist (Abb. 25) (18a):

für $\qquad 1 \lesssim l \lesssim 2$ bis $3,\ N > 100 \qquad$ wird

$$H(N, l) \approx J_0 \left(\frac{E_0}{E_j}\right)^{\gamma} e^{-\frac{0,18\,\gamma}{l^2}\left[\ln N + \frac{1}{2}\ln\ln N + 1,5\right]^2} \qquad (27)$$

b) Die Schauerhäufigkeit erreicht ein *Maximum* bei der „Gleichgewichtsschicht" (16)

$$l_m \approx 3_{10} \log N + 2,7, \qquad (28)$$

bei welcher die Schauerverteilung $H(N)$ annähernd gleich der Energie-verteilung $F(E)$ im Spektrum ist (15):

$$H(N, l_m) = H_m(N) \approx J_0 \left(\frac{E_0}{E_j}\right)^{\gamma} \left(\frac{1}{8N}\right)^{\frac{\gamma}{0,93}}. \tag{29}$$

c) Bei *dicken Schichten* fällt nach (19) die Schauerauslösung $H(l)$ wieder exponentiell ab, derart, daß sie nach etwa der doppelten Strecke des Anstiegs auf den 10. Teil des maximalen Werts gesunken ist. Die Schauerverteilung $H(N)$ ist hier nur wenig steiler als das Spektrum $F(E)$:

für $l \gtrsim 5_{10} \log N$ wird nach (19)

$$H(N, l) \approx \left(\frac{1}{N}\right)^{\frac{\gamma}{a}} \cdot e^{-\frac{\gamma}{a} bl} \cdot \left(\frac{E_0}{E_j}\right)^{\gamma} \cdot e^{-3,2\frac{\gamma}{a}}$$

mit den Koeffizienten:

$a =$	1,9	1,5
$b =$	0,40	0,25
für $N \approx$	10	1000
$l \approx$	15—20	

$$\tag{30}$$

d) Die Intensität ist, wie der Faktor $\left(\frac{E_0}{E_j}\right)^{\gamma}$ in den Formeln (27) bis (30) angibt, im schweren Material viel größer als im leichten, da z. B. in Pb die Energieteilung bis zu $E_j = 10^7$ eV herab erfolgen kann, während sie in Al schon bei der Grenze $E_j = 6 \cdot 10^7$ eV infolge der Ionisa-tionsverluste aufhört (Tabelle 2). Im Fall $\gamma = 1$ verhalten sich also die Intensitäten in verschiedenen Materialien ungefähr wie die Atom-nummern. Bei stärker abfallenden Spektren $\gamma > 1$ sind die Verhältnisse verschiedener Materialien noch extremer.

Insbesondere beschreibt die aus (25) gebildete Funktion $H(1, l)$ die Zahl der Ereignisse von mehr als einem Teilchen, also die Zahl der Koinzidenzen, die in einer Anordnung Abb. 9 von Elektronen ausgelöst werden, als Funktion der Schichtdicke.

Berechnet man für ein mittleres Spektrum (24, $\gamma \sim 1$ bis 2) diese Ab-sorptionskurve der Koinzidenzen $H(1, l)$ in verschiedenen Materialien, so findet man, wie HEITLER (H 6) (vgl. auch A 4) gezeigt hat, hinter Schichten gleicher Masse ungefähr die gleiche Intensität. Diese annähernd *massen-proportionale* Absorption der Elektronenkoinzidenzen kommt durch Kompensation zweier Wirkungen zustande: Im schweren Material ist zwar die Teilung der Anfangsenergie pro g/cm² größer als im leichten. Im schweren Material kann sie aber auch bis zu einer niedrigeren Energie herab fortgesetzt werden, so daß gleiche Massen annähernd gleiche Absorptionswirkungen haben, als ob es sich um eine Ionisationsabsorption allein handeln würde.

12. Die BHABHAschen Ionisationsschauer. Nach der Kaskaden-theorie erzeugt jedes Elektron, welches auf eine Schicht von mehr als 1 cm Pb fällt, einen Schauer. Schwere Teilchen können praktisch keine direkten Kaskadenschauer hervorrufen, da sie wegen ihrer großen Masse so wenig strahlen. Trotzdem muß man, wie BHABHA ($B\,6$) gezeigt hat, auf Grund der Kaskadentheorie erwarten, daß auch schwere Teilchen ab und zu von kleineren Schauern begleitet sind.

Denn ein schweres Teilchen gibt ja beim Durchgang durch Materie Energie durch Ionisation ab, d. h. es stößt die Elektronen der durchquerten Atomhüllen an (§ 9). Hierbei wird es ab und zu viel Energie auf ein Elektron übertragen, und ein solches stark angestoßenes Elektron erzeugt dann seinerseits eine Kaskade.

Die mittlere Zahl der Elektronen und Positronen oberhalb der für das Material charakteristischen Grenzenergie E_j (§ 6), welche auf diese Weise schwere Elektronen von 10^{10} eV hinter einer Schicht Blei begleiten, ist nach BHABHA ($B\,6$) 10%. Als mittlere Gesamtzahl kann nach (14a) etwa das Doppelte dieser Zahl angenommen werden. Die genaueren Werte dieser mittleren Elektronenzahl sind nach den Rechnungen von BHABHA für ein Teilchen der Masse $\mu = 100\,m$ (m = Elektronenmasse) bei verschiedenen Energien in Tabelle 4 angegeben ($B\,6$).

Tabelle 4. Mittlere Zahl der Elektronen, welche ein schweres Elektron der Energie E begleiten. (Nach BHABHA.)

E		Teilchenzahl $> E_j$	Gesamte Teilchenzahl
10^8 eV	Pb	—	—
	H_2O	—	—
10^{10} eV	Pb	0,09	0,19
	H_2O	0,03	0,07
10^{12} eV	Pb	0,16	0,34
	H_2O	0,07	0,15

Die nächste Tabelle (5) zeigt die Verteilung dieser Elektronenanzahl auf verschiedene Schauergrößen. Sie gibt die Wahrscheinlichkeit $\dfrac{-\partial}{\partial N}\,Q(N, E)$ an, mit welcher ein schweres Elektron der Energie E von einem Elektronenschauer mit N Teilchen begleitet ist. Unter N ist dabei wieder die Zahl der Schauerteilchen oberhalb der Ionisierungsgrenze E_j verstanden. Die wirkliche Schauergröße ist nach BHABHA schätzungsweise doppelt so groß.

Tabelle 5. Wahrscheinlichkeit $\dfrac{-\partial}{\partial N}\,Q(N, E)$, mit der ein schweres Elektron der Energie E von einem Schauer mit N Teilchen begleitet ist.

E		$N = 1$	2	4	5	10	50
10^8 eV	Pb	—	—	—	—	—	—
	H_2O	—	—	—	—	—	—
10^{10} eV	Pb	0,046	0,014	0,0057	0,0038	0,0012	$0,34 \cdot 10^{-4}$
	H_2O	0,025	0,007	0,0022	0,0011	0,0003	—
10^{12} eV	Pb	0,047	0,015	0,0060	0,0042	0,0013	$0,54 \cdot 10^{-4}$
	H_2O	0,028	0,009	0,0036	0,0024	0,0008	$0,29 \cdot 10^{-4}$

Die Tabellen können in grober Näherung durch die folgende Formel für die Wahrscheinlichkeit $Q(N, E)$ ersetzt werden, mit der ein schweres

Elektron der Energie $E > \mu c^2$ von einem Ionisationsschauer mit mehr als N Teilchen in Wasser begleitet ist:

$$\left. \begin{aligned} Q(N, E) &\approx 0{,}03 \,\frac{1}{N} \qquad \text{für} \qquad E > 8\,N E_j, \\ \dot{Q}(N, E) &\approx 0 \qquad\qquad \text{für} \qquad E < 8\,N E_j. \end{aligned} \right\} \tag{31}$$

Diese Formel ist in ihren Hauptzügen verständlich: Denn damit ein Schauer von N Teilchen entsteht, muß [nach (15)] im Mittel auf ein Atomelektron eine Energie $8\,N E_j$ übertragen werden und dazu muß das schwere Elektron mindestens die Energie $8\,N E_j$ besitzen. Ionisationsschauer von mehr als N Teilchen können also im Material mit der Ionisierungsgrenze E_j nur von schweren Teilchen oberhalb der Energie $8\,N E$ erzeugt werden. Das Verteilungsgesetz $1/N$ der Ionisationsschauer von mehr als N Teilchen (31) folgt nach (15) aus der Energieverteilung $1/E$ (§ 9 B) der Ionisationselektronen oberhalb der Energie E.

Die mittlere Zahl der Elektronen, welche mit einem schweren Elektron der Energie E im Gleichgewicht sind, wird nach (31) annähernd

$$\int_{0}^{\frac{E}{8E_j}} \frac{-\partial Q(NE)}{\partial N}\, N\, dN = 0{,}03 \ln \frac{E}{8\,E_j} \qquad (\text{in } H_2O) \tag{32}$$

in grober Übereinstimmung mit Tabelle 4.

Die Häufigkeit der Ionisationsschauer in anderen Substanzen als Wasser geht aus (31) und (32) durch Multiplikation mit dem Faktor

$$\left. \begin{aligned} \frac{0{,}8 \cdot 10^9\,\text{eV}}{E_j}\,\frac{1}{Z} \\ (Z = \text{Atomnummer } E_j = \text{Ionisierungsgrenze}) \end{aligned} \right\} \tag{33}$$

hervor.

III. Theorie der schweren Teilchen.

13. Schwere Elektronen und Kernkräfte. In der Einleitung (I) wurden bereits die Gründe auseinandergesetzt, die dafür sprechen, daß die durchdringende Komponente der Höhenstrahlung in der Hauptsache aus einer neuen Art von Elementarteilchen besteht, deren Masse etwa das 160fache der Elektronenmasse beträgt: Einerseits kann es sich bei der durchdringenden Komponente nicht um gewöhnliche Elektronen handeln, da diese nach der Theorie und nach der experimentellen Erfahrung Kaskaden erzeugen und große Schichten nicht einzeln durchdringen können. Der große Strahlungsverlust der Elektronen fällt weg bei Teilchen größerer Masse, da die Strahlungsintensität dem reziproken Quadrat der Ruhmasse der strahlenden Teilchen proportional ist. Andererseits folgt aus der gemessenen Ionisation der durchdringenden Teilchen, daß es sich zum mindesten bei den Teilchen von einem Impuls $pc < 7 \cdot 10^8$ eV nicht um Protonen handeln kann, da Protonen so geringer Impulse merklich stärker ionisieren als Elektronen. Diese Argumente zusammen mit den schon genannten Wilson-Kammeraufnahmen der schweren Elektronen geben der Annahme von der Existenz der neuartigen Teilchen einen hohen Grad von Sicherheit.

Ob alle Teilchen dieser Art die gleiche Masse haben, oder ob es in der durchdringenden Komponente Teilchen vieler verschiedener Massen gibt, darüber geben die Experimente einstweilen keine Auskunft. Auch läßt sich bisher nicht entscheiden, wie viele von den durchdringenden Teilchen sehr hoher Energie etwa Protonen sind. Die bisher vorliegenden Messungen, die eine Bestimmung der Masse der durchdringenden Teilchen versuchen, lassen sich jedoch alle mit der Annahme vereinigen, daß die durchdringende Komponente in der Hauptsache aus einer bestimmten Teilchensorte besteht, deren Ruhmasse etwa das 160fache der Elektronenmasse beträgt, und daß außer diesen Teilchen nur noch Protonen und Neutronen in relativ geringer Anzahl vorkommen.

Wenn man diese Annahme nicht macht, so gibt es einstweilen noch keine theoretischen Gesichtspunkte, die zu Aussagen über das Verhalten dieser Teilchen führen könnten. Wenn man jedoch die Existenz einer bestimmten Teilchensorte von einer Ruhmasse von etwa 160 Elektronenmassen annimmt, so liegt es nahe, diese Teilchen in Verbindung zu bringen mit einer Theorie der Kernkräfte, die im Jahre 1935 von Yukawa (Y $4, 5, 6, 7$) vorgeschlagen und von ihm und verschiedenen anderen Forschern ausgearbeitet worden ist (F 3, K 1, B 7, B 8, W 3). Yukawa hat in dieser Theorie die Existenz von Teilchen der genannten Art vorhergesagt und damit diesen Teilchen einen bestimmten Platz in dem ganzen Zusammenhang zwischen Kernkräften, β-Zerfall usw. angewiesen. Obwohl es vielleicht verfrüht wäre, von einer definitiven Bestätigung der Yukawaschen Theorie durch die Entdeckung der schweren Elektronen zu sprechen, scheint es doch natürlich, die experimentellen Ergebnisse über die durchdringende Komponente der Höhenstrahlung an Hand der Yukawaschen Theorie zu analysieren. Wir geben also zunächst eine Übersicht über die Grundgedanken und die wesentlichsten Ergebnisse der Yukawaschen Theorie.

Die Yukawasche Theorie geht von der Absicht aus, die Kräfte, die die Neutronen und Protonen im Kern zusammenhalten, in möglichst enge Analogie zu bringen zu den elektrischen Kräften. Yukawa führt also ein Kraftfeld der schweren Kernbausteine ein, das ähnlich wie das elektrische Feld durch gewisse Wellenfunktionen beschrieben wird, die einer Differentialgleichung 2. Ordnung zu genügen haben. Die Kernkräfte unterscheiden sich jedoch von den elektrischen Kräften dadurch, daß sie eine endliche Reichweite von der Größenordnung des klassischen Elektronenradius ($r_0 = 2{,}81 \cdot 10^{-13}$ cm) haben. Daher setzt Yukawa an die Stelle der elektrostatischen Potentialgleichung

$$\Delta \varphi = 0$$

die andere Gleichung

$$\Delta \varphi - k^2 \varphi = 0.$$

Diese Gleichung führt zu einem Potential der Form e^{-kr}/r. Die Konstante k bestimmt die Reichweite der Kernkräfte und hat die Größenordnung $k \sim 1/r_0$. Erweitert man nun die Potentialgleichung zur Wellengleichung, so erhält man:

$$-\frac{1}{c^2}\frac{\partial^2\varphi}{\partial t^2}+\varDelta\varphi-k^2\varphi=0. \tag{34}$$

Dies ist die DE BROGLIEsche Wellengleichung für Teilchen von der Ruhmasse $\mu = \hbar k/c$ ($P\,2$). An die Stelle der Lichtquanten in der MAXWELL-Theorie treten also wegen der endlichen Reichweite der Kernkräfte Teilchen der Ruhmasse $\hbar k/c$, und diese Ruhmasse stimmt in der Größenordnung überein mit der Ruhmasse, die aus den wenigen vorliegenden Experimenten über die Masse der schweren Elektronen folgt (etwa 160 Elektronenmassen). Ein weiterer Unterschied zwischen dem Kernfeld und dem elektrischen Feld muß darin bestehen, daß das Kernfeld zu *Austausch*kräften zwischen Neutronen und Protonen führt. Dies erreicht YUKAWA durch die Annahme, daß die dem Kernfeld entsprechenden Partikel geladen seien, daß also die Emission eines YUKAWA-Teilchens mit einer Änderung der Ladung des emittierten Teilchens verknüpft sei derart, daß die Ladung im ganzen erhalten bleibt.

Aus diesen Grundgedanken der YUKAWA-Theorie folgt, daß die in der Höhenstrahlung beobachteten schweren Elektronen, wenn sie mit den YUKAWA-Teilchen identifiziert werden können, im Gegensatz zu allen anderen geladenen Elementarteilchen den Regeln der BOSE-Statistik genügen müssen und einen ganzzahligen Spin besitzen. Bei möglichst enger Analogie der YUKAWA-Teilchen mit den Lichtquanten wird man den Spin 1 für die YUKAWAschen Teilchen erwarten. Die Theorie für diesen Fall ist von BHABHA ($B\,5, 6$) und FRÖHLICH, HEITLER und KEMMER ($F\,3$) und YUKAWA ($Y\,7$) ausgearbeitet werden. Aus Symmetriegründen muß es positiv und negativ geladene YUKAWAsche Teilchen geben, wobei der Absolutbetrag der Ladung stets ein elektrisches Elementarquantum betragen muß. Wenn die Kräfte zwischen den Kernbausteinen von der Ladung in der Weise unabhängig sind, wie dies auf Grund der Versuche von TUVE, HAFSTADT und HEYDENBURG von verschiedenen amerikanischen Forschern vermutet wurde, so müßte man annehmen, daß es auch elektrisch neutrale YUKAWAsche Teilchen gibt. Eine experimentell oder theoretisch sicher begründete Aussage hierüber ist aber wohl einstweilen nicht möglich.

14. Zerfall der schweren Elektronen. YUKAWA nimmt nun weiter eine Wechselwirkung des Kernfeldes auch mit den leichten Teilchen (Elektronen und Neutrinos) an; z. B. sollte ein sehr energiereiches Elektron bei der Ablenkung in einem Kraftfeld ein negativ geladenes YUKAWAsches Teilchen emittieren und sich dabei gleichzeitig in ein Neutrino verwandeln können. Eine solche Wechselwirkung hat zur Folge, daß das YUKAWAsche Teilchen von selbst (ohne Wechselwirkung mit anderer Materie) in ein Elektron und ein Neutrino zerfallen kann. Das

Yukawasche Teilchen besitzt also einfach eine natürliche β-Radioaktivität. Seine Lebensdauer hängt mit der Stärke jener Wechselwirkung zusammen und läßt sich dadurch abschätzen, daß man die gewöhnliche β-Aktivität der Atomkerne aus der Yukawaschen Theorie berechnet. Die natürliche β-Aktivität der Atomkerne erklärt sich in dieser Theorie durch die Annahme, daß etwa ein Neutron in ein Proton unter gleichzeitiger Erzeugung eines virtuellen Yukawaschen Teilchens übergeht und daß dieses virtuelle Teilchen gleichzeitig in Elektron und Neutrino zerfällt. Diese Vorstellung führt im wesentlichen wieder zur Fermischen Theorie des β-Zerfalls und liefert darüber hinaus die mittlere Lebensdauer des schweren Elektrons, die Yukawa zu $\tau = \frac{1}{2} \cdot 10^{-6}$ sec berechnet hat.

Da der spontane Zerfall der schweren Elektronen für die Diskussion des Verhaltens der durchdringenden Komponente wichtig ist, muß er noch etwas genauer beschrieben werden: Die Erhaltungssätze für Impuls und Energie fordern, daß ein ruhendes Yukawa-Teilchen in der Weise in Elektron und Neutrino zerfällt, daß die beiden leichten Teilchen in genau entgegengesetzter Richtung mit entgegengesetzt gleichen Impulsen davonfliegen. Die Summe der Energie der beiden leichten Teilchen muß mit der Ruhmasse des schweren Elektrons übereinstimmen. Wegen der geringen Ruhmasse des Elektrons und der verschwindenden Ruhmasse des Neutrinos folgt daraus, daß die kinetische Energie von Elektron und Neutrino beim Zerfall ziemlich genau $\mu c^2/2$ beträgt (μ bedeutet dabei die Masse des schweren Elektrons). Die Radioaktivität der schweren Elektronen unterscheidet sich also insofern von der β-Radioaktivität irgendwelcher Atomkerne, als das Elektron beim Zerfall eine scharf bestimmte Energie von etwa 40 MeV mitbekommen sollte. Die Emission des Elektrons bzw. des Neutrinos wird dabei nach allen Raumrichtungen mit gleicher Wahrscheinlichkeit erfolgen.

Der Zerfall des schweren Elektrons wird nun häufig zu einer Zeit erfolgen, wo das schwere Elektron noch eine erhebliche kinetische Energie besitzt. In diesem Fall kann man den Impuls des emittierten Elektrons durch eine Lorentz-Transformation ermitteln. Wir nehmen an, das schwere Elektron bewege sich in der Richtung x mit der Geschwindigkeit βc. Der Winkel, den die Emissionsrichtung mit der x-Achse einschließt — gemessen in dem Koordinatensystem, in dem das schwere Elektron ruht —, heiße φ. Dann erhalten wir für die Energie des Elektrons und seine Impulskomponente parallel und senkrecht zur x-Richtung im Ruhsystem des Yukawa-Teilchens die Beziehungen

$$E' = \frac{\mu c^2}{2}; \qquad p'_{\parallel} = \frac{\mu c}{2}\cos\varphi; \qquad p'_{\perp} = \frac{\mu c}{2}\sin\varphi.$$

Im Koordinatensystem des Beobachters gilt dann

$$E = \frac{\mu c^2}{2}\frac{(1+\beta\cos\varphi)}{\sqrt{1-\beta^2}}; \qquad p_{\parallel} = \frac{\mu c}{2}\frac{(\beta+\cos\varphi)}{\sqrt{1-\beta^2}}; \qquad p_{\perp} = \frac{\mu c}{2}\sin\varphi.$$

Theoretische Gesichtspunkte zur Deutung der kosmischen Strahlung.   27

Drückt man in dieser Rechnung die Geschwindigkeit des YUKAWA-Teilchens durch seinen Impuls P aus, so folgt

$$E = \frac{c}{2}\left(\sqrt{(\mu c)^2 + P^2} + P\cos\varphi\right);$$
$$p_\| = \frac{1}{2}\left(P + \sqrt{(\mu c)^2 + P^2}\cos\varphi\right); \qquad p_\perp = \frac{\mu c}{2}\sin\varphi. \qquad (35)$$

Wenn die Energie des YUKAWA-Teilchens sehr viel größer als seine Ruhmasse ist, so wird also auch das Elektron beim radioaktiven Zerfall in den meisten Fällen nahezu in der Bewegungsrichtung des YUKAWA-Teilchens ausgeschleudert. Sein Impuls in dieser Richtung kann etwa mit gleicher Wahrscheinlichkeit alle Werte von 0 bis P annehmen.

Wenn man aus der Zerfallswahrscheinlichkeit $1/\tau$ des YUKAWA-Teilchens auf die Strecke schließen will, die es im allgemeinen vor seinem Zerfall durchlaufen hat, wenn man also nach der Zerfallswahrscheinlichkeit pro cm fragt, so muß man, wie BHABHA betont hat, auf die Zeitdilatation der Relativitätstheorie achten. Man erhält für die Zerfallswahrscheinlichkeit w pro cm den Wert

$$w = \frac{\mu}{\tau P}. \qquad (36)$$

15. Sekundärprozesse der schweren Elektronen. Die YUKAWA-sche Theorie verhält sich hinsichtlich der Wechselwirkung zwischen Protonen, Neutronen und schweren Elektronen ähnlich wie die FERMI-sche Theorie hinsichtlich der Wechselwirkung zwischen leichten und schweren Teilchen ($F\,1$) und führt zu einer Folgerung, die für die Diskussion der Schauer und der HOFFMANNschen Stöße ($\S\,23$, 24) von Wichtigkeit ist. Die Wechselwirkung zwischen den drei genannten Teilchensorten wird nämlich um so größer, je größer die Energie der bei einem Zusammenstoß beteiligten Korpuskeln ist ($H\,8$). Dies hat zur Folge, daß dann, wenn bei einem Zusammenstoß eine Energie von sehr viel mehr als etwa 10^8 eV zur Verfügung steht, diese Energie im allgemeinen dazu verwendet werden wird, in einem einzigen Akt eine Reihe von Teilchen zu erzeugen, deren Energie in der Gegend von einigen 10^8 eV liegen wird ($H\,3$). Ein solcher explosionsartiger Schauer sollte nach der YUKAWAschen Theorie (im Gegensatz zur FERMIschen Theorie des β-Zerfalls) im wesentlichen nur Protonen, Neutronen und schwere Elektronen enthalten. Erst mit einer um den Faktor $e^2/\hbar c$ bzw. $(e^2/\hbar c)^2$ geringeren Wahrscheinlichkeit sollten auch Lichtquanten und gewöhnliche Elektronen entstehen. Irgendwelche genaueren Angaben über diese Explosionen lassen sich aus der YUKAWAschen Theorie deswegen nicht gewinnen, weil das Auftreten der Explosionen eben die Grenze bezeichnet, die der Anwendung der bisherigen Quantentheorie gesetzt ist ($H\,3$, $H\,5$, $H\,8$).

Entwirft man nach diesen theoretischen Vorstellungen ein Bild von den Prozessen, die sich an einem schweren Elektron bei seinem Durchgang durch Materie abspielen, so kommt man etwa zu folgendem

Resultat. Zunächst wirkt das schwere Elektron auf die umgebende Materie vermöge seiner Ladung, es verursacht also Ionisation und wird dadurch gebremst ebenso wie Protonen oder Elektronen. Die üblichen Formeln für den Energieverlust (2) pro cm und für die Reichweite als Funktion der Energie (11) wird man unbedenklich verwenden können. Beim Durchgang durch Atomkerne kann ferner das schwere Elektron in Wechselwirkung mit den Protonen und Neutronen treten. Diese Wechselwirkung kann nicht durch eine Kraft im Sinn der klassischen Theorie beschrieben werden, sondern ist zu vergleichen mit der Wechselwirkung zwischen Lichtquanten und Elektronen. Das YUKAWAsche Teilchen kann also erstens im Kern absorbiert werden, was nur unter gleichzeitiger Verwandlung eines Neutrons in ein Proton möglich ist (Analogon zum Photoeffekt). Es kann zweitens an den schweren Teilchen gestreut werden, wobei es im allgemeinen bei der Streuung auch seine Energie ändern wird (Analogon zum COMPTON-Effekt). Bei großen Energien des schweren Elektrons wird der letztere Prozeß häufiger sein als der erstere. Denn bei der Absorption des YUKAWA-Teilchens ist zur Erhaltung von Energie und Impuls die Anwesenheit von 2 schweren Teilchen in Wechselwirkung notwendig. Die Streuung kann dagegen schon an einem einzelnen schweren Teilchen im leeren Raum stattfinden. Wenn die Energie des schweren Elektrons von der Größenordnung 10^8 eV ist, so erhält der Wirkungsquerschnitt für diesen Streuprozeß etwa die Größenordnung 10^{-26} cm². Ferner gibt es Prozesse, die durch das Zusammenwirken der Kernkräfte mit den elektrischen Kräften entstehen und auf die HEITLER ($H\,8$) aufmerksam gemacht hat. Als einfachstes Beispiel hierfür sei folgender Vorgang genannt: Ein YUKAWA-sches Teilchen negativer Ladung stößt mit einem Proton zusammen. Dieses verwandelt sich in ein Neutron, und ein Lichtquant wird frei. Der Wirkungsquerschnitt für diesen Prozeß ist von HEITLER für YUKAWA-Teilchen der Energie $\sim 10^8$ eV zu 10^{-27} cm² abgeschätzt worden. Wenn ein YUKAWA-Teilchen von einer Energie, die 10^8 eV erheblich übersteigt, auf ein Proton oder Neutron trifft, so wird im allgemeinen die obengenannte Explosion stattfinden. In diesen Explosionen können außer den schweren Elektronen, wie schon erwähnt, Protonen und Neutronen und mit einer um $e^2/\hbar c$ bzw. $(e^2/\hbar c)^2$ kleineren Wahrscheinlichkeit auch Lichtquanten und Elektronen beteiligt sein.

Der Wirkungsquerschnitt für die Explosionen sollte etwa die Größenordnung 10^{-26} cm² haben; es ist aber möglich, daß dieser Wirkungsquerschnitt mit wachsender Energie der stoßenden Teilchen wieder kleiner wird. Eine bestimmte Aussage hierüber ist aus der bisherigen Theorie nicht möglich.

Die bisher genannten Prozesse geben zu einem Energieverlust Anlaß, der im wesentlichen massenproportional ist, d. h. der Energieverlust pro g pro cm² dürfte in allen Materialien ungefähr der gleiche sein.

Schließlich kann das Yukawa-Teilchen spontan in Elektron und Neutrino zerfallen. Dieser Prozeß spielt sich völlig unabhängig vom durchlaufenen Material ab, von einer Massenproportionalität dieser Absorption ist also keine Rede. Es ist zwar denkbar, daß es neben dem spontanen Zerfall auch noch einen Zerfall gibt, der durch die Wechselwirkung mit anderer Materie induziert wird. Einfache theoretische Abschätzungen machen es aber unwahrscheinlich, daß dieser induzierte Zerfall neben dem spontanen einen erheblichen Beitrag liefert.

16. Sekundärwirkungen schneller Protonen und Neutronen (H 4). Bei der durchdringenden Komponente der Höhenstrahlung werden auch Protonen und Neutronen eine Rolle spielen. Zwar ist es zur Zeit kaum möglich anzugeben, welcher Bruchteil der durchdringenden Strahlen hoher Energie aus Protonen besteht. Jedenfalls aber zeigen Experimente immer wieder das Auftreten von schweren Teilchen geringerer Energie (R 6, A 3). Die Existenz von Protonen und Neutronen in der Höhenstrahlung ist also als gesichert zu betrachten. Diese schweren Teilchen können nun nach den vorhergenannten theoretischen Überlegungen in erster Linie die folgenden Prozesse hervorrufen: Über die Ionisation der Protonen ist schon in der Einleitung das Wesentliche gesagt. Beim Zusammenstoß mit ·Atomkernen können die Protonen und Neutronen mit den Kernbausteinen durch die Austauschkraft in Wechselwirkung treten. Beim Durchgang durch Atomkerne verhalten sich also die schweren Teilchen (im Gegensatz zu den Yukawaschen Teilchen) ganz ähnlich wie die Elektronen beim Durchgang durch Materie. Durch die Kernkräfte, die eine Reichweite von der Größenordnung $2,8 \cdot 10^{-13}$ cm besitzen, ist ein durch einen Kern fliegendes energiereiches Proton imstande, Energie auf benachbarte Protonen oder Neutronen zu übertragen, also Sekundärteilchen einer bestimmten Energieverteilung zu bilden. Die hierbei abgegebene Energie wird nach der Bohrschen Theorie (B 21) zum größten Teil zu einer Erhitzung des getroffenen Atomkerns verwendet werden, der als Folge der Erhitzung nachträglich in Analogie zur Verdampfung Protonen und Neutronen emittieren wird. Zum Teil können die von den stoßenden Protonen oder Neutronen getroffenen Kernbausteine, wenn sie genügende Energie erhalten haben, den Kern auch verlassen und als Sekundärteilchen in Erscheinung treten.

Die von einem Proton oder Neutron der Geschwindigkeit βc pro cm an die Kernmaterie abgegebene Energie ist ungefähr durch die Formel (H 4)

$$\frac{\partial E}{\partial x} \approx -\frac{M c^2}{r_0 \beta^2} \cdot 6 \cdot 10^{-3} \qquad\qquad (37)$$

$(r_0 = \text{klass. Elektronenradius}; \ M \ \text{Protonenmasse})$

gegeben. Für die Reichweite R eines schweren Teilchens als Funktion der kinetischen Energie E erhält man daraus (H 4) (vgl. 2, 11):

$$R \approx r_0 \frac{E^2}{M c^2 (M c^2 + E)} \cdot 1,7 \cdot 10^2. \qquad (37\,\text{a})$$

Aus diesen Formeln folgt, daß Protonen und Neutronen, die sich nahezu mit Lichtgeschwindigkeit bewegen, im Mittel nur wenig Energie (von der Ordnung 20 MV) auf den Atomkern übertragen. Langsamere Teilchen dagegen, z. B. Protonen von $3 \cdot 10^8$ eV, können ihre ganze Energie beim Durchgang durch einen Atomkern verlieren.

Zum Teil können die von den stoßenden Protonen oder Neutronen getroffenen Kernbausteine, wenn sie genügend Energie erhalten haben, den Kern verlassen und als Sekundärteilchen in Erscheinung treten. Die Energieverteilung dieser Sekundärteilchen hängt praktisch nicht von der Energie des stoßenden Teilchens ab, sofern die letztere nur erheblich größer ist als die Energie der Sekundärteilchen. Auch hängt sie kaum ab von der Größe des getroffenen Atomkerns. Die Anzahl der Sekundärteilchen $n(E)$ einer Energie größer als E ergibt sich ungefähr $(H\ 4)$ zu

$$n(E) \approx \text{const} \cdot e^{-78 \frac{E}{M c^2}}. \tag{38}$$

Neben diesen Prozessen, die in enger Analogie zu den üblichen Ionisationsprozessen stehen, gibt es aber noch andere, die mit der Bremsstrahlung der Elektronen verglichen werden können und die insbesondere bei großen Energien der stoßenden Teilchen die Hauptrolle spielen dürften. Ein energiereiches Proton kann beim Zusammenstoß mit einem anderen schweren Teilchen ein YUKAWA-Teilchen emittieren und sich dabei in ein Neutron verwandeln. Bei hinreichend großer Energie kann es mit einem Schlage mehrere schwere Elektronen und andere schwere Teilchen nach Art der oben genannten Explosionen erzeugen. Der Wirkungsquerschnitt für diese Prozesse dürfte ähnlich wie bei den entsprechenden Prozessen der YUKAWA-Teilchen die Größenordnung 10^{-26} cm² besitzen.

IV. Die Spektren der kosmischen Strahlung und ihre Absorption.

17. Die empirischen Grundtatsachen (vgl. $M\ 3$, $S\ 6$). Es ist in der phänomenologischen Beschreibung der kosmischen Strahlung nach AUGER üblich, eine *harte und eine weiche Komponente* zu unterscheiden, wobei unter „harter Komponente" derjenige Teil der ionisierenden Strahlen verstanden wird, der mehr als 10 cm Pb durchdringt, und unter „weicher Komponente" der Teil, der von 10 cm Pb zurückgehalten wird ($A\ 5$, $A\ 6$). Diese Unterscheidung wird nahegelegt durch Koinzidenzexperimente, welche in Abb. 9 dargestellt sind. Abb. 9 gibt die Zahl der geradlinigen Zählrohr-Koinzidenzen als Funktion der Dicke einer Bleischicht an, welche zwischen die Zählrohre geschoben ist. Man sieht, daß die Häufigkeit der Koinzidenzen in den ersten 10 cm Pb stark abnimmt, um dann in einen langsamen Abfall überzugehen, und man interpretiert die zwei Äste der Absorptionskurve als Wirkung zweier Komponenten der Strahlung, deren erste schon auf einer kurzen Strecke absorbiert wird und deren zweite vom Blei weniger beeinflußt wird.

Aus den in der Einleitung diskutierten WILSON-Aufnahmen folgt nun, daß die weiche Komponente aus Elektronen und Positronen besteht, welche schon in den ersten Zentimetern Pb verwandelt und absorbiert werden (vgl. *G 1*), und daß die harte Komponente schwere Elektronen, d. h. Teilchen von 100—200 Elektronenmassen, enthält. Die Frage, wieweit der harten Komponente bei höheren Impulsen

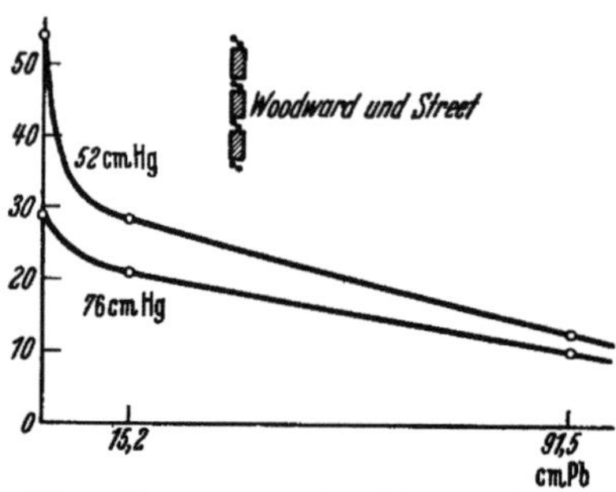

Abb. 9. Absorption der Koinzidenzen in Blei nach WOODWARD und STREET. Abszisse: Schichtdicke des Absorbers. Ordinate: Häufigkeit der Koinzidenzen nach Messungen von WOODWARD und STREET in zwei verschiedenen Höhen entsprechend 76 cm Hg und 52 cm Hg.

als 10^9 eV auch wesentliche Mengen von Protonen oder anderen schweren Bestandteilen beigemischt sind, kann bisher noch nicht entschieden werden. Wir werden hier vorläufig annehmen, daß die harte Komponente überwiegend aus schweren Elektronen besteht, und die Konsequenzen dieser Annahme verfolgen.

Der *Anteil der weichen* Komponente an der ionisierenden Gesamtstrahlung beträgt, wie man in Abb. 9 ablesen kann, im Meeresniveau etwa 30%; er sinkt nach Messungen von AUGER und Mitarbeitern unter $1^1/_2$ m Wasser auf 15% und unter 8 m Erde auf 7% (*A 5*) herab[1] und steigt in einer Höhe entsprechend 50 cm Hg auf 50% an. In größerer Höhe über dem Erdboden nimmt das Verhältnis der weichen Strahlung zur harten zunächst langsam weiter zu, wie aus den Ionisationsmessungen von BOWEN, MILLIKAN und NEHER (Abb. 10) hervorgeht.

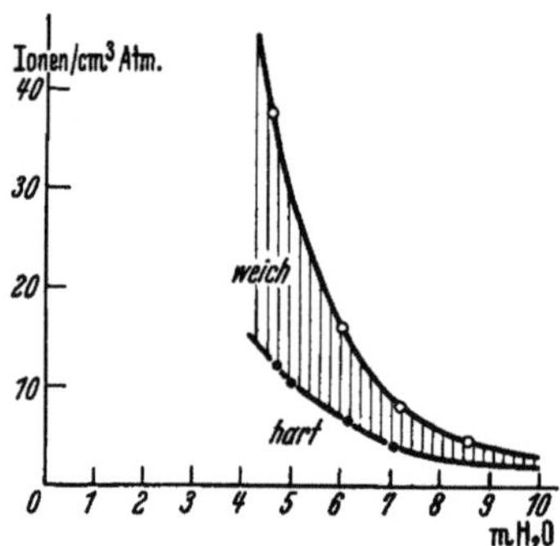

Abb. 10. Harte und weiche Komponente in der unteren Atmosphäre nach BOWEN, MILLIKAN u. NEHER: Physic. Rev. 46, 645. (1934). Abszisse: Tiefe der Atmosphäre in m Wasser. Ordinate: Ionisation ohne Panzer (o o o) und mit einem Panzer von 10 cm Blei (● ● ●) nach Messungen von MILLIKAN (March-Field, Calif.). Schraffiertes Gebiet: weicher Anteil.

Die Variation der gesamten Intensität mit der Höhe über dem Erdboden zeigen die Abbildungen 11 und 12. Abb. 11 gibt die von PFOTZER

---

[1] Nach EHMERT beträgt allerdings der Anteil, der von 5 cm Pb zurückgehalten wird, 15 bis 20% im Bodensee (*E 1*). Die Frage, wieweit dieses Ergebnis mit den Resultaten der Meßanordnungen von AUGER und Mitarbeitern verglichen werden kann, bedarf jedoch einer genaueren Untersuchung.

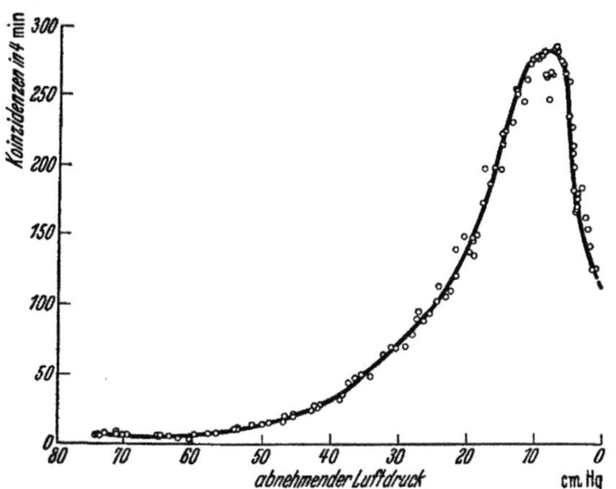

Abb. 11. Vertikalintensität der Ultrastrahlung in der Atmosphäre nach Pfotzer. [Z. Physik 102, 23 (1936).] Abszisse: Tiefe der Atmospäre in cm Hg. Ordinate: Koinzidenzzahl.

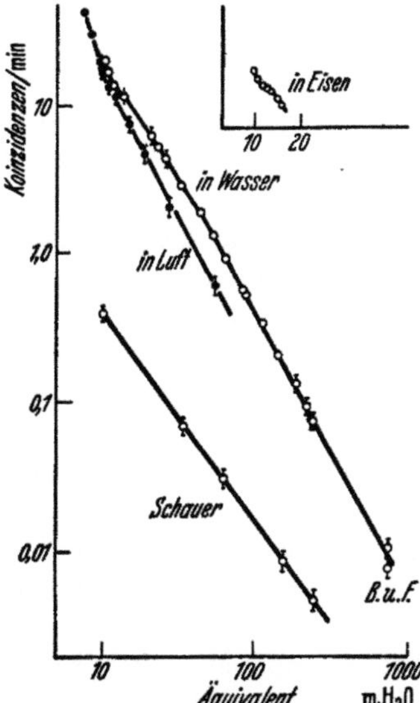

Abb. 12. Vertikalintensität im Bodensee nach Ehmert. [Z. Physik 106, 751 (1937).] Abszisse: (log) Tiefe in m Wasser von der Grenze der Atmosphäre aus. Ordinate: (log) Koinzidenzzahl. Der unterste Meßpunkt BF ist aus den Bergwerksmessungen von Barnothy und Forro übertragen. Die Kurve „in Luft" wurde von Ehmert durch schräge Koinzidenzen aufgenommen.

mit Zählrohren in Ballonaufstiegen registrierten vertikalen Koinzidenzen $(P\,1)$. Die *Vertikalintensität* nimmt zunächst in einer Höhe entsprechend 50 cm Hg auf den doppelten Wert der Erdintensität zu und erreicht schließlich in der Stratosphäre ein Maximum mit etwa dem 40fachen der Intensität in Meereshöhe. In der obersten Schicht der Atmosphäre von weniger als 8 mm Hg-Äquivalent findet dann wieder ein Absinken der Koinzidenzzahl statt.

Abb. 12 gibt die Zahl der Koinzidenzen im Bodensee nach Ehmert $(E\,1)$ an. Man erkennt, daß hier die Intensität viel langsamer pro g/cm² als in der Atmosphäre abnimmt, und daß das Gesetz der Abnahme mit der Tiefe T durch eine Potenzfunktion $T^{-\gamma}$ dargestellt werden kann, deren Exponent γ nach Ehmert in mehr als 50 m Wassertiefe $\gamma = 1{,}87$ ist $(E\,1,\ W\,7)$.

Ein etwas größeres Verhältnis der Stratosphärenintensität zur Bodenintensität als bei den Vertikalkoinzidenzen ergibt sich bei den Messungen der *Ionisation*, d. h. der allseitigen Strahlung wie die oberste Kurve der Abb. 13 (S. 35) zeigt $(B\,18)$. Vertikalintensität und Ionisation können nach Gross $(G\,7)$ unter einigen vereinfachenden Annahmen über die Absorption schräger Strahlen in der Atmosphäre ineinander umgerechnet werden (vgl. $P\,1$).

Aus dem Integral der Ionisation über die Atmosphäre ergibt sich die gesamte aus dem Weltenraum einfallende Energie $(B\,18)$. Denn diese Energie wird ja schließlich einmal zur Erzeugung von Ionen verwandt, wenn sie nur tief genug in die Erdmaterie eingedrungen ist. REGENER (*R 1*) findet eine gesamte Ionisation von $6,1 \cdot 10^9$ Ionenpaaren pro min pro cm², also (§ 9) einen Energiestrom von $1,9 \cdot 10^{11}$ eV/min · cm². BOWEN, MILLIKAN und NEHER (*B 18*) geben einen Energiestrom von $1,7 \cdot 10^{11}$ eV/min · cm² an.

Mit der Verteilung der Intensität in der Tiefe vergleichen wir die spektrale Verteilung der Intensität über verschiedene Energiebereiche. Abb. 16 (S. 41) zeigt das *Spektrum* der durchdringenden Teilchen in Meereshöhe nach BLACKETT (*B 12*) (vgl. auch *K 4, B 12, H 10, A 2*). Die Abszisse gibt die Impulse $H\varrho$, die Ordinate die Häufigkeit der Teilchen pro Impulsintervall. Die Gesamtenergie der durchdringenden Teilchen in Meereshöhe kann hieraus auf etwa $6 \cdot 10^9$ eV/min · cm² geschätzt werden.

18. Qualitative theoretische Diskussion. Der Verlauf der vertikalen Intensität in der oberen Atmosphäre kann auf Grund der Kaskadentheorie verstanden werden, wenn man annimmt, daß Elektronen oder Lichtquanten aus dem Weltenraum einfallen (*B 5, C 1*). Diese Elektronen werden sich dann in der ersten Schicht durch Multiplikation vermehren und in den späteren Schichten durch Teilung absorbieren, und sie werden auf diese Weise eine Kurve (Abb. 11) der Intensität in der Atmosphäre erzeugen, die den kaskadentheoretischen Kurven in Abb. 7 ähnlich sieht. Da die Strahlungseinheit ($l = 1$) (Tabelle 2) in Luft (275 m Normalluft) einem Druckunterschied von 2,6 cm Hg oder 0,34 m Wasser entspricht, so wird dabei eine Veränderung der Intensität um die eigene Größenordnung immer auf Strecken von einigen cm Hg zustande kommen. Aus der Lage des Maximums bei 8 cm Hg ($l \approx 3$) wird man nach Abb. 7 und Tabelle 14 auf eine mittlere Anfangsenergie von einigen 10^9 eV schließen. Den genaueren Vergleich der Strahlung in der oberen Atmosphäre mit der Kaskadentheorie werden wir in § 19 durchführen. Hierbei wird sich zeigen, daß wir zwar die weiche Strahlung in der oberen Atmosphäre den Kaskadenwirkungen zuschreiben können, daß wir aber die Elektronen in der unteren Atmosphäre auf andere Prozesse, die mit der durchdringenden Komponente in Verbindung stehen, zurückführen müssen (*B 18*).

Bei der Diskussion der durchdringenden Komponente gehen wir von der YUKAWAschen Theorie aus (Kap. III). Nach dieser Theorie sind die schweren Elektronen instabil, d. h. sie können auch im leeren Raum zerfallen. Wir müssen daher annehmen, daß sie auch im Weltenraum nicht auf längere Zeit bestehen können, und daß die schweren Elektronen, die wir an der Erdoberfläche finden, erst in der Atmosphäre aus der weichen Strahlung erzeugt wurden (*W 3, B 18*). Die Erzeugung der durchdringenden Teilchen in der Atmosphäre ist zwar noch nicht durch Ballonaufstiege erwiesen (welche in diesem Falle ein Wiederabsinken der Intensität durchdringender Teilchen in der obersten Atmosphäre ergeben müßten), aber sie kann doch als möglich betrachtet werden (*B 18*) auf Grund der Tatsache, daß die in der durchdringenden Komponente an der Erde enthaltene Energie $\sim 6 \cdot 10^9$ eV/min·cm² klein ist gegen die gesamte Energie von $2 \cdot 10^{11}$ eV/min·cm², welche an der Grenze der Atmosphäre einfällt[1].

Eine theoretische Möglichkeit für die Erzeugung schwerer Elektronen in der Atmosphäre ist innerhalb der YUKAWAschen Theorie durch die Umkehrung der in § 15 diskutierten Prozesse gegeben: Ein energiereiches Lichtquant kann beim Zusammenstoß mit einem schweren Teilchen ein oder mehrere YUKAWAsche Teilchen erzeugen (*W 3, H 8*). Da aber der Verlauf dieser Prozesse im Gebiet hoher Energien wegen der Begrenzung der heutigen Theorie noch nicht berechnet werden kann (Kap. III), werden wir die Erzeugung der schweren Teilchen hier halbempirisch behandeln. Wir werden erwarten, daß die schweren Elektronen am häufigsten in der Stratosphäre erzeugt werden, weil hier die Intensität der weichen Strahlung am größten ist, und wir werden daher etwas schematisch annehmen, daß ein Spektrum schwerer Teilchen, welches von der Energie ungefähr in der gleichen Weise abhängt, wie das erzeugende Elektronenspektrum, in der Nähe des Maximums der Ionisation ($\sim$8 cm Hg) entsteht. Ob diese einfache Annahme zulässig ist, prüfen wir dann später an dem Verlauf des BLACKETTschen Spektrums an der Erdoberfläche. Dieses Spektrum der schweren Elektronen wird nun bei seinem Durchgang durch die Atmosphäre und durch die Erde folgende Prozesse erleiden:

Erstens wird es durch Ionisation gebremst. Aus der Theorie der Ionisation kann also in gewisser Näherung ein Absorptionsgesetz der durchdringenden Teilchen abgeleitet werden (§ 20).

Zweitens werden die schweren Elektronen spontan in Elektronen und Neutrinos zerfallen. Es wird sich zeigen, daß dieser Prozeß zur Absorption der durchdringenden Komponente in der Atmosphäre erheblich beiträgt (§ 20) und daß die hierbei erzeugten Elektronen wahrscheinlich den wesentlichen Teil der in der unteren Atmosphäre gefundenen weichen Strahlung ausmachen, deren Intensität mit der YUKAWAschen Zerfallskonstanten verglichen werden kann (§ 21).

Drittens werden die schweren Teilchen beim Zusammenstoß mit Atomkernen, die in § 15 besprochenen Prozesse hervorrufen können. Wie die Diskussion des empirischen Materials zeigt, wird die bei derartigen Prozessen entstehende weiche Strahlung wahrscheinlich erst unter Wasser zusammen mit den Ionisationsschauern (§ 12) beobachtet werden können, während sie in der unteren Atmosphäre hinter den Elektronen des spontanen Zerfalls zurücktritt.

19. Das Kaskadenspektrum der Elektronen in der oberen Atmosphäre. Breiteneffekt. Zur genaueren Diskussion der Kaskadenstrahlung

---

[1] Nach einer freundlichen Mitteilung von P. M. S. BLACKETT.

in der Atmosphäre gehen wir von den Ionisationsmessungen aus, welche BOWEN, MILLIKAN und NEHER (*B 17*) in verschiedenen Breiten durchgeführt haben, und welche in Abb. 13 wiedergegeben sind. Die großen Unterschiede der Stratosphärenintensität in den verschiedenen Breiten können nach STÖRMER (*S 13*, *S 14*), LAMAÎTRE und VALLARTA (*L 1*) verstanden werden, wenn man annimmt, daß die aus dem Weltenraum kommende Strahlung in beträchtlicher Menge geladene Teilchen enthält. Geladene Teilchen können nämlich das Magnetfeld der Erde in der geomagnetischen Breite φ nur dann durchdringen, wenn ihr Impuls größer ist als ein für jede Breite charakteristischer Impuls (*S 13*):

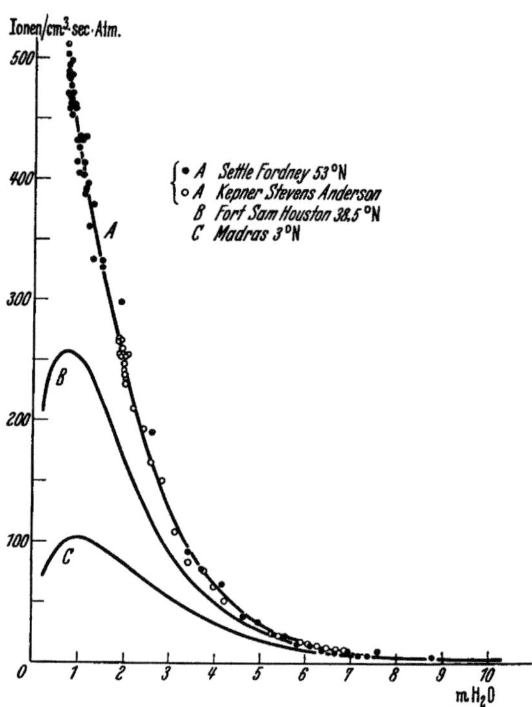

Abb. 13. Ionisation in der Atmosphäre in drei verschiedenen Breiten nach Messungen von BOWEN, MILLIKAN und NEHER. [Physic. Rev. **52**, 80 (1937).]

$$pc = E_\varphi = 1{,}9 \cdot 10^{10} \text{ eV} \cdot \cos^4\varphi.$$

$\varphi =$	0	20	40	60	80	90°	
$E_\varphi =$	1,9	1,5	0,65	0,12	0,0017	0	$\cdot\, 10^{10}$ eV

$$(39)$$

Dabei wurde angenommen, daß die Ladung der Teilchen gleich der Elementarladung ist.

Am Äquator können also einerseits weniger Elektronen einfallen als in höheren Breiten; daher ist in Abb. 13 das Maximum der Ionisation am Äquator weniger intensiv als in Polnähe. Am Äquator ist ferner die mittlere Energie der einfallenden Teilchen größer als in der Nähe der Pole; daher tritt das flachere Äquatormaximum erst bei einer größeren Schicht ein als das Maximum in höheren Breiten (*H 7*, *N 8*).

Nach (39) müssen wir den breitenabhängigen Teil der Ionisation, also die Differenz der Kurven für verschiedene Breiten 53°—38°, 38°—3° als die Wirkung von Elektronen eines schmalen Energiebereiches

$$2{,}5 \cdot 10^9\, \text{eV} < E < 6{,}7 \cdot 10^9\, \text{eV}, \qquad 6{,}7 \cdot 10^9\, \text{eV} < E < 17 \cdot 10^9\, \text{eV},$$

also als Erzeugnis von Elektronen nahezu bestimmter Energie

$$E \approx 4 \cdot 10^9\,\text{eV}, \qquad E \approx 10^{10}\,\text{eV} \tag{39a}$$

auffassen ($B\,18$). Andererseits gibt die Kaskadentheorie (Abb. 7) den Intensitätsverlauf der von diesen Energien $y = 3{,}3$, $y = 4{,}2$ erzeugten Elektronen in der Atmosphäre (14) (Abb. 7). Der Vergleich des experimentellen und theoretischen Breiteneffekts wird in Abb. 14 durchgeführt, welche der Arbeit von Bowen, Millikan und Neher ($B\,18$) entnommen ist.

Wie man sieht, kann der Verlauf der Messungen in der *oberen Atmosphäre* durch die Kaskadentheorie dargestellt werden, wenn man von einer kleinen, durch die Mittelbildung (39a) bedingten Unstimmigkeit ($N\,7$) in der Lage der Maxima absieht.

In der *unteren Atmosphäre* aber ist die kaskadentheoretische Intensität viel geringer als die wirkliche Intensität der weichen Komponente, welche nach Auger und Leprince-Riguet ($A\,7$) *in allen Breiten* denselben Bruchteil von etwa $^1/_3$ der Gesamtstrahlung in Meereshöhe beträgt. Hieraus muß gefolgert werden, daß die Elektronen in der unteren Atmosphäre sekundäre Erzeugnisse der durchdringenden Komponente sind ($B\,18$)[1]. Zum selben Resultat werden wir später auf Grund der Höhenabhängigkeit der Hoffmannschen Stöße kommen (§ 23).

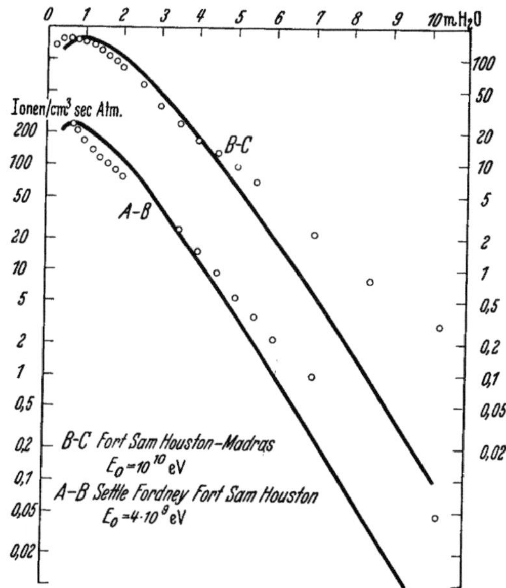

Abb. 14. Vergleich des breitenabhängigen Teils der Ionisation mit der Kaskadentheorie nach Bowen, Millikan und Neher. (Physic. Rev. **52**, 80 (1937).] Differenzen der Kurven Abb. 13 in log. Darstellung. —— Kaskadentheorie, o o o o Messung.

Wir betrachten nun die Wirkung der Elektronen oberhalb der magnetischen Grenzenergie E_φ, welche von Nordheim ($N\,6$, $N\,7$) und Heitler ($H\,7$) diskutiert wurde. Hierzu nehmen wir an, daß ein Spektrum von

$$F_1(E', 0) = J_0 \left(\frac{10^8\,\text{eV}}{E'} \right)^\gamma \tag{40}$$

Elektronen oberhalb der Energie E' aus dem Weltenraum einfällt (wobei die Konstanten γ, J zunächst unbestimmt bleiben sollen) und fragen nach dem Spektrum $F_1(E, l)$, welches sich in der Tiefe l unter der Grenze der Atmosphäre, insbesondere im Meeresniveau $l = 29$ einstellt.

---

[1] Dies geht noch deutlicher aus Abb. 14 hervor, wenn man die Grosssche Transformation berücksichtigt (§ 17).

Wie NORDHEIM ($N\,6$) gezeigt hat, ist für tiefere Schichten ($l \gtrsim 15$) die Gestalt des Spektrums $F_1(E, l)$ wieder durch das ursprüngliche Potenzgesetz (40) gegeben, und die Intensität des Spektrums $F_1(E, l)$ wird exponentiell in der Atmosphäre absorbiert mit einem Exponenten, der um so größer wird, je stärker die Potenz γ des Abfalls ist.

Denn nach § 10 ist

$$F_1(E, l) = \int\limits_{\substack{E' = E_\varphi \text{ für } E < E_\varphi \\ E' = E \text{ für } E > E_\varphi}}^{\infty} z\left(l, \ln \frac{E'}{E}\right) \frac{-\partial F_1(E', 0)}{\partial E'}\, dE', \qquad (E \geqslant E_j)$$

oder mit $E'/E = e^{y'}$ und (40)

$$F_1(E, l) = J_0 \left(\frac{10^8\,\text{eV}}{E}\right)^{\gamma} \int\limits_{\substack{y' = \ln \frac{E_\varphi}{E} \text{ für } E < E_\varphi \\ y' = 0 \quad\text{ für } E > E_\varphi}}^{\infty} z\,(l, y')\, e^{-\gamma y'} \gamma\, d y'\,.$$

Für große l ist der Wert des Integrals nahezu unabhängig von E und damit von der Breite φ und er fällt nach (19) exponentiell mit l ab; die numerische Rechnung ergibt:

$$F_1(E, l) \approx 4\, J_0 \left(\frac{10^8\,\text{eV}}{E}\right)^{\gamma} e^{-f(\gamma)\cdot(l-10)} \tag{41}$$

mit

$$f(\gamma) \approx 0{,}36\,\sqrt{\gamma - 1}\,; \quad (\gamma \lesssim 5)$$

nach NORDHEIM gilt genauer für $e \to \infty$:

$$f(\gamma) = \frac{4}{3} - \frac{1}{\gamma + 1} - \sqrt{\left(\frac{2}{3} - \frac{1}{1+\gamma}\right)^2 + \frac{4}{3(\gamma+1)\gamma}}\,. \tag{42}$$

Die Berechnung der Intensität bei kleineren Schichten führt zu einem Maximum, welches in unseren Breiten (für $\gamma = 1{,}5$ bis $2{,}5$) bei $l = 3$ bis 4 eintritt, und in welchem die Intensität $J_1 \sim 4\,J_0$ herrscht ($H\,7, N\,7$).

Die Konstante J_1 kann aus PFOTZERs Messungen in der Stratosphäre entnommen werden:

$$4\,J_0 \approx J_1 \approx 40/\text{min}\cdot\text{cm}^2\,.$$

Schwieriger ist es, die Abfallskonstante γ festzulegen. Aus der Bedingung, daß die Intensität des Kaskadenspektrums an der Erdoberfläche ($l = 29$) nicht größer als die wirkliche Intensität der Elektronen ($\sim^1/_2/\text{min cm}^2$) sein soll, folgt zunächst nur eine untere Grenze für den Exponenten: $\gamma > 1{,}3$.

Ein weiterer Anhaltspunkt kann aus dem Breiteneffekt Abb. 14 entnommen werden unter der Annahme, daß das aus dem Weltenraum einfallende Spektrum nur relativ wenig Lichtquanten und vorwiegend Elektronen und Positronen enthält. In diesem Falle müssen sich ja

$$\text{die einfallenden Energien} \left[\int\limits_{E_\varphi}^{\infty} \frac{-\partial F_1}{\partial E}\, E\, dE \text{ nach } (40)\right] \text{ verhalten wie}$$

$$\left(\frac{1}{E_\varphi}\right)^{\gamma - 1},$$

wenn E_φ die magnetische Grenzenergie (39) bedeutet.

Aus den Messungen von Bowen, Millikan und Neher folgt dann

$$\gamma \approx 1{,}8,$$

wie aus der nachstehenden Tabelle von Johnson ($J\,1$) hervorgeht:

Breite	3°	39°	52°
Ionen/cm²/min	$1{,}8 \cdot 10^9$	$3{,}2 \cdot 10^9$	$6 \;\cdot 10^9$
Energie eV/cm²/min .	$0{,}6 \cdot 10^{11}$	$1{,}0 \cdot 10^{11}$	$1{,}9 \cdot 10^{11}$
E_φ	$15 \;\cdot 10^9$	$8 \;\cdot 10^9$	$2 \;\cdot 10^9$

Bisher ist angenommen worden, daß die Gestalt des ursprünglich aus dem Weltenraum einfallenden Spektrums durch ein reines Potenzgesetz (40) dargestellt werden kann. Ein allgemeines Spektrum wird man am besten diskutieren, indem man es in Summanden der Form (40) zerlegt ($N\,6$). Es kann auch ein solches Spektrum konstruiert werden, welches die Intensität der weichen Komponente in der gesamten Atmosphäre darstellt ($N\,7$, $H\,7$). Ein solches Spektrum könnte aber, wie Heitler ($H\,7$) gezeigt hat, für die weiche Komponente in Meereshöhe nur einen Breiteneffekt von weniger als 1% ergeben, während der wirkliche Breiteneffekt der weichen Komponente nach Auger-Leprince-Riguet ($A\,7$) etwa 10% beträgt.

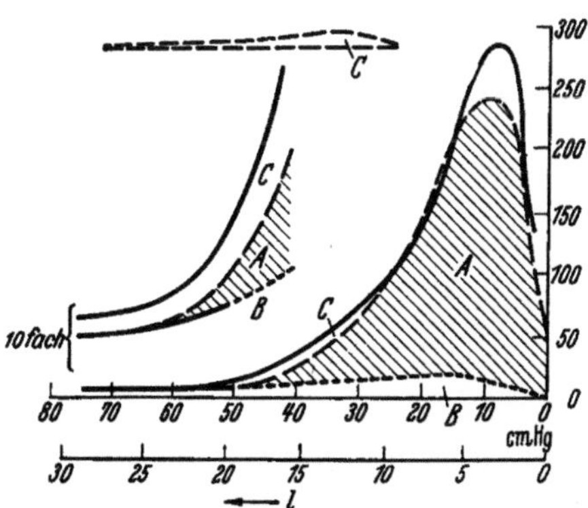

Abb. 15. Zerlegung der Strahlung in der Atmosphäre (Abb. 11). ———— Messung der Vertikalintensität von Pfotzer. - - - - - Extrapolation des durchdringenden Anteils. A (schraffiert) „Kaskadenelektronen", welche aus den Elektronen des Weltraums durch Teilung in der Atmosphäre entstehen (§ 19, $\gamma = 1{,}9$). B Schwere Elektronen, welche von A in der Atmosphäre erzeugt werden (§ 20). C „Zerfallselektronen", welche aus B entstehen (§ 21).

Hieraus muß wie auf S. 36 gefolgert werden, daß die Elektronen in Meereshöhe nicht auf das Kaskadenspektrum zurückgeführt werden können, sondern aus dem der durchdringenden Komponente entstehen.

Die Trennung der verschiedenen Anteile der kosmischen Strahlung in der Atmosphäre ist in Abb. 15 dargestellt. Die ausgezogene Kurve gibt die Vertikalintensität von Pfotzer als Funktion der Höhe (Abb. 11) noch einmal wieder. Die unterste gekreuzte Kurve B gibt die Extrapolation des Anteils an, den wir der durchdringenden Komponente zuschreiben können (§ 20), und das schraffierte Gebiet A gibt die theoretische Zahl der Kaskadenelektronen, die wir aus einem ursprünglich aus dem Weltenraum einfallenden Spektrum (40) mit der Abfallskonstanten $\gamma = 1{,}9$ erhalten ($H\,7$, $N\,7$). Man sieht, daß die Messungen in der oberen Atmosphäre durch dieses Kaskadenspektrum ganz gut dargestellt werden, daß aber in der unteren Atmosphäre ein gewisser Betrag C übrigbleibt, der aus den oben angeführten Gründen einen anderen Ursprung haben muß, und dessen Beginn durch den von REGENER und seinen Schülern ($P\,1$) berichteten „Buckel" bei 30 cm Hg bezeichnet zu sein scheint. Diesen Anteil C werden wir in § 21 berechnen.

20. Die durchdringende Komponente. Für die folgenden Betrachtungen wird angenommen, daß in der Stratosphäre (genauer bei etwa 8 cm Hg) ein Spektrum schwerer Elektronen entsteht, das in seiner Energieverteilung der Energieverteilung der weichen Komponente ähnlich sein soll. Es soll also in der Stratosphäre

$$F_s(p) = J_0\left(\frac{2\cdot 10^9\,\text{eV}}{p\,c}\right)^\gamma \tag{43}$$

schwere Elektronen oberhalb des Impulses p geben. Wir berechnen nun die Absorption eines solchen Spektrums durch Ionisation und spontanen Zerfall.

Bei der Berechnung der *Ionisation* gehen wir von der Formel (11) aus, nach der ein schweres Teilchen vom Impuls p eine Reichweite R hat, welche etwa durch

$$R(p) \approx \frac{\mu c^2}{a}\left(\frac{2+\left(\frac{p}{\mu c}\right)^2}{\sqrt{1+\left(\frac{p}{\mu c}\right)^2}} - 2\right)$$

($\mu \sim 160\,m = $ Masse des schweren Elektrons, $\mu c^2 \sim 8\cdot 10^7\,\text{eV}$)

gegeben ist.

Tabelle 6.

$a=$	Blei	Wasser	Luft	
	$1{,}2\cdot 10^7$	$2\cdot 10^6$	$2{,}5\cdot 10^3$	eV/cm

Ein Teilchen, welches ursprünglich den Impuls p_0 besaß, hat also nach Durchdringung einer Schicht T einen kleineren Impuls p, welcher durch die Gleichung gegeben ist:

$$T = R(p_0) - R(p);\ \text{ also für } p_0 \gtrsim 3\,\mu c:$$
$$T = \frac{p_0\,c - 2\,\mu c^2}{a} - R(p).$$

Das Spektrum (43) verwandelt sich daher durch Absorption nach der Tiefe T in ein energieärmeres Spektrum von

$$F_s(T,p) = J_0\left(\frac{a^{-1}\cdot 2\cdot 10^9\,\text{eV}}{T + R(p) + \frac{2\,\mu c^2}{a}}\right)^\gamma \tag{44}$$

Teilchen oberhalb des Impulses p (wegen $p_0 > 3\,\mu c$ gilt diese Formel für kleine Impulse erst von etwa $T > \mu c^2/a \sim 40$ cm H_2O ab). Dieses integrale Spektrum der durchdringenden Teilchen fällt bei hohen Impulsen $(p > 3\,\mu c)$ ab wie

$$F_s(T,p) \approx J_0\left(\frac{2\cdot 10^9\,\text{eV}}{a\,T+p\,c}\right)^{\gamma}, \quad (a\,T\approx 2\cdot 10^9\,\text{eV in Meereshöhe}); \quad (45)$$

d. h. die Zahl der schweren Elektronen oberhalb eines Impulses p in der Tiefe T ist gleich der ursprünglichen Zahl der Teilchen oberhalb eines um den Verlust in der Strecke T höheren Impulses.

Bei kleinen Impulsen kann man die Formel (43) entwickeln nach $R(p)$ und erhält für die Anzahl der Teilchen *unterhalb* des Impulses p

$$J_0\gamma\,\frac{(2\cdot 10^9\,\text{eV})^{\gamma}}{(a\,T+2\,\mu c^2)^{\gamma+1}}\cdot a\,R(p). \quad (46)$$

Da die Reichweite mit abnehmendem Impuls sehr rasch abnimmt, so macht die Formel (46) unmittelbar verständlich, warum schwere Elektronen mit geringen Impulsen so selten beobachtet werden. In der Abb. 16 ist der theoretische Verlauf des differentiellen Spektrums bei Impulsen unterhalb $p\,c = 7\cdot 10^8$ eV dargestellt.

Bisher haben wir nur von der Absorption durch Ionisation gesprochen und die Absorption durch den spontanen Zerfall vernachlässigt. Diese Vernachlässigung ist erlaubt, wenn man die Absorption der durchdringenden Komponente in Wasser oder fester Materie betrachtet, wie die Abschätzung der mittleren Zerfallszeit des schweren Elektrons durch Yukawa $(\tau \sim \frac{1}{2}\cdot 10^{-6}$ sec; vgl. § 14) zeigt. In der Atmosphäre dagegen spielt der spontane Zerfall eine erhebliche Rolle. Wir können uns dabei auf das Verhalten des Spektrums bei größeren Impulsen beschränken $(p\,c > 2\cdot 10^8$ eV), da bei kleinen Impulsen $(p < 2\cdot 10^8$ eV) der spontane Zerfall den qualitativen Verlauf (46) nur wenig ändert. Für die Änderung des differentiellen Spektrums $f(T,p) = -\partial F_s(T,p)/\partial p$ mit der Tiefe T erhält man, wenn man Absorption und spontanen Zerfall gleichzeitig berücksichtigt, die Differentialgleichung:

$$\frac{\partial f(T,p)}{\partial T} = \frac{a}{c}\,\frac{\partial f(T,p)}{\partial p} - \frac{b}{p\,T}\,f(T,p). \quad (47)$$

Hierin bedeutet T die Tiefe in cm Wasser, gemessen vom Gipfel der Atmosphäre; a den Energieverlust pro cm Wasser (vgl. Tabelle 6). Die Zerfallswahrscheinlichkeit pro cm Wasser wird in der Atmosphäre um so größer, je geringer die Dichte der Luft ist, daher enthält bei Annahme der Barometerformel das zweite Glied in (47) den Nenner T. b hängt mit der Zerfallszeit τ des schweren Elektrons zusammen durch die Formel

$$b = \frac{\mu}{\tau}\,\frac{\text{Dichte des Wassers}\cdot(T_{\text{Meereshöhe}})}{\text{Dichte der Luft in Meereshöhe}}.$$

Die Lösung der Differentialgleichung (47) lautet:

$$f(T,p) = g(p\,c + a\,T)\left(\frac{p\,c}{a\,T}\right)^{\frac{b\,c}{p\,c+a\,T}}. \quad (48)$$

Hierin stellt $g(p\,c + a\,T)$ eine willkürliche Funktion des Arguments $p\,c + a\,T$ dar. Sie bestimmt sich aus der Bedingung, daß für die

Stratosphäre, genauer für $T \approx 100$, $f(T, p)$ übergehen muß in das differentielle Spektrum zu (43). Man erhält damit für das differentielle Spektrum in Abhängigkeit von der Tiefe:

$$f(T, p) = \gamma\, J_0\, \frac{(2 \cdot 10^9\,\text{eV})^\gamma \cdot c}{[pc + a(T - 100)]^{\gamma + 1}} \left[\frac{100\,pc}{T\,(pc + a(T - 100))} \right]^{\frac{bc}{pc + aT}}. \quad (49)$$

Insbesondere wird das Spektrum in Meereshöhe $(T \sim 1000)$:

$$f(1000, p) = \gamma\, J_0\, \frac{(2 \cdot 10^9\,\text{eV})^\gamma \cdot c}{[pc + 900\,a]^{\gamma + 1}} \left[\frac{pc}{10\,(pc + 900\,a)} \right]^{\frac{bc}{pc + 1000\,a}} \quad (50)$$

Der zweite Faktor auf der rechten Seite dieser Gleichung bewirkt, daß das Spektrum langsamer abfällt als das Spektrum der weichen Komponente. Hierin drückt sich der Umstand aus, daß der spontane Zerfall auf einer gegebenen Strecke bei kleinen Impulsen häufiger eintritt als bei großen. Erst bei sehr hohen Energien $(pc \gg 900\,a \sim 1{,}8 \cdot 10^9\,\text{eV})$ wird der Abfall wieder durch den Exponenten γ allein bestimmt. Diese Folgerung wird durch die Messungen von BLACKETT $(B\,12)$ bestätigt. Abb. 16 (unterer Teil) vergleicht das theoretische Spektrum, das im Hinblick auf die weiter unten zu besprechenden Messungen von EHMERT $(E\,1)$ mit den Werten $\gamma = 1{,}87$; $bc/1000\,a = 0{,}37$ berechnet ist, mit den BLACKETTschen Messungen. Die Theorie gibt den allgemeinen Verlauf des Spektrums, allerdings nicht die Einzelheiten, befriedigend wieder.

Ferner geben die Formeln (49) und (50) nach KULENKAMPFF $(K\,5)$ eine einfache Erklärung für die auffallende Beobachtung von

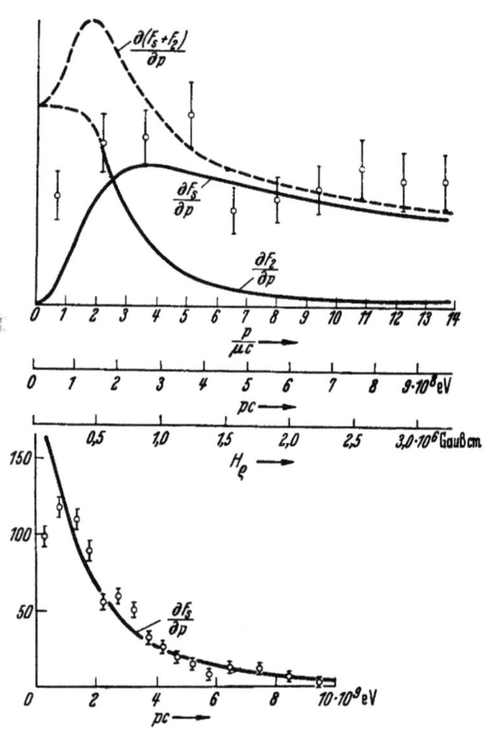

Abb. 16. Das Spektrum in Meereshöhe. Abszisse: Impuls p der Teilchen. Ordinate: Zahl $\partial F/\partial p$ der Teilchen pro Impulsintervall. Kurve $\partial F_s/\partial p$: Theoretisches Spektrum der schweren Elektronen. Kurve $\partial F_2/\partial p$: Theoretisches Spektrum der leichten Elektronen (§ 21). Kurve $\partial(F_s + F_2)/\partial p$: Theoretisches Gesamtspektrum. Meßpunkte: Gesamtspektrum nach BLACKETT. Oberer Teil: Spektren bei niedrigen Impulsen; unterer Teil: Spektrum der durchdringenden Teilchen bei hohen Impulsen.

EHMERT $(E\,1, E\,2)$ und KULENKAMPFF $(K\,5)$, daß die Absorption in Wasser zunächst langsamer erfolgt als in Luft (Abb. 12). Die Dicke der durchlaufenen Luftschicht ist dabei von EHMERT dadurch variiert worden, daß bei festgehaltenem Beobachtungsort (ungefähr auf Meereshöhe[1]) der Winkel φ der einfallenden Strahlen zur Vertikalen durch Schrägstellung der Koinzidenzanordnung verändert wurde. Man definiert dann als scheinbare Tiefe die Größe $T' = 1000/\cos\varphi$. Für die Berechnung dieser Anordnung hat man in Gl. (47) a durch $a/\cos\varphi$ und b durch $b/\cos\varphi$ zu ersetzen. Man erhält daher für die Intensitätsverteilung auf Meeresniveau in Abhängigkeit von T':

$$\underset{\text{Luft}}{f(T',p)} = \gamma J_0 \frac{(2\cdot 10^9\,\text{eV})^\gamma c}{\left(pc + \frac{9}{10}aT'\right)^{\gamma+1}} \left[\frac{pc}{10\left(pc + \frac{9}{10}aT'\right)}\right]^{\frac{bc}{pc\frac{1000}{T'}+1000a}}. \quad (51)$$

Für die Intensitätsänderung im Wasser, in dem der spontane Zerfall keine Rolle mehr spielt, folgt dagegen aus (48) $(b=0)$ und (50)

$$\underset{\text{Wasser}}{f(T,p)} = \gamma J_0 \frac{(2\cdot 10^9\,\text{eV})^\gamma c}{[pc + a(T-1000)]^{\gamma+1}} \left[\frac{pc + a(T-1000)}{10(pc + a(T-100))}\right]^{\frac{bc}{pc+aT}}. \quad (52)$$

Integriert man (51) und (52) über das ganze Spektrum, so folgt für die Abhängigkeit der Intensität von der Tiefe $(\delta = bc/1000\,a)$:

$$\underset{\text{Luft}}{F_s(T',0)} = \gamma J_0 \left(\frac{2\cdot 10^9\,\text{eV}}{\frac{9}{10}aT'}\right)^\gamma \int_0^\infty \frac{du}{(1+u)^{\gamma+1}} \left[\frac{u}{10(1+u)}\right]^{\frac{\delta}{1+\frac{9}{10}u}}, \quad (53)$$

$$\left.\begin{aligned} &\underset{\text{Wasser}}{F_s(T,0)} = \\ &= \gamma J_0 \left(\frac{2\cdot 10^9\,\text{eV}}{a(T-100)}\right)^\gamma \int_0^\infty \frac{du}{(1+u)^{\gamma+1}} \left[\frac{1}{10(1+u)}\left(u + \frac{T-1000}{T-100}\right)\right]^{\frac{\delta\cdot 1000}{u(T-100)+T}}. \end{aligned}\right\} \quad (54)$$

Hinter sehr dicken Schichten stehen nach diesen Gleichungen die Intensitäten von Wasser und Luft in einem konstanten Verhältnis, das nach Ehmert (vgl. Abb. 12) etwa $2:1$ beträgt, während der Abfall mit der Tiefe dem Gesetz $T'^{-\gamma}$ folgt. Der von Ehmert bestimmte Wert $\gamma \sim 1{,}87$ paßt gut zu den Werten, die bei der weichen Komponente gefunden wurden (§ 19). Aus dem konstanten Verhältnis kann die Größe δ und damit die mittlere Zerfallszeit des schweren Elektrons bestimmt werden. Es ergibt sich

$$\delta \sim 0{,}37; \qquad \tau \sim 2{,}7 \cdot 10^{-6}\,\text{sec}. \quad (55)$$

Dieser Wert ist etwa 5mal so groß wie der aus der Yukawaschen Theorie berechnete (§ 14). In Anbetracht der Unsicherheit mancher Einzelheiten in der Yukawaschen Theorie, insbesondere der Masse des schweren Elektrons, ist diese Übereinstimmung durchaus befriedigend. Es sei

---

[1] Die Ehmertsche Messungen sind teils in Stuttgart, teils auf der Zugspitze ausgeführt worden, ordnen sich aber nach Reduktion noch gut zu einer einheitlichen Kurve, die wir ohne allzu große Fehler auf Meereshöhe beziehen können.

auch noch hervorgehoben, daß die Experimente zunächst das Verhältnis der Zerfallszeit zur Masse des YUKAWA-Teilchens bestimmen, daß also eine Änderung der Masse auch eine entsprechende Änderung der experimentellen Zerfallszeit des schweren Elektrons nach sich zieht.

Schließlich vergleichen wir die theoretischen Formeln für die Abnahme der Intensität mit der Tiefe (49), (54) noch mit den Messungen von AUGER (*A 5*), die einen etwas langsameren Abfall als die von EHMERT ergeben. Die theoretischen Werte sind mit den nach EHMERT bestimmten Konstanten γ und δ berechnet und zeigen, daß die Theorie den verhältnismäßig langsamen Abfall der Intensität richtig wiedergibt, daß jedoch zur genauen Darstellung der AUGERschen Experimente ein etwas kleinerer Wert von γ gewählt werden müßte.

Bei der vorangehenden Berechnung des Spektrums der durchdringenden Teilchen und ihrer Absorption wurden die Prozesse (§ 15) noch nicht berücksichtigt, welche sich

Tabelle 7.

	T in m Wasser	Intensität nach AUGER	Theoretische Intensität
Jungfraujoch . .	6,6	108	120
Paris	10	75	75
Keller	30	28	17
Katakomben . .	75	5	3

aus den Kräften zwischen den schweren Teilchen und den Atomkernen ergeben. Man gewinnt jedoch aus Abb. 16 und Abb. 12 den Eindruck, daß die Mitwirkung dieser Kernprozesse das Spektrum der durchdringenden Teilchen anscheinend nur in den Einzelheiten beeinflußt. Eine genaue theoretische Verfolgung dieser Effekte, welche in § 15 besprochen wurden, ist einstweilen noch nicht möglich.

Die durchdringende Komponente hat in Meereshöhe einen *Breiteneffekt* von etwa 10% (vgl. *M 3, S 11*). Die Tatsache, daß dieser Breiteneffekt so gering ist, kann als unabhängiges Argument für die Erzeugung der durchdringenden Komponente aus der weichen Strahlung betrachtet werden: Ein durchdringendes Teilchen verliert in der Atmosphäre nur etwa $2 \cdot 10^9$ eV durch Ionisation. Wenn wir nun mit primär einfallenden Teilchen von $2 \cdot 10^9$ eV rechnen würden, so müßten wir einen Breiteneffekt der Größenordnung 100% erwarten, weil in unseren Breiten zwar schon alle Teilchen oberhalb $2 \cdot 10^9$ eV vom Magnetfeld der Erde durchgelassen werden, am Äquator aber nur die Teilchen von mehr als 10^{10} eV. Betrachtet man jedoch die durchdringende Komponente als Sekundärwirkung der weichen Komponente, so muß man annehmen, daß die Lichtquanten im Stratosphärenmaximum $(l \sim 4)$ die an der Erdoberfläche beobachteten durchdringenden Teilchen erzeugen. Diese Lichtquanten, welche selbst eine Energie von mehr als $2 \cdot 10^9$ eV besitzen müssen, werden wiederum im Mittel von solchen Elektronen des Weltenraumes erzeugt, welche mehr als etwa $2 \cdot 10^{10}$ eV Energie haben (wegen der Energieteilung in der Stratosphäre). Weltraumelektronen so hoher Energie aber werden durch das Magnetfeld der Erde nicht mehr gestört

(§ 12). Eine geringe Magnetfeldabhängigkeit wird natürlich zu erwarten sein, da nicht ausschließlich die weiche Strahlung am Stratosphärenmaximum bei der Erzeugung der durchdringenden Komponente wirksam ist, obwohl diese wegen ihrer großen Intensität sicher den Hauptbeitrag liefert.

21. Die Zerfallselektronen in der unteren Atmosphäre. Die durchdringende Komponente erzeugt nach der Yukawaschen Theorie eine weiche Sekundärstrahlung auf zweierlei Weise:

Erstens können die durchdringenden Teilchen spontan in Neutrinos und Elektronen zerfallen, die dann nachträglich durch Kaskaden multipliziert werden (§ 14).

Zweitens können die schweren Elektronen beim Zusammenstoß mit Atomkernen Lichtquanten erzeugen, die dann ihrerseits Kaskaden hervorrufen (§ 15).

Die Elektronen aus den Kernprozessen werden nahezu gleich häufig sein in Luft und hinter größeren Schichten fester Körper, da die Erzeugung der Sekundären massenproportional geschieht und die Absorption der Sekundären (nach § 11 γ) ebenfalls nahezu massenproportional erfolgt.

Die Elektronen aus dem spontanen Zerfall aber werden in Luft viel häufiger sein als hinter Wasser oder hinter anderen Schichten dichter Materie, da die Strecke, aus der die in Luft erzeugten Sekundären herauskommen können geometrisch größer ist als die entsprechende Strecke in Wasser und da auf dieser geometrisch größeren Strecke mehr Elektronen durch spontanen Zerfall erzeugt werden.

Da nun das empirische Verhältnis der weichen zur harten Komponente in Luft auf Meeresniveau größer ist als hinter festen Körpern (§ 17), so nehmen wir an, daß die Elektronen in der unteren Atmosphäre, welche ja nach § 19 Sekundärwirkungen der harten Komponente sind, vorwiegend durch *spontanen* Zerfall entstehen, und daß die Elektronen, die aus den Kernprozessen entstehen, erst unter Wasser oder hinter dickeren Schichten fester Körper neben den Ionisationsschauern merklich werden.

Wir berechnen nun das Spektrum der Elektronen des spontanen Zerfalls

$$F_2(T, p),$$

welches sich mit dem Spektrum der durchdringenden Teilchen

$$F_S(T, p)$$

ins Gleichgewicht setzt. [$F(T, p)$ = Teilchenzahl oberhalb des Impulses p in der Tiefe T unter der Grenze der Atmosphäre.]

Die Zerfallselektronen teilen wir in zweierlei Arten ein. Erstens solche, die aus dem Zerfall der durch die Ionisation zur Ruhe gebrachten schweren Elektronen entstehen (A). Zweitens solche Elektronen, die aus dem Zerfall bewegter schwerer Teilchen hervorgehen.

Die ersteren haben nur eine sehr geringe Energie von weniger als $4 \cdot 10^7$ eV, da die zur Erzeugung des Elektrons und des Neutrinos zur

Verfügung stehende Ruhenergie der schweren Teilchen $\mu c^2 = 8 \cdot 10^7$ eV beträgt. Diese Elektronen haben nach § 9 und § 11 γ etwa die Reichweite $\mu c^2/2a \approx 20$ g/cm². Ihre Gesamtmenge $\overline{A}$ ist daher ungefähr gleich der in der letzten Schicht von $\mu c^2/a \approx 20$ g/cm² steckengebliebenen schweren Elektronen, also nach Gl. (46)

$$\overline{A} \sim \frac{1}{2}\,\frac{\mu c^2}{a T} \cdot F_S{}^{\cdot}(T, 0),\qquad(56)$$

was in Meereshöhe etwa 2% der durchdringenden Teilchen ausmacht.

Einen größeren Beitrag können wir aus dem Zerfall der bewegten schweren Elektronen erhalten. Ein schweres Elektron mit dem Impuls p hat ja nach § 14 eine Zerfallswahrscheinlichkeit

$$\frac{\mu}{p\tau}$$

pro cm, wenn τ die Lebensdauer der schweren Teilchen in ihrem Ruhsystem bedeutet. Beim Zerfall eines durchdringenden Teilchens vom Impuls p wird ein Elektron erzeugt, dessen Impuls in der Bewegungsrichtung im Mittel $p/2$ beträgt (§ 14). Dieses Elektron erzeugt nun seinerseits eine Kaskade, welche nach einer weiteren Schicht l eine Zahl von

$$z\left(l, \frac{p/2}{p'}\right)$$

Elektronen oberhalb des Impulses $p' > E_j$ enthält; dabei bezeichnet z die Multiplikationsfunktion (§ 10, 11), l die Schichtdicke im Maß der Tabelle 2, deren Einheiten durch die Strecke X_0 und die Energie E_j in jedem Material charakterisiert sind.

Es entsteht also im Gleichgewicht mit dem durchdringenden Spektrum $F_S(T, p)$ ein Elektronenspektrum (für $p'c > E_j$)

$$F_2(T, p') = \int\limits_{p=2p'}^{\infty} dp\,\frac{-\partial F_s(T, p)}{\partial p} \int\limits_{l=0}^{\infty} dl\,z\left(l, \frac{p/2}{p'}\right)\frac{X_0}{\tau c}\frac{\mu c}{p} + A,$$

oder nach (17)

$$F_2(T, p') = \frac{\mu c}{2 p'}\cdot\frac{3}{4}\,\frac{X_0}{\tau c}\,F_s(T, 2p') + A.\qquad(57)$$

Wir diskutieren nun zunächst die Gesamtintensität und dann die spektrale Verteilung dieser Zerfallselektronen $F_2(T, p')$.

Die Gesamtmenge der Elektronen wird nach (14a) etwa durch $3\,F_2\!\left(T, \frac{E_j}{c}\right)$ gegeben. Die relative *Gesamtintensität*, d. h. das Verhältnis der Zahl der Elektronen zur Zahl der durchdringenden Teilchen ist dann nach (57) u. (56):

$$\varkappa \approx \frac{3\,F_2\!\left(T, \dfrac{E_j}{c}\right)}{F_s(T, 0)} \approx \frac{9}{4}\,\frac{X_0}{\tau c}\,\frac{\mu c^2}{2 E_j} + \frac{1}{2}\,\frac{\mu c^2}{a T}.\qquad(58)$$

Formel (58) gibt nach Einrechnung der Ionisationselektronen (§ 12) das empirische Verhältnis von weicher zu harter Strahlung in Meereshöhe ($X_0 = 275$ m)

$$\varkappa = \frac{1}{3}$$

dann richtig wieder, wenn wir die Zerfallszeit τ der schweren Elektronen zu

$$\tau = 2 \cdot 10^{-6} \ \text{sec} \tag{59}$$

annehmen ($E\,6$). Diese Bestimmung der Zerfallszeit schwerer Elektronen aus der Intensität der weichen Komponente in der unteren Atmosphäre steht also ungefähr in Übereinstimmung mit der Bestimmung aus den Absorptionsmessungen von Ehmert (55).

Die Strahlungsstrecke X_0 in Formel (58) bringt die oben erwähnte Abhängigkeit der Intensität der Zerfalls-Elektronen vom spezifischen Volumen des Gleichgewichtsmaterials zum Ausdruck: Nach (58) steigt mit zunehmender Höhe über dem Erdboden das Verhältnis der weichen Zerfallsstrahlung zur harten Strahlung an wie der reziproke Luftdruck, was mit den in Abb. 10, 15 dargestellten Messungen in Übereinstimmung zu sein scheint. Dieser Anstieg wird sich fortsetzen bis zu dem von Regener und seinen Schülern bei 30 mm Hg beobachteten Buckel (Abb. 11,15) in der Gesamtintensität der Vertikalkoinzidenzen, welcher den Einsatz des spontanen Zerfalls nach Formel (48) bezeichnen dürfte.

Unter Wasser ($X_0 = 34$ cm) oder hinter sonstigen Schichten dichter Materie reduziert sich die Zahl der Zerfallselektronen auf den geringen Beitrag des zweiten Gliedes in (58); hier überwiegen also die Bhabha-schen multiplizierten Ionisationselektronen (§ 12) und eventuell Elektronen, die aus Kernprozessen entstehen.

Die *Gestalt* des differentiellen Zerfallselektronenspektrums $\frac{\partial}{\partial p} F_s(T, p)$, d. h. die Zahl der Zerfallselektronen im Energieintervall dp, ist für Meeresniveau in Abb. 16 unter der Annahme der Zerfallskonstanten $\tau = 2 \cdot 10^{-6}$ eingetragen ($E\,7$). Dieses Spektrum sinkt bei höheren Energien ungefähr mit einer um 1 stärkeren Potenz ab als das Spektrum der durchdringenden Teilchen.

Das Verhältnis beider Spektren als Funktion des Impulses kann mit der Erfahrung verglichen werden, da es den mittleren Energieverlust der Teilchen verschiedener Massen vom Impuls p bestimmt. Denn schwere Teilchen erleiden in einer dünnen Schicht nur einen im allgemeinen geringen Ionisationsverlust, Elektronen aber den vollen, von der Energie unabhängigen relativen Strahlungsverlust (7): 1,72 pro cm Pb. Daher ist der über die Teilchensorten gemittelte relative Energieverlust bei dünnen Schichten pro cm Pb.:

$$\frac{1{,}72 \, \dfrac{\partial F_2(p)}{\partial p}}{\dfrac{\partial F_2(p)}{\partial p} + \dfrac{\partial F_s(p)}{\partial p}} \cdot \tag{60}$$

Dieser theoretische relative Energieverlust pro cm Pb als Funktion des Impulses ist in Abb. 3 eingetragen. Man sieht, daß er mit den Messungen von BLACKETT (*B 11*) im Einklang ist.

Während der relative Energieverlust, den die schweren Elektronen durch Ionisation in einer dünnen Schicht erleiden, nur sehr gering ist, wird er doch in einer dicken Schicht wirksam sein. In 2 cm Gold z. B. müssen nicht nur die Elektronen, sondern auch alle schweren Elektronen unterhalb eines bestimmten Impulses steckenbleiben, welcher nach (11) $pc = 1{,}3 \cdot 10^8$ eV beträgt. Dieser Wert kann sich eventuell durch Streuung und durch die Kernprozesse etwas erhöhen. In BLACKETTs Messungen (*B 11*) konnten alle Teilchen von mehr als $pc = 2{,}4 \cdot 10^8$ eV die Goldplatte von 2 cm durchsetzen, und alle Teilchen von weniger als $pc \doteq 1{,}65 \cdot 10^8$ eV blieben in der Platte stecken. Diese Ergebnisse, welche zunächst auf einen induzierten Zerfall hinzudeuten scheinen, können also mit der Theorie des spontanen Zerfalls qualitativ vereinigt werden.

V. Diskussion der Sekundärwirkungen.

22. Übersicht. Auf Grund der in den vorigen Paragraphen angegebenen Spektren müssen wir nun die folgenden Sekundärwirkungen erwarten.

Erstens wird die weiche Komponente Kaskadenschauer erzeugen (§ 10, 11). Diese Kaskaden entwickeln sich in Schichten von 1—5 cm Pb (Abb. 8), je nachdem, ob sie einige Teilchen oder einige tausend Teilchen enthalten. Sie werden im schweren Material stärker ausgeprägt sein als im leichten (§ 11 γ). Die für die Kaskadenbildung in verschiedenen Materialien charakteristischen Energien E_j und Strecken $l = 1$ sind in Tabelle 2, S. 8, gegeben.

Zweitens wird die durchdringende Komponente hinter dicken Schichten von Ionisationsschauern (§ 12) begleitet sein. Die Zahl der kleinen Ionisationsschauer ($N \lesssim 20$ in Pb, $N \lesssim 5$ in Al) muß nahezu vom Material unabhängig sein; die Häufigkeit der größeren Ionisationsschauer dagegen muß im leichten Material geringer sein als im schweren (*B 6*): Denn wenn wir die von BHABHA angegebene Häufigkeit der Ionisationsschauer (31) über das Spektrum (45) der durchdringenden Strahlung integrieren,

so erhalten wir für die relative Zahl $q(N) = \dfrac{1}{F(\mathrm{o})} \displaystyle\int_0^{\infty} Q(N,E) \dfrac{-\partial F(E)}{\partial E}\, dE$

der Schauer von mehr als N Teilchen pro durchdringendes Teilchen:

$$\left.\begin{array}{ll} q \sim \dfrac{0{,}03}{N} & \text{für} \quad N < \dfrac{aT}{8E_j} \\[2ex] q \sim \dfrac{0{,}03}{N}\left(\dfrac{aT}{8NE_j}\right)^{\gamma} & \text{für} \quad N > \dfrac{aT}{8E_j} \end{array}\right\} \quad (61)$$

$aT \approx 2 \cdot 10^9$ eV in Meereshöhe, $\gamma = 1$ bis 2 nach (50).

Drittens werden wir erwarten, daß die kosmischen Strahlen explosionsartige Schauer (§ 15) erzeugen. Auch wenn sich die Wirkungsquerschnitte zur Erzeugung von Explosionen für verschiedene Teilchen nicht sehr unterscheiden, so wird man doch in Meereshöhe vorwiegend diejenigen Explosionen beobachten, welche von der durchdringenden Komponente erzeugt werden, da die weiche Komponente meist durch die starke Absorption der viel häufigeren Strahlungsprozesse an der Ausbildung der selteneren Explosionen verhindert wird.

Da eine Explosion bereits an einem einzelnen Atomkern vor sich geht, wird die Häufigkeit der Explosionen bei dünnen Schichten unabhängig von der Größe des Schauers linear mit der Schichtdicke ansteigen im Gegensatz zu den Kaskaden, welche wegen der Verkettung vieler Atomkerne bei ihrer Erzeugung eine um so größere Strecke zu ihrer Entwicklung gebrauchen, je größer der Schauer ist (27). Bei dickeren Schichten wird die Intensität der Explosionen nur langsam mit der Schichtdicke abnehmen entsprechend der Absorption der auslösenden durchdringenden Komponente, während die Intensität der Kaskaden rasch gegen 0 geht (Abb. 8). Diese „Sättigung" der Explosionsprozesse wird nach einer Schichtdicke eintreten, die der Reichweite der in den Explosionen erzeugten Teilchen (§ 15) entspricht. Schließlich werden sich an die Schauer, welche an einem Kern erzeugt werden, häufig Kernprozesse anschließen.

23. Kleine Schauer (vgl. $G\,1$, $M\,3$, $G\,5$). A. Eine direkte Entscheidung der Frage, ob die hier diskutierten Prozesse der Schauererzeugung in der Wirklichkeit vorkommen, ist im letzten Jahre durch Wilson-Aufnahmen möglich geworden, in denen die Strahlen mehrere Absorberplatten durchsetzen mußten ($F\,8$).

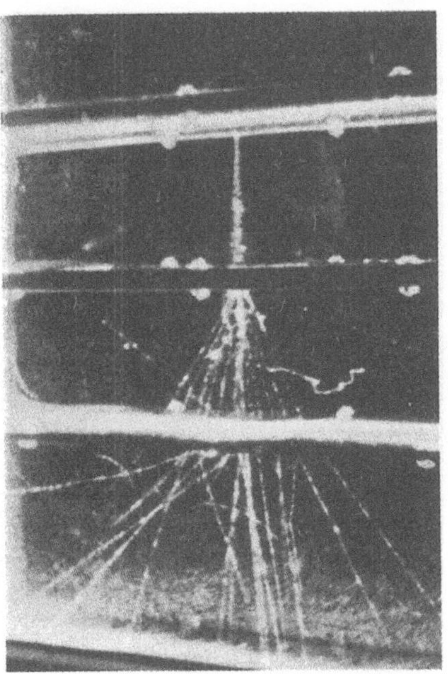

Abb. 17. Kaskadenschauer nach einer Aufnahme von Fussell vgl. $F\,8$ (Aufnahme noch nicht veröffentlicht). Die Dicke der Platten ist von oben nach unten: 6,3 mm Pb, 6,3 mm Pb, 0,7 mm Pb.

Abb. 17, welche Herr Fussell[1] freundlicherweise zur Verfügung gestellt hat, zeigt das typische Bild einer Kaskade: Der von oben ein-

---

[1] Herrn Fusell möchten wir für die freundliche Übersendung seiner Aufnahmen 17, 18, 18a und für die Erlaubnis zu ihrer Veröffentlichung herzlich danken.

dringende Strahl vervierfacht seine Zahl ionisierender Teilchen in der ersten Platte von 6 mm und vervierfacht sie weiter in einer zweiten Platte derselben Dicke. Die 16 Strahlen werden dann in der dritten dünnen Schicht von 0,7 mm nur wenig gestreut. Diese schrittweise Multiplikation der Strahlenzahl entspricht der Kaskadentheorie, wenn wir dem Anfangsstrahl eine Energie von etwa $2 \cdot 10^9$ eV zuschreiben (Abb. 7).

Die nächste Aufnahme (Abb. 18)[1] von FUSSELL zeigt eine typische Explosion. Der Explosionscharakter ergibt sich einmal aus der Tatsache, daß der erzeugende Strahl die erste Bleiplatte (6 mm) ungestört durchsetzt und den ganzen Schauer in der zweiten Platte allein (6 mm) erzeugt. Er folgt ferner aus dem Umstand, daß der Schauer einige stark ionisierende Spuren enthält. Diese stark ionisierenden Teilchen können entweder Protonen sein, welche in der Oberfläche der Bleiplatte bei einer von den Sekundären des Schauers ausgelösten Kernverdampfung ausgeschleudert werden ; oder sie sind langsame schwere Elektronen, welche direkt im Explosionsschauer erzeugt werden. Die Frage, ob der ganze Schauer aus schweren Elektronen besteht, kann aus Abb. 18 noch

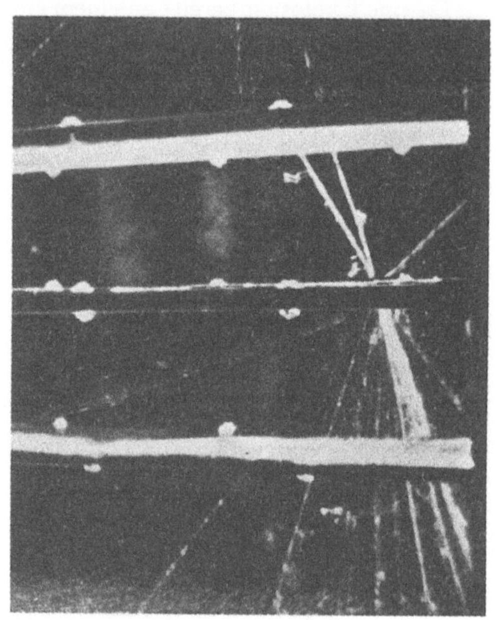

Abb. 18. Explosionsschauer nach einer Aufnahme von FUSSELL[1] vgl. *F 8* (Aufnahme noch nicht veröffentlicht). Absorber wie in Abb. 17.

nicht entschieden werden. Zur Entscheidung dieser Frage müssen weitere Aufnahmen abgewartet werden, in denen alle Schauerteilchen noch eine Platte von mehreren Millimetern Blei durchsetzen. Dabei wird sich zeigen, ob die Schauerteilchen weitere Kaskaden erzeugen, also Elektronen sind, oder ob sie ungestört durch den Absorber hindurchgehen, wie wir es bei schweren Elektronen erwarten.

Zugunsten der letzteren Annahme spricht die Aufnahme Abb. 18a[1] von FUSSELL. Der Schauer in Abb. 18a muß wiederum eine Explosion sein, weil er einige stark ionisierende Spuren enthält, und weil er in einer Bleiplatte von nur 0,7 mm entsteht, in der die Ausbildung von Kaskaden nur sehr unwahrscheinlich ist. Die Schauerteilchen scheinen aber in der Mehrzahl schwere Elektronen zu sein, weil die meisten von ihnen

---

[1] Siehe Fußnote 1 auf S. 48.

die weiteren Platten von 6 mm durchdringen können, während nur eins eine Kaskade erzeugt.

Unter 900 Aufnahmen vom Typus der Kaskaden (Abb. 17) fand Fussell (*F 8*) drei Aufnahmen vom Typus der Explosionen (Abb. 18).

Abb. 18a. Explosionsschauer nach einer Aufnahme von Fussell[1] (vgl. *F 8*, Aufnahme noch nicht veröffentlicht). Die Dicke der Bleiplatten ist von oben nach unten 0,7 mm; 6,3 mm; 6,3 mm. (Die Bahnspuren sind von Fussell durch Retusche deutlich gemacht.)

Trumpy (*T 3*) stellte ebenfalls ein Überwiegen der Kaskaden fest. Diese Verhältnisse zeigen, daß die meisten kleineren Schauer aus Pb Kaskaden sind; das ist verständlich, wenn man bedenkt, daß jedes Elektron und Lichtquant in einer Bleischicht von einigen Millimetern

---

[1] Siehe Fußnote 1 auf S. 48.

eine Kaskade erzeugt, daß aber die durchdringenden Teilchen nur mit einem sehr kleinen Wirkungsquerschnitt Explosionen hervorrufen.

B. Ein statistisches Studium der kleineren Schauer ist möglich auf Grund der Rossi-Koinzidenzen ($R\,3$, $R\,4$, vgl. $G\,1$, $M\,3$), d. h. der Beobachtung, daß mehrere Zählrohre, welche im Dreieck angeordnet sind, koinzidente Ausschläge geben, wenn sie von den Strahlen eines Schauers getroffen werden. Die Häufigkeit der Rossischen Koinzidenzen als Funktion der schauerauslösenden Schichtdicke ist in Abb. 19 nach Messungen von Morgan und Nielsen ($M\,6$) in Pb und Fe aufgetragen. Sie steigt zunächst zu einem Maximum bei 1,5 cm Pb ($l = 4$) bzw. 4,5 cm F ($l = 3$) an und sinkt dann auf einen Wert herab, der sich mit wachsender Schicht nur noch wenig ändert.

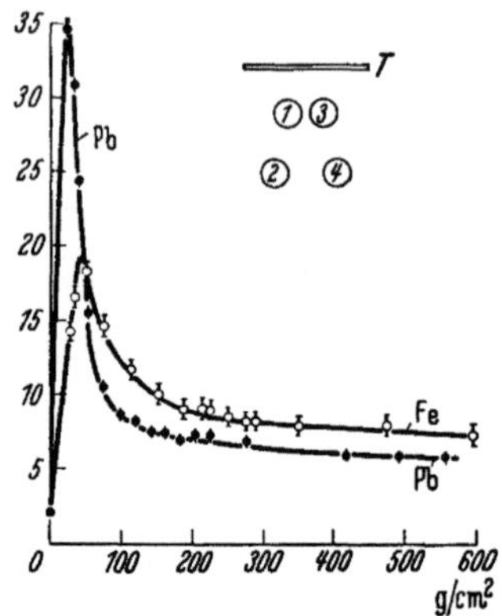

Abb. 19. Rossi-Koinzidenzen nach Morgan und Nielsen. [Physic. Rev. 52, 564 (1937).] Abszisse: Dicke der Auslöseschicht in g/cm². Ordinate: Koinzidenzen pro Stunde. ● ● ● Blei. o oo Eisen.

Es soll nun im folgenden gezeigt werden, daß die Rossi-Koinzidenzen beim Maximum von Kaskaden hervorgerufen werden, und daß sie bei größeren Schichten hauptsächlich durch die Bhabhaschen Ionisationsschauer, und wahrscheinlich nur zu geringerem Teil durch Prozesse entstehen, die mit den Explosionen verwandt sind.

Die experimentelle Trennung der reinen Kaskadenschauer von den Schauern, welche stärker durchdringende Sekundäre enthalten, wurde von Schwegler ($S\,2$) durchgeführt. Schwegler verglich die Zahl der gewöhnlichen Rossi-Koinzidenzen (oberste Kurve in Abb. 20) mit derjenigen reduzierten Zahl der Rossi-Koinzidenzen, welche er erhielt, wenn er zwischen die Zählrohre einen Block von 10 cm Pb schob (untere Kurve in Abb. 20). Hierbei zeigte sich, daß alle Koinzidenzen hinter großen Schichten auch durch den 10-cm-Pb-Block hindurchgehen. Da der Verlauf der Koinzidenzen durch den 10-cm-Bleiblock als Funktion der Dicke der Auslöseschicht (Abb. 20) den Charakter der Bhabhaschen Ionisationsschauer hat, werden wir den

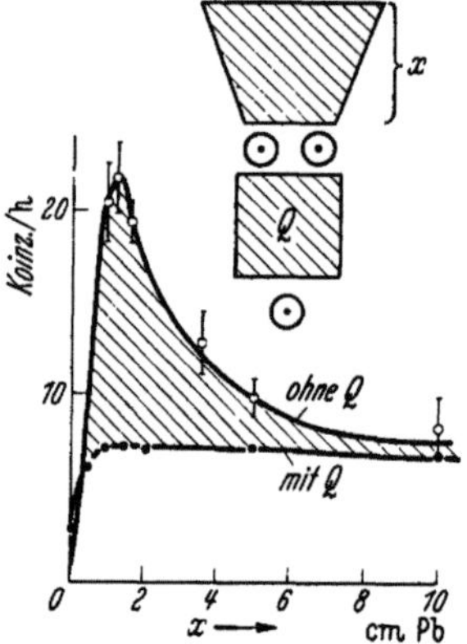

Abb. 20. Empirische Trennung der Kaskadenschauer von den übrigen nach Schwegler. [Z. Physik 96, 62 (1935).] Abszisse: Dicke der Bleischicht über den Zählrohren. Ordinate: Häufigkeit der Rossi-Koinzidenzen. Obere Kurve: ohne Bleiabsorber Q; unter Kurve: mit Bleiabsorber Q von 10 cm zwischen den Zählrohren. Schraffiert: Kaskadenanteil.

überwiegenden Teil der Rossi-Koinzidenzen hinter großen Schichten den Ionisationsschauern zuschreiben, welche von der durchdringenden

Strahlung erzeugt werden. Einen gewissen Teil der Rossi-Koinzidenzen hinter größeren Schichten müssen wir allerdings auf Grund neuerer Messungen von Schmeisser und Bothe ($S\,1$), welche in § 24 besprochen werden, auf andere Prozesse zurückführen.

Die Differenz der Schweglerschen Messungen ohne und mit Blei-block (schraffiert in Abb. 20), also die überwiegende Zahl der Koinzidenzen bei Schichten von weniger als 5 cm Pb, deutet durch ihren Verlauf auf eine kaskadenartige Entstehung hin (Abb. 8), welche durch die folgenden Gründe einen hohen Grad von Sicherheit erhält:

a) Die Materialabhängigkeit weicht von einem massenproportionalen Verhalten ab im Sinne der Tabelle 2, S. 8. Dieser Umstand wurde von Geiger und Fünfer schon vor der Kaskadentheorie als Argument für Entstehung dieser Schauer durch Bremsstrahlung und Paarbildung angeführt ($G\,3$, $G\,1$).

b) Die Schauer aus Blei sind größer als die Schauer aus Aluminium, wie aus Untersuchungen von Geiger ($G\,5$) mit mehreren Zählrohren hervorgeht. Dies kann nach der Kaskadentheorie verstanden werden, da die Multiplikation in Blei bis zu Energien von 10^7 eV herab erfolgt, während ihr in Aluminium schon bei $6 \cdot 10^7$ eV durch die Ionisation eine Grenze gesetzt ist (Tabelle 2).

c) Aus demselben Grunde wird das Ergebnis der Absorptions-messungen von Hu Chien Shan ($H\,14$) verständlich, welche zeigten, daß die Schauerteilchen aus Al energiereicher sind als die aus Blei.

Eine quantitative experimentelle Angabe der theoretisch leicht berechenbaren Häufigkeit $H(Nl)$ der Schauer von mehr als $N = 2, 3, 4, \ldots$ Teilchen als Funktion der Schichtdicke l (Abb. 8, § 11γ) ist heute bei kleinen Schauergrößen $N < 10$ noch nicht möglich. Die Größen, welche bisher quantitativ angegeben werden können, sind:

Die Häufigkeit der n-fach-Koinzidenzen in verschiedenen Anordnungen. Hieraus kann die „Ansprechwahrscheinlichkeit" ($G\,5$, $G\,2$) eines Zählrohres in einer bestimmten Anordnung ermittelt werden, d. h. die relative Zahl der Fälle, in denen dieses Zählrohr zugleich mit einer Gruppe anderer Zählrohre anspricht. Aus der Ansprechwahrscheinlich-keit kann auf die mittlere Größe eines Schauers und aus der Koinzidenz-häufigkeit kann auf die Häufigkeit des mittleren Schauers, dessen Größe mit der Schichtdicke variiert, geschlossen werden. Arley ($A\,8$) hat die Koinzidenzexperimente diskutiert unter der Voraussetzung, daß die Häufigkeit der 2fach-Koinzidenzen gleich der Häufigkeit der Schauer von mehr als 2 Teilchen sei. Es ergab sich im großen und ganzen quantitative Übereinstimmung mit der Kaskadentheorie im Sinne der Abb. 20. Eine genauere Diskussion ist auf Grund des Geiger-schen Begriffs der Ansprechwahrscheinlichkeit möglich.

Mit einem Proportionalzählrohr hat Stuhlinger ($S\,14$) die *Schauer-verteilung*, d. h. die relative Zahl $H(N, l)$ der Schauer mit mehr als $1, 2, \ldots N$ Teilchen hinter einer Schicht von 1,5 cm Pb ($l = 4$) gemessen.

Seine Resultate ließen sich annähernd durch ein Gesetz $H(N, 4) = \text{const}/N$ für $1 < N < 100$ darstellen. Sie können allerdings noch nicht unmittelbar mit der Theorie des § 11 γ, § 19 verglichen werden, solange wir die geometrischen Bedingungen für die Winkel der Strahlen nicht beherrschen. Die theoretische Schauerverteilung $H_m(N)$, welche wir im Winkelmittel bei der Gleichgewichtsschicht $l_m \sim 3 \, _{10}\log N + 2{,}7$ (§ 11 γ) erwarten, zeigt Abb. 24, welche im nächsten Paragraphen genauer besprochen wird.

Die Zahl der schauerfähigen *Lichtquanten*, d. h. der Lichtquanten oberhalb der Ionisierungsgrenze E_j (Tabelle 2), welche in der weichen Strahlung enthalten ist, wird nach der Kaskadentheorie annähernd ebenso groß wie die Zahl der Elektronen. Dies ist im Einklang mit Messungen von AUGER, LEPRINCE-RIGUET und EHRENFEST (Abb. 21) (*A 6*), nach welchen die Absorptionskurve der geradlinigen Koinzidenzen der weichen Strahlen um die eigene Größenordnung erhöht wird, wenn der Absorber seinen Platz zwischen den Zählrohren mit der Lage über den Zählrohren vertauscht, in der er alle energiereichen Lichtquanten $> E_j$ in Elektronen verwandelt.

Die Zahl der energiearmen Lichtquanten unterhalb der Ionisierungsgrenze muß dagegen nach der Strahlungstheorie viel größer sein als die Zahl der Elektronen gleicher Energie. Denn die Elektronen einer Energie

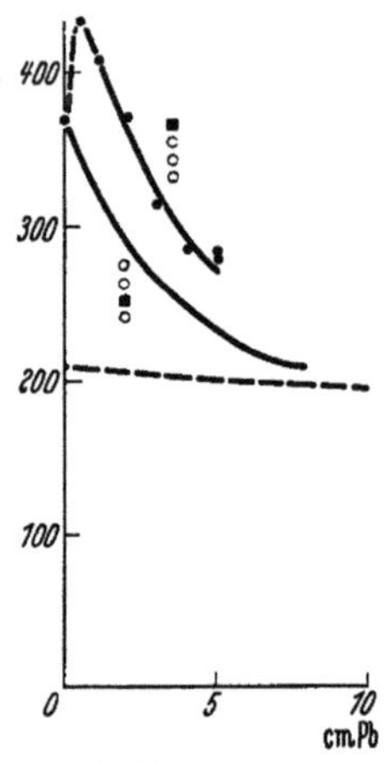

Abb. 21. Koinzidenzmessungen von AUGER, LEPRINCE-RIGUET und EHRENFEST auf dem Jungfraujoch. [P. AUGER, L. LEPRINCE-RIGUET, P. EHRENFEST: J. Physique et Radium **2**, 58 (1936).] Abszisse: Absorberdicke. Ordinate: Koinzidenzzahl. Die untere Kurve gibt die Koinzidenzen, bei denen der Absorber zwischen den Zählrohren lag, so daß nur die geladenen Teilchen wirken konnten. Bei der Aufnahme der oberen Kurve lag die Absorberschicht über den Zählrohren, so daß auch die Lichtquanten einer Energie $> E_j \sim 10^7$ eV in Elektronen verwandelt wurden und Koinzidenzen auslösten vgl. Abb. 9. — — — extrapolierter Anteil der harten Komponente.

$E < E_j$ werden durch Ionisation gehemmt; die Lichtquanten dagegen werden erst bei viel tieferen Energien der Größenordnung 10^6 eV durch COMPTON- und Photoeffekt Verluste erleiden. Während also das differentielle Spektrum $\partial F(E)/\partial E$ der Elektronen in der weichen Komponente (Abb. 16) (oder in den Sekundärstrahlen der Schauer) unterhalb der Ionisierungsgrenze E_j nahezu einen konstanten Wert behält, wird das entsprechende Spektrum der Lichtquanten bei kleinen Energien $k < E_j$ zunächst wie dk/k weiter ansteigen (§ 7). Die Strahlung in einem Schauer wird also von einer großen Menge energiearmer Lichtquanten begleitet sein, welche die Zahl der Elektronen um ein Vielfaches übertreffen kann. Nach Koinzidenzmessungen von GEIGER und ZEILLER (*G 8*) beträgt diese Zahl der Lichtquanten etwa 50 pro Elektron. Die von GEIGER und seinen Mitarbeitern gemessenen Rückstrahleffekte (vgl. *G 1*) werden von diesen Autoren ebenfalls auf energiearme Lichtquanten zurückgeführt.

24. Große Schauer (vgl. *S 6, M 3*). A. Über die großen Schauer können wir Auskunft erhalten aus den Hoffmannschen Stößen, d. h. den plötzlichen Abscheidungen großer Ionenmengen, welche in der Ionisationskammer beobachtet werden und welche aus Garben von

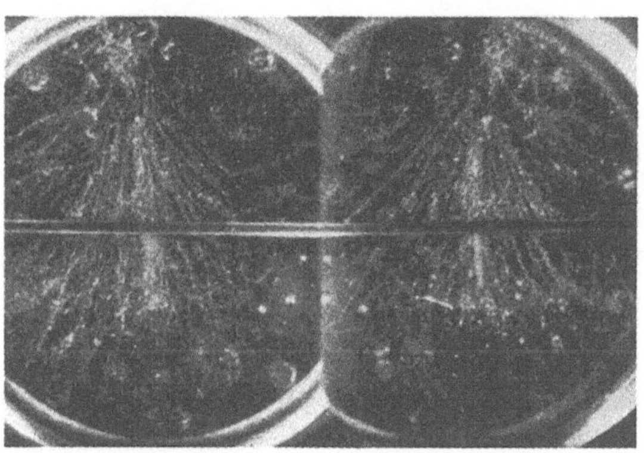

Abb. 22. Aufnahme eines stoßartigen Schauers von schätzungsweise mehr als 300 Teilchen und mehr als $1{,}5 \cdot 10^9$ eV Gesamtenergie von Anderson und Neddermeyer. [Physic. Rev. **50**, 263 (1936).]

10—1000 schwach ionisierenden Teilchen bestehen, die auch in Wilson-Aufnahmen bisweilen beobachtet wurden (Abb. 22). Es kann zwar bisher nur recht ungenau von der Ionisation auf die Schauergröße N und von der Häufigkeit der Stöße in einer Kammer auf die Häufigkeit der Schauer pro min und cm² geschlossen werden; aber innerhalb der Fehlergrenzen dieser Umrechnungen, welche eine Unbestimmtheit in der Schauergröße und Schauerhäufigkeit um einen Faktor 2 oder 3 offenlassen, können doch schon eine Reihe von Aussagen über die Statistik der großen Schauer gewonnen werden. Die hier benutzten Schätzungen für die Umrechnung zwischen den empirisch gegebenen und den theoretisch erfaßbaren Größen sind in der folgenden Tabelle zusammengestellt:

Tabelle 8.

Autor	Wirksame Fläche in cm²	Teilchenzahl pro 10^6 Ionenpaare
Böggild	700	40
Nie	1600	50
Messerschmidt . . .	500	20
Young und Street .	50	—
Carmichael	2000	—

Hierbei wurde eine Ionisation von 70 Ionenpaaren pro cm vorausgesetzt und die wirksame Fläche des Panzers wurde in allen Fällen dem Kammerquerschnitt gleichgesetzt.

Abb. 23 zeigt die Stoßauslösekurve für Stöße von mehr als 200 Teilchen in verschiedenen Materialien, d. h. die Häufigkeit der Stöße von mehr als $N = 200$ Teilchen als Funktion der Dicke des Panzers nach Nie ($N\,2$, $N\,3$). Die Kurve sieht in ihrem Verlauf der entsprechenden Kurve für die kleineren Schauer der Rossi - Koinzidenzen ähnlich (Abb. 19, 20); ihre Gestalt variiert nur wenig mit der Stoßgröße N ($M\,2$, $C\,2$, $B\,16$, $Y\,1$, $Y\,2$). Wir werden im folgenden zeigen, daß der in Abb. 22 schraffierte Anteil der Stöße beim Maximum der Kurve den Kaskaden zugeschrieben werden muß, und daß die übrigen Stöße, also vor allem die Stöße hinter dickeren und dünneren Schichten, aus Explosionen bestehen, wenn

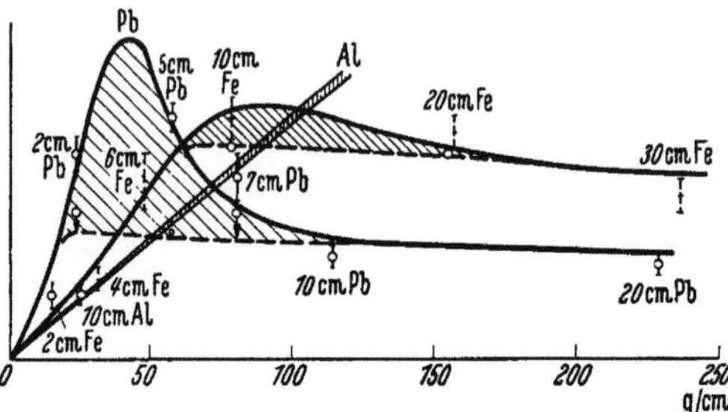

Abb. 23. Stoßauslösekurven von Nie. Abszisse: Dicke der auslösenden Schicht in g/cm². Ordinate: Häufigkeit der Stöße von mehr als $N=200$ Teilchen. Messung von Nie in Pb, Fe, Al. Schraffiert: Theoretischer Kaskadenanteil nach Abb. 8, § 11. o o o Durch Subtraktion erhaltener Anteil der Nichtkaskaden.

die bisherigen Messungen auch unter reineren Bedingungen bestätigt werden ($E\,6$). Die Trennung der beiden Arten von Stößen in Abb. 22 wurde durchgeführt, indem die theoretische Häufigkeit der großen Kaskaden (Abb. 8, § 11γ) von jedem Meßpunkt aus nach unten abgetragen wurde, wobei ein unbestimmter gemeinsamer Intensitätsfaktor geeignet gewählt wurde.

Die Argumente für die Kaskadennatur der Stöße beim Maximum der Stoßauslösekurve (schraffiert in Abb. 23) sind folgende:

a) Die Lage l_m der Maxima wird durch die Kaskadentheorie annähernd richtig wiedergegeben, wie die nachstehende Tabelle zeigt:

Tabelle 9. Lage des Maximums der Stoßauslösekurve.

Stoßgröße	$N =$	10	20	30	40	80	200	300	400
($Y\,1$, $Y\,2$) Young und Street Pb	$l_m =$	4 ± 2	5 ± 2	$5{,}5 \pm 2$					
($B\,16$) Böggild Fe	$l_m =$		$3 \pm 1{,}5$		$4 \pm 1{,}5$	$5 \pm 1{,}5$			
($N\,2$) Nie Pb .	$l_m =$						$7{,}5 \pm 4$	9 ± 4	10 ± 4
($N\,2$) Nie Fe .	$l_m =$						8 ± 2	9 ± 2	10 ± 2
(§ 11γ) theor.: $2{,}7 + 3\,{}_{10}\log N$	$N =$	5,7	6,6	7,2	7,5	8,5	9,6	10,2	10,5

Die Abweichungen zwischen Theorie und Experiment sind nur bei den Böggildschen Messungen größer, als man auf Grund der Unsicherheit in den Umrechnungsfaktoren (Tabelle 8) erwarten würde.

b) Die Breite der Maxima, welche vor allem in den Stoßmessungen von Young und Street geprüft werden kann (E 6), wird richtig dargestellt. Insbesondere wird sie in Fe dann richtig dargestellt, wenn man nur den theoretisch erwarteten, gegenüber Blei geringen Anteil der Stöße den Kaskaden zuschreibt (Abb. 23).

c) Es scheint, daß das Verhältnis der Intensitäten in den Maxima der Stoßauslösekurve richtig wiedergegeben wird durch ein Elektronenspektrum, welches wir nach § 21 und § 19 theoretisch erwarten.

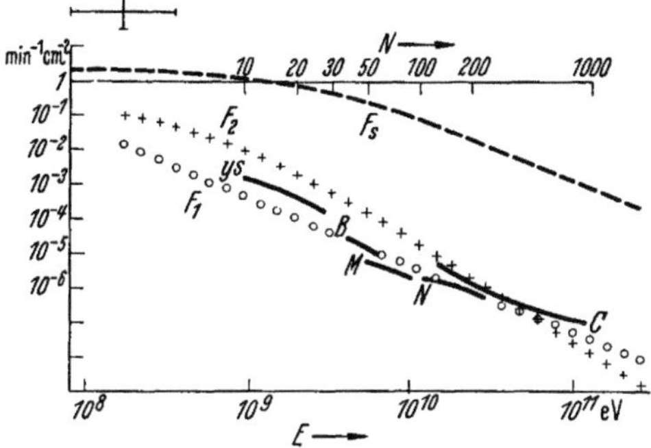

Abb. 24. Spektrum und Kaskadenhäufigkeit. Abszisse: (log) Impuls pc der Teilchen in eV. Ordinate: (log) Zahl F der Teilchen oberhalb des Impulses p pro min und cm² in Meereshöhe. o o o F_1: Kaskadenelektronen (§ 19, $\gamma = 1,85$). — — — F_s: Durchdringende Teilchen (§ 20). + + + F_2: Zerfallselektronen (§ 21). Da jedes Elektron der Energie E in Blei eine Kaskade erzeugt, deren Größe N beim Maximum l_m der Schauerauslösekurve (§ 11 γ) durch $N =$ $= \frac{1}{8} \left(\frac{E}{10^7\,eV} \right)^{0,93}$ gegeben ist, kann die Abbildung auch so gelesen werden: Abszisse: Größe N der Kaskadenschauer. Ordinate: Häufigkeit H_m der Kaskadenschauer von mehr als N Teilchen beim Maximum der Bleiauslösekurve pro min pro cm². Die ausgezogenen Kurvenstücke stellen die Stoßmessungen von YS: Young und Street, B: Böggild, M: Messerschmidt, N: Nie, C: Carmichael dar. Experimentelle Unsicherheit der absoluten Häufigkeit und Stoßgröße.

spektrum, welches wir nach § 21 und § 19 theoretisch erwarten.

Dies geht im einzelnen aus Abb. 24 hervor, in welcher die experimentelle Häufigkeit $H_m(N)$ der Kaskadenstöße beim Maximum der Stoßauslösekurve in Pb (§ 11 γ), also das Maximum der schraffierten Fläche in Abb. 23, als Funktion der Stoßgröße N dargestellt ist. Diese Funktion soll ja nach (15) ein direktes Maß für das stoßerzeugende Elektronenspektrum sein, da in dem betrachteten Fall die Häufigkeit der Stöße von mehr als N Teilchen gleich der Häufigkeit der auf die Bleischicht einfallenden Elektronen von mehr als $E = (8N)^{1,07} E_j$ eV Energie ist. Die Energieskala ist deshalb neben der Skala für die Teilchenzahl in die Abb. 24 eingezeichnet.

Wie man in der Abbildung sieht, ordnen sich die Kurvenstücke, welche die Messungen der mit den verschiedenen Ionisationskammern erreichbaren Schauergrößen darstellen ($Y\,2$, $M\,2$, $C\,2$, $B\,2$, $N\,2$, $N\,3$), zu einer Kurve zusammen, welche das Spektrum der an der Erdoberfläche einfallenden Elektronen angibt, wenn einmal absolute Stoßhäufigkeiten genau bekannt sind. Dieses empirische Elektronenspektrum scheint nun nicht sehr abzuweichen von dem theoretischen Spektrum der Elektronen, welches in § 21 und § 19 angegeben wurde. Dieses theoretische Spektrum setzt sich aus 2 Teilen zusammen: dem Spektrum der Zerfallselektronen (§ 21), welches in Abb. 24 durch Kreuze bezeichnet ist und welches bei niedrigen Energien überwiegt; und dem Spektrum der Kaskadenelektronen (§ 19), welches in Abb. 23 durch Kreise bezeichnet ist und welches in Meereshöhe erst bei hohen Energien zum

Vorschein kommt. Zu einer endgültigen Entscheidung über diese Spektren müssen aber noch genauere absolute Häufigkeitskurven der Kaskadenschauer abgewartet werden.

d) Die starke empirische Höhenabhängigkeit der HOFFMANNschen Stöße beim Maximum der Bleiauslösekurve ($Y\,3$, $M\,4$, $S\,16$) kann auf Grund der Kaskaden erzeugenden Elektronenspektren verstanden werden, während die geringere Höhenabhängigkeit der Stöße bei dicken Schichten ($Y\,1$, siehe $M\,3$) auf die Erzeugung dieser Stöße aus der durchdringenden Komponente hinweist:

Die Zunahme der Stoßhäufigkeit beim Maximum ($\sim 2{-}4$ cm Pb) beträgt auf einer Höhendifferenz von 76 auf 45 cm Hg:

Tabelle 10.

	WOODWARD ($W\,8$)	YOUNG und STREET ($Y\,2$, $Y\,3$)			MONTGOMERY ($M\,4$)
Stoßgröße N . . .	< 10	10	20	30	40
Anstiegfaktor . . .	8,5	10	17	22	26

Diese Zunahme, welche um so größer ist, je größer die Stöße sind, ist durch das Zusammenwirken der Elektronenspektren § 21 und § 19 verständlich. Denn bei kleinen Energien und damit bei kleinen Kaskadenstößen überwiegt ja in Meereshöhe das Zerfallsspektrum ($+++$ in Abb. 23) der Elektronen, welches langsam mit der Höhe ansteigt (§ 21). Bei großen Energien und damit bei großen Kaskadenstößen überwiegt aber das Kaskadenspektrum ($\circ\circ\circ$ in Abb. 24), welches sehr rasch mit der Höhe ansteigt [(41), § 19]:

Tabelle 11.

Anstiegsfaktor bei der genannten Höhendifferenz ($e = 12$)	20	40	63
Für die Abfallskonstante (§ 19) . . . . $\gamma =$	1,5	1,7	1,9

In größerer Höhe über dem Erdboden wird man allerdings bald ein Überwiegen des Kaskadenspektrums bei allen Energien erwarten. Weitere Messungen der Höhenabhängigkeit großer HOFFMANNscher Stöße können die Gestalt dieses Elektronenspektrums, die in § 19 und § 20 nur indirekt und ungenau bestimmt wurde ($\gamma = 1,8$ bis 1,9), genauer festlegen.

Aus der Höhenabhängigkeit (Tabelle 10) der HOFFMANNschen Kaskadenstöße ergibt sich noch einmal ein unabhängiges Argument für die Erzeugung der in Meereshöhe gefundenen Elektronen aus der durchdringenden Komponente: Denn wollte man ein Elektronenspektrum konstruieren, welches aus dem Weltenraum einfällt, sich dann durch Kaskaden in der Atmosphäre verwandelt und in der unteren Atmosphäre die HOFFMANNschen Kaskadenstöße erzeugt, so müßte man diesem Spektrum bei hohen Energien einen stärkeren Potenzabfall zuschreiben als bei niedrigen Energien, damit die großen Kaskadenstöße stärker mit der Höhe zunehmen als die kleinen (§ 19). Will man aber das Spektrum so konstruieren, daß es die Zahl der Elektronen in der oberen und unteren Atmosphäre darstellt, so muß man ihm bei hohen Energien einen schwächeren Abfall zuschreiben als bei kleinen Energien (§ 19) (*H 7, N 6*). Beide Bedingungen zugleich können aber nicht erfüllt werden, d. h. die weiche Strahlung in der unteren Atmosphäre muß aus der durchdringenden Strahlung entstehen.

Dem Gang der Stoßintensität mit der Höhe entspricht ihre Variation mit der Tiefe: In einem Keller ist nach Böggild (*B 16*) das scharfe Kaskadenmaximum mehr herabgesetzt als die Intensität bei großer Schichtdicke. Im Bodensee nimmt nach Weischedel (*W 1*) die Stoßhäufigkeit, welche als Häufigkeit der Stöße hinter großen Schichten aufgefaßt werden kann, ähnlich wie die durchdringende Komponente ab. Entsprechende Erfahrungen wurden mit den Rossischen Koinzidenzen gemacht (vgl. *A 5*), so daß ebenso bei den Stößen wie bei den Rossi-Koinzidenzen angenommen werden kann, daß die Maxima der Auslösekurve von der weichen Komponente erzeugt werden. Damit ist die Strahlungstheorie (§ 7) bis zu 10^{11} eV hinauf bestätigt.

B. Während also ein Teil der Stöße beim Maximum der Stoßauslösekurve alle Eigenschaften aufweist, die für Kaskaden charakteristisch sind, müssen die Stöße bei großen Schichten anderen Ursachen zugeschrieben werden.

e) Hierfür spricht erstens der Umstand, daß die kaskadentheoretische Stoßhäufigkeit bei großen Schichten schnell gegen 0 abfällt (Abb. 8), während die wirklichen Stöße bei großen Schichten eine endliche Häufigkeit behalten, die bei mittleren Stößen in Blei halb so groß ist wie die Häufigkeit beim Maximum (Abb. 23).

f) Das stärkste Argument für die Nichtkaskadennatur eines großen Teils der Stöße aber liegt in der Materialabhängigkeit: Während sich die Häufigkeit entsprechender Stöße in Pb : Fe : Al nach der Kaskadentheorie etwa wie $1 : 3^{-2} : 6^{-2}$ verhalten sollte (§ 11 γ, $\gamma \approx 2$), ist umgekehrt die Intensität der wirklichen Stöße bei dicken Schichten größer im leichten als im schweren Material (Abb. 23). Wenn wir also 50% aller Stöße beim Maximum der Bleiauslösekurve für Kaskaden halten, so können wir die Stöße beim Fe-Maximum nur zu 6% und die beim Al-Maximum nur zu etwa 1,5% auf Kaskadenwirkungen zurückführen. Die Stöße bei größeren und kleineren Schichten, vor allem die überwiegende Zahl der Stöße in den leichten Materialien muß also anderen Ursprungs sein. Die eben diskutierte Materialabhängigkeit der Hoffmannschen Stöße schließt auch die Möglichkeit aus, die Stöße bei dicken Schichten auf Bhàbhasche Ionisationsschauer zurückzuführen. Denn die großen Ionisationsschauer sollen ja im schweren Material etwa im Verhältnis der Atomnummern häufiger sein als im leichten (61), während die wirklichen Stöße eher das umgekehrte Verhalten zeigen (Abb. 23).

g) Ein Anhaltspunkt über die Natur der nichtkaskadenartigen Stöße kann aus dem Verhalten bei dünnen Schichten entnommen werden.

Abb. 25 zeigt die empirische Häufigkeit der Stöße von mehr als N Teilchen pro min und kg als Funktion der Stoßgröße N bei dünnen Schichten. Die Tatsache, daß die empirischen Kurven für verschiedene Schichtdicken nahezu zusammenfallen, bedeutet, daß hier der Anstieg

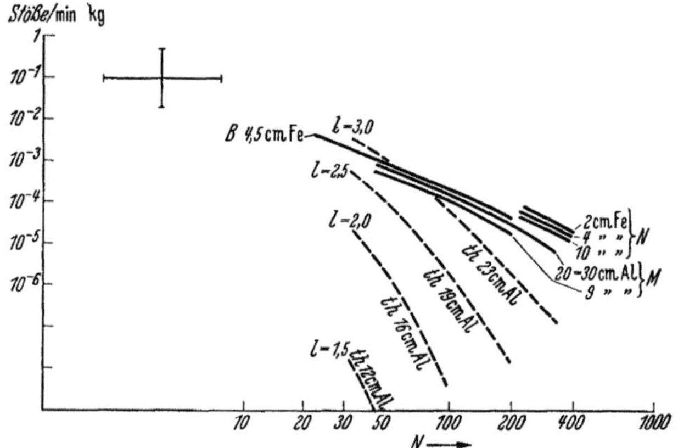

Abb. 25. Stoßverteilung bei dünnen Schichten. Abszisse: (log) Stoßgröße N. Ordinate: (log) Häufigkeit der Stöße von mehr als N Teilchen pro min pro kg. — — — th.: nach der Kaskadentheorie (§ 11 γ für $\gamma = 1,5$, Gesamtfaktor willkürlich). ———— Messung von M: MESSERSCHMIDT, N: NIE, B: BÖGGILD. ⊢—⊣ Unsicherheit in der Beurteilung des Vergleichs zwischen den Messungen M, N, B.

der Stoßhäufigkeit nahezu linear mit der Schichtdicke erfolgt; ganz im Gegensatz hierzu müßte man eine viel stärkere Abhängigkeit der Stoßhäufigkeit von der Schichtdicke erwarten, wenn die Stöße bei dünnen Schichten leichter Materialien Kaskaden wären, wie die gestrichelten Linien in Abb. 25 zeigen; z. B. sollten die Stöße pro cm² und min in 30 cm Al nach der Kaskadentheorie um einen Faktor $>10^5$ häufiger sein als die in 10 cm Al. Die wirklichen Stöße von etwa 200 Teilchen in 30 cm Al sind aber nach NIE ($N\,2$) nur um einen Faktor 3—4mal häufiger pro cm² und min als die gleich großen Stöße hinter 9 cm Al. Wenn sich also die Messungen der Stoßhäufigkeit in dünnen Schichten auch unter reineren Bedingungen, welche die Mitwirkung der Zimmerwände ausschließen, bestätigen, so muß hieraus geschlossen werden, daß die nicht-kaskadenartigen Stöße explosionsartig entstehen, da nur bei der Entstehung der Stöße an *einem* Kern ein linearer Anstieg der Häufigkeit mit der Schichtdicke verstanden werden kann.

Die Frage, ob die in den Explosionen entstehenden *Sekundären* leichte oder schwere Elektronen sind, kann durch Stoßexperimente entschieden werden. Obwohl heute eine klare Antwort noch nicht vorliegt, scheinen die bisherigen Experimente eher auf eine Erzeugung schwerer als auf die überwiegende Bildung leichter Elektronen

hinzudeuten. Die Entscheidung kann einmal getroffen werden in Stoß-
koinzidenzexperimenten. Nach Nie ($N\,3$) nimmt die Zahl der Stöße
von mehr als 100 Teilchen aus 10 cm Fe, welche gleichzeitig in 2 Kammern
auftreten, auf ihren 5. Teil ab, wenn sich zwischen den Kammern eine
Schicht von 9 cm Pb befindet. Dies ist eher zu verstehen, wenn schwere
Elektronen in den Explosionen erzeugt werden, als wenn in den Explosionen leichte Elektronen entstehen, da man für die letzteren eine
stärkere Absorption bis auf etwa 1% des anfänglichen Wertes erwarten
würde ($E\,6$). Doch ist eine sichere Entscheidung heute noch nicht möglich. Die Natur der Sekundären kann ferner aus „Übergangseffekten"
entnommen werden: Eine Schicht von einigen cm Pb müßte die Stöße
aus 10—20 cm Al multiplizieren, falls sie Elektronen enthalten, aber
nur wenig verändern, falls sie aus schweren Elektronen bestehen. Solche
Übergangsmessungen sind bisher von Böggild ($B\,16$) nur bei den
Schichten ausgeführt worden, hinter denen die Stöße zweifellos Kaskaden
sind, nämlich beim Bleimaximum. Sie bestätigten die elektronenartige
Natur und damit den Kaskadencharakter dieser Stöße beim Maximum
der Bleiauslösekurve. Entsprechende Messungen hinter solchen Schichten,
welche Explosionen hervorbringen (z. B. 10—20 cm Al), könnten über
die Frage der Sekundären in den Explosionen Auskunft geben. Hierbei
ist allerdings zu berücksichtigen, daß auch dann die Explosionsstöße
hinter größeren Schichten von beträchtlichen Mengen (etwa 20—40%)
Elektronen begleitet sein werden, wenn zunächst nur schwere Teilchen
im Schauer entstehen, da ja die in den letzten Zentimetern steckengebliebenen schweren Elektronen ihre Zerstrahlungsprodukte dem
Schauer beimengen. Diese Elektronenbeimengung wird nur dann vermieden werden können, wenn man die Stöße aus dünnen Schichten
von etwa 10—20 cm Al oder 2—4 cm Fe beobachtet, da dann nur wenig
schwere Schauerteilchen im Material steckenbleiben und nachträglich
zerstrahlen können.

Wenn wir nun vorläufig annehmen, daß in den Explosionen vorwiegend schwere Elektronen erzeugt werden, so können wir aus der
Strecke des Anstiegs der Explosionsstöße auf die Reichweite und damit
auf die Energie der schweren Schauerelektronen schließen ($E\,6$). Aus
der Anstiegstrecke von 10 cm Fe (Abb. 23) erhalten wir dann (Tabelle 6)
eine mittlere *Energie* von etwa

$$E_k \approx 10^8 \text{ eV} \tag{62}$$

für die Sekundären der Explosionen in Übereinstimmung der Größenordnung mit der kritischen Energie, welche in § 15 angegeben wurde.
Zur Erzeugung einer Explosion von N Teilchen wird also im Mittel
ein schweres Elektron von

$$E \sim N \cdot 10^8 \text{ eV} \tag{63}$$

notwendig sein.

Theoretische Gesichtspunkte zur Deutung der kosmischen Strahlung.   61

Ein schweres Elektron, welches einen Explosionsschauer von N Teilchen erzeugt, muß also ungefähr dieselbe Energie haben, wie ein leichtes Elektron, das einen Kaskadenschauer derselben Größe beim Maximum der Bleiauslösekurve hervorbringt:

$$(E = 8 \cdot N \cdot E_j = 8 \cdot 10^7 \cdot N \cdot \text{eV} \sim 10^8\, N\, \text{eV}).$$

Der *Wirkungsquerschnitt*, mit dem ein schweres Elektron einen Explosionsschauer erzeugt, kann nun angegeben werden durch Vergleich der empirischen Intensität der Explosionen bei großen Bleischichten mit der empirischen Intensität der Kaskaden beim Maximum der Bleiauslösekurve. Diese beiden Intensitäten verhalten sich ja nach Abb. 23 ungefähr wie $1:1$ und dieses Verhältnis $1:1$ scheint bemerkenswert wenig zu variieren von den kleinen Stößen von etwa 30 Teilchen ($Y\,3$, $B\,16$) an bis zu den ganz großen Stößen von 1000 Teilchen ($C\,2$). Da sich nun die Zahl $F_2(E)$ der Elektronen oberhalb der Energie $E \lesssim 10^{11}$ eV (Abb. 23) zur Zahl $F(2\,E)$ der durchdringenden Teilchen oberhalb der Energie $2\,E$ in Meereshöhe nach (57), (59) wie

$$F_2(E) : F_s(2\,E) = \frac{10^7\,\text{eV}}{E}$$

verhält, und da ferner jedes Elektron eine Kaskade erzeugt, so folgt für die Wahrscheinlichkeit, mit der ein schweres Elektron der Energie E in der Anstiegstrecke von 10 cm Fe eine Explosion erzeugt, annähernd:

$$\frac{10^7\,\text{eV}}{E} \text{ pro } 10 \text{ cm Fe.}$$

Da der Kubikzentimeter Fe $8{,}5 \cdot 10^{22}$ Atome enthält, setzt also jedes Proton und Neutron des Kerns dem ankommenden schweren Elektron der Energie $E > 10^9$ eV einen Wirkungsquerschnitt

$$Q \approx \frac{1}{2} \cdot 10^{-27}\,\text{cm}^2 \cdot \left(\frac{10^9\,\text{eV}}{E} \right) \tag{64}$$

für die Erzeugung einer Explosion entgegen.

Die Größenordnung dieses Wirkungsquerschnittes paßt ebenfalls zu den theoretischen Erwartungen des § 15. Das empirische Gesetz, nach dem der Wirkungsquerschnitt bei hohen Energien mit der ersten Potenz der Energie abzunehmen scheint, kann natürlich heute noch nicht theoretisch begründet werden.

Wenn dieses Gesetz (64) richtig ist, muß bei ganz großen Stößen $N > 1000$ allerdings wieder ein Ansteigen des Verhältnisses der maximalen Intensität zur Sättigungsintensität (Abb. 23) erwartet werden, da dann das Kaskadenspektrum hervortritt (Abb. 24).

Durch die bisherigen Betrachtungen werden die Hoffmannschen Stöße hinter großen Schichten in der Näherung verständlich, in der sie in allen Materialien gleich häufig sind und ihre Sättigung (10 cm Fe, > 30 cm Al) bei Schichten gleicher Masse erreichen. Denn die Explosionen sollen ja massenproportional erzeugt und annähernd massenproportional absorbiert werden. In Wirklichkeit sind aber die Schauer hinter großen Schichten leichter Materialien intensiver als hinter großen Schichten schweren Materials, und die Sättigungsstellen treten entsprechend im leichten Material bei etwas größeren Massen ein als im schweren (Abb. 23, Abb. 19). Ein Weg zum Verständnis dieser noch ungeklärten Materialabhängigkeit liegt vielleicht in der Möglichkeit, daß die Schauerteilchen einer Explosion noch im Kern ihrer Entstehung durch Kernionisation (§ 16, 24) gebremst werden (*E 6*). Diese Bremsung, welcher vor allem die langsamsten schweren Elektronen des Schauers unterliegen würden, könnte im großen Bleikern mehr ausmachen als im leichten Aluminiumkern, und könnte so die beobachtete Bevorzugung des leichten Materials hervorrufen. Eine theoretische Beurteilung dieses Effekts ist aber heute noch nicht möglich.

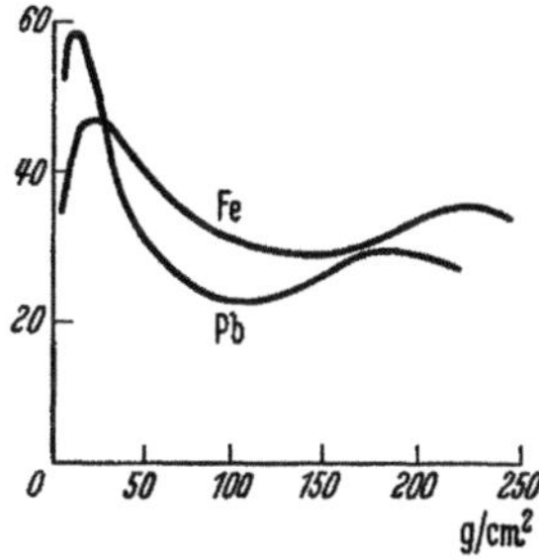

Abb. 26. Zweites Maximum der Rossischen Dreifachkoinzidenzen unter spitzen Winkeln von 7° nach Schmeisser und Bothe: Naturwiss. **25**, 833 1937). (Vgl. Abb. 19.) Abszisse: Schichtdicke. Ordinate: Zahl der Koinz./Std.

Die eben besprochene, noch ungeklärte Materialabhängigkeit ist den Hoffmannschen Stößen gemeinsam mit einer in der letzten Zeit von Schmeisser und Bothe (*S 1*) untersuchten besonderen Art von Rossi-Koinzidenzen.

Verschiedene Autoren (*K 3, M 1, D 1, H 15*) hatten bei der Registrierung der Rossischen Koinzidenzen die Andeutung eines zweiten Maximums bei 15—20 cm Pb gefunden. Schmeisser und Bothe (*S 1*) gelang es, dieses zweite Maximum in besonders starker Intensität zu erhalten durch die Beobachtung unter spitzen Winkeln. Es ergab sich (Abb. 26) ein deutlich ausgeprägtes zweites Maximum bei 17 cm Pb, 30 cm Fe; die folgenden Gründe machen nun die naheliegende Annahme wahrscheinlich, daß das zweite Maximum von den kleineren Explosionen erzeugt wird, deren große Seitenstücke wir in den nichtkaskadenartigen Hoffmannschen Stößen beobachten:

a) Das zweite Maximum wird von der durchdringenden Komponente erzeugt, wie Schmeisser und Bothe durch Messungen in einem Keller zeigen konnten, in welchem das erste Maximum durch Absorption der weichen Strahlung verkleinert wurde, während das zweite Maximum nur wenig verändert wurde.

b) Das zweite Maximum tritt bei Schichten (17 cm Pb, 30 cm Fe) ein, welche mit den Sättigungsschichten der Nichtkaskadenstöße (3—8 cm Pb, 10 cm Fe) vergleichbar sind. Sie sind etwa dreimal so groß wie die entsprechenden Schichten bei den Hoffmannschen Stößen und würden auf Sekundärenergien von etwa $3 \cdot 10^8$ eV schließen lassen, falls die erzeugten Teilchen schwere Elektronen sind. Es ist wahrscheinlich, daß in den Schmeisser-Botheschen Schauern zunächst schwere Elektronen erzeugt werden, da die großen Reichweiten von 17 cm Pb für Elektronen nicht so leicht verständlich wären. Durch nachträglichen Zerfall würden trotzdem den beobachteten Schauern wieder Elektronen beigemengt sein.

c) Für einen Explosionsschauer, dessen Gesamtenergie groß gegen die Ruhenergie 10^9 eV eines Kernbestandteils ist, wird man auf Grund der Erhaltungssätze einen spitzen Divergenzwinkel erwarten. Die Schauerteilchen werden dann nachträglich nur eine geringe weitere Winkelstreuung durch elastische Ablenkung erfahren, da sie (im Gegensatz zu den Sekundären eines Kaskadenschauers) hohe Impulse haben. Dieser Umstand dürfte mit der von SCHMEISSER und BOTHE entdeckten Spitzwinkligkeit der Schauer im 2. Maximum der ROSSI-Kurve zusammenhängen und die Erwartung nahelegen, daß auch unter den HOFFMANNschen Stößen die Explosionen durch spitze Winkel vor den Kaskaden ausgezeichnet sind.

d) Die zweiten Maxima (Abb. 26) zeigen dieselbe ungeklärte Materialabhängigkeit wie die Explosionsstöße.

Die Materialabhängigkeit des hier betrachteten Sekundäreffekts hat eine gewisse Ähnlichkeit mit den noch nicht gedeuteten „Übergangseffekten" bei dicker Schicht (S 9). Unter „Übergangseffekt" wird die schnelle Änderung in der Absorptionskurve der mit der Ionisationskammer gemessenen Intensität verstanden, welche sich beim Übergang von einem zu einem andern Absorbermaterial ergibt (S 4). Die Übergangseffekte bei dünneren Schichten von weniger als etwa 100 g/cm² zeigen eine Zunahme der Ionisation, wenn Blei hinter Aluminium gesetzt wird, und eine Abnahme, wenn Aluminium hinter Blei tritt. Die Übergangseffekte bei größeren Schichten haben gerade das umgekehrte Verhalten. Nach der Strahlungstheorie (G 1), insbesondere der Kaskadentheorie (C 1, B 3) sind die Übergangseffekte bei kleineren Schichten verständlich, da Blei die Strahlen aus Aluminium noch weiter multiplizieren kann, nicht aber Aluminium die aus Blei. Für die Übergangseffekte hinter dicken Schichten aber müssen ganz andere Sekundäreffekte verantwortlich sein, welche dieselbe Materialabhängigkeit haben wie die HOFFMANNschen Stöße bei dicken Schichten und die SCHMEISSER-BOTHEschen Koinzidenzen.

25. Kernprozesse. In der Höhenstrahlung sind häufig schwere Teilchen, die sehr viel stärker als gewöhnliche Elektronen ionisieren, nachgewiesen worden. Dieser Nachweis erfolgte einerseits mit der WILSON-Kammer, insbesondere in der schon genannten Arbeit von ANDERSON und NEDDERMEYER (A 3) sowie bei BRODE und STARR (B 19), andererseits durch die Spuren, die stark ionisierende Teilchen in der empfindlichen Schicht einer photographischen Platte hinterlassen. Durch verschiedene Forscher (HERZOG und SCHERRER (H 10), RUMBOUGH und LOCHER (R 6), FÜNFER (F 5), SCHOPPER (S 15), BLAU und WAMBACHER (B 14), TAYLOR (T 3) wurden photographische Platten teils im Laboratorium, teils in großer Höhe längere Zeit der kosmischen Strahlung ausgesetzt und wiesen nachher Spuren auf, die am einfachsten als die Bahnen von Protonen gedeutet werden konnten. Von BLAU und WAMBACHER (*B 14*) wurden ferner „Sterne" solcher Spuren beobachtet, also Punkte in der photographischen Platte, von denen offenbar mehrere Protonen gleichzeitig ausgegangen sind (Abb. 27). Schließlich hat TAYLOR (*T 3*) an einigen Stellen seiner Platten „Haufen" solcher Spuren beobachtet.

Charakteristisch für die verhältnismäßig langsamen schweren Teilchen ist zunächst die starke Zunahme ihres Auftretens mit der Höhe. ANDERSON

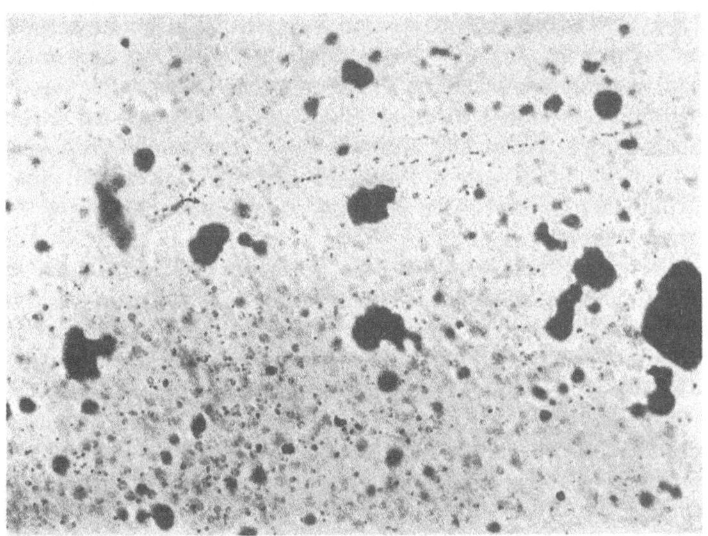

Abb. 27. Kernverdampfung nach BLAU und WAMBACHER: Nature (Lond.) 140, 585 (1937).

und NEDDERMEYER finden auf dem Pike's Peak etwa 12 mal soviel Spuren langsamer schwerer Teilchen als auf dem Meeresniveau. Ähnliche Werte folgen auch aus den Protonenspuren in photographischen Platten, während für die Sterne in den photographischen Platten von TAYLOR eine noch stärkere Zunahme mit der Höhe beobachtet wird. Daraus folgt, daß die stark ionisierenden Spuren jedenfalls durch eine Strahlung hoher Absorption hervorgerufen werden. Von den meisten Forschern wird bisher angenommen, daß es sich bei den stark ionisierenden Spuren in der Hauptsache um langsame Protonen handeln muß. Doch kann wohl nicht ausgeschlossen werden, daß auch langsame schwere Elektronen für einen Teil der starken Spuren verantwortlich sind. Ferner kann es sich in einzelnen Fällen vielleicht auch um α-Teilchen handeln. Die Spuren in den Sternen von BLAU und WAMBACHER sind wahrscheinlich zum größten Teil durch Protonen hervorgerufen.

Von der Strahlung, die diese Spuren indirekt verursacht, ist zunächst nur bekannt, daß sie in der Atmosphäre sehr stark absorbiert wird.

Theoretische Gesichtspunkte zur Deutung der kosmischen Strahlung.   65

Daraus folgt, daß schwere Elektronen und sehr energiereiche Protonen oder Neutronen nur zu einem unwesentlichen Teil an der Auslösung der langsamen Protonen beteiligt sein können. Als erzeugende Strahlung kommen also nur Elektronen und Lichtquanten, ferner nicht zu energiereiche Protonen und Neutronen in Betracht. Wenn Elektronen oder Lichtquanten in einem Kern schwere Teilchen auslösen sollen, so ist das mit einer hinreichenden Wahrscheinlichkeit wohl nur möglich durch Prozesse, die durch das Zusammenwirken der elektrischen Kräfte mit den Kernkräften entstehen (§ 15); z. B. kommt etwa die Umkehrung des in § 15 besprochenen Prozesses in Betracht: Ein hinreichend energiereiches Lichtquant erzeugt beim Zusammenstoß mit einem Neutron ein negativ geladenes schweres Elektron, wobei gleichzeitig das Neutron in ein Proton verwandelt wird und einen Teil der Energie des Lichtquants mit übernimmt. Wenn die Energie des Lichtquants groß genug ist, so wird an die Stelle dieses Prozesses ein Mehrfachprozeß vom Typus der Explosionen treten. Den Wirkungsquerschnitt für diesen Vorgang kann man wie oben zu etwa 10^{-27} cm^2 abschätzen (§ 15). Die Auslösung eines analogen Prozesses durch ein Elektron ist wohl noch um den Faktor $\sim e^2/\hbar c$ seltener. Die Aufnahmen von ANDERSON und NEDDERMEYER (*A 3*) zeigen, daß langsame Protonen häufig zugleich mit einem Kaskadenschauer beobachtet werden. Dieser Umstand deutet darauf hin, daß die langsamen Protonen tatsächlich durch Lichtquanten ausgelöst werden. Bei hohen Energien dieser Lichtquanten kann dies wohl nur durch die genannten Prozesse geschehen. Bei kleineren Lichtquantenenergien kann man auch an den direkten Photoeffekt im Kern denken, der von BOTHE und GENTNER (*B 20*) beobachtet worden ist. Doch dürften bei einem solchen Prozeß nur sehr energiearme Protonen den Kern verlassen, die in den meisten Fällen in der Materie, in der sie erzeugt sind, so schnell absorbiert werden, daß sie nicht bis in die WILSON-Kammer gelangen. Es ist also sehr fraglich, ob dieser Kernphotoeffekt bei den empirisch beobachteten Protonenspuren eine merkbare Rolle spielt. Auch wenn man den anderen obengenannten Prozeß zugrunde legt, scheint es schwierig, die relativ große beobachtete Häufigkeit der langsamen Protonen zu erklären. Doch sind die theoretischen Abschätzungen der Wirkungsquerschnitte einstweilen so unsicher, daß man die Bedeutung dieser geringen Diskrepanz noch nicht beurteilen kann. Zum Teil können die langsamen Protonen wohl auch durch Neutronen oder Protonen mittlerer Energie ausgelöst werden. Man muß hier daran denken, daß bei den Prozessen, die zur Erzeugung der durchdringenden Komponente der Höhenstrahlung führen und die ebenfalls zu dem obengenannten Typus gehören, stets auch Protonen und Neutronen mittlerer Energie erzeugt werden. Insbesondere werden bei den eigentlichen Explosionen, die durch sehr energiereiche Lichtquanten oder Elektronen ausgelöst sind, häufig schwere Teilchen aus dem Energiebereich 10^8—10^9 eV beteiligt sein. Diese Teilchen, die auf dem Weg aus der Stratosphäre, in der sie hauptsächlich entstehen, nach unten sehr rasch absorbiert werden, können vielleicht zum Teil für die Auslösung der langsamen Protonen verantwortlich sein. In der Tat glauben Anderson und Neddermeyer auf einer Aufnahme eine Kernzertrümmerung durch ein Neutron nachweisen zu können, wobei dieses Neutron allerdings wieder gleichzeitig mit einem größeren Schauer auftritt. Es ist möglich, daß es sich hier um einen Explosionsschauer gehandelt hat, doch läßt sich Sicheres hierüber aus der Photographie nicht entnehmen. Die von Blau und Wambacher beobachteten Sterne sind wohl in der natürlichsten Weise als Sekundärwirkung von Protonen oder Neutronen mittlerer Energien (10^8 bis $6 \cdot 10^8$ eV), als „Kernionisation", zu deuten. Entweder handelt es sich um Prozesse, bei denen ein schweres Teilchen dieses Energiebereichs von außen auf einen Atomkern trifft und in diesem die Sekundärteilchen auslöst; oder das schwere Teilchen wird durch ein von außen kommendes Lichtquant im Atomkern ausgelöst und erzeugt nun auf seinem Weg aus dem Atomkern heraus die Sekundärteilchen. Im letzteren Falle werden gleichzeitig mit den Protonen auch ein, oder wenn das einfallende Lichtquant energiereich genug ist, mehrere Anderson-Teilchen den Atomkern verlassen. Wir kommen damit wieder zu den explosionsartigen Prozessen, die naturgemäß meistens in einem Kern stattfinden und mit einer Emission von Sekundärteilchen aus dem betreffenden Kern verknüpft sind. Der Emission von Sekundärteilchen aus dem Kern wird auch in allen genannten Fällen noch eine Kernverdampfung von der Art, wie sie in der gewöhnlichen Kernphysik untersucht wird, folgen. Wenn man annehmen darf, daß die Explosionen der seltenere Vorgang sind, und es sich also bei den Blau-Wambacherschen Sternen meistens um Sekundärteilchen handelt, die von einem schweren Teilchen mittlerer Energie erzeugt sind, so läßt sich die Energieverteilung dieser Sekundärteilchen mit der theoretischen Energieverteilung von Formel (38) vergleichen. Die bisher vorliegenden Messungen scheinen mit der Theorie gut übereinzustimmen. Für die von Taylor beobachteten Haufen ist wohl einstweilen eine einigermaßen zuverlässige theoretische Deutung nicht möglich.

Literaturverzeichnis.

A 1. Ackemann, M.: Naturwiss. **22**, 169 (1934).
A 2. Anderson, C. D.: Physic. Rev. **44**, 406 (1933).
A 3. — and S. H. Neddermeyer: Physic. Rev. **50**, 263 (1936).
A 4. Arley, N.: Proc. roy. Soc. Lond. Im Erscheinen.
A 5. Auger, P.: Züricher Vortrag in der Sammlung „Kernphysik". Berlin: Julius Springer 1936.
A 6. — L. Leprince-Riguet et P. Ehrenfest: J. Physique et Radium **2**, 58 (1936).
A 7. — — Nature (Lond.) **133**, 138 (1934).
A 8. — C. r. Acad. Sci. Paris **206**, 346 (1938).
A 9. — et P. Ehrenfest jr.: J. Physique et Radium **5**, 204 (1937).

B 1. BAGGE, E.: Ann. Physik **30**, 72 (1937).
B 2. BARNÓTHY, J. u. M. FORRÉ: Z. Physik **104**, 744 (1937).
B 3. BETHE, H. u. W. HEITLER: Proc. roy. Soc. Lond. **146**, 83 (1934).
B 4. — Handbuch der Physik, Bd. 24, S. 1.
B 5. BHABHA, H. J. and W. HEITLER: Proc. roy. Soc. Lond. **159**, 432 (1937).
B 6. — Proc. roy. Soc. Lond. **164**, 257 (1937).
B 7. — Nature (Lond.) **141**, 117 (1938).
B 8. — Proc. roy. Soc. Lond. **166**, 501 (1938).
B 9. BLACKETT, P. M. S.: Proc. roy. Soc. Lond. **154**, 573 (1936).
B 10. — and J. G. WILSON: Proc. roy. Soc. Lond. **160**, 306 (1937).
B 11. — Proc. roy. Soc. Lond. **165**, 11 (1938).
B 12. — Proc. roy. Soc. Lond. **159**, 1 (1937).
B 13. — and J. G. WILSON: Proc. roy. Soc. Lond. **165**, 209 (1938).
B 14. BLAU, M. and H. WAMBACHER: Nature (Lond.) **140**, 585 (1937).
B 15. BLOCH, F.: Z. Physik. **81**, 363 (1933).
B 16. BÖGGILD, J. K.: Diss. Kopenhagen 1937.
B 17. BOWEN, J. S. and R. A. MILLIKAN: Physic. Rev. **53**, 217 (1938).
B 18. — — and H. V. NEHER: Physic. Rev. **52**, 80 (1937); **53**, 217 (1938).
B 19. BRODE, R. B. and M. A. STARR: Physic. Rev. **53**, 3 (1938).
B 20. BOTHE, W. u. W. GENTNER: Z. Physik **107**, 236 (1937).
B 21. BOHR, N.: Nature (Lond.) **137**, 344 (1936).
C 1. CARLSON, J. F. and J. K. OPPENHEIMER: Physic. Rev. **51**, 220 (1937).
C 2. CARMICHAEL, H.: Proc. roy. Soc. Lond. **154**, 223 (1936).
C 3. CLAY, J., A. VAN GERMERT and J. T. WIERSMA: Physica 7, 627 (1936).
C 4. CORSON, D. E. and R. B. BRODE: Physic. Rev. **53**, 773 (1938).
C 5. CRUSSARD, J. et L. LEPRINCE-RIGUET: C. r. Acad. Sci. Paris **204**, 243 (1937). — J. Physique et Radium 5, 214 (1937).
D 1. DRIGO, A.: Ricerca scient. **5**, 88 (1934).
E 1. EHMERT, A.: Z. Physik **106**, 751 (1937).
E 2. — Physik. Z. **38**, 975 (1937).
E 3. EHRENBERG, W.: Proc. roy. Soc. Lond. **155**, 532 (1936).
E 4. EHRENFEST, P. jr.: C. r. Acad. Sci. Paris **206**, 428 (1938).
E 5. EULER, H.: Physik. Z. **38**, 943 (1937).
E 6. — Naturwiss. **26**, 382 (1938). — Z. Physik. Im Erscheinen 1938.
F 1. FERMI, E.: Z. Physik **88**, 161 (1934).
F 2. FRÖHLICH, H. and W. HEITLER: Nature (Lond.) **141**, 37 (1938).
F 3. — — and N. KEMMER: Proc. roy. Soc. Lond. **166**, 154 (1938).
F 4. FÜNFER, E.: Z. Physik **83**, 92 (1933).
F 5. — Naturwiss. **25**, 235 (1937).
F 6. — Z. Physik **83**, 92 (1933).
F 7. FURRY, W. H.: Physic. Rev. **52**, 569 (1937).
F 8. FUSSELL, L.: Physic. Rev. **51**, 1005 (1936).
G 1. GEIGER, H.: Erg. exakt. Naturwiss. **14**, 42 (1935).
G 2. — u. O ZEILLER: Z. Physik **97**, 300 (1935).
G 3. — u. E. FÜNFER: Z. Physik **93**, 543 (1935).
G 4. — Züricher Vorträge, herausgeg. von E. BRETSCHER. Berlin: Julius Springer 1936.
G 5. — Physik. Z. **38**, 936 (1937).
G 6. GERBES, W.: Ann. Physik **23**, 648 (1935).
G 7. GROSS, B.: Z. Physik **83**, 214 (1933).
G 8. GEIGER, H. u. O. ZEILLER: Z. Physik **108**, 212 (1938).
H 1. HEIDEL, E.: Diss. Tübingen 1931.
H 2. HEISENBERG, W.: Ann. Physik **13**, 430 (1932).
H 3. — Z. Physik **101**, 533 (1936).
H 4. — Naturwiss. **25**, 749 (1937). — Sächs. Akad. Wiss. **89**, 369 (1937).

H 5. HEISEBBERG, W.: Ann. Physik **37**, 20 (1938).
H 6. HEITLER, W.: Nature (Lond.) **140**, 235 (1937).
H 7. — Proc. roy. Soc. Lond. **161**, 261 (1937).
H 8. — Proc. roy. Soc. Lond. **166**, 529 (1938).
H 9. — Theory of Radiation. Oxford: Clarendon-Press 1936.
H 10. HERZOG, G. u. P. SCHERRER: J. Physique et Radium **6**, 489 (1935).
H 11. — — Helvet. physic. Acta **8**, 514 (1935).
H 12. HOSEMANN, R.: Z. Physik **100**, 212 (1936).
H 13. HU CHIEN SHAN: Proc. roy. Soc. Lond. **158**, 581 (1937).
H 14. — Proc. roy. Soc. Lond. **161**, 95 (1937).
H 15. HUMMEL, J. N.: Naturwiss. **22**, 170 (1934).
J 1. JOHNSON, T. H.: Physic. Rev. **53**, 499 (1938).
J 2. ISHINO, M.: Philosophic. Mag. **32**, 202 (1916).
K 1. KEMMER, N.: Proc. roy. Soc. Lond. **166**, 127 (1938).
K 2. KOCKEL, B.: Z. Physik **107**, 153 (1937).
K 3. KULENKAMPFF, H.: Physik. Z. **35**, 996 (1934).
K 4. KUNZE, P.: Z. Physik **80**, 559 (1933).
K 5. KULENKAMPFF, H.: Verh. dtsch. physik. Ges. **1938** (Breslauer Vortrag).
L 1. LAMAÎTRE, G. and M. S. VALLARTA: Physic. Rev. **43**, 87 (1933).
L 2. LANDAU, L. and G. RUMER: Proc. roy. Soc. Lond. **166**, 213 (1938).
L 3. LEPRINCE-RIGUET, L. et J. CRUSSARD: J. Physique et Raduim **5**, 208 (1937).
M 1. MAASS, H.: Physik. Z. **35**, 858 (1934).
M 2. MESSERSCHMIDT, W.: Z. Physik **103**, 27 (1936).
M 3. MIEHLNICKEL, E.: Höhenstrahlung. Dresden u. Leipzig: Theodor Steinkopff 1938.
M 4. MONTGOMERY, C. G. and C. C. MONTGOMERY: Physic. Rev. **48**, 786 (1935).
M 5. — — Physic. Rev. **47**, 429 (1935).
M 6. MORGAN, J. E. and W. M. NIELSEN: Physic. Rev. **52**, 564 (1937).
N 1. NEDDERMEYER, S. H. and C. D. ANDERSON: Physic. Rev. **51**, 884 (1937).
N 2. NIE, H.: Z. Physik **99**, 453 (1936).
N 3. — Z. Physik **79**, 776 (1936).
N 4. NISHINA, Y., M. TAKEUCHI and T. ISHIMYA: Physic. Rev. **52**, 1198 (1937).
N 5. NORDHEIM, L. W.: Physic. Rev. **51**, 1110 (1937).
N 6. — Physic. Rev. **53**, 694 (1938).
P 1. PFOTZER, G.: Z. Physik **102**, 23 (1936).
P 2. PAULI, W. u. V. WEISSKOPF: Helvet. physic. Acta **1935**.
R 1. REGENER, E.: Nature (Lond.) **131**, 130 (1933).
R 2. — u. G. PFOTZER: Physik. Z. **35**, 779 (1934).
R 3. ROSSI, B.: Z. Physik **82**, 151 (1933).
R 4. — Z. Physik **33**, 304 (1932).
R 5. RUHLIG, A. J. and H. R. CRANE: Physic. Rev. **53**, 266 (1938).
R 6. RUMBOUGH, G. H. and G. L. LOCHER: Physic. Rev. **49**, 855 (1936).
S 1. SCHMEISSER, K. u. W. BOTHE: Naturwiss. **25**, 833 (1937). — Ann. Physik **32**, 161 (1938).
S 2. SCHWEGLER, A.: Z. Physik **96**, 62 (1935).
S 3. STARR, M. A. u. R. B. BRODE: Physic. Rev. **53**, 3 (1938).
S 4. SCHINDLER, H.: Z. Physik **72**, 625 (1931).
S 5. STEVENSON, E. C. and J. C. STREET: Physic. Rev. **49**, 26 (1936).
S 6. STEINKE, E. G.: Erg. exakt. Naturwiss. **13**, 89 (1934).
S 7. STÖRMER, C.: Z. Astrophys. **1**, 237 (1930).
S 8. — Z. Astrophys. Norwegen **2**, 4 (1937).
S 9. STREET, J. C. and E. C. STEVENSON: Physic. Rev. **47**, 891 (1935).

Theoretische Gesichtspunkte zur Deutung der kosmischen Strahlung.   69

S 10. STREET, J. C. and E. C. STEVENSON: Physic. Rev. **52**, 1003 (1937).
S 11. — and R. T. YOUNG: Physic. Rev. **46**, 823 (1934).
S 12. — — Physic. Rev. **47**, 572 (1935).
S 13. — R. H. WOODWARD and E. C. STEVENSON: Physic. Rev. **47**, 891 (1935).
S 14. STUHLINGER, E.: Z. Physik **108**, 444 (1938).
S 15. SCHOPPER, E.: Naturwiss. **25**, 557 (1937).
S 16. SWANN: Physic. Rev. **47**, 811 (1935).
T 1. TAMM, J. and D. IWANENKO: Nature (Lond.) **133**, 981 (1934).
T 2. TAYLOR: Nature (Lond.) **1938**.
T 3. TRUMPY: Norske Vidensk. Selsk. **37**, 137 (1938).
W 1. WEISCHEDEL, F.: Physik. Z. **36**, 796 (1935).
W 2. WEIZSÄCKER, C. F. VON: Z. Physik **88**, 612 (1934).
W 3. WENTZEL, G.: Naturwiss. **26**, 273 (1938).
W 4. WILLIAMS, E. J. and E. PICKUP: Nature (Lond.) **141**, 684 (1938).
W 5. — and F. R. TERROUX: Proc. roy. Soc. Lond. **126**, 289 (1930).
W 6. — Züricher Vorträge. Berlin: Julius Springer 1936.
W 7. WILSON, V. C.: Physic. Rev. **52**, 559 (1937); **53**, 337 (1938).
Y 1. YOUNG, R. T. and J. C. STREET: Physic. Rev. **46**, 823 (1934).
Y 2. — jr.: Physic. Rev. **52**, 559 (1937).
Y 3. — and J. C. STREET: Physic. Rev. **52**, 552 (1937).
Y 4. YUKAWA, H.: Proc. physic. math. Soc. Jap. **17**, 48 (1935).
Y 5. — and S. SAKATA: Proc. physic. math. Soc. Jap. **19**, 1084 (1937).
Y 6. — — Proc. physic. math. Soc. Jap. **19**, 712 (1937).
Y 7. — — and M. TAKETANI: Proc. physic. math. Soc. Jap. **20** (1938).
Z 1. ZEILLER, O.: Z. Physik **96**, 121 (1935).

hinzugefügt. Nach 4tägiger Gärung war Testosteron in einer Ausbeute von 81 % zu isolieren. Diese auf rein enzymatischem Wege verlaufene Umwandlung von Dehydroandrosteron in Testosteron bestätigt ferner die Möglichkeit der Auffassung von *Butenandt*, daß das Dehydroandrosteron eine Zwischenstufe bei der Entstehung des Testosterons im Organismus darstellen könnte.

Physikalisches Colloquium Hamburg

am 1. Dezember 1938.

Prof. Dr. W. Heisenberg, Leipzig: „*Das schwere Elektron (Mesotron) und seine Rolle in der Höhenstrahlung.*"

Einleitung. Die Höhenstrahlung läßt sich zerlegen in eine sogenannte „weiche" und „harte" Komponente von relativ geringer bzw. großer Durchdringungsfähigkeit. Die weiche Komponente wird durch einige cm Pb absorbiert, die harte dagegen vermag noch Koinzidenzen auszulösen in 2 Zählrohren, welche durch Bleischichten von 1 m und mehr getrennt sind. Die Absorption der weichen Komponente ist eng verknüpft mit dem Phänomen der „Schauer" und „Stöße". Dieser Erscheinungskomplex ist während der letzten Jahre durch enge Zusammenarbeit von Theorie und Experiment aufgeklärt worden[1]).

Die Entwicklung der Theorie führte weiter zu der Konsequenz, daß die durchdringende Komponente der Höhenstrahlung nicht aus energiereichen Elektronen oder γ-Quanten bestehen kann, denn es zeigt sich, daß die γ-Strahlung von großer Energie in der Materie eine sehr starke Absorption infolge von Pärchenbildung erfährt, — die Absorption durch *Compton*-Effekt, welche bei früheren Betrachtungen allein zugrunde gelegt worden war, ist oberhalb einiger Millionen Volt (MV) daneben ganz zu vernachlässigen —. Für schnelle Elektronen andererseits besteht beim Durchtritt durch Materie eine sehr große Wahrscheinlichkeit für Energieverluste infolge von Bremsstrahlung. Beide Teilchenarten können deshalb, auch bei extrem hohen Energien, nicht mehr als einige cm Pb durchsetzen.

Das neue Teilchen. Einen wesentlich geringeren Energieverlust beim Durchtritt durch absorbierendes Material würden Teilchen größerer Masse, etwa Protonen erleiden. Jedoch ist die Möglichkeit, die durchdringende Komponente auf Protonen zurückzuführen, dadurch ausgeschlossen, daß bei *Wilson*-Aufnahmen von Ultrastrahlungsteilchen viel zu wenig Protonen im Vergleich zur Intensität der harten Komponente gefunden werden. Das Verständnis dieser harten Komponente brachte erst eine Entdeckung *Anderssons*[2]), der in der Nebelkammer Bahnspuren von geladenen Teilchen beobachtete, die weder einem Elektron noch einem Proton zugehören konnten. Es läßt sich nämlich in der Nebelkammer einerseits durch die Krümmung im Magnetfeld der Impuls des Teilchens, andererseits durch die Tröpfchendichte (Ionenzahl) seine Geschwindigkeit feststellen. Aus beiden Daten errechnete *Andersson* die Masse zu etwa der 200fachen Elektronenmasse bzw. $^1/_{10}$ Protonenmasse[2]). Inzwischen wurden diese Teilchen auch von verschiedenen anderen Forschern beobachtet[4]). Ihre Existenz ist experimentell als gesichert anzusehen, die Ladung kann sowohl positiv als auch negativ sein. *Andersson* schlägt den Namen „Mesotron" für dieses Teilchen vor,

---

[1]) Vgl. *Euler* u. *Heisenberg*, Ergebn. exakt. Naturwiss. XVII, 1 [1938] (im Erscheinen); dort ausführliches Schrifttumverzeichnis; vgl. auch z. B. *Geiger* u. *Heyden*, diese Ztschr. **51**, 657 [1938] und die neue Arbeit von *Trumpy*, Z. Physik 111, 338 [1938].

[2]) *Andersson* u. *Neddermeyer*, Physic. Rev. **51**, 220 [1937]; **54**, 88 [1938].

[3]) Die Aufnahmen *Anderssons* gestatten daneben noch einige weitere Schätzungen der Masse, welche alle ungefähr denselben Wert liefern.

[4]) Die erste Aufnahme eines solchen Teilchens ist in der Arbeit von *Kunze*, Z. Physik 83, 1 [1933], Fig. 5 enthalten. Leider wurde damals die Konsequenz, daß es sich hier wirklich um ein neuartiges Teilchen handelt, noch nicht entschieden genug gezogen und ihre Tragweite für die Höhenstrahlung nicht verfolgt. Literatur über spätere Aufnahmen bei *Euler* u. *Heisenberg* l. c.

der weiterhin verwendet wird[5]). Partikel solcher Masse erfahren nun nach der Theorie, wie oben gesagt, einen so geringen Energieverlust beim Durchgang durch absorbierendes Material, daß sie durchaus für die durchdringende Komponente der Höhenstrahlung verantwortlich gemacht werden können. Daß sie bisher der Beobachtung entgangen waren, liegt wohl einerseits daran, daß sie in alten *Wilson*-Aufnahmen immer entweder für Protonen oder Elektronen gehalten wurden, andererseits an ihrer Instabilität, auf die wir nachher eingehen.

Zusammenhang mit der Kernphysik. Durch die Entdeckung dieses Teilchens ist für die Physik eine besonders reizvolle Situation geschaffen, als sich hier eine Verknüpfung zu ergeben scheint zwischen zwei ganz heterogenen Gebieten, nämlich der Höhenstrahlung einerseits, der Theorie der Wechselwirkungskräfte zwischen den Bausteinen des Atomkerns andererseits. Schon vor einigen Jahren hatte *Yukawa* darauf hingewiesen, daß man zu einer guten Beschreibung der Kernkräfte gelangen könnte, wenn man die Existenz eines Teilchens von etwa 100facher Elektronenmasse annimmt[6]). Sein Gedanke ist kurz folgender: Die Wechselwirkungskräfte zwischen Proton und Neutron unterscheiden sich von der elektrischen Wechselwirkung geladener Teilchen durch ihre kurze Reichweite; statt des Abfalls des elektrischen Potentials wie r^{-1} legen die Daten der Kernphysik einen Abfall des Potentials der Kernkräfte wie $r^{-1} \cdot e^{-r/\lambda}$ nahe, worin die „Reichweite" λ von der Größenordnung einiger 10^{-13} cm ist. Statt der Potentialgleichung $\Delta\varphi = 0$ genügt ein solches Kraftfeld der Gleichung $\Delta\varphi - \varphi/\lambda^2 = 0$. Zu den feineren Zügen der Kernkräfte gelangt man dann, wenn man zur Quantentheorie dieses Kernfeldes übergeht; und dabei ergibt sich, daß die Quanten dieses Feldes eine von Null verschiedene Ruhmasse haben müssen, von der Größenordnung $h/\lambda c$ — im Gegensatz zu den Quanten des elektromagnetischen Feldes, den Lichtquanten —. Setzt man für λ die experimentelle Reichweite der Kernkräfte ein, so ergibt sich als Masse des *Yukawa*schen Quants gerade etwa die 100fache Elektronenmasse. Wegen des Austauschcharakters der Kernkräfte muß außerdem das *Yukawa*sche Quant die elektrische Elementarladung (positiv oder negativ) besitzen.

Es liegt nun nahe, dies *Yukawa*-Quant mit dem *Andersson*-Mesotron zu identifizieren, und damit ergibt sich dann die Möglichkeit, aus seiner Rolle bei den Kernphänomen weitere Eigenschaften zu erschließen und diese Konsequenzen in der Höhenstrahlung zu verfolgen. Möglicherweise wird sich auch der umgekehrte Weg einschlagen lassen.

Instabilität und Lebensdauer des Mesotrons. Zunächst muß man zur Erklärung der Tatsache, daß das Mesotron nicht sonst in der Natur vorgefunden wird, wohl annehmen, daß es instabil ist[7]). *Yukawa* nahm an, daß es sich in ein gewöhnliches Elektron verwandeln könne und dabei den Massenüberschuß in kinetische Energie umsetze; aus Gründen der Erhaltung des Impulses und Drehimpulses[8]) muß dabei gleichzeitig — ganz wie beim β-Zerfall — ein ungeladenes leichtes Partikel, das Neutrino, entstehen. Mit dieser Vorstellung ergibt sich die Möglichkeit einer neuen Interpretation des β-Zerfalls: Ein β-aktiver Kern emittiert zunächst — virtuell[9]) — ein *Yukawa*-Quant und dieses zerfällt weiter in Elektron und Neutrino. Der Wert dieser Theorie liegt darin, daß sie

---

[5]) In der Literatur vorläufig gebrauchte Bezeichnungen waren: neues Teilchen, schweres Elektron, Yukon, Baritron, Penetron; der Name Mesotron (Mittellage zwischen Proton und Elektron) wurde kürzlich von *Andersson* vorgeschlagen.

[6]) *Yukawa*, Proc. physic.-math. Soc. Japan **17**, 48 [1935]; **19**, 1084 [1937]; eine leicht verständliche zusammenfassende Darstellung s. *Wentzel*, Naturwiss. **26**, 273 [1938].

[7]) Ein stabil geladenes Teilchen wäre sicherlich nicht während der letzten 30 Jahre der Beobachtung entgangen.

[8]) Aus der Natur der Kernkräfte folgt, daß das *Yukawa*sche Quant ganzzahligen Spin besitzt, beim Zerfall in ein Elektron mit halbzahligem Spin muß dann gleichzeitig ein zweites Teilchen mit halbzahligem Spin entstehen. Die Verhältnisse liegen ganz ebenso wie beim β-Zerfall.

[9]) Virtuell deshalb, weil beim β-Zerfall nur Energien von wenigen mV umgesetzt werden; zur wirklichen Erzeugung des *Yukawa*-Quants mit der Masse von etwa 100 mV reicht die Energie nicht aus. Deshalb muß die Entstehung des Quants und sein nachfolgender Zerfall bei den β-Zerfalls-Prozessen als ein einziger Akt angesehen werden. Diese Vorstellungsweise ist in der Quantentheorie geläufig.

gestattet, nunmehr aus den Daten der β-Radioaktivität auf die Lebensdauer des freien Mesotrons — wie es in der Höhenstrahlung vorkommt — zu berechnen. Sie ergibt sich zu einigen 10^{-6} s.

Nunmehr gibt es einige Experimente mit der Höhenstrahlung, die als eine direkte Messung der Lebensdauer des Mesotrons zu interpretieren sind, und die tatsächlich eine Bestätigung der *Yukawa*schen Abschätzung (etwa $3 \cdot 10^{-6}$ s) ergeben.

Die direkteste Messung ist von *P. Ehrenfest*[10]) durchgeführt worden. Er bestimmte durch Differenzmessungen die Intensität eines bestimmten Energiebereiches der harten Komponente (Anteil, welcher mindestens 40 cm Pb und höchstens 60 cm Pb durchsetzt), einmal auf dem Jungfraujoch und einmal den entsprechenden Energiebereich in Paris. Der Höhenunterschied ist etwa 3 km; die Mesotronen, die sich bei dieser Energie nahezu mit Lichtgeschwindigkeit bewegen, brauchen bis Paris herunter etwa 10^{-5} s länger. Aus dem Intensitätsunterschied ergibt sich dann die Anzahl der inzwischen zerfallenen Mesotronen und daraus ihre Lebensdauer zu $3{,}4 \cdot 10^{-6}$ s.

Eine andere Berechnungsmöglichkeit bieten die Messungen von *Ehmert*[11]). Hier mußten die Mesotronen in 2 Versuchen eine gleiche Menge Material praktisch gleicher Zusammensetzung durchlaufen, so daß der gewöhnliche Energieverlust durch Ionisation der gleiche ist. Jedoch brauchten sie in beiden Fällen verschieden lange Zeit. Nämlich eine Koinzidenzanordnung — 2 Zählrohre, dazwischen 50 cm Pb — wurde einmal 30 m tief in senkrechter Lage im Bodensee angebracht und die Koinzidenzen gezählt, und zum Vergleich wurde in Höhe des Wasserspiegels mit solcher Neigung der Apparatur die Messung wiederholt, daß die mehr durchlaufene Luftschicht der Wassersäule von 30 m entsprach. Im zweiten Fall brauchten die Mesotronen eine dem längeren Weg entsprechend längere Zeit. Aus dem Vergleich der Intensitäten läßt sich wieder die Lebensdauer bestimmen. Durch verschiedene Wahl der Wassertiefe und entsprechender Neigungswinkel ließen sich die Versuchsbedingungen abändern. Dabei ergab sich das zunächst überraschende Resultat, daß die Lebensdauer scheinbar proportional zur Energie der Mesotronen ist. Dies ist jedoch eine unmittelbare Konsequenz des Relativitätsprinzips — und zugleich eine augenfällige Bestätigung desselben. — Die Lebensdauer ist für die Mesotronen verschiedener Energie sicherlich die gleiche in dem jeweiligen Bezugssystem, in dem das Mesotron ruht. Der ruhende Beobachter, gegen den die Mesotronen bewegt sind, muß dann die relativistische Zeitdilatation in Rechnung ziehen, und diese ist gerade $= E/mc^2$, (E = Energie, m = Ruhmasse des Mesotrons). Bei Berücksichtigung dieses Umstandes[12]) berechnet sich dann die Lebensdauer des Mesotrons zu ungefähr $3 \cdot 10^{-6}$ s, in ausgezeichneter Übereinstimmung mit dem *Ehrenfest*schen Ergebnis und *Yukawa*s theoretischer Schätzung.

Auf den zitierten *Andersson*schen und *Kunze*schen Aufnahmen ist neben dem Mesotron auch die Bahnspur eines Elektrons zu sehen, dessen Energie und geometrische Lage relativ zur Spur des Mesotrons die Deutung zuläßt bzw. nahelegt, daß dieses Elektron durch Zerfall aus dem Mesotron entstanden ist.

Entstehung des Mesotrons. Da das Mesotron eine so geringe Lebensdauer besitzt, kann es natürlich nicht aus dem Weltraum kommen, sondern muß in der oberen Atmosphäre erzeugt werden, wobei die Energie aus der „primären" kosmischen Strahlung geliefert wird. Diese primäre Komponente wird wahrscheinlich zum überwiegenden Teil aus sehr energiereichen (über 10^{10} V) Elektronen und Positronen und evtl. auch aus γ-Quanten bestehen, d. h. aus energiereicher, aber weicher[13])

Strahlung. Dies ist im Einklang mit dem starken Anstieg der Gesamtstrahlung (*Regener, Pfotzer*) mit der Höhe und ihrem Maximum in der oberen Atmosphärenschicht bei etwa 10 cm Hg Luftdruck. Übrigens hatten schon *Millikan* und Mitarbeiter[14]) aus einer Reihe von Daten geschlossen, daß die harte Komponente nicht primär sei, sondern in der oberen Atmosphäre aus der weichen Komponente entstehen muß.

Die *Yukawa*sche Theorie gibt nun auch die Möglichkeit, wenigstens eine Abschätzung der Wahrscheinlichkeit für die Entstehung von Mesotronen beim Durchtritt schneller Elektronen durch Materie und gestattet damit, das Verhältnis der weichen zur harten Komponente der Höhenstrahlung in der oberen Atmosphäre und auch an der Erdoberfläche zu berechnen. Tatsächlich läßt sich darnach das gesamte experimentelle Material — einstweilen erst mehr qualitativ — verstehen. Wir dürfen deshalb nun wohl mit Recht von der quantitativen Durchführung dieser Ansätze die Bestätigung erhoffen, daß wir nunmehr — mit den Entdeckungen der letzten Jahre, vor allem der des Positrons der Pärchenerzeugung und schließlich auch des Mesotrons und seiner endlichen Lebensdauer — den Schlüssel zum vollen Verständnis der komplexen Natur der Höhenstrahlung gewonnen haben.

In der *Aussprache* betonte Heisenberg noch folgende Gesichtspunkte.

1. Ein wichtiger Zug der *Yukawa*schen Ansätze, den sie mit der alten Theorie des β-Zerfalls gemeinsam haben[15]), ist, daß bei Energien, welche groß gegen die Ruhmasse des Mesotrons (100 MV) sind, eine große Wahrscheinlichkeit dafür besteht, daß in einem einzigen Elementarakt nicht nur ein Teilchen erzeugt wird, sondern eine größere Anzahl[16]). Dieses Phänomen wird wohl mit dem von *Bothe* und *Schmeiser*[17]) beschriebenen „harten" Schauern im Zusammenhang stehen. Das gleichzeitige Entstehen mehrerer Teilchen in einem Elementarakt ergibt sich im Formalismus der Quantentheorie beim Überwiegen der höheren Näherungen gegenüber den niedrigeren, und das bedeutet, daß wir uns bei diesen Prozessen an der Grenze des durch die heutige Quantenmechanik beschreibbaren Bereichs befinden, und daß hier diejenigen Abänderungen an der Quantentheorie wesentlich werden, welche die Berücksichtigung einer universellen kleinsten Länge vermutlich mit sich bringen wird[18]). Diese Energiekonzentrationen, bei denen solche neuen Erscheinungen zu erwarten sind, lassen sich derzeit nicht entfernt im Laboratorium erzeugen, und wir sind deshalb, im Hinblick auf das experimentelle Material, an dem wir uns bei der genannten Fortentwicklung der Quantentheorie orientieren können, für lange Zeit ganz auf den betreffenden Erscheinungskomplex in der Höhenstrahlung angewiesen.

2. Die Erzeugung der Mesotronen geschieht nach der *Yukawa*schen Theorie innerhalb oder in der Nähe eines Atomkerns. Dabei wird zugleich dem Kern im allgemeinen eine hohe Anregungsenergie übertragen, der Kern wird „heiß", und es werden nachträglich mehrere Kernbausteine „verdampfen". Auf diese Prozesse werden die von *Blau* und *Wambacher*[19]) und *Andersson* beobachteten „Kernzertrümmerungen" durch die Höhenstrahlung zurückzuführen sein.

[14]) *Bowen, Millikan* u. *Neher*, Physic. Rev. **53**, 217 [1938].
[15]) *Heisenberg*, diese Ztschr. **49**, 691 [1936]. Z. Physik **101**, 533 [1936].
[16]) Im Gegensatz zur kaskadenartigen Entstehung der weichen Schauer vgl. z. B. *Geiger* u. *Heyden*, diese Ztschr., l. c.
[17]) *Bothe* u. *Schmeiser*, Ann. Physik **32**, 161 [1938].
[18]) *Heisenberg*, ebenda **32**, 20 [1938]; Z. Physik **101**, 251 [1938].
[19]) *Wambacher*, diese Ztschr.; erscheint demnächst (Bericht über den Physikertag in Baden-Baden).

Kaiser Wilhelm-Institut für physikalische Chemie und Elektrochemie

Colloquium am 8. November 1938, Berlin-Dahlem.

E. Hückel, Marburg: „*Die Mesomerievorstellung und einige ihrer Anwendungen in der organischen Chemie*"[1]).

Ausgehend von den klassischen Strukturformeln der organischen Chemie und den Versuchen, die Konstitution und das Verhalten dieser Verbindungen durch eine chemische Formel zu beschreiben (Beispiel: Benzolformeln), entwickelte der Vortragende die Vorstellungen, zu denen die neuere Quantentheorie für die Konstitution dieser Verbindungen geführt hat. Nach dieser kann ein Zustand bestimmter Energie

[1]) S. auch *E. Hückel* Z. Elektrochem. angew. physik. Chem. **43**, 752, 849 [1937] bzw. Abdruck davon: *E. Hückel*: Grundzüge der Theorie ungesättigter u. aromatischer Verbindungen, Verlag Chemie, Berlin 1938.

[10]) Noch unveröffentlichte Ergebnisse.
[11]) *Ehmert*, Z. Physik. **106**, 751 [1937]; vgl. auch *Rossi*, Nature, London **142**, 993 [1938].
[12]) Wenn man alle Überlegungen in dem Bezugssystem, in welchem das Mesotron ruht, durchführen wollte, so müßte man die entsprechende *Lorentz*-Kontraktion der Erdatmosphäre in Rechnung setzen; es ergibt sich dann natürlich das gleiche Resultat. Das zerfallende Mesotron ist also das einfachste Beispiel der im Zusammenhang mit dem Relativitätsprinzip viel diskutierten „bewegten Uhr".
[13]) Es sei nochmals betont, daß die übliche Bezeichnung harte bzw. weiche Komponente sich nicht auf die Energie der Partikeln bezieht, sondern auf ihre geringe bzw. große Absorbierbarkeit.

gehend stellte er die Blochsche Metalltheorie dar, die durch die Mitberücksichtigung der Wärmebewegung des Ionengitters eine Erklärung für die endliche freie Weglänge der Elektronen liefert. Besonders eingehend behandelte er die Fragen der elektrischen und thermischen Leitfähigkeit insbesondere das Wiedemann-Franzsche Gesetz; unter Verzicht auf eine Wiedergabe des üblichen mathematischen Apparats stellte er die anschaulichen physikalischen Ursachen des elektrischen und thermischen Widerstandes heraus. Zum Schluß berührte er noch kurz die Frage der Supraleitung.

Der letzte Vortragende, Hr. W. Heisenberg, Leipzig, sprach über die Elementarteilchen und die kosmische Strahlung. Er beschäftigte sich zunächst mit der Yukawaschen Theorie der schweren Elektronen, die ursprünglich aufgestellt wurde, um die Kernkräfte zwischen Protonen und Neutronen zu erklären. In ähnlicher Weise wie die Coulombschen Kräfte zwischen geladenen Teilchen durch das elektromagnetische Feld der Lichtquanten vermittelt werden, werden die Kernkräfte durch das Feld der schweren Elektronen oder, wie heute gebräuchlich, der Mesotronen, verursacht. Aus der Reichweite der Kernkräfte kann die Masse der Mesotronen entnommen werden. Der Nachweis der schweren Elektronen in der Höhenstrahlung wird durch die Wilson-Spuren dieser Teilchen erbracht, die bei gleicher Bahnkrümmung deutlich von den Spuren der leichten Elektronen und der Protonen zu unterscheiden sind. Die starke Absorption der Mesotronen in der Atmosphäre läßt sich nicht durch den gewöhnlichen Bremsprozeß bei der Ionisierung der Luft erklären. Vielmehr werden hierzu Zerfallsprozesse der Mesotronen benötigt, wie sie auch nach den Vorstellungen von Yukawa vorkommen sollten; und zwar erweist sich die mittlere Lebensdauer der schweren Elektronen ihrer Größenordnung von 10^{-6} sec nach als mit der Yukawaschen Theorie in Einklang. Interessant in diesem Zusammenhang ist der Umstand, daß sich hierbei die Zeitdilatation der speziellen Relativitätstheorie in einer Abhängigkeit der Lebensdauer von der Energie äußert, wie sich aus den für verschiedene Einfallsrichtungen gemessenen Energiespektren der Höhenstrahlung ergibt.

Im Anschluß an die Nachmittagssitzung fand ein festliches Abendessen statt, bei dem Hr. E. Lange dem Jubilar seine Ernennung zum Ehrenmitglied der Physikalisch-medizinischen Sozietät der Universität Erlangen überbrachte. Den Höhepunkt des Abends bildete die von musikalischen Darbietungen eingerahmte Festrede von Hrn. M. Planck, die allen Hörern im Gedächtnis bleiben wird.

Das Fest wurde beschlossen durch lustige physikalische Vorführungen, die von neuem bezeugten, daß die Physik in der Vollendung aus einer reizvollen Mischung von Wissenschaft und Kunst besteht.

wurden jedoch auch Zertrümmerungen beobachtet, bei denen α-Teilchen hoher Energie und schwerere Kerntrümmer emittiert wurden. Die auf die Trümmerteilchen übertragene Energie beträgt bei manchen Prozessen mehr als 100 Millionen Volt. Die Zahl der beobachteten Kernprozesse nimmt mit der Höhe stark zu und beträgt in der Stratosphäre (mittlere Höhe 18000 m) 0,1 pro Stunde und cm² in einer Schicht von 25 μ Dicke, hinter 1 mm Paraffin.

16. Hr. W. Heisenberg: Verwandlung der kosmischen Strahlung in der Atmosphäre.

Die kosmische Strahlung scheint primär in der Hauptsache aus Elektronen und Positronen zu bestehen, die vom Weltenraum nahezu isotrop auf die Erde fallen. Die Anzahl der im Weltenraum auftretenden Elektronen oberhalb der Energie E wird offenbar in einem weiteren Energiegebiet näherungsweise durch die Formel $const/E^{\gamma}$; $\gamma \approx 1,9$ beschrieben. Beim Eindringen in die Atmosphäre führt diese Primärstrahlung zur Ausbildung von Kaskaden, die das von Regener und Pfotzer beobachtete Maximum der Ionisation bei etwa 80 mm Quecksilberdruck erklären. In tieferen Atmosphärenschichten wird diese Strahlung schnell absorbiert, so daß ihr Beitrag zu der Ultrastrahlung auf Meeresniveau bereits verschwindend gering geworden ist. Ob dieser Primärstrahlung auch Protonen, Neutronen und Lichtquanten beigemischt sind, läßt sich aus den bisherigen Experimenten nicht mit Sicherheit entscheiden. Daß ihr Mesotronen beigemischt sind, ist nicht anzunehmen, da die Mesotronen längst radioaktiv zerfallen wären, bevor sie die Erde erreicht haben. Die durchdringende Komponente der Höhenstrahlung, die aus Mesotronen besteht, wird vielmehr aus der weichen Komponente, wahrscheinlich bevorzugt aus den Lichtquanten, in der Atmosphäre erzeugt. Aus der Ähnlichkeit des Spektrums der durchdringenden Komponente mit dem der primären Strahlung folgt, daß der Wirkungsquerschnitt für die Erzeugung der Mesotronen bei hohen Energien nahezu von der Energie unabhängig ist. Die Mesotronen werden teils durch Ionisation, teils durch radioaktiven Zerfall absorbiert Beide Prozesse führen zur Entstehung von Elektronen, die als weiche Komponente im Gleichgewicht mit der durchdringenden Komponente diese bis in große Tiefen begleitet. Die weiche Komponente auf Meeresniveau ist also in der Hauptsache Folgeprodukt der harten Komponente. Die Prozesse, in denen die Mesotronen entstehen, sind wahrscheinlich als

Explosionsprozesse aufzufassen, die mit Kernverdampfung verbunden sind. ₂₂ Die Häufigkeit der von Blau-Wambacher, Schopper und anderen gefundenen Kernzertrümmerungen paßt ungefähr zum Wirkungsquerschnitt für die Erzeugung der Mesotronen. Ein erheblicher Teil der Hoffmannschen Stöße muß ebenfalls mit diesen Explosionen in Verbindung gebracht werden, und das gleiche gilt wohl für die von Kulenkampff diskutierten häufigen Protonenspuren in Wilsonkammer-Messungen.

De atoomkern en hare samenstelling

door W. Heisenberg *)

Historische ontwikkeling.

Wanneer men uitgaat van de hypothese, dat de atoomkernen zijn opgebouwd uit eenvoudige elementaire deeltjes, nl. protonen en neutronen, dan is het opvallend hoeveel overeenkomst er bestaat tusschen de ,,kernmaterie'' en gewone materie. In de eerste plaats is het volume der atoomkernen ongeveer evenredig met hare massa, dus met het aantal elementaire bouwsteenen, zoodat de kernen ongeveer een constante dichtheid vertoonen. Verder is de energie, die benoodigd is om één elementair deeltje uit de kern te ver-

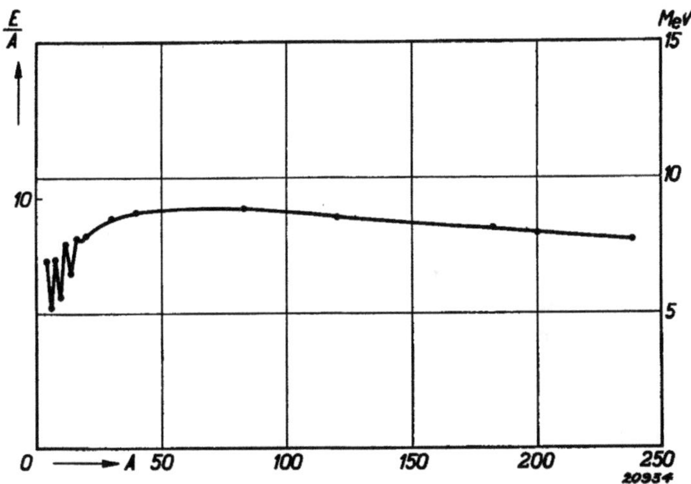

Fig. 1. Bindingsenergie, per elementair deeltje, der atoomkernen als functie van het aantal A der samenstellende deeltjes (links in 0,001 at. gew. een-heden, rechts in MeV). Door de best bekende punten is een vloeiende lijn getrokken. De periodiciteit voor A < 20 is door een zigzaglijn schematisch aangeduid.

wijderen, in eerste benadering constant (fig. 1). Men kan dus de kern vergelijken met een vloeistofdruppel, waarvan de protonen en

*) Deze voordracht, gehouden op 27 Januari 1939 voor het natuurkundig colloquium in het Laboratorium der Philipsfabrieken, werd met toestemming van Prof. H e i s e n-b e r g voor dit tijdschrift bewerkt door Dr. W. d e G r o o t.

— 89 —

neutronen de „moleculen" zijn. De constante energie, benoodigd voor het verwijderen van een elementair deeltje kan men vergelijken met een „verdampingswarmte". Een verdere conclusie, die uit het vloeistofdruppelbeeld volgt, is het bestaan van een *oppervlakte-spanning*, waardoor de geringere bindingsenergie per deeltje der lichte kernen, vergeleken met de zwaardere kernen, verklaard kan worden. Evenals bij een vaste stof of bij een vloeistof krachten tusschen de atomen en tusschen de moleculen aanwezig zijn, wier werkingssfeer ongeveer even groot is als gemiddelde afstand van 2 atomen of 2 moleculen $(2.10^{-8}$ cm) moet men aannemen, dat in de kernvloeistof tusschen de samenstellende deeltjes, dus in het bijzonder tusschen een neutron en een proton, krachten heerschen, wier werkingsfeer van dezelfde grootte-orde is als hun gemiddelde afstand in de kern $(\sim 2.10^{-13}$ cm). Deze krachten moeten t.o.v. beide deeltjes symmetrisch zijn, want de stabiele kernen bevatten ongeveer evenveel protonen als neutronen, afgezien van de verschuiving die dit evenwicht, ten gunste der neutronen, ondergaat tengevolge van de electrostatische afstooting bij de zware kernen.

Een verdere eigenschap van deze krachten moet zijn, dat ze evenals de chemische valentiekrachten *verzadiging* vertoonen, zoodat één deeltje slechts aan één ander of aan een eindig aantal andere gebonden kan zijn.

Was dit niet zoo, dan zou men als resultaat vinden dat de dichtheid van de kernvloeistof niet constant zou zijn, doch met toenemend aantal samenstellende deeltjes toe zou nemen. In gewone materie zijn weliswaar de attractiekrachten van Van der Waals tusschen de moleculen van de vloei-

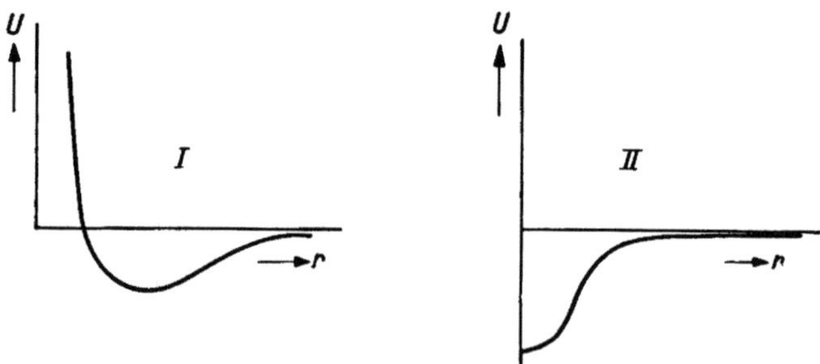

Fig. 2. I) Potentiaalkromme met aantrekkende en afstootende krachten (van der Waals-attractie- en afstootingspotentiaal tusschen de moleculen van een vloeistof).
II) Potentiaalkromme zonder afstootingstak, zooals men gewoonlijk ter verklaring van de kernkrachten aanneemt.

stof niet aan verzadiging onderhevig, doch in dat geval wordt het constant zijn van de dichtheid, onafhankelijk van de hoeveelheid der beschouwde vloeistof, veroorzaakt door de gelijktijdige aanwezigheid van afstootende krachten. Wanneer men tusschen de protonen en neutronen ook afstootende krachten toeliet (fig. 2) dan zouden de aantrekkende krachten hier evenmin verzadiging behoeven te vertoonen. Het is echter gebleken, dat de onderstelling dat er geen speciale afstootende krachten bestaan en dat de aantrekkende krachten verzadiging vertoonen, om hare eenvoudigheid de voorkeur verdient.

Daar men de chemische valentiekrachten verklaren kan door een quantummechanische plaatsruil-wisselwerkingsintegraal (der valentie-electronen), noemt men in analogie hieraan ook de verzadigbare krachten tusschen de kerndeeltjes plaatsruil-krachten (Austauschkräfte).

Een nadere studie van de krachten tusschen protonen en neutronen en tusschen protonen (en neutronen) onderling, (vooral naar aanleiding van experimenten van T u v e, H a f s t a d en H e yd e n b u r g over de strooiing van protonenstralen in waterstof, waarbij zekere afwijkingen van de wet van Coulomb werden geconstateerd) heeft aangetoond, dat deze krachten nog van de onderlinge orientatie der magnetische momenten (spin-vectoren) afhankelijk zijn.

Men vindt op deze wijze in principe vier verschillende krachtfuncties, die men op een of andere wijze met de vier mogelijkheden: identieke of niet identieke deeltjes, gecombineerd met parallele of antiparallele spinvectoren, in verband kan brengen (fig. 3). Met

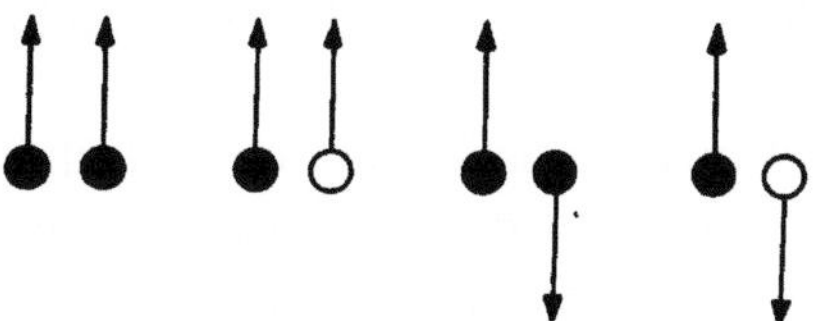

Fig. 3. De vier mogelijkheden, welke aan de vier soorten wisselwerkingskrachten tusschen elementaire deeltjes ten grondslag liggen (● proton, ○ neutron, ↑ spin).

behulp van dergelijke onderstellingen kan men de energie der eenvoudigste kernen, $_1D^2$, $_1T^3$, $_2He^4$, het deuton, het triton (waterstof-

Fig. 4. Model der vijf eenvoudigste kernen, 1 neutron, 2 proton, 3 deuton, 4 triton, 5 alphadeeltje.

isotopen) en het alphadeeltje, bevredigend weergeven (fig. 4).
Daarbij blijkt, dat de resultaten weinig afhankelijk zijn van de
vorm, die men voor de potentiaalfunctie aanneemt, zoodat het
vrijwel uitgesloten is deze vorm uit de massadefecten te bepalen.
Men kan dus van een willekeurige, met toenemende afstand snel
afnemende functie uitgaan (bijv. rechthoekige potentiaalput:
$U = - A$ voor $r \leqq a$, $U = 0$ voor $r > a$, of functie van G a u s s:
$U = - A\, e^{-r^2/a^2}$). De diameter $(2a)$ en de diepte (A) van de
potentiaalput kan men dan uit de massa-defecten der genoemde
kernen nauwkeurig bepalen.

Kernmodellen.

De behandeling van de uit een groot aantal deeltjes bestaande
,,zware'' kernen op de boven aangegeven grondslag is mathematisch een zeer moeilijk vraagstuk. Men moet zich hier voorloopig
tevreden stellen met eenvoudige modellen. De eenvoudigste onderstelling, waarbij men de gezamenlijke werking van alle andere
deeltjes op een gegeven deeltje vervangt door een enkel potentiaalveld en de kern dus denkt als een potentiaalkuil, waarin de deeltjes
zich onafhankelijk van elkaar bewegen (evenals men in de theorie
der spectra wel eens de werking van de atoomromp op de valentie-
electronen, door een enkel samengesteld potentiaalveld vervangt)
is wel in staat om sommige verschijnselen qualitatief weer te geven,
maar in andere opzichten faalt zij. In het bijzonder stelt zij niet
in staat de samenhang van de kern quantitatief te verklaren.

Eenige verbetering in de opvatting van de structuur der zware
kernen bracht de theorie van B o h r. Deze wees er op, dat ook
een relatief snel deeltje in de kernmaterie slechts een zeer beperkte
dracht heeft. Dit beteekent dus dat de energie-uitwisseling tusschen
de elementaire kerndeeltjes een zeer *snel* proces is. In het bijzonder
is uit nauwkeurige berekeningen gebleken, dat een proton om een

Fig. 5. *a*) Doordringen van een zeer energierijk deeltje door een *Pb*-kern
volgens B o h r.

b) Samentreffen tusschen een zware kern en een deeltje van relatief geringe energie.

Pb-kern $(N + Z = 207)$ geheel te doordringen, gemiddeld een energie zou moeten hebben van meer dan 200 MeV. Een deeltje van middelmatige energie zal dus, bij botsing met een kern, daarin blijven steken en zijn energie door botsingen met de aanwezige kernbestanddeelen aan deze overdragen (fig. 5). In het vloeistof-druppelbeeld kan men dit beschrijven als een toename van de *temperatuur* van de kern. Resulteert uit het samentreffen van kern- en projectiel een kernreactie, dan kan men dit als een *verdampings-proces* opvatten.

Worden dus eenerzijds door het in acht nemen der botsingen onze voorstellingen over het mechanisme der kernprocessen verhelderd, een betere verklaring van de samenhang van de kern krijgt men zoo evenmin. Men kan trachten verder te komen door, zooals door E u l e r is uitgewerkt, aan de kern een structuur toe te kennen. Blijft men voorloopig in het vloeistofdruppelbeeld, dan kan men hierop soortgelijke beschouwingen toepassen als in de electrolyt-theorie van D e b ij e, waarbij in de buurt van een positief ion de kans om een negatief ion aan te treffen grooter is dan die om een tweede positief ion te vinden en omgekeerd (evenals, volgens een populair beeld van D e b y e, in een balzaal, ondanks de op een afstand schijnbaar homogene vermenging, bij nader beschouwen rondom een dame de concentratie der heeren die der dames over-treft en omgekeerd). Men kan dit ook beschrijven als de neiging tot vorming van proton-neutron-paren en dubbele paren (alpha-deeltjes) (fig. 6).

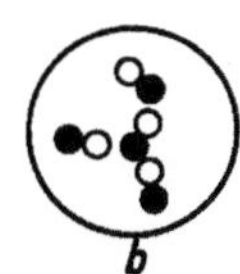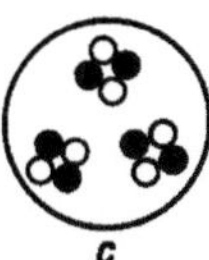

Fig. 6. Schematische voorstelling van de ontwikkelingstrappen van het vloeistofdruppel-model.
a) eenvoudige vloeistofdruppel.
b) paarvorming.
c) alphadeeltjes-vloeistof.

Een verdere soortgelijke stap is een model, waarbij men zich voor-stelt, dat de aanwezige neutronen en protonen reeds zooveel moge-lijk tot alphadeeltjes zijn gecondenseerd, zoodat deze in hoofdzaak als elementaire bouwsteenen optreden. Hierbij wordt de bindingsenergie reeds dadelijk vrij aanzienlijk, omdat men die van het alphadeeltje als gegeven kan beschouwen, maar het blijkt toch onmogelijk om uit de bestaande voorstellingen een binding *tusschen* de alphadeeltjes af te leiden, die voldoende zou zijn om de bestaande massadefecten der zware kernen te verklaren. In het bijzonder gelukt het reeds niet, zooals door W a t a n a b e is aangetoond, de bindingsenergie van He^5, waarvan de kern kan geacht worden te bestaan uit één alphadeeltje plus een neutron, op grond van de bestaande voorstellingen te berekenen.

In een nieuwe phase kwam de theorie der kernkrachten door beschouwingen van den Japanner Y u k a w a. Deze stelde zich ten doel, uitgaande van de potentiaal der kernkrachten, het *veld* te berekenen en dus voor deze krachten dezelfde stap te doen die in het electromagnetisme voert van de wet van C o u l o m b naar de vergelijkingen van M a x w e l l. Evenals het meest algemeene electromagnetische veld een stralingveld is en men daarin de energie gedragen kan denken door stralingscorpuskels (stralingsquanta, photonen), zoo werd ook Y u k a w a bij zijn berekeningen gevoerd tot een (in dit geval geladen) deeltje, dat als drager van de energie van het kernveld kon optreden. Dit deeltje zou, om de gevonden kernkrachten (werkingsstraal 2.10^{-13} cm, diepte van de potentiaalkuil 35 MeV) te kunnen verklaren, een massa moeten hebben gelijk aan ongeveer 200 electronen-massa's en een lading van één elementairlading. Zulke deeltjes zijn nu in de laatste jaren in de cosmische straling gevonden. Daardoor is deze theorie, die eerst slechts een hypothetisch karakter had, op experimenteele basis komen te staan.

Volgens de theorie van Y u k a w a is het waarschijnlijk, dat behalve de gewone plaatsruil-wisselwerking tusschen twee deeltjes, er nog krachten aanwezig zijn, die door de samenwerking van 3 of meer deeltjes ontstaan. Men kan dit ook zoo uitdrukken, dat de kracht tusschen twee deeltjes nog van de aanwezigheid van andere deeltjes in de nabijheid afhankelijk is. Men heeft tot heden toe nog geen poging gedaan, deze invloed in rekening te brengen. Zeer waarschijnlijk zou dit, vooral bij de zware kernen, zeer groote mathematische moeilijkheden meebrengen. Men is daarom bij de behandeling der zware kernen voorloopig aangewezen op benaderingsmethoden, waarbij men dan weer uitgaat van een of ander aanschouwelijk beeld.

Onlangs is het nu aan W e f e l m e i e r gelukt verschillende eigenschappen van de kern beter te verklaren, door deze als een uit alphadeeltjes opgebouwd *kristal* op te vatten. Zijn beschouwingen wijken in twee opzichten af van het boven besproken alphadeeltjes-vloeistofmodel. In de eerste plaats tracht W e f e l m e i e r niet de kracht tusschen de alphadeeltjes te berekenen, maar hij leidt deze empirisch af uit de experimenteele massadefecten en verder neemt hij de alphadeeltjes stilstaande aan. (De overtollige neutronen en eventueel protonen, denkt hij zich over tusschenruimten verdeeld). Deze voorstelling is niet alleen aantrekkelijk door hare eenvoud, maar levert bovendien quantitatief bevredigende resultaten op.

Men kan bijvoorbeeld de energie der lichte kernen, die alleen uit alphadeeltjes zijn samengesteld, weergeven, door te tellen hoeveel aanrakingen tusschen alphadeeltjes voorkomen, wanneer men zich deze voorstelt als kleine bollen, die zoo dicht mogelijk opeengepakt

Fig. 7. De kernen van Be^8, C^{12}, O^{16} en Ne^{20} volgens het alphadeeltjes-kristal-model van W e f e l m e i e r.

zijn (fig. 7), en men de (kinetische) nulpuntsenergie per deeltje als een negatief energiebedrag in rekening brengt. Men vindt dan:

kern	$_4Be^8$	$_6C^{12}$	$_8O^{16}$	$_{10}Ne^{20}$
aantal α-deeltjes	2	3	4	5
energie	a—$2c$	$3a$—$3c$	$6a$—$4c$	$9a$—$5c$

Daarbij stelt a de bindingsenergie per contactplaats voor en c de kinetische nulpuntsenergie. Stelt men nu, op grond van het experimenteele feit, dat de Be^8-kern zeer weinig stabiel is en gemakkelijk in twee alphadeeltjes uiteenvalt, $a = 2c$, dan vindt men voor de verhouding der bindingsenergieën van de andere drie kernen:

	C^{12}		O^{16}		Ne^{20}	
theorie	1,5	:	4	:	6,5	
experiment	8		16		21 $\times$ 10^{-3} at. gew. eenh.	

Verder volgt uit deze theorie, dat elementen waarbij de alpha-

deeltjes in de kern een bijzonder symmetrische configuratie vormen,
bijzonder stabiel moeten zijn. Dit is het geval bij 4 deeltjes (tetra-
ëder) en bij 13 (fig. 8). Inderdaad zijn de elementen met kernlading
$4 \times 2 = 8\ (O)$, $13 \times 2 = 26\ (Fe)$ de meest voorkomende in de
natuur.

Een ander verschijnsel, dat goed past in de theorie van W e f e l-
m e i e r en dat met de vloeistofdruppeltheorie niet zoo gemakke-

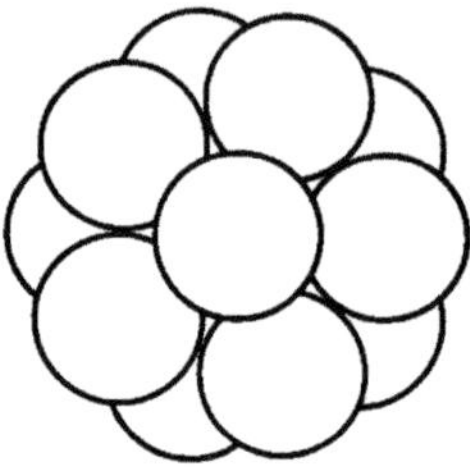

Fig. 8. *Fe*-kern volgens W e f e l m e i e r.

lijk te verklaren zou zijn, is het voorkomen van kernen met quadru-
pool-momenten. Een quadrupoolmoment beteekent een afwijking
in de ladingsverdeeling van de bolsymmetrie en wel is experimen-
teel het teeken zoo, dat men zou moeten aannemen, dat de kernen
met quadrupoolmoment langgerekt, dus eivormig zijn. Denkt men
zich aan een zeer symmetrische kern (b.v. de *Fe*-kern) een alphadeeltje toegevoegd, dan geeft dit reeds een asymmetrische ladingsverdeeling. Een tweede toegevoegd alphadeeltje zal door de vrij
aanzienlijke aantrekking de electrostatische afstooting overwinnen
en zich naast het eerste leggen, enz. Zoo groeit uit de bol een

Fig. 9. Ontstaan van een kern met quadrupoolmoment uit een symmetrische
kern, volgens W e f e l m e i e r.

eivormig lichaam (fig. 9). De neiging tot het aannemen van deze
vorm wordt door de electrostatische afstooting van de rest van de

kern nog bevorderd. Men kan zeggen, dat de uiteindelijke vorm van de kern bepaald wordt door het antagonisme tusschen de oppervlakte-spanning, die naar de bolvorm streeft en de electrostatische afstooting, die een langgerekte vorm bevoordeelt.

Een zeer sterk argument ten gunste van deze voorstelling is, zooals door W e f e l m e i e r en W e i z s ä c k e r is opgemerkt, te voorschijn gekomen uit het werk van H a h n en S t r a s s- m a n n over de transmutatie van uranium door neutronenbestraling. Het was vroeger reeds opgevallen, dat na de bestraling van uranium met neutronen een aantal radioactieve producten ontstaan, die alleen β-radioactiviteit vertoonen. Bij het uiteenvallen ontstaan dus producten met kernlading hooger dan 92. Door de β-emissie neemt het aantal protonen naar verhouding toe en daardoor zou, volgens de tot heden bestaande opvattingen, de α-instabiliteit eveneens moeten toenemen. Toch heeft men nimmer emissie van α-deeltjes gevonden. Dit resultaat bewijst min of meer, dat de electrostatische afstooting bij de transuranen een merkwaardig geringe invloed heeft. Nu is uit vele zorgvuldige proeven gebleken, dat onder de uiteenvallings-producten van de uraniumkernen een *barium*-isotoop voorkomt. De voorstelling ligt voor de hand, dat de kern U^{239} (die uit U^{238} ontstaat door opname van één neutron) en de producten, die door β-emissie hieruit ontstaan, zoo langgerekt zijn, dat ze instabiel worden en uiteen kunnen vallen tot een kern met massa ~ 139 en een andere met massa ~ 100. De mogelijkheid is dus niet uitgesloten (hoewel dit nog nader zal moeten worden onderzocht), dat niet alleen, tegelijk met de zoogenaamde transuranen, producten van veel lager atoomgewicht optreden, maar ook dat vele radioactieve producten, die tot heden voor transurane kernen werden gehouden, in werkelijkheid kernen van veel geringere massa zijn. Alleen zorgvuldige chemische proeven kunnen hier nader uitsluitsel brengen.

Indien dergelijke transmutaties mogelijk zijn, waarbij een kern in twee stukken uiteenvalt, is dit óok daarom nog belangrijk, omdat tengevolge van de kromming van het gebied der stabiele kernen in het N-Z-diagram de uiteenvallings-producten noodzakelijk ver buiten de stabiele zône komen te liggen en er dus zeer instabiele kernen kunnen ontstaan, met een korte uiteenvaltijd en een intensieve, harde β-stralen emissie.

98   W. HEISENBERG, DE ATOOMKERN EN HARE SAMENSTELLING

LITERATUUR.

H. E u l e r, Naturwissenschaften **25**, 201, 1937. Z. Phys. **105, 553**, 1937.
W. W e f e l m e i e r, Naturwissenschaften **25, 525**, 1937. Z. Phys. **107, 332**, 1938.
H. Y u k a w a, Proc. phys. math. Soc. Japan **17, 48**, 1935.
O. H a h n und F. S t r a s s m a n n, Naturwissenschaften **27, 11**, 1939.

Voor vroegere overzichten, betreffende kernstructuur, in Nederlandsche tijdschriften, zie
G. J. S i z o o,   G. E. U h l e n b e c k,   de Ingenieur **52**, E, 47, 56, 1937.
W. d e G r o o t, Philips' Tech. T. **2, 97**, 1937. Chem. weekblad **84**, 2, 1937.

Na de eerste publicatie van H a h n en S t r a s s m a n n zijn in de literatuur talrijke bevestigingen van hun ontdekking verschenen, waarvan wij noemen:
O. H a h n und F. S t r a s s m a n n, Naturwissenschaften **27, 89**, 1939.
F. J o l i o t, I. C u r i e, P. S a v i t c h, C. R. Acad. Sci., Paris, **202, 341, 343**, 1939.
O. R. F r i s c h, Nature, London, **143, 276**, 1939.

Theoretische beschouwingen verschenen van de hand van
L i s e M e i t n e r and O. R. F r i s c h, Nature, London, **143, 239**, 1939.
N. B o h r, Nature, London, **143, 330**, 1939.
C. F. v. W e i z s ä c k e r, Naturwissenschaften **27, 133**, 1939.

JULY–OCTOBER, 1939   REVIEWS OF MODERN PHYSICS   VOLUME 11

IV. Production of Secondary Radiation

On the Theory of Explosion Showers in Cosmic Rays.*

W. Heisenberg
Leipzig, Germany

THE theory of Yukawa contains interaction terms between mesotrons and nuclear particles, which lead to cross sections for the scattering of mesotrons by nuclear particles that increase with increasing energy of the mesotrons. These interaction terms, therefore, permit an application of ordinary quantum mechanics only for small energies of the particles concerned, as, for example, in interaction of heavy particles in the nucleus.

For high energies of the particles one is restricted to the use of the wave equations in the sense of a correspondence principle. Such considerations lead to the result that the collision of particles of very high energy can often lead to the production of many mesotrons in one single act. This result can be illustrated by the theory of electronic radiation given by Bloch and Nordsieck. When a particle of very high energy collides with a proton, then a certain part of the mesotron proper field of the proton is shaken off in the collision. This part of the field would correspond to a small number of mesotrons, which increases only logarithmetically with the energy of the incident particle. In the propagation of the field, however, a strong scattering of mesotrons by mesotrons and a production of mesotron pairs by mesotrons takes place, and can be treated in a way similar to that used for the scattering of photons by photons in quantum electrodynamics. In this way the production of a large number of mesotrons is to be expected.

It can also be shown that the use of the classical equations together with the correspondence principle lead to results for the scattering of mesotrons by protons that are entirely different from those of the quantum-mechanical calculation. This is due to the fact that the proper field of the nuclear particle, which is neglected in the quantum theoretical formulas, provides an additional inertia for the motion of the nuclear particle, which decreases the scattering cross section considerably.

The most general result of such correspondence considerations consists in the conclusion that in the limit of extremely high energies of the incident particles the total cross sections for the production of other particles will tend to a finite value (if no long range forces of the Coulomb type are present). This result can also be expressed in the following way: since in the collision of any two particles other particles can always be produced, the forces between these two particles always have a finite range. So, in spite of the fact that in quantum electrodynamics, for example, the interaction of photons with electrons is assumed to have an infinitely small range (δ-type of force), the virtual possibility of pair production will lead to a finite range for the force between photon and electron. Therefore, the cross sections at very high energies also remain finite and do not decrease to zero. This fact may be important in the discussion of the question as to whether the measurement of position or of distance can be performed with arbitrary accuracy.

---

* Cf. W. Heisenberg, Zeits. f. Physik **113**, 61 (1939) for complete paper.

Bericht über die allgemeinen Eigenschaften der Elementarteilchen[1]

von **W. Heisenberg**, Leipzig

II. Die Wechselwirkung der Elementarteilchen

a) Die Forderungen der Relativitätstheorie und der Quantentheorie bestimmen das Verhalten der Elementarteilchen so weitgehend, daß nur eine beschränkte Anzahl von einfachen Möglichkeiten für die Bewegungsgleichungen freier Teilchen angegeben werden kann, von denen auch ein erheblicher Teil in der Natur verwirklicht zu sein scheint. Sobald man jedoch die Wechselwirkung der Teilchen berücksichtigen will, so stehen die Aussichten für eine wirkliche Theorie dieser Wechselwirkungen viel ungünstiger. Zwar können rein formal die Forderungen von Relativitätstheorie und Quantentheorie noch befriedigt werden, aber jede derartige Theorie scheint zu divergenten Resultaten zu führen. Die Gründe hierfür lassen sich leicht angeben: Wegen der Vertauschungsrelationen für die Wellenfunktionen hat der Wert einer Wellenfunktion in einem bestimmten Raum-Zeitpunkt (im Gegensatz zum Mittelwert über ein Gebiet) unendliche Eigenwerte. Jede nichtlineare Wellengleichung, die neben einer Wellenfunktion auch das Produkt zweier oder mehrerer Wellenfunktionen am gleichen Ort enthält, ist also gewissermaßen eine Differentialgleichung mit unendlichen Koeffizienten. In einer nichtrelativistischen Wellentheorie tritt diese Schwierigkeit nicht notwendig auf. Denn dort kann die Wechselwirkung durch Fernwirkungskräfte hervorgerufen werden, so daß die Wellenfunktion an einem bestimmten Ort in der Differentialgleichung mit einem *Integral* über die Wellenfunktion (mal Potentialfunktion) über ein Raumgebiet multipliziert ist. (Vgl. die Klein-Jordan-Wignersche Theorie der Materiewellen.) In einer relativistischen Theorie jedoch, in der nur Nahewirkungen vorkommen, scheint die Divergenz unvermeidlich. Man trifft also hier auf eine Situation, die derjenigen ganz ähnlich ist, die die Physik beim Versuch einer Vereinigung von Thermodynamik und Maxwellscher Theorie antraf. Auch der Versuch, die Gibbs-Boltzmannschen Methoden auf das Strahlungsfeld anzuwenden, führte zu divergenten Resultaten. Diese Divergen-

---

[1] This review, which Heisenberg finished in the beginning of June 1939, had been intended to form the second half of a joint report with Wolfgang Pauli, entitled 'Bericht über die allgemeinen Eigenschaften der Elementarteilchen' ('Report on the General Properties of Elementary Particles'), for the 8ᵉ Conseil de Physique (8th Solvay Conference) in Brussels, 22–29 October 1939. Due to war conditions, however, the conference was cancelled. Heisenberg's two chapters 'Die Wechselwirkung der Elementarteilchen' and 'Die Grenzen der bisherigen Theorie', represented Parts II and III of the report, while Pauli treated in Part I the general properties of free particles. Pauli later published an improved form of this contribution as 'Relativistic Field Theories of Elementary Particles' in Rev. Mod. Phys. **13**, 203–232 (1941). Heisenberg's parts, published here for the first time, and Pauli's article constitute a consistent status report on elementary particle theory at the end of the thirties. (Editors)

zen beruhten, wie wir heute wissen, darauf, daß in den Naturgesetzen eine universelle Konstante von der Dimension einer Wirkung vorkommt, auf die damals noch nicht Rücksicht genommen worden war. Ähnlich dürften die Divergenzen in der Quantentheorie der Wellen darauf beruhen – wie schon von verschiedenen Forschern, z. B. Born, March, Wataghin, Heisenberg ausgesprochen wurde –, daß in den Naturgesetzen eine universelle Konstante von der Dimension einer Länge vorkommt, die bisher nicht in der richtigen Weise berücksichtigt werden kann. Für freie Teilchen kann sich aus Invarianzgründen eine solche Konstante kaum anders als im Auftreten einer Ruhmasse äußern, so daß hier ihr Einfluß klar übersehen werden kann. Bei der Wechselwirkung liegen jedoch die Verhältnisse wesentlich komplizierter.

Zunächst leuchtet ein, daß jeder Wechselwirkungsansatz nur im Sinne einer korrespondenzmäßigen Analogie gemeint sein kann. Denn er bedeutet stets: Es gibt Fragen, für die man die Wellenfunktionen durch Mittelwerte über gewisse endliche Raum-Zeitgebiete ersetzen und für die man das, was in den kleinsten Raum-Zeitgebieten geschieht, summarisch durch eine Ruhmasse der Teilchen ausdrücken kann. Die Wechselwirkung muß also auch notwendig als eine „kleine Störung" in dem Sinne aufgefaßt werden können, daß die durch sie behandelbaren Vorgänge von der Frage nach der Masse der Teilchen getrennt studiert werden können. Wo dies nicht der Fall ist, kann vielleicht durch korrespondenzmäßige Behandlung der Selbstenergie und der mit ihr zusammenhängenden Fragen noch eine qualitative Übersicht gewonnen werden, aber von einer richtigen Anwendung der Quantentheorie kann dort einstweilen wohl keine Rede sein.

Zur weiteren Untersuchung dieser Frage ist es zweckmäßig, alle in der Theorie vorkommenden Größen in „natürlichen Einheiten" zu messen. Da die Forderungen der Relativitätstheorie und Quantentheorie die Voraussetzungen für jede Theorie bilden, ist es natürlich, $\hbar$ und c als Einheiten zu benützen. Diese Annahme läuft darauf hinaus, alle Größen durch Multiplikation mit Potenzen von $\hbar$ und c auf die Dimension: Potenz einer Länge zu bringen, was stets auf eine einzige Weise möglich ist. Z. B. führt man statt der Energie ein: $\varepsilon = E/\hbar c$ (cm^{-1}), statt der Feldstärke E: $e = E/\sqrt{\hbar c}$ (cm^{-2}) usw. Man kann dann zwei Arten von Wechselwirkungen unterscheiden.

1) Das Wechselwirkungsglied in der Hamiltonfunktion enthält außer den betreffenden Wellenfunktionen nur noch einen dimensionslosen Zahlenfaktor Z. Dann muß dieser Zahlenfaktor $Z \ll 1$ sein, damit beim gegenwärtigen Stand der Theorie die Einführung dieser Wechselwirkung überhaupt mit einem klaren physikalischen Sinn verbunden werden kann [1].

2) Das Wechselwirkungsglied enthält außer den Wellenfunktionen eine Konstante von der Dimension: Potenz einer Länge. In diesem Fall wird die Wechselwirkung nur dann, wenn sie sich auf Teilchen geringer Energie bezieht, als kleine Störung betrachtet werden können. (Der Wechselwirkungsausdruck erhält also seinen physikalischen Sinn durch das Verhalten von Teilchen geringer Energie.) Für Teilchen hoher Energie kann dann zunächst aus dem Wechselwirkungsausdruck nichts entnommen werden, weil dort die Frage der Wechselwirkung von der Frage nach der Masse der Teilchen nicht mehr getrennt werden kann, weil also dort quantentheoretische Methoden einstweilen versagen.

Wenn in einer Theorie von der ersten Art Elementarteilchen mit Ruhmassen vorkommen, so muß dies immer bedeuten, daß diese Teilchen auch noch andere Wechselwirkungen der zweiten Art ausüben können (um die Sprache der klassischen Theorie zu brauchen: daß es Kräfte gibt, die Teilchen einer bestimmten Größe zusammenhalten). Auch umgekehrt haben die Ruhmassen zusammen mit den Wechselwirkungen erster Art unter Umständen zur Folge, daß in höheren Näherungen indirekt Wechselwirkungen der zweiten Art entstehen. Man kann also wohl schließen: Im allgemeinen haben wir stets mit der Existenz von Wechselwirkungen der zweiten Art zu rechnen. Es kann aber große Bereiche geben, in denen von diesen Wechselwirkungen (bis auf ihre Auswirkung in den Ruhmassen) abgesehen werden kann.

b) Das einfachste Beispiel für eine Theorie der ersten Art bietet die Quantenelektrodynamik. In ihr ist die Wechselwirkung durch die Sommerfeldsche Konstante $e^2/\hbar c$ bestimmt, die für viele Zwecke als $\ll 1$ angesehen werden kann. Diese Theorie kann wohl, soweit sie Vorgänge mit nicht zu hoher Energie- und Impulsübertragung betrifft (vgl. III), im wesentlichen als abgeschlossen betrachtet werden. Die in sie eingehenden Unklarheiten, die von den Vorgängen mit hoher Impulsübertragung oder in ganz kleinen Raum-Zeitgebieten herrühren, dürften sich, wegen der Kleinheit von $e^2/\hbar c$, in den Resultaten nur in Unsicherheiten von höchstens einigen Prozent äußern.

Ein besonders interessanter Zug der Quantenelektrodynamik besteht in der von Dirac entdeckten Möglichkeit der Paarerzeugung aus Lichtquanten und der Zerstrahlung von Positron-Elektronpaaren. Diese auch experimentell sichergestellten Vorgänge haben zur Folge, daß man zweckmäßig in der Theorie nicht mehr von der Wellenfunktion eines einzelnen Teilchens ausgeht — denn die Teilchenzahl ist ja zeitlich nicht konstant —, sondern von der Diracschen Dichtematrix, die eine Darstellung des Zustandes auch beim Vorhandensein vieler Teilchen ermöglicht. Wenn man konsequent die Diracsche Dichtematrix benützt, so kann man die physikalisch unbefriedigende Vorstellung der Diracschen „Löcher" vermeiden und die Theorie von vornherein symmetrisch in Bezug auf das Ladungsvorzeichen formulieren. (Für praktische Zwecke wird allerdings die „Löcher"vorstellung manchmal geeigneter sein.)

Die genauere Ausarbeitung dieses Programms hat zu folgenden Ergebnissen geführt [2]: Man geht aus von der Matrix $(x''t''k'' \,|\, R_S \,|\, x't'k')$, die der Diracschen Gleichung: [2]

$$\mathscr{H} R_S = \left[i\hbar \frac{\partial}{c\,\partial t'} + \frac{e}{c} A_0(x') + \alpha_s \left(i\hbar \frac{\partial}{\partial x_s'} - \frac{e}{c} A_s(x') \right) + \beta mc \right]$$

$$\cdot (x''t''k'' \,|\, R_S \,|\, x't'k') = 0 \tag{1}$$

(und entsprechend für die zweigestrichenen Indizes) genügt und die in der Löchertheorie durch

---

[2] In Eq. (1) the α_s, s = 1, 2, 3, denote the spin matrices, and A_0 and A_s the four components of the electromagnetic potential. There is a summation over twice occurring indices s. Note also that the Latin indices assume the values 1, 2, 3 and the Greek indices (used below) assume the value 0, 1, 2, 3. (Editors)

$$(x''t''k'' \,|\, R_S \,|\, x't'k') = \left(\sum_{\text{bes. Zust.}} (n) - \sum_{\text{unbes. Zust.}} (n) \right) \psi_n^*(x''t''k'') \, \psi_n(x't'k') , \quad (2)$$

in der Quantentheorie der Wellen durch

$$(x''t''k'' \,|\, R_S \,|\, x't'k') = \tfrac{1}{2} [\psi^*(x''t''k'') \, \psi(x't'k') - \psi(x't'k') \, \psi^*(x''t''k'')] \quad (3)$$

definiert wird. Aus dieser Matrix, die für $x'' \to x'$, $t'' \to t'$ singulär wird, bildet man eine neue Matrix

$$r = R_S - S , \tag{4}$$

wobei S eine bestimmte Funktion der äußeren Felder A_s, F_{rs} bedeutet. Summiert man über griechische Indizes von 0 bis 3, setzt

$$A_0 = -A^0, \quad ct = x_0 = -x^0, \quad x_\lambda = x'_\lambda - x''_\lambda, \quad \xi_\lambda = \frac{x'_\lambda + x''_\lambda}{2},$$

so wird

$$S = \exp \left(\frac{e}{\hbar c} \, \mathrm{i} \int_{P'}^{P''} A^\lambda dx_\lambda \right) \left[S_0 + \frac{a}{x_\lambda x^\lambda} + b \log \left| \frac{x_\lambda x^\lambda}{C} \right| \right] . \tag{5}$$

Hierin bedeutet S_0 den Wert von R_S im feldfreien Vakuum, ferner ist

$$a = \frac{e}{48 \pi^2 \hbar c} x_\varrho x^\sigma \alpha^\lambda \left(\frac{\partial \dot{F}_{\lambda\sigma}}{\partial \xi_\varrho} - \delta_\lambda^\varrho \frac{\partial F_{\tau\sigma}}{\partial \xi_\tau} \right) + \frac{e^2 \mathrm{i}}{96 \pi^2 \hbar^2 c^2} x_\varrho x_\sigma x^\tau \alpha^\varrho F^{\mu\sigma} F_{\mu\tau} ,$$

$$b = \frac{e}{48 \pi^2 \hbar c} \alpha^\lambda \frac{\partial F_{\tau\lambda}}{\partial \xi_\tau} - \frac{e^2 \mathrm{i}}{48 \pi^2 \hbar^2 c^2} x_\lambda \alpha^\mu (F_{\tau\mu} F^{\tau\lambda} - \tfrac{1}{4} \delta_\mu^\lambda F_{\tau\sigma} F^{\tau\sigma}) , \tag{6}$$

$$C = 4 \left(\frac{\hbar}{\gamma mc} \right)^2 .$$

γ ist die Eulersche Konstante $\gamma = 1{,}781 \dots$.

Ladungsdichte, Energiedichte usw. können dann in bekannter Weise aus der Matrix r ermittelt werden. Z. B. erhält man für die Stromdichte $s_\lambda(\xi)$:

$$s_\lambda(\xi) = -e \sum_{k'k''} \alpha_{k'k''}^\lambda (\xi k' \,|\, r \,|\, \xi k'') \quad \text{usw.} \tag{7}$$

Die durch diese Gleichungen skizzierte Theorie ist insofern formal sehr unbefriedigend, als der singuläre Charakter der Matrix S, insbesondere bei ihrer Verwendung in der Quantentheorie der Wellenfelder, stets zu umständlichen Grenzübergängen ($x' \to x''$ usw.) zwingt. Aber auch diese Unschönheit hängt offensichtlich damit zusammen, daß in der Quantenelektrodynamik die universelle Konstante von der Dimension einer Länge noch nicht in richtiger Weise berücksichtigt ist. Denn wenn durch die richtige Berücksichtigung dieser Länge die Matrix S für $x' \to x''$ regulär würde, so wäre das Auftreten der subtraktiven Matrix S nur der naturgemäße Ausdruck für die Möglichkeit der Paarerzeugung. Denn sie bewirkt, daß an Stelle der homogenen Diracgleichung $\mathscr{H} R_S = 0$ die inhomogene Gleichung

$$\mathscr{H} r = - \mathscr{H} S \tag{8}$$

tritt. Die Inhomogenität ist für die Paarerzeugung maßgebend.

Der aus der Diracschen Theorie folgende Wirkungsquerschnitt für die Paarerzeugung ist experimentell bestätigt worden: am eindrucksvollsten wohl durch die Kaskadenschauer in der kosmischen Strahlung, die von Oppenheimer, Carlson, Bhabha und Heitler auf Grund der Quantenelektrodynamik erklärt werden konnten.

Darüber hinaus liefert die oben skizzierte Theorie noch verschiedene interessante Ergebnisse, die bisher noch nicht experimentell nachgeprüft werden konnten:

1) Die Gleichung (8) führt zu einer „Polarisation des Vakuums". In der Umgebung einer elektrischen Ladung wird durch das Feld dieser Ladung neue Ladung induziert, so daß z. B. in Gebieten der Ordnung $\hbar/mc$ um die Ladung herum Abweichungen vom Coulombschen Feld auftreten. Diese Polarisationswirkungen können in der Weise ausgedrückt werden, daß man die Maxwellschen Gleichungen durch Zusatzglieder – die allerdings Raum-Zeitintegrale enthalten – modifiziert [3]. Wenn die auftretenden Felder klein sind verglichen mit einem kritischen Feld der Stärke $|E| = m^2 c^3/e\hbar$, so tritt z. B. an die Stelle der Gleichung

$$\operatorname{div} E = 4\pi s_0$$

die modifizierte Gleichung

$$\operatorname{div} E(r, t) - \frac{e^2}{\hbar c} \int dr' c \, dt' f(r, t; r', t') \cdot \square\square \operatorname{div} E(r', t') = 4\pi s_0(r, t) , \qquad (9)$$

wobei

$$f(r, t; r', t') = \frac{1}{32\pi^5} \int_0^{\pi/2} d\psi \cos^3\psi \int dk \, dk_0 \cdot \exp\{i[k \cdot (r - r') - k_0(t - t')]\}$$

$$\cdot (k^2 - k_0^2)^{-2} \cdot \log\left(1 + \frac{1}{4}\left(\frac{\hbar}{mc}\right)^2 \cos^2\psi \cdot (k^2 - k_0^2)\right) . \qquad (10)$$

Diese Abänderung hat unter anderem zur Folge, daß bei der Wechselwirkung zwischen zwei Ladungen Zusätze zum Coulombschen Kraftfeld auftreten, die kleine Termverschiebungen in den Atomspektren und ähnliche Korrektionen an den Streuformeln verursachen (Uehling l.c.).

2) Auch wenn keine Ladungen vorhanden sind, führt Gleichung (8) zu Abweichungen von der Maxwellschen Theorie [4], die dann merklich werden, wenn die Felder nicht mehr sehr klein gegen das kritische Feld $|E_k| = m^2 c^3/e\hbar$ sind. Für den Fall, daß die Felder räumlich und zeitlich langsam veränderlich sind ($|\partial E/\partial x| \ll mc/\hbar |E|$), können diese Abweichungen durch die Lagrangefunktion

$$\mathscr{L} = \frac{1}{2}(E^2 - B^2) + \frac{e^2}{\hbar c} \int_0^\infty e^{-\eta} \frac{d\eta}{\eta^3}$$

$$\cdot \left\{ i\eta^2 (E \cdot B) \frac{\cos[(\eta/|E_k|)\sqrt{E^2 - B^2 + 2i(E \cdot B)}] + \text{konj.}}{\cos[(\eta/|E_k|)\sqrt{E^2 - B^2 + 2i(E \cdot B)}] - \text{konj.}} \right.$$

$$\left. + |E_k|^2 + \frac{\eta^2}{3}(B^2 - E^2) \right\} \qquad (11)$$

dargestellt werden. Die Maxwellschen Gleichungen verlieren also durch die Möglichkeit der Paarerzeugung ihren linearen Charakter, es tritt eine gegenseitige Streuung der Lichtwellen auf, die allerdings bisher nicht experimentell nachgeprüft werden konnte. Die Wechselwirkung zwischen Strahlung und Materie, die durch die Konstante $e^2/\hbar c$ charakterisiert ist, hat also nicht nur eine Wechselwirkung zwischen Elektronen, sondern auch eine zwischen Lichtquanten zur Folge. Die Elektronen können einander auf beliebig große Abstände nach dem Coulombschen Gesetz beeinflussen. Die Lichtquanten stehen nach (11) nur in Wechselwirkung, solange sie am gleichen Ort sind. Diese letztere Form der Wechselwirkung wird aber durch die Näherungsmethode, bei der $|\partial E/\partial x| \ll mc/\hbar|E|$ angenommen wird, vorgetäuscht. Der Vergleich von (9) und (11) lehrt, daß Lichtquanten auch in endlichen Abständen eine Kraft aufeinander ausüben können, deren Reichweite (bei Energien, die nicht groß gegen mc^2 sind) ungefähr durch $\hbar/mc$ gegeben sein muß.

Dieser Sachverhalt wird in interessanter Weise beleuchtet durch ein Argument von Wick [5], nach dem die Reichweite einer Kraft, die durch Teilchen der Masse μ vermittelt wird, den Wert $\hbar/\mu c$ nicht übersteigen kann: Der Abstand der beiden Partikeln, die in Wechselwirkung treten sollen, sei r. Dann müssen also mindestens während der Zeit c/r, die ein Teilchen mit Lichtgeschwindigkeit braucht, um von der einen Partikel zur anderen zu kommen, Teilchen der Masse μ vorhanden sein. Das Vorhandensein solcher Teilchen widerspricht freilich dem Energiesatz, da ja die Energie μc^2 zur Erzeugung eines solchen Teilchens nicht verfügbar ist. Aber der Energiesatz braucht nur bis auf die durch die Ungenauigkeitsrelationen gegebenen Fehler zu gelten. Diese Fehler sind hier $\Delta E \sim \hbar/\tau = \hbar c/r$, also ist die Wechselwirkung möglich, wenn

$$\mu^2 c \lesssim \Delta E \sim \hbar c/r, \quad \text{d.h.} \quad r \lesssim \frac{\hbar}{\mu c}. \tag{12}$$

Kräfte unendlicher Reichweite, wie die Coulombschen, können also nur durch Teilchen ohne Ruhmasse vermittelt werden. Die Kräfte zwischen den Lichtquanten haben die Reichweite $r \sim \hbar/mc$.

Gleichung (11) zeigt übrigens auch, daß aus der Wechselwirkung erster Art durch die Ruhmassen der Teilchen indirekt wieder eine solche der zweiten Art hervorgerufen wird. Doch gibt es einen großen Energiebereich, in dem die letzteren Wechselwirkungen nur eine sehr geringe Rolle spielen.

c) Das interessanteste Beispiel für eine Wechselwirkung der zweiten Art bietet die Yukawasche Theorie des Mesotrons. Wenn man die aus der Erfahrung gewonnenen Ansätze für die Wechselwirkungen der Kernbausteine durch das Yukawasche Feld erklären will, so muß man dieses Feld (analog zum elektromagnetischen Feld) als vektorielles Feld einführen (d. h. Mesotronen vom Spin 1 $\hbar$ annehmen) und in die Wechselwirkung des Yukawafeldes mit den Kernbausteinen Glieder aufnehmen, die dimensionsmäßig eine Länge enthalten (im Gegensatz zur Maxwellschen Theorie). Da die Einzelheiten der Yukawaschen Theorie in einem anderen Bericht von Heitler[3] besprochen werden sollen, mag es für das Fol-

---

[3] Walter Heitler had been invited to prepare a report on mesons, or 'mesotrons', as they were called at that time. (Editors)

gende genügen, ein vereinfachtes Modell der Yukawaschen Theorie zu Grunde zu legen, das auch von Bhabha [6] behandelt worden ist: Man ersetzt die Mesotronen durch neutrale Teilchen gleicher Masse, sieht vom Austauschcharakter der Kräfte ab und behält nur die einer Länge proportionale Wechselwirkung bei.

Die Lagrangefunktion dieser vereinfachten Theorie lautet dann:

$$\mathcal{L} = \frac{1}{2}(f^2 - g^2) - \frac{\varkappa^2}{2}(u_0^2 - u^2) + \mathrm{i}\,\psi^* \frac{\partial \psi}{\partial t} - \mathrm{i}\,\psi^* \alpha_k \frac{\partial}{\partial x_k}\,\psi - \psi^* \beta \psi K, \qquad (13)$$

wobei

$$f_k = -\dot{u}_k - \frac{\partial u_0}{\partial x_k} - \mathrm{i}\,l\,\psi^* \beta \alpha_k \psi,$$

$$g_k = \operatorname{rot}_k u - l\,\psi^* \beta \sigma_k \psi \qquad (14)$$

gesetzt wird. Hierin bedeuten u_0, u bzw. f, g die in „natürlichen" Einheiten gemessenen Yukawaschen Feldgrößen; die Funktionen u_0, u und ψ sollen in (13) variiert werden, wobei die f und g aus (14) durch die u_0, u und ψ ausgedrückt werden. $\varkappa$ und K sind (in den gleichen Einheiten) die Ruhmassen von Mesotron und Proton. l ist die Wechselwirkungskonstante von der Dimension einer Länge.

In einem Ansatz von der Art der Gleichungen (13) und (14) ist die Wechselwirkung nur dann eine kleine Störung, wenn es sich um Teilchen geringer Energie handelt ($\varepsilon < 1/l$). In der Yukawaschen Theorie ist l von der Größenordnung $1/\varkappa = \hbar/\mu_{\mathrm{Mes}}c \sim 2 \cdot 10^{-13}$ cm. Der Wechselwirkungsansatz soll also Auskunft geben über die Kräfte zwischen Protonen und Mesotronenfeld, wenn die Energie der beteiligten Teilchen klein ist verglichen mit der Ruhmasse des Mesotrons, also klein gegen 10^8 eV. Diese Voraussetzung ist für die Kräfte im Atomkern erfüllt. Der Wechselwirkungsausdruck erhält also seinen physikalischen Sinn durch die Kernkräfte. Auch der β-Zerfall sollte sich nach dieser Theorie behandeln lassen.

Ebenso kann man die durch die Möglichkeit der Paarerzeugung (positives und negatives Proton) bedingte gegenseitige Streuung der Mesotronen nach Analogie von (11) berechnen [7]. Man erhält dann (für ein reines Mesotronenfeld, bei dem $|\partial f/\partial x| \ll K|f|$ sein soll), neben den aus (13) folgenden Gleichungen

$$\operatorname{div} f + \varkappa^2 u_0 = 0, \quad \operatorname{rot} g - \varkappa^2 u = 0,$$

noch die weiteren Gleichungen:

$$\dot{u}_k + \operatorname{grad}_k u_0 = -\frac{\partial \bar{\mathcal{L}}}{\partial f_k}, \qquad \operatorname{rot}_k u = -\frac{\partial \bar{\mathcal{L}}}{\partial g_k}. \qquad (15)$$

Die Funktion $\bar{\mathcal{L}}$ ergibt sich hierbei für parallele Felder $f \,\|\, g$, $|f| = f$, $|g| = g$, $w = (g + \mathrm{i}f)\,l/k$ zu:

$$\bar{\mathcal{L}} = \frac{1}{2}(f^2 - g^2) + \frac{K^4}{\pi^2} \operatorname{Re}\left\{ \frac{13}{48}\,w^2 - \frac{13}{288}\,w^4 + \frac{5}{36}\,fg \cdot \mathrm{i}w^2 + \left[-\frac{1}{16} - \frac{w^2}{8} \right. \right.$$

$$\left. \left. + \frac{w^4}{48} + fg\mathrm{i}\left(\frac{1}{4} - \frac{w^2}{12} \right) \right] \log(1 - w^2) + (-w^2 + fg\mathrm{i})\,\frac{1}{6w}\,\log\left(\frac{1+w}{1-w} \right) \right\}. \qquad (16)$$

Für allgemeinere Felder ist in diesem Ausdruck g^2-f^2 durch die Invariante g^2-f^2, f^2g^2 durch die Invariante $(f \cdot g)^2$ zu ersetzen. Die Gleichungen (15) und (16) (oder richtiger: ihr Analogon in der vollständigen Yukawatheorie) sollten die gegenseitige Streuung energiearmer Mesotronen richtig beschreiben; wenn jedoch die Energie der Mesotronen so groß ist, daß der zweite Teil in $\mathscr{L}$ gegenüber $\frac{1}{2}(f^2-g^2)$ überwiegt, so ist jedenfalls eine quantentheoretische Behandlung der Streuung nach (15) und (16) unmöglich.

Es ist lehrreich, die durch die Paarerzeugung bedingten nichtlinearen Zusätze zur Wellengleichung in der Strahlungstheorie, Gl. (11), und in der Yukawatheorie, Gl. (16), zu vergleichen. Die Zusätze zur Lagrangefunktion können in beiden Fällen geschrieben werden als Produkt dreier Faktoren. Diese sind: eine dimensionslose Konstante; ein quadratischer Ausdruck in den Feldstärken; und eine Potenzreihe in dem Verhältnis der Feldstärke zu einer kritischen Feldstärke. Der dimensionslose Faktor ist (soweit er von den universellen Konstanten abhängt) in der Strahlungstheorie $e^2/\hbar c \sim 1/137$, in der Yukawatheorie $K^2l^2 \sim 100$. Die kritische Feldstärke ist in der Strahlungstheorie $m^2c^3/e\hbar$ oder in unseren Einheiten $(mc/\hbar)^2 \cdot \sqrt{\hbar c/e^2}$, in der Yukawatheorie K/l. Der dritte Faktor, die Potenzreihe, wächst in der Strahlungstheorie bei hohen Feldstärken nur logarithmisch an; in der Yukawatheorie jedoch mit dem Quadrat der Feldstärke (multipliziert mit einem Logarithmus). Der Zusatz zur Lagrangefunktion bleibt also in der Strahlungstheorie wegen des Faktors $e^2/\hbar c$ selbst bei sehr großen Feldstärken eine kleine Störung. In der Yukawatheorie überwiegt er, sobald die Feldstärke den Wert K/l erheblich überschreitet. Dieser charakteristische Unterschied der beiden Theorien ist zweifellos durch den verschiedenen Typus der Wechselwirkung („erster" und „zweiter" Art) bedingt.

Natürlich zeigt sich der Umstand, daß die Wechselwirkung (13) bei größeren Energien der beteiligten Teilchen nicht mehr als kleine Störung betrachtet werden kann, auch noch an vielen anderen Folgerungen der Yukawaschen Theorie. Das einfachste Beispiel bildet wohl die Streuung der Mesotronen an Protonen. Der totale Wirkungsquerschnitt für die Streuung transversaler Mesotronen vom Impuls k (in cm^{-1}) und der Energie k_0 ($k_0^2 = \varkappa^2 + k^2$) an einem Proton wird auf Grund einer quantentheoretischen Störungsrechnung [8] nach Gln. (13) und (14):

$$Q = \text{const.}\ \frac{l^4 k^4}{k_0^2}, \tag{17}$$

wobei der konstante Faktor von der Richtung des Polarisationsvektors relativ zum Protonenspin abhängt. Der Wirkungsquerschnitt für die Streuung nimmt also scheinbar mit wachsender Energie der Mesotronen unbegrenzt zu, was sicher weder mit den experimentellen Tatsachen noch mit allgemeinen wellenmechanischen Überlegungen zu vereinbaren ist. Ein genaueres Studium der Störungsrechnung, die zum Ergebnis (17) führt, zeigt auch sofort, daß diese Störungsrechnung für hohe Energien $k_0 > 1/l$ divergiert, so daß Gleichung (17) für solche Energien keine Gültigkeit mehr beanspruchen kann. Es wäre aber ein Irrtum zu glauben, daß etwa nur die Störungstheorie in diesem Gebiet versagt und daß eine genaue quantentheoretische Rechnung zu vernünftigen Resultaten führen müßte. Eine solche exakte Quantentheorie nichtlinearer Wellengleichungen existiert nicht; die

Gleichungen (13) und (14) führen strenggenommen zu divergenten Resultaten. Sie verlieren also ihren Sinn, sobald ihre „korrespondenzmäßige" Verwendung durch die Störungstheorie unmöglich wird.

III. Die Grenzen der bisherigen Theorie

a) Die Notwendigkeit, die aus der Quantentheorie der Wellenfelder gewonnenen Ergebnisse in ihrem Gültigkeitsbereich zu beschränken, ergab sich zuerst bei der Berechnung der Selbstenergie des Elektrons. Zwar wird diese Selbstenergie wegen der Kleinheit von $e^2/\hbar c$ in der Quantentheorie ganz anders bestimmt als in der klassischen. Aber die Rechnungen zeigen [9], daß auch die quantentheoretische Selbstenergie unendlich wird, wenn man die auftretenden Integrale nicht durch ein Abschneiden bei gewissen Impulsen oder den entsprechenden Längen konvergent macht [10]. Man muß also eine charakteristische Länge einführen (entsprechend dem klassischen „Elektronenradius" r_0) und annehmen, daß durch neue physikalische Erscheinungen, die in so kleinen Raum-Zeitdimensionen auftreten und die in der bisherigen Theorie nicht berücksichtigt sind, für die Konvergenz des betreffenden Integrals bei kleinen Längen gesorgt werde. Es ist ein befriedigender Zug der Quantenelektrodynamik, daß der genaue Wert dieser Länge r_0 in den Ergebnissen eine sehr geringe Rolle spielt. Die Arbeit von Weisskopf hat gezeigt, daß die durch die Selbstenergie bedingte relative Ungenauigkeit der Ruhenergie des Elektrons durch die Unbestimmtheit von $e^2/\hbar c \lg (r_0/\mathrm{const})$ gegeben ist. Die Unsicherheit beträgt also kaum mehr als etwa 1 Prozent. Den gleichen Grad der relativen Unbestimmtheit haben Pauli und Fierz [11] für die Ausstrahlung eines Elektrons beim Stoß gefunden. Daß die charakteristische Länge nur logarithmisch in die Ergebnisse eingeht, scheint nach Weisskopf eine spezielle Eigenschaft der Teilchen, die dem Paulischen Ausschließungsprinzip genügen. Für die Mesotronen, die der Bosestatistik genügen, würden sich also andere Verhältnisse ergeben. Aber deren Selbstenergie wird in erster Linie durch das Yukawafeld bedingt sein.

Die relativ günstigen Ergebnisse, die in der Quantenelektrodynamik mit einer „Abschneidevorschrift" erzielt worden sind, haben zu verschiedenen Versuchen geführt, eine in sich geschlossene Quantentheorie der Wellenfelder mit relativistisch invarianten Abschneidevorschriften auszuarbeiten. Diese Vorschriften laufen darauf hinaus, die Matrixelemente der Wechselwirkung, die zum Übergang eines Teilchens von Energie und Impuls $k_0^{(1)}$, $k^{(1)}$ nach Energie und Impuls $k_0^{(2)}$, $k^{(2)}$ gehören, sehr klein zu machen, sobald

$$|(k^{(1)} - k^{(2)})^2 - (k_0^{(1)} - k_2^{(0)})^2| > \frac{1}{r_0^2} \tag{18}$$

ist [12]. Diese Gleichung läßt sich physikalisch plausibel machen. Es ist aber bisher nicht nachgewiesen worden, daß in dieser Weise eine in sich geschlossene Quantentheorie der Wellen mit sinnvollen Ergebnissen aufgebaut werden kann.

Gegen die Möglichkeit einer solchen Theorie spricht die Überlegung, daß derartige Abschneidevorschriften ja gar keine Aussagen enthalten über die doch wohl zu erwartenden neuen physikalischen Erscheinungen, die bei großen Energie- und Impulsübertragungen eintreten. Es ist kaum anzunehmen, daß eine geschlossene Quantentheorie der Wellen existiert, die gerade das Gebiet großer Energie- und Impulsübertragungen nicht enthält und die das Verhältnis der Ruhmassen der Teilchen unbestimmt läßt.

b) Aus diesem Grunde liegt es nahe, anzunehmen, daß die Quantentheorie bei Energie-Impulsübertragungen, die der Gleichung (18) genügen, grundsätzlich versagt und daß in diesem Gebiet auch durch Abschneidevorschriften kein wesentlicher Fortschritt erzielt werden kann. Dabei muß noch etwas genauer festgelegt werden, in welchem Sinne die Gleichung (18) verstanden werden soll: Wenn man annimmt, daß bei den Erscheinungen mit hohen Energieübertragungen eine universelle Konstante von der Dimension einer Länge (r_0) eine entscheidende Rolle spielt, so folgt, daß die in (18) eingehende Größe r_0 für alle Elementarteilchen den gleichen Wert haben muß und daß r_0 nicht etwa von der Ruhmasse der Teilchen abhängt. Natürlich wird man − ähnlich wie früher mit der korrespondenzmäßigen Behandlung quantenmechanischer Probleme − mit der quantentheoretischen Behandlung von Stoßvorgängen manchmal näher, manchmal weniger nahe an das durch (18) bestimmte Gebiet herankommen können. Aber es wird sich vor einer wirklichen Lösung der mit r_0 zusammenhängenden Probleme kaum genauer als durch (18) angedeutet entscheiden lassen, wie weit die nach der bisherigen Theorie gewonnenen Ergebnisse Vertrauen verdienen.

Nach Gleichung (18) könnte man zunächst vermuten, daß es überhaupt sinnlos wäre, von Längen zu sprechen, die klein gegen r_0 sind. Ein solcher Schluß wäre aber nicht berechtigt; denn auch wenn bei Prozessen von der Art (18) ganz neue physikalische Erscheinungen auftreten, so ist es doch prinzipiell stets möglich, z. B. Wellenlängen, die sehr klein gegen r_0 sind, durch Beugungsmessungen zu bestimmen, ohne daß Prozesse der Art (18) überhaupt ins Spiel kommen. Eine andere Frage ist es, ob man etwa den Ort von Teilchen genauer als bis auf Größen der Ordnung r_0 festlegen kann. Ob dies möglich ist, wird sich wohl erst entscheiden lassen, wenn die im Gebiet (18) auftretenden neuen Erscheinungen genau bekannt sind.

Die Größe von r_0 kann man nach der Yukawaschen Theorie abschätzen, indem man untersucht, bis zu welchen Energieübertragungen die Yukawasche Wechselwirkung als kleine Störung betrachtet werden kann. Man kommt dabei für r_0 auf Werte, die etwa zwischen dem klassischen Elektronenradius $e^2/mc^2 \approx 2,81 \cdot 10^{-13}$ cm und dem zehnten Teil dieser Länge liegen. Daraus dürfte folgen, daß alle Prozesse, bei denen Mesotronen oder schwerere Teilchen entstehen, schon außerhalb der Reichweite der bisherigen Theorien liegen. Dies ist insofern befremdlich, als ja die Yukawasche Wechselwirkung formal gerade eine Angabe über die Wahrscheinlichkeit der Erzeugung von Mesotronen enthält. Es wurde aber schon in Abschn. IIa) betont, daß die Yukawasche Wechselwirkung ihren physikalischen Sinn durch die Kernkräfte erhält, also durch die Erscheinungen, bei denen zwar ein Yukawasches Feld, aber keine Teilchen erzeugt werden. Die Größe und die Reichweite der Kernkräfte wird durch die Yukawasche Theorie festgelegt. Die genaue Abhängigkeit der Kernkräfte vom Abstand kann aber innerhalb dieser Theorie wohl ebensowenig wie die Häufigkeit der Mesotronenerzeugung bestimmt werden; es ist auch fraglich, welchen physikalischen Sinn man einer solchen Abhängigkeit geben könnte.

c) In der kosmischen Strahlung treten nun häufig Prozesse auf, bei denen viel Energie- und Impuls übertragen werden; also Prozesse, die der Gleichung (18) genügen. Diese Prozesse lassen sich offenbar bei unseren gegenwärtigen theoretischen Kenntnissen nicht mehr quantitativ behandeln. Man kann aber versuchen, durch Korrespondenzbetrachtungen noch einen qualitativen Einblick in die Verhältnisse im Gebiet (18) zu gewinnen. Dazu bietet sich folgende Möglichkeit dar: Wenn man von den Vertauschungsrelationen absieht, so können die nichtlinearen Wellengleichungen für die verschiedenen Feldgrößen nach den Methoden der klassischen Physik behandelt werden. Man muß dabei allerdings zunächst auf den Teilchenbegriff verzichten, erhält dafür aber auch Aussagen über Vorgänge, die sich in sehr kleinen Raum-Zeitgebieten abspielen. Den Teilchenbegriff kann man dann gewissermaßen nachträglich — allerdings nur in einer grob-qualitativen Weise — einführen, indem man etwa die Teilchen durch normierte Wellenpakete ersetzt oder die Wellen als Wahrscheinlichkeitswellen für Teilchen interpretiert. In dieser Weise erhält man qualitative Aussagen über die Prozesse mit hoher Energieübertragung.

Bei der Anwendung dieser Methode, die vielleicht eine sinngemäße Deutung der Experimente ermöglicht, erhält man von dem Zusammenstoß zweier sehr energiereicher Teilchen folgendes Bild: Die energiereichen Teilchen werden durch geeignet normierte Wellenpakete dargestellt; beim Ineinanderlaufen dieser Wellenpakete werden die nichtlinearen Glieder der Wellengleichungen für die Bewegung wesentlich und bewirken eine „turbulente" Bewegung, die zu einer weitgehenden Durchmischung des Spektrums über weite Frequenzbereiche führt. Nach einiger Zeit breitet sich der Vorgang wieder in Form einer Kugelwelle mit einem weiten kontinuierlichen Frequenzbereich aus. Deutet man diese Kugelwelle als Wahrscheinlichkeitswelle für Teilchen, so wird man auf eine große Anzahl von Teilchen geringerer Energie schließen müssen. Beim Zusammenstoß energiereicher Teilchen kann also mit der explosionsartigen Entstehung vieler Sekundärteilchen gerechnet werden.

Auch an einem spezielleren Beispiel kann man diese Schlußkette anschaulich verfolgen: Wir betrachten etwa den Zusammenstoß eines Protons mit einem anderen Teilchen. Das Proton ist von einem Yukawaschen Eigenfeld umgeben. Wenn der Stoß sehr plötzlich erfolgt, so wird sich im Moment des Stoßes — wie in der Strahlungstheorie bei Bloch und Nordsieck [13] — die Differenz des Eigenfeldes vor und nach dem Stoß als Wellenpaket selbständig machen und als Kugelwelle in den Raum wandern. Dieses Wellenpaket würde beim Beginn der Bewegung nur wenige Mesotronen relativ hoher Energie repräsentieren. Bei der weiteren Entwicklung des Wellenvorganges werden aber die nichtlinearen Glieder in Gleichung (16) eine entscheidende Rolle spielen und für eine Durchmischung des Spektrums sorgen; sie werden aus den schon vorhandenen Mesotronen neue Mesotronen geringerer Energie entstehen lassen, so daß schließlich ein Schauer von Mesotronen zu Stande kommt, wobei die Anzahl der entstehenden Teilchen wohl nur von der Energie des primären Stoßes abhängt.

Sehr große aus Mesotronen bestehende Schauer dieser Art sind bisher noch nicht mit Sicherheit beobachtet worden. Dagegen wird man einige Wilsonaufnahmen von Fussell wohl als kleinere Mesotronenschauer zu deuten haben.

Die eben skizzierte korrespondenzmäßige Verwendung der Wellengleichungen mit einer nur qualitativen Benützung des Teilchenbegriffs ergänzt die quantenmechanische Störungsrechnung in den Gebieten, in denen sie wegen der Divergenz des Störungsverfahrens versagt. Allerdings benötigt sie dabei Begriffe, die bisher in der Quantentheorie nicht vorkommen. Damit nämlich die Selbstenergie der Wellenpakete, die ein Teilchen darstellen, nicht größer wird als die Ruhmasse, darf der Durchmesser eines solchen Pakets nicht kleiner werden als eine bestimmte, durch die Ruhmasse des Teilchens gegebene untere Grenze. Die Einführung eines „Teilchenradius" ist also hier nicht zu vermeiden. Legt man z. B. die Gleichungen (13) und (14) zu Grunde, so erhält man für den Radius des Protons r_p einen Wert der Größenordnung:

$$r_p \sim \sqrt[3]{\frac{l^2}{K}} \, . \tag{19}$$

Berechnet man dann unter Benützung dieses Wertes z. B. den Wirkungsquerschnitt für die Streuung der Mesotronen an Protonen, so erhält man zwar für Yukawafelder geringer Frequenz ($k_0 < 1/l(Kl)^{-1/3}$) wieder die Gleichung (17) (solche Felder fallen räumlich stets exponentiell ab); für höhere Frequenzen (also für alle ebenen Wellen) jedoch geht die Gleichung (17) in eine andere über:

$$Q \sim \text{const.} \; l^2 \frac{k^4}{k_0^4} (lK)^{-2/3} \, , \tag{20}$$

die den Einwänden, die gegen (17) erhoben werden konnten, nicht mehr ausgesetzt ist. Für Frequenzen der Mesotronen, die den Wert K übersteigen, verliert auch Gleichung (20) ihre Gültigkeit und man kann aus anschaulichen Betrachtungen schließen, daß der Wirkungsquerschnitt dann mit wachsender Energie wieder abnehmen wird.

In einer solchen korrespondenzmäßigen Theorie können also Begriffe wie Selbstenergie und Teilchenradius nicht vermieden werden. Obwohl es in einer derartigen Theorie möglich ist, der Masse der Elementarteilchen beliebige Werte zu geben, so wird man daher doch schließen müssen, daß die Frage nach den Ruhmassen nicht mehr getrennt werden kann von der Frage, wie sich die Elementarteilchen bei Stößen im Gebiet (18) verhalten. Die Frage nach dem Wert der Ruhmassen und nach dem Wert der Wirkungsquerschnitte für Stöße mit großer Energieübertragung dürfte im Gegenteil ein einheitliches Problem darstellen, das – ähnlich wie die Quantentheorie durch die Einführung einer universellen Naturkonstanten von der Dimension einer Wirkung charakterisiert war – mit der Einführung einer universellen Naturkonstante von der Dimension einer Länge verknüpft ist.

d) Man kann hier die Frage stellen, welche experimentellen Erfahrungen am ehesten geeignet wären, um über dieses Problem neue Aufschlüsse zu vermitteln. Da bei hohen Energien häufig sehr verwickelte Stoßvorgänge mit der Entstehung vieler Teilchen zu erwarten sind, so dürften die Wirkungsquerschnitte für bestimmte Einzelvorgänge weder dem Experiment noch der Theorie einfach zugänglich sein. Dagegen sind die Gesamtwirkungsquerschnitte — etwa dafür, daß beim Stoß überhaupt neue Materie entsteht — der Messung leichter zugänglich.

Insbesondere ihre Grenzwerte für sehr hohe Energien der stoßenden Teilchen haben auch eine einfache theoretische Bedeutung. Das anschauliche Bild der Wellenpakete, deren Größe nicht unter einen bestimmten Wert herabgedrückt werden kann, legt nahe, diese Grenzwerte der Wirkungsquerschnitte mit der Vorstellung „Teilchendurchmesser" in Verbindung zu bringen. Man darf vielleicht hoffen, daß für diese Grenzwerte der Wirkungsquerschnitte ganz einfache Gesetze gelten — eben ähnlich wie sie etwa aus der anschaulichen Vorstellung „Teilchendurchmesser" folgen würden. Es ist in diesem Zusammenhang befriedigend, daß nach den Erfahrungen in der kosmischen Strahlung der Wirkungsquerschnitt für die Erzeugung von Mesotronen aus leichten Teilchen (Photonen oder Elektronen) bei hohen Energien einem Grenzwert der Ordnung $10^{-26}\,cm^2$ zuzustreben scheint. Wenn es sich herausstellen sollte, daß die Wirkungsquerschnitte für den Stoß irgendwelcher Teilchen mit wachsender Energie stets einem endlichen Grenzwert zustreben, so wäre damit auch die Frage, ob es möglich ist, Teilchenorte mit beliebiger Genauigkeit zu messen, in ein neues Licht gerückt.

Literatur [4]

1   Auf diese Bedingung hat insbesondere N. Bohr z. B. Solvay-Bericht 1933, S. 217 mehrfach hingewiesen.
2   Vgl. z. B. W. Heisenberg: Z. Phys. **90**, 209 (1934)
3   R. Serber: Phys. Rev. **48**, 49 (1935); E. A. Uehling: ebenda **48**, 55 (1935)
4   W. Heisenberg, H. Euler: Z. Phys. **98**, 714 (1936)
5   G. Wick: Nature **142**, 993 (1938)
6   H. J. Bhabha: Nature **143**, 276 (1939)
7   W. Heisenberg: Z. Phys. **113**, 61 (1939)
8   H. J. Bhabha: Proc. R. Soc. **A166**, 501 (1938); W. Heitler: ebenda **A166**, 529 (1938)
9   V. Weisskopf: Z. Phys. **89**, 27; **90**, 817 (1934)
10 Vgl. hierzu auch H. A. Bethe: Phys. Rev. **55**, 681 (1939)
11 W. Pauli, M. Fierz: Nuovo Cimento **15**, Nr. 3 (1938)
12 Vgl. z. B. G. Wataghin: Z. Phys. **88**, 92; **92**, 547 (1934)
13 F. Bloch, A. Nordsieck: Phys. Rev. **52**, 54 (1937); A. Nordsieck: ebenda **52**, 59 (1937)

---

[4] We have completed some of the references given incompletely in Heisenberg's manuscript. (Editors)

Chemisch-Physikalische Gesellschaft in Wien

**(gemeinsam mit dem Gauverein Ostmark
der Deutschen physikalischen Gesellschaft)**

Sitzung am 30. April 1941.

Prof. Dr. **W. Heisenberg,** Leipzig: *Die durchdringende Komponente der Höhenstrahlung.*

Vortr. gibt zuerst einen kurzen Überblick über die verwickelten Prozesse, die sich als Folge der aus dem Weltraum einfallenden Strahlung in der Atmosphäre abspielen, behandelt die Zerlegung der kosmischen Strahlung in die verschiedenen Komponenten und geht dann auf die Prozesse ein, durch welche die Mesonen in der Atmosphäre absorbiert werden, und auf die Vorgänge, bei denen die Mesonen in der Atmosphäre entstehen. Es wird die Ansicht vertreten, daß die von *Blau* u. *Wambacher* beobachteten Kernzertrümmerungen zugleich die Entstehungsprozesse der Mesonen und der Protonen und Neutronen seien und daß diese Kernzertrümmerungen in erster Linie von den Lichtstrahlen der primären Ultrastrahlung ausgelöst werden[10]).

Deutsche Gesellschaft für photographische Forschung

10. Tagung, 9. Mai 1941 in Berlin (Haus der Technik).

Prof. Dr. **J. Eggert,** Leipzig: *Sensitometrischer Jahresbericht.*

Es wird über die Tätigkeit des Ausschusses für Sensitometrie der Gesellschaft berichtet. Der Ausschuß hat die auf der letzten Tagung in München[11]) von englischer und amerikanischer Seite gemachten Vorschläge auf ihre Leistungsfähigkeit sowohl an sich als auch im Vergleich zum DIN-System untersucht. In Betracht gezogen wurden: 1. Das Entwicklungsverfahren von *Rawling* (reproduzierbare Zeit-Temperatur-Entwicklung); 2. Der Vorschlag von *Jones*; 3. Der Vorschlag der Amer. Standard Association. Dieser enthält die *Rawling*-Entwicklung und das *Jones*-Kriterium, verlangt aber überdies noch eine Beschränkung auf eine bestimmte Neigung der S-Kurve. Diese bestimmte Neigung der S-Kurve soll der Arbeitsweise der amerikanischen Photohändler (Photofinisher) angepaßt sein, da diese angeblich weitgehend einheitlich bis zu einer bestimmten Neigung entwickeln. Dies trifft jedoch in keiner Weise für deutsche Verhältnisse zu, wie umfangreiche Versuche bzw. Erhebungen bei deutschen Photohändlern ergaben. Es wurden bei gleichen Filmtypen an verschiedenen Stellen γ-Werte festgestellt, die sich maximal um den Faktor 4 unterschieden. — Die Vergleiche sind noch nicht endgültig abgeschlossen. Sie lassen jedoch bereits erkennen, daß die Vorschläge gegenüber dem DIN-System hinsichtlich der Erfassung der praktischen Empfindlichkeit, der Reproduzierbarkeit der Meßwerte und der Einfachheit der Ausführung der Messungen in keiner Weise grundlegende Verbesserungen darstellen.

---

[10]) Vgl. hierzu *Heisenberg,* Das schwere Elektron (Mesotron) und seine Rolle in der Höhenstrahlung, diese Ztschr. **52,** 41 [1939]; *Wambacher,* Mehrfachzertrümmerung v. Atomkernen durch kosmische Strahlung, ebenda S. 117; *Euler,* Ausgedehnte Schauer der Ultrastrahlung in der Luft, ebenda S. 702.
[11]) Vgl. diese Ztschr. **52.** 212 [1939].

Kosmische Strahlung: Vorträge gehalten im Max Planck-Institut Berlin-Dahlem, ed. by Werner Heisenberg (Springer, Berlin 1943)

ARNOLD SOMMERFELD

ZU SEINEM 75. GEBURTSTAG

AM 5. DEZEMBER 1943

GEWIDMET

KOSMISCHE STRAHLUNG

VORTRÄGE

GEHALTEN IM MAX PLANCK-INSTITUT

BERLIN-DAHLEM

VON

E. BAGGE · F. BOPP · S. FLÜGGE · W. HEISENBERG
A. KLEMM · J. MEIXNER · G. MOLIÈRE · H. VOLZ
C. F. v. WEIZSÄCKER · K. WIRTZ

HERAUSGEGEBEN VON

WERNER HEISENBERG

MIT 37 ABBILDUNGEN

BERLIN
SPRINGER-VERLAG
1943

Inhaltsverzeichnis.

Einleitung.

Kaskaden.

Mesonen.

Kernteilchen.

Geomagnetische Effekte.

Vorwort.

Die Forschungen über kosmische Strahlung werden von der Ungunst der Zeiten besonders stark betroffen. Denn einerseits müssen sie in den meisten physikalischen Laboratorien naturgemäß hinter anderen Problemen zurückstehen, andererseits ist auch die Information über Forschungsergebnisse, die im Ausland gewonnen worden sind, durch die fehlende Nachrichtenübermittlung sehr erschwert. Schließlich sind ausführliche Referate im Kriege in Deutschland einfach deswegen nicht erschienen, weil der Physiker, der im Kriegseinsatz steht, zu einer umfangreicheren Arbeit dieser Art nicht die Zeit findet. In Anbetracht der grundsätzlichen Wichtigkeit dieses Zweiges der Physik scheint es mir daher gerechtfertigt, wenn man eine Reihe von Kolloquiumsvorträgen, die einen Überblick über den jetzigen Stand der Höhenstrahlungsforschung geben sollen, zusammenfaßt und in Buchform veröffentlicht.

Die vorliegenden Vorträge wurden in den Jahren 1941 und 1942 im Kaiser Wilhelm-Institut für Physik gehalten; sie geben zum größeren Teil eine Übersicht über den Stand bestimmter Einzelfragen, wobei die amerikanische Literatur nur etwa bis zum Sommer 1941 zugänglich war. Zum kleineren Teil enthalten sie auch Ergebnisse, die noch nicht an anderer Stelle veröffentlicht worden sind. Ich erwähne die Untersuchung von MOLIÈRE über die großen Luftschauer, Rechnungen von FLÜGGE über die Neutronenverteilung in der Atmosphäre, und eine vereinfachte Kaskadentheorie, die aus einer Vorlesung stammt, die ich im Sommer 1939 in Leipzig und in Lafayette (U.S.A.) gehalten habe. Im Ganzen dürften die Vorträge ein einigermaßen zutreffendes Bild der Kenntnisse über die Ultrastrahlung etwa Ende des Jahres 1941 geben. Dieses Bild ist nicht sehr befriedigend, die Genetik der einzelnen Strahlenarten ist noch nicht genügend geklärt, und der Mechanismus der Umwandlung ist nur bei den Elektronen und Lichtquanten genauer bekannt. Man kann also nur hoffen, daß das hier gezeichnete Bild möglichst bald durch ein klareres und richtigeres ersetzt werde.

Es war der Wunsch aller Mitarbeiter, dieses Buch ARNOLD SOMMERFELD zu seinem 75. Geburtstag zu widmen. Der Dank an den Lehrer der Atomphysik in Deutschland sollte durch die Tatsache zum Ausdruck gebracht werden, daß die Art von Gemeinschaftsarbeit, die SOMMERFELD in seiner Münchener Schule begonnen hat, auch von den Jüngeren weitergeführt wird und Früchte trägt.

Für die Mühe bei der Ausarbeitung und Korrektur der Vorträge möchte ich allen Mitarbeitern danken, insbesondere auch dem Verlag, der die Herausgabe der Vorträge bereitwillig unterstützt und alle praktischen Schwierigkeiten erfolgreich überwunden hat.

Berlin-Dahlem, den 2. Juni 1943.

W. HEISENBERG.

Einleitung.

1. Übersicht über den jetzigen Stand unserer Kenntnisse von der kosmischen Strahlung.

Von **W. Heisenberg**-Berlin-Dahlem.

Mit 1 Abbildung.

Seit den Versuchen von Hess (H 11) und Kolhörster (K 3) ist bekannt, daß eine von dem Weltraum auf die Erde einfallende Strahlung in der Atmosphäre verwickelte Sekundärerscheinungen auslöst, die sich im Auftreten einer allgemeinen Ionisation, hervorgerufen durch Elementarteilchen verschiedener Art (Elektronen, Lichtquanten, Mesonen, Protonen, Neutronen) sichtbar äußern und die schließlich bei tieferem Eindringen in die Atmosphäre oder in die Erde immer schwächer werden, bis sie in etwa 1000 m Wassertiefe praktisch völlig verschwinden (vgl. auch die zusammenfassenden Darstellungen bei Miehlnickel (M 3a), Steinke (S 18a), Steinmaurer (S 18b) und in Rev. of mod. Phys. 11, 122, 1939). Zur Ordnung dieser Erscheinungen der sogenannten „Ultrastrahlung" oder „Höhenstrahlung" oder „kosmischen Strahlung" ist es seit längerer Zeit üblich, die Strahlung in der Atmosphäre in Komponenten einzuteilen. Die einzelnen Komponenten unterscheiden sich durch die verschiedenen Arten der Teilchen und ihre Entstehung.

Beim gegenwärtigen Stand unserer Kenntnisse kann man deutlich vier Komponenten auseinanderhalten. Die Existenz einer fünften Komponente, die sich in sehr großer Tiefe bemerkbar macht, ist nach einigen Versuchen wahrscheinlich. Die vier Komponenten lassen sich kurz in folgender Weise beschreiben:

1. Die weiche Strahlung (Kaskadenstrahlung). Sie besteht aus Elektronen, Positronen und Lichtquanten, die sich in der Atmosphäre nach den Gesetzen der Kaskadentheorie ineinander umwandeln, und stellt von etwa 7 km Höhe ab bis in sehr große Höhen den Hauptanteil der kosmischen Strahlung dar. Auch das bekannte Maximum der Ionisation in etwa 16 bis 20 km Höhe wird wahrscheinlich von dieser Elektronenstrahlung erzeugt. Auf Meeresniveau ist die Intensität dieser Komponente schon auf einen sehr kleinen Bruchteil der Anfangsintensität gesunken.

2. Die durchdringende Strahlung. Sie besteht aus Mesonen, die durch eine andere Strahlung (in erster Linie wohl Protonen, Lichtquanten) in der Atmosphäre erzeugt werden. Ihre Intensität nimmt nach den bisherigen Messungen bis zu den größten bisher erreichten Höhen dauernd zu, ohne ein Maximum zu erreichen — obwohl ein Maximum schließlich wohl irgendwo vorhanden sein muß, da die Mesonen als radioaktive Teilchen nicht vom Weltraum auf die Erde kommen können. Vielleicht gibt es verschiedene Arten von Mesonen, die sich durch den Wert des Drehimpulses und durch ihre Lebensdauer unterscheiden und die in verschiedenen Höhen in verschiedener relativer Häufigkeit die durchdringende Komponente bilden. Diese Komponente nimmt mit wachsender Tiefe viel langsamer ab als die weiche Komponente und kann bis zu Tiefen von 400 m Wasser und mehr nachgewiesen werden.

3. Die weiche Sekundärstrahlung der Mesonen. Sie besteht aus Elektronen, Positronen und Lichtquanten, die von den Mesonen durch Stoßprozesse und durch radioaktiven Zerfall gebildet werden. Sie steht im allgemeinen mit der Mesonenkomponente im Gleichgewicht und nimmt mit wachsender Tiefe etwas langsamer ab als die Mesonenintensität, was auf die mit der Tiefe zunehmende Härte der Mesonenstrahlung zurückzuführen ist.

4. Die Protonen- und Neutronenstrahlung. Ihre Intensität verläuft bis zu großen Höhen der Intensität der weichen Komponente parallel; diese Tatsache spricht dafür, daß die Proton-Neutron-Komponente — zum mindesten in tieferen Atmosphärenschichten — zu einem erheblichen Teil von der weichen Komponente gebildet wird. Allerdings sprechen auch starke Argumente dafür, daß Protonen vom Weltraum auf die Erde treffen — vielleicht sogar den Hauptteil der Primärstrahlung darstellen —, daß also die Protonenkomponente in großer Höhe zum größten Teil eine Primärstrahlung ist, die dort erst die Mesonenkomponente und die Kaskadenstrahlung erzeugt. — Die Intensität der Neutronen ist viel größer als die der Protonen, was offenbar mit der viel größeren Reichweite der Neutronen zusammenhängt.

Im folgenden sollen die wichtigsten der bisher bekannten Eigenschaften der vier Komponenten in einem kurzen Überblick zusammengefaßt werden.

1. Die Kaskadenstrahlung.

Wenn man das Spektrum der weichen Komponente von den Höhen, in denen Messungen vorgenommen worden sind, bis zum Gipfel der Atmosphäre extrapoliert — die Multiplikation und die Absorption dieser Komponente wird ja durch die Kaskadentheorie vollständig beherrscht —, so kommt man zu einem einfachen Spektrum, das man lange Zeit als das Primärspektrum der kosmischen Strahlung betrachtet hat. Wenn man nämlich annimmt, daß — oberhalb einer Energie von etwa 3 bis $4 \cdot 10^9$ eV — eine Anzahl $F(E)$ von Elektronen einer Energie $> E$, gegeben durch

$$F(E) \approx 0{,}05 \cdot \left(\frac{10^{10}\,\text{eV}}{E}\right)^{1,8} (\text{sec}^{-1}\,\text{cm}^{-2}) , \qquad (1)$$

vom Weltraum auf die Erde trifft, so kann man mit dieser Annahme den Verlauf der weichen Komponente mit der Höhe, das Maximum der Ionisation, und den Breiteneffekt — das Magnetfeld der Erde schneidet das Spektrum (1) je nach der geographischen Breite bei einer bestimmten Energie ab—, schließlich auch die Häufigkeit der großen Luftschauer und der HOFFMANNschen Stöße ungefähr richtig deuten.

Nach dieser Deutung sollte das Primärspektrum (1) je nach der geographischen Breite bei einer bestimmten Energie, z. B. in unseren Breiten bei etwa $5 \cdot 10^9$ eV beginnen, also keine Elektronen niedrigerer Energie enthalten. Diese Elektronen würden sich vom Gipfel der Atmosphäre ab nach der Kaskadentheorie multiplizieren und so das Maximum der Ionisation hervorrufen, dann würde das gesamte Spek-

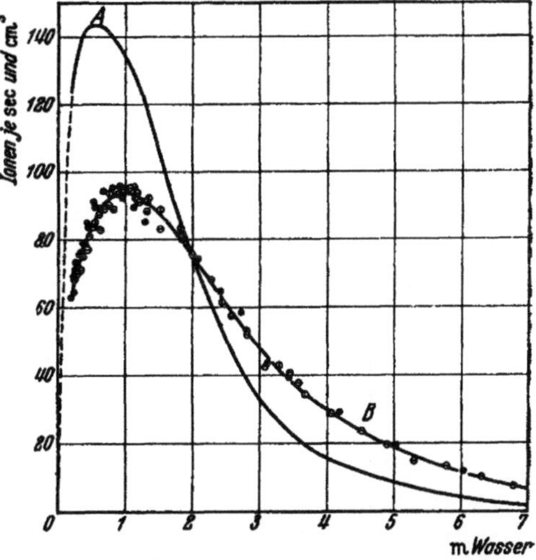

Abb. 1. Intensität der kosmischen Strahlung als Funktion der durchlaufenen Schichtdicke nach BOWEN, MILLIKAN und NEHER.
Kurve B: Intensität in Madras (3° nördl. magn. Breite).
Kurve A: Differenz der Intensitäten in Fort Sam Houston (38,5° nördl. Breite) und Madras.

trum nach dieser Theorie unter Beibehaltung seiner Form (1) exponentiell absorbiert werden. In nördlichen Breiten ist das Maximum verhältnismäßig hoch und steil, da das Primärspektrum aus vielen relativ energiearmen Teilchen besteht, am Äquator ist es erheblich niedriger, da es durch relativ wenige energiereiche Teilchen erzeugt wird (vgl. die bekannten Messungen von MILLIKAN, BOWEN und NEHER (B 32, 33, 34) und die Abb. 1). Auf Meeresniveau ist die weiche Komponente schon auf einen kleinen Bruchteil der Primärintensität gesunken, die weiche Komponente macht auf Meereshöhe wohl nur etwa 5 bis 10% der gesamten dort beobachteten kosmischen Strahlung aus.

Die weiche Strahlung löst dann, wenn sie viel Energie enthält, beim Auftreffen auf Materie stets Kaskaden aus. So bilden sich erstens in der Atmosphäre die großen Luftschauer, zweitens in festem Material die von ROSSI (R 5) beobachteten Schauer und die HOFFMANNschen (H 12)

Stöße. Die Luftschauer, die von sehr energiereichen Elektronen herrühren, reichen bis in größere Tiefen hinab und sind von Auger (A 10), Kolhörster (K 4, 5) und ihren Mitarbeitern beobachtet worden. Sie erfüllen eine Säule von etwa 50 m Radius mit Elektronen, die Dichte und die Höhe der Säule hängen von der Primärenergie ab. Die beobachteten Schauer rühren zum Teil von Elektronen von mehr als 10^{15} eV her. Die Häufigkeit dieser Schauer bietet daher eine Möglichkeit, das Spektrum (1) bis zu sehr hohen Energien zu prüfen, und bestätigt den Wert 1,8 des Exponenten in (1). Nach Euler (E 5) ergibt sich hier als Exponent $1,8 \pm 0,17$.

Die Hoffmannschen Stöße in der Ionisationskammer hinter dünnen Schichten festen Materials sind ebenfalls als Wirkung der Luftschauer anzusehen, die in den Materialschichten durch Multiplikation noch verdichtet werden. Die Stöße hinter dicken Schichten rühren jedoch nicht unmittelbar von der weichen Komponente her. Die von Rossi (R 5) beobachteten kleinen Schauer hinter verschiedenen Schichten festen Materials sind zum Teil auf die Kaskadenkomponente, zum Teil auf die Sekundärstrahlung der Mesonen zurückzuführen.

Die Deutung der weichen Komponente als Primärstrahlung wird aber neuerdings in Zweifel gezogen durch die Messungen von Schein, Jesse und Wollan (S 4, 5, 6), nach denen in sehr großen Höhen, oberhalb des Maximums der Ionisation, schnelle Elektronen nicht oder nur in sehr geringer Zahl vorhanden sind. Wenn diese Beobachtungen sich bestätigen, dann kann die weiche Komponente jedenfalls nicht die Primärstrahlung sein. Die Deutung des Breiteneffekts auf Grund eines primären Elektronenspektrums (1) muß dann fallengelassen werden, und es bleibt etwas merkwürdig, daß der Breiteneffekt so gut hatte dargestellt werden können. Freilich war auch diese Darstellung wohl nicht allzu genau gewesen.

Dagegen muß immer noch angenommen werden, daß *unterhalb* des Maximums der Ionisation die weiche Komponente im wesentlichen die Form hat, die aus (1) nach der Kaskadentheorie folgt. Das wird durch die Luftschauer, die Stöße und den Verlauf der Intensität mit der Höhe bewiesen. Man muß dann wohl annehmen, daß diese Kaskadenstrahlung durch andere Komponenten in der höchsten Atmosphäre erzeugt wird; z. B. durch die Bremsstrahlung einer primären Protonenkomponente, oder durch Bremsstrahlung und radioaktiven Zerfall einer Mesonenstrahlung, oder bei explosionsartigen Prozessen, die von energiereicheren Protonen herrühren und zur Entstehung vieler Mesonen und vieler Lichtquanten und Elektronen führen. Über die Einzelheiten dieser Prozesse ist bisher nichts bekannt.

Da bei allen derartigen Prozessen ein Teil der Primärenergie *nicht* im höchsten Teil der Atmosphäre in Kaskadenstrahlung überginge, so müßte die Primärstrahlung *mehr* Energie enthalten als das Spektrum (1). Daher müßte der Breiteneffekt beim Maximum der Ionisation, wenn dieses durch die weiche Komponente hervorgerufen wird, geringer sein als nach der bisherigen Theorie. Hierdurch entsteht eine Schwierigkeit, die die Frage aufwirft, ob das Maximum der Ionisation überhaupt durch Elektronen erzeugt wird und, wenn dies der Fall ist, durch welche Prozesse diese Elektronen entstanden sind.

2. Die Mesonenkomponente.

Die Mesonen entstehen in der Atmosphäre; sie kommen nicht vom Weltraum auf die Erde. Dies folgt aus der radioaktiven Instabilität der Mesonen, die durch eine Reihe von Experimenten sichergestellt ist. JOHNSON (J9) hat aus dem Ost-West-Effekt der durchdringenden Komponente sowie aus dem Überwiegen der positiven Mesonen geschlossen, daß die meisten Mesonen aus einer primären Protonenkomponente entstehen. Eine derart intensive Protonenkomponente ist allerdings bisher nicht beobachtet worden, und man muß daher dann annehmen, daß die energiereichen Protonen in den höchsten Schichten der Atmosphäre sehr schnell absorbiert werden. In tieferen Schichten der Atmosphäre, bis herunter zu etwa 7 km Höhe, ist allerdings auch noch die Erzeugung von Mesonen beobachtet worden. Diese Mesonenerzeugung verläuft nach den Beobachtungen parallel zur Intensität der weichen Komponente, es ist daher sehr wahrscheinlich, daß die weiche Komponente für sie verantwortlich ist. Außerdem scheinen bei diesen Prozessen vorzugsweise *langsame* Mesonen zu entstehen [vgl. die Messungen von SCHEIN, WOLLAN und GROETZINGER (S 8) und HERTZOG und BOSTICK (H 8, 9)]. Man wird daher an Prozesse denken, die der gewöhnlichen Paarerzeugung analog sind und bei denen ein Lichtquant an einem Kern ein Mesonenpaar bildet. Energiereichere Lichtquanten scheinen explosionsartige Prozesse auslösen zu können, bei denen unter Umständen viele Mesonen auf einmal entstehen. Prozesse dieser Art sind von FUSSELL (F 7), POWELL (P 6), WOLLAN (W 16), DAUDIN (D 1) in der WILSON-Kammer beobachtet worden, auch die Beobachtungen von SANTOS, POMPEJA und WATAGHIN (W 2) und JÁNOSSY und INGLEBY (J 3) über durchdringende Schauer weisen in die gleiche Richtung.

Die Mesonenkomponente wird bei ihrem Weg durch die Atmosphäre geschwächt einerseits durch die bekannte Ionisationsbremsung, andererseits durch den radioaktiven Zerfall der Mesonen. Die Bremsung kann nach den bekannten Formeln von BETHE (B 10) und BLOCH (B 26) behandelt werden; FERMI (F 1) hat neuerdings gezeigt, daß diese Formeln bei ihrer Anwendung auf feste Materialien zu verbessern sind, und daß die Bremsung pro g/cm^2 im dichten Material etwas geringer ist als im dünnen. Ferner hat KEMMER (K 1) aus der YUKAWAschen Theorie abgeleitet, daß die Übertragung sehr großer Energiebeträge von Mesonen hoher Energie an Elektronen mit einem Wirkungsquerschnitt erfolgen kann, der sehr viel größer ist als der entsprechende Wirkungsquerschnitt für den Stoß zweier Elektronen. Durch solche Stöße wird daher die Absorption energiereicher Mesonen erhöht.

Der radioaktive Zerfall des Mesons erfolgt nach der YUKAWAschen Theorie in der Weise, daß das Meson in ein Elektron und ein Neutrino zerfällt, wobei diese beiden Teilchen im Ruhsystem des Mesons ungefähr je die halbe Ruhmasse des Mesons, d. h. etwa 40 bis 50 MeV als kinetische Energie mitnehmen. Die mittlere Zerfallszeit beträgt für die Mesonen auf Meereshöhe nach den vorliegenden Experimenten im Ruhsystem des Mesons etwa 1,5 bis $2,5 \cdot 10^{-6}$ sec und nimmt für schnell bewegte Mesonen nach der Zeitdilatation der Relativitätstheorie im Verhältnis der Energie zur Ruhenergie zu. Sie kann experimentell aus der Absorption der Mesonen in Luft, aus den sich hierbei ergebenden Temperatur- und Barometereffekten, aus der Gestalt des Spektrums bei kleinen Energien und aus der Anzahl der Zerfallselektronen ermittelt werden. Der radioaktive Zerfall ist durch Aufnahmen in der WILSON-Kammer von WILLIAMS (W 12) direkt nachgewiesen worden.

Das Spektrum der Mesonen scheint bei seiner Entstehung in der höchsten Atmosphäre auch ziemlich genau einem Gesetz der Form

$$G(E) = \text{const}\, E^{-\gamma} \tag{2}$$

zu folgen [$G(E) =$ Anzahl der Mesonen einer Energie $> E$], wobei der Exponent γ innerhalb der Meßgenauigkeit ebenso groß ist wie bei der weichen Komponente ($\gamma \sim 1,8$). Dies folgt erstens aus der Abnahme der Mesonenintensität mit der Tiefe T in Schichten bis zu 300 m Wasser, die nach dem Gesetz $T^{-\gamma}$ erfolgt, zweitens aus der Energieverteilung der von den Mesonen erzeugten Kaskadenstöße. In Tiefen unter 300 m Wasser wird allerdings die Absorption stärker, doch beruht dies wohl nicht auf Abweichungen vom $E^{-\gamma}$-Gesetz, sondern auf der von KEMMER abgeleiteten zusätzlichen Absorption, die dadurch zustande kommt, daß sehr energiereiche Mesonen mit einem von der Energie unabhängigen Wirkungsquerschnitt einen erheblichen Teil ihrer Energie an ein Elektron übertragen können, vielleicht auch noch auf anderen zusätzlichen Sekundärprozessen. [Vgl. auch die Arbeiten von OPPENHEIMER, SNYDER u. SERBER (O 2) und LYONS (L 12)].

Die Ähnlichkeit der Spektren (1) und (2) läßt einen genetischen Zusammenhang zwischen beiden Spektren vermuten. Wahrscheinlich entsteht also entweder die Mesonenkomponente aus der weichen Komponente, oder die weiche Komponente aus der Mesonenkomponente, oder schließlich beide aus der gleichen Primärkomponente. Früher hat man die erste Annahme für die wahrscheinlichste gehalten. Wenn sich jedoch die Beobachtungen von SCHEIN, JESSE und WOLLAN (S 4, 5. 6) über den sekundären Charakter der weichen Komponente bestätigen, so ist die zweite oder dritte Annahme wahrscheinlicher.

Eine interessante Deutungsmöglichkeit hat sich hier aus verschiedenen Arbeiten von OPPENHEIMER, SNYDER und SERBER (O 2), MÖLLER und ROSENFELD (M 5) und ROZENTAL (R 11) ergeben: Die Theorie der Kernkräfte fordert die Existenz von Mesonen vom Spin 1 oder von zwei Mesonensorten vom Spin 1 und vom Spin 0. OPPENHEIMER hat jedoch aus der Häufigkeit der von Mesonen hervorgerufenen Kaskadenstöße geschlossen, daß die auf Meeresniveau beobachteten Mesonen wahrscheinlich den Spin 0 besitzen. Man kann daher vermuten, daß in der höchsten Atmosphäre aus der primären Protonenkomponente die beiden Mesonensorten vom Spin 1 und Spin 0 mit vergleichbarer Häufigkeit entstehen; daß jedoch die Mesonen vom Spin 1 eine sehr kurze Lebensdauer besitzen, wie es die YUKAWAsche Theorie fordert, etwa von der Größenordnung 10^{-8} sec, während die Mesonen vom Spin 0 die beobachtete größere Lebensdauer haben. Die Mesonen vom Spin 1 würden daher schon in der höchsten Atmosphäre absorbiert, ihre Energie zum großen Teil in Kaskadenstrahlung übergeführt werden — bei den energiereicheren Mesonen durch Ausstrahlung, bei den energieärmeren durch Zerfall, und nur die Mesonen vom Spin 0 könnten in die tieferen Atmosphärenschichten dringen. Ob sich diese Deutung als Arbeitshypothese bewährt, kann freilich erst die Zukunft lehren. Einstweilen ist die Genetik der einzelnen Komponenten der kosmischen Strahlung noch ganz unsicher.

3. Die weiche Sekundärstrahlung der Mesonen.

Diese Strahlung besteht eigentlich wieder aus zwei Komponenten: den Elektronen (und ihren Kaskaden), die von den Mesonen durch Stoß eine hohe Energie erhalten haben [BHABHA (B 19)], und denen, die vom Zerfall des Mesons herrühren [EULER (E 4)]. Für beide Komponenten wird die Form des Spektrums theoretisch näherungsweise durch

$$H(E) = \mathrm{const}\, E^{-2,8} \tag{3}$$

dargestellt; beide Spektren verhalten sich also gleich und fallen um eine Potenz stärker ab als das Mesonenspektrum oder die weiche Komponente. Die KEMMERschen Stoßprozesse geben allerdings noch ein schwaches zusätzliches Spektrum etwa der Form $\mathrm{const}\, E^{-1,8}$.

Die weiche Sekundärstrahlung ist im allgemeinen schon nach verhältnismäßig geringen Schichtdicken (von der Ordnung der „Strahlungseinheit" der Kaskadentheorie) mit der Mesonenstrahlung im Gleichgewicht. Die relative Gesamtmenge der Stoßelektronen, die mit der Mesonenkomponente im Gleichgewicht steht, ist in größerer Tiefe größer als auf Meeresniveau, da die mittlere Energie der Mesonen dort größer ist. Auf Meeresniveau dürften (in Luft) die Mesonen 75 bis 80%, die Stoßelektronen etwa 5%, die Zerfallselektronen 10 bis 15%, die weiche Primärstrahlung etwa 5 bis 10% der gesamten ionisierenden Strahlung ausmachen. Hinter dicken Schichten von dichtem Material verschwindet praktisch der Anteil der Zerfallselektronen, da die in Luft gebildeten Zerfallselektronen durch die dicke Schicht absorbiert worden sind, auf der relativ kurzen Strecke im dichten Material aber keine neuen gebildet werden konnten. Die Gleichgewichtsmenge der Stoßelektronen sollte theoretisch hinter Material von hoher Atomnummer größer sein als hinter Stoffen geringen Atomgewichts.

Da sich die weiche Sekundärstrahlung der Mesonen physikalisch wie die weiche Primärstrahlung verhält, bildet sie auch Schauer und Stöße wie diese. Auf Meeresniveau werden allerdings praktisch alle *größeren* Schauer oder Stöße der weichen Primärstrahlung zugeschrieben werden müssen, da das Spektrum der weichen Sekundärstrahlung viel rascher abfällt, also weniger energiereiche Elektronen enthält als das Primärspektrum. Anders ist es hinter dicken Schichten dichten Materials. Hier kann das Primärspektrum keine Rolle mehr spielen, und die auch hier noch beobachteten Stöße werden wohl nach Oppenheimer, Snyder und Serber (O 2) auf solche Stoßprozesse, wie Kemmer (K 1) sie untersucht hat, zurückzuführen sein. Die Häufigkeit der beobachteten Stöße páßt jedenfalls nach Oppenheimer, Snyder und Serber (O 2) qualitativ zu dieser Annahme. Allerdings gehören die hier angenommenen Stoßprozesse schon zu den Vorgängen, bei denen die Strahlungskräfte der Yukawaschen Theorie eine entscheidende Rolle spielen müssen, bei denen also die Anwendung der bisherigen quantentheoretischen Formeln sehr problematisch ist. Es muß daher mit der Möglichkeit gerechnet werden, daß bei diesen Prozessen gleichzeitig mit dem Stoß eine Anzahl von Mesonen erzeugt wird, wie es beim Auftreten einer starken Yukawaschen Strahlung plausibel ist. Es würde sich dann um echte explosionsartige Schauer handeln, deren Existenz, wie schon erwähnt, sowohl aus einigen Wilson-Aufnahmen von Fussell (F 7), Powell (P 6), Wollan (W 16) und Daudin (D 1), wie aus Koinzidenzmessungen von Pompeja, Santos und Wataghin (W 2), Jánossy und Ingleby (J 3) hervorzugehen scheint.

4. Die Proton-Neutron-Komponente.

Johnson (J 9) hat die Hypothese ausgesprochen und begründet, daß die Mesonen aus einer primären Protonenkomponente entstehen, und die Beobachtungen über den sekundären Charakter der weichen Komponente sprechen dann dafür, daß die ganzen Erscheinungen der kosmischen Strahlung auf eine aus Protonen bestehende Primärstrahlung aus dem Weltraum zurückzuführen sind. Wenn man diese Arbeitshypothese zugrunde legt — erst zukünftige Experimente können hier endgültig Klarheit schaffen —, so wird man über die primäre Protonenkomponente folgende Annahme machen können: Die primäre Protonenkomponente ist hinsichtlich ihrer Energieverteilung ebenfalls durch ein Spektrum der Form (1)

$$F(E) = \text{const}\, E^{-1,8} \qquad (4)$$

gegeben, wobei die Konstante sicher größer ist als in Gleichung (1). Dieses Protonenspektrum verwandelt sich in der höchsten Atmosphäre in andere Komponenten, ein großer Teil seiner Energie wird sich also schon in etwa 20 km Höhe in der weichen bzw. in der Mesonenkomponente [(1) bzw. (2)] wiederfinden. Die Gestalt (4) des Spektrums dürfte notwendig sein, um die entsprechende Form der Sekundärspektren (1) und (2) zu erklären. Dieses primäre Protonenspektrum wird offenbar in der Atmosphäre sehr stark absorbiert; denn in den höchsten Höhen, in denen bisher nach Protonen gesucht wurde, fand sich nur eine relativ schwache Protonenkomponente. Der Grund für diese starke Absorption muß wohl in Kernprozessen und in der Bremsstrahlung gesucht werden. Nebenbei sei festgestellt, daß Neutronen jedenfalls *nicht* als Primärteilchen für die kosmische Strahlung in Frage kommen, da sie nach der Theorie des β-Zerfalls radioaktiv zerfallen können. Experimentell ist der Zerfall der Neutronen bisher allerdings nicht nachgewiesen worden.

In größeren Höhen bis herunter zum Meeresniveau findet man eine verhältnismäßig schwache Protonen- und etwa hundertmal stärkere Neutronenkomponente. Diese Strahlung hat wohl nicht direkt mit der Primärstrahlung zu tun. Ihre Intensität verläuft nach FÜNFER (F 6) und KORFF (K 6) als Funktion der Höhe ganz ähnlich wie die Intensität der weichen Komponente und kann daher wahrscheinlich — ausgenommen die obersten Schichten der Atmosphäre — als Sekundärstrahlung der weichen Komponente angesehen werden. Die Anzahl der Neutronen ist dabei mit der der Elektronen dieser Komponente vergleichbar; die der langsamen Protonen ist etwa um den Faktor 100 geringer. Gleichzeitig mit einzelnen Protonenspuren werden auf photographischen Platten, die in großer Höhe exponiert sind, häufig Kernzertrümmerungen hoher Energie beobachtet [BLAU und WAMBACHER (B 25), SCHOPPER (S 12)], bei denen aus einem Kern mehrere (bis zu zwölf) Kerntrümmer herausgeschlagen werden. Hier handelt es·sich offenbar entweder um Prozesse, bei denen — etwa durch primäre Lichtquanten — die energiereichen Protonen und Neutronen aus den Atomkernen herausgeschlagen werden, oder um Sekundärprozesse, die selbst durch schnelle schwere Teilchen hervorgerufen werden [BAGGE (B 1)]. Es liegt nahe, diese Prozesse auch mit der Entstehung von Mesonen in Verbindung zu bringen und anzunehmen, daß hier im Kern durch ein Photon explosionsartige Prozesse hervorgerufen werden, bei denen gleichzeitig Mesonen und schwere Teilchen entstehen, die ihrerseits wieder im gleichen Kern weitere Sekundärprozesse hervorrufen können. Die Annahme, daß die Mesonen in mittleren Atmosphärenschichten und die schweren Teilchen durch die gleichen Prozesse und häufig in einem Akt entstehen, ist mit der Häufigkeit der Kernprozesse und der verschiedenen Teilchensorten verträglich [BAGGE (B 1)]. Auch sprechen die Aufnahmen von FUSSELL (F 7) sowie prinzipielle Gründe dafür, daß die Mesonenerzeugung meistens mit Kernprozessen verbunden sein wird. Doch reicht das bisher vorliegende experimentelle Material noch nicht aus, um diese Frage endgültig zu entscheiden.

Die Absorption der Neutron-Proton-Komponente erfolgt für die beiden Teilchensorten ganz verschieden. Die Protonen werden, wenn sie nicht eine sehr hohe Anfangsenergie haben, durch Ionisation sehr rasch gebremst. Die Neutronen dagegen werden durch Kernprozesse absorbiert oder häufiger durch Zusammenstöße mit Kernen gebremst, bis sie schließlich als langsame Neutronen in irgendwelchen Kernen (z. B. N oder H) eingefangen werden.

5. Weitere Komponenten.

Die Mesonenkomponente kann bis zu Tiefen von mehr als 400 m Wasser verfolgt werden, in größeren Tiefen wird sie verhältnismäßig rasch schwächer. Aber selbst in 1000 m Wassertiefe kann eine durch kosmische Strahlung verursachte Ionisation deutlich nachgewiesen werden [vgl. z. B. Untersuchungen von CLAY (C 4) und seinen Mitarbeitern]. Messungen von BARNOTHY und FORRO (B 6) deuten darauf hin, daß hier die Ionisation in der Hauptsache von weicher Strahlung herrührt. Wenn dies richtig ist, so kann an das Ergebnis die Vermutung geknüpft werden, daß die kosmische Strahlung in diese Tiefen durch eine neue elektrisch neutrale Komponente transportiert wird, als deren Träger man etwa neutrale YUKAWAsche Teilchen oder PAULIs Neutrinos ansetzen kann. Aber auch diese Frage kann erst durch weitere Messungen entschieden werden.

2. Die Kaskadentheorie.

Von W. HEISENBERG[1]-Berlin-Dahlem.

Elektronen und Positronen erzeugen beim Durchgang durch Materie Lichtquanten; Lichtquanten erzeugen beim Durchgang durch Materie Elektronen und Positronen. Im Wechselspiel dieser Prozesse verteilt sich die Energie eines schnellen Teilchens nach und nach auf zahlreiche

---

[1] Der vorliegende Text wurde nach einem Vortrag von W. HEISENBERG, unter Benutzung einer Ausarbeitung von D. LYONS, von C. F. v. WEIZSÄCKER abgefaßt. Dabei wurden weitergehende Rechnungen von S. FLÜGGE und G. MOLIÈRE verwendet.

langsamere Teilchen: es entsteht eine „Kaskade". Die Vermehrung und Bremsung der weichen Komponente der kosmischen Strahlung und die Bildung der Schauer und Stöße beruht auf diesem Vorgang. Der nachfolgende Bericht soll die mathematische Theorie der Kaskadenbildung deduktiv darstellen. Der Vergleich mit der Erfahrung ist den späteren Berichten überlassen.

1. Die Elementarprozesse.

a) Bremsstrahlung. Fliegt ein Elektron[1] an einem geladenen Teilchen vorbei, so wird es aus seiner Flugrichtung abgelenkt. Mit dieser Ablenkung ist eine Energieübertragung an das ablenkende Teilchen verbunden. Für Elektronen, deren Geschwindigkeit sich noch merklich von der Lichtgeschwindigkeit unterscheidet, ist dieser Vorgang die wichtigste Ursache ihrer allmählichen Bremsung beim Durchgang durch Materie. Die Hauptrolle spielt dabei die Energieübertragung an die Hüllenelektronen der Atome, welche meist zur Ionisation der Atome führt. Wir nennen diese Art der Energieabgabe kurz „Ionisation".

Bei sehr schnellen Elektronen wird ein anderer Energieabgabeprozeß wichtiger als die Ionisation: die Bremsstrahlung. Mit der Änderung der Flugrichtung des Elektrons ist eine Änderung des elektromagnetischen Feldes in der Umgebung des Elektrons verbunden. Das Elektron wird dadurch zum Ausgangspunkt einer elektromagnetischen Welle oder quantenmäßig gesprochen, es emittiert ein Lichtquant oder mehrere Lichtquanten. Die Wahrscheinlichkeit dieses Prozesses ist von SAUTER (S 1a), BETHE und HEITLER (B 16) berechnet worden. Der Wirkungsquerschnitt dQ für die Emission eines Lichtquants, dessen Energie zwischen k und $k + dk$ liegt, beim Vorübergang eines Elektrons der Energie E an einem Kern der Ladung Ze ist

$$dQ = 4\frac{Z^2 e^2}{\hbar c}\left(\frac{e^2}{mc^2}\right)^2 \frac{dk}{k}\left\{\left[\frac{4}{3}\left(1 - \frac{k}{E}\right) + \left(\frac{k}{E}\right)^2\right]\log\frac{183}{Z^{1/3}} + \text{kleinere Glieder}\right\}. \quad (1)$$

Da der Wirkungsquerschnitt proportional zu Z^2 ist, kann, außer für die allerleichtesten Elemente, die für die Höhenstrahlung keine Rolle spielen, die Wirkung des Atomkerns als groß gegen die Wirkung der Hüllenelektronen angesehen werden. Der Einfluß der Hülle ist in der obigen Formel nur in dem logarithmischen Glied summarisch berücksichtigt; dieses Glied verdankt seine Form der Berechnung der Abschirmung des Kernfeldes nach außen nach der Methode von THOMAS und FERMI.

Die eckige Klammer in (1) hat den Wert 4/3 für $k=0$, 11/12 für $k=E/2$ und 1 für $k = E$. Man begeht somit nur einen geringen Fehler, wenn man sie konstant gleich 1 setzt. Das ist die Näherung, in der wir im

---

[1] Unter „Elektronen" verstehen wir, solange nicht ausdrücklich das Gegenteil ausgesprochen wird, sowohl positive wie negative Teilchen.

folgenden rechnen wollen. Der Fehler ist geringer als die Fehler, welche durch die Annäherungen der weiteren Rechnung auftreten. Wir erhalten dann für die Anzahl der Lichtquanten im Intervall dk, die auf der Strecke dx von einem Elektron mit einer Energie $E \gg mc^2$ durch Bremsstrahlung erzeugt werden:

$$dn(k) = \frac{dx}{x_0} \cdot \frac{dk}{k} . \tag{2}$$

Dabei ist die Länge x_0, welche wir als „Strahlungseinheit" bezeichnen wollen, definiert durch

$$\frac{1}{x_0} = 4 N Z^2 \frac{e^2}{\hbar c} \cdot \left(\frac{e^2}{mc^2}\right)^2 \log \frac{183}{Z^{1/3}} . \tag{3}$$

N ist die Anzahl streuender Atomkerne im Kubikzentimeter. Wir wollen von nun an alle Längen in Strahlungseinheiten messen. Eine in Strahlungseinheiten gemessene Länge soll mit dem Buchstaben l bezeichnet werden. Gleichung (2) nimmt dann die einfache Form an:

$$dn(k) = dl \cdot \frac{dk}{k} . \tag{4}$$

Der reziproke Wert der Strahlungseinheit wächst nahezu quadratisch mit der Ordnungszahl und linear mit der Dichte der bremsenden Substanz; für Blei ist die Strahlungseinheit $1/2$ cm, für Luft über 300 m. Eine Tabelle für verschiedene Stoffe gibt der nachfolgende Bericht von G. Molière (Seite 32).

Die Strahlungseinheit ist die Strecke, auf der die Energie eines Elektrons im Mittel auf den e-ten Teil absinkt. Der mittlere Energieverlust pro cm ist

$$\frac{dE}{dx} = -\frac{1}{x_0} \int_0^E \frac{dk}{k} \cdot k = -\frac{E}{x_0} \tag{5}$$

und daher ist

$$E = E_0 e^{-x/x_0}. \tag{6}$$

Anschaulich kann man sagen: die Anzahl der Lichtquanten zwischen k und $k + dk$, die auf der Strecke x_0 erzeugt werden, ist zu $1/k$ proportional, ihre Energie also zu $k \cdot 1/k$, d. h. konstant. D. h. die Wahrscheinlichkeit für jeden Energieverlust ist ungefähr gleich groß. Da die pro Wegstrecke durch Ionisation verlorene Energie von der Primärenergie nahezu unabhängig ist, ist damit auch das Überwiegen des Energieverlustes durch Bremsstrahlung bei großen Primärenergien verständlich.

b) Paarbildung. Lichtquanten, deren Energie nicht sehr groß gegen mc^2 ist, verlieren ihre Energie beim Durchgang durch Materie vorzugsweise durch den Compton-Effekt und den Photoeffekt, also durch Zusammenstöße mit Elektronen. Die Lichtquanten der Höhenstrahlung verlieren ihre Energie hingegen überwiegend durch die Bildung von

Elektronenpaaren. Dieser Vorgang hat eine gewisse Analogie zur Bremsstrahlung; nach der DIRACschen Theorie des Positrons kann er als der inverse Prozeß zur Bremsstrahlung betrachtet werden. Bei der Bremsstrahlung geht ein Elektron anläßlich des Vorübergangs an einem Kern aus einem Zustand höherer Energie in einen Zustand geringerer Energie über. Die Umkehrung dieses Vorgangs ist, daß ein Lichtquant beim Vorübergang an einem Kern (der bei dem Prozeß notwendig ist, um den aus den Erhaltungssätzen folgenden Impulsüberschuß zu übernehmen) von einem Elektron absorbiert wird, das dabei in einen Zustand höherer Energie übergeht. Paarbildung tritt nun ein, wenn das absorbierende Elektron ursprünglich in einem der nach der DIRACschen Theorie möglichen Zustände negativer Energie war. Dieses Elektron selbst war dann nämlich, solange seine Energie negativ war, eines aus einer unendlichen Dichte von Elektronen negativer Energie, die nach der Annahme von DIRAC keine physikalischen Wirkungen ausüben. Wenn es aber positive Energie angenommen hat, so ist es erstens selbst wahrnehmbar, und zweitens tritt das zurückbleibende „Loch" in der Verteilung von Elektronen negativer Energie als Positron in Erscheinung.

Der Wirkungsquerschnitt für die Erzeugung eines Elektronenpaares, dessen eines Elektron die Energie E hat, durch ein Lichtquant der Primärenergie k (das andere Elektron hat dann die Energie $k-E$) ist für $k \gg mc^2$ nach BETHE und HEITLER (B 16):

$$dQ = 4Z^2 \frac{e^2}{\hbar c}\left(\frac{e^2}{mc^2}\right)^2 \frac{dE}{k}\left[\left(\frac{E}{k}\right)^2 + \left(1 - \frac{E}{k}\right)^2 + \frac{2}{3}\frac{E}{k}\left(1 - \frac{E}{k}\right)\right] \log \frac{183}{Z^{1/3}} \qquad (7)$$
$$+ \text{ kleinere Glieder.}$$

Die eckige Klammer ist gleich eins für $E = 0$ und $E = k$; sie hat den Wert 2/3 für $E = k/2$. Man kann sie also wieder ohne allzu großen Fehler konstant setzen, d. h. annehmen, daß die Primärenergie des Lichtquants mit gleicher Wahrscheinlichkeit in jedem Verhältnis auf die beiden Elektronen aufgeteilt wird. Integriert man die Formel über E und multipliziert mit N, so erhält man für die Wahrscheinlichkeit dafür, daß auf der Wegstrecke dx bzw. dl überhaupt ein Paar von dem Lichtquant erzeugt wird:

$$dn = \frac{7}{9}\frac{dx}{x_0} = \frac{7}{9}dl = \delta dl. \qquad (8)$$

Wir haben für 7/9 zur Abkürzung δ gesetzt.

    c) **Ionisation.** Durch das Wechselspiel von Bremsstrahlung und Paarbildung entsteht aus einem energiereichen Primärteilchen eine Kaskade, die so lange anwächst, bis die Energie der einzelnen Teilchen zu klein geworden ist. Die Grenzenergie ist definiert durch das beginnende Überwiegen des Energieverlustes der Elektronen durch Ionisation gegenüber dem Energieverlust durch Bremsstrahlung. Wenn nämlich ein soeben durch Paarbildung entstandenes Elektron nur noch so wenig Energie hat, daß es auf der nächsten Strahlungslänge x_0, die es durchläuft, den größeren Teil seiner Energie bereits durch Ionisation verliert, so ist sein weiterer Beitrag zur Kaskade in der Näherung, in der wir hier rechnen wollen, zu vernachlässigen. Wir betrachten also die Kaskadenentwicklung als abgebrochen, wenn die Energie der Elektronen gleich dem Ionisations-Energieverlust pro Strahlungslänge wird. Diesen nennen wir E_J. Wir rechnen in der einfachen Näherung, daß für $E > E_J$ der Ionisationsverlust und für $E < E_J$ die Strahlung vernachlässigt wird. Genauere Untersuchungen von Arley haben gezeigt, daß man mit dieser Näherung Fehler bis zu 50% begehen kann.

Es ist also

$$E_J = x_0 \left(\frac{dE}{dx}\right)_{\text{Ion.}} = \left(\frac{dE}{dl}\right)_{\text{Ion.}} \tag{9}$$

Der Ionisationsverlust pro cm ist proportional zur Elektronenzahl pro cm³, also etwa zu Z, x_0 hingegen ist umgekehrt proportional zu Z^2. Also ist E_J genähert proportional zu $1/Z$. Eine genäherte Interpolationsformel ist

$$E_J = \frac{1600\, m c^2}{Z}. \tag{10}$$

Eine Tabelle von E_J ist im nachfolgenden Bericht von G. Molière (Seite 32) gegeben.

2. Die Grundgleichungen der Kaskadentheorie.

Wir betrachten nun die Ausbildung einer Kaskade. Wir vereinfachen die Rechnung durch die Annahme, daß alle Teilchen genau nach vorne emittiert werden, so daß die Teilchenzahl nur als Funktion des durchlaufenen Wegs und nicht auch des Ablenkungswinkels betrachtet wird. Die Winkelstreuung ist in dem nachfolgenden Artikel von G. Molière behandelt. Unser Verfahren schließt sich eng an eine Arbeit von Landau und Rumer (L 2) an. Wir führen folgende Bezeichnungen ein:

$F(E, l) =$ Anzahl der Elektronen einer Energie $> E$ im Abstand l vom Ausgangspunkt der Kaskade.

$G(E, l) =$ Anzahl der Lichtquanten einer Energie $> E$ im Abstand l.

Die Anzahlen von Elektronen bzw. Lichtquanten zwischen den Energien E und $E + dE$ sind dann gegeben durch

$$\left. \begin{aligned} f(E, l)\, dE &= -\frac{\partial F(E, l)}{\partial E}\, dE, \\[2mm] g(E, l)\, dE &= -\frac{\partial G(E, l)}{\partial E}\, dE. \end{aligned} \right\} \tag{11}$$

Wenn diese Teilchen nun die Strecke dl weiterlaufen, ändert sich die Anzahl der Elektronen nach der Gleichung

$$F(E, l + dl) = F(E, l) - dF(E, l)_{\text{Str.}} + dF(E, l)_{\text{Paar}}, \tag{12}$$

$dF(E,l)_{\mathrm{Str}}$ ist dabei die Anzahl der Elektronen, deren Energie auf der Strecke dl durch Bremsstrahlung unter den Wert E sinkt; $dF(E,l)_{\mathrm{Paar}}$ ist die Anzahl der Elektronen oberhalb E, die auf dieser Strecke durch Paarbildung neu entstehen. Da ein Elektron der Energie $E' > E$ eine Energie $> E' - E$ mit der Wahrscheinlichkeit $dl \int_{E'-E}^{E'} dk/k$ verliert, ist

$$dF(E,l)_{\mathrm{Str}} = \int_{E}^{\infty} dE' f(E',l) \cdot dl \cdot \int_{E'-E}^{E'} \frac{dk}{k}. \tag{13}$$

Andererseits erzeugt ein Lichtquant der Energie E' auf dl nach (8) $2\delta\, dl$ Elektronen, deren Energien mit gleicher Wahrscheinlichkeit im Intervall von 0 bis E' liegen. Uns interessiert hier aber nur der Bruchteil, dessen Energie im Intervall von E bis E' liegt. Also wird

$$dF(E,l)_{\mathrm{Paar}} = \int_{E}^{\infty} dE' g(E',l) \cdot 2\delta\, dl \frac{E'-E}{E'}. \tag{14}$$

Analog erhalten wir für die Änderung der Lichtquantenzahl

$$G(E, l + dl) = G(E,l) - \delta \cdot dl\, G(E,l) + \int_{E}^{\infty} dE' f(E',l) \cdot dl \int_{E}^{E'} \frac{dk}{k}. \tag{15}$$

Setzen wir in (12) und (15) für f und g die Ableitungen $\partial F/\partial E$ und $\partial G/\partial E$ ein, dividieren durch dl und formen die entstehenden Ausdrücke durch partielle Integration um, wobei vorausgesetzt ist, daß F und G für $E' = \infty$ verschwinden, so erhalten wir die Grundgleichungen der Kaskadentheorie:

$$\frac{\partial F(E,l)}{\partial l} = -\int_{E}^{\infty}[F(E',l)-F(E,l)]\,dE'\left(\frac{1}{E'}-\frac{1}{E'-E}\right)+2\delta\int_{E}^{\infty} G(E',l)\,dE'\frac{E}{E'^2}, \tag{16}$$

$$\frac{\partial G(E,l)}{\partial l} = -\delta \cdot G(E,l) + \int_{E}^{\infty} F(E',l)\frac{dE'}{E'}.$$

3. Lösung für ein Potenzspektrum.

Diese Integro-Differentialgleichungen haben zwei Eigenschaften, die es ermöglichen, eine spezielle Lösung leicht aufzufinden: 1. sie sind linear in G und F, daher ist die Summe zweier Lösungen wieder eine Lösung, 2. sie sind homogen in E und E', daher kann man

$$\frac{E'}{E} = \xi \tag{17}$$

setzen und erhält dann Gleichungen, die nur von ξ und l abhängen.

Wir setzen

$$F(E,l) = F(l) \cdot E^{-s}, \qquad G(E,l) = G(l) \cdot E^{-s}, \qquad s > 0 \tag{18}$$

und erhalten für $F(l)$ und $G(l)$ die Differentialgleichungen

$$\left.\begin{aligned}
\frac{dF(l)}{dl} = F'(l) &= -\sigma(s)\,F(l) + 2\delta\frac{G(l)}{s+1},\\[2mm]
\frac{dG(l)}{dl} = G(l) &= -\delta G(l) + \frac{F(l)}{s}.
\end{aligned}\right\} \tag{19}$$

Diese Gleichungen hängen nicht mehr von ξ ab. Wenn also ein Potenzspektrum für $l = 0$ vorgegeben ist, so behält es für alle l diese Form bei. $\sigma(s)$ ist das folgende Integral:

$$\sigma(s) = \int\limits_1^\infty\left(\frac{1}{\xi^s} - 1\right)\left(\frac{1}{\xi} - \frac{1}{\xi-1}\right)d\xi = C + \psi(s). \tag{20}$$

$C = 0,577\ldots$ ist die Eulersche Konstante, ψ ist die logarithmische Ableitung der Gammafunktion:

$$\psi(s) = \frac{\partial}{\partial s}\log\Gamma(s+1). \tag{21}$$

Es gilt

$$\psi(s+1) = \psi(s) + \frac{1}{s}. \tag{22}$$

Dabei ist

$$\sigma(1) = 1, \quad \sigma(2) = \tfrac{3}{2}, \quad \sigma(3) = \tfrac{11}{6}$$

und

$$\lim_{s\to\infty}\sigma(s) = \log s + 0,577. \tag{23}$$

Um die Gleichungen (19) zu lösen, setzen wir

$$\left.\begin{aligned}
F(l) &= a_1 e^{-\varkappa_1 l} + a_2 e^{-\varkappa_2 l},\\
G(l) &= b_1 e^{-\varkappa_1 l} + b_2 e^{-\varkappa_2 l}.
\end{aligned}\right\} \tag{24}$$

Eingesetzt in (19) ergibt dies die Bedingungsgleichungen

$$\left.\begin{aligned}
[\varkappa - \sigma(s)]\,a + \frac{2\delta}{s+1}\,b &= 0,\\[2mm]
\frac{a}{s} + (\varkappa - \delta)\,b &= 0,
\end{aligned}\right\} \tag{25}$$

welche jeweils a_1, b_1 und $\varkappa_1$ bzw. a_2, b_2 und $\varkappa_2$ untereinander verknüpfen. Es folgt

$$\frac{a}{b} = s(\delta - \varkappa), \tag{26}$$

mit den beiden Lösungen:

$$\varkappa_{1;2} = \frac{\sigma + \delta}{2} \pm \sqrt{\left(\frac{\sigma - \delta}{2}\right)^2 + \frac{2\delta}{s(s+1)}}. \tag{27}$$

Den Verlauf der Lösungen deutet die Tabelle an:

s	0	1	∞
$\varkappa_1$	$+\infty$	$1+\delta$	$\log s + 0{,}577$
$\varkappa_2$	$-\infty$	0	δ

Drückt man a durch b und $\varkappa$ aus, so folgt

$$\left.\begin{aligned}
F(l) &= b_1 s (\delta - \varkappa_1) e^{-\varkappa_1 l} + b_2 s (\delta - \varkappa_2) e^{-\varkappa_2 l}, \\
G(l) &= b_1 e^{-\varkappa_1 l} + b_2 e^{-\varkappa_2 l}.
\end{aligned}\right\} \tag{28}$$

Anschaulich bedeutet unser Ergebnis folgendes: Die Teilchenzahl in einer Kaskade, die von vornherein ein Potenzspektrum hat, ist gegeben durch die Summe zweier Exponentialfunktionen. Von den beiden Exponenten ist der größere, $\varkappa_1$, immer positiv, der ihm entsprechende Summand fällt also jedenfalls mit wachsendem l ab, und zwar immer schneller als der andere Summand. Er entspricht den Einstellvorgängen, die auftreten, wenn des Mengenverhältnis von Elektronen und Lichtquanten anfangs nicht dasjenige ist, das sich in einer fertigen Kaskade von selbst ausbildet. Z. B. kann man eine Kaskade, in der anfangs nur Elektronen, aber keine Lichtquanten vorkommen, darstellen, indem man $b_1 = - b_2$ setzt. Dem raschen Abfall des Summanden mit $\varkappa_1$ entspricht dann zunächst ein Ansteigen von $G(l)$ mit wachsendem l, bis dieser erste Summand verschwindend klein geworden ist und der Verlauf von G nur noch durch den zweiten Summanden geregelt wird. $\varkappa_2$ ist negativ für $s < 1$, positiv für $s > 1$, für $s = 1$ hat es den Wert Null. Je nach dem Wert des Exponenten nimmt die Kaskade also schließlich ständig an Teilchen zu oder ab. Dies ist physikalisch begreiflich. Die im ganzen in einer Kaskade enthaltene Energie ist

$$E_K = \int [f(E) + g(E)] E\,dE \sim \int E^{-s}\,dE. \tag{29}$$

Dieses Integral divergiert für $s = 1$ logarithmisch, für $s < 1$ divergiert es an der oberen, für $s > 1$ an der unteren Grenze. Die Divergenz an der unteren Grenze ist physikalisch belanglos, da in Wirklichkeit das Kaskadenspektrum bei E_J aufhört. Eine Kaskade mit $s > 1$ hat also einen endlichen Energieinhalt, der sich wegen des ständigen Verlustes derjenigen Teilchen, deren Energie kleiner wird als E_J, mit wachsendem l allmählich erschöpfen muß. Eine Kaskade mit $s < 1$ hätte hingegen oberhalb des gegebenen Energiewertes noch einen unendlich großen Energievorrat, aus dem stets mehr Energie in das gerade betrachtete Energieintervall von obenher nachgeliefert wird, als es nach unten verliert. In der Natur kommen natürlich nur Kaskaden mit endlicher Gesamtenergie vor, d. h. Spektren mit $s < 1$ können nicht bis zu beliebig hohen Energien reichen. Nach anfänglichem Anwachsen der Teilchenzahl muß schließlich wieder eine Abnahme folgen.

Das primäre Spektrum der Höhenstrahlung läßt sich wahrscheinlich in guter Näherung durch ein Potenzgesetz mit $s \approx 1{,}8$ darstellen. Die Höhenstrahlung in der Atmosphäre müßte also eine Kaskade mit exponentiell nach unten abfallender Teilchenzahl sein. Nach (27) ist für $s = 1{,}8$ $\varkappa_1 = 1{,}74$ und $\varkappa_2 = 0{,}46$. Der schließliche Abfall müßte also durch $e^{-0{,}46 l}$ dargestellt sein, d. h. ein Absinken auf den e-ten Teil auf je 2 Strahlungslängen ($^3/_4$ m H_2O) zeigen. Für die weiche Komponente unterhalb des Maximums ist diese Bedingung recht gut erfüllt und bestätigt damit die Vorstellung vom Kaskadencharakter dieser Komponente. Daß ein Maximum der Intensität in hohen Luftschichten auftritt, muß von einer Abweichung des Primärspektrums vom Potenzcharakter herrühren. Eine hinreichende Erklärung würde schon die Tatsache bieten, daß geladene Teilchen unterhalb einer bestimmten Energie das erdmagnetische Feld nicht mehr durchdringen können. Nach neueren Untersuchungen scheint aber die Primärstrahlung überhaupt nicht aus Elektronen, sondern aus Protonen zu bestehen, und damit wäre die Kaskadentheorie in der hier entwickelten Form überhaupt nur auf denjenigen Teil der Strahlung anzuwenden, der sicher aus Elektronen und Lichtquanten besteht, d. h. auf die weiche Komponente etwa vom Maximum an abwärts.

4. Die Ausbildung einer Einzelkaskade.

Wir lassen nun die Voraussetzung, daß das Spektrum von vornherein den Potenzcharakter habe, fallen und berechnen die Ausbildung einer Kaskade, die von einem einzelnen Elektron bestimmter Energie ausgelöst wird. Wir folgen dabei dem von LANDAU und RUMER (L 2) eingeschlagenen Weg.

Wir führen zwei neue Funktionen einer Hilfsvariablen s ein durch die Gleichungen

$$\left. \begin{aligned} f(s,l) &= \int_0^\infty dE\, E^s f(E,l)\,, \\ g(s,l) &= \int_0^\infty dE\, E^s g(E,l)\,. \end{aligned} \right\} \tag{30}$$

Diese Funktionen haben für die Zerlegung der Funktionen $f(E,l)$ und $g(E,l)$ in Potenzspektren E^{-s} eine ähnliche Bedeutung wie die FOURIERkoeffizienten für die Zerlegung in trigonometrische Funktionen. Die neuen Funktionen $f(s,l)$ und $g(s,l)$ genügen denselben Differentialgleichungen wie unsere früheren Funktionen $F(l)$ und $G(l)$, also den Gleichungen (19). Die Lösungen können wir also nach (28) sofort hinschreiben.

$$\left. \begin{aligned} f(s,l) &= b_1 s \cdot (\delta - \varkappa_1)\, e^{-\varkappa_1 l} + b_2 s \cdot (\delta - \varkappa_2)\, e^{-\varkappa_2 l}\,, \\ g(s,l) &= b_1 e^{-\varkappa_1 l} + b_2 e^{-\varkappa_2 l}\,. \end{aligned} \right\} \tag{31}$$

Dabei ist nur zu beachten, daß b_1 und b_2 Funktionen des Parameters s sind. Sind zu Anfang ($l = 0$) keine Lichtquanten vorhanden, so ist

$$b_2 = -b_1 = b(s).\tag{32}$$

Ist zu Anfang gerade ein Elektron der Energie E_0 da, so lautet das einfallende Spektrum

$$f(E, l = 0) = \delta(E - E_0).\tag{33}$$

Die Funktion $\delta(x)$ bedeutet dabei die DIRACsche singuläre δ-Funktion, welche durch die beiden Gleichungen

$$\left.\begin{array}{ll} \delta(x) = 0 & \text{für} \quad x \neq 0, \\[2mm] \int\limits_{-\varepsilon}^{+\varepsilon} \delta(x)\,dx = 1 & \text{für jedes } \varepsilon \end{array}\right\}\tag{34}$$

definiert ist. Daher ist

$$f(s, 0) = E_0^s.\tag{35}$$

Aus (31) folgt dann

$$b(s) = \frac{E_0}{s(\varkappa_1 - \varkappa_2)}.\tag{36}$$

Dies in (31) wieder eingesetzt, ergibt

$$\left.\begin{array}{l} f(s, l) = \dfrac{E_0^s}{\varkappa_1 - \varkappa_2}\left[-(\delta - \varkappa_1)\,e^{-\varkappa_1 l} + (\delta - \varkappa_2)\,e^{-\varkappa_2 l}\right], \\[4mm] g(s, l) = \dfrac{E_0^s}{s(\varkappa_1 - \varkappa_2)}\left[-e^{-\varkappa_1 l} + e^{-\varkappa_2 l}\right]. \end{array}\right\}\tag{37}$$

In den eckigen Klammern können wir die Glieder mit $e^{-\varkappa_1 l}$ vernachlässigen. Denn da anfangs nur ein Teilchen vorhanden ist und $\varkappa_1$ stets positiv ist, sind sie immer kleiner als 1; wir interessieren uns aber nur für wirkliche Kaskaden, in denen die Teilchenzahl groß gegen eins ist.

Wir führen die folgenden Abkürzungen ein:

$$\left.\begin{array}{l} \log \dfrac{E_0}{E} = y, \\[4mm] E \cdot f(E, l) = e^{\varphi(y, l)} \end{array}\right\}\tag{38}$$

und erhalten aus (37) unter Vernachlässigung von $e^{-\varkappa_1 l}$ und aus (30):

$$\int\limits_0^\infty dy\, e^{\varphi(y, l) - ys} = \frac{\delta - \varkappa_2}{\varkappa_1 - \varkappa_2}\, e^{-\varkappa_2 l}.\tag{39}$$

Wir wollen nun zur Bestimmung von φ bzw. f aus dieser Integralgleichung eine Näherungsrechnung durchführen. Der Integrand hat bei einem y-Wert, den wir y_m nennen wollen, ein steiles Maximum. Wie wir später sehen werden, ist dieses Maximum um so steiler, je mehr Teilchen gebildet werden. y_m bestimmt sich aus der Bedingung

$$\left(\frac{\partial \varphi}{\partial y}\right)_m - s = 0.\tag{40}$$

Wir entwickeln $\varphi(y)$ in der Nähe dieses Maximums („Sattelpunktmethode"):

$$\varphi(y) - ys = \varphi(y_m) - sy_m + \tfrac{1}{2}\varphi''(y_m)(y - y_m)^2, \tag{41}$$

berücksichtigen nur die hingeschriebenen Glieder und integrieren statt von 0 von $-\infty$ an. So erhalten wir näherungsweise für das Integral

$$e^{\varphi(y_m) - sy_m}\sqrt{\frac{2\pi}{-\varphi''(y_m)}} = \frac{\delta - \varkappa_2}{\varkappa_1 - \varkappa_2}\,e^{-\varkappa_2 l} \tag{42}$$

und logarithmiert

$$\varphi(y_m) - sy_m + \varkappa_2 l = \chi(s, y_m), \tag{43}$$

mit

$$\chi(s, y_m) = \log\left[\frac{\delta - \varkappa_2}{\varkappa_1 - \varkappa_2}\sqrt{\frac{-\varphi''(y_m)}{2\pi}}\right]. \tag{44}$$

Wir lassen nun den Index m fort. Differenzieren wir (43) nach s, so erhalten wir mit (43) selbst und (40) drei Gleichungen

$$
\left.
\begin{aligned}
\text{I.} \qquad & \varphi(y, l) - sy + \varkappa_2(s)\, l = \chi(s, y), \\[2mm]
\text{II.} \qquad & -y + \varkappa_2'(s)\, l = \frac{\partial \chi(s, y)}{\partial s}, \\[2mm]
\text{III.} \qquad & \frac{\partial \varphi(y, l)}{\partial y} - s = 0.
\end{aligned}
\right\} \tag{45}
$$

Ehe wir diese Gleichungen auflösen, vergegenwärtigen wir uns ihre Bedeutung. Die Ausgangsgleichungen (30) gelten für jedes s. Die Gleichungen (45) nun verknüpfen jeden Wert von s mit einem bestimmten Wert von y. Die Gleichung (45, III) bedeutet nämlich, daß y diejenige Stelle ist, an der $F(E, l)$ sich als Funktion von E verhält wie E^{-s}. Aus den Gleichungen (II) und (III) können wir s als Funktion von y und l berechnen. Wir wissen damit, durch welche Potenz von E die Funktion F an jeder Stelle l und in der Umgebung jedes Energiewertes E approximiert wird. Setzen wir diesen Wert von s in (I) ein, so erhalten wir einen Ausdruck für $\varphi(y, l)$ selbst und können damit auch f und F direkt berechnen. Da wir schließlich nur F zu kennen wünschen, rechnen wir s, φ und f hier nicht aus, sondern bilden gleich die richtige Kombination dieser Größen.

$\chi(s, y)$ ist als Logarithmus langsam veränderlich mit s. Wir vernachlässigen daher zunächst die rechte Seite von (II). Indem wir (II) nach s differenzieren, erhalten wir

$$\frac{\partial y}{\partial s} = \varkappa_2'' l + \text{kleine Glieder.} \tag{46}$$

Durch Differentiation von (III) nach y folgt unter Berücksichtigung von (46)

$$\frac{\partial^2 \varphi}{\partial y^2} = \frac{\partial s}{\partial y} = \frac{1}{\varkappa_2'' l}. \tag{47}$$

Dies setzen wir in (44) ein und erhalten

und
$$\chi(s,y) = \log\left[\frac{\delta - \varkappa_2}{\varkappa_1 - \varkappa_2} \cdot \sqrt{\frac{1}{2\pi l\,|\varkappa_2''|}}\right] \tag{48}$$

$$\frac{\partial \chi(s,y)}{\partial s} = \frac{\partial}{\partial s}\log\frac{\delta - \varkappa_1}{\varkappa_1 - \varkappa_2} + \frac{1}{2}\frac{\varkappa_2'''}{\varkappa_2''}, \tag{49}$$

was, in (45 II) eingesetzt, eine Gleichung für s ergibt.

Wir interessieren uns im folgenden für die Gesamtzahl $n_{el}(l)$ der Elektronen in einer Kaskade nach Durchlaufung der Schichtdicke l

$$n_{el}(l) = F(E_J) = \int\limits_{E_J}^{\infty} f(E,l)\,dE = \int\limits_{0}^{y_J} e^{\varphi(y,l)}\,dy, \tag{50}$$

wo

$$y_J = \log\frac{E_0}{E_J}. \tag{51}$$

Für $E \geqq 10^{10}\,eV$ ist $y_J \geqq 5$. $\varphi(y,l)$ steigt steil mit y an; wir können daher mit guter Näherung schreiben

$$n_{el}(l) = \int\limits_{0}^{y_J} dy \cdot \exp\left[\varphi(y_J,l) + \left(\frac{\partial\varphi}{\partial y}\right)_{y_J}(y - y_J)\right] = \frac{e^{\varphi(y_J,l)}}{\left(\frac{\partial\varphi}{\partial y}\right)_{y_J}}\left(1 - e^{-\left(\frac{\partial\varphi}{\partial y}\right)_{y_J}\cdot y_J}\right). \tag{52}$$

Nach (45 III) ist $(\partial\varphi/\partial y)_{y_J} = s(y_J,l)$. Da s von der Größenordnung Eins und y_J merklich größer als Eins ist, können wir in genügender Näherung das zweite Glied in der Klammer vernachlässigen und erhalten

$$n_{el}(l) = \frac{e^{\varphi(y_J,l)}}{s(y_J,l)}. \tag{53}$$

5. Auswertung des Ergebnisses.

a) Die Lage des Kaskadenmaximums. Die Elektronenzahl einer Kaskade muß als Funktion von l zunächst zunehmen und dann wieder abnehmen. Wir bestimmen den Wert $l_{\max}$, bei dem sie ihr Maximum erreicht. $s(y_J,l)$ ist eine langsam veränderliche Funktion von l. Daher stimmt das Maximum von $n_{el}(l)$ ungefähr mit dem von $e^{\varphi(y_J,l)}$ überein. Übrigens kommt es nicht sehr auf den genauen Wert von $l_{\max}$ an, da die Funktion n_{el} am Maximum sehr flach verläuft. Wir fordern also

$$\left(\frac{\partial\varphi(y_J,l)}{\partial l}\right)_{l_{\max}} = 0. \tag{54}$$

Aus (45 I) folgt unter Vernachlässigung der rechten Seite

$$\frac{\partial\varphi(y_J,l)}{\partial l} = y\frac{\partial s}{\partial l} - \varkappa_2 - \frac{\partial\varkappa_2}{\partial s}l\frac{\partial s}{\partial l}. \tag{55}$$

Die Summe des ersten und des letzten Gliedes der rechten Seite von (55) ist nach (45 II) genähert gleich Null. Also bleibt

$$\varkappa_2(l_{\max}) \approx 0. \tag{56}$$

Nach (27) bedeutet das

$$s(y_J, l_{\max}) \approx 1. \tag{57}$$

Dies ist anschaulich einleuchtend: solange bei $E = E_J$ noch $s < 1$ ist, nimmt die Teilchenzahl dort noch zu, für $s > 1$ nimmt sie ab, nach den Ergebnissen für reine Potenzspektren. Das Gebiet nahe E_J trägt aber am meisten zur Teilchenzahl der ganzen Kaskade bei. Für $s = 1$ ist

$$\varkappa_1 = 1{,}78, \quad \varkappa_2 = 0, \quad \varkappa_2' = 0{,}98, \quad \varkappa_2'' = -1{,}41, \quad \varkappa_2''' = 2{,}83, \tag{58}$$

$$\frac{\partial \chi}{\partial s} = +0{,}6.$$

Setzen wir die obigen Werte in (45 II) ein, wobei in der verwandten Näherung die rechte Seite konsequenterweise zu vernachlässigen ist, so folgt

$$l_{\max} = \frac{1}{\varkappa_2'}\, y_J = 1{,}02 \log \frac{E_0}{E_J}. \tag{59}$$

b) Die Anzahl der Elektronen und Lichtquanten am Maximum. Aus (45 I) folgt

$$\varphi(y_J, l_{\max}) = [s y_J - \varkappa_2 l_{\max} + \chi(s)]_{s=1} = y_J + \log 0{,}13 - \log \sqrt{l_{\max}}. \tag{60}$$

Daher wird

$$n_{el}(l_{\max}) = \frac{0{,}13}{\sqrt{l_{\max}}} \cdot \frac{E_0}{E_J}. \tag{61}$$

Während sich also $l_{\max}$ nur logarithmisch mit der Anfangsenergie ändert, ist die Größe der Kaskade am Maximum zum Verhältnis der Anfangsenergie zur Ionisationsenergie proportional. Beides ist anschaulich zu erwarten. Die Entfernung des Maximums vom Ausgangspunkt ist proportional zur Anzahl von Stößen, durch die das ursprüngliche Energiequantum E_0 in Quanten der Größenordnung E_J aufgeteilt wird; da jeder Stoß die Energie ungefähr halbiert, ist diese Stoßzahl zu $\log(E_0/E_J)$ proportional. Hingegen ist die Größe der Kaskade am Maximum eben durch die Anzahl der Quanten der Größe E_J gegeben, in welche E_0 zerfällt.

Nach (37) ist das Verhältnis der Lichtquantenanzahl zur Elektronenanzahl

$$\frac{n_q}{n_{el}} \approx \frac{1}{s(\delta - \varkappa_2)}. \tag{62}$$

Am Maximum ist aber $s = 1$ und $\varkappa_2 = 0$ und daher wegen $\delta = 7/9$

$$n_q(\max) \approx \tfrac{9}{7}\, n_{el}(\max). \tag{63}$$

c) Näherungsformel für die Entwicklung einer Kaskade. Nach (53) und (45 I) ist, wenn wir für y_J kurz y schreiben:

$$n_{el}(l) = \frac{e^{\varphi(y,\, l)}}{s} \tag{64}$$

mit

$$\varphi(y, l) = s y - \varkappa_2(s)\, l + \chi(s). \tag{65}$$

Nach (45 II) ist

$$y = \varkappa_2'(s)\, l - \chi'(s). \tag{66}$$

Für $\varkappa_2$ und $\chi(s)$ machen wir nun Näherungsansätze, die in dem Gebiet $1 \leq s \leq 3$, das uns interessiert, hinreichend genau sind:

$$\varkappa_2 = 1 - \frac{1}{s}, \qquad \chi = -\frac{\alpha}{s} - \beta s - \frac{1}{2}\log l \tag{67}$$

mit

$$\alpha = 1,4, \qquad \beta = 0,56.$$

Damit folgt

$$s = \sqrt{\frac{l - \alpha}{y - \beta}}, \tag{68}$$

und

$$n_{el}(l) = \sqrt{\frac{y - \beta}{l - \alpha}} \cdot \frac{1}{\sqrt{l}} \cdot e^{-l + 2\sqrt{(l-\alpha)(y-\beta)}}. \tag{69}$$

Der wesentliche Teil dieses Ausdrucks ist der Exponentialfaktor. Der ganze Ausdruck gilt nur für größere Werte von l. Dabei überwiegt bei wachsendem l zunächst die Wurzel im Exponenten, und zwar um so stärker, je größer y ist; die Elektronenzahl wächst dann exponentiell an. Schließlich aber überwiegt das Glied $-l$, und die Kaskade wird wie e^{-l} absorbiert. Für die Lichtquantenzahl folgt nach (62) und (67)

$$n_q = \frac{n_{el}}{1 - \frac{2}{6}s}. \tag{70}$$

Im Anfang der Kaskade sind also etwa ebenso viele Lichtquanten wie Elektronen vorhanden; vom Maximum an überwiegen die Lichtquanten immer mehr.

d) Anzahl der Elektronen mit $E < E_J$. Nach ARLEY (A 5) kann man die folgende Abschätzung machen: Wir nehmen an, daß ein Elektron, das die Energie E_J erreicht hat, von nun an nicht mehr durch Strahlung, sondern nur noch durch Ionisation Energie verliert. Da E_J gerade definiert war als die Stelle, an der (nach strenger Rechnung) der Energieverlust durch Ionisation gleich dem durch Strahlung wird, muß also nunmehr das Elektron gerade noch eine Strahlungslänge weit laufen können, ehe es seine ganze Energie E_J durch Ionisation verloren hat. Die Elektronen, welche eine Energie $E < E_J$ haben, haben also, da der Energieabfall durch Ionisation proportional dem durchlaufenen Weg ist, gerade den Bruchteil $\varepsilon = 1 - E/E_J$ einer Strahlungslänge durchlaufen, seit sie die Energie E_J hatten. So folgt

$$f(E, l)\,dE \atop (E<E_J) = dE \cdot \begin{cases} f\left(E_J, l - \left[1 - \dfrac{E}{E_J}\right]\right) & \text{für} \quad l \geq 1 - \dfrac{E}{E_J} \\[2ex] 0 & \text{für} \quad l < 1 - \dfrac{E}{E_J}. \end{cases} \tag{71}$$

Entwickeln wir f in der Umgebung von l, so ergibt sich

$$f(E, l) = f(E_J, l)\left\{1 - \varepsilon \frac{\partial}{\partial l}\ln f(E_J, l)\right\}. \tag{72}$$

Wir bilden die Anzahl n_{el}^{*} aller Elektronen unter E_J:

$$n_{el}^{*} = \int_{0}^{E_J} f(E, l)\, dE = E_J f(E_J, l)\left\{1 - \frac{1}{2}\frac{\partial}{\partial l}\ln f(E_J, l)\right\} = e^{\varphi(y, l)}\left(1 + \frac{\varkappa_2}{2}\right). \quad (73)$$

Mit der oben verwendeten Näherung für $\varkappa_2$ ergibt dies die zu (53) analoge Formel

$$n_{el}^{*} = \left(\frac{3}{2} - \frac{1}{2s}\right) e^{\varphi(y, l)} = \left(\frac{3s}{2} - \frac{1}{2}\right) n_{el}. \quad (74)$$

Insbesondere ist also beim Kaskadenmaximum ($s \approx 1$) die Anzahl der Elektronen unterhalb E_J ungefähr ebenso groß wie die Anzahl der Elektronen oberhalb E_J.

3. Die großen Luftschauer.

Von G. Molière-Berlin-Dahlem.

Mit 2 Abbildungen

Im vorhergehenden Bericht wurde die Kaskadentheorie geschildert, soweit sie von der eigentlichen Entwicklung der Kaskade, d. h. der Teilchenvervielfachung und der Aufteilung der Energie handelt. Dabei konnte von der Winkelstreuung ganz abgesehen und so gerechnet werden, als ob alle Kaskadenteilchen die Richtung des ursprünglichen, die Kaskade auslösenden Teilchens unverändert beibehielten. Die Berücksichtigung dieser Winkelstreuung und die Berechnung der durch sie bedingten Winkel- und räumlichen Verteilung der Kaskadenteilchen erfordert daher eine Erweiterung der Theorie, die uns im folgenden beschäftigen soll. Die wesentlichsten Anwendungen derselben betreffen Experimente, die sich mit den „ausgedehnten Luftschauern" als der wichtigsten Kaskadenerscheinung beschäftigen (siehe den Schlußteil 7 dieses Berichtes S. 34).

1. Qualitatives Bild des Schauers.

Bevor wir uns den Rechnungen selbst zuwenden, sei ein qualitatives Bild des Kaskadenschauers entworfen: Die vollständige Vernachlässigung der Winkelablenkungen stellt in der Tat den geeigneten Ausgangspunkt dar, und der Schauer als Ganzes behält genau die Fortpflanzungsrichtung des ihn auslösenden Primärteilchens bei. Damit ist nicht gesagt, daß stets nur kleine Winkelablenkungen vorkämen; vielmehr können, wie z. B. bei Kaskaden in Blei, durchaus große Winkeldivergenzen auftreten. Für die Fortentwicklung der Kaskade von Ort zu Ort sind jedoch in erster Linie die verhältnismäßig wenigen Teilchen

6. Zusammenfassung.

Die experimentellen Untersuchungen lassen folgendes Bild der Mesonenentstehung als das wahrscheinlichste erscheinen:

Aus dem Weltraum fällt eine primäre Protonenstrahlung ein (JOHNSON), deren Energiespektrum wahrscheinlich mit dem bei der weichen Komponente und den Mesonen gefundenen übereinstimmt (vgl. 1. Bericht, HEISENBERG). Die Protonen erzeugen am äußersten Rand der Atmosphäre, d. h. jedenfalls auf einer Strecke, die kürzer als eine Strahlungseinheit der Kaskadentheorie ist, die harte Mesonenkomponente, wahrscheinlich in Mehrfachprozessen. Die weiche Strahlung, die das Maximum von PFOTZER bildet, entsteht im selben oder in einem getrennten Akt aus den Primären oder ist eine Sekundärstrahlung der Mesonen. Experimentell ist darüber nichts bekannt. Die Mesonen bilden die noch auf Seehöhe beobachtete harte Komponente der kosmischen Strahlung, deren Häufigkeitsmaximum dort bei $2 \cdot 10^9$ eV liegt. Außerdem werden Mesonen auch noch auf andere Weise, und zwar wahrscheinlich als Sekundäre der Lichtquanten der weichen Komponente in großer Anzahl, aber mit geringerer Energie ($< 5,2 \cdot 10^8$ eV) erzeugt. Auch hier werden Vielfachprozesse (Explosionen) beobachtet. In der Nähe der Erdoberfläche werden diese Mesonen sehr selten, in Übereinstimmung mit der Tatsache, daß dort auch die weiche Kaskadenstrahlung schon fast vollständig absorbiert ist. In 7 km Höhe dagegen machen sie rund 30% der gesamten Mesonen aus. Außer aus den Lichtquanten wird wahrscheinlich ein kleiner Bruchteil ($< 5\%$) dieser Mesonen aus neutralen korpuskularen Partikeln (möglicherweise neutralen Mesonen) erzeugt. Künftige Untersuchungen müssen zeigen, auf welche Weise die Mesonen- und Kaskadenstrahlung aus den Primären hervorgehen und wie die geomagnetischen Effekte und die Intensitätsverhältnisse daraus zu erklären sind.

5. Schauer mit durchdringenden Teilchen.

Von A. KLEMM und W. HEISENBERG-Berlin-Dahlem.

Mit 4 Abbildungen.

Die Frage, ob in der kosmischen Strahlung auch Schauer beobachtet werden, die mehrere durchdringende Teilchen enthalten, ist theoretisch von grundlegender Bedeutung, da aus ihrem Auftreten auf die Existenz von echten Mehrfachprozessen (oder „Explosionen") geschlossen werden kann, bei denen in einem einzigen Akt mehrere Teilchen entstehen. Derartige explosionsartige Schauer sind theoretisch nach der YUKAWASchen Theorie zu erwarten [vgl. W. HEISENBERG (H 3); auch 8. Bericht, HEISENBERG].

Von den eigentlichen Schauern sind zu unterscheiden die *Paare* durchdringender Teilchen, die gelegentlich mit zu den Schauern gerechnet werden. Im folgenden soll das Wort „Schauer" aber nur auf Prozesse mit mindestens 3 Teilchen angewendet werden, die Diskussion der Paare soll kurz vorweggenommen werden.

1. Mesonenpaare.

Die ersten Anzeichen für Paare mit durchdringenden Teilchen ergaben sich aus Versuchen von MAASS (M 1) und von SCHMEISER und BOTHE (B 30 a, S 9). Die letztgenannten Autoren untersuchen Koinzidenzen von Zählern in großem Abstand von der schauererzeugenden Schicht und finden unter solchen Bedingungen das sogenannte zweite Maximum der ROSSISchen Kurve bei 17 cm Pb stark ausgeprägt. Sie schließen hieraus auf die Existenz von Strahlen einer Reichweite von 17 cm Pb. Die verschiedentlich vorgenommenen Wiederholungen dieser Versuche mit etwas abgeänderter Geometrie haben jedoch noch nicht zu einer völligen Klarstellung der Entstehung des zweiten Maximums geführt.

Ein sicherer Beweis für das Auftreten von Paaren durchdringender Teilchen, und zwar von Mesonenpaaren, ist durch die Arbeiten von BRADDICK und HENSBY (B 35), LEISEGANG (L 3) und HERZOG und BOSTICK (B 8, 9) geliefert worden.

BRADDICK und HENSBY (B 35) nehmen zählrohrgesteuerte Nebelkammerbilder in London 30 m unter der Erde auf. Sie erhalten 1900 Aufnahmen mit Einzelspuren von Mesonen. Die Mesonen werden erkannt an ihrem glatten Durchgang durch 1,4 cm bzw. 2,5 cm Blei. Unter diesen 1900 Aufnahmen zeigen 5 Aufnahmen Doppelspuren von Mesonen mit einem offenbar gemeinsamen Ursprung in der Erdschicht oberhalb der Kammer. Bei 1900 Einzelspuren wären nur 0,057 zufällige Doppelspuren zu erwarten gewesen.

LEISEGANG (L 3) macht zählrohrgesteuerte WILSON-Aufnahmen, wobei sich über der Kammer 11 cm bzw. 16 cm Blei und in der Kammer eine 1 cm starke Bleischicht befinden. Unter 900 aufgenommenen Mesoneneinzelbahnen findet er 3 Doppelbahnen, deren Ursprung in der auslösenden Schicht liegt. Der Beweis für die Mesonennatur wird dadurch erbracht, daß die Teilchen in 1 cm Blei kaum abgelenkt werden (also energiereich sind) und keine Sekundärteilchen bilden.

HERZOG und BOSTICK (H 8, 9) schließlich haben auf einer in großer Höhe (im Flugzeug) gewonnenen Nebelkammeraufnahme die Entstehung eines Mesonenpaares beobachtet.

Es kann also wohl kein Zweifel mehr darüber sein, daß gelegentlich Paare von Mesonen gebildet werden. Es liegt nahe, anzunehmen, daß das Primärteilchen ein Photon ist, daß es sich also um Prozesse handelt,

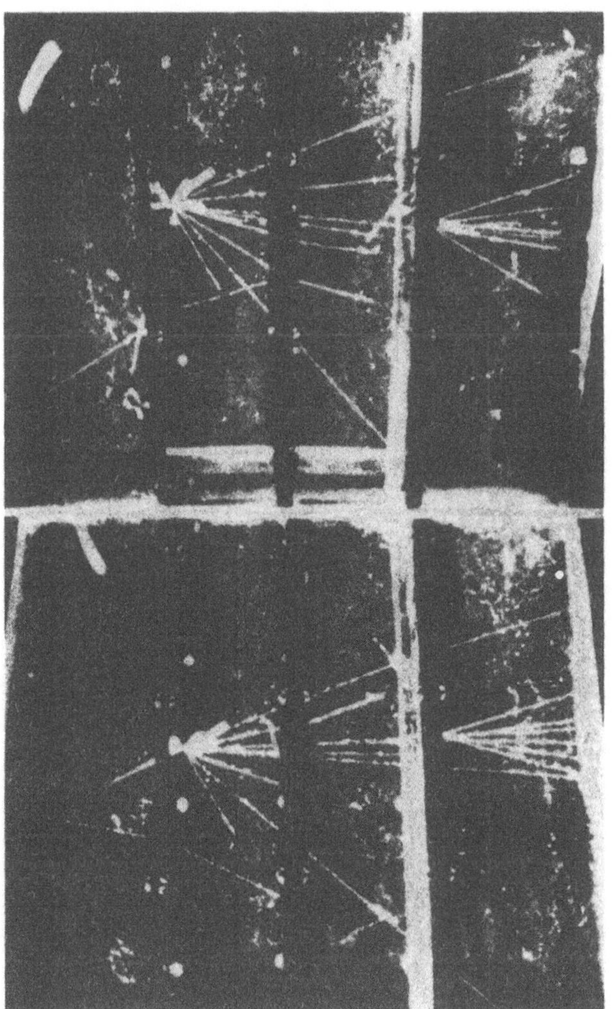

Abb. 1. Von FUSSELL beobachtete Schauer mit durchdringenden Teilchen.

die der gewöhnlichen Paarerzeugung bei den Elektronen analog sind. Die Häufigkeit der Paare läßt sich aus dem vorliegenden Material nur größenordnungsmäßig abschätzen. Der Wirkungsquerschnitt für ihre Entstehung aus Photonen scheint in der Gegend von 10^{-27} bis $10^{-26}\,\mathrm{cm}^2$

(pro Kernteilchen) zu liegen. Diese Größenordnung ist auch nach theoretischen Abschätzungen [z. B. Booth und Wilson (B 29)] nicht unplausibel.

2. Durchdringender Schauer.

Der Beweis für das Auftreten von echten Schauern mit durchdringenden Teilchen ist zuerst durch einige Nebelkammeraufnahmen von Fussell (F 7), dann durch systematische Zählrohruntersuchungen von Wataghin, Santos und Pompeja (W 2) sowie noch ausführlicher von Janossy und Ingleby (J 3) erbracht worden.

Abb. 1 zeigt das beste von Fussell beobachtete Beispiel eines solchen Prozesses. Der Schauer entsteht in einer 0,7 mm dicken Pb-Schicht. Von seinen Strahlen gehen mindestens drei ohne nennenswerte Ablenkung und ohne Sekundärspuren durch 1,2 bis 2 cm Pb, sind also als Mesonenspuren anzusehen. Die Schauererzeugung ist mit einer Kernzertrümmerung verknüpft, wie die dicken Spuren kurzer Reichweite beweisen. Der ganze Schauer tritt gleichzeitig mit Elektronen auf, die nicht vom Entstehungsort des Schauers kommen; der Prozeß scheint also im Rahmen eines größeren Luftschauers ausgelöst worden zu sein.

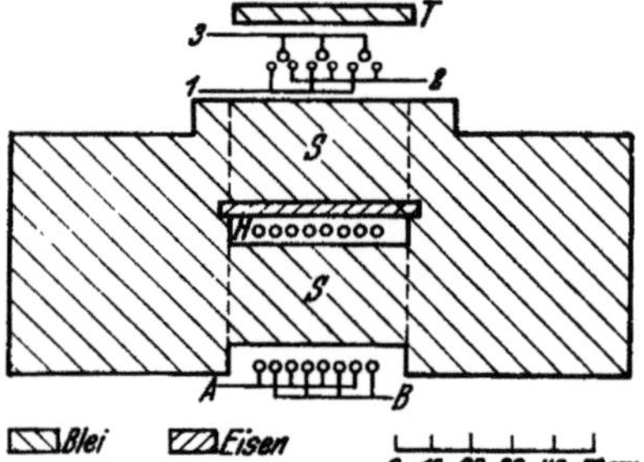

Abb. 2. Versuchsanordnung von Janossy und Ingleby.

Wataghin, Santos und Pompeja (W 2) haben in 800 m Seehöhe vier Zählrohre von je 100 cm² Fläche so aufgestellt, daß je zwei Zählrohre senkrecht untereinander stehen und die beiden Zählrohre voneinander einen horizontalen Abstand von das eine Mal 30 cm, das andere Mal 65 cm haben. Ein senkrecht von oben kommendes Teilchen, das ein Zählrohrpaar zum Ansprechen bringt, hat dabei 17 cm Blei zu durchdringen. Die Autoren beobachten Viererkoinzidenzen und finden bei 30 cm Abstand 4,5 Koinzidenzen pro Tag und bei 65 cm Abstand 3,6 Koinzidenzen pro Tag, während sie angeben, daß rein zufällig nur 0,3 Koinzidenzen pro Tag zu erwarten gewesen wären. Hier handelt es sich offenbar um Paare von durchdringenden Teilchen, wahrscheinlich Mesonen. Durch ein dazugefügtes fünftes Zählrohr wird ferner festgestellt, daß in einem erheblichen Bruchteil der Fälle mehr als zwei durchdringende Teilchen auftreten, da die fünffachen Koinzidenzen nicht viel seltener sind als die vierfachen.

Die ausführlichsten und aufschlußreichsten Untersuchungen dieser Art haben Janossy und Ingleby (J 3) angestellt.

Sie wählen die Versuchsanordnung der Abb. 2.

Bei fünffachen Koinzidenzen (1, 2, 3, A, B) wird die Zahl n der gleichzeitig ansprechenden Zählrohre H gemessen. Als „erzeugende" Schicht T wird Pb bzw. Al gewählt, die Dicke dieser Schicht wird zwischen 0 und 120 g/cm² variiert. Die absorbierende Schicht S besteht aus Blei. Die Ergebnisse sind aus der folgenden Tabelle und der Abb. 3 ersichtlich:

$n =$	0	1	2	3	4	5	6	7	8
Anzahl der Koinzidenzen (für beliebiges T)	854	911	184	139	130	108	80	85	49

Wenn $T = 0$ ist, so sorgt die experimentelle Anordnung dafür, daß nur Koinzidenzen beobachtet werden, die durch Luftschauer erzeugt werden. Denn nur dann werden die oberen drei Zählrohre 1, 2, 3 gleichzeitig ansprechen. Die Teilchen des Luftschauers lösen dann offenbar in der 25 cm dicken Bleimasse zwischen den Zählern 1, 2, 3 und H durchdringende Teilchen aus, die für das Ansprechen der anderen Zähler sorgen. Wenn nun über die ganze Anordnung das Material T gebracht wird, so scheint sich an dieser Entstehung grundsätzlich nichts

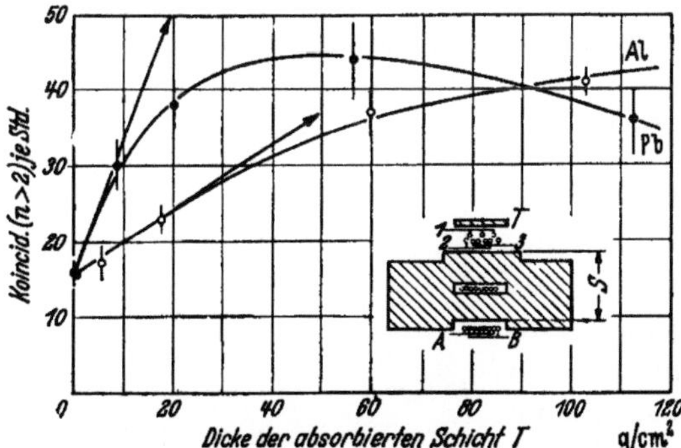

Abb. 3. Entwicklung der durchdringenden Schauer.

zu ändern. Luftschauer, die bis dahin zufällig die Zähler 1, 2, 3 *nicht* gleichzeitig auslösten, werden durch das Material T verdichtet und liefern nun Dreifachkoinzidenzen 1, 2, 3; daher der Anstieg der Häufigkeit mit T. Die Anfangsneigung der Kurven in Abb. 3 und die Sättigungsdicke bei 5 cm Pb entsprechen ganz den Verhältnissen, die nach der Kaskadentheorie zu erwarten sind. Die Auslösung der durchdringenden Teilchen wird dabei für dünnere Schichten T immer noch hauptsächlich in der großen Bleimasse unter 1, 2, 3 erfolgen. Erst bei dickeren Schichten T werden schon die durchdringenden Teilchen in T ausgelöst, der Abfall der Kurven in der Abbildung entspricht dann der Absorption der durchdringenden Teilchen.

Die Fälle $n = 0$ und $n = 1$ können durch einzelne Mesonen hervorgerufen werden, die in T und bei den Zählern A, B Sekundärelektronen auslösen. Sie sind wesentlich häufiger als die Fälle $n \geq 2$. Dagegen besteht interessanterweise kein großer Häufigkeitsunterschied mehr zwischen $n = 2$ und irgendeinem höheren Wert bis herauf zu $n = 8$. Es sieht also so aus, als ob — wenn überhaupt mehrere durchdringende

Teilchen entstehen — die Entstehung vieler solcher Teilchen nicht viel seltener ist als die von zwei oder drei.

Wenn die durchdringenden Teilchen dieser Schauer Mesonen sind — und dies ist wohl das Wahrscheinlichste —, so kann man die Häufigkeit der Schauer vergleichen mit der Häufigkeit der Mesonenentstehung in der Atmosphäre. In der Atmosphäre werden die Mesonen wahrscheinlich zum Teil von primären Protonen, zum Teil von den Lichtquanten der weichen Komponente gebildet (vgl. 4. Bericht, WIRTZ). Die langsamen Mesonen, deren Bildung man in Höhen über 7 km in großer Zahl beobachtet, entstehen wohl hauptsächlich aus den Lichtquanten der Kaskaden. Es würde sich dann also in dem Bleiblock (zwischen den Zählern 1, 2, 3 und H) bei JANOSSY und INGLEBY das gleiche abspielen, was sich in entsprechend größerer Häufigkeit auch in der oberen Atmosphäre abspielt. Daraus kann man die Häufigkeit der Mesonenschauer in diesen Versuchen sofort abschätzen: sie sollte sich zur Häufigkeit der einzelnen Mesonen etwa so verhalten wie die Kaskadenstrahlung auf Meeresniveau zu der in großer Höhe. Für ein Kaskadenspektrum der Form $E^{-1.8}$ ist nach der Theorie die Intensität auf Meeresniveau

etwa 10^4 bis 10^5 mal schwächer als in großer Höhe. Daher sollten auch die Schauer von JANOSSY und INGLEBY etwa 10^5 mal seltener sein als einzelne Mesonen, was in der Größenordnung auch ungefähr richtig ist. Diese Abschätzung spricht daher für die Deutung, daß es sich bei den Versuchen von JANOSSY und INGLEBY um Mesonenschauer (d. h. um echte explosionsartige Schauer, bei denen viele Teilchen in einem Akt erzeugt werden) handelt. Die Richtigkeit dieser Deutung könnte übrigens u. a. auch durch die Abhängigkeit der Häufigkeit von der

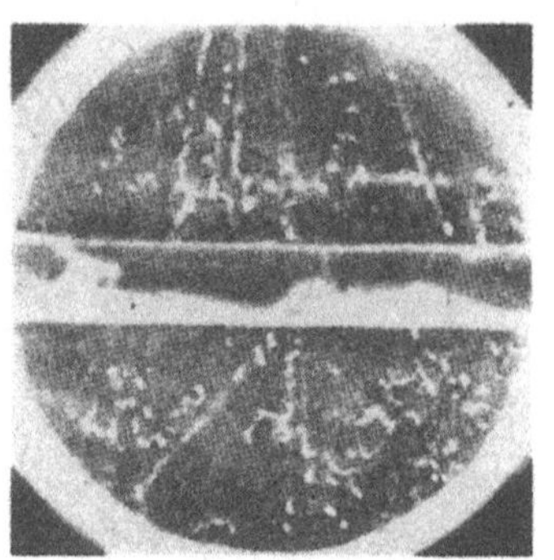

Abb. 4. Mesonenschauer nach WOLLAN.

Höhe nachgeprüft werden. Schon auf der Höhe des Jungfraujochs sollten die Schauer etwa 80 mal häufiger sein als auf Meeresniveau.

In jüngster Zeit sind noch von POWELL (P 6) und WOLLAN (W 16) Nebelkammeraufnahmen veröffentlicht worden, in denen offenbar Schauer von Mesonen, die von einem Punkt ausgehen, beobachtet werden (Abb. 4). Diese Schauer stimmen in ihrer Art wohl mit den von FUSSELL (F 7) beobachteten (s. Abb. 1) überein. Ob die von DAUDIN (D 1) beobachteten Schauer auch zur gleichen Gruppe gehören, wird wohl erst durch weitere Experimente geklärt werden können.

Schließlich sind in der letzten Zeit systematische Versuche von COCCONI, LOVERDO und TONGIORGI (C 5 a) veröffentlicht worden, die zeigen, daß die meisten Mesonenschauer in 2200 m Höhe in Verbindung mit großen Luftschauern entstehen. Diese Ergebnisse passen daher gut zu denen von Janossy und Ingleby.

Zusammenfassend kann man also feststellen: Durch die Experimente der letzten Jahre kann die Existenz von explosionsartigen Mesonenschauern mit vielen Teilchen als gesichert gelten; diese Schauer scheinen sogar nach den neueren Versuchen von Schein, Jesse und Wollan (S 4 bis 6) in der ganzen Genetik der kosmischen Strahlung in der Atmosphäre eine entscheidende Rolle zu spielen (vgl. 4. Bericht, Wirtz). Dieses experimentelle Ergebnis ist befriedigend, weil solche Schauer nach der Yukawaschen Theorie zu erwarten sind [Heisenberg (H 3)], und weil die Existenz dieser Schauer eine theoretische Begründung dafür gibt, warum die nach der gewöhnlichen Störungstheorie berechneten Wirkungsquerschnitte für Stoß- oder Entstehungsprozesse der Mesonen nicht mit den Experimenten übereinstimmen. Nur dort, wo es sich bei derartigen Rechnungen um die rein elektromagnetische Wirkung der Mesonen nach außen handelte (vgl. 10. Bericht, v. Weizsäcker), konnte eine Übereinstimmung mit dem Experiment erhofft werden.

6. Die Absorption der Mesonen.

Von H. Volz-Berlin-Charlottenburg.

Mit 1 Abbildung.

Das große Durchdringungsvermögen der kosmischen Strahlung ist, wie wir aus den Messungen von Anderson und Neddermeyer (A 4) sowie von Blackett (B 24) wissen, eine Eigenschaft der darin enthaltenen Mesonen. Diese bisher nur in der kosmischen Strahlung gefundenen Teilchen sollen nach der Theorie von Yukawa, der Teilchen solcher Masse schon vor ihrer Entdeckung zur Erklärung des β-Zerfalls und der Kernkräfte theoretisch gefordert hat, instabil sein und mit einer Lebensdauer der Größenordnung 10^{-6} Sekunden in ein Elektron und ein Neutrino zerfallen.

Ein solcher Zerfall des Mesons wird ebenso wie die normale Absorption durch Energieverlust zu einer Verminderung der Teilchenzahl beim Durchgang durch Materie führen, der gemessene Absorptionskoeffizient wird also neben dem Anteil der Bremsung noch einen weiteren Anteil enthalten, der auf den Zerfall zurückzuführen ist. Im folgenden soll zunächst die Absorption der Mesonen durch die Bremsung allein betrachtet und im Anschluß daran die Frage behandelt werden, ob die Höhenstrahlung zwei Mesonensorten enthält, und daß das eine der beiden Mesonen die lange und das andere die kurze Lebensdauer hat. Denn dann muß man beim β-Zerfall stets gerade die kurze Lebensdauer beobachten, da die kurzlebigen Mesonen den langlebigen im Zerfall zuvorkommen. In der Höhenstrahlung hingegen werden die kurzlebigen Mesonen in den hohen Atmosphärenschichten zerfallen, und auf Meeresniveau wird man nur noch die langlebigen beobachten.

Immerhin sollte man in der Höhenstrahlung noch etwas von den kurzlebigen Mesonen wahrnehmen, und eben dafür bestehen starke Indizien. So hat WEISZ (W 3) bemerkt, daß die verschiedenen Bestimmungen der Lebensdauer des Mesons aus der Höhenstrahlung eine um so längere Lebensdauer ergaben, je länger der Weg war, den die zur Messung verwendeten Mesonen in der Atmosphäre durchsetzten. WEISZ hat daraus auf eine Uneinheitlichkeit der Masse der Mesonen geschlossen; doch muß jede Uneinheitlichkeit der Zerfallszeit, auch die von uns vermutete Anwesenheit von zwei getrennten, aber in sich einheitlichen Mesonensorten, zu demselben Resultat führen, da die kurzlebigen Mesonen um so mehr ausfallen, je länger der Beboachtungsweg ist. Direkte Anzeichen für kurzlebige Mesonen in größerer Höhe hat JUILFS (J 16) aus der Richtungsverteilung der harten Komponente gefunden.

Wenn wir annehmen, daß die Mesonen vom Spin 1 kurzlebig, die vom Spin 0 langlebig sind, so ist es nunmehr verständlich, daß die Stoßmessungen den Mesonenspin 0 ergeben. Die weiteren Anwendungen, die von der Annahme der zwei Mesonensorten gemacht werden können, überschreiten den Rahmen dieses Referats.

8. Der radioaktive Zerfall der Mesonen.

Von W. HEISENBERG-Berlin-Dahlem.

Mit 1 Abbildung.

Nach der YUKAWAschen Theorie sollte ein ruhendes Meson radioaktiv in ein Elektron und ein Neutrino zerfallen können, wobei die beiden entstehenden Teilchen je ungefähr die halbe Ruhmasse des Mesons als kinetische Energie mitbekommen. Die mittlere Lebensdauer τ_0 des ruhenden Mesons sollte nach der Theorie der Größenordnung nach etwa 10^{-8} sec betragen; neuere Arbeiten von MÖLLER und ROSENFELD (M 5) und ROZENTAL (R 11) haben jedoch darauf hingewiesen, daß man vielleicht zwei Arten von Mesonen, solche vom Spin 1 und vom Spin 0, unterscheiden muß, von denen die eine Sorte diese kurze Lebensdauer haben sollte, während die andere Sorte eine (theoretisch ohne neue Annahme nicht zu bestimmende) erheblich längere Lebensdauer haben kann.

Daß die in der kosmischen Strahlung beobachteten Mesonen tatsächlich radioaktiv zerfallen, wurde zuerst von EULER (E 4) (vgl. auch K 6a) aus der Anzahl der Sekundärelektronen indirekt erschlossen, später durch Nebelkammeraufnahmen von WILLIAMS und ROBERTS (W 12) experimentell sichergestellt [Vgl. auch KUNZE (K 7)]. Eine der wichtigsten Aufgaben besteht jetzt in der experimentellen Bestimmung der mittleren Lebensdauer τ_0. Die ersten Schätzungen dieser Konstanten durch EULER auf Grund der damals vorliegenden Experimente (K 6a) gaben Werte für τ_0 zwischen $2 \cdot 10^{-6}$ und $3 \cdot 10^{-6}$ sec. Spätere Untersuchungen führten teils auf größere [POMERANTZ und JOHNSON (P 5)], teils auf kleinere Werte [KOLHÖRSTER und MATTHES (K 5)]. Die genauesten Arbeiten des letzten Jahres haben jedoch gezeigt, daß der richtige Wert von τ_0 wohl noch etwas tiefer liegt als die ursprüngliche EULERsche Schätzung, etwa bei

$$\tau_0 \sim 1,5 \text{ bis } 2,5 \cdot 10^{-6} \text{ sec} . \tag{1}$$

Von diesen Experimenten soll im folgenden die Rede sein.

Zur Bestimmung der Zerfallszeit sind bisher drei grundsätzlich verschiedene Wege beschritten worden. Man kann erstens in irgendeiner Weise die Abnahme der Mesonenintensität als Funktion der durchlaufenen Strecke untersuchen. Man erhält dann den mittleren Zerfallsweg R, der mit der mittleren Zerfallszeit τ_0 der ruhenden Mesonen, dem Impuls p und der Ruhmasse μ der Mesonen verknüpft ist durch die aus der Zeitdilatation der Relativitätstheorie folgende Beziehung:

$$R = \tau_0 c \cdot \left(\frac{p}{\mu c}\right) . \tag{2}$$

Man kann zweitens, nach EULER und WILLIAMS, aus der Anzahl der Zerfallselektronen, die mit der Mesonenkomponente im Gleichgewicht stehen, indirekt die Zerfallszeit berechnen: Von der Energie $E \sim pc$ des zerfallenden Mesons geht nämlich im Mittel die Hälfte an das Zerfallselektron über, das seinerseits nun eine Kaskade bildet. Die Energie $pc/2$ des Zerfallselektrons verteilt sich dabei schließlich irgendwie auf Elektronen niedriger Energie (10^6 bis 10^8 eV), die keine Vermehrung mehr erfahren. Es sei nun $\varkappa$ das Gleichgewichtsverhältnis der Anzahl solcher Elektronen niedriger Energie zur Anzahl der Mesonen. Dann muß im Gleichgewicht der Energieverlust der Mesonen pro cm durch Zerfall doppelt so groß sein wie der Energieverlust pro cm der Elektronen. Der Energieverlust durch Zerfall setzt sich aus zwei Anteilen zusammen: Dem Anteil $\frac{p_e}{R}$ der bewegten Mesonen und einem Anteil der zur Ruhe gekommenen Mesonen. Der letztere ist offenbar gleich der Ruheenergie des Mesons multipliziert mit der relativen Abnahme der gesamten Mesonenzahl pro cm durch Ionisierung. Die letztere ist bei einer Tiefenabhängigkeit $J \sim T^{-\gamma}$ durch $\frac{\gamma}{T}$ gegeben. Der mittlere Energieverlust a eines Elektrons zwischen 10^6 und 10^8 eV beträgt in Luft pro cm etwa:

$$a \approx 3 \cdot 10^3 \, \text{eV/cm}.$$

Es folgt also die Beziehung:

$$\frac{pc}{2R} + \frac{\mu c^2}{2}\frac{\gamma}{T} = \varkappa a, \quad \text{d. h.} \quad \varkappa = \frac{\mu c^2}{2a}\left(\frac{1}{\tau_0 c} + \frac{\gamma}{T}\right). \tag{3}$$

Allerdings wird sich später zeigen, daß auf Grund der Versuche von Rasetti (R 1) und der theoretischen Überlegungen von Tomonaga und Araki (T 2) die Gleichung (3) nur für positiv geladene Mesonen gelten kann.

Schließlich kann drittens die Zerfallszeit ruhender Mesonen direkt gemessen werden, indem man die Mesonen zur Ruhe kommen läßt und die Zeit mißt, die im Mittel bis zur Aussendung des Zerfallselektrons vergeht [Rasetti (R 1)].

Die ersten beiden Methoden geben einen experimentellen Wert für τ_0/μ, die letzte bestimmt τ_0. Aus dem Verhältnis der Resultate kann daher grundsätzlich die Masse der Mesonen erschlossen werden; doch wird sich zeigen, daß die direkten Massenbestimmungen wahrscheinlich genauer sind als die Zahlen τ_0/μ und τ_0.

Nach den ersten der drei genannten Methoden sind in der letzten Zeit drei Bestimmungen von τ_0/μ vorgenommen worden. Ageno, Bernardini, Cacciapuoti, Ferretti und Wick (A 1) haben Werte von $\tau_0\left(\dfrac{10^8\,\text{eV}}{\mu c^2}\right)$ in der Gegend von 4 bis 5 Mikrosekunden gefunden. In zwei anderen Arbeiten wurde der mittlere Zerfallsweg für Mesonen verschiedener Energien gemessen: Rossi und Hall (R 7) haben im Anschluß an eine frühere Arbeit von Rossi, Hilberry und Hoag (R 8) Absorptionsmessungen der durchdringenden Komponente in Denver (1616 m) und am Echosee (3240 m) durchgeführt; Nielsen, Ryerson, Nordheim und Morgan (N 6) haben ähnliche vergleichende Registrierungen in geringerer Höhe (125 m und 2040 m) vorgenommen. Das Prinzip der Messungen ist in allen Fällen praktisch das gleiche. Auf zwei verschiedenen Höhen werden Absorptionsmessungen der durchdringenden Komponente durch Mehrfachkoinzidenzen von 4 bis 6 Zählrohren mit dazwischen liegenden Bleischichten durchgeführt. Bei der Messung in größerer Höhe wird über der Zählrohranordnung eine bestimmte Menge Absorptionsmaterial angebracht, die so berechnet ist, daß ihre Absorption (wenn man vom Zerfall absieht) genau der Absorption der Luftschicht entspricht, die zwischen den beiden Meßorten liegt. Als Absorptionsmaterial werden Stoffe verwendet, die im Atomgewicht den Elementen Sauerstoff und Stickstoff so nahe stehen, daß man einigermaßen sicher mit den üblichen Bremsformeln umrechnen kann.

Ohne den radioaktiven Zerfall sollten nun die Meßreihen in den verschiedenen Höhen die gleichen Resultate ergeben, da die Menge des durchlaufenen Absorptionsmaterials in beiden Fällen die gleiche ist; ein kleiner Unterschied ist wegen des von FERMI gefundenen Effektes[1] zu erwarten, nach dem die Absorption in Material verschiedener Dichte nicht genau massenproportional stattfindet; der Einfluß des FERMIschen Effektes ist jedoch gering und wird als Korrektu.' berücksichtigt. Der radioaktive Zerfall bewirkt dann, daß die Intensität in geringerer Höhe kleiner ist, der Vergleich der beiden Intensitäten liefert unmittelbar den mittleren Zerfallsweg. Dabei gestattet die Intensitätsmessung hinter verschiedenen Bleidicken noch, Mesonengruppen verschiedener Energiebereiche zu unterscheiden: etwa alle Mesonen, die in einem Bleiblock bestimmter Dicke steckenbleiben, und alle, die diesen Block durchqueren können. Für jede dieser Mesonengruppen kann man dann aus der Bleidicke (und aus dem ziemlich gut bekannten Mesonenspektrum) die mittlere Energie berechnen, die Messung liefert für diese Gruppe den mittleren Zerfallsweg. Die folgende Tabelle gibt den Vergleich der mittleren Energien und der gemessenen mittleren Zerfallswege, die einander ungefähr proportional sein sollten.

Tabelle 1.

	ROSSI und HALL		NIELSEN, RYERSON, NORDHEIM und MORGAN			
pc (in 10^8 eV) .	5,1	13	3,50	4,55	5,6	14
R (in km) . . .	4,5	13,3	2,0	2,3	2,5	8

Die Tabelle zeigt, daß zwar jede einzelne Meßreihe die Proportionalität von R mit pc befriedigend zeigt, also indirekt die Zeitdilatation der Relativitätstheorie bestätigt, daß aber die beiden Meßreihen schlecht zusammenstimmen. Dementsprechend sind auch die aus den Meßreihen folgenden (und mit etwas verschiedenen Korrekturen gewonnenen) Werte für τ_0/μ recht verschieden; so finden ROSSI und HALL:

$$\tau_0\left(\frac{10^8\ \text{eV}}{\mu c^2}\right) = (3,0 \pm 0,4) \cdot 10^{-6}\,\text{sec}\,,$$

NIELSEN, RYERSON, NORDHEIM und MORGAN:

$$\tau_0\left(\frac{10^8\ \text{eV}}{\mu c^2}\right) = (1,25 \pm 0,3) \cdot 10^{-6}\,\text{sec}\,.$$

$$(4)$$

Eine direkte Bestimmung von τ_0 hat RASETTI (R 1) mit der in Abb. 1 dargestellten Anordnung vorgenommen. In der Abbildung sind parallel geschaltete Zählrohre durch eine Linie verbunden und durch einen Buchstaben bezeichnet. Die Koinzidenzapparatur 1 registriert Fünffachkoinzidenzen der Zählrohre $A\text{-}B\text{-}C\text{-}D\text{-}E$, die Apparatur 2

---

[1] Vgl. 6. Bericht, VOLZ.

Antikoinzidenzen der Apparatur *1* und der Zählrohre *F*, *G*. In *2* werden also nur die Fälle registriert, in denen ein Meson durch die Zähler *ABCD* in den Absorberblock (aus Al oder Fe) eindringt und dort steckenbleibt und ein Zerfallselektron in einen der Zähler *E* entsendet oder ein Meson vom Absorber in einen Zähler *E* hinein abgelenkt wird. Die Apparaturen *1* und *2* haben ein verhältnismäßig geringes

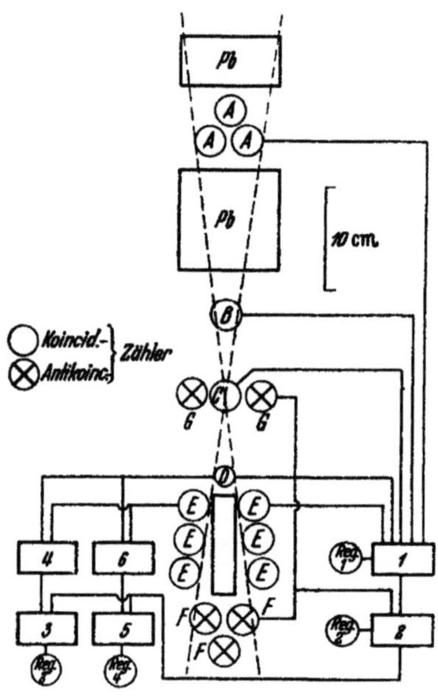

Abb. 1. Rasettis Anordnung zur Messung der Zerfallszeit der Mesonen.

Auflösungsvermögen (15 · 10^{-6} sec). Nun werden gleichzeitig durch zwei weitere Koinzidenzanordnungen *4* und *6* Zweierkoinzidenzen *D—E* mit hoher Auflösung ausgesiebt, durch die Koinzidenzzähler *3* und *5* mit der Apparatur *2* verknüpft und in der Registrieranordnung *3* und *4* gezählt. Das Auflösungsvermögen von Apparatur *4*, registriert in *Reg 3*, ist 1,95 · 10^{-6} sec, das von Apparatur *6*, registriert in *Reg 4*, ist 0,76 · 10^{-6} (bzw. in anderen Messungen 0,95 · 10^{-6}) sec. Bezeichnet man nun die in einem bestimmten, größeren Zeitraum von den Zählern *1* bis *4* gezählten Koinzidenzen mit n_1 bis n_4, so bedeutet $n_2 - n_4$ bzw. $n_2 - n_3$ die Anzahl der verspäteten Koinzidenzen, bei denen zwischen dem Ansprechen des Zählers *D* und eines Zählrohrs *E* mehr als 1,95 · 10^{-6} bzw. 0,76 · 10^{-6} sec vergangen sind. Daher sollten diese Differenzen mit der mittleren Lebensdauer τ_0 zusammenhängen durch die Gleichung:

$$\frac{n_2 - n_3}{n_2 - n_4} = e^{-\frac{(1,95 - 0,76) \cdot 10^{-6}}{\tau_0}} \tag{5}$$

Rasetti findet aus längeren Registrierungen, die teils mit Eisen, teils mit Aluminium als Absorber vorgenommen worden sind, als Mittelwert nach (5)

$$\tau_0 = (1,5 \pm 0,3) \cdot 10^{-6} \text{ sec.}$$

Dieser Wert paßt besser zu den Werten von τ_0/μ, die von Nielsen, Ryerson, Nordheim und Morgan gewonnen sind, als zu denen von Rossi und Hall; denn die Masse der Mesonen dürfte von dem Wert $\mu c^2 = 10^8$ eV höchstens um 30% abweichen. Der statistische Fehler ist aber auch bei den Messungen von Rasetti noch recht erheblich. Man kann also einstweilen nur feststellen, daß der Wert von τ_0 kleiner ist, als früher angenommen wurde, und wahrscheinlich zwischen 1 und 2,5 Mikrosekunden liegt.

Dieser geringe Wert von τ_0 und τ_0/μ ist zunächst schwer zu vereinbaren mit dem von EULER aus der Anzahl der Sekundärelektronen nach Gleichung (3) geschlossenen Wert, insbesondere da BERNARDINI, CACCIAPUOTI, FERRETTI, PICCIONI und WICK (B 9) gezeigt haben, daß die Anzahl der Zerfallselektronen noch geringer ist, als EULER früher angenommen hatte. RASETTI hat daher mit der oben beschriebenen Apparatur auch untersucht, in wievielen Fällen der radioaktive Zerfall eines im Absorberblock steckengebliebenen Mesons tatsächlich beobachtet werden kann. Er hat (nach Berücksichtigung der Absorption im Absorberblock und des von den Zählrohren E erfüllten Raumwinkels) gefunden, daß im Mittel nur etwa die Hälfte aller Mesonen radioaktiv zerfällt, während die andere Hälfte ohne Zerfall absorbiert wird.

Dieses Resultat kann befriedigend gedeutet werden durch theoretische Überlegungen, die die Japaner TOMONAGA und ARAKI (T 2) angestellt haben. Diese Forscher haben nach der YUKAWAschen Theorie die Wahrscheinlichkeit für den Einfang eines Mesons im Atomkern (unter Übertragung der Energie an einen Kernbaustein) abgeschätzt. Dabei ergibt sich ein wesentlicher Unterschied zwischen negativen und positiven Mesonen. Der Wirkungsquerschnitt für Einfang *negativer* Mesonen genügt für geringe Geschwindigkeiten dem $1/v$-Gesetz und führt zu einer Lebensdauer τ_e der negativen Mesonen für Einfang, die zwar von der Dichte des betreffenden Materials abhängt, die aber selbst in Luft noch erheblich kürzer ist als die Lebensdauer für Zerfall. (Die theoretischen Werte sind $\tau_e \sim 0,3 \cdot 10^{-7}$ sec in Luft.) Langsame positive Mesonen können dagegen wegen der COULOMBschen Abstoßung vom Atomkern praktisch überhaupt nicht eingefangen werden, für die positiven Mesonen ist der Zerfall außerordentlich viel wahrscheinlicher als der Einfang.

Aus dieser Überlegung und dem RASETTIschen experimentellen Befund kann man schließen, daß nur die positiven Mesonen radioaktiv zerfallen, daß die negativen dagegen vorher von Atomkernen eingefangen werden und dort in der Regel eine Kernumwandlung hervorrufen. Auch die Formel (3) bezieht sich daher nur auf die positiven Mesonen, $\varkappa$ bedeutet in ihr das Gleichgewichtsverhältnis der Zerfallselektronen zu den positiven Mesonen. Auf Meeresniveau beträgt nach der Diskussion im 9. Bericht von BOPP die Anzahl der Zerfallselektronen etwa 18% der Mesonen. Da etwas mehr als die Hälfte aller Mesonen auf Meeresniveau positive Ladung tragen, kann man aus der Intensität der weichen Komponente die Größe $\varkappa$ zu etwa 0,3 abschätzen. Daraus folgt nach (3):

$$\tau_0\left(\frac{10^8\,\text{eV}}{\mu c^2}\right) \sim 2 \cdot 10^{-6}\,\text{sec}.$$

Diese Zahl liegt in der Mitte zwischen den Messungen von ROSSI und HALL und denen von NIELSEN, RYERSON, NORDHEIM und MORGAN und von RASETTI.

An dieser eben geschilderten Auffassung vom radioaktiven Zerfall der Mesonen bedarf noch besonders ein Punkt der experimentellen Bestätigung. Wenn die negativen Mesonen in der Regel von Atomkernen eingefangen werden, so müssen sie dort eine Kernumwandlung hervorrufen, bei der die ganze Ruhenergie des Mesons (etwa 100 MeV) zur Erwärmung des Kerns und zur Verdampfung von Kernteilchen ausgenützt werden kann. Man müßte also am Ende der Spur eines negativen Mesons (etwa in der Nebelkammer) in der Regel eine Kernumwandlung, nicht selten mit Aussendung mehrerer Protonen, beobachten. Bisher ist noch keine derartige Beobachtung veröffentlicht worden.

9. Die Zerfallselektronen der Mesonen.

Von FRITZ BOPP-Breslau, z. Z. Berlin-Dahlem.

Mit 1 Abbildung.

Die weiche Komponente E (Elektronenkomponente) der Höhenstrahlung enthält neben einem in Meereshöhe schwachen Rest R von Kaskaden (vgl. 1. Bericht, HEISENBERG) eine von der durchdringenden Komponente M (Mesonenkomponente) herrührende Sekundärstrahlung, die sich aus mehreren Anteilen verschiedener Herkunft zusammensetzt. Erstens können die Mesonen durch Wechselwirkung mit der durchstrahlten Materie, zweitens durch β-Zerfall sekundäre Kaskaden auslösen [EULER und HEISENBERG (E 7)] (Wechselwirkungskomponente W, Zerfallskomponente Z). Im folgenden sollen einige neuere Arbeiten besprochen werden, die die Z-Komponente zu bestimmen suchen.

Grundsätzlich ist dazu jede Messung geeignet, die die Instabilität der Mesonen nachzuweisen erlaubt. Denn nach EULER und HEISENBERG (E 7) besteht ein enger Zusammenhang zwischen der Intensität der Zerfallskomponente im Verhältnis zu der der Mesonen und der Lebensdauer τ des ruhenden Mesons. WICK (B 8) und Mitarbeiter haben diesen Zusammenhang unabhängig von den beschränkenden Voraussetzungen der Kaskadentheorie unter der von WILLIAMS (W 10) hervorgehobenen Annahme abgeleitet, daß schließlich die gesamte auf die Zerfallselektronen übertragene Energie — im Mittel also die halbe Mesonenenergie — in Ionisierungsarbeit umgesetzt werden muß. Wenn man hinzufügt, daß nach RASETTI (R 1) nur die Hälfte aller in die Mesontheorie gesetzt hat, haben sich also einstweilen nicht erfüllt. Das ist auch nach allem, was in den beiden Referaten ausgeführt wurde, nicht sehr verwunderlich, da wir hier offenbar die Grenze der Leistungsfähigkeit des bisherigen Schemas bereits überschritten haben und Prozesse mit mehreren Teilchen, Nichtlinearität der Gleichungen und ähnliche Dinge hier eine gänzlich neue Situation schaffen dürften, der unser mathematischer Apparat noch in keiner Weise angepaßt ist.

12. Theorie der explosionsartigen Schauer.

Von **W. Heisenberg**-Berlin-Dahlem.

Die Wirkungsquerschnitte für irgendwelche Stoßprozesse sind bisher meistens mit Hilfe der quantenmechanischen Störungstheorie, d. h. der Bornschen Stoßtheorie, berechnet worden. Da die Matrix der Wechselwirkungsenergie im allgemeinen nur Elemente für Übergänge enthält, bei denen nur eines oder höchstens zwei Teilchen entstehen, so muß man die Störungsrechnung im allgemeinen bis zur n [bzw. $(n/2)$]-ten Näherung treiben, wenn man Prozesse erhalten will, bei denen n Teilchen gleichzeitig entstehen. Wenn die Störungsrechnung überhaupt vernünftig konvergiert, so wird daher die Wahrscheinlichkeit für die gleichzeitige Emission vieler Teilchen sehr gering. Diese allgemeine Überlegung hat zu der Vermutung geführt, daß es derartige echte Mehrfachprozesse praktisch nicht geben könne, und auch in der Yukawaschen Theorie haben Yukawa (Y 1 bis 6), Bhabha (B 19, 20) und Heitler (H 5, 6) die Wirkungsquerschnitte für Streuung und Mesonenerzeugung zunächst nach der Bornschen Methode berechnet und angenommen, daß die Konvergenz des Störungsverfahrens, die durch die bekannten Divergenzen (Selbstenergie der Teilchen) stets problematisch ist, durch eine spätere Vervollkommnung der Theorie begründet werden könne.

Die genauere Untersuchung dieser Frage [Heisenberg (H 3)] hat jedoch im Gegenteil zu dem Ergebnis geführt, daß die Yukawasche Theorie bereits zu einer Gruppe von Theorien gehört, in denen die Störungsrechnung oberhalb einer gewissen Energie der stoßenden Teilchen grundsätzlich nicht mehr konvergieren kann und in denen daher von diesen Energien ab echte explosionsartige Schauer zu erwarten sind. Die Gründe für diese Eigenschaft der Yukawaschen Theorie sollen im folgenden Referat ohne nähere mathematische Durchführung besprochen werden.

Zwei Ursachen sind an dem Zustandekommen der Mehrfachprozesse in der Yukawaschen Theorie beteiligt: erstens die enge Verwandtschaft der Yukawaschen Teilchen mit den Lichtquanten und zweitens die Besonderheiten der Wechselwirkungsenergie zwischen Mesonen und Kernbausteinen, die durch Spin und Ladung der Mesonen bedingt sind.

Die enge formale Verwandtschaft zwischen Mesonen und Lichtquanten, die von der Yukawaschen Theorie ausgesprochen wird, hat zur Folge, daß beim Zusammenstoß energiereicher Kernbausteine in ähnlicher Weise mehrere Mesonen entstehen können, wie beim Zusammenstoß elektrisch geladener Teilchen in der Regel unendlich viele Lichtquanten entstehen. Die letztere Tatsache ist durch eine Arbeit von Bloch und Nordsieck (B 27) quantentheoretisch verständlich gemacht worden: Wenn ein energiereiches geladenes Teilchen plötzlich abgelenkt wird, so wird nach der klassischen Theorie im Moment der Ablenkung die Differenz des das Teilchen umgebenden elektromagnetischen Feldes vor und nach der Ablenkung gewissermaßen abgeschüttelt und wandert als Strahlung in den Raum. Dieses Strahlungsfeld ist im Moment seiner Entstehung offenbar ein relativ kleines Wellenpaket, seine Fourier-Entwicklung ergibt dann ein Spektrum, dessen Intensität im Gebiet sehr kleiner Frequenzen einem konstanten Wert zustrebt. Dieses Spektrum stellt bei der quantentheoretischen Behandlung den Erwartungswert des entstehenden Spektrums dar. Die mittlere Anzahl dn der Lichtquanten pro Frequenzintervall dv wird daher (vgl. z. B. die bekannten Formeln für das Röntgen-Bremsspektrum)

$$dn \sim \text{const} \, \frac{dv}{hv}, \tag{1}$$

und die Integration über alle Frequenzen führt zu einer unendlichen Anzahl sehr energiearmer Lichtquanten.

Völlig analog hierzu wird bei der plötzlichen Ablenkung eines sehr energiereichen Protons oder Neutrons die Differenz des zugehörigen Yukawa-Feldes vor und nach der Ablenkung abgeschüttelt und wandert als Mesonenstrahlung in den Raum. Wieder entsteht ein Spektrum, dessen Erwartungswert näherungsweise durch (1) gegeben ist. Für die Gesamtzahl n der Mesonen erhält man jedoch hier einen endlichen Wert, da die Mesonen eine endliche Ruhmasse besitzen; das Integral über dn hat also die untere Grenze $hv = \mu c^2$. Daraus folgt, daß die mittlere Anzahl der Mesonen mit dem Logarithmus der zur Verfügung stehenden Stoßenergie anwächst, denn die obere Grenze von (1) liegt in der Gegend, wo hv mit der Primärenergie E vergleichbar wird. Es folgt daher:

$$n \sim \text{const} \cdot \lg \frac{E}{\mu c^2}. \tag{2}$$

Der konstante Faktor hat in der Lichtquantentheorie einen Wert der Ordnung $e^2/\hbar c$, also in der Yukawaschen Theorie die Größenordnung $g^2/\hbar c$, doch soll auf seine Berechnung hier verzichtet werden. Der eben genannte Effekt gibt also prinzipiell die Möglichkeit zum Verständnis der Entstehung mehrerer Mesonen in einem Akt, allerdings nur bei extrem hohen Energien des stoßenden Primärteilchens. Zur Deutung der Experimente wird dieser Effekt allein kaum ausreichen.

Die YUKAWAsche Theorie unterscheidet sich aber zweitens in einem wesentlichen Punkte von der Lichtquantentheorie: sie enthält Terme in der Wechselwirkung zwischen den schweren Teilchen und dem YUKAWA-Feld, die mit wachsender Energie der beteiligten Teilchen beliebig anwachsen. Man kann dies am einfachsten durch eine Dimensionsbetrachtung einsehen. Man kann durch Multiplikation mit geeigneten Potenzen von $\hbar$ und c jede physikalische Größe dimensionsmäßig in eine Potenz einer Länge überführen. Mißt man in dieser Weise alle Feldgrößen in Einheiten von $\hbar$ und c, so bleibt in der MAXWELLschen Theorie nur eine dimensionslose Wechselwirkungskonstante stehen $\left(\text{nämlich } \dfrac{e}{\sqrt{\hbar c}}\right)$, die Konvergenz einer Störungsrechnung hängt hier also allein vom Zahlwert dieser Konstanten ab. Die YUKAWAsche Theorie dagegen enthält wegen des Spins und der Ladung der Mesonen Wechselwirkungsterme mit einer Konstanten von der Dimension einer Länge (und der Größenordnung 10^{-13} cm). Dies hat zur Folge, daß jede Störungsrechnung schließlich divergiert, wenn die Wellenlänge der beteiligten Teilchen hinreichend klein gegen diese Konstante wird. Die Divergenz des Störungsverfahrens ist aber gleichbedeutend mit der Möglichkeit zur Erzeugung vieler Mesonen in einem einzigen Akt.

Im einzelnen kommt diese Entstehung eines Explosionsschauers wohl in folgender Weise zustande: Ähnlich wie die Erzeugung von Elektron-Positron-Paaren in der Strahlungstheorie Anlaß gibt zu nichtlinearen Gliedern in den MAXWELLschen Gleichungen, die zur Streuung von Licht an Licht führen [EULER und KOCKEL (E 6)], so gibt es auch in der YUKAWA-Theorie auf dem Umweg über Proton-Neutron-Paare nichtlineare Terme, die eine Streuung von Mesonen an Mesonen bewirken [HEISENBERG (H 3)]. Diese nichtlinearen Terme sind zwar im allgemeinen sehr klein. In den besonders dichten Wellenpaketen jedoch, die bei der Ablenkung energiereicher schwerer Teilchen abgeschüttelt werden, spielen sie bei hinreichend großer Energie eine größere Rolle als die normalen linearen Terme. Die nichtlinearen Terme werden also eine Art turbulenter Durchmischung des Wellenpaketes zur Folge haben, bis sich das Wellenpaket über so große Raumgebiete ausgebreitet hat, daß die nichtlinearen Glieder ihre Bedeutung verlieren. Das FOURIER-Spektrum des Wellenpaketes wird sich dabei nach kleineren Frequenzen hin verschieben, und das bedeutet die Entstehung vieler energiearmer Mesonen. Eine quantitative Berechnung des Mesonenspektrums innerhalb eines solchen Explosionsschauers ist beim jetzigen Stand der

Theorie noch nicht möglich. Man kann nur die ziemlich selbstverständliche Voraussage machen, daß ein solcher Schauer im allgemeinen viel mehr energiearme als energiereiche Mesonen enthalten wird und daß die mittlere Teilchenzahl um so größer wird, je größer die zur Verfügung stehende Primärenergie ist. Dagegen wird die Form des Mesonenspektrums nur wenig von der Primärenergie abhängen.

Die Grundannahme der Yukawaschen Theorie, daß nämlich die Mesonen der kosmischen Strahlung einen ganzzahligen Spin tragen und für die Kernkräfte maßgebend sind, ist einstweilen experimentell noch nicht völlig gesichert. Wigner, Critchfield und Teller (W 8) haben z. B. eine Theorie der Kernkräfte vorgeschlagen, in der die Kräfte durch Paare von Teilchen vom Spin $\frac{1}{2}\hbar$ übertragen werden, wobei es offen bleiben kann, ob diese Teilchen mit den beobachteten Mesonen identisch sind. In einer solchen Theorie würde die erste der obenerwähnten Ursachen für die Entstehung von echten Mehrfachprozessen wegfallen, dagegen würde die zweite im gleichen Maß wie in der Yukawa-Theorie gelten. Tatsächlich ist die Möglichkeit der explosionsartigen Mehrfachprozesse gerade zuerst an einer Theorie der Wigner-Critchfield-Tellerschen Art, nämlich der Fermischen β-Zerfallstheorie studiert worden [Heisenberg (H 1)].

Im Hinblick auf diese Situation in der Theorie ist es befriedigend, daß die Experimente der letzten Jahre die Existenz der explosionsartigen Schauer ziemlich sicherstellen (vgl. 5. Bericht, Klemm und Heisenberg). Für die nach der gewöhnlichen Störungstheorie berechneten Wirkungsquerschnitte kann man dann nur im Gebiet kleiner Energien Übereinstimmung mit der Erfahrung erwarten, sofern es sich um Wirkungen handelt, die durch das Yukawasche Feld übertragen werden. Die rein elektromagnetischen Wirkungen der Mesonen können allerdings wahrscheinlich unabhängig von diesen Einschränkungen nach der gewöhnlichen Theorie behandelt werden. Nur wird man mit der Möglichkeit rechnen müssen, daß bei großer Energieübertragung die so berechneten Wirkungsquerschnitte z. B. für die Bremsstrahlung eines Mesons nicht die Wahrscheinlichkeit für die Emission *nur eines* Lichtquants messen, sondern daß sie die Wahrscheinlichkeit für die Emission eines Lichtquants messen, gleichgültig, was etwa noch an energiearmen Lichtquanten und Mesonen gleichzeitig bei diesem Prozeß entsteht (vgl. die Arbeit von Bloch und Nordsieck). Man muß also mit der Möglichkeit rechnen, daß z. B. die Kaskaden, die von Mesonen durch Bremsstrahlung ausgelöst werden (vgl. 5. Bericht, v. Weizsäcker), auch noch eine Anzahl langsamer Mesonen enthalten. Hierüber liegen jedoch noch keine Experimente vor.

Literaturverzeichnis.

A 1   Ageno, M., G. Bernardini, B. N. Cacciapuoti, B. Ferretti und G. C. Wick: Physic. Rev. **57**, 945 (1940).

A 2   Alexeeva, K. I.: C. r. Acad. Sci. URSS **26**, 28 (1940).

A 3   Anderson, C. D.: Physic. Rev. **44**, 406 (1933).

A 4   Anderson, C. D., u. S. H. Neddermeyer: Physic. Rev. **50**, 263 (1936).

A 5   Arley, N.: Proc. roy. Soc. Lond. A **168**, 519 (1938).

A 6   Arley, N., u. W. Heitler: Nature (Lond.) **142**, 158 (1938).

A 7   Auger, P.: Kernphysik. (Züricher Vorträge.) Hrsg. v. Bretscher. 1936.

A 8   Auger, P., L. Leprince Ringuet u. P. Ehrenfest: J. Physique et Radium **7**, 58 (1936).

A 9   Auger, P., R. Maze u. J. Robley: C. r. Acad. Sci. Paris **208**, 1641 (1939).

A 10   Auger, P., R. Maze u. T. Grivet-Meyer: C. r. Acad. Sci. Paris **206**, 1721 (1938).

A 11   Auger, P., R. Maze, P. Ehrenfest u. A. Fréon: J. Physique et Radium **1**, 39 (1939).

A 12   Auger, P., u. A. Rosenberg: C. r. Acad. Sci. Paris **201**, 1116 (1935).

B 1   Bagge, E.: Ann. Physik **39**, 512 (1941).

B 1a   Bagge, E.; Ann. Phys. **39**, 535 (1941).

B 1b   Bagge, E.: Ann. Phys. **35**, 118 (1939).

B 2   Baños, A.: J. Math. Physics **18**, 211 (1939).

B 3   Baños, A.: J. Franklin Inst. **227**, 623 (1939).

B 4   Baños, A.: Physic. Rev. **55**, 621 (1939).

B 5   Baños, A., H. Uribe u. J. Lifshitz: Rev. Modern Physics **11**, 137 (1939).

B 6   Barnóthy, J.: Z. Physik **115**, 140 (1940).

B 7   Belenky, S.: C. r. Acad. Sci. URSS, N. s. **30**, 608 (1941).

B 8   Bernardini, G., B. N. Cacciapuoti u. B. Ferretti: Ric. Scient. **10**, 731 (1939).

B 9   Bernardini, G., B. N. Cacciapuoti, B. Ferretti, O. Piccioni u. G. C. Wick: Physic. Rev. **58**, 1017 (1940).

B 10   Bethe, H.: Ann. Physik **5**, 325 (1930).

B 11   Bethe, H.: Handb. d. Physik **24/1**, 273 (1933).

B 12   Bethe, H. A.: Physic. Rev. **55**, 1130 (1938).

B 13   Bethe, H. A.: Physic. Rev. **57**, 260 (1940).

B 14   Bethe, H. A.: Physic. Rev. **57**, 390 (1940).

B 15   Bethe, H. A.: Physic. Rev. **59**, 684 (1941).

B 16   Bethe, H., u. W. Heitler: Proc. roy. Soc. Lond. **146**, 83 (1934).

B 17   Bethe, H. A., S. A. Korff u. G. Placzek: Physic. Rev. **57**, 573 (1940)

B 18   Bethe, H. A., u. L. W. Nordheim: Physic. Rev. **57**, 998 (1940).

B 19   Bhabha, H. J.: Proc. roy. Soc. Lond. **164**, 257 (1937).

B 20   Bhabha, H. J.: Proc. roy. Soc. Lond. A **166**, 501 (1938).

B 21   Bhabha, H. J.: Nature (Lond.) **143**, 276 (1939).

B 22   Blackett, P. M. S.: Proc. roy. Soc. Lond. A **159**, 1 (1937).

B 23   Blackett, P. M. S.: Nature (Lond.) **142**, 992 (1938).

B 24   Blackett, P. M. S.: Proc. roy. Soc. Lond. **165**, 11 (1938).

B 25   Blau, M., u. H. Wambacher: Nature (Lond.) **140**, 585 (1937).

B 26   Bloch, F.: Z. Physik **81**, 363 (1933).

Literaturverzeichnis.   **169**

B 27 BLOCH, F., u. A. NORDSIECK: Physic. Rev. **52**, 54 (1937).
B 28 BOHR, N.: Philosophic. Mag. **25**, 10 (1913); **30**, 581 (1915).
B 29 BOOTH, F., u. A. H. WILSON: Proc. roy. Soc. Lond. A **175**, 483 (1940).
B 30 BOPP, F.: Im Erscheinen.
B 30a BOTHE, W.: Kernphysik, Zürich. Vorträge. 112, (1936).
B 31 BOUCKAERT, L.: Ann. Soc. Sci. Bruxelles A **54**, 174 (1934).
B 32 BOWEN, J. S., R. A. MILLIKAN u. H. V. NEHER: Physic. Rev. **52**, 80 (1937).
B 33 BOWEN, J. S., R. A. MILLIKAN u. H. V. NEHER: Physic. Rev. **53**, 217 (1938).
B 34 BOWEN, J. S., R. A. MILLIKAN u. H. V. NEHER: Physic. Rev. **53**, 855 (1938).
B 35 BRADDICK, J. J., u. G. S. HENSBY: Nature (Lond.) **144**, 1012 (1939).
C 1 CACCIAPUOTI, B. N.: Ric. Scient. **10**, 1082 (1939).
C 2 CERENKOV, P.: C. r. Acad. Sci. URSS **14**, 101 (1937).
C 3 CHRISTY, R. F., u. S. KUSAKA: Physic. Rev. **59**, 414 (1941).
C 4 CLAY, J., u. A. VAN GEMERT: Physica **6**, 497 (1939).
C 5 COCCONI, G., u. V. TONGIORGI: Z. Physik **118**, 88 (1941).
C 5a COCCONI. G., A. LOVERDO u. V. TONGIORGI: Naturwiss. **31**, 135 (1943).
C 6 COMPTON, A. H., u. I. A. GETTING: Physic. Rev. **47**, 817 (1935).
C 7 COMPTON, A. H., u. R. N. TURNER: Physic. Rev. **52**, 799 (1937).
C 8 COOPER, D. M.: Physic. Rev. **55**, 1272 (1939).
C 9 COOPER, D. M.: Physic. Rev. **57**, 68; **58**, 288 (1940).
C 10 CURRAN, S. C., u. J. E. STROTHERS: Nature (Lond.) **145**, 224 (1940),
     Nr 3667 — Proc. roy. Soc. Lond. A **172**, 72 (1939). Nr. 948.
D 1 DAUDIN, J.: Etudes sur les gerbes de rayons cosmiques. Paris 1942.
E 1 EHMERT, A.: Z. Physik **106**, 751 (1937).
E 2 EHMERT, A.: Z. Physik **115**, 326 (1940).
E 3 EULER, H.: Z. Physik **110**, 450, 692 (1938).
E 4 EULER, H.: Naturwiss. **26**, 382 (1938).
E 5 EULER, H.: Z. Physik **116**, 73 (1940).
E 6 EULER, H., u. B. KOCKEL: Naturwiss. **23**, 246 (1935).
E 7 EULER, H., u. W. HEISENBERG: Erg. exakt. Naturwiss. **17**, 1 (1938).
E 8 EULER, H., u. H. WERGELAND: Astrophysica Norwegica **3**, 165 (1940).
F 1 FERMI, E.: Physic. Rev. **57**, 485 (1940).
F 2 FERMI, E., u. B. ROSSI: Rend. R. Accad. Naz. Lincei **17**, 346 (1933).
F 3 FILIPPOV, A., I. GUREVICH u. A. ZHDANOV (A. GDANOV): J. Physic. Sowj.-
     Union **1**, 51 (1939).
F 4 FRÖHLICH, H., W. HEITLER u. N. KEMMER: Proc. roy. Soc. Lond. **166**,
     154 (1939).
F 5 FÜNFER, E.: Naturwiss. **25**, 235 (1937).
F 6 FÜNFER, E.: Z. Physik **111**, 351 (1938).
F 7 FUSSELL, L.: Physic. Rev. **51**, 1005 (1936).
G 1 GAMOW, G., u. E. TELLER: Physic. Rev. **49**, 895 (1936).
G 2 GEIGER, H., u. M. HEYDEN: Z. Physik **93**, 543 (1934).
G 3 GEIGER, H., u. W. STUBBE: Abh. preuß. Akad. Wiss., **1941**, 3, Nr 10.
G 4 GILL, P. A.: Physic. Rev. **55**, 1151 (1939).
G 5 GODART, O.: Ann. Soc. Sci. Bruxelles A **58**, 27 (1938).
G 6 GODART, O.: Physic. Rev. **55**, 875 (1939).
G 7 GODART, O.: Physic. Rev. **56**, 1074 (1939).
G 8 GODART, O.: J. Math. Physics **20**, 207 (1941).
G 9 GRIVET-MEYER, T.: C. r. Acad. Sci. Paris **206**, 833 (1938).
G 10 GRÖNBLOM, B. O.: Physic. Rev. **56**, 508 (1939).
G 11 GROSS, B.: Z. Physik **83**, 214 (1933).
H 1 HEISENBERG, W.: Z. Physik **101**, 533 (1936).

170	Literaturverzeichnis.

H 1a HEISENBERG, W.: Ber. Sächs. Akad.; Abh. math. phys. Kl. **89**, 369 (1937).
H 2 HEISENBERG, W.: Z. Physik **110**, 251 (1938).
H 3 HEISENBERG, W.: Z. Physik **113**, 61 (1939).
H 4 HEITLER, W.: Quantum-Theory of Radiation. Oxford: Clarendon-Press 1936.
H 5 HEITLER, W.: Proc. roy. Soc. Lond. **161**, 261 (1937).
H 6 HEITLER, W.: Proc. roy. Soc. Lond. A **166**, 529 (1938).
H 7 HERZOG, G.: Physic. Rev. **59**, 117 (1941).
H 8 HERZOG, G., u. W. H. BOSTICK: Physic. Rev. **58**, 278 (1940).
H 9 HERZOG, G., u. W. H. BOSTICK: Physic. Rev. **59**, 122 (1941).
H 10 HERZOG, G., u. P. SCHERRER: Helvet. phys. Acta **8**, 514 (1935).
H 11 HESS, V. F.: Physik. Z. **14**, 610 (1913).
H 12 HOFFMANN, G., u. W. S. PFORTE: Physik. Z. **31**, 347 (1930).
H 13 HUGHES, D. J.: Physic. Rev. **57**, 592 (1940).
H 14 HUTNER, R. A.: Physic. Rev. **55**, 15 (1939).
H 15 HUTNER, R. A.: Physic. Rev. **55**, 109 (1939).
H 16 HUTNER, R. A.: Physic. Rev. **55**, 614 (1939).
H 17 HALE, G. E., F. H. SEARES, A. v. MAANEN u. F. ELLERMANN: J. Astrophys. **47**, 206 (1918).
J 1 JÁNOSSY, L.: Z. Physik **104**, 430 (1937).
J 2 JÁNOSSY, L.: Proc. roy. Soc. Lond. A **179**, 361 (1942).
J 3 JÁNOSSY, L., u. P. INGLEBY: Nature (Lond.) **145**, 511 (1940); **147**, 56 (1941).
J 4 JÁNOSSY, L., u. B. C. LOVELL: Nature (Lond.) **142**, 716 (1938).
J 5 JENSEN, H.: Verh. dtsch. physik. Ges. (3) **20**, 113 (1939).
J 6 JESSE, W. P., u. P. S. GILL: Physic. Rev. **55**, 414 (1939).
J 7 JOHNSON, T. H.: Physic. Rev. **47**, 318 (1934).
J 8 JOHNSON, T. H.: Physic. Rev. **54**, 385 (1938).
J 9 JOHNSON, T. H.: Rev. Modern Physics **10**, 193 (1938).
J 10 JOHNSON, T. H.: Physic. Rev. **56**, 226 (1939).
J 11 JOHNSON, T. H.: J. Franklin Inst. **227**, 37 (1939).
J 12 JOHNSON, T. H.: Rev. Modern Physics **11**, 208 (1939).
J 13 JOHNSON, T. H., u. J. G. BARRY: Physic. Rev. **56**, 219 (1939).
J 14 JONES, H.: Rev. Modern Physics **11**, 235 (1939).
J 15 JORDAN, B. G.: Erg. exakt. Naturwiss. **16**, 47 (1937).
J 16 JUILFS, J.: Naturwiss. (im Erscheinen).
K 1 KEMMER, N.: Proc. Cambridge philos. Soc. **34**, 354 (1938).
K 2 KOENIG, H. P.: Physic. Rev. **58**, 385 (1940).
K 3 KOLHÖRSTER, W.: Physik. Z. **14**, 1153 (1913).
K 4 KOLHÖRSTER, W., J. MATTHES u. E. WEBER: Naturwiss. **26**, 576 (1938).
K 5 KOLHÖRSTER, W., u. J. MATTHES: Physik. Z. **40**, 142, 617 (1939).
K 6 KORFF, S. A.: Physic. Rev. **56**, 210 (1939) — Rev. Modern Physics **11**, 211 (1939).
K 6a KULENKAMPFF, H.: Verh. dtsch. phys. Ges. (3), **19**, 92 (1938).
K 7 KUNZE, P.: Z. Physik **80**, 559 (1933).
K 8 KUNZE, P.: Z. Physik **83**, 1 (1939).
L 1 LANDAU, L.: J. of Physics **3**, 237 (1940).
L 2 LANDAU, L., u. G. RUMER: Proc. roy. Soc. Lond. **166**, 213 (1938).
L 3 LEISEGANG, S.: Z. Physik **116**, 515 (1940).
L 4 LEMAÎTRE, G.: Ann. Soc. Sci. Bruxelles A **54**, 162 u. 194 (1934).
L 5 LEMAÎTRE, G.: Nature (Lond.) **140**, 23 (1937).
L 6 LEMAÎTRE, G., u. M. S. VALLARTA: Physic. Rev. **43**, 87 (1933).
L 7 LEMAÎTRE, G., u. M. S. VALLARTA: Ann. Soc. Sci. Bruxelles A **56**, 102 (1936).

L 8 LEMAÎTRE, G., u. M. S. VALLARTA: Physic. Rev. **49**, 719 (1936).

L 9 LEMAÎTRE, G., u. M. S. VALLARTA: Physic. Rev. **50**, 493 (1936).

L 10 LEMAÎTRE, G., M. S. VALLARTA u. L. BOUCKAERT: Physic. Rev. **47**, 434 (1935).

L 11 LEPRINCE-RINGUET, L., S. GORODETZKY, E. NAGEOTTE u. R. RICHARD-FOY: CR **211**, 382 (1940).

L 12 LYONS, D.: Physik. Z. **42**, 166 (1941).

M 1 MAASS, H.: Physik. Z. **35**, 858 (1934).

M 2 MAASS, H.: Ann. Physik (5) **27**, 507 (1936).

M 3 MATTAUCH, J., u. S. FLÜGGE: Kernphysikal. Tabellen. Berlin: Springer 1942.

M 3a MIEHLNICKEL, E.: Höhenstrahlung 1933.

M 4 MOLIÈRE, G.: Naturwiss. **30**, 87 (1942); ausführlich in Z. Physik.

M 5 MÖLLER, CHR., u. L. ROSENFELD: Danske Vid. Selsk., Math. Fys. Medd. **17**, 8 (1940).

M 6 MONTGOMERY, C. G., u. D. D. MONTGOMERY: Rev. Modern Physics **11**, 255 (1939).

N 2 NEDDERMEYER, S. H., u. C. D. ANDERSON: Rev. Modern Physics **11**, 191 (1939).

N 3 NIE, H.: Z. Physik **99**, 453 u. 776 (1936).

N 4 NIELSEN, W. M., u. K. Z. MORGAN: Physic. Rev. **54**, 245 (1938).

N 5 NIELSEN, W. M., C. M. RYERSON, L. W. NORDHEIM u. K. Z. MORGAN: Physic. Rev. **57**, 158 (1940).

N 6 NIELSEN, W. M., C. M. RYERSON, L. W. NORDHEIM u. K. Z. MORGAN: Physic. Rev. **59**, 547 (1941).

N 7 NORDHEIM, L. W.: Physic. Rev. **53**, 694 (1938).

N 8 NORDHEIM, L. W.: Physic. Rev. **56**, 502 (1939).

N 9 NORDHEIM, L. W., u. M. H. HEBB: Physic. Rev. **56**, 494 (1939).

N 10 NORDSIECK, A.: Physic. Rev. **52**, 59 (1937).

O 1 OPPENHEIMER, J. R.: Rev. Modern Physics **11**, 264 (1939).

O 2 OPPENHEIMER, J. R., H. SNYDER u. R. SERBER: Physic. Rev. **57**, 75 (1940).

P 1 PAULI, W., u. W. WEISSKOPF: Helvet. phys. Acta **7**, 709 (1934).

P 2 PFOTZER, G.: Z. Physik **102**, 23 (1936).

P 3 PFOTZER, G.: Z. Physik **102**, 41 (1936).

P 4 POMERANTZ, M. A.: Physic. Rev. **57**, 3 (1940).

P 5 POMERANTZ, M. A., u. H. JOHNSON: Physic. Rev. **55**, 104, 600, 1112 (1939).

P 6 POWELL, W. M.: Physic. Rev. **60**, 413 (1941).

P 7 PROCA, J.: J. of Physics A **7**, 347 (1936).

P 8 PRIMAKOFF, H., u. T. HOLSTEIN: Phys. Rev. **55**, 1218 (1939).

R 1 RASETTI, F.: Physic. Rev. **60**, 198 (1941).

R 2 REGENER, E., u. A. EHMERT: Z. Physik **111**, 501 (1939).

R 3 RIBNER, H. S.: Physic. Rev. **55**, 1271 (1939).

R 4 RIBNER, H. S.: Physic. Rev. **56**, 1069 (1939).

R 5 ROSSI, B.: Nature (Lond.) **132**, 173 (1933).

R 6 ROSSI, B.: Rev. Modern Physics **11**, 296 (1939).

R 7 ROSSI, B., u. D. B. HALL: Physic. Rev. **59**, 223 (1941).

R 8 ROSSI, B., N. HILBERRY u. J. B. HOAG: Physic. Rev. **56**, 837 (1939).

R 9 ROSSI, B., N. HILBERRY u. J. B. HOAG: Physic. Rev. **57**, 461 (1940).

R 10 ROSSI, B., u. V. H. REGENER: Physic. Rev. **58**, 837 (1940).

R 11 ROZENTAL, S.: Physic. Rev. **60**, 612 (1941).

R 12 RUMBAUGH, G. H., u. G. L. LOCHER: Physic. Rev. **49**, 855 (1936).

S 1 SANTAGELO, M., u. E. SCROCCO: Ric. Scient. **11**, 601 (1940).

S 1a SAUTER, F.: Ann. d. Phys. (5) **18**, 486 (1933); **20**, 404 (1934).

172   Literaturverzeichnis.

S 2 SCHEIN, M., u. P. S. GILL: Rev. Modern Physics **11**, 267 (1939).
S 3 SCHEIN, M., u. P. S. GILL: Physic. Rev. **55**, 1111 (1939).
S 4 SCHEIN, M., W. P. JESSE u. E. O. WOLLAN: Physic. Rev. **57**, 847 (1940.)
S 5 SCHEIN, M., W. P. JESSE u. E. O. WOLLAN: Physic. Rev. **59**, 615 (1941.)
S 6 SCHEIN, M., W. P. JESSE u. E. O. WOLLAN: Physic. Rev. **59**, 930 (1941.)
S 7 SCHEIN, M., u. V. C. WILSON: Rev. Modern Physics **11**, 292 (1939).
S 8 SCHEIN, M., E. O. WOLLAN u. G. GROETZINGER: Physic. Rev. **58**, 1027 (1940).
S 9 SCHMEISER, K., u. W. BOTHE: Ann. Physik **82**, 161 (1938) — SCHMEISER, K. Z. Physik **112**, 501 (1939).
S 10 SCHÖNBERG, M.: Ann. Acad. Brasil. Sci. **12**, 281 (1940).
S 11 SCHOPPER, E.: Naturwiss. **25**, 557 (1937).
S 12 SCHOPPER, E. M., u. E. SCHOPPER: Physik. Z. **40**, 22 (1939).
S 13 SCHREMP, E. J.: Physic. Rev. **54**, 153 (1938).
S 14 SCHREMP, E. J.: Physic. Rev. **54**, 158 (1939).
S 15 SCHREMP, E. J.: Physic. Rev. **57**, 1061 (1940).
S 16 SCHREMP, E. J., u. H. S. RIBNER: Rev. Modern Physics **11**, 149 (1939).
S 17 SIEGERT, B.: Z. Physik **118**, 217 (1941).
S 18 SITTKUS, A.: Z. Physik **112**, 626 (1939).
S 18a STEINKE, E. G.: Ergebn. exakt. Naturwiss. **13**, 89 (1934).
S 18b STEINMAURER, R.: Ergebn. d. kosm. Physik **3**, 38 (1938).
S 19 STETTER, G., u. H. WAMBACHER: Physik. Z. **40**, 702 (1939).
S 20 STÖRMER, C.: Z. Astrophys. **1**, 237 (1930).
S 21 STÖRMER, C.: Univ. Publ. Univ. Obs. Oslo **1934**, Nr 10, Nr 12 — Astrophysica Norvegica **1**, 115 (1936); **2**, 193 (1937).
S 22 STREET, J. C., u. R. H. WOODWARD: Physic. Rev. **49**, 198 (1936).
S 23 STUHLINGER, E.: Z. Physik **116**, 281 (1939).
S 24 SWANN, W. F. G.: Physic. Rev. **44**, 224 (1933).
S 25 SWANN, W. F. G.: Physic. Rev. **44**, 124 (1933).
S 26 SWANN, W. F. G.: Rev. Modern Physics **11**, 251 (1939).
S 27 SWANN, W. F. G.: Physic. Rev. **59**, 770 (1941); **58**, 200 (1940).
S 28 SWANN, W. F. G.: Physic. Rev. **60**, 470 (1941).
T 1 TAKEUCHI, T., T. INAI, T. SUGITA u. M. HUZISAWA: Proc. Phys.-Math. Soc. Jap. **19**, 88 (1937).
T 2 TOMONAGA, S., u. G. ARAKI: Physic. Rev. **58**, 90 (1940).
V 1 VALLARTA, M. S.: Physic. Rev. **44**, 1 (1933).
V 2 VALLARTA, M. S.: Physic. Rev. **47**, 647 (1935).
V 3 VALLARTA, M. S.: Nature (Lond.) **139**, 24 (1937).
V 4 VALLARTA, M. S.: Rev. Modern Physics **11**, 239 (1939).
V 5 VALLARTA, M. S.: Physic. Rev. **55**, 583 (1939).
V 6 VALLARTA, M. S.: J. Franklin Inst. **227**, 1 (1939).
V 7 VALLARTA, M. S., u. R. P. FEYNMAN: Physic. Rev. **55**, 506 (1939).
V 8 VALLARTA, M. S., C. GRAEF u. S. KUSAKA: Physic. Rev. **55**, 1 (1939).
V 9 VALLARTA, M. S., u. O. GODART: Rev. Modern Physics **11**, 180 (1939).
W 1 WAMBACHER, H.: Wiener Ber. **149**, 157 (1940).
W 2 WATAGHIN, G., M. D. SANTOS u. P. H. POMPEJA: Physic Rev. **57**, 61 (1940).
W 3 WEISZ, P.: Physic. Rev. **59**, 845 (1941).
W 4 WENTZEL, G.: Naturwiss. **26**, 273 (1938).
W 5 WENTZEL, G.: Helvet. phys. Acta **13**, 269 (1940).
W 6 WICK, G. C.: Unveröffentlichte Mitteilung an W. HEISENBERG.
W 7 WIDHALM, A.: Z. Physik **115**, 481 (1940).
W 8 WIGNER, E. P., C. L. CRITCHFIELD u. E. TELLER: Physic. Rev. **56**, 530 (1939).

Literaturverzeichnis.   **173**

W 9   WILLIAMS, E. J.: Proc. roy. Soc. Lond. A **169**, 531 (1939).
W 10   WILLIAMS, E. J.: Proc. roy. Soc. Lond. **172**, 194 (1939).
W 11   WILLIAMS, E. J., u. G. R. EVANS: Nature (Lond.) **145**, 818 (1940).
W 12   WILLIAMS, E. J., u. G. E. ROBERTS: Nature (Lond.) **145**, 102 (1940).
W 13   WILSON, V. C.: Physic. Rev. **53**, 337 (1938).
W 14   WILSON, V. C.: Physic. Rev. **55**, 6 (1939).
W 15   WILSON, J. G.: Proc. roy. Soc. Lond. A **172**, 517 (1939).
W 16   WOLLAN, E.: Physic. Rev. **60**, 532 (1941).
Y 1   YONG-LI, TSCHANG: Ann. Soc. Sci. Bruxelles A **59**, 285 (1939).
Y 2   YUKAWA, H.: Proc. Phys.-Math. Soc. Jap **17**, 48 (1935).
Y 3   YUKAWA, H.: Proc. Phys.-Math. Soc. Jap. **19**, 712 (1937).
Y 4   YUKAWA, H., u. S. SAKATA: Proc. Phys.-Math. Soc. Jap. **19**, 1084 (1937).
Y 5   YUKAWA, H., M. TAKETANI u. S. SAKATA: Proc. Phys.-Math. Soc. Jap. **20**, 319 (1938).
Y 6   YUKAWA, H., S. SAKATA, M. KOBAYASI n. M. TAKETANI: Proc. Phys.-Math. Soc. Jap. **20**, 720 (1938).

T a g u n g e n

Vorträge von der Physikertagung Göttingen 4.-6. 10. 46

Jensen u. *Houtermans:* Zur thermischen Dissoziation des Strahlungshohlraums (vorgetragen von *Jensen*). — Ein materiefreier Hohlraum enthält bei genügender Temperatur infolge Paarbildung zunächst eine gewisse Menge Elektronen und Positronen und zwar im Gleichgewicht mit der Strahlung. Bei noch höheren Temperaturen treten auch schwerere Teilchen (Mesonen, Protonen usw.) hinzu. Wegen der mit der Temperatur immer stärker zunehmenden Menge der Elektronen treten Unterbringungsschwierigkeiten auf. Ohne Angabe einer endgültigen Lösung wurde darauf hingewiesen, daß hier ein ernstes Problem der Quantentheorie vorliegt, dessen Lösung vielleicht durch Berücksichtigung einer invarianten kleinsten Länge möglich sein wird.

v. Laue: Anschauliches über die Supraleitungstheorie. — Bei der Supraleitung können die Kräfte auf die Materie nicht in Volumeffekten sondern nur an den Grenzschichten wirksam werden. Es wird diskutiert, wie sich diese Verhältnisse bei relativ zur Oberfläche kleiner werdendem Volumen auswirken.

Heisenberg: Zur Elektronentheorie der Supraleitung. — Zunächst wurde eine kurze Übersicht über die Vorstellungen von dem Mechanismus der Supraleitung in den letzten 10 bis 15 Jahren gegeben. Anschließend wurde im einzelnen auf die Vorstellungen eingegangen, daß die Supraleitung nur von den ganz wenigen Elektronen am Ende der Fermi-Verteilung getragen wird. Die Wechselwirkung dieser Elektronen wird diskutiert und festgestellt, daß bei hinreichend wenigen, an der Leitung teilnehmenden Elektronen eine Ordnung dieser Elektronen sich energetisch „lohnt", weil die freie Energie der geordneten Phase größer wird als die der ungeordneten. Aus diesen Vorstellungen ergibt sich zwangsläufig eine sehr verschiedene Höhe des Einsatzes der Supraleitung bei den verschiedenen Stoffen; die Ferromagentika ausgenommen, müßten sämtliche Metalle Supraleiter sein.

Oetjen: Optische Untersuchungen an elektrolytischen Lösungen. — Es wurde über Absorptionsmessungen an konzentrierten wässrigen Salzlösungen berichtet, die über den Einfluß der Umgebung Aufschluß geben sollen. Bemerkenswert an den Resultaten ist der steile Anstieg der Oszillatorenstärke für Konzentrationen oberhalb 1 molar bei den Nitraten der untersuchten Erden.

Kuss: Exakte Viskositätsmessungen bis 2000 atm. — Bei Erhöhung des Drucks in Flüssigkeiten wird vor allem die Viskosität stark verändert, weil, anstelle der gegenseitigen Beeinflussung der Moleküle durch Diffusionserscheinungen die direkte intermolekulare Beeinflussung tritt. Verschiedene Methoden der Viskositätsmessung und ihre Resultate wurden mitgeteilt.   Ko.

---

*) Von den zahlreichen Vorträgen ist hier nur eine kleine Auswahl referiert. Eine vollständige Besprechung der Tagung soll in der Zeitschrift „Chemie" erscheinen. S. auch den Vortrag Justi: Supraleitung und Periodisches System S. 207.

Nature *160*, 211–215 (1947)

RESEARCH IN GERMANY ON THE TECHNICAL APPLICATION OF ATOMIC ENERGY*

By Prof. W. HEISENBERG

Max Planck Institute of Physics, Göttingen

EVEN ten years ago, physicists were well aware that the utilization of atomic energy could not be realized without a fundamental extension of scientific knowledge. In spite of the remarkable progress in experimental nuclear physics which followed the introduction of high-voltage equipment and the invention of the cyclotron, no physical phenomenon was known, even as late as 1937, which offered the remotest possibility of exploiting the enormous quantities of energy lying latent in atomic nuclei.

It was the discovery of the fission of uranium by Hahn and Strassmann[1] in December 1938—in other words, the fact that the uranium nucleus can be split into two fragments of comparable mass when bombarded by neutrons—which brought the actual utilization of atomic energy within reach. Following this discovery, Joliot and his co-workers[2] succeeded in proving, in the spring of 1939, that in the act of fission the uranium nucleus itself emits several neutrons, thus making a chain reaction fundamentally possible. Thereafter the possibility of nuclear chain reactions was eagerly debated among physicists, particularly in the United States ; in Germany it was discussed by Flügge[3] in *Die Naturwissenschaften* in the summer of 1939. Meitner and Frisch[4] had already directed attention to the enormous quantities of energy set free in the fission process.

Public interest in the problems of atomic physics was negligibly small in Germany between the years 1933 and 1939, in comparison with that shown in other countries, notably the United States, Britain and France. Thus, while in America, previous to 1939, a whole series of modern research laboratories equipped with high-voltage plants and cyclotrons was springing up, in Germany there were only two adequately equipped laboratories ; and these were not supported by the State, but sponsored by a private body, the Kaiser Wilhelm Gesellschaft. These two institutes were the Kaiser Wilhelm Institutes at Heidelberg and Berlin-Dahlem ; each possessed a small high-voltage set suitable for nuclear research. A cyclotron for such work was altogether lacking—the Heidelberg cyclotron, again built entirely by private funds, and mainly designed for medical investigations, was started as late as 1938, and could not be tested out before 1944. Only with the outbreak of war did the awakened interest of the authorities allow of more extended facilities for nuclear research.

The following report deals with those particular investigations which had for their purpose the technical utilization of atomic energy. The many purely scientific problems which arose in more or less close connexion with the technical problem will not be discussed here ; they will be dealt with in a forthcoming FIAT* Review by Flügge and Bothe. I may, however, mention the extensive chemical investigations of Hahn and his co-workers on the fission products of uranium, carried out throughout the War in the Kaiser Wilhelm Institute for Chemistry, the great majority of which have been published.

Almost simultaneously with the outbreak of war, news reached Germany that funds were being allocated by the American military authorities for research on atomic energy†. In view of the possibility that England and the United States might undertake the development of atomic weapons, the Heereswaffenamt created a special research group, under Schumann, whose task it was to examine the possibilities of the technical exploitation of atomic energy. As early as September 1939 a number of nuclear physicists and experts in related fields were assigned to this problem, under the administrative responsibility of Diebner. I should mention the names of Bothe, Clusius, Döpel, Geiger, Hahn, Harteck, Joos and v. Weizsäcker among those so employed. At Schumann's behest, the Kaiser Wilhelm Institut für Physik in Berlin-Dahlem was nominated as the scientific centre of the new research project. The Institute came accordingly under the administration of the Heereswaffenamt ; a step which disregarded the rights of the Kaiser Wilhelm Gesellschaft, and so led to the departure of its director, Debye, who as a Dutch citizen could not continue to serve under the ægis of a German war department.

As a result of the first conferences in the autumn of 1939, it was clear that there were two lines of attack possible in the exploitation of nuclear energy. One could attempt the separation of the rare isotope U(235) from ordinary uranium. This isotope, following theoretical arguments due to Bohr, must be immediately applicable either to the controlled production of energy, using primarily the slow-neutron reaction, or directly as an explosive in bombs, using the fast-neutron reaction ; the separation of U(235) was, however, a problem which made the greatest possible demands on engineering technique. Secondly, one could mix ordinary uranium with a substance which would slow down the neutrons produced in the nuclear fission without absorbing them. These slow neutrons give rise preferentially to the fission of U(235) and thus maintain the chain reaction. A rapid deceleration of the neutrons is required, in order that the region of resonance absorption by U(238) should be rapidly traversed ; for if absorbed they are lost to the chain

---

* This article is a slightly abridged translation of a paper appearing in *Die Naturwissenschaften*.

* FIAT Reviews of German Science, 1939–1946 : a series of authoritative accounts of the progress made in both natural and applied science in Germany during the War. These reviews have been sponsored jointly by British, American and French FIAT (Field Intelligence Agency (Technical)), and a limited edition is expected to be ready for distribution before the end of 1947.

† Cf. Smyth Report, p. 27 : Initial approaches to the Government. In actual fact, U.S. Government funds were first used at the turn of the year 1939–40 (p. 28), whereas the first discussions between men of science and the American navy took place as early as March 1939.

reaction. The advantage of this arrangement is that the chain reaction can be controlled through the heat developed thereby, so that energy can be abstracted in amounts sufficient for technical applications.

Thus two lines of purely scientific investigation were marked out : first, to develop refined methods for the separation of isotopes ; secondly, by measurement of the effective cross-sections of a range of possible substances, to determine whether the alternative line of attack was at all practicable. Harteck pointed out, as early as the autumn of 1939, that it might be advantageous, in regard to the second scheme, to have the so-called moderator physically segregated from the uranium. This suggestion gave rise to theoretical investigations as to whether, with the effective cross-sections of such moderator substances as were known at the time, an arrangement having a homogeneous mixture of uranium and moderator, or one with a local separation (for example, in layers), led to the more favourable production of energy. A tentative theoretical investigation made by Heisenberg, in December 1939, led to the result that while ordinary water was unsuitable as a moderator, it should be possible with heavy water (D_2O) or very pure carbon to produce energy in positive amount, provided the moderator and uranium were arranged in layers. This arrangement, however, demanded the highest degree of purity of the substances involved. At the same time, it was evident that a certain minimum size of apparatus was necessary for the production of energy. Nevertheless, with a small set it is still possible to determine whether there would be a production of energy if the apparatus were suitably enlarged. Thus if we feed such a small plant with neutrons from an internal source, more neutrons must escape from the surface than are supplied by the source, if the layer arrangement is favourable to energy production ; if unfavourable, then fewer neutrons escape than are supplied by the source. These small model plants, which are continuously fed from a neutron source, are called 'neutron-injected piles'. The ratio k of the number of neutrons escaping from the pile to that fed in by the source can be used to characterize the pile. If $k \ll 1$, the arrangement is unsuitable for the production of energy ; if $k > 1$, energy will be produced on enlarging the pile.

In 1940, measurements of the most important effective cross-sections were carried out, especially by Bothe and his collaborators, and by Döpel and Heisenberg. At the same time, investigations on the masses and energies of the fission products were being pursued by Jentschke and Prankl[5] and by Flammersfeld, P. Jensen and Gentner[6], and on the spectrum of the neutrons produced by Kirchner and v. Droste and by Bothe and Gentner[7]. The theory of neutron absorption in the U(238) resonance line was laid down by Flügge and Heisenberg. On the technical side the following results were the most important : the absorption cross-section of heavy water proved to be so low that this substance was certainly usable in the construction of a uranium pile (Döpel and Heisenberg). The work of v. Droste on large quantities of sodium uranate, and of Harteck and the Hamburg group—Groth, H. Jensen, Knauer and Süss—on U_3O_8 in solid carbon dioxide, furnished the first criteria for the distribution of neutron density in certain arrangements of uranium and moderator. In the autumn of 1940, the first pile, constructed of layers of U_3O_8 and light paraffin, was built at Berlin-Dahlem and its characteristics measured (Wirtz,

Fischer, Bopp). This model pile gave, as expected, $k < 1$, that is, the arrangement was not suitable for the production of energy. Nevertheless, it yielded valuable data for further piles using alternate layers of U_3O_8 and heavy water.

In the summer of 1940, v. Weizsäcker pointed out that a uranium pile, besides generating the fission products of uranium, will constantly produce the uranium isotope U(239) and its transformation series ; and that theoretically these transformation products should show the same properties as U(235) in regard to neutron fission. It remained thereby an open question whether β-decay ended at element No. 93 or 94 or even later ; for, since no cyclotron was available in Germany, these elements could not be prepared in sufficient quantity for the examination either of their nuclear properties or chemical characteristics. Nevertheless, it appeared likely, from v. Weizsäcker's work, that an energy-producing pile might be used for the production of an atomic explosive, even though the practical details involved remained uncertain. In fact, this method had been employed on the grand scale in America. The American piles deliver as a transformation product of U(239) the element plutonium $_{94}^{239}$Pu, which is used in the manufacture of atomic bombs.

Important technical problems immediately arose out of these scientific investigations. The production of U_3O_8 of the highest purity was assigned to the Auer-Gesellschaft by the Heereswaffenamt. The casting of the corresponding metal powder was afterwards allotted to Degussa in Frankfurt. The production of heavy water, which was obviously of the greatest importance for the construction of a uranium pile, was planned at the Norsk Hydro factory at Rjukan in Norway. Harteck, in conjunction with Süss, H. Jensen and Wirtz, developed a number of projects which resulted in an increase of heavy water production at Rjukan far beyond the former output of 10–20 litres a month. Moreover, Harteck and Clusius put forward detailed plans for the production of heavy water in Germany. The improvements in the Norsk Hydro factory finally increased production in the summer of 1942 to about 200 litres per month. Further, steps were taken by order of the Heereswaffenamt for the production of very pure carbon. The attempts to exceed the degree of purity afforded by the best technical electrographite failed, however, for the time being.

The most important progress in the uranium project was achieved during the year 1941. Initially some negative results were recorded. Thus, the enrichment of U(235) by the Clausius–Dickel thermal diffusion method, using gaseous UF_6, proved impossible (Fleischmann, Harteck and Groth). The absorption cross-section for neutrons of the highest purity electrographite was determined in the Kaiser Wilhelm Institute at Heidelberg (Bothe and Jensen[4]), and the behaviour of pure carbon estimated from the results. It appeared, according to the information then available, that even the purest possible carbon was unsuitable for the construction of a uranium pile ; whereas, as is well known, carbon has been used in the United States with complete success. Whether the Heidelberg conclusions were falsified by insufficient consideration of the chemical impurities present in commercial graphite (for example, hydrogen or nitrogen), or by deficiencies in the theory, can scarcely be assessed for the moment. In any event, the Heidelberg experiments on graphite and beryllium (Fünfer and Bothe[5]) made it clear, in connexion with

Nature *160*, 211–215 (1947)

later experiments in the Berlin–Dahlem Institute, that both pure carbon and pure beryllium were highly suitable for use as an outer cover for a uranium pile, inasmuch as their low absorption cross-section and high reflecting power restrict the spread of the neutrons escaping from the pile, thus reducing its minimum dimensions.

In the summer of 1941, 150 litres of heavy water were available for the first experiments on a neutron-injected pile built up of uranium and heavy water (Döpel and Heisenberg, Leipzig). The uranium and heavy water were arranged in alternate spherical layers with the neutron source at the centre. The oxide U_3O_8 which was first employed produced only a slight increase in the number of escaping neutrons, which could scarcely be considered a clear proof that $k > 1$. The use of pure uranium metal, however, gave such a decided improvement that no further doubt of a real increase in the number of neutrons ($k > 1$) was possible (about February or March 1942). Here then was the proof that the technical utilization of atomic energy was possible, and that the mere enlargement of the Leipzig apparatus must furnish an energy-producing uranium pile.

At the same time, important administrative changes were taking place. At a meeting held in the building of the Reichsforschungsrat in Berlin, on February 26, 1942, the results to date were reported to the Minister of Education, Rust, and several directors of war research. The uranium project was transferred from the Heereswaffenamt to the control of the Reichsforschungsrat; and the then president of the Physikalisch-Technische Reichsanstalt, Esau, was made responsible for the project. On June 6, 1942, there was a second meeting at Harnack House in Berlin, when the results of the uranium project were reported to Speer, as Minister for War Production, and to the armament staff.

The facts reported were as follows: definite proof had been obtained that the technical utilization of atomic energy in a uranium pile was possible. Moreover, it was to be expected on theoretical grounds that an explosive for atomic bombs could be produced in such a pile. Investigation of the technical sides of the atomic bomb problem—for example, of the so-called critical size—was, however, not undertaken. More weight was given to the fact that the energy developed in a uranium pile could be used as a prime mover, since this aim appeared to be capable of achievement more easily and with less outlay. As to the separation of the uranium isotopes, no method was known which would have allowed of the production of an atomic explosive without an enormous and therefore impossible technical equipment. Incidentally, the use of protoactinium as an atomic explosive was also considered, since its nucleus is fissionable by neutrons with energies down to 10^5 eV., with the consequent possibility of a fast chain reaction. It was, however, considered to be impracticable to prepare the necessary quantities of the element.

Following this meeting, which was decisive for the future of the project, Speer ruled that the work was to go forward as before on a comparatively small scale. Thus the only goal attainable was the development of a uranium pile producing energy as a prime mover—in fact, future work was directed entirely towards this one aim. Again during the summer of 1942, discussions were held with heat experts on the technical problems of heat transfer from the uranium to the working material (that is, water or steam). Technical experts from the Navy attended the meeting with a view to the possible use of a uranium power unit in warships. The Kaiser Wilhelm Institut für Physik was restored to the Kaiser Wilhelm Gesellschaft, with the author as director. In preparation for investigations on the larger uranium piles planned in the Institute, a spacious underground laboratory was added (Wirtz).

About this time, however, the strain of the War on the already overloaded German industry was making itself felt. Uranium and uranium slugs were produced in such small quantities that deliveries were late and the larger-scale experiments were repeatedly postponed. Nevertheless, important progress was made. As early as 1941, a research group at the Heereswaffenamt (Diebner, Pose, Czulius) had made measurements on a large pile built up as a lattice of uranium cubes in a paraffin matrix; the subsequent theoretical investigation (Höcker) demonstrated that the lattice construction could in certain circumstances show advantages over the layer arrangement. An experiment made by this group with a model pile of uranium cubes in D_2O ice did, in fact, yield a larger increase in the number of neutrons than the Leipzig pile. In a later experiment, using 500 litres of heavy water, a further increase in the number of neutrons was recorded. Measurements made in the Heidelberg Institute with a small model pile defined the relation between the increase in the number of neutrons on one hand, and the thickness of the layers on the other; while experiments undertaken by Bothe and Flammersfeld at Heidelberg, and by Stetter and Lintner in Vienna, threw new light on the fission processes occurring in U(238). A theoretical investigation by Bothe stressed the importance of the 'stopping distance' (*Bremslänge*) for the minimum size of a self-sustaining pile.

In preparation for further experiments with larger quantities of heavy water and uranium metal, the Kaiser Wilhelm Institut für Physik in Berlin began a study of the effect of graphite and water as an outer cover for the pile. Resonance absorption in uranium had been studied by Volz and Haxel and by Sauerwein. Further, the absorption cross-sections of a series of different substances were measured by Ramm, as well as by Volz and Haxel at the Berlin Technical High School. As regards the question of thermal stabilization of the energy production resulting from the temperature broadening of the resonance lines, experiments by Sauerwein and Ramm with the Berlin-Dahlem high-voltage plant were significant.

In the spring of 1943, the Norsk Hydro electrolytic plant was put out of action in a Commando raid. Its reconstruction was begun, but finally the responsible army command in Norway reported that effective protection of the plant, particularly against air raids, was impossible. In October 1943 the plant was completely destroyed in a heavy air raid. Nevertheless, about two tons of heavy water were available in Germany at the time: a quantity which, according to our calculations, was just enough for the construction of an energy-producing pile. The Reichsforschungsrat had made no effective provision for the construction of a new heavy water factory in Germany, and the pilot plant at I.G. Leuna made slow progress. It was proving, in fact, barely possible, in view of air raids and the overall strain on German production, to undertake such big building projects. The production of uranium slugs came to a temporary standstill after the raids on Frankfurt in the spring of 1944.

Even then, some progress was achieved by the Harteck–Groth–Beyerle group. As early as 1942, this group had succeeded in developing an ultracentrifuge for the enrichment of the isotope U(235). It was planned to use uranium enriched with the rare isotope in the construction of improved uranium piles, possibly in conjunction with ordinary water. At about this time, the direction of the uranium project was transferred from Esau to Gerlach. Gerlach had taken over the physics section of the Reichsforschungsrat, and he strove to promote more particularly the scientific side of the uranium problem ; and at that, not only the physical, but also the medical aspect. In connexion with the medical applications, the construction of a low-temperature pile, in liquid carbon dioxide, was undertaken on Harteck's suggestion. Such a pile, even of small dimensions, could be expected to yield profitable amounts of radioactive elements for tracer research, in view of the decreased absorption in the resonance lines at low temperatures.

In the winter of 1943–44, a model pile of 1·5 tons of heavy water and about the same weight of uranium plates was constructed in the Dahlem air-raid shelter through the co-operative efforts of the Kaiser Wilhelm Institutes for Physics in Berlin and Heidelberg (Wirtz, Fischer, Bopp, P. Jensen, Ritter). The number of neutrons injected from the internal source was multiplied by the factor 3, a performance approaching considerably nearer to what we. called the *Labilitätspunkt*, at which the ratio k increases beyond limit and at which the uranium pile begins to radiate independently of the neutron source and thus to produce energy. The relation between neutron increase and layer thickness fulfilled expectations. Further, the stopping distances of the fission neutrons in carbon and heavy water were redetermined (Wirtz) and the former inexact measurements considerably improved. These experiments were made in the air-raid shelter of the Institute during the heaviest air raids on Berlin, and were naturally to some extent hindered by the raids. On February 15, 1944, the Kaiser Wilhelm Institut für Chemie received a direct hit. In the meantime, the Kaiser Wilhelm Institut für Physik had been partly evacuated to Hechingen. On the instructions of Gerlach, a cellar cut out of the solid rock, situated in the village of Haigerloch, was equipped for the rebuilding of the uranium pile. It was not until February 1945, however, that the greater part of the necessary material (about 1·5 tons of heavy water, 1·5 tons of uranium, 10 tons of graphite, cadmium for the regulating rods, etc.) was finally assembled at Haigerloch, and a new pile, this time built up of uranium cubes in heavy water, with an outer cover of graphite, constructed (Wirtz, Fischer, Bopp, Jensen, Ritter). A branch of the Reichsforschungsrat at Stadtilm was allotted the remaining quantity of heavy water and a great part of the available uranium. The Haigerloch pile yielded a sevenfold neutron increase. The material available at Haigerloch, however, was just insufficient to attain $k = \infty$. A relatively small amount of uranium would in all probability have sufficed ; but it was no longer possible to obtain it, since transport from Berlin or Stadtilm could no longer reach Hechingen. On April 22, Haigerloch was occupied, and the material confiscated by the Americans.

When we compare the German work reported here with the corresponding Anglo-American effort, so far as it has been made known, then the beginning of 1942 seems to be the turning point. Up to that time, both sides were dealing predominantly with the scientific problem as to whether nuclear energy could be utilized at all, and what fundamental methods had to be employed to that end. Both sides had arrived almost simultaneously at very similar results, if one excludes the field of isotope separation, in which the Anglo-Americans had made much greater progress. Furthermore, in the United States far more attention had been given to laying the groundwork for subsequent full-scale development of the uranium project ; so that the first self-supporting pile was functioning as early as December 1942.

It remained to determine the technical sequel to these results. In the United States, the final decision was taken to go for the production of atomic bombs, with an outlay that must have amounted to a considerable fraction of the total American war expenditure ; in Germany an attempt was made to solve the problem of the prime mover driven by nuclear energy, with an outlay of perhaps a thousandth part of the American. We have often been asked, not only by Germans but also by Britons and Americans, why. Germany made no attempt to produce atomic bombs. The simplest answer one can give to this question is this : because the project could not have succeeded under German war conditions. It could not have succeeded on technical grounds alone : for even in America, with its much greater resources in scientific men, technicians and industrial potential, and with an economy undisturbed by enemy action, the bomb was not ready until after the conclusion of the war with Germany. In particular, a German atomic bomb project could not have succeeded because of the military situation. In 1942, German industry was already stretched to the limit, the German Army had suffered serious reverses in Russia in the winter of 1941–42, and enemy air superiority was beginning to make itself felt. The immediate production of armaments could be robbed neither of personnel nor of raw materials, nor could the enormous plants required have been effectively protected against air attack. Finally— and this is a most important fact—the undertaking could not even be initiated against the psychological background of the men responsible for German war policy. These men expected an early decision of the War, even in 1942, and any major project which did not promise quick returns was specifically forbidden. To obtain the necessary support, the experts would have been obliged to promise early results, knowing that these promises could not be kept. Faced with this situation, the experts did not attempt to advocate with the supreme command a great industrial effort for the production of atomic bombs.

From the very beginning, German physicists had consciously striven to keep control of the project, and had used their influence as experts to direct the work into the channels which have been mapped in the foregoing report. In the upshot they were spared the decision as to whether or not they should aim at producing atomic bombs. The circumstances shaping policy in the critical year of 1942 guided their work automatically towards the problem of the utilization of nuclear energy in prime movers. To a German physicist, this task seemed important enough. The mere possibility of solving the problem had been rendered possible by the discovery of the German scientific workers Hahn and Strassmann; and so we could feel satisfied with the hope that the important technical developments, with a peace-time application, which must eventually grow out of their discovery, would likewise find their beginning in Germany, and in due course bear fruit there.

[1] *Naturwiss.*, **27**, 11 (1939).
[2] *Nature*, **143**, 470 (1939).
[3] *Naturwiss.*, **27**, 402 (1939).
[4] *Nature*, **143**, 239 (1939).
[5] *Z. Phys.* **119**, 696 (1942).
[6] *Z. Phys.*, **120**, 450 (1943).
[7] *Z. Phys.*, **119**, 568 (1942).
[8] *Z. Phys.*, **122**, 749 (1944).
[9] *Z. Phys.*, **122**, 769 (1944).

Publication of results for which no source is cited was prohibited during the War.

A LONG-TERM PLAN FOR THE NILE BASIN

EGYPT, an agricultural country with practically no rainfall, is entirely dependent on the annual flooding of the Nile. Under natural conditions only that part of the country could be cultivated which was reached by the flood waters, and as the bulk of the population was generally little above the subsistence level, a poor flood meant widespread famine. The area of cultivation could only be extended by irrigation of land not reached by the natural flood : that is, by constructing dams to raise the level of the water, without interfering with the maximum level of the flood further downstream. The study of this possibility was begun about 1890 by Sir William Willcocks, as a result of which a relatively small dam was built at Aswan. This was completed in 1902, but was soon found to be inadequate ; no other suitable site could be found at the time and the Aswan dam was heightened in 1912. Since 1920 the study of the Nile Basin has been assigned to the Physical Department of the Ministry of Public Works, Egypt, under the director-general, Dr. H. E. Hurst, who has gradually pushed investigations farther and farther upstream to the head-waters of the river in Lake Victoria in Uganda and Lake Tana in Abyssinia. From 1931 onwards a comprehensive series of reports has been issued under the title "The Nile Basin" ; the present volume, by H. E. Hurst, R. P. Black and Y. M. Simaika, is the seventh of the series*.

The earlier proposals considered only how to make the best use of the flood water year by year. Storage from one year to another on the Nile itself presented very great difficulties, partly because of the immense volume of storage required to allow for a run of poor floods, and partly because of the enormous loss by evaporation in Egypt. Consequently, a poor flood still caused very great hardship ; the situation was not so bad as it might have been because the Nile draws its water from a very large basin, and the fluctuations of rainfall were not in general similar over the whole area : but this advantage is largely offset by the losses in the extensive marshes of the Sudd. The Bahr el Jebel, which carries the whole of the discharge of Lake Albert, averaging 25 milliards (thousand millions) of cubic metres a year, flows through the flat Sudan plain, where it loses nearly half its water by evaporation and transpiration. Owing to this loss, the contribution of the White Nile to the main stream is relatively unimportant, and the bulk of the water supply of Egypt has to be met by the Blue Nile alone.

* The Nile Basin. Vol. 7: The Future Conservation of the Nile. By Dr. H. E. Hurst, R. P. Black and Y. M. Simaika. (Ministry of Public Works, Egypt : Physical Department Paper No. 51.) Pp. xi + 178 + 27 plates. (Cairo : S.O.P. Press, 1946.) P.T. 125 : 25s.

The population of Egypt is increasing rapidly, and the only way to guard against future famine is to accumulate a reserve of water so large that it can provide against the worst succession of bad years to be expected on any reasonable basis. This reserve must be in regions where evaporation is relatively small compared with that in Egypt, where the volume of water can be increased to a sufficient extent without greatly increasing the evaporating surface, and far enough apart for the variations from year to year to be practically independent, so that runs of bad years would not be likely to coincide in all areas. The authors find that these requirements can be met by dams at Nimule below Lake Albert and at the outlet of Lake Tana, with possibly at a later stage a further reserve in Lake Victoria. The amounts of storage considered necessary, after allowing for the loss of some 40 per cent on the long journey to Aswan, are enormous : they vary slightly according to different details in the plans but roughly amount to 140 milliards of cubic metres in Lake Albert and Lake Victoria together, with afterwards another 40 milliards, and anything up to 17 milliards in Lake Tana. Compared with these figures the capacity of the first Aswan dam, one milliard cubic metres, is indeed "a trifling amount". In order to reduce the losses in the passage of the Sudd, it will also be necessary either to dig a diversion canal or to embank the Bahr el Jebel. The scale of the problem is shown by the length of the proposed canal—about 300 km, or nearly 200 miles. The ramifications are endless, ranging from consideration of the disturbance of dock and harbour facilities to the need to remove and rebuild churches. The cost would be very high, but against it must be set not only the livelihood of many more people in Egypt but also the prevention of flooding and the supply of hydro-electric power. These details are matters for a full engineering inquiry : but the authors have given a lead in what must be one of the most gigantic plans ever seriously put forward.

There is one detail which requires to be mentioned here. The authors repeatedly emphasize their idea of "Century storage", which briefly means that the reserve must be sufficient to deal with the greatest accumulated deficit to be expected in a hundred years. To obtain this figure, they examined a collection of about sixty long records of discharges, river-levels, rainfall, temperature and pressure, ranging from 10 up to as many as 200 years (with one river-level record of 400 years), and plotted the greatest accumulated departures from the mean values against the square root of the number of years. They write : "If we have N observations of an element whose standard deviation is σ, then if R is the maximum accumulated excess or deficit from the mean, on the average

$$R = 1 \cdot 65 \, \sigma \, \sqrt{N}$$

. . . The form of this result is confirmed by theoretical investigations based on the theory of probability applied to the tossing of coins." Unfortunately, they do not give the theory, and the reviewer has not been able to find it anywhere in the literature of statistics ; but the form of the expression seems to him to be incorrect. It is easily shown that in a series of values obeying the Gaussian distribution, the maximum possible accumulated departure is $0 \cdot 4 \, \sigma \, N$, but the value of $0 \cdot 4 \, N$ is less than $1 \cdot 65 \, \sqrt{N}$ when N is less than 17. Thus, for a series of less than seventeen years, the authors' expression

In *FIAT Review of German Science 1939–1945: Nuclear Physics and Cosmic Rays. Part II,* ed. by W. Bothe, S. Flügge (Dieterichsche Verlagsbuchhandlung, Wiesbaden 1948) pp. 143–165

7.1 GROSSVERSUCHE ZUR VORBEREITUNG DER KONSTRUKTION EINES URANBRENNERS

von

W. HEISENBERG und K. WIRTZ

Max Planck-Institut für Physik, Göttingen

7.1.1 Theoretische Grundlagen

Unter einem Uranbrenner* versteht man eine Anordnung aus Uran und Bremsmittel, in der eine steuerbare Kettenreaktion von Spaltungsprozessen an den Atomkernen des Urans abläuft. Das Bremsmittel ist dabei notwendig, um die Spaltneutronen möglichst schnell an den Resonanzlinien des ^{238}U vorbeizuführen, in denen sie sonst absorbiert würden und für die Reaktionskette verloren gingen. Es ist daher auch zweckmäßig, Uran und Bremsmittel räumlich zu trennen, damit die Neutronen erst, wenn sie thermisch geworden sind, wieder ins Uran zurückdiffundieren. In einer solchen Anordnung kann die Anzahl der Neutronen im wesentlichen durch die folgenden Prozesse verändert werden:

1. Ein Neutron kann bei höheren Energien im ^{238}U und bei beliebiger Energie im ^{235}U einen Spaltungsprozeß hervorrufen und dabei Neutronen freimachen.

2. Es kann an einer Resonanzstelle von ^{238}U eingefangen werden und den Kern ^{239}U bilden, der dann weiter durch β-Zerfall in ^{239}Np und ^{239}Pu übergeht.

3. Es kann im Bremsmittel eingefangen werden.

4. Es kann durch die Oberfläche des Brenners entweichen.

Die Kettenreaktion kann nur stattfinden, wenn die Prozesse 1 ebenso häufig oder häufiger sind als die Prozesse 2, 3, 4 zusammengenommen. Prozeß 4 spielt eine um so geringere Rolle, je größer die Substanzmenge ist, um die es sich handelt. Bei der Konstruktion eines Uranbrenners besteht also die erste Aufgabe darin, eine Anordnung zu finden, bei der die Prozesse 1 häufiger sind als 2 und 3 zusammen.

Die im folgenden beschriebenen Großversuche dienten dazu, nachzuprüfen, ob dieses Ziel bei bestimmten Anordnungen erreicht werden kann; auch kann man dabei die wichtigsten Materialkonstanten in der Kombination bestimmen, in der sie für die Konstruktion des Uranbrenners gebraucht werden.

Alle Versuche dieser Art wurden nach dem folgenden Schema aufgebaut: Die gewählte Anordnung aus Uran und Bremssubstanz befindet sich in einem großen, meist zylindrischen oder kugelförmigen Gefäß. Im Zentrum dieses Gefäßes liegt eine Neutronenquelle. Das Gefäß ist umgeben von einer Hülle aus einer geeigneten Bremssubstanz, in den meisten Fällen einfach Wasser.

---

* In Amerika wird hier das Wort ,,Uranium Pile" verwendet, das insbesondere für die kleineren Anlagen, die noch keine merkbare Erwärmung zeigen, besser ist als Uranbrenner; trotzdem wird dieser Ausdruck im folgenden für alle Apparaturen verwendet, da er sich im Kriege eingebürgert hat.

Um die Bilanz zwischen den Prozessen 1, 2 und 3 experimentell festzulegen, genügen dann die folgenden Messungen: Man bestimmt die Anzahl der pro Zeiteinheit von der Neutronenquelle ausgehenden Neutronen, etwa dadurch, daß man das Gefäß zunächst leer läßt und die Dichteverteilung der thermischen Neutronen im Bremsmittel des Außenraums (also meistens im Wasser) mit Hilfe irgendwelcher Indikatoren ausmißt. Die Gesamtmenge der Neutronen außen sei N_0. Dann bringt man die gewählte Anordnung von Uran und Bremsmittel in das Gefäß und bestimmt die Gesamt-Neutronenmenge innen (N_i) und außen im Bremsmittel des Mantels (N_a). Bezeichnet man den Absorptions-Koeffizienten für thermische Neutronen im äußeren Bremsmittel mit ν_0, so ist die Anzahl der pro Zeiteinheit von der Quelle ausgesandten Neutronen offenbar $N_0 \nu_0$. Da beim Vollversuch $N_a \nu_0$ Neutronen absorbiert werden, müssen im Innern des Gefäßes $\nu_0 (N_a - N_0)$ erzeugt worden sein; der effektive Neutronenproduktions-Koeffizient $\bar{\nu}$ der Anordnungen beträgt also:

$$\bar{\nu} = \frac{N_a - N_0}{N_i}\, \nu_0 . \tag{1}$$

Wenn $\bar{\nu} > 0$ ist, so überwiegen die Prozesse 1 über 2 und 3. Wenn jedoch $\bar{\nu} < 0$ ist, überwiegen 2 und 3.

Die Formel (1) bedarf bei der praktischen Anwendung noch gewisser Korrekturen. Erstens gehen von der Quelle ja nicht thermische, sondern energiereiche Neutronen aus. Diese energiereichen Neutronen erfahren eine gewisse Vermehrung dadurch, daß sie Spaltungen in ^{238}U-Kernen hervorrufen, die von ihnen getroffen werden. Für diese Vermehrung ist zu korrigieren. Zweitens wird der Überschuß von Neutronen im Innern in Form von schnellen Neutronen produziert. Man muß ihn also mit der Anzahl der schnellen, nicht der thermischen Neutronen im Innern vergleichen.

Um den ersten Effekt zu berücksichtigen, multipliziert man N_0 mit einem Faktor Y, der angibt, um wieviel die Anzahl der primären Neutronen durch Spaltung von ^{238}U erhöht wird. Dieser Faktor muß durch besondere Messungen oder durch Berechnungen auf Grund der Untersuchungen von STETTER und LINTNER[2] (vgl. auch Ziff. 7. 2) bestimmt werden. Den zweiten Effekt kann man dadurch in die Formel (1) aufnehmen, daß man N_i mit einem Faktor e^w multipliziert, wobei w die Wahrscheinlichkeit dafür bedeutet, daß ein schnelles Neutron während des Bremsvorgangs in ^{238}U eingefangen wird (Prozeß 2). Dieser Faktor ist meistens rechnerisch auf Grund der Messungen

---

[1] W. HEISENBERG, Die Möglichkeit technischer Energiegewinnung aus der Uranspaltung. I. Forschungsbericht 1939. Unveröffentlicht. - W. HEISENBERG, Bericht über die Möglichkeit technischer Energiegewinnung aus der Uranspaltung. II. Forschungsbericht 1940. Unveröffentlicht.

[2] W. HEISENBERG, Über die Möglichkeit der Energieerzeugung mit Hilfe des Isotops 238. Forschungsbericht. Unveröffentlicht. — W. BOTHE u. A. FLAMMERS-FELD, Über die Vermehrung thermischer und schneller Neutronen in Uran. Forschungsbericht 1941. Unveröffentlicht. — G. STETTER u. K. LINTNER, Schnelle Neutronen im Uran. Der Zuwachs durch den Spaltprozeß und der Abfall durch unelastische Streuung. Forschungsbericht 1941. Unveröffentlicht. — G. STETTER, Schnelle Neutronen im Uran. Forschungsbericht 1942. Unveröffentlicht.

von VOLZ, HAXEL[3] und SAUERWEIN[4] und theoretischer Überlegungen von FLÜGGE[4] (vgl. auch Ziff. 4. 5) bestimmt worden. Die verbesserte Formel lautet schließlich:

$$\bar{\nu} = \frac{N_a - N_0 Y}{N_i}\, e^{-w} \nu_0 \,. \tag{2}$$

Der in dieser Weise bestimmte Wert von $\bar{\nu}$ ist ein Maß für die Güte der Anordnung. Eine Anordnung ist offenbar um so besser, je größer $\bar{\nu}$ ist. Dabei hängt $\bar{\nu}$ natürlich auch noch von der Gesamtdichte der Anordnung ab, d. h. eine reine Erhöhung der Dichte würde eine Erhöhung von $\bar{\nu}$ ergeben, ohne daß damit grundsätzlich in der Bilanz der Prozesse 1, 2, 3 etwas gebessert worden wäre. Man wird daher am besten etwa $\bar{\nu}/\varrho$ ($\varrho=$ Dichte) als geeignetes Maß für die Güte einer Anordnung betrachten. Die im folgenden beschriebenen Großversuche geben daher in ihrer historischen Reihenfolge geordnet im wesentlichen eine Reihe nach wachsenden Werten von $\bar{\nu}/\varrho$ (Tab. 4, S. 157).

In USA (Smyth-Report) wird anstelle von $\bar{\nu}$ der sog. Multiplikationsfaktor k als Kriterium für die Güte der Anordnung verwendet. k gibt an, um wieviel sich im Mittel eine Neutronengeneration von der Entstehung bis zur Absorption vermehrt. $k > 1$ bedeutet vermehrungsfähige, $k < 1$ vermehrungsunfähige Substanzanordnung. Die Umrechnung von $\bar{\nu}$ auf k ist im Anhang angegeben.

Der gemessene Wert von $\bar{\nu}$ kann dann auch mit dem theoretischen Wert verglichen werden, der nach den in Ziff. 7. 3 erwähnten Methoden berechnet wurde. In diese Theorie gehen als wichtigste und experimentell relativ schwer zugängliche Konstanten die folgenden beiden Größen ein: Die Anzahl X der Neutronen, die im gewöhnlichen Uran pro eingefangenes thermisches Neutron durch Spaltung frei werden (ein Uranbrenner aus gewöhnlichem Uran ist natürlich nur möglich, wenn X größer als 1 ist. Dabei hängt X auch vom Reinheitsgrad des benutzten Urans ab, wird also durch Verunreinigungen herabgesetzt). Ferner eine Konstante, die die Resonanzabsorption in ^{238}U mißt. Die Resonanzabsorption war als Funktion der Schichtdicke von SAUERWEIN[4] gemessen worden; dabei waren diese Messungen in bezug auf den Gang mit der Schichtdicke genauer als hinsichtlich der Absolutwerte. Der Neutronenproduktions-Koeffizient $\bar{\nu}$ hängt aber so empfindlich von X und dem absoluten Wert der Resonanzabsorption ab, daß man für die experimentelle Bestimmung dieser Größen am besten die Messungen von $\bar{\nu}$ benutzt. Jeder Großversuch ergibt eine Beziehung zwischen X und der Resonanzabsorption. Trägt man diese

---

[3] H. VOLZ, Über die Absorption des Urans im Resonanzgebiet. Forschungsbericht 1941. Unveröffentlicht. — H. VOLZ u. O. HAXEL, Über die Absorption von Neutronen im Uran. Forschungsbericht 1941. Unveröffentlicht.

[4] S. FLÜGGE, Berechnung des Bruchteils von Neutronen, der in einem Uran-Wasserstoffgemisch in der Resonanzlinie absorbiert wird. I., II., III. Forschungs-. bericht 1940. Unveröffentlicht. — S. FLÜGGE u. K. SAUERWEIN, Untersuchungen über den Resonanzeinfang von Neutronen beim Uran. I., II. Forschungsbericht 1942. Unveröffentlicht. — K. SAUERWEIN, Untersuchungen über den Resonanzeinfang von Neutronen beim Uran. Forschungsbericht 1941. Unveröffentlicht.

Beziehung in einem Diagramm als Kurve auf, so müssen sich die aus den verschiedenen Großversuchen gewonnenen Kurven alle in einem Punkt schneiden, aus dem der Wert von X und der Betrag der Resonanzabsorption zu bestimmen ist. Praktisch schneiden sich die Kurven wegen der Unvollkommenheit der Theorie und der Messungen nicht genau in einem Punkt. Man kann aber als Ergebnis der Messungen als besten Wert für X für das in Deutschland verwendete U-Metall etwa

$$X = 1{,}18 \tag{3}$$

ansehen.

Wenn man für die thermischen Neutronen den Gesamt-Absorptionsquerschnitt in gewöhnlichem Uran und den Anteil kennt, der davon auf die Spaltung in ^{235}U entfällt (vgl. Ziff. 4. 4), so kann man aus dem Verhältnis die Anzahl der Neutronen berechnen, die im Mittel bei einem Spaltungsprozeß an ^{235}U ausgesandt werden. Die Absorptionsmessungen in Uran ergeben für den effektiven Absorptionsquerschnitt pro Uranatom[5]:

$$\sigma_{\mathrm{abs}} = 6{,}2 \cdot 10^{-24}\ \mathrm{cm}^2. \tag{4}$$

Für den Spaltungsquerschnitt haben Messungen von v. DROSTE und anderen etwa den Wert

$$\sigma_{\mathrm{sp}} = 3{,}3 \cdot 10^{-24}\ \mathrm{cm}^2 \tag{5}$$

geliefert. Daraus folgt als Neutronenzahl X_{235} pro Spaltungsprozeß in ^{235}U:

$$X_{235} = 1{,}18 \cdot \frac{6{,}2}{3{,}3} = 2{,}2. \tag{6}$$

Diese Zahl bezieht sich auf die Spaltung durch thermische Neutronen. Bei schnellen Neutronen könnte die Zahl etwas höher sein.

Wenn man Anordnungen mit $\bar{\nu} > 0$ konstruiert hat, so kann man einen energieliefernden Uranbrenner bauen, sobald man nur die Anordnung hinreichend stark vergrößert. Der Bau eines Uranbrenners ist also von diesem Augenblick an eine reine Materialfrage. Da es im Hinblick auf die Kostbarkeit der Materialien zunächst erwünscht erscheint, den Uranbrenner möglichst klein zu halten, entsteht als nächstes Problem die Frage, wie groß die Anordnung mindestens gemacht werden muß, damit sie selbständig brennt. Diese „kritische Größe" hängt außer von $\bar{\nu}$ in erster Linie von der sogenannten effektiven Diffusionskonstante $\bar{D}$ ab. Man kann nämlich nach BOTHE[6] das Verhalten der Neutronen im Innern des Uranbrenners in grober Näherung durch die effektive Diffusionsgleichung beschreiben:

$$\frac{\partial n}{\partial t} = \bar{D}\,\Delta n + \bar{\nu} n. \tag{7}$$

n bedeutet dabei die Anzahl der Neutronen pro ccm; die Konstante $\bar{D}$ oder die Länge $l = \sqrt{\dfrac{\bar{D}}{\bar{\nu}}}$ gibt ein Maß dafür, wie schnell die Spaltungsneutronen

[5] W. BOTHE u. A. FLAMMERSFELD, Forschungsbericht 1940, Unveröffentlicht. R. u. K. DÖPEL u. W. HEISENBERG, Bestimmung der Diffusionslänge thermischer Neutronen im Präparat. 38. Forschungsbericht 1940. Unveröffentlicht.
[6] W. BOTHE, Forschungsbericht. Unveröffentlicht.

im Uranbrenner diffundieren können und ist physikalisch in erster Linie durch die Strecke bestimmt, die ein Spaltungsneutron zurücklegen kann, bevor es thermisch geworden ist. Der Verlauf der Neutronendichte im Innern des Brenners gestattet die Bestimmung dieser effektiven Diffusionskonstante und muß daher gemessen werden. Bei dem Versuch, die Ergebnisse solcher Messungen theoretisch darzustellen, ergab sich allerdings, daß die Diffusionsgleichung (7) die Verhältnisse nur sehr ungenau wiedergibt. Eine wesentlich bessere Darstellung erhält man nach BOPP[7], wenn man das grundsätzlich recht komplizierte Problem in folgender Weise vereinfacht. Man teilt die Neutronen auf in zwei Sorten: in „thermische" und in „schnelle" Neutronen. Dann setzt man für jede dieser Neutronensorten eine Diffusionsgleichung an und bestimmt die eingehenden „Konstanten" aus dem Experiment. In dieser (rechnerisch schon recht komplizierten) Weise kann man tatsächlich den experimentellen Dichteverlauf befriedigend darstellen und damit auch die kritische Größe ziemlich genau bestimmen.

Wenn man eine bestimmte Anordnung aus U und Bremsmittel vergrößert, so steigt das Verhältnis $Z = \dfrac{N_a}{N_0}$ an und geht bei der kritischen Größe gegen unendlich. Der Wert von Z ist daher auch ein bequemes Maß für die Güte einer Apparatur, d. h. dafür, wie weit man noch von der kritischen Größe entfernt ist. Z ist in der Tab. 4 (S. 157) für die verschiedenen Großversuche eingetragen, die in ihrer historischen Reihenfolge auch im wesentlichen eine Reihe von wachsenden Z-Werten ergeben.

Eine weitere Verbesserung des Uranbrenners kann dadurch erreicht werden, daß man den Brenner mit einer Substanz umgibt, die die Neutronen gut reflektiert, aber möglichst wenig absorbiert[8]. Bei den ersten Großversuchen wurde der Brenner fast stets mit Wasser oder Paraffin umgeben, jedoch ist natürlich eine Umgebung mit einer schwächer absorbierenden Substanz günstiger, wobei die Bremsung schneller Neutronen in diesen Substanzen auch eine wesentliche Rolle spielt. Daher wurden die letzten Uranbrenner in einen Graphitmantel eingebaut. Dieser Graphitmantel hat auch tatsächlich die Neutronenvermehrung erheblich heraufgesetzt.

Bekanntlich sind in Amerika große Uranbrenner gebaut worden, die kein schweres Wasser, sondern nur Graphit als Bremsmittel enthalten. Dazu war der in Deutschland benutzte Graphit nicht geeignet, da er nach Messungen von BOTHE und JENSEN[9] einen zu hohen Absorptionskoeffizienten besaß. Allerdings wurde damals aus diesen Messungen geschlossen, daß auch reine Kohle ungeeignet sei. Als Mantel für den Uranbrenner aus D_2O konnte der Graphit jedoch gut verwendet werden[10].

[7] F. BOPP, Forschungsbericht. Unveröffentlicht.
[8] F. BOPP u. E. FISCHER, Der Einfluß des Rückstreumantels auf die Neutronenausbeute des U-Brenners. Forschungsbericht 1944. Unveröffentlicht.
[9] W. BOTHE u. P. JENSEN, Die Absorption thermischer Neutronen in Elektrographit. Forschungsbericht 1941. Z. Physik **122**, 749 [1944].
[10] Im folgenden sind noch einige Arbeiten genannt, die theoretische Probleme des

7.1.2 Übersicht über die in Deutschland durchgeführten Großversuche

Die Reihenfolge der Großversuche wurde durch physikalische Gesichtspunkte und zu einem erheblichen Teil durch äußere, mit der Beschaffung des Materials zusammenhängende Umstände bestimmt. Wir geben im folgenden eine allgemeine historische Übersicht über die Versuche und beschreiben als Beispiel in Ziff. 7.1.3 einen solchen Versuch in den Einzelheiten.

Man kann etwa 4 Gruppen unterscheiden: einige Großversuche von mehr vorbereitendem Charakter; dann die Versuche des Physikalischen Instituts der Universität Leipzig (L_1 bis L_4, DÖPEL und HEISENBERG); die der Versuchsstelle Gottow des Heereswaffenamtes (G_1 bis G_3, BERKEI, CZULIUS, DIEBNER, HARTWIG und HERRMANN) und die umfangreicheren Versuche des Kaiser-Wilhelm-Instituts für Physik (B_1 bis B_8, BOPP, FISCHER, HEISENBERG, WIRTZ), von denen die letzten im Zusammenhang mit dem Heidelberger KWI (BOTHE, JENSEN, RITTER) ausgeführt wurden.

Zu den vorbereitenden Großversuchen gehört der „CO_2-Versuch" von HARTECK, JENSEN, KNAUER und SUESS[11] in Hamburg vom August 1940,

Uranbrenners behandeln. Die Titel einer Reihe weiterer Arbeiten sind nicht mehr verfügbar.

F. BOPP u. E. FISCHER, Der Einfluß des Rückstreumantels auf die Neutronenausbeute des U-Brenners. Forschungsbericht. Unveröffentlicht. — W. BOTHE, Strenge Behandlung der Diffusion in absorbierenden Mitteln. Forschungsbericht 1940. Z. Physik **118**, 401, 1941; **119**, 493, 1942. — K. H. HÖCKER, Berechnung der Energieerzeugung in der Uranmaschine. II. Kohle als Bremssubstanz. Forschungsbericht 1940. Unveröffentlicht. — K. H. HÖCKER, Berechnung der Energieerzeugung in der Uranmaschine. IV. Wasser als Bremssubstanz. Forschungsbericht 1940. Unveröffentlicht. — K. H. HÖCKER, Über die Anordnung von Uran und Streusubstanz in der U-Maschine. Forschungsbericht. Unveröffentlicht. — K. H. HÖCKER, Über die Abmessungen von Uran und schwerem Wasser in einer Kugelstrukturmaschine. Forschungsbericht 1943. Unveröffentlicht. — P. O. MÜLLER, Die Neutronenabsorption in Kugelschalen aus Uran. Forschungsbericht 1940. Unveröffentlicht. — P. O. MÜLLER, Berechnung der Energieerzeugung in der Uranmaschine. III. Schweres Wasser. Forschungsbericht 1940. Unveröffentlicht. — P. O. MÜLLER, Über die Temperaturabhängigkeit der Uranmaschine. Forschungsbericht 1940. Unveröffentlicht. — H. VOLZ, Über die Geschwindigkeitsverteilung der Neutronen in Gemisch von schwerem Wasser und Uran. Forschungsbericht 1941. Unveröffentlicht. — C. F. v. WEIZSÄCKER, P. O. MÜLLER u. K. H. HÖCKER, Berechnung der Energieerzeugung in der Uranmaschine. Forschungsbericht 1940. Unveröffentlicht. — C. F. v. WEIZSÄCKER, Die Energiegewinnung aus dem Uranspaltprozeß durch schnelle Neutronen. Forschungsbericht 1940. Unveröffentlicht. — C. F. v. WEIZSÄCKER, Eine Möglichkeit der Energiegewinnung aus ^{238}U. Forschungsbericht 1940. Unveröffentlicht. — C. F. v. WEIZSÄCKER, Bemerkungen zur Berechnung von Schichtenanordnungen. Forschungsbericht 1941. Unveröffentlicht. — C. F. v. WEIZSÄCKER, Über den Temperatureffekt der Schichtenmaschine. Forschungsbericht 1941. Unveröffentlicht. — C. F. v. WEIZSÄCKER, Verbesserte Theorie der Resonanzabsorption in der Maschine. Forschungsbericht 1942. Unveröffentlicht.

[11] P. HARTECK, H. JENSEN, FR. KNAUER u. H. SUESS, Über die Bremsung, die Diffusion und den Einfang von Neutronen in fester Kohlensäure und über ihren Einfang in Uran. Forschungsbericht 1940. Unveröffentlicht. — P. HARTECK, Versuche mit dem Kohlensäureblock; Methoden zur Anreicherung von schwerem Wasser; die experimentellen Möglichkeiten zur Trennung der Uranisotope. Forschungsbericht 1941. Unveröffentlicht.

bei dem 15 t festes Kohlensäureeis und eine relativ geringe Menge Uranoxyd (185 kg) diskontinuierlich zu einem etwa $2 \times 2 \times 2$ m großen Block zusammengebaut wurden, in dem die Diffusion, die Bremsung und der Einfang von Neutronen einer (Ra + Be)-Quelle studiert wurde. Ein Einfluß des Urans auf die Neutronenvermehrung wurde nicht beobachtet, dagegen die Bremslänge und die Diffusionslänge der Neutronen in CO_2-Eis bestimmt. Ähnliche Messungen an einem Würfel von etwa 1 m Kante, bestehend aus 2000 Stück Paketen Natrium-Uranat ($Na_2U_2O_7 + 6H_2O$) je 1 kg, eingepackt in Papier, das gewissermaßen zusammen mit dem Hydratwasser als Bremssubstanz diente, führte v. DROSTE[12] in Berlin (September 1940) aus. Wesentlich genauere Ergebnisse für die theoretische Auswertung lieferte ein Versuch von BOTHE und FLAMMERSFELD[13] (BF, Mai 1941), der die Neutronenvermehrung in einem homogenen Gemisch von 4432 kg U_3O_8 und 435 kg Wasser in einem Zylinder von 992 l, der sich in einem H_2O-Streumantel befand, untersuchte und den Neutronenproduktions-Koeffizienten $\bar{\nu}/\varrho$ bestimmte (vgl. Tab. 4, S. 157).

Die Leipziger Versuche[14] L_1 bis L_4 (1941—42) wurden dadurch besonders wichtig, daß sie schließlich zum erstenmal positive Neutronenproduktions-Koeffizienten $\bar{\nu}/\varrho$ und Vermehrungsfaktoren $Z > 1$ ergaben und damit die Möglichkeit selbständig arbeitender energieliefernder Uranbrenner bewiesen. Der erste sichere Nachweis gelang im Februar 1942; in USA (vgl. Smyth-Report) etwa um dieselbe Zeit. In USA ging von diesem Zeitpunkt ab die Entwicklung sehr rasch. Schon am 2. Dezember 1942 wurde in Chicago von FERMI und Mitarbeitern der erste selbständige Pile errichtet, in dem zum erstenmal die Kern-Kettenreaktion vor sich ging. Dies ist in Deutschland bekanntlich nie erreicht worden.

L_1 bestand aus kugelschalenförmig angeordneten Schichten von Paraffin (~ 2 cm) und Uranoxyd ($\sim 7{,}5$ cm). Den Kern bildete Uranoxyd. Ihn umgaben abwechselnd zwei Schichten Paraffin und zwei weitere Schichten U_3O_8. Diese Schichtung ähnelt der gleich zu beschreibenden von B_1 und führte wie diese nicht zur Neutronenproduktion. Die einzelnen Schichten waren durch Halbkugeln aus Aluminiumblech voneinander getrennt.

In L_2 (Abb. 45) wurden zum erstenmal 150 l schweres Wasser als Bremssubstanz verwendet. Um einen D_2O-Kern von $\sim 12{,}5$ cm Radius lag eine Schicht von $\sim 3{,}5$ cm Uranoxyd, dann folgte wieder eine D_2O-Schicht von $\sim 15{,}5$ cm

[12] G. v. DROSTE, Bericht über einen Versuch mit 2 t Natriumuranat. Forschungsbericht 1940. Unveröffentlicht.

[13] W. BOTHE u. A. FLAMMERSFELD, Messungen an einem Gemisch von U-Oxyd und Wasser; der Vermehrungsfaktor X und der Resonanzeinfang w. Forschungsbericht 1941. Unveröffentlicht.

[14] R. u. K. DÖPEL u. W. HEISENBERG, Versuche mit einer Schichtenanordnung von Wasser und Präparat 38. Forschungbericht 1941. Unveröffentlicht. — R. u. K. DÖPEL u. W. HEISENBERG, Versuche mit Schichtenanordnungen von D_2O u. 38. Forschungsbericht 1941. Unveröffentlicht. R. u. K. DÖPEL u. W. HEISENBERG, Die Neutronenvermehrung in einem D_2O-38-Metallschichtensystem. Forschungsbericht 1942. Unveröffentlicht. — R. u. K. DÖPEL u. W. HEISENBERG, Die Neutronenvermehrung in 38-Metall durch rasche Neutronen. Forschungsbericht 1942. Unveröffentlicht. — W. BOTHE, Bemerkungen zum Leipziger D_2O-Versuch. Forschungsbericht 1942 unveröffentlicht.

und dann eine weitere U_3O_8-Schicht von $\sim 3{,}5$ cm Dicke. Die ganze Kugel befand sich ebenso wie L_1 in einem Rückstreumantel von H_2O. L_3 verwendete an Stelle von U_3O_8 U-Metallpulver der Schüttdichte 12. Die Anordnung war dieselbe wie bei L_1, jedoch fehlte die äußere Uranschicht. Beim Versuch L_4 war sie vorhanden und bestand ebenfalls aus Uranmetallpulver. Auch bei L_2, L_3 und L_4

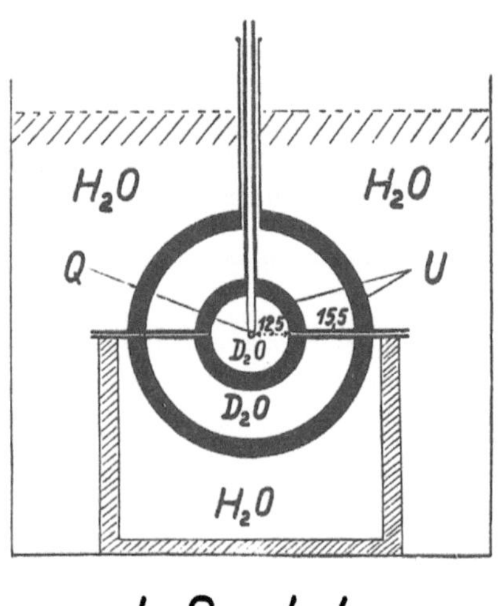

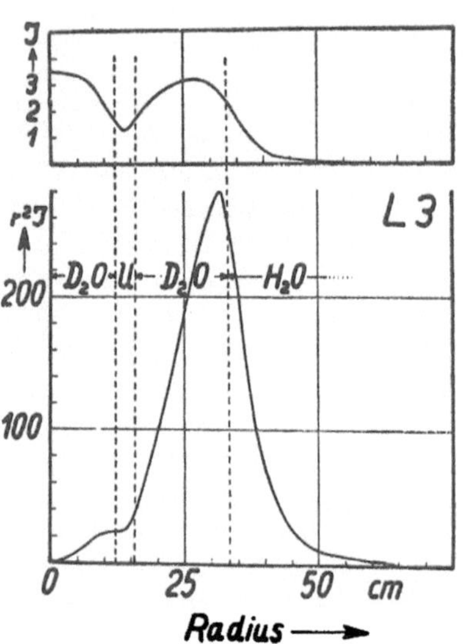

Abb. 45. Schema der Versuchsanordnung. Dicke der Urankugelschalen 3,5 cm. Uran und D_2O sind durch nicht eingezeichnete Wände aus Aluminiumblech getrennt. Abstandsangaben in der Figur in cm. Einzelheiten im Text.

Abb. 46. Verlauf der Neutronenintensität bei L_3 im Innen- und Außenraum in Abhängigkeit vom Radius. Bis 33 cm: Innenraum; ab 33 cm: Außenraum.

waren die einzelnen Schichten durch Kugelschalen aus Aluminiumblech getrennt. Durch einen vertikal ins Innere führenden Meßschacht konnte die Neutronenquelle ins Zentrum der Kugel gebracht werden und außerdem der Verlauf der Neutronenintensität in der Schichtung verfolgt werden. Die Intensität (Abb. 46) der austretenden Neutronen wurde im umgebenden Wasserstreumantel bestimmt (vgl. die ausführliche Beschreibung derartiger Messungen im folgenden Abschnitt 7. 1. 3). L_4 ergab positives $\bar{\nu}/\varrho$ und zum erstenmal ein Z, das deutlich > 1 war. Daten und Ergebnisse sind in der folgenden Tabelle 1 zusammengestellt, soweit sie noch bekannt sind.

Die Arbeitsgruppe in Gottow[15] führte 3 Großversuche (G_1 bis G_3) durch. G_1 (1941) war eine Packung aus 6800 Stück Würfeln von 9,7 cm Kante aus U_3O_8-Pulver (insgesamt ca. 25 to), die in Kästen aus Paraffin (insgesamt 4,4 to)

[15] Versuchsstelle Gottow des Heereswaffenamtes (Mitarbeiter: F. BERKEI, W. BORRMANN, W. CZULIUS, K. DIEBNER, G. HARTWIG, K. H. HÖCKER, W. HERRMANN, H. POSE u. W. REXER). Bericht über einen Würfelversuch mit

Tabelle 1. Daten und Ergebnisse der Leipziger Kugelschichtversuche L_2 bis L_4.

	Innenkugel $R = 12,5$ cm	1. Schicht Dicke $= 3,5$ cm	2. Schicht Dicke $=15,5$ cm	3. Schicht Dicke $= 3,5$ cm	Gesamt-volumen	Mittlere Dichte ρ	Prod.-Koeff. $\bar{\nu}$	Massenprod.-Koeff. $\bar{\nu}/\rho$	Vermehrungs-faktor Z
L_2	9 kg D_2O	37 kg U_3O_8	155 kg D_2O	105 kg U_3O_8	212 l	1,5	< 0 $+ 65*$	< 0 $+ 44*$	~ 1
L_3	9 kg D_2O	108 kg U-Pulver	155 kg D_2O	keine	159 l	1,7	0 $+ 88*$	0 $+ 50*$	~ 1
L_4	9 kg D_2O	175 kg U-Pulver	155 kg D_2O	580 kg U-Pulver	212 l	3,8	$+ 90$	$+ 23,5$	1,1

gefüllt wurden. Die Wandstärke des Paraffins zwischen den Würfeln betrug 2 cm. Die ganze Anordnung war in einen zylindrischen Al-Kessel von 2,50 m Querschnitt und 2,30 m Höhe eingebaut, der wiederum in Wasser als Streumantel stand. Geeignete Schächte gestatteten das Einführen des Neutronenpräparates und der Meßindikatoren. Dieser Versuch brachte den ersten Hinweis darauf, daß Würfelanordnung günstiger als Schichtenanordnungen sein könnte. Die Neutronenproduktion war negativ (vgl. Tabelle 4).

Der Versuch G_2 (1942—43) war von ganz anderer Art. Die theoretische Auswertung der bisherigen Großversuche hatte nahegelegt, U-Metallwürfel von 5 cm Kante (2,2 kg) zu versuchen. Es wurden etwa 100 Stück dieser Würfel in etwa 220 l festes D_2O-Eis (bei — 10° C) in einer Paraffin-Kugel von 37,5 cm Radius ohne sonstiges Halterungsmaterial eingefroren; äußerer Streumantel war Paraffin. Es ergab sich eine Neutronenvermehrung, die erheblich stärker war als bei dem früheren Leipziger Versuch L_4 (vgl. Tab. 4).

G_3 (1942/43) stellt einen entsprechenden Versuch in flüssigem D_2O dar. G_{3a} bestand aus derselben Kugel vom Radius 37,5 cm, die zunächst einen Streumantel von 13,5 cm D_2O besaß, der schließlich wiederum von Paraffin umgeben war. G_{3b} enthielt dagegen in einer Al-Kugel von 51 cm Radius 240 U-Würfel von 5 cm Kante (insgesamt 540 kg U-Metall) und 525 l D_2O. Das Ganze war umgeben von Paraffin. Dieser letzte Versuch brachte eine Vermehrung $Z = 2,1$ und bewies damit die Vorzüge der Würfel. In all diesen Versuchen waren die Würfel in dichtester Kugelpackung mit 8 cm gegenseitigem Abstand aufgehängt (Ergebnisse in Tabelle 4).

Uranoxyd und Paraffin (G_1). Forschungsbericht. Unveröffentlicht; sowie 2 weitere Berichte über G_2 und G_3. — F. BOPP, E. FISCHER, W. HEISENBERG, C. F. v. WEIZSÄCKER u. K. WIRTZ, Untersuchungen mit neuen Schichtenanordnungen aus U-Metall und Paraffin. Forschungsbericht. Unveröffentlicht. — W. HEISENBERG, Auswertung der Gottower Versuche ($G_{2,\,3}$). Forschungsbericht. Unveröffentlicht.

* Die Produktionskoeffizienten > 0 ergaben sich, wenn eine Korrektur derart angebracht wurde, daß das in dem Brenner befindliche Halterungsmaterial zum Außenraum gerechnet wurde.

Die systematischen Versuche im Kaiser-Wilhelm-Institut für Physik in Dahlem (B_1 bis B_8)[16] begannen im Dezember 1940 mit horizontalen Schichtungen von Uranoxyd und Paraffin (B_1 und B_2) in einem $\sim 1{,}4$ m hohen Al-Zylinder von $\sim 1{,}4$ m $\varnothing$. Geringste Neutronenabsorption ergab in Übereinstimmung mit der damaligen theoretischen Erwartung der Versuch B_1 mit 27 g/cm² U_3O_8 in jeder Schicht, entsprechend ~ 7 cm Schichtdicke, die mit 2,1 cm dicken Paraffin-Schichten abwechselten (Abb. 47). Rund 16 solcher Doppelschichten mit insgesamt rund 6800 kg U_3O_8 lagen aufeinander, zum Teil stabilisiert und geebnet durch eingelegte Al-Bleche. Die ganze Anordnung war von einem Wasserstreumantel umgeben. Durch einen axialen Schacht konnten eine Neutronenquelle von 500 mg Ra + Be-Pulver sowie Indikatoren zur Intensitätsmessung ins Innere gebracht werden.

Sobald (1941/42) größere Mengen U-Metall, zunächst in Pulverform, zur Verfügung standen, wurden damit entsprechende Versuche (B_3 bis B_5) zunächst ebenfalls mit Paraffin als Bremsmaterial ausgeführt. Die horizontalen Schichten wurden in eine Al-Kugel von 28 cm Radius (Abb. 48a und 48b) eingebaut. Wiederum konnten Sonden und Neutronenquelle durch einen Schacht ins Innere gebracht werden. Drei Variationen der Schichtdicken wurden durchgemessen, deren Daten und Neutronen-

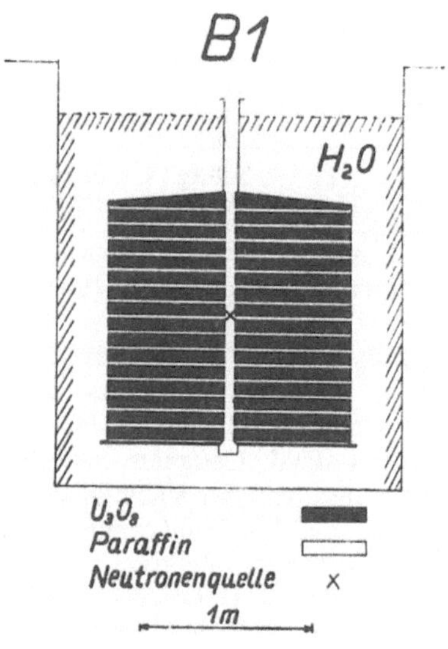

Abb. 47. Vertikalschnitt durch B_1; Kessel: Aluminiumblech.

[16] W. HEISENBERG (unter Mitarbeit von A. DEUBNER, E. FISCHER, C. F. v. WEIZSÄCKER u. K. WIRTZ). Bericht über die ersten Versuche an der im Kaiser-Wilhelm-Institut f. Physik aufgebauten Apparatur. Forschungsbericht 1940. Unveröffentlicht. — W. HEISENBERG (unter Mitarbeit von A. DEUBNER, E. FISCHER, C. F. v. WEIZSÄCKER u. K. WIRTZ). Bericht über Versuche mit Schichtenanordnungen von Präparat 38 und Paraffin am Kaiser-Wilhelm-Institut f. Physik, Berlin-Dahlem. Forschungsbericht 1941. Unveröffentlicht. — F. BOPP, E. FISCHER, W. HEISENBERG, C. F. v. WEIZSÄCKER u. K. WIRTZ. Vorläufiger Bericht über Ergebnisse an einer Schichtenkugel aus 38-Metall und Paraffin. Forschungsbericht 1942. Unveröffentlicht. — W. HEISENBERG, Bemerkungen zu dem geplanten halbtechnischen Versuch (B_6 und B_7) mit 1,5 to D_2O und 3 to 38-Metall. Forschungsbericht 1942. Unveröffentlicht. — W. HEISENBERG, F. BOPP, E. FISCHER, C. F. v. WEIZSÄCKER u. K. WIRTZ, Messungen an Schichtenanordnungen aus 38-Metall und Paraffin. Forschungsbericht 1942. Unveröffentlicht. — F. BOPP, E. FISCHER, W. HEISENBERG, K. WIRTZ, W. BOTHE, P. JENSEN, O. RITTER: Schichtenversuche mit Uranmetall in D_2O (B_6). Forschungsbericht. Unveröffentlicht. — F. BOPP, E. FISCHER, W. HEISENBERG, K. WIRTZ, Kaiser-Wilhelm-Institut Berlin; W. BOTHE, P. JENSEN, O. RITTER, Kaiser-Wilhelm-Institut Heidelberg: 3 Berichte über B_6, B_7. Forschungsberichte 1944 und 1945. Unveröffentlicht.

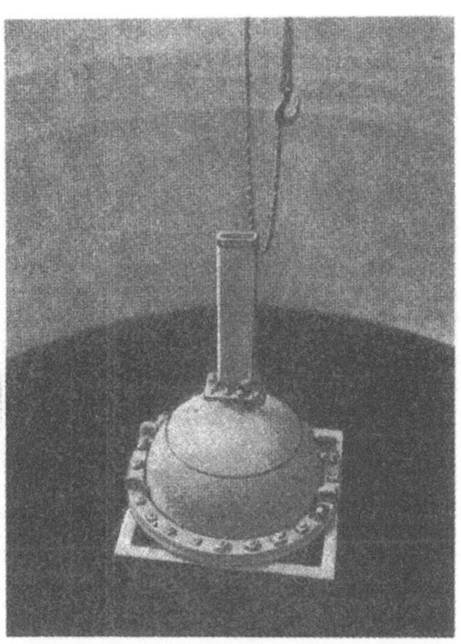

Abb. 48a. Äußere Ansicht von B_3. Die Aluminiumkugel steht auf einem Traggerüst in einem Wasserbassin. Das Wasser des Streumantels ist nur zum Teil eingelassen. Oben auf der Kugel der Kamin zum Einführen der Neutronenquelle.

produktions-Koeffizienten in der folgenden Tabelle 2 zusammengestellt sind (vgl. auch Abb. 49 u. 50). In Übereinstimmung mit der theoretischen Erwartung ergab B_3 die geringste Neutronenabsorption. Keiner der Versuche ergab Neutronenvermehrung.

Die Versuche B_6, B_7 (1944) und B_8 (1945) benutzten die Erfahrungen von L_{2-4} und $G_{2,3}$ und sollten die Verhältnisse bei Kombinationen von D_2O mit U-Metall systematisch klären. Als Bremssubstanz diente also jetzt schweres Wasser, von dem rd. 1,5 to dem Kaiser-Wilhelm-Institut für Physik zur Verfügung standen. Obwohl die Vorzüge der Würfel damals schon bekannt waren, wurden um der Systematik willen zunächst wiederum Schichten untersucht (B_6 und B_7). In einem zylindrischen Kessel aus „Elektron"-Metall (Magnesium mit geringen Al-Zusätzen, eine Legierung mit sehr geringer Neutronenabsorption) von 124 cm Querschnitt und 124 cm Höhe, umgeben von einem H_2O-Streumantel, wurden horizontale Schichten aus U-Metall-Platten übereinander angeordnet. Die Schichten waren durch geeignet geformte Magnesiumblechstücke distanziert. Eine U-Schicht be-

Tabelle 2. Daten der Schichtenkugeln B_3 bis B_5.

Versuch	B_3	B_4	B_5
Zahl der U-Schichten	19	12	7
Schichtdicke: U in g/cm² .... (in Klammern: Dicke in cm)	18 (1,7)	39 (3,8)	75 (7,3)
Paraffin in g/cm² .... (in Klammern: Dicke in cm)	1,44 (1,6)	1,44 (1,6)	1,44 (1,6)
Gesamtmasse U in kg	551	740	864
Gesamtmasse Paraffin in kg	44	37	12,5
Mittelere Dichte	5,7	7,2	8,4
Neutronenproduktions-Koeffizient	— 209	— 213	— 344

Schüttdichte des U-Pulvers im Brenner: 10—11.

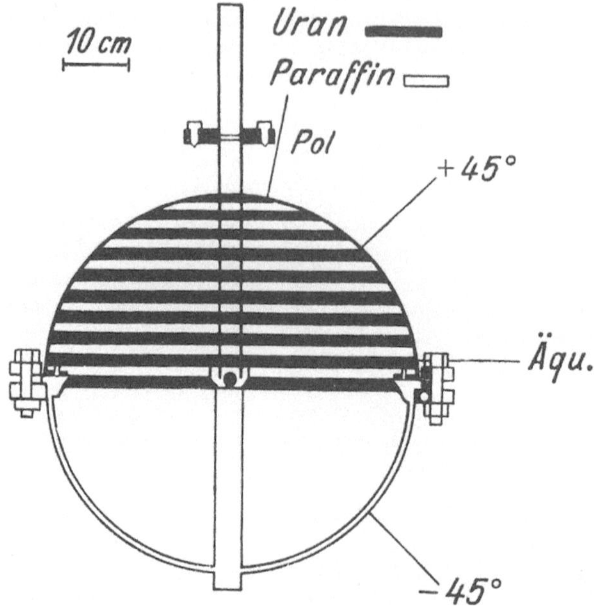

Abb. 48 b. Schema des Innenaufbaus von B_3. Schnitt durch die Schichten von Uranmetallpulver (schwarz) und Paraffin (weiß). Die Schichtung setzt sich in die untere Halbkugel fort, ist dort jedoch nicht eingezeichnet. Im Zentrum die Neutronenquelle.

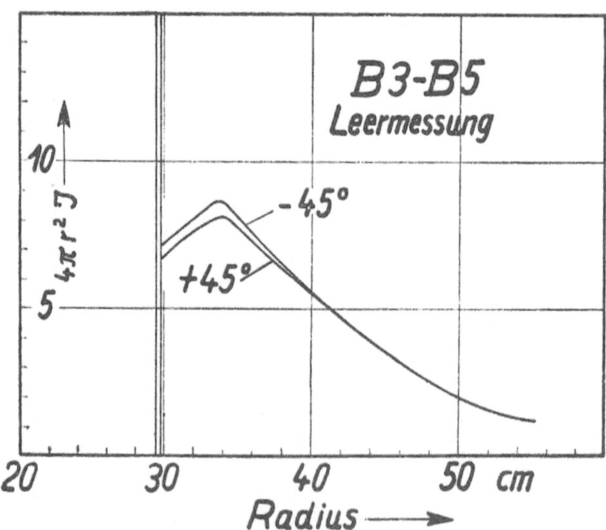

Abb. 49. Verlauf der Neutronenintensität im Außenraum (H_2O) bei leerer Anordnung (B_3 bis B_5).

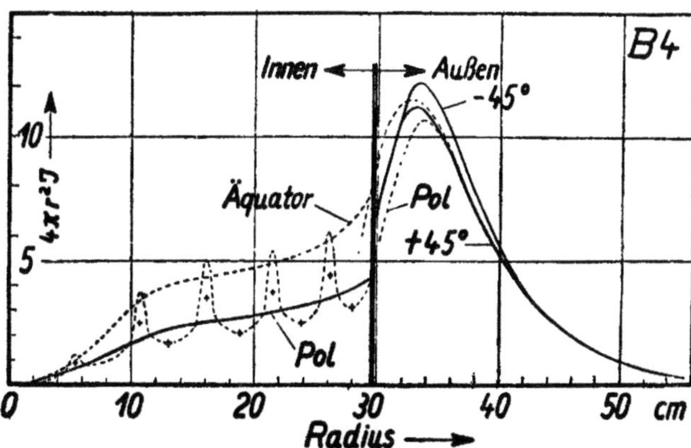

Abb. 50. Verlauf der Neutronenintensität im Innen- und Außenraum bei B_4. In den Paraffinschichten ($\downarrow$) erhöhte Intensität der thermischen Neutronen.

stand aus 9 zusammengelegten Standard-Platten von $30\times30\times1$ cm und von etwa 16 kg Gewicht und 8 Platten von $^1/_4$ dieser Größe. Die Abb. 51 zeigt eine Aufsicht auf eine Schicht solcher Platten, wie sie horizontal im Kessel lag. Der Versuch B_6 bestand aus 4 Variationen, die in der folgenden Tabelle 3 zusammengestellt sind.

Tabelle 3. Daten der Versuche B_6 und B_7

	B_{6a}	B_{6b}	B_{6c}	B_{6d}	B_7
Schichtdicke U in cm	1	1	1	2	1
Schichtdicke D_2O in cm ..	10	18	26	26	18
Anzahl der U-Schichten ..	12	7	5	5	7
Gesamtmenge U in to	2,12	1,25	0,89	1,78	1,25
Gesamtmenge D_2O in to .	$\sim 1,5$	$\sim 1,5$	$\sim 1,5$	$\sim 1,5$	$\sim 1,5$
Neutronenproduktions-Koeffizienten $\bar{\nu}/\rho$	$+\,91$	$+\,145$	$+\,118$	$+\,96$	$+\,145$
Vermehrungsfaktor Z	1,56	2,35	2,12	2,06	3,6

Es ergaben sich erwartungsgemäß positive Neutronenproduktions-Koeffizienten $\dfrac{\bar{\nu}}{\rho}$ und dementsprechend Vermehrungsfaktoren $Z > 1$. Die günstigste Schichtung war B_{6b} (Abb. 52), wie BOTHE und FÜNFER[17] schon vorher in Modellversuchen gefunden hatten. Sie wurde in dem ebenfalls in der Tabelle angeführten Versuch B_7 wiederholt, mit dem Unterschied, daß hier zunächst ein Streumantel von Graphitkohle in einer Schicht von ~ 40 cm Dicke (insgesamt ~ 10 to Kohle) den Magnesiumkessel umgab. Die Kohle war umgeben von einem zweiten

[17] W. BOTHE u. E. FÜNFER, Schichtenversuche mit Variation der U- und D_2O-Dicken. Forschungsbericht Dezember 1943. Unveröffentlicht.

Aluminiumkessel (vgl. auch Ziff. 7. 1. 3), der sich wiederum in H_2O befand· Die Neutronenvermehrung stieg durch diese Maßnahme auf $Z = 3,6$, was z. T· daran liegt, daß Kohle als Streusubstanz Neutronen sehr viel weniger als H_2O absorbiert[8].

Bei diesen Schichtenversuchen war eine hinreichende Übereinstimmung mit der im Laufe der Zeit immer weiter verfeinerten Theorie (HEISENBERG,

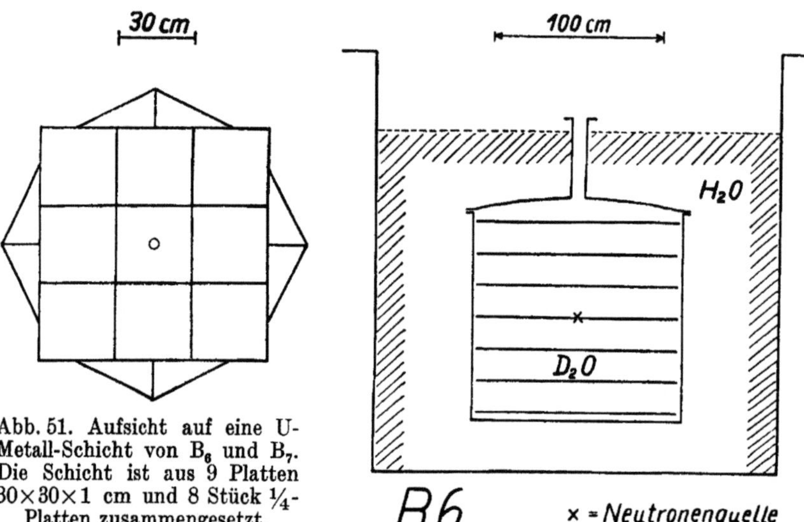

Abb. 51. Aufsicht auf eine U-Metall-Schicht von B_6 und B_7. Die Schicht ist aus 9 Platten $30 \times 30 \times 1$ cm und 8 Stück $\frac{1}{4}$-Platten zusammengesetzt.

Abb. 52. Vertikalschnitt durch B_{8b}. In dem mit D_2O gefüllten Kessel liegen 1 cm dicke horizontale Schichten aus U-Metallplatten; die Schichten sind durch nicht eingezeichnete Distanzstücke aus Magnesiumblech von einander getrennt. Abstand der Schichten 18 cm.

v. WEIZSÄCKER, HÖCKER) erreicht worden. Als bester Wert für die Größe X, die Anzahl der pro eingefangenes Neutron neu entstehenden Neutronen, ergab sich, wie schon erwähnt, $X = 1,18$. Da sich vor allem durch die Gottower Versuche herausgestellt hatte, daß Uranwürfel eine höhere Neutronenvermehrung liefern würden, wurde als nächster Versuch des Dahlemer Bunkers ein Raumgitter aus Würfeln im gleichen Magnesiumkessel mit derselben Menge D_2O und in demselben Kohlemantel geplant. Dieser Versuch (B_8), der schließlich in Haigerloch kurz vor Kriegsende durchgeführt wurde, wird im folgenden Abschnitt als Beispiel eines Großversuchs ausführlicher beschrieben.

Zur technischen Durchführung der Versuche sei noch vermerkt:

Die Versuche B_1 bis B_5 des Kaiser-Wilhelm-Instituts für Physik wurden abseits vom Institut in einer kleinen Baracke ausgeführt, die für diesen Zweck gebaut und eingerichtet war. Sie enthielt im wesentlichen eine im Fußboden eingelassene „Brunnengrube", in der die Uranbrenner aufgestellt wurden und die mit Wasser als Streumantel gefüllt werden konnte. Außerdem befanden sich darin Pumpanlagen, Laboreinrichtungen und ein Aufbewahrungsort für die Neutronenquelle. Die folgenden Versuche B_{6a-d} und B_7 wurden in einem inzwischen (1943) am Institut selbst unterirdisch neu errichteten Bunkerlaboratorium durchgeführt, das sowohl zur Sicherung gegen Luftangriffe als auch im Hinblick auf den Schutz gegen die Strahlung selbsttätiger Uranbrenner mit 2 m dicken Eisenbetonwänden umgeben war. Es enthielt ein Hauptlabor mit „Brunnengrube", Pumpen, Laboreinrichtungen, Belüftungs- und Heizanlagen sowie einen Lagerraum für die D_2O-Tanks und die D_2O-Aufbereitungsanlage. Für den Betrieb selbsttätiger Uranbrenner waren Luftabsaugeinrichtungen für radioaktive Gase, ein ferngesteuerter Präparattransport, Fernbeobachtung durch wassergefüllte Glasdoppelfenster usw. vorgesehen. Das Bunkerlaboratorium umfaßte noch eine Werkstatt und mehrere Laboratorien für die Bearbeitung von Uranmetall, die Untersuchung des D_2O usw. Auch in Gottow war ein Speziallaboratorium in einer Baracke eingerichtet worden, und ebenfalls in Haigerloch in einer Felsenhöhle.

Schon bei Beginn der Großversuche war klar, daß schweres Wasser eine äußerst günstige Bremssubstanz für Uranbrenner ist. Die rasche Bremsung der Neutronen würde kleine Anordnungen mit verhältnismäßig geringen Substanzmengen ermöglichen. Deshalb wurde von Beginn der Arbeiten ab versucht, D_2O in größeren Mengen zu beschaffen. Die einzige Stelle, die es produzierte, war damals die Firma Norsk Hydro in Rjukan in Norwegen. Bei Kriegsbeginn betrug ihre Leistung etwa 10 l im Monat, die bis 1941 zunächst auf 120 l/Monat und später bis auf über 300 l/Monat mit Hilfe verschiedener Zusatzanlagen (vgl. Ziff. 7. 4 HARTECK, Isotopentrennung in technischem Maßstab) gesteigert wurde*. 1943 wurde die Anlage zuerst durch ein Sprengkommando, dann durch einen Luftangriff zerstört. Das gelieferte D_2O (D-Gehalt etwa $99^0/_0$) verunreinigte sich im Laufe der Versuche mit H_2O. Deshalb wurde im Kaiser-Wilhelm-Institut für Physik in Berlin-Dahlem eine elektrolytische D_2O-Aufbereitungsanlage errichtet, die in der Lage gewesen wäre, aus 1 to D_2O im Laufe von etwa 2 Monaten einige Prozent H_2O zu entfernen. Diese Anlage war bei Kriegsende kurz vor

Tabelle 4. Massenproduktions-Koeffizienten $\bar{\nu}/\rho$ und Vermehrungsfaktoren Z der Großversuche (ρ = mittlere Dichte von Uran + Bremssubstanz; $\bar{\nu}$ = Neutronenproduktions-Koeffizient in sec^{-1}). Anordnung etwa in historischer Reihenfolge

	B_1	L_1	BF	B_2	G_1	B_3	B_4	B_5	L_2	L_3
$\bar{\nu}/\rho$:	—295	—378	—770	—575	—522	—209	—217	—344	<0	<0
Z:	—	—	—	—	—	—	—	—	~1	~1

	L_4	G_2	G_{3a}	G_{3b}	B_{6a}	B_{6b}	B_{6c}	B_{6d}	B_7	B_8
$\bar{\nu}/\rho$:	+23,5	+180	+215	+182	+91	+145	+118	+96	+145	+215
Z:	1,1	1,37	1,65	2,1	1,65	2,35	2,12	2,06	3,06	6,7

* vgl. auch: K. WIRTZ, Die elektrolytische Schwerwassergewinnung in Norwegen; zusammenfassender Vortrag 1942, unveröff.

ihrer Inbetriebnahme. Schließlich sei bemerkt, daß in Dahlem eine besondere Tankanlage zur Aufbewahrung von D_2O in Gestalt von glasierten Stahlkesseln eingerichtet worden war.

Die Gesamtergebnisse der Großversuche sind in der Tabelle 4 (Seite 158) zusammengefaßt, die deutlich die historische Entwicklung der Uranarbeiten wiedergibt.

7.1.3 Der Versuch B_8 in Haigerloch[18]

Die früheren Versuche hatten in Übereinstimmung mit der Theorie ergeben, daß Würfel bzw. Kugeln die besten Formen des Uranmetalls im D_2O-Brenner sind. Die günstigste Dimension der Würfel sollte nach der Theorie etwa 6—7 cm Kante sein. Von den Gottower Versuchen war jedoch eine größere Anzahl Würfel von 5 cm Kante vorhanden. Es wurde deshalb im Hinblick auf die Unmöglichkeit, genügend rasch andere Würfel zu produzieren, beschlossen, weitere Würfel von 5 cm herzustellen, um die vorhandenen zu ergänzen. Um die Versuche mit den früheren, speziell mit B_6 und B_7, vergleichen zu können, sollten sie, wie gesagt, in demselben Magnesiumzylinder mit Kohlestreumantel wie B_7 ausgeführt werden. Hierzu waren, wenn ein möglichst günstiges räumliches Würfelgitter erzielt werden sollte, rund 680 Stück Uranwürfel erforderlich, insgesamt etwa 1,5 to. Sie mußten an Al-Drähten am Deckel des Zylinders in Ketten in das D_2O hineingehängt werden. Der Deckel des Kessels mußte umkonstruiert werden, um dieses Gewicht tragen zu können. Der Kohlestreumantel sollte sich wiederum in einem äußeren Al-Kessel befinden, der vom Wasserstreumantel umgeben war. Alle diese Gesichtspunkte führten schließlich u folgendem Aufbau des Versuchs:

a) **Technischer Aufbau des Versuchs B_8.** Die Abb. 53 gibt einen vertikalen Schnitt durch den Aufbau. In einem größeren, nicht näher gezeichneten Wasserbassin im Fußboden des Labors steht auf Holzrosten der äußere zylindrische Al-Kessel von 210 cm Querschnitt, 210 cm Höhe und 5 mm Wandstärke. Oben wird ein gewölbter Deckel wasserdicht aufgeschraubt, der einen aus dem Wasser herausragenden Kamin trägt, durch den Präparat und Innensonden und das D_2O eingeführt werden. Dann kommt innen die ca. 40 cm dicke Schicht aus Graphitkohle vom Gesamtgewicht von 10 to. Sie besteht aus rechteckigen Blöcken von $5\times10\times50$ cm der Dichte 1,7. Es handelt sich um eine verhältnismäßig reine Kohle, die jedoch nicht speziell für die vorliegenden Bedürfnisse hergestellt war und deshalb wohl bei weitem nicht die Reinheit besessen haben kann, die die von den Amerikanern für die Piles hergestellte Spezialkohle besitzen dürfte. In diesem Kohlemantel eingebaut befand sich der zylindrische Magnesiumkessel von 124 cm Breite und 164 cm Höhe und 3 mm Wandstärke.

[18] F. BOPP, E. FISCHER, W. HEISENBERG, K. WIRTZ, Kaiser-Wilhelm-Institut f. Physik in Berlin bzw. Hechingen, und W. BOTHE, P. JENSEN und RITTER, Kaiser-Wilhelm-Institut, Inst. f. Physik in Heidelberg, Bericht über den Versuch B_8 in Haigerloch. Forschungsbericht 1945. Unveröffentlicht.

In diesen wurde oben ein Deckeleinsatz ca. 40 cm tief eingelassen, dessen Boden zum Aufhängen der „Ketten" von U-Würfeln diente. Dieser Boden bestand aus einer 1 cm dicken Magnesiumplatte, die nach oben durch vier Eisenstangen

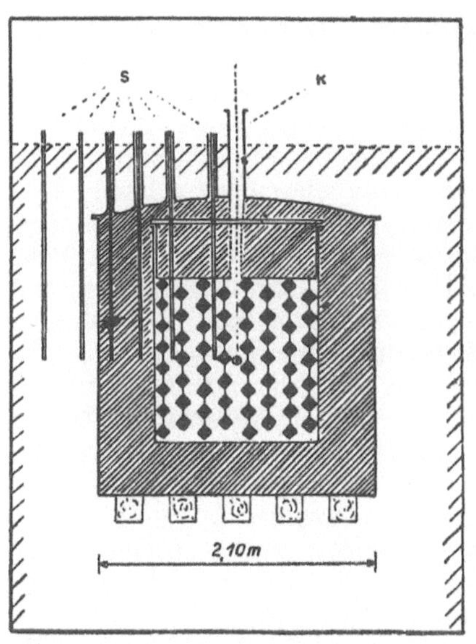

Abb. 53. Schema des Großversuchs B_8 in Haigerloch. Im Zentrum die Neutronenquelle ⊙ umgeben von den Uranwürfeln, die kettenförmig in schwerem Wasser hängen. Eng schraffiert der Kohlemantel, der wiederum von gewöhnlichem Wasser (weit schraffiert) umgeben ist. In der linken oberen Hälfte die Sonden S zur Bestimmung der Neutronenintensität, K = „Kamin" zum Einfüllen des D_2O. Würfelzahl und -größe sind willkürlich gezeichnet. Ergebnisse im Text.

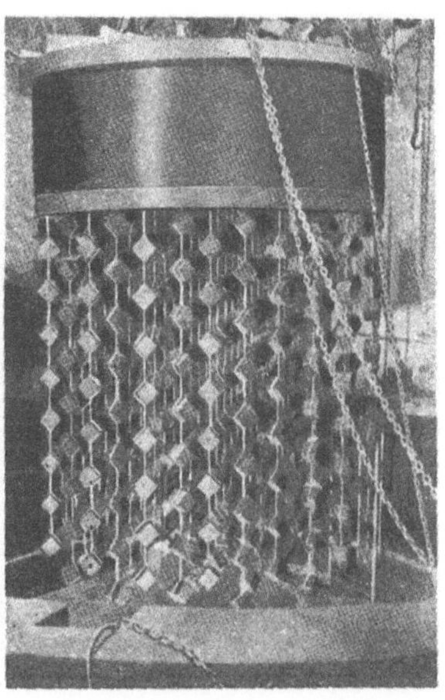

Abb. 54. Raumgitter der U-Würfel in B_8. Würfeldimensionen $5 \times 5 \times 5$ cm; Abstand nächster Nachbarn 14 cm. Die Würfel sind an 78 Ketten am Boden des auf dem Bild sichtbaren Deckeleinsatzes über der Kesselanlage aufgehängt, 40 Ketten mit 9 Würfeln, 38 Ketten mit 8 Würfeln. (Dies sind weder die günstigsten Würfeldimensionen noch die günstigsten Abstände für maximales $\bar{\nu}$; vgl. Text.)

zusätzlich mit einem eisernen Traggerüst verbunden war, das quer über dem Einsatz lag. Durch den Boden des Einsatzes wurde im Innern des Kessels ein Raum von 124 cm Höhe begrenzt, in den das schwere Wasser eingefüllt wurde, und in den die Würfelketten hineinhingen. In diesen Ketten — insgesamt 78 Stück — waren die U-Würfel durch Al-Draht verbunden, der durch eine Bohrung im Deckelboden hindurchgeführt und dort befestigt wurde. In den Deckeleinsatz selbst wurde ebenfalls Kohle eingefüllt und dadurch der Kohlestreumantel in den verlängerten Magnesiumkessel hinein fortgesetzt. Axial ist ein Meßkanal eingebaut, der gestattet, die Neutronenquelle in das Zentrum des Brenners zu bringen. Der komplizierte Deckeleinsatz, der notwendig war, um die Uranketten aufhängen zu können, die Kohleschicht darauf zu bringen und zugleich das D_2O sicher einzufüllen, erforderte eine ziemlich große Materialmenge, nämlich insgesamt 32 kg Magnesiummetall und 75 kg Eisen im Traggerüst. Das Eisen war von der Uran-D_2O-Mischung mehr als 40 cm entfernt, störte die Vermehrung also nicht mehr allzu sehr, mußte aber bei der Auswertung der Messungen berücksichtigt werden. Durch den Kamin sollten später auch die Cd-Schieber eingeführt werden, die dann, wenn die kritische Größe erreicht werden kann, zweckmäßig zur Stabilisierung der Anlage verwendet werden.

Der Zusammenbau erfolgte derart, daß zuerst in den großen Al-Kessel (1) die Bodenschicht Kohle kam, darauf der Magnesiumkessel (3) zentrisch aufgestellt und dann die seitliche Kohlefüllung bis zu seinem oberen Rand geführt wurde. Dann wurde dieser Teil der Apparatur abgedeckt und darüber an einem Kran der Deckeleinsatz aufgehängt und an seinem Boden die vorbereiteten U-Metall-Würfelketten befestigt. Abb. 54 zeigt die am Einsatz fertig aufgehängten Ketten, die eine räumlich dichteste Kugelpackung der Würfel ergeben, kurz ehe sie vom Kran in den Kessel eingelassen wurden.

Nach dem Einlassen wurde der Deckeleinsatz mit Kohle gefüllt und der Rest der äußeren Kohleschicht vollendet. Nach Aufbringen des äußeren Kesseldeckels wurde das mit Antikorrosionsmitteln versehene Wasser des Streumantels eingelassen, und erst wenn sich die Apparatur als wasserdicht erwies, wurde durch den Kamin und den zentralen Kanal der Schichtung das D_2O in das Innere gepumpt. Dabei wurde die Zunahme der Neutronenintensität I durch Silberindikatoren verfolgt. Die Neigung der Kurve $1/I$, aufgetragen gegen die Menge des eingefüllten D_2O, läßt sofort erkennen, bei welcher Füllung die Intensität ∞, d. h. selbständige Reaktion des Brenners, voraussichtlich überschritten würde. Bei dem vorliegenden Versuch B_8 wurde dieser Labilitätspunkt nicht erreicht.

b) Messung der Neutronenintensität. Wie in Ziff. 7. 1. 1 erläutert, verlangt die Auswertung des Uranbrenners die Angabe des Integrals N_0 der Gesamtneutronenzahl im Außenraum, die von der Quelle bei leerer Anlage geliefert wird, d. h. bei nicht eingefülltem Uranmetall und D_2O; ferner dasselbe Integral N_a bei zusammengebautem Brenner und das Integral N_i der Neutronenzahl über den ganzen Innenraum, d. h. die Füllung Uranwürfel + D_2O. $N_a : N_0$ ergibt den Vermehrungsfaktor Z. Ursprünglich wurden die Neutronenintensitäten von Punkt zu Punkt durch Dysprosiumindikatoren ($\sim$ 300 mg Dysprosiumoxyd auf einer Al-Unterlage) ausgemessen, der Intensitätsverlauf in Kurven aufgetragen und diese Kurven entsprechend der Symmetrie der Anordnung über den ganzen Raum integriert. Später war das Ziel, durch geeignete Form des Indikators einen Teil der Integration gleich bei der Messung vorzunehmen. Dieser Weg wurde zuerst von BOTHE und FÜNFER[17] beschritten. Für zylindrische Brenner führte dies zu folgendem Verfahren: Dysprosiumoxyd wurde in verd. HNO_3 bis zur Sättigung aufgelöst. (Etwa 10 g Dy_2O_3 in 100 ccm.) Diese Lösung wurde in Glasröhren gefüllt, die parallel zur Zylinderachse, also senkrecht, in den äußeren und Innenraum der Anordnung eingeführt wurden (vgl. Abb. 53). Entsprechend dem Aufbau des Versuchs B_8 gab es folgende Integrationsräume: im äußeren Wasser den Wassermantelraum und das über dem gewölbten Al-Deckel stehende Wasser, den Wasserdeckelraum; entsprechend den Kohlemantel und den Kohledeckel; außerdem den Innenraum. Jeder

Raum wurde einzeln ausgemessen. Im Wassermantelraum z. B. reichten die vertikalen Flüssigkeitssonden von der Wasseroberfläche bis zum „Äquator" des Brenners, d. h. in die Höhe, in der innen die Neutronenquelle hing. Die Dysprosiumlösung wurde bis zur Sättigung aktiviert. Dann wurde die Sonde entleert, die Lösung umgeschüttelt und in einem Glaszählrohr von etwa 0,2 mm Wandstärke mit Flüssigkeitsmantel (Konstruktion BOTHE und FÜNFER) gemessen. Derartige Sonden wurden in jeden einzelnen Raum in verschiedenen Abständen R von der Zylinderachse eingebracht. Sie ergaben in jedem Raum einen bereits in Richtung der Zylinderachse integrierten Intensitätsverlauf in Abhängigkeit von R, der dann nur noch über $2\pi R$ integriert zu werden brauchte, um die Gesamtintensität der oberen Hälfte der Anordnung zu liefern. Die Beiträge der einzelnen Räume wurden mit gewissen Korrekturen versehen und mit den jeweiligen Produktionskoeffizienten multipliziert. Ihre Summe ergibt das Gesamtintegral. Für den Innenraum wurde entsprechend verfahren. Für jeden Integrationsraum mußten demnach beim Aufbau des Brenners Indikatorenkanäle vorgesehen werden. Sie bestanden für die Innenräume Kohle und D_2O aus dünnen, wasserdicht eingeführten Al-Röhren, in die die Glassonden gesteckt wurden. Von BOPP wurde außerdem eine weitere Vereinfachung der Integration dadurch erzielt, daß durch geeignete Wahl der Sondenorte in Abhängigkeit von R und der Sondenquerschnitte für einen zusammenhängenden Integrationsraum alle Sonden vor der Messung zusammengemischt werden konnten und mit Hilfe einer einzigen Messung sofort das Integral über die Neutronenzahl dieses Raumes erhalten wurde; hierbei wurde das bekannte Gaußsche Verfahren zur Integration ausgenutzt, bei dem man das Integral durch die mit geeigneten Gewichten versehenen Funktionswerte an einigen ausgezeichneten Punkten ausdrückt. Alle Intensitäten wurden auf eine Eichstrahlenquelle reduziert.

c) Ergebnisse der Messungen. Es ergaben sich folgende Werte in willkürlichen Einheiten. $N_0 = 16{,}9$; $N_a = 113{,}4$ und $N_i = 846$. Der Quotient $N_a : N_0$ liefert den Vermehrungsfaktor von B_8:

$$Z = 6{,}7.$$

Der Neutronenproduktions-Koeffizient ergibt sich nach Ziff. 7. 1. 1 aus folgender Beziehung:

$$\bar{\nu} = \frac{N_a - N_0 Y}{N_i} \cdot e^{-w} \cdot \nu_{H_2O} = \frac{9{,}65}{846} \cdot 0{,}96 \cdot 3920 = 430 \, \text{sec}^{-1}.$$

Daraus folgt der Massenproduktions-Koeffizient von B_8 (vgl. auch Ziff. 7. 1. 2, Tab. 4):

$$\frac{\bar{\nu}}{\varrho} = 215.$$

162   W. HEISENBERG UND K. WIRTZ · URANBRENNER

7.1.4 Folgerungen für den energieliefernden Uranbrenner

Die Übersicht über die im Vorhergehenden beschriebenen Großversuche zeigt, daß man durch Vergrößerung der beschriebenen Apparaturen — bei dem letzten Haigerlocher Versuch würde eine relativ geringe Vergrößerung genügen — den kritischen Punkt erreichen und damit zu einem energieliefernden Brenner kommen kann. Es entstand daher die Frage, welche Vorgänge eintreten, wenn man den kritischen Punkt erreicht und wie man dafür sorgen kann, daß man den Ablauf der Kettenreaktion stets in der Gewalt behält. Das, was beim Überschreiten des kritischen Punktes geschieht, kann man qualitativ am einfachsten aus der Diffusionsgleichung (7) ablesen mit dem Vorbehalt, daß diese Gleichung (7) die tatsächlichen Verhältnisse nur recht ungenau darstellt. Die Beschreibung der genaueren und zum Teil schon recht komplizierten Theorie würde den Rahmen dieses Berichtes überschreiten. Wenn man annimmt, daß die Gleichung (7) im Innern des Brenners gilt, und daß die Eigenschaften des Bremsmittels im Außenraum dafür sorgen, daß am Rande der Anordnung (die zur Vereinfachung als kugelförmig angenommen wird) eine Grenzbedingung von der Form

$$\frac{d}{dr}\,(nr) = -\,\frac{\gamma}{l}\,(nr) \qquad (\gamma = \mathrm{const})$$

gilt — was auch nur eine grobe Annäherung ist —, so erhält man die folgenden Typen von Lösungen: Wenn der Radius R der Anordnung kleiner ist als die kritische Größe R_k: $\left(\mathrm{ctg}\,\dfrac{R_k}{l} = -\gamma\right)$, so ist eine Neutronenquelle von der Stärke Q im Zentrum umgeben von der Neutronendichte im Brenner:

$$n\,(r) = \frac{Q}{4\,\pi\,Dn}\left(\cos\frac{r}{l} + \alpha\,\sin\frac{r}{l}\right) ; \;\; \alpha = \frac{\sin\dfrac{R}{l} - \gamma\cos\dfrac{R}{l}}{\cos\dfrac{R}{l} + \gamma\sin\dfrac{R}{l}}.$$

Der Vermehrungsfaktor Z wird dann:

$$Z = \frac{1 + \gamma\dfrac{R}{l}}{\cos\dfrac{R}{l} + \gamma\sin\dfrac{R}{l}}.$$

Ist jedoch der Radius R größer als der kritische Wert, so wächst schließlich die Dichte zeitlich über alle Grenzen, und zwar nach der Formel:

$$n\,(r) \sim e^{\beta t}\,\frac{\sin r/l'}{r}\,, \;\; \text{wobei } \mathrm{ctg}\,\frac{R}{l'} = -\gamma \text{ und } \beta = \frac{\overline{D}}{l^2} - \frac{\overline{D}}{l'^2}.$$

Der für das exponentielle Anwachsen charakteristische Koeffizient β hängt dabei von der Differenz zwischen dem tatsächlichen Radius und dem kritischen Radius in der Form ab:

$$\beta \approx \overline{v}\,\frac{2\,(R - R_k)}{R_k}.$$

FOLGERUNGEN FÜR DEN ENERGIELIEFERNDEN URANBRENNER 163

Das „Anlaufen" der Neutronendichte geschieht also um so langsamer, je weniger die tatsächliche Größe die kritische Größe überschreitet. Außerdem wird, wie aus amerikanischen Arbeiten hervorgeht, die Anlaufszeit noch dadurch erheblich erhöht, daß zum Teil die beim Spaltungsprozeß freiwerdenden Neutronen die Atomkerne nicht sofort, sondern mit einer gewissen zeitlichen Verzögerung verlassen. Obwohl die Anzahl dieser zeitlich verzögerten Neutronen sehr gering ist, verglichen mit der Anzahl der im Mittel bei der Spaltung entstehenden Neutronen, kann wegen der empfindlichen Abhängigkeit des $\bar{\nu}$ von X der Einfluß dieses Effekts auf die Anlaufszeit sehr groß sein.

Das exponentielle Anwachsen der Neutronendichte findet praktisch dadurch ein Ende, daß mit diesem Anwachsen eine Erwärmung der Apparatur Hand in Hand geht. Diese Erwärmung kann in zweierlei Weise zur Stabilisierung des Brenners bei einer bestimmten Temperatur führen: Zunächst bewirkt die Erwärmung des Urans eine Verbreiterung der Resonanzlinien durch Dopplereffekt und damit eine Erhöhung der Resonanzabsorption. Die Temperaturerhöhung und damit die verstärkte Resonanzabsorption wird so lange fortschreiten, bis die tatsächliche Größe des Brenners mit der kritischen Größe bei der betreffenden Temperatur übereinstimmt. Der Brenner wird sich also von selbst auf einer bestimmten Temperatur stabilisieren, wobei diese Temperatur durch die Größe des Brenners gegeben ist. Eine weitere Stabilisierung wird man zweckmäßig nach einem Vorschlag von JOLIOT dadurch erreichen können, daß man Cadmiumschieber in die Apparatur bringt, die die thermischen Neutronen und insbesondere auch die Neutronen etwas höherer Temperaturen stark absorbieren. Bei einer Temperaturerhöhung der in das Uran eindringenden thermischen Neutronen wird das Häufigkeitsverhältnis der im Cadmium absorbierten Neutronen, verglichen mit den im Uran absorbierten Neutronen, zugunsten der ersteren verschoben. Dies führt zu einer Verminderung der Spaltungsprozesse und damit zur Stabilisierung. Bei dieser Form der Stabilisierung nutzt man die Erwärmung des schweren Wassers aus, die allerdings zunächst sehr viel geringer ist als die Erwärmung des Urans. (Der Hauptteil der Spaltungsenergie wird im Uran in Wärme umgesetzt, während nur der kleinere Teil der Energie, der auf die Neutronen entfällt, zur Erwärmung des schweren Wassers verwendet wird.) Erst durch Wärmeleitung vom Uran her wird ein größerer Teil der bei der Spaltung erzeugten Wärme nachträglich auch ins D_2O übergeführt. Die Temperatur der thermischen Neutronen wird sich nach der Temperatur des schweren Wassers, nicht nach der des Urans richten.

Wie groß die Cadmiumschieber sein müssen, um eine zuverlässige Stabilisierung der Anlage zu geben, bedarf im einzelnen Fall noch einer näheren Untersuchung. Rein technisch sind die Cadmiumschieber schon deswegen besonders zweckmäßig, weil man durch Heraus- oder Hineinschieben des Cadmiums die kritische Größe und damit die Temperatur, auf der sich der Brenner stabilisiert, willkürlich einstellen kann.

Wendet man die allgemeinen Überlegungen auf die speziellen numerischen Verhältnisse des Haigerlocher Brenners (B_8) an, so erhält man etwa folgendes Bild: Nach den Messungen ist die Diffusionslänge l der inneren Anordnung aus U-Würfeln und D_2O etwa 35 cm. Der Kohlemantel sorgt dafür, daß — wie aus dem gemessenen Werte $Z = 6{,}7$ hervorgeht $-\gamma = 0{,}824$ ist. (Dabei wird die Haigerlocher Anordnung als Kugel vom Radius 71 cm betrachtet.) Der kritische Punkt $(Z \to \infty)$ würde also erreicht worden sein bei $\operatorname{ctg}\dfrac{R}{35} = -0{,}824$, d. h. $R \approx 80$ cm, was einer Vergrößerung des Volumens um nicht ganz die Hälfte entspricht. Praktisch war allerdings nicht die direkte Vergrößerung geplant, da mehr D_2O nicht zur Verfügung stand, sondern es sollten auch in den Kohlemantel U-Stücke gebracht werden, was wahrscheinlich auch zur Erreichung des kritischen Punktes genügt hätte.

Hinsichtlich der Energieerzeugung des Uranbrenners muß noch folgender Punkt beachtet werden: Die Temperatur, auf die sich der Brenner einstellt, hängt, wie gesagt, von der kritischen Größe, also etwa von der Stellung des Cadmiumschiebers ab. Diese Temperatur wird von dem Brenner eingehalten, unabhängig davon, wieviel Energie dem Brenner entzogen wird. Die Frage der Energielieferung hängt also allein davon ab, wieviel Energie aus dem Brenner etwa durch Wärmeleitung, Konvektion usw. abgeführt werden kann. Wenn viel Energie entzogen wird, so sinkt die Temperatur im Innern um einen ganz geringen Betrag. Damit sinkt die kritische Größe, was zu einer so starken Steigerung der Neutronenproduktion und der Spaltungsprozesse führt, daß die entzogene Energie sofort nachgeliefert wird. Bei der Frage der technischen Verwendung des Uranbrenners zur Energieerzeugung, etwa für Dampfmaschinen und dgl., müssen also zunächst Fragen des Wärmeübergangs[18a], dann auch reine Materialprobleme wie Korrosionsschutz des Urans u. ä. behandelt werden. Die Darstellung dieser Probleme würde aber den Rahmen dieses Berichtes überschreiten.

ANHANG

Daten für die in Deutschland verwendeten Substanzen

1. Einfangsquerschnitte der Bremsmaterialien für thermische Neutronen.

Substanz:	H	D	He	Be	C	O
$\sigma \cdot 10^{24}$ cm²:	0,24	0,0015	0	fast 0	0,006$_4$	0,002

2. Einfangsquerschnitte der Halterungsmaterialien für thermische Neutronen

Substanz:	Al	Mg	Fe
$\sigma \cdot 10^{24}$ cm²:	0,42	0,31	1,6

3. Wirkungsquerschnitte des Urans, $\sigma \cdot 10^{24}$ cm² für thermische Neutronen

    a) Absorption 0,1 bis 0,2
    b) Spaltung 3,3
    c) unbekannte Absorption 2,8

    totaler Absorptionsquerschnitt 6,2

[18a] W. Fritz u. E. Justi, Bericht über die Leistung der Uranmaschine. Forschungsbericht 1942. Unveröffentlicht.

4. Zusammenstellung der Werte von ν

Substanz	ν in sec^{-1}
Uranmetall	69 000
U-Pulver	46 600
U_3O_8-Pulver	13 100
Paraffin	4 570
H_2O	3 920
Kohle	143
D_2O	24

5. Umrechnung vom Produktions-Koeffizienten $\bar{\nu}$ auf den Multiplikationsfaktor k des Smyth-Reports. Es ist

$$k = \frac{\nu_i + \bar{\nu}}{\nu_i},$$

wobei ν_i = mittlerer Absorptionskoeffizient für thermische Neutronen im Brenner. Es ist $k > 1$ für $\bar{\nu} > 0$; $k < 1$ für $\bar{\nu} < 0$. Daraus wurden die Multiplikationsfaktoren einiger Versuche bestimmt:

Versuch	B_1	L_4	G_2	B_6b	B_7	B_8
$\bar{\nu}$ (sec^{-1})	-950	90	360	266	266	430
k	0,92	1,01	1,09	1,08	1,08	1,11

7.2 DER BEITRAG DER SCHNELLEN NEUTRONEN ZUR NEUTRONENVERMEHRUNG IM URAN

von

O. HAXEL,

Max Planck-Institut für Physik, Göttingen

7.2.1 Einleitung

In einem „Uranbrenner", der Deuterium oder andere leichtatomige Substanzen zur Abbremsung der Neutronen enthält, wird die Kern-Kettenreaktion überwiegend von der Spaltung des ^{235}U durch thermische Neutronen getragen. Die Spaltung des ^{235}U durch schnelle Neutronen spielt keine Rolle, da der Spaltquerschnitt für schnelle Neutronen mehrere hundertmal kleiner ist als der für thermische Neutronen. Dagegen bringt die Spaltung des 139mal häufigeren Isotops ^{238}U einen Beitrag, der insbesondere dann ins Gewicht fällt, wenn der Brenner aus Schichten aufgebaut ist, die so dick sind, daß die darin entstehenden schnellen Neutronen eine merkliche Wahrscheinlichkeit haben, auf Uranatome zu treffen, bevor sie in die Bremssubstanz austreten. Da man den Brenner aus einem anderen Grunde, nämlich zur Verringerung des Neutronenverlustes durch Resonanzabsorption sowieso aus Schichten aufbauen muß, wird man den hierbei

TWO LECTURES

BY

W. HEISENBERG

**DIRECTOR OF THE
MAX PLANCK INSTITUTE FOR PHYSICS
IN GÖTTINGEN**

1

The Present Situation in the Theory of
Elementary Particles

2

Electron Theory of Superconductivity

CAMBRIDGE
AT THE UNIVERSITY PRESS
1949

CONTENTS 5

LECTURE 1

The Present Situation in the Theory of
Elementary Particles

page 9

LECTURE 2

Electron Theory of Superconductivity

page 27

PREFACE 7

The following two lectures were delivered in December 1947 at the Cavendish Laboratory, as an introduction to discussions on two different topics in atomic theory.

The theory of elementary particles belongs to a part of atomic physics in which the fundamental laws are still unknown, and rather different views have been expressed in recent years about the next steps to be taken in this field. The first lecture tries to explain why the well-known divergencies in meson theory and nuclear physics may be considered, not as a difficulty, but rather as a natural and therefore satisfactory feature of the present 'correspondence' theory. It is only in quantum electrodynamics that a clear separation of the divergent from the other terms leaves open a wide field of application of standard quantum-theoretical methods.

The second lecture deals with an application of quantum mechanics, the theory of superconductivity. The views presented in it may be considered as an attempt to combine different features of the previous rather discordant theories into a consistent picture of the phenomenon. Many discussions on the wide range of experimental work on superconductivity done in the Royal Society Mond Laboratory have been of great help to the author.

Finally the author wishes to express his gratitude to Professor Sir Lawrence Bragg and the members of the Cavendish Laboratory for the invitation to Cambridge, and for many interesting discussions.

W. H.

Göttingen
28 May 1948

9

LECTURE 1

THE PRESENT SITUATION IN THE THEORY OF ELEMENTARY PARTICLES

1. The problem of the elementary particles has been treated in recent years in many papers by standard methods of quantum theory [1]. One has, in fact, set up equations for a system described in classical terms in space and time—for instance, a system of particles interacting with each other or a system of wave-fields— and one has then 'quantized' that set of equations. This process of 'quantization' is defined as the attempt to find new equations within the mathematical framework of quantum theory, which in the limit of large quantum numbers will go over into the classical equations from which one started. Obviously this process is not unique. There may be many quantum-theoretical equations corresponding to the classical picture (e.g. the Klein-Gordon equation and the Dirac equation for the electron), therefore the process of quantization is just a method of guessing the right quantum theoretical description for the system concerned. The process is not even unique in the reverse direction. For one set of quantum theoretical equations one may find several classical pictures of the system, which are obtained by going to the limit of macroscopic dimensions in different ways. So, for example, the well-known quantum-mechanical equations for the atom may be considered

as describing electrons moving around the nucleus or standing waves of negative electronic matter.

In spite of all this the process of quantization leaves only little freedom when applied to a simple classical system, and it has been applied with equal success to systems of particles and to wave-fields. This success has, however, been limited to the case of unrelativistic theories. So long as we have to deal with the normal interactions of matter that act instantaneously at large distances the quantized equations lead to consistent results and describe the experiments with fair accuracy. When, on the other hand, we have to deal with relativistic systems, serious difficulties arise, which have been analysed in many papers in recent years [2].

The reason for these difficulties may be stated in several ways. When starting from a particle description one would have to introduce a force of interaction between particles in a relativistically invariant manner. This force should then be connected with the two invariant functions of the co-ordinates x_ν of the distance between the two particles concerned $x_\nu = x'_\nu - x''_\nu$, i.e. the square of the space-time distance

$$\Sigma x_\nu^2 = r^2 - c^2 t^2, \tag{1}$$

and the relativistic δ-function, defined by

$$\int_\Omega dx_1 \ldots dx_4\, \delta(x) = \left\{ \begin{array}{l} 1 \text{ when } \Omega \text{ contains the point} \\ \quad x_\nu = 0, \\ 0 \text{ otherwise.} \end{array} \right\} \tag{2}$$

The introduction of Σx_ν^2, however, is equivalent to the assumption of a wave-field propagated with the

velocity of light from the one particle to the other (since Σx_ν^2 vanishes even at large distances if $r = \pm ct$). Therefore the use of x_ν^2 would effectively mean that one has turned from a particle picture to a wave picture, which we have to consider later. The δ-type of interaction, on the other hand, is quite meaningless in classical physics, since two mass-points never get infinitely close to each other. One may, however, try to attach some meaning to a force of the type (2) in quantum theory. Actually a repulsion of two particles with a δ-type interaction energy would seem to lead to sensible results, since such an interaction would just require that the wave-function of the system vanishes in the region $x_\nu = 0$. A more careful investigation shows, however, that the δ-type interaction would only lead to the boundary condition $r\phi \to 0$ for $r \to 0$, and therefore would not lead to any real interaction. An attraction of the δ-type on the other hand leads to inconsistent results, for the system would have infinite negative eigenvalues for the energy. Therefore the particle picture is not very useful in relativistic theory; even more, as it offers no reasonable description of the creation or annihilation of particles.

In classical physics the wave picture gives an adequate description of all actions propagated from a point to its immediate surroundings and therefore is generally used in the mathematical formulation of relativistic phenomena. The quantization of wave-fields, however, leads to the conclusion that every wave-field is automatically connected with its corresponding type of particle, and that any wave equation which contains interaction

terms means effectively δ-type interactions between particles. Therefore it is not unexpected that every relativistic quantum theory of a system with interactions so far has given divergent results. Mathematically the divergency results as a rule from interference of the waves concerned with other waves of infinitely short wave-length.

2. This situation would present a real difficulty if we had reason to believe that the correspondence between the classical wave equations and their quantum theoretical analogue should be close even at the shortest wave-lengths, or, in the particle picture, at the shortest distances between the particles. This, however, is certainly not so—for two reasons:

(a) Any future theory of the elementary particles must contain, besides the fundamental constants $\hbar$ and c, a third fundamental constant of the dimensions of a length or a mass [3] (or a combination of such with $\hbar$ and c). This follows simply from the fact that, on account of purely dimensional reasons, one cannot derive the mass of an elementary particle from the constants $\hbar$ and c. The actual mass of the main elementary particles suggests that this new constant may be considered as a length l of the order of magnitude $l \sim 10^{-13}$ cm. If the future theory contains such a constant in whatever form, it is natural to assume that the usual correspondence between the classical wave description and its quantum-theoretical analogue only holds for distances very much greater than l, but fails in the region of smaller distances.

(b) Quite apart from this argument, we know from experiment that at high energies all particles are related;

13 that is to say, in a collision between any two very energetic particles any other particles can be created. In such a general form this may not follow directly from experiment; but the experiments prove directly that there are transitions between nucleons and mesons (the π-, μ- and σ-particles of Powell[4]); mesons, electrons and neutrinos; protons, electrons and light quanta, etc. Therefore one can go from any particle to any other particle at least by intermediate steps, and it is thereby very probable that in a collision process of sufficiently large energy particles of any type can be created.

It has always seemed to me an especially interesting question whether at high energies many particles of various types can be created simultaneously in an explosion-like process. From the earlier forms of the theory of nuclear forces it had been concluded that such explosion phenomena could occur[5], and in the beginning there seemed to be direct evidence for it from cosmic-ray showers. Then by the theory of the cascade process, developed by Carlson and Oppenheimer[6], Bhabha and Heitler[7], at least a large part of these showers could be explained satisfactorily, and consequently the theoretical evidence for explosions of the type mentioned was disputed. Recently, however, definite experimental evidence has been obtained that explosion processes, in which many penetrating particles are created in one single act[8], do occur in cosmic radiation, and more detailed theoretical analysis of the usual theories (Lewis, Oppenheimer and Wouthuysen[9]) confirmed the earlier prediction. Therefore, the existence of such multiple processes can now be considered as well esta-

14 blished. All this shows at once that any theory which does not contain all particles, or their corresponding wave-fields, simultaneously, must necessarily fail at high energies.

(c) The real situation in the theory of elementary particles can be made clearer through a comparison of the elementary particles with the compound system of unrelativistic quantum theory[10]—and not, as it is usually done, with the electrons of this theory. Consider the interaction of, say, a sodium and a chlorine atom. This interaction can, for small velocities of the atoms, indeed, be represented in a good approximation by a Hamiltonian containing the kinetic energies of these two atoms and a suitable potential energy. Such a Hamiltonian, however, would be useless for calculating the effect of very energetic collisions of the two atoms. Because in such collisions the atoms may part as Na^+ and Cl^- ions, or as Na^+ ion, Cl atom and an electron, or even as Na nucleus, Cl nucleus and 28 electrons. Obviously, any conclusions, drawn from the Hamiltonian mentioned above, which require the recurrence to matrix elements involving high-energy terms, would lead to completely wrong results.

The analogy can be carried somewhat further by discussing such questions as: Should we picture the neutron as consisting of a proton and a negative meson, or should we rather assume that the proton consists of a neutron and a positive meson? Such questions have an answer only in the same sense as we can give an answer to the question: Does the HCl molecule consist of an H atom and a Cl atom, or rather of an H^+ ion and a

Cl$^-$ ion? Such questions can only refer to a qualitative description of the wave-function and may be answered as such. A quantitative answer can only be given by writing down the wave-function or some equivalent to this in the future theory. **15**

The real description of the behaviour and the interaction between a sodium and a chlorine atom can only be given by the Hamiltonian referring to the two nuclei and all the 28 electrons, the solutions of which comprise all possible particles such as Na^+, Na^{++}, etc. In the same way the behaviour of elementary particles at very high energies can only be described by a mathematical formalism which comprises all elementary particles and which does not refer only to a special type such as protons or neutrons or electrons. Therefore, if the whole group of elementary particles can be represented by a Hamiltonian function at all, this primary Hamiltonian will probably not be expressed by the nucleon or the meson wave-field or any other of the well-known wave-functions, and will certainly have no similarity to any of the Hamiltonians hitherto derived from correspondence to classical pictures.

3. (a) It is instructive to follow this possibility of a primary Hamiltonian a little further. Let us therefore suppose that by means of certain operators for which commutation rules are given, we may construct a four-vector J_ν, representing the total momentum and energy of the system 'matter'. $H = icJ_4$ may then be regarded as the Hamiltonian of the system. Every state of this system can be represented by a vector in Hilbert space (Becker and Leibfried[11]), and we may call ϕ_0 the vector which represents free space without matter or radiation. Then **16**

$$J_\nu \phi_0 = 0. \tag{3}$$

If we confine ourselves to solutions, for which $J_k = 0$ ($k = 1, 2, 3$), then there will be discrete eigenvalues of $H = icJ_4 = \kappa_n c\hbar$, which we may take as defining a particle at rest. From any such solution we may by means of a Lorentz transformation go over to another solution representing the same particle with a momentum say $\mathbf{k}_n \hbar$ ($\mathbf{k}_n$ being the wave number, $k_n^0 = \sqrt{(\kappa_n^2 + k_n^2)}$ and $\kappa_n \hbar/c$ the rest mass of the particle). The operator through which one can change the Hilbert vector ϕ_0 into ϕ_{k_n}, representing this particle, may be called O_{k_n}. Therefore

$$\phi_{k_n} = O_{k_n} \phi_0. \tag{4}$$

(In unrelativistic wave theory the operators O_{k_n} for the electrons are just the Fourier components of the Schroedinger wave ψ; the operators for, say, a hydrogen atom, however, are much more complicated expressions.) There is no reason to assume that for any two particles of different type the corresponding operator O_{k_n} of the one particle must commute with the conjugate $O_{k_m}^+$ of the other. To each pair of discrete eigenvalues κ_n and κ_m (assuming $J_k = 0$ for $k = 1, 2, 3$) there is also a continuous spectrum, going from $\kappa_n + \kappa_m$ to infinity and representing the coexistence of two such particles with opposite momentum. It may often happen that a discrete eigenvalue lies in the range of one or several of the continuous spectra. In this case there may exist some interaction between the two solutions which would lead to a finite lifetime of the discrete state. The discrete

17   eigenvalue would no longer be a really sharp eigenvalue, but rather a sharp peak in the continuous spectrum of a finite breadth, the breadth being given by the lifetime in the usual way.

Empirically we already know a great deal about the lowest stationary states of our system (4, 12). The following diagram gives a qualitative and certainly not final representation of the energy spectrum, omitting the continuous parts, and states, which in a very good approximation may be considered as compound states.

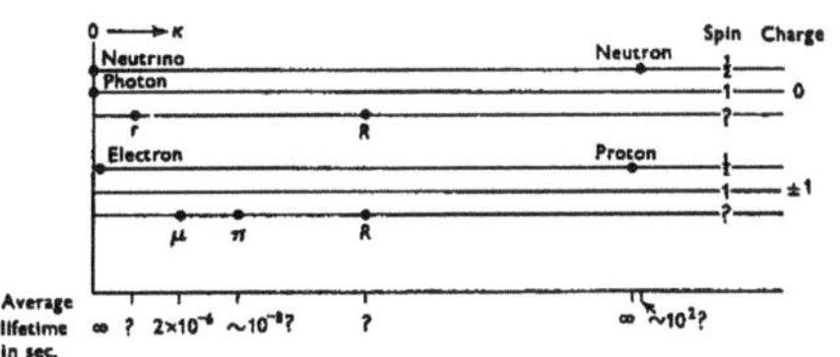

The most outstanding feature of this spectrum is the extreme stability of states that lie as high as the proton mass. This stability must mean that transitions from these states to continuous states are forbidden by some quantum rule which cannot yet be formulated. It is quite likely that in future other stationary states, probably with still shorter lifetimes, will be found.

(b) For small energies of the interacting particles, the primary Hamiltonian will give rise to sub-Hamiltonians, which in a certain approximation can be used

18   to account for the interaction at small energies. Consider, for example, the state described by the Hilbert vector

$$O_{k_n} O_{k_m} \phi_0, \tag{5}$$

meaning the simultaneous existence of two particles with momenta k_n and k_m. This state will not be a stationary state; but in a certain approximation one may try to find a Schroedinger function $\chi(k_n, k_m)$ such that

$$\phi = \int \chi(k_n k_m)\, dk_n\, dk_m\, O_{k_n} O_{k_m} \phi_0 \tag{6}$$

becomes a stationary state, at least approximately. This implies that

$$H\phi = E\phi. \tag{7}$$

If one takes the spur of this equation in Hilbert space by multiplying by the vector

$$O_{k'_n} O_{k'_m} \phi_0, \tag{8}$$

one finds the Schroedinger equation

$$\int dk_n dk_m \left[(k'_n k'_m \mid H \mid k_n k_m) \right.$$
$$\left. - E(k'_n k'_m \mid U \mid k_n k_m) \right] \chi(k_n k_m) = 0. \tag{9}$$

Here $(k'_n k'_m \mid U \mid k_n k_m)$ is defined as the spur

$$(O_{k'_n} O_{k'_m} \phi_0 \mid O_{k_n} O_{k_m} \phi_0) = (k'_n k'_m \mid U \mid k_n k_m). \tag{10}$$

This spur is in general not the unit matrix. Equation (9) is, of course, only an approximation which it may be possible to use at low energies. It is this sub-Hamiltonian which generally will correspond to a classical picture of particles or waves and which ought to be similar to the Hamiltonians that have been studied hitherto.

For higher energies one will have to go back to the   19 exact solutions of (7) which for a collision problem may be described as (13)

$$\phi = O_{k'_1}, O_{k'_2} \ldots \phi_0$$
$$+ \int \delta_+(E' - E'') (k'_1 k'_2 \ldots \mid r \mid k''_1 k''_2 \ldots)\, dk''_1 \ldots O_{k''_1} O_{k''_2} \ldots \phi_0. \tag{11}$$

The second term on the right-hand side contains the outgoing waves and may refer to a different number of particles from the first term. By putting

$$(k'_1 \ldots \mid R \mid k''_1 \ldots) = \delta(E' - E'') (k'_1 \ldots \mid r \mid k''_1 \ldots) \tag{12}$$

and

$$S = 1 + R, \tag{13}$$

one obtains the S-matrix, which represents the asymptotic behaviour of the wave-function at large distances. The matrix r is in general *not* uniquely determined by the wave-function, since the non-singular parts of the wave-function may be distributed in different ways among the matrix elements of r. The matrices R and S, however, are uniquely determined by the wave-functions.

The general properties of the matrices S and R have been discussed in a number of papers (13). The most important properties are

$$S^\dagger S = SS^\dagger = 1, \tag{14}$$

and the fact that S must be relativistically invariant. The stationary states of the system can be derived from the poles of the matrix elements in S with some exceptions or restrictions. But in connexion with bound states in many-body problems a closer inspection of the properties of the S-matrix is still necessary.

(c) It has been emphasized before that in regions of   20 the order of magnitude of $l \sim 10^{-13}$ cm. no correspondence between quantum theory and any classical picture can be expected. It is, for instance, very doubtful whether any significance can be attached to the question what the exact form of the interaction potential is between neutron and proton (14). If one combines this fact with the experience that every Hamiltonian that contains interactions has so far given divergent results, one is led to discuss the possibility that for such problems no Hamiltonian exists and only the S-matrix has any significance (13). This would mean that the final theory of matter would not be given by a certain Hamiltonian, as was suggested in §3 (a) and (b), but by defining a certain S-matrix; the concept of potential energy which proves so useful in macroscopic dimensions would be dropped in favour of the S-matrix at very small dimensions.

This assumption would in the first instance seem to provide us only with an empty frame for the future theory into which the picture has still to be painted (2). This objection does, however, not describe the real situation, for two reasons. First, the alternative assumption of a primary Hamiltonian is an empty frame in exactly the same sense. Secondly, the tentative solutions which have been written down in relativistic wave-mechanics so far are actually not defined by a Hamiltonian, but by an S-matrix. One has, in fact, as a rule started with a Hamiltonian and realized that it has no solutions; one has then used certain matrix elements out of this Hamiltonian to define, arbitrarily, scattering processes for the system concerned; in other

21 words, one has chosen a certain S-matrix for the system instead of the Hamiltonian which does not give consistent solutions.

(d) It may be useful to compare this situation with a similar situation in the theory of general relativity. When curved surfaces were first studied by the mathematicians some two hundred years ago, it was considered obvious that such surfaces should be defined by equations of the type $f(x, y, z) = 0$, that they should be pictured as embedded in a three-dimensional Euclidian space. Gauss then introduced the metric tensor g_{ik}. In general relativity we now know that the question whether our Riemann geometry world is embedded in a Euclidian world of more dimensions is of no importance. What we need is a determination of the g_{ik}, and the determination is not obtained from an equation of the type

$$f(x, y, z, t, w) = 0$$

referring to a Euclidian world of five dimensions, but from the Einstein field equations which have an entirely different origin. In the same way it may not be very important to know whether the final matrix S is embedded in a primary Hamiltonian as described in § 3 (a) and (b), but what we need is some entirely new way of determining S, analogous to the Einstein field equations. It may turn out that there is no primary Hamiltonian, or it may turn out that there are infinitely many such Hamiltonians giving the same S; in both cases the Hamiltonians are not very important.

Discussing this problem from the point of view of the experiments one may also argue against the exist-

22 ence of a Hamiltonian and thereby of a Schroedinger probability function in the following way. In quantum mechanics the Schroedinger function defines the probability of finding the particle at a certain point, irrespective of the means by which the position of the particle is measured; whether one uses very fast electrons or protons or γ-rays, the probability distribution which results from the experiment will be the same. If we come to a region of the order 10^{-13} cm., however, it is quite improbable that the result of a measurement of such accuracy could be independent of the means by which it is carried out, since every particle used for the measurement will itself have a 'size' of the order 10^{-13} cm. Therefore it will not be possible to translate the results of such collision experiments into an objective probability distribution for the one particle which one has tried to measure.

4. (a) Keeping this in mind the question still arises in what way a theory defined from an S-matrix goes over into a theory of the usual type in the limit of dimensions large compared to $l \sim 10^{-13}$ cm. This problem would not occur if only forces of finite range of the type of nuclear forces were found in nature. Because, then, there would be no classical picture to which any correspondence would be needed. The whole problem of correspondence comes in through the long-range forces, that is, the electromagnetic and the gravitational forces. It may be that once the final theory of matter has been found one can look for a rough approximation to the solutions by omitting electromagnetic effects ($e^2/\hbar c \ll 1$) and gravitational effects. In this approximation the

23 spectrum (Fig. 1) divides itself up in non-combining parts, and the solutions of this first approximation do not correspond to any classical picture. It is only through the long-range forces that the concept of potential energy gains its value, and only there a correspondence to the classical description is to be postulated. It seems not improbable that the value of $e^2/\hbar c$ will finally be fixed just through the condition that the formalism determining the S-matrix must, for larger dimensions, go over into the usual quantum-mechanical formalism, because it is only through the quantity e that such a correspondence is needed. The numerical value of $e^2/\hbar c$ plays, indeed, a decisive role in this correspondence, as was pointed out by Bohr (15) many years ago. If $e^2/\hbar c$ was of the order of unity or larger, the size of atoms would probably shrink to the size of the nuclei, and the energy spectra of the atoms would be replaced by something entirely different, the radiation forces being of the same order as all other forces. It is difficult to see how any correspondence could be formulated in this case. On the other hand, $e^2/\hbar c$ must also be different from zero for long-range forces to appear. Similar considerations apply to gravitation, but it will certainly be a very much better approximation to neglect gravitation than to neglect electromagnetic forces. It has been noticed before that β-decay and the neutrino may possibly drop out of the picture of matter if gravitation is neglected.

(b) In conclusion to this analysis of the present situation it may be permissible to discuss possible future steps towards a solution of these problems. It seems to

24 me that first some purely mathematical problems have to be solved before even new experimental results can help us any further. The properties of S-matrices for many-body problems with bound states must be analysed from the usual scheme of unrelativistic quantum theory and must be formulated by a few simple equations. These equations must, then, together with $S^\dagger S = 1$ and other conditions, be considered as determining equations for the S-matrix (as a first—perhaps still rather 'empty' —analogue to the Einstein field equations for the g_{ik}). The very fact that so far relativistic Hamiltonians have given inconsistent results may suggest that the combination of the conditions mentioned with the relativistic invariance of S already leads to rather narrow restrictions. Within this mathematical framework one may, probably only led by principles of simplicity and invariance, try to find an S-matrix that represents the lowest states of the spectrum (Fig. 1) in the approximation $e^2/\hbar c \ll 1$, that is, the π-, ν- and μ-particles of Powell, the proton and neutron and all atomic nuclei. It may be that this can only be achieved through the formulation of some further restrictions on the S-matrix which cannot be derived from unrelativistic quantum theory. Only after this first step has been taken may one try to connect up the theory with the present unrelativistic quantum theory of electrostatic forces, and one may perhaps hope at this stage to find the reason for the numerical value of $e^2/\hbar c$.

Among the other purely mathematical problems that have to be solved there is still left the important question whether any relativistic Hamiltonians can be defined

25 which contain interaction terms and still give consistent results. One may look for such Hamiltonians among the integral operators of the type used in recent field theories of the electron. But it may also turn out, as has been discussed, that such Hamiltonians do not exist.

LITERATURE

(1) Compare, for example, G. WENTZEL, *Quantumtheorie der Wellenfelder*, Wien, Deuticke, 1943.

(2) W. PAULI. *Report on an International Conference on Fundamental Particles and Low Temperatures*, 1, 5, 1947, London, Taylor & Francis.

(3) W. HEISENBERG. *Ann. d. Phys., Lpz.*, (5), **32**, 20, 1938.

(4) C. M. G. LATTES, G. P. S. OCHIALINI and C. F. POWELL. *Nature, Lond.*, **160**, 453, 486, 1947.

(5) W. HEISENBERG. *Z. Phys.* **101**, 533, 1936.

(6) J. F. CARLSON and J. R. OPPENHEIMER. *Phys. Rev.* **51**, 220, 1937.

(7) H. BHABHA and W. HEITLER. *Proc. Roy. Soc.* A, **159**, 432, 1937.

(8) The most recent paper is that by W. H. FRETTER, *Phys. Rev.* **73**, 41, 1948, in which numerous quotations on the earlier work are given.

(9) H. W. LEWIS, J. R. OPPENHEIMER and S. A. WOUTHUYSEN. *Phys. Rev.* **73**, 127, 1948.

(10) Compare J. FRENKEL. *J. Phys. U.S.S.R.* **9**, 465, 1945.

(11) R. BECKER and G. LEIBFRIED. *Phys. Rev.* **69**, 34, 1946.

(12) G. D. ROCHESTER and C. C. BUTLER. *Nature, Lond.*, **160**, 855, 1947.

(13) W. HEISENBERG. *Z. Phys.* **120**, 513, and 673, 1943; *Z. Naturforsch.* **1**, 608, 1946.
CH. MØLLER. *K. danske vidensk. Selsk.* **23**, no. 1, 1945; **24**, no. 19, 1946; *Nature, Lond.*, **158**, 403, 1946.
S. T. MA. *Phys. Rev.* **69**, 668, 1946; **71**, 195, 1947.
D. TER HAAR. *Physica.*

(14) C. F. VON WEIZSAECKER. *Z. Phys.* **118**, 489, 1941.

(15) N. BOHR. *Report on the International Congress on Physics in Rome*, 1931.

27

LECTURE 2

THE ELECTRON THEORY OF SUPERCONDUCTIVITY

The phenomenon of superconductivity, discovered by Kamerlingh-Onnes as long ago as 1911, was given a satisfactory phenomenological description ten years ago through the work of F. and H. London[1], v. Laue[2] and others. Hitherto it has seemed very difficult, however, to get an explanation of the phenomenon in terms of the well-known quantum theory of electrons in metals, and several earlier attempts, such as Kronig's[3] theory of electron lattices, or Welker's[4] theory of magnetic interactions, have met with serious difficulties. Recently a new attempt[5] has been made, and the following discussion is meant as an analysis of the physical aspects of this theory.

The basis of the new survey is the assumption that at low temperatures a phenomenon of condensation can take place among the free electrons in the metal. Since the free electrons obey Fermi statistics, which is essential in deriving the conduction properties of metals, they fill approximately a sphere in momentum space whose radius is determined by the density of the free electrons. At very low temperatures only the few electrons near the surface of this sphere can undergo any changes, and therefore also the condensation can take place only on the surface of this sphere. The condensed phase may be

pictured as an electron lattice of very low density, the lattice constant being of the order of 10^{-7} cm. It is assumed that the condensation occurs through the ordinary Coulomb interactions between the electrons, not by means of a magnetic interaction.

This condensation causes spontaneous currents consisting of moving electron lattices. This can be easily visualized by remembering that all electrons on the surface of the Fermi sphere have the same absolute value of the velocity, but different directions. Any aggregation of these electrons, unsymmetrical in momentum space, will lead to an ordered state moving with a velocity of the same order. Therefore the condensation phenomenon provides a picture for a state in the metal, in which there is an electric current, but no heat transfer or resistance. Normally these spontaneous currents will be distributed at random in a superconductor like the magnetized domains in a ferromagnet. It is only by the action of an electric or magnetic field that they can be ordered and thereby produce a total current.

The condensation phenomenon gives also a reasonable explanation of the specific-heat anomaly and of the fact that heat conduction is smaller in the superconducting than in the normal state.

The action of an electric field consists mainly in changing the momentum distribution of the free electrons which in turn will transfer momentum to the electron lattices, so that indirectly a connexion between supercurrent and electric field is established of the well-known form

$$\lambda \dot{\mathbf{J}}_s = \mathscr{E} \tag{1}$$

29 (the first equation of London). At the same time the orientation of the spontaneous current domains will lead to stresses and to a potential energy in the condensed electron phase; according to the theory of London this can again be expressed in terms of the constant λ of equation (1).

Finally, the Meissner effect, the fact that magnetic fields cannot penetrate into a superconductor, is explained as the result of the equilibrium between the Lorentz force of the magnetic field upon the supercurrents and the action of the stresses.

1. These various assumptions will now be discussed in detail, and we begin with the phenomenon of condensation itself. Let us assume that the radius of the sphere in momentum space filled with electrons is P. Upon the surface of this sphere we can build wave-packets of thickness ΔP (in momentum space), either symmetrical (covering the whole surface of the sphere), or unsymmetrical (either covering only a part of the surface, or covering the surface with varying intensity). These wave-packets, translated into co-ordinate space, can be described as symmetrical or unsymmetrical distributions of charge which fall off as a function of the radius r as r^{-2} between the radii $\hbar/P$ and $\hbar/\Delta P$ ($\hbar$ is Planck's constant, divided by 2π). Inside the radius $\hbar/P$ the charge distribution is nearly constant, outside of $\hbar/\Delta P$ it falls off very rapidly. If one tries to describe the condensed phase as a lattice of such wave-packets, then the price in kinetic energy which has to be paid for the localization of one electron in the wave-packet is of the order $P\Delta P/m$, where m is the effective mass of the

30 electron in the energy band. On the other hand, one can hope to gain an amount of Coulomb energy which can be estimated by assuming that around the electron the other electrons are distributed in a nearly uniform charge cloud from which only the wave-packet is lacking. Therefore the gain in Coulomb energy is essentially the Coulomb interaction of the electron in the centre with the wave-packet which turns out to be of the order

$$\frac{e^2\Delta P}{\hbar}\ln\frac{P}{\Delta P}$$

(e is the electronic charge). Hence for sufficiently small values of ΔP the gain in Coulomb energy will be greater than the loss in kinetic energy, and at very low temperatures the condensation will take place.

While this rough estimate shows the physical reason for the condensation, the real calculation is much more complicated for several reasons. The Pauli principle ensures that the electrons of like spin are kept apart in any case; therefore there is practically no gain in Coulomb energy by establishing an ordering of the type mentioned between them. The ordering will only change the Coulomb energy between electrons of opposite spin direction. At the same time the Coulomb exchange energy between the electrons in the two phases has to be considered and will affect the results. Finally, a proper evaluation of the Coulomb energy term would require the solution of a typical many-body problem, which seems to be almost impossible at present. The best approximation to the energy of condensation so far has been obtained by Koppe, but even there the evaluation

of the Coulomb interaction needs refinement. As a final formula for the energy A of condensation (per unit volume) at zero temperature (wave-packet of central symmetry) Koppe (5) finds

$$A=\frac{P^5}{64\pi^2 mh^3 Z}e^{-8.9\,-16Z},\qquad (2)$$

where

$$Z=\frac{\pi}{4}\frac{\hbar P}{mc^2}.\qquad (3)$$

The characteristic feature of this formula is the exponential which is a direct consequence of the logarithmic term in the Coulomb energy. It is this exponential that reduces the condensation energy from a value of ordinary atomic order to a value of about 10^{-8} times smaller, and gives rise to a very sensitive dependence of this energy upon P and the effective mass m of the electron.

If the primary assumption is valid, that the condensation takes place through the Coulomb forces, then one can scarcely think of any other mechanism which reduces the ordinary Coulomb energies to so small values, but one described by a formula of type (2). On the other hand, it has been suggested by Welker that it is not the Coulomb forces but magnetic interactions between the electrons that cause superconductivity, possibly indirectly through the Meissner effect. In this case the order of magnitude for A would be given by atomic energies multiplied by $(v/c)^2$, v being the velocity of the electrons $(v=P/m)$; no exponential function would then be required to explain the observed smallness of A. It should be emphasized in connexion with later discussions of the Meissner effect that a theory of

32 this type, suggested by Welker, may be described as an essentially relativistic theory, since, if we go to the limit $c\to\infty$, A would vanish and the phenomenon of superconductivity would disappear. A theory of the type (2), however, may be called a non-relativistic theory, since A does not depend on c.

Empirically two strong arguments can be given in favour of a formula of type (2). First, the transition temperatures of different superconductors vary by a factor up to a hundred, which means a variation of A by a factor up to 10^4; such a variation can be easily explained by an exponential dependence on the constants of the metal, but would be difficult to explain otherwise. Secondly, it is known experimentally (6) that the transition temperatures depend very strongly upon the pressure to which the metal is subjected. If l is a linear dimension in the metal, T_s the transition temperature, the experiments give

$$\frac{\Delta T_s}{T_s}\approx 27\,\frac{\Delta l}{l}.\qquad (4)$$

Such a strong dependence could scarcely be explained by a magnetic theory of superconductivity but follows naturally from a theory of the type (2).

2. We shall now consider how this phenomenon of condensation gives rise to currents without resistance. To make this point clear we simplify our model by assuming that the wave-packet covers a certain fraction ω of the sphere in the momentum space with equal thickness, as shown in Fig. 1. We further assume that the energy of condensation is then just $A\omega$, A being the value of (2), and we confine the calculation to very

33   low temperatures so that there will be only a small hole in the wave-packet. The state of lowest energy would, of course, always be the one with complete condensation. But at finite temperatures there may be a gain in entropy if a hole is made in the wave-packet. Without the condensation the free energy of the electrons would be proportional to T^2; at first sight one might imagine that when condensation takes place this free energy would just be reduced by a factor $1 - \omega$. But this is not to be expected, since only the number of states above the radius to which electrons can be excited is reduced by $1 - \omega$, while the number of states below P, from which the electrons are taken, remains unchanged. The correct calculation gives, according to Koppe, something near the geometric mean between 1 and $1 - \omega$, namely, $\sqrt{(1 - \omega)}$, so that the free energy has a value of the order const. $T^2 \sqrt{(1 - \omega)}$. For the total free energy f_s (per unit volume) one finds accordingly

$$f_s = -A\omega - BT^2\sqrt{(1 - \omega)}, \qquad (5)$$

and $\partial f_s/\partial \omega = 0$ leads to

$$A = \frac{BT^2}{2\sqrt{(1 - \omega)}}. \qquad (6)$$

If, for simplicity, one assumes (6) to be valid even for higher temperatures up to the point $\omega = 0$, $T = T_s$ (transition point), one finds

$$\omega = 1 - \left(\frac{T}{T_s}\right)^4, \quad f_s = -A\left[1 + \left(\frac{T}{T_s}\right)^4\right], \qquad (7)$$

and the energy per unit volume becomes

$$u_s = -A\left[1 - 3\left(\frac{T}{T_s}\right)^4\right], \qquad (8)$$

34   while the specific heat

$$c_v = \frac{du_s}{dT} = 12A\frac{T^3}{T_s^4}. \qquad (9)$$

These formulae, which had already been used long ago by Gorter and Casimir (7), describe the experiments more accurately than would be expected from the rather rough assumptions made in our calculation.

These calculations were, however, carried out not primarily for the deduction of the specific heat, but in order to show why, even at the lowest temperatures, a hole will be produced in the wave-packet which will cause a dissymmetry and therefore a current.

Against this assumption of a spontaneous current the argument has often been used that the lowest state of the electronic phase must always be a state without current, and that for any given state with current one can always find—by a simple translation in momentum space—another state with lower energy and without current. It is indeed possible, by pushing the electronic distribution of Fig. 1 slightly to the left, to reach a state without current and with an energy smaller than (8) by an amount of the order $(1 - \omega)^2$ const. This new state would, however, not be a thermodynamically possible state because the distribution of the free electrons would be different from a Fermi distribution. If, on the other hand, the electronic distribution is rearranged according to Fermi statistics, only electrons on the side of the hole in the wave-packet will be excited (the excitation of electrons from the right side will be very improbable on account of the displacement of the wave-packet as indicated in a somewhat exaggerated way in Fig. 2).

Therefore the gain in entropy will be less and the free energy will be higher than in the state with current, in spite of the fact that the energy itself is lower. In this way a state with current can be thermodynamically stable, that is to say, such a current produces no Joule's heat and therefore shows no resistance.   35

This statement seems to contradict an old rule derived by Peierls, that the thermodynamically stable state of a system of electrons can never have a current.

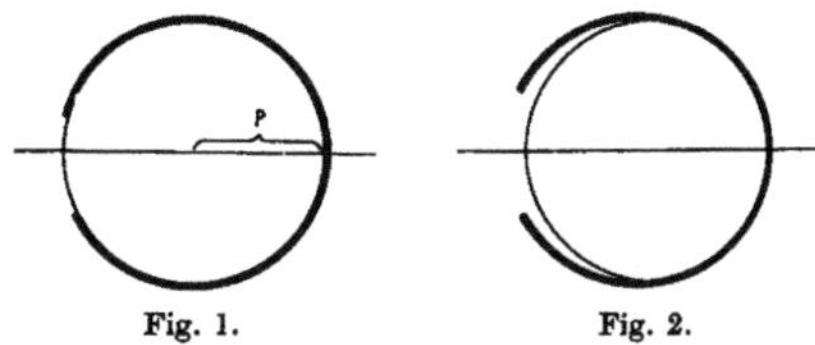

Fig. 1.   Fig. 2.

The contradiction disappears, however, when one distinguishes carefully between the theoretical case of a metal without resistance, i.e. a perfect ionic lattice, and the case of the real metal with impurities and thermal vibrations. If there were no perturbations by impurities and thermal vibrations, then the total momentum of the electrons would be conserved. It would then be possible for the electronic distribution to be translated in momentum space (without change in shape) and any distribution reached in that way would again represent thermodynamical equilibrium. Therefore in this case the state of lowest free energy would certainly be the one in which there is no current—in which the current

of the condensed phase is just compensated by the current of the free electrons. In this state the centre of the 'Fermi sphere' would, however, not coincide with a state of zero momentum relative to the ionic lattice.   36

In the real metal, on the other hand, the irregularities in the ionic lattice affect the free electrons and the condensed phase in different ways. The Fermi sphere of the free electrons is now forced into such a position, that its centre coincides with the zero momentum state relative to the ionic lattice, and the condensed phase is at the same time split into small domains. Here again the total current vanishes in agreement with Peierls's rule, if one averages over regions which are so large, that thermodynamical equilibrium is established, i.e. regions of an order of magnitude larger than the electronic mean free path. But it vanishes on account of the statistical distribution of the domains, while in the single domain the current of the condensed phase is not compensated by the current of the free electrons and produces together with it no Joule's heat.

Concerning the problem of the specific heat, it should be mentioned that a somewhat more accurate determination has been carried out by Koppe. A really satisfactory treatment, however, would require a detailed theory of the shape of the wave-packet which has not yet been given; it may also be necessary to consider the possibility of excited states even in the condensed phase which so far has been entirely neglected.

Within the approximation of the formulae (5)–(9) one can easily derive the change in heat conduction which takes place when condensation occurs. In normal

37 metals the coefficient κ_n for heat conduction can, to a good approximation, be obtained from the classical formula of the kinetic theory of gases

$$\kappa_n = c_v l \frac{v}{3}, \qquad (10)$$

c_v being the specific heat, l the mean free path, v the velocity of the electrons. In a superconductor the condensed phase does not carry heat; therefore we have to replace c_v by the specific heat of the uncondensed electrons alone, considering the value ω as given at each point. According to (5) c_v is then simply to be multiplied by the factor $\sqrt{(1-\omega)}$. The probability for elastic collisions of the electrons below the radius P of the Fermi sphere is not changed by the condensation; for electrons above P it is reduced by $1-\omega$. As an average one may assume that the probability decreases, and therefore the mean free path increases, by the factor $1-\tfrac{1}{2}\omega$. The velocity $v = P/m$ remains unchanged. Finally, we get for the coefficient κ_s of heat conduction in the superconductor

$$\kappa_s = \frac{\sqrt{(1-\omega)}}{1-\tfrac{1}{2}\omega}\,\kappa_n = 2\,\frac{(T/T_s)^2}{1+(T/T_s')^4}\,\kappa_n. \qquad (11)$$

For a metal with high electrical residual resistance κ_n varies linearly with T. In this case, plotting the heat resistance $R = 1/\kappa$ against temperature, we get the curves of Fig. 3. This picture is in good agreement with earlier measurements of de Haas and Bremmer on tin [8]. Formula (11) does not agree equally well with recent measurements of de Haas and Rademakers [9] on a sample of very pure lead, where the main part of the normal

38 electric resistance was still due to thermal vibrations. It may be that in such a case some refinement concerning the definition of a mean free path is required.

3. The fact that by means of condensation spontaneous currents are formed produces in the metal a state which should in some way resemble the state in a ferromagnet. The distortions in the ionic lattice will cause stresses also in the electronic lattice and will tend to break it up into relatively small domains among which the direction of the current is distributed at random. Since the mean free path of the normal electrons is usually very large at low temperatures, it seems probable that the average size of a domain of uniform supercurrent is small compared to the mean free path. On the other hand, the current can only be stationary in the metal if it is closed, or if it reaches the surface from which it may continue as normal electric current. Therefore the most probable shape for the domains will be that of thin threads either finally closed in a ring, or reaching the surface. The thickness of the threads will usually be small compared to the mean free path, and within a region of the order of the mean free path these threads will be distributed at random.

If this general picture is correct, the question arises: In what way can an external electric field change the

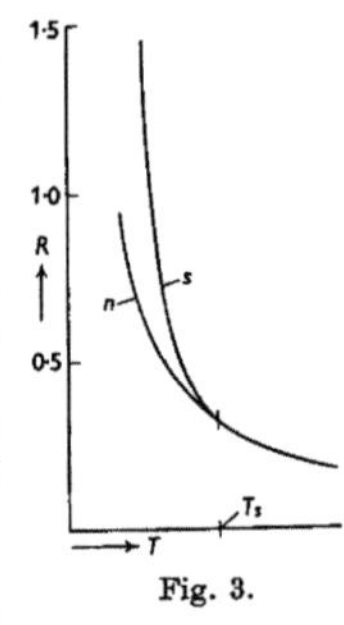

Fig. 3.

statistical distribution to produce an average current, and what other changes in the metal may be expected simultaneously? 39

In a normal metal an external electric field displaces the Fermi sphere in the direction of the field by such an amount that the average momentum transferred to the electrons per unit time from the field is finally transferred to the ionic lattice by means of collisions. The amount of the displacement directly measures the current and thereby the conductivity.

In a superconductor the electric field will act in the same way, but it will also affect the condensed electron phase. The electric field will transfer a certain amount of momentum per unit time directly upon the condensed phase; but simultaneously there will be a still larger continuous transfer of momentum from the free electrons to the condensed phase through the collisions between the free electrons and the electronic lattices. The free electrons can collide both with the ionic and the electronic lattices, therefore only a certain fraction of the momentum per unit time, which they get from the field, will go to the ionic lattice; the remaining fraction will go to the electronic lattices and will thereby change the statistical distribution of the spontaneous currents. It turns out that this second type of momentum transfer is much larger than the direct action of the field on the lattices.

Since this momentum transfer is due to collisions, it is possible to describe it quantitatively by introducing a mean free path for the collision of a free electron with the electronic lattices. We can define l_e and l_i as mean free paths for collisions with the electronic or ionic lattice respectively. Then for the total mean free path we have 40

$$\frac{1}{l} = \frac{1}{l_i} + \frac{1}{l_e}. \qquad (12)$$

A detailed calculation of the quantity l_e is, however, very difficult and has not yet been carried out. In a first approximation it is justifiable to assume that the collision probability is proportional to the density of the electronic lattice. Therefore we may put

$$l_e = \frac{L}{\omega}, \qquad (13)$$

and consider L as independent of the lattice density. L would essentially depend upon the irregularities in the electronic lattices. These deviations from the perfect lattice are again caused by the irregularities in the ionic lattice and are therefore connected with the value of l_i. The irregularities of the ionic lattice are, in fact, responsible for the splitting of the electronic lattice into many small domains, and their size and shape will essentially influence L.

In calculating this interaction between free and bound electrons one must also keep in mind another important point. The probability for a certain momentum transfer from a free electron either to an ion or to an electron depends, on account of the Pauli principle, on the direction of the spontaneous current in that region. Therefore, averaging over a larger area, containing many domains, the force which the 'Fermi liquid' of free electrons exerts upon the ionic lattice and the

41 supercurrent phase depends upon the supercurrent flowing in that region. Any deviation from the random distribution of domains, or, in other words, any average supercurrent, will therefore lead to a change in the equilibrium of forces between the three phases (free electrons, bound electrons, ions) and will be connected with stresses and a corresponding potential energy in the electronic lattice phase. Stresses and potential energy will be directly connected with the mean free path l_e.

It seems strange at first sight that a typical kinetic quantity, such as a mean free path, should be connected with a typical static quantity such as a potential energy or a stress. But one can give examples of such a connexion also in classical physics. If, for example, one rotates a disk by means of an external force in a vessel filled with gas, then the stresses and the potential energy in the disk depend directly on the mean free path of the molecules in the gas.

After these general remarks we have to account for the electrodynamical behaviour of a superconductor in a more quantitative way. If, in a first approximation, we disregard magnetic effects or, as discussed in § 1, go to the limit $c \to \infty$, then the action of an electric field produces the following mechanism. An external electric field $\mathscr{E}$ transfers per unit time and volume the momentum $ne\mathscr{E}$ upon the electrons, n being the density of conduction electrons in the metal. This momentum is divided into the fraction l/l_e going into the electronic lattices and the fraction l/l_i going into the ionic lattice ($l/l_i + l/l_e = 1$). Therefore the momentum transferred

42 upon the superconducting lattices per unit time and volume is $ne\mathscr{E}l/l_e$. Hence

$$\frac{d}{dt}\left(\frac{m}{e}\mathbf{J}_s\right) = ne\mathscr{E}\,\frac{l}{l_e}, \tag{14}$$

or

$$\lambda \dot{\mathbf{J}}_s = \mathscr{E}, \tag{15}$$

where

$$\lambda = \frac{m}{ne^2}\,\frac{l_e}{l} = \frac{m}{ne^2}\left(1 + \frac{L}{l_i\omega}\right). \tag{16}$$

According to measurements of Shoenberg[10] and Pippard[11] on Hg and Sn at low temperatures, λ seems to be about twelve times larger than the value m/ne^2, previously given by London (this would mean $L/l_i \sim 11$). But one has to remember that neither n nor m are well known in a metal, and that all the complications caused by the crystal structure, overlapping of energy bands, stresses and impurities in the ionic lattice, etc., enter in the determination of m and n. For small values of ω, λ should, according to (16), increase as ω^{-1}, in agreement with experiment. This simple connexion may, however, be modified through the Meissner effect and an effect recently measured and discussed by London[12], Pippard[13] and Reuter and Sondheimer[14]. Since these effects are connected with the magnetic field and disappear in the limit $c \to \infty$, they will be discussed further in § 4.

When an electric field produces a supercurrent according to (15), the work done per unit time and volume is $\mathbf{J}_s\mathscr{E}$, and consequently the free energy in the electronic lattices is increased by

$$f_J = \tfrac{1}{2}\lambda J_s^2. \tag{17}$$

43 At the same time, according to London[1], a stress in the superconducting phase appears, which can be described as a pressure in the direction of the current, and a tension perpendicular to the current, both of the amount $\tfrac{1}{2}\lambda J_s^2$. Formally, one can express the stresses by the tensor

$$T_{ik} = \lambda(J_i^s J_k^s - \tfrac{1}{2}\delta_{ik}J_s^2). \tag{18}$$

It is obvious that the simple equations (15), (17) and (18) only apply when the size, or rather the thickness, of the superconducting domains is small compared to the mean free path l of the electrons. The equations (15), (17) and (18) refer to average values taken over many domains. The immediate connexion (18) between the current J_s and the stresses applies also in the reverse direction. If it were possible to produce by external strains on the ordered superconducting phase the stresses (18), then the current J_s would appear as the only stable distribution of spontaneous currents belonging to these stresses[15]. The current is determined by the stresses except for its sign. The superconductor has, therefore, a certain similarity with a piezo quartz crystal. If through strains one produces stresses in the quartz crystal, then the random distribution of electric dipole moments in the crystal is replaced by an ordering from which an average electric moment results.

It is well known that only a limited amount of current can be sent through a superconductor. Formerly it was believed that the current was limited through the action of the magnetic field which the current produces. Now we know, however, that the limitation can be expressed by the equation

$$f_n \geqslant f_s + \tfrac{1}{2}\lambda J_s^2, \tag{19}$$

44 where f_s is given by (5) and (7); f_n is the free energy (per unit volume) in the normal metal. Equation (19) contains simply the thermodynamical statement that superconductivity is only possible when the total free energy in any part of the superconducting state is smaller than it would be if this part were in the normal state. One can easily see from (16) that the limitation (19) would persist even in the limit $c \to \infty$ in which there is no magnetic field; therefore the magnetic field has no immediate connexion with the limitation of the current.

In the approximation $c \to \infty$ we can discuss the question whether it would be possible in thermodynamical equilibrium to have supercurrents with $\operatorname{rot}\mathbf{J}_s \neq 0$. Let us assume that λ depends on the co-ordinates x, y, z; in other words, let us consider an inhomogeneous superconductor consisting of various materials. Furthermore, the sources of the currents $\mathbf{J}_s$ which are usually at the surface of the superconductor may be given. We can then divide the current $\mathbf{J}_s$ into two parts, $\mathbf{J}_1^s + \mathbf{J}_2^s$, obeying the equations

$$\operatorname{div}\mathbf{J}_2^s = 0, \quad \operatorname{rot}(\lambda\mathbf{J}_1^s) = 0. \tag{20·1}$$

Then the energy of the supercurrents becomes

$$\int \frac{\lambda}{2}J_s^2 d\tau = \int \frac{\lambda}{2}J_1^{s^2} d\tau + \int \frac{\lambda}{2}J_2^{s^2}d\tau, \tag{20·2}$$

since the contribution $\int \lambda\mathbf{J}_1^s\mathbf{J}_2^s d\tau$ vanishes on account of (20·1). Therefore to each distribution $\mathbf{J}_1 + \mathbf{J}_2$ would exist another distribution $\mathbf{J}_1 - \mathbf{J}_2$, which belongs to the same distribution of sources and leads to the same energy. Hence, from purely statistical reasons, we

45 would expect the racemic mixture of both states, that is, a distribution with $J_2 = 0$, which also gives the smallest value of the energy possible for the given sources. In the normal case $\lambda = \text{const.}$, the assumption $J_2 = 0$ reduces to $\operatorname{rot} \mathbf{J}_s = 0$.

It may be emphasized that it is in just this respect that a superconductor behaves in an essentially different way from a perfect conductor. In a perfect conductor, e.g. a gas of electrons in a vessel of axial symmetry (assuming again $c \to \infty$, i.e. no magnetic field), a state with $\operatorname{rot} \mathbf{J}_s \neq 0$ would be possible because the electronic gas may rotate as a whole, and the conservation of angular momentum would then prevent any change in $\operatorname{rot} \mathbf{J}_s$. In a superconductor, however, there is no conservation of angular momentum for the electronic phase alone, since the electronic phase interacts with the ionic lattice; therefore (for $\lambda = \text{const.}$, $B = 0$) the state $\operatorname{rot} \mathbf{J}_s \neq 0$ is not possible in thermodynamical equilibrium.

4. When c is taken as finite, then the magnetic field will influence the distribution of current in the superconductor. We will discuss this influence first under the assumption that c is still extremely large, so that the magnetic field certainly penetrates into the whole superconductor. This assumption is, of course, equivalent to the assumption that we have to deal with superconductors of a size small compared to the penetration depth.

In this case, the magnetic field, which penetrates practically without change into the superconductor, will exert the usual Lorentz forces upon the supercurrents in the same way as, say, upon the electrons in a helium atom. Therefore these Lorentz forces must be balanced

46 by some other forces; either, as in the helium atom, by some uniform rotation of the current system producing Coriolis forces, or by stresses produced in the superconducting phase through some change in the current distribution. Actually, these two alternative possibilities reduce to one, since in the average over many domains a rotating current distribution can be considered as a stationary current distribution different from the original one.

Treating the magnetic influence from the first point of view, we would perhaps (for the local supercurrent, i.e. the domains) try to set up an equation of the type

$$\frac{d}{dt}\mathbf{j}_s = [\mathbf{j}_s, \mathbf{B}]\,\text{const.} \tag{21}$$

It would, however, be difficult to determine the constant on the right-hand side, since the force on the domains will again depend largely on the disturbance of the motion of the free electrons. But quite apart from the special numerical value of this constant, which may be derived in connexion with a rigorous theory of the constant λ, one sees that in a region where the average current vanishes, there still remains an average rotation of the supercurrents which must lead to a value of $\operatorname{rot} \mathbf{J}_s$ different from zero.

We therefore turn to the alternative point of view and try, for a homogeneous superconductor ($\lambda = \text{const.}$), to find a stationary distribution of current for which the forces produced by the stresses just cancel the magnetic forces. The equation

$$\frac{1}{c}[\mathbf{J}_s \mathbf{B}] - \operatorname{div} T = 0, \tag{22}$$

together with (18), then leads to

$$\operatorname{rot}(\lambda \mathbf{J}_s) = -\frac{1}{c}\mathbf{B}. \tag{23}$$

This derivation of the second London equation from the stress tensor may seem somewhat artificial, since the form of the stress tensor was originally derived from this second equation. But from our point of view, the essential part of the superconducting mechanism is the balance of forces between the ionic lattice, the gas of free electrons, and the electronic lattices with their stresses. We consider the Meissner effect as a secondary consequence of this balance of forces. Therefore only a rigorous theory of the constants λ and L will give a complete insight into the mechanism of the Meissner effect.

These arguments can also be expressed in a way which shows the relation with previous discussions by London[1]. The magnetic potential $\mathbf{A}$, belonging to a constant value $\mathbf{B}$, can be defined as

$$\mathbf{A} = -\tfrac{1}{2}[\mathbf{rB}]. \tag{24}$$

If one expresses the total electronic current by means of the wave-functions ψ of all the electrons one finds

$$\mathbf{J} = -\frac{i\hbar e}{2m}\Sigma(\psi^* \operatorname{grad}\psi - \psi \operatorname{grad}\psi^*) - \frac{e^2}{mc}\mathbf{A}\Sigma\psi^*\psi \tag{25}$$

(Σ to be taken over all electrons). Since the magnetic field is considered as a small perturbation, we may (neglecting the electronic spin) put

$$\psi_k = \psi_k^0 + \frac{e}{2mc}\sum_{l \neq k}\frac{(k\,|\,\mathbf{B}\,[\mathbf{rp}]\,|\,l)}{E_l - E_k}\psi_l^0, \tag{26}$$

48 where ψ_k is the perturbed, ψ_k^0 the unperturbed wave-function of an electron; in this approximation we treat the electrons as independent particles, moving under the influence of the field of the ions. We shall first apply equations (25) and (26) to the theory of diamagnetism of a normal insulator. Then we can consider in particular the wave-function of an electron moving around an ion at the centre of the co-ordinate system. For this electron the angular momentum around the ion is at least approximately conserved if the interaction with neighbouring ions is weak. Therefore the first term in (25) vanishes in this approximation, since it depends, according to (26), upon the transition matrix elements of the angular momentum. Hence the diamagnetic current is given by the last term in (25). If we consider, on the other hand, an electron moving around an ion at some distance from the centre of the co-ordinate system, then the second term in (25) is very much larger. This second term, however, is practically compensated by the first term. Because there is now no conservation of angular momentum, the matrix elements in (26) are different from zero. As van Vleck[20] has pointed out in detail, what remains after the compensation is just a term of the same size as the one for the electron in the centre of the co-ordinate system. Therefore a normal insulator shows only a small diamagnetism.

If we now go over to the electrons in the superconducting phase, essentially the same formulae apply, only we have to take into account the interaction of the electrons. Therefore the wave-function ψ has to be replaced by a function of all the electrons, the matrix

49 elements refer to the change of the total angular momentum of all electrons, and in the calculation of J one has to integrate over the co-ordinates of all electrons but one. If the effect of the electronic lattices could be considered as 'external forces', like those from the ionic lattice, then again the first term in (25) would compensate the second term almost completely for any value of $\mathbf{r}$, so that only the ordinary diamagnetism of the metal would remain. But since the electronic lattices are a part of all the electrons, all the matrix elements in (26), referring to collisions between electrons and electron lattices, will disappear because such collisions cannot change the total angular momentum of the electrons. Hence the first term in (25) is, roughly speaking, reduced by the factor l/l_i, since only the collisions with the ions contribute. If one disregards the small diamagnetism of the normal metal, we are, therefore, left with the current

$$\mathbf{J} = -\frac{e^2}{mc}\mathbf{A}\Sigma\psi^*\psi\left(1-\frac{l}{l_i}\right) = \frac{e^2}{2mc}[\mathbf{rB}]n\frac{l}{l_e} = \frac{1}{2c\lambda}[\mathbf{rB}], \quad (27)$$

which is identical with (23).

As the next step we drop the assumption that c should be considered as very large, or, in other words, that our superconducting body should be small compared to the penetration depth. If we drop this assumption then B can no longer be constant in the superconductor, since the current itself produces a magnetic field. It will still be justifiable to apply locally the previous arguments and therefore to consider the second London equation (23) as valid.

50 One essential change, however, is introduced by the fact that the penetration depth, for the actual value of c, turns out to be much smaller than the mean free path l of the electrons. The consequences of this effect have been worked out for the skin effect of normal metals at high frequencies by Pippard[13] and by Reuter and Sondheimer[14]. Their results can be essentially stated by saying that the effective number of free electrons contributing to conductivity is reduced by the ratio of the penetration depth to l, as compared with the case of small frequencies. This change may also affect equation (16) for the constant λ. Only a detailed theory of the mean free paths l_i and L will allow a satisfactory calculation of the value of λ. It would only be through such a detailed theory that one would gain insight into the balance of forces between the gas of free electrons, the ionic lattice and the electronic lattices and their stresses; on the other hand, the calculation of the mean free paths will be still more complicated than in the theory of the normal metals.

In conclusion, it may be useful to compare the present theory with previous theories of superconductivity. The essential difference from several of the more recent attempts is the assumption that the perfect conductivity rather than the diamagnetism is the primary feature of the phenomenon; that superconductivity is brought about by the Coulomb interaction of the electrons, not by magnetic forces. In these points we follow the lines of the old theory of Kronig[3]. With regard to the specific heat the present theory does not assume a special gap in the electronic level system, but it assumes condensa-

51 tion by means of interaction which indirectly produces a gap of varying size. (Compare the older work by Gorter and Casimir[7], Daunt and Mendelssohn[16], Welker[4] and Sommerfeld[17].) It is just this gap which by its unsymmetrical position on the Fermi sphere produces the current domains which also were discussed long ago (compare Sommerfeld and Bethe[18], and recent experiments by Justi[19]). In the present theory the London stresses play a more important role than hitherto accepted[15]. With regard to the destruction of superconductivity by means of a magnetic field, we consider as the primary phenomenon the existence of a maximum current density which is simply due to the fact that the presence of a current increases the stress in the lattices and thereby the free energy of the superconducting phase. It is only through the Meissner effect that the magnetic field produces currents and thereby indirectly destroys superconductivity. The Meissner effect itself is interpreted along the lines previously described by London[1], but a detailed theory of the two London equations and the constant λ has not yet been worked out. As a whole the present theory should be considered as an attempt to reconcile different features of previous theories in a consistent picture rather than as an entirely new description of the phenomenon.

LITERATURE **52**

(1) F. and H. London. *Proc. Roy. Soc. A*, **149**, 71, 1935.
F. London. *Une conception nouvelle de la superconductibilité*, Paris, 1937; *Nature, Lond.*, **140**, 796, 834, 1937.
(2) M. v. Laue. *Ann. Phys., Lpz.*, (5), **32**, 71, 253, 1938; *Theorie der Supraleitung*, Springer, Berlin, 1946.
(3) R. de L. Kronig. *Z. Phys.* **78**, 744, 1932; **80**, 203, 1932.
(4) H. Welker. *Phys. Z.* **39**, 920, 1938.
(5) W. Heisenberg. *Z. Naturforsch.* **2a**, 185, 1947; **3a**, 65, 1948.
H. Koppe. *Ann. Phys., Lpz.*, (6), **1**, 405, 1947; *Z. Naturforsch.* **3a**, 1, 1948.
(6) B. Lasarev and L. Kan. *J. Phys. U.S.S.R.* **8**, no. 6, 361, 1944.
(7) C. S. Gorter and H. Casimir. *Z. techn. Phys.* **15**, 539, 1934; *Phys. Z.* **35**, 963, 1934.
(8) W. J. de Haas and H. Bremmer. *Commun. Phys. Lab. Univ. Leiden*, no. 214d, 1931.
(9) W. J. De Haas and A. Rademakers. *Physica*, **7**, 992, 1940.
(10) D. Shoenberg. *Nature, Lond.*, **143**, 434, 1939; *Proc. Roy. Soc. A*, **175**, 49, 1940.
M. Désirant and D. Shoenberg. *Nature, Lond.*, **159**, 201, 1947.
(11) A. B. Pippard. *Nature, Lond.*, **159**, 434, 1947.
(12) H. London. *Proc. Roy. Soc. A*, **176**, 522, 1940.
(13) A. B. Pippard. *Proc. Roy. Soc. A*, **191**, 385, 399, 1947.
(14) G. E. H. Reuter and E. H. Sondheimer. *Nature, Lond.*, **161**, 394, 1948.
(15) W. Heisenberg. *Nachr. Ges. Wiss. Göttingen*, 1946.
(16) J. G. Daunt and K. Mendelssohn. *Proc. Roy. Soc. A*, **185**, 225, 1946; *Nature, Lond.*, **150**, 604, 1942.
(17) A. Sommerfeld. *Z. Phys.* **118**, 467, 1941.
(18) A. Sommerfeld and H. Bethe. *Handb. Phys.* **xxiv**, 2, 555, Berlin, 1933.
(19) E. Justi. *Ann. Phys., Lpz.*, **42**, 84, 1942; *Phys. Z.* **43**, 130, 1942.
(20) J. H. van Vleck. *Theory of Electric and Magnetic Susceptibilities*, Oxford, 1932, especially p. 276.

Il Nuovo Cimento (9) 6, Supplemento, 493–497, 503–508 (1949)

Die Erzeugung von Mesonen in Vielfachprozessen.

W. Heisenberg

Göttingen

1. – Die Wechselwirkung zwischen den Nukleonen im Kern ist möglicherweise von einer solchen Art, daß bei großer Feldstärke des Mesonfeldes die nichtlinearen Glieder eine entscheidende Rolle spielen, und daß dadurch bei einem energiereichen Stoß zweier Nukleonen viele π-Mesonen mit einem Schlag erzeugt werden [1]. Diese Annahme, deren Berechtigung durch das Experiment entschieden werden muß, hat zu den folgenden praktisch prüfbaren Folgerungen geführt [2]:

Sei P_0 die Energie, M die Ruhmasse des stoßenden Nukleons, k die Wellenzahl, k_0 die Energie, $\varkappa$ die Ruhmasse der ausgesandten Mesonen (die Lichtgeschwindigkeit wird gleich eins gesetzt), so ist die maximale Energie ε, die beim Stoß in Mesonenenergie umgesetzt werden kann

$$1) \qquad \varepsilon_{\mathrm{max}} = \sqrt{2M(P_0 + M)} - 2M \; .$$

Die dabei zu erwartende Anzahl von Mesonen ist

$$n \sim \frac{\varepsilon}{k} \Big/ \lg \frac{\varepsilon(4M + \varepsilon)}{\varkappa(4M + 2\varepsilon)} \, ,$$

wenn für das Mesonenspektrum im Schwerpunktssystem die plausible Form $\mathrm{d}\varepsilon = a k_0^{-1} \mathrm{d}k_0$ vorausgesetzt wird. Die Aussendung erfolgt im Schwerpunktssystem isotrop. In grober Annäherung folgt also für die Mesonenanzahl etwa

$$n \sim \sqrt{5 \frac{P_0}{M}} \; .$$

Im System des Beobachters ergibt sich für die Winkelverteilung

$$\overline{\cos \vartheta} \sim 1 - \frac{8}{n^2} \, ,$$

                     W. HEISENBERG

wobei ϑ den Winkel zwischen Primärteilchen und dem ausgesandten Meson bedeutet. Die wahrscheinlichste Energie der ausgesandten Mesonen beträgt etwa

$$k_0 \sim \frac{1}{10} \sqrt{P_0 M}\ .$$

Wenn die primären Protonen nach einem Spektrum $P_0^{-1,7}$ verteilt sind, so sollte nach diesen Formeln die Anzahl der Schauer mit mindestens n Mesonen wie $(n \lg n)^{-3,4}$ abnehmen.

Findet ein solcher Stoß zwischen Nukleon und Nukleon im Innern eines schweren Atomkerns statt, so wird das erzeugte Mesonenfeld mit den übrigen Nukleonen des Kerns in Wechselwirkung treten und einen Teil der Nukleonen mit sich reißen. Dabei kann sich die Zahl der Mesonen vergrößern oder verkleinern; ihre Winkelverteilung wird sich jedenfalls erheblich verbreitern. Die Gesamtzahl der aus dem Kern mit Lichtgeschwindigkeit emittierten Teilchen wird also jedenfalls um einen Faktor höher sein als in einem leichten Kern, und für den Vermehrungsfaktor Z kann man als Abschätzung etwa setzen:

$$Z = \exp\left[\beta(A^{1/3} - 1)\right],$$

wobei β aus dem Experiment zu bestimmen ist (A ist das Atomgewicht).

2. - *a*) Dieses allgemeine Bild führt zu der Vorstellung, daß es grundsätzlich zwei verschiedene Sorten von Schauern (und alle möglichen Zwischenfälle) geben wird, wie BERNARDINI auch vom experimentellen Material her betont hat. Die einen finden in leichten Kernen (C, O) oder am Rande eines schweren Atomkerns statt; bei ihnen spielen die Sekundärprozesse nur eine geringe Rolle. Man wird bei ihnen auch nur wenige langsame Protonen oder α-Teilchen finden und sie daran erkennen können. Die anderen Prozesse sind mehr oder weniger zentrale Stöße auf schwere Kerne (Ag, Br, Pb), bei denen die sekundäre Multiplikation den ganzen Vorgang erheblich verändert. Bei diesen Sternen wird man auch eine große Zahl von langsamen Protonen oder α-Teilchen aus dem Kern beobachten [3]. Ein Vergleich zwischen der beobachteten und der theoretischen Winkelverteilung der Mesonen läßt sich nur bei den Stößen der ersten Sorte sinnvoll durchführen. Aber da liegt bisher wenig Erfahrungsmaterial vor, sodaß ein zuverlässiger Vergleich erst in einiger Zeit vorgenommen werden kann.

b) Die Häufigkeit der Mesonenschauer als Funktion der Schauergröße kann jedoch an den neuen Ergebnissen von POWELL und seinen Mitarbeitern studiert werden [4]. Die Abbildung gibt die Anzahl der Schauer mit mindestens n dünnen Spuren als Funktion von n in doppelt logarithmischem Maßstab an (schwarze Punkte); die gerade Linie stellt das Gesetz $(n \lg n)^{-3,4}$ dar.

Man erkennt, daß der Abfall bei großer Teilchenzahl tatsächlich dem theoretischen Gesetz gut folgt. Der Knick bei etwa $n = 8$, — wenn er sich als reell erweist, — kann vielleicht mit dem Abschneiden des primären Spektrums bei einigen 10^9 eV durch das erdmagnetische Feld in Verbindung gebracht werden. Wählt man von allen Sternen in der Statistik von POWELL nur diejenigen aus, die höchstens 6 dicke Spuren enthalten, so erhält man die Kreuze der Abbildung. Man kann annehmen, daß es sich hierbei um Zusammenstöße

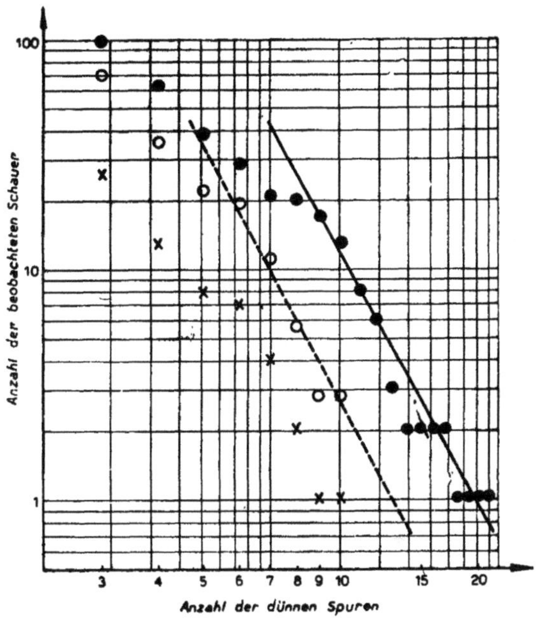

Fig. 1.

mit leichten Atomkernen (C, O) oder um Stöße am Rande eines schweren Atomkerns (Ag, Br) handelt. Um einen Vergleich hinsichtlich der Mesonenzahl zu erhalten, bezogen auf die gleiche Anzahl von Zusammenstößen, muß man die Kreuze der Abbildung noch um einen Faktor erhöhen, der durch das Verhältnis der wirksamen Oberfläche aller Kerne zur Oberfläche der leichten Kerne (und der Randgebiete der schweren Kerne) gegeben ist. Eine Abschätzung auf Grund der chemischen Zusammensetzung der empfindlichen Schicht gibt für den Faktor etwa 2,8. So entstehen aus den Kreuzen der Abbildung die als Ringe angegebenen Punkte. Die Verteilung der Mesonenschauer hat für leichte Kerne also ungefähr die gleiche Form wie für schwere (man muß dabei bedenken, daß die als Ringe angegebenen Punkte aus einer sehr kleinen Statistik gewonnen sind), nur ist die Anzahl der dünnen Spuren in Ag und Br offenbar etwa 1,5 mal größer als für C und O. Daraus schließt

Il Nuovo Cimento (9) 6, Supplemento, 493–497, 503–508 (1949)

man .für die Konstante β etwa $\beta = 1/6$, und für den Vermehrungsfaktor Z erhält man:

	H	C	Ag	Pb
$Z=$	1	1,2	1,8	2,3

Die Anzahl der dünnen Spuren in Blei sollte also etwa doppelt so groß sein wie in Kohle, was auch gut zu dem experimentellen Befund in Nebelkammermessungen [5] paßt.

Wenn durch die Sekundärprozesse hauptsächlich Nukleonen mitgerißen werden, so sollte in Blei etwa die Hälfte aller dünnen Spuren von Protonen herrühren, in Kohle dagegen nur etwa 20 %. Diese Deutung wird durch die Untersuchungen von BLACKETT, BUTLER und ROCHESTER und ihren Mitarbeitern [6] über den positiven Exzeß der durchdringenden Teilchen nahegelegt.

Mit der genannten Annahme über den Vermehrungsfaktor Z würde der von POWELL beobachtete Knick in der Verteilungskurve der Schauer in reinem Wasserstoff bei Schauern von etwa 5 dünnen Spuren liegen. Dieser Wert entspricht auch ungefähr der Anzahl, die gelten sollte für ein Primärteilchen von der durch das erdmagnetische Feld zugelassenen Minimalenergie. Bei photographischen Platten, die in der Nähe des erdmagnetischen Äquators exponiert sind, sollte der Knick also bei größeren Teilchenzahlen liegen.

c) Die genannten Ergebnisse stehen auch in engem Zusammenhang mit Messungen über lokale durchdringende Schauer, wie sie früher von JANOSSY und WATAGHIN, neuerdings von verschiedenen anderen Forschern [7] ausgeführt worden sind. Nach diesen Messungen ist die Anzahl der Schauer, die man aus bestimmten Schichtdicken, gemessen in Gramm pro Quadratzentimeter, bekommt, in leichtem Material (C, Al) größer als in schwerem (Pb). Bei geringen Schichtdicken ist dies ohne weiteres verständlich, da bei gleicher Masse die wirksame Atomoberfläche in leichtem Atommaterial größer ist als in schwerem. Bei großer Schichtdicke dürften geometrische Effekte mit eine Rolle spielen, insbesondere dürfte die größere Winkelausdehnung der Schauer aus schwerem Material zu einer Verringerung der beobachteten Schauerhäufigkeit führen. Nach den genannten Versuchen von HAXEL [7] erhält man aus Paraffin mehr Schauer als aus Kohle, was wieder darauf hindeutet, daß schon beim Zusammenstoß zweier isolierter Nukleonen Mesonenschauer gebildet werden.

3. - Gegen die Annahme der Vielfachentstehung von Mesonen war früher von HEITLER und seinen Mitarbeitern eingewendet worden, daß die Matrixelemente der Wechselwirkungsenergie, die zu Vielfacherzeugung führen, unter Umständen auch eine große Dämpfung hervorrufen können, die die Vielfacherzeugung wieder verhindert. Dieser Einwand kann jedoch durch den Nachweis widerlegt werden, daß man S-Matrizen konstruieren kann, die zur Vielfacherzeugung führen. HEITLER und JANOSSY schlagen die Entstehung der Schauer durch kaskadenartige Prozesse im Atomkern vor und glauben, daß die bisherigen Experimente auch mit dieser Hypothese in Einklang stehen. Die Frage, ob beim Stoß zweier Nukleonen in der Regel viele Mesonen erzeugt werden, bedarf daher noch der experimentellen Entscheidung.

LITERATUR

[1] W. HEISENBERG: *Zeits. f. Phys.*, **113**, 61 (1939). *Vorträge über Kosmische Strahlung*, Berlin, Springer, 1943 (S. 57 u. 115); G. WATAGHIN: *Akad. Brasileira de Cienças* (1943); *Phys. Rev.*, **74**, 975 (1948).

[2] W. HEISENBERG: *Nature*, **164**, 65 (1949); *Zeits. f. Phys.*, **126**, 569 (1949).

[3] L. LEPRINCE-RINGUET, F. BOUSSER, HOANG-TCHANG-FONG, L. JAUNEAU u. D. MORELLET: *Compt. Rend.*, **229**, 163 (1949).

[4] R. K. BROWN, U. CAMERINI, P. H. FOWLER, H. HEITLER, D. T. KING u. C. F. POWELL: *Phil. Mag.* (7), **40**, 862 (1949).

[5] A. LOVATI, A. MURA, G. SALVINI, G. TAGLIAFERRI: *Vortrag in Como*.

[6] P. M. S. BLACKETT, G. D. ROCHESTER, C. C. BUTLER u. Mitarbeiter: *Vortrag in Como*.

[7] J. E. CLAY: *Physica*, **14**, 594 (1948); H. A. MEYER, G. SCHWACHHEIM u. G. WATAGHIN: *Phys. Rev.*, **74**, 846 (1948); E. P. GEORGE: *Vortrag in Como*; O. HAXEL: (wird demnächst veröffentlicht).

Il Nuovo Cimento (9) 6, Supplemento, 493–497, 503–508 (1949)

case of thin absorbers over the top tray of counters it is most likely that the main shower is produced inside the thick layers of lead *below* the top tray and the only function of the top absorber is to give rise to one or two additional particles aiding the setting off of the top counters.

In this way the difference between carbon and paraffin can also be understood. The relatively frequent collisions with hydrogen increase slightly the sizes of showers without very great loss of energy to the primary.

Thus there is no difficulty in understanding the first slopes of the transition effects in terms of plural production. The interpretation of records under thick absorbers (saturation) is complicated involving arguments of geometry and such arguments can hardly be relied on. Thus it does not seem worth while to go into any detail regarding the exact interpretation of shower intensities under greater thicknesses of material.

DISCUSSIONE E OSSERVAZIONI SEGUITE ALLE RELAZIONI

di W. HEISENBERG, *Göttingen*, (pag. 493)

e di W. HEITLER e L. JANOSSY, *Dublin*, (pag. 499).

— B. FERRETTI, *Roma*:

1 should like to make some remarks about the experiment of Prof. HAXEL. In this experiment the production of penetrating showers in paraffin and carbon is compared. Apparently he has found a considerably greater number of penetrating showers in paraffin than in carbon for the same thickness in g/cm^2 of the two substances. I do not think that this is really a proof of the multiple production of mesons in collisions against the hydrogen of the paraffin for the following reasons: the amount of hydrogen in paraffin is about 15 % of the amount of carbon, so that even taking into account the « shadow effects » for production of penetrating showers in carbon (these effects cannot certainly be greater than 50 % in carbon) one should expect an increase in the number of penetrating showers due to the hydrogen, of not more than 30 % when comparing two sheets of carbon and paraffin having the same number of g/cm^2 of carbon only. Comparing two sheets having the same number of g/cm^2, the effect that one would expect should not be greater than about 15 %. However, taking the background into account, the effect observed by Prof. HAXEL is greater than 100 % and the statistical errors are greater than 15 %. I do not think therefore that the result of HAXEL's experiment can be an argument in favour of multiple production.

I agree completely with Prof. HEISENBERG's idea that in a sufficiently small volume it is impossible to talk of the number and of the kind of particles which are present. However what I do not know is what is meant by this volume: is it the volume of a heavy nucleus, or the classical volume of an electron, or even of the order of magnitude of the cube of the COMPTON wave-length of the proton? The interpretation of experiments made on complex nuclei, like carbon, lead and so on, will depend on this circumstance. Consequently only a clear experiment in which nucleon-proton collisions are involved may be really decisive for the question of the multiplicity of production of mesons. Reciprocally, once the question of the multiplicity of production of mesons is solved, the results of the experiments on complex nuclei may give information about the volume in which the ordinary concepts of number of particles, cascade production and so on, lose their meaning.

— W. HEISENBERG, *Göttingen*:

Concerning the difference between Carbon and Paraffine one should of course take into account the rather large statistical curve involved and consider the experiment only as an indication in the direction of the multiple process, not as a proof. One may perhaps hope to get the final decision in the near future from three types of experiments: 1) Frequency of showers of larger size in Carbon; 2) Direct observation of collisions with single hydrogen-atoms in photographic plates; 3) Observations on the production of nuclear explosions by negative particles, as mentioned by Prof. BLACKETT.

— R. BRODE: *Berkeley*:

Prof. FRETTER has observed showers of 15 minimum ionizing particles from Carbon plates.

— L. JANOSSY, *Dublin*:

The distribution of relativistic particles in stars when plotted on an ordinary scale shows a knee opposite to that which appears in a logarithmic plot. This knee is in agreement with the predictions of HEITLER and JANOSSY in terms of plural production. The physical significance of the knee is the relatively large contribution of the light elements in the emulsion to small stars. HEISENBERG's interpretation of the knee in terms of the latitude cut off seems to me artificial. It does not seem likely that the cut off of the primary nucleon spectrum at the top of the atmosphere could possibly persist down to the levels where the stars have been observed. The knee of the size-frequency curve seems therefore to be rather an argument in favour of plural production and not an argument in favour of multiple production.

— W. HEISENBERG, *Göttingen*:

A decision between the two conflicting explanations of the knee can probably be reached, when one collects experimental data from photographic plates exposed near the equator at great altitude. I hope that such material will be available in the future. If the hypothesis of multiple production is correct, then the knee should change to showers of larger size in such plates.

— M. G. E. COSYNS, *Bruxelles*:

How can the pair-production observed in high energy nuclear explosions be explained by the theory of multiple production?

— W. HEISENBERG, *Göttingen*:

As to the production of light-quanta and electrons in nuclear explosions I would like to say, that if the explosion can be considered as a temperature equilibrium at

Il Nuovo Cimento (9) 6, Supplemento, 493–497, 503–508 (1949)

extremely high temperature ($\sim 3 \cdot 10^8$ eV) in a very narrow region, then about one third of the energy could easily be contained in the electron-photon-component, and thereby one could understand the multiple production of photons and electrons.

— G. P. THOMSON, *London*:

It is well known that a number of single tracks can be seen in photographic plates. These tracks are too numerous to be due to stars outside the emulsion and they are strongly concentrated in a vertical direction. It seems likely that these tracks proceed from an event which at its origin is composed wholly of lightly-ionising particles. Such events are hard to find, but they are being searched and when found their multiplicity should give valuable information.

— J. E. CLAY, *Amsterdam*:

Attention was asked for an experiment on the production of penetrating pairs of particles in different materials for layers of the same number of atoms (in 10 cm water). Paraffine, H_2O, Al, Fe, Pb, P, 0,27, 0,29, 0,26, 0,23 and 0,27 respectively. *Rev. of Mod. Phys.*, **21**, 82 (1949).

— O. PICCIONI, *Brookhaven, L. I., N. Y.*:

In connection with the question of the meson emission, whether multiple or plural, I would like to mention an experiment on local penetrating showers [1] at two different latitudes (22º and 44º geom. lat.) and two altitudes (300 g cm^{-2} and 400 g cm^{-2}). The latitude effect turned out to be 10 % at both altitudes, which shows that the particles triggering the apparatus arise from primaries having the same energy whether the apparatus is working at 300 g cm^{-2} or at 400 g cm^{-2}. This would be more in favour of the hypothesis that protons lose a large amount of energy in a single collision, rather than a gradual loss of energy through the atmosphere, as suggested by the scheme of emission in several nucleon-nucleon collisions.

— A. ROGOZINSKY, *Meudon*:

In connection with the counter experiments described by Professor HEISENBERG, I should like to mention similar ones, carried out at the Observatoire de Meudon by M. A. VOISIN. In the experiments a penetrating shower detecting apparatus was put *a*) outside and *b*) inside a room, surrounded by a quasi-spherical screen of water 1 meter thick. An increase of the order of 100 % of the frequency of the penetrating showers was observed in the latter case.

— L. MEZZETTI, *Roma*:

In connection with the discussion on the type of the processes responsible for the production of penetrating showers, I point out the preliminary results obtained by Dr. QUERZOLI and myself at the L.T.G. and already reported in the morning session. We have found that the ratio of the mean free path for production of showers containing at least two penetrating particles in lead and carbon is equal to the ratio of the mean free paths corresponding to the geometrical cross-sections and independent

---

[1] Performed by T. WALSH and O. PICCIONI.

from the total multiplicity of the shower particles, for not too high multiplicities. For high multiplicities, on the contrary, lead appears to be more efficient than carbon in shower production; this is quite reasonable, owing to the fact that the multiplicity measured by us includes also comparatively slow protons and electrons eventually produced in the nuclear collision.

— G. BERNARDINI, *Roma*:

In my opinion the problem largely discussed today — i.e. in what type of process the mesons are produced — cannot be separated from certain information concerning the energy of the nucleons involved in the process itself. At extremely high energies, for instance like that producing the Rochester star, it is quite reasonable to think that the process shows the aspect of a *multiple* process. It should be a completely different matter to observe that *usually* a nucleon of energy around 10 GeV produces simultaneously more then one meson. That would be actually a test of the *multiple* creation of mesons. In this direction it might be pointed out that the data reported by BLACKETT, which represent an argument in favour of the *plural* production, could be considered consistent with the several highly energetic meson stars observed by COSYNS, LEPRINCE-RINGUET and POWELL, and supporting the *multiple* hypothesis. As a matter of fact, because of the nucleonic cascade, the nucleon spectrum near sea level must be composed largely of nucleons which are still able to produce mesons, but which have a quite low average energy in comparison with the energies found at high altitudes.

— G. BERNARDINI, *Roma*:

[Information comunicated in a letter received from Dr. R. E. MARSHAK (*Rochester*, N. Y., U.S.A.) concerning a spectacular and extremely interesting star observed in a β-sensitive plate (Kodak NTB3) by the Rochester group].

The star is due to a collision of an extremely energetic primary cosmic alpha particle. The collision gives rise to a very narrow penetrating shower of 23 relativistic singly charged particles together with a more diffuse shower of 33 relativistic particles. This star consists of 75 prongs representing a minimum of 81 charges of which about 56 correspond to minimum ionization for relativistic singly charged particles and constitute the shower part of the star.

There are 18 non-relativistic heavy particles emitted in the nuclear explosion corresponding to at least 23 charges. The energy in the non-shower part of the explosion can be estimated to ~ 2.8 GeV. The star contains 81 charges, or 77 after subtracting 2 fast alpha particles. Thus there is an excess of 30 charges over that of Ag(47) the heaviest nucleus in the emulsion. It is to be concluded that at least one-half of the 56 shower particles are mesons (or possibly electrons) and not nucleons.

The shower part of the star consists of a sharp well-defined very dense core (C) whose total angular spread projected on the plane of the emulsion is 2.2°. The axis of this core is an exact continuation of the direction of the incident alpha particle. The core contains 23 ± 2 tracks, assumed to be tracks of relativistic particles, as the number was determined by grain counting the core over a known distance and dividing by the number of grains produced by a minimum ionization track over that distance. The core is so dense up to the region where the main part of it leaves the emulsion, that not all individual tracks are resolvable. All the resolvable tracks, some of which continue in the emulsion for at least 2000 μ correspond to minimum ionization for singly charged particles.

The maximum deviation of any of the 33 tracks from minimum ionization is $\pm 10 \%$. (Minimum ionization for singly charged particles is 16.5 grains/75 μ).

Il Nuovo Cimento (9) 6, Supplemento, 493–497, 503–508 (1949)

It was possible to trace the core C to the next plate in the stack. The region where the narrow shower core was expected to penetrate into the next plate (an area of 25 mm²) was surveyed for tracks whose projected length and direction of travel in the emulsion were such that they could have been caused by the explosion. One would expect that if none of the charged particles of the core are lost we should find again 23 ± 2 tracks satisfying the criterion. One finds 44 tracks of relativistic particles in or near the cone; 38 tracks lie inside the cone, 6 more lie near the edge of the cone or just outside of it. All but 9 of the tracks have such orientations that if projected back they pass near to but not through the center of the star. A small uncertainty in location due to the recording of coordinates on the mechanical stage of the observing microscope as well as multiple scattering of the particles (they pass through at least 2 cm of glass for which the mean free path for multiple scattering is .327 μ) are assumed to be responsible for this deviation. In surveying an additional area of 29.2 mm² for tracks with the same criterion we find only 7 tracks, all but one of which lie very close to the core and could easily be thought to have been scattered out of it. We thus conclude that considerable multiplication of charged particles has occured in the core.

This conclusion is supported by the following observations. In the original plate we find a track in the outer fringe of the narrow core C that originates in the emulsion, has the direction of the core and corresponds to exactly twice minimum ionization over a length of 2000 μ. This track is resolvable after 2000 μ into two minimum ionization tracks and therefore can be considered as due to pair production by a gamma ray (or another non-ionizing particle) which must be associated with the star. A first estimate of the energy of this pair (neglecting scattering effects) is given by the relation: $E_\gamma \simeq mc^2/\Delta\theta \simeq 2$ GeV (a separation between the members of the pair of .5 μ after 2300 μ). Actually multiple scattering of the electrons produced by a 2 GeV gamma ray would give a separation of the tracks of 3-5 μ after 2000 μ. Since the observed separation is only 0.5 μ the divergence due to multiple scattering is more important than the mc^2/E spread, and gives us an estimate of the gamma energy of approximately 15 GeV.

In the cone another pair production is observed in the emulsion. After traversing 700 μ, one of the members of this pair creates another pair, suffering only very slight scattering in the process. On the same basis as above (neglecting scattering) approximately 700 MeV was estimated for the original pair and 200 MeV for the pair produced by the electron.

The number of tracks in the cone, excluding those that do not lie strictly in it is 38. If we subtract 6 (the number determined by scanning the control area) which is certainly an exaggerated correction we obtain 32 minimum ionization tracks. As we originally had 23 ± 2 this means that we have 9 extra particles in the next plate plus the pair found in the original plate giving us 6 ± 1 pairs created in the core over a mean distance of 2.7 cm of glass and 1500 μ of emulsion. Considering the radiation lengths of glass and emulsion it was found that the original number of γ-rays was $N_0 = 25 \pm 4$.

If one assumes that the observed electron pairs are produced by gamma rays, then if the gamma rays are the result of the decay of a neutral meson with a rest mass corresponding to 150 MeV [1], by assuming that the gamma ray is converted immediately into pairs, one can obtain an upper limit on the life-time of the neutral

---

[1] Cfr. Berkeley evidence on γ-rays from neutral meson.

DISCUSSIONE E OSSERVAZIONI

meson from the pair observed to originate in the emulsion 500 μ from the star. $T_{\text{lab}} < 500 \cdot 10^{-4}/3 \cdot 10^{10} = 1.7 \cdot 10^{-12}$ s. For the estimate of gamma ray energy of 15 GeV the neutral meson giving rise to this must have an energy of at least 15 GeV. In the rest system of the neutral meson this corresponds to a life-time of $T_0 < 1.7 \cdot 10^{-14}$ s. Hence on the above assumptions, the upper limit for the life-time of the neutral mesons is $1.7 \cdot 10^{-14}$ s.

The total energy release in the star can be estimated to be ~ 200 GeV. To summarize we conclude that the event described (which may be the initial stage of development of an AUGER shower) gives direct evidence in favour of the multiple production of high energy gamma rays in stars (either directly or through some intermediate process such as the decay of a neutral meson).

Progress of Theoretical Physics, Vol. 5, No. 4, July~August, 1950

The Yukawa Theory of Nuclear Forces in the Light of Present Quantum Theory of Wave Fields.

W. Heisenberg

Max Planck-Institut für Physik, Göttingen.

(Received June 5, 1950)

The Yukawa theory of nuclear forces[1] has led to many successes and, owing to the present state of quantum theory, to some difficulties. Among the successes one remembers first the existence of the π-meson and the possibility of describing the spin dependency and the saturation of nuclear forces by means of simple vector fields or pseudo-scalar fields. Among the difficulties we mention the divergence of the interaction at small distances of the nucleons and the impossibility of getting the correct mass defect for heavy nuclei when one takes the constants of the Yukawa field from the mass defect of light nuclei.[2] Furthermore, the existence of closed neutron and proton shells in the nucleus[3] and the behaviour of the cross section for elastic collisions of nucleons at very high energies indicate, that the Yukawa potential is not correct at small distances of the nucleons.

These difficulties cannot be really solved yet; but the recent progress in quantum theory of wave fields[4] shows so clearly the way towards the solution of these problems, that it may be worth while to discuss this way, even if it is still too early to work it out in the mathematical details.

In the relativistic quantum theory of wave fields we have learned, that the divergent results arise from the singularities in the commutation function. Therefore the correct theory will have to start with a *regular* commutation function. This starting point leads to a number of problems, which have been dealt with recently in many papers.[5] We mention the most important results: The wave function that obeys a regular commutation rule, corresponds necessarily to several different types of elementary particles, not only to one type. This implies, that nucleons interact, as Bhabha[6] has suggested, not only by means of π-mesons but also by other types of particles, in such a way, that the singularity of the force at small distances will disappear. Furthermore, the two coordinated wave functions, that obey the regular commutation rule, cannot be hermitian conjugates in the ordinary sense.[5] This leads probably to a change in the hamiltonian formalism in the range of the " smallest length " l_0 ($l_0 \sim 10^{-13}$ cm), which corresponds to a lack of point-to-point causality, again tending to wash out singularities of the field.

W. Heisenberg

Thereby already many of the difficulties may have disappeared. The potential inside a nucleus will now be rather smooth, certainly much smoother than one would expect from potentials of the type $\frac{1}{r}e^{-\kappa r}$ or the corresponding tensor force potential. As a result there will be only small forces acting upon a nucleon inside a nucleus; it is only at the surface of the nucleus that the nucleons will be pulled back into the nucleus by strong forces. This explains quite naturally the existence of separated neutron and proton orbits and closed shells in the nucleus.

The order of the closed shells can be understood, according to Haxel, Jensen, Suess[3] and Göppert-Mayer,[3] from a strong spin-orbit coupling of every nucleon. Gaus[7] has shown that this strong spin-orbit coupling results under certain conditions immediately from the vector-meson theory of Yukawa. Therefore one may at this point conclude from the experiments, that at larger distances of two nucleons the symmetrical vector meson theory with the mass of the π-meson will probably give a fairly good approximation, while at smaller distances the higher masses will come into play and the deviations from the hamiltonian formalism will make the definition of a potential rather doubtful, as it was expected long ago from the concept of the " smallest length."[8]

The existence of neutral mesons, possibly of the scalar type, may produce forces without the property of saturation. This would explain naturally the rather large mass defects of heavy nuclei as compared to the mass defects of light nuclei. The observed saturation would then, as Teller[9] has suggested, probably be brought about by the non-linear interaction terms in the field equation, which prevent the Yukawa field to increase above a certain value.

Another difficulty for the vector-meson theory was the sign of the quadrupole-moment of the deuteron, which seemed to favour the pseudo-scalar rather than the vector theory. The quadrupole moment of the deuteron is determined by the tensor force, which depends strongly on the potential at small distances; the mass defect and the spin-orbit coupling depend more strongly on the outer part of the potential function. Therefore the higher masses and the deviation from Hamilton formalism may be decisive for the quadrupole moment of the deuteron, while the mass defect and the spin-orbit coupling are mainly produced by the vector field of the normal π-mesons.

Finally the cross-section for elastic collision of very fast nucleons will decrease more rapidly with increasing energy than one would expect from the Yukawa potential $\frac{1}{r}e^{-\kappa r}$ and the corresponding tensor potential. One may express this mathematical result by stating, that the introduction of the " smallest length " l_0 in the primary commutation function leads also to a " largest force " of the order l_0^{-2} (or in ordinary units $\frac{\hbar c}{l_0^2}$), so that a momentum transfer of much more than $\frac{\hbar}{l_0}$ in an elastic collision will be a rather rare event. The collision of very

Progress of Theoretical Physics 5, 523–525 (1950)

energetic nucleons will instead as a rule lead to the creation of new particles, first of π-mesons and at still higher energies of other masses. The quantitative question at which energies the deviations from the simple Yukawa potential appear cannot yet be solved.

References.

1) H. Yukawa, Proc. Phys.-Math. Soc. Jap. **17** (1935), 48.

2) Compare f. i. H. Euler, ZS f. Phys. **105** (1937), 553, or H. Primakoff and T. Holstein, Phys. Rev. **55** (1938), 1218.

3) O. Haxel, J. H. D. Jensen u. H. E. Suess, Naturwiss. **35** (1948), 376; **36** (1949), 153; Phys. Rev. **75** (1949), 1766, and M. Goeppert-Mayer, Phys. Rev. **75** (1949), 1969.

4) S. Tomonaga, Prog. Theor. Phys. **1** (1946), 27; Phys. Rev. **74** (1948), 224; R. P. Feynman, Phys. Rev. **74** (1948), 939, 1430; J. Schwinger, Phys. Rev. **74** (1948), 1439; **75** (1949) 651; **75** (1945), 790; F. J. Dyson, Phys. Rev. **75** (1949), 486, 1736.

5) Compare W. Heisenberg, Zur Quantentheorie der Elementarteilchen, ZS. f. Naturforschung, to appear shortly, which contains a number of references of the most important papers.

6) H. J. Bhabha, Phys. Rev. **77** (1950), 665.

7) H. Gaus, ZS f. Naturforschung **4a** (1949), 721.

8) W. Heisenberg, Ann. d. Phys. **32** (1938), 20.

9) E. Teller, Private communication.

In *Proceedings of the International Congress of Mathematicians, Cambridge, Massachusetts, U.S.A., August 30–September 6, 1950, Volume II* (American Mathematical Society, Providence, R. I., 1952) pp. 292–296

ON THE STABILITY OF LAMINAR FLOW[1]

W. HEISENBERG

The stability of laminar flow has for a long time been a subject of considerable dispute, and it is only recently that clarity has been achieved in the essential points. Recently two surveys of the problem, by Lin [3] and by Tollmien [13] have been published. Since these surveys and the calculations contained in them have contributed essentially to the clarity in this problem, I need not here repeat the survey in detail. I would like however to go shortly through the history of the problem and to add a few remarks at those points where I think that a further clarification is necessary. At the same time I would like to discuss the physical interpretation of the mathematical results and to compare it with the physical interpretation of the statistical isotropic turbulence.

1. We confine ourselves in the usual way to the two-dimensional flow between parallel walls, and we know from the work of Squire that the extension to three-dimensional perturbations would not alter the problem essentially. Let us call the original velocity distribution $w(y)$, the perturbation in velocity u and v, with

$$u = \frac{\partial \psi}{\partial y}, \qquad v = -\frac{\partial \psi}{\partial x},$$

and put

$$\psi(x,y) = \varphi(y)e^{i\alpha(x-ct)},$$

then

$$(1) \qquad (w - c)(\varphi'' - \alpha^2 \varphi) - w''\varphi = -\frac{i}{\alpha R}\left(\varphi'''' - 2\alpha^2 \varphi'' + \alpha^4 \varphi\right)$$

according to Orr [6] and Sommerfeld [9].

This differential equation represents an eigenvalue problem of a similar type to those one meets in wave mechanics. It is therefore natural to use similar methods for solution in both cases, and I would like to mention that the asymptotic method [1] which was used long ago in treating equation (1) was essentially the same method which later in wave mechanics has been worked out by Wentzel, Kramers, and Brillouin.

If one wants to know the stability at very large Reynolds numbers, it is natural to put the right side in the Orr-Sommerfeld-equation equal to zero and to omit two boundary conditions. The liquid then is allowed to slip along the walls and one gets the equation used by Rayleigh:

$$(2) \qquad (w - c)(\varphi'' - \alpha^2\varphi) - w''\varphi = 0.$$

---

[1] This address was listed on the printed program under the title *Die Stabilitätsfragen der Flüssigkeitsdynamik im Zusammenhang mit der statischen Turbulenztheorie.*

292

ON THE STABILITY OF LAMINAR FLOW 293

This equation is sufficient to decide the instability in all cases where damped or amplified solutions of (2) exist, because then (2) contains no singular point in which viscosity could come into play. It should also be emphasized—and this is a point in which I cannot agree entirely with the mathematical formulation in Lin's analysis—that whenever there is a damped solution of (2), there must also

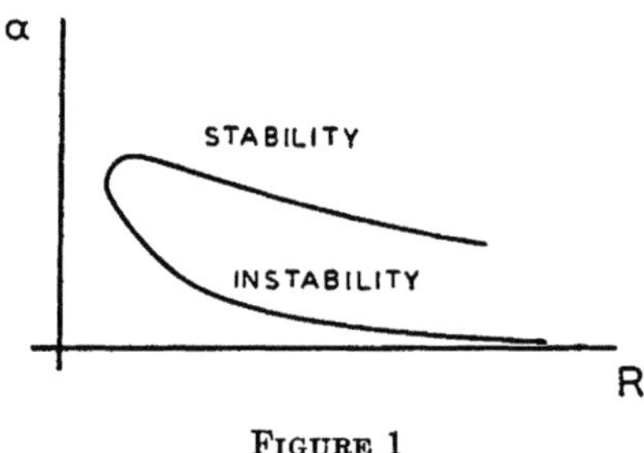

FIGURE 1

be an amplified one with the same wave length, because one may simply change the sign of α in (2). Physically you may say that one may always simply reverse the sign of time in any mechanical problem in which viscosity plays no role. It does play no role here for damped or amplified solutions since the inertia forces (left side of (1)) are at every point much stronger than the viscosity (right side of (1)). Therefore for every damped solution one can find an amplified one, and every profile of such type is unstable at very high Reynolds numbers.

The real problem, however, arises from the fact that equation (2) allows such solutions only for very special profiles, namely profiles with a point of inflection ($w'' = 0$ for $y = y_0$). As early as 1880 Lord Rayleigh has shown that one can have amplified solutions for such profiles, and starting from the complete equation (1) the limits of instability for finite Reynolds numbers are given by a curve of the type of Fig. 1. (Compare [14], [3], and [13].)

2. But some of the most important profiles have no point of inflection, and it is for those that one must from the beginning go back to the complete equation (1). This is already necessary in any case where one considers neutral disturbances; because for real values of c, equation (2) contains the singular point $w = c$ where viscosity becomes important. The simplest profile of this type is the Couette motion $w(y) = y$, where (2) has no solution, and v. Mises [4] and Hopf [2] accordingly had shown very early that (1) leads only to damped vibrations. The linear profile has been therefore known to be stable long ago. It should be emphasized in this connection, that a damped solution of equation (1) does generally *not* go over into a solution of equation (2) in the limit $R \to \infty$, since equation (2) has generally no solution. In fact, as Lin has pointed out, there is a radical difference between the damped and the amplified solutions of equation (1), contrary to the behaviour of equation (2). Generally one should expect that

294   W. HEISENBERG

in the limit $R \to \infty$ a solution of (1) will not go over into a solution of (2) which fulfills the boundary conditions; but for every solution of (2) there will be a solution of (1) which approaches it in the limit $R \to \infty$.

The next simplest profile is the Poiseuille motion between parallel walls

$$w = 1 - y^2 \qquad \text{(walls at } y = \pm 1\text{)}.$$

For this distribution equation (2) has one solution corresponding to a somewhat

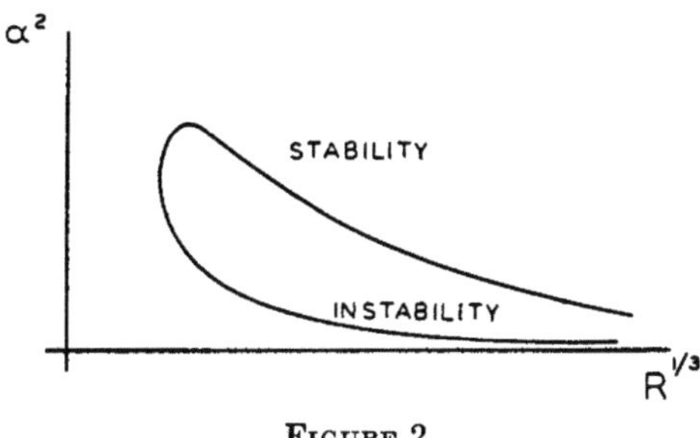

Figure 2

degenerate neutral disturbance: $\varphi(y) = w$, $\alpha = c = 0$. It looked natural to try whether this solution would change into an amplified solution when viscosity was taken into account. When this problem was attacked in 1924 by the method of asymptotic expansion [1], it turned out that one actually did get instability, and recently the much more accurate calculations of Lin [3] have confirmed this view. One gets an instability range of the type shown in Fig. 2. Naturally both branches of the neutral curve go to $\alpha \to 0$ for very large R, because they must tend towards the solution $\varphi(y) = w$, $\alpha = c = 0$. (In the paper of 1924 only one branch was calculated, the other one was very roughly estimated; the curve above has been given by Lin [3].)

Thus the calculations seemed to give the very natural result that whenever equation (2) has any solution at all (amplified, damped, or neutral), then the profile becomes unstable at sufficiently large Reynolds numbers. It should be mentioned, however, that not all recent calculations on the Poiseuille motion have led to this result; e.g., Pekeris [7] seems to get stability, at least for very small values of α; so that it would certainly be worthwhile to extend the calculations, possibly also on the higher harmonics.

The result described would probably have been generally accepted as plausible if shortly after 1924 a paper of Noether [5] had not appeared, in which he claimed to show that a curve of neutral disturbance can never exist for any continuous velocity distribution. The mathematical methods of Noether were better than those in the earlier papers, and the whole situation became rather obscure. Then in 1929 Tollmien [14] took up the question of stability for profiles of the boundary layer type. He got instability for such profiles and was able to calculate the limiting neutral curve. Quite recently, in 1944, Tollmien [13] has improved

Proceedings of the International Congress of Mathematicians II (1952) pp. 292-296

the mathematical methods considerably and got similar results, and in 1947 Schubauer and Skramstad [8] could prove by careful experiments that Tollmien's calculation of the stability limits was correct.

Therefore the paper of Noether, which in his time had made the whole theory of instability suspicious, seems to contain some mistake, but this mistake has not yet been found.

Concerning the experimental side of the problem one should add that in any given experiment the limit of stability may be quite different from the theoretical one as soon as the experiments introduce new sources of instability which are not accounted for by the simple theory.

For instance the primary turbulence of the wind tunnel in which the experiment is done can according to Taylor [11] produce instability, and the same may be true for any perturbations of the stream at the entrance into the region of measurement. But this does not change the validity of the simple theory.

3. Taking now for granted that instability arises generally even in those cases in which the inviscid equation allows only a neutral solution, the question arises how viscosity can cause instability. From simple arguments one would expect damping rather than amplifying. But here one should remember that an inviscid fluid is a system of an infinite number of degrees of freedom, which normally interact so that the energy is dissipated among all modes of vibration. It is only for very special geometrical conditions that this transfer of energy does not take place. Therefore, if a neutral disturbance is possible in the inviscid fluid, the viscosity may easily change the phases of the vibration in such a manner that the transfer of energy begins, which then means amplification of the vibration.

There it is very probable, though not certain, that the amplification of the perturbation leads at once to complete turbulence, that is, to the statistical distribution of energy among all degrees of freedom. This is not quite certain since we know from the work of Taylor [10] on the cylindrical Couette case that special modes of vibration may be developed which have been described as cellular motions and which do not mean real turbulence. But this seems to be a special consequence of the centrifugal forces. In most cases the instability can safely be assumed to lead directly to turbulence. Such turbulence need not be isotropic since the walls introduce deviations from isotropy, but it will be a statistical distribution of energy among many degrees of freedom.

Finally the question remains: what happens to those profiles where the inviscid equation (2) has *no* solution at very high Reynolds numbers? These profiles certainly will be stable at extremely high Reynolds numbers if one can keep away all outer perturbations, e.g. at the entrance of the flow. But for a finite perturbation one should expect that there will be a Reynolds number from which on these perturbations will start the exchange of energy between the different degrees of freedom going, and thereby will cause turbulence. This has, however, never been followed mathematically.

 W. HEISENBERG

REFERENCES

1. W. HEISENBERG, Annalen der Physik vol. 74 (1924) p. 577.
2. L. HOPF, Annalen der Physik vol. 44 (1914) p. 1.
3. C. C. LIN, Quarterly of Applied Mathematics vol. 3 (1946) pp. 117, 217, 277.
4. R. v. MISES, Heinrich Weber-Festschrift, 1912, p. 252.
5. F. NOETHER, Zeitschrift für Angewandte Mathematik und Mechanik vol. 6 (1926) pp. 232, 339, 428, 497.
6. W. M. F. ORR, Proceedings of the Royal Irish Academy vol. 27 (1906–1907) pp. 9–26, 69.
7. C. L. PEKERIS, The Physical Review vol. 74 (1948) p. 191.
8. G. B. SCHUBAUER and H. K. SKRAMSTAD, Journal of the Aeronautical Sciences vol. 14 (1947) p. 69.
9. A. SOMMERFELD, Atti del IV Congresso internazionale dei Matematici, Rome, 1909.
10. G. J. TAYLOR, Proc. Roy. Soc. London Ser. A vol. 102 (1923) p. 541.
11. ——, Proc. Roy. Soc. London Ser. A vol. 156 (1936) p. 307.
12. ——, Proceedings of the Fifth International Congress of Applied Mechanics, Cambridge, 1938.
13. W. TOLLMIEN, Fiat Review of German Science 1939–1946 vol. 11 Hydro- and Aerodynamics (1948) pp. 21–53.
14. ——, N. A. C. A. Technical Memoire no. 609 (1931), no. 792 (1936).

MAX PLANCK INSTITUTE FOR PHYSICS,
GÖTTINGEN, GERMANY.

KOSMISCHE STRAHLUNG

VORTRÄGE

GEHALTEN IM MAX-PLANCK-INSTITUT FÜR PHYSIK
GÖTTINGEN

ZWEITE AUFLAGE

VON

L. BIERMANN · P. BUDINI · J. BUSCHMANN · M. DEUTSCHMANN
E. FREESE · K. GOTTSTEIN · R. HAGEDORN · W. HEISENBERG
F. G. HOUTERMANS · H. JAHN · G. LÜDERS · R. LÜST · W. MACKE
H. M. MAYER · P. MEYER · G. MOLIÈRE · R. OEHME · K. OTT · W. PAUL
F. SAUTER · A. SCHLÜTER · K. SYMANZIK · M. TEUCHER
C. F. v. WEIZSÄCKER · K. WIRTZ · B. ZUMINO

HERAUSGEGEBEN VON

WERNER HEISENBERG

MIT 256 ABBILDUNGEN
UND 1 TAFEL

SPRINGER-VERLAG
BERLIN · GÖTTINGEN · HEIDELBERG
1953

Vorwort zur 2. Auflage.

In den zehn Jahren, die seit dem Erscheinen der 1. Auflage verstrichen sind, haben die Kenntnisse von der kosmischen Strahlung eine außerordentliche Erweiterung und Vertiefung erfahren. Nach dem Ende des Krieges sind in jedem Jahr Hunderte von Arbeiten veröffentlicht worden, die über neue Erfahrungen oder neu entwickelte experimentelle Methoden berichten oder die das theoretische Verständnis der Zusammenhänge erweitern. So ist es notwendig geworden, das vorliegende Buch praktisch neu zu schreiben, um der Fülle des hinzugekommenen Materials einigermaßen gerecht zu werden. Das Bild, das die 1. Auflage vom Gesamtgebiet der kosmischen Strahlung entworfen hatte, mußte an vielen Stellen ergänzt und in einigen Punkten grundsätzlich geändert werden. Die wichtigsten Verbesserungen sind durch die Entdeckung der neuen Mesonenarten bedingt, die, wie wir jetzt wissen, in der Genetik der kosmischen Strahlung eine entscheidende Rolle spielen. Ferner ist neu hinzugekommen das Kapitel I von der Entstehung der kosmischen Strahlung, für die es erst seit einigen Jahren im Zusammenhang mit anderen astrophysikalischen Erscheinungen eine plausible Erklärung gibt. Die Reihenfolge der übrigen Kapitel ist so gewählt worden, wie es den heutigen Vorstellungen von der Genetik der Strahlenarten entspricht. Das Schlußkapitel V schildert das Zusammenwirken der verschiedenen Komponenten nach einer verallgemeinerten Kaskadentheorie. Ein Anhang enthält Ergänzungen, die teils theoretische oder mathematische Einzelfragen, teils experimentelle Methoden zum Gegenstand haben.

Ebenso wie die 1. Auflage ist auch die 2. aus einem Kolloquium des Max Planck-Instituts für Physik (des früheren Kaiser Wilhelm-Instituts für Physik) hervorgegangen. Die einzelnen Abschnitte entsprechen den Kolloquiumsvorträgen, die durch die ihnen folgenden Diskussionen ergänzt und verbessert wurden. Durch die gemeinsamen Diskussionen konnte auch versucht werden, die Ungleichmäßigkeiten, die sich bei der Zusammenarbeit vieler verschiedener Autoren zwangsläufig ergeben, nachträglich wieder auszugleichen. Es handelt sich also ebenso wie früher um eine Gemeinschaftsarbeit des Instituts, an der sich freundlicherweise auch die Herren Prof. HOUTERMANS, Prof. PAUL und Prof. SAUTER vom Physikalischen Institut der Universität Göttingen und Dr. BUDINI vom Physikalischen Institut der Universität Triest beteiligt haben.

Allen Mitarbeitern, insbesondere denen, die die mühevolle Redaktion der einzelnen Kapitel übernommen haben, sei an dieser Stelle noch einmal der Dank für die aufgewendete große Arbeit ausgesprochen; dieser Dank gilt auch besonders den Herren LÜDERS und LÜST, die mich bei den praktischen Redaktionsaufgaben unermüdlich unterstützt haben, sowie dem Verlag für verständnisvolle Zusammenarbeit.

Göttingen, den 10. November 1952.   W. HEISENBERG.

Inhaltsverzeichnis.

Einführung.

I. Die Herkunft der kosmischen Strahlung.

Redaktion: L. Biermann.

II. Nukleonen und π-Mesonen.

Redaktion: W. Heisenberg und K. Wirtz.

III. μ-Mesonen.

Redaktion: K. Wirtz.

Inhaltsverzeichnis.   VII

VIII   Inhaltsverzeichnis.

Übersicht über den heutigen Stand
der Kenntnisse von der kosmischen Strahlung.

Von W. Heisenberg.

Seit den Versuchen von Hess [*He 13*] und Kolhörster [*Ko 13*] ist bekannt, daß eine vom Weltraum auf die Erde einfallende Strahlung in der Atmosphäre verwickelte Sekundärerscheinungen auslöst; diese äußern sich im Auftreten einer allgemeinen Ionisation, hervorgerufen durch Elementarteilchen verschiedener Art (Elektronen, Lichtquanten, μ-, π- und andere Mesonen, Protonen und Neutronen) und werden schließlich bei tieferem Eindringen in die Atmosphäre oder in die Erde immer schwächer, bis sie nach etwa 1000 m Wassertiefe praktisch völlig verschwinden. Über den Ursprung dieser Strahlung konnte man noch vor wenigen Jahren kaum begründete Aussagen machen. Da man aber seit 1946 aus den Versuchen von Forbush [*Fo 46*] und Ehmert [*Eh 48*] weiß, daß auch die Sonne von einzelnen Punkten ihrer Oberfläche gelegentlich kosmische Strahlung emittiert, da man ferner durch die Untersuchungen von Freier, Lofgren, Ney, Oppenheimer, Bradt und Peters [*Fr 48a, b*] erfahren hat, daß die Strahlung aus schnellbewegten Atomkernen besteht, in ähnlicher Elementverteilung wie bei der gewöhnlichen Sternmaterie, kann man kaum daran zweifeln, daß die kosmische Strahlung zu einem großen Teil einfach von den Sternen ausgesandt wird.

a) Genetik der kosmischen Strahlung.

In den turbulenten Vorgängen des Plasmas der Sternoberflächen können sich offenbar gelegentlich für kurze Zeit riesige elektrische Kraftfelder (vgl. I, 4) ausbilden und zwar unter Umständen auch in Gebieten, in denen die Materiedichte gerade sehr gering ist. In solchen Gebieten werden dann alle geladenen Teilchen zu sehr hohen Energien beschleunigt. Schnelle Elektronen verlieren ihre Energie später verhältnismäßig rasch wieder durch Bremsstrahlung. Die schnellbewegten schweren Teilchen aber, unter denen die Protonen am häufigsten sind und ganz allgemein die Ionen der verschiedenen Atomsorten etwa nach dem Häufigkeitsverhältnis der Elemente in der Sternmaterie vorkommen, verlassen den Bereich des Sterns und beschreiben in den turbulenten Magnetfeldern des interstellaren Raums im galaktischen System vielfach verschlungene Bahnen, bis sie endlich auf ein Hindernis auftreffen und dabei absorbiert werden, oder — wohl viel seltener — das galaktische System verlassen. Richtmyer und Teller [*Ri 49*] haben die spezielle Hypothese untersucht, daß die auf der Erde beobachtete kosmische Strahlung zum größten Teil von der Sonne stamme, daß also die von der Sonne emittierten Teilchen durch ein im Gebiet unseres Planetensystems liegendes Magnetfeld immer wieder in dieses Gebiet zurückgeführt würden. Dazu wären jedoch recht große Magnetfelder sehr spezieller Struktur notwendig, und es ist daher wahrscheinlicher, daß die kosmische Strahlung ein Phänomen unseres Milchstraßensystems, nicht unseres Planetensystems ist, wobei die Sonne offenbar viel weniger kosmische Strahlung emittiert als viele andere Sterne. UNSÖLD [*Un 49b, 51*] hat auf die Ähnlichkeit hingewiesen, die in dieser Beziehung zwischen der Emissionsstärke der Sterne für kosmische Strahlung und für Ultrakurzwellenstrahlung (Radiowellen mit einer Wellenlänge in der Größenordnung 1 m) besteht, die beide wahrscheinlich von Vorgängen der gleichen Art herrühren. FERMI [*Fe 49a*] hat gezeigt, daß die geladenen Teilchen im interstellaren Raum durch die zeitlich veränderlichen Magnetfelder weiter beschleunigt werden können; es ist aber noch unsicher, wie wichtig dieser Vorgang für die Energiebilanz der kosmischen Strahlung ist. Jedenfalls bildet sich schließlich durch die vielfache Umlenkung der geladenen Teilchen in den statistisch verteilten Magnetfeldern des galaktischen Raumes eine weitgehend isotrope Richtungsverteilung der kosmischen Strahlung aus; nur in der Nähe der emittierenden Sterne oder bei sehr hohen Energien mögen starke Abweichungen von der Isotropie auftreten.

Auf die Atmosphäre der Erde treffen also von außen vor allem Protonen, und in geringerer Zahl Atomkerne und positive Ionen schwererer Elemente (vgl. I, 2). Die Energieverteilung der Strahlung, die in irgendeiner Weise durch den Entstehungsprozeß auf den verschiedenen Sternen bestimmt wird, ist am oberen Rand der Atmosphäre nicht mehr die gleiche wie im Weltraum (vgl. I, 1). Denn Teilchen kleiner Energie werden vom erdmagnetischen Feld abgelenkt und können unter Umständen die Erdatmosphäre nicht mehr erreichen. Die untere Grenze der Teilchenenergie hängt dabei außer von der Teilchensorte noch von der geographischen Breite und der Richtung der Teilchen ab. In mittleren Breiten liegt sie für Protonen in der Größenordnung $3 \cdot 10^9$ eV. Für höhere Energien entspricht die Energieverteilung jener im interstellaren Raum. Man kann die letztere näherungsweise durch ein Potenzgesetz darstellen; d.h. die Anzahl der Protonen einer Energie $>E$ ist ungefähr durch

$$N(E) = \text{const } E^{-\gamma} \tag{1}$$

gegeben. Allerdings variiert nach den Experimenten der Exponent γ noch langsam mit der Energie. Bis zu etwa 10^{10} eV gibt $\gamma \approx 1$ eine brauchbare Annährung, für größere Energien hat man $\gamma \approx 1,8$ gemessen, oberhalb von 10^{15} eV dürfte auch dieser Wert noch zu klein sein.

Die aus dem Kosmos kommenden Protonen und schwereren Atomkerne treffen in der Erdatmosphäre auf Stickstoff- und Sauerstoffatomkerne. Wenn die Energie der ankommenden Protonen noch relativ niedrig ist ($\lesssim 10^{10}$ eV), schlagen sie aus dem getroffenen Atomkern zunächst nur einzelne Nukleonen heraus, und zwar um so mehr, je höher die Energie ist; außerdem erwärmen sie dabei den Atomkern, so daß von der Oberfläche des Atomkerns noch einige langsame Nukleonen wegverdampfen (vgl. II, 1, 2, 9 und 10).

Bei hoher Primärenergie spielt jedoch die Bremsstrahlung der Nukleonen eine entscheidende Rolle. Diese Bremsstrahlung ist beim energiereichen Stoß zweier Nukleonen nicht, wie bei den Elektronen, γ-Strahlung, sondern vor allem Mesonenstrahlung. Die γ-Strahlung spielt nur eine ganz untergeordnete Rolle. Der Hauptteil der Bremsstrahlung besteht für Energien bis zu etwa 10^{11} eV aus den geladenen und neutralen π-Mesonen (vgl. II, 2 und 7). Bei höheren Energien kommt noch ein steigender Anteil von anderen Mesonenarten und wohl auch von Nukleon-Antinukleonpaaren hinzu (vgl. II, 12).

Die geladenen π-Mesonen zerfallen in der Atmosphäre nach einer mittleren Lebensdauer von etwa $2,5 \cdot 10^{-8}$ sec radioaktiv in ein μ-Meson gleicher Ladung und ein neutrales Teilchen ohne Ruhmasse, das Neutrino (vgl. II, 5). Die μ-Mesonen (vgl. III) stellen die durchdringende Komponente der kosmischen Strahlung dar; ihre Wechselwirkung mit anderen Elementarteilchen ist so klein, daß sie einen Teil der Energie der kosmischen Strahlung bis zur Erdoberfläche und mit abnehmender Intensität noch weiter bis zu 1000 m Wassertiefe tragen können.

Die neutralen π-Mesonen zerfallen nach einer mittleren Lebensdauer von etwa $6 \cdot 10^{-15}$ sec radioaktiv in je zwei γ-Quanten, die nun ihrerseits Elektron-Lichtquantenkaskaden, d. h. die sog. weiche Komponente der kosmischen Strahlung auslösen (vgl. II, 6, IV und V, 4 bis 8).

Das μ-Meson zerfällt nach einer mittleren Lebensdauer von $2,15 \cdot 10^{-6}$ sec in ein Elektron und zwei Neutrinos. Außerdem kann es durch Stöße energiereiche Elektronen und Lichtquanten hervorbringen. Die durchdringende Komponente erzeugt daher als Sekundärstrahlung ebenfalls einen Teil der weichen Komponente aus Elektronen und γ-Quanten (vgl. III und V, 5).

Man teilt ganz allgemein die kosmische Strahlung in der Atmosphäre zweckmäßig in folgende Komponenten ein: Die Nukleonen- und

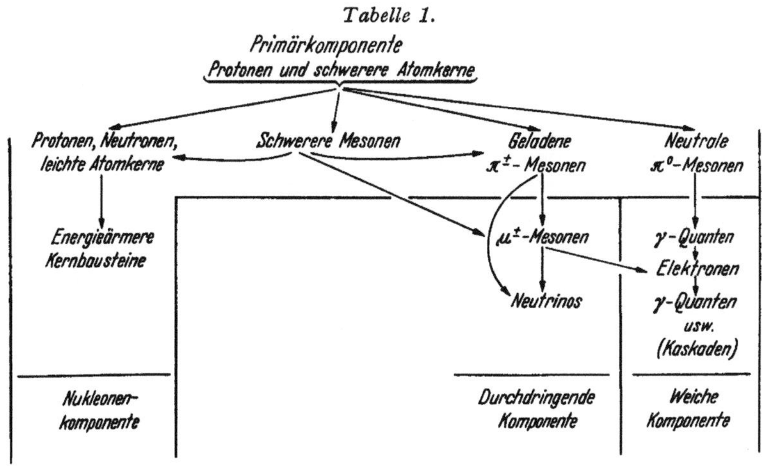
Tabelle 1.

4   Allgemeine Übersicht.

π-Mesonenkomponente, die in der Atmosphäre schnell absorbiert wird (zu dieser Komponente gehören auch die schwereren Mesonenarten), die von den μ-Mesonen gebildete durchdringende Komponente, und schließlich die weiche Komponente, die aus Kaskaden von Elektronen und γ-Quanten besteht.

Die Genetik der verschiedenen Strahlenarten ist schematisch in der Tabelle 1 dargestellt, die Intensität der verschiedenen Komponenten als Funktion der Höhe oder richtiger der durchlaufenen Materiemenge ist für eine geomagnetische Breite von etwa 50° auf Grund des vorliegenden experimentellen Materials in Abb. 1 ohne großen Genauigkeitsanspruch wiedergegeben.

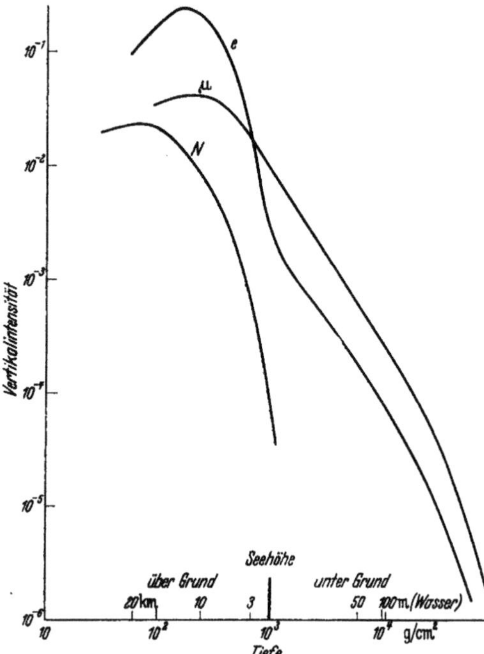

Abb. 1. Qualitative Darstellung der Höhenabhängigkeit der verschiedenen Komponenten der kosmischen Strahlung für rund 50° geomagnetische Breite. (N Nukleonenkomponente; μ durchdringende Komponente; e weiche Komponente.)

b) Die Nukleonenkomponente und ihre Sekundärstrahlung.

Die Nukleonenkomponente besteht in sehr großer Höhe (Materiemenge über dem Beobachtungsort $< 20\,\mathrm{g/cm^2}$) zum größten Teil aus der Primärstrahlung; mit tieferem Eindringen in die Atmosphäre kommen in steigendem Maße die Bruchstücke (Protonen, Neutronen, α-Teilchen usw.) der von der Primärstrahlung getroffenen Stickstoff- und Sauerstoffkerne hinzu. Der Wirkungsquerschnitt für den Stoß zweier Atomkerne ist im wesentlichen durch den geometrischen Querschnitt der beiden Kerne bestimmt. Die mittlere Absorptionsschicht beträgt daher für Protonen in Luft etwa 60 g/cm², für schwerere Atomkerne entsprechend weniger. Die Zahl der schwereren Atomkerne nimmt folglich beim Eindringen in die Atmosphäre sehr rasch ab; schon bei einem Druck von 100 g/cm² ist ihr Anteil relativ zur Nukleonenkomponente sehr klein geworden. In dieser Höhe ist andererseits die Anzahl der (sekundären!) Neutronen schon der Zahl der Protonen vergleichbar. Die Nukleonenkomponente nimmt mit zunehmender Tiefe weniger rasch ab als die Primärstrahlung, ihre Intensität sinkt etwa nach Durchqueren von je 125 g/cm² auf den e-ten Teil. Da man in größerer Tiefe in der Hauptsache nicht mehr die primären, sondern die sekundären, tertiären usw. Kernbruchstücke beobachtet, verschiebt sich die Energieverteilung der

Nukleonen mit zunehmender Tiefe nach geringeren Energien. Protonen geringer Energie (< 100 MeV) werden dann durch Ionisationsbremsung schnell zur Ruhe gebracht und scheiden damit aus der Strahlung aus. Bei geringeren Energien gibt es daher in der kosmischen Strahlung viel mehr Neutronen als Protonen. Die Neutronen werden weiter durch elastische Zusammenstöße mit den Stickstoff- und Sauerstoffkernen bis fast zu thermischen Geschwindigkeiten verlangsamt und in diesem Energiegebiet von den Stickstoffkernen schließlich eingefangen.

Beim Stoß energiereicher Nukleonen entstehen als Bremsstrahlung verschiedene Mesonenarten, am häufigsten die zuerst von POWELL und seinen Mitarbeitern [*La 47a, 47b*] beobachteten π-Mesonen. Diese Teilchen kommen als geladene (Masse etwa 276 Elektronenmassen) und als neutrale (Masse etwa 265 Elektronenmassen) vor und können offenbar bei einem einzelnen Stoß bereits in größerer Zahl [*He 49b*] (vgl. auch [*He 39a*]) emittiert werden. Die Zahl der beim Zusammenstoß zweier Nukleonen emittierten Mesonen nimmt mit der im Schwerpunktsystem zur Verfügung stehenden Energie zu, hängt aber außerdem offenbar noch stark vom Stoßabstand ab. Bei größeren Energien ($> 10^{10}$ eV im Schwerpunktsystem) können neben den π-Mesonen noch schwerere Elementarteilchen erzeugt werden. Man hat in den letzten Jahren eine ganze Reihe solcher Teilchen gefunden, die man nach ihren Eigenschaften als V-, τ-, ζ-, χ- und $\varkappa$-Mesonen unterscheidet, und wahrscheinlich werden in der Folgezeit noch weitere Arten entdeckt werden. Alle beim Stoß unmittelbar erzeugten Mesonen stehen mit den Nukleonen in starker Wechselwirkung. Sie verursachen daher, wenn sie nicht vorher schon zerfallen, beim Auftreffen auf Atomkerne die gleichen Prozesse wie Nukleonen, auch ihr Wirkungsquerschnitt für den Stoß mit Atomkernen ist von dem der Nukleonen nicht sehr verschieden. Die Bremsstrahlung der Nukleonen steht daher mit dem energiereichen Teil der Nukleonenkomponente im Gleichgewicht und nimmt beim Durchgang durch die Atmosphäre nach dem gleichen Gesetz ab wie dieser. Auch die Energieverteilung der π-Mesonen dürfte bei höheren Energien der der Nukleonen und damit bei sehr hohen Energien der der Primärkomponente entsprechen; sie wird also auch näherungsweise durch ein Potenzgesetz von der Form (1) mit dem gleichen Exponenten γ gegeben sein, der in einem weiten Energiebereich ungefähr den Wert 1,8 hat.

c) Die durchdringende Komponente.

Ein geladenes π-Meson zerfällt nach einer (im Ruhsystem des Mesons gemessenen) mittleren Lebensdauer von $2{,}5 \cdot 10^{-8}$ sec in ein μ-Meson und (wahrscheinlich) ein Neutrino. Die μ-Mesonen entstehen daher hauptsächlich in einer Höhe von etwa 20 km aus den in der hohen Atmosphäre gebildeten π-Mesonen. Die Masse des μ-Mesons ist etwa 210mal größer als die des Elektrons. Im Ruhsystem des π-Mesons hat das beim Zerfall entstehende μ-Meson eine kinetische Energie von etwa 4,1 MeV. Ein schnellbewegtes π-Meson erzeugt also beim Zerfall ein schnellbewegtes μ-Meson,

6   Allgemeine Übersicht.

dessen Energie sich zu der des π-Mesons ungefähr verhält wie die Masse des μ-Mesons zu der des π-Mesons. Das Spektrum der μ-Mesonen entspricht daher in einem weiten Energiegebiet dem der π-Mesonen, fällt also ebenfalls in diesem Gebiet etwa wie $E^{-1,8}$ ab. Nur π-Mesonen sehr hoher Energie ($\gtrsim 10^{11}$ eV) durchlaufen wegen der relativistischen Zeitdilatation eine so große Wegstrecke bis zum Zerfall, daß sie meist in der Atmosphäre einen Atomkern treffen, bevor sie zerfallen können. Oberhalb von 10^{11} eV nimmt das μ-Mesonenspektrum daher schneller ab als das π-Mesonenspektrum.

Das μ-Meson kann Wirkungen auf andere Elementarteilchen praktisch nur durch elektrische Kräfte, d. h. vermöge seiner Ladung, ausüben. Es verliert daher beim Durchgang durch Materie fast nur durch Ionisationsbremsung an Energie und kann, wenn seine Anfangsenergie hoch genug ist, sehr große Materieschichten durchsetzen. μ-Mesonen einer Energie von 10^{11} eV können schon etwa 500 m Wasser durchdringen. Daher bilden die μ-Mesonen die sog. durchdringende Komponente, deren Intensität mit der Tiefe nur langsam abnimmt. In der Art des Abfalls spiegelt sich das Mesonenspektrum, daher sinkt die Intensität als Funktion der Tiefe T bis zu einigen 100 m Wasser wie $T^{-1,8}$ und von da noch steiler ab.

Bei geringeren Energien spielt auch der radioaktive Zerfall der μ-Mesonen für die Intensitätsabnahme der durchdringenden Komponente eine Rolle. Der recht genau gemessenen Lebensdauer von $2,15 \cdot 10^{-6}$ sec entspricht bei einem Meson der Energie 10^9 eV eine Wegstrecke von 6 km, zur Energie 10^{11} eV gehört aber schon ein Zerfallsweg von 600 km. Nur unterhalb von 10^{10} eV kann sich daher der Zerfall in der Intensitätsabnahme bemerkbar machen.

Da das μ-Meson in ein Elektron und zwei Neutrinos zerfällt, liefert der Zerfall einen Beitrag zur weichen, aus Elektronen und Lichtquanten gebildeten Komponente und zur Neutrinokomponente. Die letztere muß im Prinzip zur durchdringenden Komponente gerechnet werden; die Wechselwirkung der Neutrinos mit anderen Elementarteilchen ist aber so gering, daß sie praktisch nicht beobachtet werden können. Die Intensität der direkt von der Sonne kommenden Neutrinostrahlung mag sogar außerordentlich hoch sein, zu beobachten ist sie einstweilen nicht.

d) Die weiche Komponente.

Die in der hohen Atmosphäre gebildeten neutralen π-Mesonen zerfallen nach einer (im Ruhsystem des Mesons gemessenen) mittleren Lebensdauer von etwa $6 \cdot 10^{-15}$ sec in je zwei γ-Quanten und erzeugen damit — wieder hauptsächlich in etwa 20 km Höhe — den Anfang der aus Lichtquanten und Elektronen bestehenden weichen Komponente. Die Energieverteilung der γ-Quanten entspricht wieder der der neutralen π-Mesonen und folgt daher in einem weiten Bereich dem $E^{-1,8}$-Gesetz. Da jedes γ-Quant beim Vorbeigang an einem Atomkern ein Elektronenpaar, und jedes Elektron bei der Ablenkung durch einen Kern wieder ein γ-Quant erzeugen kann, teilt sich die Primärenergie nach kurzer Wegstrecke in viele Elektronen und γ-Quanten auf, was zur Bildung der sog. weichen Kaskadenschauer führt. Je nach der Energie des einzelnen γ-Quants schließt sich also beim Eindringen in die tiefere Atmosphäre eine größere oder kleinere Kaskade von Elektronen und Lichtquanten an. Ein Paar von γ-Quanten sehr hoher Energie ($> 10^{13}$ eV) erzeugt eine Kaskade, die die Atmosphäre teilweise oder ganz durchdringt und in der Gegend des Maximums aus vielen Tausenden von Elektronen und Lichtquanten besteht. Häufig werden sich auch die von den verschiedenen neutralen π-Mesonen eines oder mehrerer aufeinanderfolgender Nukleonenstöße gebildeten Kaskaden zu einem großen Bündel vereinigen. In dieser Weise kommen die über einige hundert Quadratmeter ausgedehnten „Luftschauer" zustande, die zuerst von AUGER [*Au 38*] in etwa 3000 m Höhe beobachtet worden sind. Diese Luftschauer enthalten auch einen kleinen Anteil durchdringender Partikeln, die wohl indirekt noch von den Nukleonenstößen herrühren, die die neutralen π-Mesonen und damit die Kaskaden erzeugt hatten.

Die Gesamtintensität der weichen Komponente nimmt mit der Tiefe rasch ab, sie sinkt etwa mit je 100 g/cm² durchquerter Luftschicht auf den e-ten Teil. Ihr Abfall wird allerdings in tieferen Schichten wieder dadurch verzögert, daß hier durch den Zerfall der μ-Mesonen und durch Stoßionisation und Ausstrahlung dieser Mesonen Energie von der durchdringenden Komponente in die weiche übergeht. Schließlich stellt sich in größerer Tiefe ein Gleichgewicht zwischen der durchdringenden Komponente und der von ihr erzeugten weichen Komponente her.

Da das Verhalten der Elektronen und Lichtquanten durch die Quantenelektrodynamik mathematisch genau dargestellt wird, läßt sich die weiche Komponente theoretisch viel genauer verfolgen als die übrigen Komponenten. Insbesondere liegen über die Ausbildung der Kaskaden, ihre räumliche Ausbreitung, Energie- und Winkelverteilung sehr ausführliche mathematische Untersuchungen vor, die einen eingehenden Vergleich mit der Erfahrung zulassen.

Die kosmische Strahlung erweckt heutzutage vor allem aus zwei Gründen Interesse: Sie erlaubt, mit Elementarteilchen höchster Energie zu experimentieren und bringt Kunde von Vorgängen auf den Sternen und im interstellaren Raum. Das Interesse stammt also zum Teil aus der Atomphysik, zum Teil aus der Astronomie.

Die Atomphysik hat die atomaren Vorgänge bei niedrigen Energien durchforscht, und die eigentliche terra incognita ist heute die Physik der Elementarteilchen selbst. Um hier vorwärts zu kommen, muß man mit Teilchenenergien experimentieren, die groß gegen die Ruhmasse der schwersten Elementarteilchen sind, und das ist mit technischen Mitteln einstweilen noch nicht möglich; nur in der kosmischen Strahlung findet man Teilchen so hoher Energie in der Natur vor. Daher sind auch in den letzten 20 Jahren viele der wichtigsten Fortschritte in der Physik der Elementarteilchen mit Hilfe der kosmischen Strahlung erzielt worden.

                              Allgemeine Übersicht.

Für den Astrophysiker andererseits bringt die kosmische Strahlung
Informationen einer ganz neuen Art. Viele Jahrhunderte hindurch haben
die Astronomen nur durch die Lichtstrahlen Kunde von den Sternen
erhalten. Auch wenn im Lauf der Zeit die spektrale Zerlegung des Lichtes
oder der Bau immer besserer optischer Instrumente erstaunlich viele
Einzelheiten über die Vorgänge auf den Sternen vermittelt hat, muß die
Beobachtung einer neuen Strahlenart, die von den Sternen auf die Erde
kommt, die Kenntnisse von den fernen Weltkörpern und dem Raum
zwischen ihnen wesentlich fördern. Daher ist auch die Hydrodynamik
bewegter ionisierter Gase, die man zum Studium der Vorgänge auf den
Sternoberflächen oder im interstellaren Raum braucht, in den letzten
Jahren im Zusammenhang mit der Theorie der Entstehung der kosmischen Strahlung weiter entwickelt worden.

4. Die mittlere Lebensdauer τ bezüglich des Übergangs in zwei γ-Quanten ist $< 5 \cdot 10^{-14}$ sec (wohl etwa $6 \cdot 10^{-15}$ sec s. u.).

Wie schon erwähnt ist neben dem Übergang des π°-Mesons in zwei γ-Quanten auch noch der Prozeß $\pi^\circ \to e^+ + e^- + \gamma$ zu erwarten, allerdings mit einer etwa hundertmal geringeren Wahrscheinlichkeit. Neuere Messungen von Energie und Öffnungswinkel der Elektron-Positronpaare, die in der Platte in unmittelbarer Nähe von Sternen entstehen, machen es sehr wahrscheinlich, daß dieser Prozeß wirklich vorkommt [*Da 52b*, *An 52c*]. Auch neue Experimente mit künstlichen Mesonen deuten darauf hin [*An 52d*]. Die Ausmessung von 62 derartigen „direkten Paaren" liefert $\tau = (6 \, {}^{+6}_{-3}) \cdot 10^{-15}$ sec [*An 53*].

7. Theorie der Explosionsschauer.

Von W. Heisenberg.

Schon im Abschn. 2 und 3 dieses Kapitels wurde auseinandergesetzt, daß beim Stoß eines Nukleons auf einen Kern häufig als Bremsstrahlung ein Bündel von Mesonen (bei nicht zu hoher Energie in erster Linie π-Mesonen) emittiert wird. Bei der Deutung solcher Prozesse haben sich lange Zeit zwei Auffassungen gegenübergestanden, die man (in etwas unklarer Bezeichnung) als Theorie der „Mehrfach"- und „Vielfach"-Erzeugung unterscheidet. Nach der ersteren Theorie [*He 42, 49f, 50b, Ja 43b*] entsteht beim Stoß zweier Nukleonen in der Regel nur ein einziges Meson, die beobachtete größere Zahl kommt durch hintereinander ablaufende Stoßprozesse im gleichen Atomkern (Kaskadenbildung) zustande. Nach der älteren zweiten Theorie [*He 36, 39a, 43b, 49a, 49b, Le 48, Fe 50a, 51b, Wa 48b*] entstehen beim energiereichen Stoß zweier Nukleonen in der Regel viele Mesonen; die in einem schwereren Atomkern nachfolgende Kaskadenbildung vermehrt die Mesonenzahl nur um einen Faktor der Größenordnung 1 (nach halbempirischen Abschätzungen beträgt der Faktor für Schauer mittlerer Größe bei Kohle etwa 1,2, bei Blei etwa 2 bis 3 [vgl. *He 49c*]).

Die experimentellen Ergebnisse der letzten Jahre haben die zweite Auffassung sehr wahrscheinlich gemacht, und zwar aus folgenden Gründen: 1. Da bei „Mehrfacherzeugung" im Kern mindestens ebenso viele energiereiche Nukleonenstöße stattgefunden haben müssen, wie Mesonen auftreten, sollte die Zahl der grauen Spuren (Protonen höherer Energie) proportional zur Zahl der Mesonen anwachsen. Tatsächlich wächst die Zahl der grauen Spuren aber schwächer an, so wie nach der Theorie der Vielfacherzeugung zu erwarten ist (vgl. Abschn. 2c, Abb. 6). 2. Die Häufigkeit der Schauer als Funktion der Mesonenzahl sollte bei Mehrfacherzeugung in leichten Kernen schneller abfallen als in schweren. Tatsächlich ist die Häufigkeitsverteilung der Schauer in leichtem und schwerem Material gerade bei großen Schauern nicht allzu verschieden (vgl. Abschn. 2c, Abb. 13). 3. Es sind verschiedentlich Schauer beobachtet worden, bei denen neben einer größeren Zahl von dünnen Spuren weder graue noch schwarze Spuren auftreten. Hier kann es sich nur um Ereignisse handeln, bei denen das ankommende Nukleon

7. Theorie der Explosionsschauer.

149

entweder einen Wasserstoffkern getroffen hat, oder auf ein Nukleon am Rande eines größeren Kerns gestoßen ist, wobei sich dieser Kern nicht weiter am Stoßvorgang beteiligt hat. In die letztere Gruppe von Prozessen gehören auch die Schauer, bei denen neben einer größeren Mesonenzahl nur ganz wenige schwarze Spuren auftreten; der Kern ist dabei offenbar nur wenig erhitzt worden, so daß durch nachfolgende

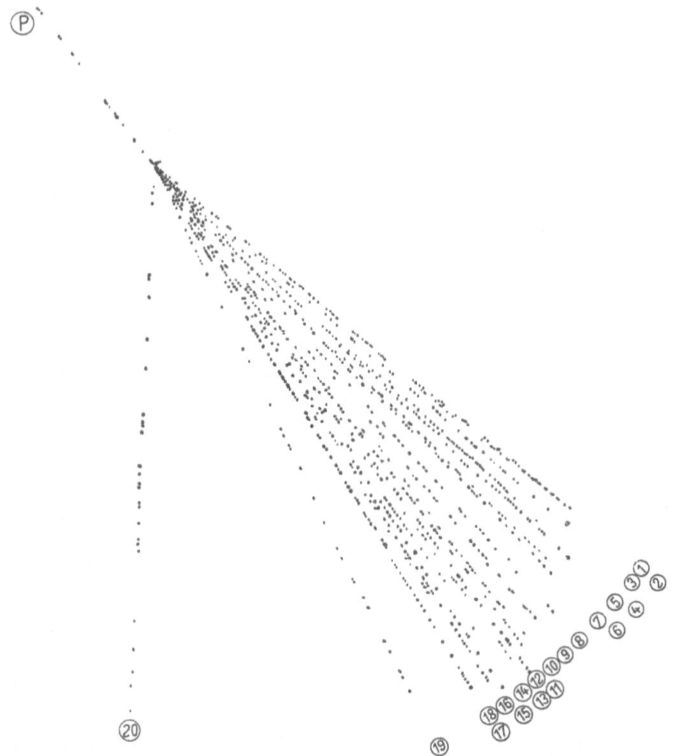

Abb. 1. Ein Proton von etwa 40 GeV erzeugt einen Mesonenschauer (wahrscheinlich $18\pi^{\pm}$-Mesonen und 2 Protonen, ferner etwa $9\pi^{\circ}$-Mesonen) (Aufnahme aus [*Te 52 b*]).

Verdampfung nur einige langsame Protonen emittiert wurden. Auch die Beobachtung von durchdringenden Schauern aus Wasserstoff [*Vi 51b*] oder der Vergleich zwischen Paraffin und Kohle (O. HAXEL, unveröffentlicht), vgl. Abschn. 3 insbesondere Abb. 3, beweist das Vorkommen der Vielfacherzeugung.

Die Abb. 1—4 zeigen einige typische Fälle von Vielfacherzeugung. Die Energie des Primärteilchens kann bei solchen Vorgängen nach einem Vorschlag von FERMI [*Fe 50a, 51b*] leicht aus der Winkelverteilung der Mesonen abgeschätzt werden. Im Schwerpunktsystem (S-System) müssen die Mesonen ungefähr symmetrisch um die Mittelebene verteilt sein. Bei der LORENTZ-Transformation in das System des Laboratoriums (L-System) geht diese Mittelebene in einen Kegel über, der also etwa

150   II. Nukleonen und π-Mesonen.

die Hälfte der vorhandenen Mesonen enthält und damit experimentell leicht zu bestimmen ist. Der Öffnungswinkel ϑ des Kegels hängt dann, wenn die Mesonen im Schwerpunktsystem nahezu mit Lichtgeschwindigkeit emittiert werden, mit der Geschwindigkeit β des Schwerpunktsystems und der kinetischen Energie E des Primärteilchens nach den Formeln zusammen (M Nukleonenmasse):

$$\left.\begin{array}{c} \sin\vartheta = \sqrt{1-\beta^2} \\[2mm] \text{und} \qquad E = 2Mc^2\left(\frac{1}{\sin^2\vartheta} - 1\right). \end{array}\right\} \tag{1}$$

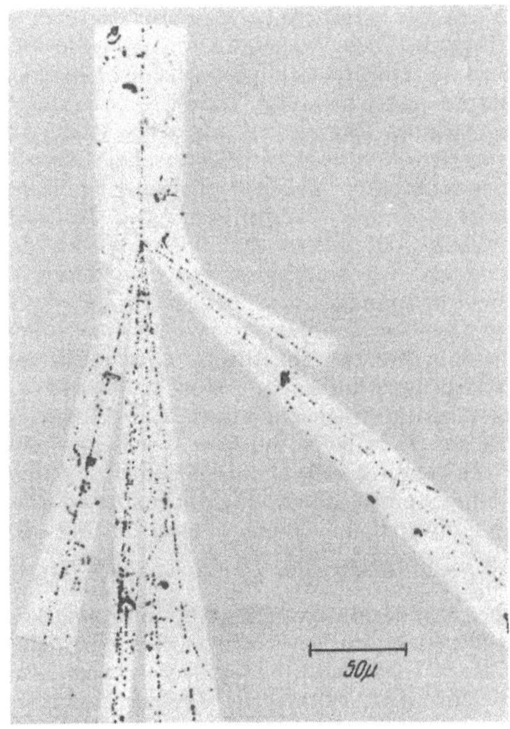

Abb. 2. Primärproton von etwa 2000 GeV, 7 geladene Teilchen (Aufnahme aus [*Ca 50 a*]).

Abb. 3. Primärteilchen von etwa 20 GeV, 9 geladene Teilchen (Aufnahme M.P.I. f. Physik).

Allerdings trifft die Voraussetzung über die Geschwindigkeit der Mesonen im allgemeinen nicht zu. Zur genaueren Bestimmung der Primärenergie muß man daher andere Verfahren verwenden, die im Anh. 7d beschrieben sind. Die rechte Seite der ersten Gl. (1) kann sich dabei um einen Faktor zwischen 0,5 und 1 ändern.

Bei dem Vorgang von Abb. 1 z. B. stößt ein Proton von etwa 40 GeV auf ein anderes Proton und erzeugt dabei 18 geladene und wohl noch etwa 9 neutrale Mesonen, die in einem Bündel von etwa 20° Öffnung emittiert werden. ϑ beträgt ungefähr 10°. Bei dem von SCHEIN [*Sche 51*] beobachteten Vorgang von Abb. 4 ist die Primärenergie noch sehr viel größer, wie aus dem sehr kleinen Öffnungswinkel hervorgeht, jedoch die Mesonenzahl geringer. Man erkennt schon aus den wenigen Beispielen der Abb. 1—4, daß offenbar kein eindeutiger Zusammenhang zwischen Primärenergie und Teilchenzahl besteht.

a) Vielfacherzeugung als Folge starker Wechselwirkung.

α) Die Wechselwirkung zwischen Teilchen bzw. den ihnen zugeordneten Wellenfeldern kann grundsätzlich einer der beiden folgenden Arten angehören (vgl. [*He 36*]). Wir wollen von einer Wechselwirkung der ersten Art sprechen, wenn sie mit zunehmender Energie der beteiligten Teilchen entweder abnimmt oder konstant bleibt, von einer Wechselwirkung der zweiten Art, wenn sie mit der Energie der beteiligten Teilchen zunimmt. Benützt man natürliche Einheiten ($\hbar = c = 1$), so hat der Kopplungsparameter, nach dem man in der Störungstheorie entwickelt, bei Wechselwirkungen der ersten Art die Dimension einer reinen Zahl oder einer negativen Potenz einer Länge, bei einer Wechselwirkung der zweiten Art die Dimension einer positiven Potenz einer Länge. Die Quantenelektrodynamik ist eine Theorie der ersten Art, der Kopplungsparameter ist die Zahl $\dfrac{e}{\sqrt{\hbar c}}$; auch die normale pseudoskalare Kopplung zwischen π-Mesonen und Nukleonen, ferner der nichtlineare Term mit φ^3 in der Mesonenwellengleichung ist erster Art. Dagegen sind die pseudovektorielle Kopplung der π-Mesonen mit den Nukleonen, oder die FERMISsche Kopplung für die Theorie des β-Zerfalls Wechselwirkungen zweiter Art. Treibt man Störungstheorie und entwickelt nach Potenzen des Kopplungsparameters, so erhält man Matrixelemente für die gleichzeitige Erzeugung von n Teilchen im allgemeinen erst in der n-ten Näherung. Sofern die Störungstheorie konvergiert, wird die Wahrscheinlichkeit für die Erzeugung von vielen Teilchen daher stets sehr klein. Sie kann also bei Theorien der ersten Art auch für hohe Primärenergie stets klein bleiben. Bei Theorien der zweiten Art

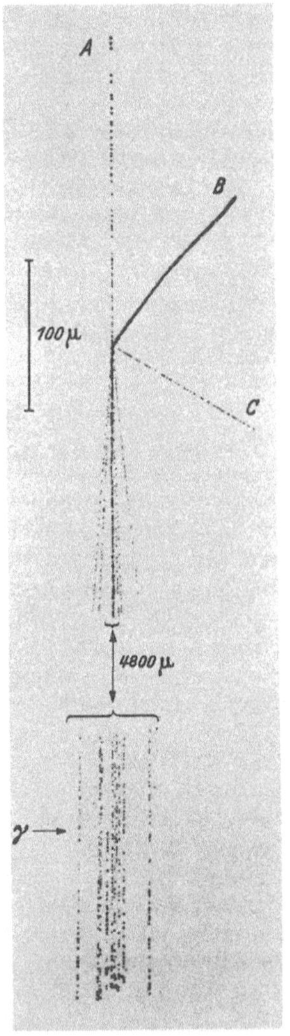

Abb. 4. Primärproton von etwa 30000 GeV, 15 geladene Teilchen (Aufnahme aus [*Sche 51*]).

                    II. Nukleonen und π-Mesonen.

dagegen divergiert die Störungstheorie bei hinreichend hoher Energie der beteiligten Teilchen; hier ist also mit der Erzeugung vieler Teilchen in einem Akt zu rechnen.

Die Unterscheidung zwischen den beiden Arten der Wechselwirkung ist in der mathematischen Behandlung der Quantentheorie der Wellenfelder neuerdings prinzipiell wichtig geworden. Theorien der ersten Art können durch Renormalisierung zu endlichen und physikalisch sinnvollen Resultaten führen. Theorien der zweiten Art lassen sich nicht renormalisieren und müssen im Sinne einer „nichtlokalisierbaren" Wechselwirkung verändert werden, um einen physikalischen Sinn zu erhalten.

Die Tatsache, daß häufig viele Mesonen in einem einzigen Akt erzeugt werden, macht also wahrscheinlich, daß für die Mesonen eine Wechselwirkung der zweiten Art existiert. Über ihre spezielle Form ist einstweilen nichts bekannt.

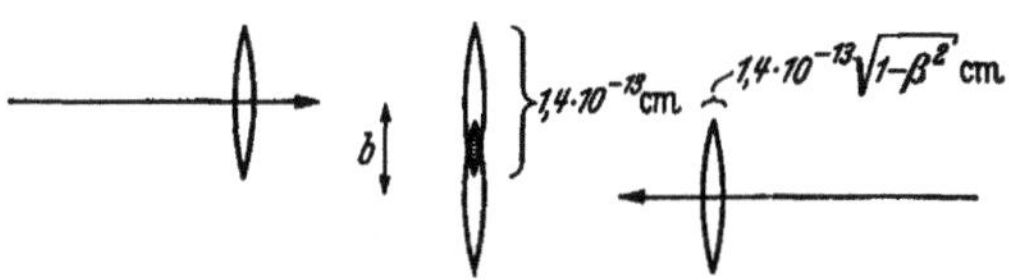

Abb. 5. Zusammenstoß der Nukleonen (schematisch).

Da es sich bei der Erzeugung vieler Mesonen um einen Vorgang handelt, bei dem wegen der hohen Quantenzahl die quantentheoretischen Züge des Mesonenfeldes nur eine untergeordnete Rolle spielen, kann man den Vorgang zunächst ganz anschaulich beschreiben. Im Schwerpunktsystem erscheinen die beiden aufeinanderzufliegenden Nukleonen, die wir uns ursprünglich als Kugeln von $1{,}4 \cdot 10^{-13}$ cm Durchmesser vorstellen wollen, als flache Scheiben vom gleichen Durchmesser, deren Dicke (durch Lorentz-Kontraktion) um den Faktor $\sqrt{1-\beta^2}$ geringer ist als ihr Durchmesser (Abb. 5). Im Moment des Zusammenstoßes überlappen diese Scheiben in einem gewissen Gebiet, dessen Größe vom Stoßabstand b abhängt. In diesem Gebiet entsteht eine starke Wechselwirkung, die die Nukleonen bremst, so daß im ganzen Gebiet der beiden Scheiben in diesem Moment ein erheblicher Teil der Energie in das Mesonenfeld übergeht und sich von dort — nach kurzer Zeit unabhängig von den beiden Nukleonen — in den Raum ausbreitet.

Der Anfangszustand des Mesonenfeldes ist also dadurch charakterisiert, daß im Gebiet der beiden Scheiben im Augenblick des Stoßes eine sehr hohe Energiedichte herrscht, während außen das Mesonenfeld verschwindet. Wenn sich das Feld von diesem Augenblick an bereits nach der linearen Wellengleichung

$$\Box\varphi - \varkappa^2\varphi = 0 \tag{2}$$

ausbreitete $\left(\varkappa = \dfrac{m_\pi \cdot c}{\hbar}\right)$, so wäre seine spektrale Zusammensetzung von der Zeit unabhängig und daher identisch mit der im ersten Augenblick. Wie die spätere Rechnung zeigt und wie auch anschaulich wegen der geringen Ausdehnung des Ursprungsgebietes einleuchtet, ist hier die Energie einigermaßen gleichmäßig über alle Wellenzahlen bis zu einer maximalen hin verteilt, die etwa durch die reziproke Dicke

7. Theorie der Explosionsschauer.

153

der Scheiben, d.h. durch

$$p_{0\,\mathrm{max}} \sim \frac{\varkappa}{\sqrt{1-\beta^2}} \qquad (3)$$

gegeben ist. Die Anzahl der Mesonen zwischen der Energie (incl. Ruhenergie) p_0 und $p_0 + dp_0$ beträgt also

$$dn = \mathrm{const}\,\frac{dp_0}{p_0} \; ; \qquad (4)$$

die Anzahl aller Mesonen wächst nur logarithmisch mit der im S-System zur Verfügung stehenden Energie an.

Dies genügt aber kaum, um die tatsächlich beobachteten Mesonenzahlen zu erklären. Vielmehr muß man annehmen, daß die innere Wechselwirkung des Mesonenfeldes (und vielleicht in geringerem Maß die Wechselwirkung mit den Nukleonen) den Ausbreitungsvorgang weiter beeinflußt, so daß er sich nicht einfach nach der Wellengleichung (2) sondern einer nichtlinearen Gleichung vollzieht. Dann verändert sich die spektrale Zusammensetzung noch während des Ausbreitungsvorgangs, die Energie im Spektrum verschiebt sich nach den kleineren Wellenzahlen; sie verteilt sich also nicht mehr auf einige wenige energiereiche, sondern auf viele energiearme Mesonen. Erst dieser Vorgang führt zu hohen Mesonenzahlen, so wie sie z.B. in Abb. 1 beobachtet sind.

Zur Abschätzung der Mesonenzahl sind verschiedene Wege beschritten worden. Grundsätzlich ist eine zuverlässige Berechnung natürlich erst dann möglich, wenn über die Natur der Wechselwirkung Klarheit besteht. Man kann aber hoffen, daß — ähnlich wie in der Wärmelehre oder der Theorie der Turbulenz — die spezielle Form und Größe der Wechselwirkung dann nicht mehr wichtig ist, wenn man weiß, daß es sich um starke Energiedissipation handelt. Auf diesen Grenzfall starker Energiedissipation beziehen sich die folgenden Überlegungen.

β) Zunächst hat man die Wirkung der nichtlinearen Terme mit den entsprechenden Wirkungen in der Theorie der Turbulenz verglichen [*He 49 a, 49 b*]. Da sich dort im Gleichgewicht eine spektrale Verteilung einstellt, die einem Potenzgesetz folgt, lag es nahe, als Endzustand für das ausgesandte Mesonenfeld ebenfalls statt (4) ein Gesetz der Form

$$dn = a\,\frac{dp_0}{p_0^{\alpha+1}} \qquad (5)$$

mit $\alpha > 0$ anzunehmen. Allerdings ist eine so einfache Form des Gesetzes nur für $p_0 \gg \varkappa$ zu erwarten, da ja für $p_0 \approx \varkappa$ die Wechselwirkung nicht mehr „stark", ihre spezielle Form also wichtig werden wird. Schon wegen des Phasenraumvolumens müßte z.B. die rechte Seite von (5) den Faktor p/p_0 enthalten. Auf solche Verbesserungen soll aber hier verzichtet werden, da das Spektrum im Gebiet $p_0 \approx \varkappa$ doch von der speziellen Form der Wechselwirkung abhängt.

Speziell wurde in den erwähnten Arbeiten die Annahme $\alpha = 1$ studiert, die, wie weiter unten besprochen wird, als Grenzgesetz für starke Wechselwirkung eine besondere Bedeutung besitzt. Mit $\alpha = 1$ kann man die Mesonenzahl in folgender Weise abschätzen. Wir nehmen an, daß

154   II. Nukleonen und π-Mesonen.

der Bruchteil γ der maximal im S-System zur Verfügung stehenden Energie ins Mesonenfeld übergeht:

$$\varepsilon = \gamma \left[\sqrt{2M(E+2M)} - 2M \right].$$
(6)

Dann ist offenbar

$$\varepsilon = a \int_{\varkappa}^{p_{0\,max}} \frac{dp_0}{p_0} = a \lg \frac{p_{0\,max}}{\varkappa},$$
(7a)

also

$$a = \frac{\varepsilon}{\lg \dfrac{p_{0\,max}}{\varkappa}}$$
(7b)

und

$$n = a \int_{\varkappa}^{p_{0\,max}} \frac{dp_0}{p_0^2} = \frac{a}{\varkappa} \left(1 - \frac{\varkappa}{p_{0\,max}} \right) = \frac{\varepsilon \left(1 - \dfrac{\varkappa}{p_{0\,max}} \right)}{\varkappa \lg \dfrac{p_{0\,max}}{\varkappa}}.$$
(7c)

Für die mittlere Energie pro Meson im S-System folgt:

$$\bar{p}_0 = \frac{\varepsilon}{n} = \varkappa \frac{\lg \dfrac{p_{0\,max}}{\varkappa}}{1 - \dfrac{\varkappa}{p_{0\,max}}}.$$
(8)

Diese mittlere Energie ist also unabhängig von γ und wächst auch mit der Primärenergie E nur logarithmisch an. Eine etwas genauere Abschätzung von $\bar{p}_0$ ist in [*He 52a*] gegeben worden.

Der Erwartungswert $\bar{n}$ der Mesonenzahl ist dagegen proportional zu γ. Dieser Bruchteil γ wird in irgendeiner Weise mit dem Stoßparameter zusammenhängen, und man kann nur qualitativ voraussagen, daß γ bei streifenden Zusammenstößen klein, bei zentralen Stößen groß sein wird. Wenn man „starke" Wechselwirkung voraussetzt, kann man für zentrale Stöße wohl $\gamma = 1$ ansetzen. Es liegt nahe, zur Abschätzung von γ diese Größe dem Überlappungsintegral der ursprünglichen Mesonenfelder der beiden Nukleonen proportional zu setzen; diese Felder haben die Form $\dfrac{e^{-\varkappa r_1}}{r_1}$ und $\dfrac{e^{-\varkappa r_2}}{r_2}$; der Minimalabstand der Nukleonenmittelpunkte (d.h. der Stoßparameter) sei b; dann ist also

$$\gamma = \text{const} \int \frac{e^{-\varkappa r_1}}{r_1} \frac{e^{-\varkappa r_2}}{r_2} d\mathfrak{r} \quad \text{für } r_{12} = b.$$
(9)

Daraus folgt:

$$\gamma = e^{-b\varkappa}$$
(10)

und der Wirkungsquerschnitt für einen Wert von γ zwischen γ und $\gamma + d\gamma$ wird

$$d\sigma = \frac{2\pi}{\varkappa^2} \frac{d\gamma}{\gamma} (-\lg \gamma).$$
(11)

7. Theorie der Explosionsschauer.

155

Kleine Werte von γ sind also stark bevorzugt. Will man aus (11) einen totalen Wirkungsquerschnitt ableiten, so muß man berücksichtigen, daß zu kleine Werte von γ überhaupt keine Vielfachprozesse mehr ergeben. Man kann etwa fordern, daß mindestens zwei Mesonen entstehen sollen und erhält dann für $p_{0\,\max} \gg \varkappa$ nach (6) und (7)

$$\gamma_{\min} \approx \frac{2\bar{p}_0}{\left[\sqrt{2M(E+2M)} - 2M\right]} \tag{12}$$

und für den totalen Wirkungsquerschnitt

$$\sigma \approx \frac{\pi}{\varkappa^2} \lg^2 \gamma_{\min} \cdot \tag{13}$$

Ebenso ergibt sich für die Erwartungswerte von γ und n für $p_{0\,\max} \gg \varkappa$:

$$\bar{\gamma} \approx \frac{2}{\lg^2 \gamma_{\min}} \ ; \qquad \bar{n} \approx \frac{2}{\lg^2 \gamma_{\min}} \frac{\sqrt{2M(E+2M)} - 2M}{\bar{p}_0} \tag{14}$$

Die folgende Tabelle 1 gibt nach (6) bis (14) für einige Energiewerte des Primärteilchens den totalen Wirkungsquerschnitt und die Erwartungswerte von γ, n und p_0. Bei den Erwartungswerten für γ und n ist allerdings zu beachten, daß die Tabelle sich auf Stöße zwischen einzelnen Nukleonen, also auf Schauer im Wasserstoff bezieht. Bei schwereren Atomkernen kommen streifende Zusammenstöße nach einer Bemerkung von Le Couteur [Le 52b] relativ seltener vor, da sie nur für Nukleonen am Rand der Kerne möglich sind. Die Erwartungswerte für γ und n werden hier also größer. Da die Schwankungen um den Wert $\bar{n}$ nach (11) sehr groß sind — es werden ja gelegentlich auch zentrale Stöße mit $\gamma \approx 1$ auftreten —, ist in der vorletzten Zeile der Wert von n für $\gamma = 1$ als $n_{\max}$ eingetragen; die letzte Zeile gibt die (von γ unabhängige) mittlere kinetische Energie der Mesonen im S-System. Die Zahlen der ersten Spalte sind eingeklammert, da hier die Voraussetzungen der Abschätzung, nämlich starke Energiedissipation, kaum zutreffen.

Tabelle 1.

E	10	10^2	10^3	10^4 GeV
σ	(0,22)	0,53	0,90	1,4 $\cdot 10^{-24}$ cm^2
$\bar{\gamma}$	(0,56)	0,235	0,138	0,086
$\bar{n}$	(7,4)	8,7	12,5	19,1
$n_{\max}$	(13)	37	90	222
$\bar{p}_0$	(0,214)	0,322	0,458	0,608 GeV

Vergleicht man die in den Abb. 1—4 dargestellten Schauer mit den Werten der Tabelle, so erkennt man, daß für die Schauer in Abb. 1 und 3 ein ziemlich zentraler Stoß vorgelegen haben muß, während für die beiden anderen Schauer die Teilchenzahl in der Gegend von $\bar{n}$ liegt. Allerdings ist bei diesem Vergleich zu bedenken, daß bei hohen Primärenergien neben den π-Mesonen noch schwerere Teilchenarten erzeugt werden, die in der Tabelle 1 nicht berücksichtigt sind. Die Korrektur dieser Tabelle für die anderen Teilchenarten soll aber erst später erörtert werden (vgl. auch [He 52a]).

γ) Fermi [Fe 50a, 51b] hat eine etwas andere anschauliche Vorstellung für die Abschätzung der Mesonenzahlen zugrunde gelegt. Er nimmt an, daß die starke Wechselwirkung nur im Moment des Stoßes im Gebiet der beiden Scheiben (vgl. Abb. 5) wirksam sei und daß sie dort ein echtes thermodynamisches Gleichgewicht zur Folge habe, das mit der Ausdehnung des Mesonenwellenpakets sofort einfriert. Die Zahl der Mesonen kann man dann in folgender Weise abschätzen: Die Energiedichte dieses Wellenpakets im Moment des Stoßes ist von der Größenordnung $\varkappa^3 E$, die Temperatur erhält also nach dem Stefan-Boltzmannschen Gesetz die Größenordnung $(\varkappa^3 E)^{+\frac{1}{4}}$. Fermi gibt daher für die mittlere Energie der emittierten Mesonen im S-System

$$\bar{p}_0 \sim \varkappa^{\frac{3}{4}} E^{\frac{1}{4}} \tag{15}$$

an. Wenn man nun weiter mit $\gamma = 1$ rechnet, so erhält man nach Fermi Werte für $\bar{n}$, die bei größeren Energien etwa denen von Tabelle 1 entsprechen. Wenn man jedoch erheblich kleinere Werte von $\bar{\gamma}$ einsetzt, etwa die der Tabelle 1, so würde man für $\bar{n}$ viel kleinere Werte als die von Tabelle 1 erhalten und die Experimente kaum darstellen können. Es ist daher nach den Experimenten wahrscheinlich, daß die Formel (15) für große E zu hohe Werte für $\bar{p}_0$ und damit bei Berücksichtigung des richtigen γ-Wertes zu kleine Teilchenzahlen liefert. Tatsächlich haben z. B. Hopper, Biswas und Derby [Ho 51e] einen Schauer von $E = 1000$ GeV mit neun Mesonen im einzelnen ausmessen können und festgestellt, daß dort $\gamma = 0,1$ und $\bar{p}_0 = 0,44$ GeV (nach Tabelle 1 ergibt sich $\bar{p}_0 = 0,458$ GeV) gewesen ist, während die Fermische Formel (15) für $\bar{p}_0$ einige GeV ergäbe. Andererseits berichtet Perkins [Pe 52b] über Messungen von $\bar{p}_0$, die Werte etwa in der Mitte zwischen denen von Gl. (8) und (15) liefern.

Man kann bei konsequenter Durchführung der Fermischen Überlegung aber auch einsehen, daß die Energiedissipation noch weiter gehen wird, als Gl. (15) dies annimmt. Denn wenn die Mesonen die Energie (15) erhalten haben, so muß sich das Wellenpaket schon mindestens zu einer Dicke von der Ordnung der zugehörigen Wellenlänge, d. h. $\bar{p}^{-1}$ ausgedehnt haben. Dann ist aber die Energiedichte schon geringer als $E \varkappa^3$, die Temperatur niedriger als $E^{\frac{1}{4}} \varkappa^{\frac{3}{4}}$ geworden, und es besteht kein Grund gegen die Annahme, daß sich nun ein neues Gleichgewicht mit kleinerem $\bar{p}_0$ einstellt. Das Gleichgewicht kann erst dann einfrieren, wenn die Energiedichte so gering geworden ist, daß die nichtlinearen Wechselwirkungsglieder in der Wellengleichung nicht mehr angreifen. Das dürfte aber erst bei Energiedichten $\sim \varkappa^4$ der Fall sein.

Die Fermische Überlegung gibt auch eine einfache Abschätzung für die relative Häufigkeit verschiedener Mesonensorten in energiereichen Schauern. Mesonen, deren Ruhmasse größer ist, als die durch die primäre Energiedichte definierte Temperatur, werden bei Annahme eines Temperaturgleichgewichts nur selten emittiert werden; die relative

7. Theorie der Explosionsschauer.   **157**

Häufigkeit der Mesonen, deren Ruhmasse erheblich kleiner ist als diese Temperatur, sollte zu Anfang einfach durch die statistischen Gewichte gegeben sein, wenn für die betreffende Mesonensorte thermodynamisches Gleichgewicht besteht. Allerdings könnte dieses Verhältnis der Anzahlen der verschiedenen Mesonen während der Ausdehnung des Wellenpakets noch dadurch verschoben werden, daß sich etwa infolge der Wechselwirkung schwerere Mesonen in leichtere, z.B. χ- oder τ-Mesonen in π-Mesonen verwandeln; hier hängt offenbar alles von der speziellen Form der Wechselwirkung ab. Nach den π-Mesonen scheinen am häufigsten die χ- und $\varkappa$-Mesonen, deren Ruhmasse etwa 0,76 bzw. 0,55 GeV betragen soll, in energiereichen Schauern aufzutreten. Sie sollten (wegen $E^{\frac{1}{3}}\varkappa^{\frac{1}{3}} > 0{,}5$ GeV) etwa von einer Primärenergie von 100 GeV ab in ähnlicher Zahl wie die π-Mesonen erzeugt werden. Nega-tive Protonen dagegen sind erst bei Schauern von 10^4 GeV oder mehr zu erwarten.

FERMI hat ferner darauf hingewiesen [Fe 51 b], daß bei Schauern mit größeren Stoß-abständen vom Mesonenwel-lenpaket nicht nur Energie,

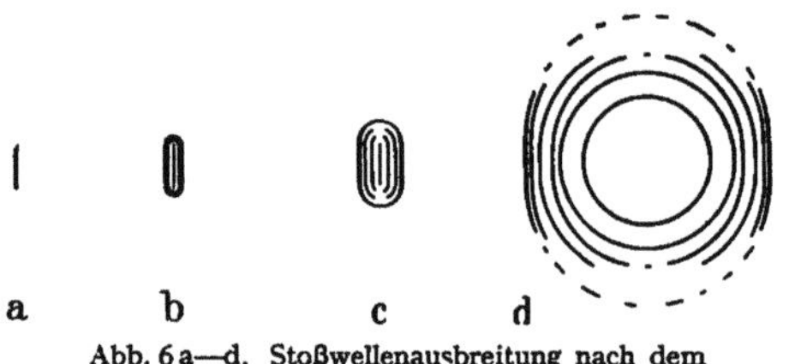

Abb. 6 a—d. Stoßwellenausbreitung nach dem Zusammenstoß (s. Text).

sondern auch Drehimpuls aufgenommen werden muß. Dies hat auch bei der thermodynamischen Deutung zur Folge, daß die Mesonen im S-System nicht in allen Richtungen gleichmäßig ausgesandt werden können; vielmehr werden die Richtungen der beiden einfallenden Teil-chen und die Ebene senkrecht zur Drehimpulsachse bevorzugt sein. Die Winkelverteilung wird in Abs. b, β noch ausführlicher behandelt werden.

δ) Den wirklichen Verhältnissen wird man (vgl. [He 52 a]) wahrschein-lich am besten gerecht, wenn man die Mesonenerzeugung mit einem Stoßwellenvorgang vergleicht, so wie er etwa bei den Schallwellen von Explosionen in der angewandten Physik bekannt ist. Tatsächlich kann man für energiereiche Stöße, bei denen viele Mesonen erzeugt werden, in erster Näherung von den quantentheoretischen Zügen des Vorgangs absehen, da es sich ja um den Grenzfall großer Quantenzahlen handelt, und kann die quantentheoretische Interpretation nachträglich nach den Methoden des Korrespondenzprinzips vornehmen. Man muß dann nach den Verfahren der klassischen Physik die Ausbreitung der Mesonenwellen mit Hilfe einer nichtlinearen Wellengleichung studieren.

Wir beginnen mit der anschaulichen Beschreibung des Vorgangs. Die obigen Abbildungen (Abb. 6a—d) stellen zu aufeinanderfolgenden Zeiten qualitativ die Energiedichte in einer Ebene (im S-System) dar, die die Bahnen der Mittelpunkte der Primärteilchen enthält. Im Stoß-moment ist die ganze Energie in der Scheibe der beiden Nukleonen konzentriert (a). Dann entfernen sich nach rechts und links zwei Stoß-fronten. Der Hauptteil der Energie sitzt noch in den beiden Stoßfronten, aber auch im Raum zwischen ihnen ist Wellenerregung vorhanden, die den Rest der Energie enthält (b). Die Stoßfronten schreiten nun

158   II. Nukleonen und π-Mesonen.

weiter fort, die Erregung in ihrem Kielwasser erstreckt sich über einen weiteren Raum, der zum Ausgangspunkt neuer Wellenausbreitung wird. Die Energie in den Stoßfronten ist kleiner geworden, sie hat sich in das übrige Wellengebiet (und damit zu größeren Wellenlängen) verlagert (c). Bei weiterem Fortschreiten sinkt die Erregung im Mittelpunkt ab, es bildet sich eine richtige Welle aus, die in der Richtung der Stoßfronten schneller fortschreitet als senkrecht dazu, da Wellen kürzerer Wellenlänge eine höhere Fortpflanzungsgeschwindigkeit (Gruppengeschwindigkeit) besitzen; nur mit sehr geringer Intensität wird sich nach allen Seiten eine Erregung auch schon mit Lichtgeschwindigkeit ausbreiten. Die Energie in den Stoßfronten ist jetzt so klein geworden, daß auch hier die Nichtlinearitäten keine wichtige Rolle mehr spielen; die weitere Fortpflanzung geschieht also nach der gewöhnlichen linearen Wellengleichung (d).

Übersetzt man den Wellenvorgang in diesem letzten Stadium nach dem Korrespondenzprinzip in die Sprache der Korpuskularvorstellung, so erkennt man, daß ein großer Teil der Energie in Form von Mesonen, deren Wellenlänge mit dem Durchmesser der „Scheibe" (a), d.h. mit $1/\varkappa$ vergleichbar ist, nach allen Richtungen ausgestrahlt wird; daß aber energiereichere Mesonen senkrecht zur Scheibe, d.h. näher zur Bewegungsrichtung der Primärteilchen emittiert werden. Die energiereichsten Mesonen, deren Wellenlänge etwa der Dicke der Scheibe entspricht, werden in Winkeln von der Größenordnung $\vartheta \sim \varkappa/p_{0\,\mathrm{max}}$ relativ zur Primärrichtung ausgesandt (Mesonen der Energie p_0 entsprechend in Winkeln der Ordnung $\varkappa/p_0$), da der Impuls senkrecht zur Primärrichtung nur von der Größenordnung $\varkappa$ sein kann; denn (diese Voraussetzung muß hier ausdrücklich gemacht werden) das ursprüngliche Wellenpaket kann in der Scheibenebene als einigermaßen gleichförmig angesehen werden.

Das bisher gezeichnete Bild entspricht etwa den Verhältnissen bei zentralem Stoß. Bei größerem Stoßabstand ist, wie schon erwähnt, zu beachten, daß die Stoßwelle Drehimpuls enthalten muß. Das geschieht etwa in der Weise, daß die rechte Stoßfront im unteren, die linke im oberen Teil intensiver ist. Der ganze Wellenvorgang wird dann weniger symmetrisch (er verliert die Spiegelungssymmetrie um die beiden Achsen, behält aber die Invarianz gegenüber einer Drehung um 180°), verläuft aber sonst qualitativ ähnlich.

b) Die Stoßwellengleichung.

α) Die Ausbreitung der Stoßwellen kann auch mathematisch im einzelnen verfolgt werden, wenn die Wechselwirkung der Mesonen in Form eines nichtlinearen Gliedes in der Wellengleichung gegeben ist. Es läge nahe, hier zunächst die Wirkung eines Gliedes der Form φ^3 in der Wellengleichung zu untersuchen, wie es nach SCHIFF und THIRRING [*Schi 51, Th 52*] aus der Absättigung der Kernkräfte und der Forderung der Renormalisierbarkeit der Mesonentheorie folgt. Es stellt sich aber heraus, daß ein solches Glied keine „starke" Wechselwirkung zur Folge hat und nicht zu einer eigentlichen Vielfacherzeugung führt. Als

7. Theorie der Explosionsschauer. **159**

Beispiel für eine „starke" Wechselwirkung wählen wir daher nach dem Vorbild von Born ([*Bo 33*], s. auch [*He 39a*]) die Lagrange-Funktion:

$$L = l^{-4}\left(\sqrt{1 + l^4\left[\left(\frac{\partial \varphi}{\partial x_\nu}\right)^2 + \varkappa^2\varphi^2\right]} - 1\right), \tag{16}$$

die für kleine Werte von l in die übliche Lagrange-Funktion übergeht.

Um die Stoßwellengleichung lösen zu können, vereinfachen wir die Anfangsbedingungen: die Scheibe in Abb. 6a sei unendlich ausgedehnt und unendlich dünn. Die Koordinate senkrecht zur Scheibenachse sei x; φ hängt dann nur von x und t ab. Da die Erregung der Welle nur in dem Weltpunkt $x = 0$ und $t = 0$ erfolgt, kann φ wegen der Lorentz-Invarianz von L nur von $s = t^2 - x^2$ abhängen. Wir suchen also Lösungen $\varphi = \varphi(s)$, die L zum Extremum machen und für $s < 0$ verschwinden. Man erhält für φ unter diesen Voraussetzungen die Wellengleichung

$$\frac{d}{ds}(\varphi' s) + \frac{\varkappa^2}{4}\varphi = 2l^4 s \varphi'^2 \frac{\varphi' + \varkappa^2\varphi}{1 + l^4\varkappa^2\varphi^2}. \tag{17}$$

Für $l = 0$ kommt man zur linearen Wellengleichung (2) für den Spezialfall $\varphi = \varphi(s)$ zurück; die Lösung lautet dann

$$\left.\begin{array}{ll} \varphi = a\, J_0\!\left(\varkappa\sqrt{s}\right) & \text{für}\quad s > 0, \\ = 0 & \text{für}\quad s < 0. \end{array}\right\} \tag{18}$$

(a Integrationskonstante.)

Daß diese Lösung die Wellengleichung auch an der Stelle $s = 0$ befriedigt, erkennt man, indem man (17) über die Stelle $s = 0$ hinwegintegriert. Die Lösung (18) ist das eindimensionale Analogon zur Schwingerschen $\varDelta$-Funktion; sie stellt eine Stoßwelle dar, bei der unendlich viel Energie in der Singularität an der Stelle $t = x$ oder $s = 0$ enthalten ist (die Ableitung hat dort den Charakter einer δ-Funktion). Die Fourier-Analyse gibt das Spektrum (4) der gewöhnlichen Bremsstrahlung; man erhält also wenige energiereiche Mesonen, keine eigentliche Vielfacherzeugung.

Setzt man $l \neq 0$ aber $\varkappa = 0$ (verschwindende Ruhmasse der Mesonen), so lautet die Lösung der nichtlinearen Wellengleichung (17):

$$\left.\begin{array}{l} \varphi = \dfrac{1}{a}\lg\left(1 + \dfrac{a^2}{2l^4}s + \dfrac{a}{2l^4}\sqrt{4\,l^4 s + a^2 s^2}\right) \quad \text{für}\quad s \geq 0 \\ = 0 \quad \text{für}\quad s \leq 0. \end{array}\right\} \tag{19}$$

Für $l \neq 0$ und $\varkappa \neq 0$ läßt sich die Lösung nur noch als Reihenentwicklung angeben. Man kann setzen:

$$\varphi(s) = \frac{1}{\varkappa l^2} f(\zeta); \qquad \zeta = s\varkappa^2 \tag{20}$$

und erhält die Näherungsformeln:

$$\left.\begin{array}{ll} f(\zeta) = \sqrt{\zeta}\left(1 + a\zeta + \dfrac{27a^2 + 2a - 1}{10}\zeta^2 + \cdots\right) & \text{für}\quad \zeta \ll 1, \\ \approx \eta\,\zeta^{-\frac{1}{4}}\cos\left(\sqrt{\zeta} + \delta\right) & \text{für}\quad \zeta \gg 1, \\ = 0 & \text{für}\quad \zeta \leq 0, \end{array}\right\} \tag{21}$$

160   II. Nukleonen und π-Mesonen.

wobei die Konstanten η und δ durch die Integrationskonstante a bestimmt sind.

Entwickelt man $\varphi(s)$ nach FOURIER, so erhält man bei den beiden Lösungen (19) und (20) für große p-Werte $(p_0 \gg \varkappa)$ ein Spektrum der Form (5) mit $\alpha = 1$, so wie es in Abschn. a, β angenommen wurde; denn der Verlauf des Spektrums für große p-Werte ist durch den Verlauf der Stoßwelle bei sehr kleinen s-Werten gegeben. In (19) und (20) verhält sich die Wellenfunktion dort wie $\sqrt{s}$, die Ableitung $\varphi'(s)$ also wie $s^{-\frac{1}{2}}$ und die Energiedichte wie $1/s$; das Integral über die Energiedichte divergiert logarithmisch. Man erhält jetzt eine echte Vielfacherzeugung der Mesonen und zwar genau die Verhältnisse, die in Abs. a, β und Abs. a, δ geschildert wurden.

Man kann hier auch klar erkennen, wodurch ein Potenzgesetz der Form (5) speziell mit $\alpha = 1$ ausgezeichnet ist. Die Annahme eines Potenzgesetzes bedeutet nur, daß $\varphi(s)$ für $s \geq 0$ in der Nähe von $s = 0$ nach Potenzen von s entwickelbar sei, d.h., daß dort keine wesentliche Singularität liegen soll. Dies ist in der Tat nötig, wenn $\varphi'(s)$ für $s = 0$ unendlich werden, und die Wellengleichung an der Stelle $s = 0$ doch gelten soll. $\alpha = 1$ ist aber dann der höchste Wert des Exponenten, der überhaupt auftreten kann. Denn wenn $\alpha > 1$ wäre, so wäre die Gesamtenergie in der Stoßwelle unter den idealisierten Anfangsbedingungen, die zu $\varphi = \varphi(s)$ führen, endlich. Dies kann aber nicht sein, da zu einem relativistisch invarianten Wellenvorgang auch ein invarianter Vierervektor von Energie und Impuls gehört. Jeder endliche Vierervektor zeichnet aber eine Richtung aus; daher muß der Vierervektor hier unendlich sein.

Man erkennt aus dieser Überlegung, daß das Potenzgesetz

$$dn = a\,\frac{dp_0}{p_0^2} \tag{22}$$

gerade dem Grenzfall starker Wechselwirkung unabhängig von ihrer speziellen Form entspricht, und unabhängig von der Frage, ob es sich um skalare, vektorielle oder irgendwelche anderen Teilchen handelt. Für verschwindende Wechselwirkung wird der Exponent α in Gl. (5) gleich 0; aber auch ein Wechselwirkungsglied der Form φ^3 in der Wellengleichung führt nur zu $\alpha = 0$. Denn da φ schon bei der Stoßwelle ohne Wechselwirkung im ganzen Raum endlich ist [Gl. (18)], stellt eine Wechselwirkung mit φ^3 keine starke Störung dar. Erst wenn φ' im nichtlinearen Wechselwirkungsglied vorkommt, wird der Charakter der Lösung grundsätzlich geändert, da φ' bei der Lösung (18) an der Stelle $s = 0$ unendlich wird. Eine Wechselwirkung der „zweiten Art" wird bei hohen Primärenergien stets zu einer starken Wechselwirkung werden.

Wenn man von den idealisierten Anfangsbedingungen (unendlich dünne und unendlich ausgedehnte ebene Scheibe) zu den realen in Abs. a beschriebenen Anfangsbedingungen übergeht, so werden die Verhältnisse mathematisch sehr verwickelt, dürften aber grundsätzlich der anschaulichen Beschreibung von Abs. a, δ entsprechen.

β) Für den Fall starker Wechselwirkung kann man die Winkelverteilung der auftretenden Mesonen daher einstweilen auch nur mit Hilfe dieser anschaulichen Beschreibung abschätzen. In Abs. a, δ wurde darauf hingewiesen, daß die Impulse der Mesonen senkrecht zur Primärrichtung den Wert $\varkappa$ größenordnungsmäßig nicht überschreiten können. Daran ändert sich auch grundsätzlich nichts, wenn der Stoßabstand von der Größenordnung $1/\varkappa$ war und die Stoßwelle daher einen erheblichen Drehimpuls enthält. Nur die Intensitäten der Stoßwelle oben und unten müssen dann verschieden sein (etwa rechts unten und links oben groß, rechts oben und links unten klein). Der Drehimpuls wird zu einem erheblichen Teil in den energiereichen Mesonen stecken.

Man wird demnach nicht annehmen können, daß die Winkelverteilung relativ zur Primärrichtung in Stößen mit Drehimpuls qualitativ anders aussieht als in Stößen ohne Drehimpuls. Wohl aber folgt aus der anschaulichen Überlegung, daß die Winkelverteilung stark von der Primärenergie abhängt.

In Stößen kleinerer Primärenergie wird die Anzahl der energiereichen Mesonen mit anisotroper Winkelverteilung klein sein; die Winkelverteilung ist daher im ganzen ungefähr isotrop. Erst bei wirklich großer Primärenergie nimmt auch die Zahl der (im S-System) energiereichen Mesonen zu, und deren Winkel relativ zur Primärrichtung sollte die Größenordnung $\varkappa/p_0$ nicht überschreiten, die Verteilung im ganzen also anisotrop sein. Bei sehr energiereichen Schauern sollte man also in der Regel im L-System zwei Kegel unterscheiden können, einen mit sehr kleinem, einen mit größerem Öffnungswinkel, in die die Mesonen emittiert werden; und zwar sollte dies unabhängig vom Wert von γ gelten.

γ) Bisher war angenommen worden, daß es sich um die Stoßwellengleichung für eine skalare Wellenfunktion φ handelt, die bei schwacher Wechselwirkung einfach in die Wellengleichung (2) übergeht. Die richtige Gleichung für die Mesonenentstehung dürfte aber erheblich komplizierter sein und im Grenzfall schwacher Wechselwirkung in eine verwickeltere lineare Wellengleichung übergehen, zu deren Lösung nicht nur die Wellen der π-Mesonen, sondern auch die der schwereren Mesonenarten, insbesondere z. B. der χ- und $\varkappa$-Mesonen gehören.

An der Form (22) der Energieverteilung wird sich hierdurch nichts ändern. Bei hohen Werten von p_0 wird die Verteilung auf die verschiedenen Mesonenarten von p_0 unabhängig und durch die Stoßwellengleichung bestimmt sein. Obwohl die verschiedenen Mesonenarten im allgemeinen in vergleichbarer Häufigkeit auftreten dürften, besteht kein Grund mehr zu der aus der FERMIschen Theorie folgenden Annahme, daß das Häufigkeitsverhältnis einfach durch die statistischen Gewichte gegeben sei; vielmehr enthält dieses Häufigkeitsverhältnis indirekt Informationen über die Form der Stoßwellengleichung. Bei kleineren Werten von p_0 bleiben natürlich nur noch die leichteren Mesonen übrig, und man kann eine grobe Abschätzung über das Verhältnis der Mesonenarten als Funktion der Gesamtenergie erhalten, wenn man für jede Mesonensorte das Spektrum (22) einfach bis zu $p_0 = \varkappa_i$ ($\varkappa_i =$ Ruhmasse der betreffenden Sorte) herunter fortsetzt.

162   II. Nukleonen und π-Mesonen.

Aus dieser Überlegung folgt auch, daß die schwereren Mesonen fast nur in der Stoßwellenfront, d.h. unter kleinen Winkeln zur Primärrichtung (oder genau entgegengesetzter Richtung) im S-System ausgesandt werden; denn p-Werte, die klein gegen $\varkappa_i$ sind, werden jeweils nur selten vorkommen. Die Winkelverteilung der schwereren Mesonen im S-System ist also stark anisotrop.

c) Vergleich mit der Erfahrung.

Für den Vergleich der Theorie mit der Erfahrung kann man nach den Überlegungen der letzten Abschnitte die Tabelle 1 mit folgender Einschränkung zugrunde legen. In Tabelle 1 war angenommen worden, daß nur π-Mesonen entstehen. In Wirklichkeit werden bei hohen Primärenergien auch χ- und $\varkappa$-Mesonen und andere Teilchen entstehen. Ein Teil der Energie wird also den π-Mesonen entzogen und für die schwereren Mesonen verwandt werden; da der Erwartungswert von p_0 für die letzteren wegen der größeren Ruhmasse sicher größer sein wird, als für die ersteren, wird die Gesamtzahl bei sehr hoher Primärenergie ($> 10^{12}$ eV) etwas kleiner sein als nach Tabelle 1. Da andererseits gerade in der Teilchenzahl eine sehr große Streuung zu erwarten ist (wegen der starken Streuung der γ-Werte), da ferner die $\bar{\gamma}$-Werte für schwerere Atomkerne größer sind als in Tabelle 1, fällt diese Abänderung kaum ins Gewicht (genauere Angaben in [He 52a]).

Wir haben für einige Schauer der hier betrachteten Art, bei denen Teilchenzahl und Primärenergie bekannt sind, die Zahlen in Tabelle 2 zusammengestellt. Dabei ist die Anzahl der neutralen Mesonen als etwa gleich der Hälfte der geladenen geschätzt und einbezogen worden. Die Primärenergien sind aus der Winkelverteilung abgeschätzt und zum Teil recht unsicher. Das Material stammt aus Beobachtungen von SCHEIN und Mitarbeitern [Sche 51], POWELL und Mitarbeitern [Ca 50a, 52a], TEUCHER [Te 52b], PICKUP und VOYVODIC [Pi 51b] (die Auswertung ist hier besonders ungenau, da die genauen Winkelmessungen nicht veröffentlicht sind; die hier geschätzten Energiewerte sind erheblich geringer, als sie von den Autoren selbst auf Grund eines speziellen Vergleichs mit der FERMIschen Theorie abgeschätzt wurden), HOPPER, BISWAS und DERBY [Ho 51e] und aus neueren Beobachtungen am M.P.I. f. Physik. Der aus der beobachteten Teilchenzahl mit dem theoretischen Wert von $\bar{p}_\iota$ berechnete γ-Wert ist in der dritten Zeile angegeben. Man erkennt, daß die Teilchenzahlen tatsächlich keine einfache Funktion der Energie sind, sondern offenbar stark streuen in einem Bereich, wie er

Tabelle 2.

E	20	30	40	40	90	130	1000	2000	30000 GeV
n	13	9	18	25	10	18	9	12	21
γ	0,7	0,4	0,75	≈ 1	0,29	0,42	0,10	0,095	0,055
Beob.	[1]	[Pi 51b]	[Pi 51b]	[Te 52b]	[Pi 51b]	[Pi 51b]	[Ho 51e]	[Ca 50a, 52a]	[Sche 51]

[1] Schauer von Abb. 3.

7. Theorie der Explosionsschauer.   **163**

nach Tabelle 1 zu erwarten ist. Der Wert von γ und damit der Wert von $\bar{p}_0$ ist nur bei zwei Schauern auch experimentell bekannt; nur in diesen beiden Fällen (TEUCHER [*Te 52b*] und HOPPER [*Ho 51e*]) konnte die mittlere Energie der Mesonen im Schwerpunktsystem bestimmt werden. Tabelle 3 gibt die Ergebnisse:

Tabelle 3.

E	40	1000 GeV
$\bar{p}_0$ beobachtet	0,29	0,44 GeV
$\bar{p}_0$ berechnet nach Tabelle 1	0,27	0,46 GeV

PERKINS [*Pe 52b*] berichtet über Messungen an Schauern von 50 bis 1000 GeV Primärenergie, die für p_0 etwa 1,5 GeV ergeben haben. Allerdings sind sie zum Teil an Schauern mit einigen dicken Spuren durchgeführt, bei denen die Messungen sehr unsicher werden. Leider liegen für noch energiereichere Schauer bisher keine Messungen vor. Man kann aber aus der Form des allgemeinen Energiespektrums der μ-Mesonen noch indirekt einen Schluß auf die Energieverteilung der π-Mesonen im S-System beim Einzelstoß ziehen. Potenzgesetze von der Form (5) haben für $\alpha \leq 1$ die Eigenschaft, daß ein Potenzgesetz für das primäre Spektrum von der Form $E^{-\gamma}$ sich auch im allgemeinen Spektrum der π-Mesonen im L-System mit dem gleichen Exponenten reproduziert, worauf schon vor langer Zeit EULER hingewiesen hat. (Für $\alpha = 1$ kommt allerdings noch ein Faktor hinzu, der von der Energie langsam, d.h. logarithmisch abhängt.) Für $\alpha > 1$ oder für ein exponentiell abfallendes Temperaturspektrum würde jedoch das gesamte π-Mesonenspektrum im L-System stärker abfallen als $E^{-\gamma}$. Da tatsächlich das μ-Mesonenspektrum und damit das erzeugende π-Mesonenspektrum in einem weiten Bereich wie $E^{-\gamma}$ abzufallen scheint mit dem gleichen Exponenten γ wie das Primärspektrum, kann man auf die Gültigkeit eines Potenzgesetzes der Form (5) mit $\alpha \leq 1$ schließen.

Endlich muß noch der Vergleich hinsichtlich der Winkelverteilung durchgeführt werden. Bei Schauern bis zu 100 GeV sollte die Verteilung der Mesonen im Schwerpunktsystem noch ziemlich isotrop sein; dies ist z.B. bei dem großen Schauer in Abb. 1 unmittelbar zu sehen. Bei hohen Primärenergien dagegen sollte die Verteilung anisotrop werden, da der Winkel im S-System, in dem Mesonen der Energie p_0 ausgesandt werden, die Größenordnung $\varkappa/p_0$ nicht überschreiten sollte. Bei dem Schauer von 1000 GeV sind zwei energiereichere Mesonen beobachtet worden, für die im S-System p_0 den Wert 0,86 bzw. 0,67 GeV hat. Das erste wird in einem Winkel von 4° zur Rückwärtsrichtung, das zweite unter 23° zur Primärrichtung ausgesandt. Bei dem Schauer von 30000 GeV sollte der Mittelwert von p_0 etwa 0,7 GeV betragen (sofern es sich dabei um π-Mesonen handelt), aber es sollten Mesonen bis zu etwa 10 GeV im S-System vorkommen; der größere Teil der Mesonen sollte also in Kegeln mit Öffnungswinkeln von etwa 15° (im S-System) um Vorwärts- und Rückwärtsrichtung liegen. Auch dies scheint der

Fall zu sein. Ferner wird von PERKINS [*Pe 52 b*] berichtet, daß die Winkelverteilung der χ- und $\varkappa$-Mesonen im S-System durchweg stark anisotrop sei, wie die Theorie es fordert. Schließlich kann man für Schauer von 10^{15} bis 10^{17} eV Schlüsse über die Winkelverteilung aus der Verteilung der Kerne der großen Luftschauer ziehen. Die Tatsache, daß diese Kerne, die ja von verschiedenen π°-Mesonen des gleichen Primärschauers herrühren, erfahrungsgemäß nur um Abstände der Ordnung 10 cm voneinander getrennt sind, läßt sich nur einfach erklären aus einer Winkelverteilung der π°-Mesonen im S-System, bei der $\vartheta \approx \varkappa/p_0$ ist, sie kann also als unmittelbare Bestätigung der theoretischen Winkelverteilung gelten. Die Einzelheiten der großen Luftschauer werden in Kap. V, 6 und 7 ausführlich behandelt.

8. Kaskaden im Atomkern.

Von K. SYMANZIK.

Über Kernzertrümmerungen in Photoplatten liegen, wie in Abschn. 2 gezeigt ist, schon umfangreiche und eingehende experimentelle Untersuchungen vor. Im Vergleich hierzu ist die theoretische Beschreibung dieser Erscheinungen noch sehr unbefriedigend. Dies liegt sowohl an der begrenzten Leistungsfähigkeit der gegenwärtigen experimentellen Methoden als auch daran, daß die Entstehung eines „Sterns" im allgemeinen sicher ein sehr verwickelter Prozeß ist.

In theoretischer Hinsicht wäre die Beobachtung von Sternen in reinem Wasserstoff der einfachen Bedingungen wegen besonders erwünscht. In der Emulsion erfolgen zwar auch Zusammenstöße der einfallenden Teilchen mit Wasserstoffkernen, doch sind diese Ereignisse selten und nicht immer mit Sicherheit als solche zu erkennen. Bei Sternen von schweren Kernen hat man es aber mit der Wechselwirkung vieler Elementarteilchen zu tun. Selbst wenn man sich zur Vereinfachung den Sternerzeugungsprozeß weitgehend in Einzelprozesse zwischen je zwei Elementarteilchen auflösbar denkt, so sind doch die Charakteristika dieser Prozesse (Wirkungsquerschnitte, Multiplizitäten, Winkelverteilung usw.) bei den in Betracht kommenden hohen Energien noch fast unbekannt. Der umgekehrte Schluß von dem Stern auf die zu ihm führenden Einzelprozesse ist namentlich dadurch erschwert, daß die Beziehung zwischen der Primärenergie und den Sternparametern (Schauerteilchenzahlen, graue Teilchen usw.) nur bis zu etwa 10 GeV hinauf empirisch bekannt und insbesondere auch die Zuammensetzung der schnellen Teilchen von Sternen (π, p-Teilchen) noch sehr unsicher ist. So scheint es noch nicht möglich zu sein, aus den Sternen über die verschiedenen Einzelprozesse bei hohen Energien wirklich eindeutige Schlüsse zu ziehen (vgl. [*Ca 52 a*] und Abs. c, β).

Bei den Sternen lassen sich einerseits einzelne Ereignisse studieren (Abschn. 7), andererseits statistische Betrachtungen anstellen (Abschn. 2). Die Analyse von Einzelereignissen verspricht Erfolg nur in den Fällen, wo das Ereignis genügend bestimmt ist (z. B. wenn aus dem Fehlen grauer

Literaturverzeichnis.*

Die Zahlen hinter den Anfangsbuchstaben bedeuten das Erscheinungsjahr, nach dem innerhalb eines Anfangsbuchstabens geordnet ist. Kursivzahlen geben die Seiten im Buch an, auf der die Arbeit zitiert ist.

Au 38   AUGER, P., R. MAZE u. T. GRIVET-MEYER: C. r. Acad. Sci. Paris **206**, 1721 (1938). *7.*

Bo 33   BORN, M.: Proc. Roy. Soc. London, A **143**, 410 (1933). *159.*

Ca 50a   CAMERINI, U., P. H. FOWLER, W. O. LOCK u. H. MUIRHEAD: Phil. Mag. **41**, 413 (1950). *150, 162.*

Ca 52a   CAMERINI, U., W. O. LOCK u. D. H. PERKINS: Progress in Cosmic Ray Physics, herausgeg. von J. G. WILSON, S. 1 ff. Amsterdam 1952. *162.*

Eh 48   EHMERT, A.: Z. Naturforsch. **3a**, 264 (1948). *1.*

Fe 49a   FERMI, E.: Phys. Rev. **75**, 1169 (1949). *2.*

Fe 50a   FERMI, E.: Progr. Theor. Phys. **5**, 570 (1950). *148, 149, 156.*

Fe 51b   FERMI, E.: Phys. Rev. **81**, 683 (1951). *148, 149, 156, 157.*

Fo 46   FORBUSH, S. E.: Phys. Rev. **70**, 771 (1946). *1.*

Fr 48a   FREIER, P., E. J. LOFGREN, E. P. NEY, F. OPPENHEIMER, H. L. BRADT u. P. PETERS: Phys. Rev. **74**, 213 (1948). *1.*

Fr 48b   FREIER, P., E. J. LOFGREN, E. P. NEY u. F. OPPENHEIMER: Phys. Rev. **74**, 1818 (1948). *1.*

He 13   HESS, V. F.: Phys. Z. **14**, 610 (1913). *1.*

He 36   HEISENBERG, W.: Z. Phys. **101**, 533 (1936). *148, 151.*

He 39a   HEISENBERG, W.: Z. Phys. **113**, 61 (1939). *5, 148, 159.*

He 42   HEITLER, W., u. H. W. PENG: Proc. Cambridge Phil. Soc. **38**, 296 (1942). *148.*

He 43b   HEISENBERG, W.: Vorträge über Kosmische Strahlung, herausgeg. von W. HEISENBERG, 1. Aufl., S. 115 ff. Berlin 1943. *148.*

He 49a   HEISENBERG, W.: Nature **164**, 65 (1949). *148, 153.*

He 49b   HEISENBERG, W.: Z. Phys. **126**, 569 (1949). *5, 148, 153.*

He 49c   HEISENBERG, W.: Nuovo Cim. Suppl. (9) **6**, 493 (1949). *148.*

He 49f   HEITLER, W.: Rev. Mod. Phys. **21**, 113 (1949). *148.*

He 50b   HEITLER, W., u. L. JÁNOSSY: Helv. Phys. Acta **23**, 417 (1950). *148.*

He 52a   HEISENBERG, W.: Z. Phys. **133**, 65 (1952). *154, 156, 157, 162.*

Ho 51e   HOPPER, V. D., S. BISWAS u. J. F. DERBY: Phys. Rev. **84**, 457 (1951). *156, 162, 163.*

Ja 43b   JÁNOSSY, L., C. B. MCCUSKER u. G. D. ROCHESTER: Phys. Rev. **64**, 345 (1943). *148.*

Ko 13   KOLHÖRSTER, W.: Phys. Z. **14**, 1153 (1913). *1.*

La 47a   LATTES, C. M. G., H. MUIRHEAD, G. P. S. OCCHIALINI u. C. F. POWELL: Nature **159**, 694 (1947). *5.*

La 47b   LATTES, C. M. G., G. P. S. OCCHIALINI u. C. F. POWELL: Nature **160**, 453, 486 (1947). *5.*

Le 52b   LECOUTEUR, K. J.: Diskussionsbemerkung auf der Konferenz in Kopenhagen, Juni 1952. *155.*

Pe 52b   PERKINS, D. H.: Vortrag auf der Konferenz in Kopenhagen, Juni 1952. *156, 163, 164.*

Pi 51b   PICKUP, E., u. L. VOYVODIC: Phys. Rev. **82**, 265 (1951). *162.*

Ri 49   RICHTMYER, R. D., u. E. TELLER: Phys. Rev. **75**, 1729 (1949). *1.*

Sche 51 SCHEIN, M., J. J. LORD u. J. FAINBERG: Phys. Rev. **81**, 313 (1951). *151, 162.*

Schi 51 SCHIFF, L. I.: Phys. Rev. **84**, 1 (1951). *158.*

Te 52b   TEUCHER, M.: Naturwiss. **39**, 68 (1952). *149, 162, 163.*

Th 52   THIRRING, W.: Z. Naturforsch. **7a**, 63 (1952). *158.*

Un 49b   UNSÖLD, A.: Z. Astrophys. **26**, 176 (1949). *2.*

Un 51   UNSÖLD, A.: Phys. Rev. **82**, 857 (1951). *2.*

Vi 51b   VIDALE, M., u. M. SCHEIN: Phys. Rev. **84**, 593 (1951). *149.*

Wa 48b   WATAGHIN, G.: Phys. Rev. **74**, 975 (1948). *148.*

---

* In reprinting the references, we have restricted ourselves to those quoted in Heisenberg's articles. (Editors)

Physica *19*, 897-908 (1953)

Heisenberg, Werner
1953

Physica XIX
897-908
Lorentz
Kamerlingh Onnes
Conference

DOUBTS AND HOPES IN QUANTUM-ELECTRODYNAMICS

by WERNER HEISENBERG *)

When I have to sum up the results of our conference on quantum-electrodynamics I feel this is an especially difficult task, because here the situation is still worse than in the case of superconductivity. In superconductivity at least the foundations of the theory are safe because we know that the problem is finally to solve the Schrödinger equation for electrons in a cristal. In quantum-electrodynamics however even the foundations are quite uncertain. When I remember the many discussions on electrodynamics there were in the old times with L o r e n t z, E h r e n f e s t and K r a m e r s, I feel that it is difficult to keep up with this tradition, and I am almost tempted to quote a sentence by E h r e n f e s t which was not meant too seriously. He emphasized how difficult it always is to explain a subject to an audience, so that the audience can really understand it and he added: "but if one has not understood the problem oneself then to explain it to the audience in such a way that the audience can understand it — that is really difficult". This being now more or less the case, I will start with the optimictic side of the problem and later on come to the difficulties and finally I will try to say what hopes one may have in the present situation.

At the beginning of the conference C a s i m i r told us about the electrodynamics of L o r e n t z and of the history of our present problems. Since I have to talk about the quantum-theoretical side of the problems I would like to mention that when one tried in the beginning of the thirties to apply quantum theory to electrodynamics the result was rather disappointing. Let me start to write down those two formulae, which are the basis of quantum-electrodynamics:

---

*) Summarizing address in the final session of the Conference at Leiden, June 26th, 1953, taken from the recording magnetophone tape.

Physica *19*, 897–908 (1953)

$$\left[\gamma_\mu\left(\frac{\partial}{\partial x_\mu} - iA_\mu\right) + m\right]\psi(x) = 0 \tag{1}$$

$$\frac{\partial}{\partial x_\mu}\left(\frac{\partial A_\nu}{\partial x_\mu} - \frac{\partial A_\mu}{\partial x_\nu}\right) = -\frac{ie}{2}\left[\psi^\dagger(x)\,\gamma_\nu\psi(x)\right] \tag{2}$$

The first equation is simply the Dirac equation which, as we know, describes the electron with all its properties very well. Equation (2) is the still better known Maxwell equation, the right-hand side being the current, A is the electromagnetic potential and $\psi(x)$ the Dirac wave function. When one tried to quantize these equations one first of all realized that the mathematics become rather complicated, secondly that the results one gets after much labour are already well known from classical theory and finally that one meets a number of divergencies. One gets infinite terms and at that time on did not know what to do with them. One invented a kind of "subtraction physics", one cancelled the infinite terms and tried to get some sensible results out of the remaining finite terms. This situation was extremely unsatisfactory; for as has been said by some physicist — I do not know by whom — the fact that a term is infinite is not a sufficient reason to drop it. A decisive progress has however been made after 1946 by an idea which, as far as I know, has been first mentioned by K r a m e r s in discussions, and was carried out by B e t h e [1]) in a still unrelativistic treatment. Later came to this a very important development of relativistic quantum-electrodynamics by T o m o n a g o [2]), S c h w i n g e r [3]), D y s o n [4]), F e y n m a n [5]) and many others.

The idea was simply that the quantities e and m which enter into the equations (1) and (2) and which were meant to be the charge and the mass of the electron are really something different. When one starts from these equations one finds that the state which describes one single electron in space has not the energy mc^2 but a different energy and also that the electron reacts upon the field not with the charge e but with a different charge. This is of course a very old result because also at the time of L o r e n t z when one had no quantum theory yet one knew that an electron would not only have its mechanical mass but also an electromagnetic mass. Therefore it was natural to use other equations with e' and m' instead of e and m and to choose these new quantities in such a way that the electron gets its real experimental mass and charge. Then it could be shown, and that was the great result after 1946, that actually after the introduction of the experimental values of e and m the infinities disappeared and that one could get definite results. In the mathematics it was important that one could expand in perturbation theory with respect to the fine structure constant $e^2/\hbar c$. This quantity being about $^1/_{137}$ the expansion converges sufficiently rapidly to give accurate results. In our discussions in the last days of the conference, we heard from the lectures of L a m b and B l o c h how well the results fit with the experiments. Actually there was an old difficulty that the level spacing in the fine structure of hydrogen was not accurately given by the S o m m e r f e l d formula or by the D i r a c theory of the electron. These level spacings have now been remeasured very accurately by radar technique by L a m b and R e t h e r f o r d [6]). On the other hand they have been calculated by the new scheme of quantum-electrodynamics and the results fit extraordinarily well with the experiments. Perhaps one of the most beautiful examples how well this new scheme works has been given by B l o c h who told us about the most recent values of the magnetic moment of the electron. When one calculates the "g-value" of the electron from the Dirac equations alone one finds exactly two. Now the new mathematical scheme of quantum-electrodynamics gives a different value because it includes the influence on the magnetic moment of the electron from the virtual creation of electron-positon pairs around this electron, the result being

$$g_{el} = 2\left(1 + \frac{a}{2\pi} - 2{,}973\,\frac{a^2}{\pi^2} + \ldots\right) = 2 \times 1{,}001\ 145$$

The accuracy of recent experiments is so high that even the term $- 2{,}973\,a^2/\pi^2$ has now been verified or at least been made probable by the experiments. Therefore in this respect one can say that the situation is very satisactory.

One should perhaps mention in this connection that there are still some non-trivial results of quantum-electrodynamics which have not yet been checked experimentally. If one is interested in the behaviour of the electromagnetic field in such an energy region where the energy of the light quanta is not sufficient to create pairs, one can introduce the virtual creation and annihilation of pairs into the Maxwell equations which accordingly are to be modified to a

Physica *19*, 897–908 (1953)

certain extent and one gets equations which describe the behaviour of the field alone. The Maxwell equations have to be changed in two very important respects. Firstly one gets instead of differential equations integro-differential equations or, as one says nowadays, a non-local theory. One finds that the expression on the left side of equation (2) is modified by multiplication with a certain form-factor and integration over the four-dimensional space, which can be interpreted to mean that the light-quanta have a finite extension of the order of the Compton wavelength of the electron. This form-factor has been calculated long ago by U e h l i n g [7]) and S e r b e r [8]). Furthermore one gets not only such a non-local character of the theory but also a non-linearity into the theory which means that there is an interaction between light-quanta. Two beams of light which cross each other should not go just through each other as in Maxwells theory but there should be an actual scattering of light by light. Also this effect has been calculated long ago by E u l e r and K o c k e l [9]).

Nowadays we have good reason to believe that all these results of quantum-electrodynamics are correct. It would be very interesting if some time they could be checked, but such experiments are probably very difficult. But if we suppose for the moment that even this side of quantum-electrodynamics is an agreement with the facts, which we believe, then from this point of view the situation looks extremely satisfactory.

We come now to the pessimistic side of the picture. If we want to calculate some effect, for instance the magnetic moment of the the electron or the level-splitting in hydrogen by means of an expansion to powers of a, we have to renormalize the theory. That means that the quantities e' and m' in the equations differ from the real charge and mass of the electron by a certain amount. This amount turns out to be infinite in this approximation. So actually e' and m' are infinite quantities and strictly speaking the equations have no sense at all, although they give the correct results. At the conference we had a very imporant contribution from K ä l l é n who looked into this matter very carefully without using an expansion into powers of a. Then apperently there are three possibilities:

 a) If we add up all the terms with different powers of a the infinities disappear, so that actually m' and e' are finite, although they will probably be different from the experimental mass and charge of the electron, the differences being finite quantities. In this case we would have a consistent theory.

b) The situation resembles the one we get by expanding with respect to a. This would mean that those quantities which can be observed are always finite but m' and e' are infinite From the formal equations (1) and (2) however one could derive equations which only contain finite quantities and which settle the experimental questions.

c) The series with respect to a will diverge and so the theory has no meaning as long as we try to confine our attention to electrons and light-quanta only. The theory can only be closed if we remember that there are other particles besides electrons and photons and therefore these other particles will have to be connected with the whole scheme. This latter possibility has specially been emphasized by Thirring[10]) and Hurst[11]). Rosenfeld has pointed out to us at the conference that one should again subdivide this possibility into two cases — but I will not enter upon these details. One result of the paper of Källén is that possibility *a)* is really ruled out. It is thus certainly impossible that m' and e' in (1)and (2) are finite quantities and that we have a finite and simple mathematical scheme.

The two possibilities which are left can simply be described in the following way: The theory is either "closed" or it is "open"; "closed" means that one can define quantum-electrodynamics in a consistent mathematical scheme without referring to other particles, and "open" means that one has to take the other particles, mesons, protons, *etc.* into account to make the theory consistent.

Now I will try to discuss what are the consequences of these two possible assumptions. Let us assume first that the theory is closed. Then it should be possible to formulate it without any infinities by going over from the original equations with e' and m' to other equations in which the experimental mass and the experimental charge of the electron occur. These equations however would probably be very complicated, which is not very satisfactory. Another consequence would be that the famous factor $e^2/\hbar c$, the fine structure constant, should be defined within this scheme. One idea to this has been presented by Casimir at the conference. He discussed the

Physica *19*, 897–908 (1953)

assumption that the Coulomb repulsion which tries to break the electron into pieces is compensated by the zero-point pressure of the light-quantum field. This might lead to a value for $e^2/\hbar c$. The electron was supposed to be a sphere with a radius of the order of e^2/mc^2, like in the old times. I would like to point out however that this assumption about the radius can scarcely be justified. If quantum-electrodynamics is a closed scheme then one must take the equations (1) and (2) seriously as they stand, and analysing them one finds that one can describe the electron either as a kind of jellyfish with an extension of 10^{-11} cm or as a point charge. There is really nothing in this picture which would say that the electron radius is of the order of 10^{-13} cm.

I will make one final remark about the assumption of a "closed" theory. Imagine that we want to describe the following experiment. An electron of a very high energy say 100 BeV hits another electron. We know that other particles such as mesons or may be pairs of nucleons will be created in such a collision and it is obvious that this can never come out of the closed theory. So if a closed theory would exist, it would be a theory which is undoubtedly wrong for very high energies.

Let us now turn to the assumption that the theory is open, which I think is the opinion of most physicists to-day. Then the situation looks very much better. Our equations can now quite well be understood from correspondence arguments because in this case the quantities m' and e' are almost the experimental mass and charge of the electron. There are only very small corrections of the order $e^2/\hbar c$ which distinguish m' and e' from the experimental values e and m. The renormalization is now only a small correction, not something fundamental and that is perhaps more satisfactory from the point of view of correspondence. In the second place one would not expect that the quantity $e^2/\hbar c$ would be fixed within quantum-electrodynamics. On the contrary, one would expect that this quantity will only be fixed much later, namely at the same time as the ratios of the masses of the different particles are fixed. If we take this point of view we can conclude that it is quite impossible to improve quantum-electrodynamics in any way except by going into the question of heavy particles like mesons, protons, *etc.*, because it is only there that any changes as compared with the present theory will occur.

Because it seems generally to be very difficult to write down theories which contain no infinities, it has been suggested long ago by W a t a g h i n that one should replace the fundamental equations by integral equations or "non-local" equations with a non-locality of the order of 10^{-13} cm. According to S t ü c k e l b e r g and F i e r z it would then follow, that in such small regions there are deviations from causality, where the term causality is used in the sense of the theory of special relativity. Also other extensions of the theory have been suggested. D i r a c spoke this morning at the conference on an entirely different possibility of extending the theory by introducing the concepts of ether and of absolute time without spoiling the Lorentz-invariance of the theory.

I would like to say now a few words about non-local theories. One should distinguish two kinds of non-locality. The first kind is really not a serious kind and it is probably better to describe it as due to the finite size of the elementary particles. One gets this kind of non-locality already, as we have seen, if one goes over from the original Maxwell equations to modified equations which contain the effect of virtual creation and annihilation of pairs. We have here no real deviations from the causality. There is however a much more serious kind of non-locality, and especially F i e r z [12]) has pointed out that as soon as one introduces integro-differential equations instead of differential equations there is a great danger that one loses causality altogether.

I should perhaps now say what one means with causality in this respect. We are not concerned with the deviations of causality which we always have in quantum mechanics, that is to say with the statistical behaviour of radiation and matter. We know however from the special theory of relativity that any action which is caused by some event at a certain point in space-time should only spread out with the velocity of light and that two events which are distant from each other by a spacelike vector should have nothing to do with each other. It is this kind of causality which could be spoiled if we introduce integro-differentiale equations, and actually it will be spoiled in many cases as was shown by F i e r z. There is still a chance that one can find theories which spoil this concept of causality only within very small regions, say again of the order of 10^{-13} cm, and especially the papers of M ø l l e r and K r i s t e n s e n [13]) have tried to construct theories of this type, but it is not certain that it is

Physica *19*, 897–908 (1953)

possible to do this in a satisfactory way. It may be that whenever one introduces the non-local character in such a way that the theory converges it always gives strong deviations from causality which we cannot accept.

As a second point in this connection I would like to emphasize that this problem of local or non-local theory must be settled by experiments, by research about collisions of particles with very high energies. From the uncertainty relations we know that we must have very big momenta at our disposal when we try to find a very good localization, and therefore it is only from collision experiments with very high energies that one can get some information about this question of non-locality. Here again there are three possibilities.

Les us first suppose we have a non-local theory with a length of the order of 10^{-13} cm which converges. Then I feel the result for the high energy collisions would be that any energy or momentum much bigger than the restmass of the π-meson becomes very improbable because this length of 10^{-13} cm corresponds to the restmass of the π-meson in the uncertainty relation. This however does not seem to be the case, for the experiments seem to show that sometimes very much higher energies are transferred.

The second case would be that we have a purely local theory with small interaction like in quantum-electrodynamics. Then the wave-functions ψ are more or less identical with those operators which create particles and that amounts to the particles having a strong point-centre; therefore we should expect a very strong transfer of energy and momentum, but only very rarely the creation of several particles. This again seems definitely not in agreement with the experiments.

The final possibility would be like this. The interaction is still of the local type, at least in the limiting case, where one can neglect quantisation, but the local interaction is very strong. Then of course the operator which creates a particle is very different from the wave function which obeys the local commutation relation and therefore the particle is much more washed out although it is still possible to transfer a large amount of energy. I think that this third possibility has the biggest chance to survive, but certainly a definite answer cannot be given now.

So you see that at the present moment all these questions are still completely unsettled. What can we do to improve the situation?

We can work from two sides, either from the experiments or from the mathematical methods. If one starts from the experiments, I think, the most useful attempt will be of the kind which has been presented to us by P a i s. The most important experimental results are not the masses of the particles, because it is always difficult to derive a theory from quantitative features, but the most important properties are selection rules and symmetry properties. P a i s has tried to use several conservation laws which so far have no adequate place in the theory, namely the conservation of the number of nucleons, the long lifetime of the V-particle and some similar features to derive a mathematical scheme which represents these properties of the particles by means of some invariance properties of the equations. P a i s introduces new variables and shows that one gets new invariance properties which will be sufficient to represent the selection rules which are observed. The other way would be to study what kind of mathematical schemes one can hope for to represent the theory of the elementary particles [14]). Here I would like to be more optimistic than many other physicists. Usually one hears sentences like this: It will be extremely difficult to derive later on the value of $e^2/\hbar c$ because then we will have to show that a certain theory only exists for a special value of this quantity. Or sometimes people say: if we have so many different particles and if we have to introduce a wave-function for every kind of particle, then the theory becomes so complicated that we can never hope to get a consistent scheme. Now I think it is not at all as difficult as that and I will now write down a formula which will certainly not give the correct theory of the elementary particles but which is only meant as a kind of foundation for optimism in this respect:

$$\gamma_\mu \frac{\partial \psi}{\partial x_\mu} = A\gamma_\lambda\psi(\psi^\dagger\gamma_\lambda\psi). \tag{3}$$

Here the constant A has the dimension of the square of a length. Let us assume (what is rather doubtful) that this equation can be quantized and gives convergent results. If that would be the case, the results of this equation would qualitatively (not quantitatively) agree with all what we know about the elementary particles. Say we have a state called vacuum, the relativistically invariant state of lowest energy; let us further assume that we have one discrete

Physica *19*, 897–908 (1953)

level. This discrete level would mean an elementary particle of a given mass. So we can write down the following equation

$$\Psi_{K,0} = O_{K,0}\Psi_0$$

The operator $O_{K,0}$ makes from the Hilbert-vector of the vacuum state Ψ_0 a one-particle state with energy K and momentum zero. From this one particle state we can by Lorentz-transformation form another state where the same particle is now moving with a certain velocity. We can superpose such Hilbert vectors and get a new Hilbert vector which represents that particle in a certain region of space, a kind of wave-packet in the region R. Then we can write

$$\Psi_{R_1} = O_{R_1}\Psi$$

where Ψ_{R_1} corresponds to a particle in the region R_1. Of course we can do that for any region in space. Finally we can have two particles of the same kind, one in region R_1 and the other in region R_2 with the Hilbert vector

$$\Psi_{R_1 R_2} = O_{R_1}.O_{R_2}.\Psi_0$$

If R_1 en R_2 are sufficiently distant from eachother the operators O_{R_1} and O_{R_2} commute; therefore there is no difficulty in writing down the state-vector and we see at once that we have not only a solution corresponding to one particle but to infinitely many of these particles. Furthermore we see that the equation (3) will probably have solutions with different masses, because we will probably have different discrete levels. The equation will lead to Fermi particles and Bose particles because whenever the operator $O_{K,0}$ is an odd function of ψ it will give a Fermi particle, when it is an even function it will give a Bose particle. So it appears that such a simple equation may lead to quite a number of different particles and obviously the ratio of the masses of these particles will be determined just as well as the different stationary states of the H-atom are determined in ordinary quantum mechanics. Also the quantity $e^2/\hbar c$ would be determined by such a theory for the equation would describe the interactions between the different particles in the same way as the Schrödinger equation of a molecule determines the interactions between different atoms in the molecule. It is also important to note that $\psi(x)$ itself, the wave function, is definitely not the operator $O_{K,0}$ but that this operator $O_{K,0}$ is a complicated function of ψ. Then one final remark. If somebody would say that ψ must be something like an operator creating a particle because one can assume first that the constant A is very small, this would be altogether wrong. Namely if we change the value of A we change really nothing in the equation. A change of A just means a change of the specific length, therefore the whole matter of the world would be changed simultaneously, all the eigenvalues would be changed by the same constant factor and so really nothing would be changed. So we see that it has no meaning in such an equation to expand with respect to the coupling parameter A.

I realize of course that one must be very careful with making statements which start with "If". Equation (3) may be too simple, and it is quite doubtful whether local theories of this type can be quantised. I only wanted to point out that so far physicists have looked for too complicated schemes. We have been accustomed to introduce one wave-function for the proton, one for the neutron, one for the π-meson, one for the electron, *etc.* and so we got hopelessly complicated schemes and have made things much more difficult than they are. I think there is a good chance that a wave equation of a very simple kind may be so rich as to comprise quite a number or perhaps all of the elementary particles. One should not try to extend the present foundations of physics and to study very complicated schemes unless one is absolutely forced to do so. The only reason to go over to more complicated systems than equation (3) may be the experimental evidence about selection rules, and here I would like to quote again the paper of P a i s. The number of conservation laws which we can derive from an equation is given by the number of invariance properties of the equation. It may quite well be that in this respect our equation (3) is too simple and therefore such an extension as tried by P a i s is natural. Generally one should not try too many extensions. It would however be extremely valuable if somebody succeeded in finding at least one mathematical example which really converges. This of course is the central problem again and here I can certainly not offer any solution. I wanted only to say that especially with regard to the complication of our picture of elementary particles one should not be too pessimistic.

Received 8-8-53.

REFERENCES

1) B e t h e, H. A., Phys. Rev. **72** (1947) 339.
2) T o m o n a g a, S., Progr. theor. Phys. **1** (1946) 27.
3) S c h w i n g e r, J., Phys. Rev. **74** (1948) 1439; **75** (1949) 651; **76** (1949) 790.
4) D y s o n, F. J., Phys. Rev. **75** (1949) 486; **75** (1949) 1736.
5) F e y n m a n n, R. P., Phys. Rev. **76** (1949) 749; **76** (1949) 769.
6) L a m b Jr., W. E. and R e t h e r f o r d, R. C., Phys. Rev. **72** (1947) 241.
7) U e h l i n g, E. A., Phys. Rev. **48** (1935) 55.
8) S e r b e r, R., Phys. Rev. **48** (1935) 49.
9) E u l e r, H. and K o c k e l, B., Naturwiss. **23** (1935) 246.
10) T h i r r i n g, W., Helv. phys. Acta **26** (1953) 33.
11) H u r s t, Proc. Camb. phil. Soc. **48** (1952) 625.
12) F i e r z, M., Helv. phys. Acta **23** (1950) 731.
13) K r i s t e n s e n, P. and M ø l l e r, C., Dan. mat. fys. Medd. **27** (1952) no. 7.
14) H e i s e n b e r g, W., Z. Naturf. **5a** (1950) 251; (1950) 367.

In *Louis de Broglie: Physicien et Penseur* (Albin Michel, Paris 1953) pp. 283–286

REMARQUES SUR LA THÉORIE NEUTRINIENNE DE LA LUMIÈRE

par

W. HEISENBERG

Il y a longtemps, Louis de Broglie [1] a indiqué la possibilité de représenter les propriétés d'une onde lumineuse, en interprétant cette onde lumineuse comme une combinaison d'ondes de spineurs. Le quantum de lumière individuel doit être considéré ainsi comme composé de la sorte, de deux neutrinos, ou plus généralement de deux particules spinorielles. La relation mathématique, qui sert de fondement à une telle théorie, a été, depuis, étudiée de divers côtés [2]. Elle intervient d'une façon générale, si l'on étudie par exemple les ondes sonores, qui peuvent se propager dans un gaz de Fermi.

Cette idée, que le quantum de lumière est aussi un système composé, a, ces dernières années, vu croître sa probabilité d'être justifiée grâce à la découverte de nombreuses nouvelles particules élémentaires. En effet, l'on doit manifestement considérer toutes les particules élémentaires comme des états stationnaires du même système, la matière. Les fonctions d'ondes spinorielles doivent nécessairement intervenir dans la représentation du champ

de la matière, car il existe de nombreuses particules élémentaires de spin demi-entier. Ainsi, le quantum de lumière serait aussi composé d'un nombre pair de particules spinorielles. Cette description, maintenant, soulève quelques difficultés caractéristiques, qui proviennent de cette circonstance que le quantum de lumière n'a pas de masse au repos ; ces difficultés doivent être discutées dans ce qui suit.

Pour préciser la représentation intuitive, nous voulons supposer que le quantum de lumière est formé de deux neutrinos. Si l'on peut admettre que ces neutrinos et le

(1) L. de Broglie, *C. R.* **213**, 473 (1936).
(2) P. Jordan et R. de L. Kronig, *Z. f. Physik*, **100**, 569 (1936). P. Jordan, *Z. f. Physik*, **102**, 243 (1936), **105**, 114 (1937). S. Tomonaga, *Progr. theor. phys.*, **5**, 544 (1950).

quantum de lumière ont une masse au repos finie, comme l'a fait L. de Broglie, alors l'on pourrait considérer le système à partir du centre de gravité du quantum de lumière. Les deux neutrinos devraient se trouver dans un état stationnaire, d'extension finie dans l'espace, où ils seraient liés l'un à l'autre comme l'électron et le proton dans l'atome d'hydrogène. Si l'on se ramène maintenant, par transformation, au système du laboratoire et si l'on passe au cas limite de la masse au repos tendant vers zéro, alors l'extension dans l'espace du quantum de lumière dans la direction du mouvement se réduit à zéro, par contraction de Lorentz, quelqu'ait pu être primitivement l'extension de la fonction d'onde dans cette direction. Dans le système de référence du laboratoire, le quantum de lumière serait ainsi un élément plan, dont le plan serait perpendiculaire à la direction de propagation de la lumière.

C'est à ce caractère bidimensionnel du quantum de lumière qu'est lié ceci : il est apparu beaucoup plus simple de former une théorie neutrinienne de la lumière à une seule dimension, par rapport à la direction de propagation, plutôt qu'une théorie à trois dimensions. Si les neutrinos et le quantum de lumière ne peuvent, fondamentalement, se propager que dans une direction (ou la direction exactement opposée) il ne serait besoin d'aucune force pour maintenir les deux particules ensemble ; car alors les deux

neutrinos ne pourraient être séparés par une perturbation extérieure (par exemple, dispersion) que si l'un des neutrinos n'inverse complètement sa direction de propagation ; et cela ne pourrait se produire que sous l'influence de grosses forces. Il en va différemment dans l'espace à trois dimensions. Ici, deux neutrinos, qui chemineraient en quelque sorte, par hasard, l'un près de l'autre et dans la même direction, pourraient être séparés par dispersion à l'aide d'un très petit échange d'impulsion. Ici, une force est ainsi nécessaire pour maintenir ensemble les deux neutrinos. Cette force doit relier ensemble les deux particules, de façon que la fonction d'onde du quantum de lumière n'ait qu'une extension finie dans le plan perpendiculaire à la direction de propagation ; en ce cas, il est indifférent que l'on considère chacune des deux particules spinorielles avec ou sans masse au repos, mais il est important, cependant, que la masse au repos totale s'annule.

Si l'on cherche maintenant à introduire dans la théorie des interactions entre les deux neutrinos répondant aux exigences de la théorie de la Relativité, alors surgit la difficulté fondamentale suivante : d'après Stueckelberg ([3]) et Fierz ([4]), dans une théorie qui maintient la condition de causalité de la théorie de la Relativité restreinte, ne peuvent survenir que des interactions du type des fonctions S_F ou Δ_F de Feynman (c'est-à-dire d'un potentiel retardé pour l'émission, avancé pour l'absorption). Des interactions de cette espèce ne peuvent, en effet, entraîner, fondamentalement, aucune liaison des deux neutrinos ; en effet une action émise à partir d'un neutrino avec la vitesse de la lumière ne peut rattraper l'autre neutrino, car celui-ci se trouve dans une autre position du plan perpendiculaire à la direction de propagation, et se déplace lui aussi avec la vitesse de la lumière. Si, malgré cela, on veut admettre des inter-

286 actions, il faut introduire des actions à distance, qui vérifient les conditions d'invariance de la théorie de la Relativité. De telles actions à distance peuvent être construites ; mais elles ne vérifient pas la condition de causalité et ne peuvent plus être exclusivement représentées par les fonctions S_F et Δ_F. Ceci signifie que dans de très petits domaines d'espace-temps, la succession dans le temps de la cause à l'effet peut être permutée.

Si l'on veut ainsi maintenir l'idée que les quanta de lumière sont composés de particules spinorielles — et en faveur de cette conception militent des arguments généraux de poids — il faut alors abandonner une théorie causale dans le sens de la théorie de la Relativité, et arriver à une théorie généralisée, où l'espace et le temps, dans de très petits domaines, seraient estompés d'une manière particulière. Ces dernières années, l'on a développé, de diverses parts, les bases et les débuts de telles théories, mais nous ne nous en occuperons pas ici. Nous nous bornerons à souligner ceci : la pensée exprimée par L. de Broglie en 1936, que les quanta de lumière doivent être aussi considérés comme des édifices composés, conduit à des problèmes de principe de la même importance que ceux que souleva la découverte célèbre des ondes de matière.

Werner HEISENBERG.

(Prix Nobel)

Max Planck-Institut für Physik,
Göttingen (Allemagne).

(3) E. STUECKELBERG et D. RIVIER. *Helv. phys. Acta*, **23**, 215 (1950).
(4) M. FIERZ. *Helv. phys. Acta*, **23**, 731 (1950).

Traduit de l'allemand par Jacques WINTER (École Polytechnique).

In *Proceedings of the 1954 Glasgow Conference on Nuclear and Meson Physics,* ed. by E. H. Bellamy, R. G. Moorhouse (Pergamon, London New York 1955) pp. 293–295

PART VI

FIELD THEORY

Quantization of non-linear wave equations

W. Heisenberg

Max-Planck-Institut für Physik, Göttingen

MANY new elementary particles have been discovered in recent years, and it seems likely that similar new experimental results will be obtained in the near future. Still, the qualitative picture of the elementary particles and their behaviour will probably not be changed appreciably. Therefore it seems reasonable already now to look for a theoretical description of the elementary particles which agrees qualitatively, not quantitatively, with what we know about their properties.

For this purpose it will certainly not be useful to introduce a new wave function for each new elementary particle, since such a mathematical scheme would be extremely complicated and could scarcely lead to an explanation of the masses and other properties of the particles. The simplest alternative is the introduction of only one wave function for matter. This wave function would have to obey a non-linear field equation in order to represent the interaction of matter, and it has to be studied whether such an equation can lead to a number of different elementary particles with reasonable properties (HEISENBERG).

As an example, the wave equation

$$\gamma_\nu \frac{\partial \psi}{\partial x_\nu} + l^2 \psi(\psi^+\psi) = 0 \tag{1}$$

will be discussed, $\psi(x)$ being a spinor wave function, considered as an operator in a Hilbert-space which is to be defined later. It is necessary to start with a spinor wave function, because otherwise it would be impossible to get Fermi-particles as discrete eigenvalues of the system. Equation (1) cannot be renormalized, therefore the normal rules of quantization can possibly not be applied.

In order to get qualitative information about the commutation rules, the following expression can be studied:

$$\chi_\alpha(x,x') = e^{-i[a_\nu\psi_\nu^+(x') + \psi_\nu(x')a_\nu^+]} \psi_\alpha(x) e^{i[a_\nu\psi_\nu^+(x') + \psi_\nu(x')a_\nu^+]} \tag{2}$$

a_ν being an arbitrary constant spinor that anticommutes with $\psi(x)$ and $\psi^+(x)$. $\chi_\alpha(x,x')$ obeys the wave equation (1), if $\psi_\alpha(x)$ obeys it. If a_ν is chosen as very small, $\chi_\alpha(x,x')$ can be expanded with respect to a_ν:

$$\chi_\alpha(x,x') = \psi_\alpha(x) - ia_\nu[\psi_\alpha(x)\psi_\nu^+(x') + \psi^+(x')\psi_\alpha(x)] + ia_\nu^+[\ldots] + \tag{3}$$

Since it is natural to assume, that the commutators vanish for space-like distances $x - x'$, one sees that $\chi_\alpha(x,x')$ corresponds to solutions of the wave equation (1) which differ from ordinary solutions by a disturbance which

293

W. Heisenberg

starts at $x = x'$ and expands into the future and the past within the light-cone $s = (x_\nu - x'_\nu)^2 = 0$. $\chi_\alpha(x,x')$ cannot be analytical at the light-cone.

In conventional quantum theory one would expect, that $\chi_\alpha(x,x')$ can be divided into two parts

$$\chi_\alpha(x,x') = \chi_\alpha^0(x,x') + c_\alpha(x - x') \tag{4}$$

where $c_\alpha(x - x')$ is a c-number (except for a sign-function) and contains the singular part near the light-cone $s = (x_\nu - x_\nu')^2 = 0$, whereas $\chi_\alpha^0(x,x')$ is a q-number function, which vanishes at the light-cone. $c_\alpha(x - x')$ may for instance be taken as the vacuum-expectation value of $\chi_\alpha(x,x')$. In this case, $c_\alpha(x - x')$ would have to obey a wave equation of the type

$$\gamma_\nu \frac{\partial c}{\partial x_\nu} + l^2 c(c^+ c) + c\kappa(s) = 0 \tag{5}$$

where $\kappa(s)$ is a function of $s = (x_\nu - x'_\nu)^2$, which results from a vacuum expectation value of the type $(\chi_\alpha^0(x,x'))^2$; for the discussion of the behaviour of c near the light-cone $\kappa(s)$ may be replaced by a constant value κ.

The solutions of (5) that are important for the determination of the commutation rules behave for large values of s like Bessel-functions, or more accurately, like the ordinary Schwinger S-function of the linear-wave theory.

For small values of s, however, $c_\alpha(x - x')$ oscillates more and more quickly, so that at $s = 0$ c_α has an essential singularity where the oscillation becomes infinitely fast. If a_ν in (2) decreases, the region of fast oscillations is limited to the immediate neighbourhood of $s = 0$. In the limiting case $a_\nu \to 0$, $c_\alpha(x - x')$ behaves everywhere outside of $s = 0$ like the ordinary S-function in the linear theory. At the light-cone $s = 0$, however, it does not behave like the Dirac δ-function $\delta(s)$ or its derivative; it remains undetermined for $s = 0$, its integral over a region near $s = 0$ being infinitely small when the region tends to zero.

These results allow us to determine the qualitative behaviour of the vacuum expectation value of the commutator

$$S_{\alpha\nu}(x - x') = \langle 0\,|i[\psi_\alpha(x)\psi_\nu^+(x') + \psi_\nu^+(x')\psi_\alpha(x)]|\,0\rangle \tag{6}$$

$S_{\alpha\nu}(x - x')$ behaves for $s \neq 0$ qualitatively like the Schwinger-function, but for $s = 0$ it is undetermined without leading to a function of the δ-type. This holds, if equation (4) is accepted, and shows, that even on a surface $t - t' = 0$ the commutation rules are different from the usual ones.

A change in the commutation rules of this kind involves great changes in the whole structure of the theory. The most important one concerns the Hilbert-space, on which $\psi(x)$ operates. If one characterizes the states of the system by their state vector and the corresponding mass, then one may divide the Hilbert-space into two parts.

Hilbert-space I contains all states up to a very large but finite mass M, Hilbert-space II all others. The states of Hilbert-space II contribute in (6) only to the behaviour of $S(x - x')$ in the immediate neighbourhood of $s = 0$. The states of Hilbert-space I contribute to (6) ordinary S-functions and give rise to δ-functions at $s = 0$. A regular Hilbert-space II, in which the usual rules of quantum theory hold, could not subtract these δ-functions to give the function (6) described. Therefore one is forced to assume, if one wants to keep

294

Quantization of non-linear wave equations

the relation (4), that Hilbert-space II is to be replaced by a symbolic Hilbert-space, which does subtract the δ-functions and which therefore does not contain states which can actually occur in nature. Therefore in Hilbert-space II the usual rules of quantum theory do not apply.

This means radical changes in the whole theory as compared to the usual quantum theory of fields, and further calculations will have to show whether a theory of this type can qualitatively lead to results corresponding to what one observes.

For the calculation of eigenvalues the so-called new Tamm-Dancoff-method has been used. For the mass of the lowest Fermion this method gives in the second approximation

$$\kappa_{1/2} \approx 7{\cdot}45/l$$

The mass of the lowest Boson has not yet been calculated. Calculations on the forces between two Fermions, however, indicate that there are long-range forces of the Coulomb type, the exact nature of which has not yet been worked out; such long-range forces should be connected with Bosons of rest-mass zero, but definite results cannot yet be given.

It should be emphasized that in a theory of the type described particles without interaction do not exist; the interaction is determined together with the mass and other properties of the particles.

REFERENCE

[1] For the literature compare W. HEISENBERG; *Z. Naturforschung* **9a,** 292, 1954 where the necessary quotations are given.

DISCUSSION

B. F. X. Touschek: Is the mass determined from the behaviour of the S_1-function?

W. Heisenberg: Yes. A Tamm-Dancoff method is used.

R. E. Marshak: There are two types of matter, Fermi-Dirac and Einstein-Bose particles. For example, Fermi interactions are weak, and should not this provide a very distinct separation of Bose and Fermi interactions, which should appear in a general theory?

W. Heisenberg: This is perhaps true; however, you do obtain Bose particles as well as Fermi particles correctly in my method.

G. Breit: Do the successes of the electromagnetic-field theory of radiative corrections persist in this method?

W. Heisenberg: I have not yet investigated this.

C. Möller: Is the mass determination a convergent process?

W. Heisenberg: In the Tamm-Dancoff method there are an infinite number of equations which are broken off and the convergence of the mass determination process as the break-off proceeds to infinity is not yet proved, but it seems likely. One gets good results in the first approximation.

Il Nuovo Cimento (10) 2, Supplemento, 96–103 (1955)

The Production of Mesons in Very High Energy Collisions (*).

W. HEISENBERG

Max-Planck-Institut für Physik - Göttingen

When nuclear encounters take place at very high energies, say of 100 GeV or more, then usually many mesons are produced. This does not necessarily mean that already in a nucleon-nucleon collision many mesons are produced because it might be that a production of many mesons takes place through a cascade process inside a nucleus. This was the view which was for a long time held by HEITLER. On the other hand, I think there has recently been pretty good evidence that also in a single nucleon-nucleon collision many mesons can be produced. Best evidence for this are perhaps the jets which are occasionally observed and in which there are great numbers of shower tracks without the appearance of any heavy tracks. Although it is in most cases not probable that these events are actually simple nucleon-nucleon collisions one may say that even if they constitute reactions with nucleons at the surface of a nucleus the nucleus cannot have participated in the process very much since else one would have observed the evaporation of particles. That such events are possible has been emphasized by the theory long ago and the attempts have also started very early to connect the theoretical estimates with experiment. I remind you of the papers by WATAGHIN — but it was only in recent years that one has tried to give a more detailed theoretical description. Such description was given by FERMI and myself and recently by LANDAU. In the following I shall not repeat the details of FERMI's and my own papers but I will give a survey of LANDAU's work which is perhaps not generally accessible since it was published in Russian. At the end I will compare the results of the three theories with recent experiments. LANDAU's paper appeared after those of FERMI and myself and tries to improve on the angular distribution of the particles as found by FERMI.

We describe everything in center of mass system. To begin with we have two

---

(*) Reproduced from the magnetophone-tape after the lecture of the|Author.

Il Nuovo Cimento (10) *2*, Supplemento, 96–103 (1955)

nucleons which are pictured in Fig. 1 as very thin disks (Lorentz-contraction). Where the two disks collide we have a region of very high energy density and of extremely strong interaction and there- fore one can assume that a large proportion of the proton energy goes into the mesonic field. As well in Fermi's as in Landau's theory the assumption is made that the whole kinetic energy is transferred into the mesonic field, whereas I assumed that this proportion depends on the degree of interact- ion. It is small when the impact parameter is large, whereas the total energy goes into the mesonic field when the particles collide head-on.

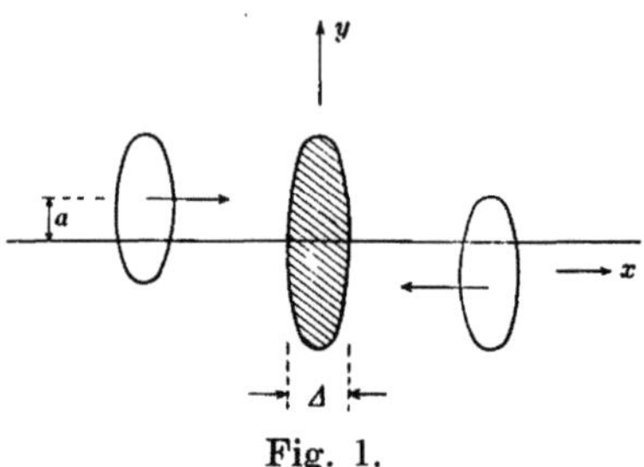

Fig. 1.

Now at the beginning we have a very high energy density of the mesonic field concentrated in a small region and we must ask what happens. In Fermi's theory the interaction in the meson field in supposed to be so strong that we can consider this as a state of temperature equilibrium, the temperature simply being calculated from the energy density. It is assumed that the mesons are a kind of Bose gas, the energy density is given by the initial energy of the process, and thereby one can calculate the temperature from Boltzmann's law. FERMI then assumes that when the region of the meson field starts ex- panding the temperature equilibrium is at once frozen in, the energy distri- bution and the number of particles stay constant.

In my paper I did not consider the mesons as a Bose gas but rather as a wave-field which obeyes a non-linear wave-equation, the non-linearity represent- ing the strong interaction. So I simply had to calculate how such a nonlinear- wave equation would lead to an expansion and to what kind of expansion.

LANDAU tries to improve the Fermi theory. He is not satisfied with the assumption that the particles spread out without any further interactions iso- tropically into space. For some time, he says, there will still be interaction because the energy density is still very high. When the energy density has fallen below a certain critical value, only then one may expect the interactions to become so weak that they do not affect the energy distribution. Now LANDAU has defined this point as the one at which the temperature reaches the order of $\mu c^2/k$ (where μ is the rest mass of π-meson and k is the Boltzmann constant). In the region of strong interaction the particles will be accelerated by hydro- dynamic pressure almost only in the x-direction (Fig. 1) and therefore the angular distribution will be much less uniform than in Fermi's theory. LANDAU distinguishes three steps of the whole development:

1) A hydrodynamic state of affairs where the gas which is first con- centrated in a small region, expands nearly linearly to the sides. Therefore

                                    W. HEISENBERG

we can treat this as a two-dimensional problem in x and t. The process is still non-adiabatic, i.e. the interaction is so strong that heat-transfer will take place from one element to another with the total entropy slightly increasing. When the linear expansion has reached a certain degree than also the expansion sidewards will come into play and then he considers the expansion as an adiabatic process. Finally a state is reached in which the temperature in that part of the volume which is considered goes down to the order of $\mu c^2/k$. From this moment on the interaction is broken off, LANDAU assumes that then the energy and angular distribution remain constant. He arrives in this way at angular and energy distributions which are different from FERMI's.

He assumes, however, that the total number of particles is the same as in the Fermi theory, he actually normalized his number of particles so as to fit the Fermi theory. We now consider the steps in detail:

First, the period from t_0 to t_1 (Fig. 1). LANDAU starts from the energy-momentum tensor

$$T^{ik} = pg^{ik} + (\varepsilon + p)u^i u^k \; ;$$

p (hydrodynamic pressure in the system where the gas is at rest) equal to, $\varepsilon/3$ for pure Bose gas, neglecting the rest mass,

ε energy density in the rest system,

$\vec{u}$ four dimensional velocity with which a volume-element in the gas is moving

$$g^{ik} = \begin{pmatrix} -1 & & & 0 \\ & 1 & & \\ & & 1 & \\ 0 & & & 1 \end{pmatrix} \qquad u = \left\{ \frac{1}{\sqrt{1-\beta^2}} , \; \frac{\vec{\beta}}{\sqrt{1-\beta^2}} \right\} .$$

In this bose-gas the temperature T is proportional to $\varepsilon^{\frac{1}{4}}$, therefore the entropy goes as $\varepsilon^{\frac{3}{4}}$.

He now applies to the tensor T^{ik} the ordinary rules

$$\frac{\partial T^{ik}}{\partial x^k} = 0 \; ,$$

For our two-dimensional problem this reduces to two equations. If one considers only those parts of the gas which move approximately with the velocity of light ($u = u^0 \gg 1$) these equations can be replaced by the following:

$$\frac{\partial}{\partial t}(\varepsilon u^2) = -\frac{1}{4}\frac{\partial \varepsilon}{\partial \zeta} , \qquad \frac{\partial \varepsilon}{\partial t} = -\frac{\partial}{\partial \zeta}\left(\frac{\varepsilon}{u^2}\right) ; \qquad \zeta = t - x$$

of course such hydrodynamic equations have very many solutions and it is not quite easy to pick out such solutions which represent the phenomenon well;

Il Nuovo Cimento (10) 2, Supplemento, 96–103 (1955)

therefore I feel that the solution LANDAU picked out is perhaps somewhat arbitrary, but at least one can check that it is actually an approximate solution. Here it is

$$\varepsilon = \varepsilon_0 \exp\left[-\tfrac{4}{3}(\eta + \tau - \sqrt{\tau\eta})\right] \quad \text{where} \quad \tau = \lg(t/\Delta) \quad \eta = \lg(\zeta/\Delta).$$

For the above ε, u^2 must be of the following form:

$$u^2 \approx \frac{t}{2\zeta}\sqrt{\frac{\tau}{\eta}},$$

ε and u^2 are solutions only if we neglect the differential quotient of τ and η at this point.

Because of $T^{00} = \varepsilon u^2$ and $u^2 \sim t/\zeta$ we obtain for the energy dE in a layer of thickness $d\zeta = \zeta \, d\eta$

$$dE = \varepsilon a^2 u^2 d\zeta,$$

where a is the radius of the disk. Therefore

$$dE \sim \exp\left[-\tfrac{1}{3}((\sqrt{\tau} - 2\sqrt{\eta})^2\right] d\eta.$$

In this manner one finds for the entropy

$$dS \sim \exp\left[-\tfrac{1}{2}(\sqrt{\tau} - \sqrt{\eta})^2\right] d\eta.$$

If you plot the energy density as a function of ζ you obtain qualitatively Fig. 2.

2) We now come to the period $t_1 \to t_k$. The instant t_1 depends on the position of the volume element in the gas, it is not the same all over the gas.

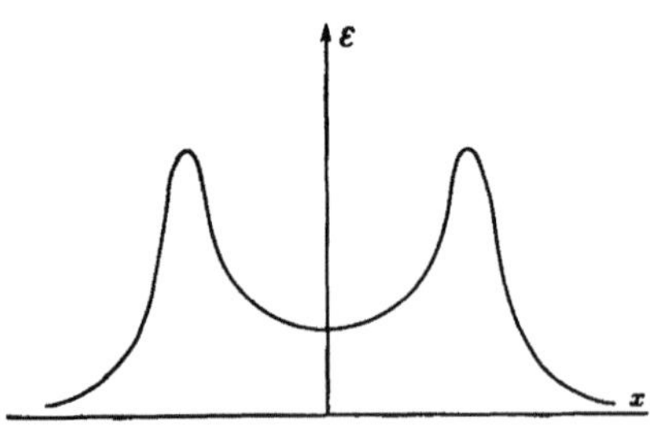

Fig. 2.

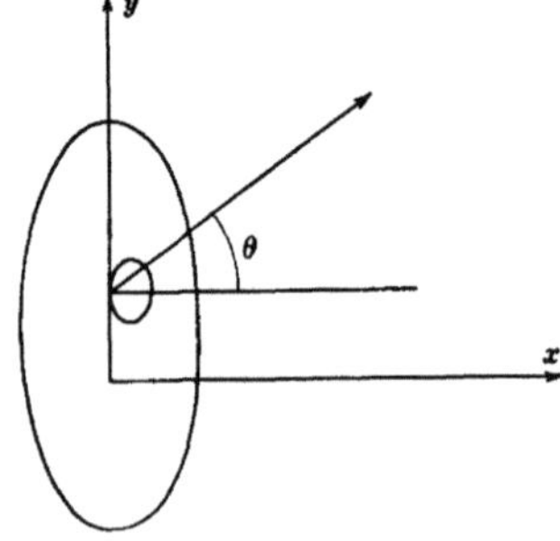

Fig. 3.

Let us now assume in this step $(t_1 \to t_k)$ that the expansion can take place in all directions. The question is under what direction the different volume

                                  W. HEISENBERG

elements are emitted. LANDAU says that the velocity of any volume element
is caused by the acceleration due to hydrodynamic pressure.

The angle θ (compare Fig, 3) is then determined by,

$$\theta = \left(\frac{\partial T_{yy}}{\partial y}\right) \Big/ \left(\frac{\partial T_{xx}}{\partial x}\right),$$

which is the ratio of force in y-direction to that in x-direction. In y-direction
the energy density decreases smoothly and the derivative $\partial T_{yy}/\partial y$ will be
of the order of magnitude ε/a. While $\partial T_{xx}/\partial x$ will be of the order ε/ζ. Therefore we have

$$\theta \sim \zeta/a \ .$$

Now it seems easy to calculate the value of t_1. The one-dimensional solution is valid if $t\zeta \ll a$; therefore it has the time limit $t_1 = a^2/\zeta$.

Thus in the middle the volume elements start to spread out to all directions
soon, but at the outer edge they are always pressed in the forward direction
by hydrodynamic pressure.

Beyond the time t_1 Landau assumes that the process is adiabatic.

$$\varepsilon u^2 t^2 = \text{const} \ .$$

This means that the energy density times volume is constant. Whereas the
relation $s u t^2 \sim \text{const}$ follows from the fact that the entropy density times
volume is constant. From these two equations we have

$$u \sim t$$

$$\varepsilon \sim 1/t^4 \ .$$

Just at the time t_1, we introduce $\tau_1 = L + \lambda$ and $\eta_1 = L - \lambda$, where

$$\lambda = \lg (a/\zeta) \qquad \text{and} \qquad L = \lg (a/\Delta) = \lg (E/2Mc^2) \ .$$

Therefore one obtains for the entropy

$$ds \sim \exp \left[\sqrt{L^2 - \lambda^2}\right] d\lambda \ .$$

Because of the proportionality between the entropy and the number of particles
in the volume elements $(dN \sim ds)$ this is equivalent to

$$dN \sim C \exp \left[\sqrt{L^2 - \lambda^2}\right] d\lambda \ .$$

Il Nuovo Cimento (10) *2*, Supplemento, 96–103 (1955)

Having introduced $\lambda = -\lg \mathrm{tg}\,(\theta/2)$ one obtains the angular distribution in the center of mass system:

$$\mathrm{d}N \sim \exp\left[\sqrt{L^2 - \lambda^2}\right](\mathrm{d}\theta/\sin\theta)\,.$$

The constant factor is gained by normalization:

$$\int \mathrm{d}N = N\,.$$

3) *Determination of time t_k*. The solution having the required properties for $t \geqslant t_1$, is

$$\varepsilon = \varepsilon_0 \left(\frac{t_1}{t}\right)^4 \exp\left[-\frac{4}{3}\,(2L - \sqrt{L^2 - \lambda^2})\right].$$

the instant t_k of the free spreading-out of the particles is given by

$$t_k = t_1 \left(\frac{\varepsilon_0}{\varepsilon_k}\right)^{\frac{1}{4}} \exp\left[-\tfrac{1}{3}(2L - \sqrt{L^2 - \lambda^2})\right],$$

where

$$\varepsilon_k \sim \frac{\mu c^2}{(\hbar/\mu c)^3}\,,$$

ε_k being the energy density for which $T_k = \mu c^2/k$. t_k again is different for the different volume elements.

Let us now calculate the velocity u of the volume element at the instant t_k, i.e. the velocity with which the mesons are left when the interaction has stopped

$$u = \frac{t_k}{t_1}\,u_{t_1} = \frac{t_k}{t_1}\sqrt{\frac{t_1}{\zeta}} = \frac{t_k}{a}\,.$$

The energy-distribution is therefore:

$$\mu u \sim M \exp\left[-\frac{L}{6} + \lambda + \tfrac{1}{3}\sqrt{L^2 - \lambda^2}\right],$$

$$(\mu,\ \text{meson mass};\ \ M,\ \text{proton mass}).$$

Passing over to the angles one gets

$$\mu u \sim \frac{\text{const}}{\theta} \qquad \text{for small } \theta.$$

Having finished this short report on Landau's calculations, let us now go into the physics of these equations. One important result is that mesons of high energy are emitted only under small angles θ. I.e., the angular distribution is very anisotropic in the sense that only slow mesons can be emitted in all directions.

The energy distribution is not very different from Fermi's and is essentially given by

$$\mathrm{d}E \sim \exp\left[-\tfrac{1}{3}(\sqrt{\tau} - 2\sqrt{\eta})^2\right].$$

This energy distribution has however a high energy tail in such a manner that the number of particles of high energy E decreases as $\mathrm{d}E/E$.

This high energy tail is longer than in Fermi's or my theory. We shall now make a detailed comparison of the three theories.

In both the Fermi and Landau theories we have in the initial state a Bose gas with a fixed number of particles and very strong interacton. In my theory we have a mesonic field of very high energy concentration. It is a well defined field, not a statistical assembly like a Bose gas, which will obey a nonlinear wave equation. In the Fermi and Landau theory practically all kinetic energy of the nucleons goes into the meson field, whereas in my theory the possibility of glancing collisions is taken into account where only a small part of the kinetic energy goes into the meson field.

For the transition $t_0 \to t_1$ all the 3 theories are rather different. Perhaps Landau's theory here resembles mine more than Fermi's since it takes into account the interaction between the particles at this stage. Both theories also give a strong dependency of the number of high energy mesons on the angle of emission, that is a strong anisotropy, contrary to Fermi's theory.

From t_k onwards the 3 theories make the same assumption of no further interactions taking place.

The energy distributions resulting from the theories are qualitatively given in Fig. 4.

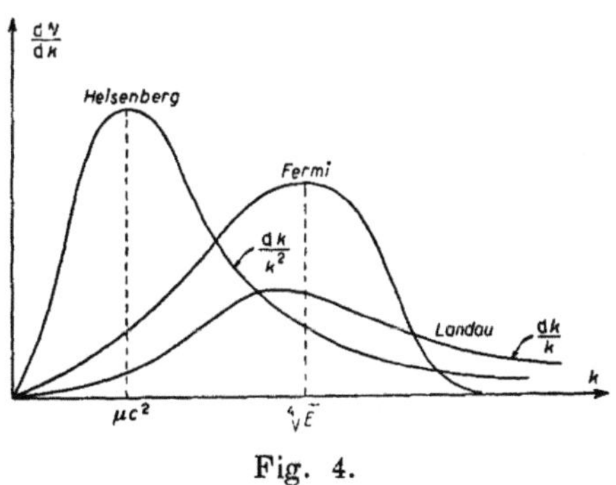

Fig. 4.

As to the comparison with experiment there seems to be good evidence that the multiplicities are rather low at high energies. If one assumes the energy distribution of my theory to be correct it is possible to get the right multiplicities only by making plausible assumptions on the fraction of the energy going into the meson field. If, on the other hand, the glancing collisions should not play an important role one would have to conclude from the small number of mesons that the energy distribution as given by Fermi or Landau is better.

Il Nuovo Cimento (10) *2*, Supplemento, 96–103 (1955)

Which of the two alternatives is true has not yet been decided by the experiments. I might mention that in the one case which so far still seems to be the most useful one (20 shower particles being spread out over an angle of about 50 degrees) the average measured energy of the secondaries is much smaller than given by the Fermi theory and seems to fit quite well my theory. But one case is really no case and one has to wait for further experimental results.

Now with regard to the angular distribution the famous shower of Schein shows a strong anisotropy in the center of mass system. This has been interpreted by Fermi by taking into account angular momentum conservation.

Still the anisotropy is so marked, that it may be taken as an argument in favour of Landau's or my theory.

There is another argument which comes from the observation of large air showers. There, too, seems to be the indication of a very strong anisotropy

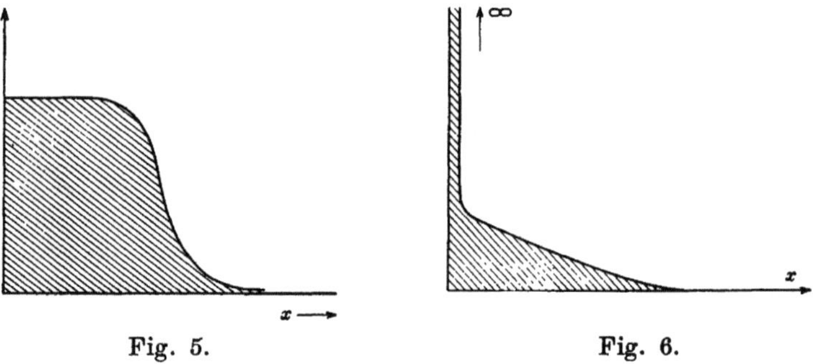

Fig. 5.   Fig. 6.

since the shower cones are always distributed within distances only of the order of 10 or 20 cm.

In conclusion I should like to mention that the angular distribution is very important for the study of the structure of elementary particles.

It is easy to see that the angular distribution depends on the distribution in space of the mass density of the nucleon. In Landau's and my theory a smooth distribution (Fig. 5) was assumed which gives a strong anisotropy whereas in current theories like quantum electrodynamics one uses distribution functions of δ-function type (with a tail) (Fig. 6) which would result in rather isotropic angular distributions. For these reasons it is important to study the angular distributions in order to obtain information about the structure of elementary particles.

Science *122*, 1215–1220 (1955)

Theory of Elementary Particles

W. Heisenberg

It is obvious that at the present state of our knowledge it would be hopeless to try to find the correct theory of the elementary particles. On the other hand, one may try to form some kind of picture of how such a future theory of elementary particles will look, because, even if we realize that we know only very few details about the elementary particles, we have already quite a good qualitative picture of them, and we feel that even if the experiments go on for 5 or 10 or even more years, this qualitative picture will scarcely change.

Perhaps the best way to start this subject is to give a short review of what we know about the elementary particles, and then the problem of the theory will not be to find the correct theory, but rather it will be to find a model of such a theory. That is, one can, even at the present time, make the attempt to construct theories that at least qualitatively give something very similar to the elementary particles that we see now in nature. Only at a much later stage can we hope to find the correct theory. What I shall try to tell here is in some ways quite ambitious, because it is a model for the real theory of elementary particles—that is, a theory that comprises all knowledge about atomic events, in one single mathematical scheme. On the other hand, it is not so very ambitious, because it is *not* an attempt to find such a theory but only to find a theory that qualitatively resembles it—in other words, a kind of model of such a theory.

Our Knowledge of
Elementary Particles

What do we know about the elementary particles? First of all, we know that there is a great number of different elementary particles. We know a mass spectrum of such particles and the masses of many. For instance, the mass of the proton is 1836 times larger than the mass of the electron, so the electron seems to be an especially light particle. Most other particles seem to be heavier by at least a factor of 100. If we consider these masses as something similar to the stationary states in the hydrogen atom, then we see that some of these masses are stable states, and others are unstable states.

The electron apparently is a stable particle and has, therefore, a very sharply defined mass. The proton also seems to be a stable particle, but the neutron is not stable. The neutron can decay, emitting a proton and electron and neutrino. The neutron has a lifetime of roughly a quarter of an hour. Then there are the mesons, which are still much more unstable. Their lifetime is very much shorter. The μ meson has a lifetime of 2×10^{-6} second. The π meson has a lifetime of 2.5×10^{-8} second. The neutral π has a lifetime of only 10^{-15} second. So we see that all different degrees of stability may occur, and as a rule one can assume that when the particles get heavier and heavier, then the chances that they are stable are smaller and smaller, so that probably above a certain mass value all particles will have only an extremely short lifetime. Therefore they will not have a well-defined mass, and then it is of no use to speak about elementary particles.

Dr. Heisenberg is director of the Max Planck Institute of Physics, Gottingen. This article is based on a tape recording of a lecture given before the Indiana Academy at Purdue University, Lafayette, Ind., 15 Oct. 1954.

We know still more about the particles, or, I should say, about the results of any future theory of elementary particles. For instance, we know that any such future theory of elementary particles must contain some invariance properties. It must obey, for instance, all the properties of invariance that are involved in the Lorentz transformation. So they must be invariant for what one calls the inhomogeneous Lorentz group. These invariance properties will lead to a number of conservation laws—conservation of energy, momentum, angular momentum. In connection with quantization, they will also lead to the fact that the angular momentum always is either an integer multiple of $\hbar$ or a half quantum integral of $\hbar$. So all these results must come out of such a theory of elementary particles.

For experiments, these conservation rules mean, for example, that we have selection rules that some particles can only decay into certain other ones. And sometimes we even do not know yet what the selection rules are that apparently are present. For instance, according to all known conservation laws, we think that a proton could disintegrate into a positron and one or several light quanta. But we see that this is not the case; so there must be new conservation laws and, therefore, new invariance properties that have not been accounted for in the present theories.

If we take all this qualitative knowledge together, it seems reasonable to believe that even in 5 or 10 years from now the general picture 'of this knowledge will probably not have been changed. In 5 or 10 years from now we will certainly know a number of new particles beyond those that we know already. We will have better knowledge of the cross sections and of the production probability of these particles. We will know in what number these particles are created in high-energy collisions, and so forth. But still qualitatively this picture will not have been changed. One important feature of this picture also is that all these particles are connected. By *connected* I mean that when we have a sufficient amount of energy at our disposal—when, for instance, two elementary particles collide at very high energies—then apparently any other type of particle can be created, either directly in the collision or some time after the collision through radioactive decay. So we cannot divide

all existing elementary particles into different groups that have nothing to do with one another. Such a division is in principle certainly not possible. All the elementary particles are connected.

A Wave Equation for Matter

Let us take this qualitative picture of matter, of the behavior of matter, and ask: How can the theory of elementary particles possibly look? We can say: Since we must have in this theory the invariance for the inhomogeneous Lorentz group, it is very natural that such a theory will in some way be connected with a wave function depending on x, y, z, and t. Because, if we write the wave equation for such a wave function, then it is easy to do it in such a way that the invariance for the inhomogeneous Lorentz group actually is present. This wave equation that we want to write, however, will certainly not be a wave equation for a special kind of waves—light waves or meson waves—or a wave equation for nucleons, or anything like that, because the mesons, light quanta, and nucleons must come out of the equation; they cannot be put into it. So this wave equation, if it exists at all, will be an equation for matter, not for any special kind of elementary particles.

What kind of wave function do we have to introduce to represent matter? We may think, just because we have no other mathematical tools, of functions that are scalars, or spinors, or vectors, or tensors—in any case some of these relativistic functions or operators. It would certainly not be convenient or sensible to start by assuming that this wave function of matter is a scalar or a vector, because then this wave equation could never lead to spinor particles. On the other hand, if we assume that this wave function of matter is a spinor, then there is a chance to represent not only the spinor particles but also the scalar and the vector particles, because, if we start with half integers for spin quantum numbers, we can also get spins that are integers by taking several of these half quanta together; but if we start with integral spin numbers, we can never get the half spin quantum numbers. So it looks natural to assume that, if someday we can write a wave equation for matter, this should be a spinor equation.

Then again, one could think of a spinor equation that is just a linear wave equation, like the Dirac equation. This, however, could certainly not represent the facts, because we know that all elementary particles interact. A linear wave equation, however, will never lead to any interaction, and therefore one cannot expect a linear equation to represent the experimental situation for the elementary particles. So we have to start with a nonlinear equation for a spinor wave function, and we shall see whether we can in this way get a model for a theory of the elementary particles.

What is the simplest nonlinear wave equation for a spinor wave? I think I can quickly write it.

$$\gamma_\mu \partial \psi / \partial x_\mu + l^2 \psi (\psi^+ \psi) = 0 \qquad (1)$$

Here, $\psi^+_{\text{Heisenberg}} = \overline{\psi}_{\text{Schwinger}} = \psi^\dagger \beta = -i \psi^+_{\text{Pauli}}$, where $\psi^\dagger$ is the hermitian conjugate of ψ and β is the Dirac matrix operator for $(1 - v^2/c^2)^{1/2}$. One can argue that other equations are just as simple or perhaps slightly simpler, but essentially this is a very simple equation. As I said before, I do not believe that this is necessarily the correct wave equation. I just want to see whether such a wave equation can lead to a picture of the elementary particles, which at least qualitatively represents what we know about them.

This wave equation has two parts. The first part is just part of the ordinary Dirac wave equation for a spinor function, $\gamma_\mu \partial \psi / \partial x_\mu = 0$. That would be a Dirac equation for neutrinos. Then there

is added a term where l represents a constant of the dimension of a length and $l^2 \psi (\psi^+ \psi)$ is an interaction term. It is the simplest interaction term one can write. It must be a term of the third order, because with a spinor function it would not be possible to have a second-order term with the correct transformation properties. One could imagine other terms of the fifth order and the seventh order, but this seems to be the simplest one. Also, instead of this term of the third order, one could take other terms with some γ operators in them, but this would not essentially change the situation.

I think qualitatively such an equation seems to be a reasonable starting point for a theory of matter. The question is: Is there any chance that the quantization of such an equation will lead to an ensemble of elementary particles, some stable, others unstable, from which one can then calculate other interactions, and so forth? The next question is: Can such an equation be quantized according to the methods that we know for the quantization of wave fields?

The answer to this latter question is: No, because we know now from the theories of Schwinger and Tomonaga, Feynman, and others that in the quantization of fields, one will always run into the so-called "divergency difficulties," and this can be overcome only in some cases by a formalism, which is called the process of renormalization. Not all equations can be renormalized. On the contrary, we can divide all possible interactions into two types: one type can be renormalized and shows what can be called weak interaction; the other type has what we may call strong interaction, and for strong interactions this process of renormalization does not work. This interaction here, however, belongs to the strong-interaction type, and regardless of what kind of nonlinear wave equation we would write for spinor waves, we would always get the strong-interaction type, which cannot be renormalized. Therefore, we have to invent a new scheme of quantization. We have to change the rules of quantization in such a way that on one side we still preserve those fea-

Science *122*, 1215–1220 (1955)

tures of quantum theory which we know must be true and still avoid the divergence difficulties and get to mathematical schemes that really work.

Commutation Relationships

The next and most difficult problem in connection with such a wave equation is the question: What assumptions can we make about the commutation relationships? So, we will now be interested in a commutator between ψ at one point and ψ or ψ^+ at another point. Let me write this commutator.

$$\{\psi_a(x) ; \psi_\nu^+(x')\}_+ = - iS_{a\nu}(x, x') \quad (2)$$

This commutator is, in this case, written with a + sign between the two expressions on the left side, because we expect for a spinor wave the anticommutation rules that we know from Fermi statistics. The sum $\psi_a(x)\psi_v^+(x') + \psi_v^+(x')\psi_a(x)$ is, in the ordinary theory, 0 for any nonvanishing spacelike distances between x and x' and becomes a delta function when the points are close together. This anticommutator (multiplied by i) is usually called the S-function, after Schwinger who made much use of it. The problem is: Can we for this nonlinear theory define a new S-function which in a linear theory would be the Schwinger function?

Let me first state some of its general properties. In the linear theory we know that the anticommutator must be 0 whenever the distance between x and x' is a spacelike distance. This is a necessary condition if we want to preserve the properties of causality that follow from the theory of special relativity. From the theory of special relativity, we learn that all action can be propagated only with a velocity less than or equal to the velocity of light. This means that when two points in a four-dimensional world have a spacelike distance, then no action can go from one point to the other, and vice versa. Therefore, at two such points the wave function must always commute, or in this scheme anticommute, because otherwise it would mean that we would

have a deviation from ordinary causality. Therefore, we can form the picture shown in Fig. 1. By "0" we indicate the

regions where the anticommutator shall be 0. It shall be different from 0 in what one calls the future cone and the cone of the past. The dividing lines between this future cone and those parts where the commutator is 0 form the so-called "light cone." These are the points to which a wave can be propagated with the velocity of light.

Such functions as S in the linear theory are called propagation functions, because they really represent only waves that obey the normal wave equation and are propagated as perturbations from a certain point. There is a singular point $x = x'$, $t = t'$, and from this point a wave propagates into the future or into the past. The function that represents the anticommutator is just such a propagation function. This is so in a linear theory. This is quite understandable, because in a linear theory the commutator itself must obey the wave equation. In a nonlinear case, however, this is not true, and we have first to find the connection between the "propagator" on the one side and the commutator on the other side. Now we have to invent some kind of mathematical trick to see the connection between the propagation functions and the commutator. To find this mathematical connection I will have to write a few formulas. Consider the equation

$$\chi_a(x, x') = \exp\{ - i[a_\nu\psi_\nu^+(x') + \text{conj}]\}$$
$$\psi_a(x) \exp\{i[a_\mu\psi_\mu^+(x') + \text{conj}]\} \quad (3)$$

Right in the middle we find the operator $\psi_a(x)$. Of course, our ψ's are not only functions now, they are also operators, and they shall be noncommuting quantities. And this ψ_a is multiplied on the left side and on the right side with certain factors, which are each other's reciprocal.

The quantities a_ν or a_μ appearing in the exponents shall be the components of an arbitrary spinor with the property of anticommuting with all the wave functions ψ_a. This a_ν is just introduced as a mathematical tool to get the right connection between the propagation functions and the commutator. The factors on both sides of $\psi_a(x)$ depend on x' but not on x. What we have introduced is nothing but a canonical transformation of $\psi_a(x)$ independent of x or α, and

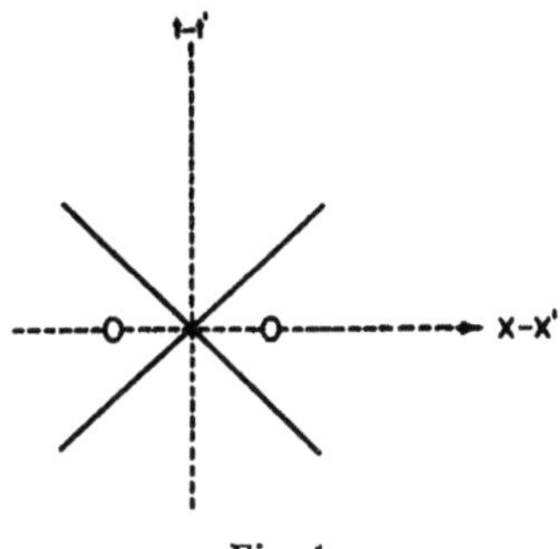

Fig. 1.

therefore one can easily see that the new function $\chi_a\ (x,\ x')$ also obeys the wave equation

$$\gamma_\mu \partial\chi/\partial x_\mu + l^2\chi(\chi^+\chi) = 0 \qquad (1a)$$

Now we can study what the χ will be like. If we assume that this arbitrary spinor a_ν we have introduced is a very small quantity—and since it is arbitrary we can *take* it as very small—then the expansion with respect to a_ν is the following:

$$\chi_a(x, x') = \psi_a(x) - ia_\nu$$
$$\{\psi_\nu^+(x')\,;\psi_a(x)\}_+ + \ldots \qquad (4)$$

The first nonlinear expression appearing in the expansion of χ with respect to a_ν is just the anticommutator. And now we can see the relationship between commutator and propagation function. We see that χ apparently corresponds to a solution of the wave equation, which is not a smooth solution but is a solution where superimposed on a smooth solution there is such a perturbation as we have seen from our picture (Fig. 1)—that is, a perturbation that starts from a point $x = x'$. So the χ is a kind of propagation function. It represents a solution of the wave equation that has a perturbation starting from one point. If we assume that the a_ν is very small, then it is a very small per-

turbation at the point $x = x'$. And the commutator is then the difference between the original smooth solution and this perturbed solution. So now we know qualitatively at least what the connection between commutator, on the one side, and propagation function, on the other side, must be. The commutator must correspond to the difference between two solutions of the original wave equation, one of which is smooth and the other has this perturbation at the point $x = x'$.

Knowing this we can go further and write our function χ_a, the operator that we have defined, in the following way. We can say

$$\chi_a(x, x') = \chi_a{}^0(x, x') + c_a(x - x') \qquad (5)$$

and this definition is to be understood in the following way. The singularity of the function χ_a on the light cone will be contained completely in the c-number function $c_a(x - x')$ (more correctly, the products of the amplitudes a_ν and a c-number). Such a division is actually possible in any present-day quantum theory, because in present-day quantum theory we always assume that the anticommutator at the origin, at the point $t = t'$, is a c-number. For instance, we usually write the anticommutator as a delta function. Here we assume that we do not know which kind of c-number function c_a $(x - x')$ is, but it is an ordinary function and not a field operator creating or annihilating particles. Therefore we may say that we split the χ_a up into one part $\chi_a{}^0$ which is an operator, but which is smooth at the light cone, and another part c_a which contains the main singularities and is a c-number.

As I said before, this splitting into two parts has always been possible in present-day quantum theory and will, of course, in a linear theory lead to the ordinary commutation relationships. For instance, we can assume that $c_a(x - x')$ is just the vacuum expectation value of the operator $\pm \chi_a$. Then we can put the χ into the wave equation and, if we have done so, we can take the vacuum expectation value of the wave equation, and we find the wave equation for c. Actually it turns out that this function ς is a solution of the original wave equation with only slight modifications. Let me write it:

$$\gamma_\mu \partial c / \partial x_\mu + l^2 c (c^+ c) + c \kappa(s) = 0;$$
$$s = \Sigma_\mu (x_\mu - x_\mu')^2 \quad (6)$$

For the c the wave equation that one gets is just the same as the ordinary wave equation that is on the left side and is the one from which we started; however, there is one term added, which is the function c times a function of the space-time distance $(x - x')^2$. Since this last term will not affect the behavior of the c near the light cone, and we are interested only in the behavior of c near the light cone, we can just as well assume that this is approximately constant and say c times a certain constant κ, which we can adjust according to what is convenient in the equation.

This gives us a qualitative picture of how the commutator will look near the light cone and how we can derive these properties from a solution of the wave equation. I should mention the following point before I go on. What we need for all further discussion is just the behavior of the commutation relationship near the light cone, because in ordinary theory we already know that we can derive the whole theory if we know only the commutator in the immediate surroundings of the point $x = x'$, $t = t'$. So also here we can be quite satisfied with knowing the commutation relationship very near to this point, because all the rest can be derived by integrating the wave equation. That is, we can then proceed from the time t to time $t + dt$, and so forth, and thereby we can get the whole solution. We are interested only in the behavior near the light cone, and this behavior we can get from solving Eq. 6.

The Solutions

Now I do not want to go into the mathematics of the solution, but I would like to write the solutions in the form of a few pictures. If we solve the same problem for the linear wave equation, then,

of course, we would also find for the function c just a linear wave equation. The term with the third power of c would be left out, and we would get as a solution for the anticommutator the well-known propagation function of Schwinger. Let us for a moment assume that we are not dealing with spinor particles but are dealing with scalar particles, and then we do not have to deal with the S-function of Schwinger but with the Δ-function of Schwinger, which is a function of the distance between x and x' only. This makes it easier to draw pictures, because then we have one single function of only one variable s as in the function shown in Fig. 2.

Vertically we have the Δ-function of Schwinger, and horizontally we plot the space-time distances between the two points. In the case of the linear theory this commutation function of Schwinger, the so-called Schwinger Δ-function, has the following property. It is a Bessel function for all finite distances s, and it is a Dirac δ-function just at the point $s = 0$. So at the Δ-axis we have drawn the δ-function as s that would go to infinity, and the oscillating function for positive s is the Bessel function. (Fig. 2.)

This would be the solution of our wave equation for c, if the nonlinear terms were not present. Now we have to study the behavior in the case of the nonlinear theory, where we will first draw a picture for those cases in which the constant a_ν is still finite and not infinitely small. Then, of course, what we get is not the commutator but actually a sum of terms, the first of which is the commutator, and then there are higher terms, which, of course, somewhat change the picture, so that only in the limit for $a_\nu \rightarrow 0$ it will become the commutator. The picture will then look like Fig. 3.

For large space-time distances the function again will be nearly a Bessel function, because then the nonlinear terms are very small and do not have strong influence. But for small space-time distances the influence of the nonlinear terms is felt, and then there are some deviations from the old picture. Hence, out at the right we have the Bessel function again, but in the inner part it turns out that there are very fast oscillations so that the function starts oscillating quite rapidly near $s = 0$, resulting in an infinitely frequent oscillation with infinite amplitude very near the origin. It is readily apparent that such a function can easily be integrated over the whole distance from $s = 0$ to any finite value of s. If we now go to the case where the $a\nu$ is exceedingly small, then this region of very fast oscillation moves always closer and closer to the origin, so finally we are left with a Bessel function for all finite values of s, and only in the origin do we have the fast oscillations (Fig. 4). This means that

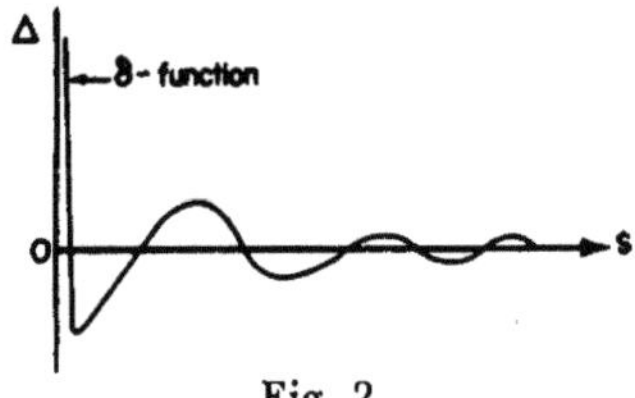

Fig. 2.

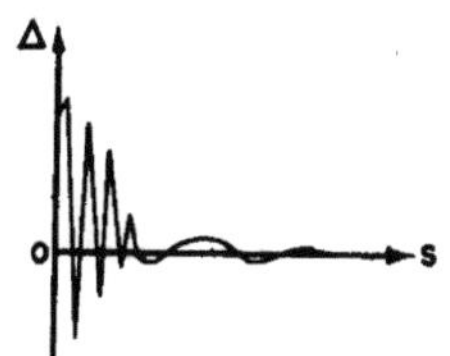

Fig. 3.

now our commutator in the case of the nonlinear theory does almost look like the commutator in the linear theory. The only difference is that the delta function at the origin disappears, and instead of the delta function we have an infinitely fast oscillation; that is, we have an essential singularity, and therefore the value of the function is not defined at this point, but the integral is defined and is always 0 if we only go close enough to the origin. Really *the only difference is the disappearance of the delta function.*

Now this actually helps a lot for the whole divergence problem, because as soon as one starts with the commutation function that has no delta function at the origin, all the divergencies disappear, and we do get a convergent theory.

Before going on, I must write one function that can be derived from the Schwinger propagation function S, which Schwinger calls the S_1-function. This S_1-function is derived from the propagation function S in the following way. One writes a Fourier expansion of the S_1-function, changes the sign of all the Fourier components of which the frequency has a negative sign, and gets a new function, which can be derived from the old one by an integral operation. This function, multiplied by i, is called the S_1-function, and theoreticians know the properties of this function. Here I have a special reason for writing it, and I will just do so in spite of the fact that at this moment it cannot be understood very well why it is useful:

$$S_1(x - x') = i\,[S^{(+)}(x, x') - S^{(-)}(x, x')] \doteq$$
$$\frac{\kappa^2}{4\pi}\,\gamma_\mu\,\frac{\partial}{\partial x_\mu}\,\mathrm{Im}\left[\frac{H_1^{(1)}(u)}{u} + \frac{2i}{\pi u^2} - \frac{i}{\pi}\ln\left|\frac{\gamma u}{2}\right|\right] - \frac{\kappa^3}{4\pi}\,\mathrm{Im}\left[\frac{H_1^{(1)}(u)}{u} + \frac{2i}{\pi u^2}\right]$$
$$\text{with } u = s^{1/2}\kappa; \quad s \lessgtr 0 \qquad (7)$$

The point is that this function of u or s has the one property that is important: namely, it contains terms that fall off slowly with distance. The terms in Eq. 7 not containing Hankel functions do *not* appear in Schwinger's work and are added here because of the omission of the Dirac δ-function from our S-function.

The next problem is: If we introduce such commutation relationships, have we any hope that this can lead to a consistent mathematical scheme of quantization? In order to explain why I believe that this can lead to such a scheme, I must go back to some of the mathematical fundaments of present-day quantum theory. I think one can understand these fundaments even if one does not go into the details.

In ordinary quantum theory, the commutation relationships would not be the ones that I put down here. As a matter of fact, in ordinary quantization of wave fields, one does start with rather similar commutation relationships; however, one includes the delta function at the origin,

Science *122*, 1215–1220 (1955)

and thereby one gets into all the divergency difficulties. If one omits the delta function, as we are inclined to do here, one ruins the theory completely. That is, one makes in this way a complete change, and the problem is: What price has to be paid for it? Certainly one does not get such a change for nothing. There must be some very serious deviation from ordinary quantum theory. And this I can explain in the following way.

The fundamental difference between quantum theory and classical theory is that in quantum theory not only the actual state of a system is important but at the same time all possible states of a system. For instance, when one is calculating the normal state of a hydrogen atom, it is not sufficient to know that the electron moves in an orbit of radius 10^{-8} centimeter, but it makes a difference whether the hydrogen atom is in a very small volume or in a very big volume. The eigen-states really are different, depending on whether the box in which the hydrogen atom is contained is small or big. That is, the possibility of the atom to get to very great distances is involved in the calculation of the eigenvalue. Or in calculating the scattering of a particle, we usually calculate it with the help of so-called "intermediate virtual" states. These states actually never are occupied, and yet for the scattering it is a problem to know what are the intermediate virtual—that is, possible—states. Therefore, contrary to classical theory, all possible values for a certain quantity are important in the mathematical formulas.

Coming back to the quantization of waves, we say that not only such wave functions as actually do occur in nature are important for quantization of waves but also all "possible" wave functions. From this aspect one comes very easily to an almost absurd conclusion. If, one says, space and time are really continuous in a mathematical sense, then the wave functions of the following type also

belong to the possible wave functions. Assume a wave function that has the

value 1 at every point where the coordinates have rational values and has the value 0 at every other point. Such a wave function is pure nonsense from the point of view of the physicist. Still it would be difficult in normal quantum theory to exclude any possible wave function from the mathematical scheme, even if it has some absurd properties, for instance, infinitely fast oscillations. This situation is probably the root of the so-called "divergency difficulties."

How do these divergency difficulties occur in the ordinary mathematical scheme? We usually say that all the stationary states of a quantum theoretical system define a certain Hilbert space. For instance the states of the hydrogen atom can be defined as the vectors in the Hilbert space, and we use it in quantum theory. Here I think it is reasonable to divide the Hilbert space into two parts. We can say that all existing stationary states up to a certain maximum energy, or rather mass, of the total system may be called Hilbert space No. 1. All other states may be called Hilbert space No. 2. The limiting mass may be extremely big. Let us assume the whole mass of the universe. Then it is obvious that only rather smooth functions can be expanded by using the states of Hilbert space No. 1 only, and for the infinitely many other wave functions one would either need Hilbert space No. 2 for expansion or one could not represent them at all. On the other hand, the states of the Hilbert space No. 2 do not occur in nature, and therefore it may be possible to change the rules of quantum theory with respect to these states of the second kind. That is what I do when I omit the δ-function. This change is actually necessary for the following reason: One can calculate the commutator by first going from the vacuum to the first group of excited states and then back to vacuum. Then I go from vacuum to the second group of excited states and back to vacuum, and so on. In any of these cases from every transition I get a function of this type:

$$<\Omega|\psi_a(x)|\Phi><\Phi|\psi_\nu{}^+(x')|\Omega>$$
$$\div <\Omega|\psi_\nu{}^+(x')|\Phi><\Phi|\psi_a(x)|\Omega>$$

where Ω is the vacuum and Φ the intermediate state. This is to be summed over Φ for getting the vacuum value of the anticommutator of Eq. 2. Now it has been shown in papers by Gell-Mann and by Low and by Källen and by Lehmann that each group contributes a δ-function at the origin and that all these δ-functions at the origin do not cancel, but they add up. If one says that the δ-functions at the origin cancel, it means that one has given up quantum theory for Hilbert space No. 2. One has sacrificed this

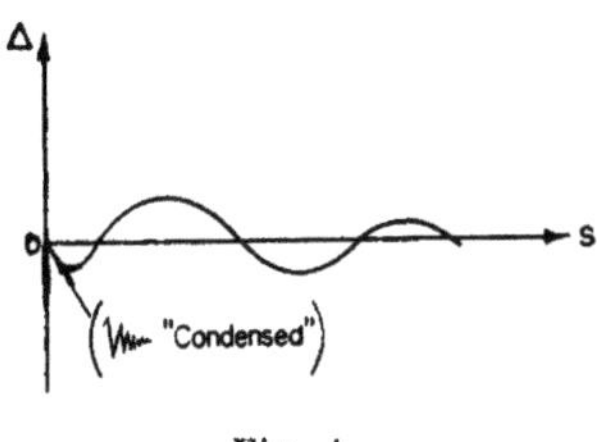

Fig. 4.

and, instead, has got some new mathematical scheme in which one has replaced the total Hilbert space by a thing which one may call a Hilbert space with a roof on top of it. I do not know whether this picture helps the mathematicians, but it shows the purpose of Hilbert space No. 2.

Having introduced this kind of Hilbert space, I am now far away from ordinary quantum theory but perhaps not too far from the experimental situation. So we replace Hilbert space No. 2 by a kind of imaginary Hilbert space, or by what I have called the roof on top of the Hilbert space.

The rest is just straightforward mathematics. So far we have had physical assumptions formulated in a mathematical language—assumptions about the physics of the problem—but from here on we have only mathematics. I shall not go into any calculations but shall just speak about the method and then give the results.

The method that can be used most conveniently is the so-called "new" Tamm-Dancoff method. It is a method that has been developed by Schwinger, Gell-Mann and Low, Freese and Zim-

mermann of Göttingen, and Goldberger, following an old paper of Tamm and Dancoff in 1941. So it is a rather well-known mathematical frame nowadays, and the great advantage of this frame is that one can work out a mathematical scheme in which one is interested only in matrix elements for those operators or products of operators that lead from the vacuum to a state of a finite energy. That is, one has to do only with matrix elements in Hilbert space No. 1, in which ordinary quantum theory shall be true. The whole contribution from the states of Hilbert space No. 2 comes only in the form of the commutation relationship. So it is quite sufficient for the calculation to know the behavior of the commutator near the point $x = x'$. This we have defined by means of the function given in Eq. 7. So we have actually a mathematical scheme by which we can calculate the energy eigenvalues, and it turns out that no divergency difficulties occur.

Now I will tell the results. One can ask: Are there stationary states; and, are there stationary states, say, with the spin $\frac{1}{2}$, so that the angular momentum is $(\frac{1}{2})\hbar$? The result is that there is a lowest stationary state with spin $\frac{1}{2}$, and the eigenvalue is given by putting κ (the energy or the rest mass of the system) equal to $7.45/l$. That it must have the factor $1/l$ in it is obvious, because l is a constant of the dimension of a length, and if $\hbar$ and c are made equal to 1, which is always done in these calculations, $1/l$ is the same as the dimension of a mass, and therefore this $7.45/l$ is just the eigenvalue of a mass.

We see that this equation leads to one particle, a fermion, which has this mass. If we assume that l is of the order of the Compton wave length of a π meson, which is a sensible assumption for this kind of a theory, then the mass of this particle turns out to be roughly that of the proton. Then one can also ask whether there are particles of Bose statistics, and of integer spin number. One can write the conditions for it, and actually one does get a kind of Bethe-Salpeter equation which leads to the existence of Bose particles. The mass for these Bose particles will again contain the factor $1/l$; and then it will depend on the numerical coefficient of $1/l$ for this Bose particle, whether it is stable or unstable. If the factor would turn out to be of the order of 20, then, of course, it would be unstable, because it could disintegrate into two of these fermions. If, however, the mass turns out to be only, say $1/l$, then it would be a stable particle and could correspond to the π mesons. (*Note added in proof*: Later calculations by Kortel, Mitter, and me have led to the eigenvalues $0.95/l$; $3.32/l$; $0.33/l$ and $1.74/l$ for the masses of Bose particles.) Strangely it turns out in the calculations, as far as we could see (but the calculations are not quite finished yet), that apparently one solution for the rest mass of the Bose particle is just 0. That is, one gets something like light quanta out of this calculation. The reason can be seen as follows. One can start with a different question and then it is easier to judge the answer. What is the interaction between two of the fermions we found? That, of course, follows again from the wave equation. All the interactions are defined by this wave equation. So one has just to calculate what happens when two such fermions are scattered by each other. And it is calculated by normal application of the Tamm-Dancoff method. Then it turns out that between two Fermi particles we have a long-range force of the Coulomb type —a force where the potential energy drops off as $1/r$. This long-range force is connected with the term

$$\frac{\kappa^2}{4\pi}\, \gamma_\mu\, \frac{\partial}{\partial x_\mu}\, \mathrm{Im}\left(-\frac{i}{\pi}\ln\left|\frac{\gamma u}{2}\right|\right)$$

$$\text{wit } u = s^{1/2}\kappa$$

appearing in our new S_1-function as given by Eq. 7. That is, just the very fact that one has omitted the δ-function in the

1220

commutator produces in the S_1-function an additive term that only very slowly decreases as the distance increases. For as I said before, these terms in Eq. 7 added to the Hankel function are the direct consequence of the omission of the δ-function at the origin. The scattering can be calculated in a very rough approximation with the Feynman graph shown in Fig. 5. Say we have two such fermions coming in, then we have to assume interaction through two more such fermions, and finally two come out. If one calculates this Feynman graph, then between these vertex points one has to put in the S_1-function, and therefore one gets an interaction of the Coulomb type. Really this Feynman graph is not a good approximation, so the calculation has to be done more carefully.

So, it seems that this equation has Bose particles of the rest mass 0 as eigenvalues, and this is, of course, a very interesting contribution to the problem of the elementary particles, because it shows that also the light quanta in the real theory of elementary particles may have to do with just this singularity of the S-functions at the origin. If the model of the elementary particles that is formed by the theory is correct, it would mean that the Coulomb forces—the electromagnetic forces—are for nature that method by which nature avoids divergency difficulties, which otherwise are always met in the theory.

Next Steps

The next problem is to calculate the higher boson states and also to calculate for them the corresponding value of $g^2/\hbar c$ and to find whether these quanta are scalar or vector quanta, and so forth. All this is just now in progress, so I cannot report the result. I just want to mention a few problems that are important and should be solved before one can take such a model quite seriously.

One of the most important questions will be: Does there exist some invariance property that corresponds to the gage invariance in electrodynamics? Only if this gage invariance is actually present, does one have a real analogy to the experimental situation. This gage invariance is, of course, decisive for the conservation of charge, the determination of $\dfrac{e^2}{\hbar c}$, and so forth. It may be that the gage invariance comes out by itself in the theory. This would be extremely interesting, if it were so. It may also be that it restricts the possible assumptions about the main interaction term. This would also be a very interesting result.

The next step would be to calculate the masses of the different Fermi particles and the different Bose particles and see whether that has any resemblance to the actual elementary particles.

This is the general picture of what I wanted to tell. I would like to add a few remarks about the difference between such a scheme and what one has hitherto done in the theory of elementary particles, especially in the quantization of wave fields. Usually in the quantization of wave fields one says: We have free particles, and there is an interaction. We first assume that the interaction is small, and then later we try also to account for strong interactions. Here we see that such an assumption would be complete nonsense. There is absolutely no meaning in saying: Let us first assume that the coupling, the nonlinear term, is small. If we would assume that this constant l, which has the dimension of a length, is small, this would not change anything in the theory at all, because it would just make a similarity transformation in the whole theory. That is, all masses would become bigger proportional to $1/l$, but the spectrum of these masses would not be changed. So it just means that the dimension of the whole world would change, but the eigenvalues and the whole spectrum—all that—would not be changed. Therefore in such a theory the idea of small interaction is just nonsense. Also the idea of free particles that have no interaction in a first approximation is nonsense in such a theory, because the particles are found in exactly the same mathematical frame in which all the interactions are found. That is, in such a theory, not only all the masses of the particles would be determined, but at the same time all the interactions would be determined. Therefore, in such a theory it is quite obvious that one would get a definite value, for instance, for the fine-structure constant $e^2/\hbar c$. Then, one may say: Is there not a danger that one still

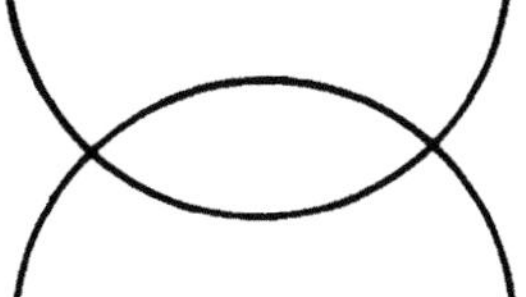

Fig. 5.

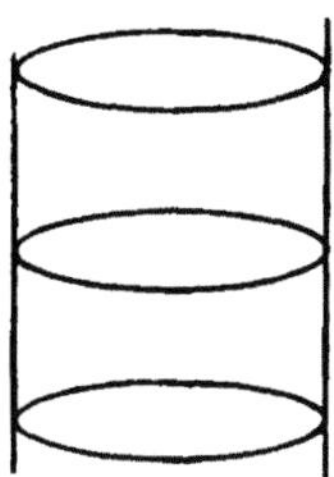

Fig. 6.

will come to some contradiction in such a theory, for instance, with respect to the law of causality? May it not be that, on account of introducing this rather strange commutation function, one gets deviations from causality, which then lead to a scheme that we cannot accept from the experiments? On the other hand, we have introduced the commutator from the closest possible analogy with the propagation functions, and the propagation functions, since they are calculated classically, are functions that obey the ordinary rules of causality. Therefore I put it that it seems as least unlikely that one gets into trouble with deviation from the causality, although one must admit that the theory is not yet so well studied that one can be absolutely certain.

There may also be some difficulties with the convergence of the mathematical scheme, but in any case one can say that all those divergency difficulties, which one knows normally from the quantization of waves, do not occur here. Whether other difficulties may occur—say, the question whether the Tamm-Dancoff method converges—is a different matter and remains to be seen.

So, generally speaking, I certainly do not say that this is already a good and sensible model for the theory of the elementary particles, but I would like to say

Science *122*, 1215-1220 (1955)

that, whatever one will in the future do for obtaining a theory of elementary particles, one will have to look in a similar direction as here. That is, one will have to look for a theory in which one does not start from the wave equation for the mesons or the nucleons or anything like that, but one will have to start with an equation for matter only, and one will have to try to derive all the different masses of the elementary particles from just one wave equation, of which these masses shall be the eigenvalues. So I think this tendency toward such a theory is almost necessary, but as to the exact form in which one will gradually be induced to give it, one will have to be led by what will come out of it.

Discussion. In what sense can one say that "fermion states" or "bosom states" are found from Eq. 1? Answer: Such states are obtained from the vacuum state Ω by operating on it by an odd or an even number of factors ψ or ψ^+. Maybe a meson interaction between fermions will look, as a Feynman diagram, like a ladder with loops. (Fig. 6). Again let me stress that the limit $l \to 0$ is meaningless; the theory goes over into "ordinary" quantum theory in problems of low energy where one does not consider transitions to virtual states of high energy, because one stays inside Hilbert space No. 1 anyhow.

Der gegenwärtige Stand der Theorie der Elementarteilchen *).

Von W. HEISENBERG, Göttingen.

In den ersten Jahrzehnten der Entwicklung der Quantentheorie betrachtete man die Elementarteilchen als etwas Gegebenes. Es handelte sich damals in erster Linie um die Elektronen und eventuell noch um die ganzen Atomkerne, deren Massen, Ladungen, magnetische Momente experimentell bestimmt, aber nicht als Gegenstand von theoretischen Untersuchungen betrachtet wurden. Dies änderte sich später, als neben den Elektronen und Protonen noch eine Reihe anderer Elementarteilchen entdeckt wurden: die Neutronen, μ-Mesonen, π-Mesonen, Neutrinos und viele andere, erst in der letzten Zeit genauer untersuchte Gebilde. Erst diese Vielzahl der Elementarteilchen wies die Physiker darauf hin, daß auch die Elementarteilchen einer Erklärung bedürfen, und daß es wohl eines Tages eine Theorie geben müßte, aus der man die Verhältnisse der Massen aller Elementarteilchen, ihre Ladungen, Drehmomente und magnetischen Momente würde verstehen und deduktiv ableiten können. Im folgenden soll zunächst eine historische Übersicht gegeben werden über die Versuche, die in den vergangenen Jahrzehnten unternommen wurden, um die Gesetzmäßigkeiten für einzelne Elementarteilchen oder für Gruppen von ihnen mathematisch zu formulieren. Die Perioden vom Abschluß der Quantenmechanik bis zum Beginn des Krieges und nach dem Krieg bis heute werden getrennt behandelt. Erst im Anschluß daran wird eine genauere Analyse der heutigen Situation versucht werden.

Schon relativ früh hatte sich die Erkenntnis durchgesetzt, daß zum Verständnis der Elementarteilchen die Quantelung der Wellenfelder die Voraussetzung sei. Denn schon die Untersuchungen EINSTEINs zur Strahlungstheorie hatten gezeigt, daß man die Lichtquanten durch Quantisierung der einzelnen FOURIER-Komponenten eines MAXWELLschen Feldes erhalten kann. Ferner konnten JORDAN, KLEIN und WIGNER kurz nach der Entstehung der Quantenmechanik nachweisen, daß die konsequente Quantisierung des unrelativistischen DE BROGLIEschen Materiewellenfeldes unter Berücksichtigung der elektrostatischen Wechselwirkung genau zur SCHRÖDINGER-Gleichung des Mehrkörperproblems der Quantenmechanik führt. Damit war die weitere Entwicklung grundsätzlich vorgezeichnet. Man mußte zunächst die allgemeinen mathematischen Methoden für die Quantelung der Felder entwickeln, sie auf die bekannten Materie- und Strahlungsfelder anwenden und konnte dann hoffen, vom Bekannten zum Unbekannten fortschreitend, zu einem Verständnis der verschiedenen Felder und ihrer Elementarteilchen zu gelangen.

Der Durchführung dieses Programms stellten sich aber schon bald sehr ernste Schwierigkeiten in den Weg. Zwar schien der mathematische Prozeß der Quantelung der Wellenfelder formal eindeutig und unproblematisch zu sein, aber bei seiner Anwendung auf die Wechselwirkung zwischen dem Elektronenfeld und dem MAXWELLschen Feld stellte sich heraus, daß sich eine unendliche Selbstenergie der Elektronen und andere Divergenzen ergaben, die eine physikalische Interpretation des vollständigen Gleichungssystems unmöglich machten. Nur wenn man sich bei der störungstheoretischen Behandlung mit einer niedrigen Näherung begnügt, erhält man noch konvergente und physikalisch brauchbare Ergebnisse.

Tatsächlich hatte diese erste Form der Quantenelektrodynamik erhebliche Erfolge aufzuweisen. Die DIRACsche Strahlungstheorie erklärte Streuung und Absorption des Lichtes durch die Atome, ebenso die spontane Emission und die Resonanzfluoreszenz. Es gelang in der Folgezeit sogar, Effekte vorherzusagen, die in der klassischen Theorie völlig fehlten und für die Quantenelektrodynamik charakteristisch waren: Die Polarisation des Vakuums in der Umgebung einer elektrischen Ladung und die Streuung von Licht an Licht; mathematisch gesprochen also nichtlokale und nichtlineare Abweichungen von den MAXWELLschen Gleichungen. Diese Effekte sind auch experimentell später nachgewiesen worden. Alle diese Erfolge erschienen aber von Anfang an bedroht durch die Tatsache, daß die Quantenelektrodynamik die schon erwähnten Unendlichkeiten enthält, die es unmöglich machen, der Theorie eine feste mathematische Grundlage zu geben. Natürlich setzten schon sehr bald die Bemühungen ein, die Ursachen für die Unendlichkeiten im mathematischen Formalismus aufzuspüren. Insbesondere tauchte·die Frage auf, ob die Unendlichkeiten etwa durch die Störungstheorie nur vorgetäuscht werden und bei einer strengen Behandlung der Gleichungen vermieden werden können. Für die Behandlung dieser Frage fehlten aber zunächst die mathematischen Hilfsmittel.

Als im Lauf der Zeit neben den Elektronen und Lichtquanten auch andere Elementarteilchen, Neutronen und Neutrinos, μ-Mesonen usw. in gewissen Experimenten in Erscheinung getreten waren und damit in das Blickfeld der theoretischen Physik gerieten, wurden auch die Wechselwirkungen der neuen Teilchen mit den alten nach den mathematischen Methoden der Quantentheorie der Wellenfelder behandelt. Die FERMIsche Theorie des β-Zerfalls aus dem Jahre 1934 ist ein besonders erfolgreiches Beispiel für eine derartige Anwendung der neuen Methoden.

Gelegentlich einer etwas späteren Untersuchung der FERMIschen Wechselwirkung stellte sich dabei eine wichtige Klassifizierung der verschiedenen Wechselwirkungen heraus, die für die weitere Entwicklung der Quantentheorie der Wellenfelder eine erhebliche prinzipielle Bedeutung erlangt hat. Die Wechselwirkung zwischen Elementarteilchen, die etwa in einem Stoßvorgang mit einer gewissen im Schwerpunktsystem gemessenen kinetischen Energie zusammenstoßen, kann entweder mit wachsender Energie zunehmen oder abnehmen oder konstant bleiben. Je nachdem, welcher dieser drei Fälle vorliegt, ergeben sich sehr verschiedene Folgerungen für die Brauchbarkeit einer

---

*) Vortrag, gehalten am 23. 9. 1955 auf der Physikertagung in Wiesbaden.

störungstheoretischen Entwicklung und für die physikalischen Vorgänge. Wenn die Wechselwirkung klein ist und mit wachsender Energie abnimmt oder konstant bleibt, so kann die Störungstheorie auch bei hohen Energien ausreichend konvergieren. Bei Stoßprozessen wird dann die Erzeugung eines einzigen neuen Teilchens immer sehr viel häufiger sein als die gleichzeitige Erzeugung mehrerer Teilchen, die ja in den höheren störungstheoretischen Näherungen auch vorkommt. Wenn jedoch die Wechselwirkung mit wachsender Energie zunimmt, so muß die Konvergenz der Störungstheorie bei einer gewissen Energie aufhören. Es können dann beim Stoß zweier sehr energiereicher Teilchen viele neue Teilchen auf einmal erzeugt werden, aber die Störungstheorie ist zur Behandlung solcher Vorgänge prinzipiell ungeeignet. Die Wechselwirkung zwischen Elektronen und Lichtquanten in der Quantenelektrodynamik gehört zur ersten Gruppe der Wechselwirkungen, die zwischen Elektronen, Neutrinos und Nukleonen in der Fermischen Theorie des β-Zerfalls gehört zur zweiten. Aus der Fermischen Theorie würde sich also eine Vielfacherzeugung von Elektronen und Neutrinos bei hinreichend energiereichen Stößen ergeben. Daß die Vielfacherzeugung von Elementarteilchen tatsächlich eine wichtige Rolle spielen kann, wurde experimentell erst viel später, und zwar an den π-Mesonenschauern, dann auch in der jüngsten Zeit bei γ-Strahlenschauern nachgewiesen. Zunächst aber konnte man feststellen, daß die Kopplungskonstante in der Fermischen Wechselwirkung beim β-Zerfall außerordentlich klein ist. Für die Erklärung der Lichtquanten- und Elektronenschauer in der kosmischen Strahlung genügte also zunächst die Kaskadentheorie von Bhabha und Heitler, Oppenheimer und Carlson, die nur die quantenelektrodynamischen Wechselwirkungen berücksichtigte.

Bei diesem Stand der Erkenntnis von den Elementarteilchen trat die durch den zweiten Weltkrieg bedingte große Pause ein, die in vielen Ländern für eine Reihe von Jahren die Forschung auf unserem Gebiet unterbrach. Das Interesse wandte sich den Anwendungen der Kernphysik zu. Doch sei aus der Zeit des Krieges ein Ergebnis erwähnt, das in den theoretischen Erörterungen der letzten Jahre eine gewisse Rolle spielt. Von den in der Quantentheorie der Wellen auftretenden Divergenzen war schon gesprochen worden. Da es sich damals in den mehr als zehn Jahren seit den Anfängen dieser Theorie als unmöglich herausgestellt hatte, die Divergenzen zu beseitigen, lag der Schluß nahe, daß die Grundlagen der Quantentheorie noch einer Abänderung bedürfen, wenn man sie auf die Wellenfelder anwenden will. Über die Natur dieser Abänderungen war einstweilen nichts bekannt. Wohl aber konnte man die Frage stellen, welche Teile der bisherigen Theorie bei einer solchen Abänderung wahrscheinlich unangetastet bleiben sollten. Dabei stellte sich heraus, daß die mathematische Größe, die das asymptotische Verhalten der Wellenfunktionen bei Stößen in weitem Abstand vom Stoßzentrum darstellt, die sog. S-Matrix, auch in einer zukünftigen Theorie existieren und ähnliche Eigenschaften aufweisen müßte wie in der bisherigen Theorie. Die Eigenschaften der S-Matrix wurden dann sorgfältig studiert, es sei insbesondere auf die Untersuchungen von Møller verwiesen, und aus dieser Analyse ergaben sich wichtige Rückschlüsse auf die allgemeine Struktur einer Quantentheorie der Elementarteilchen. Damit hatte man eine feste Grundlage für spätere Erweiterungen der Theorie gewonnen. Man konnte für diese spätere Theorie der Elementarteilchen z. B. an mathematische Formulierungen denken, bei denen das lokale Verhalten der Wellenfunktionen der Elementarteilchen nicht definiert wird, bei denen aber trotzdem eine unitäre S-Matrix abgeleitet werden kann.

In der Zeit nach dem Kriege gab es zunächst wichtige experimentelle Fortschritte, die eine Menge neuen Materials über die Elementarteilchen zu Tage förderten: die π-Mesonen, V-Teilchen und τ-Mesonen wurden entdeckt, ihre Masse und sonstigen Eigenschaften experimentell bestimmt. Aber über die experimentellen Fortschritte soll hier nicht berichtet werden.

Ganz unabhängig von dieser experimentellen Entwicklung aber gab es auch eine wichtige theoretische Entdeckung. Schon die Untersuchungen der dreißiger Jahre über die Quantenelektrodynamik hatten gezeigt, daß die Massen und Ladungen, die man für die Elektronen schließlich aus den Gleichungen erhält, verschieden sind von den Konstanten m_0 und e_0, die man in die Grundgleichungen hineinsteckt. Für Masse und Ladung ergeben sich sogar bei Anwendung der Störungstheorie unendliche Werte, wenn man mit endlichen Werten m_0 und e_0 begonnen hat. Soviel ich weiß, hat Kramers schon früh gelegentlich den Gedanken geäußert, man solle die Ausgangswerte m_0 und e_0 zunächst unbestimmt lassen und fordern, daß die endgültigen Werte e und m mit den experimentellen übereinstimmen. Dieses Programm der Renormierung der Konstanten ist zum erstenmal konsequent durchgeführt worden in einer Arbeit von Bethe aus dem Jahr 1947, in der Bethe den Einfluß der Strahlungskräfte auf die Energieniveaus des Wasserstoffatoms untersucht. Dabei stellte sich heraus, daß nach einer Renormierung der Elektronenladung und -masse die vorher genannten Divergenzen in der Störungstheorie verschwinden, und daß die Rückwirkung des Strahlungsfeldes auf die Elektronen gerade die Verschiebungen der Energieniveaus im Wasserstoffatom gegenüber der Sommerfeldschen Formel bewirkt, die experimentell schon seit langem vermutet und von Lamb und Retherford durch die Ultrakurzwellenmethode definitiv nachgewiesen worden waren. Die große Genauigkeit, mit der die Verschiebung durch die Theorie dargestellt wurde, ließ an der Echtheit dieser Erklärung keinen Zweifel mehr zu. Das war ein wichtiger Fortschritt, weil damit gezeigt war, daß die Quantenelektrodynamik einen viel höheren Wahrheitsgehalt hatte, als man bis dahin vermuten konnte. Insbesondere entstand nun die Hoffnung, daß die Quantenelektrodynamik durch den Prozeß der Renormierung zu einer mathematisch geschlossenen Theorie umgeformt werden könnte, in der Divergenzen überhaupt nicht mehr auftreten. Bald wurde auch gefunden, daß der Prozeß der Renormierung bei allen den Theorien möglich ist, die zu jener ersten Gruppe gehören, bei der die Wechselwirkung mit wachsender Energie entweder abnimmt oder konstant bleibt; während für jene andere Gruppe von Theorien, bei denen die Wechselwirkung mit wachsender Teilchenenergie wächst, der Renormierungsprozeß nicht durchgeführt werden kann. Es entstand also die Vorstellung, es gebe grundsätzlich zwei Arten von Quantenfeldtheorien, die renormierbaren und die nichtrenormierbaren. Für die renormierbaren Theorien ließen sich alle Divergenzen beseitigen, in den nichtrenormierbaren Theorien blieben sie bestehen. Nur die renormierbaren Theorien seien also mathematisch widerspruchsfrei, nur sie kämen für die Quantentheorie der Elementarteilchen in Frage. Daher gehöre auch die Theorie der π-Mesonen in ihrer Wechselwirkung mit den Nukleonen zu dieser ersten Gruppe. Die Experimente hatten in der Tat den Spin 0 für die π-Mesonen nachgewiesen, was die Renormierbarkeit ermöglicht. Nun muß ich gleich an dieser Stelle hervorheben, daß die eben geschilderten Vorstellungen, so verlockend sie scheinen, zweifellos falsch sind. Die renormierbaren Theorien lassen sich im allgemeinen nicht mathematisch geschlossen so formulieren, daß sie den Anforderungen der Quantentheorie entsprechen, und die Experimente über elektromagnetische Vorgänge großer Energie in der kosmischen Strahlung haben gezeigt, daß die Quantenelektrodynamik bei großen Energien falsch wird, und zwar im Sinne eines steigenden Anteils nichtrenormierbarer Wechselwirkungen. Dies hat sich allerdings erst in der allerletzten Zeit herausgestellt, und daher muß vorher noch die weitere Entwicklung der Theorie vor dieser letzten Phase besprochen werden.

Schon in den ersten Jahren nach dem Kriege wurde der mathematische Apparat der Quantenfeldtheorie in einer Reihe von wichtigen Arbeiten durch Tomonaga, Schwinger, Feynman und Dyson u. a. sehr erfolgreich weiterentwickelt. Die relativistische Invarianz der Theorie wurde in diesen neuen Darstellungen voll ausgenützt, was einen sehr viel klareren Einblick in die Struktur der Theorie ermöglichte. Die neuen Methoden gestatteten auch eine genauere Analyse der Frage, unter welchen Bedingungen eine derartige Quantenfeldtheorie den Kausalitätsforderungen der speziellen Relativitätstheorie genügt. An dieser Stelle haben Stueckelberg und Fierz einen wichtigen Fortschritt erzielt. Sie konnten nachweisen, daß eine Theorie, in der eine S-Matrix für die Streuwellen definiert wird, die also in einem gewissen Sinne den Bedingungen der Quantentheorie genügt, im allgemeinen die Anforderungen der Kausalität im Sinne der speziellen Relativitätstheorie noch nicht erfüllt. Vielmehr müssen noch gewisse zusätzliche Bedingungen befriedigt werden, auf die hier nicht näher eingegangen werden kann. Daher haben auch die Versuche, sog. nichtlokale Theorien zu entwickeln, in denen relativistisch invariante Fernwirkungen angenommen werden, bisher noch nicht zu einem größeren Erfolge geführt.

Mit Hilfe der neuen Methode wurde auch die Theorie der π-Mesonen und ihrer Wechselwirkung mit den Nukleonen entwickelt. Allerdings waren hier die Ergebnisse viel weniger befriedigend als in der Quantenelektrodynamik. Die störungstheoretische Entwicklung, die in der Quantenelektrodynamik im wesentlichen nach Potenzen von $e^2/\hbar c$ fortschreitet, konvergiert zwar dort sehr gut; in der Theorie der π-Mesonen aber ist die entsprechende Kopplungskonstante groß. Obwohl die Theorie renormiert werden kann, ist daher die erste störungstheoretische Näherung unbrauchbar, und man ist auf die Verwendung anderer Näherungsmethoden angewiesen. Ich erinnere an die Arbeiten über starke Kopplung von Wentzel und über intermediäre Kopplung von Tomonaga. Die Ergebnisse geben daher die experi-

mentellen Erfahrungen auch nur sehr ungenau wieder. Von einer quantitativen Übereinstimmung, wie in der Quantenelektrodynamik bei der Verschiebung der Feinstrukturniveaus, kann hier keine Rede sein.

Die eigentlichen Einwände gegen die uneingeschränkte Verwendung der renormierbaren Theorien kamen aber im letzten Jahr von einer anderen Seite. Lee hatte ein Modell einer renormierbaren Quantenfeldtheorie entwickelt, das gegenüber der Quantenelektrodynamik den Vorteil aufwies, ohne Verwendung der Störungstheorie integrierbar zu sein, sonst aber viel Ähnlichkeit mit der Quantenelektrodynamik hatte. Die Analyse dieses Modells durch Källén und Pauli zeigte, daß infolge der Renormierung die Hamilton-Funktion des Ausgangssystems nicht mehr hermitisch ist. Das hat zur Folge, daß als Lösung dieser Hamilton-Funktion stationäre Zustände auftreten, die Pauli und Källén „Geister“-Zustände nennen, und deren Existenz bewirkt, daß die Metrik im Hilbert-Raum nicht mehr positiv definit und die S-Matrix nicht mehr unitär ist. Das würde physikalisch bedeuten, daß man negative Wahrscheinlichkeiten zulassen müßte, was natürlich logisch unsinnig ist. In anderen Worten: die renormierte Theorie ist keine physikalisch interpretierbare Quantentheorie mehr.

Ich möchte diesen Sachverhalt an der Quantenelektrodynamik noch etwas ausführlicher darstellen, obwohl hier die mathematischen Beweise vielleicht noch nicht über jeden Zweifel erhaben sind. Wenn man in den Ausgangsgleichungen der Quantenelektrodynamik mit einer Ladung e_0 beginnt, so wird nach Durchführung der Rechnungen die wahre Ladung des Elektrons e eine Funktion von e_0 sein. Macht man, um alle Divergenzen zu vermeiden, die Theorie zunächst durch die Einführung eines Maximalimpulses P (eines „Abschneideimpulses“) konvergent (mit der Absicht, am Schluß den Grenzübergang $P \to \infty$ durchzuführen), so kann man die Beziehung zwischen e und e_0 in der Form schreiben:

$$e^2 = \frac{e_0^2}{1 + e_0^2\, F(P,\, e_0^2)}.$$

Man kann nachweisen, daß

$$F(P,\, e_0^2) > 0.$$

Nach den Überlegungen von Pauli und Källén und nach neueren Arbeiten von Landau und Thirring scheint man ferner nachweisen zu können, daß es für F eine untere Schranke gibt mit der Eigenschaft:

$$F(P,\, e_0^2) > \text{const} \cdot \lg P \quad \text{für jedes } e_0^2.$$

Das bedeutet aber, daß für jedes endliche e_0 $e^2 \to 0$ für $P \to \infty$, d.h., daß zu jeder endlichen Ausgangsladung eine wahre Ladung 0 gehört; in anderen Worten, es gibt überhaupt keine elektromagnetische Wechselwirkung. Setzt man aber willkürlich e gleich der wahren Ladung des Elektrons, so folgen für endliche P oberhalb einer gewissen Grenze negative Werte von e_0^2, d.h. eine imaginäre Kopplungskonstante e_0 und eine nichthermitische Hamilton-Funktion. Als Folge der imaginären Kopplung treten die schon besprochenen „Geister“-Zustände auf, die dem Hilbert-Raum eine indefinite Metrik geben. Ihre Energie E liegt, wenn die genannte Vermutung richtig ist, sehr hoch; es ist $\log (E/mc^2) \approx 137$. Praktisch spielen diese Zustände also bei niedrigen Energien keine Rolle, und wenn man einen endlichen und nicht zu großen

Die Naturwissenschaften *42*, 637–641 (1955)

Abschneideimpuls beibehält, fallen die Geister-Zustände weg. Die Quantenelektrodynamik läßt sich aber ohne willkürliche Abschneidevorschrift nicht mehr zu einer geschlossenen, relativistisch invarianten mathematischen Theorie umformen.

Diese wichtigen Feststellungen, die dem Glauben an die prinzipielle Bedeutung der renormierbaren Theorien den Boden entziehen, haben in der Quantenelektrodynamik noch nicht allzu viel praktische Bedeutung. Für Theorien mit großer Kopplungskonstante aber, z.B. für die Wechselwirkung der π-Mesonen mit den Nukleonen, sind sie auch praktisch sehr einschneidend. Denn die Geister-Zustände werden in der π-Mesonen-Theorie bei experimentell durchaus zugänglichen Energien, sagen wir in der Größenordnung 1 GeV, liegen. Man kann also mit der π-Mesonen-Theorie nur dann zu interpretierbaren Resultaten kommen, wenn man die Theorie durch einen relativ kleinen Abschneideimpuls so stark verändert, daß sie bestenfalls die Wirklichkeit bei kleinen Energien ungenau und bei großen Energien gar nicht mehr darstellt.

Im ganzen kann man die Ergebnisse dieser neuesten Entwicklung also dahin zusammenfassen, daß der Prozeß der Renormierung allgemein nicht genügt, um aus einer Quantenfeldtheorie mit ihren bekannten Divergenzen eine mathematisch brauchbare Theorie zu machen. Man muß also in der Quantentheorie der Wellenfelder doch noch grundsätzliche Änderungen vornehmen, bevor man sie als eine tragfähige Basis für eine Theorie der Elementarteilchen benützen kann.

Es ist bei dieser Sachlage nicht unbefriedigend, daß neuerdings auch die Experimente die Unzulänglichkeit der bisherigen Quantenelektrodynamik zeigen. In den letzten Jahren haben sich durch die Beobachtungen von Schein, Kaplon und Ritson und der Turiner Gruppe gezeigt, daß die elektromagnetischen Kaskadenprozesse höchster Energie in der kosmischen Strahlung anders ablaufen, als man nach der Kaskadentheorie der bisherigen Quantenelektrodynamik erwarten sollte. Man beobachtet Schauer mit einer unverhältnismäßig großen Anzahl von Dreiergabeln, d.h. von Paarerzeugung durch Elektronen, und gelegentlich findet offenbar eine Vielfacherzeugung von γ-Strahlen statt, d.h., in einem einzigen Stoßprozeß werden gelegentlich bis zu 20 γ-Strahlen auf einmal erzeugt. Es gibt also jedenfalls Teilchen, die bei hohen Energien erheblich stärker strahlen, als es Elektronen nach der bisherigen Quantenelektrodynamik tun sollten, die sich aber sonst ähnlich wie Elektronen zu verhalten scheinen. Diese Experimente lassen offenbar noch zwei mögliche Deutungen zu: Entweder die stärker strahlenden Teilchen sind einfach Elektronen; das würde bedeuten, daß die Quantenelektrodynamik bei höheren Energien der Elektronen, etwa von 10 GeV ab, versagt, und daß bei diesen Energien eine Wechselwirkung mit dem Strahlungsfeld ins Spiel kommt, die mit wachsender Energie zunimmt, also eine nichtrenormierbare Wechselwirkung. Oder aber die stärker strahlenden Teilchen sind keine Elektronen; sie sind z.B. Teilchen mit dem Spin 1, die bekanntlich bei hohen Energien stärker strahlen sollen. Dann beweist dies ebenfalls die Existenz von nichtrenormierbaren elektromagnetischen Wechselwirkungen. Wir besitzen einstweilen noch nicht genügend experimentelles Material, um zwischen den beiden Alternativen sicher entscheiden zu können.

Die Unzulänglichkeit der renormierbaren Theorien und die Notwendigkeit, an den Grundlagen der Quantenfeldtheorie etwas zu ändern, ergibt sich also in gleicher Weise aus der theoretischen Analyse und aus den Beobachtungen der elektromagnetischen Schauer. Bei dieser Sachlage kann man erneut die Frage aufwerfen, ob denn, abgesehen von den eben genannten Schwierigkeiten, die Renormierungstheorien eine befriedigende Darstellung der Elementarteilchen versprochen hätten. Hier fällt nun sogleich ein grundsätzlicher Mangel auf. Die Renormierungstheorien in ihrer bisherigen Form beziehen sich auf bestimmte Gruppen von Elementarteilchen, die als gegeben betrachtet werden, z.B. auf Elektronen und Lichtquanten oder auf Nukleonen und π-Mesonen. Wie sollte man von einem solchen Ausgangspunkt aus zu einer Theorie kommen, die die Massen und sonstigen Eigenschaften aller Elementarteilchen verständlich macht? Es ist gelegentlich die ganz vage Hoffnung ausgesprochen worden, man werde eben später die Massen und Wechselwirkungen aller Elementarteilchen in eine riesige Gesamt-Hamilton-Funktion einsetzen und damit alle Elementarteilchen zusammenfassen müssen und dann vielleicht nachweisen können, daß die Gleichung nur für ganz bestimmte Werte der Massenverhältnisse und Kopplungskonstanten renormierbar wird und dann konvergente Lösungen zuläßt. Daraus könne man die tatsächlichen Massenverhältnisse usw. bestimmen. Wenn man an die sehr große Zahl der beobachteten verschiedenartigen Elementarteilchen denkt, erkennt man aber sofort, daß diese Hoffnung nur auf dem Umweg über eine ganz monströse Mathematik verwirklicht werden könnte. Wenn man gewissermaßen als Trost hinzufügt, daß diese besondere Mathematik vielleicht auf eine einfache Wurzel ganz anderer Struktur wird zurückgeführt werden können, so muß man jedenfalls zugeben, daß diese einfache Wurzel dann nicht eine renormierbare Gleichung für eine bestimmte Gruppe vorhandener Elementarteilchen sein wird.

Als obersten Grundsatz beim Suchen nach einer Theorie der Elementarteilchen wird man also aufstellen müssen: Es wird sich bei einer solchen Theorie um eine Grundgleichung für die Materie, nicht aber für einzelne Elementarteilchen handeln, und aus dieser Grundgleichung für die Materie müssen sich alle Elementarteilchen etwa als Eigenlösungen ergeben. Daraus geht auch sofort hervor, daß die Elementarteilchen in dieser Theorie eine ähnliche Rolle spielen werden wie die stationären Zustände eines komplizierteren atomaren Systems in der Quantenmechanik, nicht wie die Elektronen in der alten Theorie. Ferner entstehen hier die Elementarteilchen sozusagen gleichzeitig mit ihrer Wechselwirkung; der Begriff des freien Teilchens ohne Wechselwirkung wird also sinnlos. Alle Versuche, die Massen der Elementarteilchen aus Gleichungssystemen zu bestimmen, die noch keine Wechselwirkung enthalten, sind also von vornherein zum Scheitern verurteilt. Während in den bisherigen Untersuchungen über die S-Matrix, insbesondere bei Dyson, die gebundenen Zustände als ein- oder auslaufende Wellen meist vernachlässigt worden sind, wird man also umgekehrt in Zukunft das Interesse vor allem auf die gebundenen Zustände konzentrieren müssen, da in einer richtigen Theorie alle Elementarteilchen sozusagen gebundene Zustände sind. Ansätze

Heft 24
1955 (Jg. 42)   Kurze Originalmitteilungen.   **641**

in dieser Richtung finden sich in der japanischen Literatur bei NISHIJIMA. Die Hauptschwierigkeit wird zuerst darin bestehen, überhaupt einmal eine Quantenfeldtheorie, und sei es auch nur als Modell, anzuschreiben, die keine Unendlichkeiten oder sonstige mathematische Widersprüche mehr enthält.

Von den bisherigen Versuchen in dieser Richtung soll nur einer erwähnt werden, den der Verfasser zusammen mit KORTEL und MITTER genauer ausgearbeitet hat. Dort wird von einer einfachen Wellengleichung ausgegangen

$$\gamma_\nu \frac{\partial \psi}{\partial x_\nu} + l^2 \psi(\psi^+ \psi) = 0,$$

in der ψ das Wellenfeld der Materie, aber nicht einer bestimmten Sorte von Elementarteilchen bedeutet. Die Quantisierungsvorschrift wird gegenüber der gewöhnlichen Quantenfeldtheorie abgeändert, denn eine Renormierung ist hier nicht möglich; die übliche Quantisierung würde also zu mathematischen Ungereimtheiten führen. Die Abänderung wird dadurch vollzogen, daß neben dem gewöhnlichen HILBERT-Raum noch ein zweiter HILBERT-Raum oder HILBERT-Teilraum postuliert wird mit Zuständen, die physikalisch nicht realisiert werden können. Dieser HILBERT-Raum II spielt nur bei den virtuellen Zwischenzuständen eine Rolle und bewirkt, daß in den Vertauschungsfunktionen die δ-Funktionen auf dem Lichtkegel, die sonst die Divergenzen verursachen, hier verschwinden. Als ein- oder auslaufende Wellen aber sollen die Zustände dieses HILBERT-Raums II nicht vorkommen. Der HILBERT-Raum II steht, wie sich später herausgestellt hat, in engem Zusammenhang mit dem Raum der PAULI-KÄLLÉNschen Geister-Zustände in den Renormierungstheorien. Die Zustände dieses HILBERT-Raums II machen ebenfalls die Metrik im gesamten HILBERT-Raum indefinit und führen damit formal zu negativen Wahrscheinlichkeiten, die nicht physikalisch interpretiert werden können. Ferner sind sie jenen Geister-Zuständen auch insofern ähnlich, als sie ebenfalls die gröbsten Singularitäten in der Vertauschungsfunktion beseitigen. Aber sie sind nicht mathematisch identisch mit den Geister-Zuständen der PAULI-KÄLLÉNschen Arbeit und unterscheiden sich von ihnen in einem wesentlichen Punkt. Die Zustände des HILBERT-Raums II verhalten sich zu den Geister-Zuständen etwa wie ein Dipol zu einem Pol. Auf Einzelheiten kann hier nicht eingegangen werden. Aber dieser Unterschied hat zur Folge, daß in der S-Matrix, im Gegensatz zu den Renormierungstheorien, Übergänge von den normalen Zuständen zu den Geister-Dipolen nicht vorkommen können, daß also die S-Matrix innerhalb der normalen Zustände unitär bleibt. Die negative Wahrscheinlichkeit spielt also in der S-Matrix keine Rolle und kommt

bei der physikalischen Interpretation nicht vor. Das bedeutet, daß — wenigstens in jeder endlichen Näherung — die Theorie qualitativ zu unseren heutigen Kenntnissen von den Elementarteilchen paßt. Daraus folgt allerdings noch nicht, daß die Modelltheorie mathematisch konsistent ist, da noch nicht nachgewiesen ist, daß das ganze Verfahren beim Übergang von niedrigen zu höheren Näherungen schließlich gegen endliche Grenzwerte konvergiert. In den niedrigen Näherungen ist das Spektrum der Elementarteilchen, das sich aus der genannten einfachen Wellengleichung ergibt, dem experimentellen Spektrum erstaunlich ähnlich. Insbesondere ist ein erheblicher Teil der Elektrodynamik in ihr enthalten; aber das mag noch Zufall sein.

Die Modelltheorie sollte nur als Beispiel dafür erwähnt werden, wie eine Theorie der Elementarteilchen grundsätzlich aussehen kann. Sie ist auch ein Beispiel für ein Schema, in dem zwar eine unitäre S-Matrix, ähnlich wie in der gewöhnlichen Quantenfeldtheorie, definiert werden kann, aber eine physikalische Interpretation des lokalen Verhaltens der Wellenfunktionen wegen der negativen Wahrscheinlichkeit nicht mehr möglich ist. Ferner zeigt sie, wie die Masse der Elementarteilchen eine Folge der schon in der Grundgleichung auftretenden Wechselwirkung der Materie ist, wie die Elementarteilchen also immer von vornherein schon miteinander in Wechselwirkung stehen.

Im ganzen wird man also den heutigen Stand der Theorie der Elementarteilchen etwa in folgender Weise zusammenfassen können: Die Renormierungstheorien vom Typus der Quantenelektrodynamik mögen als angenäherte Darstellungen gewisser Gruppen von Elementarteilchen und ihrer Wechselwirkungen gute Dienste leisten; für die Formulierung einer geschlossenen Theorie der Elementarteilchen sind sie aber ungeeignet. Für eine solche geschlossene Theorie wird man die bisherigen Methoden der Quantisierung der Felder erweitern müssen. Es muß aber auch in dem erweiterten Schema diskrete Eigenwerte für die Massen und eine unitäre und relativistisch invariante S-Matrix für die Stoßprozesse geben. Wahrscheinlich spielt eine Erweiterung des HILBERT-Raums in dem vorhin besprochenen Sinne für die Verbesserung der Quantenfeldtheorie eine wesentliche Rolle. Sie bietet vielleicht die Möglichkeit, zwar die unitäre S-Matrix beizubehalten, aber auf die lokale Beschreibung der Wellenfunktion zu verzichten, und das würde für die Beschreibung der Experimente grundsätzlich ausreichen. Es besteht durchaus die Hoffnung, aus einer recht einfachen Grundgleichung einmal alle Elementarteilchen als Eigenlösungen herzuleiten.

Max-Planck-Institut für Physik, Göttingen.
Eingegangen am 28. September 1955.

Kurze Originalmitteilungen.

Für die Kurzen Originalmitteilungen sind ausschließlich die Verfasser verantwortlich.

Hagelwachstum eine Quelle der Gewitterelektrizität.

Die Ballonaufstiege bei Kew haben gezeigt, daß ein aktives Gewitter *drei* Raumladungsgebiete verschiedenen Vorzeichens enthält: Ein weit ausgebreitetes positives in größten Höhen, darunter noch mächtiger an Ausdehnung und Ladung ein negatives und schließlich ganz unten, eng konzentriert auf das Gebiet des ausfallenden Starkniederschlags, noch ein kleineres positives. Gerade dieses scheint von besonderer Bedeutung zu

sein, da von den zur Erde gehenden Blitzen die überwiegende Mehrzahl ihrer Polarität nach von dem elektrischen Feld zwischen dieser untersten positiven und der darüberliegenden negativen Ladung ausgeht.

Trotz vieler Bemühungen ist es noch nicht gelungen, die Vorgänge ausfindig zu machen, die zur Erzeugung und Aufrechterhaltung der recht beträchtlichen Raumladungen von vielen Coulomb dienen könnten. Fast unbeachtet blieb

Reviews of Modern Physics *29*, 269–278 (1957)

REVIEWS OF MODERN PHYSICS   VOLUME 29, NUMBER 3   JULY, 1957

Quantum Theory of Fields and Elementary Particles

W. Heisenberg

Max Planck Institut für Physik, Göttingen, Germany

THE present article gives a general discussion of the problems arising in a theory of elementary particles, together with a survey of papers[1-6] that have been published on this subject in German periodicals. These papers deal with a special model for the theory of elementary particles that has been constructed to show some of the main features of such a theory; the author believes that the model does in fact represent a system of elementary particles and their interactions in a manner qualitatively suitable also for the real system of particles.

1. GENERAL REMARKS ON FIELD THEORY AND ELEMENTARY PARTICLES

Any attempt to construct a field theory of elementary particles must from the outset meet the well-known difficulties arising from the combination of the space-time structure of special relativity with quantum theory. Whenever one applies the normal rules of quantization to a differential equation that is Lorentz-invariant and represents interaction between fields, one seems to get divergent results. It may be difficult in the present state of the theory to give a rigorous mathematical proof that these difficulties cannot be completely avoided; but hitherto no solution has been presented. For a time the process of renormalization seemed to offer such a solution. But in the only case where the mathematical structure could be analyzed completely, the Lee model,[7] Källén and Pauli[8] showed that the process of renormalization leads to the implicit introduction of so-called "ghost-states" which destroy the unitarity of the S matrix and thereby violate the rules of quantum theory. To get convergent schemes within the framework of quantum theory one has therefore been forced to introduce the interaction as a nonlocal one,[9] for instance by means

of a so-called "cutoff-factor."[10] This however implies deviations from the kind of causality[11] that follows from the space-time structure of special relativity. It is still an open question how serious the deviations from relativistic causality must be in order to ensure a convergent mathematical scheme. But complete local causality seems not to be compatible with quantization. Therefore any field theory of elementary particles or of matter must start by offering some solution to the central mathematical problem: how to combine quantization with a certain greater or lesser degree of relativistic causality. Judging from the experiments, the deviations from causality can scarcely exceed space-time regions of the order of 10^{-13} cm.

The next important problem concerns the fundamental quantities that appear in the mathematical formulation of the theory of elementary particles. Nearly all conventional theories start by introducing some field operators representing the wave fields connected with some specified elementary particles, e.g., meson field or electromagnetic field operators. This procedure, however, requires a definition of the concept "elementary particle." Is there any criterion by which we can distinguish between an elementary particle and a compound system? Is it any more justified to introduce a meson field into the fundamental equations than, e.g., a hydrogen field or an oxygen field?

The author believes that it is essential for any real progress in the theory of matter to recognize that such criteria do not exist.[12] If one asks physicists how they would like to define the nature of an elementary particle as distinct from a compound system, one frequently gets the following answer: an elementary particle is a particle for which one introduces a separate wave function in the fundamental equations. Sometimes one gets the different answer: any particle with spin <1, charge <2, isotopic spin $<\frac{3}{2}$ is elementary; all other particles are compound systems. Obviously both definitions contain elements of complete arbitrariness. Any careful investigation into the properties of atomic particles shows that there may be quantitative differ-

[1] W. Heisenberg, Nachr. Akad. Wiss. Göttingen **1953**, p. 111.
[2] W. Heisenberg, Z. Naturforsch. **9a**, 292 (1954).
[3] Heisenberg, Kortel, and Mitter, Z. Naturforsch. **10a**, 425 (1955).
[4] W. Heisenberg, Z. Physik **144**, 1 (1956).
[5] W. Heisenberg, Nachr. Akad. Wiss. Göttingen **1956**, p. 27.
[6] R. Ascoli and W. Heisenberg, Z. Naturforsch. **12a**, 177 (1957).
[7] T. D. Lee, Phys. Rev. **95**, 1329 (1954).
[8] G. Källén and W. Pauli, Dan. Mat. Fys. Medd. **30**, No. 7 (1955).
[9] Compare H. Yukawa, Phys. Rev. **76**, 300 (1950) and **77**, 219 (1950); P. Kristensen and C. Møller, Dan. Mat. Fys. Medd. **27**, No. 7 (1952); C. Bloch, Dan. Mat. Fys. Medd. **27**, No. 8 (1952).

[10] Compare G. Wataghin, Z. Physik **88**, 92 (1934) and **92**, 547 (1934), and G. Wentzel, Helv. Phys. Acta **13**, 269 (1940).
[11] Compare M. Fierz, Helv. Phys. Acta **23**, 731 (1950), and E. C. G. Stueckelberg and G. Wanders, Helv. Phys. Acta **27**, 667 (1954).
[12] Compare W. Heisenberg, Naturwissenschaften **42**, 637 (1955).

ences between different particles which suggest that it might be convenient in a given experiment to call one particle elementary and the other a compound system; but no qualitative distinction between elementary and compound can be made. A proton certainly looks like an elementary particle for energies <100 Mev, but it may be considered as composed of a Λ^0 particle and a θ^+ particle in collisions of much higher energies. One might argue that the Λ^0 particle and the θ^+ particle are unstable while the proton is stable, that therefore the proton cannot be composed of Λ^0 and θ^+. The fallacy of the argument is, however, seen at once from the case of the deuteron, which is stable but usually considered as composed of proton and neutron, the latter being unstable. For an understanding of matter and of the atomic particles it is essential to realize that the question whether the proton is elementary or a compound system has no answer. This result should not prevent us from using the term "elementary particle" whenever it is convenient to disregard its inner structure. But it should not be misunderstood as making some specific statement about the nature of the particle.

To avoid the two fundamental difficulties which have been put here at the beginning of the discussion, the efforts of many physicists have in recent years been concentrated on the S matrix.[13] The S matrix is the quantity immediately given by the experiments. It is not difficult to construct S matrices which fulfill the requirements of Lorentz-invariance and unitarity without encountering any divergent terms. At the same time the problem of the "elementary" particles does not occur immediately in the S matrix, since any incoming or outgoing particles are here characterized by their wavefunctions ψ_{in} or ψ_{out}, irrespective of whether they are compound or not.

The S-matrix formalism does not by itself guarantee the requirements of relativistic causality. Therefore many recent investigations have dealt with supplementary conditions to be put on the S matrix to ensure relativistic causality. The best known example is the treatment of the dispersion relations.[14] Insofar as these conditions state relations between observable quantities they will serve as a very useful tool for the interpretation of the experiments. If, however, their mathematical formulation makes use of an extrapolation of the S matrix into regions of momentum space, where the relations $\mathbf{p}^2+\kappa^2=0$ for the respective particles are not fulfilled, their value might be very limited, since it is difficult to see what the momentum vector $p\mu$ of, say, a hydrogen atom in its normal state can mean when the relation $\mathbf{p}^2+\kappa^2=0$ (where κ is the total mass of the atom in its normal state) is not fulfilled. In other words:

the extrapolation into these regions of momentum space requires eventually the distinction between elementary particles and compound systems which cannot be more than a more or less suitable convention. It is perhaps not exaggerated to say that the study of the S matrix is a very useful method for deriving relevant results for collision processes by going around the fundamental problems. But these problems must be solved some day and one will then have to look for a mathematical formalism that allows one to calculate the masses of the particles and the S matrix at the same time. The S matrix is an important but very complicated mathematical quantity that should be derived from the fundamental field equations; but it can scarcely serve for formulating these equations.

As a result of the foregoing discussion we can try to state some general principles for a theory which attacks the problem of the fundamental field equations for matter.

1. The field operators necessary for formulating the equations shall not refer to any specified particle like proton, meson, etc.; they shall simply refer to matter in general.

2. The particles (elementary or compound) should be derived as eigensolutions of the field equations.

3. The fundamental field equations must be nonlinear in order to represent interaction. The masses of the particles should be a consequence of this interaction. Therefore the concept of a "bare particle" has no meaning.

4. Selection rules for creation and decay of particles should follow from symmetry properties of the fundamental equations. Therefore the empirical selection rules should provide the most detailed information on the structure of the equations.

5. Besides the selection rules and the invariance properties, the only other guiding principle available seems to be the simplicity of the equations.

The empirical spectrum of elementary particles looks very complex. All hitherto observed particles have lifetimes $>10^{-15}$ sec; for most of them the lifetime is $>10^{-11}$ sec. The natural lifetime for elementary particles that can decay into others would, however, be of the order 10^{-22} to 10^{-23} sec. This comparison shows that the observed particles represent those rare cases where the selection rules provide an exceptionally long lifetime. When one compares in the optical spectra of atoms the number of levels with the natural lifetime of the order 10^{-8} sec with the small number of metastable levels where the lifetime is longer, say by a factor of 10^8, one gets an idea of how complicated the real spectrum of elementary particles may possibly be.

Under these circumstances it seems advisable first to study a simplified model, which may be constructed in accordance with the foregoing principles. The model to be discussed in the following sections is certainly

[13] Compare J. A. Wheeler, Phys. Rev. 52, 1107 (1937), and W. Heisenberg, Z. Physik 120, 513, 673 (1943); Z. Naturforsch. 1, 608 (1946); C. Möller, Kgl. Danske Videnskab. Selskab. 23, No. 1 (1945); 24, No. 19 (1946).

[14] Compare Gell-Mann, Goldberger, and Thirring, Phys. Rev. 95, 1612 (1954).

too simple to represent the real spectrum of elementary particles. But it shows the main features of an adequate theory inasmuch as it represents a world consisting of elementary particles with qualitatively similar properties to our own.

2. MODEL FOR A THEORY OF MATTER

(a) Wave Equation and Quantization

The model which has been studied in detail in the series of papers mentioned above[1-6] starts from the equation*

$$\gamma_\nu \frac{\partial \psi}{\partial x_\nu} - l^2 \psi(\psi^+\psi) = 0. \qquad (1)$$

$\psi(x)$ is defined as a spinor wave function representing matter. The equation has been chosen according to the principles 1, 3, 5 of the foregoing section. Equation (1) is certainly too simple to represent the real particles since it does not contain the isobaric spin variable.

Equation (1) defines a quantum theory for matter only if commutation relations for the operator $\psi(x)$ are added. At this point one meets the difficulty mentioned in the beginning. This difficulty will be discussed in some detail.

To get some information on the possible assumptions for the commutation relations and on the eventual connection between "commutator" and "propagator," it is convenient to consider the operator:

$$\chi_\alpha(x,x') = \exp\{-i[a_\nu \psi_\nu{}^+(x') + \text{conj.}]\}\psi_\alpha(x)$$
$$\times \exp\{+i[a_\nu \psi_\nu{}^+(x') + \text{conj.}]\}. \qquad (2)$$

("Conj." in the exponent means the Hermitian conjugate.) a_ν is an arbitrary constant spinor that anticommutes with $\psi(x)$ and $\psi^+(x)$ everywhere. $\chi_\alpha(x,x')$, considered as a function of x, satisfies the wave Eq. (1). If the components a_ν are chosen as very small, $\chi_\alpha(x,x')$ can be expanded as

$$\chi_\alpha(x,x') = \psi_\alpha(x) - ia_\nu[\psi_\alpha(x)\psi_\nu{}^+(x') + \psi_\nu{}^+(x')\psi_\alpha(x)]$$
$$- ia_\nu{}^*[\psi_\alpha(x)(\psi_\nu{}^+(x'))^* + (\psi_\nu{}^+(x'))^*\psi_\alpha(x)]$$
$$+ \cdots. \quad (3)$$

(The star * denotes Hermitian conjugation.) To preserve relativistic causality, one usually assumes the anticommutators to be zero for space-like distances $(x-x')^2 > 0$. This assumption means that $\chi_\alpha(x,x')$ corresponds to a solution of (1), in which a secondary wave starts from the singular point $x=x'$ and fills the cones of future and past. The amplitude of this wave will be very small everywhere except in the immediate neighborhood of the lightcone, if the a_ν are very small. In all conventional forms of quantum theory it has been

---

* The original papers start from an equation that differs from Eq. (1) by the sign of the second term. But—as has been pointed out[6]—if one wants to use conventional formalism, one should say that the calculations actually refer to Eq. (1).

assumed that the anticommutators (or commutators) are singular c-number functions in the immediate neighborhood of the lightcone. If this assumption is taken over into Eqs. (2) and (3), one can put

$$\chi_\alpha(x,x') = \chi_\alpha{}^0(x,x') + c_\alpha(x-x'), \qquad (4)$$

where $c_\alpha(x-x')$ is a c number function (except for a sign function occurring in the a_ν) and contains the singularities at the lightcone, while $\chi_\alpha{}^0(x,x')$ is smooth at the light cone. One may further assume in defining $c_\alpha(x-x')$ that the vacuum expectation value of $\chi_\alpha{}^0(xx')$ vanishes. By inserting (4) into (1) and taking the vacuum expectation value, one finds for c_α as function of x an equation of the type

$$\gamma_\mu \frac{\partial c}{\partial x_\mu} - l^2 c(c^+c) - \kappa(s) \cdot c = l^2 \langle \Omega | \chi^0(\chi^{0+}\chi^0) | \Omega \rangle, \quad (5)$$

where $s = (x-x')^2$, and $\kappa(s)$ is essentially given by the vacuum expectation value of $|\chi^0(x,x')|^2$. The right side of (5) vanishes in the neighborhood of the lightcone.

To study the possible assumptions for the anticommutator

$$S_{\alpha\nu}(xx') = \psi_\alpha(x)\psi_\nu{}^+(x') + \psi_\nu{}^+(x')\psi_\alpha(x), \qquad (6)$$

one has therefore to look for the solutions of the classical nonlinear wave equation (5) where the right-hand side can be put equal to zero, since we are only interested in the neighborhood of the lightcone. In this region $S_{\alpha\nu}(x-x')$ should behave as $c_\alpha(x-x')$ for small a_ν; or more accurately: one should study the continuous group of solutions of (5) that corresponds to different values of a_ν and should finally assume

$$S_{\alpha\nu}(x-x') \sim - \lim_{a_\nu \to 0} \left(\frac{\partial c_\alpha(x-x')}{\partial a_\nu} \right) \qquad (7a)$$

near the lightcone. For the second anticommutator it is plausible to assume as usual

$$\psi_\alpha(x)\psi_\nu(x') + \psi_\nu(x')\psi_\alpha(x) = 0 \qquad (7b)$$

near the lightcone.

For the definition of a quantum theory for matter it will be sufficient to state, besides (1), the commutation rules only near the lightcone, since their value in other parts of space and time will then follow from (1). The equations (5) and (7a) state the required connection between "anticommutator" and "propagator."

Mathematical analysis of (5) shows that its solutions exhibit infinite oscillations in the neighborhood of the light cone.[2] This is entirely different from the behavior of $S_{\alpha\nu}$ in a linear wave theory, where $S_{\alpha\nu}(x-x')$ behaves like the derivative of the Dirac δ function.

At this point we meet the difficulty mentioned in the beginning of Sec. 1. If we are primarily interested in the vacuum expectation value

$$S_{\alpha\nu}(xx') = \langle\Omega|\psi_\alpha(x)\psi_\nu^+(x')+\psi_\nu^+(x')\psi_\alpha(x)|\Omega\rangle$$
$$= \sum_\Phi \langle\Omega|\psi_\alpha(x)|\Phi\rangle\langle\Phi|\psi_\nu^+(x')|\Omega\rangle$$
$$+ \sum_\Phi \langle\Omega|\psi_\nu^+(x')|\Phi\rangle\langle\Phi|\psi_\alpha(x)|\Omega\rangle, \quad (8)$$

each group of intermediate states Φ, belonging to a definite mass eigenvalue κ, contributes an ordinary Schwinger function $S_{\alpha\nu}{}^\kappa(xx')$ belonging to that mass; and the δ and δ' functions of the different intermediate masses all add up with the same sign. Therefore the sum over Φ in (8) can never lead to the behavior of $c_\alpha(x-x')$, the infinite oscillations near the light cone, since the integral contribution of the oscillations vanishes near the lightcone, while that of the δ functions does not vanish.

At a finite distance from the lightcone the sum over Φ in (8) can very well represent a solution of (5) for small a_ν, since here the nonlinear term $c(c^+c)$ can be neglected and the variable coefficient $\kappa(s)$ in (5) corresponds to the different mass values of the states Φ. But near the light cone the rules of quantization must be changed in order to avoid the contradiction between (5) and (8).

(b) Hilbert Space II and the Unitarity of the S Matrix

The only feasible way of getting rid of the δ and δ' functions on the lightcone in (8) seems to be an extension of Hilbert space by the introduction of new intermediate states Φ, which change the metric of Hilbert space into an indefinite one. This extra group of states, called Hilbert space II in the papers mentioned, is chosen so as to cancel the δ and δ' functions on the lightcone in (8). They do not contribute to $S(xx')$ in other parts of space-time. But they will then contribute to the functions $S_1(xx')$ or $S_P(xx')$ also in other parts of space-time, if the normal Schwinger relation

$$S_1(\mathbf{r},t) = \frac{1}{\pi}\int_{-\infty}^{+\infty} S(\mathbf{r},t-t')dt'/t' \quad (9)$$

is taken over from conventional theory.

If, in a first approximation, one considers in Hilbert space I (which comprises all physical states of the system) only one group of intermediate states Φ, belonging to a single mass κ, the corresponding part of the S function in momentum space would be

$$\tfrac{1}{2}S(p) = \frac{p\mu\gamma\mu+i\kappa}{\mathbf{p}^2+\kappa^2}. \quad (10)$$

The subtraction of the δ and δ' functions on the light cone can then be performed by putting, in this same approximation,

$$\tfrac{1}{2}S(p) = \frac{p\mu\gamma\mu+i\kappa}{\mathbf{p}^2+\kappa^2} - \frac{p\mu\gamma\mu+i\kappa}{\mathbf{p}^2} + \frac{p\mu\gamma\mu\kappa^2}{(\mathbf{p}^2)^2}. \quad (11)$$

The contributions from Hilbert space II appear to belong to a mass value 0. But by writing the contributions in a different way, namely,

$$\frac{\kappa^2 p\mu\gamma\mu}{(\mathbf{p}^2)^2} - \frac{p\mu\gamma\mu+i\kappa}{\mathbf{p}^2} = \lim_{\epsilon\to0} \frac{p\mu\gamma\mu\cdot\kappa^2}{\epsilon}\left(\frac{1}{\mathbf{p}^2} - \frac{1}{\mathbf{p}^2+\epsilon}\right)$$
$$\times\left(1 - \frac{ip\mu\gamma\mu}{\kappa} - \frac{\mathbf{p}^2}{\kappa^2}\right), \quad (12)$$

one sees that the extra states of Hilbert space II can be considered as "dipole" states, composed of one normal state with mass 0 and a "ghost state" with mass $\sqrt{\epsilon}\to0$. ("Ghost state" on account of the negative sign in the metric.) As a result of the dipole character of these states, their representatives do not depend simply exponentially on space-time as do the representatives of states in Hilbert state I. A simple calculation of the representatives on the basis of (11) and (12) shows that either the covariant or the contravariant representative has the space-time dependence (in a suitable coordinate system)

$$(t+\text{const})e^{i\mathbf{p}\mathbf{x}} \quad (13)$$

while the other representative has the usual exponential form $e^{i\mathbf{p}\mathbf{x}}$.

This is an important result, because it shows that the states of Hilbert space II—contrary to the "ghost states" of Källén and Pauli[8]—do not destroy the unitarity of the S matrix. In fact if in a collision problem all incoming waves belong to Hilbert space I, the total wave function depends exponentially on space-time in both representations. This behavior cannot be changed by collision [on account of the invariance of Eq. (1) for the inhomogeneous Lorentz group], therefore also the outgoing waves cannot contain contributions from Hilbert space II, since they would destroy the exponential form of the total wave function. Therefore the extension of Hilbert space, which was necessary in order to avoid contradiction between (5) and (8), does not destroy the unitarity of the S matrix and may therefore be compatible with the experimental results. (Introduction of Hilbert space II has in this respect some resemblance to the method of Gupta[16] in quantum electrodynamics.) Of course this extension does mean a certain deviation from local relativistic causality. The local behavior of $\psi(x)$ cannot be interpreted in the usual manner, since $\psi(x)$ contains contributions from Hilbert space II, and their conven-

---

[16] S. N. Gupta, Proc. Roy. Soc. (London) **63**, 681 (1950); **64**, 850 (1951); K. Bleuler, Helv. Phys. Acta **23**, 567 (1950). Compare P. A. M. Dirac, Proc. Roy. Soc. (London) **A180**, 1, (1942).

tional interpretation would, on account of the negative sign in the metric, lead to negative probabilities, which have no meaning.[†]

Actually these contributions from Hilbert space II change the form of $S_1(xx')$ and $S_F(xx')$ fundamentally. For large values of $(x-x')^2>0$ these functions decrease more slowly (like $(x-x')_\nu\gamma_\nu/(x-x')^2$) than the usual Schwinger functions and this behavior leads to long range forces between the particles. It will be shown later that it is through these contributions that Eq. (1) contains quantum electrodynamics with a specific value of the Sommerfeld fine-structure constant.

(c) Methods of Integration

The equations (1) and (11) should be sufficient to define a quantum theory for matter. Since Eq. (11) is only a first approximation, it should generally be replaced by its precise form:

$$\tfrac{1}{2}S(p)=\int\rho(\kappa)d\kappa\left[\frac{p\mu\gamma\mu+i\kappa}{\mathbf{p}^2+\kappa^2}-\frac{p\mu\gamma\mu+i\kappa}{\mathbf{p}^2}+\frac{p\mu\gamma\mu\kappa^2}{(\mathbf{p}^2)^2}\right] \quad (14a)$$

with

$$\int\rho(\kappa)d\kappa=1,$$

where $\rho(\kappa)$ is the mass spectrum of fermions. $\rho(\kappa)$ should be derived from Eqs. (1) and (14).

On account of the infinite oscillations on the light-cone, one can simply put

$$S_{a\nu}(xx')=0 \quad \text{for} \quad (x-x')^2=0. \quad (14b)$$

The only method of integration that has been used so far for Eqs. (1) and (14) is what is sometimes called the new Tamm-Dancoff method.[16] One considers τ functions of the general type

$$\tau_{\alpha\beta\gamma}(x_1x_2|x_3)=\langle\Omega|T\psi_\alpha(x_1)\psi_\beta(x_2)\psi_\gamma{}^+(x_3)|\Phi\rangle \quad (15)$$

as covariant representatives of the states Φ. (T means the time-ordered product.) By means of Eq. (1) one can find differential equations, which connect the derivatives of one τ function with another τ function with a number of variables larger by two. This differential equation can be integrated by means of the Green function $G_F(xx')$ of the Dirac equation for mass zero. (The Feynman functions G_F are chosen in order to satisfy the boundary conditions.) By repeating this process a connection can be established between the

original τ function and another one, where the number of variables is larger by any even number. This latter τ function is then expressed in terms of the so-called φ functions through the process of contraction:

$$\tau(x_1x_2\cdots|y_1y_2\cdots)=\varphi(x_1x_2\cdots|y_1y_2\cdots)-\tfrac{1}{2}S_F(x_1y_1)$$
$$\times\varphi(x_2\cdots|y_2\cdots)-\cdots+\tfrac{1}{4}S_F(x_1y_1)S_F(x_2y_2)$$
$$\times\varphi(x_3\cdots|y_3\cdots)+\cdots-\cdots, \quad (16)$$

and finally all those φ's are neglected, the number of variables of which is larger than that of the original τ function.

In this way a linear integral equation for the τ function is established that can be used for the determination of the mass eigenvalues or the S matrix. This integral equation can be represented by a Feynman graph, consisting of two types of lines, one representing the G_F function, the other representing the S_F function. From (1) and (16) one can easily derive the following rules for these graphs:

1. In every point of the graph four lines meet. The initial and the final points of the graph are considered as identical, or the graph is repeated as a periodic pattern.

2. The numbers of S_F and G_F lines in the graph are equal.

3. Every point is connected with one of the final points through one sequence of G lines.

4. The connection of two lines at a point means matrix multiplication of the respective operators.

5. The kernel of the integral equation is given, when the points of the graph and the pattern of G lines are given. One has to sum over the contributions from all possible S line patterns belonging to the given G line pattern.

This scheme differs from conventional formalism by two characteristic features. The contraction is performed by the function S_F, which is *not* identical here with the Green function G_F; and there are no δ terms in the equations, which in the usual theory arise from exchange of ψ-factors $\psi(x_1)$ and $\psi^+(x_2)$ at the point $t_1=t_2$. Here all $\psi(x)$ functions anticommute on a given subspace $t=$ const. Therefore, quantization is introduced only by the contractions.

This last difference is a significant consequence of the nonlinearity of the equations. If in a linear theory the commutator vanished on a given subspace $t=$ const, it would, by virtue of the wave equation, vanish everywhere. Therefore one must start with a δ function at the point $x=x'$ on the subspace $t=$ const. In a nonlinear theory the commutator can vanish everywhere on $t=$ const and still be different from zero at other times. It is a well-known property of nonlinear equations—the fact that a solution sometimes cannot be continued—which has to be used here in the definition of quantization.

[†] *Note added in proof.*—A mathematical analysis of this method of quantitization can be obtained from the Lee model. The constants g_0 and m_v of the Lee model can be adjusted to let the energies of the normal V particle and the "ghost-state" coincide. In this case the two states form a dipole as in (12), the renormalized wave functions ψ_v commute on the subspace $t=$ const and the two parts of Hilbert space are distinguished by the asymptotic behavior of the wave functions. The details will be published in *Nuclear Physics*.

[16] M. Gell-Mann and F. Low, Phys. Rev. **84**, 350 (1951); Freese, Z. Naturforsch. **8a**, 776 (1953).

The contraction function S_F can, in a low approximation, be represented by (11), and the eigenvalue κ is to be obtained from the integral equation. In higher approximations one may consider several eigenvalues or, finally, the form (14).

Whether this whole procedure will, by going from low to high approximations, converge to a final solution, is still an open question. It does give finite results in any approximation. But it may be necessary to define the integral equations by averaging over certain groups of G-line graphs, in order to obtain convergence. This problem has been treated in detail by Matthews and Salam for conventional formalism.[17] It may also be necessary to replace the φ functions by slightly different groups of functions, as suggested by Maki.[18] The procedure has been studied in detail in the example of the anharmonic oscillator. But whatever the results of such mathematical investigations will be, one will probably get some—though inaccurate—information on the solutions of (1) and (14) by using the rules described in this section for the low approximations.

(d) The Lowest Eigenvalues

The lowest eigenvalues of (1) and (14) have been calculated along the lines described in the foregoing section.[3] For the eigenvalues of the fermion type all graphs of the type of Fig. 1 have been combined to

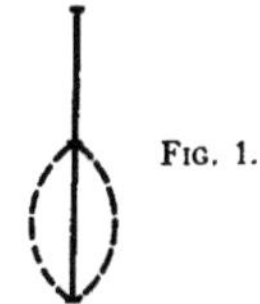

Fig. 1.

define the integral equation. (The G lines are given as full, the S lines as dotted lines.) Only τ functions with one or three variables have been used.

The result was that there exists—in this approximation—only one eigenvalue for the mass κ of a fermion:

$$\kappa = 7.426/l. \qquad (17)$$

The spin of this particle is 1/2.

The bosons have been calculated from the graph given in Fig. 2. Only τ functions for two variables $[\psi_\alpha(x)$

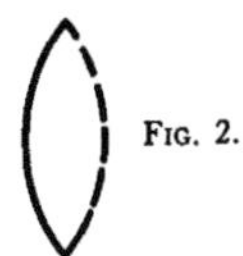

Fig. 2.

and $\psi_\beta^+(x)$ with equal space-time coordinates] have been used. Four different bosons with nonvanishing

[17] P. T. Matthews and A. Salam, Proc. Roy. Soc. (London) A221, 128 (1954).
[18] Z. Maki, Progr. Theoret. Phys. 15, 237 (1956).

masses were found, with mass values, spins, and parities as shown in Table I.

TABLE I.

Mass κl	Spin	Parity
0.33	1	−1
0.95	0	1
1.74	0	−1
3.32	0	1

It is interesting that (1) and (14) give for the boson masses values which are considerably smaller than the mass of the fermion. This fits well with the empirical fact that the masses of π meson and θ meson are considerably smaller than the mass of the nucleon.

It may be useful at this point to compare the properties of the model, given by (1) and (14), with the general principles laid down at the end of Sec. 1.

The wave function $\psi(x)$ in (1) refers to matter in general, not to a specified particle. The particles do actually come out as the eigensolutions of the equations. Since $\psi(x)$ has been chosen as spinor, while the commutation relation states the value of the anticommutator, all particles with half quantum spin obey Fermi statistics, all particles with integer spin obey Bose statistics. States of the first kind are obtained when one applies an odd number of $\psi(x)$ operators on the vacuum state, the Bose states are created by applying an even number of ψ's.

The nonlinear term in (1) is multiplied with a coupling constant with the dimension of a length. A variation of l will simply change all mass eigenvalues by a constant factor. The ratio of the eigenvalues depends only on the general form of the nonlinear term. The masses of the particles are entirely a product of interaction, namely of the nonlinear term. Therefore their interactions are given simultaneously with their masses; the concept of a bare particle has no meaning in this theory.

Besides the bosons given in Table I analysis of (1) reveals the existence of still another group of boson states belonging to the rest mass 0 (which had not been expected in the papers[1,2]). Their existence can be seen as follows.[3] When one applies the integral operator represented by the graph of Fig. 2 on a τ function belonging to a total momentum vector J_μ with $J_\mu^2 = 0$, one gets generally an infinite result. But there are special forms of the τ function

$$\tau_{\mu\beta}(x|x) = \langle\Omega|\psi_\alpha(x)\psi_\beta^+(x)|\Phi\rangle = e^{\iota J_\mu x_\mu}\tau_{\alpha\beta}, \qquad (18)$$

for which the divergencies disappear. These special τ functions can be used to satisfy the integral equations.

As conditions for the cancellation of the infinite terms one finds

$$J_\mu\gamma_\mu(\tau - \mathrm{spur}\,\tau) = 0,$$
$$(\tau - \mathrm{spur}\,\tau)J_\mu\gamma_\mu = 0, \qquad (19)$$
$$\gamma_\mu\tau\gamma_\mu = 0.$$

Reviews of Modern Physics *29*, 269–278 (1957)

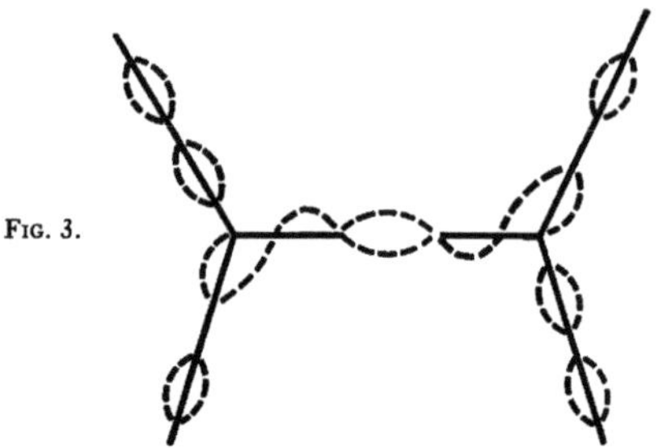

FIG. 3.

If one introduces two four-vectors A_ν, which satisfy $A_\mu J_\mu = 0$ and are linearly independant of each other and of J_μ (there are just two such vectors), one can solve the equations (19) for τ by putting

$$\tau = J_\mu A_\nu \gamma_{\mu\nu} \tag{20}$$

where

$$\gamma_{\mu\nu} = (i/2)(\gamma_\mu \gamma_\nu - \gamma_\nu \gamma_\mu). \tag{21}$$

There are two independent solutions, since there are two independent vectors A_ν. Another equivalent solution would be

$$\tau = J_\mu A_\nu \gamma_{\mu\nu} \gamma_5 \tag{22}$$

and actually this solution has been discussed.[3] But it is easily seen that the solutions (22) are not linearly independent of (20); they can be obtained from (20) by a linear transformation. Therefore there are just two independent solutions for each vector J, characterized by the "vectors of polarization" A_ν. It has been shown[3] that these bosons of rest mass 0 have all the transformation properties of light quanta and belong to the spin spin values ± 1 (the axis of the spin being parallel or antiparallel to the direction of propagation).

Existence of these bosons is closely connected with the existence of long range interactions between particles, which in turn are a consequence of the dipole states in Hilbert space II.

(e) Interaction between Particles. Electrodynamics and the Fine Structure Constant

The interaction between particles in collision processes can be treated by essentially the same method as the mass eigenvalues. To give a graph-picture of the interaction one can represent the incoming and outgoing particles by infinite tails, which are just periodic repetitions of the elementary graphs used for the calculation of their masses.[3] The interaction is ·then represented by a pattern of S and G lines connecting the periodic tails. A typical example is shown in Fig. 3.

It follows from rule 3 in section 2(c) for the graph patterns that there must be a break in the G lines connecting the two particles, which can be filled by two S lines, as is shown in Fig. 3. The two S_F functions vary for large space-like distances of x and x' as

$(x-x')_\nu \gamma_\nu/(x-x')^2$, their product varies roughly as $(x-x')^{-2}$. Therefore the two S functions, connecting the inner endpoints of the G lines, show together a behavior similar to that of the D_F function of quantum electrodynamics, and produce long range forces between the particles qualitatively similar to Coulomb forces.

Of course the multitude of graphs of the general type of Fig. 3 contains many different interactions, produced by all the different fields that belong to the different particles, bosons, and fermions. Actually the interaction will be a mixture of contributions from all particles that cannot be disentangled. Only in the case of bosons of rest-mass 0 can one separate their influence by studying the long range forces, since all other particles would produce only short range forces.

Calculation of the exact form of the long range forces is, however, rather complicated. One may for instance try to derive the cross section for collisions in which very little momentum J_μ is transferred from one fermion to the other:

$$P_\mu^{(1)} - P_\mu^{(1)\prime} = J_\mu = P_\mu^{(2)\prime} - P_\mu^{(2)}; \quad |J_\mu^2| \ll \kappa^2. \tag{23}$$

In this case it is necessary to take into account very long tails of the type of Fig. 4 in between the fermions,

FIG. 4. 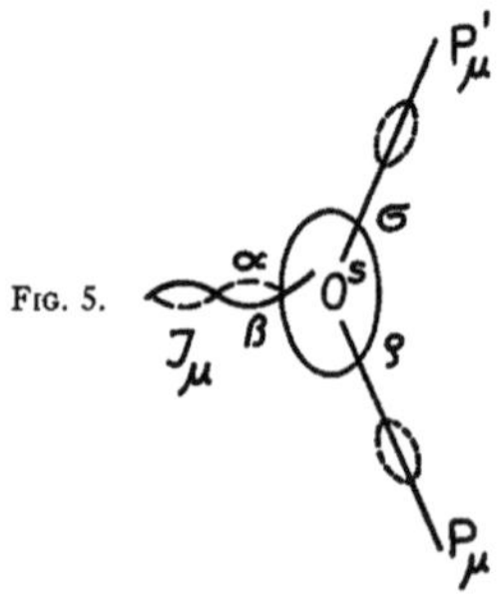

because for very small J_μ the tail means bosons (of rest-mass 0) that are nearly free; and for free bosons the tail could be infinitely long. Therefore the calculation has been carried out in two steps. In the paper cited in reference 3, the summation over all tails of different lengths was performed and led to an operator connecting the two fermions, which was composed of the eigenfunctions (20) or (22) of the light quanta [(104) of the paper cited], thereby showing that the long range forces are actually transferred by means of the field corresponding to the light quanta of Sec. 2(d). Then the operator O^S (page 440 of the paper), which connects as a kind of vertex part the former operator with either of the fermions, had to be calculated. O^S is represented by a graph of the general type of

FIG. 5.

Fig. 5. From arguments of symmetry and invariance[6] O^S has the general form

$$O_{\alpha\beta|\rho\sigma}\nu = \text{const}\frac{J_\mu}{\sqrt{J^2}}(\gamma_{\mu\nu})_{\alpha\beta}(\gamma_\nu)_{\rho\sigma}. \qquad (24)$$

The constant factor could be calculated only rather inaccurately, since its determination required the use of special methods of approximation the accuracy of which is somewhat doubtful. Finally the transition matrix element for the collision turns out to be

$$\frac{\text{const}}{J^2}[u^+(P^{(1)})\gamma_\mu u(P^{(1)})][u^+(P^{(2)})\gamma_\mu u(P^{(2)})]$$

$$\cdot\delta(P^{(1)}-P^{(1)\prime}+P^{(2)}-P^{(2)\prime})-\text{exchange term}. \qquad (25)$$

[$u(P)$ is the eigenfunction belonging to a fermion of momentum vector P.] This is exactly the form for the transition element in quantum electrodynamics. The value of the fine structure constant α_F was determined [using the inaccurate value of the constant in (24)] with the result

$$\alpha_F \approx 1/267. \qquad (26)$$

This shows that quantum electrodynamics with a special value of the fine structure constant is contained in the Eqs. (1) and (14) of our model for a theory of matter. One could not expect that the value of the fine structure constant should just be the empirical value 1/137, since the model theory is not yet the correct theory. But the fact that one gets a definite value of this constant of the right order of magnitude, seems to indicate that the model is in this respect not too far from the truth.

Equation (25) is perhaps not sufficient to show that (1) and (14) contain the complete scheme of quantum electrodynamics. Actually (25) has been supplemented[6] by proof that the conservation of charge holds generally and that one can construct field operators $F_{\mu\nu}$, which obey the Maxwell equations. These field operators are closely connected with the operators $\psi(x)\gamma_\mu\psi^+(x)$ which have the same transformation properties under the Lorentz group, but they are not identical with them. The operators $\psi(x)\gamma_\mu\psi^+(x)$ represent, besides the fields $F_{\mu\nu}$, other properties of matter, for instance the spin density of the fermions. Conservation of charge follows essentially from the identity $\partial^2 F_{\mu\nu}/\partial x_\mu\partial x_\nu=0$ and is not primarily connected with the invariance of the fundamental equations (1) and (14) against the transformation $\psi\rightarrow\psi e^{i\alpha}$. Since all particles can be considered as combinations of fermions, their charge can only be an integral multiple of the elementary charge of the fermion. The charge of the antifermion is opposite to that of the fermion. The bosons calculated from the graph [Fig. 2 in Sec. 2(d)] are neutral.

A few remarks may be added concerning the smallness of the fine structure constant. This smallness certainly has nothing to do with the value of the coupling parameter l. It is primarily a consequence of the mathematical form of the two functions G_F and S_F which are combined in the graph of Fig. 2. Their form leads to the conditions (19) which are very restrictive with respect to the possible solutions for $\tau_{\alpha\beta}$. In calculating the operator connecting the two fermions one has to make use of the relation:

$$\delta_{\alpha\beta}\delta_{\rho\sigma}=\tfrac{1}{4}\sum_N \gamma_{\alpha\sigma}{}^N\gamma_{\beta\rho}{}^N, \qquad (27)$$

where the sum is to be taken over all 16 elements of the Dirac algebra. Of this sum only the tensor terms $(\gamma_{\mu\nu})_{\alpha\sigma}(\gamma_{\mu\nu})_{\beta\rho}$ contribute to the interaction on account of the conditions (19), and only three of the six tensor terms contribute to the electromagnetic forces on account of the properties of the operator O^S in (24). These reasons for the smallness of the fine structure constant would therefore remain even if the form of the nonlinear term in (1) were altered.

If one were to define a corresponding coupling constant for the short range forces, its value should be of the order unity, since there are many bosons of different symmetries and the restrictive conditions (19) do not appear. This result fits well with the empirical fact of a large coupling constant for the Yukawa interaction. But one should remember that it should not, according to the graphs of the type of Fig. 3, be possible to single out a special short range interaction for a given boson from the rest of interactions.

(f) Nonlinear Integral Equations for S_F

The form of the function S_F has so far been derived from the assumed existence of fermions together with the assumption of the states of Hilbert space II. Since S_F is identical with a τ function of two variables it should also be possible to derive S_F from integral equations in the same manner as is done with the other τ functions. There is only the one essential difference

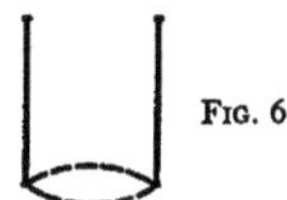

Fig. 6.

that the integral equations for S_F would be nonlinear, since S_F enters twice into these equations, once as the initial τ function and again as the function of contraction. For instance, a graph of the type of Fig. 6 would lead to an integral equation of the general type

$$\gamma_\mu\frac{\partial S_F}{\partial x_\mu}\sim G_F S_F{}^3. \qquad (28)$$

An integral equation of this type would probably, on account of its similarity to (1), lead to oscillations near the light cone, which do not show up in the

Reviews of Modern Physics 29, 269–278 (1957)

approximate solution (11). Therefore one could not expect to derive the solution (11) from an approximate integral equation like (28); still the exact form (14) could quite well be a solution of the exact integral equations. Actually it should be possible, at least in principle, to determine the spectrum $\rho(\kappa)$ of the fermions from this integral equation.

An attempt[3] was made to use differential equations corresponding to the integral equations of Fig. 6 for a determination of the asymptotic behavior of S_F at large space-like values of x and x'. There were two differential equations, one connecting $S_F(xx')$ with $\tau(xx|xx')$, the other connecting $\tau(xx|xx')$ with $\tau(xxx'|xx'x')$ and by means of three contractions again with $S_F(xx')$. Kita[19] has pointed out that these equations[3] contained an error of sign and that, after the correction of the error, the equations do not lead to the correct asymptotic behavior of S_F. But a closer investigation has since shown that actually the whole procedure for the calculation cannot be justified, since $\tau(xx|xx')$ is identical with $\varphi(xx|xx')$, on account of (14b), $S_F(0)=0$. The evaluation of $\tau(xxx'|xx'x')$ by means of contractions, however, is only possible if the φ functions of four variables can be neglected. The second differential equation would therefore use the φ function of four variables on the left-hand side while it would neglect it on the right-hand side. This cannot lead to any reasonable approximation.

It has so far not been possible to improve this calculation and to get information on the form of the S_F function from the nonlinear integral equations, since the construction of the φ functions with four variables would require very complicated mathematical investigations.

3. EXTENSION OF THE MODEL TOWARDS A REALISTIC THEORY OF MATTER

(a) Introduction of the Isobaric Spin

The model of (1) and (14) cannot represent the real system of elementary particles since the isobaric spin has not yet been introduced into the equations. (One could of course argue that the isobaric spin should not be put into the equations, but that it should come out of them, since the isobaric spin is closely connected with the charge. It may be that a careful study of the way in which the charge is attached to the particles will lead to a deeper insight into the nature of isobaric spin. For the time being however it seems necessary to introduce the isobaric spin into the equations.) To extend the model by its introduction to a more realistic one, it will not be sufficient to define the wave function ψ as spinor both in ordinary space and in isobaric spin space; for then all particles with half quantum spin would also have half-quantum isobaric spin, and integer spin values would be connected with integer isobaric

spin. But the selection rules put forward by Pais,[20] Gell-Mann,[21] Nakano and Nishijima[22] and others seem to show that, e.g., the Λ^0 particle has the spin 1/2 and isobaric spin 0. Therefore one needs at least two fundamental fields (as has been suggested by Goldhaber[23]), say ψ and χ, the one of which is a spinor in ordinary and in iso-space, while the other, χ, may be a spinor in ordinary space, but a scalar in iso-space.

One may of course choose other combinations for the two fields, but these assumptions are the simplest ones and allow one to construct a theory very similar to the model (1) and (14). We have now to investigate how far one can come with two such functions in describing the real system of elementary particles.

In the system of the real particles one usually distinguishes three types of interactions: the strong interaction, for instance between nucleons, π mesons, and hyperons; the electromagnetic interaction; and the weak interaction which for instance produces the decay of the Λ^0 particle or other radioactive processes.

The coupling constant for the weak interactions is extremely small compared to normal nuclear constants of the same dimension. They will certainly not play any appreciable role in determination of the masses. Therefore it will be convenient first to neglect the weak interactions all together and to introduce them later as a very small perturbation.

The electromagnetic interaction cannot, if we follow the model of (1) and (14), be separated from the strong interaction, since it is one of its consequences. On the other hand from the empirical selection rules one is inclined to think that the strong interactions are invariant against any rotations in isobaric spin space, while the electromagnetic interaction is only invariant against rotations around the z axis of this space.

Such a result may possibly be achieved by making the fundamental wave equation invariant against all rotations in iso-space, but assuming a commutation rule which is only invariant for rotations around the z axis of this space. This procedure may possibly lead to the intended result, since the electromagnetic interactions are connected with the constant κ in the commutator, which does not appear in the wave equation. If the sign of κ is coupled with the z direction of iso-space, this direction would have been introduced into the electromagnetic interactions, but the influence on the mass values might be comparatively small.

The program indicated has not yet been followed in detail. Just as an example of how it might possibly be carried out we quote the two equations[6]:

$$L = \psi^+\gamma\mu\frac{\partial\psi}{\partial x_\mu} + \chi^+\gamma\mu\frac{\partial}{\partial x_\mu}\chi + l^2(\psi^+\chi)(\chi^+\psi) \qquad (29)$$

[19] H. Kita, Progr. Theoret. Phys. 15, 83 (1956).

[20] A. Pais, Phys. Rev. 86, 663 (1952).
[21] M. Gell-Mann, Phys. Rev. 92, 833 (1953).
[22] T. Nakano, K. Nishijima, Progr. Theoret. Phys. 10, 581 (1953).
[23] M. Goldhaber, Phys. Rev. 92, 1279 (1953) and 101, 433 (1956).

and

$$\tfrac{1}{2}S^{\psi}(p)\approx\frac{p\mu\gamma\mu+i\kappa\tau_3}{p^2+\kappa^2}-\frac{p\mu\gamma\mu+i\kappa\tau_3}{p^2}+\frac{p\mu\gamma\mu\kappa^2}{(p^2)^2}. \qquad (30)$$

L is the Lagrangian for the wave equation and τ_3 is the z component of isobaric spin. The brackets in (21) are spinors in iso-space, but their product is an iso-scalar. The quantitative consequences of (29) and (30) have not yet been worked out. But it is easy to see what the qualitative consequences of these two equations would be with respect to the selection rules (compare Goldhaber[23]).

Besides the usual conservation laws of energy, momentum, and parity one has in (29) and (30) invariance for the transformations $\psi\rightarrow\psi e^{i\alpha}$ and $\chi\rightarrow\chi e^{i\beta}$. One of these invariances may be interpreted as conservation of the baryonic number (which is 1 for nucleons and hyperons, -1 for their antiparticles, and 0 for mesons and leptons). The rotational symmetry of (29) and (30) around the z axis of iso-space provides conservation of the z component of the isobaric spin. There will be no conservation of the total isobaric spin, since τ_3 appears in (30). But the transitions involving changes of the total isobaric spin may be somewhat less frequent than the others. Finally one has the conservation of electromagnetic charge which is connected with the introduction of the states of Hilbert space II and the conditions (19).

These selection rules together provide the existence of the "strangeness-quantum number" and its conservation. It is not necessary to invoke any new symmetry for the interpretation of this quantum number, as had been suggested by Racah[24] and Espagnat and Prentki.[25]

Therefore two rather simple equations such as (29) and (30) do actually account for all the known selection rules between baryons and mesons. Before we discuss the role of the leptons in this scheme it may be useful to form a general idea of the possible form of the weak interactions.

(b) Weak Interactions

The most characteristic feature of weak interaction between baryons and mesons seems to be the fact that it changes the z component of the isobaric spin by 1/2 without changing any of the other quantum numbers. Therefore one might think of interactions of the type

$$\text{const}\,[(\psi^+\chi)(\psi^+\psi)+\text{conj.}]. \qquad (31)$$

Such expressions in the Lagrangian are scalars in ordinary space but spinors in iso-space. The constants would therefore play the role of a spinor in iso-space. This situation would lead to an interesting consequence. If one performs a rotation in iso-space by 360°, the expressions (31) change their sign, since a spinor changes its sign under this transformation. Therefore one should have written (31) with an indefinite sign $\pm$, because one cannot distinguish between the two signs. If this is true, however, one could also allow instead of (31) interactions of the different type

$$\pm\,\text{const}\,[(\psi^+\chi)(\psi^+\gamma_5\psi)+\text{conj.}]. \qquad (32)$$

Such an expression would not violate the invariance of the Lagrangian for reflections in ordinary space. Still, if one combines the expressions (31) and (32), one would not expect the conservation of parity in weak interactions. Actually one of the two expressions could produce radioactive decay of the τ meson into three π mesons, while the other could cause the decay into two π mesons. Therefore this discussion favors the idea of Yang and Lee[26] that parity is actually not conserved in weak interaction. For interactions of spinor type in isobaric space [as in (31) and (32)] the invariance of the Lagrangian under reflections in ordinary space does not guarantee the conservation of parity.

(c) Role of the Leptons in the Scheme

While it is comparatively easy to combine all qualitative features of baryons and mesons in two simple equations like (29) and (30), it is difficult to see what place could be occupied by the leptons in such a scheme. The leptons are connected with the other particles both by electromagnetic interaction and by the weak interaction. Even if the weak interaction is neglected the mass spectrum of leptons should not be changed; but in this case all transitions from π mesons into leptons, β decay, etc., should be stopped. Therefore one should have a selection rule which forbids all transitions from the heavier particles into leptons except pair creation. So far no reason for such a selection rule can be given. It may be necessary to introduce a third wave function into the fundamental equation in order to account for the leptons or it may be that a closer investigation of the states in Hilbert space II would lead to a basis for such a selection rule without new wave functions. For the time being no solution for the problem of the leptons can be given.

[24] G. Racah, Nuclear Phys. 1, 301 (1956).
[25] B. d'Espagnat and J. Prentki, Nuclear Phys. 1, 33 (1956).
[26] T. D. Lee and C. N. Yang, Phys. Rev. 102, 290 (1956).

In *International Conference on Mesons and Recently Discovered Particles e 43° Congresso Nazionale di Fisica: Comunicazioni* (Ciclogrfia Borghero, Padova 1958) p. IX–I

GENERAL REMARKS CONCERNING THE THEORY OF ELEMENTARY PARTICLES

W. HEISENBERG
Göttingen

As basis for a theory of the elementary particles thefollowing principles are discussed:

1) The field operators neccessary for formulating the fundamental equations do not refer to any specified particle like proton,meson etc..they shall simply refer to matter in general.

2) The particles-elementary or compound-should be derived as eigensolutions of the field equations.

3) The fundamental field equations must be nonlinear to represent interaction. The masses of the particles should be a consequence of this interaction. The concept of a 'bare' particle has no meaning.

4) The conservation laws and selection rules for the creation and the decay of particles follow from the symmetry and invariance properties of the fundamental equations. Therefore the empirical selection rules should provide the most detailed information on the structure of the equations. -- The nonlinear spinor equation $\gamma_\nu \frac{\delta \psi}{\delta x_\nu} = 1^2 \psi(\psi^*\psi)$ together with an unconventional method of quantisation represent a model in general agreement with the foregoing principles. A mathematical analysis of the method of quantisation can be given by means of the Lee model and shows that the S-matrix is unitary. The spinor model leads to a reasonable mass spectrum of Fermions and Bosons and contains electrodynamics for a specified value of the Sommerfeld fine-structure constant ($\sim \frac{1}{267}$). For a more realistic theory of matter the isobaric spin has to be taken into account;the representation of the selection rules for the strange particles seems to be rather obvious,while for the leptons a satisfactory answer has not yet been given.

In *1958 Annual International Conference on High Energy Physics at CERN: Proceedings*, ed. by B. Ferretti (CERN, Geneva 1958) pp. 119–126

RESEARCH ON THE NON-LINEAR SPINOR THEORY WITH INDEFINITE METRIC IN HILBERT SPACE

W. HEISENBERG

Max-Planck Institut für Physik, Göttingen

The present report will be confined to recent developments in a non-linear spinor theory of matter, that had been treated in several previous papers [1]. The theory is different from the conventional theories especially by the use of an indefinite metric in Hilbert space and by the introduction of dipole-ghost states, the rôle of which has been analysed in some detail with the help of the Lee model [2].

The theory in the form that has been suggested [3] in connection with Pauli's transformation [4] starts from two characteristic invariants :

$$J = \int \bar{\psi}\,\gamma_\nu\,\frac{\partial}{\partial x_\nu}\,\psi\,d\tau \quad \text{and} \quad I = \int (\psi\gamma_\mu\gamma_5\psi)^2\,d\tau. \quad (1)$$

Both expressions are invariant with respect to the Lorentz group, the Pauli - Gürsey group [5] and the PCT-transformations; J is the conventional Lagrangian of the neutrino theory.

If one chooses any function of J and I as Lagrangian, the variation of $\mathfrak{L}$: $\delta\mathfrak{L} = 0$ leads to the wave equation :

$$\gamma_\mu\,\frac{\partial}{\partial x_\mu}\,\psi \pm l^2\,\gamma_\mu\gamma_5\,\psi\,(\bar{\psi}\,\gamma_\mu\gamma_5\,\psi) = 0. \quad (2)$$

(The constant $\dfrac{\partial\mathfrak{L}}{\partial I}\Big/\dfrac{\partial\mathfrak{L}}{\partial J}$ has the dimension of a square of

a length and may be put equal to $\pm\,l^2$.)

The determination of some of the simplest mass eigenvalues has been carried out recently by Mitter and Schlieder [*] in a first approximation by means of the Tamm - Dancoff method. This method seems for the time being to be the only method available for treating problems of a relativistic non-linear field theory with indefinite metric, in which the commutator cannot be given from the beginning but must come out as a result of the regularity conditions. While the mathematical basis of a non-linear spinor theory with indefinite metric is at present still uncertain in a similar manner as the more conventional theories involving interactions, there seems to be no reason to consider the Tamm - Dancoff method (when applied to

a mathematically well-defined eigenvalue problem) as less reliable than other approximation methods.

The Tamm - Dancoff method needs to begin with an assumption about the form of the commutator, the consistency of which has to be checked later in the course of the calculations. This process has been described in detail in some of the previous papers [6]. In a first approximation it is convenient to restrict the calculation to the " strong interaction " terms where one has full invariance for the isospin group, and to look apart from such complications as ̂ -conjugation and degeneration of the vacuum discussed in the preprint [3].

The only assumption for the commutator showing the complete isospin invariance seems to be

$$\langle \psi_\alpha(x)\,\overline{\psi}_\beta(x') \rangle = -\gamma_\nu{}^{\alpha\beta}\,\frac{\partial}{\partial x_\nu}\,F(s)$$

$$= 2\int e^{\,ip\,(x-x')}\,dp\,dK\,\frac{p_\nu\,\gamma_\nu{}^{\alpha\beta}}{p^2+K^2}\cdot\varrho\,(K)$$

$$\approx 2\int e^{\,ip(x-x')}\,dp\,\left[\frac{p_\nu\,\gamma_\nu{}^{\alpha\beta}}{p^2+K^2} - \frac{p_\nu\,\gamma_\nu{}^{\alpha\beta}}{p^2} + \frac{p_\nu\gamma_\nu{}^{\alpha\beta}\cdot K^2}{(p^2)^3}\right].$$

$$(3)$$

In the last line, which is meant as a first approximation, only one discrete mass eigenvalue is considered; the last two terms represent the contributions from the " dipole-ghost ". Other discrete eigenvalues and the continuous spectrum have been neglected.

If (3) is used as a " contraction function " in the Tamm - Dancoff method, the graph leads for a spinor particle of

momentum vector $\mathfrak{J}_\mu$ to an eigenvalue equation of the form

$$f(\mathfrak{J}^2)\cdot\mathfrak{J}_\mu\,\gamma_\mu{}^{\alpha\beta}\varphi_\beta = 0\,. \quad (4)$$

The function $f(\mathfrak{J}^2)$ has been determined numerically by Mitter and Schlieder and has only one zero defining a mass value [6].

(*) A detailed account of these calculations and other discussions connected with the non-linear spinor theory will be published by the Göttingen group in the Zeitschr. f. Naturforschung.

The solutions of (4) can be divided into two groups. The one belongs to the Dirac equation $\Im_\nu \gamma_\nu \varphi = 0$ and the mass value zero. It can be interpreted as defining four two-component leptons of mass zero (e^+, e^-, ν $\bar{\nu}$).

The other group belongs to a Klein - Gordon equation for a spinor wave function of the type:

$$(\Im_\mu{}^2 + K^2)\, \varphi_a = 0 \,. \tag{5}$$

It describes four baryons of equal mass ($P\ N\ \bar{P}\ \bar{N}$). Equation (5) can be decomposed into two Dirac equations

$$\Im_\mu \gamma_\mu \varphi = iK\, \hat{\varphi},$$
$$\Im_\mu \gamma_\mu \hat{\varphi} = iK\varphi, \tag{6}$$

as had been shown earlier by Gürsey. The symmetry between the matrix elements (φ-functions) φ and $\hat{\varphi}$ suggests the introduction of the $\hat{\ }$-conjugation and the use of a spinor

$$\Psi = \begin{pmatrix} \psi \\ \hat{\psi} \end{pmatrix}. \tag{7}$$

If one introduces Pauli-matrices Σ in this spinor space, one may write $\hat{\psi} = \Sigma_1 \psi$ and replace (5) by

$$(\Im_\mu \gamma_\mu - iK\, \Sigma_1)\, \varphi = 0. \tag{8}$$

The group theoretical problems connected with the $\hat{\ }$-conjugation and the reformulation of the commutator with the help of the Σ-matrices have been treated in detail by Duerr. We mention as one of the results the form of the Pauli-transformations, which must — as one sees from (6) — be written as

$$\Psi \to a\, \Psi + b\gamma_5\, \Sigma_3\, C^{-1}\, \overline{\Psi},$$
$$\Psi \to e^{\,ia\,\gamma_5 \Sigma_3}\, \Psi. \tag{9}$$

Before coming to the symmetry properties of the solutions we have to discuss the three transformations P, C, T (parity, charge conjugation and time reversal). Since the Lagrangian is invariant for these transformations one would expect corresponding quantum numbers. In the present theory, however, neither P nor C commute with the Pauli-transformations describing charge and baryonic number, if P is written in the conventional form

$$P: \quad x_k \to -x_k\,, \qquad \Psi \to \gamma_4\, \Psi. \tag{10}$$

Therefore, this kind of parity cannot be defined for particles having a definite baryonic number; the only exception being particles with baryonic number zero, e.g. π mesons, since their wave function is not affected by the transformation $e^{ia\gamma_5}$.

One may, however, define a second kind of parity by the relation

$$P: \quad x_k \to -x_k\,, \qquad \Psi \to \gamma_4\, \Sigma_1\, \Psi. \tag{11}$$

This operation commutes with the Pauli-transformations, since $\gamma_4\, \Sigma_1$ commutes with $\gamma_5\, \Sigma_3$. Therefore, one can define a parity for heavy particles by means of (11) and it is this second kind of parity that can be used for deriving selection rules. At this point it may be added that in the present theory the transition from particle to antiparticle is not given by the operation C, but by PC (first kind of P) or by $C\, \Sigma_1$.

Similar calculations as for the nucleons have been carried out by Mitter and Schlieder for the π meson group, i.e. for particles of baryonic number zero. The method used was in both cases indentical with the method applied in [6]. The calculations have so far been restricted to particles of spin zero. In this first approximation which is defined by the graph $\bigcirc$, the integral equation for the eigenvalues refers to eigenfunctions in which the two particles are at the same point. This restriction means that one has to do only with s-states. For particles of spin 1 one could scarcely hope to get an approximation without taking the p-states into account.

Table I gives the representation of the different particles in the form of the operators which annihilate (A) or create (C) the particles in question. If one wants to construct operators that only annihilate or only create, one has to combine the ψ- and the $\hat{\psi}$-term. (The explanation of the signs is given at the bottom of the table.)

Since, on account of the complexity of the calculations, it has so far not been possible to go to a higher approximation in the Tamm - Dancoff method, i.e. to an approximation defined by a more complicated graph, one may try to get some check on the reliability of the calculations by making use of the $\hat{\psi}$-functions. The symmetry between the ψ- and the $\hat{\psi}$-function suggests that one might get a better approximation, when one replaces $\bar{\psi}\, \gamma_\mu \gamma_5 \psi$ in the calculations by $\frac{1}{2}\, (\bar{\psi}\gamma_\mu \gamma_5 \psi + \hat{\bar{\psi}}\gamma_\mu \gamma_5 \hat{\psi})$. The numbers given in Table II as second approximation refer to this procedure. The parity of the neutron comes out as opposite to the parity of the proton, while the charged pions become scalar particles. This result is not identical with the usual description in the text books (where proton and neutron have the same parity and the $\pi^\pm$ particle is pseudoscalar), but it is equivalent to the usual description, since one can arbitrarily attach an odd parity to the charge. Therefore the result of the calculations is in agreement with the existing experimental evidence. The $\pi_1{}^0$ of the isospin-triplet comes out as pseudo-scalar with respect to the parity of the second kind in agreement with the experiments. However, its parity of the first kind is even, like that of the charged pions.

The isosinglet state $\pi_0{}^0$ represents a pseudoscalar particle with respect to both definitions of parity. Its mass comes out as considerably higher than that of the $\pi_1{}^0$ particles and it should therefore be highly unstable [*]. Hitherto there seems to be no good experimental evidence for an isosinglet $\pi_0{}^0$ particle.

The determination of the mass value leads to a satisfactory agreement between the first and the second approximation for the nucleons. Therefore, one may conclude that the correct mass value of the nucleons lies somewhere in the neighbourhood of $K\ell \sim 7$, which means that ℓ should be roughly equal to the Compton radius of the π meson $\left(\ell \sim \dfrac{\hbar}{\mu_\pi c}\right)$, in order to yield the correct nucleon mass. For the π mesons the two approximations give very different values. This unsatisfactory result seems to

TABLE I

	P	N	$\bar{P}$	$\bar{N}$
A	$\psi_r,\ \hat{\psi}_l$	$D^{-1}\bar{\psi}_r,\ -D^{-1}\hat{\bar{\psi}}_l$	$D^{-1}\bar{\psi}_l,\ -D^{-1}\hat{\bar{\psi}}_r$	$\psi_l,\ \hat{\psi}_r$
C	$\bar{\psi}_l,\ \hat{\bar{\psi}}_r$	$\psi_l D,\ -\hat{\psi}_r D$	$\psi_r D,\ -\hat{\psi}_l D$	$\bar{\psi}_r,\ \hat{\bar{\psi}}_l$

	π^+	π^0	π^-
A	$\psi_{r_1}\psi_{l_2}-\psi_{r_2}\psi_{l_1},\ \hat{\psi}_{l_1}\hat{\psi}_{r_2}-\hat{\psi}_{l_2}\hat{\psi}_{r_1}$	$\psi_{r_1}\bar{\psi}_{l_1}+\psi_{r_2}\bar{\psi}_{l_2},\ -\hat{\psi}_{l_1}\hat{\bar{\psi}}_{r_1}-\hat{\psi}_{l_2}\hat{\bar{\psi}}_{r_2}$	$\bar{\psi}_{r_1}\bar{\psi}_{l_2}-\bar{\psi}_{r_2}\bar{\psi}_{l_1},\ \hat{\bar{\psi}}_{l_1}\hat{\bar{\psi}}_{r_2}-\hat{\bar{\psi}}_{l_2}\hat{\bar{\psi}}_{r_1}$

$$\pm\left\{\psi_{l_1}\bar{\psi}_{r_1}+\psi_{l_2}\bar{\psi}_{r_2},\ -\hat{\psi}_{r_1}\hat{\bar{\psi}}_{l_1}-\hat{\psi}_{r_2}\hat{\bar{\psi}}_{l_2}\right\}$$

$$\bar{\psi}=\psi^*\gamma_4;\ \ D=C\gamma_5;\ \ D^{-1}=\gamma_5 C^{-1}.$$

$$C\gamma_\mu C^{-1}=-\tilde{\gamma}_\mu.\ \ \ D\gamma_\mu D^{-1}=\tilde{\gamma}_\mu.$$

$$\text{,,}r\text{''}:\gamma_5=+1;\ \ \text{,,}l\text{''}:\gamma_5=-1.$$

$$\text{,,}1\text{''}:\sigma_3=+1;\ \ \text{,,}2\text{''}:\sigma_3=-1.$$

TABLE II

Particles	Nucleons	πmesons		?	
	$P\ \ N\ \ \bar{P}\ \ \bar{N}$	$\pi_1{}^+\ \ \pi_1{}^0\ \ \pi_1{}^-$	$\pi_0{}^0$	$+\ \ \ 0\ \ \ -$	0
Parity 1st kind	not defined	$+\ \ +\ \ +$	$-$	$+\ \ +\ \ +$	$-$
Partiy 2nd kind	$+\ \ -\ \ -\ \ +$	$+\ \ -\ \ +$	$-$	$+\ \ -\ \ +$	$-$
Graph					
Mass $K\ell$ 1st approx.	7.08	3.0	7.0	—	
2nd approx.	6.67	0.41	5.9	2.5	2.5

(*) At the Conference the author had reported that the mass of the isosinglet state came out as equal to the mass of the isotriplet. A renewed controlling calculation has, however, revealed an error in the earlier computations of the neutral π-states, which did not affect the mass of the $\pi_1{}^0$-state but did affect strongly the mass of the $\pi_0{}^0$-state. The author regrets the error in his verbal report and wants to emphasize that an independent repetition of these rather complicated calculations by some other group — if possible by different methods — would be very desirable.

be connected with the very high sensitivity of the meson mass in the calculations for changes in the nucleon mass. The π eigenvalue depends on the nucleon mass and the calculation shows that a change of the nucleon mass by 10% produces a change of the meson mass by more than a factor 2. On the other hand, the nucleon mass can easily be inaccurate by 10-15%. Therefore, a somewhat accurate determination of the π meson mass has so far not been possible. But it may be possible in this special case to calculate later a genuine second Tamm - Dancoff approximation. The singlet π_0^0 mass is not equally sensitive.

The wave equation (2) contains an indefinite sign in the non-linear term. A change of this sign is equivalent to a change of the sign of the commutator (3). For instance the transformation $x_\nu \rightarrow -x_\nu$ will change simultaneously the sign in (2) and in (3). But there is no simple transformation that leads from one sign in (2) to the other, without simultaneously also changing the sign in (3).

The only transformation that can produce this effect is a rather radical change made possible by the indefinite metric. If one changes the norm of all physical states of half-integer spin from $+1$ to -1 (keeping the norm of the states of integer spin as $+1$), one changes the sign of (3) without changing (2), and if one simultaneously replaces x_ν by $-x_\nu$, one changes the sign in (2) without changing (3).

By this transformation one gets into an entirely new set of states, which in relation to the first set can be considered as a non-combining term system. The assumption of a norm -1 of all states with half-integer spin does not inter-fere with the assumption of a unitary S-matrix. In this new system the particles corresponding to the nucleons get in the first approximation the same mass as those of the first system. The particles corresponding to the pions, however, get a different mass (the value is given in the third column in Table II). It is tempting to connect this non-combining term-system with the existence of the " strange particles " and especially the third column of Table II with the θ-particles. But the analysis of the solutions has not yet been carried far enough to justify such an identification.

The mathematical analysis of the indefinite metric has not been carried further by the Göttingen group since the research connected with the Lee model. But I might mention at this point two recent papers by Ascoli and Minardi [7], in which these authors try to get a more rigorous axiomatic basis for the use of the indefinite metric. One of their results concerns the equivalence between a local and causal theory with indefinite metric and non-local theory with definite metric. But I understand that Pauli will speak in the discussion about such problems, therefore I need not go into these questions.

In conclusion, I would like to state the present situation concerning the non-linear spinor theory as follows : the question whether this attempt can lead to a theory of the elementary particles is still completely open. But I think that the possibilities offered by the indefinite metric and the results reported here are interesting enough to make it worth while to look still more deeply into the conse-quences of such an attempt.

LIST OF REFERENCES

1. Heisenberg, W. Rev. mod. Phys., *29*, p. 269, 1957.
2. Heisenberg, W. Nuclear Physics, *4*, p. 532, 1957.
3. Heisenberg, W. and Pauli, W. On the isospin group in the theory of the elementary particles. (preprint)
4. Pauli, W. Nuov. Cim., *6*, p. 204, 1957.
5. Gürsey, F. Relation of charge independance and baryon conservation to Pauli's transformations. (preprint)
6. Heisenberg, W. Nachr. Akad. Wiss. Göttingen, Math. phys. Kl. 2, p. 111, 1953.
    Heisenberg, W. Kortel, F. and Mitter, H. Z. Naturforsch., *10a*, p. 425, 1955.
7. Ascoli, R. and Minardi, E. On quantum theories with indefinite metric. (preprint)

DISCUSSION

Pauli : I am glad that during the last month I had occasion to discuss anew with many colleagues the problems of the elementary particles with respect to group theory and to field quantization. But the logical connection between field equations and commutation relations has not been sufficiently clarified neither by rigorous, nor by more clumsy intuitive methods. Therefore, reliable methods for calculating mass ratios of elementary particles are not yet available.

Regarding the papers of Heisenberg and collaborators on the spinor model, which appeared in the last years, after various discussions I reached the conclusion that they are mathematically objectionable.

To illustrate this, consider a spinor field $\psi(x)$ obeying a differential equation of first order

$$\gamma^\nu \frac{\partial}{\partial x_\nu} \psi(x) + F\big(\psi(x)\big) = 0 \qquad (1)$$

where the argument of the function F depends only on the value of $\psi(x)$ at a particular point. (We assume that F and its derivatives have no singularities; for example, imagine F to be a polynomial.) Assume, further, that the field not only anticommutes for space-like points,

$$\{\psi_a{}^*(x), \psi_\beta(x')\} = 0 \quad \text{and} \quad \{\psi_a(x), \psi_\beta(x')\} = 0 \qquad (2a)$$
$$\text{for } (x_\nu - x'_\nu)^2 > 0,$$

but that this also holds for $x = x'$

$$\{\psi_a{}^*(x), \psi_\beta(x)\} = 0, \qquad (2b)$$

where in the usual theory a singularity of the δ-type appears.

As one knows from the Lee model, the commutation rules (2a) and (2b) can be fulfilled for an indefinite metric. They are, however, in contradiction to the assumption of a differential equation of the first order (Eq. (1)). Indeed, from (1) and (2a, b) it follows for all derivatives

$$\left\{ \psi_a(x), \ \frac{\partial^n}{\partial t'^n}\psi_\beta(x') \right\} = 0 \quad \text{for} \quad t = t'. \qquad (3)$$

This is in disagreement with the usually assumed mass spectra. The same contradiction still holds for several spinor fields obeying first order differential equations if the anti-commutativity of the field at the same point is supposed for all these fields.

Now the question arises whether in the quoted papers the contradicting assumptions (1, 2a, b) are actually used. In spite of the introduction of time-ordered products (τ- and φ-functions) instead of the field operators themselves, it is my opinion that an assumption fully equivalent to the discussed one has been made in the course of the integration with the help of the Tamm - Dancoff method. Therefore, I do not trust the excuses which have been made for the contradicting results in question and I stress my opinion that the latter is still existing.

In this connection, I also wish to point out that I do not overlook the possibility of replacing the regularity of the anti-commutator at the same point by an oscillating behaviour with an essential singularity. This would give rise to an essentially new situation if this oscillating behaviour were explicity applied in the course of the integration of the equations for the τ- and φ-functions. But as soon as the oscillations are replaced by an average zero, which seems to me not admissible, the equivalence with the assumptions (1) (2a,b) and this contradiction reappears.

I disbelieve all the more in the possible excuses for such a contradiction as it can very easily be avoided. For instance, it is absent in the Lee model, where instead of (1) an integral equation appears. Indeed we know that the commutation relations (3) are not fulfilled in the Lee model with an indefinite metric.

My remarks do not pretend to say anything new. I only wanted to clarify my own position with regard to these problems. The general problem of whether an indefinite metric can be used in physics is not yet covered by my present remarks.

Heisenberg [*]: Pauli's remark contains three different questions which I would like to answer separately and at some length in order to make this side of the theory completely clear. The three questions are:

1. Is the assumption that $\psi(x)$ and $\bar{\psi}(x')$ anticommute everywhere on the subspace $t - t' = 0$, but not for time-like $x - x'$, compatible with the non-linear differential equation?

2. Are the assumptions made about the anticommutator used in a consistent way when they are applied in the Tamm - Dancoff method in connection with the equation $S(0) = S_F(0) = 0$?

3. Should not the differential equation be replaced by an integral equation in the time-variable and is not the Lee-model an example for such a replacement?

For the first question we have to remember that — as had always been assumed — the anticommutator should behave near the light-cone like that classical solution of the non-linear differential equation that vanishes for space-like distances. Such a solution has been constructed (the calculations had been carried out for the older spinor model, but not yet for the new wave equation) and it shows the following behaviour: it has infinitely frequent oscillations in the immediate neighbourhood of the light-cone for time-like distances. It vanishes for space-like distances. The statement that it vanishes everywhere for $t = t'$ has, however, to be taken with some precaution: at the point $x = x'$ its value is not defined, on account of the infinite oscillations. But its space integral at $t = t'$ including the point $x = x'$ definitely vanishes. There is no δ-function of the space co-ordinates. Therefore, we see that this behaviour is actually compatible with the differential equation and there is no inconsistency in using an anticommutator of this (oscillating) type, which is, of course, not analytic at the light-cone.

With regard to the second question, we note that the non-linear wave equation of the operators can be considered as a compact way of writing the infinitely many differential equations between the τ-functions of any number of variables. Actually, for the calculations it would be sufficient to assume that the field equation means nothing else but this infinite set of τ-equations. (Instead of using the time-ordered products one could, of course, also use other kinds of products of field operators.) Now we know that the τ-function

$$\tau(x_1 x_2 \dots \mid y_1 y_2 \dots)$$

behaves in the neighbourhood of the point, say $x_i - y_k = 0$, like $\frac{1}{2} S_F(x_i y_k) \cdot \tau(x_1 .. x_{i-1}, x_{i+1} ... \mid y_1 .. y_{k-1} y_{k+1} ...)$ plus a

---

[*] This part of the discussion has not been reported verbatim but is based on a paper sent by Heisenberg after the Conference.

function which is less singular near the light-cone. Therefore, the value of a τ-function at the point $x_i = y_k$ is undefined in the same way as the value of $S_F(0)$.

On the other hand, those τ-equations that result from the wave equation contain always one τ-function (that with the higher number of variables) in which three co-ordinates are equal, e.g. either

$$x_i = x_k = y_l \quad \text{or} \quad x_i = y_k = y_l \, .$$

This is a necessary consequence of the postulate of micro-causality. Therefore these τ-equations have no definite meaning unless one defines explicitly what is meant in the equation by the value of the τ-function at this point, e.g. $x_i = x_k = y_l$.

The simplest definition is given by the assumption that in the equations one shall put

$$\tau(x_1 \ldots | y_1 \ldots) | x_i = x_k = y_l = \lim_{\substack{x_i \to y_l \\ x_k \to y_l}} [\tau(x_1 \ldots | y_1 \ldots)$$
$$- \tfrac{1}{2} S_F(x_i \, y_l) \, \tau(x_1 \ldots x_{i-1} \, x_{i+1} \ldots | y_1 \ldots y_{l-1} \, y_{l+1} \ldots)$$
$$- \tfrac{1}{2} S_F(x_k \, y_l) \, \tau(x_i \ldots x_{k-1} \, x_{k+1} \ldots | y_1 \ldots y_{l-1} \, y_{l+1} \ldots)] \, .$$

This limiting value exists on account of the behaviour of the τ-functions and it gives a possible interpretation of the τ-equations. The definition can be replaced by the simpler formula $S_F(0) = 0$, or more accurately by the statement: "... in the infinite set of τ-equations one can calculate as if $S_F(0) = 0$ was generally correct". It is only this definition of the τ-functions with three equal variables which gives a unique meaning to the wave equation. This meaning is compatible both with the oscillatory behaviour of the anticommutator and the postulate of microcausality. The equation $S_F(0) = 0$ should be considered as a statement about the wave equation (or the τ-equations) rather than as one about the anticommutator.

In momentum space our assumptions concerning the function $S(xx')$ or $S_F(xx')$ will probably mean the following. The mass spectrum $\varrho(K^2)$, from which $S(xx')$ can be constructed, will contain discrete states and a continuous part. For very large values of K^2 the continuous part will show an oscillating behaviour in order to represent the oscillating behaviour of $S(xx')$ near the light-cone. The discrete part will contain the discrete states like nucleons, electrons, etc. and the regularizing dipole ghosts. In the lowest approximations of the Tamm - Dancoff method only the lowest discrete states and the dipole ghost can be taken into account; the higher discrete states and the continuum have to be neglected. Therefore, one obtains a reasonable approximation only if the contribution from the lowest states is considerably bigger than that of the higher ones and the continuum. This situation is well known from the optical spectra of atoms. The conditions, that the lowest lines contain the main part of the total sum of f-values ($\Sigma f_\nu = 1$ is the Thomas - Kuhn sum rule) is frequently, but not always,

fulfilled. The uncertainty of the Tamm - Dancoff method lies just in the uncertainty about the fulfilment of this condition.

The answer to the third question arises from the well-known axiom of quantum field theory, according to which the field operators belonging to an arbitrarily small time-interval Δt must be sufficient to construct from vacuum the complete Hilbert space. Therefore, it must be possible to establish a unique relation between the operators referring to the interval between t and $t + \Delta t$, and those belonging to the interval between $t + \Delta t$ and $t + 2\Delta t$. Since Δt can be taken as arbitrarily small, this relation can be written as a differential equation in the time co-ordinate, which may possibly contain integrations over space co-ordinates.

However, it has to be noted that in a relativistic theory the occurrence of integrations over the space co-ordinates at this point would lead to a violation of the condition of microcausality. Even if the anticommutator vanishes for space-like distances within the time interval t, $t + \Delta t$, it would not fulfil this condition in the intervals between $t + \Delta t$ and $t + 2\Delta t$, if the connection would involve integrations over the space co-ordinates. Of course one could try to give up the condition of microcausality; but so far it has not been possible to drop this condition without violating at the same time the principle of macro-causality. It was one of the main results of the theory of special relativity, that causal connections have to be expressed as differential equations that are invariant for the Lorentz transformation. I cannot see how one could get away from this necessity.

Perhaps I should summarize the results of this discussion as follows: if one wants to keep microcausality, one has to connect the field operators at different times by means of a differential equation that is Lorentz invariant, or by other equations that are equivalent to such differential equations. The anticommutator between $\psi(x)$ and $\bar\psi(x')$ is zero for space-like distances and is indefinite at the point $x = x'$; the space integral of the anticommutator between $\psi(x)$ and $\bar\psi(x')$ at the same time $t = t'$ vanishes. Since the wave equation contains the product of three field operators at the same point one can give a definite meaning to the wave equation by defining this product or the corresponding τ-functions by a limiting process, or more simply by $S(0) = S_F(0) = 0$. Without such a definition the wave equation would not contain any statement at all.

Pauli: It is probably not true that the operators in one time-plane define the whole Hilbert space. You will have to take into account that ψ_ν is not a measurable field.

Heisenberg: That does not matter. It belongs to the Hilbert space; I mean, I can use any operators to define the Hilbert space, but I only need the assumption that the operators in a Hilbert space defined in a certain time-interval are sufficient to construct the complete Hilbert space, and I think that this assumption is completely true also in the Lee model.

Chew: I do not understand.

Heisenberg: Oh yes! There is an apparent contradiction, because in the Lee model the renormalized wave equation takes the form of an integral equation in the time co-ordinate. In spite of this, it should be possible, on account of the axiom mentioned above, to reformulate it into a differential equation, which will probably involve integrations over space co-ordinates.

Pauli: I do not think we will reach an agreement on this point. There are many other points to discuss in Heisenberg's paper.

Stueckelberg: I just want to ask a rather trivial question. There is the D_C (in Europe) or D_F causality defined in terms of perturbation theory. Its basis is that we distinguish between positive and negative frequencies, and define D_C or D_F functions by positive frequencies. On the other hand, there is the causality, defined by dispersion relations, saying that the vacuum expectation value of the commutator $[\psi(x), \psi(y)]_\pm$ is only different from zero for time-like $x-y$. The question I am asking is whether Heisenberg or anybody in this group can answer the question: is there any relation between the condition of positive frequencies in the D_C, (D_F) definition and the positive energy equation in dispersion relations theory?

Oppenheimer: I think I can answer the question. If you have vanishing commutators, then you may construct either a causal or an anti-causal D-function. If you do not have vanishing commutators, it is very doubtful that you could construct either.

Gell-Mann: There is just one more comment in answer to Stueckelberg's question. If the causality in the sense of dispersion relations exists and you talk about the matrix elements of the commutators, then you can also talk, if you like, about the D_C or D_F functions for the same problem, and for positive frequencies in the Fourier transform these will agree and for negative frequencies they will disagree only in the sign of the imaginary part. So a statement about one is effectively a statement about the other, and the people working on the dispersion theories conventionally work with the commutator rather than with the D_C and D_F function just for greater convenience.

Stueckelberg: Does this mean that if you speak of commutators you do not really need somehow a series development?

Gell-Mann: Well, one hopes in dispersion theory not to make use of perturbation theory. Of course, nobody really knows how to work with field theory aside from perturbation theory but people try, and they have produced a certain number of results especially in dispersion theory.

Oppenheimer: The answer is the following. Of course, to know the commutators in full you need to know the theory and know whether it is linear or not, but to

know that something vanishes will be true of all functions. Thus this statement does not restrict you to any linear theory.

Klein: I would like to ask a very simple question which can be actually answered. Will Heisenberg please remind us of what is the meaning of the graphs for the calculations of the masses.

Heisenberg: When you work with the Tamm - Dancoff approximation you deal with the τ-function $\langle \Omega \,|\, \psi(x) \,|\, \Phi \rangle$. Then from this τ-function you go over to some function with more variables, by means of a Green's function, using the wave equation which has been written. So this is something like, say

$$\langle \Omega \,|\, \psi(x) \,|\, \Phi \rangle = \langle \Omega \,|\, dx' \, G(x, x') \, \psi(x') \, \bar{\psi}(x') \, \psi(x') \,|\, \Phi \rangle.$$

Then you can again go over to the next step and you introduce two Green's functions,

$$G(x, x') \cdot G(x', x'').$$

In this way you get higher τ-functions and then at the end you make the so-called contractions. Thereby these τ-functions can be expressed approximately by the τ-functions of a lower number of variables. The contraction function is essentially the commutator, or more accurately the Schwinger S_F-function. The graph states which points x or x' or x'' are to be connected by G_F- or S_F-functions in the final integral equation.

Thereby one can get a simple formal picture for such a rather complicated integral equation, just as in the conventional Feynman-graphs.

Johnston: I should like to ask Heisenberg one point about the masses of the $\pi_0{}^0$ and $\pi_1{}^0$ in the scheme he drew up on the board. He said that in the first approximation the masses became equal, and he said that perhaps if other corrections were included, mentioning particularly electromagnetic corrections, this situation would possibly change. Does the electromagnetic interaction come from the first wave equation written on the board, or is this something extra that has to be added?

Heisenberg: That part of the calculation concerning the electromagnetic fields and particles has not yet been carried out for this wave equation, but it had, in a former paper by Ascoli and myself, been done for the old non-linear spinor equation. At that time it turned out that the electromagnetic forces can come out of such an equation; actually they are contained in it. So at this point we do not yet have any definite results for the new equation, but I think it is very probable that one can just repeat the old calculations for this model here and we will get very similar results.

Wentzel: There are so many questions to be asked that I can hardly start, but let me ask what has become of the problem that you once connected with the degeneracy of the vacuum; the necessity to explain the large number of quantum numbers one needs, for instance, in baryon physics

Heisenberg: I think this question of a duplication of the vacuum is more a matter of calculation technique than of fundamentals. For instance, when we have the solutions for the leptons and for the baryons then, of course, we can separate them in two ways; one can either say that the operator ψ is the sum of two operators, one referring to changes of the baryonic number, and the other referring to changes of the leptonic number, but one can also introduce two kinds of vacuum for the matrix element between a state and the vacuum, so that instead of doubling the operators we double the vacuum, and thereby one can perform transformations not only on the operators but also on the vacuum. If one has to calculate a vacuum expectation value of a product of not only two but $2n$ ψ-functions, then, of course, this value would become a matrix with 2^n rows and columns. But perhaps I should state it more clearly in the following way. In the old theory of the atom, when Dirac introduced the spin, his argument was as follows. In the normal theory one has momentum and co-ordinate p and q and we get an equation of the second order on account of the Lorentz-invariance. But now, instead of taking a square root, one can also introduce a new degree of freedom which can be put on the wave functions, which means on the Hilbert vectors. In this way, Dirac introduced the duplication of states in Hilbert space by saying: we have not only one state vector, namely a function ψ, but we have ψ_1 and ψ_2 and thereby we have a duplication of the states. In the same way you can say here that we start from a Lagrangian in which, of course, there is nothing to be seen of any duplication; but when we now try to find a commutator which corresponds to this Lagrangian, it may turn out that a square root is to be taken (e.g. as for the mass in the Klein - Gordon-equation), and therefore a duplication may occur in the vacuum or in the state vectors.

Wentzel: This partly answers my question, but looking at reality we have a quantum number that is called strangeness. Is the multiplicity that is inherent in your formalism sufficient to explain or to incorporate all the quantum numbers which we need nowadays? Of course you can go on and introduce a σ and a Σ_1, and by always taking another power of two you can get there. But is this a natural procedure?

Heisenberg: I think it is characteristic for these quantum numbers — you referred to the quantum numbers called l and l_N in the preprint — that they are apparently defined only by cyclic groups. We need no space group, no rotation in space, to explain these quantum numbers. It seemed natural to us that just the cyclic groups should be put into the Hilbert space. It was especially emphasized by Dürr that actually cyclic groups may be transferred to the vacuum while space rotations can not. The reason is the following: if the operators transform like spinors, then if one tries to replace their transformation by transformations in Hilbert space, $\Lambda' = T^{-1} \Lambda T$, one needs for the corresponding transformations in Hilbert space not half-integral representations of the rotation group but quarter-integral representations which do not exist. Therefore, three-dimensional rotations can certainly not be put on the vacuum, while one-dimensional rotations can be transferred on the Hilbert space. It seemed natural to connect this result with the other empirical fact that these extra groups are actually only cyclic groups. So I think it is not unnatural to say that these two cyclic groups belonging to l and l_N are actually properties of the operators which can be expressed by a duplication of the vacuum or of the state vectors.

Pauli: I completely disagree with the answer of Heisenberg. I think this is not only unnatural but this is mathematically impossible. You cannot use the multiplicity of the vacuum in order to obtain the strangeness, because then the charge of two protons is not any longer two, if the charge of one proton is one. This I discussed already in April and I wonder that you again repeat it all.

Heisenberg: Of course I again disagree completely with what Pauli said, because I do not see the slightest reason why one should not take out of the vacuum as many charges as one has protons; that is the same as in the Dirac theory. Dirac has introcuded the doubling for each new electron, and for ten electrons he had also ten spins.

Pauli: The point is that the additivity of charges is different from the additivity of other operators, so that the answer was again completely false: the additivity of charges is different from the additive behaviour of other operators like spin and isotopic spins, and masses. This is the point. The number of the vacua does not multiply if you have two protons. I think this was disproved in April.

Gell-Mann: May I attempt to re-state the situation? I am not sure if I understand all points but perhaps it will come out. As I understand the idea, it would be something like this. You take a twin vacuum and an operator, which you would ordinarily call a neutron operator. You act on the twin vacuum with the neutron operator. On one state it makes a neutron and on another state it makes a Λ. So you get two states, neutron and Λ. I think the difficulty is, that if you take two neutron operators, the two vacua will give you only double the number of states but you need four times as many states now, namely neutron-neutron, neutron-Λ, Λ-neutron, and Λ-Λ. So, depending on how many particles you want to describe, N, you need 2^N vacua. This must vary with the number of particles we wish to describe. I suppose one can do it but it seems to be complicated and very much like adding another field.

Heisenberg: I agree completely with what Gell-Mann just said about this point; but at the same time, I propose to postpone the discussion for half a year and then we will know more about it.

Pauli: Well, I think that it is superfluous. I think that in half a year the answer would be the same as Gell-Mann gave just now.

isungen. Hierbei ist der enzy-
artialabbau von besonderem
tehenden Oligopeptide werden
ind Elektrophorese, meist auf
ster einer solchen kombinierten
de sind für die Proteine charak-
en, wie Fingerabdrücke, ihre
scheidung, was bei Milchsäure-
en Ursprungs praktiziert wurde.
bau mit Phenylisothiocyanat
säurereihenfolge in den Peptid-
rden. Die Verwendung von
stattet die Bearbeitung aller-

sich auch antibiotische oder
nlichen Baus, cyklische Ver-
ren enthalten, die in den ver-
vorkommen. So enthalten die
olierten Giftkomponenten aus
itterpilz mehrere verwandte

peptidsynthesen befinden sich
er Entwicklung. Für den Auf-
id zu den von E. FISCHER und
ten Methoden zahlreiche neue
Teil das Arbeiten in wäßriger
chemischen Medium, erlauben.
iren der lebenden Zelle glaubt
sein, doch ist der Mechanismus
am Protein noch immer unge-
Polypeptide entstehen bei der
tischen Enzymen auf gewisse
itischen Proteinabbaus (Plas-
auf synthetischem Weg aus
hydriden der α-Aminosäuren
ivaten von niederen Peptiden.

g der Nucleinsäuren
iehrung der Viren

RAMM, Tübingen

iren hat man eine Reihe von
nmengefaßt, die in ihrer Größe
ien und anderer Mikroorganis-
nehrung nur in lebenden Zellen
en der Biochemie geht dahin,
nzeichen auf gemeinsame Prin-
ückzuführen. Die chemische
Viren Nucleinsäuren enthalten,
ucleinsäure- oder dem Desoxy-
:hören. Dieses Merkmal haben
ermehrungsfähigen Zellelemen-
cleinsäure oder Nucleinsäure-
len sich stets im Innern der
Hülle besteht bei den einfachen
ei den komplizierter gebauten
, Kohlenhydrate und Lipoide.
rsuchungen spricht dafür, daß
nehrungsvorgang die Nuclein-
während die Hülle dem Schutz

sich die Möglichkeit, nähere Einblicke in den Vermehrungs-
prozeß zu gewinnen. Durch die Erweiterung unserer
Kenntnisse über die Virusvermehrung lassen sich auch
wichtige Aufschlüsse über die Biosynthese der normalen
Zellproteine erhalten.

Am Dienstag, dem **30. September,** führt Herr Professor Dr. H. KOPFERMANN den Tagesvorsitz. Der Tag
steht unter dem Zeichen der Elementarteilchen und
bringt Vorträge zum Geophysikalischen Jahr.

Allgemeiner Überblick und Fortschritte in der Theorie der Elementarteilchen

Von W. HEISENBERG, Göttingen

Die experimentellen Ergebnisse über die Elementarteilchen geben Veranlassung, zwischen drei Arten von
Wechselwirkungen zu unterscheiden: Den starken Wechselwirkungen, die z.B. als Kernkräfte zwischen den
Nukleonen in Erscheinung treten, den elektromagnetischen Kräften und shließlich den schwachen Wechselwirkungen, die sich im radioaktiven Zerfall äußern. Die
Symmetrieeigenschaften der Elementarteilchen werden
durch Quantenzahlen beschrieben, deren wichtigste die
Baryonenzahl, die elektrische Ladung, der Isotopenspin
und der gewöhnliche Spin sind. Mit Hilfe dieser Quantenzahlen ist phänomenologisch eine weitgehende Ordnung
der Elementarteilchen und ihrer Auswahlregeln bei
Übergängen erreicht worden. Für die starken Wechselwirkungen gelten die schärfsten Auswahlregeln, bei den
elektromagnetischen ist die Isospin-Symmetrie gestört,
bei den schwachen auch die Parität.

Die Gleichung

$$\gamma_\nu \frac{\partial}{\partial x_\nu} \psi \pm l^2 \gamma_\mu \gamma_5 \psi (\overline{\psi} \gamma_\mu \gamma_5 \psi) = 0$$

— aufgefaßt als eine Gleichung für den Materiefeld-
Operator in einem Hilbert-Raum mit indefiniter Metrik —
ist invariant gegen die Lorentz-Gruppe, die Isospingruppe (in Pauli-Gürseyscher Darstellung) und die drei
Operationen: Ladungsumkehr, Parität, Zeitumkehr. Da
diese Gleichung auch die Lichtquanten als Eigenlösungen
zu enthalten scheint und — auf dem Umweg über die
Existenz von Teilchen der Ruhmasse Null und damit
von Kräften unendlich langer Reichweite — zu Symme-
trieverminderung Anlaß geben kann, scheint die Gleichung
zum mindesten qualitativ geeignet für die Darstellung
des wirklichen Systems der Elementarteilchen. Es wird
über die Ergebnisse der Rechnungen berichtet, die zur
Durchführung des eben angedeuteten Programms bisher
unternommen worden sind.

Erzeugung von Elementarteilchen im Laboratorium

Von W. PAUL, Bonn

Es werden die experimentellen Bedingungen diskutiert, unter denen Elementarteilchen entstehen können
und damit gezeigt, warum Protonen und Elektronen-
beschleuniger höchster Energie benötigt werden. Das

In *Proceedings of the 1960 Annual International Conference on High Energy Physics at Rochester*, ed. by E. C. G. Sudarshan, J. H. Tinlot, A. C. Melissinos (University of Rochester, Rochester, N.Y., 1960) pp. 851–858

DISCUSSION

Noyes: I wonder if there is an experimentalist here who is familiar with the details of the K_{e3} spectrum? As I recall, most methods have a very strong experimental bias against measurements of high energy electrons in this decay. I wonder if someone can make an experimental comment on that point, as the evidence given for a 350 MeV $\pi^{0\prime}$ depends entirely on whether or not such bias is present.

Heisenberg: There seems to be no answer.

Feinberg: I would like to make a comment on another point. The $\pi^{0\prime}$ could be looked for by the same type of experiment that was described the other

tion, in particular by the reaction $d+d \to \alpha+\pi^{0\prime}$. If the $\pi^{0\prime}$ interacts strongly, then one would expect that the cross section for this, well above threshold, would be, maybe, 10^{-29} cm^2, which is much larger than the cross section for production of α + photon, so I think finding this particle, if it exists, might be possible by such an experiment.

(The following remark by Dr. Chamberlain was added after the discussion.)

Chamberlain: If the $\pi^{0\prime}$ mass is greater than two pion masses, it will be difficult to distinguish its production from double π^0 production in the $d+d$

RECENT RESEARCH ON THE NONLINEAR SPINOR THEORY OF ELEMENTARY PARTICLES

W. Heisenberg

Max-Planck-Institut für Physik und Astrophysik, München, Germany

During the last year research on the nonlinear spinor theory has been carried out in Munich mainly along three lines which may be characterized by the three topics: indefinite metric in Hilbert space, group theory, and approximation methods for the calculation of eigenvalues; some work has also been done on the analytical behavior of matrix-elements.

1. INDEFINITE METRIC IN HILBERT SPACE

(a) The question of whether the probability interpretation of quantum theory is compatible with the use of an indefinite metric in Hilbert space has been taken up from a general mathematical viewpoint in three papers by Schlieder. Starting from given symmetry groups Schlieder studies bilinear forms

which are invariant under the given transformations, and constructs the corresponding fundamental metric tensor. He then states supplementary conditions which are sufficient for the probability interpretation of the asymptotic waves. While for finite and compact groups one comes back in this way essentially to the conventional theory, infinite and non-compact groups may lead to more general representations in a space with indefinite metric which still are compatible with the probability interpretation of the asymptotic waves. Besides the special cases that have been studied in connection with Quantum electrodynamics by Bleuler and Gupta, and in connection with the Lee model by the author of the present report, Schlieder mentions the case of systems in which superselection rules exist. These rules divide the space of states into different incoherent sectors. It is easily seen that one would not run into any difficulties with the probability interpretation if e.g., the metric in one sector is positive definite, while it is negative definite in another sector.

Schlieder discusses finally the representation of the scale transformation in a Hilbert space with indefinite metric. This transformation may actually be used here for defining a quantum number, while in a definite metric it would only lead to trivial results. The reason is seen from the following argument: if one wants to introduce a quantum number for the scale transformation, one has to say that the Hilbert vector of a state with quantum number λ is multiplied by η^λ in a scale transformation (η being the scale factor). In a Hilbert space with definite metric the relation

$$\langle \phi_i | AB | \phi_f \rangle = \sum_n \langle \phi_i | A | \chi_n \rangle \langle \chi_n | B | \phi_f \rangle \qquad (1)$$

shows that $|\chi_n\rangle\langle\chi_n|$ must be invariant under the scale transformation. This would only be possible for $\lambda = 0$, since $|\chi_n\rangle$ may be considered as the complex conjugate of $\langle\chi_n|$; the scale transformation would therefore become trivial. In an indefinite metric however, relation (1) is to be replaced by

$$\langle \phi_i | AB | \phi_f \rangle = \sum_n \langle \phi_i | A | \chi^n \rangle \langle \chi_n | B | \phi_f \rangle \qquad (2)$$

This equation is invariant under the scale transformation, if only $|\chi^n\rangle$ transforms like $\eta^{-\lambda}$ when $\langle\chi_n|$ transforms like η^λ. This is a possible, and in fact necessary, assumption, since $|\chi^n\rangle$ is the inverse vector to $|\chi_n\rangle$ defined by

$$\langle \chi^m | \chi_n \rangle = \delta_{nm} . \qquad (3)$$

Therefore λ can be used as a quantum number.

(b) The possible presence of a dipole ghost of mass zero in the mass-spectrum of the vacuum-expectation values has been the object of an investigation by Mitter. By means of the Tamm-Dancoff method one can derive for the S_F function a nonlinear integro-differential equation, which in the lowest approximations reduces to a nonlinear differential equation. From a discussion of the solutions of this nonlinear equation Mitter tries to check whether the conditions

$$\int_0^\infty \xi |\rho| d\rho = 0 \quad \text{and} \quad \int_0^\infty \xi |\rho| \rho d\rho = 0 \qquad (4)$$

are actually enforced by the nonlinear character of the equations. The calculations have so far been carried out only for a somewhat simplified version of the theory, in which for symmetry reasons one could not expect the occurrence of mass eigenvalues different from zero. In this case Mitter gets as the only solution obeying the boundary conditions:

$$S_F = \int d^4 p\, e^{ip(x-x')} \frac{p_\nu \gamma_\nu}{(p^2)^2} \qquad (5)$$

which corresponds to a pure dipole ghost. In this solution the condition $\int_0^\infty \xi |\rho| d\rho = 0$ is satisfied; the other condition $\int_0^\infty \xi |\rho| \rho d\rho = 0$ however is not satisfied. The extension of the calculations to the realistic case where mass eigenvalues different from zero are possible has not yet been carried out.

2. GROUP THEORY

(a) Conservation of parity.

The fundamental equation

$$\gamma_\nu \frac{\partial \psi}{\partial x_\nu} \pm l^2 \gamma_\mu \gamma_5 \psi (\overline{\psi} \gamma_\mu \gamma_5 \psi) = 0 \qquad (6)$$

is invariant under the transformation

$$\psi(\mathbf{r}, t, l) = \gamma_4 \psi(-\mathbf{r}, t, l) .$$

But this transformation cannot be used as space reflection parity, since it does not commute with the Touschek transformation applied here for defining the baryonic number. Therefore a "parity of the second kind" had been introduced:

$$\psi(\mathbf{r}, t, l) = -i \frac{p_\nu \gamma_\nu}{|p|} \gamma_4 \psi(-\mathbf{r}, t, l) \qquad (7)$$

which came out as a natural consequence of the fact that the fermion wave functions obey a Klein-Gordon equation instead of a Dirac equation. This transformation is primarily defined on the energy shell of the nucleon. If one generalizes it as in Eq. (7) then Eq. (6) seems however not to be invariant under the non-local operation (Eq. (7)), therefore the parity of the second kind defined in this way can at most be approximately conserved. A similar situation had been reported by Thirring for his theory based upon the use of two Weyl spinors for the nucleons.

Duerr has studied the question whether one cannot for a theory starting from Eq. (6) introduce space reflection by an operation different from Eq. (7) and coinciding with it on the nucleon energy shell so that the equation should be strictly invariant under this operation. If this is possible parity should be strictly conserved for strong and electromagnetic interactions. The non-conservation of parity for the weak interactions should then be interpreted as a lack of symmetry in the ground state " world ", from which the particles are produced. It seemed convenient for this investigation to rewrite Eq. (6) in a different form closely related to a form suggested by Schremp and Gürsey. Instead of the spinor ψ_α one introduces a new spinor $\chi_{\alpha\tau}$ by the relation

$$\chi_{11} = \psi_1; \quad \chi_{21} = \psi_2; \quad \chi_{12} = -\psi_4{}^*; \quad \chi_{22} = \psi_3{}^* \quad (8)$$

The first index of χ refers to the Dirac spin, the second to the isospin. In this notation Eq. (6) takes the form

$$\sigma_v \frac{\partial}{\partial x_v} \chi \pm l^2 \sigma^\mu \chi(\chi^* \sigma_\mu \chi) = 0 \quad (9)$$

where $\sigma_v = (\sigma_k, 1)$ and $\sigma^v = (\sigma_k, -1)$. The bracket is to be interpreted as $\chi^*{}_{\alpha\tau}\sigma_{\mu,\,\alpha\beta}\chi_{\beta\tau}$. The advantage of the form (9) as against Eq. (6) lies in the fact that the isospin-group is seen more clearly in Eq. (9) than in Eq. (6); the σ_μ-matrices in Eq. (9) indicate that we have to do with a two-dimensional representation of the proper Lorentz-group. If one introduces the transposition matrices C_σ and C_τ in the conventional way by

$$\sigma_k^T = -C_\sigma^{-1}\sigma_k C_\sigma; \qquad \tau_k^T = -C_\tau^{-1}\tau_k C_\tau \quad (10)$$

the operation $\psi(\mathbf{r}, t, l) \to \gamma_4\psi(-\mathbf{r}, t, l)$ can be expressed for the χ spinor as

$$\chi(\mathbf{r}, t, l) = C_\sigma^{-1}C_\tau^{-1}\chi(\mathbf{r}-, t, l) \quad (11)$$

Eq. (9) is invariant under this operation.

This is essentially a CP transformation.

It had been suggested in an earlier paper that the field operators should be considered as functions not only of x_μ but also of l and that the operators at different values of l may be connected by the com-

mutator $\{\chi(x, l), \chi^*(x', l')\}$. If this is done, one realizes that Eqs. (6) and (9) are invariant under the transformation

$$\chi(\mathbf{r}, t, l) \to \chi(\mathbf{r}, t, -l) \quad (11')$$

This transformation can be connected with space reflection in the following manner : the states of the system are usually characterized by the momenta $p_v = (p_k, p_0)$ where by definition the energy $p_0 > 0$. What one measures in an experiment is, however, not the values x_v or the momenta p_v but the ratios x_v/l and the momenta $\pi_k = p_k l$. Since it is necessary to keep the energy positive one may define :

$$\rho_k = \frac{x_k}{l}, \qquad \rho_0 = \frac{x_0}{|l|}; \qquad \pi_k = p_k l, \quad \pi_0 = p_0|l|.$$

A change of sign in l reserves the sign of the space coordinates ρ_k but not of the time ρ_0. Therefore the operation (11') represents reflection in space, connected with a reflection of l :

$$\chi(\rho_k, \rho_0, l) \to \chi(-\rho_k, \rho_0, -l) \quad (12)$$

To bring out this symmetry more clearly, one may define a new 4×2 component spinor by

$$\mathbf{X}(\rho_k, \rho_0, l) = \begin{pmatrix} \chi(\rho_k, \rho_0, l) \\ \chi(\rho_k, \rho_0, -l) \end{pmatrix} \quad (13)$$

and at the same time go over from the Pauli-matrices σ_k to Dirac-matrices Γ_v by which one represents the complete Lorentz-group in the conventional manner. The Γ_v therefore have a different physical meaning from the γ_v in Eq. (6) :

$$\Gamma_4\mathbf{X}(\rho_k, \rho_0, l) = \mathbf{X}(\rho_k, \rho_0, -l) \quad (14)$$

For this operator $\mathbf{X}$ which now can be restricted to positive values of l, Duerr obtains the differential equation :

$$\Gamma_v\frac{\partial}{\partial x_v}\mathbf{X} \pm \frac{l^2}{2}\{\Gamma_\mu\mathbf{X}(\overline{\mathbf{X}}\Gamma_\mu\mathbf{X}) + \Gamma_\mu\Gamma_5\mathbf{X}(\overline{\mathbf{X}}\Gamma_\mu\Gamma_5\mathbf{X})\} \quad (15)$$

The operation P of space reflection (11') or (12) may now be written in the conventional form

$$\chi(\rho_k, \rho_0) \to \Gamma_4\mathbf{X}(-\rho_k, \rho_0) \quad (16)$$

Eq. (15) is identical with Eq. (9), written once for $\chi(\rho_k, \rho_0, |l|)$, and once in the corresponding form for $\chi(\rho_k, \rho_0, -|l|)$. The brackets in Eq. (15) contain, as in Eq. (9), the unit-matrix in isospin space. The isospin group takes the conventional form

$$\mathbf{X} \to e^{i\alpha_\bullet \tau_\bullet} \mathbf{X} \tag{17}$$

The introduction of the operator $\chi(\rho_v, -l)$ besides $\chi(\rho_v, l)$ would in itself lead to a doubling of the number of states, if the states produced by $\chi(\rho_v, -l)$ were not by definition identified with the states produced by $\chi(\rho_v, l)$. This identification is formally carried out by introducing the commutator between $\chi(\rho_v, l)$ and $\chi(\rho_v, -l)$, which appears as the mass term in the commutator of $\chi(\rho_v, l)$. The sign of the mass term is arbitrary in the beginning: it decides the parity of the nucleon, which can be chosen arbitrarily. But when the choice has been made, the number of states is the same as before the introduction of $\chi(\rho_v, -l)$.

Finally one gets for the vacuum expectation value of $\mathbf{X}\overline{\mathbf{X}}$:

$$\langle \Omega | \mathbf{X}_{\alpha\sigma}(x) \overline{\mathbf{X}}_{\beta\rho}(x') | \Omega \rangle = \int d^4 p\, e^{ip(x-x')} \times$$

$$\times \left[\delta_{\sigma\rho} \left(\frac{\Gamma^v_{\alpha\beta} p_v + i\kappa\delta_{\alpha\beta}}{p^2 + \kappa^2} \right) + \frac{\text{regularizing}}{\text{terms}} + \frac{\text{contributions from}}{\text{continuous spectrum}} \right] \tag{18}$$

In this way one sees that Eq. (16) defines a space reflection parity which is strictly conserved and coincides on the energy shell of the nucleons with the parity of the second kind defined by Eq. (7). The non-conservation of parity for the weak interactions must be due to a lack of symmetry in the ground-state " world ", which would show up in the higher approximations of some vacuum-(or rather " world ") expectation values.

(b) Representation of the strange particles.

It has been emphasized in an earlier paper that the non-conservation of isospin in the electromagnetic forces must be due to an asymmetry of the ground state " world ". The " world " possesses a very big total isospin, and this ground state is, therefore, highly degenerate. Nambu has in a recent paper pointed to the analogy of this assumption with the situation in superconductivity. The analogy is actual-ly quite close, since the lowest state of a superconductor has lost the invariance under the gauge-transformation; the gauge-transformation, however, belongs to the isospin group in the representation of Eq. (6). In this connection I might mention a paper of Yamazaki in which he tries to apply the Tamm-Dancoff method to the problem of superconductivity in the form given to it by Bogolubov. Yamazaki intends to check in this way the validity of the Tamm-Dancoff method. He has been able to show that in this case the Tamm-Dancoff method leads exactly to the results of Bogolubov.

If the ground state of the world has a big isospin, the strange particles may be produced by attaching some part of this isospin to a nucleon or a π-meson. In a similar manner the outer valency electron of the calcium atom produces a triplet spectrum instead of a doublet spectrum, since a spin $\frac{1}{2}$ of the inner shells has been attached to the valency electron.

For a mathematical representation of this possibility one has to express the degeneracy of the ground state. If particles of strangeness 1 are considered, it is sufficient to label only the last isospin $\frac{1}{2}$ of the ground state by an index having the values ± 1. This isospin can be attached to the particles if an isospin inter-action exists of the same general type as the spin-spin interaction in the outer atomic shells. From Eq. (9) one would not immediately expect such an interaction. But the large mass difference between the triplet and the singlet π meson derived from this equation shows that such an interaction is indirectly produced by exchange terms in a manner similar to the spin-spin interaction in the outer atomic shells; e.g. the energy difference between the orthohelium and the parahelium terms. Therefore, the isospin interaction may also show up in the vacuum expecta-tion values.

If one introduces the notation $|\Omega_\alpha\rangle$ for the two different ground states ($\alpha = \pm 1$), one expects for the vacuum expectation value of the product of two field operators the general form:

$$\langle \Omega_\alpha | \chi_\mu(x, l)\, \chi_v{}^*(x', l) | \Omega_\beta \rangle = a\delta_{\alpha\beta}\delta_{\mu v} + b\tau^i_{\mu v}\tau^i_{\beta\alpha} \tag{19}$$

where the indices refer to the isospin and a, b are functions of the space-time coordinates and the Dirac spin. Eq. (19) is invariant under rotations in isospin-space if the rotation is applied on the operator χ_μ and the ground state simultaneously. The second term on the right hand side is due to the isospin interaction; its size should be determined later on by a condition of consistency.

Since Eq. (9) is invariant under the CP transformation (11) and since there is no experimental evidence against this symmetry in the ground state, Eq. (19) should also be invariant under the operation (11) without any corresponding transformation in the ground state. If χ and χ^* in Eq. (19) refer to the same value of l this condition however can only be fulfilled by $b = 0$, since the term $\tau_{\mu\nu}^i \tau_{\alpha\beta}^i$ would change sign under the CP transformation. On the other hand if χ and χ^* refer to opposite values of l, the second term in (19) can be made invariant under the CP transformation. Therefore, using again the spinor $\mathbf{X}$ instead of χ, the general form of the vacuum expectation value could, in analogy to (18), be expressed by

$$\langle \Omega_\lambda | \mathbf{X}_{\alpha\sigma}(x, l) \overline{\mathbf{X}}_{\beta\rho}(x', l) | \Omega_\mu \rangle = \int \rho(\kappa^2) d\kappa^2 \int d^4 p\, e^{ip(x-x')} \times$$

$$\times \left[\frac{\delta_{\lambda\mu}\delta_{\sigma\rho}\Gamma_{\alpha\beta}^\nu p_\nu + i\kappa(a\delta_{\sigma\rho}\delta_{\lambda\mu}\delta_{\alpha\beta} + b\Gamma_{\alpha\beta}^5 \tau_{\rho\sigma}^i \tau_{\mu\lambda}^i)}{p^2 + \kappa^2} + \right.$$

$$\left. + \text{regularizing terms} \vphantom{\frac{\Gamma}{p^2}} \right] \qquad (20)$$

The change of sign in the expression $\tau_{\sigma\rho}^i \tau_{\mu\lambda}^i$ under the operation C is compensated for by the change of Γ^5 under P in the complete transformation CP.

Eq. (20) contains two important items of information:

i) If one carries out calculations in an approximation in which one uses only operators for one value of l and as contraction functions the vacuum expectation values of products of only two field operators with the same l, (13), then an interaction between the isospin of the particle and of the ground state cannot be expressed, and the strange particles cannot appear as eigenvalues. In order to get the strange particles one must either use the operators $\chi(\rho_\nu, l)$ and $\chi(\rho_\nu, -l)$ simultaneously or one must include vacuum expectation values of products of at least four field operators. The vacuum expectation values of products of four χ-operators of the same l may in fact express an interaction in a CP invariant manner:

$$\langle \Omega_\alpha | \chi_\lambda(x, l) \chi_\mu^*(x', l) \chi_\rho(x'', l) \chi_\sigma^*(x''', l) | \Omega_\beta \rangle =$$

$$= c\delta_{\alpha\beta}\delta_{\lambda\mu}\delta_{\rho\sigma} + id\tau_{\beta\alpha} \cdot [\tau_{\lambda\mu} \times \tau_{\rho\sigma}] \qquad (21)$$

where c and d are functions of x, x', x'', x'''

ii) While Eq. (20) is invariant under the CP transformation, applied only on the field operators, it is not invariant under P or C separately. Therefore, a single strange particle cannot have well defined space reflection parity. Only a pair of such particles of opposite strangeness may have a parity. This seems to be the reason for the well-known fact that a K-meson can disintegrate both into two or three π-mesons.

A calculation of the mass eigenvalues for the strange particles has so far not been carried out, since such a calculation would, according to i), either require a very substantial extension of the Tamm-Dancoff method or an entirely different, new approach.

3. APPROXIMATION METHODS FOR THE CALCULATION OF EIGENVALUES

In a theory with indefinite metric in Hilbert space, where the field operators obey a nonlinear differential equation, the anticommutator, say, of $\chi(x)$ and $\chi^*(x')$ will generally not behave like a δ-function at the point $x = x'$. Therefore, the operator representing translation in time (the Hamiltonian) cannot generally be expressed by the field variables at a given instant of time. Consequently, the conventional methods of calculating eigenvalues, as the Ritz method, fail. This situation occurs already in the Lee model after renormalization; therefore, Sekine has studied approximation methods in the Lee model.

If one introduces a finite interval of time, Δt, the equations of motion in the Lee model can be expressed as integro-differential equations for the renormalized operators without any infinite constants. The anticommutator between $\psi^*(\mathbf{r}, t)$ and, say, $\int_{t-\Delta t}^{t} dt' \psi(\mathbf{r}', t')$ behaves approximately like $\delta(\mathbf{r}, \mathbf{r}') \dfrac{\text{const}}{\log|\Delta t|}$ and vanishes for $\Delta t \to 0$.

Sekine introduces an approximate Hamiltonian H' with the following properties: the matrix-elements of H' between any two states of energy, E_i and E_f, are equal to those of the exact Hamiltonian H, if only

$$E_i \ll \frac{1}{\Delta t} \quad \text{and} \quad E_f \ll \frac{1}{\Delta t}.$$

H' depends only on the operators at a given instant of time $t = t_0$. The commutation of H' with $\int_{t-\Delta t}^{t} \psi(\mathbf{r}', t') dt'$ produces the equations of motion in a sufficient approximation for all matrix elements in the region of small energies $(E \ll 1/\Delta t)$.

When these conditions can be fulfilled, the conventional methods of variation may be applied on H' and lead to eigenvalues which for $E \ll 1/\Delta t$ are identical with the corrrect eigenvalues of the system. Formally, the calculations of Sekine are closely related to a conventional cut-off procedure, but this procedure is used only as a mathematical tool, while the eigenvalues and matrix elements are independent of it.

The apparent success of this method in the Lee model suggests the following procedure in the nonlinear spinor theory: the wave equation and the corresponding local Hamiltonian may be modified by some cut-off process which can be non-relativistic (e.g. limitation of integrals in momentum space by a maximum value K) and which changes the theory into a non-local one. For the non-local theory the anticommutator will not vanish everywhere for a given time $t = t_0$; it will still contain δ-like terms which, however, go to zero in the limit $K \to \infty$. In this modified theory the conventional methods of variation may be applied; they will — so one may hope — lead to eigenvalues which converge towards the correct eigenvalues in the limit $K \to \infty$. The theory would be relativistically invariant only in this limit.

This extension of the results from the Lee model to the nonlinear spinor theory has, however, not yet been investigated in detail; therefore it is still unknown whether a useful method of approximation can be constructed in this way.

4. VACUUM EXPECTATION VALUES OF THE PRODUCT OF FOUR FIELD OPERATORS

In analogy to the investigations by Lehmann of the vacuum expectation value of the product of two field operators, Montaldi has studied the vacuum expectation value of the product of four field operators of the type:

$$\langle \Omega | \psi_\alpha(x) \overline{\psi}_\beta(x') \psi_\gamma(x'') \overline{\psi}_\delta(x''') | \Omega \rangle \tag{22}$$

Generally the analytic form of such expressions may be very complicated. But in the special case of Eq. (6), where the invariance under the Pauli-Gürsey group and the Touschek group are required, as well as the Lorentz invariance, then on account of the postulate of microcausality, Eq. (22) can be reduced to only two functions of 6 mass-variables. These two functions must fulfill a number of symmetry conditions resulting from the underlying groups. Besides that, through Eq. (6), they can be connected with the mass-spectrum $\xi(\kappa^2)$ in the vacuum expectation value of two field operators. Finally these functions must obey restrictive conditions of the type of Eq. (4), in order to avoid divergencies due to the nonlinear character of the fundamental field equation.

The results of Montaldi can be stated in the following formula:

$$\langle \Omega | T \psi_\alpha(x_1) \psi_\beta(x_2) \overline{\psi}_\lambda(x_3) \overline{\psi}_\mu(x_4) | \Omega \rangle =$$

$$= \int_0^\infty d\kappa_1^2 \dots \int_0^\infty d\kappa_6^2 K^F_{-1,\alpha\beta}(x_1 - x_2, \kappa_1) K^F_{1,\lambda\mu}(x_3 - x_4, \kappa_6) \Delta_F(x_1 - x_3, \kappa_2) \Delta_F(x_1 - x_4, \kappa_3) \Delta_F(x_2 - x_3, \kappa_4) \times$$
$$\times \Delta_F(x_2 - x_4, \kappa_5) M_1(\kappa_1^2, \dots \kappa_6^2) +$$

$$+ \int_0^\infty d\kappa_1^2 \dots \int_0^\infty d\kappa_6^2 K^F_{0,\alpha\lambda}(x_1 - x_3, \kappa_2) K^F_{0,\beta\mu}(x_2 - x_4, \kappa_5) \Delta_F(x_1 - x_2, \kappa_1) \Delta_F(x_1 - x_4, \kappa_3) \Delta_F(x_2 - x_3, \kappa_4) \times$$
$$\times \Delta_F(x_3 - x_4, \kappa_6) M_2(\kappa_1^2 \dots \kappa_6^2) +$$

$$+ \int_0^\infty d\kappa_1^2 \dots \int_0^\infty d\kappa_6^2 K^F_{0,\alpha\mu}(x_1 - x_4, \kappa_3) K^F_{0,\beta\lambda}(x_2 - x_3, \kappa_4) \Delta_F(x_1 - x_2, \kappa_1) \Delta_F(x_1 - x_3, \kappa_2) \Delta_F(x_2 - x_4, \kappa_5) \times$$
$$\times \Delta_F(x_3 - x_4, \kappa_6) M_3(\kappa_1^2 \dots \kappa_6^2) +$$

$$+ \quad \text{three similar terms, with the } K \text{ functions replaced by } H \text{ functions and } M_1, M_2, M_3 \text{ replaced by } M_4, M_5, M_6 \text{ respectively.} \tag{23}$$

The K and H functions are defined as follows: (C = charge conjugation matrix)

$$K_0^{\pm}(x) = \gamma_\nu \frac{\partial}{\partial x_\nu} \Delta^{\pm}(x) \qquad H_0^{\pm}(x) = \gamma_5 K_0^{\pm}(x)$$

$$K_{-1}^{\pm}(x) = \gamma_\nu C^{-1} \frac{\partial}{\partial x_\nu} \Delta^{\pm}(x) \qquad H_{-1}^{\pm}(x) = \gamma_5 K_{-1}^{\pm}(x)$$

$$K_1^{\pm}(x) = C\gamma_\nu \frac{\partial}{\partial x_\nu} \Delta^{\pm}(x) \qquad H_1^{\pm}(x) = K_1^{\pm}(x)\gamma_5 \tag{24}$$

furthermore,

$$Z_F(x) = \begin{cases} Z^+(x) & \text{for } x_0 > 0 \\ -Z^-(x) & \text{for } x_0 < 0 \end{cases}$$

where $Z_F = K^F$, H^F or Δ_F.

For simplicity, we do not quote here the symmetry conditions that must be satisfied by the spectral functions $M_i(\kappa_1^2 \ldots \kappa_6^2)$ due to the above mentioned requirements. We shall limit ourselves to giving the conditions analogous to Eq. (4) for the two independent mass spectra (which we shall denote by N and Q), in terms of which the functions M_i can be explicitly constructed: ($A = N$ or Q):

$$\int_0^\infty d\kappa_1^2 \ldots \int_0^\infty d\kappa_6^2 A(\kappa_1^2 \ldots \kappa_6^2) = 0$$

$$\int_0^\infty d\kappa_1^2 \ldots \int_0^\infty d\kappa_6^2 \, \kappa_j^2 \, A(\kappa_1^2 \ldots \kappa_6^2) = 0$$

$$\int_0^\infty d\kappa_1^2 \ldots \int_0^\infty d\kappa_6^2 \, \kappa_j^2 \, \kappa_l^2 \, A(\kappa_1^2 \ldots \kappa_6^2) = 0$$

$$\vdots$$

$$\int_0^\infty d\kappa_1^2 \ldots \int_0^\infty d\kappa_6^2 \, \kappa_1^2 \ldots \kappa_6^2 \, A(\kappa_1^2 \ldots \kappa_6^2) = 0$$

The vacuum expectation value of products of four field operators should contain terms that can be interpreted as due to the creation and annihilation of π-mesons; therefore, the spectra N and Q should in some way contain not only the mass of the nucleon but also the mass of the π-meson.

DISCUSSION

BLOKHINTSEV: Concerning the work of Mitter on these commutation rules, I should like to know what kind of initial conditions are taken for the commutation rules?

HEISENBERG: As far as I know, in ordinary theory, the initial condition never has any effect on the commutation rules.

DUERR: May I comment on this point? What Mitter actually did was to derive, in an approximate fashion, not really the anticommutator but rather the vacuum expectation value of the T-product, that is the propagation function, under the following conditions: first, that the anticommutator has the symmetry properties connected with the theory; second, that the anticommutator has the property of being zero for space-like distances; third, that the energies are positive; forth, that you have invariance with respect to time reversal. Under these conditions, he then derived the stated form of the propagation function.

BREIT: I want to ask whether any of these theories have consequences for quantum electrodynamics? Presumably, if there are consequences they would be in the nature of some interaction of another field with the electron-positron and electromagnetic fields. If there were consequences, perhaps they could be tested, because some quantum electrodynamics tests are carried out very precisely. Of course, in the same connection, there is the question of the muon. I wonder if any of these theories would have bearing on the behavior of the muon?

HEISENBERG: May I answer very briefly: I am convinced there are consequences but they have not been worked out. Actually, one can probably only approach this problem in the following order: the most trivial thing one can do is to treat the nucleons and pions and the strong interactions. The next step must be the strange particles. After that, one may come to quantum electrodynamics, namely, to the existence of the photon and only at this stage one may discuss these consequences.

DUERR: I want to make a short comment on this l-parity, i.e., the parity which is connected with the reversal of l. I want to give a " nonmystical " interpretation of this. In order to incorporate parity in a strict fashion, you are forced, in a way, to double the number of components of the field operator. If you double the number of components of the field operator, however, you immediately run into the difficulty that you can write down five different invariant fourth order expressions, i.e., you have five Fermi interaction terms which will involve five different coupling constants. So you look for a method of doubling the components which does not increase the number of coupling constants. Now, in this special case which we have investigated, we introduce the doubling of the components in a very special fashion in the following sense: in the original equation l enters only quadratically. We now could slightly modify this theory in stating that the theory may also depend on the absolute value of l, i.e., it may depend on the sign of the square root of l^2. If you use this degree of freedom in that way, then of course, this is equivalent to doubling the number of components, and parity can be introduced, but you do not introduce new kinds of interactions.

DYNAMICAL THEORY OF ELEMENTARY PARTICLES SUGGESTED BY SUPERCONDUCTIVITY

Y. Nambu

The Enrico Fermi Institute for Nuclear Studies, University of Chicago, Chicago, Illinois

My talk is based on some work done in collaboration with Dr. Jona-Lasinio.

We would like to propose here a theory of elementary particles which is based on a mathematical analogy between the dynamics of relativistic particles and that of superconductors in the theory of Bardeen, Cooper and Schrieffer [1]. That there can exist such an analogy is not surprising. Both relativistic quantum field theory and solid state physics deal with many body problems of large media, and in fact we already know many instances where field theoretical techniques have been successfully applied to problems of solid state physics. We shall see presently that this interaction of the two branches of physics can be reciprocal, and that solid state physics can provide us with useful models which help us understand the dynamics of elementary particles.

I. We start with the comparison of the Dirac equation for a nucleon, say, and the Bogolubov-Valatin relation [2,3] for an elementary excitation (quasi-particle) in a superconducting medium. They are given respectively by

$$E\psi_1 = \sigma \cdot \mathbf{p}\psi_1 + m\psi_2$$

$$E\psi_2 = -\sigma \cdot \mathbf{p}\psi_2 + m\psi_1 \qquad (1)$$

$$E = \pm\sqrt{p^2 + m^2}$$

and

$$E\psi_{p+} = \varepsilon_p\psi_{p+} + \phi\psi^{\dagger}_{-p-}$$

$$E\psi^{\dagger}_{-p-} = -\varepsilon_p\psi^{\dagger}_{-p-} + \phi\psi_{p+} \qquad (2)$$

$$E = \sqrt{\varepsilon_p^2 + \phi^2}$$

Here the Weyl representation is used for the Dirac equation: ψ_1, ψ_2 correspond to the eigenstates of chirality $\gamma_5 = \pm 1$. In Eq. (2), $\psi_{p\pm}$ is the wave function of an electron with momentum p and spin $\pm$ (up or down), so that $\psi^{\dagger}_{-p-}$ effectively represents

Acta Physica Austriaca *14*, 328–339 (1961)

Der derzeitige Stand der nichtlinearen Spinortheorie der Elementarteilchen*

Von

W. Heisenberg

Max-Planck-Institut für Physik und Astrophysik, München

(Eingegangen am 19. Dezember 1960)

Unter dem Terminus „nichtlineare Spinortheorie der Elementarteilchen" wird im folgenden der Versuch verstanden, aus der Gleichung

$$\gamma_\nu \frac{\partial \psi}{\partial x_\nu} \pm l^2 \gamma_\mu \gamma_5 \psi \left(\bar{\psi} \gamma_\mu \gamma_5 \psi\right) = 0$$

die Eigenschaften der Elementarteilchen, also ihre Massen, ihre Wechselwirkung, die Art ihres Entstehens usw., herzuleiten.

In den letzten Jahren ist eine Reihe von Untersuchungen an dieser Theorie durchgeführt worden, es sind verschiedene mathematische Probleme aufgeklärt, Querverbindungen zu anderen Gedankenkreisen hergestellt und die physikalischen Konsequenzen der Theorie genauer als früher durchgearbeitet worden. Daher lohnt es sich, einen Überblick über den derzeitigen Stand zu geben.

Zunächst sollen die Grundtatsachen nochmals genannt werden, aus denen die Theorie entstanden ist, und die Grundgedanken, die als Folgerungen aus diesen Grundtatsachen gezogen worden sind.

Von den vielen Erfahrungen, die in den letzten Jahren an verschiedenen Stellen, etwa in der kosmischen Strahlung oder mit den großen Beschleunigungsmaschinen in Kalifornien, in Dubna, in Genf, gewonnen worden sind, ist vielleicht die wichtigste, daß die Elementarteilchen ineinander umgewandelt werden können. Wenn Elementarteilchen mit sehr hoher Energie aufeinandertreffen, so entstehen dabei neue Elementarteilchen, teils derselben, teils anderer Art. Wenn man die Ergebnisse der Experimente qualitativ zu einem einfachen Bild zusammenordnen will, kann man sagen: Die Elementarteilchen sind nicht letzte, unzerstörbare Einheiten der Materie, sondern verschiedene Formen ein und derselben Substanz. Man kann den Grundstoff, aus dem alle Elementarteilchen bestehen, einfach Energie nennen. Wir können sagen, aus Energie entsteht die Materie, indem sie sich in die

---

* Vortrag, gehalten am 3. November 1960 auf der Tagung der Österreichischen Physikalischen Gesellschaft in Graz.

Form der Elementarteilchen begibt. Die Elementarteilchen sind also nur verschiedene Formen, stationäre Zustände, in denen dieser Grundstoff Energie sich manifestieren kann.

Neben dieser ersten und wichtigsten experimentellen Tatsache gibt es eine zweite von ähnlichem Gewicht, nämlich die Feststellung, daß die Wechselwirkung zwischen den Elementarteilchen der Forderung der Kausalität genügt. Damit ist folgendes gemeint: Bei allen Prozessen, bei denen Elementarteilchen aufeinander wirken, bei denen neue Teilchen entstehen oder wieder absorbiert werden, immer gilt der aus der Relativitätstheorie bekannte Satz, daß Wirkungen sich nie schneller als mit Lichtgeschwindigkeit fortpflanzen können und daß die Ursache immer früher ist als die Wirkung. Diese Tatsache allein hat schon wesentliche Einschränkungen in den möglichen Theorien zur Folge, die mathematisch zu untersuchen sind.

Schließlich liefern die Experimente noch ein drittes, wichtiges Erfahrungsmaterial: Es hat sich herausgestellt, daß bei den Umwandlungen der Elementarteilchen viele verschiedene Erhaltungssätze gelten; zum Teil die schon aus der alten Physik bekannten Erhaltungssätze für Energie und Impuls usw., zum Teil neue, die durch gewisse Quantenzahlen beschrieben werden können. Man muß also annehmen, daß es in den grundlegenden Naturgesetzen die den Erhaltungssätzen entsprechenden Symmetrie-Eigenschaften gibt. Denn wir wissen ja aus der Quantentheorie, oder ganz allgemein aus der gesamten früheren Physik, daß ein Erhaltungssatz immer dann auftritt, wenn die grundlegenden Naturgesetze gewisse Symmetrien, das heißt Invarianzeigenschaften aufweisen.

Diesen drei Grundtatsachen: Verwandlung der Elementarteilchen, Kausalität und Erhaltungssätze, entsprechen drei Grundannahmen der nichtlinearen Spinortheorie, die zwar kaum als mathematisch zwangsläufige Folgen jener Grundtatsachen, wohl aber als ihre natürliche Interpretation angesehen werden können.

1. Wenn es wahr ist, daß alle Elementarteilchen aus demselben Stoff „Energie" bestehen, dann ist es vernünftig anzunehmen, daß man alle Elementarteilchen mittels eines einzigen Feldes, nämlich eines einheitlichen Materiefeldes beschreiben kann. Es wird daher versucht, nur einen einzigen Feldoperator einzuführen, von dem aus alle Elementarteilchen in irgend einer Weise hergeleitet werden können. Da es Elementarteilchen mit halbzahligem Spin gibt, kann dieses zugrundeliegende universelle Feld nur ein Spinorfeld sein. Man wird also annehmen, daß, wenn es ein Grundgesetz der Materie gibt, dies durch eine Spinorgleichung dargestellt werden muß. Da es Wechselwirkungen gibt, muß die Spinorgleichung nichtlinear sein; denn in der Physik wird in der Regel eine Wechselwirkung durch die Abweichung von der Linearität beschrieben.

2. Die zweite Grundtatsache war, daß die Wechselwirkung kausal verläuft. Es ist in den letzten Jahren viel darüber nachgedacht worden, ob die Kausalität als eine, wie man sagt, streng lokale Forderung ver-

standen werden muß, oder ob es verallgemeinerte, weniger scharfe Begriffe von Kausalität gibt, bei denen zwar die Kausalität nicht in den kleinsten Bereichen streng gilt, aber in weiteren Bereichen doch näherungsweise erhalten bleibt. Bei den bisherigen Untersuchungen hat sich immer wieder herausgestellt, daß die einzig eindeutige und befriedigende Art, die Kausalität zu definieren, doch wahrscheinlich darin besteht, sie wenigstens formal als lokale Eigenschaft, also sozusagen im unendlich Kleinen zu fordern. Man hat einstweilen jedenfalls keinen Grund, von dieser Forderung abzuweichen. Wenn man die Forderung so stellt, dann muß die grundlegende Materiegleichung eine Differentialgleichung sein. Diese Aussage ist nur mit Vorbehalt richtig, da eventuell noch ein etwas allgemeinerer Gleichungstypus in Betracht käme, nämlich eine Integrodifferentialgleichung, in der nur ein beliebig kleines, jedoch endliches Zeitintervall Δt als Integrationsbereich vorkommt. Solche Integrodifferentialgleichungen kann man in einem übertragenen Sinne auch noch als Differentialgleichungen bezeichnen, sie entsprechen nämlich Differentialgleichungen mit unendlichen Koeffizienten. Mit diesem etwas erweiterten Sprachgebrauch kann man sagen: wenn die Forderung der Kausalität in strengem Sinne gilt, dann muß die zugrundeliegende Materiegleichung eine nichtlineare Differentialgleichung sein.

3. Schließlich kann man aus den vielen Erhaltungssätzen, die man empirisch beobachtet, schließen, daß die Differentialgleichung eine sehr hohe Symmetrie besitzen muß; denn die vielen Erhaltungssätze und die zugehörigen Quantenzahlen können nur durch entsprechend viele Symmetriegruppen gedeutet werden.

Beim Versuch, eine solche einheitliche Materietheorie zu formulieren, ergab sich die überraschende Erfahrung, daß schon die einfachste nichtlineare Spinorgleichung, die man in Anlehnung an die bisherige Feldtheorie anschreiben kann, praktisch alle die Gruppen und zugehörigen Symmetrie-Eigenschaften enthält, die man für die Deutung der Erfahrung braucht.

$$\gamma_\nu \frac{\partial \psi}{\partial x_\nu} \pm l^2 \gamma_\mu \gamma_5 \psi \left(\bar\psi \gamma_\mu \gamma_5 \psi\right) = 0$$

	Gruppe:		Deutung:
(1)	Lorentzgruppe	$\rightarrow$	Erhaltung von Energie, Impuls, Drehimpuls etc.
(2)	$\psi \rightarrow a\,\psi + b\,\gamma_5\,c^{-1}\,\bar\psi$ $\lvert a\rvert^2 + \lvert b\rvert^2 = 1$	$\rightarrow$	Isospin J
(3)	$\psi \rightarrow e^{i\alpha\gamma_5}\,\psi$	$\rightarrow$	J_N
(4)	$\psi \rightarrow \eta^{3/2}\,\psi\,(x\,\eta,\,l\,\eta)$	$\rightarrow$	Λ
(5)	$\psi \rightarrow \gamma_4\,c^{-1}\,\bar\psi\,(-\,\vec{\mathfrak{r}},\,t,\,l)$	$\rightarrow$	PC (Parität + Ladungskonjugation)
(6)	$\psi \rightarrow \psi\,(x,\,-\,l)$	$\rightarrow$	P (Parität)

Der derzeitige Stand der nichtlinearen Spinortheorie der Elementarteilchen   331

$$\text{Nukleonenzahl} \qquad N = J_N + \Lambda$$

$$(7) \quad \text{Leptonenzahl} \qquad L = J_N - \Lambda$$

$$\text{Ladung} \qquad Q = J_3 + \Lambda - s/2$$

$[s = \text{,,Seltsamkeit`` (strangeness)}]$

Die vorstehende Tabelle gibt links die Gruppeneigenschaft, rechts die zugehörige physikalische Deutung.

An erster Stelle steht die LORENTZ-Gruppe (1); sie bedeutet, daß in der zugehörigen Theorie Energie, Impuls und Drehimpuls erhalten sind, wie in der üblichen Physik. Die Transformation (2), die von PAULI und GÜRSEY studiert wurde, ist der räumlichen Drehungsgruppe isomorph, sie kann daher mit dem sogenannten Isospin, einer an den Elementarteilchen seit etwa 30 Jahren bekannten Eigenschaft, in Verbindung gebracht werden. Die Transformation (3), die von TOUSCHEK u. a. untersucht worden ist, kann zur Deutung einer Quantenzahl J_N benützt werden, die nach (7) mit Baryonen-Zahl und Leptonen-Zahl verknüpft ist. Der Skalentransformation (4), über die nachher noch gesprochen werden soll, kann man eine Quantenzahl Λ zuordnen, die ebenfalls für Baryonen- und Leptonenzahl wichtig ist. Schließlich gibt es noch die Transformation (5), die man mit dem Produkt aus Parität und Ladungskonjugation in Verbindung bringen kann und die Transformation (6), nämlich die Umkehr von $+ l$ in $- l$, die, wie nachher noch diskutiert werden soll, mit der Parität zusammenhängen kann. (7) zeigt noch die Verbindung dieser Zahlen mit der Nukleonenzahl, der Leptonenzahl und der Ladung. Die einzige Quantenzahl, die in diesem Schema noch nicht gedeutet werden kann, ist die sogenannte Seltsamkeit (strangeness); sie mag entweder mit noch nicht untersuchten diskreten Gruppen der Gleichung zusammenhängen, oder durch Verhältnisse erklärt werden, wie man sie aus den optischen Spektren der Atome, z. B. in der Beziehung zwischen Orthohelium- und Parheliumspektrum kennt.

Nachdem die Symmetrie-Eigenschaften der Gleichung mit gewissen empirischen Bestimmungsstücken der Elementarteilchen in Verbindung gebracht sind, muß man versuchen, aus der Gleichung und den zugehörigen Symmetrien die physikalischen Eigenschaften der Elementarteilchen auszurechnen, also z. B. zu zeigen, daß das Massenverhältnis zwischen Nukleonen und π-Mesonen gerade den beobachteten Wert hat.

Es hatte sich schon am Anfang dieser Theorie herausgestellt, daß man den Rahmen der früheren Quantenmechanik an verschiedenen Stellen erheblich erweitern muß, um der Ausgangsgleichung einen Sinn zu geben. Diese Erweiterungen sind der Gegenstand von vielen Diskussionen geworden, weil man sich bei jeder Erweiterung natürlich immer fragen muß, ob die Erweiterung erstens nötig und zweitens möglich ist.

Ich möchte diese Probleme, die gerade in den letzten Jahren oft erörtert worden sind, vorerst kurz besprechen und dann später auf die Fortschritte eingehen, die in ihrer Behandlung neuerdings erzielt worden sind.

1. Zunächst hat sich herausgestellt, daß man, um die Ausgangsgleichung zu quantisieren, nicht mit einem Hilbertraum auskommt,

in dem wie üblich eine positive Metrik gilt; sondern man muß einen Hilbertraum mit indefiniter Metrik benutzen. Zur Vorgeschichte möchte ich erwähnen, daß der Vorschlag, eine indefinite Metrik im Hilbertraum zu verwenden, schon 1942 von Dirac gemacht worden ist. Pauli hatte aber ein Jahr später gezeigt, daß man bei dieser indefiniten Metrik in Schwierigkeiten mit der Wahrscheinlichkeitsinterpretation der Quantentheorie geraten kann. Wir hatten bei der nichtlinearen Spinortheorie zunächst einige Hinweise dafür gefunden, daß man die Schwierigkeiten vielleicht doch überwinden kann. Später hatten Pauli und Källén am Lee-Modell gezeigt, daß es Fälle gibt, in denen anscheinend die Wahrscheinlichkeitsinterpretation nicht durchgeführt werden kann. An einem speziellen Fall des Lee-Modells konnte andererseits wieder nachgewiesen werden, daß die Wahrscheinlichkeitsdeutung doch mit der indefiniten Metrik vereinbar sein kann. Mit anderen Worten: Die Frage, ob diese Benützung der indefiniten Metrik möglich und notwendig ist, bedarf noch ausführlicher Untersuchungen.

Das war aber nicht die einzige Erweiterung des früheren quantentheoretischen Rahmens, sondern es waren noch mehrere andere Erweiterungen nötig.

2. Man hat in dieser Theorie zur Deutung der experimentellen Quantenzahlen auch Transformationen im Raum des Parameters l, des Längenparameters in der Ausgangsgleichung, in Betracht gezogen. Es erscheint zunächst sehr ungewöhnlich, daß man einen Parameter, der so etwas wie eine Kopplungskonstante zu bedeuten scheint, auf die gleiche Stufe stellt mit den Koordinaten x, y, z und t, und Transformationen dieses Parameters zuläßt.

Man kann die Verwendung von Transformationen im Längenparameter auch in der Weise ausdrücken, daß man nicht von einer vierdimensionalen, sondern einer fünfdimensionalen Theorie spricht. Man betrachtet nämlich die Operatoren ψ als Funktionen der fünf Variablen x, y, z, t und l. Das ist in der früheren Quantentheorie weder üblich noch möglich gewesen, und daher handelt es sich auch hier um eine Neuerung, deren Sinn erst diskutiert werden muß.

3. Ein Punkt, der bisher weniger Beachtung gefunden hat, der aber auch von größter Wichtigkeit im Zusammenhang mit der Theorie zu sein scheint, ist die Frage nach den Randbedingungen. In der alten Quantenmechanik war mit der Formulierung der Wellengleichung (etwa in der Schrödingerschen Form) das Problem praktisch vollständig definiert. Zwar ist bekannt, daß man die Eigenwerte nur erhält, wenn man die Schrödingergleichung durch Randbedingungen ergänzt, also etwa fordert, daß die Wellenfunktion im Unendlichen verschwinden soll, keine Singularitäten auftreten und dergleichen. In der unrelativistischen Quantenmechanik waren die Randbedingungen aber immer von so einfacher Art, daß keine Schwierigkeit entstand. In der nichtlinearen Spinortheorie sind die Verhältnisse erheblich komplizierter, weil es gewisse „Unendlichkeiten“ gibt, die in der gewöhnlichen Quantenmechanik nicht vorkommen. Dort genügte es, über das Verhalten der Wellen-

funktion im räumlich Unendlichen Aussagen zu machen. Hier kann aber auch die Teilchenzahl unendlich werden. Die Experimente zeigen, daß beim Stoß neue Teilchen erzeugt werden können. Beim Zusammenstoß von zwei sehr energiereichen Teilchen können unter Umständen sogar viele neue entstehen. Im Prinzip ist daher die Teilchenzahl unendlich. Und da es Teilchen mit der Ruhemasse Null gibt, kommen auch praktisch Fälle vor, bei denen selbst in einem einzigen Experiment unendlich viele Teilchen erzeugt werden. Wenn man eine unendliche Teilchenzahl in Betracht ziehen muß, so muß man auch sagen, wie sich die Wellenfunktionen bei unendlichen Teilchenzahlen verhalten. Dadurch kommen neue Arten von Randbedingungen ins Spiel, die es im ersten Augenblick überhaupt als fraglich erscheinen lassen, ob das Problem durch die Differentialgleichung allein schon eindeutig beschrieben ist. Es wäre ja denkbar, daß die Differentialgleichung zwar gilt, aber das Problem noch nicht vollständig bestimmt, daß vielmehr zur Festlegung noch neue, bisher unbekannte Randbedingungen notwendig sind.

4. Eine andere Erweiterung bezieht sich auf die Symmetrie des Grundzustandes: In der gewöhnlichen Quantenfeldtheorie, etwa der Quanten-Elektrodynamik, hat man immer angenommen, daß das Vakuum die höchste mögliche Symmetrie besitzt; das heißt das Vakuum soll invariant sein gegen alle die Transformationen, gegen die auch die Gleichung selbst invariant ist. Man kann auch sagen, das Vakuum sollte keinerlei physikalische Eigenschaften besitzen, außer etwa der, leer und damit symmetrisch zu sein.

In der neuen Theorie mußte jedoch, um die Erfahrung darzustellen, angenommen werden, daß auch der Grundzustand unsymmetrisch ist, daß er nicht die volle Symmetrie der Gleichung besitzt und damit zum Träger physikalischer Eigenschaften wird; er sollte also „Welt" und nicht „Vakuum" genannt werden.

Diese Annahme war deswegen notwendig, weil man ja gewisse Abweichungen von der Symmetrie erklären muß. Es ist z. B. empirisch bekannt, daß die Isospininvarianz bei den elektromagnetischen Kräften nicht mehr gilt. Wenn der Grundzustand nicht die volle Symmetrie besitzt, muß er entartet sein, weil man durch Anwendung der Transformation zu einem anderen Grundzustand kommen kann, der physikalisch die gleichen Eigenschaften besitzt wie der ursprüngliche. Auf dieser Entartung des Grundzustandes beruht dann auch die Existenz der „strange particles", der „seltsamen" Teilchen. Die Hyperonen z. B. kommen dadurch zustande, daß sich die Nukleonen sozusagen vom Vakuum einen Teil des dort vorhandenen Isospins abzweigen und anlagern können. Solche Verhältnisse kommen schon bei speziellen Systemen der gewöhnlichen Quantenmechanik vor und können dort studiert werden. Sie führen jetzt zu grundsätzlichen Unterschieden zwischen der nichtlinearen Spinortheorie und der üblichen Quantenfeldtheorie.

5. Schließlich will ich als letzten Punkt die Schwierigkeiten des Näherungsverfahrens nennen: Man hat zur Berechnung von Eigenwerten in den ersten Arbeiten über die nichtlineare Spinortheorie mmer eine Näherungsmethode benützt, die als das sogenannte Tamm-Dancoff-Verfahren bekannt ist. Sie ist recht umstritten. Es handelt sich dabei um eine Näherungsmethode, deren Genauigkeitsgrenzen schwer abzuschätzen sind und bei der auch die Beziehung zu anderen Näherungsmethoden nicht allzu klar ist.

Jedenfalls wäre es für die Beurteilung dieses letzten Punktes wichtig, wenn man andere Näherungsverfahren zum Vergleich benützen und auswerten könnte, um zu sehen, wie gut oder schlecht die Ergebnisse der anderen Verfahren mit den Tamm-Dancoff-Ergebnissen übereinstimmen.

Damit sind die wichtigsten Gebiete bezeichnet, in denen die neue Theorie von der früheren Quantentheorie abgewichen ist, und es soll jetzt diskutiert werden, bei welchen Problemen im letzten Jahr Fortschritte erzielt worden sind.

Der Stand vor einem Jahr läßt sich etwa in folgender Weise kurz zusammenfassen:

Wenn man voraussetzt, daß die genannten Erweiterungen der Quantentheorie mathematisch einwandfrei durchgeführt werden können, dann stellen sich für die einfachsten Teilchen, nämlich für die Nukleonen und für die Bosonen in der untersten Näherung der Tamm-Dancoff-Methode Massen heraus, die qualitativ mit den beobachteten übereinstimmen. Die Symmetrie-Eigenschaften der Teilchen stimmen mit den beobachteten überein. Nur bei der Parität war noch eine Unklarheit übriggeblieben: Es ließ sich zwar die sogenannte „Parität 2. Art" definieren, aber es konnte nicht gezeigt werden, daß sie bei den starken Wechselwirkungen streng erhalten bleibt und erst bei den ganz schwachen Wechselwirkungen, nämlich bei den radioaktiven, verletzt wird.

Diese Schwierigkeit war auch in anderen, ähnlichen Theorien aufgetreten, z. B. in einer Theorie von Thirring, der vorgeschlagen hatte, alle Teilchen durch drei Felder darzustellen.

Abgesehen von dieser Problematik hatte sich aber bei der ersten Übersicht herausgestellt, daß man für die Massenverhältnisse einigermaßen vernünftige Werte und für die Streuung zwischen Nukleonen und π-Mesonen etwas qualitativ Richtiges bekommt; daß sich ferner auch für die β-Wechselwirkung die richtigen Symmetrie-Eigenschaften ergeben.

Die seitdem erzielten Fortschritte sollen nun an den vorhin aufgezählten Problemen im einzelnen besprochen werden.

Die wichtigste Erweiterung gegenüber der früheren Quantentheorie war die Verwendung der indefiniten Metrik im Hilbertraum. Zunächst waren zwei Beispiele bekannt, in denen es gelungen war, die indefinite Metrik mit der üblichen Wahrscheinlichkeits-Deutung der Quantentheorie zu vereinigen. Bleuler und Gupta hatten die Quantenelektrodynamik in einem Hilbertraum mit indefiniter Metrik formuliert und gezeigt, daß in diesem speziellen Fall die indefinite Metrik nicht zu Schwierigkeiten mit der Wahrscheinlichkeits-Deutung führt. Sehr ähnlich konnte ein spezieller Fall des sogenannten LEE-Modells behandelt werden, der den Annahmen in der nichtlinearen Spinortheorie möglichst genau nachgebildet war. Auch hier konnte die Wahrscheinlichkeits-Deutung beibehalten werden. Allerdings gibt es andere Fälle des LEE-Modells, insbesondere den von PAULI und KÄLLÉN untersuchten, wo das nicht möglich schien.

Neuerdings hat SCHLIEDER die allgemeinen gruppentheoretischen Voraussetzungen dafür untersucht, daß die indefinite Metrik mit der Wahrscheinlichkeits-Deutung verträglich ist. Er ist zu einer Reihe von neuen Beispielen gekommen, hat aber noch nicht den allgemeinen Nachweis erbringen können, daß es immer möglich ist, Wahrscheinlichkeits-Deutung und indefinite Metrik zu vereinigen. Ferner hat eine Untersuchung von MITTER nachgewiesen, daß sich für die sogenannten Kontraktionsfunktionen (Erwartungswerte für das Produkt von zwei Wellenfunktionen) aus der zugrundegelegten Differentialgleichung jedenfalls in der niedrigsten. Näherung eine Form ergibt, die mit einer definiten Metrik im Hilbertraum nicht vereinbar ist, sondern die zwangsläufig zu einer indefiniten Metrik führt.

Bei der Einführung der indefiniten Metrik kann man ja immer die zwei Fragen stellen: Ist sie möglich und ist sie notwendig? Die SCHLIEDERsche Arbeit gab einen Beitrag zu der Frage, ob sie möglich ist, indem sie eine Anzahl von Beispielen aufzeigte, in denen tatsächlich die Wahrscheinlichkeits-Deutung mit der indefiniten Metrik vereinbart werden kann. Die MITTERsche Arbeit gibt einen Beitrag zu der anderen Frage, ob sie nötig ist. Es scheint sich herauszustellen, daß die Vereinigung von Kausalität, Lorentzinvarianz und Wechselwirkung zumindest in diesem speziellen Fall die Einführung der indefiniten Metrik erzwingt.

Inzwischen sind aber auf diesem Gebiet auch noch andere, interessante Untersuchungen erschienen, und zwar möchte ich vor allem die Arbeiten von MAKSIMOW, SUDERSHAN und von FERRETTI erwähnen. Dort wird gezeigt, daß auch in dem PAULI-KÄLLÉNschen Fall des LEE-Modells die Wahrscheinlichkeits-Deutung, die man bisher für unmöglich gehalten hatte, durchgeführt werden kann, wenn man die Formeln anders interpretiert. Dieses Ergebnis legt den Gedanken nahe, daß es einen allgemeinen gruppentheoretischen Grund dafür gibt, daß die Wahrscheinlichkeitsinterpretation bei den asymptotischen Zuständen durchgeführt werden kann. Die Mathematiker unterscheiden kompakte und nicht-kompakte kontinuierliche Gruppen. Eine kompakte Gruppe ist eine Gruppe, bei der man nicht durch Limes-Bildung zu einem Element außerhalb der Gruppe kommen kann, in einer nichtkompakten dagegen kann die Limes-Bildung aus der Gruppe herausführen. Wenn man die Gesamtheit der Transformationen untersucht, die in dem betreffenden Hilbertraum möglich sind, dann handelt es sich sicher um eine nichtkompakte Gruppe, denn es gehört ja z. B. die

Lorentzgruppe dazu, die nichtkompakt ist. Wenn man aber nur das asymptotische Verhalten der Wellen in Betracht zieht und „auf der Energieschale bleibt", das heißt nur diejenigen Transformationen zuläßt, bei denen die gesamte Energie und der gesamte Impuls festgehalten werden, dann ist die so definierte Gruppe eine kompakte Gruppe. Von den Mathematikern ist gezeigt worden, daß für kompakte Gruppen die gewöhnliche unitäre Darstellung der Transformationen naturgemäß ist und daß man jede andere auf sie zurückführen kann. Für die nichtkompakten Gruppen ist dagegen umgekehrt die indefinite Metrik (als Darstellung die nichtunitäre Transformation) naturgemäß. Daher ist es begreiflich, daß man für die aus- und einlaufenden Wellen doch die Wahrscheinlichkeitsinterpretation durchführen kann, selbst wenn die Wahrscheinlichkeits-Deutung für die lokalen Verhältnisse auf Schwierigkeiten stößt. Aus den Untersuchungen von Sudershan, Maksimow und Ferretti wird also die Vermutung nahegelegt, daß der Gebrauch der indefiniten Metrik im gesamten Hilbertraum nie zu Schwierigkeiten mit der Wahrscheinlichkeitsdeutung bei den asymptotischen Zuständen führt. Wenn sich dieser Schluß zu einem mathematischen Beweis verschärfen läßt, wäre also an dieser Stelle eine wesentliche Schwierigkeit der Theorie beseitigt.

Die zweite Erweiterung betraf die Verwendung von Transformationen in l. Gegen eine Verwendung dieser Transformationen könnte man zunächst etwa folgendermaßen argumentieren: Die Differentialgleichung enthält nur einen bestimmten Wert von l, also wird der ganze zeitliche Ablauf des Geschehens innerhalb dieses einen l-Wertes entschieden. Es muß also völlig gleichgültig sein, ob noch andere Werte daneben in Betracht gezogen werden, sie liefern nur verschiedene, gleichberechtigte Bilder der wirklichen Welt, aber nichts Neues.

Das wäre richtig, wenn es nicht das Problem der Randbedingungen gäbe, von dem vorhin gesprochen wurde. Der wirkliche zeitliche Ablauf wird eben nicht nur durch die Differentialgleichung, sondern auch durch die Randbedingungen bestimmt. Eine neue Untersuchung von Dürr gibt ein Beispiel dafür, wie eine Transformation im l-Raum tatsächlich entscheidend zur Vervollständigung der Theorie benützt werden kann, wenn man die Randbedingung bei Unendlichwerden der Teilchenzahl so stellt, daß sie einer Symmetrieforderung im l-Raum entspricht.

Es wurde vorhin schon erwähnt, daß in der Theorie zunächst die strenge Erhaltung der Parität für Nukleon und π-Meson problematisch war. Man weiß andererseits aus vielen Experimenten, daß für Nukleon und π-Meson offenbar die Parität bei den starken Wechselwirkungen in Strenge erhalten bleibt. In der Arbeit von Dürr wird zunächst gezeigt, daß die Ausgangsgleichung unmittelbar keine Darstellung der vollständigen Lorentzgruppe enthält, sondern nur eine solche der eigentlichen Lorentzgruppe, die die kontinuierlichen Transformationen ohne Spiegelung umfaßt. Die Spiegelungsoperation ist also ursprünglich in der Gleichung nicht enthalten, jedenfalls so lange nicht, als man nur von *einem* Wert des Parameters l spricht. Man kann zwar eine „Parität zweiter Art" durch die Operation

$$\psi(p) \to \frac{p_\nu \gamma_\nu}{|p|} \cdot \psi(\vec{p}, p_0)$$

definieren. Es besteht aber zunächst kein Anlaß dafür anzunehmen, daß diese Parität zweiter Art streng erhalten bleibt. Die folgenden Formeln können den Sachverhalt erläutern:

$$\chi_{11} = \psi_1; \qquad \chi_{21} = \psi_2; \qquad \chi_{12} = -\psi_4^*; \qquad \chi_{22} = \psi_3^* \tag{1}$$

$$-i\, \sigma_\nu \frac{\partial}{\partial x_\nu} \chi \pm l^2 \sigma^\mu \chi \,(\chi^* \sigma_\mu I \chi) = 0 \tag{2}$$

$$\sigma_\nu = (1, \sigma_k) \qquad \sigma^\nu = (-1, \sigma_k) \tag{3}$$

$$\chi\,(\vec{\mathfrak{r}}, t, l) \to c_\sigma^{-1} c_\tau^{-1} \chi^*\,(-\vec{\mathfrak{r}}, t, l) \tag{4}$$

$$\vec{\xi}_k = \frac{\vec{x}_k}{l}, \qquad \xi_0 = \frac{x_0}{|l|}; \qquad \pi_k = p_k \cdot l, \qquad \pi_0 = p \cdot |l| \tag{5}$$

$$X(\vec{\xi}_k, \xi_0, l) = \begin{pmatrix} X\,(\vec{\xi}_{,k}\, \xi_0,\, l) \\ X\,(\vec{\xi}_k,\, \xi_0,\, -l) \end{pmatrix} \tag{6}$$

$$\Gamma_\nu \frac{\partial}{\partial x_\nu} X \pm \frac{l^2}{2} \{\Gamma_\mu X'\,(\bar{X}\, \Gamma_\mu X) + \Gamma_\mu \Gamma_5 X\,(\bar{X}\, \Gamma_\mu \Gamma_5 X)\} = 0 \tag{7}$$

$$\langle \Omega\,|X_{\alpha\sigma}\,(x, l)\, \bar{X}_{\beta\varrho}\,(x', l)|\,\Omega\rangle' = \int d^4 p\; e^{ip(x-x')} \delta_{\sigma\varrho} \left(\frac{\Gamma_{\alpha\beta}^\nu\, p_\nu + i\,\varkappa\, \delta_{\alpha\beta}}{p^2 + \varkappa^2} + \dots \right) \tag{8}$$

DÜRR hat zunächst die Ausgangsgleichung umgeformt, indem er nicht die DIRACschen Spinorindizes (mit vier Komponenten) verwendet, sondern neue Spinoren (1) benützt. Er führt statt des ursprünglichen Operators ψ mit den Komponenten ψ_1, ψ_2, ψ_3, ψ_4 einen neuen Operator χ ein, der zwei Indizes hat, die aber immer nur die Werte 1 und 2 annehmen [siehe (1)]. Der erste Index bezieht sich auf den Spin, der zweite auf den Isospin. Dann erhält die ursprüngliche Gleichung die Form (2). Man sieht hier, daß nur die PAULIsche Spinmatrix (3) und damit eine zweidimensionale Darstellung der Lorentzgruppe vorkommt. Die Gleichung ist zwar gegen die Isospingruppe invariant, es ist in ihr aber die Spiegelung gar nicht enthalten, da ja nur die eigentliche Lorentzgruppe dargestellt ist. Das bedeutet, daß man zunächst gar nicht sagen kann, was eine Spiegelung hier heißen sollte.

Die Invarianzeigenschaft (4) kann mit dem Produkt aus Spiegelung und Ladungskonjugation in Zusammenhang gebracht werden, sie entspricht also der sogenannten PC-Transformation.

Läßt man verschiedene l-Werte gleichzeitig zu, so kann man eine weitere Symmetrie-Eigenschaft feststellen: Die Gleichung ist invariant gegenüber einer Vorzeichenumkehr des Parameters l. Man kann nach (5) neue Raum- und Zeitkoordinaten definieren, die aus Dimensionsgründen immer Beziehungen zwischen x und l sein müssen. Man führt dann statt der Ausgangsfunktion χ, die in (2) vorkommt, die beiden Operatoren $\chi\,(x, + l)$ und $\chi\,(x, - l)$ ein und definiert etwa einen Spinor (6). Dies bedeutet eine Verdoppelung, indem $+ l$ und $- l$ für den Parameter zugelassen werden. Wenn man durch diese Verdoppelung von den σ_ν zu neuen Diracmatrizen übergeht, die jetzt mit Γ_μ bezeichnet werden sollen, so erhält man die Gleichung (7). Sie ist nichts anderes als die Ausgangsgleichung, die hier zweimal angeschrieben wurde, nämlich einmal für $+ l$ und einmal für $- l$. Verknüpft man die verschiedenen l-Werte miteinander, indem man die Vertauschungsrelation zwischen $\chi\,(x, + l)$ und $\chi\,(x, - l)$ nach Gleichung (8) verschieden von Null annimmt, dann wird die Verdopplung, die man ursprünglich durch die zwei l-Werte bewirkt hatte, wieder aufgehoben und man hat eine Darstellung der Parität gewonnen. Es läßt sich leicht zeigen, daß die so formulierte Gleichung (7) tatsächlich gegenüber der üblichen Paritätstransformation $X\,(\vec{\xi}, \xi_0) \to \Gamma_4\,X\,(-\,\vec{\xi}, \xi_0)$ invariant ist. Das Verfahren von Dürr läuft darauf hinaus, daß man die Randbedingungen für große Teilchenzahl gegenüber der üblichen Formulierung verändert, nicht aber die Feldgleichung (2). Man fordert zusätzlich eine Symmetrie zwischen einer n-Punktfunktion und einer $(n + 2)$-Punktfunktion, die schließlich die Darstellung der Parität (durch eine Erweiterung des Begriffs „Parität zweiter Art") gestattet.

Die vor einem Jahr noch ungeklärte Frage, ob die Parität zweiter Art sich zu einer strengen Erhaltungsgröße erweitern läßt, kann also jetzt mit Hilfe der $\pm\,l$-Verdopplung positiv entschieden werden. Die Arbeit von Dürr zeigt weiterhin, in welcher Weise grundsätzlich die Benutzung einer l-Transformation möglich ist. Die Transformation im l-Raum kann als Ersatz oder als eine vereinfachte Schreibweise für eine Aussage über die Randbedingung bei großer Teilchenzahl gelten; eine Vermehrung der Zahl der Eigenzustände ist mit der Erweiterung in den l-Raum nicht verbunden. In diesem Sinne ist die Ausnützung der Transformationen im l-Raum möglich und wohl auch notwendig.

Eine weitere Besonderheit der nichtlinearen Spinortheorie bestand in der Annahme, daß der Grundzustand entartet sei und nicht die volle Symmetrie der Ausgangsgleichung besitze. Im letzten Jahr ist eine sehr interessante Verbindung zwischen der Theorie der Elementarteilchen und der Theorie der Supraleitung hergestellt worden. Nambu hat darauf hingewiesen, daß die Hamiltonfunktion, die der Theorie der Supraleitung zugrundeliegt, mathematisch ähnlich aussieht, wie die Differentialgleichung der nichtlinearen Spinortheorie. Bogoljubov hat gezeigt, daß der Grundzustand in der Theorie der Supraleitung auch ein entarteter Zustand ist, also ein Zustand, der eine geringere Symmetrie aufweist, als die Differentialgleichung, von der ausgegangen wird. Bogoljubov hat in diesem Zusammenhang auf eine ganze Klasse von Problemen hingewiesen, bei denen ähnliche Verhältnisse herrschen. Außer der Supraleitung nannte er den Ferromagnetismus, die Suprafluidität, die Bildung von Kristallen, die Kondensation von Flüssigkeiten usw. Er hat damit nachgewiesen, daß die Annahme von der Entartung des Grundzustandes kein Novum ist, sondern durchaus in den Rahmen der bisherigen Quantentheorie gehört. NAMBU konnte speziell zeigen, daß die Entartung bei der Supraleitung in ihrer formalen Struktur der Entartung in der Elementarteilchentheorie weitgehend entspricht, wobei die letztere schon durch die Forderung einer Parität für die Nukleonen zustandekommt. Die Entartung des Grundzustandes kann also im wesentlichen als geklärt gelten.

Eng mit diesem Punkt verknüpft ist die Frage nach der Darstellung der „seltsamen" Teilchen. Die Annahme einer Entartung des Grundzustandes ermöglicht eine solche Darstellung und läßt Schlüsse auf die Symmetrie-Eigenschaften dieser Teilchen zu. Die Schwierigkeit bestand zunächst darin, daß ψ sowohl im gewöhnlichen als auch im Isospinraum ein Spinor ist, daß also eine ungerade Zahl von ψ-Operatoren einen halbzahligen Spin und Isospin, eine gerade Anzahl dagegen einen ganzzahligen Spin und Isospin hervorruft. Die „seltsamen" Teilchen sind andererseits gerade dadurch charakterisiert, daß sie halbzahligen Spin und ganzzahligen Isospin oder umgekehrt besitzen. Die Darstellung dieser Teilchen wird aber möglich, wenn das Vakuum selbst einen Isospin besitzt, da dann z. B. ein Nukleon einen Isospin $\frac{1}{2}$ vom Vakuum abzweigen und damit zu einem Hyperon werden kann. Führt man diesen Gedanken mathematisch durch, so stellt sich interessanterweise heraus, daß für ein einzelnes seltsames Teilchen die Parität nur unter Einbeziehung des vom Grundzustand übernommenen Isospins (des „Spurions") definiert werden kann; daß also diese Parität nur für ein Paar solcher Teilchen mit der gewöhnlichen Parität verglichen werden kann. Dazu paßt die experimentell gefundene Tatsache, daß beim Zerfall eines Λ-Teilchens in ein Nukleon und ein π-Meson oder eines K-Teilchens in zwei oder drei π-Mesonen die Parität sicher nicht erhalten bleibt. — Die Untersuchungen über die seltsamen Teilchen sind aber noch nicht abgeschlossen.

Schließlich sei noch die Frage nach der Brauchbarkeit des TAMM-DANCOFF-Verfahrens kurz besprochen. Die mathematische Ähnlichkeit der Theorie der Supraleitung mit der Theorie der Elementarteilchen legte den Gedanken nahe, das TAMM-DANCOFF-Verfahren auf die Supraleitung anzuwenden, da man hier das Resultat ja aus anderen Untersuchungen kennt. YAMAZAKI hat diese Rechnung durchgeführt und dabei die schon bekannten Resultate exakt wiedergewonnen. Das TAMM-DANCOFF-Verfahren hat also im Fall der Supraleitung die Probe völlig bestanden, und das kann als Stütze für die bisherigen Rechnungen in der Elementarteilchentheorie gelten. Trotzdem sollte man versuchen, noch neue Approximationsverfahren zu entwickeln. Auch hier wird noch viel mathematische Arbeit geleistet werden müssen.

In *Conférence Internationale d'Aix-en-Provence sur les Particules Élementaries – The Aix-en-Provence International Conference on Elementary Particles. Volume 1,* ed. by E. Cremieu-Alcan, P. Falk-Variant, O. Lebey (C. E. N. Saclay, Gif-sur-Yvette 1961) pp. 163–168

THEORY OF THE STRANGE PARTICLES

H.-P. DURR and W. HEISENBERG

(presented by W. HEISENBERG)

In the following report I would like to comment on a paper by Dürr and myself concerning the theory of the strange particles. It will not be possible in a brief talk to give a full account of the mathematical details of the paper, nor will it be necessary, since the preprints of the paper have been circulated some four months ago and Dürr has reported on the paper at the la Jolla-conference. But I would like to discuss some of the essential points of the paper which can be understood without going into the details.

The calculations of the paper had led to some definite predictions concerning the properties of hyperons and K-mesons of strangeness 1, and concerning the existence of unstable excited states, so called "resonance" states, of such particles. These predictions are essentially different from those of other theories, for instance of the theory of "global symmetry". We hope that the experiments will decide about these points in a near future. While we have to wait for these decisions, we have frequently been asked, to what extent the predictions of this paper are a rigorous and necessary consequence of the underlying grouptheoretical assumptions and to what extent they might depend on the approximation methods used, which are necessarilly more uncertain and controversial.

In order to answer this question I shall start with the grouptheoretical assumptions, a large part of which had been stated in a paper published two years ago by Dürr, Mitter, Schlieder, Yamazaki and myself.

We start from the nonlinear spinor equation

$$\gamma_\nu \frac{\partial \Psi}{\partial x_\nu} \pm l^2 : \gamma^\mu \gamma^5 \Psi (\Psi \gamma_\mu \gamma_5 \Psi) : = 0$$

with
$$\Psi = \Psi_\alpha \quad \alpha = 1 \ldots 4$$

which may, according to Dürr, also be rewritten in the form

$$-i \sigma_\nu \frac{\partial}{\partial x_\nu} \chi \pm l^2 : \sigma^\mu \chi (\chi^* \sigma_\mu \chi) : = 0,$$

with
$$\chi = \chi_{\alpha\beta} \quad \begin{matrix} \alpha = 1, 2 \ \text{(spin)} \\ \beta = 1, 2 \ \text{(isospin)} \end{matrix}$$

Connection :
$$\chi_{11} = \Psi_1; \ \chi_{21} = \Psi_2 ; \ \chi_{12} = - \Psi_4^* ; \ \chi_{22} = \Psi_3^*$$

where χ is a Weyl-spinor and an isospinor. Parity is trated along the lines explained in the paper by Dürr, as an approximate symmetry which is valid only if the effect of the leptons can be neglected.

When one applies a field operator Ψ or χ, or products of such fieldoperators on a vacuum which is completely symmetrical and non degenerate, the resulting states have either half-quantum spin and isospin, if the number of field operators is odd, or integer spin and isospin, if this number is even. Therefore one cannot get strange particles in this way, some of which have half quantum spin and integer isospin ; this fact had been considered as a serious difficulty in the early stages of the theory. On the other hand the non-conservation of isospin in the electromagnetic interactions shows that the ground state can be symmetrical under isospin rotation only up to a certain approximation, it cannot be rigorously symmetrical. In fact the number of neutrons and protons in the world seems to be very different, therefore the groundstate "world", from which the

163

particles are created, has probably a very high isospin, it is strongly degenerate. This assumption had been suggested in the old paper as explanation for the lack of symmetry in the electromagnetic forces. It has recently found some support from the theory of superconductivity - as pointed out by Nambu -, where one has learned that a degenerate groundstate is a quite normal situation in quantum theory. As a matter of fact, concerning the problem of isospin the nonlinear spinor theory is still more closely related to the case of ferromagnetic bodies, where the degeneracy refers to the direction of the total spin, than to superconductivity. While the lack of symmetry produced by the big isospin of the world will be essential for the understanding of electromagnetic forces, the analogy to ferromagnetism can be used still in a different way to explain the existence of strange particles. It had been suggested in the old paper, that the strange particles are created by taking an isospin $1/2$ or 1 off the total isospin of the world and attaching it to a nucleon or π - meson. The possibility of such couplings in the case of ordinary spin is demonstrated by the Russel-Saunders coupling say in the Calcium atom.

In this way the nonlinear spinor theory gives simply a definite interpretation to the concept of a "spurion", introduced by Wentzel several years ago, and we will discuss this interpretation first in connection with the hyperons. The "spurion" was assumed by Wentzel to have no properties with respect to the Lorentz-group, I.e. no energy or momentum or position, but it should have just the isospin $1/2$. In the nonlinear spinor theory the spurion is to be represented by a degeneracy of the vacuum. If the spurion would have no other properties but the isospin $1/2$, it would be sufficient to consider only two vacuum states transforming as an isospinor ; these states should be completely symmetrical under all possible transformations of the theory except isospin rotation. The only possible interaction term between nucleon and spurion would then be the scalar product of the isospin of the spurion and the isospin of the nucleon. This term however is too simple and could not be used alone for introducing an interaction, for the following reason :

The operation of charge conjugation C has in the case of particles with isospin to be supplemented to the operation G = CJ, introduced by Lee and Yang, in order to lead to a particle with the same quantum numbers. (The operation J means essentially the reversal of isospin). Instead of PCT one might then write PG. (JT). If the transformation G is applied on the spurion alone, the scalar product of spurion isospin and nucleon-isospin changes sign. Therefore the interaction of the antispurion must have opposite sign to that of the spurion, and if one would identify spurion and antispurion, the interaction term would necessarily vanish.

Therefore one must distinguish between spurion and antispurion and introduce four vacuum states. Formally one has then the choice between two possibilities : Since the spurion must be symmetrical under PCT, it can either be symmetrical under PG and JT or under G(JT) = CT and P. But the original field equation is symmetrical under PG and not under CT, therefore within the frame of the nonlinear spinor theory the spurion should be symmetrical under PG. Spurion and antispurion can then be distinguished by the parity operation P, which may be represented by a Pauli-matrix Σ_1. Σ_1 represents the parity of the spurion in the same sense, as Γ_4 représents the parity of the nucleon at rest ; Σ_2 corresponds to the "helicity" Γ_5 of the nucleon.

Thereby the group-theoretical assumptions of the theory have been stated precisely, and we will now turn to the consequences with regard to the properties of the hyperons.

Any theory which can account for the existence of nucleons will, quite independently of the approximations used, lead to a Dirac equation for the nucleons :

$$(\Gamma^\mu p_\mu - i\varkappa)\; \varphi = 0$$

In order to get the Dirac equation for the hyperons, we must supplement it with interaction terms between spurion and nucleon. The only interaction terms that can be added are now the two underlined expressions in the following Dirac equation of the hyperons :

$$\left\{\Gamma^\mu p_\mu \,(1 + \alpha\; \Sigma_2\Gamma_5\,) - i\varkappa\,[1 + i\eta\; \underline{\Sigma_3\,\Gamma_5\;(\rho\,\tau)}\,]\right\}\; \varphi = 0.$$

$$\varphi = \varphi_{\alpha\beta\gamma\delta}(p)$$

α = Parity of "Spurion"	$\Sigma_2\,\Sigma_3$ acting on α
β = Isospin of "Spurion"	ρ acting on β
γ = Dirac spin	Γ^μ acting on γ
δ = Isospin of field operator	τ acting on δ
p = Energy-momentum	

164

For simple grouptheoretical reasons these are the only terms that are not symmetrical if the transformations P or G or rotation in isospace are applied on the spurion alone or the nucleon alone, but are symmetrical, if they are applied on both simultaneously. They are symmetrical with respect to GP or JT, wether applied alone or simultaneously.

The main terms which are fully symmetrical could indirectly from the interaction receive contributions that contain the operator Σ_1 ; but such contributions could only appear in rather high order terms and should be negligible, if the constants α and η are not too big.

Therefore the general form of the Dirac equation of the hyperons does not depend on any special method of approximation. It is only in the numerical determination of α and η (which will generally still depend on p^2), that the approximation method comes in. Fortunately for the qualitative picture of the spectrum of hyperons these numerical values [or rather : the numerical behaviour of the functions $\varkappa\,(p^2)$, $\alpha\,(p^2)$ and $\eta\,(p^2)$] are not too important. Since the masses of the hyperons are not very different from those of the nucleons, it should be sufficient in a first approximation to treat $\varkappa$, α and η as constants. For the energy in the rest-system one obtains :

$$P_0 = -\varkappa\,\frac{\alpha\,\eta\,\Sigma_1\Gamma_4(\rho\tau)}{1-\alpha^2} \pm \frac{\varkappa}{1-\alpha^2}\,\sqrt{1-\alpha^2+\eta^2\,(\rho\tau)^2}$$

$\Sigma_1\Gamma_4$ is the parity of the hyperon.

$$(\rho\tau) = \begin{cases} +1 & \text{for the triplet } (\Sigma) \\ -3 & \text{''\quad''\quad singlet } (\Lambda). \end{cases}$$

In the Tamm-Dancoff-method the value $\alpha = 0.63$ was calculated. This figure might be changed considerably by using a different method of approximation ; still we think, that a value of α between 0.5 and 0.8 should be a reasonable estimate. The constant η should be considerably smaller than

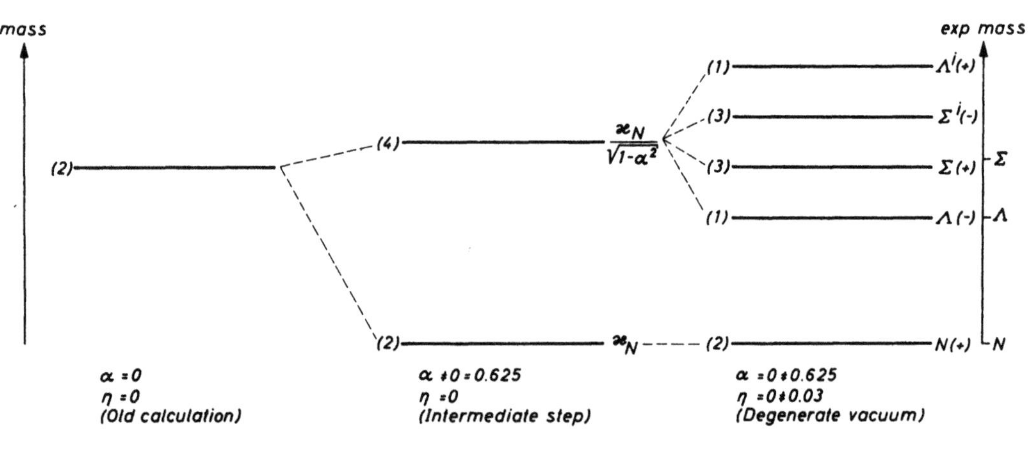

Scheme of the mass eigenvalues of the fermions ($I_N = 1$)

mass splittings:
$$\Lambda^i - \Lambda \approx 2\varkappa\,\frac{3\alpha}{1-\alpha^2}\,\eta\,\varkappa_N$$
$$\Sigma^i - \Sigma \approx 2\varkappa\,\frac{\alpha}{1-\alpha^2}\,\eta\,\varkappa_N$$

Figure 1

165

α ; but the Tamm-Dancoff-method did not give reliable results for η, since small changes of α could lead to very big changes in η. Therefore η was taken from the experiments, to give the best fit (at $\alpha = 0.63$) with the observed masses.

If $|\alpha|$ is of this order of magnitude and $|\eta|$ very much smaller, one gets always the qualitative picture of figure 1. It should be emphasized, that the spectrum does not depend on the sign of α or η. If the sign of the product $\alpha\eta$ is reversed, one should also reverse the sign of the parity $\Sigma_1 \Gamma_4$ to get the same terms. But in the experiments one can only compare the parity of systems having the same strangeness. Therefore a common factor, given by the sign of $\alpha\eta$, would not show up in the experiments and may be omitted.

The masses of the hyperons are bigger than that of the nucleon ; one may say : simply because the interaction with the spurion increases the mass. There are four different hyperons, according to the values $(\rho\tau) = 1$ and $- 3$, $\Sigma_1\Gamma_4 = \pm 1$. The lowest and highest hyperon must be isosinglets or Λ - particles, the two intermediate ones are isotriplets or Σ - particles. The parity of the lower Λ - particle must be opposite to that of the lower Σ - particle. If the two highest states are sufficiently far above the two other states, they can disintegrate into the lower ones by π - meson emission, therefore they may be called then "excited states" or "resonance states". All these results are immediate consequences of the group structure and would not be changed within wide limits of the constants or functions $\alpha\ (p^2)$ and $\eta\ (p^2)$.

It may be that the two higher terms can be identified with the resonance states recently found by the Berkeley-group (Alston et al.). This identification is possible only if the empirical resonance states have spin 1/2, isospin 1 and 0, and the calculated parity ; I would like to mention that Nogami has in a recent paper given an interpretation of the lower Y^* state from the dispersion theory which agrees with regard to spin and parity with our interpretation.

Even if these figures agree, there remains one very important property of the particles which has not been claculated from the theory, their electric charge. Electromagnetic interaction has not yet been investigated in the nonlinear spinor theory ; therefore neither the concept of electric charge nor the quantum number strangeness, closely related to the charge, could be discussed in the paper on which I am reporting.

Furthermore, the nonlinear spinor theory gives so far no information concerning hyperons of spin 3/2. Such states have been discussed in the theory of "global symmetry", but have not yet been investigated in the nonlinear spinor theory.

Turning now from the hyperons to the bosons, it is easily seen that even the simplest case of bosons of spin 0 and baryon nomber 0 is more complicated than the case of hyperons of spin 1/2, since the bosons may have isospin 0 and 1. The existence of these two kinds of " π - mesons" follows from the group theoretical assumptions alone ; their calculated mass eigenvalues however depend on the approximation method used. In the Tamm-Dancoff-method for the bosons only the terms of lowest order have been taken, while for the fermions terms of the second order could be included. Therefore the calculation for the bosons is still more uncertain than that for the fermions. In this rough estimate the mass of the singlet π - meson comes out somewhat below the nucleon mass, that of the triplet roughly five times smaller, while the empirical mass of the triplet π - meson is roughly one seventh of the nucleon mass. The singlet π - meson may possibly be identified with the ω - meson recently announced by the Berkeley-group. The large massdifference between singlet - and triplet - π's plays a decisive rôle also for the mass spectrum of strange bosons. Therefore one cannot easily distinguish between results from the grouptheory alone and results depending upon the approximations. Attaching a "Spurion" to a singlet π - meson should produce two doublets, one of which may possibly be identified with the excited K -meson found recently by the Berkeley-group ; attaching it to the triplet - π's should lead to two doublets and two quartets. The mass eigenvalues of these terms, as estimated from the Tamm-Dancoff-method, are seen in figure 2. One quartet and one doublet term coming from the triplet π - meson appears at a very low mass value comparable to leptonic masses. Whether this low eignevalue would remain in higher approximations is quite doubtful, and could be decided only in connexion with the lepton-problem. The other doublet from the triplet - π and the lower doublet from the singlet - π would coincide for $\alpha = 1/2$ and $\eta = 0$ and could also coincide at other values of α, if η is suitably chosen. We believe that these two doublets should be identified with the K-mesons. From the transformation properties under GP and isospin-rotation one would then conclude (figure 3).

If the theory in a later stage would lead to electrodynamics and to the conservation of electric charge, the two doublets must necessarily coincide in that approximation in which one neglects electromagnetic interactions. Because, if there would be a mass difference, the state $K^+ >$ would

166

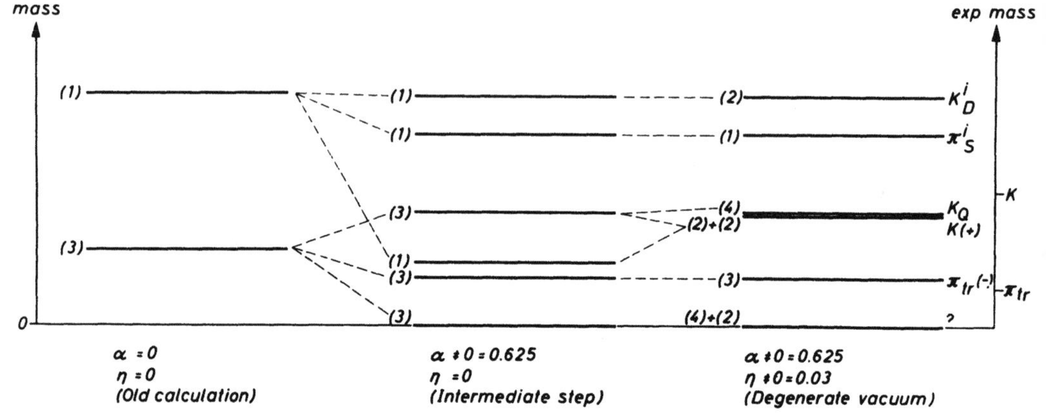

Figure 2

K-mesons as combinations of π-mesons with "spurion".

Isospin	$\frac{1}{2}$	$-\frac{1}{2}$
Doublet from singlet	$\frac{1}{\sqrt{2}}(K^+ \rangle - \overline{K}^\circ \rangle)$	$\frac{1}{\sqrt{2}}(K^\circ \rangle - K^- \rangle)$
Doublet from triplet	$\frac{1}{\sqrt{2}}(K^+ \rangle + \overline{K}^\circ \rangle)$	$\frac{1}{\sqrt{2}}(K^\circ \rangle + K^- \rangle)$

Figure 3

not be stationary, it would go over into $\overline{K}^\circ \rangle$ after a time interval given by the invers mass difference. The coincidence of the two doublets means a relation between those terms that are responsible for the isosinglet - iso-triplet - separation (the isospin - isospin interaction) and the coupling between the spurion-isospin and the particle isospin.

The parity of the boson states could be determined only by the postulate, that the different Tamm-Dancoff equations - the number of which is higher than the number of unknown wave functions - should be fulfilled with the smallest possible deviations. This condition leads to the result, that - so far as parity is concerned - the K, $\overline{K}$ - meson should be considered rather as a compound of Λ and $\overline{N}$ or $\overline{\Lambda}$ and N, than as a compound of Σ and $\overline{N}$ or $\overline{\Sigma}$ and N. This result is perhaps plausible in itself ; but I have to emphasize that it does not follow from grouptheory alone ; therefore it is less certain than the result of opposite parity of Λ - and Σ - hyperons.

The final results of the paper are combined in the table of figure 4. When the final experimental results will be available some day, their comparison with the thoeretical results on the hyperons will offer a direct check on the underlying grouptheoretical assumptions. Their comparison with the calculated properties of the bosons will give less definite information but might allow some conclusions on the usefulness of Tamm-Dancoff approximations.

167

Mass eigenvalues

Particles		$N \ \bar{N}$	$\Lambda \ \bar{\Lambda}$	$\Sigma \ \bar{\Sigma}$	$\Lambda^{\mathrm{I}} \ \bar{\Lambda}^{\mathrm{I}}$	$\Sigma^{\mathrm{I}} \ \bar{\Sigma}^{\mathrm{I}}$	π_{tr}	π^{I}_{s}	$K \ \bar{K}$	K^{I}_{qu}	K^{I}_{D}	?
Mass	theor.	1	1, 195	1, 250	1, 365	1, 312	0, 194	0, 76	0, 437	0, 448	0, 915	0, 006
	exp.	1	1, 19	1, 265			0, 149		0, 532			
Relative parity	theor.	+ -	- +	+ -	+ -	- +	-	-	+	-		
	exp.	+ -					-					

After this table had been published (March 61), several new experimental data have been published or announced at meetings or communicated ot us, which give very interesting information on the figures of the table. I am not in a position to judge the reliability of these data ; but if one takes the data tentatively as correct, one comes to the comparison of figure 5. The agreement between experiment (new data in circles) and theory seems to be as good as could be expected in a Tamm-Dancoff approximation.

Particles		N	Λ	Σ	Σ^{*}	Λ^{*}	π_{Tr}	$\pi^{*}_{3} = \Omega$	K	K^{*}_{Qu}	K^{*}_{D}
Mass	theor.	1	1, 195	1, 250	1, 312	1, 365	0, 194	0, 76	0, 437	0, 448	0, 915
	obs.	1	1, 190	1, 265	(1, 47)	(1, 505)	0, 1485	(0, 80)	0, 532		(0, 95)
Parity	theor.	+	-	+	-	+	-	-	+	-	
	obs.	+	(-)	(+)	(-)	(+)	-		(+)		

$\bigcirc$: More recent experimental results, still <u>uncertain</u>.

In *1962 International Conference on High-Energy Physics at CERN: Proceedings*, ed. by J. Prentki (CERN, Geneva 1962) pp. 675–680

ELECTRODYNAMICS IN THE NON-LINEAR SPINOR THEORY

W. Heisenberg

Max-Planck-Institut für Physik und Astrophysik, München

(invited paper presented by W. Heisenberg)

The interpretation of electrodynamics in the non-linear spinor theory requires the solution of three rather different problems. 1. The eigenvalue equation for the bosons may lead to solutions with mass zero ($J^2 = 0$). One has to investigate, whether such solutions exist and whether they exhibit the correct transformation properties of the photon with respect to the Lorentz group; Maxwell's equations would appear as a consequence of the Lorentz-properties of the photon. 2. The behaviour of the photons with respect to the isospin-group must be treated on the basis of a degeneracy of the vacuum, which may explain the lack of isosymmetry in electrodynamics. From this discussion the interaction of the photon with the particles (the " charge " of the particles) should follow. If the theory is correct, it should give the well known rule of Gell-Mann and Nishijima. 3. The coupling constant $e^2/\hbar c$ should be derived numerically from the theory. For each of these problems the present lecture can only describe the main steps in the mathematical treatment without giving any details or proofs. The details of the investigation, which has been carried out by Dürr and the author, will be published later [1].

1. BOSON EIGENSTATES WITH REST MASS ZERO

If the fundamental field equation in the non-linear spinor theory is written in the form given by Dürr [2]:

$$-i\sigma^\mu\frac{\partial\chi}{\partial x_\mu} \pm l^2\sigma^\mu\chi(\chi^*\sigma_\mu\chi) = 0 \tag{1}$$

the vacuum expectation value of the product of two field operators may in a first approximation be expressed as

$$\langle 0|\chi(x)\chi^*(x')|0\rangle$$

$$= (2\pi)^{-4}\int d^4p\, e^{ip(x-x')}p_\nu\sigma^\nu\left[\frac{1}{p^2+\kappa^2}-\frac{1}{p^2}+\frac{\kappa^2}{(p^2)^2}\right] \tag{2}$$

The corresponding Hilbert space can be interpreted as consisting of nucleon-states of mass κ and " dipole-ghost "-states of mass zero. For the ghost states the situation is similar to that in the Lee model. There are states $\Phi_0\rangle$ obeying the neutrino equation and states $\Phi_{\text{dip}}\rangle$ not obeying this equation. Both kinds of states have the norm zero but $\Phi_0\rangle$ and $\Phi_{\text{dip}}\rangle$ belonging to the same momentum are not orthogonal to each other. These ghost states have an important consequence for the bosons.

For any mass of the boson, the state seems to " disintegrate " into two ghost states of the $\Phi_0\rangle$ type, moving with suitable momentum in opposite directions. Therefore the wave function of the boson must contain an in-going or out-going spherical wave. This wave however has the norm zero; therefore it does not prevent the boson state from having a finite norm and being a discrete state (under suitable boundary conditions). In momentum space consequently the wave function of the boson has a pole at the relative momentum of the fermions corresponding to the rest mass of the boson (compare earlier calculations on the π-mesons). For a boson of rest mass zero however the situation is radically different. For such a boson there is no rest system in which the two ghost fermions could move in opposite directions. Asymptotically this boson may be composed only of two ghost states moving in exactly the same direction as the boson itself, in such a way that their momenta add up to the momentum of the boson.

These two ghost fermions move with the velocity of light in the same direction, therefore their distance does not change, there is no out-going or in-going wave. At large distances between them any solution of the neutrino-equation for one of the particles would describe a possible asymptotic behaviour. Since no restriction from a boundary condition occurs, one should expect a solution of the eigenvalue equation of the boson for rest mass zero only if the application of the characteristic integral operator in this equation on the wave function can produce a state of finite norm. This however is not at all trivial, since the singularities of the integral operator corresponding to the production of the ghost states would as a rule prevent a convergent Fourier transformation and lead to an infinite norm. Therefore it is only by the compensation of several terms in the eigenfunction of the boson that the singularities at $J^2 = 0$ can be avoided and a solution of finite norm established. This condition distinguishes between different symmetries of the boson state (a simplified version of this criterion has been used in an earlier paper by Mitter, Kortel and the author [3]).

In the following table the symmetry is characterized by the main terms in the matrix element $\langle 0|\chi(x)\chi^*(x')|J_\nu,B_\mu\rangle$ leading from vacuum to a boson state of energy-momentum J_ν and, in case of spin 1, of polarization B_μ (with $J_\nu B^\nu = 0$). Only states with spin $s = 0$ and spin $s = 1$ are considered. PC in the table refers to the intrinsic parity of the boson state. ($z = x - x'$; $\varepsilon_{\mu\nu\kappa\lambda}$ is the completely antisymmetric tensor.)

PC \ s	0	1
$+$	$\sigma^\nu z_\nu$	$\sigma^\nu B_\nu$; $\sigma^\nu z_\nu B^\mu z_\mu$; $\varepsilon_{\mu\nu\kappa\lambda}\sigma^\mu J^\nu z^\kappa B^\lambda$
$-$	$\sigma^\nu J_\nu$	$\sigma^\nu J_\nu z^\mu B_\mu$

Each term in the wave function is multiplied by a function of z^2 and $(J^\nu z_\nu)^2$ which has been omitted in the table. Terms containing the factor $J^\nu z_\nu$ have been omitted too, since they would vanish for $J^\nu z_\nu = 0$, i.e. for equal times of the two fermions in the rest system (or what corresponds to it for $J^2 = 0$) and would not give rise to singularities. In the three cases $s = 0$, PC $= +$ or $-$, and $s = 1$, PC $= -$ only

one term occurs; this term would for $J^2 = 0$ give rise to the mentioned singularity unless it is multiplied by some power of $J^\nu z_\nu$ which would make it vanish at " equal times " i.e. for $J^\nu z_\nu = 0$. Since the norm of the state could be calculated from the plane $J^\nu z_\nu = 0$, it cannot be a state of finite norm.

Therefore only the symmetry $s = 1$, PC $= +$ remains for a boson of rest mass zero, and this is just the symmetry of the real photon. This particle may be pictured as consisting of two fermions of parallel spin being in an s- or d-state with regard to their orbital angular momentum. For a given form of the integral-operator in the eigenvalue equation of the boson the different terms belonging to this symmetry can actually be combined in a way to make the singularity at $J^2 = 0$ vanish.

If one wants to get information about the symmetry of this state under the operations P and C separately, one has to replace the form (1) of the fundamental equation by a form given by Dürr (l.c.), containing the Dirac matrices Γ_μ instead of the σ_μ:

$$\Gamma^\mu\frac{\partial\chi}{\partial x_\mu} \pm \frac{l^2}{2}\left[\Gamma^\mu\chi(\overline{\chi}\Gamma_\mu\chi)+\Gamma^\mu\Gamma_5\chi(\overline{\chi}\Gamma_\mu\Gamma_5\chi)\right] = 0 \qquad (3)$$

Using the same technique as has been applied for the determination of the parity of the π-meson, one finds P $= -$, C $= -$ for the photon. By defining

$$\langle 0|A_\mu(x)|h\nu\rangle = \text{const } \langle 0|\chi^*(x)\sigma_\mu\chi(x)|h\nu\rangle \quad \text{or}$$

$$= \text{const } \langle 0|\overline{\chi}(x)\Gamma_\mu\chi(x)|h\nu\rangle \qquad (4)$$

and

$$F_{\mu\nu} = \frac{\partial A_\mu}{\partial x_\nu} - \frac{\partial A_\nu}{\partial x_\mu}, \qquad (5)$$

and disregarding for a moment the isospin dependence of the $\chi(x)$, which will be discussed below, Maxwell's equations may be derived from the symmetry of the photon state.

2. *a*) PROPERTIES OF THE PHOTON IN ISOSPIN-SPACE

If the field operators, as in (1), are considered as spinors also in isospin space and if the vacuum is assumed to be symmetric in this space, one should expect two different kinds of photons belonging to isospin 0 and 1, like η-meson (which has probably isospin 0) and π-meson.

The real world however is not symmetrical under rotations in isospace, the number of neutrons and protons is very different. This lack of symmetry makes itself felt in the mass difference between proton and neutron and thereby introduces a term proportional to τ_3 in the 2-point functions of the vacuum. The big isospin of the world acts in some way like an outer " magnetic " field in the z-direction of isospace. For the photons the first consequence of this perturbation is, that the photon-eigenfunctions containing τ_1 and τ_2 can no longer contribute to electrodynamics and cannot produce long range forces. Because such a photon would, when producing an interaction between two nucleons, change proton into neutron and vice versa, it should therefore transfer mass from the one nucleon to the other (the mass-difference proton-neutron); the interaction must then decrease exponentially with the distance.

If τ_3 appears in the 2-point function, it must be multiplied there by another operator referring to the vacuum. From the symmetry of the field equation one would first expect an expression $\tau_k \Lambda_k$, where Λ_k marks the direction of the isospin of the world. In the theory of strange particles such products have been discussed [4] with the idea that some finite isospin may be taken from the total isospin of the world and attached to the particles. For these strange particles the complete symmetry of the field equation was preserved.   Now we have to discuss the possibility, that a definite direction in isospace makes itself felt on the particles. Since this direction shall not be changed—the whole world should not take part in any transformation—we can be satisfied with the expression $\tau_3 \Lambda_3$ and thereby fix the z-axis as the direction of the big isospin of the world. We should however not exclude the possibility that the property of the vacuum represented by Λ_3 could belong to those properties which can be attached to the particles, like in the case of the strange particles. This remains to be seen in connection with the theory of strange particles.   For the moment it is sufficient to state which properties are represented by Λ_3 . Since the rotation in isospace has been dropped, only the transformation PG remains, under which $\tau_3 \Lambda_3$ should be invariant. Λ_3 like τ_3 should be invariant under PG and all other transformations of the Lorentz group. Therefore one may put $\Lambda_3^2 = 1, \Lambda_3 = \pm 1$ and introduce the rule, that Λ_3 (like τ_3) changes sign under PG.

The connection of Λ_3 with the properties of the strange particles will be discussed later.

The eigenvalue equation of the photon may now be indicated by

$$f(p, J) = \int K(p, q, J, \tau_3 \Lambda_3) dq f(q, J) , \qquad (6)$$

p and q signifying the relative momentum of the fermions and J the total momentum of the photon. From (6) one sees immediately, that the solution must contain the projection operator $\tfrac{1}{2}(\Lambda_3 \pm \tau_3)$ as factor in order to make $\tau_3 \Lambda_3$ a number. The sign is arbitrary and may, by suitable definition of Λ_3, simply be fixed as $\tfrac{1}{2}(\Lambda_3 + \tau_3)$. It is important to note that the projection operator could not be $\tfrac{1}{2}(1 + \tau_3 \Lambda_3)$. This can be seen from the structure of the integral operator $K(p, q, J; \tau_3 \Lambda_3)$, which contains terms belonging to graphs of type a) or b).

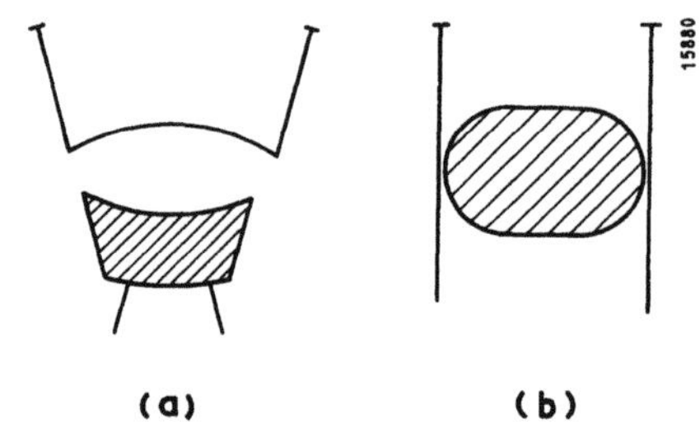

(a)   **(b)**

The lines signify successive matrix multiplication in Dirac spin space, τ_3 and Λ_3 space.   If the projection operator would be $\tfrac{1}{2}(1 + \tau_3 \Lambda_3)$, the unit in the bracket would lead to contributions of the graphs a) which are lacking for $\tau_3 \Lambda_3$; therefore the application of K would not reproduce the factor $\tfrac{1}{2}(1 + \tau_3 \Lambda_3)$. For the operator $\tfrac{1}{2}(\Lambda_3 + \tau_3)$ however only the graphs b) will contribute, on both terms in the same way. Therefore the eigenfunction of the photon can be written in the form

$$f(q, J) = \tfrac{1}{2}(\Lambda_3 + \tau_3) g(q, J) . \qquad (7)$$

If this particle interacts with a nucleon, the vertex part will also contain the projection operator as factor, as can be seen from the corresponding graph. Therefore $\tfrac{1}{2}(\Lambda_3 + \tau_3)$ can simply be interpreted as the charge operator. For a system consisting of several fermions the additivity of charge follows from Maxwell's equations. Therefore the charge may generally be represented by

$$Q = \sum_i \tfrac{1}{2}(\Lambda_3^i + \tau_3^i) , \tag{8}$$

where the sum $\sum_i$ is to be taken over all units carrying isospin. From (8) one sees, that charge appears to be composed of two parts, the isospin $\tfrac{1}{2}\tau_3$ and the " hypercharge " $\tfrac{1}{2}\Lambda_3$ [*]. The latter part is caused by the asymmetry of the world with respect to rotations in isospace; it would not occur in a symmetrical world and is in its structure more closely related to the isotriplet—than to the isosinglet-photon of the symmetrical world. In spite of this relation to the asymmetry of the world, Λ_3 appears in electrodynamics as a property attached to the particles, not to the vacuum; it is only the index 3 which is fixed to the vacuum. Therefore one should expect a close connection of the property Λ_3 with the " spurions " discussed in the theory of strange particles.

2. b) CONNECTION BETWEEN HYPERCHARGE AND STRANGENESS

In an earlier paper on the strange particles the vacuum had been replaced by a system of four degenerate states (" spurion "), which have isospin $\pm\tfrac{1}{2}$ and ± 1 and are invariant under PG. In order to represent this PG invariance the states had been explained as a mixture of " spurion " and " antispurion " (s_α and $\bar{s}_\alpha$), the latter being the PG-transformed spurion. Formally the PG invariance led to the result, that for any state represented by a wave function ϕ_α obeying the eigenvalue equation

$$O_{\alpha\beta}\phi_\beta = 0 , \tag{9}$$

there exists a representation by another wave function

$$\chi_\gamma^* = (i\rho_2\Sigma_1)_{\gamma\delta}\phi_\delta$$

obeying the equation

$$O_{\alpha\beta}\chi_\alpha^* = 0 \tag{10}$$

with the same eigenvalue.

If one wants to interpret these functions ϕ_α and χ_α^* as matrix elements, one may in the case of creation of a Σ-hyperon write:

$$\langle\Sigma|\chi_\lambda^*|\bar{s}_\alpha\rangle = \chi_{\alpha\lambda}^* , \qquad \langle s_\alpha\Sigma|\chi_\lambda^*|0\rangle = \phi_{\alpha\lambda} . \tag{11}$$

The operator χ_λ^* may either create from an antispurion in the vacuum a Σ-hyperon, or from vacuum create a Σ-particle and a spurion together. The PG invariance of the vacuum implies that these two possibilities are equivalent, that the two matrix elements belong necessarily to the same eigenvalue. As for the normal particles the same eigenvalue-equation will also hold for matrix elements referring to the annihilation of $\bar{\Sigma}$:

$$\langle 0|\chi_\lambda^*|\bar{s}_\alpha\bar{\Sigma}\rangle \quad \text{and} \quad \langle s_\alpha|\chi_\lambda^*|\bar{\Sigma}\rangle , \tag{12}$$

only the sign of the energy for these solutions will be reversed.

From these representations one sees at once how the quantum number strangeness can be defined. A state or particle has strangeness s, if the matrix element ϕ representing its creation, varies like $\phi \to e^{i\eta s}\phi$ under the gauge-transformation of the " spurion " s_α:

$$s_\alpha\rangle \to e^{i\eta}s_\alpha\rangle ; \qquad \langle s_\alpha \to e^{-i\eta}\langle s_\alpha \tag{13}$$

$$\bar{s}_\alpha\rangle \to e^{-i\eta}\bar{s}_\alpha\rangle ; \qquad \langle \bar{s}_\alpha \to e^{i\eta}\langle \bar{s}_\alpha . \tag{14}$$

In this way strangeness can be defined in spite of the PG invariance of the vacuum states used for the strange particles. But the PG invariant mixture of s_α and $\bar{s}_\alpha$ must for this purpose be separated into s_α and $\bar{s}_\alpha$. This separation cannot be represented in the space of the index α, on which the operators Σ_1, Σ_2, $\Sigma_3\rho_k$ act in the way described in the earlier paper. It is necessary to distinguish s_α and $\bar{s}_\alpha$, e.g. by a quantum number λ_3 which is $\lambda_3 = +1$ for s_α and $\lambda_3 = -1$ for $\bar{s}_\alpha$. The strangeness is then $s = \sum_i \lambda_3$.

For the mixture this quantum number could be dropped. Since λ_3 has the same transformation properties as Λ_3 in (8), it must be identified with the hypercharge of the spurion and we may simply write $\lambda_3 = \Lambda_3^s$, the index s referring to the

---

[*] The word " hypercharge " here is used in its actual sense, while in current literature frequently not $\tfrac{1}{2}\Lambda_3$, but Λ_3 is called hypercharge.

spurion. The total charge of the spurion is then $\frac{1}{2}(\lambda_3+\rho_3) = \frac{1}{2}(\Lambda_3^s+\rho_3)$ in analogy to (8).

From this point of view one can well understand that in the eigenvalue equation for a system consisting of one baryon and two spurions, both the Ξ-particle and the nucleon appear as solutions (paper by Dürr and Géhéniau, to be published in Nuovo Cimento). The nucleon belongs to combinations of s_α and $\bar{s}_\alpha$, the Ξ or anti-Ξ to combinations s_α, s_α or $\bar{s}_\alpha$, $\bar{s}_\alpha$. Therefore the transformation property represented by the operator Λ_3 can actually be attached to the particles; it may also be represented by a combination of two spurions. The expression $s_\alpha(i\rho_2\Sigma_1)_{\alpha\beta}s_\beta$ is invariant under isospin rotation and space reflexion, but belongs to $\sum_i \Lambda_3^i = +2$, the expression $\bar{s}_\alpha(i\rho_2\Sigma_1)_{\alpha\beta}\bar{s}_\beta$ belongs to $\sum_i \Lambda_3^i = -2$. The hypercharge of the pair of spurions is $+1$, of the pair of antispurions -1. Hence the difference between hypercharge and $\frac{1}{2}$ baryonic number is always an integer.

From the equations

$$Q = \tfrac{1}{2}\sum_i (\Lambda_3^i+\tau_3^i)$$

and   (15)

$$s = \sum_i{}' \Lambda_3^i$$

(where $\sum_i{}'$ only runs over the spurions), the rule of Gell-Mann and Nishijima follows. Charge is the sum of the z-component of isospin and of hypercharge; hypercharge is $\frac{1}{2}$ times the sum of baryonic number and strangeness. The contribution to charge from the lepton has not yet been considered.

From a purely group theoretical point of view the structure of the theory with respect to charge and electrodynamics may finally be described as follows.

The non-linear spinor theory starts from one field operator depending on x and on two (two-valued) spin-indices. The interpretation would seem simpler if the theory started from two field operators each depending on one (two-valued) spin index, one of them also depending on x. In this case the one field operator, depending on x, could belong to a representation of the proper inhomogeneous Lorentz group, including a gauge group; the other operator could quite independently belong to the representation of the isospin group, including another gauge group. In the actual theory however there is only one field-operator carrying both spin indices. This would, if no degeneracy of the vacuum is considered, introduce two connections between the Lorentz group and the isospin group. Any state with half integer angular momentum would also have half integer isospin; integer momentum would belong to integer isospin. And the two gauge groups would be reduced to one, they would become identical. If however the world is unsymmetrical and the vacuum degenerate, the situation is different again. The inhomogeneous part of the Lorentz group allows the decomposition of 4-point, 6-point-functions etc. at large distances into products of complementary transformation properties. The resulting degeneracy of the vacuum with respect to isospin transformations destroys the connections between the Lorentz group and the isospin group. Angular momentum and isospin become quite independent, the two gauge groups are again separated. The big isospin of the world introduces an asymmetry e.g. of the 2-point function which on account of the PG invariance can be expressed only by the product of two operators in isospin space, the one referring to the isospin of the fermions in the direction of the big isospin of the world, the other referring to the isospin gauge property of the fermions. The boson of rest mass zero and spin 1, the photon, therefore contains these two operators in its eigenfunction in the form of a projection operator. Hypercharge refers to the isospin gauge group, baryonic charge to the Lorentz gauge group; for the nucleons hypercharge and baryonic charge are identical (except for the trivial factor 2 from our definition of hypercharge). The contribution of the leptons cannot be discussed at the present stage, since the leptons have not yet been investigated in the framework of this theory.

3. THE NUMERICAL VALUE OF THE COUPLING CONSTANT

A numerical calculation on the value of $e^2/\hbar c$ has not yet been attempted; but the essential steps of such a calculation may be briefly described. The eigenfunction $f(q, J)$ of the photon in (7) can at least in principle be determined from the integral equation (6), which in turn may be derived from a low order Tamm-Dancoff approximation. Which degree of accuracy can be obtained in this way remains to be seen. The most important step is then the determination of the norm of this state which requires the solution of some mathematical problems concerning the structure of the underlying Hilbert space (of indefinite metric !). When the norm has been determined, the matrix element

$$\langle h\nu|\chi^*(x)\tfrac{1}{2}(\Lambda_3+\tau_3)\sigma_\mu\chi(x)|0\rangle \qquad (16)$$

—with $\langle h\nu$ being a normalized state of the photon— can be calculated and compared with the matrix element

$$\langle h\nu|A_\mu(x)|0\rangle \qquad (17)$$

of quantum electrodynamics. This comparison would enable one to rewrite the field equation (1)

$$-i\sigma_\mu\frac{\partial\chi}{\partial x_\mu} \pm l^2\sigma^\mu\chi(\chi^*\sigma_\mu\chi) = 0$$

in the form

$$-i\sigma_\mu\frac{\partial\chi}{\partial x_\mu} \pm \text{const } \sigma^\mu A_\mu\chi + \ldots = 0 \qquad (18)$$

and thereby to determine the coupling constant. This procedure shows that the theory leads to a definite value of $e^2/\hbar c$, but so far no numerical estimate can be given. It should be emphasized that the value does not depend on special assumptions on the ground-state " world ", except the fact that it has a big isospin. Here we meet a similar situation as in the theory of general relativity: the value of the centrifugal force does not depend on special features of the cosmological model except the existence of very big masses at very large distances. In electrodynamics the existence of a mass difference between neutron and proton produces the unsymmetrical projection operator in the photon eigenfunction $\frac{1}{2}(\Lambda_3+\tau_3)$. The resulting difference of charge in proton and neutron in turn causes the mass-difference between proton and neutron. Therefore the assumed asymmetry is self-consistent, the numerical value of $e^2/\hbar c$ depends only on the existence of the asymmetry, not on special properties of the groundstate.

LIST OF REFERENCES

1. H.P. Dürr and W. Heisenberg, to be published in Zs.f.Naturf.
2. H.P. Dürr, Zs.f.Naturf. *16a*. 327 (1961).
3a) W. Heisenberg, F. Kortel and H. Mitter, Zs.f.Naturf. *10a*, 425 (1955).
 b) R. Ascoli u. W. Heisenberg, Zs.f.Naturf. *12a*, 177 (1957).
4. H.P. Dürr and W. Heisenberg, Zs.f.Naturf. *16a*, 726 (1961).

In *Physikertagung Stuttgart: Hauptvorträge der Jahrestagung 1962 des Verbandes Deutscher Physikalischer Gesellschaften*, ed. by E. Brüche, H. Franke (Physik Verlag, Mosbach/Baden 1963) p. 93

8. Elektrodynamik in der nichtlinearen Spinortheorie

(Kurzfassung)

Von W. Heisenberg, München

Die Ableitung der Elektrodynamik aus der nichtlinearen Spinortheorie erfordert die Lösung dreier ziemlich unabhängiger Probleme. 1. Es ist zu zeigen, daß es Bosonenzustände der Ruhmasse 0 als Eigenlösungen gibt, die die Symmetrieeigenschaften der Photonen in bezug auf die *Lorentz*gruppe aufweisen. Dabei ergeben sich die *Maxwell*schen Gleichungen. 2. Der Grundzustand der Welt muß in bezug auf die Isospingruppe unsymmetrisch sein. Aus dieser Unsymmetrie muß die bekannte Unsymmetrie der Elektrodynamik hergeleitet werden. Dabei ergibt sich für die Ladung der Teilchen die Regel von *Gell-Mann* und *Nishijima*. 3. Die Kopplungskonstante, d.h. die *Sommerfeld*sche Feinstrukturkonstante, kann aus der grundlegenden Feldgleichung ermittelt werden, wenn es gelingt, die Normierung des Photonenzustandes durchzuführen. Das mathematische Problem der Normbestimmung ist bisher noch nicht befriedigend gelöst.

Die Entwicklung der einheitlichen Feldtheorie der Elementarteilchen

Von W. Heisenberg, München

Der in den letzten zehn Jahren ausgearbeitete Entwurf einer einheitlichen Feldtheorie der Elementarteilchen ([1] bis [5]) geht von folgenden allgemeinen Vorstellungen aus:

Die Elementarteilchen sind nicht, wie man früher etwa angenommen hätte, unveränderliche, unteilbare Grundbausteine der Materie. Sie können vielmehr ineinander umgewandelt werden; bei einem energiereichen Stoß zwischen irgendwelchen Elementarteilchen können Teilchen jeder anderen Art entstehen. Daher gibt es auch keinen grundsätzlichen Unterschied zwischen „elementaren" und zusammengesetzten Teilchen. Alle Teilchen sind gewissermaßen nur Formen einer Grundsubstanz, die man Materie oder Energie nennen kann. Die Energie wird zur Materie, indem sie sich in die Form eines Elementarteilchens begibt. Diese Formen müssen durch ein einheitliches Naturgesetz bestimmt sein, sich aus ihm herleiten lassen. Die Elementarteilchen sind also stationäre Zustände eines physikalischen Systems „Materie" in ähnlicher Weise, wie etwa die vielen verschiedenen Energiezustände des Eisenatoms stationäre Zustände eben des Systems „Eisenatom" sind. In der Quantenmechanik der Atomhüllen kann der gesetzmäßige Zusammenhang nach Jordan, Klein und Wigner durch eine einfache nichtlineare Feldgleichung mathematisch dargestellt werden. Dadurch wird die Frage nahegelegt, ob etwa auch das Naturgesetz, das die Elementarteilchen bestimmt, mathematisch in der Form einer Feldgleichung ausgedrückt werden kann. Grundsätzlich könnte man sich auch allgemeinere, aber dann wohl erheblich abstraktere Formulierungen für ein solches Gesetz vorstellen. Zu konkreten mathematischen Aussagen, die durch den Vergleich mit dem Experiment auf ihre Richtigkeit geprüft werden können, kommt man aber wohl am ehesten, wenn man — sozusagen versuchsweise — eine bestimmte Feldgleichung postuliert, die sich als natürlicher Ausdruck für die beobachteten Gesetzmäßigkeiten anbietet, und ihre Folgerungen untersucht. Von dieser Überlegung ausgehend ist die Feldgleichung

$$\gamma_\mu \frac{\partial \psi}{\partial x_\mu} \pm l^2 \gamma_\mu \gamma_5 \psi \left(\bar{\psi} \gamma_\mu \gamma_5 \psi \right) = 0 \qquad (1)$$

vorgeschlagen worden. $\psi = \psi(x)$ ist hierbei der Feldoperator der Materie (ein sogenannter Dirac-Spinor), x steht für die Raum- und Zeitkoordinaten, die γ_μ sind die bekannten Diracschen Matrizen. Die verschiedenen Symmetrieeigenschaften der Elementarteilchen sollen aus den Invarianzeigenschaften der Feldgleichung (1) entspringen. Für die Wahl dieser speziellen Gleichung lassen sich als empirische Gründe drei Erfahrungskomplexe anführen: Die bei den Vorgängen an Elementarteilchen beobachteten Erhaltungssätze scheinen durch die Invarianzeigenschaften eben dieser Gleichung richtig dargestellt zu werden. Die Existenz der Wechselwirkung wird durch den nichtlinearen Charakter der Gleichung beschrieben. Die aus der relativistischen Struktur von Raum und Zeit folgende Form der Kausalität (Fortpflanzung von Wirkungen nur mit Lichtgeschwindigkeit) wird durch den Charakter der Gleichung als Differentialgleichung gewährleistet. Jeden-

falls dürfte die Feldgleichung (1) die einfachste Gleichung sein, die die genannten Forderungen befriedigt.

1. Die mathematische Axiomatik der Quantenfeldtheorie

In Anbetracht der mathematischen Schwierigkeiten der Quantenfeldtheorie, die in den vergangenen Jahrzehnten gründlich studiert worden sind, ist es von vornherein keineswegs sicher, daß eine Feldgleichung von der Art (1), zusammen mit der Forderung der Antikommutativität der Feldoperatoren bei raumartigen Abständen, ein sinnvolles mathematisches Problem definiert. Man kann aber auf Grund der Diskussionen der letzten Jahre die hier zugrunde gelegte Axiomatik gegen andere Axiomatisierungsversuche der Theorie der Elementarteilchen klar abgrenzen. Es lassen sich etwa drei Stufen der Axiomatik (I, II, III) unterscheiden, von denen jede folgende die axiomatischen Voraussetzungen der vorhergehenden durch Zusatzforderungen weiter einengt.

I. Das Verhalten der Elementarteilchen in weitem gegenseitigem Abstand (ihre Wirkungsquerschnitte bei Stößen usw.) soll durch eine unitäre S-Matrix beschrieben werden können, die die empirisch bekannten Gruppen (Lorentzgruppe, Isospingruppe usw.) darstellt, und die soweit analytisch ist, wie die oben genannte Forderung der relativistischen Kausalität dies verlangt.

II. Es wird die weitere Annahme hinzugefügt, daß ein lokaler Feldoperator $\psi(x)$ existiert, der bei raumartigen Abständen (anti)kommutiert und in einem geeigneten Hilbert-Raum Transformationen bewirkt. Wegen der Kausalitätsforderung muß $\psi(x)$ dann einer Differentialgleichung (oder einer Integrodifferentialgleichung mit beliebig kleinem Zeitintegrationsintervall Δt) genügen.

III. Zu I und II wird noch die Forderung hinzugefügt, daß die Metrik im Hilbert-Raum positiv definit sein soll; ferner sollen die asymptotischen Zustände (d. h. die Zustände der freien Teilchen) genügen, den ganzen Hilbert-Raum aufzuspannen.

Daß die Forderung I mathematisch erfüllbar sei, wird heutzutage wohl allgemein angenommen. Chew [6] hat neuerdings sogar den Standpunkt vertreten, daß man auch nicht wesentlich mehr brauche, um eine Theorie der Elementarteilchen zu definieren, daß die Annahmen II und III also vielleicht unerfüllbar, jedenfalls aber unnötig seien. Immerhin ist der axiomatische Rahmen I sehr weit, und man wird kaum umhin können, ihn durch Zusatzforderungen einzuengen (z. B. die der „maximalen Analytizität"), bevor man zu eindeutigen mathematischen Ergebnissen kommt. Dann aber entsteht natürlich das Problem, ob die Zusatzforderungen erfüllt werden können.

Die Axiomatik III ist am engsten und kommt der bisherigen Quantenmechanik am nächsten; sie ist von Wightman [7] und anderen präzis formuliert und zur Ableitung vieler wichtiger Resultate z. B. über Dispersionsrelationen benützt worden. Allerdings ist es bisher nicht möglich gewesen, ein mathematisches Beispiel einer lorentzinvarianten Theorie mit kausaler Wechselwirkung anzugeben, das dieser Axiomatik genügt.

Die mit der Feldgleichung (1) beginnende Theorie geht von der Vermutung aus, daß die Axiome III nicht erfüllt werden können, da die definite Metrik im Hilbert-Raum eine mindestens δ-funktionsartige Singularität der sogenannten Propagatoren (Vakuumerwartungswerte des Produktes zweier Feldoperatoren) auf dem Lichtkegel erzwingen würde, die mit der kausalen Wechselwirkung wahrscheinlich nicht verträglich ist. Jedoch wird angenommen, daß die Axiome II erfüllbar sind und daß die vorausgesetzte Wechselwirkung für eine geringere Singularität auf dem Lichtkegel sorgt, die dann eine indefinite Metrik im Hilbert-Raum notwendig macht. Einen zwingenden Beweis für die Unerfüllbarkeit von III und die Erfüllbarkeit von II gibt es bisher nicht. Die Theorie steht also einstweilen mathematisch noch auf schwankendem Boden.

Auch wenn man die Axiomatik II als erfüllbar voraussetzt, genügt die Feldgleichung (1) noch nicht zur eindeutigen Bestimmung der stationären Zustände. Man muß sie durch Randbedingungen und durch eine Aussage über den Grundzustand („Vakuum" oder „Welt") ergänzen. Zu den beiden aus der Quantenmechanik bekannten Randbedingungen bei räumlich unendlichem oder verschwindendem Abstand zweier Teilchen kommen in der Quantenfeldtheorie noch die Randbedingungen bei unendlich hoher Teilchenzahl, die bisher nur im Rahmen der sogenannten Tamm-Dancoff-Methode formuliert worden sind. Wieviel Spielraum die Theorie an dieser Stelle enthält, ist bisher noch nicht klargestellt worden. Eine Untersuchung von Dürr [4] über die Definition der Raumspiegelungsparität scheint zu zeigen, daß dieser „Rand" (bei unendlich hoher Teilchenzahl) auch für die Symmetrieeigenschaften der Lösungen von Bedeutung ist. Noch mehr Spielraum aber dürfte die Theorie bei der Wahl des Grundzustandes besitzen, dessen Definition eine Aussage über die Welt im Großen, über das kosmologische Modell der Welt enthält.

2. Die Entartung des Grundzustandes und die „seltsamen" Teilchen

Die Feldgleichung (1) ist invariant gegenüber Drehungen im Isospinraum, wenn man diese mit den Pauli-Gürseyschen Transformationen identifiziert. In der wirklichen Welt gelten aber die Erhaltungssätze für den Isospin nur näherungsweise, sie versagen in der Elektrodynamik. Daher kann die auf (1) beruhende Theorie die wirklichen Verhältnisse nur dann richtig darstellen, wenn man annimmt, daß der Grundzustand nicht die volle Symmetrie bei Drehungen im Isospinraum besitzt, daß er also im quantenmechanischen Sinne entartet ist. Empirisch läßt sich diese Annahme damit begründen, daß ja auch die Anzahl der Neutronen in der Welt von der der Protonen wesentlich verschieden ist, daß also die Welt im Ganzen einen sehr großen Isospin zu besitzen scheint. Diese Annahme von der Entartung des Grundzustandes führt zu einer Reihe interessanter Schlußfolgerungen, deren genauere Analyse in den letzten Jahren vor allem durch den Vergleich mit analogen Verhältnissen in der Theorie der Supraleitung und des Ferromagnetismus möglich geworden ist.

Zunächst gestattet der große Isospin des Grundzustandes das prinzipielle Verständnis der „seltsamen" Teilchen (strange particles [5]). Da der Feldoperator ein Spinor sowohl im Raum des Drehimpulses als

auch im Isospinraum ist, können durch Anwendung eines Produktes solcher Operatoren auf ein symmetrisches, nichtentartetes Vakuum nur Teilchen entstehen, die bei halbzahligem Spin auch halbzahligen Isospin, oder bei ganzzahligem Spin auch ganzzahligen Isospin besitzen, wie z.B. Proton oder π-Meson. Seltsame Teilchen wie Λ- oder Σ-Hyperonen oder K-Mesonen können so nicht beschrieben werden. Bei Anwendung der Feldoperatoren auf ein unsymmetrisches Vakuum aber kann vom Isospin des Vakuums ein Teil abgezweigt und an die durch die Operatoren erzeugten Partikeln angehängt werden. Diesen abzweigbaren Teil kann man im Anschluß an frühere Überlegungen von Wentzel [8], d'Espagnat und Prentki [9] u. a. als „Spurion" bezeichnen und damit eine Art von Gebilden andeuten (das Wort „Teilchen" wäre hier unberechtigt), die keine Lorentzeigenschaften — also weder Energie oder Impuls noch Ort und Zeit — besitzen, wohl aber über Isospin verfügen. Die Invarianz der Ausgangsgleichung (1) gegenüber Zeitspiegelung und die Erfahrung, daß diese Invarianz selbst bei den schwachen Wechselwirkungen nicht durchbrochen wird, können dann dazu benützt werden, die Eigenschaften der Spurionen bei den Transformationen der übrigen diskreten Gruppen festzulegen. Dabei hat sich herausgestellt, daß die Spurionen noch eine Raumspiegelungseigenschaft, d.h. Parität besitzen müssen [5] und daß man Spurionen und Antispurionen unterscheiden und damit eine Quantenzahl „Seltsamkeit" (strangeness) definieren kann [13], [14]. Die Frage, ob diese Annahmen über den Grundzustand zwangsläufig aus der Feldgleichung (1) folgen, wird man allerdings wohl verneinen sollen. Denn man kann zwar plausibel machen, daß die Annahmen konsistent sind, daß sie keine inneren Widersprüche enthalten. Aber diese Konsistenz wäre wohl auch bei anderen Annahmen über den Grundzustand zu erreichen, d.h., die Grundgleichung läßt wahrscheinlich noch verschiedene kosmologische Modelle der Welt zu. Wenn dies zutrifft, kann die Richtigkeit der über den Grundzustand der Welt gemachten Annahmen nur durch den Vergleich mit der Erfahrung (etwa hinsichtlich der Eigenschaften der seltsamen Teilchen oder der Elektrodynamik) nachgeprüft werden.

Bisher sind Rechnungen über stationäre Zustände und Abschätzungen der zugehörigen Massenwerte aus Gründen der Einfachheit nur für Teilchen vom Drehimpuls 0 oder $\frac{1}{2}$ durchgeführt worden. Dabei wurde die sogenannte neue Tamm-Dancoff-Methode benützt; andere Methoden stehen in einer derartigen nichtrenormierbaren Quantenfeldtheorie bisher nicht zur Verfügung. Da praktisch nur die niedrigste Näherung gerechnet werden kann, kommt man über eine recht ungenaue Abschätzung der Eigenwerte nicht hinaus. Der Vergleich mit Tamm-Dancoff-Rechnungen am anharmonischen Oszillator, bei dem man die exakten Eigenwerte kennt, läßt bestenfalls eine Ungenauigkeit von etwa $\pm 15\%$ erwarten [1].

Für die Behandlung der seltsamen Teilchen war es dabei eine wesentliche Hilfe, daß, wie Nambu [10] hervorgehoben hat, die mathematischen Probleme hier denen der Bardeen-Bogoljubowschen [11] Theorie der Supraleitung sehr ähnlich sind. In der Theorie der Supraleitung können die grundsätzlichen Schwierigkeiten der Quantenfeldtheorie keine Rolle spielen; die Entartung des Grundzustandes und die sich aus

Heft 1
1963 (Jg. 50)   W. Heisenberg: Die Entwicklung der einheitlichen Feldtheorie der Elementarteilchen   5

ihr ergebenden Folgerungen können also hier ohne die Problematik der relativistischen Feldtheorie studiert werden. Yamazaki [12] hat festgestellt, daß auch die neue Tamm-Dancoff-Methode bei der Anwendung auf das Problem der Supraleitung die richtigen Resultate liefert.

Für die Teilchen mit der Seltsamkeit (strangeness) 0 oder 1 sind die Abschätzungen mit dieser Methode vorgenommen worden; für Teilchen der Seltsamkeit 2 haben Dürr und Géhéniau [13] mit Hilfe gruppentheoretischer Überlegungen Eigenwertgleichungen abgeleitet, in denen unbestimmte Konstanten vorkommen, die bisher nur aus der Erfahrung bestimmt wurden. Im ganzen hat man mit solchen Methoden ein theoretisches Massenspektrum für Teilchen vom Drehimpuls 0 und $\frac{1}{2}$ erhalten, das gut zu dem empirischen paßt. Über den Vergleich mit der Erfahrung wird aber weiter unten noch ausführlicher zu berichten sein.

3. Elektrodynamik [14]

Die Existenz von Elementarteilchen der Ruhmasse Null hängt eng mit der auf Grund der kausalen Wechselwirkung angenommenen indefiniten Metrik im Hilbert-Raum zusammen. Die Beseitigung der δ-funktionsartigen Singularitäten auf dem Lichtkegel der „Propagatoren" wird bei der Entwicklung dieser Propagatoren nach einem Massenspektrum zum mindesten formal dadurch bewerkstelligt, daß „Geisterzustände" (speziell: „Dipolgeister") zur Ruhmasse Null auftreten, die in der niedrigsten Näherung auch die Norm Null besitzen [3]. Solche Geisterzustände verschwindender Ruhmasse und Norm gibt es sowohl bei den Fermionen als auch bei den Bosonen. Vermöge der Wechselwirkung können diese Zustände teilweise — sozusagen nachträglich — wieder eine endliche Norm erhalten und damit zu echten Teilchen der Ruhmasse Null werden. Bei den Bosonen kann, wie die genauere Untersuchung zeigt, eine endliche Norm zwar nicht für Spin 0, wohl aber bei Zuständen vom Spin 1 auftreten, die die Symmetrieeigenschaften der empirischen Photonen aufweisen. Insofern folgen die Existenz und die Eigenschaften der Photonen aus der Feldgleichung (1), und die Maxwellschen Gleichungen für das elektromagnetische Feld sind wiederum eine Konsequenz des Verhaltens der Photonen bei den Transformationen der Lorentzgruppe.

Für die Eigenschaften der Photonen bei den Transformationen der Isospingruppe aber wird die Entartung des Vakuums entscheidend wichtig. Man würde zunächst Photonen vom Isospin 0 und 1 erwarten, ähnlich wie es bei den Bosonen endlicher Masse die η- und π-Mesonen bzw. die ω- und ϱ-Mesonen gibt. Der große Isospin der Welt aber wirkt wie ein „äußeres Feld" im Isospinraum und verursacht einen Massenunterschied zwischen Proton und Neutron; d.h., die Masse des Nukleons wird etwas verschieden, je nachdem sein Isospin dem der Welt parallel oder antiparallel gerichtet ist. Das hat zur Folge, daß es Photonen der Ruhmasse 0 mit einem Isospin parallel oder antiparallel zu dem der Welt nicht geben kann. Denn solche Photonen müßten, wenn sie etwa eine Kraft langer Reichweite zwischen Neutronen und Protonen bewirken sollten, auf der einen Seite das Neutron in ein Proton, auf der anderen das Proton in ein Neutron verwandeln. Dabei müßte auch Masse übertragen werden (nämlich die Massendifferenz zwi-

schen Proton und Neutron), und das kann wegen der Unbestimmtheitsrelationen nur bei Kräften endlicher Reichweite vorkommen. Es bleiben dann zunächst noch zwei mögliche Photonenzustände übrig, deren Isospinkomponente in Richtung des Gesamtisospins der Welt verschwindet. Die durch die Massendifferenz Proton-Neutron ausgedrückte unsymmetrische Wirkung des Gesamtisospins erzeugt Übergänge zwischen diesen beiden Zuständen, so daß nur Summe oder Differenz als mögliche stationäre Zustände vorkommen können. Nur einer von ihnen (z. B. die Summe) kann zur Ruhemasse Null gehören und Kräfte langer Reichweite hervorrufen. Daß in einer unsymmetrischen Welt die Zustände der anderen Art dann keine Kräfte langer Reichweite erzeugen können, läßt sich auch in folgender Weise anschaulich einsehen. Die Isospin-Unsymmetrie der Welt im Großen kann nur dadurch zustande kommen, daß es neutrale Teilchen gibt, die dann im Überschuß (gegenüber sich in der Ladung kompensierenden geladenen Teilchen) vorhanden sind. Ihre Ladung kann zwar für die Photonen der einen Sorte verschwinden, nicht aber zugleich auch für die der anderen. In bezug auf diese anderen Photonen — wenn es sie als Teilchen der Ruhmasse 0 gäbe — würde dann die Welt im Ganzen geladen sein, eine mittlere Ladungsdichte besitzen, was mit der Translationsinvarianz nicht verträglich wäre.

Die genauere Untersuchung des eben geschilderten Sachverhalts hat gezeigt, daß die Eigenfunktion des Photons einen Projektionsoperator im Isospinraum als Faktor enthält, dessen einer Summand der Isospin in Richtung des Gesamtisospins der Welt, dessen anderer Summand eine Quantenzahl ist, die mit der Eichgruppe im Isospinraum verknüpft ist. Da dieser Projektionsoperator auch die Wechselwirkung des Photons mit anderen Teilchen bestimmt, kann man ihn als Ladungsoperator bezeichnen; die Ladung erscheint als Summe von Isospin und „Zusatzladung" (hypercharge). Die letztere ist durch die Eichgruppe im Isospinraum gegeben. Dadurch wird die bekannte von Gell-Mann und Nishijima gefundene empirische Regel für die Ladung verständlich. Durch die Existenz der Spurionen wird die Eichgruppe im Isospinraum von der mit den Lorentztransformationen verknüpften Eichgruppe unabhängig. Die Zusatzladung setzt sich additiv zusammen aus Baryonenzahl (Eichtransformation am Feldoperator) und Seltsamkeit (Eichtransformation an den Spurionen, die keine Lorentzeigenschaften besitzen).

Der Beitrag der Leptonen zur Ladung konnte bisher nicht diskutiert werden, da die Theorie der Leptonen im Rahmen der einheitlichen Feldtheorie noch nicht ausgearbeitet worden ist. Eine Berechnung der Kopplungskonstante $e^2/\hbar c$, die etwa durch eine Normierung des Photonzustandes mit Hilfe der Tamm-Dancoff-Methode vorgenommen werden könnte, ist bisher noch nicht ausgeführt worden.

4. Schwache Wechselwirkungen und Leptonen

Für die Deutung der schwachen Wechselwirkungen im Rahmen der einheitlichen Feldtheorie existiert bisher nicht viel mehr als ein allgemeines Programm [3]. Es wird erwartet, daß von den Geisterzuständen der Fermionen einige vermöge der Wechselwirkung eine von Null verschiedene Norm, einige — nämlich solche mit elektrischer Ladung — sogar eine von Null verschiedene Masse erhalten. Ihr Anteil am Feldoperator bleibt aber klein, was zur Folge hat, daß es sich um „schwache" Wechselwirkung handelt. Mit dieser Vorstellung wird die experimentell beobachtete Symmetrie der β-Wechselwirkung verständlich, da die β-Wechselwirkung dann einfach ein Abbild der in der Feldgleichung (1) ausgedrückten starken Wechselwirkung wird, die in der Tat dem beim β-Zerfall beobachteten $(A\text{-}V)$-Typus entspricht. Das Versagen der Erhaltungssätze für Isospin, Parität und Seltsamkeit bei den schwachen Wechselwirkungen ist damit aber noch nicht erklärt. Man wird es wohl wieder mit Symmetrieabweichungen im Grundzustand in Verbindung bringen müssen.

5. Vergleich mit der Erfahrung

Die großen Teilchenbeschleuniger in Berkeley, Dubna, Genf, Brookhaven und anderen Orten haben in jüngster Zeit ein so umfangreiches experimentelles Material über Elementarteilchen zu Tage gefördert, daß ein Vergleich mit dem Entwurf der einheitlichen Feldtheorie nun an vielen Stellen möglich geworden ist. Während noch vor einigen Jahren nur die wenigen metastabilen Zustände der Materie bekannt waren, die wegen ihrer relativ langen Lebensdauer sichtbare Teilchenspuren in photographischen Platten oder Nebel- oder Blasenkammern hinterlassen, kennt man heute schon einen Teil der sicher viel zahlreicheren stationären Zustände mit kurzen Lebensdauern (bis herunter zu etwa 10^{-23} sec), die als Analogon der normalen stationären Zustände etwa in den Atomspektren zu betrachten sind. Die Fülle dieser Zustände, die nach bestimmten, empirisch schon recht genau bekannten Gesetzen ineinander umgewandelt werden können, zeigt ganz unmittelbar, daß es sich hier um ein zwar kompliziertes, aber aus einem einheitlichen Naturgesetz entspringendes Massenspektrum handeln muß. Das allgemeine Bild, von dem die einheitliche Feldtheorie ausgegangen ist, paßt also gut zu den inzwischen neu gewonnenen Erfahrungen.

Die Übereinstimmung betrifft aber auch schon speziellere Züge dieser Theorie. Wenn wie in (1) das fundamentale Materiefeld als Spinorfeld angenommen wird, so sind die Bosonen der Baryonenzahl 0 in erster Näherung als zusammengesetzt aus Nukleon und Antinukleon aufzufassen, wie Fermi und Yang [15] schon vor längerer Zeit vorgeschlagen haben. Tatsächlich können die vier einfachsten und am besten bekannten Mesonen der Seltsamkeit Null, die π-, η-, ϱ- und ω-Mesonen nach ihren Symmetrieeigenschaften gerade als S-Zustände (d.h. Zustände ohne Bahndrehimpuls) eines Nukleon-Antinukleonsystems gedeutet werden.

Ferner ist neuerdings die Streuung sehr energiereicher Nukleonen an anderen Nukleonen untersucht worden. Das empirische Verhalten der S-Matrixelemente wurde verglichen mit jenem analytischen Verhalten, das sich ergibt, wenn man die den Nukleonen und anderen Resonanzzuständen entsprechenden Singularitäten als sogenannte Regge-Pole [16] auffaßt. Regge-Pole treten dann auf, wenn es sich um gebundene Zustände aus anderen Teilchen, nicht um eigentliche Elementarteilchen im ursprünglichen Sinne handelt. Die befriedigende Übereinstimmung mit dieser Vorstellung der Regge-Pole scheint zu zeigen, daß die Nukleonen sich jedenfalls in dieser Beziehung wie gebundene Zustände verhalten, daß sie keinen harten Kern „nackter" Nukleonen besitzen. Es gehört gerade zu den Grundannahmen der einheitlichen Feldtheorie, daß alle Elementarteilchen gewissermaßen zusammengesetzte Gebilde sind, daß es „nackte" Teilchen schon deshalb nicht gibt, weil bei den Propagatoren keine δ-funktionsartigen Singularitäten auf dem Lichtkegel auftreten.

Diese allgemeine Übereinstimmung beweist aber natürlich noch nicht, daß die Feldgleichung (1), wenn man sie im Sinne der Axiomatik II interpretiert, schon eine brauchbare Grundlage für die einheitliche Feldtheorie abgibt. Um diese spezielle mathematische Form (1) nachzuprüfen, wird man die numerischen Ergebnisse für Masseneigenwerte und Kopplungskonstanten, die etwa mit Hilfe des Tamm-Dancoff-Verfahrens gewonnen worden sind, mit den empirischen Daten vergleichen müssen. Dabei macht die Ungenauigkeit der Tamm-Dancoff-Methode, bei der man schon im günstigsten Fall Fehler von $\pm 15\%$ erwarten muß [1], den Vergleich reichlich unsicher. Die Rechnungen, die unter Einbeziehung der seltsamen Teilchen von der Seltsamkeit (strangeness) 1 für Zustände vom Drehimpuls 0 und $\tfrac{1}{2}$ ausgeführt worden sind, ergaben qualitativ für Nukleon, π-Meson, Λ-, Σ-Hyperon und K-Meson das richtige Massenspektrum. Für das Massenverhältnis zwischen Proton und π-Meson z. B. lieferte die Rechnung ungefähr den Wert 5, während der empirische Wert 6,7 beträgt. Beim K-Meson liegt die Abweichung in der entgegengesetzten Richtung, bleibt aber auch im Rahmen der für Tamm-Dancoff-Abschätzungen zu erwartenden Ungenauigkeit.

Ein erheblich höheres Gewicht für den Vergleich zwischen Theorie und Experiment dürften die Voraussagen haben, die in diesem Zusammenhang für die Existenz von unstabilen Resonanzzuständen gemacht worden sind, über die zu jener Zeit experimentell noch nichts bekannt war. Zu solchen Voraussagen sei zunächst allgemein festgestellt, daß es bei einer gegebenen Symmetrie, die durch Angabe aller Quantenzahlen (einschließlich Isospin, Parität, G-Parität usw.) festgelegt werden kann, jedenfalls ein Kontinuum von Zuständen geben muß, dessen untere Grenze aus den empirischen Massen der bekannten Elementarteilchen leicht berechnet werden kann. Daneben wird es im allgemeinen höchstens einen relativ langlebigen stationären Zustand zu dieser Symmetrie geben, der dann, wenn er unterhalb des Kontinuums liegt, ein echter diskreter Zustand, sonst nur ein Resonanzzustand größerer oder geringerer Schärfe ist. Die Existenz mehrerer solcher diskreter Zustände ist zwar grundsätzlich möglich, aber wegen der kurzen Reichweite der starken Wechselwirkung nicht wahrscheinlich. Empirisch scheint im Bereich der starken Wechselwirkungen bisher kein Fall bekannt zu sein, bei dem es mehr als einen stationären Zustand gleicher Symmetrie gäbe. Eine theoretische Voraussage bezieht sich also bei gegebener Symmetrie im allgemeinen auf das Vorhandensein und die Lage dieses einen diskreten stationären Zustands.

Die Rechnungen hatten fünf solche Resonanzzustände verschiedener Symmetrien ergeben [5]: ein Boson der Seltsamkeit 0 und je zwei Fermionen und zwei Bosonen der Seltsamkeit 1. Das Boson der Seltsamkeit 0, für das auch Spin und Isospin den Wert 0 haben sollen, ist inzwischen experimentell gefunden [17] und als η-Meson bezeichnet, die Identität in allen Quantenzahlen nachgewiesen worden. Die theoretisch berechnete Masse war etwa 550 bis 700 MeV[1]), die experimentelle liegt bei 550 MeV; der Spielraum des theoretischen Werts entspricht der zu erwartenden Ungenauigkeit einer Tamm-Dancoff-Abschätzung.

Bei den Fermionen handelt es sich um zwei Hyperonen mit der Baryonenzahl 1 und dem Isospin 1 und 0. Das Hyperon mit dem Isospin 1 unterscheidet sich vom Σ-Teilchen durch die Parität, seine Masse sollte nach der Rechnung etwa bei 1250 MeV liegen. Für das andere Hyperon vom Isospin 0, dessen Parität der des Λ-Teilchens entgegengesetzt ist, ergibt die Theorie eine Masse von etwa 1300 MeV. Das erstere kann möglicherweise mit dem experimentell beobachteten Zustand Y_1^* (1385 MeV), das letztere mit Y_0^* (1405 MeV) identifiziert werden; jedoch ist der Drehimpuls dieser Zustände noch umstritten, die Messungen schwanken zwischen den Werten $\frac{1}{2}$ und $\frac{3}{2}$. Die Identifikation ist nur möglich, wenn der Drehimpuls $\frac{1}{2}$ ist.

Dann ergab die Rechnung ein angeregtes K-Meson vom Isospin $\frac{1}{2}$, Drehimpuls 0, Seltsamkeit 1 mit einer Masse von etwa 650 bis 850 MeV[1]), das mit beiden Paritäten vorkommen sollte. Das Auftreten eines Paritätsdubletts an dieser Stelle ist besonders charakteristisch für die „Spurion"-Theorie der seltsamen Teilchen. Denn es beruht darauf, daß dieser Zustand auf Grund seiner Symmetrieeigenschaften durch Anlagerung eines Spurions an ein η-Meson entstehen kann. Die Parität tritt ganz allgemein nur in der Wechselwirkung des Spurionisospins mit einem Teilchenisospin auf. Da das η-Meson keinen Isospin besitzt, kann die Wechselwirkung zwischen η-Meson und Spurion nicht mehr von der Parität abhängen; d.h., für beide Paritäten muß sich wenigstens in dieser Näherung die gleiche Masse ergeben. Ob in höheren Näherungen eine (dann wahrscheinlich geringfügige) Aufspaltung eintritt, ist noch nicht untersucht worden.

Außerdem ist bemerkenswert, daß dieser Zustand ähnlich wie das η-Meson nur durch elektromagnetische (oder schwache) Wechselwirkungen zerfallen sollte. Empirisch ist ein Resonanzzustand K^* bei 885 MeV bekannt [18], dessen Drehimpuls noch nicht sicher festgestellt ist ([19] und [20]). Die Messungen schwanken zwischen den Spinwerten 0 und 1. Ferner gibt es Andeutungen eines weiteren Resonanzzustandes [21] vom Typus K^* bei 730 MeV. Ob das von der Theorie geforderte angeregte K-Meson mit einem dieser beiden Zustände identifiziert werden kann, wird durch eine genauere Analyse der Resonanzen und ihrer Zerfallsprozesse entschieden werden.

Schließlich ergab die Theorie einen Bosonenzustand mit der Seltsamkeit 1 und dem Isospin $\frac{3}{2}$, für den empirisch bisher keine Anhaltspunkte gefunden sind; seine Masse sollte etwas größer als die des K-Mesons sein. Wenn seine wirkliche Masse größer als die von $K+\pi$ ist, so würde die Tatsache, daß der Zerfall dieses Teilchens in $K+\pi$ ein erlaubter Übergang wäre, vielleicht eine hinreichende Erklärung dafür bieten, daß die Resonanz so verwaschen ist, daß sie nicht mehr beobachtet wird.

Die theoretischen Rechnungen haben sich bisher im wesentlichen auf Zustände vom Spin 0 oder $\frac{1}{2}$, Bahndrehimpuls 0, und Seltsamkeit 0 oder 1 beschränkt. Im Bereich dieser Quantenzahlen scheinen experi-

mentell einstweilen auch keine anderen Zustände als die aus der Theorie folgenden bekannt zu sein.

Für den Vergleich von Theorie und Experiment sind auch noch neuere Abschätzungen der Kopplungskonstante des π-Mesons durch Dhar [22] zu erwähnen, in denen die Dispersionsrelationen mit Tamm-Dancoff-Rechnungen verbunden werden; diese Abschätzungen ergeben empirisch befriedigende Resultate.

Ferner sei hervorgehoben, daß die relative Parität von Λ- und Σ-Teilchen nach der Theorie ungerade sein muß; experimentell ist dieser Wert noch nicht sicher bekannt [23].

Im ganzen kann man hoffen, daß die rasche Erweiterung des experimentellen Materials, die in den kommenden Jahren zu erwarten ist, bald eine sehr viel zuverlässigere Beurteilung der Brauchbarkeit der in der Entwicklung begriffenen einheitlichen Feldtheorie der Elementarteilchen gestatten wird.

Anmerkung bei der Korrektur (19. 12. 1962): Die neuesten dem Verfasser bekanntgewordenen experimentellen Ergebnisse machen das folgende Bild wahrscheinlich: Der beobachtete Resonanzzustand Y_0^* (1405 MeV) kann mit dem angeregten Λ-Zustand der Theorie bei 1300 MeV identifiziert werden. Dafür spricht insbesondere die Interferenz mit dem $S_{1/2}$-Zustand des Systems $K+N$ (Ber. v. d. Rochester-Konf., S. 322). Der Zustand Y_1^* (1385 MeV) scheint dagegen ein unaufgelöstes Dublett aus den Zuständen $P_{1/2}$ und $P_{3/2}$ des $\Lambda\pi$-Systems zu sein; denn die hohe beobachtete Polarisation des Λ bei dem Zerfall des Y_1^* (Ber. v. d. Rochester-Konf., S. 318) deutet auf eine sehr geringe Spin-Bahn-Kopplung hin. Die $P_{1/2}$-Komponente dieses Dubletts kann mit dem angeregten Σ-Zustand der Theorie identifiziert werden. Es sieht danach so aus, als seien wenigstens drei der aus der Theorie folgenden Zustände experimentell an der erwarteten Stelle im Massenspektrum gefunden worden.

Literatur

[1] Heisenberg, W.: Nachr. d. Gött. Akad. d. Wiss. 1953, S. 111. — [2] Heisenberg, W., F. Kortel u. H. Mitter: Z. Naturforsch. 10a, 425 (1955). — [3] Dürr, H.P., W. Heisenberg, H. Mitter, S. Schlieder u. K. Yamazaki: Z. Naturforsch. 14a, 441 (1959). — [4] Dürr, H.P.: Z. Naturforsch. 16a, 327 (1961). — [5] Dürr, H.P., u. W. Heisenberg: Z. Naturforsch. 16a, 726 (1961). — [6] Chew, G.F.: Vortrag auf der Konferenz in La Jolla, Juni 1961. — [7] Wightman, A.S.: Physic. Rev. 101, 860 (1956). — Lehmann, H., K. Symanzik u. W. Zimmermann: Nuovo Cim. 1, 205 (1955); 6, 319 (1957). — [8] Wentzel, G.: 6th Ann. Roch. Conf. 1956, p. VIII. — Physic. Rev. 101, 1214 (1956). — [9] d'Espagnat, B., u. J. Prentki: Nuovo Cim. 3, 845 (1956). — [10] Nambu, Y.: Phys. Rev. Letters 4, 380 (1960) u. Bericht v. d. Rochesterkonf. 1960, S. 858. — [11] Bogoljubow, N.N.: J. Exp. Theor. Phys. USSR. 34 (7), 41 (1958). — [12] Yamazaki, K.: Nucl. Physics 23, 139 (1961). — [13] Dürr, H.P., u. J. Géhéniau: Nuovo Cim. (im Erscheinen) und Bericht v. d. Rochesterkonf. 1962, S. 681. — [14] Heisenberg, W.: Bericht v. d. Rochesterkonf. 1962, S. 675. — [15] Fermi, E., u. C.N. Yang: Physic. Rev. 76, 1739 (1949). — [16] Regge, T.: Nuovo Cim. 14, 951 (1959); 18, 947 (1960). — [17] Pevsner, A., R. Kraemer, M. Nussbaum, C. Richardson, P. Schlein, R. Strand, T. Toohig, M. Block, A. Engler, R. Gessaroli u. C. Meltzer: Phys. Rev. Letters 7, 421 (1961). — [18] Alston, M., L.W. Alvarez, P. Eberhard, M.L. Good, W. Graziano, H.K. Ticho u. S.G. Wojcicki: Phys. Rev. Letters 6, 300 (1961). — [19] Alston, M., G.R. Kalbfleisch, H.K. Ticho u. S.G. Wojcicki: UCRL 10232 (1962). — Ticho, H.K.: Bericht v. d. Rochesterkonf., Genf 1962, S. 291. — [20] Armenteros, R.: Bericht v. d. Rochesterkonf., Genf, 1962, S. 295 und Kurs XXVI der Varenna-Sommerschule 1962. — [21] Alexander, G., G.R. Kalbfleisch, D.H. Miller u. G.A. Smitz: Phys. Rev. Letters 8, 447 (1962). — [22] Dhar, J.: Diss. München 1962. — [23] Tripp, R.D., M.B. Watson u. M. Ferro-Luzzi: Phys. Rev. Letters 8, 175 (1962). — Adair, R.K.: Private Mitteilung vom 20. 2. 1962; siehe auch Snow, G.A.: Bericht v. d. Rochesterkonf., Genf, 1962, S. 795.

München, Max-Planck-Institut für Physik und Astrophysik

Eingegangen am 4. Oktober 1962

---

[1]) Der untere Wert ergibt sich, wenn man das theoretische Massenverhältnis auf das π-Meson bezieht, der obere, wenn man mit dem Proton vergleicht.

In *Commemoration of the Fiftieth Anniversary of Niels Bohr's First Papers on Atomic Constitution Held in Copenhagen on 8–15 July 1963. Part 2: Session on Elementary Particles* (Institute for Theoretical Physics, Copenhagen 1963) pp. 1–21

The present situation in the theory of elementary particles

W. Heisenberg

Max-Planck-Institut für Physik und Astrophysik, München

When in the old times we listened to our teacher Niels Bohr, we noticed that he usually started very early in the history of quantum theory; and the listeners, being young and critical, sometimes asked (using a German phrase): Why does he always begin with Adam and Eve? Today, since I am getting older, I feel how difficult it is not to start with Adam and Eve; therefore I will begin with the question: What is an elementary particle? Some ten or twenty years ago most physicists would have answered without hesitation: The elementary particles are the smallest, ultimate units of matter. Such a statement however leads to the unavoidable question, why it should not be possible to split the elementary particles. The experimenting physicists have been able to divide the atoms into electrons and nuclei, and to split the nuclei into protons and neutrons. Why should they not be able to split protons or neutrons or electrons into still smaller units?

Such experiments are being carried out nowadays on a large scale by means of the big accelerators in Berkeley, Dubna, Geneva, Brookhaven. Very energetic collisions between elementary particles are the only processes where a splitting of the particles could be expected. These experiments have led to a most remarkable result, which had been discussed theoretically already more than twentyfive years ago. From the centre of such a collision process actually many particles emerge; but these particles are not smaller units of matter. They are in fact always the same kinds of elementary particles, independent of the colliding particles and of the energy in the collision. The emerging particles need not be smaller than (or different from) the colliding ones, since the large kinetic energy of the colliding particles can be transmuted into matter in the collision (multiple production of particles). Therefore the most natural description of these processes consists in stating, that the elementary particles are just different forms of the same 'substance' which may be called matter or energy, and that they are produced from energy in the collision.

If this general picture is correct, the different elementary particles appear as different stationary states of a system 'matter' in the same sense as the different energy levels of an atomic system, e.g. the iron atom, appear as the stationary states of this system. The spectrum of elementary particles may be just as complicated as that of the iron atom, and it is in fact so complicated. The different states can be distinguished by their quantum numbers, their energy or mass, their angular momentum, their lifetime etc., just like in atomic physics forty years ago. No distinction can be made in principle between an elementary particle and a compound system.

The experimental results concerning collision processes suggest an interesting but still uncertain extrapolation. It may be that even in collisions of most extreme energies (10 000 GeV or more) nothing very sensational will happen. The number of particles created may in the average be larger than at smaller energies, it will probably increase slowly with increasing primary energy; but the mixture of different kinds of secondary particles will be roughly always the same, the cross sections will be smooth functions of the energy. The jet-showers in cosmic radiation, with energies up to 10^5 GeV, seem to agree with this general picture, but the statistical fluctuations are still rather large.

If these results will be confirmed, it would mean that at the elementary particles we have actually come to an end in dividing matter. Any further 'division' or 'splitting' of elementary particles would not lead to new or smaller particles; it would be pointless to use higher and higher energies in collisions, because nothing new will happen.

This final result is not yet certain, but it looks rather probable. If it turns out to be correct, it would not mean that physics has been closed. Physics would be closed only at the limit of highest energies or smallest spatial dimensions. It would still be open in the limit of very large spatial dimensions (cosmology) or very large particle numbers (biology).

1. The mathematical description

The axiomatic basis for a mathematical description of the phenomena in high energy physics is at present still a controversial subject. One may distinguish mainly three possible frames for an axiomatic foundation of the theory, the following one always being narrower than the foregoing.

- 3 -

I. One assumes the existence of a unitary S-matrix, representing the groups of the system and representing causality by means of the analytical behaviour of its matrix elements.

II. One assumes, besides I, the existence of one or several local field operators $\psi(x)$ acting in a Hilbert space and obeying a differential equation, which represents (is invariant under) the groups and causality.

III. The Hilbert-space is (besides I and II) assumed to have a positive metric and to be constructed by the asymptotic operators only.

The first frame (I) seems to be wide enough for describing the experiments. Many interesting results have been derived by using mainly this wide frame. The dispersion relations have clarified the connection between the poles (corresponding to discrete stationary states) and the cuts (i.e. the continuous spectra) in the S-matrix-elements. The Mandelstam-relations have given a deeper insight into the interplay between particles of one kind acting as potentials between particles of another kind and thereby forming particles (compound systems) of a third kind, which may possibly be again identical with the first kind. The idea that "all particles are composed of all other particles" may in this way find an adequate but approximate mathematical formulation. The study of Regge poles has led to interesting possibilities concerning the asymptotic behaviour of matrix elements at very high energies.

On the other hand the S-matrix elements for complicated processes will be functions of many variables, the analytical behaviour of which may be extremely complicated. It will be very difficult to predict which behaviour follows from the postulate of causality. In any case the general axiomatic frame I has to be supplemented at this point by mathematical assumptions concerning the representation of causality in the analytical behaviour. One has the impression that by far the easiest way of formulating these assumptions is the introduction of a local field operator. The great success of quantum-electrodynamics is another strong argument in favour of this step.

The axiomatic frame III on the other hand is the narrowest, and comes closest to the frame of old unrelativistic quantum mechanics. It has been studied carefully in many details, and it leads - if nontrivial solutions exist at all - to satisfactory results concerning the representation of causality (dispersion relations). But so far no example is known of a Lorentz-invariant theory with interaction, fulfilling the axioms III. Therefore these axioms may be too narrow.

- 4 -

Under these circumstances I am inclined to believe that the final mathematical description will belong roughly to frame II i.e. that it will make use of a local field operator which however operates in a Hilbert space with indefinite metric. Recent investigations have shown that such a scheme with indefinite metric in Hilbert-space may very well be suited to define a unitary S-matrix. The Bleuler-Gupta version of quantum electrodynamics or the Lee model are good examples for a theory of this type.

2. Group theory

When, forty years ago, we learned atomic physics from Niels Bohr, the problem was not, to explain a special spectral line and its properties, but to explain the line spectrum. In the same way, in elementary particle physics nowadays, our problem is not to explain a special particle and its properties, but to explain the spectrum of elementary particles. The most characteristic features of a spectrum are the multiplets and the selection rules which control the transitions between different states. These selection rules and multiplets are the immediate expression of the group properties (or symmetries) of the underlying natural law and of the corresponding conservation laws.

The observations suggest that in elementary particle physics there should be strict invariance under the complete Lorentz group (including PC and T) and under three gauge groups (baryonic and leptonic number, electromagnetic charge), approximate invariance under the isospingroup (isomorphic to SU_2) and under P. Recent experimental results seem to give some significance to the SU_3-group (with the isospin group and isospin-gauge group U_2 as subgroup), which however can hold only in a very rough approximation. Masses which should be equal in a SU_3-theory, are actually different at some places by a factor up to 4 or a mass difference up to 400 MeV. Therefore this group can scarcely be a genuine property of the underlying natural law, but it may very well be valid in the same sense as one may say that a cube is almost a sphere; i.e. as an approximate symmetry reflecting the properties of special representations of the exact lower symmetries.

If one believes that a mathematical description can be found within the axiomatic frame II, one has to look for a field operator representing (in a nontrivial way) these groups (excluding SU_3) and for an invariant nonlinear differential field equation. If one adds the postulate that the field equation and the commutator should have the maximum symmetry possible for this

- 5 -

operator and that it should represent no other groups than those observed, operator and field equation seem to be uniquely determined. The field operator must then be a double Weyl-spinor (in Lorentz-space and isospinspace) and the equation is (in the form given by <u>Dürr</u>):

$$ i\sigma^\mu \frac{\partial \chi}{\partial x_\mu} \pm \sigma^\mu \chi \, (\chi^* \sigma_\mu \chi) := 0. \tag{1} $$

This equation is invariant under Lorentzgroup and isospingroup and may be just sufficient to explain the observed groups. (The violation of the isospingroup will be discussed later.) Since the fundamental fieldoperator in (1) is a spinor, equation (1) seems to take the fermions as 'more fundamental' than the bosons. The bosons appear already in the lowest approximation as composed of fermion and antifermion; it is encouraging from this point of view, that the four lowest possible "S-states" of the fermion - antifermion system have just the symmetries of the four most important bosons π, η, ρ, ω. In order to obtain the three gauge groups from (1) one has to include the conformal transformation (dilatation) $x \rightarrow \eta x$ (with η real) and

$$ \chi(x) \rightarrow \eta^{\frac{3}{2}} \chi(\eta x). \tag{2} $$

This interesting group has recently been investigated also independently from (1). Since the rest mass of the particles is not invariant under (2), the physical significance of the group is seen most clearly at extremely high energies, where the rest masses of the particles may be neglected. If one considers the collision of two particles with the very high energy E ($E \gg mc^2$) in the c.m. system, the cross section for the creation of a shower of secondaries of total energy $\geq \mathcal{E}$ in the c.m. system ($\mathcal{E} \gg mc^2$) should then from dimensional reasons have the form

$$ \sigma = \mathcal{E}^{-2} f\!\left(\frac{E}{\mathcal{E}}\right). \tag{3} $$

- 6 -

Since the unknown function $f\left(\dfrac{E}{\mathcal{E}}\right)$ seems to depend only very slightly on E, one would expect a formula of the type:

$$\sigma = const. \; \mathcal{E}^{-2} \; ln^{\alpha}\left(\frac{E}{\mathcal{E}}\right).\tag{4}$$

Actually the frequency of jet showers in cosmic radiation with primary energies between 10^2 and 10^5 GeV seems to agree with a formula of this type with $\alpha \sim 1$. But with the existing statistical material no definite conclusions can yet be drawn.

For the commutator the dilatation group would (according to **Mitter**) lead to

$$\left\{\chi(x)\,\chi^{*}(x')\right\} = \frac{\vec{\sigma}^{\,\nu}(x-x')_{\nu}}{(x-x')^2} \; f_{per}\left(ln\,\frac{(x-x')^2}{x_0^2}\right)\tag{5}$$

$$\text{for } \left|(x-x')^2\right| \to 0,$$

$$\text{and } 0 \text{ for } (x-x')^2 > 0$$

$$\text{(space like distances),}$$

where f_{per} is a periodic function (of the logarithm), the average of which vanishes, and x_0 is a constant. The validity of the dilatation group at very high energies would emphasize the point of view, that at extremely high energies nothing new will happen.

The transformation properties of the field operator and the definition of the groups are actually sufficient for determining the anticommutator and the field-equation, at least in the special case which is discussed here. In fact a field theory of the type (1) may be interpreted in two rather different ways. One may either take the fundamental field equation as a mathematical object defining by its symmetries the groups relevant in nature and the law of relativistic causality. Or one may take the abstract groups and causality as the real substance of the natural law underlying the system of elementary particles, and take the field operator and the field equation as simple re-

- 7 -

presentations of these groups. From the second point of view one would expect, that essentially the same mathematical structure could be expressed also by other types of fieldoperators or fieldequations which are just more complicated representations of the same groups; this may very well be true.

In any case, the possibility of expressing the underlying group structure and causality in the simple form (1) should be explored very thoroughly, because it would provide a rigid frame for a theory for which else it would be very difficult to find any precise starting point.

3. Broken symmetries and degeneracy of the vacuum

The isospin group is broken by the electromagnetic interactions, the space reflexion parity by the weak (radioactive) interactions. In both cases particles of rest mass zero come in, lightquanta and neutrinos, and seem to be somehow connected with the reduction of symmetry. Particles of zero mass are equivalent to long range forces. These facts would find a rather natural explanation, if the long range forces would produce an interaction with the world at large and if the groundstate ('world' rather than 'vacuum') was itself not symmetrical under the groups concerned. The lack of symmetry in the ground state is in fact a very natural phenomenon, it is known from many systems in ordinary quantum mechanics, e.g. the hydrogen atom in Dirac's theory, ferromagnetism, superconductivity, superfluidity, crystal structure etc. There is no reason why in field theory the groundstate should be without properties i.e. without deviations from complete symmetry. In the mathematical formalism of quantum field theory the groundstate would in this case introduce new operators representing those groups, for which the groundstate is not symmetrical, it would however not introduce any new groups.

If these general ideas are applied to the isospin-group, one would expect that the groundstate in elementary particle physics carries a large isospin. In fact the number of protons in the world is very different from the number of neutrons, therefore the groundstate may easily carry a very large isospin responsible for the deviations from symmetry. Such an assumption could on the one hand explain, why the electromagnetic forces violate the isospin group, it could on the other hand offer a simple description of the strange particles. These particles can be interpreted as consisting of a 'normal' particle (created by the field operator in (1)) and a 'spurion', i.e. an isospin cut from the big isospin of the world and attached to the particle. The spurion operator belongs to the description of the groundstate. Its symmetry properties are not uniquely determined by the field equation, e..g. by (1), even if the groundstate is assumed to be symmetrical under the complete Lorentzgroup. But there are only very few, essentially just three, possible assumptions for the properties of the spurions. In an earlier paper it had been assumed, that the spurions carry the isospin $1/2$, strangeness and parity. In this case the detailed calculations concerning the properties of strange particles led to a negative relative $\Lambda\Sigma$-parity, contrary to recent experiments in CERN. Therefore this assumption had to be dropped. More recent investigations have been carried out on the different assumption, that the spurions have only isospin and strangeness (the latter connected with the isospin gauge group), but no parity. In this case the formalism is much simpler than in the former case. It leads to even $\Lambda\Sigma$-parity and in a first approximation to a mass spectrum of the form

$$m^2 = a + b \cdot Bu + c \left[T(T+1) - u^2 \right], \qquad (6)$$

where B is the baryonic number, T the isospin and u the hypercharge (u = 1/2 for the proton) of the system, a, b, c being constants depending only on the other properties (e.g. the Lorentz properties) of the particles. The mass formula (6), agreeing fairly well with the observations, can be derived equally from the assumption, that the isospingroup (isomorphic to U_2) is to be considered as a subgroup of SU_3, the latter referring to an approximate symmetry of the system. The connection between these two apparently different assumptions has not yet been analysed in detail.

The lack of symmetry in the groundstate is also conspicuous with respect to the dilatation (conformal) group, where the existence of a mass scale, e.g. the mass of the nucleon, can be understood only if the groundstate introduces this scale e.g. by means of the 'vacuum expectation values'. This mass may then be considered as an operator taken from the groundstate, like the spurion, which may be 'attached' to the particles. There is still complete invariance under the dilatation group, if the operator takes part in the transformation. Therefore, the introduction of such spurion-like operators which may be attached to the particles does not prevent the strict validity of the corresponding conservation law. It is only the interaction with macroscopic properties of the groundstate e.g. a big total isospin of the world, which may formally destroy the validity of the conservation law.

4. Electrodynamics

If the lack of iso-invariance in electrodynamics can be interpreted in this way, the lightquantum may, with respect to its isospin-properties, be compared with a sound quantum (phonon) in a crystal with an axis. Such a phonon cannot possibly carry a well defined total angular (spin) momentum (or a component of this momentum perpendicular to the axis), since the axis of the crystal prevents the simple application of the rotation group. In the same way the lightquantum may be considered as a mixture of states with isospin 0 and 1 ; only the component of the isospin in the direction of the (isospin-)axis, the charge, is well defined.

It has been emphasized by Goldstone on the basis of a mathematical argument that an unsymmetrical groundstate enforces the existence of particles of rest mass zero which in some way restore the conservation law concerned. The arguments of Goldstone as applied in the theory of elementary particles of Nambu refer to the assumed conservation of the pseudovector current and seem to show, that mesons of the symmetry of $\widetilde{\pi}$- and η-mesons, but of rest mass zero should exist. These arguments however are not convincing when viewed in the light of recent investigations of Johnson on the Thirring model. On the other hand they are related to a genuine physical connection between broken symmetries and particles of rest mass zero. It is true that in a crystal with an axis a collision process violating apparently the conservation of angular momentum can transfer this residual angular momentum to the crystal only by means of sound waves. On the other hand this sound wave is a collective motion with a continuous exchange of angular momentum between the phonon and the crystal. The phonon cannot be used to restore the conservation law in its original form, since it cannot possibly have a well defined total angular momentum. In the same way the lightquanta (including the longitudinal photons in the Coulomb force) cannot be used to restore formally the conservation of (total) isospin; still the existence of the photon is necessary on account of the conservation law, if the groundstate is unsymmetrical in isospace.

10

The double nature of the photon (isospin 1 and 0) is closely related to
the two kinds of electric charge, isospin-charge and hypercharge, which are
well known since many years. Isospin-charge refers (in its grouptheoretical
aspect) to the rotation around the preferred axis in isospace, hypercharge
to the gauge group in this space. The gauge group in isospace could not be
separated from the gauge group in Lorentzspace if the particles were described
by the fieldoperators only. But since the spurion operators of the groundstate
(having a gauge property in isospace) may be included, the two gauges are
actually separated. In this way electrodynamics fits very well into the general grouptheoretical picture of elementary particle physics and strong
interactions.

The rest mass zero of the photon implies the gauge invariance of the
second kind, i.e. invariance for a gauge transformation where the gauge factor depends on the space time coordinates. Consequently the vector current
$j_\mu(x)$ interacting with the photon must obey the local conservation law
$\frac{\partial j_\mu(x)}{\partial x_\mu} = 0$. A similar relation does not hold for the pseudovector
currents interacting with the $\widetilde{\pi}$- and η-meson fields, since these pseudoscalar particles have a finite rest mass. The equation $\frac{\partial j_\mu(x)}{\partial x_\mu} = 0$
holds separately for the two parts of the vector current which refer to the
isospincharge or the hypercharge respectively, if weak interactions are neglected. It is only the weak interaction, that mixes the two types of vector
current. The first of these currents represents the third component of a
vector in isospace. The two other components of the same vector in isospace
mean two other kinds of vector currents which play an important rôle in weak
interactions. The local conservation law for these two currents must necessarily hold if only strong interactions are considered, since there is complete
symmetry in isospace for these interactions. Electromagnetic and weak interactions may however interfere with this law.

The rest mass zero of the photon and the gauge invariance of the second
kind are the essential conditions for making a 'classical' field out of the
photonfield. It is only on account of the long range and of the additivity
of the forces from different sources that a reasonable 'classical limit' of
the photon field operator exists. The existence of this limit has been of
great heuristic value in studying the mathematical properties of a relativistic field theory. In quantum electrodynamics the postulate of relativistic

causality was clearly expressed by the differential character of the underlying field equations and by the commutativity (or anticommutativity) of the
field operators at space like distances. At the same time the mathematical
formalism required the introduction of an indefinite metric in Hilbert space,
if one wanted to make Lorentzinvariance and causality immediately visible in
the equations (<u>Bleuler</u> and <u>Gupta</u>). In this way quantum electrodynamics gave
the first example, that a field theory with an indefinite metric in Hilbert-
space can very well lead to the definition of a unitary S-matrix necessary
for describing the observations. The great success of quantum electrodynamics
in the description of the experimental situation emphasizes the importance
of such a mathematical scheme.

5. Weak interactions

The well known phenomenological description of the weak interactions by
means of a Fermi interaction term reveals on the one hand the universal nature of this interaction; it demonstrates on the other hand, that this universal and fundamental interaction does not contain the space reflexion parity, it is not invariant under P. The universality is shown by the - at
least approximate - equality of the Fermi coupling constant for many different particles, baryons, mesons and leptons. The observed A-V-type interaction
indicates that primarily only particles of one chirality interact. This situation fits extremely well with the hypothesis of an underlying universal field
equation of the form (1), since the field operator in this equation is only
a Weyl (two component-) spinor, it does not represent parity and refers only
to one chirality. Therefore one may say that the fundamental interaction in
(1) is just an (A - V)-type Fermi interaction. It is only by means of the
finite mass-eigenvalues, that the parity concept is introduced indirectly.
The field operator $\widetilde{\chi}(x) = \sigma^{\mu} : \chi(x)\left(\chi^{*}(x)\,\sigma_{\mu}\,\chi(x)\right):$ supplements
in a certain way the original field operator $\chi(x)$ to form a pair of opera-
tors $\chi(x)$, $\widetilde{\chi}(x)$ which represents the space reflexion parity (<u>Dürr</u>). The
parity concept however breaks down completely, when Fermions of rest mass
zero, the neutrinos, come into play.

In isospin space the universal Fermi interaction refers to currents
(vector- or axial-vector currents), which are iso-vectors perpendical to the

preferred axis. For the vectorcurrents of this kind it follows from electro-
dynamics that $\dfrac{\partial j_\mu(x)}{\partial x_\mu} = 0$ if only strong interactions are considered.
Therefore the Fermi coupling of the vector currents from baryons and mesons
is not changed by renormalization i.e. by the dressing of the particles due
to strong interactions. For the pseudovector (or axial vector-) currents
however the law $\dfrac{\partial j_\mu(x)}{\partial x_\mu} = 0$ does not hold, and a certain change of the
coupling constant by the 'dressing' is observed.

For the theory of leptons and weak interactions it is extremely important
to find out, whether the known conservation laws (primarily for baryonic and
leptonic number, electric charge) are sufficient to explain all the selection
rules observed. In this connection the recent discovery of the two kinds of
neutrinos has given a most valuable information. A distinction between these
two neutrinos by means of known quantum numbers seems possible only, if the
leptonic number of the negative μ^--meson is opposite to the leptonic number
of the (negative) electron. Whether this attachment of quantum numbers is
sufficient for describing all experiments, is still an open question. Therefore one cannot decide at the present moment whether the leptons and the weak
interactions fit completely into the grouptheoretical scheme expressed by (1),
but it may well be that they do.

Some theoreticians have discussed the possibility that the weak interactions of the Fermi-type are produced indirectly by means of an 'intermediate
boson' (Lee, Yang, Phys.Rev. 119, 1410 (1960)). If this could be demonstrated
experimentally, the intermediate boson could possibly establish a most interesting connecting link between the electromagnetic and the weak interactions.

Comparison with quantum mechanics

This very brief and incomplete survey of our main present problems in
elementary particle physics was primarily intended to show how similar our
present situation is to that we met and had to discuss with Niels Bohr forty
years ago. There was a large experimental material consisting in tables of
spectral lines or stationary states, selection rules, transition probabilities
etc., just as we have it now for the elementary particles. One could see
groupings of levels and other regularities, as we do now, and one could try
to understand it by suitable quantum numbers or underlying symmetries. Thus
the phenomenological part of the work was very similar. Besides that there

- 13 -

were definite indications - but unfortunately no proof - that the mathematical frame of older physics was insufficient for describing the new facts, and that one had to widen this older frame. In the same way nowadays we have indications but no proof that a quantum field theory based on the traditional axioms will not be sufficient, that we will have to widen the older axiomatic frame. Just as in the old times, the single fact - as important it may be - cannot be separated from the other facts; we have learned from Niels Bohr, that a real understanding can be achieved only if all the different facts can be combined in a consistent picture. Such a picture can be formed only after a long and thorough analysis of the many difficulties occurring in the interpretation of the observations, there is no short cut from a special detail to a real theory of the phenomena. Already the existing experimental evidence on elementary particles forces us now to say that we will either some day have a unified theory of the whole system of elementary particles, or we will not understand it at all. Keeping this in mind we may only hope that the young physicists working nowadays on these problems will have as much pleasure from their work as we had in the old times, when we worked here in Copenhagen under the guidance of Niels Bohr.

- 14 -

<u>Discussion</u>

<u>Dirac</u>:

Heisenberg has given us an excellent transition from Adam and Eve to present day-physics. It always seems to me that the $\underline{S}$-matrix does not apply to the whole of physics but just to high energy physics; it has its limitation somewhere for lower energy physics. I would like to have your view of that.

<u>Heisenberg</u>:

Well, I could not agree more. I have never really liked the idea that one should explain everything by means of the $\underline{S}$-matrix; but one had to emphasize this point perhaps to some extent for a time in order to gain some freedom from the old Hamiltonian scheme. But certainly the $\underline{S}$-matrix does not contain everything.

<u>Pais</u>:

There are many things in the talk of Professor Heisenberg which I have not understood, but there are only a few things about which I am able to ask concrete questions. I want to ask a question about the conformal group. You come to the conclusion, or conjecture, that cross sections should vary with the energy as $1/\underline{E}^2$. Now, of course cross sections do not only depend on energy, but also on the energy-momentum transfer. When you go to extremely high energies, and you want to liberate yourself from the masses (which is what you want to do), it is not only the energy which has to be large compared with the masses, but also the momentum transfers. On the other hand, high energy cross sections have the tendency to be peaked very much forward, and low momentum transfers play a very important role at the

- 15 -

very highest energies. Therefore the question arises
whether one can really ever hope to have an experimental
quantity for which one is totally free from the masses.
One can of course, but these quantities are not total
cross sections, they are partial cross sections at very
high energies and at large angles. I do not know how
this is connected with the conformal group, but my
question is: do you have anything to say about high
momentum transfers as well?

Heisenberg:

I quite agree that it is of course difficult to
define experimentally what one means if one neglects the
masses. Then of course the total cross sections would
probably be infinite like in electrodynamics. Therefore
I do not look for the total cross section, but I do look
for a cross section, say, for the creation of one or
several secondary particles having together more than
1/10 of the primary energy in the c.m. system. By
stating 1/10 I have provided for that all the energies
are changed according to the same scale, and then I
should say that such a cross section, which is well
defined, should vary as $1/\underline{E}^2$ at very high energies. I
think that is perhaps at present the best definition one
can give, and one should be able to measure such a cross
section experimentally. It is not the total cross
section which counts, it is the cross section for something definite to happen. Besides that anything else
may happen; but we postulate that at least one secondary
particle should be created and that more than 1/10 of
the primary energy should go over into the secondaries.

Casimir:

Well, just a silly question, perhaps. You were
emphasizing the analogy between the situations today

and 30 years ago; of course there are always differences
and similarities, but it seems to me there is in our way
of looking at these things one rather striking difference. If I interpret it rightly we had in the early
days of quantum mechanics all these experimental data,
we had also a lot of make-shift procedures like vector
models and so on to say something about states and
spectral lines. What did then happen? In some way or
other mathematical formalisms were found to describe
particles: matrix calculus, Schrödinger equation; then
it was found a little bit later that properties of
symmetry and transformation groups and all that sort of
things accounted for these general rules of the vector
model. So to say, the Hamiltonian and the description
and the equation were there first, and these happened to
have certain symmetry and transformation properties and
they gave the general interpretation of these rules.
For instance, one knew there was an electron spin, Pauli
wrote down an equation to describe this spin non-
relativistically, and then it was found that it was
connected with two-valued representations of the
rotation group, not the other way round. It seems to
me that that is the difference between what we are
trying to do today, and what you did in those days.
Today we have again a whole set of experimental rules,
and we think that they will in some way or other have
to be connected with properties of groups, invariance,
representations and what not, and we try to get at the
Hamiltonian in that way. So it seems to me that it is
a rather different way of approach from this point of
view. Well, it is a very trivial remark, but it seems to
me it is a somewhat fundamental difference.

- 17 -

<u>Weisskopf</u>:

I would like to make a few remarks. First to
the beginning of Heisenberg's talk, when he said that
we now so to speak come to an end in a line of new
phenomena: after all, in processes starting with
elementary particles, we always get elementary particles
back. I am, as Heisenberg knows, not very convinced by
this argument, because I always feel that it defines
the phenomena too narrowly. It is true that we shall
not get situations like we had in atomic physics and
nuclear physics, namely that we get to smaller particles
so to speak, to the constituents; but do we not already
now see that nature presents its own variety in a
different way: for example we find excited states of
the nucleon, we find many-nucleon states, and we might
find - let me just now use my fantasy, because you said
it is logically impossibe - for example an ω , and
excited states of this ω , and so on. Well, we don't
know either whether the list of states of nucleons is
even finite, although there might be some indication
for that; but if this list of excited states of nucleons
is finite, we have the ω ; and after all we have also
certain phenomena such as gravity, and I would say the
universe that presents us with possibilities that we
just don't know of. Therefore I think if one restricts
the possible phenomena by this analogy, one runs a
danger; in fact one cannot do this in my opinion. Now
there is, however, if I interpret rightly what you say,
perhaps a hope that the list of new phenomena is at an
end; now this hope of course is an extremely important
thing, and here I have always the feeling, when I look
at the results and the papers that come from your group,
that we cannot but admire this hopeful spirit, because
you people are really the only ones in the whole world
of physicists who at least try to find a solution of

- 18 -

the kind you are indicating. Whether this is possible
or not is another question, but I always think we should
deeply admire the group that tries it.

<u>Heisenberg</u>:

May I say a few words about this question of
'closure'. I think that this is a very important
question, and I also feel that I sometimes have been
misunderstood in this respect. I certainly do not mean
that physics will be closed; I will comment on this somewhat later. What I mean here is only that it looks as
if at very high energies nothing very new and exciting
will happen. At the present stage, of course, we shall
get very many new phenomena, we shall certainly find
many new elementary particles, may be the ω , may be
other resonant states; that is quite obvious, here we
agree completely. But the problem is whether at an
energy which is still much higher - say above 10 or 20
GeV in the c.m. system - there are any very interesting
new phenomena. I could really imagine that at this
point physics can be closed. We may just get the normal
continuous spectrum, which we should expect, and nothing
very exciting will happen at extremely high energies.
An equivalent expression for the same idea would be the
statement that the structure of elementary particles in
the sense of Hofstadter's experiments would not show any
central hard core of very small radius. Especially if
this conformal group would be valid then of course we
could simply calculate at any energy what will happen,
because it would always be similar to what happens at
lower energies.

Let us for the moment assume that this were so.
Even then I certainly do not mean that physics would be
closed and we could have no new phenomena. I would
feel, however, that physics is open at entirely different
points. I could imagine that physics is definitely open

- 19 -

at the very low energy region, namely where physics goes
over into biology. Now I am coming to the subject of
Delbrück's lecture which he has not given yet; so I
think that I would better not say anything about it.
But I only feel that at the boundary between physics and
biology there may be a situation which will force us
again to introduce entirely new concepts. Actually new
concepts, not known in physics so far, have been in use
already in biology for many years. Therefore I could
well imagine that our present schemes are not sufficient
to connect physics with biology in a satisfactory manner.
At the extremely high energies, however, I doubt whether
physics would still be open. Another open boundary is
of course the boundary to gravitation and to the cosmological questions. So really the opinion I expressed
was meant as a very modest remark; I only felt that at
energies much higher than the mass of the nucleon there
may not be very interesting new phenomena. Of course I
may be wrong; it is quite obvious that we just don't
know the answer, yet.

<u>Rabi</u>:

Heisenberg's lecture reminded me of a time when
Bohr and Einstein were together in Princeton. Many of
you have seen and heard these confrontations that never
failed to be interesting not only scientifically but
personally. At that time Einstein made a remark: "When
I was a young man", he said, "I thought all a physicist
had to know of mathematics was multiplication and
addition". He added: "I since changed my mind". It is
interesting to note, however, that this was a period
when Einstein's success was almost minimal in spite of
tremendous efforts, and I could not help feeling that
the wonders of mathematics had been substituted for the
wonders of physics. This is something which tends to
happen in the course of time, and I am impelled to make

- 20 -

this remark because there are some young people present
and I want to warn them against knowing too much
mathematics and losing the actual taste for the facts.
It is part of the greatness of Niels Bohr, that that
is one thing he never lost.

<u>Wheeler</u>:

I only wanted to bring up the existence of one
area of elementary particle physics which would seem
to me to pose issues that I at any rate don't have the
faintest idea how we are going to approach. It is not
so much the question of so very high energies that I am
concerned with as the question of very large numbers of
nucleons; when one gets to the point of 10^{57} nucleons,
the number of nucleons to make up the mass of the sun,
and deals with the condition in which one has catalyzed
matter to the end point of thermonuclear combustion and
has cooled it down to the point where temperature does
not really contribute at all to the pressure that
sustains the star. Then one has at the centre of the
star conditions of such very high pressure that the
object no longer can have any condition of stable
equilibrium; so one comes to a point when if we add one
more kilo of matter to the system it is bound to go over
to some new very unfamiliar condition. Of course the
equations themselves predict that the pressure at the
centre becomes infinite, the separation between positive
and negative energies energy states becomes zero, so
that one is forced to conclude that one has here a state
of affairs very different from what one normally deals
with. It would appear that one has come to the place
where there is no escape from saying that the new kilo
of matter that one has added up to the outside must
somehow be counter-balanced by a kilo of matter that gets
squeezed out of existence at the centre; or if it is not
squeezed out of existence the concept of conservation of
baryons must have some very much more sophisticated

- 21 -

interpretation than we are accustomed to give it today.
Therefore I would suppose that we don't have really to
go to the realm of unbelievable energies, or unbelievable
extensions in space and issues of cosmology: really on
the very modest scale of a star we are at the border line
of quite a new issue in physics, where we shall have to
be prepared for quite new concepts. I wonder whether you
have any comments on this border line area between
elementary particle physics and gravitation physics.

Heisenberg:

No real opinion of my own. I remember, I should
say with pleasure, the idea that gravitation could
possibly be connected with the weak interactions in a
similar way as on the other hand weak interactions may
be connected with the electromagnetic ones, so that the
square of the electromagnetic coupling constant could
roughly be the weak coupling constant and the square of
the weak coupling constant would be the gravitational
coupling constant. I remember that there have been some
papers in the early years; have you, Dirac, not written
about these problems once? Ah, it was Gamow and Teller.
Yes, well, but I am not very familiar with these ideas,
and I find it simply too early in the present state to
think about these problems; but some day these problems
will certainly come up again.

Kronig:

I just want to make a remark on the historical
analogy which Professor Heisenberg brought forward between
fifty years ago and now. Fifty years ago of course the
correspondence principle was a very strong guide in leading
out of difficulties and was always stressed by Bohr as
the basis for every discussion. I don't think we have in
our situation today anything which is as strong as the
correspondence principle of Bohr was then.

In *Rendiconti della Conferenza Internazionale di Siena sulle Particelle Elementari - Proceedings of the Sienna International Conference on Elementary Particles. Volume 1* (Societa Italiana di Fisica, Bologna 1963) pp. 445-447

$$\frac{\dfrac{f(\lambda_p, k, R)}{\partial f(\lambda_p, -k, R)}}{\partial \lambda_p} = \frac{u_+}{v_-} (kR)^{1/3} K_p e^{i(5\pi/6)} + \ldots \qquad (13)$$

The relations (12), (13) and (5) are now to be introduced in (6).

REFERENCES -

(1) - E. Predazzi and T. Regge, Nuovo Cimento 14, 518 (1962).
(2) - O. Brander, Phys. Letters 4, 218 (1963); G. Berendt, Nuovo Cimento 27, 551 (1963).
(3) - S. Ciulli, Gr. Ghika and M. Stihi, "Hard core and Regge poles", to be published in Phys. Rev. Letters.
(4) - R. G. Newton, J. Math. Phys. 3, 867 (1962).

REMARKS ON THE ‹SPURION› - THEORY OF STRANGE PARTICLES

W. Heisenberg
Max-Planck-Institut für Physik und Astrophysik, München

In the non linear spinor theory of elementary particles the strange particles can be re presented by making use of a degenerate groundstate which may attach a "spurion" to a nucleon or a boson.

Dürr and Géhéniau[1] have pointed out, that for the symmetry properties of these spurions one has to choose between essentially three different possibilities, listed in table I. The fundamental field equation does not decide between these three possibilities. The table gives the symmetry operations under which the spurions should be invariant or covariant (L proper Lorentzgroup, I isospin rotation, P space reflexion parity, T time reversal, C charge conjugation, G = C J particle-antiparticle-conjugation = C x reversal of isospin).

TABLE I

Invariant under	Covariant under	Spurion operators
1) L GP TJ	I P G	Σ_1 Σ_2 $\Sigma_{3\rho k}$
2) L P CT	I G	ρk
3) L PCT	I P PG	$\Sigma_2 \, \rho k$

The third column contains the spurion operators needed for expressing these properties.

In the earlier papers[2] only the first possibility had been considered, since it seemed most plausible in connection with the weak interactions, and its consequences had been studied.

446

In this first scheme the $\Lambda\Sigma$-parity comes out to be odd. Recent results obtained at CERN[3] however seem to show that the $\Lambda\Sigma$-parity is even. This would disprove the first possibility. Therefore recently the second possibility has been studied in some detail by Dürr and the author.

In this second scheme the spurions have not got a space reflexion property, therefore the operators Σ can be dropped, and the number of strange energy levels is roughly reduced to one half compared with the first possibility. This number is further reduced, if one makes the assumption compatible with the fundamental field equation, that the wave function is symmetrical under the permutation of all isospinvariables of the particles bound together in the system (nucleon and spurions), and likewise symmetrical under permutation of the isospin variables of the antiparticles (antinucleon and antispurions). The distinction between spurions and antispurions allow the introduction of a quantum-number hypercharge or strangeness corresponding to the behaviour under an isospin-gauge group (compare Dürr and Géhéniau[1]). In this way one finally arrives at a set of assumptions about the ground state which leads to the result that the mass of a state depends only on the isospin and the hypercharge (and the Lorentzproperties) of the system.

If we call ρ_k^P the isospin-operators (Pauli-matrices) of all particles (nucleon and spurions), ρ_k^a the corresponding operators of the antiparticles (antinucleon and antispurions), n_p resp. n_a the number of particles resp. antiparticles, $n = 1/2\,(n_p - n_a)$ the hypercharge, then the only expression in the eigenvalue-equation which is invariant under isospin rotation and PCT, and which depends neither on the Lorentzproperties nor on the total number $n_p + n_a$, has the form

$$a\left[(\sum_1^{n_p}\vec{\rho}^P - \sum_1^{n_a}\vec{\rho}^a)^2 - (n_p + n_a + 2)^2\right] =$$

$$= a\left[-(\sum_1^{n_p}\vec{\rho}^P + \sum_1^{n_a}\vec{\rho}^a)^2 + 2n_p(n_p + 2) + 2n_a(n_a + 2) - (n_p + n_a + 2)^2\right] \right\} \quad (1)$$

$$= -4a\left[\bar{I}(I + 1) - n^2\right] \qquad (a = \text{const.}) \qquad (2)$$

where
$$I(I + 1) = \frac{1}{4}\,(\sum_1^{n_p}\vec{\rho}^P + \sum_1^{n_a}\vec{\rho}^a)^2 \qquad (3)$$

characterizes the total isospin of the system. The minus sign for the operators of the antiparticles in the first bracket of (1) is necessary for the PCT-invariance of the theory. This expression (2) has the same form as the expression which had been derived from a symmetry reducing perturbation in the SU_3-group[4], and seems to agree well with the experimental situation. Especially in the case of the bosons, the eigenvalue equation for π - and $\quad$ -meson in the earlier papers[5] $(x = \sqrt{|J^2/k^2|}\,)$

$$\pi: \quad 1 - 2Q(x) = 0$$
$$\text{and} \qquad \eta: \quad 1 - 6Q(x) = 0 \qquad (4)$$

can now be supplemented by a corresponding equation determining the mass of the Kaon:

$$K: \; 1 - 5\,Q(x) = 0. \qquad (5)$$

For the fermions the eigenvalue equation may, besides (2), contain an other term depending on the Lorentzgauge

$$b \cdot B \cdot n, \qquad (6)$$

where B is the baryonic number and b a constant. This term is responsible for the mass difference between nucleon and Ξ-particle. The details of the mass calculations will be published in a later paper by Dürr and the author.

In comparing the two possibilities 1) and 2) of table I with the experimental results on the strange particle states, the most important point on the experimental side is the number of levels with Dirac spin 1/2 or 0. If all the fermion levels of the type Y_0^x and Y_1^x have Dirac spin values 3/2 or higher, only the states Λ and Σ remain with spin 1/2 and only the second possibility offers a description by means of the spurion picture. In this case the $\Lambda\Sigma$-parity must be even. If however the states Y_0^x (1405 MeV) and Y_1^x (1385 MeV) should have the Dirac spin 1/2, one would expect that their parity is opposite to the parity of Λ and Σ; because they should differ by at least one quantum number from Λ and Σ. This situation could be represented by the first (or third) possibility of table I. But an even $\Lambda\Sigma$-parity could then scarcely be understood, unless some very strange kind of resonance effect should have interchanged the order of the two levels Σ and Y_1^x. Therefore the experimental results favouring an even $\Lambda\Sigma$-parity strongly suggest that the spin of all Y_0^x and Y_1^x states should be $\geq$ 3/2.

REFERENCES -

(1) - H. P. Dürr and J. Géhéniau, Nuovo Cimento 28, 132 (1962).
(2) - H. P. Dürr and W. Heisenberg, Z. Naturforsch. 16a, 726 (1961).
    W. Heisenberg, Report on the High energy Conference in Geneva 1962, pg. 675.
(3) - H. Courant, H. Filthuth et al., Phys. Rev. Letters 10, 409 (1963).
(4) - e. g. B. d'Espagnat, Report on the High energy Conf. in Geneva 1962, pg. 917.
(5) - H. P. Dürr, W. Heisenberg, H. Mitter, S. Schlieder and K. Yamazaki, Z. Naturforsch.
    14a, 441 (1959).
    H. P. Dürr, Z. Naturforsch. 16a, 327 (1961).

DIFFRACTION SCATTERING AND THE FORCES AT SMALL DISTANCES

R. Oehme[x]
The Enrico Fermi Institute for Nuclear Studies and Department of Physics
The University of Chicago, Chicago, Illinois[+]
CERN - Geneva

In this report I would like to discuss some ideas concerning the relation between forces at small distances and the properties of high energy diffraction peaks in crossed channels.

We consider scattering amplitudes within the framework of dispersion theory, i. e., we assume that the amplitude $F(s, t)$ has analytic properties as indicated by Feynman graphs. A very important assumption is the existence of a polynomial bound in one variable for finite values of the other. Under these conditions we can discuss the high-energy limit of $F(s, t)$ for $t \longrightarrow \infty$ in terms of the leading singularities of the interpolating partial wave amplitudes $F_{\pm}(s, \lambda)$ in the s-channel[1].

As is well known, a shrinking diffraction peak can be obtained by introducing a Pomeranchuk trajectory. It seems to be more difficult to produce a non-shrinking diffraction pattern together with a constant total cross-section and a pure imaginary amplitude in the asymptotic region such that for s = 0:

(x) - Guggenheim Fellow
(+) - Permanent address. Present address: Dept. of Physics, Imperial College, London.

In *XII International Conference on High Energy Physics, Dubna, August 5–15, 1964. Volume 2,* ed. by Ya. Smorodinski, V. G. Grishin, A. A. Komar, V. A. Nikitin, L. D. Solovyev, S. M. Korenchenko (Atomizdat, Moscow 1966) pp. 245–247

STRANGE PARTICLES IN THE NONLINEAR SPINOR THEORY

H. P. Dürr, W. Heisenberg

Max-Planck-Institut für Physik und Astrophysik, München
(Presented by H. P. DÜRR)

Strange particles are represented in the nonlinear spinor theory of elementary particles [1] by making use of the isospin property of the groundstate (vacuum): «spurions» carrying isospin 1/2 may be detached from the groundstate and attached to systems constructed from local field operators to form strange particles. In addition to isospin further properties have to be attributed to the spurions to be consistent with the requirement of Lorentz- and CPT-invariance of the vacuum. According to investigations by Dürr and Géhéniau [2] three such possibilities present themselves:

The first and simplest possibility from a group theoretical point of view adds a parity property and was reported earlier. Here the spurions transform according to the simplest, nontrivial representation of $P \times SU_2$ (discrete reflection group × isospin group). Some immediate implications for the baryons, e. g. odd $\Lambda\Sigma$-parity, however, are in contradiction to present experimental evidence. Therefore this possibility is ruled out.

Hence the second possibility was investigated which is the subject of the present paper (the third possibility which we do not consider is a combination with the first one). Here a «spurion number» (gauge group) is added which is identified with the hypercharge $u = {}^1/_2 Y$. The spurions in this case transform according to the simplest, nontrivial representation of $U_2 = g \times SU_2$ in which 4 different states (spurions and antispurions with $I_3 = \pm {}^1/_2$) can be distinguished. Spurions and antispurions are transformed into each other under G-conjugation. Because of the absence of spin they obey Bose statistics.

At the first step we consider a general system composed of n_s spurions and n_a antispurions, and construct the most general 2-spurion correlation operator, i. e. the operator which for ordinary particles corresponds to the pair interaction operator. One finds:

$$O(n_s, n_a) = \Big(\sum_{s=1}^{n_s} \vec{\tau}_s - \sum_{a=1}^{n_a} \vec{\tau}_a \Big)^2 - f(n_s + n_a) =$$

$$= 4(\vec{I}_s - \vec{I}_a)^2 - f(n_s + n_a) =$$

$$= 2 \sum_{i<j}^{n_s+n_a} \varepsilon_{ij} \vec{\tau}_i \cdot \vec{\tau}_j + [3(n_s + n_a) - f(n_s + n_a)]$$

where $\vec{\tau}_s$, $\vec{\tau}_a$ are the isospin matrices acting on the spurions and antispurions, respectively; $f(n_s + n_a)$ is some arbitrary functions of the total number of spurions and antispurions; ε_{ij} is a sign function which is ± 1 depending whether the indices refer to like or unlike «particles». The operator is determined by the requirements of isospin-, gauge — and G-invariance and the condition that $O(n_s, n_a)$ under C-conjugation of a s i n g l e spurion or antispurion transforms into $O(n_s - 1, n_a + 1)$ or $O(n_s + 1, n_a - 1)$, respectively, which can be deduced from the fact that the annihilation of a spurion is equivalent to a creation of an antispurion. The eigenvalue of the operator for a system with isospin I and hypercharge $u = 1/2 \, Y = n_s - n_a$ then is given by

$$O(n_s, n_a) = -4[I(I+1) - u^2] + \\ + [(n_s + n_a)(n_s + n_a + 4) - f(n_s + n_a)].$$

I.e. the I, u combination is exactly of the form which occurs in connection with broken SU_3 theories. However, above formula does not contain any of the features particular for SU_3, which — depending on the representation — express themselves in certain limitations on the possible values I, u leading to characteristic multiplets. I and u in above formula are only limited by $I \geqslant u$.

To connect above results with the actual problem of constructing single particle states and determine the isospin-hypercharge dependence of their mass eigenvalue operators, one has to consider systems which are constructed from spurions and the spinor-isospinor field operators $\psi(x)$ of the theory. In the field operators the isospin-hypercharge properties seem inseparably connected with the Lorentz properties. In particular, the isospins of two field operators will not be symmetrical, in general, as required above. However, the situation is different if only field operators at the **same** space-time point are considered. Due to the particular field equation

$$N^{i\sigma\nu} \frac{\partial}{\partial x^\nu} \psi(x) = l^2 \sigma_\mu \cdot \psi(x) [\psi^*(x) \sigma^\mu \psi(x)]:$$

the interaction $\sim \sigma_\mu \sigma^\mu = I \cdot I - \vec{\sigma} \cdot \vec{\sigma}$ will vanish whenever the spins in $\psi(x)\,\psi(x)$ are symmetrical $(\vec{\sigma} \cdot \vec{\sigma} = +1)$. Because of the antisymmetry and the equal coordinates of ψ this implies that there is no interaction if the isospins are antisymmetrical, i. e. whenever they do not behave like spurions. Hence above spurion considerations can be equally applied for states constructed from field operators at the same point and spurions. Besides the operator $O(n_s, n_a,)$ however, also the operator Bu may now occur, where B is the baryon number, which is connected with the field operators. Particles constructed from field operators at the same space-time point we may call «primary particles».

For the actual calculations of the masses of the particles we proceed according to the following programm:

1.) The existence of a dominating baryon pole at some mass in the $\psi(x)$ — contraction function is assumed. The eigenvalue equation for $\varphi(x) = \langle 0 \,|\, (x) \,|\, |p\rangle$ is derived in lowest New-Tamm-Dancoff approximation under this assumption and examined with respect to a discrete baryon solution [3]. Such a solution exists; the mass of the baryon is fixed by the selfconsistency requirement (baryon boot-strap calculation). The inclusion of spurions in the matrix elements does not alter this calculation, i.e. there is no splitting of the baryon levels at this point.

2) The eigenvalue equations for the primary strangeness zero bosons with «eigenfunctions» $\varphi_B(x) = \langle 0 \,|\, \psi(x)\,\psi(cx) \,|\, p\rangle$ are derived in lowest NTD-approximation using above ψ — contractions with the selfconsistent baryon pole. They appear as «S-states» of the baryon-antibaryon glued together by the contact interaction of the nonlinear term, and correspond to the η, π, ω, ζ mesons. However, only the spin 0 states η, π were considered [3]. In case of ω, ϱ also D-states have to be included which is more difficult. For η, π the eigenvalue equation is of the form:

$$1 + [3 - I(I+1)]\, 2Q\left(\frac{\mu_I^2}{m^2}\right) = 0$$

with $Q\left(\frac{\mu_I^2}{m^2}\right)$ a known function, which depends on the mass ratio of meson to baryon mass.

3) The boson eigenvalue equations are extended to include the K-mesons by considering the wave functions $\langle 0 \,|\, S\,\psi(x)\,\psi^*(x)\,|\,p\rangle$. The isospin-wave function of S and ψ must be symmetrical. Because of the well-known relationship (change of coupling transformation)

$$(\overline{123}) = \sqrt{\frac{3}{4}}\,(\widetilde{123}) + \sqrt{\frac{1}{4}}\,(12\overline{3})$$

where — designates a symmetrical, ⌢ an antisymmetrical combination, one immediately obtains the results that the K behaves like 75% and 25% π, and hence above eigenvalue equation can be generalized to

$$1 + [3 - [I(I+1) - u^2]]\, 2Q\left(\frac{\mu_{I,u}^2}{m^2}\right) = 0$$

to include the K. According to the above derivation this formula should **not** be used for other states than η, π, K. E.g. a K-quartet state $(I = 3/2,\ u = 1/2)$ can only occur if also a $I = 2$, $u = 0$ — state is used from the beginning. Without such a state the K-quartet will just obey the same eigenvalue equation as the π, and hence the spurion will not be actually «bound». On the other hand the consideration of a $I = 2$ state will introduce field operators at different space-time points, and therefore such particles are not «primary» particles anymore in the above sense. Hence there exist only 8 «primary» boson states of spin 0: Their eigenvalue operator contains as a factor the operator $I(I+1) - u^2$. Since the function $Q^{-1}(x^2)$ behaves essentially like $\ln^{-1} x^2$ with $x^2 = \mu^2/m^2$, it can be well approximated in the relevant region by $(a + bx^2)$ which leads to the well-known boson mass formulas. The same procedure would lead to a spin-1 «octet» if the spin-1 states were predominantly ^3S-states which does, however, not seem likely from our calculations.

4) The baryon eigenvalue calculations are repeated with the inclusion of not only the baryon contraction functions but also the boson contraction functions (boson pole terms in the ψ — 4-point functions); i.e. baryon mass corrections due to virtual η, π, K, as calculated in 3) are included. Due to the K-dressing the eigenvalue operators of baryons with zero, one- and two spurions now differ from each other. Because of the symmetry requirements in the definition of the zero, one- and two-spu-

rion baryons, the additional term has the characteristic factor: $a\,Bu + b\,[I\,(I+1) - u^2]$. The inclusion of the boson-contraction terms improves the mass ratios between baryons and bosons. The signs of the constants a and b are such that the empirical mass sequence of the baryons results, however only if the zerospurion system (the original system) is identified with the Ξ, the twospurion system with the nucleon. The one-spurion system corresponds to the and Λ, Σ. The numerical values of a and b are still somewhat too small. However, it seems plausible that the inclusion of the spin-1 mesons may eventually improve the situation. The baryon calculation is not necessary limited to the «octet» states but may also include other states, e.g. quartet statet. To exclude these «unwanted» states the hypothesis was used that spurions can only be «bound» to a s i n g l e field operator ψ (x) such that the resulting electric charge never exceeds the value one. Such an assumption seems necessary

in view of the local conservation of charge. It this case there would be only 8 primary baryons.

5.) As a future step one may consider all elementary particles which cannot be constructed from field operators at the same space-time point. These particles may be termed «secondary particles». It seems rather plausible that for these more loosely bound particles, e.g. the 3/2, 3/2 resonance state, the interaction through exchange of primary particles is the most important one. Since the primary particles roughly allow a broken SU_3 terminology, also the secondary particles may allow a similar description.

REFERENCES

1. D ü r r H. P., H e i s e n b e r g Z. W. Naturforsch., 16a, 726 (1961).
2. D ü r r H. P. Géhéniau J. Nuovo cimento, 28, 132 (1963).
3. D ü r r H. P. et al. Z. Naturforsch., 14a, 441 (1959).

Progress of Theoretical Physics, Supplement, 298-303 (1965)

Supplement of the Progress of Theoretical Physics, Commemoration Issue for
the 30th Anniversary of the Meson Theory by Dr. H. Yukawa, 1965

Particles as Collective Stationary States

W. HEISENBERG

*Max-Planck Institute for Physics and Astrophysics
Munich, Germany*

(Received June 23, 1965)

The hypothesis of *Yukawa* which connected the nuclear forces with some kind of strongly interacting particles, has led in a rather straight line of reasoning to the modern concept of an elementary particle. The atomic nucleus may be considered as a compound system of nucleons and a field of force; this field is according to Yukawa equivalent to virtual pions (or as we now know: ρ mesons etc.). This 'virtual' existence of pions in the nucleus is a consequence of the localisation of the nuclear field which—on account of the uncertainty relations—allows to consider the pions 'off their mass shell'. If this situation is described by saying that the nucleus potentially consists of nucleons and pions, it should be equally permitted to say that any system (atom, nucleus or elementary particle) potentially consists of any other particles producing together the same symmetry as the system concerned. In this way the particles appear generally as collective states[1] resembling the excited states of a solid body or a liquid, like phonons, excitons, polarons etc. The solid body in quantum mechanics should in elementary particle physics be replaced by the ground state 'world' or 'universe', but else the analogy should be rather close.

This picture of the elementary particles which seems to be generally accepted nowadays suggests the application of mathematical methods which have proved useful in solid state physics or more generally in many body problems. In the following lines the method of the Green-functions[2] shall be briefly discussed, and applied to the theory of coupling constants, recently given by Dhar and Katayama.[3] The term Green-function is used here in the generalised sense as vacuum expectation value of products of some fundamental field operators describing the system. In quantum mechanics of solid bodies the connection between the different Greenfunctions is ultimately established by means of the Schrödinger equation or a Klein-Jordan-Wigner wave equation or a corresponding Hamiltonian. A similar procedure can be adopted in elementary particle physics, if a fundamental field operator of 'matter' is introduced obeying a non-linear differential equation which by its group structure defines the physical properties of objects within this world and thereby of the elementary particles.

When the connection between the different Greenfunctions has in this way been established one gets an infinite set of non-linear integral equations leading to Greenfunctions of more and more variables. Such an infinite set must be cut off in order to define a solution. For this purpose it is convenient to separate the Greenfunctions into products of Greenfunctions with a smaller number of variables and the irreducible rest which is usually called the η function.[4] The η function (or 'correlation function') of a certain high number of variables may be neglected in order to define an approximation. Thereby the infinite set of equations is cut off, the number of equations is then equal to the number of the unknown η functions, and an approximative solution may be found. This general method shall be applied here to the non-linear spinor theory of elementary particles.

In this theory one starts from the equation

$$i\sigma_\nu \frac{\partial \chi}{\partial x_\nu} + l^2 : \sigma_\nu \chi (\chi^* \sigma^\nu \chi) := 0 \,, \tag{1}$$

where $\chi(x)$ is a Weylspinor in Lorentzspace and a spinor in isospace. In Feynman's notation the ordinary Greenfunction of the linear Weyl equation will be denoted by a single line:

$$\underline{\qquad} \qquad \text{Greenfunction} \quad \frac{p_\nu \bar\sigma^\nu}{p^2} \,, \tag{2}$$

the η functions (the irreducible parts of 2-point, 4-point, etc. functions) by circles with outgoing lines marking the variables:

$$\text{2-point } \eta \text{ function,}$$
$$\text{4-point } \eta \text{ function.} \tag{3}$$

Then the two lowest relations between η functions are

$$\tag{4}$$

$$\tag{5}$$

(The graphs are always meant as properly symmetrized with respect to the variables. If the symmetrization would explicitly be taken into account in the notation, the Feynman picture of (5) would be considerably

 W. Heisenberg

more complicated.)   It is only the local vertex of four lines, which is a characteristical feature of the fundamental equation (1).

If in Eq. (5) the 6-point η function is omitted, one thereby defines an approximation which may be used for determining the two lowest η functions and hence indirectly masses and coupling constants of the lowest fermion- and boson-states.

We first consider Eq. (5), which in this approximation may be written in the form

$$\left[\,1-\,\rlap{\times}\bigcirc\,\right]\cdot\rlap{\times}\bigcirc \;=\; \rlap{\times}\bigcirc\!\bigcirc \;+\cdots \tag{6}$$

(where $+\cdots$ means terms connected with symmetrization).
If the 2-point η function $-\!\bigcirc\!-$ is known, this is an inhomogeneous linear integral equation for $\rlap{\times}\bigcirc$.   For the determination of the boson eigenvalues it is sufficient to consider the term due to the creation and annihilation of bosons in $\rlap{\times}\bigcirc$:

$$\rlap{\times}\bigcirc \;\approx\; \rlap{\times}\bigcirc\!\!-\!\!\rlap{\times}\bigcirc \tag{7}$$

and to discuss only the homogeneous equation

$$\left[\,1-\,\rlap{\times}\bigcirc\,\right]\cdot \rlap{\text{|}}\bigcirc \;=\;0. \tag{8}$$

If the two variables on the left-hand side are put equal, Eq. (8) is simplified to

$$\left[\,1-\,\bigcirc\!\!-\!\!\bigcirc\,\right]\,\rlap{\text{|}}\bigcirc \;=\;0, \tag{9}$$

which is just an algebraic equation for calculating the boson masses. This equation has been discussed in earlier papers.[5]

For the calculation of coupling constants in scattering processes the inhomogeneous equation (6) can be integrated by

$$\rlap{\times}\bigcirc \;=\; \dfrac{1}{1-\bigcirc\!\!-\!\!\bigcirc} \;+\cdots. \tag{10}$$

(The terms $+\cdots$ are unimportant in the neighbourhood of the poles.)
Equation (10) may be verified immediately by inserting (10) into (6).

Particles as Collective Stationary States   301

This expression (10) is to be compared with the expression

$$\text{(diagram)} \tag{11}$$

in a phenomenological theory determining the scattering of two fermions by means of an intermediate boson, the propagator of which is marked by $\sim\!\sim\!\sim$ in (11). The pole of the propagator at the boson mass is produced in (10) by the denominator which vanishes according to (9) at the boson eigenvalues.

Before we follow the line of Dhar and Katayama (l.c.) for getting numerical values of the coupling constants by comparing (10) and (11), we will go back to Eq. (4) in order to get a first information about the 2-point function $\text{---}\!\circ\!\text{---}$. Inserting (5) into (4) and omitting again terms with the 6-point η function gives

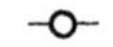

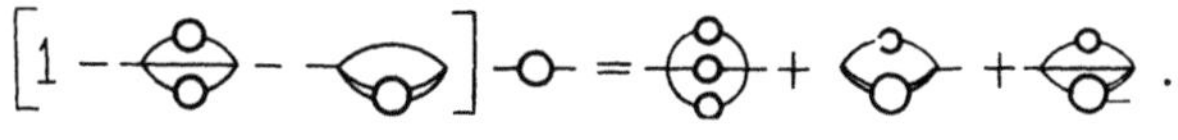

$$\left[1 - \text{(diagram)} - \text{(diagram)} \right]\!\text{---}\!\circ\!\text{---} = \text{(diagram)} + \text{(diagram)} + \text{(diagram)} . \tag{12}$$

This complicated non-linear equation could again be considered an in-homogeneous linear equation for $\text{---}\!\circ\!\text{---}$, if the operators in the bracket [ ] and on the right-hand side would be given. In this case the fermion eigenvalues could be calculated from the homogeneous linear equation

$$\text{---}\!\circ\!\text{---} \;\approx\; \text{(diagram)} \tag{13}$$

and

$$\left[1 - \text{(diagram)} - \text{(diagram)} \right]\!\text{(diagram)} = 0 . \tag{14}$$

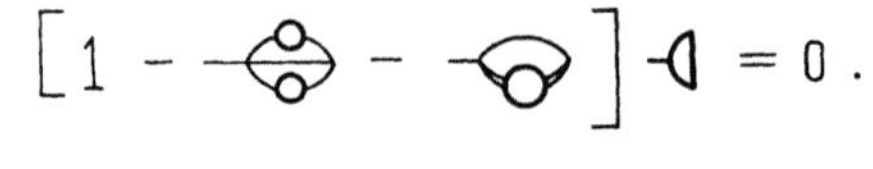

This is in fact the way in which the fermion eigenvalues have been calculated in the earlier papers.[5] The second operator (diagram) had been neglected in the first calculations, but parts of it were later included in a paper by Dürr and the author.[6] In these investigations the operators in the bracket were obtained by replacing $\text{---}\!\circ\!\text{---}$ by the trial function

$$\text{---}\!\circ\!\text{---} \sim \frac{p_\nu \bar{\sigma}^\nu \cdot \kappa^4}{(p^2)^2 (p^2 + \kappa^2)} \tag{15}$$

and determining the mass κ of the fermions by equating it with the lowest eigenvalue of (14).

Equation (12) can now be used for checking and improving the trial function (15). If (12) is solved by putting

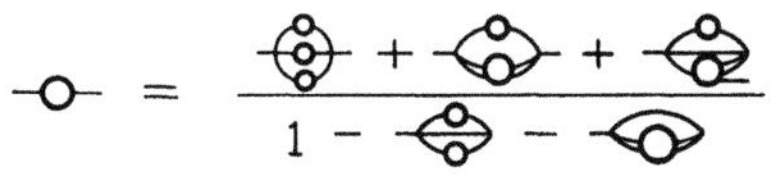

302   W. Heisenberg

$$-\!\circ\!- \;=\; \frac{\text{(diagram)} \;+\; \text{(diagram)} \;+\; \text{(diagram)}}{1 \;-\; \text{(diagram)} \;-\; \text{(diagram)}} \tag{16}$$

and by inserting (15) and (10) on the right-hand side, one gets a new 2-point function $-\!\circ\!-$, which, like (15), contains a pole at $p^2+\kappa^2=0$ and — as can be seen by studying the behaviour of $-\!\circ\!-$ at large distances $(x-y)^2$— a double pole at $p^2=0$. But else it will be different from (15); it will in fact be a much more complicated analytical function. This procedure can be repeated by inserting now the new function $-\!\circ\!-$ on the right-hand side of (16) and redetermining κ by postulating, that the pole of the new function $-\!\circ\!-$ coincides with the pole on the new right-hand side of (16), etc. One may hope that this procedure converges and leads to a final form of $-\!\circ\!-$ in the approximation defined by the omission of the 6-point η function.

The coupling constant between fermions and bosons can be calculated by comparing the graphs (10) and (11), when the graph $\succ\!\!\prec$ refers to a phenomenological theory with vector coupling between the fermion and the boson field. In this theory the wave equation for the fermion field could be written as

$$i\sigma_\nu \frac{\partial \chi}{\partial x_\nu} + \frac{f}{\kappa_B}\, \sigma_\nu \chi \cdot \frac{\partial \phi}{\partial x_\nu} + \cdots = 0 \tag{17}$$

in case of scalar bosons, or

$$i\sigma_\nu \frac{\partial \chi}{\partial x_\nu} + f\sigma_\nu \chi \cdot A_\nu + \cdots = 0 \tag{18}$$

for vector bosons.

In field theory the operator $\bigcirc$ can be evaluated, when $-\!\circ\!-$ is approximated by the trial function (15). The result is

$$\bigcirc = -\left(\frac{\kappa l}{2\pi}\right)^2 \left[\frac{3}{4}P_{I=0} + \frac{1}{4}P_{I=1}\right]\left[P_{s=0}\cdot q_0(\lambda) + P_{s=1}\cdot q_1(\lambda)\right], \tag{19}$$

where $\lambda = -\dfrac{J^2}{\kappa^2}$, J_μ is the momentum vector of $\bigcirc$, and

$$q_0(\lambda) = \ln\lambda - \frac{1}{\lambda} - \frac{(1-\lambda)^2}{\lambda^2}\ln|1-\lambda|,$$

$$q_1(\lambda) = \left(1 - \frac{2}{3}\lambda\right)\ln\lambda + \frac{2}{3} + \frac{1}{3\lambda} + \frac{(1-\lambda)^2(1+2\lambda)}{3\lambda^2}\ln|1-\lambda|. \tag{20}$$

The projection operators P determining the isospin or the Dirac spin

Particles as Collective Stationary States 303

of the operator $\langle\!\!\!\!\bigcirc\!\!\!\!\rangle$, are

$$P_{I=0}=\frac{1}{2}\cdot\mathbf{1}\cdot\mathbf{1} \; ; \quad P_{I=1}=\frac{1}{2}\cdot\boldsymbol{\tau}_k\cdot\boldsymbol{\tau}_k \; ; \tag{21}$$

$$P_{S=0}=\frac{J_\mu J_\nu}{J^2} \; ; \quad P_{S=1}=g_{\mu\nu}-\frac{J_\mu J_\nu}{J^2} \; . \tag{22}$$

For the interaction term in the fundamental equation (1) it is important to notice, that—on account of the anticommutativity of the field operators—this term is Fierz symmetrical; therefore the expression $\mathbf{1}\cdot\mathbf{1}$ in isospace may be replaced by its Fierz symmetrical version

$$\mathbf{1}\cdot\mathbf{1}\rightarrow\frac{3}{4}\cdot\mathbf{1}\cdot\mathbf{1}+\frac{1}{4}\cdot\boldsymbol{\tau}_k\cdot\boldsymbol{\tau}_k \; . \tag{23}$$

Comparing now the two graphs (10) and (11) and neglecting in this comparison the distinction between —— and —o— of the fermion lines, one gets at once

$$f^2=\frac{4\pi^2}{\partial q_0/\partial\lambda} \qquad \text{for } S=0 \text{ (scalar bosons),}$$

$$f^2=-\frac{4\pi^2}{\partial q_1/\partial\lambda} \qquad \text{// } S=1 \text{ (vector bosons).}$$

These formulae have been used by Dhar and Katayama (l.c.) for determining the π-N-coupling constant $(f^2/4\pi\approx0.07)$ and by Dürr, Yamamoto, Yamazaki and the author[7] for calculating the Sommerfeld fine structure constant $(e^2/4\pi\approx1/120)$.

References

1) W. Heisenberg, The present situation in the theory of elementary particles, in: Two lectures, Cambridge Univ. Press 1949; Nachr. d. Gött. Akad. Wiss. IIa, **8** (1953), 111; Physica **19** (1953), 897.

2) Compare A. A. Abrikosov, L. P. Gorkov and I. E. Dzyaloshinski, *Quantum Field Theory in Statistical Physics*, (Prentice Hall 1963), and P. Nozières, *Theory of Interacting Fermi Systems* (Benjamin New York-Amsterdam 1964).

3) J. Dhar and Y. Katayama, Nuovo Cim. **36** (1965), 533.

4) K. Symanzik, Green Functions, etc., Hercegovni lectures. W. Zimmermann, Nuovo Cim. **13** (1959), 503.

5) H. P. Dürr, W. Heisenberg, H. Mitter, S. Schlieder and K. Yamazaki. Zs. f. Naturf. **14a** (1959), 441.

6) H. P. Dürr and W. Heisenberg, On the 'Spurion' theory of strange particles (to appear in Nuovo Cim.).

7) H. P. Dürr, W. Heisenberg, H. Yamamoto and K. Yamazaki, Quantum electrodynamics in the non-linear spinor theory and the value of Sommerfeld's fine structure constant (to appear in Nuovo Cim.).

In *Proceedings of Seminar on Unified Theories of Elementary Particles, July 1965,* ed. by H. Rechenberg
(MPI für Physik und Astrophysik, Munich 1965) pp. 1-15

Survey on the Present Situation

Concerning Unified Theories of Elementary Particles

W. Heisenberg

Max-Planck-Institut für Physik und Astrophysik
München (Germany)

Introduction

May I welcome you all to this seminar on unified theories of elementary
particles. As you know the seminar goes back to a tradition of about five
years, and the ideas of the seminar were initiated by the Rochester physicists, I think especially Marshak and Sudarshan. Contrary to the big
Rochester conference the plan was that there should be ample time for discussions in a kind of leisure and vacation (of course depending on the
weather conditions!). Perhaps one could say that the main Rochester conference is intended for the exchange of informations, while this seminar is
intended for the exchange of views and opinions on rather difficult theoretical problems.

The theme of the conference, the 'unified theory of elementary particles' suggests that we start at least from some common hope considering
the theory of elementary particles: namely that it should be possible to
understand the spectrum of the particles and their properties from some
underlying unifying principle. Or we could say that we start from a common
concept of elementary particles: the elementary particles are just different forms of a fundamental substance which we may call matter or energy;
energy becomes matter by assuming the form of an elementary particle. Some
physicists like the phrase: Every elementary particle consists of all
other elementary particles - which is of course a rather paradoxical way
of talking about the situation. But anyway, when we take this general
view, then we compare the elementary particles not with the electrons of
quantum mechanics but rather with the stationary states of a complicated
atom like the iron atom or - and this might be a better comparison still -
with the excited states of a solid body, like excitons and polarons. By
this comparison of course we mean that the interaction always plays a

- 2 -

very essential role even for the single objects, which are 'dressed up' by the interaction. If we start from this general picture we see that our present problems are really very similar to the problems which we had in quantum mechanics fourty years ago. Perhaps you allow me to say a few words about this similarity which I feel is extremely striking. At that time we started from a very large experimental material on spectral lines, energy levels, selection rules, transition probabilities etc. Nowadays we start from a similar experimental material. We know many elementary particles, excited states of short life time; we know selection rules and transition probabilities. In the old period of physics we had at the same time very serious difficulties. The difficulties could be expressed by a few hard questions. For instance the famous question: Is light a wave motion or does it consist of particles? Or the other problem: What is the orbit of an electron in an atom?

In our time we also have very serious difficulties. They can be expressed for instance by the question: Does the probability concept of quantum mechanics apply to the interior of an atomic nucleus? Or perhaps better: Does it apply to the interior of a proton, of an electron, of a pion? And again like in the old times the answer to these questions can be given only when the new theory has been found. I remember quite well that even in the early twenties there were physicists who believed that one could use Newtonian mechanics in quantum theory and could calculate atomic levels by such methods. It was only afterwards when one knew how the new theory worked that the necessity of abandoning the old concepts was really accepted. This is of course exactly our present situation. We do not know yet what finally the structure of the new theory will be.

Since our present situation is so similar to that fourty years ago it is natural to believe that also similar methods should lead to progress. Therefore I would like to think of this seminar in the same way as I think of the old discussions in Bohr's institute in Copenhagen. There we obeyed the general rule that whatever detail was discussed or analysed, it was always discussed in the light of certain very hard questions. Not with the hope that one could answer the questions from the detail; but with the idea that all these different details are connected in some hidden background of physics which one should try to throw light upon; and only by insisting upon always to discuss the

- 3 -

details in the light of the hard questions one really could make progress.
It was especially Niels Bohr who insisted upon considering every detail
as a kind of comment on the hard questions in the background.

Therefore I would like to start the present conference by writing on
the blackboard some hard questions which I consider as the most important
ones. Of course you may already disagree about the hard questions, which
therefore could be an object of discussions. I will select mainly three
topics for which I can just give the title:

Group Theory, Mathematical Scheme and Approximation Methods.

Let us start with group theory, though I do not believe that the group
theoretical problems are really the hardest ones. But group theory has made
much progress in recent years, many physicists have been interested in the
group theoretical side of the unified theory of elementary particles. Therefore it may be worthwhile to start with the group structure of the natural
law responsible for elementary particle physics.

<u>Group structure</u>

The first question should then be

1) What are the <u>exact</u> symmetries of the underlying natural law?

The emphasis is here on the word 'exact'. Most of us believe that there
must be some exact symmetries in the underlying natural law. I think it
is reasonable also to believe that those symmetries which are observed
in nature as holding with very high precision should fundamentally be
exact symmetries - in the sense that if deviations are observed, these
deviations should be caused by actions from outside, say from the structure of the whole world or from some outer forces which are artificially
applied etc. So there must be a number of exact symmetries, and it will
be very important to know what are the exact symmetries of the underlying natural law.

These exact symmetries have to be contrasted to the <u>approximate</u>
symmetries which are now so well known, and there we may look for
different causes of the approximate symmetries. Therefore I would as
a second question put here

2) Are the observed approximate symmetries due to

a) an <u>exact</u> symmetry broken by an asymmetry of the groundstate, or

- 4 -

b)   secondary dynamical effects.

In the second case they would not be intrinsic symmetries of nature at all, but just be produced approximately by secondary effects within the dynamics of the system. And then of course there may be

c)   other causes.

Dirac: Is there a rigorous difference between a) and b) ?

Heisenberg: Yes, there is a very marked difference between the two cases, and I will explain it in the case of quantum mechanics. How would we answer these questions in quantum mechanics? We would for instance say that quantum mechanics is exactly symmetrical under the Galilei group. Still we have sometimes only approximate symmetries. Suppose we have a crystal, and we ask for the behaviour of excitons in the crystal; then for the interaction between two excitons the group of rotations would not be fulfilled exactly, because that collision process takes place in a surrounding where the rotational symmetry is destroyed by the axis of the crystal. Therefore, in spite of the fact that the quantum mechanical equations are exactly invariant under the rotation group, we would not expect to see conservation of angular momentum in a collision process; a part of the angular momentum could be transferred upon the crystal. Of course in a more general sense we can say that angular momentum will always be strictly conserved. When the whole crystal is taken as part of the system, the total angular momentum is constant. But the motion of an infinite crystal could not be observed. Therefore we would say in this case, that we have an exact symmetry broken by the asymmetry of the groundstate, which is the situation described under a) .

As an example for the second possibility b) we take the multiplet structure in the optical spectra of atoms. This multiplet structure is brought about by the

- 5 -

smallness of the spin-orbit interaction. In a certain
approximation it is allowed to rotate the spins of the
electrons alone or the orbits alone, while the theory is
rigorously invariant only when both are rotated simultaneously. Another example of such an approximate symmetry
of type b) is the seniority number in atomic nuclei.
The dynamical symmetries are approximate symmetries from
the very beginning, while in case a) we have exact symmetries broken by an asymmetry of the groundstate. I
should also mention that the Goldstone theorem, which will
be a major object of our discussions, allows to distinguish between the two cases: in case a) there will be
Goldstone particles, and in case b) there will not be
Goldstone particles. So there is a clear distinction between the two kinds of approximate symmetries. The possibility c) 'other causes' has been included, not because
I would know of any other causes but only because we can't
exclude them.

In quantum mechanics a clear answer can be given to
the questions listed above: The Schrödinger equation is
strictly invariant under the Galilei group; we have case a)
in the crystals, and case b) in the multiple structure
of optical spectra.

But can we give the answers in elementary particle
physics? There are a few answers about which most physicists will agree, and there are other answers which are
most controversial. I hope that there will be many discussions in our seminar just about these grouptheoretical
problems.

I think most physicists would agree that we should
count the Lorentzgroup L as an exact symmetry, and then
probably also parity P , charge conjugation C and
time reversal T ; because these groups, if disturbed at
all are only broken by very small effects. P and C are
broken by the weak interactions, PC seems to be broken
by a still weaker interaction in K_2-decay, and the proper

- 6 -

Lorentzgroup so far does not seem to be broken at all. Therefore most physicists would agree that these are good groups of the underlying natural law.

Sudarshan: You include P, C and T under the exact ones. And on the other hand you list secondary approximate symmetries under a different group. Are you thinking of the weak interactions as being somewhat less fundamental than the other interactions?

Heisenberg: No. I would list P and C both under 1) and 2a). P and C are exact symmetries in my opinion, but they are broken by some asymmetry of the groundstate. The weak interactions would be caused by a very weak asymmetry of the groundstate, which could be a consequence of the cosmological structure of the world. The weak interactions should not be an intrinsic property of the underlying natural law.

Generally speaking I would find it difficult to believe that the fundamental natural law should contain a symmetry which is almost perfect but not quite perfect. Therefore I would prefer to think of these very small deviations as due to an asymmetry of the groundstate.

I would also like to include under 1) and 2a) the SU2-group of isospin or rather U2 of isospin and strangeness.

Dirac: Are the same symmetries in both 1) and 2) or have you different symmetries in 1) and 2a) ?

Heisenberg: Any approximate symmetry listed under 2a) must also appear under 1), since it is due to an exact symmetry. There may however be symmetries which occur under 1) but not under 2a), namely those where we do not observe any deviation from symmetry; for instance Lorentzgroup and PCT.

Then we have secondary dynamical effects, and I would believe that the famous groups which have been so successful in recent work: SU3, SU6 etc. should be listed under 2b) . This problem is of course open for discussion. But my own reasons for classifying SU3 etc. under 2b) are the following: First of all there are very strong deviations

- 7 -

from these symmetries. While for SU2 or U2 the deviations
are really very small say of the order of $\frac{1}{1000}$ in the
mass defects, here for SU3 the effects are of the order
of say $\frac{1}{3}$; hence the deviations from these symmetries are
very big. Furthermore we are here induced to include with
SU3 the spin of the particles and thereby to come to SU6.
On the other hand the spin is a representation of the
Lorentzgroup. And the Lorentzgroup can never be included
in these unitary symmetries because it is a noncompact
group. Therefore one has the impression that these groups
are by their very nature only approximate groups, secondary
dynamical effects.

We could also apply the criterion of the Goldstone
theorem. In connection with U2, the isospin group, we have
probably a Goldstone particle namely the photon, while we
have certainly no Goldstone particle connected with SU3
or SU6. If we had one then this Goldstone particle should
have a very big interaction, it would have to produce the
large deviations from the SU3-symmetry. Such a particle
would certainly have been observed, if it existed, and it
has not been observed.

Besides that the quark particles as simplest representation of SU3 would probably appear if SU3 would be a
fundamental group; but they have not been observed. As I
said before, this is a problem which should be discussed
carefully and I hope that during our seminar we will hear
many arguments in favour or against the proposed classification.

With respect to P and C, I would like to mention that
here the Goldstone theorem would probably not apply in
its conventional form. P and C are discrete groups, and the
Goldstone theorem has so far been formulated only for a
continuous group; one does not know what the mathematical
conclusions are when we start from discrete groups; they
are again a matter for further investigation.

- 8 -

This is already more or less what I wanted to say about the group
structure; I just wanted to write down a kind of program for the
discussions of our conference. I hope that when we go into the details - and everybody knows how successful the work with these
higher groups has been - that we do find arguments in favour or
against such a classification of the higher groups.

As I said before, the grouptheoretical problems are really not
the hardest ones. I hope that it will be comparatively simple to
get to reasonable answers to the questions listed above. But now
I come to the really hard questions, which have to do with the

Mathematical Scheme.

What will be the mathematical tools of the unified theory?
I would like to classify the different possibilities in the following way: Probably most physicists will agree that we should have an
S-matrix, that an S-matrix is a good instrument, by which we can
describe the results of experiments. Therefore the widest frame is a

1) Unitary S-matrix, representing groups and causality.

Of course this means already that we have asymptotic states,
states that can be described by the motion of particles without
interactions. Furthermore the S-matrix must be invariant under
the groups listed under 1) and 2a), not invariant under the
groups 2b); and by its analytical properties it must somehow
express causality. This 'somehow' of course involves already a
lot of difficulties, as everybody knows. And we hope to learn
about these problems again in the discussions.

In the end it may turn out that the S-matrix is not really
sufficient. The S-matrix is a useful tool; but if we speak of
causality, we try to guess the properties of the S-matrix elements from field theory. Therefore one may argue that we have
to add some part of field theory to the existence of an S-matrix,
and we add for instance the existence of field operators. Our
second frame, somewhat narrower than the first, is therefore

- 9 -

2) <u>S-matrix + field operators, Hilbert space, local commutation rules.</u>

The Hilbert space in which the field operators act, will now contain not only the asymptotic states, but possibly also other states. This problem has not been decided. In the field operators causality should be expressed by <u>local commutation rules</u> and the field operators should fulfill a nonlinear differential equation.

Whether this is possible, whether these mathematical tools can really be used in quantum field theory, is one of the most critical questions and I might perhaps just give a few arguments concerning this problem. In favour of the field operators one can say that the analytical properties of the S-matrix have so far always been derived from analogy with field theory. Field theory has undoubtedly been very successful in quantum electrodynamics; and whenever we study the analytical properties of the S-matrix we first ask how are the corresponding relations in a field theory which we understand. Therefore all the dispersion relations and connected problems have been discussed first in the light of a field theory which one believed to exist or where one just made reasonable approximations in perturbation theory. And even if the approximations should not be valid at all, we believe that the analytical properties of the S-matrix elements have been derived correctly.

On the other hand there is already from the very beginning a serious difficulty in a field theory with interaction, connected with the concept of Hilbert space and with the concept of stationary states. This point has been emphasized recently in a lecture by Dirac in Lindau. Let us think of a simple Schrödinger representation of a state in quantum mechanics. For instance the normal state of the hydrogen atom can be expressed by a wave function of three space coordinates. Any stationary state can be expressed by a simple analytical function of a finite number of variables. But when we think of quantum field theory with interaction then the problem becomes very serious at once. Because, as I said it before, every particle consists of all other particles. That means when we have a particle defined by some symmetry then of course we can produce the same symmetry by having several particles in some suitable configuration. In fact we can have any number of particles because we can always add pairs of particles and antiparticles.

- 10 -

Therefore if we want to express the state of a proton, say, in the same Schrödinger way, as we express the state of a hydrogen atom, then we would have to introduce functions of not only one variable but of three, five, seven and so on, of any finite number of variables, even of an infinite number of variables, because one proton may be considered as consisting of a proton plus one pair, a proton plus two pairs, proton plus several bosons and pairs, and so on. Therefore a Schrödinger representation of a state would actually consist of an infinite number of functions, and among these functions there are infinitely many functions with a number of variables which is larger than any given finite quantity. So this is a rather difficult concept of a stationary state, and therefore one may well doubt whether such a representation of states is possible at all. On the other hand I would not be too sceptical at this point, since it should be possible to replace the Schrödinger representation by what may be called an approximate representation. This approximate representation of a state vector may be obtained by omitting all functions with more than n variables, where n is a given finite number. Finally one would have to prove that if one makes n larger and larger then the results converge towards finite limits. Therefore I hope that the mathematicians someday will be able to introduce in a rigorous way this concept of an approximate state vector, and the approximate state vector should be defined by a finite number of functions of a finite number of variables. These finite sets of functions with a finite number of variables should then be handled in principle just like the old Schrödinger vectors. But I have to emphasize again: Whether such a representation is possible or not, that is a completely open question.

If we believe that Hilbert space can be defined in a reasonable way, and that we actually can work with field operators, then of course there is a well known connection between the field operators and the S-matrix. Then and only then can we extend the S-matrix elements to the so called Green's functions, we can speak about the v.e.v. of products of field operators, and as everybody knows, these v.e.v. then are very closely connected with the S-matrix elements. The Green's functions could in momentum space contain poles, and whenever they

- 11 -

contain poles, that means in physics, that we have particles of the
corresponding masses, and the behaviour of the Green's functions near
the poles just defines the S-matrix. But as I said before: it is an
open question whether such a mathematical description is possible at
all.

Finally we can in our axiomatic foundation try to make a third
restriction. Many physicists believe, that to the axioms listed above
one can add the assumption, that the metric in the Hilbert space
should be a positive definite metric, and that the total Hilbert space
can be defined by the asymptotic states alone:

3) <u>S-matrix + field operators + definite metric, Hilbert space defined
by· asymptotic states only.</u>

In this case we come more or less to the axiomatic frame which
has been studied by Wightman, Lehmann and Symanzik, Källen and others.
This axiomatic frame is very close to that of quantum mechanics. If
this third possibility could be realized then we would undoubtedly
have the smallest deviations from old quantum mechanics. But of course
this means now postulating still more than under point 1 or 2, and I
think that there are definite reasons to believe that we will get
into difficulties at this point.

Again I know that this is a very controversial question, and I
hope to learn from the discussions in our present seminar. But again
I would like to give some reasons why we in Munich mostly like to
believe that 3) is not possible, while 2) is possible. The most important argument comes from the problem of causality. If a field theory
with interactions is possible, then this means that we must introduce
a local interaction of the type: product of several field operators.
There should be some nonlinear differential equation or a group of
differential equation like in quantum electrodynamics, and in these
differential equations the product of field operators at the same
point in space and time should occur. For instance in quantum electrodynamics we introduce the interaction term as product of the vector
potential and the Dirac wave function at the same point in space and
time. Such an interaction may be called an interaction of the δ-func-
tion type, and it is a very difficult concept, because normally
- if the field operators obey the rules of canonical quantization -

- 12 -

the product of two field operators at the same point is mathematically
just not defined. Such a term would mean an enormously violent interaction, and this violent interaction really breakes the conventional
rules of quantization. Mathematically the difficulty appears in the
following way: when we consider a two-point-function of the type
$\langle 0 | \chi(x)\, \chi^*(y) | 0 \rangle$ as function of the difference x-y ,
then for a theory without interaction at the light cone we have a
δ -function. But if this δ -function occurs, then the product of
two field operators at the same point has no meaning.

Therefore I would feel that there is only the one way out which
Dirac has suggested some twenty years ago. We simply assume that there
is no δ -function on the light cone in the two-point-function or in
higher functions in case of strong local interaction. In this case we
know from the papers of Källen, Lehmann and Symanzik that the metric
in Hilbert space cannot be definite. Of course there may be other
ways out of these difficulties, which have not been found yet; but I
doubt whether one can really give a good new answer to these rather
difficult questions. So we may just stop at this point and say: the
problem of the mathematical axioms of field theory with interaction
is still open. Most of us will agree that frame 1) is possible, probably a few will also agree that frame 2) is possible, and there will
be, I hope, many discussions on the problems of 3).

Finally there is a third group of problems which I can just write
down because I feel that they are minor problems, though perhaps you
will not quite agree to this view. What are the

Mathematical Methods of Approximation

which we can use? So we have to speak about the <u>reliability of
approximations</u>. I mention this topic here especially because I feel that
in the past the problems of no 2, of the 'Mathematical Scheme' and the
problem of the reliability of approximations have frequently been mixed
together; and this should not be done. Methods of approximations are generally not suited to solve such general problems as those of the mathematical axioms of field theory. The methods of approximation can be discussed
only after one has come to a wellfounded opinion concerning the mathematical
frame. But when one has an opinion on these problems then and only then

- 13 -

the problem of approximation comes in.

There have been discussions in the literature on the different methods of approximation. I think everybody agrees that the methods of <u>perturbation theory</u> cannot be used for the strong interactions. Therefore I do not even list it here in these lines.

<u>Dürr</u>:   But they are good in quantum electrodynamics.

<u>Heisenberg</u>:   Yes,definitely perturbation theory is good in some special cases, I agree.

Then there is a method which has recently been used with considerable success. I will describe it in very general terms as <u>relations between Green's functions</u>. These relations between Green's functions have been successful in the problems of many body physics; e.g. for calculating the excited states of crystals or treating the problems of superconductivity, superfluidity etc. I think that these methods are also very well suited to solve problems in elementary particle physics; if one believes in the axiomatical scheme 2), then the Green's functions can be easily defined as v.e.v. of some products of operators. Even within the frame 1) the concept of Green's functions can be used as a kind of analytical continuation of S-matrix elements which is necessary already for representing causality. One could probably list the <u>bootstrap calculations</u> under this heading, and I understand there will be some lectures in our seminar which will deal with this problem. The bootstrap calculations pick out from the many possible relations between Green's functions certain special ones, and try from these special relations to get information about stationary states, masses and coupling constants.

Finally I will list here the <u>Tamm-Dancoff method</u> which has been somewhat discredited for many years because there have been problems in which the Tamm-Dancoff method did not work. Still I believe that these were problems in which the mathematical basis was not defined well enough and did contain contradictions. At first sight it looks as if this method could be well suited for the real situation in elementary particle physics. We know that we cannot define state vectors in the same way as in old quantum mechanics; we must always consider a state vector as an infinite set of functions of which we can only take a finite part. But if only a finite part of this infinite set can be used then the TD method is

- 14 -

natural, since it just gives an advice how to omit the higher
parts which shall be neglected. Therefore I hope that the talk
of Stumpf on convergence of the TD method will be a good start
for discussions. We know now, that in some cases where the
foundations are well defined, the TD method actually does converge and thus gives reasonable results.

Perhaps other methods will be discussed in our seminar.
I would like to emphasize again that one can probably learn a
lot about approximation methods from many body physics. I said
in the beginning that elementary particles ought to be compared
with the excited states of solid bodies in quantum mechanics,
not with the electrons of quantum mechanics. These objects can
be treated very well by the methods of the Green's functions
and therefore I feel it would be very useful if many physicists
could work on the connection between many body physics and
elementary particle physics. Generally we should not be too
afraid of using any of these methods if only we start from a
well defined axiomatic basis. Probably the most controversal
question will be whether we should stop at frame 2) or at 3),
but the method of approximation is not too important.

<u>Conclusion</u>

This is more or less all I wanted to tell you as a kind of introduction to this seminar. I might perhaps add a few - you may
say philosophical - remarks about the methods one should apply
here, even if they are not meant too seriously. In the old times
we frequently spoke about the difficult situation we had to
cope with compared to physics in earlier times. In normal peri-
ods of physics the fundamental laws are known. Then the task of
the theoretical physicist is to derive by rigorous mathematical
methods starting from the fundamental laws the interpretation
of the phenomena. In this normal state of physics a rigorous
mathematical proof is an essential result. We know from old
Newtonian astronomy how complicated the problems may be, and
how difficult it is sometimes to find solutions; but finally it
is more or less a mathematical problem to come from the well
known fundamental laws to the complicated results. But then in

- 15 -

the early twenties we had a very different situation, and I
think that our present situation resembles this other one. At
that time the fundamental laws were not known; therefore a
mathematical proof was not always a very useful result, because
perhaps it only meant that the assumptions had been wrong. In
such a state of affairs it is always a little difficult to
start from one single phenomenon and to look for the explanation of one single detail, because whether one should believe
such an interpretation or not can not depend on a rigorous
mathematical proof. The rigorous mathematical proof means
little, because we don't know whether the assumptions have
been right. The interpretation can be tested only by comparing
it with other details, with other problems, and it is only the
equilibrium between the interpretations of many different phenomena, by which finally one can acquire a feeling how the fundaments of the theory probably are. Well, you may argue that
this is a very disagreeable and disgusting state of affairs,
since we always have to walk in the mud rather than on a good
solid road. But on the other hand I must say that I found this
time 40 years ago most fascinating and I would hope that you
also find this period now, and especially of course our discussions here at Feldafing most fascinating.

In *Proceedings of Seminar on Unified Theories of Elementary Particles, July 1965,* ed. by H. Rechenberg
(MPI für Physik und Astrophysik, Munich 1965) pp. 281-299

Quantum Electrodynamics in the Nonlinear Spinor Theory

W. Heisenberg

Max-Planck-Institut für Physik und Astrophysik

München (Germany)

In our Munich institute some calculations have recently been carried
out (by Dürr, Yamamoto, Yamazaki and Heisenberg) concerning the connection
between quantum electrodynamics (q.e.d.) and the general theory of elementary particles. The first part of my talk will be devoted to a general
method for calculating coupling constants, applicable equally well to
strong or electromagnetic interactions. The second part will deal with
coupling in electrodynamics and will lead to a numerical value of the fine
structure constant.

Concerning the coupling between say π and N, or η and N I refer to
a paper of Dhar and Katayama and another paper of Yamazaki. The mathematical method used in these papers is closely related to a method which is
frequently used in many body physics: the <u>method of the so called Green's
functions,</u> i.e. the vacuum expectation values (v.e.v.) of products of field
operators.

We start from the hypothesis, that a field equation which contains
<u>all exact</u> (and no other) symmetries of the underlying natural law provides
a reasonable basis for a theory of elementary particles. Under this condition the correct equation seems to be

$$-i\sigma_\nu \frac{\partial \chi}{\partial x_\nu} + \ell^2 \, 1\cdot\sigma^\nu \; : \chi(x)\left(\chi^*(x)\sigma_\nu \cdot 1 \, \chi(x)\right): \; = 0 .$$

$\chi(x)$ is a Weyl spinor in Lorentz space and also an isospinor. By : : we
mean the Wick product - i.e. in our case the product of three field operators at the same point, but after subtracting the 2-point contractions.
And the $\mathbf{1}$ means a unit matrix in the isospace.

Let me now explain the <u>Feynman notation</u> which I use in connection
with Green's functions. We identify a Green's function $\frac{\text{Pr}\,\sigma^\nu}{p^2}$ with ⸺

a two-point function   with ─○─

a four-point function   with ⋈

By the symbols $-\!\!\bigcirc\!\!-$ and $\bigtimes\!\!\bigcirc$ and so on I do not mean the complete v.e.v. but the irreducible parts of these expressions which are sometimes called η -functions; - i.e. we first subtract from the ordinary Green's functions all products of lower Green's functions, included in them. To express the connection between a two-point function and a four-point function we take the v.e.v. of the corresponding products on both sides of the fundamental equation, and integrate it. The two-point function is then given by

$$\langle 0|\overline{T}\chi(x)\,\chi^*(y)|0\rangle \;=\; \langle 0|\overline{T}\int G(x\,x')\,dx'\,\sigma^\nu_:\chi(x')\big(\chi^*(x')\,\sigma_\nu\chi(x')\big):\chi^*(y)|0\rangle$$

with a Green's function $G(x\,x')$. The product of the four χ we combine to a four-point function, in which three coordinates are equal. Symbolically we express this equation now by

$$-\!\!\bigcirc\!\!-\;=\;-\!\!\bigcirc\!\!\bigcirc\!\!-\quad;$$

the vertex part $-\!\!\!\prec$ is the characteristical feature of the fundamental equation. The next equation combines the four-point function with a six-point function, but besides that we get so called 'contraction' terms

$$\bigtimes\!\!\bigcirc\;=\;-\!\!\prec\!\!\bigcirc\!\!\prec\;+\;\bigtimes\!\!\bigcirc\!\!\prec\;+\;\bigtimes$$

The second term on the right hand side consists of a product of a four-point function and a two-point function, and the third term contains three two-point functions. This short hand writing (in which the symmetrisation with respect to the different lines is understood, but not explicitly written) establishes the graphs that will be used.

Now we get a <u>first approximation</u> by taking the two equations together and dropping the six-point function. Then we have two equations for two unknown functions, for the two-point function and the four-point function

$$-\!\!\bigcirc\!\!-\;=\;-\!\!\bigcirc\!\!\bigcirc\!\!-$$

$$\big[\,1\,-\,\bigtimes\,\big]\,\bigtimes\!\!\bigcirc\;=\;\bigtimes$$

Proc. Seminar on Unified Theories of Elementary Particles (1965) pp. 281–299

These two equations are very complicated nonlinear integral equations. The second equation however is an inhomogeneous linear equation which determines both the poles of the four-point function and the residues at the poles, if the two-point function is considered as given. If we are first interested in the poles corresponding to boson states, this four-point function may be considered as composed of matrix elements for creation and annihilation of bosons

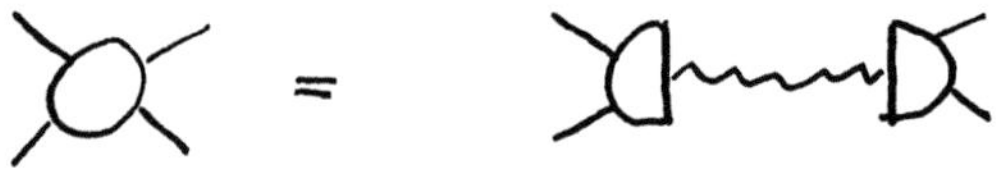

creation of a boson — annihilation of a boson

The homogeneous equation which determines the eigenvalues of the bosons is therefore

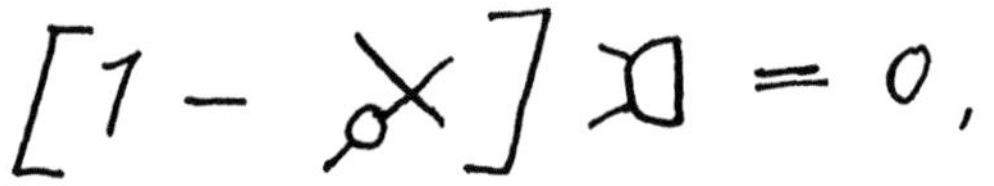

which we can simplify by putting the coordinates x and y together

So one can say that the masses of the bosons are just determined by putting this integral operator  equal to 1. The inhomogeneous part now gives us the residue of the boson poles which is connected with the coupling constants. We can formally integrate the second equation by

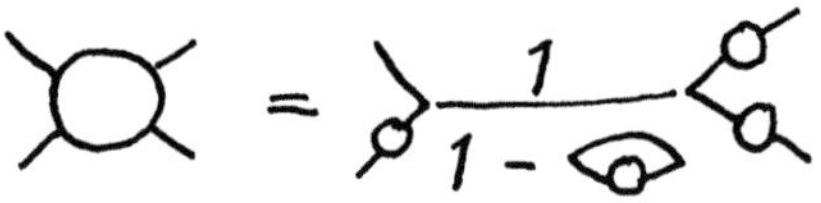

Inserting this expression gives in fact

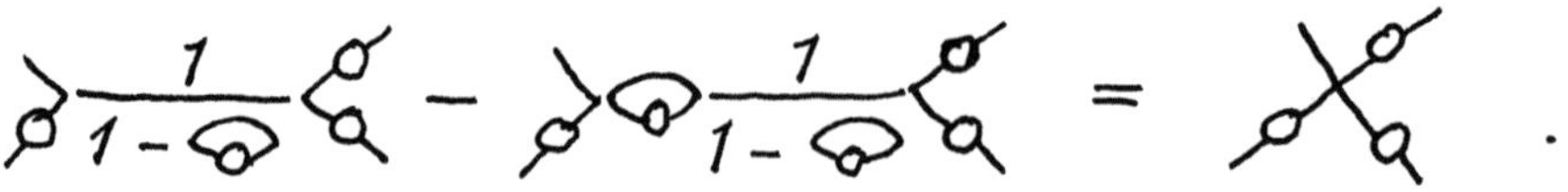

This whole procedure, of course, only works when we already know the two-point function. We may get it by putting the approximation of the second

equation into the first equation

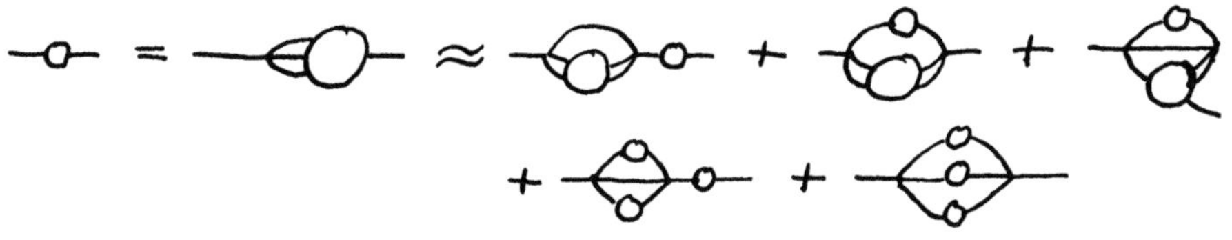

or 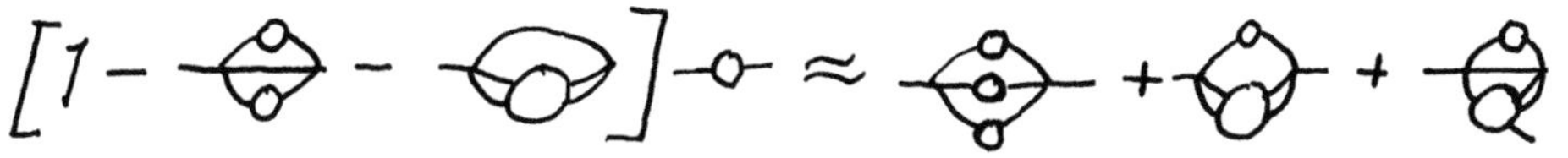

which is again a very complicated nonlinear equation. Again one has for the fermion eigenfunctions the homogeneous linear eigenvalue equation

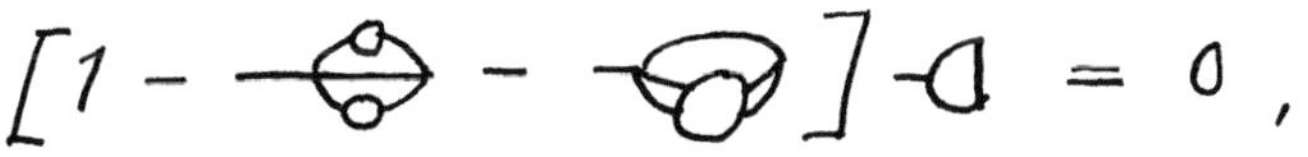

if the integral operators in the bracket are considered as given. The general procedure then should be: first to take a suitable trial function for the two-point function; then determine the four-point function, put the calculated four-point function and the trial two-point function into the equation for the two-point function, and compute the second approximation for the two-point function, and so on. This formalism of the bootstrap type should finally - at least in the approximation where we have neglected the six-point function - lead to reasonable values for the two-point function and the four-point function. It is true that practically this program cannot be carried out very far; but one can try to get some simple parametrization of the two-point function. For this purpose we write down the Lehmann-Källen representation of the two-point function

 $$= \int d(\kappa^2)\, \rho(\kappa^2)\, \frac{p_\nu \overline{\sigma}^\nu}{p^2+\kappa^2} \; .$$

In a normal field theory we expect in this function δ- and δ'-functions on the light cone. Such a singular behaviour, however, would not lead to a solution of our equations as one can easily see e.g. from the term  which is the product of three two-point functions, and the third power of a δ-function (or a δ'-function) is not a sensible quantity. Therefore we conclude that we cannot have δ- and δ'- functions on the light

- 285 -

cone, and that means that we must have the following conditions in the
Lehmann-Källen representation

$$\int \rho(\kappa^2)\, d(\kappa^2) = \int \rho(\kappa^2) \cdot \kappa^2\, d(\kappa^2) = 0 \ ,$$

which can only be fulfilled when we assume an <u>indefinite metric</u>. So we
must take an indefinite metric, if we want to give our equations a meaning.
With these two conditions we can rewrite the Lehmann-Källen representation
into the form

$$-\!\bigcirc\!- \ = \ \int d(\kappa^2)\rho(\kappa^2) \frac{p_\nu \bar{\sigma}^\nu \cdot \kappa^4}{(p^2)^2(p^2 + \kappa^2)} \ .$$

The regularization is carried out by means of a double pole at the mass
zero.

As a first approximation we try to use just one single mass

$$-\!\bigcirc\!- \ = \ \frac{p_\nu \bar{\sigma}^\nu \cdot \kappa^4}{(p^2)^2(p^2 + \kappa^2)}$$

Physically this equation can be interpreted by saying that we have one pole
at the mass κ which should be identified with the centre of gravity of
the baryons, and a double pole at mass zero which should be connected with
the leptons for following reasons: From the Lee model we have learned that
such a double pole does not take part in the interactions, i.e. it does
not contribute to the S-matrix. Assuming that we have a similar situation
here, we may say that there is an approximation in which the leptons
don't take part in any interaction and have the mass zero. In higher approximations this situation may be changed; the double pole of the charged
leptons will then not be at mass zero, and it may be changed into something more complicated.

With the first approximation of the two-point function we calculate
the four-point function, and from there we may go through the 'bootstrap'-
method and see whether we get finally something convergent. The first
condition is that the pole of the new approximation must be the same as
the pole of the original two-point function. This condition determines
the average mass of the baryons.

With this general scheme we can calculate boson eigenvalues - that
we have done in earlier papers - and coupling constants. The first

step is to study the behaviour of the integral operator ⌬ . When you
do the integration with our approximate two-point function you find the
following expression

$$\text{⌬} = -\tfrac{1}{2}\left(\tfrac{3}{4}\cdot 1\cdot 1 + \tfrac{1}{4}\,\tau_k\,\tau_k\right)\left(\tfrac{\kappa\ell}{2\pi}\right)^2\left[\tfrac{\gamma_\mu\gamma_\nu}{\gamma^2}\,q_0(\lambda) + \left(g_{\mu\nu} - \tfrac{\gamma_\mu\gamma_\nu}{\gamma^2}\right)q_1(\lambda)\right]$$

$$= \left(\tfrac{\kappa\ell}{2\pi}\right)^2\left(\tfrac{3}{4}\cdot P_{T=0} + \tfrac{1}{4}\,P_{T=1}\right)\left(P_{S=0}\,q_0(\lambda) + P_{S=1}\,q_1(\lambda)\right),$$

with $\lambda = -\dfrac{\gamma^2}{k^2}$. The $P_{T=0}$ and $P_{T=1}$ are projection operators for
isospin 0 and 1 respectively, and the $P_{S=0}$ and $P_{S=1}$ the projection
operators for Dirac spin 0 and 1. These are the four possibilities, if we
start from a two-point function, where the two coordinates are equal. We
begin with an s-state of a fermion-antifermion system, and we can add
only the two spins to give 1 or 0, and the two isospins to give 1 or 0
respectively. The projection operator with isospin 0 is multiplied by $\tfrac{3}{4}$
and that with isospin 1 is multiplied by $\tfrac{1}{4}$, because in the interaction
vertex one has to take the Fierz symmetrized product, i.e.

$$\tfrac{1}{2}\left(1_{\alpha\beta}\cdot 1_{\gamma\delta} + 1_{\alpha\delta}\cdot 1_{\gamma\beta}\right) = \tfrac{3}{4}\,1_{\alpha\beta}\,1_{\gamma\delta} + \tfrac{1}{4}\,\tau^k_{\alpha\beta}\,\tau^k_{\gamma\delta}\ .$$

In a similar fashion one can discuss the Dirac spins. Finally one gets by
simple Feynman integration two different analytical functions of

$$\lambda = \frac{\text{mass of the boson}^2}{\text{mass of the baryon}^2}\ ,$$

$$q_0(\lambda) = \ln\lambda - \tfrac{1}{\lambda} - \frac{(1-\lambda)^2}{\lambda^2}\ln|1-\lambda| \approx \ln\lambda - \tfrac{3}{2}\ \text{for}\ \lambda \ll 1,$$

$$q_1(\lambda) = \left(1 - \tfrac{2}{3}\lambda\right)\ln\lambda + \tfrac{2}{3} + \tfrac{1}{3\lambda} + \frac{(1-\lambda)^2(1+2\lambda)}{3\lambda^2}\ln|1-\lambda|$$

$$\approx \ln\lambda + \tfrac{1}{2}\ \text{for}\ \lambda \ll 1.$$

In order to calculate coupling constants we proceed in the following way:
Take the equation for the four-point function and integrate it formally

- 287 -

This expression is to be compared with the expression for the scattering
amplitude in a phenomenological theory

If we have a phenomenological theory with a vector coupling - coupling
constant f - , or with derivative coupling, if the intermediate boson is
a scalar particle, then the calculated amplitude is essentially a product
of the two vertex parts, both containing the coupling constant f, and of
a boson propagator. For scalar bosons one gets

$$f^2 \; \frac{\dfrac{\gamma_\mu \gamma_\nu}{\kappa_B^2}}{\gamma^2 + \kappa_B^2} \qquad\qquad (\;\kappa_B \text{ mass of the boson}).$$

To this expression corresponds in our field theory

$$\ell^2 Z \; \frac{\dfrac{\gamma_\mu \gamma_\nu}{\gamma^2}}{1 - 1 - Z\left(\frac{\kappa \ell}{2\pi}\right)^2 \dfrac{\partial q_0}{\partial \lambda} \dfrac{\gamma^2 + \kappa_B^2}{\kappa^2}} \qquad \text{with} \qquad
\begin{aligned}
Z &= \tfrac{3}{4} \text{ for } T = 0, \\[1mm]
Z &= \tfrac{1}{4} \text{ for } T = 1.
\end{aligned}$$

For vector bosons the phenomenological theory gives

$$f^2 \; \frac{g_{\mu\nu} - \dfrac{\gamma_\mu \gamma_\nu}{\gamma^2}}{\gamma^2 + \kappa_B^2}$$

the unified field theory

$$\ell^2 Z \; \frac{g_{\mu\nu} - \dfrac{\gamma_\mu \gamma_\nu}{\gamma^2}}{1 - 1 - Z\left(\frac{\kappa \ell}{2\pi}\right)^2 \dfrac{\partial q_1}{\partial \lambda} \dfrac{\gamma^2 + \kappa_B^2}{\kappa^2}} \qquad .$$

By comparison one finally gets

$$f_0^2 = \frac{4\pi^2}{\dfrac{\partial q_0}{\partial \lambda}} \quad \text{for } s = 0, \text{ and} \qquad f_1^2 = -\frac{4\pi^2}{\dfrac{\partial q_1}{\partial \lambda}} \quad \text{for } s = 1.$$

Dhar and Katayama applied this formalism to the pion and η-meson. Since
λ is small you have $\dfrac{\partial q_0}{\partial \lambda} \approx \dfrac{1}{\lambda}$ and therefore

$$\left(\frac{f^2}{4\pi}\right)_{pion} = \pi \lambda = \pi \left(\frac{\kappa_{pion}}{\kappa_N}\right)^2 = 0,07 \quad ,$$

which is to be compared with the experimental value 0.08. You see, that the calculation is really a straightforward procedure, and the only step that is definitely different from many body physics or similar schemes, is, that we do not hesitate to fulfill the Green's function equations, and therefore we do go into the indefinite metric.

<u>Zumino:</u>   What value of ℓ enters?

<u>Heisenberg:</u> ℓ is just a parameter of the dimension of a length. You may put it simply equal to unity, and then of course it would fix the unit of length. ℓ enters in the combination $\frac{\kappa\ell}{2\pi}$ which is computed by the fermion equation. From the empirical mass of the nucleon ℓ is then fixed as a length of the order of magnitude of 10^{-13} cm; the best value is at present $\ell \approx 1.2 \cdot 10^{-13}$ cm.

(Lecture continued)

Coming to electrodynamics I might first make a few preliminary remarks concerning the <u>Goldstone theorem</u> (G.th.). We know from this theorem that, whenever we explain a broken symmetry by means of a degeneracy of the vacuum then we expect a boson of mass zero. Let me specify the statement a little bit further. First of all we should clearly distinguish between a <u>real</u> breaking of symmetry and an only <u>formal</u> breaking of symmetry. By 'real breaking' I mean, that afterwards in our experiments we do see violations of the conservation law. We suppose that our fundamental laws are exactly invariant under certain groups, e.g. our nonlinear field equation is strictly invariant under the isospin group. Then we assume that the ground state has a big isospin - like a ferromagnet has a big magnetic moment - and that thereby the symmetry is broken. This case should be distinguished very sharply from the other case, where one formally may invent some mathematical scheme to make the ground state not invariant under the group concerned, but still one tries to preserve the conservation law. In this latter case it may formally be possible to establish a particle of mass zero, too, and to give it certain properties. But in such a case the particle would have well defined properties with respect to the group, and therefore we would never see any violation of the conservation laws. Hence a manifest violation of the conservation law does not only require that we have a massless particle, but it also requires that this particle has partly undefined properties with respect to the group under question.

Concerning the <u>physical content</u> of the G.th. I have to add a warning.
Some physicists argue that whenever we have a group broken by a degeneracy
of the ground state then we can with the help of the G.th. derive immediately the particle of mass zero and its symmetry properties with respect
to all other groups. This statement is not quite correct, and I want to
express the physical content of the G.th. in this way: If we have strict
symmetry and therefore strict conservation, and if the theory is at least
approximately local, then in a single collision process the quantity under
question need not be conserved among the colliding particles; some amount
of the quantity may be transferred up on the ground state at this point.
The G.th. then states e.g. that we cannot at once transfer angular momentum on the whole ferromagnet, but we have to transfer it locally. I.e.,
this transferred quantity will only gradually spread over the whole body,
and this spreading out is connected with a 'Goldstone' wave. The larger
the area in which the transferred quantity, e.g. the angular momentum is
spread out the smaller the energy connected with it. Because after we have
smeared out it completely then we know that we just get from one vacuum
to another vacuum which requires no energy. Therefore we must have a particle of mass zero. So the Goldstone theorem just states: If we have non-
conservation of some quantity due to a degenerate unsymmetrical vacuum,
then there must exist a particle of mass zero which will carry away the
conserved quantity in an unknown amount. Formally, of course you can say
that, taking everything into account, the quantity is conserved again;
but it may be impossible to check this conservation, because the particle
in question has no well defined property with respect to the transformation concerned. Therefore one has to distinguish three objects in connection with the G.th.:

1) <u>Spurions</u> or <u>spurion states</u> refer to the case $p_\mu = 0$. Generalized
operators belonging to $p_\mu = 0$ must necessarily exist in this mathematical
scheme, because it must be possible to go from one vacuum to a neighbouring
vacuum. This may already be a rather controversial statement among the
mathematicians, and therefore I will enlarge a little bit up on this statement. One usually says, that we have to start from one vacuum; and all
the other vacua or ground states which could possibly be created from this
one ground state by the transformation concerned need not be considered
at all because we would need an integration over the whole infinite space

to construct such an operator from the field operators. We may study this problem from a simple model. Let us start e.g. with a finite ferromagnet where we have N spins; if we turn around just one spin, that can certainly be done by one operator. Smearing out this single spin over the whole space will lead to a state which is different from the original one with respect to spin direction by an angle of the order $\frac{1}{N}$. We will need in the formalism operators which correspond to turning around a finite number of spins or isospins which are not localized. The process therefore corresponds to turning around the whole infinite ground state by an angle which is infinitely small. Operators of this type will be called spurion operators, the resulting states spurion states.

2) <u>Goldstone states</u> have $p^2 = 0$ but $p_\mu \neq 0$. As soon as the quantity under question can be localized in the original field equation then always we will have such Goldstone states.

3) <u>Goldstone particles</u> have $p^2 = 0$ and $p_\mu \neq 0$, but in addition have a definite symmetry with respect to all other transformations. The characteristic feature of the Goldstone particle is, that it fulfills an <u>eigenvalue equation</u> - it can come out as a pole of a Green's function - and therefore also can appear in nature as a real particle. When there is one kind of Goldstone particle then there may be many different Goldstone states, because whenever one adds spurions to a Goldstone particle, one gets a Goldstone state. Thus a Goldstone state is something more general than a Goldstone particle, and that is the reason why one cannot from the formal proof of the G.th. by Salam and Weinberg conclude, what the symmetry of the Goldstone particles must be. One can only say what the symmetry properties of those Goldstone states are, which appear in the formal proof.

So by analyzing this situation in detail one sees that all the proofs of the G.th. - that of Salam and Weinberg, that of Nambu - say nothing else, but that it must be possible to transfer the quantity under question to the ground state in form of some wave emission, which has the property, that for very big wave length the energy must be very small or, in other words, the particle therefore must have zero mass.

<u>Bludman:</u>   Do you see in a ferromagnet the difference between a Goldstone particle and a Goldstone state?

Heisenberg: Yes. In the ferromagnet the Goldstone particle is definitely the Bloch spin wave. But I can, of course, also construct Goldstone states which are different from the spin waves by just attaching a spurion to it, in other words, I add the spin wave to a vacuum which is slightly different from the original one.

Bludman: Do the Goldstone states play a physical rôle in the ferromagnet?

Heisenberg: In a finite ferromagnet, of course, these states play a rôle because I really can distinguish the different possible ground states. But I would say, even for an infinite ferromagnet it would be reasonable to consider states, where this infinite ferromagnet would have changed the direction of its magnetic moment by an infinitely small angle.

Sudarshan: Do you expect Goldstone particles and Goldstone states both be present? Or are there Goldstone particles without spurions attached?

Heisenberg: The point is that the Goldstone states can physically be scarcely distinguished from the Goldstone particles, because the Goldstone state is different from a Goldstone particle only in so far, as somewhere in the infinite space one spin has been turned around, so that my ground state is really different from the original one. So it perhaps would be a better way of saying: I take a mixture of all these possible ground states together with the spin wave and call that Goldstone particle.

There is another point I want to mention in connection with the vacuum degeneracy: If the ground state is degenerate, e.g. with respect to isospin then you have to decide what happens to all other groups. One might perhaps say, the best thing would just be to take the groundstate to be symmetric under all other groups and only asymmetric under this one. This, however, may be impossible on account of the discrete groups. E.g., we had the mathematical difficulty to decide in the case of the

	isospin degeneracy whether our ground state has to be invariant under particle-antiparticle reflection or not. The preservation of the other groups is a problem in itself, and for every special case you have to do quite a lot of labour to find out, what is actually the symmetry you can consistently assume for your vacuum, for your spurions, and for your Goldstone particles.
Bludman:	Do you think of the ground state as a pure state or as a coherent state?
Heisenberg:	I would definitely prefer to think about a 'mixture' of several ground states which I have to consider at the same time; states, that are transformed into each other by these infinitely small rotations.

(Lecture continued)

I want to make another remark. The scheme in which e.g. isospin is not conserved, is not determined as the only possible solution by the fundamental equation. The equation will allow quite a number of assumptions about the ground state, e.g. it will certainly allow that the ground state is really symmetrical under all groups. On the other hand it may also be possible to destroy isospin invariance together with some other properties than particle-antiparticle conjugation. So, for instance, the scheme which we tried three years ago - where P was not conserved - is logically and also mathematically consistent. The situation here is really similar to that in general relativity, where you may have many worlds - a de Sitter world, an Einstein world, and so on. So I cannot prove from general principles which ground state should be realized. I have rather the feeling that the ground state is a kind of boundary condition, and so you may have the same differential equation, and many possible boundary conditions, or at least several possible boundary conditions. But in assuming a special ground state, the proof of consistency is, of course, a very important step in the whole theory.

What we actually did in our theory was not to start from the non-conservation of isospin to construct the corresponding Goldstone particle. We rather tried the other way around, i.e. we asked: Can we construct a theory which contains a Goldstone particle 'photon', and then see whether we get a consistent scheme. If one starts from the Weinberg-Salam proof then one could for the Goldstone particle or rather for the Goldstone state only get states with no Lorentz properties, because one assumes that the ground state

- 293 -

is invariant under Lorentz transformations. Therefore one never could get
vector particles but only get scalar ones. On the other hand we know from
nature that we have photons with a scalar part, namely the Coulomb force
or some part of the Coulomb force, and therefore the scalar part of the
photon may well act as that part of the Goldstone state which appears in
the proof of Weinberg and Salam.

Now the most important part of this question is: Do we get in that
scheme Goldstone particles which obey an <u>eigenvalue equation with mass zero?</u>
This particle at the same time must have an ambiguous nature with respect
to isospin - it must be a 'schizon' in the Lee sense - because it should
break isospin, and a real asymmetry does only occur, if there are particles
of which one does not know the nature with respect to that group. Therefore
we expect a state which is a mixture of isospin 0 and 1 for the photon
- we exclude other isospins because we want to have a simple object - , and
look if we can find a consistent mathematical scheme.

The equation

will, of course, as a rule not give the mass zero. On the other hand we
know that it must give mass zero, if an asymmetry is consistent. This problem seems to be solved by a remark of Johnson, who found out that in Q.e.d.
we have to start with mass zero of the electron, and then the mass is
established as a result of interaction only by the mathematical process of
consistency. If we assume a two-point function with a mass in it then we
can reproduce the mass by means of electromagnetic forces, and consistency
requires that we should get the same mass out that we have put into the
scheme. If we, however, have established a q.e.d. scheme with a certain
mass then we can quite easily construct some other scheme with a different
mass, because we have just to apply <u>scale transformation</u>. All equations of
q.e.d. with a rest mass zero of the photon are invariant under this transformation, therefore e.d. forces do not determine the mass of the lepton;
- possibly they determine the mass ratio of the leptons.

Can we apply the same scale transformation to our fundamental equation?
The equation - with constant ℓ - allows for the following scale transformation

$$\chi(x) \longrightarrow \sqrt{\eta'}\, \chi(x\eta) \ .$$

So $\chi^{*}\chi$ transforms like ($1/$ length), and that means that $\chi^{*}\sigma_{r}\chi$

really transforms like $A_\mu(x)$ both with respect to Lorentz properties and also to the scale transformation. This remark solves our problem because the leptons occur as a double pole in the two-point function. In a first approximation this double pole is located at mass zero, but we may as well put the charged leptons or the average mass of the charged leptons to some other value arranged in such a way, that now the eigenvalue equation for the photon is satisfied. Then we have a consistent scheme and also the average mass of the lepton is not overdetermined, because the electrodynamic side alone does not fix it.

The next step is to see whether we get by such an assumption the connection between this mathematical scheme and ordinary q.e.d. In ordinary q.e.d. we calculate a matrix element from vacuum to a photon with momentum J and polarization B in perturbation theory

$$\langle 0 | \chi(x) \chi^*(y) | {}^{\hbar\nu}_{J,B} \rangle =$$

$$\mathrm{const.} \int d^4p\, e^{ip(x-y)} \frac{p_\rho \gamma^\nu B_\rho \gamma^\sigma (J-p)_\tau \gamma^\tau}{(p^2+\mu^2)[(J-p)^2+\mu^2]}\, e^{iyJ}$$

(with $\chi(x)$ and $\chi^*(x)$ electron or lepton operators). The selfenergy of the photon diverges quadratically, and in order to get rid of this divergence one usually argues as follows: The photon must have the mass zero and therefore we do have gauge invariance. So we have to subtract from this term the same term for momentum zero, i.e. under the integral

$$\frac{p_\rho \gamma^\nu B_\rho \gamma^\sigma (J-p)_\tau \gamma^\tau}{(p^2+\mu^2)^2}$$

and thereby we can change the quadratic singularity into a logarithmic singularity. Following the simple procedure in the textbook of Jauch and Rohrlich you get essentially

$$\mathrm{const.} \int d^4p\, e^{ip(x-y)} \frac{p_\rho \gamma^\nu B_\rho \gamma^\sigma (J-p)_\tau \gamma^\tau}{(p^2+\mu^2)^2 [(J-p)^2+\mu^2]}\, e^{iyJ} \, .$$

By renormalizing further, i.e. by replacing the original bare operators by the renormalized one will finally get rid of the logarithmic divergency.

So we end up with a matrix element which contains a double pole at lepton mass μ , and another single pole at the point $(J - p)^2 = -\mu^2$, and some convergence factor. In our theory we find on the other hand

$$\langle 0 | \chi(x) \chi^*(y) | {}^{h\,\nu}_{J,\beta} \rangle = \text{const.} \cdot \int d^4p \; e^{i\,p(x-y)\,+\,i\,J y}$$

$$\frac{p_\nu \gamma^\nu \; \mathcal{B}_\delta \gamma^\delta \; (J-p)_\tau \gamma^\tau}{(p^2)^2 (p^2 + \kappa^2)(J-p)^2} \cdot \frac{1}{2}(\lambda_3 + \tau_3)$$

One recognizes at once the part of the two-point function and the part of the Green's function. So the two expressions are really quite similar except for the point, that in q.e.d. the mass of the charged leptons is taken into account, while in the nonlinear spinor theory in this approximation the mass is taken equal to zero.

The projection operator $\frac{1}{2}(\lambda_3 + \tau_3)$ means that a light quantum consists only of the charged leptons and not of the uncharged ones, therefore in this theory it would be reasonable to take the leptons not at the mass zero, but at the real mass of the charged leptons. Because of the $e - \mu$ mass difference we have to take an average mass of the leptons just in the same way as we have taken an average mass of the nucleons. So we simply improve the expression by writing in the average lepton mass μ , i.e. we replace the term $(p^2)^2$ in the denominator of the two-point function by $(p^2 + \mu^2)^2$. Then the two expressions for the matrix element are similar in all essential points.

Calculating the <u>coupling constant</u> is now a straightforward procedure. At first we solve the eigenvalue equation, and determine the average mass of the leptons

$$1 + \frac{1}{4}\left(\frac{\kappa \ell}{2\pi}\right)^2 q_1(\lambda,\varepsilon) = 0 \qquad , \text{ with}$$

$$\left(\lambda = \frac{\kappa \beta^2}{\kappa^2} , \quad \varepsilon = \frac{\mu^2}{\kappa^2} \right)$$

$$q_1(\lambda,\varepsilon) = \frac{1}{3}\Bigg[\frac{1-\varepsilon}{\lambda}(1 - 2\varepsilon + 2\lambda) + \frac{(1-\lambda)^2(1+2\lambda)}{\lambda^2}\ln|1-\lambda|$$

$$- \frac{\varepsilon^2}{\lambda^2}(3 - 2\varepsilon)\ln\left|\frac{\varepsilon-\lambda}{\varepsilon}\right| + (3 - 2\lambda)\ln|\varepsilon - \lambda| \Bigg]$$

$$\approx \ln|\varepsilon - \lambda| + 1 - \frac{\lambda}{3\varepsilon} \qquad \text{for} \quad \lambda \ll \varepsilon \ll 1.$$

For the photon ($\lambda = 0$) this gives $q_1(0, \varepsilon) = \ln \varepsilon + 1$.

Comparing this result with the eigenvalue equation of the pion, which contains the same numerical factor from the isospin, we find

$$\ln \varepsilon_{lepton} = \ln \lambda_{pion} - \frac{5}{2}$$

$$\mu_{lepton} = \kappa_{pion} \cdot e^{-\frac{5}{4}} = \frac{\kappa_{pion}}{3.5} \approx 40 \, MeV.$$

For the coupling constant now we have to differentiate q_1 with respect to λ:

$$\left(\frac{\partial q_1(\lambda, \varepsilon)}{\partial \lambda} \right)_{\lambda = 0} = -\frac{2}{3\varepsilon}$$

So finally we get

$$\alpha \approx \frac{\pi}{-\frac{\partial q_1}{\partial \lambda}} = \frac{3\pi}{2} \varepsilon = \frac{3\pi}{2} \frac{\mu_{lepton}^2}{\kappa_N^2} \approx \frac{1}{120} \, ,$$

which is in reasonable agreement with the value $\frac{1}{137}$ of the fine structure constant.

In so far the calculation of the coupling constant was a straightforward procedure. But to begin with we had to establish that the photon is a particle with mass zero, which was only possible with the help of the lepton spectrum. So if in a theory of all elementary particles one wants to have a degeneracy with respect to the isospin - which involves a Goldstone particle photon - then that can only be done by having a kind of equilibrium between the baryon and the lepton spectrum. This equilibrium - so far as it is calculated here - only connects the average mass of the baryons with the average mass of the leptons.

May I come back for a moment to the behaviour of <u>photons with respect to the isospin group</u>. The photon has to be a mixture of states of isospin 1 and 0. In the eigenvalue equation we have to take an asymmetrical two-point function, so it must contain a term τ_3 which has, of course, the consequence that the masses of proton and neutron are not equal. Without any strangeness one would expect a projection operator $\frac{1}{2}(1 + \tau_3)$ in the photon. The term $\frac{1}{2}(\lambda_3 + \tau_3)$ comes out from the existence of the strange particles and from the necessity to obtain a Fierz symmetrization not only in isospace but also in hypercharge space. (Indeed, $\lambda_3 = 1$ if

- 297 -

we have no strangeness!) Such a structure goes through the vertex of the eigenvalue equation unchanged. So our assumption about the eigenfunction of the photon is consistent, and our procedure leading to a reasonable value for the fine structure constant agrees with the procedure used for the (π-N)-coupling.

Discussion

Barut:
Why do you use an average mass for the leptons, and in the case of the baryons you just take the nucleon mass?

Heisenberg:
We could just as well put in the average mass of the baryon octet. Then we would get a slightly different numerical value, e.g. for the fine structure constant $\frac{1}{170}$, but that is not essential in this approximation.

There is another point I should like to mention, namely that we must have, for all particles in the world, always an integer multiple of the same charge. That comes about by the Ward identity which is valid, since the mass of the photon is exactly zero. As soon as the mass would be different from zero, the electromagnetic coupling constants of different particles could be different. For particles which are produced by several field operators we can have charges 2, 3 or else.

Zumino:
I don't see where the violation of the symmetry comes in. You use the same formulas in connection with symmetry breaking ground state which you used for the pion.

Heisenberg:
If we would start with a symmetrical ground state - and such a solution should be possible - we would never find a Goldstone particle with mass zero. But nevertheless this would give a consistent scheme. Then we can ask differently: Can we construct from the same equation a ground state which is degenerate under isospin rotation, and there are several possibilities to that. One of them allows to include the photon as the Goldstone particle. I don't think that's problematic. You must assume that you have strict conservation of isospin of the

- 298 -

	fundamental equation, and also that this theory has some degree of locality in it. Then the G.th. follows, if the ground state is degenerated.
Zumino:	How do you choose the solution?
Heisenberg:	We are satisfied when we find a mathematical scheme which resembles the real situation in nature, and which is consistent. There may be other schemes which do not correspond to nature, other cosmological models of the world, which do not correspond to the real world.
Bludman:	Among the possible solutions one must be more stable than the other one.
Heisenberg:	Here we compare different vacua which by definition have no energy and no momentum, since they are invariant under the Lorentz group; therefore the question of stability in the dynamical sense can scarcely come in.
Sudarshan:	On the one hand you discuss the physical ideas underlying the theory of the origin of the photon in terms of long range correlations. And yet you assert that there must be a Goldstone boson, because it has been proved. Why is it necessary to refer to this proof without taking all the implications of the proof? Your physical argument is more plausible.
Zumino:	Would your vacuum also be degenerate with respect to the Lorentz group?
Heisenberg:	No. It could perhaps be made formally degenerate as Bjørken did. But this is quite unnecessary.
Sudarshan:	This theory has the nice aspect that it ties photons baryons leptons, and mesons together. But it has one apparently very dangerous feature which might kill it, and that is the question of conservation of baryon number. In this theory it looks, as if it is possible to get the transition from two baryons to two leptons. Or are there some extra selection rules which forbid those processes?
Heisenberg:	We hope that when we have really understood the leptons we will have lepton conservation.

- 299 -

Dürr: Baryons are poles, leptons are double poles. So if one
         finds a rule that a pole can never become a double pole,
         then it is alright. The double poles are really inter-
         esting objects, because they don't take part in strong
         interactions.

Introduction to the Unified Field Theory of Elementary Particles

W. HEISENBERG

Max-Planck-Institut für Physik und Astrophysik

1966

INTERSCIENCE PUBLISHERS
a division of John Wiley & Sons London New York Sydney

Preface

The present book contains a series of lectures which have been given at the University of Munich during the summer term 1965, with the hope to draw the attention of the younger generation of physicists to the present development of the unified field theory of elementary particles.

The picture in which the elementary particles appear as dynamical systems, comparable to the stationary states of complicated atoms or molecules, and resulting, as in quantum mechanics, from an underlying universal law, has for a long time not appealed to the majority of physicists. It was in fact in obvious disagreement with the term 'elementary' used for these particles and with the resulting hope for a simple understanding of their nature. On the other hand, in the nearly twenty years that have elapsed since its first explicit formulation, this picture has found the support of many more recent experiments. New elementary particles have been discovered in great number. The spectrum of particles and the selection rules in transition processes have been analysed successfully with the help of quantum numbers in a similar way as the optical spectra of atoms. The special mathematical form suggested in 1958 for the underlying natural law has been supported by the later observation of the η-meson, by the determination of coupling constants and, more recently, by the possibility of connecting the photon (and thereby electrodynamics) with the degeneracy of the ground state, the theorem of Goldstone, and the existence of 'strange' particles. All these results seemed to justify the attempt to present the main ideas of the unified field theory in a simple form understandable even for those physicists who are not yet specialists in the field of elementary particle physics.

In all lectures the emphasis has therefore been on the physical ideas of the theory, and only to a lesser degree on the mathematical scheme. In the present state of the theory it would be premature to start from a set of precise mathematical axioms and to proceed by rigorous mathematical deductions. What is required is a mathematical description adapted to the

v

experimental situation which does not seem to contain contradictions and which therefore may be developed into a rigorous mathematical scheme in the future. The history of physics teaches that as a rule a new theory can be formulated in a precise mathematical language only after all essential physical problems have been solved. The lectures do give a general outline of the expected mathematical scheme without proofs of existence or convergence; besides that they contain a few rather elementary calculations of the simplest mass eigenvalues and coupling constants. A much more elaborate description of the mathematical details will be given in a larger book by H. P. Dürr who has taken the most active part in the development of the theory. The dynamical aspects of the theory have been reviewed recently by Rampacher, Stumpf and Wagner.[76]

The mathematical formalism contains some unconventional features which formerly have rendered its understanding somewhat difficult: the indefinite metric in Hilbert space and the degeneracy of the ground state. But in recent years the indefinite metric has been studied in connexion with the Bleuler–Gupta version of quantum electrodynamics and with the Lee-model, the degeneracy of the ground state plays an important part in modern solid state physics. Therefore these extensions of conventional quantum mechanics should not be too problematic; they have been discussed carefully in the lectures.

The phenomenological theories which have successfully brought some order into the complicated experimental material: dispersion relations, Regge poles, groups of unitary transformations like SU_3, SU_6, etc., have not been presented in these lectures in any detail. They can be studied in several good textbooks. The way in which these theories can possibly be connected with the results of the unified field theory has been described in the last part of the present book. At this point again one recognizes that the unified field theory is merely in the first stage of its development and that many difficult problems are open for further investigation. In the space between strong interactions, electromagnetism, weak interactions and gravitation, seen from the viewpoint of this theory, there is still a wide field of research.

The author is indebted to Dr. Bui-Duy-Quang for much help in preparing the manuscript, to the members of the Max-Planck-Institute for Physics and Astrophysics for many valuable discussions and suggestions, and to the computing group of the institute for calculating the numerical table.

Contents

Contents

Contents

ix

ERRATA

Page 14,   formula 2(12), line 2.   *For $S^{-1}O'|\text{in}\rangle$, read $S^{-1}O'_i|\text{in}\rangle$.*

Page 26,   formula 3(4).   *For $\dfrac{\chi\partial}{\partial x^\nu}$, read $\dfrac{\partial\chi}{\partial x^\nu}$.*

formula 3(5).   *For $\dfrac{\chi\partial_{\beta\gamma}}{\partial x^\nu}$, read $\dfrac{\partial\chi_{\beta\gamma}}{\partial\chi^\nu}$*

Page 101, line 8 from bottom.   *For §7-2, read §7-3.*
Page 102, line 22 from top.   *For §7-2, read §7-3.*
Page 114, line 14 from top.   *For §7-2, read §7-3.*
Page 142, formula A(44).   *For $g(z)$, read $G(z)$.*
Page 143, line 4 from top.   *For $g(z)$, read $G(z)$.*
Page 154, line 5 from bottom   *For A(89), read A(88).*
line 2 from bottom.   *For K read $K_{\mu\nu}$.*
Page 160, formula A(110).   *For $K\kappa^2$, read $\kappa^2 K_{\mu\nu}$.*
Page 168, line 14 from top.   *For Y_2^*, read Y_1^*.*

1

The Important Experimental Facts

1-1 Definition of the elementary particle

Some fifty years ago only three kinds of 'elementary particles' were known. Electron and proton were considered to be the smallest units of matter, the photon or light quantum the smallest unit of energy in radiation. This concept of the 'smallest unit' had its origin in the problem of the divisibility of matter. A piece of matter can, by mechanical or chemical means, be decomposed into the single atoms. The atom may be separated into its nucleus and the surrounding electrons. The atomic nucleus can be split into protons and neutrons. It would be justified to speak about 'smallest units' only if these units could not be divided again into smaller parts.

The only processes in which a splitting of elementary particles could possibly be expected, are collisions between particles of extremely high energy. The big accelerators in Berkeley, Dubna, Geneva, Brookhaven have been designed just to study these collision processes. Following their construction many experiments have led to the result that from such a high energy collision between two particles actually many particles may emerge; but the emerging particles are not necessarily smaller than the colliding particles. On the contrary one finds always the same various kinds of particles, whatever the nature of the colliding particles may have been. The phenomena can be described more adequately by saying, that the big kinetic energy of the colliding particles has been transmuted into matter, into the emerging particles ('multiple production of particles'[1,2]). The photograph on the next page, taken from the bubble-chamber of the Proton-Synchrotron at CERN, Geneva, shows a typical example of multiple production of particles. Energy can become matter by assuming the form of an elementary particle. The different elementary particles may be considered as different forms in which the fundamental substance, matter or energy, can exist.[3] They may be called 'smallest units' only in the sense that when they are split, their parts are not smaller units but units of roughly the same size.

1

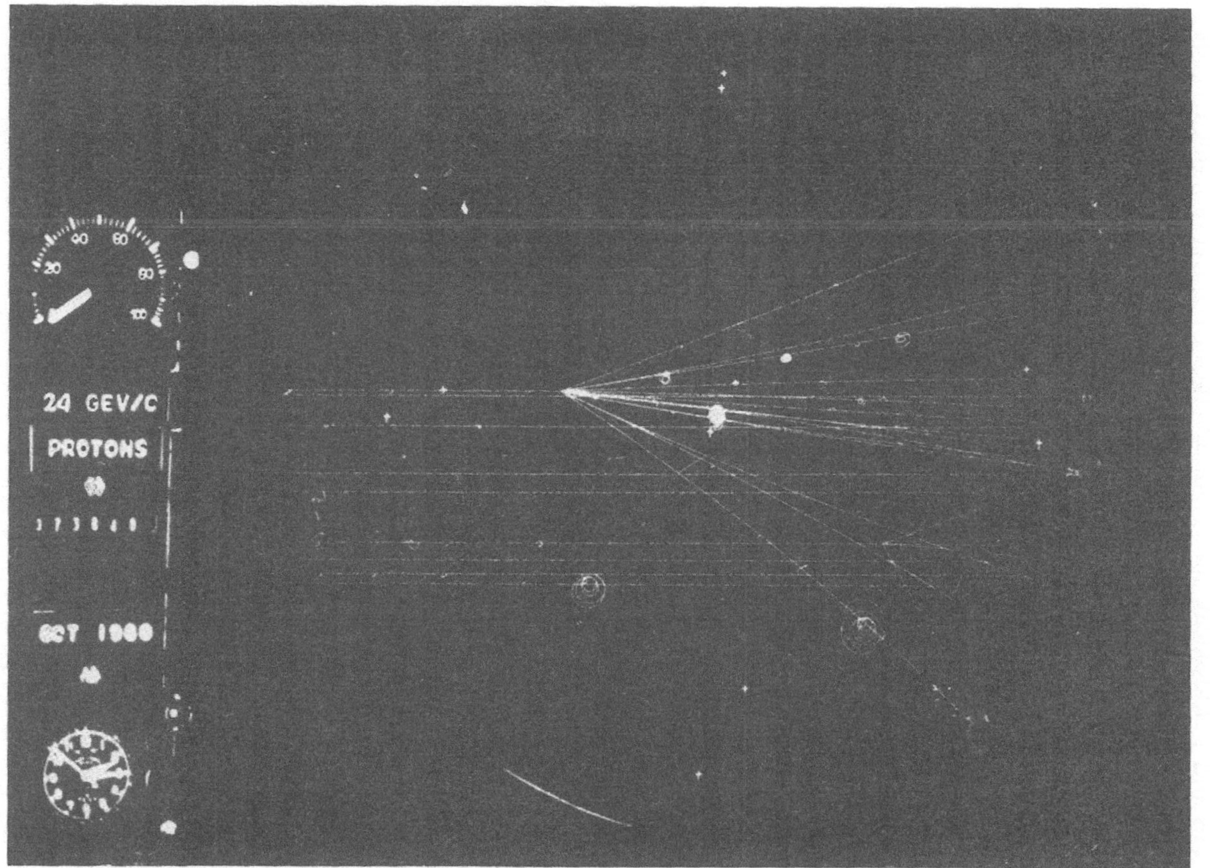

A proton of 24 GeV from the Proton-Synchrotron at CERN hits a proton in the hydrogen bubble chamber and produces a number of secondaries, mostly pions.

2   *Unified Field Theory of Elementary Particles*

This solution of the problem of the 'smallest units of matter' is somewhat surprising and paradoxical, and leads to another question which needs careful consideration. In earlier times the atoms of the chemists or the atomic nuclei were pictured as compound systems, consisting of many elementary particles, while the electron or the proton were taken as indivisible units and therefore 'elementary'. In the present situation such a distinction would be artificial; it would in fact scarcely be possible to find any good definition which could distinguish an 'elementary' particle from a compound system. A pion, for example, could be considered as a system consisting of one or several pairs of nucleon and antinucleon. The nucleon could be built from Λ-hyperon and K-meson, the photon from muon and antimuon, etc. The well-known formula, 'every elementary particle consists of all other elementary particles', seems to be a good description of the paradoxical situation, with which we are confronted in the experiments. For practical purposes it may still be convenient to speak of a compound system, if the energy necessary for dividing the system is small compared with the rest masses of the resulting parts. But this is a rather vague quantitative definition, not a qualitative one. Therefore it is reasonable to drop this distinction altogether.

One of the main reasons for this new situation in elementary particle physics is the process of pair creation, the existence of antiparticles and antimatter.[4, 5] Wave functions as representatives of states cannot be separated completely unless they have different symmetries. Any wave function will as a rule contain contributions from any other wave function of exactly the same symmetry. In unrelativistic quantum mechanics states with different numbers of particles have different symmetries on account of the gauge groups, therefore they can be separated. In relativistic field theory this is no longer true, since we have pair creation.

Another aspect of this situation is the complexity of elementary particle physics. The theory of elementary particles can scarcely be less complicated than quantum chemistry; there will be no counterpart to the optical spectrum of the hydrogen atom and its simple mathematical laws. If in quantum chemistry we want to calculate the binding energy, for example, of the two oxygen atoms in the oxygen molecule, we cannot ignore the fact, that in collisions of higher energy the atoms may be ionized and eventually disintegrated into nucleus and electrons. Therefore we cannot hope for any simple analytical function describing accurately the potential between the two atoms and thereby defining the binding energy. In the same way the mass of the pion cannot be calculated from a simple potential between nucleon and antinucleon, because the analytical behaviour of such a function should reflect the experimental fact, that in high energy collisions between nucleon and antinucleon many particles may be created. Therefore in the theory of elementary particles the problem of mathematical approximation will be as important and as difficult as in quantum chemistry.

If the different elementary particles are taken as different forms of the fundamental substance 'energy' or 'matter', it should be emphasized that they are frequently only transient forms of very short lifetime. A few of them are stable like proton, electron or deuteron, a few others have rather 'long' lifetimes between 10^{-6} and 10^{-16} sec (like pion or Ω^{-}-hyperon), the vast majority have lifetimes below 10^{-16} sec. (In the context of the present book the terms 'mass' and 'lifetime' refer always to the rest system of the particle concerned.) Hence there will be a continuous transition between states of long lifetime, the properties of which can be determined with great accuracy, and others of very short lifetime, the properties of which are not well defined. This situation is clearly seen in the experiments. A continuous spectrum of states does not belong to definite symmetries, for example, to definite values of the angular momentum or of isospin. A discrete stationary state does belong to definite symmetries. A maximum in the continuous spectrum may, depending on its width, be called a resonance state or just a maximum and may or may not be seen in several symmetries. All intermediate cases should be possible. The term 'elementary particle' will be used quite freely for any of these forms, independent of their stability.

1-2 Classification of elementary particles

From the foregoing discussion it should be clear already, that there is a close analogy between the spectrum of different elementary particles and the spectrum of different stationary states in the quantum theory of the atom. Therefore in the classification of the different elementary particles one should be guided by those principles of classification that have proved useful in the theory of the atom.

The obvious starting point for such a classification is an analysis of those features of the observations and of the underlying natural law that in classical physics have been called 'laws of conservation'; for instance conservation of energy, momentum, angular momentum, charge, etc. The physicists have learned from the mathematicians that conservation laws are connected with symmetries, with the group structure of the underlying natural law. If the underlying law is invariant under a certain transformation, the operator performing this transformation may be used for constructing an observable, which stays constant during the processes described by the natural law. The different states of the system may be considered as different representations of the group. These representations can, from a mathematical point of view, be characterized by discrete numbers (in case of compact groups) or continuous parameters (in case of non compact groups), the discrete numbers being called quantum numbers from a phenomenological point of view. The conservation law may be stated by saying that the sum of the quantum numbers or the sum of the continuous parameters of the interacting systems stays constant during the collision process, or, in the case of a discrete group, by saying that the product of the quantum numbers (for example, parity) stays constant. All these relations are well known from unrelativistic quantum mechanics; but it is a very important result of all recent observations on elementary particle processes, that in this respect no fundamental difference has been found between elementary particle physics and quantum mechanics. The elementary particles can be classified by quantum numbers, mass-values and lifetimes just like the stationary states in quantum mechanics.

A difference comes in however, by the relevant groups. The most important group in elementary particle physics is undoubtedly the Lorentz group,[6] which replaces the Galilean group of unrelativistic quantum mechanics and produces the conservation laws of energy, momentum and angular momentum. The conservation laws for baryonic number, leptonic number and electromagnetic charge should be interpreted as the result of corresponding gauge groups. Another important group is connected with the 'isospin' of the particles and with charge symmetry;[7] this group seems to be isomorphic with the group of rotations in threedimensional space or with the group SU_2 (group of unitary and unimodular transformations of two complex variables) and produces the isospin-multiplets which will be discussed in more detail below. The empirical fact that electromagnetic charge is composed of two parts, isospin charge and hypercharge,[8] and that the latter is connected with the concept of strangeness,[9] suggests a connexion between the isospin group and the gauge group producing strangeness. These two groups should therefore probably be combined to the group U_2 (group of unitary transformations of two complex variables), which is just the direct product of SU_2 and one gauge group. Finally the observations show the relevance of several discrete groups, space reflection parity (P),[10] charge conjugation (C), time reversal (T) and G-conjugation[11] (product of C and isospin reversal).

While the existence of these groups contains the most important information concerning the structure of the underlying natural law, it is somewhat disturbing, that some of these symmetries seem to be only approximately valid. In fact, the existence of multiplets in the spectrum of elementary particles shows already that there are weaker interactions which are less

The Important Experimental Facts   5

symmetrical. In case of complete symmetry any n-dimensional representation of the group should lead to n identical eigenvalues for the mass. A splitting of these eigenvalues can be produced only by an interaction which is less symmetrical. In quantum mechanics it is well known that, for example, an outer electric field destroys the rotational symmetry of the original system and thereby leads to a splitting of the optical lines (Stark effect). The symmetry can be reduced by actions from outside. On the other hand the multiplet structure of the optical spectra rests upon the fact, that the introduction of the magnetic spin–orbit interaction destroys the independence of rotational symmetry of the spin and the orbital motion. In this case it is not an action from outside which breaks the symmetry; by neglecting the magnetic forces one had artificially produced a higher symmetry (the independence of rotation of spin and orbit), which is broken afterwards. One learns from these examples, that one should distinguish between two kinds of symmetries or invariances. Symmetries of the first kind are inherent in the underlying natural law (for example, invariance under rotation in quantum mechanics). Symmetries of the second kind rest upon approximations which may be more or less justified in the different states of the system concerned. Since symmetries of the first kind may be broken by action from outside, the term 'outside' needs some specification. In many cases the outside action can be arranged arbitrarily by the experimenting physicist; the system can be put into an electric or magnetic field, enclosed in a box, etc. In other cases the action may come from the world as a whole and cannot be influenced by the physicist. In Einstein's theory of gravitation the centrifugal force results from the outside action of the groundstate of the world or from the very distant masses. It cannot be changed arbitrarily.

Therefore the observation of mass-multiplets in the spectrum of elementary particles, not caused by artificial outer fields, leaves the choice between two different interpretations. The symmetries may be inherent, i.e. of the first kind, but broken by the ground state 'world'; or they may belong to the second kind, caused by suitable approximations, i.e. approximate by nature. The decision between these two possibilities is obviously very important for the formulation of the theory.

Besides the groups that have been mentioned so far, the mass spectrum seems to contain some rather wide multiplets of a higher order suggesting the relevance of the group SU_3,[12] with the U_2 of isospin and strangeness as a subgroup. Two octets of bosons, one octet and one decuplet of fermions correspond to this assumption. The mass differences between different components of the multiplets are large (roughly a hundred times larger than in the case of the isospin-multiplets), but follow rather closely a formula resulting from a broken SU_3-group.[13] The mass of the Ω^--hyperon had been

6   *Unified Field Theory of Elementary Particles*

predicted (in 1962) from this formula.[14] Again the question arises whether the SU_3-symmetry belongs to the first or the second kind.

1-3 The different types of interactions

The appearance of approximate symmetries in the mass spectrum and in the interactions between elementary particles suggests a division of the interactions into different types according to the symmetry which they obey. Historically the 'strong interactions' were defined as those which are responsible for the binding between nucleons in the atomic nucleus and for the scattering of pions from nucleons. The electromagnetic interactions were known already from classical physics. The 'weak interactions' were observed in the radioactive decay. With the present knowledge it is probably more convenient to define as 'strong interactions' that part of the total interaction which is strictly symmetrical under all groups mentioned above, except SU_3; as 'electromagnetic' interactions those which violate the iso-spin invariance and G-conjugation, but which are symmetrical under the other groups; as weak (radioactive) interactions those which violate conservation of strangeness and iso-charge, P, C and G, but are still in-variant under the proper Lorentz group, PC and PCT and obey conserva-tion of baryonic and leptonic number and electric charge. There may be other weak interactions, which are less symmetrical. Recent experiments suggest the existence of a small interaction violating PC. Finally there is gravitation which plays a very important rôle for the structure of the world in large dimensions, but seems to have practically no immediate influence on the structure of the elementary particles.

The strong interactions could possibly again be divided into two parts, those which are symmetrical under SU_3 and those which are not. But the mass splitting caused by the second part is only slightly smaller than that caused by the first part. (1 GeV may be a characteristic energy for the first part, 0·2 to 0·4 GeV for the second part.) The mass formulas within the SU_3-multiplets however seem to hold with a higher accuracy than what would be expected from this ratio.

The electromagnetic forces which cause the splitting of the isospin-multiplets (U_2-multiplets) are very much weaker. The corresponding mass differences are of the order of 1 MeV, i.e. roughly the electrostatic energy of a sphere with the electronic charge and a radius of the order 10^{-13} cm. The violation of the isospin symmetry is seen immediately from the fact, that the electromagnetic properties of protons and neutrons are quite different, while for strong interactions they are the same particles, seen from a different frame of reference, they differ just by the 'direction of isospin'.

The Important Experimental Facts **7**

The strength of the weak (radioactive) interactions can be estimated by comparing, for example, the lifetime for the radioactive decay of charged pions ($\sim 10^{-8}$ sec) with the electromagnetic decay of neutral pions ($\sim 10^{-16}$ sec). The weak interactions are by a factor $\sim 10^{-8}$ smaller than the electromagnetic ones. The mass splitting caused by weak interactions is negligible compared with the present accuracy of mass measurement. The breakdown of space-reflection parity in weak interactions can be recognized immediately from the polarization of particles emitted in radioactive decay.

Finally the gravitation between electron and proton is roughly by a factor 10^{-39} smaller than the electrostatic force between them.

1-4 The spectrum of elementary particles

The discussions concerning the different symmetries or groups relevant in elementary particle physics and the different types of interaction define the general framework within which the complicated spectrum of elementary particles can be analysed. In Fig. 1 the observed stationary states are given as functions of their charge as abscissa and their mass as ordinate. To each state there exists a corresponding antistate of equal mass and opposite charge (which in a few cases of charge zero is identical with the state). These antistates are as a rule not listed separately in the figure. Only states with baryonic numbers 1, 0, -1 and leptonic numbers 0, or leptonic numbers 1, 0, -1 and baryonic number 0 are taken as 'elementary particles'.

The first division of the spectrum is into the two types: fermions and bosons, i.e. particles which obey Fermi statistics or Bose statistics. On account of Lorentz- and *PCT*-invariance all particles with half integer Dirac-spin must be fermions, all particles with integer Dirac-spin bosons. The bosons belong to baryonic and leptonic number zero. The term 'meson' has frequently been used for particles of intermediate mass (100 to 800 MeV), mostly bosons, but also for muons (μ-mesons) and is not very convenient for classification.

The fermions can be divided again into two very different classes of particles, the baryons and the leptons. The lepton group comprises only a few different particles (electrons, muons and two kinds of neutrinos) with an interesting common property: the absence of any strong interaction. The strongest interaction for the leptons is electromagnetic (for electrons and muons) or 'weak' interaction (for neutrinos). All other fermions belong to the baryon group. It is only within the baryon group that one can recognize isospin-multiplets and SU_3-multiplets. The symmetries relevant in the main interactions are responsible for the structure of the mass spectrum as well. In Fig. 1 the lowest SU_3-octet and the decuplet for Dirac spin 3/2 are

8   *Unified Field Theory of Elementary Particles*

marked by a bracket. The many isospin-multiplets are easily recognized by
the near equality of the masses for different charges. From the figure one
sees at once that the electromagnetic interactions can in many cases be

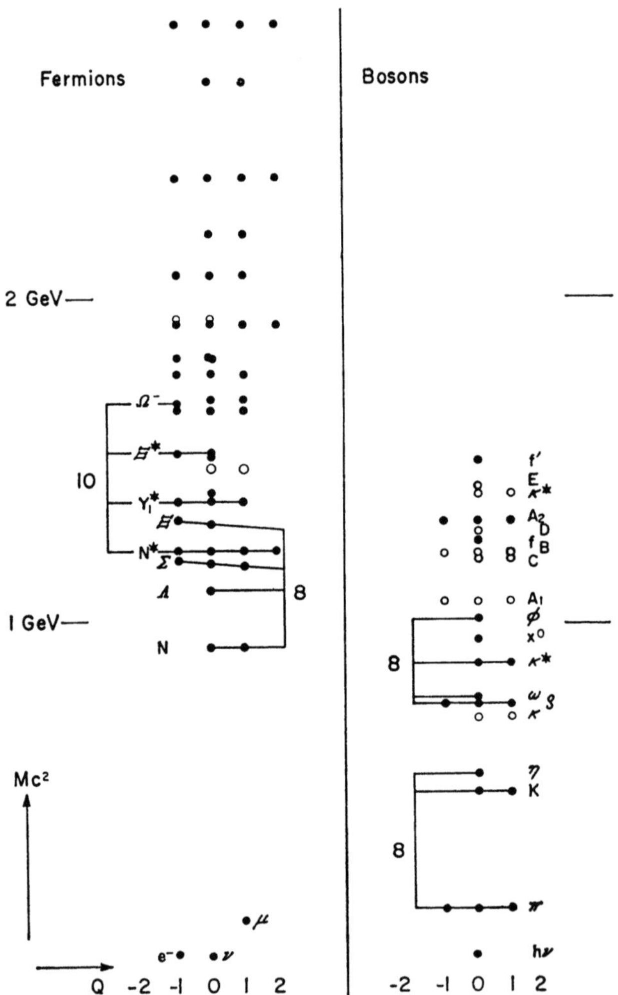

FIG. 1. Mass spectrum of elementary particles. Q is the charge. States given by
open circles are less certain than the others.

considered as a small perturbation, while there is only little difference
between the SU_3-symmetric and the SU_3-asymmetric part of the strong
interactions.

On the boson side again the SU_3-octets are marked by a bracket; the width of the different isospin- or SU_3-multiplets is roughly the same as on the fermion side. The photon has a special position in the scheme comparable to that of the leptons among the fermions, since it does not take part in the strong interactions. The charge of the photon is zero, but its isospin is not well defined.

1-5 Interaction and causality

The interaction between different elementary particles has been studied in experiments on collision processes. In scattering or in the production of new particles the laws of conservation have been checked and thereby the group structure of the underlying natural law established. Besides this information which is essentially equivalent to the information drawn also from the mass spectrum, the most important result of observations on collision processes concerns the problem of causality. This term 'causality' is meant here in the sense of the theory of special relativity: Actions are propagated from point to point along the light cone (Nahewirkung), the time order of cause and effect is never questioned. It is well known that this kind of local causality does not interfere with the statistical character of quantum mechanics; and it would be very important to know to what extent the real events in nature follow this scheme.

Two important conclusions can be drawn from the experiments.

(1) An unambiguous localization of events is not possible beyond an accuracy of $\sim 0.5 \times 10^{-13}$ cm. This follows from the fact, that in scattering processes the larger transversal momenta seem to be distributed, even for highest primary energies up to 10^5 GeV, according to a Gaussian $\exp(-p_\perp^2/p_0^2)$ with $p_0 \sim 400$ MeV/c.[15] Then, on account of the uncertainty relations, p_0 limits the accuracy of measurements of the position.

(2) In spite of this limitation, the predictions derived from the assumption of strict local causality in the relativistic sense (dispersion relations, etc.) have never been disproved. One may argue that on account of the limitation the experiments do not prove local causality. Still, from a mathematical point of view, local causality is a rather simple concept, and it is important to realize that it has not been contradicted by any experiment.

The experimental results mentioned under (1) should be taken with the precaution, that high energy scattering between leptons has not yet been observed. 'High energy' in this context is defined by the condition, that in the centre of mass system the kinetic energy of the two colliding leptons should exceed several GeV. At present one cannot exclude the possibility that high energy scattering of leptons could lead to a distribution of large transversal momenta totally different from $\exp\left(-p_\perp^2/p_0^2\right)$. The complete absence of strong interactions may create a new situation, and the mathematical scheme of quantum electrodynamics may be taken as an indication in this direction. With the existing experimental evidence however, this problem must be left completely open.

Experimental results on high-energy scattering, for example, of electrons from nuclei, may also be interpreted as informations concerning the charge distribution, or mass distribution, or more generally the structure of the particles concerned. This interpretation is somewhat problematic, since it requires assumptions about the mechanism of interaction, which are correct within conventional quantum mechanics, but may not be equally correct in the final relativistic theory. Still, it gives a reasonable picture, for example, of the nucleon which should be as good as any picture of these smallest objects can be. The nucleon appears as a rather soft cloud of matter, with a radius of roughly 10^{-13} cm. The density has a maximum but apparently no strong singularity in the centre and falls off rapidly at radii $\sim 10^{-13}$ cm. A similar picture should be true for all particles taking part in the strong interactions. For leptons however the picture could be quite different.

2

The Mathematical Tools

2-1 Mathematical description of free particles

Since most experiments on elementary particles begin with and end up with free particles moving through a cloud-chamber, bubble-chamber or emulsion, possibly interacting still with an outer electric or magnetic field, a mathematical description of the phenomena will conveniently start with a description of the free particles. For this purpose the well-known methods of the conventional linear field theory are sufficient.[16] For each kind of elementary particle a field operator $\psi_\alpha(x)$ is introduced, obeying the linear differential equation

$$(\Box + \kappa^2)\,\psi_\alpha(x) = 0, \qquad\qquad 2(1)$$

where κ represents the mass of the particle, x the space–time co-ordinates. The index α is used for characterizing the transformation properties of the particle under the different groups and thereby stating its spin, isospin and other quantum numbers. Even x may simply be considered as a continuous parameter necessary for defining the transformation properties under the inhomogeneous part of the Lorentz group, the translations. Mathematically the free particle states may be considered as special representations of the underlying groups. A free particle state can then be expressed by applying the field operator $\psi_\alpha(x)$ on the vacuum state:

$$\psi_f\rangle = \int d^4x\, f_\alpha(x)\,\psi_\alpha^*(x)|0\rangle, \qquad\qquad 2(2)$$

2(2) may be called the 'contravariant representation' of the state $\psi_f\rangle$ by means of the function $f_\alpha(x)$. The function $g_\alpha(x)$ defined by

$$g_\alpha(x) = \langle 0|\psi_\alpha(x)|\psi_f\rangle \qquad\qquad 2(3)$$

will be interpreted as 'covariant representation' of $\psi_f\rangle$. As one sees from 2(1), the function $g_\alpha(x)$ fulfills the same wave equation as $\psi_\alpha(x)$:

$$(\Box + \kappa^2)\,g_\alpha(x) = 0.$$

This is generally not true for the function $f_\alpha(x)$.

11

                           Unified Field Theory of Elementary Particles

An important mathematical quantity is the so called 2-point function, the vacuum expectation value of the product of the free field operator and its conjugate

$$\langle 0|\psi_\alpha(x)\,\psi^*_\beta(y)|0\rangle = \int d^4p\, e^{ip(x-y)}\frac{c_{\alpha\beta}(p)}{p^2-\kappa^2},\qquad 2(4)$$

where the numerator $c_{\alpha\beta}(p)$ depends on the special representation concerned. It is convenient on the right-hand side of 2(4) to integrate on the real axis in p_1, p_2, p_3, but in the complex p_0-plane on a path of integration depending on the order of factors on the left-hand side. For a different order of factors, for example, for the time ordered product or for the commutator or anticommutator a different path of integration must be chosen. The details of these relations, for which the existence of a pole at $p^2-\kappa^2=0$ is essential, are described in the textbooks on quantum field theory and are briefly summarized in Appendix A I.

When the particles move under the influence of an outer electromagnetic field, their behaviour can be described by a linear wave equation differing from 2(1) by supplementary terms which depend on the charge, the magnetic moment and other electromagnetic properties of the particles.

The many different field operators for the many different free particles and their corresponding wave equations do in this way contain a great number of important informations concerning the properties of the particles. They are suited for defining the state of affairs at the beginning and at the end of an experiment, for describing the incoming and the outgoing particles. Therefore they are frequently called the 'in' or 'out' operators or asymptotic operators.

On the other hand they are not very useful as mathematical tools for formulating a theory of the elementary particles. They are secondary quantities, derived from the asymptotic behaviour of other more fundamental mathematical expressions describing the inner mechanism of the world of elementary particles. The asymptotic operators can be constructed only after the solutions of the mathematical problems of elementary particle physics have been found.

The initial or the final states are, by definition, states in which the interaction between the particles can be neglected. Such states, and only such states, can be constructed by applying sums of products of asymptotic operators on the vacuum. They are usually supposed to form a complete set of states in the sense that any process of interaction will finally, when the interaction has become negligible, lead to a state within this set. Even if this is true, the state of interaction itself may belong to a wider group of states which cannot be represented by the asymptotic states alone. Besides that this idea of a complete set of asymptotic states encounters the difficulty that any such state will as a rule contain an unknown number of infra-red light quanta[17] which cannot be observed. Therefore the actual state is not a 'pure case' in the quantum theoretical sense but rather a 'mixture'. These difficulties are, however, of a more formal nature and shall not be discussed here in detail. In many cases the infra-red light quanta are not important and can simply be neglected.

2-2 The S-matrix

The S-matrix or scattering matrix relates the initial to the final states, the incoming cluster of particles to the outgoing cluster.[18] Therefore it is a first and very important mathematical tool for describing—at least pheno-menologically—the interaction. Mathematically the S-matrix is a unitary transformation matrix, which transforms from the initial to the final asymptotic states:

$$|\text{in}\rangle = S|\text{out}\rangle, \qquad |\text{out}\rangle = S^{-1}|\text{in}\rangle \qquad 2(5)$$

or, in slightly different notation

$$S = |\text{in}\rangle\langle\text{out}|. \qquad 2(6)$$

The unitarity of the S-matrix is essential for its physical interpretation. From

$$S^*S = SS^* = 1 \qquad 2(7)$$

one gets for the diagonal elements

$$\sum_{\text{out}} |\langle\text{out}|\text{in}\rangle|^2 = \sum_{\text{in}} |\langle\text{out}|\text{in}\rangle|^2 = 1. \qquad 2(8)$$

This equation allows to assume, that $|\langle\text{out}|\text{in}\rangle|^2$ is the probability for finding the state $|\text{out}\rangle$ in an experiment which starts with the state $|\text{in}\rangle$, or for finding the initial state $|\text{in}\rangle$ in an experiment, where we know that the final state is the special state $|\text{out}\rangle$. In this way the S-matrix can be compared immediately with the observations in a given experiment; in fact it seems to comprise everything that can actually be observed.

Besides the postulate of unitarity the S-matrix has to fulfill other conditions resulting from the fundamental group structure of the underlying natural law. If this law is invariant under a certain group, the S-matrix must be invariant as well. The operators O_i performing given transformations within the group can be represented as unitary operators in the space of the asymptotic states. For these operators the invariance of the S-matrix can be formulated by

$$O_i^{-1} S O_i = S \qquad\qquad 2(9)$$

or

$$O_i S - S O_i = 0. \qquad\qquad 2(10)$$

The generating operators must commute with the S-matrix.

If the initial and final states are influenced by outer fields, for example, electromagnetic fields, the invariance of the S-matrix will be broken, unless either these fields themselves are invariant under the transformation, or they take part in the transformation. In the second case the invariance of the S-matrix can be verified experimentally only when the fields can be chosen arbitrarily. A field, however, which is produced by the world as a whole, or by very distant masses (like the centrifugal force in the theory of general relativity) cannot be changed arbitrarily. Therefore in this case the invariance of the S-matrix is practically broken; its invariance, which would still be true 'in principle', and the corresponding law of conservation cannot be tested experimentally. This is the reason why, as has been pointed out in §1-2, one cannot distinguish experimentally in any simple way between an intrinsic lack of symmetry in the fundamental natural law and a perturbation of this symmetry by an outer field which comes from the whole world or from very distant masses.

The invariance of S produces immediately the corresponding laws of conservation. If, for example, the initial state belongs to an eigenvalue O_i' of the transformation operator O_i:

$$O_i|\text{in}\rangle = O_i'|\text{in}\rangle, \qquad\qquad 2(11)$$

the final state belongs to the same eigenvalue O_i':

$$O_i|\text{out}\rangle = O_i S^{-1}|\text{in}\rangle = O_i S^{-1} O_i^{-1} O_i|\text{in}\rangle \qquad\qquad 2(12)$$
$$= S^{-1} O'|\text{in}\rangle = O_i'|\text{out}\rangle.$$

Therefore the quantity O_i' has been 'conserved' during the interaction process. This conservation would not be true, if there were outer fields in the initial or final state, which are not invariant under O_i. In this case one might say, that some of the quantity O_i' has been exchanged between the outer field and the elementary particles. The conservation law may then be restored formally by including in the balance the quantity connected with the outer fields and their sources. Practically this conservation could however not be verified experimentally when the outer fields come from the whole world or from very distant masses.

Besides unitarity and invariance under the fundamental groups, the

S-matrix must in some way contain a mathematical formulation of causality, i.e., of that relation between cause and effect which follows from special relativity (compare §1-5). This is necessary since the S-matrix permits a complete construction of the outgoing states from the incoming states. Many investigations carried out in recent years have given a limited insight into the mathematical representation of causality in the analytical behaviour of S-matrix elements.[19] The simplest example is the elastic scattering of two interacting particles. In this case the analytical behaviour of the matrix elements corresponding to local causality has been analysed thoroughly. It leads to various 'dispersion relations', which seem to be a very valuable tool for a theoretical analysis of the experimental data. For more complicated cases however, like the production of new particles in a collision, only little is known about the analytical structure of the S-matrix elements representing causality. A first survey can probably be obtained by treating problems of this kind by a first-order perturbation calculation in a local field theory. Whether the analytical behaviour derived in this way should be generally valid in an exact theory, is however an open question. On the one hand one cannot exclude the possibility that the real interaction has some features of non-locality, on the other hand a completely local field theory with interaction could possibly contain mathematical inconsistencies. Therefore, with the present knowledge of quantum field theory a definite answer to the problem of causality cannot be given.

The postulate of causality in the relativistic sense has still another interesting consequence for the asymptotic operators and the S-matrix. If interaction can be carried only from a point to a neighbouring point, it is expected that for the asymptotic operators such solutions of 2(1) will play a rôle, which for a very short time interval are concentrated in a very small space region. The Fourier transform of such solutions (of the type of Schwinger's Δ-function) contains positive and negative frequencies. Therefore the asymptotic operators are conventionally interpreted as both creation and annihilation operators, i.e. as operators which create antiparticles and annihilate particles, while their conjugate operators create particles and annihilate antiparticles. An S-matrix element may be constructed formally by a certain number of in- and out-operators:

$$\langle \text{out}|\text{in}\rangle = \langle 0|\psi_{\text{out}}^{(1)}\,\psi_{\text{out}}^{(2)}\ldots\psi_{\text{in}}^{*(1)}\,\psi_{\text{in}}^{*(2)}\ldots|0\rangle. \qquad 2(13)$$

By changing one of these asymptotic operators from a creation to an annihilation operator or vice versa, and thereby changing it from an in- to an out-operator, one arrives at a different S-matrix element containing the same total number of asymptotic operators, i.e. referring to the same total number of in- and out-going particles. In case of strict local causality

             Unified Field Theory of Elementary Particles

these two matrix elements, considered as analytical functions of the co-ordinates in the asymptotic operators, should be connected by analytical continuation. This relation is usually called 'crossing symmetry'[20] and seems to agree well with the existing observations. Its mathematical analysis can be given only after establishing a mathematical scheme for calculating S-matrix elements from a more general theoretical basis. Eqn. 2(13) is just a formal definition, not a basis for a theoretical calculation.

In this way the postulate of causality finds an adequate mathematical representation in the analytical character of the S-matrix elements, considered as functions of the co-ordinates of the incoming and outgoing particles. On the other hand the theory of analytical functions of many complex variables is very complicated. Therefore it would probably be difficult to reach in this way general mathematical statements concerning the behaviour of S-matrix elements depending on many variables.

The three main features of the S-matrix: unitarity, invariance under the fundamental groups and analytical representation of causality, are possibly sufficient in principle, to determine its structure completely. Still, on account of its mathematical complexity, the S-matrix can scarcely be a good mathematical tool for defining the natural law underlying elementary particle physics. A (possibly incomplete) description of the phenomena by means of an S-matrix will certainly be possible. But the S-matrix should be the result, not the definition of the theory, it should be the outcome of a mathematical process starting from axioms which can be formulated in a much simpler way. This is certainly the rôle which the S-matrix plays in unrelativistic quantum mechanics; and since quantum mechanics is just a special part of elementary particle physics, one should expect that the rôle of the S-matrix in this more general field is essentially the same.

2-3 The field operators

The next mathematical tool which possibly may be used for a mathematical description of elementary particle phenomena is a local field operator $\chi(x)$. There are mainly two arguments for an applicability of such operators. In unrelativistic quantum mechanics Jordan, Klein and Wigner[21] have shown that the fundamental equation can be formulated by means of a field operator. Besides that, the success of the postulate of local causality (dispersion relations and crossing symmetry) in elementary particle physics suggests the existence of some fundamental field operator and a corresponding differential equation ('field equation') which could possibly provide the definition of the mathematical structure of the theory.

Field operators which can be used in this way for defining the theory are

The Mathematical Tools **17**

of a very different nature from the asymptotic field operators of free particles discussed in §2-1. Such fundamental field operators are not connected with only one special type of particle; they might be interpreted as creating and annihilating matter in general, not only special stationary states of matter.

If the fundamental field operators $\chi_\alpha(x)$ constitute a simple non-trivial representation of the underlying symmetry groups, one may, by building expressions of the type

$$O_\lambda(x) = \int \chi_\alpha(x)\,\chi_\beta(x+\xi_1)\,\chi_\gamma^*(x+\xi_2)\ldots f_{\alpha\beta\gamma\ldots\lambda}(\xi_1,\xi_2\ldots)\,d\xi_1\,d\xi_2\ldots \quad 2(14)$$

construct many other more general representations of the same groups. Operators of this general type may again be interpreted as creating and annihilating matter. Their connexion with the 'asymptotic' operators (free particle operators) can be seen in the following way. The 2-point functions constructed from $O_\lambda(x)$ may contain (in their momentum representation) poles at certain mass values:

$$\langle 0|O_\lambda(x)\,O_\mu^*(y)|0\rangle = \int d^4p\,e^{ip(x-y)}\left(\frac{c_{\lambda\mu}(p)}{p^2-\kappa^2}+\ldots\right). \qquad 2(15)$$

In this case an asymptotic operator $\psi_\lambda(x)$ could be defined, which in its 2-point function gives just the same pole term as $O_\lambda(x)$, but no other pole or cut. Then one may say, that the operator $O_\lambda(x)$ 'contains' the free particle operator $\psi_\lambda(x)$, that $O_\lambda(x)$ creates or annihilates particles of mass κ, characterized with respect to their quantum numbers by the transformation properties of $O_\lambda(x)$. In this sense $O_\lambda(x)$ will generally contain several free particle operators; its 2-point function will, besides several pole terms, generally also contain cuts representing continuous spectra belonging to this symmetry. The knowledge of the analytical properties of the 2-point function 2(15) would already imply the complete knowledge of all particles of this symmetry. For a given symmetry many operators $O_\lambda(x)$ may be constructed, containing different numbers of $\chi(x)$ and $\chi^*(x)$ operators; still the poles in their 2-point functions 2(15) will be identical (not the residua), since the pole belongs to a definite symmetry. The different operators $O_\lambda(x)$ belonging to the same symmetry will presumably be connected by some fundamental field equation, which defines interaction and, by its transformation properties, the symmetries of the underlying natural law.

If fundamental field operators can in this way be applied for a mathematical description of the phenomena, the postulate of local causality should be expressed by the simple assumption, that the commutator (or,

                    Unified Field Theory of Elementary Particles

in case of spinor field, the anticommutator) vanishes for space like distances of the co-ordinates. This assumption is probably sufficient for the definition of local causality, if it is combined with the axiom, that the field operators taken within the arbitrarily small (but finite) time interval between t and $t+\Delta t$ should be sufficient[22] for constructing the complete Hilbert space from the ground state $|0\rangle$. We will not go into the mathematical details of these assumptions, since a complete analysis of the axiomatic situation in this kind of field theory has not yet been given; nor will we try to specify too closely the Hilbert space of states, in which the operators $\chi_\alpha(x)$ act. The structure and the metric of the Hilbert space should finally be decided upon by the requirement of consistency of the whole mathematical scheme. The Hilbert space must definitely contain the asymptotic states, which can be constructed from the groundstate by the asymptotic operators; because these are the physical states which can be immediately interpreted in terms of observations. But it may contain other states as well,[23] as a kind of mathematical supplement necessary—or at least useful—for a simple formulation of the theory.

If the postulate of local causality holds and if the operators of the time interval between t and $t+\Delta t$ are sufficient for constructing the complete Hilbert space, the operators taken at other times must be related to the operators of the time interval by a mathematical process compatible with the assumption of commutability (or anticommutability) at space like distances. The simplest definition of such a process could be given by a differential equation for the field operators. The differential equation should in the case of local causality be invariant under Lorentz-transformations and would therefore be hyperbolic; hence it would be compatible with commutability of observables at space-like distances. Still a differential equation for the field operators raises problems which do not exist in a classical field theory.

This equation should define interaction, therefore it cannot be a linear equation. The interaction must be local, hence the corresponding term in the equation should contain the product of field operators at the same space time point. As a rule, however, such a product is not well defined. It would certainly not be defined if the field operators would obey a canonical commutation rule. The commutator (or anticommutator) is probably singular near the origin. The problem of this singularity cannot be separated from the other problem of defining what the interaction term actually means.[23] In fact the singularity of the commutator would be largely determined by the interaction term, since the 2-point function and the commutator must be compatible with the differential equation.

Therefore, any attempt of defining a quantum field theory with interaction by a differential equation requires a careful study of the singularity of the 2-point function near the origin.

A differential equation for the field operators constitutes an infinite set of differential equations for infinitely many matrix elements of products of field operators. The solutions of this set are not sufficiently defined by the equations. The differential equations must be supplemented by boundary conditions in order to construct a Hilbert space from their solutions. In conventional quantum mechanics there are essentially two kinds of boundaries: infinite space-like distances, and very small distances at singular points. The behaviour of the wave functions at these boundaries is restricted by the condition, that the norm of the state represented by the wave function should be finite. In a relativistic quantum field theory there may be a third boundary.[24] The interaction defined by the differential equation should allow the creation of pairs. In this case the infinite set of matrix elements connected by differential equations contains matrix elements with an arbitrarily high number of variables. The third boundary therefore refers to an infinitely high number of variables. The condition of finite norm requires first of all a definition of the norm in this case and raises difficult mathematical problems, which will be discussed later only in connexion with the special field theory described in the present book.

2-4 Connexion between field operators and S-matrix

The general procedure 2(14) for constructing an operator $O_\lambda(x)$ from the field operator $\chi(x)$ permits—as has been stated before—the construction of many different operators $O_\lambda(x)$ belonging to the same symmetry and also of different operators belonging to different symmetries. If any of these operators of equal symmetry leads to a pole in the corresponding 2-point function 2(15), the same pole will as a rule occur in the 2-point function of any of the other operators. The existence of the pole in 2(15) will in fact prove, that there exists a discrete stationary state $|\psi\rangle$ of mass κ and of the symmetry given by $O_\lambda(x)$, for which the matrix element

$$\phi_\lambda(x) = \langle 0|O_\lambda(x)|\psi\rangle \qquad\qquad 2(16)$$

obeys the wave equation

$$(\Box + \kappa^2)\,\phi_\lambda(x) = 0. \qquad\qquad 2(17)$$

$\phi_\lambda(x)$ may, except for a constant factor, be identified with the wave function $g_\lambda(x)$ of 2(3) for the 'free particle' state $|\psi\rangle$. In the same way any of the other operators $O_\lambda(x)$ of equal symmetry will lead to similar matrix elements which as a rule are different from zero and differ only by the constant factor from $g_\lambda(x)$. This was meant by the statement, that the operators $O_\lambda(x)$ 'contain the free particle field' $\psi_\lambda(x)$.

The vacuum expectation value

$$\langle 0 | T O^{(1)}_{\lambda_1}(x_1)\, O^{(2)}_{\lambda_2}(x_2) \dots O^{(n)}_{\lambda_n}(x_n) | 0 \rangle \qquad 2(18)$$

(where the operators $O^{(i)}_{\lambda_i}(x_i)$ may belong to different or equal symmetries) is now closely related to an S-matrix element. (The letter T means the time ordered product in the conventional sense, in which operators with larger time values are on the left-hand side of those with smaller time values.) If some of the time variables in 2(18) go to very large positive values $(t_1, t_2 \dots t_s \to +\infty)$, the others to very large negative values $(t_{s+1} t_{s+2} \dots t_n \to -\infty)$, the matrix element may, in this limit, be interpreted as an S-matrix element $\langle \text{out} | \text{in} \rangle$ in 2(6). The operators $O^{(s+1)}$, $O^{(s+2)} \dots O^{(n)}$ create the incoming particles at $t \to -\infty$, while $O^{(1)}$, $O^{(2)} \dots O^{(s)}$ annihilate at $t \to +\infty$ the outgoing particles. Therefore this asymptotic value of 2(18) defines an S-matrix element.

In momentum space 2(18) will have the general form

$$\delta(p_1 + p_2 + \dots + p_n) \frac{f(p_1, p_2 \dots p_n)}{(p_1^2 - \kappa_1^2)(p_2^2 - \kappa_2^2) \dots (p_n^2 - \kappa_n^2)}. \qquad 2(19)$$

For the asymptotic value of 2(18) in ordinary space $(t \to \pm\infty)$ only the infinitesimal surrounding of the poles in 2(19) is important. This is true, independent of the choice between $t \to +\infty$ or $t \to -\infty$ in any one of the operators. If in a special operator one goes over from the limit $t \to -\infty$ to $t \to +\infty$, one thereby goes over from the first S-matrix element to a different one referring to the same number of particles. The two S-matrix elements are connected by analytical continuation in 2(18). This connexion is the immediate expression of 'crossing symmetry'.[20]

2-5 The axiomatic formulation

The S-matrix description of elementary particle phenomena (§2-2) seems to be a rather wide frame, which, judging from previous mathematical analysis, should be consistent, unless very special assumptions would be added concerning the analytical representation of causality. The concept of a local field operator (§2-3) is much more problematic. The condition of (anti-) commutativity at space-like distances and the assumption of a differential equation for the field operator impose severe restrictions on the mathematical scheme. If they can be fulfilled, the differential equation should allow a simple formulation of the natural law underlying elementary particle physics. Even this frame is considerably wider than the axiomatic

The Mathematical Tools   **21**

frame which has frequently been discussed in connexion with quantum electrodynamics and has been constructed in close analogy to the axiomatic frame of conventional unrelativistic quantum mechanics.

This rather narrow frame, which has been studied in detail by Wightman,[25] Lehmann, Symanzik and Zimmermann[26] imposes two further restrictions (beyond the assumptions of §2-3) on the mathematical scheme. It assumes that the metric in the Hilbert space on which the field operators act should be definite and positive, and that the asymptotic field operators alone should be sufficient for constructing the complete Hilbert space. The second condition gives the decisive restriction. Together with the unitarity of the S-matrix it enforces already the definite metric for all those subspaces of the total Hilbert space, which are separated from the rest by a superselection rule. The definite metric sets sharp limits to the possible singularities of the 2-point functions in the neighbourhood of the origin and the light-cone. The singularity must be at least of the type of a δ-function. If the interaction does not allow a singularity of this type, the conditions cannot be fulfilled. Hitherto it has not been possible to give any example of a theory with local interaction fitting consistently into this narrow axiomatic frame.

This result is somewhat strange since—given the unitarity of the S-matrix —the postulate of positive metric for the Hilbert space does not seem to be a very severe restriction. One cannot see easily why the addition of unphysical states to the (asymptotic) Hilbert space would make it much simpler to define a local field operator and to construct a non-linear differential equation for its change in time. (As a special example of this situation we mention the Bleuler–Gupta[27] version of quantum electrodynamics.) Therefore it is important to notice that from a group theoretical point of view the unitarity of the S-matrix is a very natural postulate, while the assumption of a definite metric in the complete Hilbert space is not. This can be seen in the following way.

The group of all transformations which can be represented in the complete Hilbert space is necessarily a non-compact group. This follows from the fact, that the Lorentz group must belong to this group as a subgroup, and is non-compact. A finite representation of a non-compact group can be given only in space with indefinite metric. It is true that representations of a non-compact group are possible in a space of an infinite number of dimensions with definite metric. But the finite representations which could refer to a subspace of the total infinite Hilbert space are simpler, and in so far, more natural, and they require the indefinite metric. The S-matrix, on the other hand, refers to specified values of energy and momentum, it is defined 'on the energy-shell'. When the total energy and momentum are given, the group of remaining transformations is a compact group. Compact groups can be represented in a space with positive definite metric and a finite number of dimensions. Hence for the S-matrix unitarity is a natural requirement.

From the point of view of physical interpretation only the unitarity of the S-matrix is necessary. This unitarity is sufficient for permitting the conventional probability interpretation of the S-matrix elements. But a corresponding interpretation of local events within space regions of 10^{-13} cm is not required; one may well doubt whether it is possible. Therefore, the reduction of the total Hilbert space to the space of the asymptotic states and the postulate of definite metric seem to be an unnecessary constraint which may rather complicate than simplify the mathematical structure of field theory with interaction.

If one tries to formulate a field theory within the wider frame of §2-3, the concept of 'asymptotic states' may become problematic. Even in the limit $t \to \pm \infty$, states may remain which cannot be composed of eigenstates of energy or momentum. Still the transformation from the incoming to the outgoing states, expressed by the S-matrix, should exist. By a further transformation within the incoming and the outgoing states the S-matrix can be brought into its diagonal form. Among the diagonal elements belonging to eigenstates of energy and momentum some will be of the form exp $i\alpha$ (α real), the others will not. Only the states belonging to the first group should be called physical states; they form a subspace of the total Hilbert space, and the process of interaction can never lead from a state within this subspace to a state outside of it. Within this subspace the conventional definition of probability can be taken over from ordinary quantum mechanics.

2-6 Hilbert space with indefinite metric

A quantum field theory of the general type §2-3 seems to require a Hilbert space with indefinite metric. For such a space the mathematical formalism is somewhat different from the conventional formalism in the textbooks, which refers to definite metric. Therefore the essential concepts concerning the Hilbert space with indefinite metric should be outlined briefly.[28]

If the Hilbert space is constructed by a number of fundamental state vectors $|\Phi_n\rangle$, then every state vector $|\psi\rangle$ in this space can be defined by two sets of components.

We can either write

$$|\psi\rangle = |\Phi_l\rangle \cdot \psi^l, \qquad\qquad 2(20)$$

where it is understood that the summation shall be carried out if the same index occurs as lower and upper index. Then the ψ^l are called the contravariant components of $|\psi\rangle$.

Or one can put

$$\psi_l = \langle\Phi_l|\psi\rangle. \qquad 2(21)$$

The ψ_l are the covariant components of $|\psi\rangle$.

The metric is defined by the relation

$$g_{ik} = \langle\Phi_i|\Phi_k\rangle. \qquad 2(22)$$

In connexion with 2(22) we require

$$g_{ik} = g_{ki}^* \qquad 2(23)$$

and from 2(20,21,22):

$$\psi_l = g_{lk}\psi^k. \qquad 2(24)$$

From the $|\Phi_l\rangle$ one can generally construct the 'inverse' system of state vectors $\Phi_l\rangle$, defined by

$$\langle\Phi^k|\Phi_l\rangle = \delta_l^k. \qquad 2(25)$$

From 2(20) and 2(25) one gets

$$\psi^l = \langle\Phi^l|\psi\rangle. \qquad 2(26)$$

The norm of $|\psi\rangle$ can be calculated from the components

$$\langle\psi|\psi\rangle = \psi_l^*\psi^l = \psi^{*l}\psi_l = \psi^{l*}\psi^k g_{lk} = \psi_l^*\psi_k g^{lk}. \qquad 2(27)$$

The matrix elements of an operator A can be introduced by

$$\langle\Phi_l|A|\Phi_k\rangle = {}_lA_k; \quad \langle\Phi_l|A|\Phi^k\rangle = {}_lA^k, \text{ etc.}$$

and matrix multiplication then follows the rule

$$_k(AB)_l = {}_kA_m{}^mB_l = {}_kA^m{}_mB_l = {}_kA_m g^{mn}{}_nB_l.$$

A selfadjoint operator is defined by

$$\langle\Phi_k|A|\Phi_l\rangle = \langle\Phi_l|A|\Phi_k\rangle^* \quad \text{or} \quad {}_kA_l = {}_lA_k^*.$$

In the indefinite metric one may also use the term 'pseudohermitian' for this property. In this sense the adjoint or pseudohermitian conjugate B^* to the operator B is given by

$$\langle\Phi_k|B^*|\Phi_l\rangle = \langle\Phi_l|B|\Phi_k\rangle^*$$

or

$$_kB_l^* = ({}_lB_k)^*. \qquad 2(28)$$

The expectation value of an operator B in the state $|\psi\rangle$ is defined by

$$\bar{B} = \frac{\langle\psi|B|\psi\rangle}{\langle\psi|\psi\rangle}; \qquad\qquad 2(29)$$

it is real for a selfadjoint operator.

Observable quantities are generally represented by selfadjoint operators. The eigenvalues of these operators need not necessarily be real. Let $|\psi\rangle$ be an eigenstate of the selfadjoint operator O, and O' the corresponding eigenvalue:

$$O|\psi\rangle = O'|\psi\rangle, \qquad\qquad 2(30)$$

then multiplication from the left by $\langle\psi|$ leads to

$$\langle\psi|O|\psi\rangle = O'\langle\psi|\psi\rangle. \qquad\qquad 2(31)$$

The expression on the left-hand side is real according to 2(28). If the norm $\langle\psi|\psi\rangle$ is different from zero, then O' must be real and is in fact identical with the expectation value 2(29) of O in the state $\psi\rangle$. If O' is imaginary or complex, the norm $\langle\psi|\psi\rangle$ must necessarily vanish. Since physical states are states with a finite norm, the eigenvalues and the expectation values of selfadjoint operators, representing observables, are always real in these states. Complex eigenvalues of such operators can belong only to 'ghost'-states of norm zero.

If $|\psi_1\rangle$ and $|\psi_2\rangle$ are two different eigenstates of O, belonging to the two eigenvalues O_1' and O_2', then from

$$O|\psi_1\rangle = O_1'|\psi_1\rangle \quad \text{and} \quad O|\psi_2\rangle = O_2'|\psi_2\rangle \qquad 2(32)$$

or

$$\langle\psi_2|O = O_2'^*\langle\psi_2| \qquad\qquad 2(33)$$

one concludes

$$\left.\begin{array}{l}\langle\psi_2|O|\psi_1\rangle = \langle\psi_2|\psi_1\rangle O_1' = O_2'^*\langle\psi_2|\psi_1\rangle, \\ \text{or} \\ (O_1' - O_2'^*)\langle\psi_2|\psi_1\rangle = 0.\end{array}\right\} \qquad 2(34)$$

Therefore the two states must be orthogonal to each other, unless $O_1' = O_2'^*$. Two physical eigenstates with the different real eigenvalues O_1' and O_2' are necessarily orthogonal.

The eigenstates of an operator O will as a rule not be sufficient for constructing the complete Hilbert space.

These relations holding for a Hilbert space with indefinite metric replace the corresponding relations known from conventional quantum mechanics. They may be sufficient as an explanation for the use which is made of the Hilbert space with indefinite metric in the later chapters.

3

The Fundamental Field Equation

3-1 The mathematical form of the field equation

In the attempt to formulate a unified field theory of elementary particles the guiding principles discussed in §1 are the empirical group structure, i.e. the laws of conservation, and relativistic causality. The empirical fact that some of the groups are only approximately valid (compare §1-2), requires a distinction between fundamental groups, characterizing the underlying natural law, and approximate symmetries arising from the dynamics of the system. The non-linear spinor theory described in the present book takes the Lorentz group and, for the isospin, the group U_2 as fundamental, but the group SU_3 or higher groups as approximate dynamical symmetries*. The arguments for this choice will be discussed in §3-2 and §3-5.

Starting from these assumptions there seems to be only one simple equation which can possibly represent the observed elementary particles and shall be taken as the basis of the present book:[29]

$$i\sigma^\nu \frac{\partial \chi(x)}{\partial x^\nu} + l^2 \, \sigma^\nu : \chi(x) \, (\chi^*(x) \, \sigma_\nu \chi(x)) : \, = 0. \qquad 3(1)$$

l is an arbitrary constant of the dimension of a length, if the 2-point function $\langle 0|\chi(x)\chi^*(y)|0\rangle$ is in the conventional way taken as the minus third power of a length. l may simply be taken as the unit of length.

The field operator $\chi(x)$ is defined as a 2-component spinor (Weyl-spinor) under Lorentz transformations and a 2-component spinor in isospace. If the first index of $\chi(x)=\chi_{\alpha\beta}(x)$ refers to Lorentz space, the second to isospace, the explicit form of 3(1) should be

$$i\sigma^\nu_{\alpha\beta} \frac{\partial \chi_{\beta\gamma}(x)}{\partial x^\nu} + l^2 \, \sigma^\nu_{\alpha\beta} : \chi_{\beta\gamma}(x) \, \chi^*_{\delta\epsilon}(x) \, \sigma_{\nu,\delta\eta} \, \chi_{\eta\epsilon}(x) : \, = 0. \qquad 3(2)$$

The $\sigma^\nu = (1, \boldsymbol{\sigma})$ are the conventional Pauli matrices.

*Different fundamental assumptions have been discussed in various papers[80]

25

                      Unified Field Theory of Elementary Particles

The dots : : in the second term refer to the definition of a product of three field operators at the same space–time point. This product is meant here as a 'Wick product' and is given by the following process:

$$: \chi(x)\,\chi^*(x)\,\chi(x): \; = \lim_{\delta\to 0,\,\epsilon\to 0} [\chi(x)\,\chi^*(x+\delta)\,\chi(x+\epsilon+\delta)$$

$$- \langle 0|\chi(x)\,\chi^*(x+\delta)|0\rangle\,\chi(x+\epsilon+\delta)$$

$$- \chi(x)\,\langle 0|\chi^*(x+\delta)\,\chi(x+\epsilon+\delta)|0\rangle]; \qquad 3(3)$$

(ϵ and δ are space-like vectors; therefore $\langle 0|\chi(x)\chi^*(x+\delta)|0\rangle$ could also be replaced by $\langle 0|T\chi(x)\chi^*(x+\delta)|0\rangle$). Eqn. 3(3) requires, that the limit on the right-hand side ($\delta \to 0$, $\epsilon \to 0$) exists for each matrix element of the product. Thereby some restrictions are assumed for the singularity of a product of the type $\chi(x)\chi^*(x+\delta)$ in the neighbourhood of $\delta \to 0$, which will be discussed in detail in §3-3.

Eqn. 3(1) can be written in many different forms. It could for instance be replaced by:

$$+ i\sigma^\nu \frac{\chi\partial}{\partial x^\nu} + l^2 \tau_k \sigma^\nu : \chi(\chi^* \sigma_\nu \tau_k \chi): \; = 0 \qquad 3(4)$$

or

$$i\sigma^\nu_{\alpha\beta} \frac{\chi\partial_{\beta\gamma}}{\partial x^\nu} + l^2 \tau_k^{\gamma\eta} \sigma^\nu_{\alpha\beta} : \chi_{\beta\eta}\,\chi^*_{\delta\epsilon}\,\sigma_{\nu,\delta\zeta}\,\tau_k^{\epsilon\xi}\,\chi_{\zeta\xi}: \; = 0. \qquad 3(5)$$

The τ_k ($k=1,2,3$) are the Pauli matrices in isospin space.

These equations are equivalent to 3(1) and 3(2) on account of the anti-commutation rule for $\chi(x)$, which will be discussed below (§3-3).

Another possibility is the introduction of a Dirac-spinor (4-component spinor) $\psi_\alpha(x)$ with the field equation:

$$\gamma_\nu \frac{\partial\psi}{\partial x_\nu} + l^2 \gamma_\nu \gamma_5 : \psi(\bar\psi \gamma^\nu \gamma_5 \psi): \; = 0. \qquad 3(6)$$

This equation 3(6) is equivalent to 3(1), if—in the conventional representation of the matrices σ_ν and γ_ν (compare p. 166)—one puts

$$\psi_1 = \chi_{11}, \qquad \psi_2 = \chi_{21}, \qquad \psi_3^* = \chi_{22}, \qquad \psi_4^* = -\chi_{12}. \qquad 3(7)$$

Eqn. 3(6) was the first form of the non-linear spinor equation, which had been recognized as a possible basis for the theory of elementary particles. From a group theoretical point of view, however, this form is not very convenient, since the isospin transformations would be represented in an unusual way.

It is natural that the mathematical structure of a theory of this general type can be expressed in many different ways. The fundamental structure of quantum chemistry for instance can be expressed by the Schrödinger equation of a system with many electrons, or by a quantum field theory of matter waves as developed by Klein, Jordan and Wigner,[21] or by a Hamiltonian in matrix mechanics, etc. The mathematical formalism of quantum theory is extremely flexible and allows for a great number of different but equivalent representations. This is certainly true also for the structure described by 3(1) or 3(4) or 3(6), which, however, will be used mostly in its simplest form 3(1):

$$i\sigma^\nu \frac{\partial \chi}{\partial x^\nu} + l^2 \sigma^\nu : \chi(\chi^* \sigma_\nu \chi) : = 0.$$

3-2 Group structure of the fundamental equation

Eqn. 3(1) is invariant under a great number of linear transformations. The complete analysis of these symmetries establishes both the properties by which the elementary particles or stationary states following from the theory can be classified, and the laws of conservation holding for these properties. It should be remembered in this connexion, that the symmetry of the fundamental equation is, in itself, not sufficient for the symmetry of the theory. It is only when the commutations rules and the ground state obey the same symmetry, that the symmetry of the theory and the laws of conservation follow without restriction. The symmetry of the commutator and the ground state will be discussed in §3-3 and §3-5. By comparing the results of the group theoretical analysis of the theory with the empirical facts concerning multiplets and conservation laws, one gets a first insight into the possibilities of 3(1) as a basis for a theory of elementary particles.

Among the continuous groups of 3(1) the most important is undoubtedly the proper Lorentz group. Its inhomogeneous part is represented in 3(1) by the translations:

$$x_\nu \rightarrow x_\nu + a_\nu; \qquad \chi(x) \rightarrow \chi(x+a) \tag{3(8)}$$

with four continuous parameters a_ν. The invariance of 3(1) under these transformations provides the conservation of energy and momentum.

The homogeneous part of the Lorentz transformations is represented by

$$\chi_{\gamma\lambda}(x) \rightarrow e^{-(i/2)(\alpha_k + i\beta_k)\sigma_{k,\gamma\delta}} \chi_{\delta\lambda}(x') \tag{3(9)}$$

with six real, infinitesimal continuous parameters α_k, β_k; the x' are the co-ordinates after the Lorentz transformation defined by α_k and β_k. For infinitesimal α_k and β_k, the value α_k can simply be identified with the angle of rotation around the k-axis, the value β_k with the velocity in the k-

direction. The invariance of 3(1) under 3(9) leads to the conservation of angular momentum and to the three integrals concerning the motion of the centre of gravity.

The next important continuous group is the isospin group, which is isomorphic to the group SU_2 or to the group of rotations in three dimensional space. Eqn. 3(1) is invariant under the transformations

$$\chi_{\alpha\lambda}(x) \;\rightarrow\; e^{(i/2)\eta_k \tau_k \lambda\mu}\chi_{\alpha\mu}(x) \tag{3(10)}$$

with the three real infinitesimal continuous parameters η_k. Therefore the three components of the total isospin should be conserved. Empirically the conservation of isospin is only an approximation. It is violated by the electromagnetic interactions. This fact can be interpreted in a theory starting from 3(1) by assuming that the ground state is not completely symmetrical; that the ground state is not just vacuum, but 'the world'. The electromagnetic forces provide an exchange of isospin between the elementary particles and the world. This asymmetry of the ground state will be discussed in more detail in §3-5 and §7.

The fundamental equation 3(1) is moreover invariant under the scale transformation or dilatation

$$\chi(x) \;\rightarrow\; \eta^{1/2}\chi(x\eta). \tag{3(11)}$$

This important one-parameter group can play a rôle in connexion with particles of rest mass zero, compare, for example, Ref. 63. The existence of particles with a finite rest mass violates the scale invariance, unless the masses participate in the transformation, and may be considered as result of an asymmetry of the ground state with respect to the dilatation group. Possibly the asymmetry could even be attributed to the commutator. The recent development of the theory in the direction of quantum electrodynamics suggests a connexion between the conservation of the leptonic number and the scale transformation. This question can be settled however only after the theory of the leptons has been completed.

Finally 3(1) is invariant under the gauge transformation

$$\chi(x) \;\rightarrow\; e^{i\alpha}\chi(x). \tag{3(12)}$$

This invariance may be used for explaining the conservation of the baryonic number. Thereby the list of continuous groups of 3(1) has been exhausted. The complete group is characterized by fifteen continuous parameters.

Comparing these results with the empirical symmetries, multiplets and conservation laws—and disregarding for the moment the discrete groups—one sees that most of the empirical group structure can be accounted for by 3(1). At the present state of the analysis there might still be some doubt, whether the four quantum numbers charge, hypercharge, baryonic number and leptonic number can be explained by the three gauge (or gauge-like) transformations of 3(1) (the transformations 3(10) for $\eta_k = (0,0,\eta_3)$ and 3(11) and 3(12)). But this question will find a satisfactory answer in connexion with the theory of strange particles (§7). The possibility of connecting an isospin property with the ground state 'world' leads to the separation of the gauge group 3(12) into two gauge groups, the one referring to Lorentz space, the other one to isospin space. The only observed symmetry which is certainly not contained in 3(1) as a fundamental property, is the group SU_3. This result goes well together with the empirical fact (compare §1-2) that the deviations from SU_3 are very big, roughly a hundred times bigger than the deviations from the isospin group SU_2. Still the approximate validity of SU_3 in some parts of the spectrum of elementary particles needs an explanation. It will be found later on by some approximate symmetries resulting from the degeneracy of the ground state (§7-1) and the corresponding dynamical consequences of 3(1).

With regard to the discrete groups it is well known that for a relativistic wave equation such groups are related to the Lorentz group. Generally one expects the three discrete operations:

C particle–antiparticle conjugation

P space inversion

T time reversal.

If the isospin is a good quantum number, the operation C should be replaced by the operation G, which is different from C by a 180°-rotation J in isospin space:

$$G = CJ. \qquad\qquad 3(13)$$

It is then only this operation G which leads from a particle to an antiparticle of the same quantum numbers. While G, C and P are linear operations, the time reversal T is an antilinear operation.

Eqn. 3(1) is invariant under the proper Lorentz group. Therefore it is, according to a theorem of Lüders[30] and Pauli,[31] also invariant under $PCT = PGJT$.

Moreover it is invariant under PG, PC or T. The transformation PG is represented by

$$\chi(\mathbf{r}, t) \; \to \; R\chi^*(-\mathbf{r}, t), \qquad\qquad 3(14)$$

where R is a matrix in spin- and isospin space defined by the relations:

$$\sigma_k = -R\sigma_k^T R^{-1}; \qquad \tau_k = -R\tau_k^T R^{-1}. \qquad\qquad 3(15)$$

σ_k^T and τ_k^T are the transposed matrices σ_k, τ_k. In the conventional representation

$$\left(\sigma = \begin{vmatrix} 0 & 1 \\ 1 & 0 \end{vmatrix}, \quad \begin{vmatrix} 0 & -i \\ i & 0 \end{vmatrix}, \quad \begin{vmatrix} 1 & 0 \\ 0 & -1 \end{vmatrix}\right)$$

one may put

$$R = \sigma_2 \tau_2. \qquad\qquad 3(16)$$

Eqn. 3(1) is, however, not invariant under the operations P or G or C separately. In fact a 2-component spinor (Weyl spinor) $\chi(x)$ can represent the proper Lorentz group, but it cannot represent the complete Lorentz group which includes the space reflection. If in the differentials in 3(1) one reverts the space co-ordinates, one gets a new equation, which can be fulfilled by a new operator $\tilde{\chi}(x)$:

$$i\overline{\sigma}^{\nu} \frac{\partial \tilde{\chi}}{\partial x^{\nu}} + \overline{\sigma}^{\nu} \tilde{\chi}(\tilde{\chi}^* \, \overline{\sigma}_{\nu} \tilde{\chi}) = 0 \qquad\qquad 3(17)$$

with $\overline{\sigma}^{\nu} = (1, -\sigma)$. This new operator $\tilde{\chi}$ has different transformation properties from χ under the proper Lorentz group. It transforms like:

$$\tilde{\chi}_{\alpha\lambda} \;\rightarrow\; e^{-(i/2)(\alpha_k - i\beta_k)\sigma_{k,\alpha\beta}} \tilde{\chi}_{\beta\lambda}. \qquad\qquad 3(18)$$

The spinor $\sigma^{\nu}\chi(\chi^* \sigma_{\nu}\chi)$ transforms in the same way as $\tilde{\chi}$.

If the ground state and the commutator would have the complete symmetry of the fundamental equation 3(1), the parity operation would not belong to the equation, all mass eigenvalues would be zero on account of the scale invariance and all particles would have a definite helicity. The situation is different however, if the scale invariance is broken by the ground state, if discrete mass eigenvalues of finite mass exist. In this case the Klein–Gordon equation for the fermion

$$(\Box + \kappa^2)\langle 0|\chi(x)|N\rangle = 0 \qquad\qquad 3(19)$$

leads automatically to a doubling of the number of components, as has been pointed out by Dürr.[29] 3(19) can then be replaced by the two equations:

$$i\sigma_{\nu} \frac{\partial}{\partial x_{\nu}} \langle 0|\chi(x)|N\rangle = \kappa \langle 0|\tilde{\chi}(x)|N\rangle \qquad\qquad 3(20)$$

and

$$i\overline{\sigma}_{\nu} \frac{\partial}{\partial x_{\nu}} \langle 0|\tilde{\chi}(x)|N\rangle = \kappa \langle 0|\chi(x)|N\rangle.$$

The operator $\tilde{\chi}(x)$ formally entering in these matrix elements has the same

The Fundamental Field Equation 31

transformation properties as $\tilde{\chi}(x)$ in 3(17), and may be identified with it, since the mass spectrum of 3(17) could be assumed to be identical with that of 3(1). By suitable assumptions about the ground state (compare 3-5) then the parity operation can be established.

Formally the two spinors χ and $\tilde{\chi}$ may be combined to one 4-component spinor

$$\psi = \begin{pmatrix} \chi \\ \tilde{\chi} \end{pmatrix} \qquad 3(21)$$

and the two equations 3(1) and 3(17) can be expressed by one equation for ψ:

$$\gamma_\mu \frac{\partial \psi}{\partial x_\mu} + \frac{l^2}{2} [\gamma_\mu \psi(\bar{\psi}\gamma_\mu \psi) + \gamma_\mu \gamma_5 \psi(\bar{\psi}\gamma_\mu \gamma_5 \psi)] = 0 \qquad 3(22)$$

with $\bar{\psi} = \psi^* \gamma_4$. The γ_μ are the Dirac matrices.

The parity operation then is represented by

$$\psi(\mathbf{r}, t) \rightarrow \gamma_4 \psi(-\mathbf{r}, t) \qquad 3(23)$$

and 3(22) is invariant under this transformation. The final conclusion is, that the parity operation would not exist for a theory defined by 3(1), if ground state and commutator had the complete symmetry of 3(1). If, however, the scale invariance is broken by the ground state or the commutator, if finite mass eigenvalues of fermions exist, the parity operation can be established as a kind of complementary remnant of the broken scale transformation. In this case the theory will be strictly invariant under the space reflexion parity if the commutator and the ground state have this property.

3-3 The 2-point Green's function and the anticommutator

When a local field operator $\chi(x)$ has been defined (as in 3(1)), the most natural way of satisfying the postulate of relativistic causality is the assumption

$$\{\chi(x)\chi^*(y)\} = 0 \quad \text{for space-like} \quad (x-y). \qquad 3(24)$$

This condition imposes the well-known restrictions on the analytic behaviour of the 2-point Green's functions, which have been studied in axiomatic field theory. Following the arguments of this theory we will assume, that a spectral representation of the 2-point function can be given in the way suggested by Källén and Lehmann.[32] Intuitively we suppose that the

                   Unified Field Theory of Elementary Particles

operator $\chi(x)$ annihilates or creates fermions or antifermions of many different masses κ, therefore we expect that the 2-point function can be represented by an integral over the mass-spectrum

$$F(xx') = \langle 0|T\chi_{\alpha\lambda}(x)\,\chi^*_{\beta\mu}(x')|0\rangle$$

$$= i(2\pi)^{-4} \int \rho(\kappa^2)\,d(\kappa^2) \int d^4p \,\frac{p_\nu\,\overline{\sigma}^\nu_{\alpha\beta}}{-p^2+\kappa^2}\,\delta_{\lambda\mu}\,e^{-ip(x-x')}. \quad 3(25)$$

The spectral function $\rho(\kappa^2)$ contains δ-functions for the discrete eigenvalues and continuous functions for the continuous spectra. The integration over p_1, p_2, p_3 should be carried out on the real axis, the integration over p_0 in the complex plane along a path of integration which depends on the order of factors on the left-hand side in the way given in the textbooks and in the appendix A I; then eqn. 3(24) is fulfilled.

In deriving 3(25) it has been assumed that the 2-point function should be invariant under the proper Lorentz group, the isospin group and 3(12) and 3(14). If moreover 3(25) should be invariant under the scale transformation 3(11), it would reduce to

$$F(xx') = \langle 0|T\chi_{\alpha\lambda}(x)\,\chi^*_{\beta\mu}(x')|0\rangle = \mathrm{const}\,l^{-2} \int \frac{d^4p}{(p^2)^2}\,p_\nu\,\overline{\sigma}^\nu_{\alpha\beta}\cdot e^{-ip(x-x')}\,\delta_{\lambda\mu}.$$

$$3(26)$$

In this case the 2-point function would be completely invariant under the full group of the equation 3(1). But a closer analysis shows that a solution of this type is impossible, since 3(26) is incompatible with the interaction term in 3(1). This can be seen in the following way.

Eqn. 3(1) requires that the operator $\sigma^\nu:\chi(x)(\chi^*(x)\sigma_\nu\chi(x)):$ and matrix elements containing this operator should exist. In the Wick product the strongest singularities are taken away by the process 3(3). Still by means of 3(1) one operator, for example, $\chi(x)$ could be replaced by using the integrated form of 3(1):

$$\chi(x) = \chi_0(x)-l^2 \int d^4x'\, G(xx')\,\sigma^\nu:\chi(x')\,(\chi^*(x')\,\sigma_\nu\chi(x')):, \quad 3(27)$$

where $G(xx')$ is the Green's function belonging to the Weyl equation $i\sigma^\nu(\partial\chi/\partial x^\nu)=0$. $\chi_0(x)$ obeys the Weyl equation.

$$G(xx') = (2\pi)^{-4} \int d^4p \,\frac{p_\nu\,\overline{\sigma}^\nu}{p^2}\,e^{-ip(x-x')}. \quad 3(28)$$

The operator then goes over into

$$\sigma^\nu : \chi(x)(\chi^*(x)\,\sigma_\nu\,\chi(x)) := -l^2 \int d^4x'\,\sigma^\nu\,G(xx')\,\sigma^\mu : \chi(x')(\chi^*(x')\,\sigma_\mu\,\chi(x')):$$
$$\times : (\chi^*(x)\,\sigma_\nu\,\chi(x)) : + \sigma^\nu : \chi_0(x)(\chi^*(x)\,\sigma_\nu\,\chi(x)) :.$$
$$3(29)$$

The first term contains as the most singular part expressions of the type

$$\int d^4x'\,G(xx')\,F(xx')\,F(x'\,x), \qquad\qquad 3(30)$$

which are not taken away by the procedure of the Wick product. The singularity of $F(xx')$ at the point $x = x'$ is therefore limited by the condition, that 3(30) should exist. Similarly an operator of the form

$$\sigma^\nu : \chi(x)(\chi^*(x)\,\sigma_\nu\,\chi(x)) : \sigma^\mu : \chi(x')(\chi^*(x')\,\sigma_\mu\,\chi(x')) : \qquad 3(31)$$

has as most singular part terms of the form

$$F(xx')\,F(x'\,x)\,F(xx'). \qquad\qquad 3(32)$$

The integral of this product over x or x' should exist, therefore again the singularity of $F(xx')$ is limited.

The expression 3(26) for $F(xx')$, based on the scale invariance of $F(xx')$, would contain a term of the form $(x-x')_\nu\overline{\sigma^\nu}.\delta[(x-x')^2]$ which makes the integral 3(30) or the product 3(32) meaningless. Therefore the assumption 3(26) of a scale invariant $F(xx')$ is incompatible with the interaction term in 3(1).

This result can be interpreted as stating that interaction without mass is impossible, or more accurately that there must be at least some fermions of finite mass, if there is an interaction. It is not surprising, that interaction and mass are closely connected. Still it is somewhat surprising that a fundamental equation with a certain symmetry should not allow a symmetrical commutator or a symmetrical ground state, and should therefore not lead to the corresponding law of conservation. But one can easily find a counterpart to this situation in classical physics or in unrelativistic quantum mechanics. If 3(1) is interpreted as a classical wave equation derived from a Lagrangian, this underlying Lagrangian is not invariant under 3(11). Therefore even in the classical theory there would be no corresponding law of conservation. Or in quantum mechanics: the equation of motion for the anharmonic oscillator $\ddot{q} = -\alpha q^3$ is invariant under the transformation $q(t) \to \eta q(t\eta)$. But neither the classical Lagrangian nor the quantum theoretical commutator are invariant; hence there is no corresponding law of conservation.

                          Unified Field Theory of Elementary Particles

In spite of this general situation the scale transformation 3(11) will play an important rôle in connexion with the theory of particles of rest mass zero, especially of the photons. Such particles are not excluded by the existence of interaction, they are in fact a necessary consequence of a degeneracy of the ground state according to a theorem of Goldstone, which will be discussed in §7 and §8.

In the special case of the scale transformation it should be emphasized, that the symmetry here is broken not only by the ground state but by the commutator. Even the commutator in the immediate neighbourhood of the light cone cannot contain the term $(x-x')_\nu \overline{\sigma^\nu}.\delta[(x-x')^2]$, hence cannot be symmetrical. Therefore one cannot find a unitary or pseudo-unitary transformation in Hilbert space performing the operation 3(11).

This result raises the general question, to what extent the commutator and the 2-point function are determined by the fundamental wave equation. In the present state of the mathematical analysis an answer to this question cannot be given. The analogy with unrelativistic quantum mechanics suggests, that the 2-point function as representative of the ground state cannot be chosen arbitrarily, nor can it be uniquely determined; the same substance, for example, can form different crystals which are dynamically (though not thermodynamically) stable. Therefore the problem of the ground state plays a similar rôle in the present theory as the problem of cosmology in general relativity, where one can assume different 'cosmological models'. The question whether a special kind of ground state is possible, can be answered only by proving its consistency. There may be many different assumptions about the ground state that lead to consistent solutions.

Coming back to the possible scale invariance of the ground state one may ask whether solutions can be found which at least for the commutator in the immediate neighbourhood of the light cone are 'nearly' scale invariant. An analysis of the conditions which connect the wave-equation 3(1) with the 2-point function has led Mitter[34] to the conjecture, that the 2-point function in the neighbourhood of the light cone could be of the form

$$F(xx') \sim \frac{(x-x')_\nu \overline{\sigma^\nu}}{l^2(x-x')^2} \cdot f_{\text{per}}\left(\ln \frac{(x-x')^2}{x_0^2}\right) \qquad 3(33)$$

where f_{per} is a periodic function with the average value zero, possibly an elliptic function. Such a behaviour would be compatible with the existence of 3(30) and 3(32); i.e. with the interaction. At the same time it would be invariant under 3(11) for discrete values of η, namely for those values, which change the argument of f_{per} by an integer multiple of the period

($\eta = \eta_0^n$ with integer n). If not only the vacuum expectation value of the commutator, but quite generally the commutator would have a behaviour near the light cone corresponding to this 2-point function 3(33), then the discrete transformation 3(11) would define a real symmetry and it should be possible to find a pseudounitary transformation in Hilbert space fulfilling

$$U\chi(x)\,U^{-1} = \eta^{1/2}\chi(\eta x) \qquad\qquad 3(34)$$

for the specified values of $\eta = \eta_0^n$. For the present moment we will not follow this possibility, which might be important in connexion with the theory of collisions with extremely high energy (compare §9-5).

Quite independent of the problem of scale invariance the existence of 3(30) and 3(32) imposes limitations on the possible mass spectrum $\rho(\kappa^2)$ in 3(25). The 2-point function must not contain terms of the type of a δ-function or its derivatives in the neighbourhood of the light cone. On the other hand, the ordinary Schwinger function belonging to the mass κ does contain a term proportional to $\delta'[(x-x')^2]$ and another term proportional to $\kappa^2\delta[(x-x')^2]$. Therefore the corresponding terms in 3(25) can vanish only when [35]

$$\int \rho(\kappa^2)\,d(\kappa^2) = 0, \qquad\qquad 3(35)$$

$$\int \rho(\kappa^2)\,\kappa^2\,d(\kappa^2) = 0. \qquad\qquad 3(36)$$

These conditions imply the indefinite metric in Hilbert space, since $\rho(\kappa^2)$ must assume positive and negative values in order to fulfill 3(35) and 3(36). If the conditions 3(35) and 3(36) are fulfilled, the expression 3(25) can be rewritten in a regularized form:

$$F(xx') = i(2\pi)^{-4}\int \rho(\kappa^2)\,d(\kappa^2)\int d^4p\left(\frac{1}{-p^2+\kappa^2}+\frac{1}{p^2}+\frac{\kappa^2}{(p^2)^2}\right)p_\nu\overline{\sigma^\nu}\,e^{-ip(x-x')}$$

$$= i(2\pi)^{-4}\int \rho(\kappa^2)\,d(\kappa^2)\int d^4p\,\frac{p_\nu\overline{\sigma^\nu}\cdot\kappa^4}{(p^2)^2(-p^2+\kappa^2)}\,e^{-ip(x-x')}. \qquad 3(37)$$

In this expression the contribution from each mass value is formally regularized by a pole and a dipole at the mass value zero. Therefore the conditions 3(35) and 3(36) may be dropped if the form 3(37) is used and if the integral over κ^2 converges. In fact, if $\rho(\kappa^2)$ in 3(37) does not fulfill the conditions 3(35), 3(36), it may without change in 3(37) be replaced by $\bar\rho(\kappa^2) = \rho(\kappa^2) - \delta(\kappa^2)\int\rho(\kappa^2)\,d(\kappa^2) + \delta'(\kappa^2)\int\rho(\kappa^2)\,\kappa^2\,d(\kappa^2)$, which does fulfill 3(35) and 3(36). In this case the double pole at the mass zero would be physically significant.

 Unified Field Theory of Elementary Particles

The special form 3(33) of the 2-point function near the light cone could be obtained from 3(37), if the mass spectrum at high values of κ^2 would be given by

$$\rho(\kappa^2) \underset{\kappa^2\to\infty}{\to} \frac{\text{const}}{l^2\,\kappa^4}\, g_{\text{per}}\left(\ln\frac{\kappa^2}{\kappa_0^2}\right) \qquad 3(38)$$

where again the average value of the periodic function g_{per} should vanish. In the present state of the theory the conjecture 3(38) is an interesting possibility for the asymptotic behaviour of $\rho(\kappa^2)$, but a definite statement concerning this behaviour cannot be made.

Since an exact knowledge of $\rho(\kappa^2)$ in 3(25) or 3(37) would be practically impossible before the whole theory is completed, one has for the time being to be satisfied with approximate expressions for $\rho(\kappa^2)$ which may in the best case represent $\rho(\kappa^2)$ reasonably well in a certain important region of κ^2-values.

Whether such a representation is sufficient for other calculations will depend upon the special problem concerned. The simplest approximation of this type would be the assumption[23] of a single fermion mass in 3(37):

$$F(xx') \approx i(2\pi)^{-4} \int d^4p\, \frac{p_\nu\,\overline{\sigma^\nu}\cdot\kappa^4}{(p^2)^2(-p^2+\kappa^2)}\, e^{-ip(x-x')}. \qquad 3(39)$$

The mass κ could possibly in a first step be identified with the nucleon mass or with the average mass of the lowest baryon octet, if the fundamental equation 3(1) leads to fermion solutions of this type. In this case the 2-point function would actually contain a double pole at the mass zero; the consequences of this double pole will be investigated in §3-4. The behaviour of $F(xx')$ in the neighbourhood of the light cone in 3(39) is less singular than in 3(33). The approximate 2-point function 3(39) is not compatible with the scale invariance 3(34) and the corresponding behaviour of the commutator. Still it may be a sufficient approximation for many problems in which processes of extremely high energy play only a less important rôle.

When the scale invariance is broken by the ground state, the parity operation may be established in the way described in 3(20)–3(23). In this case the 2-point function belonging to 3(22) could be written as

$$\langle 0|\psi_{\alpha\lambda}(x)\,\bar\psi_{\beta\mu}(x')|0\rangle = (2\pi)^{-4} \int d(\kappa^2)\,\delta_{\lambda\mu}\cdot\int d^4p\,e^{-ip(x-x')}$$

$$\times\frac{\rho_1(\kappa^2)(-p_\nu\gamma^\nu_{\alpha\beta}+i\kappa)+\rho_2(\kappa^2)(-p_\nu\gamma^\nu_{\alpha\beta}-i\kappa)}{-p^2+\kappa^2}.$$

$$3(40)$$

The Fundamental Field Equation 37

The conditions 3(35) and 3(36) can be generalized to

$$\int [\rho_1(\kappa^2)+\rho_2(\kappa^2)]\, d(\kappa^2) = 0,$$

$$\int [\rho_1(\kappa^2)+\rho_2(\kappa^2)]\, \kappa^2\, d(\kappa^2) = 0, \qquad\qquad 3(41)$$

$$\int [\rho_1(\kappa^2)-\rho_2(\kappa^2)]\, \kappa\, d(\kappa^2) = 0.$$

When these conditions are fulfilled, 3(40) may be rewritten in the regularized form

$$\langle 0|\psi_{\alpha\lambda}(x)\,\bar{\psi}_{\beta\mu}(x')|0\rangle = (2\pi)^{-4} \int d(\kappa^2)\, \delta_{\lambda\mu} \int d^4p\, e^{-ip(x-x')}$$

$$\times \left\{ [\rho_1(\kappa^2)+\rho_2(\kappa^2)]\frac{p_\nu \gamma^\nu_{\alpha\beta}\cdot \kappa^4}{(p^2)^2\,(p^2-\kappa^2)} \right.$$

$$\left. + [\rho_1(\kappa^2)-\rho_2(\kappa^2)]\frac{i\kappa^3}{p^2(p^2-\kappa^2)} \right\} \qquad 3(42)$$

and the conditions 3(41) can be dropped again, if the regularized form 3(42) is used. The 2-point functions 3(40) or 3(42) are invariant under the space reflexion parity.

It should be emphasized in this connexion that the transformation $\psi(x) \to \gamma_5\psi(x)$ or more generally $\psi(x) \to \exp(i\alpha\gamma_5)\psi(x)$ does not belong to the 4-component version 3(22) of the fundamental equation, since 3(20) does not allow an independent gauge transformation of $\chi(x)$ and $\tilde{\chi}(x)$. Therefore the two mass spectra $\rho_1(\kappa^2)$ and $\rho_2(\kappa^2)$ should not be equal.

In the formulae 3(25) to (42) it has been assumed, that the 2-point function should be invariant under the isospin group. In the real world, however, the isospin group is just approximately valid. Therefore the fundamental equation 3(1) can describe the real world only, if the ground state is not completely symmetrical under the isospin transformations. Hence the equations 3(25) to (42) may be useful approximations; but in the final theory the 2-point functions must contain smaller terms which are not invariant under rotations in isospin space. The effect of these terms is seen in the electromagnetic interactions which will be discussed in chapter 8.

Quite generally there is no reason why the ground state 'world' should be symmetrical with respect to any of those transformations under which the fundamental equation 3(1) is invariant. On the contrary, the world is

            Unified Field Theory of Elementary Particles

probably quite asymmetrical. But this asymmetry should be felt in the properties of the elementary particles only when long range forces prevent the complete separation of the particles from the rest of the world. Therefore the degree of asymmetry in the 2-point functions should be a measure for the lack of independence between the particles and the world.

The 2-point function contains the vacuum expectation value of the (anti) commutator; but a general statement concerning the (anti) commutator, of the type used in quantum mechanics, could not be given. Besides the causality condition 3(24), stating that $\{\chi(x)\chi^*(x')\}$ should vanish for space-like distances $x-x'$, the expression $\{\chi(x)\chi^*(x')\}$ will be an operator or q-number, not a c-number. In this respect $\{\chi(x)\chi^*(x')\}$ should be compared with an expression of the type $[q(t)q(t')]$ in ordinary quantum mechanics, which is generally a q-number (except in the special case of the harmonic oscillator). Still the quantum mechanical expression approaches a c-number in the limit $|t-t'| \to 0$. In the same way it may be that $\{\chi(x)\chi^*(x')\}$ approaches a c-number in the limit $|(x-x')^2| \to 0$. Since the latter expression is probably singular in this limit, the c-number property would mean that the strongest singularity would be identical for all diagonal terms of the operator and that the off-diagonal terms would be less singular. This would be the closest approach imaginable of quantum field theory (with interaction) to ordinary quantum mechanics. For the time being one does not know whether the mathematical scheme does actually work that way.

3-4 Ghost- and dipole-ghost states in the 2-point function

The approximation 3(39) of the 2-point function would, if interpreted physically, suggest the existence of fermions of mass κ (possibly the nucleons or, more generally, baryons) and of a singularity at the mass zero which cannot simply be called a 'particle', since 3(39) contains a double pole at the mass zero. Such a singularity has a characteristic influence on the S-matrix and should therefore be studied before 3(39) can be considered as a useful approximation.

If in a scattering process a fermion can be emitted, the 2-point function should determine the fermion spectrum; this function would appear as a factor responsible for the emission, but would be multiplied (in momentum space) with a function of momentum arising from the 'vertex', at which the emission takes place. Hence the fermions possibly emitted can be derived from the 2-point function by considering it first as Green's function of a system of particles or rather states without interaction.

The approximation 3(39) then indicates the existence of three fermion states belonging to mass zero, two ghost-states and one dipole-ghost state.[37]

The Fundamental Field Equation 39

This can be seen by performing the integration over p_0 in 3(39). Thereby one gets (with $|p| = |\mathbf{p}|$ and $E = \sqrt{(\mathbf{p}^2 + \kappa^2)} = \sqrt{(|p|^2 + \kappa^2)}$)

$$F(xx') = -(2\pi)^{-3} \int d^3p \, e^{i(\mathbf{p},\mathbf{x}-\mathbf{x}')} \cdot \left\{ \frac{E + \boldsymbol{\sigma}\mathbf{p}}{2E} e^{-iE(x-x')_0} \right.$$

$$\left. - \frac{d}{dp_0} \left[\frac{(\kappa^2 + p^2)(p_0 + \boldsymbol{\sigma}\mathbf{p})}{(p_0 + |p|)^2} e^{-ip_0(x-x')_0} \right]_{p_0 = |p|} \right\}. \qquad 3(43)$$

If one divides into two terms, in which $\boldsymbol{\sigma}$ and $\mathbf{p}$ are parallel or antiparallel, 3(43) becomes

$$F(xx') = -(2\pi)^{-3} \int d^3p \, e^{i(\mathbf{p},\mathbf{x}-\mathbf{x}')} \cdot \left\{ \frac{|p| + \boldsymbol{\sigma}\mathbf{p}}{2|p|} a_+ + \frac{|p| - \boldsymbol{\sigma}\mathbf{p}}{2|p|} a_- \right\} \qquad 3(44)$$

with

$$a_+ = \frac{E + |p|}{2E} e^{-iE(x-x')_0}$$

$$- \left[-i\kappa x_0 + \frac{|p|}{\kappa} - \frac{\kappa}{4|p|} \right] e^{-i|p|x_0} \cdot \frac{\kappa}{2|p|} e^{i|p|x_0}$$

$$- \frac{\kappa}{2|p|} e^{-i|p|x_0} \left[+i\kappa x_0' + \frac{|p|}{\kappa} - \frac{\kappa}{4|p|} \right] e^{i|p|x_0'},$$

$$a_- = \frac{E - |p|}{2E} e^{-iE(x-x')_0} \qquad 3(45)$$

$$- \frac{\kappa^2}{4|p|^2} e^{-i|p|(x-x')_0}.$$

The five parts of this expression 3(44), (45) can be interpreted as

$$F(xx') = \int d^3p \{ \langle 0|\chi(x)|N_{\mathbf{p}}^+ \rangle \langle N_{\mathbf{p}}^+|\chi^*(x')|0 \rangle$$

$$+ \langle 0|\chi(x)|D^+ \rangle \langle G^+|\chi^*(x')|0 \rangle$$

$$+ \langle 0|\chi(x)|G^+ \rangle \langle D^+|\chi^*(x')|0 \rangle \qquad 3(46)$$

$$+ \langle 0|\chi(x)|N^- \rangle \langle N^-|\chi^*(x')|0 \rangle$$

$$+ \langle 0|\chi(x)|G^- \rangle \langle G^-|\chi^*(x')|0 \rangle \}.$$

The letters N, D, G refer to nucleon states, dipole ghost and ghost states. Since according to §2–6

$$\langle 0|\chi(x)\chi^*(x')|0 \rangle = {}_0\chi(x)_n g^{nl} {}_l\chi^*(x')_0, \qquad 3(47)$$

one learns from 3(46), that the metric in the space of the three states D^+, G^+, G^- is given by

$$g^{nl} = \begin{array}{c|ccc} & D^+ & G^+ & G^- \\ \hline D^+ & 0 & 1 & 0 \\ G^+ & 1 & 0 & 0 \\ G^- & 0 & 0 & -1 \end{array} \qquad\qquad 3(48)$$

The dipole ghost state D^+ is not an eigenstate of energy and momentum, as is seen from the term containing the factor x_0 in the matrix element $\langle 0|\chi(x)|D^+\rangle$ in 3(45). It had been emphasized in §2-6, that the eigenstates of energy and momentum are generally not sufficient to construct the whole Hilbert space.

A situation similar to this has been studied in detail in the Lee-model[36, 37] which allows a complete mathematical analysis. The essential mathematical features of the model are discussed in the Appendix A II. The results of this analysis cannot be applied automatically to the present problem which is much more complicated; but they can be used as a guide for the further mathematical development. If one is led by the analogy to the Lee-model, one would expect, that a physical interpretation cannot, and should not, be given for solutions of the scattering problem, in which all three states D^+, G^+ and G^- appear as incoming or outgoing waves. On the other hand one could, on the side of the incoming waves, add to the actual particles a state containing G^+ in arbitrary amount without changing the physical interpretation of the incoming group of particles, since the norm of G^+ is zero. This freedom could be used to change the factor coming from the vertex for the outgoing fermion waves in such a way that it vanishes at the point $p^2=0$. Thereby the double pole of the 2-point function would be changed into a single pole, and only the state G^+ would remain at $p^2=0$ for the outgoing waves as well. This state would again be irrelevant for the physical interpretation. In this way a special group of solutions of the scattering problem can be found which permits construction of a unitary S-matrix. It is characteristic for this group of solutions, that in spite of the singularity of the 2-point function at $p^2=0$ no emission of fermions of mass zero should be observed; the unitarity of the S-matrix is preserved without any contribution from fermions of zero mass.

These mathematical properties of the approximation 3(39) for the 2-point function agree surprisingly well with the real situation in the experiments. The singularity at $p^2=0$ could in some way represent the leptons. Then the leptons would on the one hand be used for regularizing the 2-point function, on the other hand they would in this approximation not take part in any scattering processes, they would not participate in the strong interaction of the heavy particles, in agreement with the observations.

Therefore we expect that the approximation 3(39) for the 2-point function can actually be a very useful approximation which will permit the construction of a unitary S-matrix for the strong interactions and, what is closely connected with it, stationary states of reasonable properties. This will be discussed in detail in Chapters 5 to 7.

3-5 Asymptotic properties of the 4-point function and the degeneracy of the ground state

The 4-point function of the type

$$F(xx'yy') = \langle 0|\chi(x)\chi^*(x')\chi(y)\chi^*(y')|0\rangle \qquad 3(49)$$

contains a great number of informations concerning the elastic scattering of two fermions, the existence and the masses of bosons, etc., which cannot be summarized by simple approximations like those of the 2-point functions. There are limiting cases however, in which the 4-point function is closely related to the 2-point function. We will briefly discuss the special limiting case where the two variables x, x' are separated by a very large, space-like distance from the two variables y, y'.

If the 4-point function 3(49) is written in the form

$$F(xx'yy') = \langle 0|\chi(x)\chi^*(x')|J\rangle\langle J|\chi(y)\chi^*(y')|0\rangle, \qquad 3(50)$$

then for very large distances only those intermediate states $|J\rangle$ are important, whose momentum or energy are very small. Hence the most important intermediate state is the ground state $|0\rangle$ itself, and one would expect that the 4-point function in the limit approaches the product of the two 2-point functions, replacing $|J\rangle$ in 3(50) simply by $|0\rangle$.

This statement is to be modified however, if the ground state is degenerate, if it is less symmetrical than the fundamental equation 3(1). It is well known from unrelativistic quantum mechanics, that such a situation can occur. The ground state of a ferromagnet, of a crystal or of a super-conductor are obvious examples for a degeneracy of this kind. In the first two cases a rotation in space, in the third a gauge transformation will lead from the assumed ground state to a different one. In elementary particle physics this possibility of a degenerate ground state has first been discussed in connexion with the isospin group,[29] later also in connexion with other groups.[38, 39] If the ground state is invariant under the Lorentz group, but degenerate under another continuous group, there will be a continuum of intermediate states $|J\rangle$ of energy and momentum zero and of arbitrarily small energies and momenta. In this case the 4-point function 3(50) will in the limit of a large space-like distance between x, x' and y, y' not only contain the product of two 2-point functions of the type $\langle 0|\chi(x)\chi^*(x')|0\rangle$, but also products of the more general 2-point functions $\langle 0|\chi(x)\chi^*(x')|J\rangle$, where $|J\rangle$ is different from $|0\rangle$ by the transformation concerned; possibly the term is multiplied by a factor decreasing slowly with the distance as a long-range force. If these generalized 2-point functions contain poles in momentum space, they would indicate the existence of 'strange' particles, namely of particles which have normal transformation properties with respect to the Lorentz group, but 'strange' transformation properties with respect to the group, under which the groundstate is degenerate. In this way the existence of the observed strange particles can be understood as a natural consequence of the degeneracy of the groundstate under the isospin group.[40]

The theory of the strange particles will be discussed in detail in Chapter 7. The degeneracy of the ground state has still another aspect which should be mentioned already now. An operator performing the transformation concerned in the whole ground state 'world' cannot be constructed from the field operator by finite processes, since space and time are infinite. But it should be possible to construct operators performing this transformation in finite regions; the amount of energy necessary for such a transformation should become smaller when the region becomes larger, on account of the degeneracy.

These operators will not commute with the field operators in this region and will produce changes in the 2-point function; they will thereby reveal the existence of matrix elements of the type of the 2-point function leading from the vacuum to a boson state of mass zero. (The mass zero follows from the fact, that the energy becomes smaller with increasing wave-length.)

The existence of such boson states has been emphasized in a theorem by Goldstone.[41] In a crystal or in a superfluid liquid the sound quanta (phonons) play the rôle of the (somewhat generalized) Goldstone particles. In a ferromagnet the spin-waves of Bloch[43] fill this place. In elementary particle physics when the isospin group is broken by an asymmetry of the ground state, the light quanta or photons (primarily their longitudinal and scalar part in the Coulomb field) are the corresponding Goldstone particles. The fields connected with these particles have all the properties necessary for 'classical' fields. They are of the boson type, are represented by scalars, may be related to others which are vectors or tensors, and have long range. The connexion of quantum electrodynamics with the Goldstone theorem will be discussed in detail in Chapter 8.

3-6 The problem of local conservation laws

If a classical field theory would be defined by 3(1), there would be a number of local conservation laws, closely connected with the invariance properties of the equation. In fact, defining

$$j_\mu = \chi^* \sigma_\mu \chi \quad \text{and} \quad j_{\mu,k} = \chi^* \sigma_\mu \tau_k \chi, \qquad 3(51)$$

one arrives in the classical theory at the continuity equations

$$\frac{\partial j_\mu(x)}{\partial x_\mu} = 0 \quad \text{and} \quad \frac{\partial j_{\mu,k}(x)}{\partial x_\mu} = 0. \qquad 3(52)$$

The quantities

$$Q = \int j_0 \, d^3x \quad \text{and} \quad T_k = \int j_{0,k} \, d^3x \qquad 3(53)$$

would be conserved, i.e. constant, in agreement with the invariance of 3(1) under the gauge transformation 3(12) and under the isospin transformation 3(10).

In a quantum theory based upon 3(1) and 3(22) the situation is quite different. While the operators performing the transformations commute with the S-matrix and therefore define quantities which are conserved, they can as a rule not be represented by simple expressions of the type 3(53). Local conservation laws, like 3(52) do generally not exist.

It is easily seen that the arguments leading to 3(52) in a classical field theory fail in quantum field theory, since the product of two field operators at the same point is a singular quantity. Starting from the expression

$$j_\mu(x, \epsilon) = \chi^*(x+\epsilon) \, \sigma_\mu \chi(x), \qquad 3(54)$$

we see that 3(1) leads to

$$\frac{\partial j_\mu(x, \epsilon)}{\partial x_\mu} = il^2\{ : (\chi^*(x+\epsilon) \, \sigma_\nu \chi(x+\epsilon)) \, \chi^*(x+\epsilon) : \sigma^\nu \chi(x) \qquad 3(55)$$

$$-\chi^*(x+\epsilon) \, \sigma^\nu : \chi(x) \, (\chi^*(x) \, \sigma_\nu \chi(x)) : \}.$$

The most singular part (in the limit $\epsilon \to 0$) of the expression on the right-hand side comes from the vacuum expectation value of products like $\chi^*(x+\epsilon) \sigma_\nu \chi(x)$. Therefore we get

$$\frac{\partial j_\mu(x, \epsilon)}{\partial x_\mu} = il^2 \langle 0|\chi^*(x+\epsilon) \, \sigma^\nu \chi(x)|0\rangle \cdot [: \chi^*(x+\epsilon) \, \sigma_\nu \chi(x+\epsilon) :$$

$$- : \chi^*(x) \, \sigma_\nu \chi(x) :] + \ldots \qquad 3(56)$$

where $+\ldots$ means other similar terms and terms which are less singular. The behaviour of this vacuum expectation value for small ϵ is not yet known.

If we follow the conjecture 3(33) of Mitter,[34] 3(56) becomes

$$\frac{\partial j_\mu(x,\epsilon)}{\partial x_\mu} = \text{const}\,\frac{\epsilon^\nu}{\epsilon^2}f_{\text{per}}(\ln \epsilon^2)\,[:\chi^*(x+\epsilon)\,\sigma_\nu\,\chi(x+\epsilon):$$

$$-:\chi^*(x)\,\sigma_\nu\,\chi(x):].\qquad\qquad 3(57)$$

In the limit $\epsilon \to 0$ the value of the right-hand side is not well defined, there is no reason why it should be zero. It may be zero for special matrix-elements of $\partial j_\mu(x,\epsilon)/\partial x_\mu$, and different from zero for others. For the currents $j_{\mu,k}(x,\epsilon)$ the situation is essentially the same.

The operators $j_\mu(x,\epsilon)$ and $j_{\mu,k}(x,\epsilon)$ will contain matrix elements describing the creation or annihilation of bosons; these matrix elements should be well defined even for $\epsilon=0$. If the boson has the spin zero and momentum J_μ, the matrix elements should have the general form

$$\langle 0|\,j_\mu(x,0)|B\rangle = \text{const}\,J_\mu.\,e^{iJx}.\qquad\qquad 3(58)$$

The local conservation law 3(52) would at once lead to the conclusion $J_\mu^2=0$, the mass of the boson would be zero. Eqn. 3(57) shows however, that this is not necessary; bosons of spin zero can very well have a mass different from zero.

In spite of this lack of local conservation laws the conserved quantities like isospin, momentum, charge, etc. should not be considered as spread out over the whole space. In fact the free particle operators or asymptotic operators discussed in §2-1 do permit the construction of local vector currents fulfilling the continuity equation 3(52), as is well known from conventional field theory. They do not allow the construction of local pseudovector currents, if the masses of the free particles are different from zero. It had been emphasized in §3-3, in connexion with the 4-component version 3(22) of the theory, that the transformation $\psi \to \exp(i\alpha\gamma_5)\psi$ does not belong to the theory. But the vector currents, corresponding in their isospin transformation properties to the four currents 3(51), do fulfill the continuity equation.

Therefore the uncertainty concerning the localization of the conserved quantities is limited to the range of interaction. If one disregards for the moment electromagnetic and weak interactions, the conserved quantities may be pictured as localized within a range of 10^{-13} cm around the particles concerned. This degree of localization is certainly sufficient for the interpretation of the experiments.

From a mathematical point of view one should expect that the operators Q and T_k in 3(53) can be represented by the field operators within the finite but arbitrarily small time-interval between t and $t+\Delta t$, since these operators are probably sufficient to construct the complete Hilbert space according to an axiom of Haag.[22] Therefore one should be able to write

$$Q = \int J_0(x)\, d^3x \quad \text{and} \quad T_k = \int J_{0,k}(x)\, d^3x, \qquad 3(59)$$

where $J_0(x)$ and $J_{0,k}(x)$ are obtained from the field operators within t and $t + \Delta t$ by a possibly non-local operation. Whatever the nature of this operation may be, one should expect that the densities J_0 and $J_{0,k}$ can be considered as time-components of vectors J_μ and $J_{\mu,k}$ fulfilling the equations

$$\frac{\partial J_\mu(x)}{\partial x_\mu} = 0 \quad \text{and} \quad \frac{\partial J_{\mu,k}(x)}{\partial x_\mu} = 0, \qquad 3(60)$$

since we know that

$$\frac{\partial Q}{\partial t} = 0 \quad \text{and} \quad \frac{\partial T_k}{\partial t} = 0.$$

One plausible way for constructing $J_\mu(x)$ and $J_{\mu,k}(x)$ would be the assumption, that these currents can be obtained from the limiting process:

$$J_\mu(x) = \lim_{\epsilon \to 0} \tfrac{1}{2}\{\chi^*(x+\epsilon)\,\sigma_\mu\,\chi(x) - \chi(x)\,\sigma_\mu^T\,\chi^*(x+\epsilon)$$

$$+ \tilde{\chi}^*(x+\epsilon)\,\bar{\sigma}_\mu\,\tilde{\chi}(x) - \tilde{\chi}(x)\,\bar{\sigma}_\mu^T\,\tilde{\chi}^*(x+\epsilon)\}, \qquad 3(61)$$

while the corresponding process for the pseudovector currents would not converge. Similar relations should hold for $J_{\mu,k}(x)$. In this case the only non-local operation entering into the construction of $J_\mu(x)$, $J_{\mu,k}(x)$ would be space reflection, connecting $\chi(x)$ and $\tilde{\chi}(x)$ in Eqn. 3(20). Whether the limiting process 3(61) does actually lead to convergent results could be decided only after a more thorough investigation of the singularity of the 2-point function at the light cone.

4

Methods of Approximation

4-1 General remarks concerning approximation methods in quantum field theory

The mathematical difficulties which one encounters in the attempt to find solutions of a problem in quantum field theory, are very much greater than in quantum mechanics. They arise primarily from the degree of complication, which is an immediate consequence of the experimental fact, that—using the well-known paradoxical phrase—'every elementary particle consists of all other elementary particles'. A hydrogen atom in quantum mechanics consists of electron and proton; a wave function depending on electron and proton co-ordinates is therefore sufficient for its description. In elementary particle physics a proton potentially consists of nucleon and boson or nucleon and two bosons, etc. or nucleon and a pair of nucleon and antinucleon, or nucleon and two pairs, etc. Hence in principle one should write down an infinite set of wave functions, depending on one, two, or any number of variables in order to describe the properties of a single proton. In the same way the mathematical relations from which the masses or scattering cross-sections should be determined, can be defined only as infinite sets of equations containing functions with any number of variables. Consequently one cannot hope to find exact solutions making use of simple analytical functions. One has to look for approximate solutions, and for this purpose the first question must be, what can be neglected in order to reach a certain degree of accuracy?

Intuitively it should be expected that the probability of finding, for example, a proton as consisting of very many particles, is very small. Therefore one should neglect those parts of the total wave function which mean that the particle in question consists of very many others. The calculations should be confined to functions of not more than n variables, and the degree of accuracy obtainable would then improve with increasing n. One should select, from the infinite set of equations, a finite number of equations referring to functions of not more than n variables. In order to

46

get a sufficient number of equations for the unknown functions, one will have to include equations which in their exact form contain functions with more than n variables. These latter ones should then be approximated by connecting them with other functions of not more than n variables. It is this connexion which decides about the kind of approximation that shall be used. By this general procedure one of the boundary conditions which had been discussed in §2-3, will be fulfilled automatically. In principle every state will, in its representing wave function, contain parts depending on an arbitrarily high number of variables, and it is a serious mathematical problem to determine the contribution to the norm of these parts. In the approximative wave function these parts are simply neglected above the number n of variables, their contributions to the norm are neglected as well. Therefore the finiteness of the norm is not endangered by the parts depending on very many variables, and the boundary condition referring to the region of very many variables is automatically fulfilled.

Whether this procedure leads to a convergent mathematical process will, in most cases, remain an open question. As a rule the convergence of a special method of approximation can be tested only in the case of simple models, which have some important features in common with quantum field theory. But the proof of convergence for the simplified model does not necessarily imply the convergence of the field theory. Moreover the convergence of the approximative eigenvalues to the real eigenvalues does possibly not imply the convergence of the approximative wave functions to the real wave functions.

The older methods of perturbation theory which have been so successful in quantum mechanics are apparently quite useless for the theory defined by 3(1), since in this equation no term can be considered as much smaller than the others. What one needs would be approximation methods of the type of the Ritz method of quantum mechanics in which one first has to guess the general shape of the wave function; then by changing parameters or introducing other improvements one can try to get closer and closer to the real wave function. Frequently the equations are relativistic integral equations of the Bethe–Salpeter[44] type. The methods developed by the mathematicians for general integral equations of the Fredholm[45] type may be useful in this case. For a first guess of the shape of the wave function the results of perturbation theory may sometimes be suggestive.

The degree of complication which is the main cause of the difficulties in quantum field theory has led to the introduction of a kind of shorthand writing for integral expressions or integral equations which has the great advantage of suggesting a physical interpretation of the mathematical procedure. By means of Feynman[46] diagrams complicated integrals can be represented as simple pictures giving some insight into the mechanism of the interaction to be described. In the following text we will frequently use the technique of Feynman graphs in a somewhat generalized form for writing an eigenvalue equation or a relation for determining coupling constants.

4-2 Approximative representations of state vectors by truncated sets of functions

Several approximation methods in quantum field theory start from the matrix elements for time ordered products of field operators. These matrix elements are especially useful for calculations concerning the S-matrix. A state $|\psi\rangle$ of the system characterized by the fundamental field equation 3(1) may be defined by the infinite set of functions

$$\tau(x_1 x_2 \ldots | y_1 y_2 \ldots) = \langle 0|T\chi(x_1)\chi(x_2)\ldots\chi^*(y_1)\chi^*(y_2)\ldots|\psi\rangle. \quad 4(1)$$

These τ-functions may be called a covariant representation of $|\psi\rangle$. The letter T means, that the operators with a larger value of the time variable should be on the left of the operators with smaller time values.

From this set of τ-functions a second set of functions[47] can be derived by putting

$$\begin{aligned}
\tau(x_1 x_2 \ldots | y_1 y_2 \ldots) = {}&\phi(x_1 x_2 \ldots | y_1 y_2 \ldots)\\
&+(-1)^z \phi(x_2 \ldots | y_2 \ldots)\langle 0|T\chi(x_1)\chi^*(y_1)|0\rangle\\
&+(-1)^z \phi(x_2 \ldots | y_1 y_3 \ldots)\langle 0|T\chi(x_1)\chi^*(y_2)|0\rangle\\
&+\ldots\\
&+(-1)^z \phi(x_3 \ldots | y_3 \ldots)\langle 0|T\chi(x_1)\chi^*(y_1)|0\rangle\\
&\times \langle 0|T\chi(x_2)\chi^*(y_2)|0\rangle\\
&+\ldots
\end{aligned}$$

$$4(2)$$

where z means in every expression the number of transpositions necessary for getting the operators in the vacuum expectation values together. $(-1)^z$ will in the following equations always be used with this interpretation. This set of ϕ-functions can be considered as a second covariant representation of $|\psi\rangle$, making use of a second set of basic vectors. The main singularities of the τ-functions on the light cones $(x_n - y_m)^2 = 0$ are presumably contained in the 2-point functions; therefore the ϕ-functions should be less singular on the light cones than the τ-functions. The ϕ-functions can be constructed from the τ-functions, when the 2-point function is known.

The main assumption of the so-called 'new' Tamm-Dancoff[48] method consists in taking a 'truncated' set of ϕ-functions, in which all ϕ's of more than n variables are put equal to zero, as an approximation for the real wave function of the state $|\psi\rangle$. This assumption, justified or not, enables one to construct from the field equation a sufficient set of equations for determining the finite set of ϕ-functions for $|\psi\rangle$. In this way approximate eigenvalue equations can be derived, as in conventional quantum mechanics, if the 2-point function is known. Unfortunately the 2-point function cannot be known exactly, unless the whole problem has been solved. Therefore it is necessary, to replace the 2-point functions in 4(2) by an approximative 2-point function. The problem of convergence of this new Tamm–Dancoff method then presents itself in the following form.

Let us assume that the mass eigenvalue of a certain state is calculated by using a certain approximative 2-point function and by going successively to higher numbers of n in the truncated set of ϕ-functions. Then the question arises whether the eigenvalues, calculated in this way, converge with increasing n to a limiting value and whether this value depends on the approximative 2-point function. In the present state of quantum field theory we are still far from an answer to this question. But an answer can be given for a strongly simplified model. The anharmonic oscillator defined by the equation of motion $\ddot{q} = -\alpha q^3$ has some resemblance with 3(1) and can be treated by the new Tamm–Dancoff method. The τ-functions then depend only on the time-variables instead of space–time variables, the 2-point function may be approximated by the 2-point function of the harmonic oscillator, multiplied by an arbitrary factor. In this case (specialized to a one-time formalism) Stumpf, Wagner and Wahl[49] have been able to prove, that the eigenvalues do converge with increasing n to a limiting value, that this value is independent of the approximative 2-point function within a rather wide range of values of the arbitrary factor and that the limiting value coincides with the exact quantum mechanical solution. In the numerical calculation already for $n = 10$ the relative deviations in the lowest eigenvalues are of the order 10^{-5}. This result may be taken as some justification for the new Tamm–Dancoff method even in the field theoretical case 3(1), but it does not contain any mathematical proof for its convergence.

Other approximation methods which are mainly intended for the calculation of S-matrix elements make use primarily of τ-functions concerning transitions from vacuum to vacuum, i.e. of vacuum expectation values of time ordered products. These functions are frequently called Green's functions,[50] or 2-point functions, 4-point functions, etc. The term 'Green's functions' however, is somewhat misleading. A Green's function belongs to a linear differential equation and therefore to a well-defined mass.

The 2-point functions or 4-point functions can be composed of Green's functions and possibly other terms by integrating over mass spectra (with indefinite mass density), but they are not Green's functions themselves. From this set of τ-functions of the vacuum a third[51, 52] set may be constructed by the following relations:

$$\tau(x_5 x_2 \ldots x_n | y_1 y_2 \ldots y_n) = (-1)^z \eta(x_1 y_1) \eta(x_2 y_2) \ldots \eta(x_n y_n)$$

$$+ \text{permutations}$$

$$+ (-1)^z \eta(x_1 x_2 | y_1 y_2) \eta(x_3 y_3) \ldots \eta(x_n y_n)$$

$$+ \text{permutations}$$

$$+ (-1)^z \eta(x_1 x_2 x_3 | y_1 y_2 y_3) \eta(x_4 y_4) \ldots$$

$$\eta(x_n y_n) + \text{permutations}$$

$$+ (-1)^z \eta(x_1 x_2 | y_1 y_2) \eta(x_3 x_4 | y_3 y_4)$$

$$\times \eta(x_5 y_5) \ldots + \text{permutations}$$

$$+ \ldots$$

$$4(3)$$

For the simplest η-functions the relations are:

$$\tau(x_1 | y_1) = \eta(x_1 | y_1)$$

$$\tau(x_1 x_2 | y_1 y_2) = -\eta(x_1 | y_1) \eta(x_2 | y_2) + \eta(x_1 | y_2) \eta(x_2 | y_1)$$

$$+ \eta(x_1 x_2 | y_1 y_2)$$

$$\tau(x_1 x_2 x_3 | y_1 y_2 y_3) = -\eta(x_1 | y_1) \eta(x_2 | y_2) \eta(x_3 | y_3) \pm \text{permutations}$$

$$+ \eta(x_1 x_2 | y_1 y_2) \eta(x_3 y_3) \pm \text{permutations}$$

$$+ \eta(x_1 x_2 x_3 | y_1 y_2 y_3).$$

$$4(4)$$

These η-functions are frequently called truncated Green's functions, which however again would be a somewhat misleading term. The η-functions have the important property of going to zero when any variable or group of variables is separated from the rest by a large space-like distance. The η-functions can be considered as a representation of the ground state $|0\rangle$. Again the field equation 3(1) may be used for establishing relations between the different η-functions. If all η-functions with more than n variables are neglected, one thereby defines a certain approximation, for which only a finite number of these relations can be postulated. Unlike the case of the ϕ-functions, for which the resulting equations are linear, the relations between the η-functions are non-linear.

Finally a fourth set of functions[52] may be defined by

$$\tau(x_1 x_2 \ldots | y_1 y_2 \ldots) = \zeta(x_1 x_2 \ldots | y_1 y_2 \ldots)$$
$$+ (-1)^z \zeta(x_2 \ldots | y_2 \ldots)\, \eta(x_1|y_1) + \text{permutations}$$
$$+ (-1)^z \zeta(x_3 \ldots | y_3 \ldots)\, \eta(x_1|y_1)\, \eta(x_2|y_2)$$
$$+ \text{permutations}$$
$$+ (-1)^z \zeta(x_3 \ldots | y_3 \ldots)\, \eta(x_1 x_2 | y_1 y_2)$$
$$+ \text{permutations}$$
$$+ (-1)^z \zeta(x_4 \ldots | y_4 \ldots)\, \eta(x_1|y_1)\, \eta(x_2|y_2)$$
$$\times \eta(x_3|y_3) + \text{permutations}$$
$$+ (-1)^z \zeta(x_4 \ldots | y_4 \ldots)\, \eta(x_1 x_2 | y_1 y_2)$$
$$\times \eta(x_3|y_3) + \text{permutations}$$
$$+ (-1)^z \zeta(x_4 \ldots | y_4 \ldots)\, \eta(x_1 x_2 x_3 | y_1 y_2 y_3)$$
$$+ \text{permutations}$$
$$+ \ldots$$

$$4(5)$$

This set of ζ-functions is a representation of the state concerned like the set of τ-functions, or of ϕ-functions. For the vacuum, i.e. for matrix elements $\langle 0| \ldots |0\rangle$, the ζ-functions are zero. The ζ-functions can be considered as the matrix elements of a generalized Wick product:

$$\zeta(x_1 x_2 \ldots | y_1 y_2 \ldots) = \langle 0| : \chi(x_1)\, \chi(x_2) \ldots \chi^*(y_1)\, \chi^*(y_2) \ldots : |\psi\rangle, \quad 4(6)$$

in which all creation operators are on the left-hand side of all annihilation operators (including those for bosons, bound states, etc.).

Again one may define an approximation by putting all ζ-functions with more than n variables equal to zero. Possibly all these different approximations lead to a convergent mathematical process. It will then be a matter of convenience which approximation shall be used in a special problem. The four different sets of functions (τ-, ϕ-, η-, ζ-functions) which have been defined in this chapter, can be obtained in a rather elegant way from simple functionals defined by means of some auxiliary field. This more formal aspect of the infinite sets will be discussed in the mathematical appendix A IV.

While it would certainly be premature to speak of a solid mathematical basis for the approximation methods in quantum field theory, it should be emphasized that no fundamental difficulty concerning these methods has been found in the investigations carried out hitherto. The difficulties which have been discussed so frequently in earlier years had their origin in the futile attempt to apply the methods of canonical quantization to interacting local fields or to use methods of perturbation theory in problems where this was not justified, or in restricting unnecessarily the Hilbert space of field theory to a space of positive metric. But the approximation methods derived from the approximative representation of states by truncated sets of functions did not as such introduce serious difficulties.

4-3 The new Tamm–Dancoff method

The new Tamm–Dancoff method[48] represents state vectors approximately by truncated sets of ϕ-functions according to 4(2). In order to calculate mass eigenvalues or cross-sections in scattering processes from the fundamental field equation 3(1), it will be necessary to derive from the field equation relations between τ-functions which then in turn can be translated into relations between ϕ-functions.

The general procedure for doing this can be described in the following way. Let us assume a τ-function of the type 4(1)

$$\tau(x_1, x_2 \ldots | y_1, y_2 \ldots) = \langle 0 | T\chi(x_1)\,\chi(x_2) \ldots \chi^*(y_1)\,\chi^*(y_2) \ldots | \psi \rangle.$$

By applying the field equation 3(1) to $\chi(x_1)$, we get

$$
\begin{aligned}
i\sigma_\nu \frac{\partial}{\partial x_{1\nu}} \tau(x_1 \ldots | y_1 \ldots) = {} & -l^2 \langle 0 | T : \sigma^\nu \chi(x_1)\,(\chi^*(x_1)\,\sigma_\nu \chi(x_1)) : \chi(x_2) \ldots \\
& \chi^*(y_1) \ldots | \psi \rangle \\
& + (-1)^z . c . \delta(x_1 - y_1) . \langle 0 | T\chi(x_2) \ldots \chi^*(y_2) \ldots | \psi \rangle \\
& + (-1)^z . c . \delta(x_1 - y_2) . \langle 0 | T\chi(x_2) \ldots \\
& \chi^*(y_1)\,\chi^*(y_3) \ldots | \psi \rangle \\
& + \ldots.
\end{aligned}
$$

$$4(7)$$

The terms containing the δ-functions arise from the rearrangement of factors in the time-ordered T-product, when the time co-ordinate of x_1 passes the time co-ordinate of y_1, y_2, etc. The factor c depends on the behaviour of the commutator in the neighbourhood of the origin and is not known *a priori*. If the Källén–Lehmann representation 3(25) is assumed for the 2-point function, c is essentially given by $\int \rho(\kappa^2)\, d\,(\kappa^2)$. Therefore in canonical quantization this factor would simply be a number. In a theory with interaction of the type 3(1) however, the condition 3(35) enforces $\int \rho(\kappa^2)\, d\,(\kappa^2) = 0$ and therefore $c = 0$. In spite of this result we have included the terms $c . \delta(x-y)$ in 4(7), because it may be convenient in certain approximation methods to retain the terms in any finite approximation and to go to the limit $c \to 0$ only together with the limit $n \to \infty$. Therefore we will leave these terms undefined for the moment and will integrate 4(7) by means of the causal Green's function belonging to the equation

$$i\sigma_\nu \frac{\partial G}{\partial x_\nu} = \delta^4(x)$$

$$G(x) = (2\pi)^{-4} \int_c d^4p\, e^{-ipx} \frac{p_\nu \overline{\sigma^\nu}}{p^2} \qquad 4(8)$$

in the following form:

$$\tau(x_1 \ldots | y_1 \ldots) = -l^2 \int d^4x'\, G(x_1, x') \langle 0|T\sigma^\nu : \chi(x') (\chi^*(x')\, \sigma_\nu \chi(x')):$$

$$\times \chi(x_2) \ldots \chi^*(y_1) \ldots |\psi\rangle$$

$$+ \langle 0|T\chi_0(x_1)\, \chi(x_2) \ldots \chi^*(y_1) \ldots |\psi\rangle \qquad 4(9)$$

The operator $\chi_0(x_1)$ introduced in 4(9) fulfills the equation

$$i\sigma_\nu \frac{\partial \chi_0}{\partial x_\nu} = 0. \qquad 4(10)$$

The commutator of $\chi_0(x)$ with $\chi^*(y)$ is supposed to approach the commutator of $\chi(x)$ and $\chi^*(y)$ in the neighbourhood of the origin. In this way, by applying $i\sigma^\nu(\partial/\partial x_1^\nu)$ on 4(9), one comes back to 4(7). Still one need not specify this commutator for the later calculations. The second line in 4(9) can be simplified further by introducing

$$F_0(xy) = \langle 0|T\chi_0(x)\, \chi^*(y)|0\rangle. \qquad 4(11)$$

Eqn. 4(10) then leads to

$$\langle 0|T\chi_0(x_1)\, \chi(x_2) \ldots \chi^*(y_1) \ldots |\psi\rangle = F_0(x_1 y_1)\, (-1)^z \langle 0|T\chi(x_2) \ldots$$

$$\chi^*(y_2) \ldots |\psi\rangle$$

$$+ F_0(x_1 y_2)\, (-1)^z \langle 0|T\chi(x_2) \ldots$$

$$\chi^*(y_1)\, \chi^*(y_3) \ldots |\psi\rangle$$

$$+ \ldots. \qquad 4(12)$$

Inserting 4(12) into 4(9) gives the required relation between τ-functions of different numbers of variables. This relation contains the function $F_0(xy)$, just as the relation between τ-functions and ϕ-functions contains the 2-point function $F(xy)$. $F_0(xy)$ is uniquely determined by $F(xy)$ on account of 4(10) and of the postulate, that its behaviour at the origin should be the same as that of $F(xy)$. If $\int \rho(\kappa^\nu)\,d(\kappa^2)=0$, then also $F_0(xy)=0$. Moreover one concludes from 4(9)

$$F(xy) = F_0(xy) - l^2 \int G(xx')\,dx'\langle 0|T\sigma^\nu : \chi(x')\,(\chi^*(x')\,\sigma_\nu\chi(x')) : \chi^*(y)|0\rangle.$$

$$4(13)$$

If the τ-functions are replaced by the ϕ-functions according to 4(2), the equations 4(9) and 4(12) can be reformulated as integral relations for the ϕ-functions. For this purpose it is convenient first to translate the differential equation 4(7) into an equation for the ϕ-functions. The terms with $c\,\delta(x-y)$ in 4(7) then drop out, since they occur in the same way on the left-hand side from the differentiation of the functions $F(xy)$ in 4(2). The equation 4(7) can be rewritten as

$$i\sigma_\nu^{\alpha\beta}\frac{\partial}{\partial x_{1\nu}}\phi(x_1\ldots|y_1\ldots) = -\,l^2\,\sigma_\nu^{\alpha\beta}\,\sigma^{\nu,\gamma\delta}\,\phi(x_1\underset{\beta}{x_1}x_2\ldots|\underset{\gamma}{x_1}y_1y_2\ldots)$$

$$+(-1)^z l^2\,\sigma_\nu^{\alpha\beta}\,\sigma^{\nu,\gamma\delta}\,F(\underset{\beta}{x_1}y_1)\cdot\phi(x_1 x_2\ldots$$

$$|\underset{\gamma}{x_1}y_2\ldots)+\text{permutations}$$

$$+(-1)^z l^2\,\sigma_\nu^{\alpha\beta}\,\sigma^{\nu,\gamma\delta}\,F(\underset{\gamma}{x_2}x_1)\,\phi(x_1\underset{\beta}{x_1}\underset{\delta}{x_3}\ldots$$

$$|y_1 y_2\ldots)+\text{permutations}$$

$$+(-1)^z l^2\,\sigma_\nu^{\alpha\beta}\,\sigma^{\nu,\gamma\delta}\,F(\underset{\beta}{x_1}y_1)\,F(\underset{\gamma}{x_2}x_1)$$

$$\times\phi(\underset{\delta}{x_1}x_3\ldots|y_2\ldots)+\text{permutations}$$

$$+(-1)^z l^2\,\sigma_\nu^{\alpha\beta}\,\sigma^{\nu,\gamma\delta}\,F(\underset{\beta}{x_1}y_1)\,F(\underset{\gamma}{x_2}x_1)$$

$$\times F(\underset{\delta}{x_1}y_2)\cdot\phi(x_3\ldots|y_3\ldots)+\text{permutations}$$

$$+i\sigma_\nu^{\alpha\beta}\frac{\partial}{\partial x_{1\nu}}\{(F_0(\underset{\beta}{x_1}y_1)-F(\underset{\beta}{x_1}y_1))\,(-1)^z$$

$$\times\phi(x_2\ldots|y_2\ldots)+\text{permutations}\}.$$

$$4(14)$$

This relation (which can be derived most conveniently by the methods described in Appendix A IV) can be transformed immediately into an integral equation. Starting from a truncated set of ϕ-functions one thereby arrives at a set of homogeneous linear differential or integral equations for determining the unknown functions ϕ, if the function $F(xy)$ is given. This set fixes the mass eigenvalues for the state $|\psi\rangle$ in just the same way, as the Schrödinger equation determines the eigenvalues of the energy in quantum mechanics. The degree of approximation is limited by the accuracy with which the truncated set of ϕ-functions can represent the state vector. It is a characteristic feature of the new Tamm–Dancoff method that 4(14) does not contain the singularity of $F_0(xy)$ at the light cone, which is compensated by the singularity of $F(xy)$. Quantization is introduced in 4(14) essentially through the function $F(xy)$, which will have a singularity at the light-cone different from the δ-like singularity of a linear theory.

It should be noticed that the equations 4(14) are exact equations even if the functions $F(xy)$ occurring in 4(14) are very different from the exact 2-point functions; only in the neighbourhood of the origin their behaviour should be similar, i.e. they should give rise to the same terms $c\delta(x-y)$ in 4(7) as the real 2-point functions. If the set 4(14) of equations for the ϕ-functions is broken off at a certain number of variables, then and only then the exact form of $F(xy)$ will become important. In the following calculations we will (according to 3(35) and 3(36)) always assume $F_0(xy)=0$.

While the calculation of the mass eigenvalues from homogeneous linear integral equations is simple in principle (not in the actual calculation), the determination of $F(xy)$ requires the solution of non-linear equations. Let us assume, that the process described in 4(2), 4(9) and 4(14) leads to an eigenvalue equation for a fermion state which after elimination of all higher ϕ-functions can for the lowest function $\tau(x)=\phi(x)=\langle 0|\chi(x)|\psi\rangle$ be written in the (somewhat simplified) form

$$i\sigma_\nu \frac{\partial}{\partial x_\nu} \phi(x) = i\sigma_\nu \frac{\partial}{\partial x_\nu} \int d^4x' \, \mathscr{L}(xx') \, \phi(x'). \qquad 4(15)$$

Then for $F(xy)$ we expect an equation

$$i\sigma_\nu \frac{\partial}{\partial x_\nu} F(xy) = i\sigma_\nu \frac{\partial}{\partial x_\nu} \int d^4x' \, \mathscr{L}(xx') \, F(x'y) + \mathscr{M}(xy), \qquad 4(16)$$

where both $\mathscr{L}(xx')$ and $\mathscr{M}(xy)$ depend on $F(xy)$ in a very complicated manner. If the functions $F(xy)$, etc. are written in momentum space

$$F(xy) = (2\pi)^{-4} \int d^4p e^{ip(x-y)} F(p),$$

the equations 4(15) and 4(16) read

$$[1 - \mathcal{L}(p^2)]\,\sigma^\nu p_\nu\,\phi(p) = 0 \qquad\qquad 4(17)$$

$$\left.\begin{aligned}[1 - \mathcal{L}(p^2)]\,\sigma^\nu p_\nu\,F(p) &= \mathcal{M}(p^2) \\[2mm] F(p) &= \frac{\overline{\sigma^\nu} p_\nu\,\mathcal{M}(p^2)}{p^2[1 - \mathcal{L}(p^2)]}\end{aligned}\right\} \qquad 4(18)$$

$\mathcal{L}(p^2)$ and $\mathcal{M}(p^2)$ are again complicated functionals of $F(p)$.

The actual derivation of an equation of the type 4(18) will be discussed in §4-5. For the present purpose we are satisfied with stating that there should be a close connexion between the procedure of calculating the mass eigenvalues for fermions and the determination of the 2-point function; formally the connexion is established by the 'inhomogeneous term' $\mathcal{M}(p^2)$. It should be emphasized however, already at this point, that in a non-linear equation the distinction between 'homogeneous' and 'inhomogeneous' term is purely formal and has no mathematical significance. The connexion should only ascertain that the poles of the 2-point function coincide with the discrete solutions of the eigenvalue equation. On the other hand the first procedure requires the solution of a set of homogeneous linear integral equations, the second problem the solution of complicated non-linear integral equations, mathematically a very different problem. Only the first procedure can be considered as closely analogous to the procedure in unrelativistic quantum mechanics or quantum field theory. In unrelativistic quantum field theory the commutator of the fields is given as δ-function, the rest follows by solving complicated linear equations, essentially the Schrödinger equations. If in the relativistic field theory the 2-point function is given, again the rest follows from solving linear equations. Also the 2-point function in relativistic field theory could be replaced by the commutator in the neighbourhood of the light cone, since the rest of the 2-point function could then be derived by analytic continuation. Hence in both cases, besides the fundamental field equation, the knowledge of the commutator in the neighbourhood of the origin is the essential information needed for formulating the set of homogeneous linear equations and deriving its eigenvalues.

With regard to the second part of the problem however, the analogy between relativistic and unrelativistic theory fails. In unrelativistic field theory the commutator is simply given as δ-function; in relativistic theory it is to be calculated from a complicated non-linear equation. This second part of the problem will be discussed in §4-5.

The first part of the problem still requires a prescription concerning the selection of equations for the truncated set of ϕ-functions from the infinite set of equations 4(2), 4(14). This infinite set would hold only for an exact representation of the state vector $|\psi\rangle$. From a mathematical point of view it would be most natural to postulate, that for the truncated set of ϕ-functions all equations of the infinite set 4(2), 4(14) should be fulfilled approximately with the smallest possible error. By a suitable definition of this latter term 'smallest error' the ϕ-functions could be well defined. On the other hand, from a practical point of view, such a procedure would be rather complicated. Therefore one will usually try to pick out a suitable set of simple equations truncated in the same way as the set of ϕ-functions, and not to consider the other equations. Such a procedure should, however, be used with caution, the degree of approximation obtainable in this way should be investigated carefully.

4-4 Description of integral equations by Feynman diagrams

The integral equations which in the Tamm–Dancoff method arise from combining the equations 4(2), 4(9), 4(14), can be represented by Feynman diagrams [46] in the following way. The coefficients in the linear equations are always given by combinations of the two functions $F(xy)$ and $G(xy)$. Therefore it is convenient, to represent these functions by lines, leading from the point x to the point y. We will use the following notation:

$$G(xy) \qquad | \qquad\qquad\qquad 4(19)$$

$$F(xy) \qquad \phi \quad .$$

τ-functions which are vacuum expectation values, for example, 2-point or 4-point functions shall be represented by bubbles, the variables in the $2n$-point function being indicated by lines starting from the bubble:

$$\tau(xyzu) \qquad \bigcirc \qquad\qquad\qquad 4(20)$$

η-functions will be indicated in the same way, but with a letter η inside the bubble

$$\eta(xyzu) \qquad \bigcirc\!\!\eta \quad .$$

ϕ-functions belonging to transitions from the vacuum to a particle state will be described by half bubbles, indicating thereby that a vacuum expectation value can be produced by creation and annihilation of intermediate states:

$$\bowtie = \vdash\!\!\!\sim\!\!\!\dashv + \cdot \; \cdot \; \cdot \; \cdot \; \cdot$$

$$\phi(xyz) \qquad \dashv$$

$$\phi(xxy) \qquad \dashv \; .$$

ζ-functions will be marked by a letter ζ in the half bubble

$$\zeta(xyz) \qquad \dashv\boxed{\zeta} \; .$$

If a distinction of $\chi(x)$ and $\chi^*(x)$ is necessary, it can be indicated by the direction of an arrow on the corresponding line:

$$\phi(xy|z) \qquad \leftarrow\!\dashv \; .$$

Finally the diagrams will contain vertex points where four lines meet, as can be seen from 4(9). In this equation x' is the vertex point. One of the four lines meeting there is necessarily a G-line, the others may be F- or G-lines or may represent variables in a ϕ-function. All vertex points define the same connexion of the Dirac indices and isospin indices of the meeting lines, since they all arise from the same interaction term in the fundamental equation. If necessary the Dirac or isospin matrices can be indicated by writing them into the corner of the two connected lines in the vertex: for example

$$\sigma_\mu \times \sigma'^\mu \qquad \text{or} \qquad \times \; .$$

The integrations should be carried out over all vertex points, and only there.

As an example we mention the lowest approximation for the eigenvalue equation of a boson state $|B\rangle$; in this approximation all ϕ-functions with more than two variables shall be omitted. From 4(2), 4(9) and 4(12) we get:

$$\langle 0|T\chi^*(x)\,\sigma_\mu\,\chi(x)|B\rangle = -l^2 \int d^4x' \langle 0|T\chi^*(x)\,\sigma_\mu\,G(xx')\,\sigma^\nu:\chi(x')$$

$$\times (\chi^*(x')\,\sigma_\nu\,\chi(x')):|B\rangle$$

$$\approx -l^2 \int d^4x'\, tr(\sigma_\mu\,G(xx')\,\sigma^\nu\,F(x'\,x))$$

$$\times \langle 0|T\chi^*(x')\,\sigma_\nu\,\chi(x')|B\rangle. \qquad 4(21)$$

(The sign *tr* means 'trace' with respect to the Dirac indices.) This integral equation will be expressed as Feynman diagram by

$$\langle\sigma_\mu \rangle = \langle \sigma_\mu\ \sigma_\nu \rangle\langle \sigma^\nu \rangle \quad \text{or simply by} \quad \langle\ \rangle = \langle\ \rangle\langle\ \rangle \qquad 4(22)$$

which is a specialization of the more general equation $\quad = \quad$

The kernel of the integral equation can be written as

$$tr[\sigma_\mu\,l^2\,G(xx')\,\sigma^\nu\,F(x'\,x)]+\ldots = \bigcirc \qquad 4(23)$$

As a second example we consider the eigenvalue equation for a fermion in an approximation, in which all ϕ-functions with more than three variables are omitted. The two relations between the ϕ-functions of one or of three variables can be symbolized by:

$$\qquad 4(24)$$

In this case it would already be very complicated to write out all the integrations explicitly in the way it has been done in 4(21); the Feynman diagram is a convenient method of describing the general structure of the equations.

 Unified Field Theory of Elementary Particles

4-5 The non-linear problems; η-functions and S-matrix

For the representation of the ground state itself the set of ϕ-functions is not very convenient; the ϕ-function $\phi(x|y)$ vanishes in this case by definition. The τ-functions are identical with the 2-point functions, 4-point functions, etc. and are the most adequate representation. If a truncated set of functions should be used for an approximate representation of the ground state, the truncated set of η-functions 4(3) and 4(4) is probably the best choice, since an η-function of $2n$ variables results mainly from the creation and annihilation of states which can be represented by not less than n variables. If n is large, this contribution should be comparatively unimportant.

The equations for the η-functions obtained from 4(9) are non-linear, since 4(3), 4(4) define already non-linear relations between different η-functions. The first two equations of the resulting infinite set can be indicated by Feynman graphs in the following way:

$$\qquad\qquad 4(25)$$

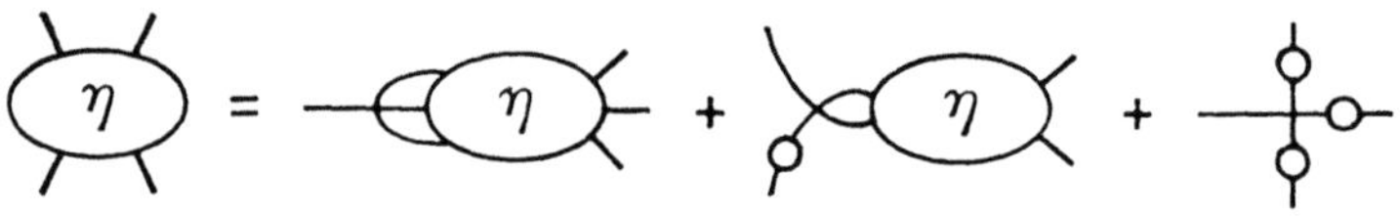

$$\qquad\qquad 4(26)$$

If the term with the η-function of six variables is omitted in 4(26) in order to get a first approximation, we obtain

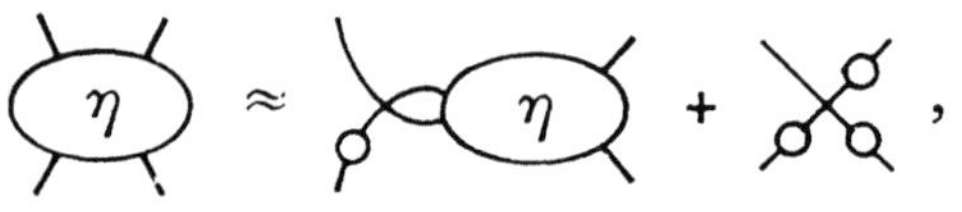

$$\qquad\qquad 4(27)$$

and by inserting 4(27) into 4(25)

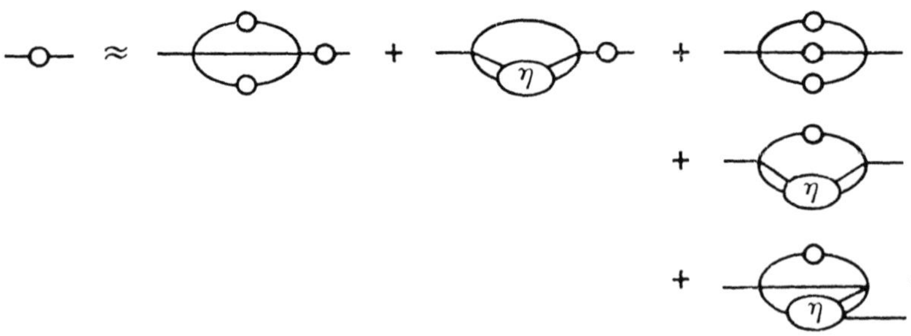

$$\qquad\qquad 4(28)$$

This latter equation has in fact the form of 4(18) required by general considerations. The operators $\mathscr{L}(xx')$ or $\mathscr{L}(p^2)$ and $\mathscr{M}(xx')$ or $\mathscr{M}(p^2)$ of 4(18) are here represented by

Methods of Approximation 61

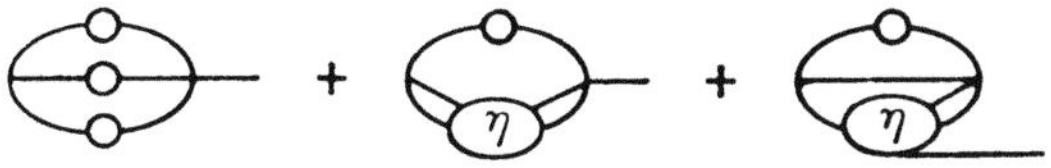

$$\mathscr{L} = \quad + \quad , \qquad 4(29)$$

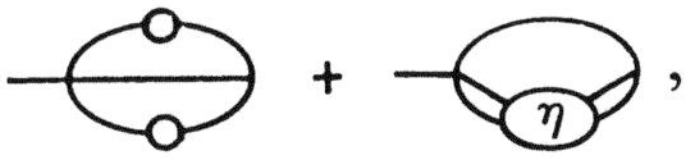

$$\mathscr{M} = \quad + \quad + \quad 4(30)$$

If the functions $-\!\!\circ\!\!- = F(xy)$ and  inside the integral operators could be considered as fixed, 4(28) could be taken as an inhomogeneous linear equation for $-\!\!\circ\!\!-$. In this case the corresponding homogeneous linear equation would be

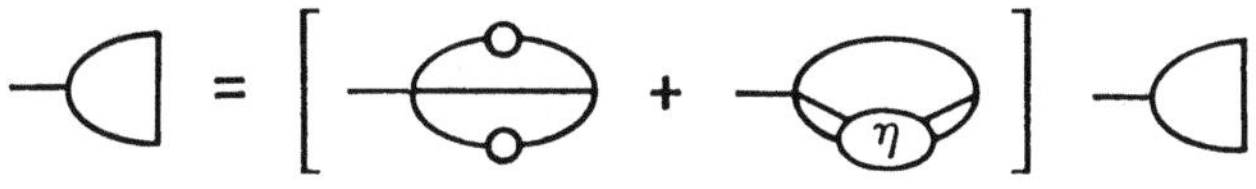

$$= \left[\quad + \quad \right] , \qquad 4(31)$$

where $-\!\!\circ\!\!-$ can be constructed by creation and annihilation of fermions

$$-\!\!\circ\!\!- = \qquad 4(32)$$

The 'homogeneous' equation 4(31) is in fact equivalent to an eigenvalue equation for the fermions and determines the position of the poles; it will be used later on (in a somewhat simplified form) for the calculation of the fermion eigenvalues. The 'inhomogeneous' equation 4(28) would determine the residue at the poles and the complete behaviour of $-\!\!\circ\!\!-$.

On the other hand $-\!\!\circ\!\!-$ and  inside the integral operators 4(29) and 4(30) cannot be considered as fixed. 4(28) is just a very complicated non-linear integral equation for $-\!\!\circ\!\!-$ and  , and it will scarcely be possible to find any general procedure for solving non-linear equations of this degree of complication.

From a practical point of view the only way of handling such equations as 4(28) will be a parametrization of $-\!\!\circ\!\!-$. In the following calculations we will use the general form 3(37). The mass spectrum $\rho(\kappa^2)$ will be simplified by assuming just a few discrete masses, which should—as a postulate of consistency—coincide with the eigenvalues of 4(31); the residua at the poles should be determined by fulfilling 4(28) with the highest possible accuracy. In the calculations of the present book only the simplest case of just one discrete mass value will be considered; $-\!\!\circ\!\!-$ will be represented by 3(39). The mass value κ in 3(39) should then lie in the neighbourhood of the lowest eigenvalue of 4(31).

In a similar way as in 4(28), eqn. 4(27) can be considered as an inhomogeneous linear equation which—if —o— is considered as fixed—may be used for the calculation of boson eigenvalues and the corresponding residua in ⧅. The homogeneous linear equation corresponding to 4(27) is in fact

$$\text{[diagram]} \qquad 4(33)$$

and has been mentioned in §4-4 as basis for the determination of the boson eigenvalues. Practically it will be used in the form

$$\text{[diagram]} \qquad 4(34)$$

mentioned already in 4(22).

The τ-functions of the vacuum, $2n$-point functions or η-functions, are closely related to the S-matrix. If some time variables in a $2n$-point function are taken as tending to $+\infty$, the others to $-\infty$, the asymptotic behaviour of the function will determine elements of the S-matrix. In fact, combinations of field operators of the type 2(14) may be used for defining operators $O_\lambda(x)$ of a given symmetry; if the 2-point function 2(15), constructed from these operators $O_\lambda(x)$, contains poles, the τ-functions of the vacuum will as a rule have the same poles, the operators $O_\lambda(t \to -\infty)$ will contain the corresponding free field operator ψ_{in}, or $O_\lambda(t \to +\infty)$ the operator ψ_{out}. Eqn. 2(13) then defines the S-matrix element from the asymptotic behaviour of the $2n$-point function. Many different τ-functions can lead to the same S-matrix element, if operators O_λ of the same symmetries are taken out of the Green's functions. In this way a complete determination of the τ- or η-functions of the ground state is equivalent to a complete determination of the S-matrix.

It should be noticed at this point that the τ- or η-functions or their asymptotic parts contain contributions from unphysical states connected with the indefinite metric in Hilbert space. The physical states can be extracted according to §2-5 from the asymptotic part by diagonalizing the S-matrix and by picking out those states for which the diagonal elements have the absolute value unity. It may be that by this process some element of non-locality (like the Coulomb force in quantum electrodynamics) is introduced in the theory.

As a practical example the scattering of two fermions may be calculated from the η-function ⊙. In the approximation 4(27) the scattering amplitude can be determined by integrating 4(27) formally [53]:

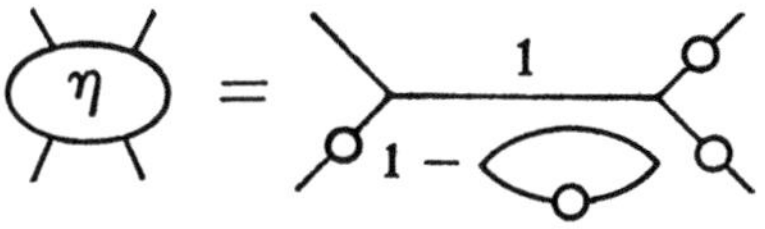

$$\text{(diagram)} \qquad 4(35)$$

Inserting 4(35) into 4(27) gives in fact:

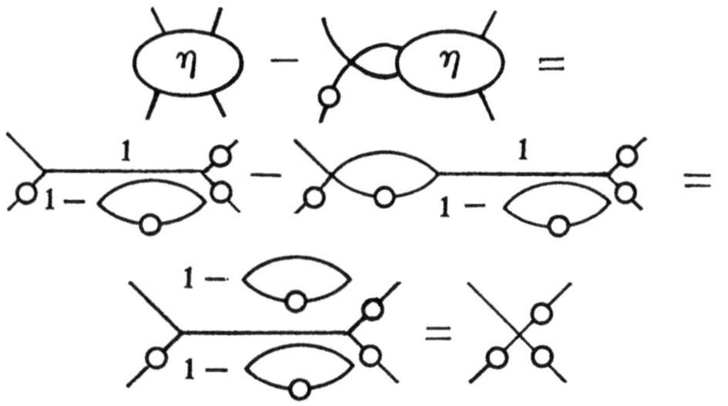

$$\text{(diagram)} \qquad 4(36)$$

The four outgoing lines can be taken as representing the incoming and outgoing fermions. One of these lines is not an $F(xy)$-line —o—, as it should be, but a G-line ———; this is a consequence of the low approximation which cannot take into account completely the dressing of the fermion-lines. The connecting term $\dfrac{1}{1-\text{⟨o⟩}}$ represents the boson fields acting as scattering force between the fermions. Its poles are determined by the equation $1 = \text{⟨o⟩}$, which is just the eigenvalue equation of the bosons. The residue of the poles may be used for calculating the coupling constants between fermions and bosons.

Special examples for the calculation of coupling constants will be given in the Chapters 6 and 8.

The methods which have been described here for the determination of the η-functions are closely related to recent methods used in quantum mechanics of many body problems. A number of results can be taken over without much change from many body physics to relativistic field theory. The only fundamental difference consists in the fact, that many body physics starts with canonical quantization and long-range forces, while relativistic field theory starts from local interaction and therefore non-canonical quantization. In all problems in which this difference is un-important the methods of many body physics may be transferred into relativistic field theory.

One remark shall be added concerning the homogeneous linear equations determining the eigenvalues. If by an investigation of the η-functions of the vacuum some information has been obtained concerning not only the 2-point function, but also the η-function of four or even more variables these informations can be used for improving the set of linear equations. In this case it will be convenient to employ a truncated set of ζ-functions (compare 4(5)) as an approximate representation of the state. The η-functions, or what is known about them, may be used in connecting the ζ-functions with the τ-functions and in constructing thereby the linear integral equations for the ζ-functions.

This whole mathematical scheme is consistent only if the different methods using truncated sets of ϕ-functions or η-functions or ζ-functions converge towards the same limit when the number of variables, at which the set is cut off, tends to infinity. In the present state of quantum field theory a proof of convergence cannot be given. The convergence may possibly depend on the way how the truncated set of equations is selected, for example, by choosing just as many equations as unknown functions, or by taking a larger number of equations to be fulfilled with the smallest possible errors. The simple example of the anharmonic oscillator seems to show that the convergence may depend on such details. Still one may hope that by a suitable selection of equations for the various truncated sets of functions a convergent scheme emerges which can be considered as a consistent solution of the field equation 3(1).

5

Solution of the Simplest Eigenvalue Equations

5-1 The lowest boson states [54]

The simplest group of bosons $|B\rangle$ can be represented by ϕ- or τ-functions of the type

$$\phi_\mu(x|x) = \tau_\mu(x|x) = \langle 0|T\chi^*(x)\,\sigma_\mu\chi(x)|B\rangle. \qquad 5(1)$$

Bosons of this group can be considered as composed of a fermion and an antifermion in an S-state (orbital angular momentum zero). The Dirac spin of these bosons may be zero or one, depending on whether the spins of the two fermions are parallel or antiparallel. In the first case the four-vector ϕ_μ must have the direction of the energy-momentum vector J_μ of the boson. In the second case its direction will be given by a vector of polarization B_μ fulfilling the condition

$$B_\mu J^\mu = 0. \qquad 5(2)$$

In the first case $\phi_\mu(x|x)$ corresponds to the derivative with respect to x_μ of the scalar wave function of the boson in the phenomenological theory. In case of spin 1 $\phi_\mu(x|x)$ can be compared with the vector potential $A_\mu(x)$ which fulfills the Lorentz condition

$$\frac{\partial A_\mu}{\partial x_\mu} = 0. \qquad 5(3)$$

The equations 5(2) and 5(3) are then equivalent. The isospin of the boson can be fixed in the same way. With the definition

$$\phi_{\mu\nu}(x|x) = \langle 0|T\chi^*(x)\,\sigma_\mu\tau_\nu\chi(x)|B\rangle \qquad 5(4)$$

and

$$\tau_\nu = (1, \boldsymbol{\tau}) \quad \text{as isospin matrix} \qquad 5(5)$$

65

 Unified Field Theory of Elementary Particles

$\phi_{\mu,0}$ represents bosons of isospin zero, $\phi_{\mu,i}$ or $\phi_{\mu i}P_i$ (P_i polarization vector in isospace) bosons of isospin 1.

The eigenvalue equation shall be established in the lowest approximation, in which all ϕ-functions of more than two variables will be neglected. The eigenfunction can be derived from ⟨⟨⟩ = ⟩⟨⟩, and the eigenvalue equations can be written as Feynman diagram:

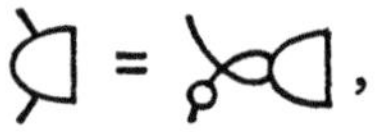

$$\langle\!\langle\;=\;\bigcirc\!\bowtie\!\langle\!\langle \tag{5(6)}$$

The simplest term was already mentioned in 4(21). For an exact evaluation it is convenient, to make use of the equivalence between the two forms 3(1) and 3(4) of the fundamental field equation, and to replace in the interaction term the product of two unit-matrices in isospace by its Fierz symmetrical version:

$$1 \cdot 1 \;\rightarrow\; \tfrac{3}{4}\cdot 1\cdot 1 + \tfrac{1}{4}\cdot \tau_k\tau_k. \tag{5(7)}$$

This replacement simplifies the process of 'contraction', i.e. the reduction of the τ-function of four variables to the ϕ-function of two variables by means of 4(2). Moreover the kernel ⬡ should be made symmetrical with respect to the two variables in $\phi_\mu(x|x)$:

$$\bigcirc\;\longrightarrow\;\tfrac{1}{2}\left[\,\bigcirc\;+\;\bigcirc\,\right], \tag{5(8)}$$

in order to preserve the *GP*-invariance of the fundamental equation in the approximation. In our special case 5(6) the kernel is already symmetrical beforehand. In selecting the equations for the truncated set of ϕ-functions the ideal method would be to postulate the whole infinite set of equations with the smallest possible errors. In practical calculations this method can scarcely be applied. But in the selection of special equations one should at least carefully avoid the violation of symmetries existing in the underlying field equation.

After these preliminary remarks the complete kernel of the eigenvalue equation for $\phi_\mu(x|x)$ can be written as

$$\bigcirc = K_{\mu\nu}(xx') = -\tfrac{1}{2}tr[2\sigma_\mu l^2 G(xx')(\tfrac{3}{4}\cdot 1\cdot 1 + \tfrac{1}{4}\tau_k\tau_k)\,\sigma_\nu F(x'\,x)]$$

$$-\tfrac{1}{2}tr[2\sigma_\mu F(xx')(\tfrac{3}{4}\cdot 1\cdot 1 + \tfrac{1}{4}\tau_k\tau_k)\,\sigma_\nu l^2 G(x'\,x)], \tag{5(9)}$$

the trace being taken over the Dirac indices only. The factor 2 in the bracket comes from the two modes of 'contraction'. For $F(xx')$ we use the approximation 3(39), depending only on the one parameter κ; $G(xx')$ is the Green's function 4(8). From 5(9), 3(39) and 4(8) we get

$$K_{\mu\nu}(xx') = (\tfrac{3}{4}\cdot 1 \cdot 1 + \tfrac{1}{4}\tau_k\tau_k)(2\pi)^{-8}l^2 \int d^4p\, d^4q\, e^{-i(p-q)(x-x')}$$

$$\times \left[\frac{tr(\sigma_\mu p_\rho\,\overline{\sigma^\rho}\,\sigma_\nu q_\tau\,\overline{\sigma^\tau})}{p^2(q^2)^2(q^2-\kappa^2)} + \frac{tr(\sigma_\mu p_\rho\,\overline{\sigma^\rho}\,\sigma_\nu q_\tau\,\overline{\sigma^\tau})}{(p^2)^2(p^2-\kappa^2)q^2}\right] \qquad 5(10)$$

By Fourier transformation:

$$K_{\mu\nu}(J) = \int d^4x K_{\mu\nu}(x-x')\,e^{+iJ(x-x')}$$

$$= -(\tfrac{3}{4}\cdot 1 \cdot 1 + \tfrac{1}{4}\tau_k\tau_k)(2\pi)^{-4}l^2{}_i \int d^4p\,\kappa^4\, tr[\sigma_\mu p_\rho\,\overline{\sigma^\rho}\,\sigma_\nu(J-p)_\tau\,\overline{\sigma^\tau}]$$

$$\times\left[\frac{1}{(p^2)^2(p^2-\kappa^2)(J-p)^2} + \frac{1}{((J-p)^2)^2((J-p)^2-\kappa^2)p^2}\right]$$

$$= -(\tfrac{3}{4}\cdot 1 \cdot 1 + \tfrac{1}{4}\tau_k\tau_k)\left(\frac{\kappa}{2\pi}\right)^4 l^2{}_i \int d^4p\,\cdot$$

$$\times 2[-g_{\mu\nu}(p,J-p) + p_\mu(J-p)_\nu + p_\nu(J-p)_\mu]$$

$$\times\left[\frac{1}{(p^2)^2(p^2-\kappa^2)(J-p)^2} + \frac{1}{[(J-p)^2]^2[(J-p)^2-\kappa^2]p^2}\right]. \qquad 5(11)$$

The integration in momentum space requires a definition of the path of integration in the complex p_0-plane. Since $F(xy)$ has been defined as vacuum expectation value of the time ordered product and since the Green's function $G(xy)$ is, on account of the boundary conditions, meant in the same way, the path can according to well-known rules be taken along the real p_0-axis, if every pole term of the type $1/(-p^2+\kappa^2)$ is replaced by $\lim_{\delta\to 0} 1/(-p^2+\kappa^2-i\delta)$. The integrations in p_1, p_2, p_3 should then be carried out along the real axis, unless a remaining singularity requires a new discussion of the boundary conditions. Actually, when the integration in p_0 has been carried out, there remains a singularity in **p**-space, which in the special co-ordinate system, where the boson is at rest, i.e. $J_\mu=(J_0,0,0,0)$ is located at the point $|\mathbf{p}|=\tfrac{1}{2}J_0$. This singularity is closely connected with the assumption of a dipole-ghost state in the approximate 2-point function and will be discussed in detail in §5-2. For the time being we will be satisfied with defining that at this pole $1/(J_0-2|\mathbf{p}|)$ the principal value shall be taken. Under this condition the integration in 5(11) can be carried out in a rather elementary way by the methods developed by Feynman. The details of the calculations are given in the Appendix A III. The result is:

$$K_{\mu\nu}(J) = -\tfrac{1}{2}(\tfrac{3}{4}\cdot 1\cdot 1 + \tfrac{1}{4}\tau_k\tau_k)\left(\frac{\kappa l}{2\pi}\right)^2$$

$$\times\left[\frac{J_\mu J_\nu}{J^2}q_0(\lambda) + \left(g_{\mu\nu} - \frac{J_\mu J_\nu}{J^2}\right)q_1(\lambda)\right], \qquad 5(12)$$

where $\lambda = \dfrac{J^2}{\kappa^2} = \dfrac{\kappa_B^2}{\kappa^2}$ (κ_B is the boson mass),

$$q_0(\lambda) = \ln\lambda - \frac{1}{\lambda} - \frac{(1-\lambda)^2}{\lambda^2}\ln|1-\lambda| \qquad 5(13)$$

$$q_1(\lambda) = (1-\tfrac{2}{3}\lambda)\ln\lambda + \tfrac{2}{3} + \frac{1}{3\lambda} + \frac{(1-\lambda)^2(1+2\lambda)}{3\lambda^2}\ln|1-\lambda|. \qquad 5(14)$$

Before going into a numerical discussion it should be noticed, that the expressions $\tfrac{1}{2}\cdot 1\cdot 1$ and $\tfrac{1}{2}\cdot\tau_k\tau_k$ are projection operators in isospace, while $J_\mu J_\nu/J^2$ and $g_{\mu\nu}-(J_\mu J_\nu)/J^2$ are projection operators in Dirac spin space. This can be seen in the following way. A state of isospin 0 is characterized by a ϕ-function, which in isospace has the form $1_{\alpha\beta}$. For this ϕ-function we have

$$1_{\alpha\beta}\cdot(\tfrac{1}{2}\cdot 1_{\beta\alpha}\cdot 1_{\gamma\delta}) = 1_{\gamma\delta}; \qquad 1_{\alpha\beta}\cdot(\tfrac{1}{2}\tau^k_{\beta\alpha}\cdot\tau^k_{\gamma\delta}) = 0. \qquad 5(15)$$

For a ϕ-function of isospin 1, for example, $\tau^l_{\alpha\beta}$ we have

$$\tau^l_{\alpha\beta}(\tfrac{1}{2}\cdot 1_{\beta\alpha}\cdot 1_{\gamma\delta}) = 0, \qquad \tau^l_{\alpha\beta}(\tfrac{1}{2}\tau^k_{\beta\alpha}\tau^k_{\gamma\delta}) = \tau^l_{\gamma\delta}. \qquad 5(16)$$

Therefore $\tfrac{1}{2}.1.1$ is a projection operator for isospin 0, $\tfrac{1}{2}\tau_k\tau_k$ a projection operator for isospin 1. With respect to Dirac spin we have already mentioned, that the four vector ϕ_μ is parallel to J_μ for Dirac spin 0, and parallel to B_μ ($B_\mu J^\mu = 0$) for Dirac spin 1. In the first case we have

$$J^\mu\frac{J_\mu J_\nu}{J^2} = J_\nu; \qquad J^\mu\left(g_{\mu\nu} - \frac{J_\mu J_\nu}{J^2}\right) = 0. \qquad 5(17)$$

In the second case:

$$B^\mu\frac{J_\mu J_\nu}{J^2} = 0; \qquad B^\mu\left(g_{\mu\nu} - \frac{J_\mu J_\nu}{J^2}\right) = B_\nu. \qquad 5(18)$$

Hence $J_\mu J_\nu/J^2$ is the projection operator for Dirac spin 0, $g_{\mu\nu} - J_\mu J_\nu/J^2$ the projection operator for Dirac spin 1.

Solution of the Simplest Eigenvalue Equations 69

This result enables us to write down immediately the eigenvalue equations for the four simplest cases: Dirac spin $S=0,1$ and isospin $T=0,1$.

$$S = 0 \quad T = 0 \qquad 1+\tfrac{3}{4}\left(\frac{\kappa l}{2\pi}\right)^2 q_0(\lambda) = 0, \qquad 5(19)$$

$$T = 1 \qquad 1+\tfrac{1}{4}\left(\frac{\kappa l}{2\pi}\right)^2 q_0(\lambda) = 0, \qquad 5(20)$$

$$S = 1 \quad T = 0 \qquad 1+\tfrac{3}{4}\left(\frac{\kappa l}{2\pi}\right)^2 q_1(\lambda) = 0, \qquad 5(21)$$

$$T = 1 \qquad 1+\tfrac{1}{4}\left(\frac{\kappa l}{2\pi}\right)^2 q_1(\lambda) = 0. \qquad 5(22)$$

The eigenvalues appear as functions of the parameter κ in the 2-point function; the numerical values for $\lambda^{1/2}=(|J|/\kappa)=(\kappa_B/\kappa)$ (κ_B is the boson mass) can be seen from Fig. 2.

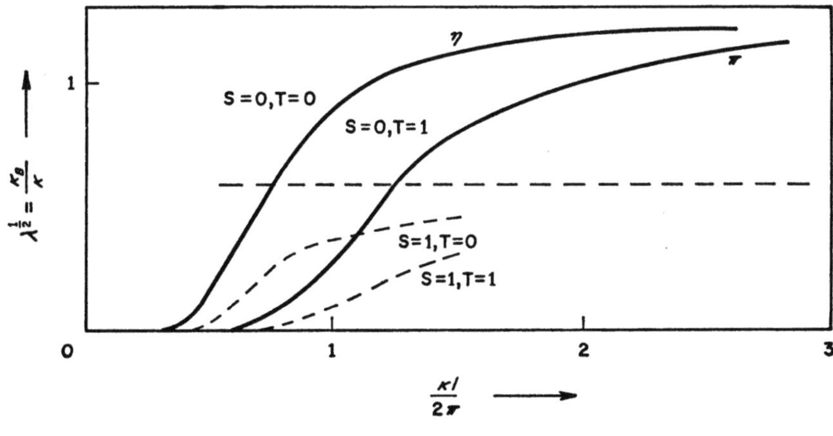

Fig. 2.

The solutions with spin 0 represent those particles which experimentally are known as η- and π-mesons. For values of $\kappa l/2\pi$ between 0·8 and 1·0 the ratio of the masses of η- and π-meson agrees roughly with the experimental value, and if κ is identified with the mass of the nucleon—as may be allowed in the lowest approximation—the ratio of the boson masses to the nucleon mass again agrees qualitatively with the observed values. Actually the η-meson had been predicted (in 1958) from 5(19).[29] It should be emphasized that the quantity $\kappa l/2\pi$ is not a free parameter. It will be determined by postulating—as had been pointed out in §4-5—that the characteristic pole $p^2 - \kappa^2 = 0$ in the 2-point function agrees with the eigenvalue resulting from the fermion equation. The theory does not contain any free parameter. Actually the values of $\kappa l / 2\pi$ calculated in low approximations by this method lie in the region of 0·8 to 1·05, as will be discussed in §5-3, §5-4 and Chapter 7.

The two other eigenvalues of Dirac spin 1 would seem to correspond, with regard to their symmetries, to particles like ω- and ρ-meson. This holds however only with one exception. When in later calculations the norm of the boson states will be discussed it will turn out that the norm of the two states with $S = 0$ is positive, while the norm of the two other states with $S = 1$ is negative. Hence, with the boundary conditions discussed briefly in §3-4, the states with $S = 1$ cannot appear as outgoing waves (they 'cannot be created'), while the states with $S = 0$ can. Therefore the eigenvalues defined by 5(21) and 5(22) cannot be identified with existing vector mesons, they are in a certain sense 'ghost' particles which cannot appear as real particles. This can be seen more clearly by studying the singularity of the wave function of the bosons at the point $|\mathbf{p}| = J_0/2$.

5-2 The singularity in momentum space of the boson wave function

The essential part of the boson wave function

$$\phi_\mu(x|y) = \langle 0 | T\chi^*(x)\, \sigma_\mu\, \chi(y) | B \rangle \qquad\qquad 5(23)$$

calculated in first approximation from the graph can be written in the form

$$\phi_\mu(x|y) = e^{iJy} \int d^4p\, e^{-ip(x-y)} \frac{-g_{\mu\nu}(p, J-p) + p_\mu(J-p)_\nu + p_\nu(J-p)_\mu}{(p^2)^2\,(p^2 - \kappa^2)\,(J-p)^2}\, \phi^\nu(0|0).$$

$$5(24)$$

The integration in p_0 shall be carried out along the real axis, if the poles like $1/p^2$ are meant as limiting values of $1/(p^2 + i\delta)$. We will discuss the behaviour of the wave function $\phi(x|y)$ at the time $x_0 - y_0 = 0$, and will assume $J_\mu = (J_0, 0, 0, 0)$. The p_0-integration then can be performed by shifting the path of integration, which goes along the real axis from $-\infty$ to $+\infty$, to large positive or large negative imaginary values, leaving only residual integrations around the poles, which are on both sides in the neighbourhood ($\sim \delta$) of the real axis. The integration will give a finite result, even in the limit $\delta \to 0$, unless the path of integration is squeezed in between two poles which coincide in the limit $\delta \to 0$. This is the case for the values $|\mathbf{p}| = (J_0/2)$, where the double pole from $(p^2)^2 = 0$ at the point $p_0 = \sqrt{(|\mathbf{p}|^2 - i\delta)}$ and the

Solution of the Simplest Eigenvalue Equations 71

single pole from $(J-p)^2=0$ at the point $p_0=J_0-\sqrt{(|\mathbf{p}|^2-i\delta)}$ coincide. After the integration a double pole of the form $(J_0-2|\mathbf{p}|)^{-2}$ will remain, unless the numerator of 5(24) vanishes at the point $J_0-2|\mathbf{p}|=0$.

This mathematical behaviour can be interpreted by noticing, that the introduction of the ghost and dipole ghost states 3(48) at mass zero in the approximate 2-point function must at least formally lead to processes, in which a boson can disintegrate into a fermion–antifermion pair of this type 3(48). Therefore we expect from the centre of the boson a spherical wave, the fermion and the antifermion leaving the boson state in opposite directions.

From this point onwards the analysis leads to different results for the $S=0$ and $S=1$ states. For the $S=0$ states $\phi^\nu(0|0)$ is proportional to J^ν, therefore the numerator of 5(24) is

$$-J_\mu(p,J-p)+(J-2p)_\mu(pJ)+p_\mu J^2 = J_\mu \cdot p^2 + p_\mu(J^2-2pJ). \quad 5(25)$$

This expression vanishes at the point $|\mathbf{p}|=(J_0/2)=p_0$. Then, after the p_0-integration, only a single pole of the form $(J_0-2|\mathbf{p}|)^{-1}$ will remain. This pole will produce a spherical wave of the general type $r^{-1}\exp[\pm i(J_0/2)r]$ (where $r=|\mathbf{x}-\mathbf{y}|$), after the $\mathbf{p}$-integration has been performed. At large distances r, i.e. when fermion and antifermion are very far from each other, we get nearly plane waves for each of the particles fulfilling (for at least one of the particles on account of the G-line in the graph $\rangle\!\!\circ\!\!\prec\!\!\Box$) the Weyl equation $p_\nu\sigma^\nu\phi=0$. Therefore asymptotically only the state G^+ of 3(48) is present, in agreement with the postulated boundary conditions. The state G^+ has the norm zero; therefore the spherical wave $r^{-1}\exp[\pm i(J_0/2)r]$ does not prevent the norm of the boson state from becoming finite. The boson states with $S=0$ may appear as outgoing or incoming particles, since their asymptotic behaviour does not interfere with the boundary condition, that only G^+ should be present asymptotically among the zero mass fermions.

For the $S=1$ states however the situation is different. Here $\phi^\nu(x|x)$ is proportional to B^ν and the numerator of 5(24) becomes

$$-B_\nu(p,J-p)+(Bp)(J-2p)_\nu. \quad\quad\quad 5(26)$$

This expression does generally not vanish at $|\mathbf{p}|=(J_0/2)=p_0$. Hence after the p_0-integration has been performed a double pole of the form $(J_0-2|\mathbf{p}|)^{-2}$ will remain, which produces a spherical wave of the type $r^{-1}(a+br)\exp[\pm i(J_0/2)r]$. This means that asymptotically, for large r, not only the state G^+, but also the dipole-ghost state D^+ will appear, in disagreement with the postulated boundary conditions. Therefore under the given boundary conditions the boson states 5(21) and 5(22) with $S=1$ cannot appear as outgoing or incoming particles, as 'free' states; they cannot represent any observed vector mesons. The theory of the real vector mesons will (with the exception of the photon, which shall be discussed in Chapter 8) require higher approximations, in which the contact interaction described by  does not play the decisive rôle.

Coming back to the $S=0$ bosons we notice that for any value of J_0 a spherical wave of the general type $r^{-1}\exp[\pm irJ_0/2]$ or more explicitly

$$\frac{a\cos(rJ_0/2)+b\sin(rJ_0/2)}{r}$$

will appear and that the condition  =  fixes at given J_0 only the relation between the two constants a and b. A solution containing the term $r^{-1}[\sin(rJ_0/2)]$ however would mean a scattering state in which a ghost wave of the G^+-type would be scattered by an antifermion. Solutions, in which a finite wave packet of the G^+-type is scattered, may exist for any values of J_0, but they have the norm zero and therefore do not represent any physical state. Hence the discrete real boson state with a finite norm is defined by the condition $b=0$; only the wave $r^{-1}[\cos(rJ_0/2)]$ shall appear asymptotically. This condition fixes the $|\mathbf{p}|$ integration to the principal value, as stated in §5-1, and thereby fixes J_0 or κ_B.

5-3 The simplest fermion states

For the calculation of the mass eigenvalues of fermion states[29] the necessary integral operations in the lowest approximation have been described by Feynman graphs in 4(24); in this approximation all ϕ-functions with more than three variables have been neglected. In the following calculations the integral equations shall be simplified further by neglecting all contributions to 4(24) except

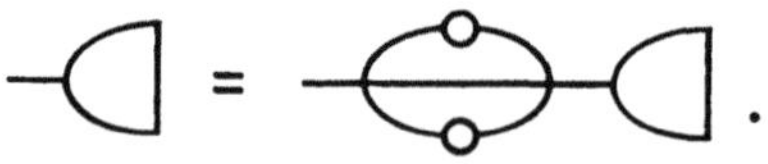 5(27)

This approximation is definitely more primitive than 4(24) and may be even less accurate than the boson calculation 5(6); but it has the advantage of changing the integral equation into an algebraic equation in momentum space and may be sufficient for a very first survey. The calculation of the mass eigenvalues thereby is reduced to the evaluation of the kernel 

with the methods developed by Feynman. Again, like in the boson case, the kernel should, if necessary, be made symmetrical in the variables:

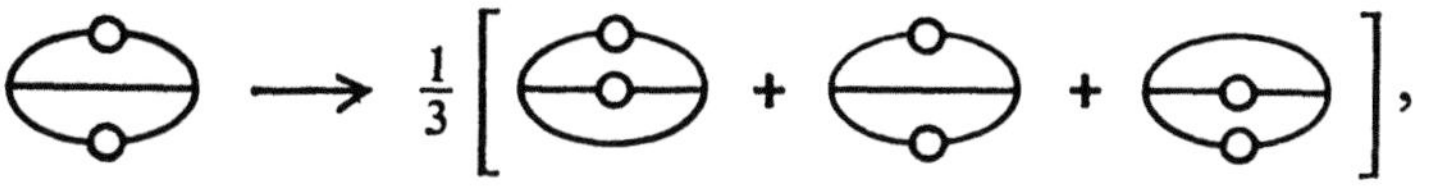

5(28)

but again it is already symmetrical beforehand.

When by Fourier transformation the kernel $K(x-x')$ has been changed into $K(J)$, this expression $K(J)$ is defined as integral over two momentum variables $d^4p\,d^4q$. The integrations in p_0 and q_0 shall as usually be carried out along the real axis, after the pole terms have been defined as

$$\lim_{\delta\to 0}\frac{1}{p^2+i\delta},$$

etc. When the p_0, q_0 integrations have been performed, the remaining function of momenta is singular at those points which can be interpreted as leading to a disintegration of the initial fermion into three fermions of mass zero. Therefore again, like in the boson case, in- or outgoing spherical waves will result depending, for example, on the distance between one fermion and the centre of mass of the two others. In this case only waves of the normal type $r^{-1}\exp(\pm ipr)$ will appear. This can be seen from the qualitative argument, that at large distances $|x-x'|$ the function $F(xx')$ of 3(39) decreases as $1/(|x-x'|)$, while $G(xx')$ decreases as $|x-x'|^{-3}$. The integral $\int d^4x' F(xx')G(x'y)\exp(ix'J)$ of the boson case therefore may contain terms of the type

$$\frac{1}{|x-y|}(a+b|x-y|)\,e^{\pm ip|x-y|}$$

while the integral $\int d^4x' F(xx')F(yx')G(zx')\exp(ix'J)$ will asymptotically contain only terms of the type

$$\frac{1}{|z-(x+y)/(2)|}\,e^{\pm ip|z-(x+y)/(2)|}.$$

Hence the existence of free fermions of finite mass is compatible with the boundary condition, that only G_+ states shall appear asymptotically as fermions of mass zero.

Again it must be postulated, that the fermion state is not a scattering state of G_+; therefore the integrations in momentum space (after the p_0 and q_0 integration have been performed) are defined as principal values. Thereby the mass of the fermion is fixed.

 Unified Field Theory of Elementary Particles

The actual evaluation of the kernel ⊜ by the methods developed by Feynman, is considerably more complicated than the evaluation of ⟨⊙⟩ in the boson case. As result of the calculations which are completely analogous to 5(9)–5(12) the eigenvalue equation of the fermion becomes:

$$\left[1+\frac{3}{2}\left(\frac{\kappa l}{2\pi}\right)^4 L(\lambda)\right]\sigma^\nu J_\nu \cdot \phi(J) = 0 \qquad 5(29)$$

with $\lambda = +(J^2/\kappa^2) = (\kappa_F^2/\kappa^2)$. The analytical function $L(\lambda)$ is discussed in detail in Appendix A V. Its behaviour for real positive values of λ is given in Fig. 3.

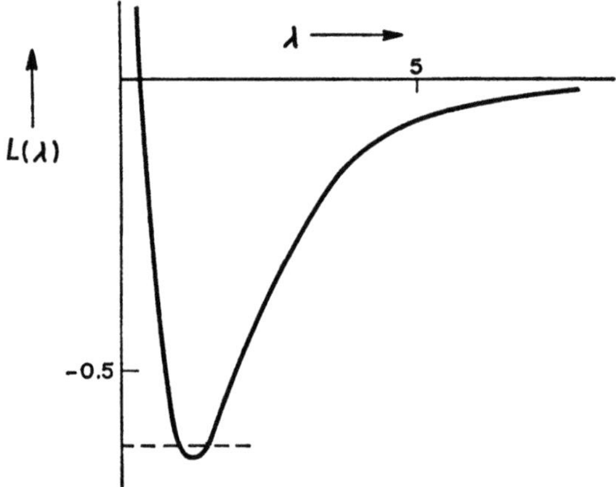

Fig. 3.

The equation 5(29) has several solutions. One solution is defined by $\sigma_\nu J^\nu \phi = 0$ or $J^2 = 0$, i.e. the fermions may be neutrinos with mass zero. Finite masses are possible, if there are values of λ, for which the bracket

$$\left[1+\frac{3}{2}\left(\frac{\kappa l}{2\pi}\right)^4 L(\lambda)\right]$$

is equal to zero. Fig. 3 shows that in this case there will be as a rule two solutions, given by the two intersections of the L-curve with a line parallel to the abscissa. Only the lower of the two values of λ should finally lead to a discrete stationary state, because in higher approximations the state with higher mass could probably disintegrate into the lower state and a pion. Therefore it is natural to take the lower of the two states as representative of the stable baryons (of spin $\frac{1}{2}$ and isospin $\frac{1}{2}$) and to postulate, that this mass should be identical with the mass κ defined by the pole of the approximate 2-point function 3(39). This identification would in a very first approximation solve the non-linear problem discussed in §4-5.

In this case the lower point of intersection should be at $\lambda = 1$. The value of $L(1)$ is $-0\cdot6247$, hence $(\kappa l/2\pi) = 1\cdot02$. From an empirical point of view, the mass ratios between pion, η-meson and nucleon would, according to §5-1, require a somewhat lower value around $(\kappa l/2\pi) \approx 0\cdot9$; still the approximate agreement should be considered as quite sufficient, since we have to do here with the lowest approximation.

If in this approximation $F(p)$ is taken from 3(39) and $\mathscr{L}(p^2)$ in the non-linear equation 4(18) is identified with $-\frac{3}{2}(\kappa l/2\pi)^4 L(\lambda)$ of 5(29), then $\mathscr{M}(p^2)$ of 4(18) becomes:

$$\mathscr{M}(p^2) = \frac{i\kappa^4\left[1+\frac{3}{2}\left(\frac{\kappa l}{2\pi}\right)^4 L\left(\frac{p^2}{\kappa^2}\right)\right]}{p^2(-p^2+\kappa^2)} \qquad 5(30)$$

This function (which has no pole at the point $p^2 = \kappa^2$) does not agree with the right-hand side of 4(30), but it may be qualitatively similar. The present approximation 5(27) and 5(30) is more primitive than the non-linear equations 4(29) and 4(30).

5-4 Higher approximations

The problem of the higher approximations presents itself in two different forms. On the one hand one may ask, how the calculations 5(19), 5(20) and 5(29) for the simplest eigenstates could be improved. On the other hand there are many eigenstates which cannot be calculated at all by approximations of the type shown in §5-1 or §5-3. The first question may be answered by arguing that at least, in principle, the higher Tamm–Dancoff approximations—consisting of sets of linear integral equations—can be solved by the generalized Fredholm-formalism and that the approximate eigenvalues will probably converge to the correct values, like in the case of the anharmonic oscillator. If the new Tamm–Dancoff method in its present form should not converge, one could probably find other truncated sets of functions or other sets of linear integral equations leading to a convergent procedure. But practically in the present state of the theory these higher approximations cannot be calculated on account of their complication. Therefore it is not very important from a practical point of view to know whether the Tamm-Dancoff method converges or other methods should be preferred. The lowest approximations are nearly identical for many different procedures, and higher approximations cannot be calculated so far.

The only realistic way of improving the eigenvalue equations 5(19), 5(20) and 5(29) seems to be the addition of special terms resulting from higher graphs which for mathematical or physical reasons should be especially important. Such an addition can scarcely be extended to a systematic procedure; but it may be useful for explaining special features in the observations. Actually such calculations have been carried out both for the bosons of §5-1 and the fermions of §5-3. For the bosons, eigenvalue equations of the type

$$5(32)$$

have been investigated by Stumpf and Yamamoto.[55] While the first term is essentially equivalent to 5(6), the second term has a small influence on the mass of the pion, and lowers considerably (by 20–30 per cent) the mass of the η-meson. In the case of the fermions, terms have been included (in a paper by Dürr and Heisenberg)[40] which correspond to the creation and annihilation of virtual bosons; such terms would have appeared automatically, if the calculations had been started from the operator $\mathscr{L}$ of 4(29) instead of that of 5(27), since $\mathscr{L}$ in 4(29) contains the term . Graphically they may be represented by

$$5(33)$$

where $\sim\!\!\sim\!\!\sim$ stands for the regularized boson propagator. These terms produce two important changes. The value of $\kappa l/2\pi$ is lowered from 1·02 to about 0·9; and the creation and annihilation of K-mesons causes a splitting of the masses of the baryon pole into the N-, Λ-, Σ- and Ξ-eigenvalues. This splitting will be discussed in detail in Chapter 7 concerning the theory of strange particles. In both cases, for fermions and bosons, the corrections improve the agreement of the calculated masses with their observed values considerably.

There exist many other states which on account of their transformation properties cannot be treated by simple eigenvalue equations of the type 5(19), 5(20) or 5(29). Bosons in which the constituting fermions are not in an S-state (in which the orbital angular momentum is not zero) or fermions with a spin $\frac{3}{2}$ or more require a more complicated ϕ-function from the very beginning. Already vector mesons like ω, ϕ, ρ, K^* may to a large part consist of a d-state (orbital momentum 2) of the fermion–antifermion system and a spin 1 in opposite direction and may therefore belong to this group. In these cases already the lowest relevant eigenvalue equation resulting from the Tamm–Dancoff formalism is a genuine integral equation which cannot be reduced to an algebraic equation for the mass. Numerical methods may some day lead to approximate eigenvalues; but so far no serious attempts have been made to solve such integral equations.

5-5 Parity of the eigenstates

The calculations of this chapter have been carried out with the 2-dimensional representation of the Lorentz group in 3(1). Therefore in this scheme only the operation GP is defined, but not G and P separately. The transformation properties of the calculated eigenstates under GP can be easily determined. The fermions are changed into the corresponding antifermions by GP, hence for the fermions the question of a symmetry under this operation does not arise. For the bosons (π and η) one concludes from 3(14) and 3(15), that under GP, applied on the eigenfunction of the boson at rest $[J^\mu = (\kappa_B, 0, 0, 0)]$

$$\langle 0|T\chi^*(0)\,\sigma^0\chi(0)|B\rangle \;\rightarrow\; \langle 0|T\chi(0)\,R^{-1}\sigma^0 R\chi^*(0)|B\rangle$$

$$= -\langle 0|T\chi^*(0)\,\sigma^0\chi(0)|B\rangle \quad 5(34)$$

$$\langle 0|T\chi^*(0)\,\sigma^0\tau_k\chi(0)|B\rangle \;\rightarrow\; \langle 0|T\chi(0)\,R^{-1}\sigma^0\tau_k R\chi^*(0)|B\rangle$$

$$= +\langle 0|T\chi^*(0)\,\sigma^0\tau_k\chi(0)|B\rangle. \quad 5(35)$$

Therefore the η-meson has $GP = -1$, the pion $GP = +1$.

The separate operations P and G can be defined only by using the 4-dimensional representation of the Lorentz group in 3(22). If finite mass eigenvalues exist for 3(1), one can—as has been pointed out in §3-2—construct the operator $\tilde{\chi}(x)$, fulfill the equation 3(17) with the same mass eigenvalues and connect $\chi(x)$ and $\tilde{\chi}(x)$ by 3(20). The two sets of eigenvalue equations, using $\chi(x)$ or $\tilde{\chi}(x)$, can be combined into one set by starting from 3(22) and the 2-point function 3(40) or (42). It is then a matter of definition—and, as we see, of consistent definition—to assume, that the baryon pole occurs in 3(40) only once, namely in $\rho_1(\kappa^2)$, but not in $\rho_2(\kappa^2)$. The corresponding baryon can be defined as having $P = +1$; then the corresponding antibaryon must have $P = -1$.

The combination of the two sets of eigenvalue equations in one set by means of 3(20), 3(40) or 3(42) does not yet establish a connexion between $\langle 0|T\chi^*(x)\,\sigma^\nu\chi(x)|B\rangle$ and $\langle 0|T\tilde{\chi}^*(x)\,\overline{\sigma^\nu}\tilde{\chi}(x)|B\rangle$, which would decide about the parity of the bosons.

Using a physical argument one may, however, say that an eigenfunction of the type 5(1) with two equal co-ordinates describes a system consisting of a baryon and an antibaryon in an S-state (orbital angular momentum zero); hence its parity must be negative $P = -1$. Mathematically this result can be derived from connecting the matrix element $\langle 0|T\chi^*(x)\tilde{\chi}(x)|B\rangle$ both with $\langle 0|T\chi^*(x)\sigma^0\chi(x)|B\rangle$ and $\langle 0|T\tilde{\chi}^*(x)\sigma^0\tilde{\chi}(x)|B\rangle$ through the 2-point function 3(40) or 3(42). Following the procedure of 4(21) we can replace in $\langle 0|T\chi^*(x)\tilde{\chi}(x)|B\rangle$ either $\chi^*(x)$ or $\tilde{\chi}(x)$ by an expression of the type $\int d^4x'\, G(xx'):\sigma^\nu\chi(\chi^*\sigma_\nu\chi):$.

In a first approximation the resulting expressions can be reduced to matrix elements with only two field operators by using the 2-point function

$$\hat{F}(xy) = \langle 0|T\chi^*(x)\,\tilde{\chi}(y)|0\rangle \qquad\qquad 5(36)$$

which can be taken from 3(40) or 3(42). Then for the boson at rest one gets the two equations

$$\langle 0|T\chi^*(x)\,\tilde{\chi}(x)|B\rangle = \int d^4x'\, K_0(xx')\langle 0|T\chi^*(x')\,\sigma^0\chi(x')|B\rangle,$$

$$\langle 0|T\chi^*(x)\,\tilde{\chi}(x)|B\rangle = \int d^4x'\, \tilde{K}_0(xx')\langle 0|T\tilde{\chi}^*(x')\,\sigma^0\tilde{\chi}(x')|B\rangle. \quad 5(37)$$

The integral operators K_0 and $\tilde{K}_0$ are essentially given by

$$K_0 = \text{const}\, tr(\sigma_0\, G(xx')\, \hat{F}(x'\,x)\ldots)$$

$$\tilde{K}_0 = \text{const}\, tr(\sigma_0\, G(x'\,x)\, \hat{F}(xx')\ldots) \qquad\qquad 5(38)$$

with the same constant. From the general properties of G and $\hat{F}$ (compare A(13) in the appendix) one concludes $G(xx') = -G(x'\,x)$; $\hat{F}(xx') = \hat{F}(x'\,x)$ and $\tilde{K}_0 = -K_0$. Therefore the equations 5(37) finally require

$$\langle 0|T\chi^*(x)\,\sigma^0\chi(x)|B\rangle = -\langle 0|T\tilde{\chi}^*(x)\,\sigma^0\tilde{\chi}(x)|B\rangle \qquad 5(39)$$

which is equivalent to $P = -1$. The essential point in this derivation is the assumption, that in the expressions 3(40) or 3(42)

$$\hat{F}(xy) \neq 0$$

or

$$\rho_1(\kappa^2) \neq \rho_2(\kappa^2). \qquad\qquad 5(40)$$

It is only when the fermions of finite mass have a definite parity by definition, that the parity of the bosons can be determined. Finally the parity of π- and η-meson and, as will be seen later, also of the K-meson comes out as $P = -1$.

6

Theory of Coupling Constants

6-1 Preliminary remarks and definition of coupling constants

In a collision between two particles many different processes may be observed, like scattering, creation of new particles, transmutation of particles into others, etc. The only restrictions are the laws of conservation; i.e. the total symmetry of the incoming particles must be identical with the total symmetry of the outgoing particles. The probability of a given process can be determined from the asymptotic behaviour of Green's functions, as discussed in §4-5. From a practical point of view it has frequently been found convenient to subdivide the process concerned into less complicated virtual processes, which may be said to happen one after the other. As an example we mention the creation of three pions in a collision between a pion and a proton. A possible division would be:

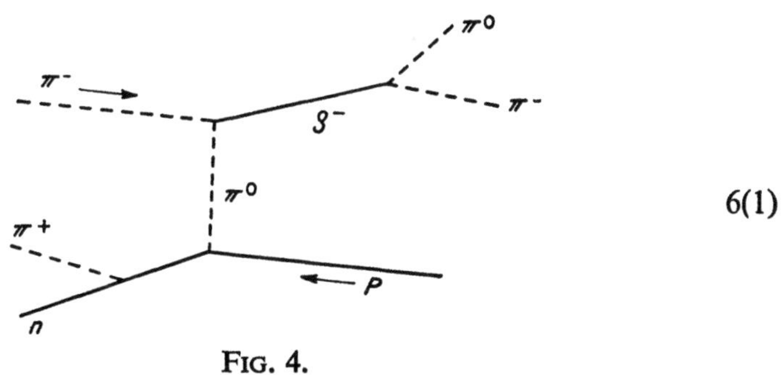

$$6(1)$$

FIG. 4.

This subdivision is by no means unique. Another possibility is shown in 6(2) on p. 80.

There may be as a rule many 'channels' by which the process can take place. The advantage of the subdivision lies in the fact, that one (or a few) of these patterns may play the dominant rôle in the complete S-matrix element. 'Dominant rôle' means, that one gets a reasonable approximation to the

79

 Unified Field Theory of Elementary Particles

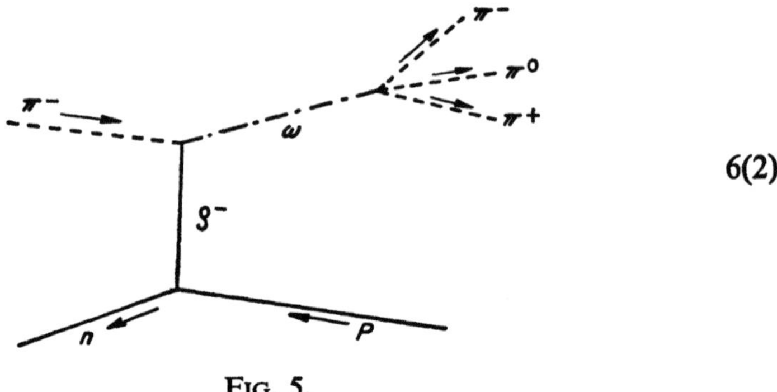

$$6(2)$$

FIG. 5.

S-matrix element by calculating only the 'contribution' of this one pattern. This contribution can be estimated, if the S-matrix elements of the more primitive processes contained in the pattern are known. On the other hand, the assumption that the S-matrix is the sum of contributions of single patterns, is certainly not correct. In the process which had been used as an example the pattern

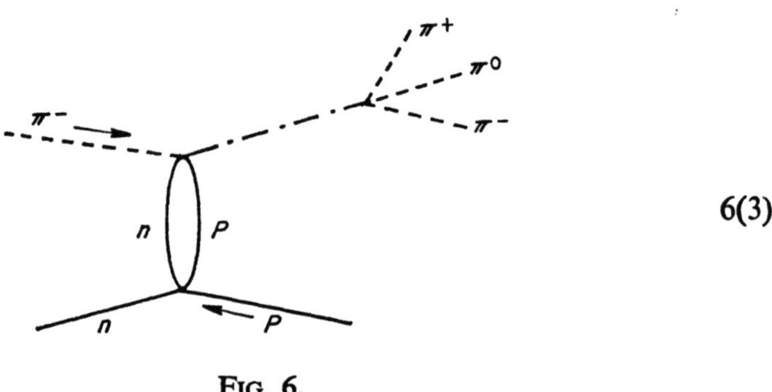

$$6(3)$$

FIG. 6.

would already contain the pattern 6(2), its contribution could not simply be added to the contribution of 6(2). Therefore the concept of a 'dominant' contribution is justified only in the neighbourhood of the poles, i.e. if the energy-momentum vector for a given line is so close to the pole in the corresponding 2-point function (if it is 'nearly on the mass-shell'), that this term is much bigger than any other contribution. Still, this pole-approximation of S-matrix elements is a useful tool for the phenomenological analysis of the experiments. Therefore some knowledge about the simplest S-matrix elements can be very helpful for the investigation of the more complicated processes. The most primitive S-matrix element would be one containing only three 'free field'-operators, for example, one incoming and two outgoing fields like

$$\langle 0|\psi_{\text{in}}^{(1)}(x_1)\,\psi_{\text{out}}^{(2)}(x_2)\,\psi_{\text{out}}^{(3)}(x_3)|0\rangle \qquad 6(4)$$

Such S-matrix elements do not exist in a strict mathematical sense. Their physical interpretation would be the desintegration of particle (1) into the two particles (2) and (3). If this process is possible, i.e. is compatible with all conservation laws, the particle (1) could not appear as asymptotic incoming field, since it has only a finite life time; if not, the matrix element would vanish. Still one can give a meaning to an expression of the type 6(4) by referring not to the S-matrix but to the τ-functions of the vacuum, from which the S-matrix elements can be derived by asymptotic processes; not to the free fields $\psi_{\text{in}}^{(1)}$, $\psi_{\text{out}}^{(2)}\ldots$ but to the operators $O_\lambda(x)$ discussed in §2-3. The behaviour of the matrix element

$$\langle 0|O^{(1)}(x_1)\,O^{(2)}(x_2)\,O^{(3)}(x_3)|0\rangle \qquad 6(5)$$

(where $O^{(1)}$ has the same transformation properties as $\psi_{\text{in}}^{(1)}$, etc.) may be studied in the neighbourhood of the poles, corresponding to the free fields $\psi_{\text{in}}^{(1)}$, $\psi_{\text{out}}^{(2)}$, $\psi_{\text{out}}^{(3)}$. If the mass κ_i characterizes a pole belonging to $O^{(i)}(x)$ [in the sense of 2(15)], the matrix element 6(5) will in momentum space contain a term of the form

$$\langle 0|O^{(1)}(p_1)\,O^{(2)}(p_2)\,O^{(3)}(p_3)|0\rangle$$

$$= \frac{c^{(123)}}{(p_1^2-\kappa_1^2)\,(p_2^2-\kappa_2^2)\,(p_3^2-\kappa_3^2)}\cdot\delta^{(4)}(p_1+p_2+p_3)+\ldots, \qquad 6(6)$$

the other terms being less singular. The non-existence of the S-matrix elements 6(4) is seen from the fact, that one cannot find three real four-vectors p_1, p_2, p_3 fulfilling the equations

$$p_1^2-\kappa_1^2 = 0, \qquad p_2^2-\kappa_2^2 = 0, \qquad p_3^2-\kappa_3^2 = 0, \qquad p_1+p_2+p_3 = 0. \qquad 6(7)$$

(It would seem that the equations could be fulfilled, if particle (1) can actually decay into the particles (2) and (3); but in this case the pole $p_1^2-\kappa_1^2=0$ would not be on the real axis, κ_1 would contain an imaginary part.) The equations 6(7) can, however, be satisfied by complex four vectors which may be reached by analytical continuation. Therefore $c^{(123)}$ is a well-defined number, depending still upon the symmetries and the masses of the three particles. The constant $c^{(123)}$ is proportional to the amounts of free field operators which are contained in the operators $O^{(i)}$. If the operator $O^{(i)}(x)$ obeys the equation 2(15), this means that it tends asymptotically to

$c_i \psi^{(i)}_{\text{in/out}}(x)$, where the value of c_i can be deduced immediately from 2(15); the free fields $\psi^{(i)}$ are defined as conventionally normalized. Therefore

$$f^{(123)} = \frac{c^{(123)}}{c_1 c_2 c_3} \tag{6(8)}$$

is essentially an information about the interaction of the three particles, which is independent of the special operators $O^{(i)}$ used for the representation of the particles. $f^{(123)}$ may be called the coupling constant between the three fields and is (except for trivial numerical factors) equivalent to the coupling constant in older phenomenological theories.

Historically the coupling constant had been defined by starting from a Lagrangian for the free fields, to which in a somewhat rash manner an interaction term had been added as 'small perturbation'. The first order term in the perturbation would then lead to an expression of the type 6(6), as can be seen in the following way:

The Lagrangian may (for real fields ϕ_i) be defined by

$$\mathscr{L} = \int d^4x \left[\tfrac{1}{2} \sum_1^3 \phi_i (\Box + \kappa_i^2) \phi_i + f \phi_1 \phi_2 \phi_3 \right] \tag{6(9)}$$

From $(\Box + \kappa_1^2)\phi_1 = -f\phi_2\phi_3$ one gets

$$\phi_1(x) = \phi_1(x)_0 - f \int d^4x'\, G(xx')\, \phi_2(x')\, \phi_3(x'), \tag{6(10)}$$

where $G(xx')$ is the Green's function belonging to $(\Box + \kappa_1^2)\phi_1 = 0$, and

$$\langle 0 | \phi_1(x_1)\, \phi_2(x_2)\, \phi_3(x_3) | 0 \rangle$$

$$= -f \int d^4x'\, G(xx') \langle 0 | \phi_2(x')\, \phi_3(x')\, \phi_2(x_2)\, \phi_3(x_3) | 0 \rangle + \dots$$

$$\approx -f \int d^4x'\, G(xx') \langle 0 | \phi_2(x')\, \phi_2(x_2) | 0 \rangle$$

$$\times \langle 0 | \phi_3(x')\, \phi_3(x_3) | 0 \rangle + \dots \tag{6(11)}$$

which in momentum space is essentially

$$\frac{f}{(p_1^2 - \kappa_1^2)(p_2^2 - \kappa_2^2)(p_3^2 - \kappa_3^2)} \cdot \delta^{(4)}(p_1 + p_2 + p_3). \tag{6(12)}$$

However, this whole procedure is not justified in quantum field theory, since for the operators of free fields there cannot be any interaction; an interaction term $f\phi_1\phi_2\phi_3$ would change the masses, it could certainly not be considered as small perturbation. Formulae of the type 6(9)–6(12) do *not* follow from a fundamental field equation like 3(1). But a coupling constant may be defined by 6(6) and 6(8), if an analytical continuation of the matrix elements to complex momentum variables is possible.

If coupling constants shall be determined experimentally, it can be done by extrapolating the measured curves for certain cross-sections to the pole. The extrapolation replaces the analytical continuation. This procedure will lead to reliable values only if one can come rather close to momenta fulfilling 6(7); for example, if two masses are equal and the last mass is very small. This is the main reason why at present only the coupling of light quanta to all kinds of elementary particles and the pion–nucleon coupling are well known.

6-2 Coupling of π- and η-mesons to baryons [53, 56]

A simple way of calculating the coupling constant of π- or η-mesons to nucleons (or $\varXi$-particles) consists in comparing the 4-point function of the fermions in the neighbourhood of the boson pole with the corresponding quantity in a phenomenological theory.

In a theory of this latter type the free fermions could be represented by two Weyl spinors $\chi(x)$ and $\tilde{\chi}(x)$ (since the fermions have a finite mass) or by one Dirac spinor $\psi(x)$, the free boson by a pseudoscalar isoscalar $\Phi(x)$, or a pseudoscalar isovector $\Phi_i(\lambda)$. Since we are in this connexion not interested in the mass of the fermions or the 'dressing' of the fermion lines, it will be sufficient to discuss a 2-component Weyl equation of the type

$$i\sigma^\nu \frac{\partial \chi}{\partial x^\nu} + \frac{f}{\kappa_B}\sigma^\nu \frac{\partial \Phi}{\partial x^\nu}\cdot\chi + \ldots = 0 \quad \text{for isospin 0} \qquad 6(13)$$

or,

$$i\sigma^\nu \frac{\partial \chi}{\partial x^\nu} + \frac{f}{\kappa_B}\cdot\tau_k \sigma^\nu \frac{\partial \Phi_k}{\partial x^\nu}\chi + \ldots = 0 \quad \text{for isospin 1.} \qquad 6(14)$$

κ_B is the boson mass, f the phenomenological (pseudo) vector coupling constant. The relevant part of the 4-point function

$$\langle 0|T\chi(x_1)\,\chi(x_2)\,\chi^*(x_3)\,\chi^*(x_4)|0\rangle$$

of the phenomenological theory could then in the lowest Tamm–Dancoff approximation and in the neighbourhood of the boson pole be represented by

$$6(15)$$

                           Unified Field Theory of Elementary Particles

where $\sim\!\sim\!\sim$ stands for the boson propagator. This expression is to be compared with the corresponding expression 4(27) or 4(35) of the unified field theory

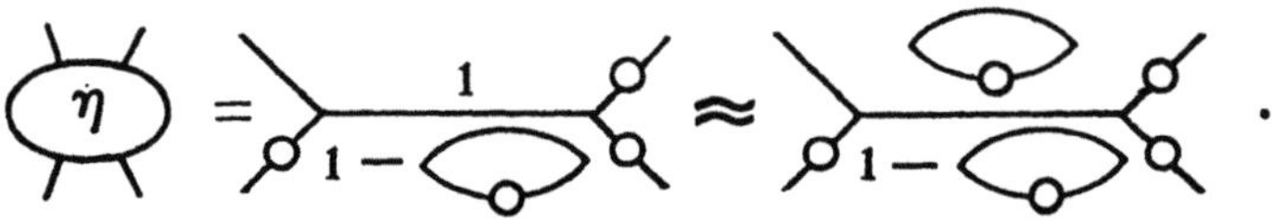

$$ \left(\eta\right) = \cdots \approx \cdots \qquad . \qquad 6(16) $$

The two pictures 6(15) and 6(16) are nearly equivalent with regard to the fermion lines, with the only difference that one G-line in 6(15) is replaced by an F-line in 6(16). Since the 'dressing' of the fermions (which has no meaning in the phenomenological formulation) shall not be discussed in this connexion, the difference is unimportant. The boson line $\sim\!\sim\!\sim$ is replaced by $\dfrac{1}{1-\bigcirc}$ in the non-linear theory, and the vertices in this theory are those of 3(1) or more explicitly 5(7).

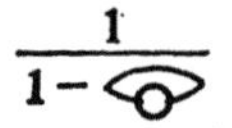

The relevant part of the graph 6(15) can therefore be written as

$$ \frac{f.(iJ_\mu/\kappa_B).(-iJ_\nu/\kappa_B)f}{J^2-\kappa_B^2} = f^2\frac{(J_\mu J_\nu)/(\kappa_B^2)}{J^2-\kappa_B^2} \qquad 6(17) $$

The corresponding part of the graph 6(16) is

$$ \frac{Z.l^2.(J_\mu J_\nu)/(J^2)}{1+Z(\kappa l/2\pi)^2 q_0(\lambda)}, \qquad 6(18) $$

where $Z=\tfrac{3}{4}$ for isospin 0 and $Z=\tfrac{1}{4}$ for isospin 1. $(J_\mu J_\nu)/(J^2)$ is the projection operator for Dirac spin 0 (compare 5(18)). By expanding the denominator in the neighbourhood of the pole one gets

$$ \frac{Zl^2.(J_\mu J_\nu)/(J^2)}{1-1+Z(\kappa l/2\pi)^2(\partial q_0/\partial\lambda)(J^2-\kappa_B^2)/(\kappa^2)} = 4\pi^2\frac{(J_\mu J_\nu)/(\kappa_B^2)}{(J^2-\kappa_B^2)(\partial q_0/\partial\lambda)}. \quad 6(19) $$

Comparing 6(17) and 6(19) gives

$$ f^2 = \frac{4\pi^2}{\partial q_0/\partial\lambda}. \qquad 6(20) $$

The quantity $\partial q_0/\partial\lambda$ is positive for the values of λ corresponding to the boson masses. This result confirms, that the norm of the boson states is positive, as had been discussed in §5-1, and as had been assumed in the phenomenological theory 6(13). For the numerical evaluation it is convenient to expand $q_0(\lambda)$ in 5(13) for small values of λ:

$$ q_0(\lambda) \approx \ln\lambda - \frac{3}{2} + \frac{\lambda}{3} + \dots \quad \text{for} \quad \lambda \ll 1. \qquad 6(21) $$

Theory of Coupling Constants 85

Hence

$$\frac{\partial q_0(\lambda)}{\partial \lambda} \approx \frac{1}{\lambda}+\frac{1}{3}+\ldots \quad \text{for} \quad \lambda \ll 1,$$

6(22)

and

$$\frac{f^2}{4\pi} \approx \pi\lambda\left(1-\frac{\lambda}{3}\right) = \pi\frac{\kappa_B^2}{\kappa^2}\left(1-\frac{\kappa_B^2}{3\kappa^2}\right).$$

6(23)

If the empirical mass of the pion is inserted for κ_B, the empirical mass of the nucleon for κ, one gets

$$\frac{f^2}{4\pi} = 0.07$$

6(24)

in satisfactory agreement with the empirical value 0·08.

In the phenomenological theory one could have assumed a pseudoscalar coupling instead of a pseudovector coupling. In this case the pseudoscalar coupling constant g is connected with the pseudovector coupling constant by the relation

$$g = \frac{2\kappa}{\kappa_B}\cdot f.$$

6(25)

Therefore the theory gives

$$\frac{g^2}{4\pi} = 4\pi\left(1-\frac{\kappa_B^2}{3\kappa^2}\right),$$

6(26)

$$\frac{g^2}{4\pi} = 12.5 \quad \text{for pions}$$

$$\frac{g^2}{4\pi} = 11.1 \quad \text{for } \eta\text{-mesons.}$$

(For purely historical reasons we have registered $f^2/4\pi$ and $g^2/4\pi$ instead of g or f.) The empirical values are 14·4 for the pion–nucleon coupling and 12·1 for the η-meson. The latter value is still very uncertain, since it has been derived rather indirectly from the empirical nucleon–nucleon scattering (Scotti and Wong).[57]

Similar calculations could be carried out for the eigenvalues 5(21) and 5(22) belonging to Dirac spin 1. In this case the phenomenological equations would read

$$i\sigma^\nu \frac{\partial \chi}{\partial x^\nu}+f\sigma^\nu \chi A_\nu+\ldots = 0$$

6(27)

 Unified Field Theory of Elementary Particles

and 6(15) should be replaced by

$$f^2 \frac{-g_{\mu\nu}+(J_\mu J_\nu)/(J^2)}{J^2-\kappa_B^2}. \qquad 6(28)$$

The comparison with the non-linear theory 3(1) would then imply

$$f^2 = -\frac{4\pi^2}{\partial q_1/\partial\lambda}. \qquad 6(29)$$

By 6(28) the sign has been reversed against 6(20). Since $\partial q_1/\partial\lambda$ is again positive for the eigenvalues 5(21) and 5(22), 6(29) cannot be fulfilled; this result indicates that the norm of such bosons with spin 1 would be negative. If in the phenomenological theory one would have assumed a negative norm of the bosons, the sign in 6(28) and 6(29) would be the reverse and 6(29) could be satisfied. The fact, that the norm of these states 5(21) and 5(22) is negative, had been mentioned already in §5-1. The discussion of the boundary conditions in §5-2 has led to the result, that these states cannot appear as incoming or outgoing particles, that they are in fact ghost states, not real particles.

6-3 The norm of the boson states

The expressions $\chi^*(x)\sigma_\mu\chi(x)$ contain matrix elements belonging to the creation or annihilation of bosons and correspond in this respect to the phenomenological operators $\partial\Phi/\partial x_\mu$; therefore one may put

$$\langle 0|\chi^*(x)\,\sigma^\mu\chi(x)|B\rangle = \frac{a}{l}\left\langle 0\left|\frac{\partial\Phi(x)}{\partial x_\mu}\right|B\right\rangle \qquad 6(30)$$

$$\langle 0|\chi^*(x)\,\sigma^\mu\tau_k\chi(x)|B\rangle = \frac{a}{l}\left\langle 0\left|\frac{\partial\Phi_k(x)}{\partial x_\mu}\right|B\right\rangle \qquad 6(31)$$

and ask what the value of the dimensionless constant a is, if $\Phi(x)$ is normalized in the conventional way. If in the interaction term of the fundamental equation 3(1) two operators $\chi^*(x)$ and $\chi(x)$ are combined to form a boson operator (this can be done in two ways) the comparison of 6(30) and 6(31) with 6(13) and 6(14) determines at once the constant a:

$$2Zl.a = \frac{f}{\kappa_B}, \qquad 6(32)$$

where again $Z=\frac{3}{4}$ for isospin 0, $Z=\frac{1}{4}$ for isospin 1. The factor 2 on the left-hand side is due to the two equivalent ways of combining $\chi^*(x)$ with $\chi(x)$.

From 6(20) and 6(32):

$$a^2 = \frac{\pi^2}{Z^2 l^2 \kappa_B^2 (\partial q_0/\partial\lambda)}.$$

6(33)

With the definition

$$\langle 0|T(\chi^*(x)\sigma_\mu\tau_\rho\chi(x)).(\chi^*(y)\sigma_\nu\tau_\sigma\chi(y))|0\rangle = \delta_{\rho\sigma}(2\pi)^{-4}\int d^4J\, e^{iJ(x-y)} G_{\mu\nu}(J)$$

6(34)

6(30), 6(31) and 6(33) lead to the pole approximation:

$$G_{\mu\nu}(J) \approx \frac{J_\mu J_\nu}{J^2}\frac{1}{4Zl^2(1-\langle\!\!\bigcirc\!\!\rangle)}.$$

6(35)

The relations 6(33) and (35) are essentially equivalent to similar relations derived in the literature (for example, Cutkosky and Leon)[58] as consequence of the Bethe–Salpeter equation. 6(35) can be obtained immediately from 6(16) by referring the outer lines on both sides to the same point:

6(36)

The parts  and  are similar to the kernel  except for the following differences: The two ends are not connected to form a vertex of 3(1); therefore one would have to add a factor Zl^2, if one wanted to replace the two parts  and  by factors  . Also one should multiply by a factor 2 on both sides for this purpose, since in  one has taken out only one combination $\chi^*(x)$ and $\chi(x)$, while in the kernel  one always adds the two equivalent combinations. Therefore the complete factor is $4Zl^2$, which appears in the denominator of 6(35); the kernel  can in the neighbourhood of the pole be replaced by unity. There is still the difference between the lines and on the right-hand side, which presumably should become unimportant in higher approximations.

The equations 4(27) and 6(36) together establish a connexion between the 2-point functions of the fermions, i.e. the normalization of the fermion wave functions, and the normalization of the boson wave functions. In unrelativistic quantum mechanics such a connexion can be seen immediately. The complete separation between particles and forces in this theory allows replacement of the boson at a certain time by a wave packet consisting of

 Unified Field Theory of Elementary Particles

two fermions without any interactions. The interaction influences the motion of the fermions later on, it binds them together, but it is irrelevant for the norm of the boson, which must be identical with the norm of the wave packet. In relativistic field theory a separation between particles and forces is impossible, therefore the connexion is more complicated. Still the method of calculating the boson norm in quantum mechanics should have its counterpart in relativistic field theory, at least within a certain approximation. Calculations of this kind have been carried out successfully by Yamazaki[56] in the case of π- and η-mesons.

6-4 Some general remarks about the 'pole'-approximation of S-matrix elements

The calculations of the previous sections allow some generalizations concerning more complicated S-matrix elements. Let us consider the case of two incoming and n outgoing particles. The S-matrix element for this process can be derived from the vacuum expectation value

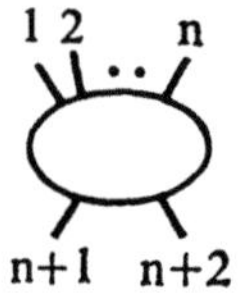

$$= \langle 0|TO_1(x_1)\,O_2(x_2)\ldots O_n(x_n)\cdot O_{n+1}(x_{n+1})\cdot O_{n+2}(x_{n+2})|0\rangle.$$

$$6(37)$$

The operators $O_{n+1}(x_{n+1})$ and $O_{n+2}(x_{n+2})$ must have the symmetries of the two incoming particles, the operators $O_1(x_1)\ldots O_n(x_n)$ those of the outgoing particles. The S-matrix is obtained by studying the asymptotic behaviour of  in the limit $t_{n+1},\,t_{n+2} \to -\infty$, $t_1,\,t_2\ldots t_n \to +\infty$. The normalization of the operators can be taken from their 2-point functions 2(15). If one goes over from co-ordinate to momentum space and hence from $O_n(x_n)$ to its Fourier transform $O_n(p_n)$, 6(37) becomes in the neighbourhood of the relevant poles

$$= \frac{f(p_1\ldots p_{n+2})}{(p_1^2-\kappa_1^2)\,(p_2^2-\kappa_2^2)\ldots(p_{n+2}^2-\kappa_{n+2}^2)}\,\delta^{(4)}(p_1+p_2\ldots+p_{n+2}).\quad 6(38)$$

For the asymptotic behaviour, only the neighbourhood of the poles is decisive, irrespective of the direction of the limit $t \to +\infty$ or $t \to -\infty$. In this way the 'crossing symmetry' can be understood. For the form of 6(38) it makes no difference whether, for example, $O_n(x_n)$ represents asymptotically an outgoing particle of mass κ_n or an incoming antiparticle

Theory of Coupling Constants 89

of the same mass. The direction of the limit $t \to +\infty$ or $t \to -\infty$ will be important only for the path of integration in the complex $(p_n)_0$ space. Hence there will be a simple analytical connexion between the two S-matrix elements, defined by the outgoing particle κ_n or the incoming antiparticle κ_n.

The collision process described by ⬡ in 6(38) may partly be due to a force of rather long range, for example, a one-pion exchange or a one-κ-meson or one-η-meson exchange. This would mean, that the expression contains a pole term, for example, of the type

$$\text{⬡} \approx \text{⟩⟨} \; , \qquad\qquad 6(39)$$

where the line $\sim\!\sim\!\sim$ represents the exchanged boson. In this case the behaviour of ⬡ in the neighbourhood of the boson pole could again be derived from an inhomogeneous integral equation

$$\text{⬡} = \text{inhom. part} + \text{⬡⬡}_\kappa \; , \qquad\qquad 6(40)$$

where the kernel ⊘$_\kappa$ could in a first approximation again be replaced by ⤫ of 4(27), and the inhomogeneous part would be the simplest Feynman diagram leading to ⬡.

Experimentally, the pole term can be seen[59] only if the mass of the exchanged particle and the exchange of momentum are rather small and if other neighbouring masses can be excluded by conditions of symmetry. In other cases it will scarcely be possible to single out the contributions of one pole from the statistical mixture of the contributions of all other possible channels. Hence for the analysis of the experiments the pole approximation or 'one particle exchange model' has only a limited value.

7

Strange Particles

7-1 Degeneracy of the ground state

Following the discussions of the Chapters 1 and 3, the deviations from isospin-symmetry in electrodynamics shall be taken as indication for an asymmetry of the ground state.[29] In fact the number of protons in the world seems to be very different from the number of neutrons, the number of electrons is very different from the number of neutrinos. Even if matter and antimatter should be distributed in the universe with equal average density—many galaxies might consist of matter, equally many of anti-matter—and if the total isospin should be small in this way, the big asymmetry would remain, since in matter the total isospin would point in one direction, in antimatter in the opposite direction. Hence there would be a macroscopic deviation from symmetry in isospace.

An asymmetry of the ground state and therefore a degeneracy of this state is a well-known phenomenon in many systems discussed in conventional quantum mechanics. Ferromagnetism, superfluidity, superconductivity, crystal structure are obvious examples. In such cases two important new phenomena appear[29] which shall be discussed first qualitatively from a physical point of view before going into the mathematical analysis: The degeneracy of the ground state enforces the existence of bosons of rest mass zero, as has been pointed out in a mathematical form by Goldstone.[41] Some property of the ground state can be attached to the particles thereby changing normal particles into strange particles.[40]

The theorem of Goldstone which will be discussed in a somewhat generalized form can be understood from the following argument. If the ground state is degenerate, this means that there must be a transformation which leaves the Hamiltonian invariant, but changes the ground state into a different possible ground state. The generator of this transformation is an observable for which the invariance of the Hamiltonian (or fundamental field equation) provides a conservation law. If a process, for example, a collision between two particles, takes place within the system, the con-

90

servation law may seem to be violated, since part of the quantity measuring the observable may be transferred upon the system as a whole, i.e. finally upon the ground state. If the theory is local—a very general kind of locality is sufficient—the transferred quantity will immediately after the process be located in the neighbourhood of the point, where the collision has taken place, and will be spread afterwards by a kind of spherical wave, comparable to the waves on the surface of a pond after a stone has been dropped into the water. These waves resemble 'classical' waves, since the conserved quantity is an observable, and therefore the waves are of the Bose type. In the limit of extremely large wave-length the motion produced by the wave goes over into the assumed transformation of the ground state. Since this transformation does not require energy, the frequency of the wave will decrease as the wave-length increases. This behaviour corresponds to a wave motion which in a relativistic theory is connected with particles of rest mass zero.

The waves or bosons of mass zero will physically restore the conservation law; but they will not necessarily appear to do it, since they will as a rule not have a well-defined transformation property under the transformation concerned. The asymmetry of the ground state will produce these bosons as mixed states ('schizons')[60] composed of parts with different symmetries. Therefore in all experiments in which these bosons are absorbed or emitted the conservation law may seem to be violated.

On the other hand any particle will now be surrounded by a cloud of virtual Goldstone bosons (for example, any charged particle is surrounded by a Coulomb field), therefore any emission of a particle will as a rule be connected with the emission of 'infra-red' Goldstone bosons, a phenomenon described in detail by Bloch and Nordsieck[17] for the case of electrodynamics. It is only in an approximation in which all these effects can be considered as negligible, the masses of the infra-red bosons be included in the masses of the particles, that the conservation law can be checked experimentally.

The appearance of the strange particles can be understood from the interaction between single particles and the physical attributes of the ground state. This interaction will first lead to a splitting of mass eigenvalues which had been equal without the interaction. In a ferromagnet, for example, the energy of an excited electron will depend on the direction of its spin with respect to the direction of the total magnetic moment. If the interaction becomes big, it may lead to a strong coupling of the Russel–Saunders[61] type between the spin of the excited electron and the spin of the nearest electron in the ground state. In this case the excited electron would appear to have the spin 1 or 0 instead of its natural spin $\frac{1}{2}$. In a similar way strange particles may be created by attaching an isospin $\frac{1}{2}$ or 1 from the ground state to a normal particle.

Finally the degeneracy of the ground state may produce some change in the group theoretical structure of the theory by means of the new degrees of freedom introduced in the ground state. In the present theory this change can be seen at two points: the breaking of the scale transformation by the introduction of a finite mass pole in the 2-point function of the ground state gives room for the space reflection parity as a new operation. The breaking of the isospin group by the photons and spurions allows the separation of the gauge group 3(12) into two gauge groups, one referring to Lorentz space, the other one to isospace.

It may be convenient to discuss the two essential features of a degeneracy of the ground state, the Goldstone-bosons and the strange particles, with the help of a simple model of conventional quantum mechanics. Therefore we will, in the next section, treat a simplified model of a ferromagnetic body for demonstrating and analysing these features.

7-2 Comparison with a simplified model of ferromagnetism [62]

Let us consider a linear chain of N atoms, the chain being bent to a ring so that atom no. N and atom no. 1 are neighbours. The electrons are assumed to be fixed at their atoms, one electron at each atom; the spins of two neighbouring electrons may have an interaction tending to put these spins parallel. Such a system would not yet be a ferromagnet in the thermodynamical sense, but in its lowest state all magnets would be parallel, and that is sufficient for the present purpose. The ground state is $(N+1)$-fold degenerate, the projection of the total spin on the z-axis may have the values

$$-\frac{N}{2}, -\frac{N}{2}+1, \ldots +\frac{N}{2}-1, \frac{N}{2}.$$

The state of the system can so far be characterized by notifying the spin-direction of each single atom. Besides these electrons in their lowest states there may be an excited electron in a higher state (no holes are assumed in the lowest states); some overlapping of its wave functions at atom n and atom $n+1$ is assumed, the excited electron may therefore move around in the ring. The spin of this electron interacts with the spin of the electron in the lowest state at the same atom. For a description of the presence of this excited electron we introduce the operators a_n^* for the creation, a_n for the annihilation of such an electron at the nth atom $(a_m^{\lambda} a_n^{*\mu} + a_n^{*\mu} a_m^{\lambda} = \delta_{nm} \cdot \delta_{\lambda\mu})$; $\lambda = +$ or $-$ denotes the direction of its spin.

The Hamiltonian of the system may then be written as

$$H = N\alpha - \alpha \sum_{1}^{N} \sigma_n \sigma_{n+1} + \beta \sum_{1}^{N} a_n^{*\lambda} a_n^{\lambda}$$

$$- \sum_{1}^{N} (a_n^{*\lambda} a_{n+1}^{\lambda} + a_{n+1}^{*\lambda} a_n^{\lambda})(\gamma + \delta \sigma_n \sigma_{n+1}) \qquad 7(1)$$

$$- \epsilon \sum_{1}^{N} a_n^{*\lambda} \sigma_{\lambda\mu} a_n^{\mu} \cdot \sigma_n.$$

σ_n are the Pauli matrices acting on the spin of the electron in the ground state of atom n, σ refers to the excited electron. Among the $N+1$ degenerate possible ground states we define as $|0\rangle$ the state, where the projection on the z-axis of the total spin is $N/2$.

We first consider those states of the system, in which there is no excited electron. The state vector

$$|n\rangle = \tfrac{1}{2}(\sigma_{n,x} - i\sigma_{n,y})|0\rangle \qquad 7(2)$$

would describe a non-stationary state of the system in which all spins are in the $+z$-direction except the one at atom n which has the opposite direction. A stationary state can then be obtained by putting

$$|k\rangle = \frac{1}{\sqrt{N}} \sum_{1}^{N} e^{ikn}|n\rangle; \qquad k = \frac{2\pi}{N}r \quad r = 1, 2, \ldots \qquad 7(3)$$

The energy of this state can be evaluated from 7(1):

$$E_k = 4\alpha(1 - \cos k), \qquad 7(4)$$

while the energy of the ground state is zero: $E=0$ for $|0\rangle$. $|k\rangle$ describes the presence of a spin wave of the type first discussed by Bloch.[43] k may be called the momentum of the spin wave; the energy goes to zero for $k \to 0$, as is required by the Goldstone theorem. Since our model is essentially unrelativistic, we cannot expect a boson of rest mass zero, but we have to expect a state fulfilling the condition $E_k \to 0$ for $k \to 0$. The spin waves of Bloch hence play the rôle of the Goldstone waves. The state $|k\rangle$ with $k=0$ is *not* identical with $|0\rangle$; but it is one of the other possible ground states, where the projection of the total spin momentum on the z-axis is $(N/2)-1$. One may call it a 'spurion' state, if one uses this term in a general way for any state not different from the ground state in energy and momentum, but different from it with respect to other properties, in our special case the projection of the spin momentum on the z-axis. The Bloch waves may seem to carry the spin 1 according to 7(2), but it is important to notice that this spin must always have a direction opposite to the total magnetic moment, which has been assumed to point in the z-direction. The Bloch wave state $|k\rangle$ may also be written in the form

$$|k\rangle = \sum_1^N e^{ikn} \cdot \tfrac{1}{2}(1-\sigma_z)_n|0'\rangle, \qquad 7(5)$$

where $|0'\rangle$ is the 'spurion' state $|k=0\rangle$. In this form the Bloch wave does not seem to carry z-spin, but to be a mixture of a state of spin 0 and spin 1, the projection of the spin on the z-axis being zero.

If an electron in an excited state is present, the two cases of small spin interaction ($\epsilon \ll \alpha, \gamma, \delta$) and of strong spin interaction ($\epsilon \gg \alpha, \gamma, \delta$) should be treated separately.

If $\epsilon \ll \alpha, \gamma, \delta$, the state vector of a stationary state can be written approximately as

$$|e_k^\lambda\rangle \approx \frac{1}{\sqrt{N}} \sum_1^N e^{ikn} a_n^{*\lambda}|0\rangle. \qquad 7(6)$$

The energy is different for the two states $\lambda = +$ and $-$ and is given by $7(1)$:

$$E_k^\pm \approx \beta - 2\gamma \cos k \mp \epsilon. \qquad 7(7)$$

Hence there is a doublet energy band, the excited electron has the spin $\tfrac{1}{2}$ and the two levels are split by the energy difference 2ϵ, thereby indicating that the symmetry of the Hamiltonian under the rotation of all spins seems to be broken by the ground state. Besides the excited electron there could also be Bloch spin waves; and in higher approximations one could see that the excited electron is already dressed up by spin waves, even if no separate spin waves are present; but these higher approximations shall not be discussed in this context.

If $\epsilon \gg \alpha, \gamma, \delta$ one gets approximate state vectors of stationary states for a triplet and a singlet.

Triplet:

$$|e_k^{+1}\rangle \approx \frac{1}{\sqrt{N}} \sum_1^N e^{ikn} a_n^{*+}|0\rangle,$$

$$|e_k^0\rangle \approx \frac{1}{\sqrt{2N}} \sum_1^N e^{ikn}(a_n^{*-}|0\rangle + a_n^{*+}|n\rangle) \qquad 7(8)$$

$$|e_k^{-1}\rangle \approx \frac{1}{\sqrt{N}} \sum_1^N e^{ikn} a_n^{*-}|n\rangle.$$

Singlet:

$$|e_k^s\rangle \approx \frac{1}{\sqrt{2N}} \sum_1^N e^{ikn}(a_n^{*-}|0\rangle - a_n^{*+}|n\rangle).$$

7(9)

The corresponding energies are, according to 7(1):
Triplet:

$$E_k^{+1} \approx \beta - (2\gamma + 2\delta)\cos k - \epsilon$$

$$E_k^0 \approx \beta + 2\alpha - (\gamma + 3\delta)\cos k - \epsilon$$

$$E_k^{-1} \approx \beta + 4\alpha - 4\delta\cos k - \epsilon.$$

7(10)

Singlet:

$$E_k^s \approx \beta + 2\alpha - (\gamma + 3\delta)\cos k + 3\epsilon.$$

7(11)

In this approximation resonance effects between overlapping or neighbouring energy bands have been neglected. Therefore the formulae are valid only for certain ranges of the constants α, β, γ; for the triplet states the condition

$$|4\alpha^2 + 4\alpha(\gamma - \delta)\cos k + (\gamma - \delta)^2| \ll |4\alpha^2 - (\gamma + 3\delta)^2 \cos^2 k|$$

is required.

The resulting situation may be described by saying that there is a strange electron of spin 1 and another strange electron of spin 0, the two energy bands are separated by 4ϵ. Within the triplet band there is a splitting of the levels depending on the angle between the spin 1 of the strange electron and the magnetic moment of the ground state.

Again, besides the excited electron, Bloch spin waves could be present, and in higher approximations the electron would be dressed up by Bloch spin waves. It should also be emphasized that the excited strange electron carries a normal charge; the electrons in the ground state do not participate in the transport of charge even if their spin seems to be carried with the excited electron.

The present model could also be used for discussing vacuum expectation values of products of operators; and if the operators a_n, a_n^* would be considered as time dependent operators $a_n(t)$ and $a_n^*(t')$ (changing over from the Schrödinger picture to the picture of time independent state vectors), the analogy to field theory would be rather close. These possibilities will, however, be explored only briefly and qualitatively in the present lectures, since they would lead into a considerable degree of complication. The difficulty arises from the question, which ground state should be considered for defining these 'ground state expectation values'. If only the state $|0\rangle$ would be used, the expectation value $\langle 0|a_m(t)a_n^*(t')|0\rangle$ would get contributions from the intermediate states $|e_k^{\pm}\rangle$ and $|e_k^{+1}\rangle$, $|e_k^0\rangle$, but not from $|e_k^{-1}\rangle$

in the first approximation, which therefore seems somewhat artificial. There would, however, be contributions to $\langle k|a_m(t)a_n^*(t')|k'\rangle$ from the two intermediate states $|e_l^0\rangle$, $|e_l^{-1}\rangle$. This would also be true for $k=0$ or $k'=0$. In higher approximations there would be contributions to all these expressions $\langle 0|\ldots|0\rangle$ or $\langle k|\ldots|k'\rangle$ from all the excited states, but it may be that each single contribution would tend to zero for $N\to\infty$. From a physical point of view this difficulty can be well understood. Physically one cannot well distinguish between an empty space and a space in which there are infra-red light quanta or Goldstone particles, the wave-length of which is larger than the space in which the experiment is carried out. Therefore it would seem reasonable to consider as ground state a mixture of states with an arbitrary number of infra-red light quanta or Goldstone particles. The poles in the 2-point functions would be changed into the starting points of cuts when the dressing of the particles is taken into account properly. The limit for the wave-length of the permitted infra-red Goldstone particles would, however, depend on the size of the region in space which one wants to consider. All these complications could probably be worked out on a solid mathematical basis for the present simple model, but they will not be discussed in this context.

Comparing this model with the real situation in the theory of elementary particles, one should first notice that the deviations from symmetry in isospace for the elementary particles are of a somewhat different nature compared with the deviations from rotational symmetry in the spin space of the model. The symmetry under particle–antiparticle conjugation is not present in the model. The ground state in elementary particle physics should therefore be pictured rather as consisting of matter with a big total isospin pointing in one direction, and of a nearly equal amount of anti-matter with an isospin pointing in the opposite direction. In some way anti-ferromagnetism might be a better model for this situation than ferro-magnetism. Therefore also the Goldstone particles should have a different symmetry in isospace from that of the Bloch waves in spin space.

Quite aside from this problem of a suitable model, the general question arises as to what extent the assumptions concerning the symmetry or asymmetry of the ground state are arbitrary and to what extent they are determined by the underlying fundamental equation. If the analogy with quantum mechanics of solid bodies is permitted for elementary particle physics, it seems to show, that in principle, several stable ground states may appear as solutions of the quantum mechanical equations. 'Stable' in this content means 'dynamically stable', not 'thermodynamically stable'. Therefore, the ground state cannot always be uniquely determined by the fundamental wave equation. On the other hand, the ground state can certainly not be arbitrary. The requirements of consistency may allow a few solutions of different symmetries for the ground state 'world', just as the quantum mechanical equations for the solid body allow a few stable solutions of different symmetries. But one would scarcely expect a continuous manifold of solutions. For a solid body it may be different if one allows forces from outside, for example, tensions exerted upon the boundaries of the body. But for the universe one could scarcely connect any meaning with the concept 'forces from outside'. Therefore the different possible ground states may in some way correspond to the different cosmological models in the theory of general relativity. Boundary conditions are essential for the determination of a cosmological model, and the same may be true for the ground state of elementary particle physics. Aside from these it can only be the postulate of internal consistency which decides about the possibility of a certain ground state.

7-3 The 'spurion'-formalism [40]

The discussions on the ferromagnetic model have, among other results, led to the conclusion that the concept of vacuum expectation values and generalized Green's functions becomes somewhat problematic, when the ground state is degenerate and when, therefore, Goldstone particles exist which may be added to the ground state in arbitrary number if only their energy is sufficiently small.

For calculating mass eigenvalues in a first approximation it should be possible to neglect these secondary effects of the photons and to use the technique of generalized Green's functions or of matrix elements $\langle 0|T\chi(x)\dots|\psi\rangle$, even for the strange particles. If in a vacuum expectation value of the type

$$\langle 0|\chi(x_1)\chi^*(x_2)\chi(y_1)\chi^*(y_2)\chi(y_3)\dots|0\rangle \qquad 7(12)$$

the co-ordinates x_1, x_2 are separated from the other co-ordinates y_1, $y_2,\dots$ by large space-like distances $(|x-y|\gg|x_1-x_2|)$, the remaining part will contain terms not only with the factor

$$\langle 0|\chi_\lambda(x_1)\chi^*_\mu(x_2)|0\rangle = \delta_{\lambda\mu}f(x_1-x_2), \qquad 7(13)$$

but also with a factor of the type:

$$\langle 0|\chi_\lambda(x_1)\chi^*_\mu(x_2)|J\rangle = g(x_1-x_2)\,\tau^k_{\lambda\mu}T^k. \qquad 7(14)$$

(λ,μ are isospin indices). The intermediate states $|J\rangle$ may in this approximation be counted as having energy and momentum zero and as being symmetric under the proper Lorentz group; T^k is connected with the isospin properties of this intermediate state.

T^k is a representation of the isospin group referring to some of the isospin indices of the operators $\chi(y_1)\chi^*(y_2)\ldots$, if the two vacua on the left and the right-hand side of 7(12) are identical and if the deviations from symmetry in isospace are neglected for these vacua. In this case the contribution of terms with the factor 7(14) would seem to establish a connexion between the isospin indices in the region x_1, x_2 and in the region y_1, y_2,…. But the contribution would decrease with the distance between the x's and y's as a long-range force, probably as $1/(|x-y|^2)$, since the energy and momentum range of the contributing intermediate states $|J\rangle$ would decrease as $|x-y|$ increases. These terms should therefore not interfere with causality.

There is no *a priori* reason for a special value of the total isospin $T(T+1)$ connected with T^k. Therefore, several terms with different values of $T(T+1)$ may occur. If such terms contain poles in momentum space, they may be considered as the contributions of strange particles to the 2-point function. The description of these particles require, besides the operator $\chi_\lambda(x)$, a reference to the isospin indices on which the T^k act. Therefore it is convenient to introduce the concept of a 'spurion' which has no Lorentz properties, which, however, possesses properties in isospace, and may be 'attached' to a normal particle to form a strange particle, like in the model of §7-2.

Using this spurion concept, one would picture the ground state as a 'sea' of infinitely many spurions and antispurions, most of which are connected in spurion–antispurion pairs of isospin zero. The asymmetry of the ground state is an essential feature of this picture, but for many problems, even in the theory of strange particles, the deviations from rotational symmetry in isospace may play only an unimportant rôle and may therefore be neglected in a first approximation.

A mathematical formalism corresponding to this picture requires first the definition of the properties—or more precisely the transformation properties—of the spurions which thereby acquire some properties of a particle. For the interpretation of the observations it seems sufficient to assume that a single spurion has the isospin $\frac{1}{2}$, and that it has a gauge property, i.e. that one can distinguish between spurion and antispurion. Hence altogether four states must be introduced, two spurion states s_1, s_2, and two antispurion states a_1, a_2. Following the picture of the spurion–antispurion sea, a spurion 'hole' should be equivalent to an antispurion. Therefore one may define

$$a_\lambda = c_{\lambda\mu} s_\mu^*, \qquad\qquad 7(15)$$

where c is given by

$$-\tau_k = c^{-1}\tau_k^T c. \qquad\qquad 7(16)$$

For the matrix element of any operator a spurion s_λ on the bra side (in Dirac's notation) would be equivalent to a spurion hole s_λ^* on the ket side:

$$\langle \ldots s_\alpha \ldots |O| \ldots s_\beta \ldots \rangle = \langle \ldots s_\beta^* \ldots |O| \ldots s_\alpha^* \ldots \rangle. \qquad 7(17)$$

The four states s_1, s_2, a_1, a_2, transform according to the 2-dimensional representation of the group $U_2 = G \cdot SU_2$, where G refers to a spurion number gauge group, SU_2 to the isospin rotations. The spurions have no properties with respect to the Lorentz group, i.e. they are completely invariant under this group, both under the proper Lorentz group and the discrete operations P and T.

If one starts from this definition of spurions, the strange particle is pictured as consisting of a normal particle, created by the field operator $\chi(x)$ (or a product of field operators), and of one or several spurions which have been attached to it. Besides that, the field operator $\chi(x)$ itself has an isospin index; therefore one may say that already $\chi(x)$ consists of two parts, namely an operator creating a particle without isospin, and another operator creating a spurion; these two parts are combined in the one field operator. Finally, the properties of a particle should depend only on the total isospin of the particle and on its gauge property in isospace, i.e. on the difference between the numbers of spurions and antispurions in the particle. The formal addition of spurion–antispurion pairs without isospin should not change the system. Hence the spurions from the ground state and those from the field operators should be treated upon the same footing.

Spurions belonging to the same point in space–time, i.e. attached to the same particle cannot be distinguished. Therefore, the rules of Bose statistics should be applied to the spurions connected with one or several field operators at the same space–time point. The wave function should be symmetrical under the permutation of the isospin indices connected with the same point. Therefore the isospins of all spurions (or of all antispurions) at the same point should be parallel. This assumption is in agreement with the properties of the interaction term in the fundamental field equation 3(1). This term contains, in its Fierz-symmetrical version 5(7), the expressions

$$1.1 - \sigma_k \sigma_k \qquad \text{in Dirac spin space,}$$

$$\tfrac{3}{4}.1.1 + \tfrac{1}{4}\tau_k \tau_k \qquad \text{in isospin space.}$$

The first line vanishes if the Dirac spins are parallel ($\sigma_k \cdot \sigma_k = +1$)—two 'bare' particles can reach the same point only if their Dirac spins are anti-parallel—the second line vanishes if the isospins are antiparallel ($\tau_k \cdot \tau_k = -3$). The colliding particles must have parallel isospin. If, at a given time, the isospins of spurions at the same space–time point are parallel, this state

  Unified Field Theory of Elementary Particles

of affairs will persist at any time (on account of 3(1)). Spurions at different points in space–time are not related to each other by conditions of symmetry.

If a strange particle is considered as consisting of n_s spurions and n_a antispurions referring to the same point (including those from the field operators), its mass should depend on its total isospin T and on its gauge property in isospace

$$u = \tfrac{1}{2}(n_s - n_a), \qquad\qquad 7(18)$$

which later will be identified with the hypercharge (in the literature hypercharge is frequently denoted by $Y = 2u$). It should not depend upon $n_s + n_a$. The total isospin T cannot be smaller than $|u| : T \geqslant |u|$. The eigenvalue operator O_{eig} determining the mass of the system can be written in terms of isospin matrices referring to spurions and antispurions. If, in a first approximation, terms with products of more than three different isospin matrices are neglected, the dependence on isospin of the operator O_{eig} must be given by

$$O_{\mathrm{eig}} = C\left(\sum_1^{n_s} \tau_s - \sum_1^{n_a} \tau_a\right)^2 - f(n_s + n_a) \qquad\qquad 7(19)$$

in order to comply with the condition 7(17); the function $f(n_s + n_a)$ should compensate any dependence on $n_s + n_a$ of the first term. Since the Bose statistics of spurions and antispurions implies

$$\left(\sum_1^{n_s} \tau_s\right)^2 = n_s(n_s + 2), \qquad \left(\sum_1^{n_a} \tau_a\right)^2 = n_a(n_a + 2), \qquad 7(20)$$

we get from 7(19):

$$O_{\mathrm{eig}} = -C\left(\sum_1^{n_s} \tau_s + \sum_1^{n_a} \tau_a\right)^2 + 2n_s(n_s + 2)\,C + 2n_a(n_a + 2)\,C - f(n_s + n_a)$$

$$= -4CT(T+1) + C(n_s - n_a)^2 + C(n_s + n_a)^2 + 4C(n_s + n_a)$$

$$\qquad -f(n_s + n_a)$$

$$= -4C[T(T+1) - u^2] + \mathrm{const.} \qquad\qquad 7(21)$$

Besides this dependence on isospin the operator may depend on the gauge property u, but only in the form $u.B$, where B is the baryonic number. This is necessary since O_{eig} must be invariant under G. Therefore one gets in a first approximation for the eigenvalue operator

$$O_{\mathrm{eig}} = a + b[T(T+1) - u^2] + c.Bu. \qquad\qquad 7(22)$$

where a, b and c will be analytical functions of the square of the mass, as in 5(12) or 5(29). A formula of this type 7(22) for the square of the mass has been derived by Okubo[13] and Gell–Mann[14] from the assumption of broken

SU_3 multiplets in the mass spectrum; it seems to agree well with the observed masses. The present theory however does not leave any free space for such an assumption. Therefore, it will be a question of the number of levels within a 'multiplet' of strange particles, whether the theory derived from 3(1) contains something like an approximate SU_3 symmetry.

7-4 Strange bosons

For bosons of baryonic number $B=0$ the third term of 7(22) vanishes. If the Dirac spin of the bosons is zero, as for π-, K- and η-meson, the eigen-value equation for the K-mesons can be obtained immediately from the relations 5(19) and 5(20) holding for the non-strange particles π and η. These two equations can be used to determine the two quantities a and b in 7(22). The result is:

$$1+\frac{3-T(T+1)+u^2}{4}\left(\frac{\kappa l}{2\pi}\right)^2 q_0(\lambda) = 0 \qquad 7(23)$$

as eigenvalue equation for π-, K- and η-meson. The mass values calculated from 7(23) agree reasonably well with the observed masses. A value $\kappa l/2\pi \approx 0.92$ which seems to fulfill approximately the requirements of consistency in the fermion spectrum gives:

$$7(24)$$

	π	K	η
$\dfrac{\kappa_B}{\kappa}$	0·198	0·737	0·827

The absolute values of κ_β/κ are somewhat too high, but their relation is nearly correct.

The equations 7(22) and 7(23) prove the existence of K-mesons of roughly the correct mass from 3(1), if the spurion formalism described in §7-2 can be applied consistently to the ground state of the present theory. They do not prove, that a direct calculation of the mass of the K-meson does lead to 7(23) and that, besides the kaon, there are no other strange bosons of Dirac spin 0 belonging to the π- and η-group. This proof can however, be supplied in the following lines.

The eigenfunction of the K-meson depends on 3 isospin indices, two from the field operators and one from the additive spurion. One of the indices from the field operators (let us assume the last one) refers to an antispurion, the two others to spurions. Then the wave function $\phi_{\alpha\beta\gamma}$ must be symmetrical in the two spurion indices α and β:

$$\phi_{\alpha\beta\gamma} = \phi_{\overline{\alpha\beta}\gamma} \qquad\qquad 7(25)$$

(— means symmetry, ⌢ antisymmetry). This wave function can be composed of two wave functions which are symmetrical or antimetrical in β and γ:

$$\phi_{\overline{\alpha\beta}\gamma} = \frac{\sqrt{3}}{2}\,\phi_{\alpha\widehat{\beta\gamma}} + \tfrac{1}{2}\phi_{\alpha\overline{\beta\gamma}}. \qquad\qquad 7(26)$$

The factors $\sqrt{3}/2$ and $\tfrac{1}{2}$ are the Clebsch–Gordan coefficients belonging to a state of isospin $T=\tfrac{1}{2}$. Eqn. 7(26) can be interpreted as stating, that a kaon consists with a probability $\tfrac{3}{4}$ of an η-meson combined with a spurion, with a probability $\tfrac{1}{4}$ of a pion combined with a spurion. Since the additive spurion does not appear explicitly in the interaction term of 3(1), the factor of $q_0(\lambda)$ in the eigenvalue equation of $\phi_{\alpha\overline{\beta\gamma}}$ should be a weighted average of the factors for η-meson and pion, the weights being $\tfrac{3}{4}$ resp. $\tfrac{1}{4}$. This leads immediately back to 7(23).

If the same calculation would be carried out for an assumed kaon state of isospin $T=3/2$, the wave function would be completely symmetrical $\phi_{\overline{\alpha\beta\gamma}}$ and would therefore consist only of a pion combined with a spurion. The eigenvalue would be identical with that of the pion; hence the spurion would not really be attached to the pion, it would be a free spurion in the 'sea' of the ground state and the particle would remain a pion. A consistency between the general spurion formalism §7-2 and the direct calculation of the mass eigenvalues can be obtained only if the kaon and no other strange particle state is added to the π- and η-group. In this way the octet of the spin 0 bosons π-, K- and η-meson finds a natural explanation in the combination of 3(1) with a degenerate ground state. The SU_3-symmetry is not assumed at any point in this argument, nor is there any approximation, in which the masses of π-, K- and η-meson could be considered as equal. Still the final results, namely the octet and the Okubo[13] mass formula, do resemble the consequences of a broken SU_3-group, to some extent.

7-5 Strange fermions

If the masses of the simplest fermions are calculated from the approximation 5(27), a formation of strange particles cannot be expected. The mass is produced here by the creation and annihilation of fermion–antifermion pairs, the Lorentz properties of which do not depend upon the isospin. It would be consistent to assume, that N, Λ, Σ, Ξ-particles all have the same mass, i.e. that the spurions would not really be attached to the particles. The situation is different however, when the creation and annihilation of bosons is taken into account in the improved eigenvalue equation 5(33). The contributions from the boson poles represent minor corrections to the main mass term which comes from the continuous spectrum of fermion–antifermion pairs; the mass of the boson does depend on the isospin. The empirical fact that the relative mass splitting in the fermion-octet N, Λ, Σ, Ξ is considerably smaller than in the lowest boson octet agrees well with the general idea of a secondary correction caused by the boson poles.

Therefore we will try to estimate the contributions to the eigenvalue equation of the fermions from the creation and annihilation of bosons represented by the graph —⁓—◁ in 5(33). For this purpose it will be sufficient, to use the phenomenological description 6(13), 6(14) of the creation of bosons, taking for the coupling constant the value of f 6(20) calculated from the theory. Only the contributions from the pseudoscalar bosons π, K, η will be considered, since hitherto no calculations about the vector mesons have been carried out on the basis of the present theory.

If in the interaction term of 6(13) or 6(14) the wave function $\chi(x)$ is replaced by the Green's function integral derived from the same wave equation, one gets at once a mass term of the following form:

$$i\sigma^\nu \frac{\partial \chi(x)}{\partial x^\nu} = \sigma^\nu \int d^4x'\, G(xx')\, \sigma_\mu\, \chi(x') \frac{f^2}{\kappa_B^2} \left\langle 0 \left| \frac{\partial \Phi(x')}{\partial x'_\mu} \cdot \frac{\partial \Phi(x)}{\partial x^\nu} \right| 0 \right\rangle + \ldots \quad 7(27)$$

or in case of isospin 1 of the boson

$$i\sigma^\nu \frac{\partial \chi(x)}{\partial x^\nu} = \sigma^\nu \int d^4x'\, G(xx')\, \sigma_\mu\, \chi(x') \frac{f^2}{\kappa_B^2} \left\langle 0 \left| \frac{\partial \Phi_i(x')}{\partial x'_\mu} \cdot \frac{\partial \Phi_i(x)}{\partial x^\nu} \right| 0 \right\rangle. \quad 7(28)$$

The contribution 7(27) could in momentum space be written as

$$\sigma^\nu J_\nu \chi = \frac{f^2}{\kappa_B^2} (2\pi)^{-4} \cdot i \int d^4p\, \frac{\sigma^\nu p_\nu (J-p)_\rho \overline{\sigma^\rho}\, \sigma_\mu p^\mu}{(J-p)^2 (p^2 - \kappa_B^2)} \cdot \chi + \ldots \quad 7(29)$$

The integral in this expression diverges at $p \to \infty$.

In the non-linear spinor theory the pole term $(p_\mu p_\nu)/(p^2 - \kappa_B^2)$ is actually produced in the way indicated in 6(34) and 6(35). The expression 6(36) can however not contain δ-functions at the point $|x-y| = 0$. Since the 2-point function of the fermions has been regularized in the first approximation 3(39) by a double pole at the mass zero, it is natural to assume the same kind of regularization for the boson 2-point function 6(34). Therefore 7(29) should be replaced by

$$\sigma^\nu J_\nu \chi = f^2 (2\pi)^{-4} i\kappa_B^2 \int d^4p \, \frac{\sigma^\nu p_\nu (J-p)_\rho \, \overline{\sigma^\rho} \, \sigma_\mu p^\mu}{(J-p)^2 (p^2)^2 (p^2 - \kappa_B^2)} \chi + \dots \qquad 7(30)$$

This expression agrees at the pole $p^2 - \kappa_B^2 = 0$ with 7(29) and gives the required behaviour of 6(34) in the neighbourhood of $|x-y| = 0$. It should be emphasized at this point, that the regularization of the 4-point function 6(34) does not necessarily require an indefinite metric in Hilbert space or ghost states with negative norm. Actually this term does not only contain the contribution from the bound state κ_B, but should also comprise the correction of the contribution from the continuous spectrum due to the binding forces between fermion and antifermion. This correction will probably go in the opposite direction compared with the pole contribution. The regularizations in 6(34) and in 3(39) should be carried out in the same way.

The integral in 7(30) is finite. For the numerator we have

$$\sigma^\nu p_\nu (J-p)_\rho \, \bar\sigma^\rho \, \sigma_\mu p^\mu = -2p_\mu \sigma^\mu(p, J-p) + (J-p)_\rho \sigma^\rho . p^2. \qquad 7(31)$$

In the integration, p in $p_\mu \sigma^\mu$ can be replaced by its component parallel to J

$$p_{\nu\|} = J_\nu \frac{(Jp)}{J^2}, \qquad 7(32)$$

for the numerator we can therefore substitute

$$\frac{J_\nu \sigma^\nu}{J^2} [p^2 J^2 - 2(Jp)^2 + p^2(Jp)]. \qquad 7(33)$$

Finally the contribution 7(30) becomes

$$\sigma^\nu J_\nu \chi = \frac{1}{8\pi} \left(\frac{f^2}{4\pi}\right) r\left(\frac{J^2}{\kappa_B^2}\right) . \sigma^\nu J_\nu \chi + \dots \qquad 7(34)$$

where

$$r\left(\frac{J^2}{\kappa_B^2}\right) = \frac{2i\kappa_B^2}{\pi^2} \int d^4p \, \frac{p^2 J^2 + p^2(Jp) - 2(Jp)^2}{(p^2)^2 (p^2 - \kappa_B^2)(J-p)^2 . J^2}. \qquad 7(35)$$

The evaluation of $r(\lambda)$ by means of Feynman's methods gives $(\lambda = J^2/\kappa_B^2)$

$$r(\lambda) = (2-\lambda)\ln\lambda + 1 + \frac{(1-\lambda)^2}{\lambda}\ln|1-\lambda|. \qquad 7(36)$$

The contributions to the eigenvalue equation from π- and η-mesons, calculated in this way, cause some change in the mass, but do not lead to any splitting of the masses of N-, Λ-, Σ-, Ξ-fermions. π- and η-meson do not contain spurions; hence the spurion will in the expression simply stay with the fermion, the contribution is not influenced by the presence of a spurion in the fermion.

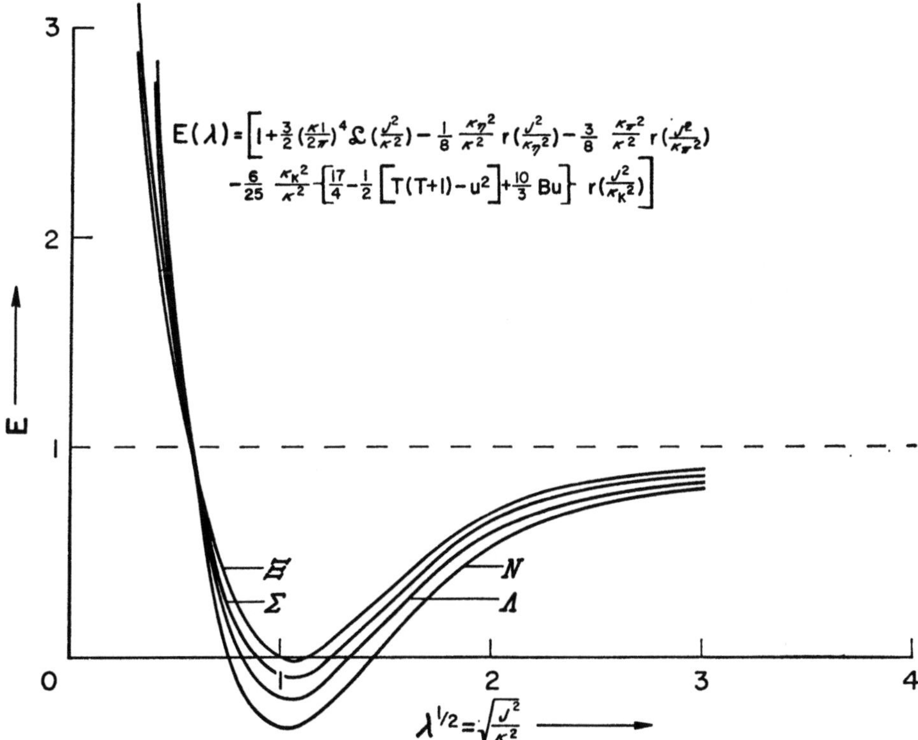

$$E(\lambda) = \left[1 + \frac{3}{2} \left(\frac{\kappa_1}{2\pi} \right)^4 \mathcal{L} \left(\frac{J^2}{\kappa^2} \right) - \frac{1}{8} \frac{\kappa_\eta^2}{\kappa^2} r \left(\frac{J^2}{\kappa_\eta^2} \right) - \frac{3}{8} \frac{\kappa_\pi^2}{\kappa^2} r \left(\frac{J^2}{\kappa_\pi^2} \right) \right.$$
$$\left. - \frac{6}{25} \frac{\kappa_K^2}{\kappa^2} \left\{ \frac{17}{4} - \frac{1}{2} \left[T(T+1) - u^2 \right] + \frac{10}{3} Bu \right\} r \left(\frac{J^2}{\kappa_K^2} \right) \right]$$

FIG. 7. The characteristic factor $E(\lambda)$ in the eigenvalue equation 7(38) of the strange particles.

A difference however, comes in with the contribution from the kaons. The spurion may leave the fermion line and go into the kaon line, therefore the presence of spurions changes the contribution from the production and annihilation of kaons. This term then should contain a factor of the general form 7(22), where the constants a, b and c are determined by Clebsch–Gordon coefficients. The calculations which shall not be carried out here do actually lead to a factor

$$\tfrac{17}{4} - \tfrac{1}{2}[T(T+1) - u^2] + \tfrac{10}{3}Bu. \qquad 7(37)$$

            Unified Field Theory of Elementary Particles

Finally the eigenvalue equation of the fermions becomes:

$$\sigma^\nu J_\nu \left[1 + \frac{3}{2}\left(\frac{\kappa l}{2\pi}\right)^4 L\left(\frac{J^2}{\kappa^2}\right) - \frac{1}{8}\frac{\kappa_\eta^2}{\kappa^2} r\left(\frac{J^2}{\kappa_\eta^2}\right) - \frac{3}{8}\frac{\kappa_\pi^2}{\kappa^2} r\left(\frac{J^2}{\kappa_\pi^2}\right) \right.$$

$$\left. - \frac{6}{25}\frac{\kappa_K^2}{\kappa^2}\left\{\frac{17}{4} - \tfrac{1}{2}[T(T+1)-u^2] + \frac{10}{3} Bu\right\} r\left(\frac{J^2}{\kappa_K^2}\right) \right]\phi = 0 \quad 7(38)$$

With a value $\kappa l/2\pi = 0\cdot 92$ the resulting masses are

	N	Λ	Σ	Ξ
κ_F/κ	0·74	0·81	0·88	1·04
number of additive spurions	2	1	1	0

$$ 7(39)$$

The absolute values of κ_F/κ in 7(39) are somewhat too low, but their relation is qualitatively in agreement with the observations. The contributions from the vector mesons have not been included in 7(27) or (28), they would modify the numerical values to some extent, but would probably not change the picture qualitatively. It follows from 7(39), that the field operator alone creates Ξ-particles, not nucleons, and that the nucleon is a two-spurion system in the sense of the spurion formalism. The limitation of this lowest fermion–spurion multiplet to the octet N, Λ, Σ, Ξ cannot be understood as simply as in the boson case. But the discussion of quantum electrodynamics in Chapter 8 will show that the octet is required for the consistency of the asymmetrical ground state. Again, as in the boson case, the SU_3 symmetry has not been assumed. But the Okubo[13] mass formula has its counterpart in 7(22), and there is an approximation in which the masses of the octet can be considered as equal, namely 5(27).

The two octets η, K, π and N, Λ, Σ, Ξ comprise essentially all those strongly interacting states which can be represented by field operators at the same point. Already the vector bosons ρ, ω, K^*, ϕ seem to be primarily d-states of the fermion–antifermion system, without much contribution from the contact interaction $\diagdown\!\!\!\!\diagup$. If these states η, K, π and N, Λ, Σ, Ξ, in which the contact interaction plays the dominant rôle, do simulate a broken SU_3 symmetry, the same should be true to some extent—on account of general requirements of consistency—for all higher states. These excited

Strange Particles 107

states can be considered as compound systems of N, Λ, Σ, Ξ and η, K, π, where finite range forces produced by virtual creation and annihilation of particles are more important than the SU_3-asymmetrical contact interaction. Calculations on the decuplet N^*, Y_1^*, Ξ^*, Ω^-, for example, have been carried out on this basis, without any immediate connexion with the present theory and without any reference to contact interaction.[77]

8

Quantum Electrodynamics[64]

8-1 The Goldstone theorem[41]

The asymmetry of the ground state with respect to the isospin group has been used in Chapter 7 as explanation for the strange particle poles in the Green's functions and as basis for the spurion formalism. In the following sections the asymmetry itself will be the object of the investigations. It has been emphasized already in early papers on this subject, that empirically the asymmetry of the ground state seems to be closely connected with the existence of long-range forces, i.e. of particles with rest mass zero.[29] The asymmetry with respect to the isospin group comes in through the long-range forces of electrodynamics, the asymmetry with respect to the space reflection parity appears in the weak interactions, and this is the first interaction which affects the neutrinos. It can be well understood that short-range forces allow a clear separation of the particles from the rest of the world, while long-range forces may lead to a dependence of the properties of the particles on the state of the world in large dimensions.

This connexion has found a mathematical expression in the theorem of Goldstone, which has been discussed qualitatively and in a somewhat generalized form in §7-1 and §7-2. In the framework of the present theory the Goldstone theorem is the basis for an understanding of quantum electrodynamics. Therefore the general mathematical proof of the theorem given by Goldstone, Salam and Weinberg[41] and Nambu[42] will be presented here for the special case of an asymmetry of the ground state with respect to the isospin group.

The conservation of isospin resulting from the group properties of 3(1) will allow the construction of the operators T_k, performing the rotations in isospace. For infinitesimal α_k:

$$e^{(i/2)\alpha_k T_k}\chi = e^{i\alpha_k T_k}\chi e^{-i\alpha_l T_l}. \qquad 8(1)$$

These operators can, according to §3-6, be connected with vector currents $J_{\mu,k}(x)$:

108

$$T_k = \int J_{0,k}(x)\, d^3 x; \qquad \frac{\partial J_{\mu,k}(x)}{\partial x_\mu} = 0. \qquad\qquad 8(2)$$

In conventional field theories these currents would be local. But it has been pointed out in §3-6, that a non-linear theory does generally not allow the construction of local currents. Hence $J_{\mu,k}(x)$ will probably need for its construction the field operators of a finite region of the order 10^{-13} cm (or 10^{-23} sec) around x. The construction of a pseudovector, fulfilling the equation of continuity 8(2) will not be possible at all, according to §3-6.

The asymmetry of the ground state may be expressed by the fact, that the mass of the fermions depends on the direction of their isospin, for example, that the $\varXi_0$ and $\varXi_-$ baryons have different masses. If we take the third axis in isospace as direction of the big isospin of matter in the world, we get for the 2-point function

$$\langle 0|\chi_\alpha^*(y+\epsilon)\,\sigma_\nu\,\chi_\beta(y)|0\rangle = \epsilon_\nu[1_{\beta\alpha}+d(\epsilon^2)\cdot\tau_{3,\beta\alpha}]\cdot c(\epsilon^2), \qquad 8(3)$$

where ϵ is a finite difference between the two points in 8(4). $d(\epsilon^2)$ is determined by the mass difference between $\varXi_0$ and $\varXi_-$, and α and β are isospin indices.

If we define the boson operators

$$\varPhi_{\nu,k}(y,\epsilon) = \chi^*(y+\epsilon)\,\sigma_\nu\,\tau_k\,\chi(y), \qquad\qquad 8(4)$$

the equation 8(1) leads to

$$\langle 0|e^{i\alpha_k T_k}\,\varPhi_{\nu,l}(y,\epsilon)\,e^{-i\alpha_m T_m}|0\rangle$$

$$= \langle 0|\chi^*(y+\epsilon)\,\sigma_\nu\,e^{-(i/2)\alpha_k \tau_k}\,\tau_l\,e^{+(i/2)\alpha_m \tau_m}\,\chi(y)|0\rangle \qquad 8(5)$$

Taking especially $\alpha_k=(\alpha,0,0)$ and comparing the two terms prop. α and α^2 of the expansion on both sides, we get the two equations (using the abbreviation $ab-ba=[a,b]$):

$$\langle 0|[T_1,\varPhi_{\nu,l}]|0\rangle = \langle 0|\chi^*(y+\epsilon)\,\sigma_\nu\,\tfrac{1}{2}[\tau_1,\tau_l]\,\chi(y)|0\rangle, \qquad 8(6)$$

$$\langle 0|[T_1[T_1,\varPhi_{\nu,l}]]|0\rangle = \langle 0|\chi^*(y+\epsilon)\,\sigma_\nu\,\tfrac{1}{4}[\tau_1[\tau_1,\tau_l]]\,\chi(y)|0\rangle. \qquad 8(7)$$

The right-hand side of 8(6) is different from zero, according to 8(3), if $l=2$. The right-hand side of 8(7) is different from zero for $l=3$.

Therefore it is convenient, to study the Källén–Lehmann representation of the two expressions:

$$\langle 0|[J_{\mu,1}(x),\varPhi_{\nu,2}(y,\epsilon)]|0\rangle \quad \text{and} \quad \langle 0|[J_{\mu,1}(x)[T_1,\varPhi_{\nu,3}(y,\epsilon)]]|0\rangle$$

in the limit of very small values of ϵ. If in these expressions only the term with the highest singularity for $\epsilon \to 0$ is written explicitly, we get from 8(3):

$$\langle 0|[J_{\mu,1}(x), \Phi_{\nu,2}(y, \epsilon)]|0\rangle$$

$$= \int d(\kappa^2)\,\rho(\kappa^2)\,\frac{\partial}{\partial x^\mu}\,\Delta(x-y, \dot{\kappa})\,.\,\epsilon_\nu\,.\,2id(\epsilon^2)\,c(\epsilon^2)+\ldots \qquad 8(8)$$

and

$$\langle 0|[J_{\mu,1}(x)\,[T_1, \Phi_{\nu,3}(y, \epsilon)]]|0\rangle$$

$$= \int d(\kappa^2)\,\rho(\kappa^2)\,\frac{\partial}{\partial x^\mu}\,\Delta(x-y, \kappa)\,.\,\epsilon_\nu\,.\,2d(\epsilon^2)\,c(\epsilon^2)+\ldots \qquad 8(9)$$

where $\Delta(x-y, \kappa)$ are the Schwinger functions, and in both cases we have

$$\int d(\kappa^2)\,\rho(\kappa^2) = 1. \qquad\qquad 8(10)$$

On the other hand the equation of continuity 8(2) gives

$$\kappa^2\,\rho(\kappa^2) = 0 \quad \text{and therefore} \quad \rho(\kappa^2) = \delta(\kappa^2). \qquad 8(11)$$

The mass spectrum of bosons created by the operators $\Phi_{\nu,2}(y, \epsilon)$ or $\Phi_{\nu,3}(y, \epsilon)$, contains bosons of rest mass zero; this is the content of the Goldstone theorem.

The operator T_1 occurring in 8(9) commutes with the total energy and momentum; it does not contain matrix elements belonging to transitions between different values of the energy–momentum vector. One may say, that this operator induces transitions from the ground state to a 'spurion' state in the sense of 7(5). Therefore the question whether $\Phi_{\nu,2}(y, \epsilon)$ or $\Phi_{\nu,3}(y, \epsilon)$ should be taken as the operator creating a Goldstone particle, depends, as in the model §7-2, on the choice of the ground state, on which this operator is supposed to act. If the ground state is taken as a mixture of several possible states, which seems to be the most natural assumption from a physical point of view, the representation of the Goldstone particles by means of a projection operator of the type 7(5) seems most adequate, since it shows most clearly, that no transport of charge takes place in the ground state together with the transport of the Goldstone particle.

If it is permitted to use the analogy with the model of §7-2 for the isospin case, one would expect that the photon operator could in isospace be represented by a projection operator analogous to $\frac{1}{2}(1-\sigma_{z,n})$ in 7(5). σ_z should then be replaced by τ_3, the sign in the bracket can be chosen arbitrarily. The unit operator in 7(5) refers to the spin indices of the electrons in the ground state, it refers to the same indices as σ_z. Therefore in isospace the unit operator must refer to the gauge property in isospace, to a property connected also with the ground state, not the gauge property in Lorentz space. If strange particles appear in the way indicated in Chapter 7, the gauge property in Lorentz space is separated from the gauge property in isospace; the latter one is expressed by the hypercharge quantum number $u = \frac{1}{2} Y$. Therefore we expect the projection operator $\frac{1}{2}(Y + \tau_3)$ in the wave function of the photon. This operator also permits the symmetry of the ground state under charge conjugation—a feature not present in the ferromagnetic model §7-2.

If the photon operator contains this term $\frac{1}{2}(Y + \tau_3)$, the photon appears as a one-to-one mixture of a state of isospin 0 and a state of isospin 1, the third component of the isospin being zero. If such a state shall be the result of the eigenvalue equation of the photon, there must initially be a degeneracy between the states with isospin 0 and 1, so that a small perturbation will produce this one-to-one mixture. Therefore the strange baryons must be distributed in such a way, that the degeneracy follows, and it is apparently this condition which produces the well-known baryon octet.

The one-to-one mixture is necessary for the consistency of the degenerate ground state: The asymmetry of the photon is a result of the asymmetry of the 2-point function. The asymmetry of the masses is in turn due to the asymmetry of the electromagnetic forces, to production and annihilation of photons. If the asymmetry of the photon would become smaller when the asymmetrical term in its eigenvalue equation becomes smaller, one would, in the attempt to get a consistent solution, finally end up with a symmetrical ground state. Only if the photon state is completely asymmetrical i.e. a one-to-one mixture, it may produce a consistent asymmetry of the ground state. This will be discussed in more detail in §8-3.

The formulae 8(8)–8(11) do not give any definite evidence concerning the symmetry of the Goldstone particle under Lorentz transformation. Since $\Phi_{\nu,3}$ is a vector, the Dirac spin of the particle may be 0 or 1. Judging from the analogy to the model §7-2, one would be inclined to consider the Coulomb part of the electromagnetic field, which is a mixture of a scalar and a longitudinal field, as the primary Goldstone field; this field is then supplemented by the other components.

The theorem of Goldstone has another interesting consequence with respect to the approximate SU_3-symmetry in the mass spectrum of elementary particles. If the underlying natural law would be invariant under SU_3 and if this symmetry would be broken by the ground state—as we have assumed for the isospin group U_2 but not for SU_3—there should exist a Goldstone particle of mass zero responsible for the breaking of the SU_3-group. The interaction of this particle should be rather strong in order

 Unified Field Theory of Elementary Particles

to produce the large deviation from the SU_3-symmetry. Empirically such a particle does not exist. Therefore the underlying natural law cannot be invariant under SU_3 or higher groups of this type (SU_6, U_{12}, etc.). This result is one of the main arguments for starting with a fundamental field equation, which does not represent SU_3. The approximate SU_3 symmetry can appear afterwards as result of dynamics, especially as consequence of the possibility of detaching spurions from the big isospin in the ground state.

8-2 The scale transformation

The photon, like any other elementary particle, should appear as eigen-solution of 3(1), or more specifically of an equation of the type 5(6). Such an equation will, however, generally not lead to the mass value zero. If the mass zero is required by the Goldstone theorem, this must mean that the asymmetry of the ground state together with the rest mass zero of the photon creates some new freedom which then may be used for fulfilling the eigen-value equation of the photon at zero mass. Actually this freedom consists in the free choice of the average mass of the charged leptons.

In quantum electrodynamics the mass of the leptons seems to be of purely electromagnetic origin. In recent papers on electrodynamics, especially by Johnson,[63] it has been emphasized, that the mass of the bare electron should be taken as zero and that the equations of electrodynamics are therefore invariant under a scale transformation:

$$A_\nu(x) \rightarrow \eta A_\nu(x\eta), \qquad\qquad 8(12)$$

$$\psi(x) \rightarrow \eta^{3/2} \psi(x\eta). \qquad\qquad 8(13)$$

This scale invariance is broken by the ground state, which introduces a finite mass of the electron. For consistency it is necessary that the mass resulting from the eigenvalue equation of the electron is identical with the mass, that has been put into the 2-point function of the ground state. This mass cannot, however, be determined by electrodynamics, since one could always, by a scale transformation, go over from one mass value to another. If all charged leptons, electrons and muons, are taken into the calculation, the ratio between the masses of the different leptons could possibly be determined by some weak interactions resulting from higher approximations in electro-dynamics. Still all the masses, and therefore also the average mass, could be changed arbitrarily by a common scale factor. Hence it is the rest mass zero and in this sense the scale invariance of the photon, which is res-ponsible for the free choice of the average lepton mass. It is true that in the present theory afterwards the average lepton mass will be determined by the eigenvalue equation of the photon, i.e. by postulating the rest mass zero of the photon. But within electrodynamics alone such a determination is impossible. The eigenvalue equation of the photon will finally relate the average mass of the charged leptons to the mass spectrum of the strongly interacting particles.

In the present theory based upon 3(1) the situation with respect to the scale transformation is essentially the same as in quantum electrodynamics. The fundamental equation 3(1) is invariant under 3(11) and the commutator may be of the form 3(33). The operator $\chi^*(x)\,\sigma_\nu\chi(x)$ behaves like $A_\nu(x)$ in 8(12) under Lorentz and under scale transformation. Still the ground state destroys the scale invariance by introducing finite mass eigenvalues.

In the approximation 3(39) of the 2-point function the double pole at the mass zero (compare §3-4) represents the leptons and ghost states, the pole at $p^2-\kappa^2=0$ the baryons. One may say that in this approximation the leptons are separated from the baryons by the scale transformation. The lepton part alone would in fact be invariant under this transformation, as one learns from 3(26). The baryon part is not invariant, unless κ would take part in the transformation. The separation would indicate that the number of baryons and the number of leptons are conserved independently. At this point, however, the result is rather trivial, since according to §3-4 leptons are neither created nor absorbed in the approximation 3(39), the leptons do not take part in the strong interactions. The electromagnetic mass of the charged leptons has been neglected in 3(39), and it will be necessary to improve the 2-point function in this respect before the eigenvalue equation of the photon can be formulated.

8-3 The eigenfunction of the photon

In the framework of the present theory one would define, according to §5-2, the τ- or ϕ-function

$$\phi^\mu_{\alpha\beta}(x|y) = \langle 0|T\chi^*_\alpha(x)\,\sigma^\mu\chi_\beta(y)|\gamma\rangle \qquad 8(14)$$

as 'eigenfunction' of the photon γ, characterized by its momentum $J_\nu(J^2=0)$ and its polarization B, $(B_\nu J^\nu=0)$. α and β are isospin indices. We have to study the behaviour of this eigenfunction in isospace and in Lorentz space.

The behaviour of the photon eigenfunction in isospace has been discussed briefly in §8-1. The analogy with the model of ferromagnetism suggests the appearance of the projection operator $\frac{1}{2}(Y+\tau_3)$ in isospace:

$$\phi^\mu_{\alpha\beta} = a^\mu \cdot \tfrac{1}{2}(Y+\tau_3)_{\alpha\beta} \qquad 8(15)$$

(with a^μ independent of isospin). The photon should be a one-to-one mixture of states of isospin 0 and isospin 1. Again at this point one can see, that a projection operator $\frac{1}{2}(1+\tau_3)_{\alpha\beta}$, which at first sight might appear as natural analogon to $\frac{1}{2}(1-\sigma_z)$ in 7(5), would not fulfill the necessary requirements. For negligible deviation from symmetry there would be no degeneracy between the two parts of isospin 0 and isospin 1, therefore a state with the projection operator $\frac{1}{2}(1+\tau_3)$ could not be an eigen solution. On the other hand, the photon should refer to the gauge property in iso-space, which is defined also for the spurions in the ground state, not to the gauge property in Lorentz space. Therefore, the 1 must be replaced by $Y=2u$ in 8(15). If one denotes more explicitly the intermediate states and their spurions in 8(14), it is convenient on the right-hand side of 8(14) to write the quantum numbers u and T_3 (third component of isospin) in their spurion representation (compare §7-2):

$$\phi^\mu_{\alpha\beta}(x|x) = \langle 0|\chi^*_\alpha(x)\,\sigma^\mu|s_\delta s_\epsilon\ldots J\rangle\langle Js_\delta s_\epsilon\ldots|\chi_\beta(x)|\gamma\rangle \qquad 8(16)$$

$$= a^\mu\cdot{}_{\beta\delta\epsilon}(u+T_3)_{\alpha\delta\epsilon}.$$

The physical interpretation of this equation is simple. The photon 'consists of' fermion–antifermion pairs, each contribution containing a factor $u+T_3$ determined by the symmetry of the fermion concerned. Within the octet N, Λ, Σ, Ξ the factor has the eigenvalues $\pm 1, 0$. The neutral fermions, with $u+T_3=0$ do not contribute to the photon eigenfunction.

With the assumption 8(16) the two contributions to the photon eigen-function coming from the symmetries u (isospin 0) and T_3 (isospin 1) are actually degenerate from the beginning, if the values of u are distributed symmetrically around zero like those of T_3, as can be seen by inserting 8(16) into the eigenvalue equation of the photon. The symmetry is satisfied for the octet N, Λ, Σ, Ξ, but would not be satisfied if other fermions for example of isospin $\frac{3}{2}$ would be added. Therefore it is just the fermion octet together with the special symmetry 8(16) of the photon that makes a consistent solution possible, in which the ground state is asymmetrical under the isospin group.

The factor

$$q = u+T_3 \qquad\qquad 8(17)$$

decides about the participation of the fermion–antifermion pair in the eigenfunction of the photon, and hence also about the interaction of the fermion (or antifermion) with the photon. We are therefore entitled to call it the electromagnetic charge quantum number of the fermion. We conclude from 8(17), that the charge q is composed of two parts, the hypercharge u and the isospin charge T_3, in agreement with the well-known empirical rule derived by Gell-Mann and Nishijima.[8] The additivity of charge in Maxwell's equations will then provide the correctness of this rule for any more complicated system. It is interesting to see that already in the ferromagnetic model there exists [in Eqn. 7(5)] a relation analogous to the Gell-Mann–Nishijima rule.

With respect to its Lorentz properties the eigenfunction of the photon (of Dirac spin 1) would, according to 5(11) and the general formula

$$\bowtie = \bowtie\!\!\!\bowtie \;,$$ be given essentially by

$$\phi^\mu(x|y) \sim \text{const}\, e^{-iyJ} \int d^4p\, e^{-ip(x-y)}(u+T_3) \frac{tr[\sigma^\mu p_\rho \overline{\sigma^\rho}\sigma_\nu B^\nu(J-p)_\tau \overline{\sigma^\tau}]}{(p^2)^2\,(p^2-\kappa^2)\,(J-p)^2}\;;$$

$$8(18)$$

but it may be necessary, as had been mentioned in §8-2, to improve the position of the poles by taking the electromagnetic masses, especially of the leptons, into account. This expression, Eqn. 8(18), may be compared with the corresponding expression in conventional quantum electrodynamics, where the spinor $\chi(x)$ is replaced by a Dirac spinor $\psi(x)$, creating electrons or muons, and where the heavy particles are neglected. In first-order perturbation theory one would get for the terms with the highest powers in p:

$$\phi^\mu(xy) \sim \text{const}\, e^{-iyJ} \int d^4p\, \frac{tr[\gamma^\mu p_\rho \gamma^\rho \gamma_\nu B^\nu(J-p)_\tau \gamma^\tau]}{(p^2-\mu^2)\,[(J-p)^2-\mu^2]}\, e^{-ip(x-y)}, \quad 8(19)$$

where μ is the mass of the electron (or muon). For a comparison with 8(18) we are, however, not interested in results of perturbation theory, but in exact expressions holding for the renormalized operators. For $x=y$ the expression 8(19) reduces to the self-energy integral of the photon, which as is well known diverges quadratically. In order to get rid of this infinite self-energy of the photon one usually argues that on account of the gauge invariance of the theory one should subtract in 8(19) a term of the same form for $J=0$, i.e. of the form

$$\frac{tr[\gamma^\mu p_\nu \gamma^\nu B_\rho \gamma^\rho p_\tau \gamma^\tau]}{(p^2-\mu^2)^2}\,. \qquad 8(20)$$

The quadratic divergence is thereby changed into a logarithmic divergence, which later in the process of renormalization would be changed into a convergent expression, and the main term (with the highest powers of p) could be written as

$$\frac{tr[\gamma^\mu p_\nu \gamma^\nu B_\rho \gamma^\rho p_\tau \gamma^\tau]}{(p^2-\mu^2)^2\,[(J-p)^2-\mu^2]}\,. \qquad 8(21)$$

Comparing 8(18) and 8(21) one sees that the γ^μ of conventional quantum electrodynamics are, in the present theory, replaced by the σ^μ, on account of the two-dimensional representation of the Lorentz group. This difference is unimportant; it would disappear, if we had started from 3(22) instead of 3(1). Besides that, the double pole ($p^2=0$) at mass zero in 8(18), coming from ϕ in ⟩⊂⟨ , has been changed to the mass of the charged leptons in 8(21), and the same change has been performed for the other pole $(J-p)^2=0$, coming from the Green's function in ⟩⊂⟨ . Finally the denominator of 8(18) contains the factor $p^2-\kappa^2$, which makes the integral convergent, while 8(21) is logarithmically divergent (and would be made convergent in the further process of renormalization). It is interesting to see that the double pole, which in 8(18) appears as consequence of the approximation 3(39) for the 2-point function, arises in conventional quantum electrodynamics from the process of renormalization.

If for the photon the same degree of approximation is chosen as for the other bosons in §5-1, it will be consistent to improve the 2-point function ϕ in 8(18) by taking the electromagnetic mass of the charged leptons into account. Since we have in this approximation no way of calculating the mass ratio of the different charged leptons, we will replace their mass by an average lepton mass μ. By moving the double pole to this finite mass value, the singularity discussed in §5-2 disappears, since the light quantum cannot, under conservation of energy and momentum, disintegrate into a pair of particles of which at least one has a finite mass. In the Green's function the Tamm–Dancoff method does not introduce any finite mass values, therefore we will follow this procedure here as well. The resulting numerical values would be changed only slightly, if the pole $(J-p)^2=0$ would be transferred to the average lepton mass.

Finally 8(18) should be replaced by

$$\phi^\mu(x|y) \sim \text{const}\, e^{-iJy} \int d^4p\, \frac{tr[\sigma^\mu p_\rho\, \overline{\sigma^\rho}\, B_\nu\, \sigma^\nu (J-p)_\tau\, \overline{\sigma^\tau}]}{(p^2-\mu^2)^2\,(p^2-\kappa^2)\,(J-p)^2}\, e^{-ip(x-y)} \cdot (u+T_3).$$

$$8(22)$$

8-4 The eigenvalue equation of the photon and the coupling constant

The initial degeneracy between the u- and T_3-parts of the photon eigenfunction shows, that in the first approximation the eigenvalue equation can simply be taken from 5(22), if only the function $q_1(\lambda)$ is replaced by $q_1(\lambda, \epsilon)$ (with $\epsilon=\mu^2/\kappa^2$) which differs from $q_1(\lambda)$ by the position of the double pole in the 2-point function; this double pole has been moved from zero to the average lepton mass μ. The analytic expression for $q_1(\lambda, \epsilon)$ is

$$q_1(\lambda, \epsilon) = q_1(\lambda) - q_1\left(\frac{\lambda}{\epsilon}\right) + \frac{\lambda}{\epsilon}(1-\epsilon)\frac{dq_1(\lambda/\epsilon)}{d(\lambda/\epsilon)} \qquad 8(23)$$

$$= \frac{1}{3}\left[\frac{1-\epsilon}{\lambda}(1-2\epsilon+2\lambda) + \frac{(1-\lambda)^2(1+2\lambda)}{\lambda^2}\ln|1-\lambda|\right.$$

$$\left. - \frac{\epsilon^2}{\lambda^2}(3-2\epsilon)\ln\left|\frac{\epsilon-\lambda}{\epsilon}\right| + (3-2\lambda)\ln|\epsilon-\lambda|\right].$$

In the limiting case $\lambda \ll \epsilon \ll 1$ it can be simplified to

$$q_1(\lambda, \epsilon) \approx \ln|\epsilon-\lambda| + 1 + \frac{\lambda}{3\epsilon} + \ldots \qquad 8(24)$$

Finally the eigenvalue equation has the form:

$$1 + \frac{1}{4}\left(\frac{\kappa l}{2\pi}\right)^2 q_1(\lambda, \epsilon) = 0; \qquad 8(25)$$

it should be sufficient for determining the average mass μ of the charged leptons and the coupling constant, which for the photons is identical with Sommerfeld's fine structure constant.

For the numerical calculation it is convenient to eliminate the constant $\kappa l/2\pi$ (which is not very accurately known) by comparing 8(25) with the eigenvalue equation 5(20) of the pion. From 5(20) and 8(25) we get, since $\lambda=0$ for the photon:

$$q_0(\lambda)_{\text{pion}} = q_1(0, \epsilon). \qquad 8(26)$$

Using the expansion

$$q_0(\lambda) \approx \ln\lambda - \tfrac{3}{2} \quad \text{for } \lambda \ll 1 \qquad 8(27)$$

and 8(24) we derive

$$\ln\lambda_{\text{pion}} - \tfrac{3}{2} \approx \ln\epsilon + 1, \qquad 8(28)$$

$$\epsilon = \lambda_{\text{pion}} \cdot e^{-5/2} \qquad 8(29)$$

and, using the empirical value $\kappa_{\text{pion}} \sim 140$ MeV:

$$\mu = \frac{\kappa_{\text{pion}}}{3 \cdot 5} \approx 40 \text{ MeV}. \qquad 8(30)$$

This value is in reasonable agreement with the observations since one does not know so far which kind of average value should be taken between electron and muon mass.

The theory of the coupling constant can be taken directly from 6-2.

        Unified Field Theory of Elementary Particles

Denoting the coupling constant of the photon by e instead of f, we get from 6(29):

$$\frac{e^2}{4\pi} = -\frac{\pi}{\{\partial q_1(\lambda, \epsilon)/\partial\lambda\}_{\lambda=0}} = \frac{3\pi}{2}\epsilon = 0\cdot386\,\frac{\kappa_{\text{pion}}^2}{\kappa^2}. \qquad 8(31)$$

$e^2/4\pi$ (corresponding to $e^2/\hbar c$ in the older notation) is identical with Sommerfeld's[65] fine structure constant. Inserting in 8(31) the empirical values of the pion mass and the nucleon mass, gives

$$\frac{e^2}{4\pi} = \frac{1}{120} \qquad 8(32)$$

in satisfactory agreement with the empirical value $\frac{1}{137}$.

The norm of the photon state is positive on account of the negative sign of $\{\partial q_1(\lambda, \epsilon)/\partial\lambda\}_{\lambda=0}$ (compare §6-2). Therefore the photon—contrary to the spin 1 bosons of Chapter 5—is a real particle and not a ghost state.

8-5 Special properties of the photon and the electromagnetic field

The rest mass zero of the photons introduces some simplifications which are missing for the other bosons and which constitute characteristic features of electrodynamics.

The fundamental field equation 3(1) contains in its interaction term matrix elements representing the interaction of the field $\chi(x)$ with the photon field 8(14). These terms could be written in the form of the phenomenological theory

$$-i\sigma_\nu\frac{\partial\chi}{\partial x_\nu} + e.q.\sigma_\nu\chi A^\nu + \ldots = 0, \qquad 8(33)$$

where q is the charge quantum number of 8(17). If one considers matrix elements of 8(33) belonging to the creation of a fermion with momentum p_μ, one may take as intermediate state in the interaction term a state of momentum $p_\mu - J_\mu$; then A^ν produces a photon of momentum J_μ. If p_μ is nearly on the mass shell of an existing fermion with mass κ_F ($p^2 - \kappa_F^2 = 0$), $p_\mu - J_\mu$ can be chosen to be very nearly on this mass shell as well, since on account of $J^2 = 0$ the J_μ can be chosen arbitrarily small. Therefore also χ in the interaction term refers to the same particle. Hence the coupling does not depend on the amount of the 'free field' with mass κ_F, which is present in the field operator $\chi(x)$, since the field operators $\chi(x)$ in both terms of 8(33) can refer to the same infinitesimal surrounding of the mass shell κ_F. The dressing of the fermion lines in 6(16) is completely unimportant for the calculation of the coupling. This fact is in conventional quantum

Quantum Electrodynamics **119**

electrodynamics expressed by the Ward [66] identity. For the present purpose it is sufficient to conclude from 8(33) that the electromagnetic interaction is universal in the following sense. Except for the quantum number q, which has the eigenvalues ± 1, 0 for the fermions contained in $\chi(x)$ and which enters as factor, the interaction is the same for all these fermions. Since all operators creating stationary states can be constructed from products of $\chi(x)$ and $\chi^*(x)$ according to 2(14), and since the additivity of charge follows from Maxwell's equations, it is seen that all electric charges must be integer multiples of the elementary charge e. Even the leptons which on account of their double-pole nature in 3(39) have no immediate interaction with bosons like η- or π-meson, have the same interactions with the photons as the baryons. The universality of electromagnetic interaction is the consequence of the rest mass zero of the photons.

The electromagnetic current can be taken as a consequence of Maxwell's equations, which follow from the symmetry of the photon and the structure of the photon interaction vertex

$$q = \int d^3 x J_{0,q}(x) \quad \text{and} \quad \frac{\partial J_{\mu,q}(x)}{\partial x_\mu} = 0. \qquad 8(34)$$

$J_{\mu,q}(x)$ is a vector current like $J_{\mu,k}(x)$, which in the same way may depend on the field operators $\chi(x)$ in a finite surrounding of x. Again a representation of the type 3(61) may be possible; but this possibility depends upon the singularity of the 2-point function near the origin which is not sufficiently known at present.

8-6 The lepton masses

The leptons appearing as double poles in the approximative 2-point function 3(39) do not take part in the strong interaction. Therefore their mass, in this approximation, is due to their interaction with the photons. The electromagnetic interaction on the other hand leaves a scale factor undetermined, and it is essentially the eigenvalue equation of the photon which contains the information concerning the masses of the leptons. When in the first approximation the baryon octet is represented by one single pole in the 2-point function, one cannot expect more in this approximation than a determination of the average lepton mass.

If one wants to go further, one should first notice that the form 8(16) of the photon eigenfunction in isospace means that this eigenfunction is composed of equal contributions from the charged baryons in the octet according to the formula

$$\phi \sim \text{const}\,(P\bar{P} + \Sigma^+ \bar{\Sigma}^+ - \Sigma^- \bar{\Sigma}^- - \Xi^- \bar{\Xi}^-). \qquad 8(35)$$

The eigenvalue equation 4(22) of the photon

may be written as

$$\phi = (FG).\text{vertex}.\phi. \qquad 8(36)$$

The vertex part from the interaction term in 3(1) acting within the space of baryon–antibaryon pairs in the octet multiplies ϕ with a constant factor and leaves the form 8(35) unchanged, as one can see by direct calculation. The factors (FG) however depend on the masses and would therefore destroy the form 8(35), if F would refer to different baryon masses regularized by the same lepton mass. The simultaneous validity of 8(35) and 8(36) requires that the four factors FG belonging to the four masses of P, Σ^+, Σ^-, Ξ^- are equal. This can be achieved only by assuming at least two different lepton masses for regularization, one belonging to P, the other to Ξ^-. The terms FG for Σ^+ and Σ^- can then be brought to the correct value by using a mixture (roughly half and half) of the two leptons in the function F.

The masses of the two leptons are finally determined by the requirement, that the factor FG for proton or Ξ^- has the same value as the factor FG calculated in §8-4 for the average baryon mass and the average lepton mass. For carrying out this calculation we will assume the same analytical form of F for both baryons, proton and Ξ^-:

$$F_{\text{Ba}} = i(2\pi)^{-4}.w_{\text{Ba}}.\int d^4p \left(\frac{\kappa_{\text{Ba}}^2-\mu^2}{p^2-\mu^2}\right)^2 \frac{p_\nu \overline{\sigma^\nu}}{-p^2+\kappa_{\text{Ba}}^2}.e^{-ip(x-x')}, \qquad 8(37)$$

where the weight w_{Ba} should be determined by the residue of the pole at $p^2=\kappa_{\text{Ba}}^2$ in the complete 2-point function 4(18). From 4(18) we get

$$w_{\text{Ba}} = \left[\frac{\mathcal{M}(p^2)}{p^2\,\mathcal{L}'(p^2)}\right]_{p^2\,=\,\kappa_{\text{Ba}}^2}. \qquad 8(38)$$

The eigenvalue equation 8(25) for the determination of the lepton mass is then to be replaced by

$$1+\frac{1}{4}\left(\frac{l\kappa_{\text{Ba}}}{2\pi}\right)^2 w_{\text{Ba}}.q_1(0,\epsilon) = 1+\frac{1}{4}\left(\frac{l}{2\pi}\right)^2 \frac{\mathcal{M}(\kappa_{\text{Ba}}^2)}{\mathcal{L}'(\kappa_{\text{Ba}}^2)}.q_1(0,\epsilon) = 0, \qquad 8(39)$$

where $\epsilon=(\mu/\kappa_{\text{Ba}})^2$, and μ is the mass of the lepton belonging to this baryon (proton or Ξ^-). The values of $\mathcal{L}'$ can be taken from Fig. 3 and 7(38)

$$\mathcal{L}' = \frac{\partial\mathcal{L}(p^2)}{\partial(p^2)} = -\frac{1}{\kappa^2}\frac{\partial E}{\partial\lambda}; \qquad 8(40)$$

κ in 8(40) is the average mass of the octet (~ 1150 MeV). The function $\mathcal{M}(p^2)$ is not well known in the present stage of the theory. The complicated integrals in 4(30) have not yet been evaluated, and 5(30) can be considered as a reasonable approximation only, if the right-hand side of 5(30) agrees at least qualitatively with 4(30). Still two rather different simplifications suggest themselves; and the true behaviour of $\mathcal{M}(p^2)$ may lie between these limits. One may either argue that the values of $\mathcal{M}(p^2)$ for the different baryon masses in the octet should be approximately equal. Or one may conclude from 5(30) that $\mathcal{M}(p^2)$ should vanish at the second zero of $E(\lambda)$, since there should be no pole of $F(p^2)$ at this second zero. Hence $\mathcal{M}(p^2)$ at the baryon pole of $F(p^2)$ should be approximately proportional to the difference between p^2 at the pole and p^2 at the second zero of $E(\lambda)$. The two cases will be distinguished as I and II in the table. The 'average' values of the table have been calculated as average values between Λ- and Σ-baryon, since the values for Ξ are very uncertain on account of the neighbourhood of the minimum of $E(\lambda)$ in Fig. 3 to the point of intersection for the Ξ. The table gives the calculated weights w_{Ba} multiplied with κ_{Ba}^2 except for a constant factor, which is unimportant, if the masses of the leptons are determined from a comparison with the mass of the pion, as in 8(26).

		Ξ	Σ	Average	Λ	N
$w_{\text{Ba}} \cdot \kappa_{\text{Ba}}^2 \cdot \text{const}$	I	11·35	1·6	1·22	0·848	0·421
	II	1·395	1·27	1·12	0·974	0·636

The masses of the two leptons can now be calculated from the equation

$$(w_{\text{Ba}} \cdot \kappa_{\text{Ba}}^2)_{\text{Average}} \cdot q_0(\lambda_{\text{pion}}) = (w_{\text{Ba}} \cdot \kappa_{\text{Ba}}^2) \cdot q_1(0, \epsilon), \qquad 8(41)$$

which replaces 8(26).

Using the numerical values of the table we get for the masses of the two leptons (in MeV):

	Ξ^-	P
I	590	0·14
II	80	3·6
Observed	105·66	0·511

The lepton regularizing Ξ^- can be identified with the muon, the other lepton regularizing the proton can be identified with the electron. The theory gives a qualitative understanding of the lepton masses in spite of the rather widely diverging results of the two extreme simplifications I and II. The terms FG in the eigenvalue equation of the photon belonging to Σ^+ and Σ^- can be regularized by a 'mixture' of the two leptons.

In the present stage of the theory one could not exclude the existence of other leptons. The leptonic number is conserved for each kind of leptons separately, since pair creation or annihilation by photons is the only process producing or destroying leptons, if weak interaction is neglected. For a more complete understanding of the leptons a more thorough analysis of the functions $\mathcal{L}(p^2)$, $\mathcal{M}(p^2)$ and $E(\lambda)$ along the lines of 4(29), 4(30) and 7(38) would be necessary.

9

Relations of the Unified Field Theory to Phenomenological Theories

9-1 The so-called bootstrap method [67]

The term 'phenomenological theory' will, in the context of the present survey, be used for a theory which tries to connect various empirical data on a theoretical basis without attempting to formulate explicitly an underlying natural law. Theories of this kind can lead to very useful representations of the observed phenomena and may appear later as result of a certain approximation method applied on the complete theory. As a famous historical example we mention the representation of the planetary orbits by cycles and epicycles in the astronomy of Ptolemäus. The cycles and epicycles could later be interpreted as the first terms in a Fourier expansion of the true Newtonian orbits.

In elementary particle physics, the 'bootstrap' method makes use of the intuitive idea that an elementary particle can be considered as composed of others and that in favourable cases one special composition out of the many possible compositions can be taken as the most important one; that furthermore, among all the interactions between the particles concerned, one special interaction may have the strongest influence. A study of the analytical behaviour of the S-matrix elements belonging to these most important contributions may lead between the various masses to relations which step by step may be used for improving the values of the constants, and which can be checked by experiments. In favourable cases they might be sufficient for determining the mass ratios and coupling constants.

A well-known example for this bootstrap method is the interaction between pions and ρ-mesons. It may be a good approximation to consider the ρ-meson as composed of just two pions. Among the forces binding the two pions together, the exchange of one ρ-meson may be the most important one. In this case there are essentially only two unknown constants, the

123

 Unified Field Theory of Elementary Particles

ratio of the masses of pion and ρ-meson and the coupling constant of the $\pi\pi\rho$-vertex which should be determined by relations between S-matrix elements.

This scheme can, in the framework of the unified field theory 3(1), be understood as the result of some special Feynman graphs appearing in high Tamm–Dancoff approximations. In fact one could for the ρ-meson derive from 3(1) an eigenvalue equation of the type

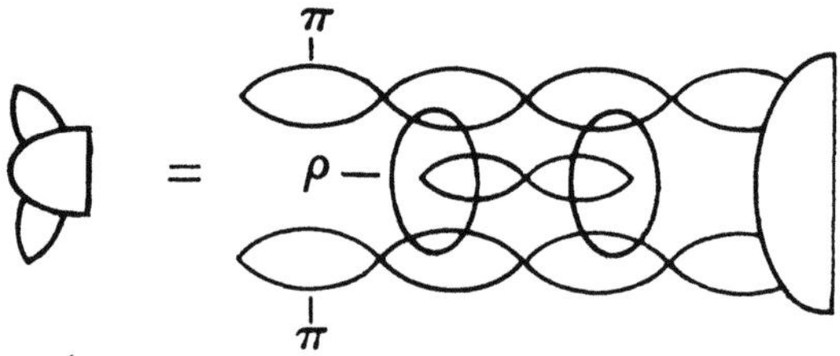

$$9(1)$$

which corresponds to the prescription of the bootstrap method. This picture however does not answer the question why the graph 9(1) should be so much more important than many other possible graphs. At this point one is left with the general argument that poles at low mass values are presumably more important for the behaviour of the functions than poles and cuts at high mass values; and one may argue that, within the two boson octets, only the ρ-meson can decay into two pions. It seems to be quite impossible to know beforehand what degree of approximation can be obtained by picking out just one such graph of the type 9(1).

In principle of course one could include in the bootstrap formalism a great number of possible compositions and processes compatible with the assumed symmetry. The degree of complication would then correspond rather closely to the degree of complication in a similar high-order Tamm–Dancoff approximation. Whether calculations on such a basis could actually be carried out is, however, another question.

A bootstrap formalism interpreted in this very general way may be fundamentally equivalent to the unified field theory 3(1), if (and only if) it starts from the same group structure. In fact, besides the group structure both 3(1) and the bootstrap formalism contain just the principle of relativistic causality. Eqn. 3(1) formulates this principle by means of a local differential equation, the bootstrap method formulates it by statements concerning the analytical behaviour of S-matrix elements which are again indirectly taken from field theory. It may be somewhat complicated to include in the bootstrap formalism electrodynamics and the breaking of symmetries by the ground state, but even this should not be impossible in principle.

Unified Field Theory and Phenomenological Theories **125**

9-2 Analytic behaviour of S-matrix elements

The unified field theory supposes the existence of asymptotic states and of a unitary S-matrix. The analytic behaviour of the S-matrix elements is restricted by the condition that it must be compatible with that degree of relativistic causality which follows from the local differential equation 3(1). This degree of causality seems to be sufficient for the description of the experiments. It may give slightly more freedom for the analytic behaviour than the *Mandelstam* relations and the other conventional dispersion relations,[19] but this freedom is probably not important. Therefore we expect that the major part of the consequences of conventional dispersion theory will also follow from the unified field theory. There may be special features of the unified field theory coming in through the special way of regularization for the approximate 2-point function 3(39). The regularization is carried out by introducing a double pole at the lepton masses, in agreement with conventional quantum electrodynamics. In the usual treatment of dispersion problems, poles or double poles at the lepton masses and poles or cuts corresponding to the existence of the photons are not considered. In that approximation in which the problems of electrodynamics can be neglected the difference is probably unimportant.

With respect to the problem of the Regge [68] poles it should be emphasized, that in the unified field theory all poles corresponding to stationary states should be of the Regge type, since all stationary states can be interpreted as bound states consisting of other (not necessarily smaller) units. The only possible exception could be the neutrinos which are excluded from all normal interactions. If the weak (radio-active) interactions are taken into account even the neutrinos may appear as bound states.

9-3 The approximate symmetries (SU_3, SU_6, etc.)[12]

The empirical spectrum of elementary particles contains a number of wide multiplets (compare §1-4 and §7) which can be interpreted as result of an approximate invariance under the group SU_3 of the underlying natural law. If the Dirac spin of the particles is included in the group theoretical analysis, some of these multiplets can be combined to even larger families of particles which may be considered as big multiplets produced by higher approximate symmetries of the type SU_6, U_{12}, etc. These symmetries seem to be approximate from the very beginning, not exact symmetries perturbed by an asymmetrical ground state; because no 'Goldstone-particles' (bosons of rest mass zero) are seen which could be responsible for the large deviations from the supposed symmetry. Therefore the general question arises whether such approximate symmetries can be understood from the special equation 3(1) which does not seem to contain any immediate reference to higher groups as SU_3, SU_6, etc.

For a general understanding of this situation one should perhaps start with a statement which is meant not as a mathematical law but as a general rule allowing exceptions: 'Any simple object which is strictly symmetrical under a group g has a good chance of being approximately symmetrical under larger groups G containing g as a subgroup, if the number of degrees of freedom is sufficient'. The meaning of this statement can be explained for example by the symmetries of the patterns produced in a kaleidoscope. Let us assume that the mirrors of the kaleidoscope guarantee the exact hexagonal symmetry of the pattern. Besides that the pieces of broken glass etc. in the kaleidoscope are distributed more or less at random. Still frequently we see higher approximate symmetries; for example approximate circles in nice colours or dodecagonal patterns. The original hexagonal symmetry is a subgroup of the circular symmetry or of the dodecagonal symmetry. We may mention another example: a vase which is constructed as a cylindrical vessel, for which height and diameter are not too different, may easily have an approximate spherical symmetry. There may be large deviations from the higher symmetry, still it is easily recognized in spite of these deviations. It should therefore be emphasized that the statement of such an approximate symmetry is a rather weak statement, if the deviations are large. A third example may be taken from atomic physics. The Schrödinger equation of a heavy atom is strictly invariant under rotations in space. It is not invariant under independent rotations of either the spins or the orbits of the electrons. Still in some parts of the spectrum the spin-orbit interaction is relatively small; therefore the resulting approximate symmetry can be recognized through the existence of the optical multiplets.

Hence it is certainly not surprising that a simple mathematical object like Eqn. 3(1) leads to approximate symmetries under higher groups containing the exact groups of 3(1) as subgroups. It is also understandable that the groups of unitary transformations of a number of complex variables play an important rôle, since in the Hilbert space of the asymptotic states the unitary transformations correspond to the canonical transformations of classical mechanics. If by chance a few stationary states have the same mass, the unitary transformations between these states may easily produce an approximate symmetry. Therefore it is rather the vagueness of the concept of approximate symmetry to which we attribute the apparent validity of an approximate SU_3, SU_6, etc. invariance.

If one wants to follow the appearance of these symmetries in detail, one

Unified Field Theory and Phenomenological Theories   **127**

should first notice that the stationary states in the spectrum of 3(1) can be divided into two distinct groups (with possibly a few intermediate cases): on the one hand the states which can be produced by field operators at only one point in space and time (namely by $\chi(x)$, $\chi(x)\chi^*(x)$ and $\chi(x)\chi^*(x)\chi(x)$), S-states, for which the contact interaction plays the decisive rôle; on the other hand all other states, for which—if considered as composed of the states of the first kind—long-range forces determine the binding and thereby, indirectly, the masses (compare §7-5). There may be intermediate cases like the vector mesons ρ, ω, K^*, ϕ, which should be a mixture of S- and D-states of the fermion–antifermion system. As S-states they would belong to the first kind, as D-states to the second. If they are predominantly D-states (which seems to be more likely) they should be counted with the second kind. Finally only the two octets N, Λ, Σ, Ξ and π, K, η, and in some way the photon, belong to the first kind. These two octets, as they are derived from 3(1), do in fact simulate the SU_3-symmetry in the special form of the 'eight-fold way';[12] the *Okubo* mass formula follows from the arguments described in §§7-2 and 7-4. This happens in spite of the asymmetry with respect to SU_3 of the fundamental contact interaction $\underset{\rho}{\times}$. The summation over the contributions from various virtual processes to the masses seems to be responsible in the case of the fermions for the mass formula simulating the approximate SU_3-symmetry.

The stationary states of the second kind can be considered as 'composed' of those of the first kind, the long-range binding forces being due to fields belonging to the first or second kind. It is plausible that also parts of this second spectrum simulate the approximate SU_3-symmetry, if the spectrum of the composing particles has this property. Whether this general explanation is sufficient for an interpretation of the spectrum can be decided only by explicit calculations concerning the higher states which have not been carried out so far.

The classification of the approximate symmetries SU_3, SU_6, U_{12}, etc. as secondary structures should not be misunderstood as an undervaluation of their heuristic value. It may be that in elementary particle physics these symmetries play a similar rôle as the concept of valency in chemistry. The heuristic value of the valency concept in chemistry is probably greater than that of quantum mechanics or of the Schrödinger equation in this field. It would be difficult to derive from quantum mechanics definite limitations for the applicability of the valency concept or to understand in detail why it is so successful. But considering the degree of complication in quantum chemistry this concept is probably the most efficient tool for predicting the results of chemical experiments. In a similar way the approximate symmetries may be the most efficient tool in elementary particle physics for predicting new resonances or similar phenomena.

9-4 The weak interactions

The weak or radio-active interactions are distinguished from the strong and the electromagnetic interactions by the violation of several symmetries and their resulting selection rules. Isospin charge (the third component of the isospin) and hypercharge are not always conserved separately, only the total charge is conserved; the space reflection parity is broken and neutrinos may be created in radio-active decay. Since all these selection rules did follow from the fundamental field equation 3(1), when the ground state was assumed to be symmetrical under the corresponding transformations, the weak interactions should be caused by some small additional asymmetry of the ground state. In the framework of the present theory the weak interactions should therefore be interpreted on a similar basis as the electromagnetic forces. The additional deviation from symmetry must be very much smaller than the asymmetry in isospace responsible for electromagnetism, since the lifetimes for radio-active decay are larger by a factor 10^6 or more than the lifetimes for electromagnetic decay. The 2-point or 4-point functions characterizing the ground state hence seem to contain very small terms not invariant under space reflexion or under the single gauge groups connected with isocharge or hypercharge. If this assumption is correct, the general arguments of §7-1 apply. The theorem of Goldstone would again lead from the asymmetry of the ground state to the existence of particles of rest mass zero. In this case the experimental results suggest that neutrinos or pairs of neutrinos might play the rôle of the Goldstone particles. The existence of the neutrinos would then in turn be responsible for the breaking of the parity. This statement does not in principle contradict the fact that parity violating weak decays can occur without the emission of neutrinos (e.g. $\Lambda \to \pi + N$). Even here the analogy to electromagnetism is complete. Isospin- or G-parity violating decays can occur through the action of Coulomb forces, without the emission of photons (e.g. $\eta \to 3\pi$). The usual Goldstone argument would, for the radio-active decay, require some modification, since it is primarily a discrete group, the space reflection parity, which is broken by the 'weak' asymmetry of the ground state. Parity is not a classical observable, therefore the argument, that the Goldstone particles must be bosons, could possibly fail. The neutrino with a definite helicity is in the same way a one-to-one mixture of two states with opposite parity, as the photon is a one-to-one mixture of two states with isospin 0 and 1. From this point of view the interpretation of the neutrino as Gold-

Unified Field Theory and Phenomenological Theories   129

stone particle is plausible. On the other hand a theory of the weak interactions on this basis has not yet been worked out. The numerical value of
the Sommerfeld fine structure constant, which is about ten times smaller
than that of the corresponding π–N coupling constant, could be understood
[8(31) and (32)] as a result of the usual relations for coupling constants
applied on a boson of spin 1 and rest mass zero. One cannot see at present,
how the extremely small value of the coupling constant in the Fermi interaction for radio-active decay could follow from similar relations.

Still there are mainly two empirical facts which suggest a close connexion
of the fundamental equation 3(1) with the phenomena of weak interaction;
these are the universality of weak interactions and their transformation
properties in Lorentz space.

In a phenomenological theory the weak interactions[69] can in a first
approximation be represented by a Lagrangian

$$\mathscr{L} = G j_\mu^*(x) j_\mu(x), \qquad\qquad 9(2)$$

where G is the coupling constant and the 'weak current' $j_\mu(x)$ is in Lorentz
space given by

$$j_\mu(x) = \sum_{a,b} \bar{\psi}_a \gamma_\mu (1+\gamma_5) \psi_b. \qquad\qquad 9(3)$$

The indices a, b refer to the particles and (using the abbreviation $p=$ proton,
$n=$ neutron, $\Lambda = \Lambda$-particle, $\nu, \nu' =$ neutrinos, $e=$ electron, $\mu=$ muon) the
current contains the terms:

$$j(x) = \bar{\psi}_\nu \ldots \psi_{e^-} + \bar{\psi}_{\nu'} \ldots \psi_{\mu^-} + \bar{\psi}_p \ldots \psi_n + \bar{\psi}_p \ldots \psi_\Lambda. \qquad 9(4)$$

The current creates or annihilates electric charge and is in so far a 'charged'
counterpart to the 'neutral' electromagnetic current.

The interaction 9(2) is in fact very similar to the fundamental interaction
in 3(1). The factor $1+\gamma_5$ in 9(3) indicates, that 9(2) actually represents an
interaction between Weyl-spinors of the type $(\chi^* \sigma_\mu \chi)(\chi^* \sigma^\mu \chi)$; the parity
symmetrical terms $(\tilde{\chi}^* \overline{\sigma_\mu} \tilde{\chi})(\tilde{\chi}^* \overline{\sigma^\mu} \tilde{\chi})$ are omitted. For the baryon currents
the expressions 9(3) are later slightly modified by the 'dressing' of the
baryons, but 9(3) seems to be the primary interaction. The fact that all
currents in 9(4) appear with approximately the same numerical factor,
suggests that all particles are created by the same fundamental spinor $\chi(x)$
and that the universality of the coupling is produced in the same way as the
universality of the electromagnetic coupling. On account of this universality
the weak interaction can be understood only from a 'unified' theory of
elementary particles, a theory in which all elementary particles appear as
solutions of the same underlying natural law. The form 3(1) of the fundamental field equation was in fact first suggested by the empirical form 9(2)

and (3) of the weak interactions. On the other hand, a theory of the weak interaction could scarcely be built on the basis of 3(1) before the spectrum of the leptons has been understood completely from this theory.

Beyond the radioactive interactions there seems to be another small interaction asymmetrical even under PC, which can be seen in the decay of the K_2-meson. Again this interaction should be interpreted as the result of small PC-asymmetrical terms in the generalized Green's functions characterizing the properties of the ground state. The smallest interaction finally is gravitation.[78] Since the gravitational field is a boson field of long range corresponding to particles of zero mass, it has many properties of a genuine Goldstone field and should therefore eventually find an interpretation on this basis. The universality of gravitation is another argument in favour of this view.

As can be seen from these discussions, the classical fields, electromagnetism and gravitation, appear in the present theory as a result of certain features of the actual universe which may be called contingent in the sense that they do not follow alone from the fundamental field equation 3(1). They are connected with special assumptions about the ground state, about the cosmological model of the world, which are permitted but not enforced by the underlying natural law 3(1). Again one should stress at this point the analogy to the theory of general relativity. The normal forces of inertia, like the centrifugal force, appear in general relativity as consequence of the boundary conditions or possibly—if Mach's principle can be applied —as a reaction from the very distant masses; they do not follow from the field equations alone and are, in this sense, contingent features of the actual world.

9-5 The high-energy jet-showers

The opposite limiting case to the weak interactions is realized in the interaction during a collision process of extremely high energy. In cosmic radiation showers have been observed produced by protons of energies up to 10^5 GeV. If such a proton collides with another proton in a hydrogen atom the energy in the c.m. system may be as high as a few hundred GeV. This energy would be very much bigger than the mass of any known particle or resonance state. Therefore one should expect that in these showers at least in the first stage of their development the special properties of the ground state play only a secondary rôle, the symmetries of the underlying natural law should be clearly visible. If 3(1) contains the correct formulation of this law, the conservation laws, not only for energy, momentum, angular momentum, baryonic and leptonic number, but also for isospin,

Unified Field Theory and Phenomenological Theories 131

strangeness should be valid with great accuracy; but there should be no good conservation law corresponding to the groups SU_3, SU_6, etc.

In these showers, as a rule, many secondary particles are produced; the average energy of a secondary particle in the c.m. system seems to increase only slowly (perhaps logarithmically) with the primary energy, therefore the number of secondaries increases with the total energy transferred into the secondaries slightly less than linearly. The total cross-section for a collision producing at least one pion seems to be nearly the geometrical one, given by the size of the colliding particles. One may however ask about the cross-section σ_ϵ for a process, in which a total energy $\geqslant \epsilon$ (in the c.m. system) is transferred into secondary particles. If the primary energy (in the c.m. system) is E, and if we consider only those cases where both E and ϵ are large compared with the masses of the important resonances, the invariance of 3(1) under the scale transformation would lead to the formula [70]

$$\sigma_\epsilon \approx \frac{\text{const}}{\epsilon^2} f\left(\frac{\epsilon}{E}\right).$$
9(5)

For $\epsilon \ll E$ the function $f(\epsilon/E)$ would be nearly constant, σ_ϵ would depend only weakly on the primary energy E. The formula 9(5) seems to be compatible with all the experimental material[15] on jet-showers up to 10^5 GeV, thereby indicating the importance of the scale transformation.

The jet-showers contain still another important information. The secondaries are frequently produced in several intermediate steps: In the first act some high resonance states are produced, which then decay very quickly into lower resonance states of longer lifetime, they decay again etc. and only at the end of this process many stable particles have been created. The transversal momenta of the secondaries can be taken as an indication concerning the masses of the resonance states involved (the term 'resonance states' in this context refers to states with a lifetime longer than the time necessary for their creation, say roughly 10^{-24} sec). Resonance states with large masses would, as a rule, produce secondaries with transversal momenta comparable to say one-half of their masses. Empirically the transversal momenta, even for the jet-showers of highest energy, seem to be distributed according to a Gaussian $\sim \exp(-p^2/p_0^2)$, where $p_0 \sim 0.4$ GeV. Transversal momenta > 1.5 GeV have never been observed with certainty. Therefore resonances with masses > 3 GeV virtually play no rôle in high-energy collisions. If they should exist, they are created only with a very small probability. This result is another argument against a fundamental SU_3-structure in the underlying natural law. If SU_3 would be fundamental one would expect that the simplest non-trivial representation of this group, the three-dimensional representation, plays a rôle in physics; the corresponding particles have not been found in the mass region up to 2·5 GeV; therefore they should have still larger masses, if they existed. The distribution of the transversal momenta in the jet-showers makes their existence very unlikely.

The general behaviour of the jet-showers goes well together with the other empirical fact, that the characteristic length in elementary particle physics, for example in the experiments of Hofstadter[71] on the electro-magnetic structure of nucleons, is of the order $0·5 . 10^{-13}$ cm, not very much smaller. The existing experiments may not definitely exclude much smaller structures, but they certainly do not give any evidence for structures below 10^{-14} cm.

10

Concluding Remarks

The unified field theory presented in this book is an attempt to formulate the natural law which is visible behind the complicated spectrum of elementary particles, their interactions and symmetries, and behind the connexions formulated in the existing phenomenological theories. The great wealth of experimental data on elementary particles collected during the last fifteen years may be considered as a sufficient reason for believing that such an attempt is not premature in the present state of physics, that actually all the important facts are known which are needed for a formulation of the theory.

Taking this point of view, it may be useful to recollect briefly the arguments for choosing the field equation 3(1) as basis of the theory. Any attempt to formulate the underlying natural law must necessarily start from answering a number of critical questions concerning two main topics: the group theoretical structure of this law, and the mathematical tools for its formulation.

In group theory the critical questions may be formulated by asking: What are the exact symmetries of the underlying natural law? Are the approximate symmetries which are found in nature due to exact symmetries broken by an asymmetry of the ground state 'world', or are they the result of dynamics, valid in a rough approximation only for a limited range of phenomena? The answers to these questions are decisive for the mathematical formulation of the unified theory.

In the present attempt the dividing line between the exact symmetries and the approximate symmetries of dynamical origin has been drawn between the groups U_2 and SU_3. The reasons for this decision have been discussed extensively in the text of this book. Perhaps the most striking argument in favour of this dividing line is the existence of a Goldstone particle for U_2, the photon, and the non-existence of such a particle for SU_3.

With respect to the mathematical tools the critical questions are: Is the S-matrix alone a sufficient mathematical instrument for formulating the theory, including its causal structure? Or can one use the concept of a local

133

 Unified Field Theory of Elementary Particles

field operator, commuting (or anticommuting) at space-like distances and obeying a differential equation? Can the Hilbert space on which these field operators act be constructed from the asymptotic operators alone?

The present theory gives a negative answer to the last, but a positive answer to the second question. The resulting indefinite metric in Hilbert space introduces a marked extension of the conventional mathematical scheme of quantum mechanics. This extension can be interpreted physically as stating, that the concept of probability cannot be applied to the 'local' region, to dimensions in space of the order 10^{-14} cm or less. The universal length entering into the theory through the constant l or through the mass of the proton may be defined as that length below which the quantum theoretical concept of probability fails.

These answers to the critical questions may not be the only possible answers. But they seem to be the simplest, considering the fact that it has not been possible so far to construct a relativistic theory with interaction within the narrow framework of the Wightman axioms.

If the answers to the critical questions are accepted, both with respect to group theory and to the mathematical tools, the fundamental equation 3(1) seems to be uniquely determined.

The concept of a unified field theory of elementary particles has another aspect which needs some discussion; that is its relation to other parts of physics. In the first instant it would seem that all laws in other parts of physics should simply be considered as consequences of the one fundamental law; that this universal law should comprise all possible phenomena. It is mainly this feature of the unified theory which has raised much criticism, since it seems too ambitious a programme to formulate all physical laws in one simple equation.

Two answers can be given to this criticism. First one may say that the claim of universality is not due to any ambitious programme but is an unavoidable consequence of the fact that the elementary particles are the smallest units of matter. The relation of quantum mechanics to chemistry is similar: since all atoms and molecules consist of nuclei and electrons, it is unavoidable that the fundamental equations of quantum mechanics must comprise all chemistry. This does not mean that all chemical problems have been solved or could be solved by quantum mechanics. On the contrary the heuristic value of quantum mechanics in chemistry has so far not been too impressive. Still we cannot doubt that the chemical phenomena take place within the framework of quantum mechanics. In a similar way the unified field theory of elementary particles must give the framework for all physical phenomena.

Secondly it should be emphasized that the fundamental equation does not determine completely the laws in all other parts of physics. For instance the electromagnetic laws do not follow from the equation unless one adds specific assumptions about the asymmetry of the ground state, i.e. about the cosmological model of the world. In a similar way radioactivity and gravitation are probably connected with the structure of the world in large dimensions. The boundary condition concerning the ground state is still flexible to some extent, it has to be adapted to the real world; and this adaptation is by no means trivial.

When the essential problems of high-energy physics have been solved some day, physics will, even aside from the question of boundary conditions, still be open at the side of the low energies. We know from experience that the complicated phenomena connected with the existence of life, of living organisms require for their description new concepts beyond those of conventional physics and chemistry. In the present state of biology it is an open question whether those biological concepts could find a place within the mathematical framework of quantum theory. One may hope that the great wealth of forms expressible, for example, by the statistical matrices in quantum theory, should be sufficient for describing even biological phenomena. But it may also be that new mathematical forms, different even in principle from those of quantum theory, are needed. Whatever the answer to this question may be, the answer is irrelevant for high-energy physics. And in turn the final formulation of the natural law underlying high-energy physics does not answer any one of the problems arising at the border-line between physics and biology. Therefore the apparent universality of the natural law underlying elementary particle phenomena should not lead to a misinterpretation of its relevance in other parts of natural science. The possibility and necessity of a unified field theory of matter should be judged primarily on its value for understanding the complicated phenomena of elementary particle physics.

Mathematical Appendix

AI.   The 2-Point Functions of Linear Field Theories

The Källén–Lehmann representation 3(25) of 2-point functions is based upon the 2-point functions of linear field theories. In the simplest case of a scalar field obeying a Klein–Gordon equation all the different 2-point functions like $\langle 0|\phi(x)\phi^*(y)|0\rangle$ or $\langle 0|T\phi(x)\phi^*(y)|0\rangle$ or $\langle 0|[\phi(x),\phi^*(y)]|0\rangle$ etc. can be expressed by an integral of the form

$$\Delta_C = (2\pi)^{-4} \int_C d^4p \, \frac{e^{-ipx}}{-p^2+\kappa^2}, \qquad\qquad \text{A(1)}$$

where C characterizes the paths of integration in the complex p_0-plane. These functions have first been introduced by Jordon and Pauli[72] and have later been studied in detail by Schwinger.[73]

If C defines a closed path in a finite region of the complex p_0-plane, the function Δ_C obeys the homogeneous Klein–Gordon equation

$$(\square+\kappa^2)\Delta_C = 0. \qquad\qquad \text{A(2)}$$

This follows from A(1), since the poles in the integral vanish in A(2) by the operation $\square+\kappa^2$. Four different solutions of the homogeneous equation A(2) are frequently used in the literature, specified as Δ, Δ_+, Δ_- and Δ_1. They are defined by

$$[\phi(x),\phi^*(y)] \qquad\qquad\qquad = -i\Delta(x-y) \qquad\qquad \text{A(3)}$$

$$\langle 0|\phi(x)\phi^*(y)+\phi^*(y)\phi(x)|0\rangle = \Delta_1(x-y) \qquad\qquad \text{A(4)}$$

$$\langle 0|\phi(x)\phi^*(y)|0\rangle \qquad\qquad = -i\Delta_+(x-y) \qquad\qquad \text{A(5)}$$

$$\langle 0|\phi^*(y)\phi(x)|0\rangle \qquad\qquad = i\Delta_-(x-y) \qquad\qquad \text{A(6)}$$

The corresponding paths of integration in the complex p_0-plane are given in the figure:

136

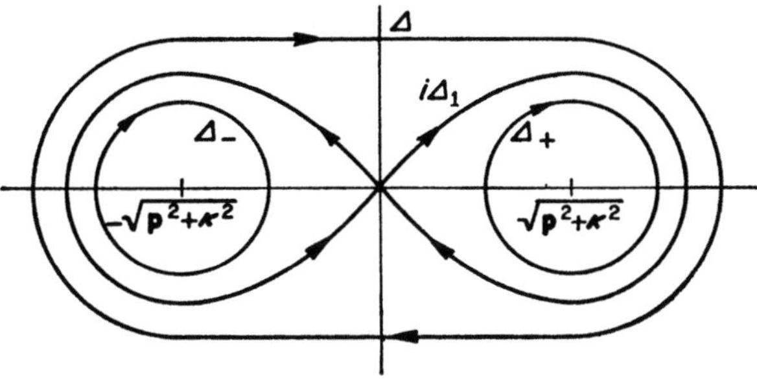

Fɪɢ. 8.

The function $\Delta(x-y)$ is zero for all finite space-like distances $(x-y)$, as can be seen by expanding (for $(x-y)_0=0$) the path of integration to a circle at infinity.

If a path of integration is chosen going from $p_0=-\infty$ to $p_0=+\infty$, the function A(1) does not obey the homogeneous Klein–Gordon equation; on account of

$$(2\pi)^{-4}\int_{-\infty}^{+\infty} d^4p\, e^{-ipx} = \delta^4(x) \qquad\qquad \text{A(7)}$$

Δ_C now fulfills the inhomogeneous equation

$$(\Box+\kappa^2)\Delta_C = \delta^4(x) \qquad\qquad \text{A(8)}$$

and corresponds in this way to a Green's function.

Five functions of this type are used in the literature: $\Delta_R, \Delta_A, \Delta_{1R}, \Delta_{1A}, \Delta_P$. The corresponding paths of integration are:

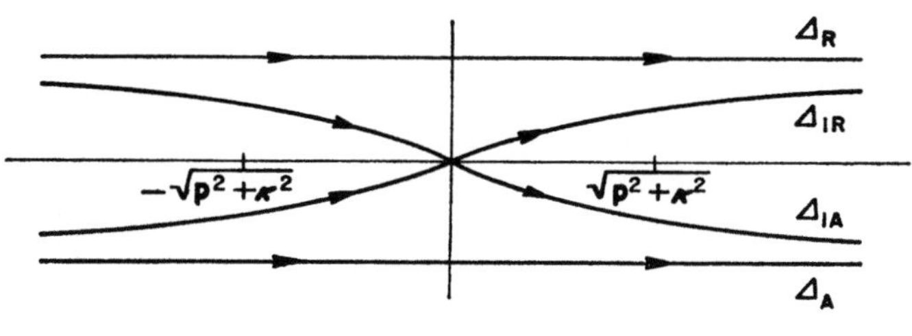

Fɪɢ. 9.

For $\Delta_P=\tfrac{1}{2}(\Delta_A+\Delta_R)$ the principal value has to be taken at the two poles. The retarded function Δ_R vanishes for negative values of x_0, the advanced function Δ_A vanishes for positive values of x_0.

The important time-ordered 2-point function is given by

$$\langle 0|T\phi(x)\,\phi^*(y)|0\rangle = -i\Delta_{1R}(x-y). \qquad \mathrm{A(9)}$$

The nine different Δ_C-functions are connected by a number of linear relations:

$$\left. \begin{aligned}
\Delta &= \Delta_+ + \Delta_- = \Delta_R - \Delta_A \\
i\Delta_1 &= \Delta_+ - \Delta_- = \Delta_{1R} - \Delta_{1A} \\
\Delta_P &= \tfrac{1}{2}(\Delta_R + \Delta_A) = \tfrac{1}{2}(\Delta_{1R} + \Delta_{1A})
\end{aligned} \right\} \qquad \mathrm{A(10)}$$

$$\left. \begin{aligned}
\Delta_{1R} + \Delta_- &= \Delta_R & \Delta_{1A} - \Delta_- &= \Delta_A \\
\Delta_{1R} - \Delta_+ &= \Delta_A & \Delta_{1A} + \Delta_+ &= \Delta_R \\
\Delta_P - \tfrac{1}{2}\Delta &= \Delta_A & \Delta_P + \tfrac{1}{2}\Delta &= \Delta_R \\
\Delta_P - \tfrac{i}{2}\Delta_1 &= \Delta_{1A} & \Delta_P + \tfrac{i}{2}\Delta_1 &= \Delta_{1R}
\end{aligned} \right\} \qquad \mathrm{A(11)}$$

Finally we denote the reality conditions:

$$\left. \begin{aligned}
\Delta_+^* &= \Delta_-, & \Delta_R^* &= \Delta_R \\
\Delta_A^* &= \Delta_A, & \Delta_{1R}^* &= \Delta_{1A}
\end{aligned} \right\} \qquad \mathrm{A(12)}$$

and the symmetry properties:

$$\left. \begin{aligned}
\Delta(-x) &= -\Delta(x), & \Delta_1(-x) &= \Delta_1(x), & \Delta_P(-x) &= \Delta_P(x) \\
\Delta_R(-x) &= \Delta_A(x), & \Delta_{1R}(-x) &= \Delta_{1R}(x), & & \\
\Delta_{1A}(-x) &= \Delta_{1A}(x), & \Delta_+(-x) &= -\Delta_-(x). & &
\end{aligned} \right\} \qquad \mathrm{A(13)}$$

If, in a linear field theory, the field operator is not a scalar, but a spinor or vector, the integrand in the formula corresponding to A(1) will contain in the numerator matrices referring to the spinor or vector indices of the field operator, which may be connected with the components of the momentum p to form invariant expressions. The denominator will, however, be the same, therefore the definition of different 2-point functions by means of the different paths of integration will be essentially the same.

For the calculations of the text the most important 2-point function is that of the time-ordered product; the special connexion of this function with the principle of relativistic causality has first been emphasized by Feynman.[46, 74] The function expresses in a natural way the fact that a particle can be annihilated only after it has been created.

AII.　The Lee Model [36, 28]

A II-1　Definition of the model

In the approximative 2-point function 3(39) the regularization is performed by means of a double pole at the average lepton mass; this mass is in 3(39) simply put equal to zero. The consequences of such a double pole have first been studied with the help of a model proposed by Lee as a simplified analogon to quantum electrodynamics. Therefore the analysis of this model shall be repeated here.

The model assumes the existence of a real boson field ϕ_θ, obeying the Klein–Gordon equation

$$(\Box + m_\theta^2)\,\phi_\theta = 0; \qquad\qquad \text{A(14)}$$

the corresponding particles are called θ-particles. Besides this boson field the model contains, in its simplest version, a particle of infinite mass localized at the origin, which may exist in two states, called N-state and V-state. The boson field interacts with this particle in such a way, that the transition from the V-state to the N-state occurs only by emission of a θ-particle, the transition from N-state to V-state only by absorption of a θ-particle.

$$V \to N+\theta, \qquad N+\theta \to V. \qquad\qquad \text{A(15)}$$

For a mathematical representation of this situation in the Schrödinger picture we use the field operator at the time $t=0$:

$$\phi_\theta(\mathbf{x}, 0) = \phi_\theta(\mathbf{x}) \qquad\qquad \text{A(16)}$$

and its Fourier transformed operator $\phi(\mathbf{k})$:

$$\phi_\theta(\mathbf{x}) = (2\pi)^{-3/2} \int\limits_{-\infty}^{+\infty} d^3 k\, e^{i\mathbf{kx}} \phi_\theta(\mathbf{k}). \qquad\qquad \text{A(17)}$$

The conjugated variable is

$$\dot\phi_\theta(\mathbf{x}) = \pi(\mathbf{x}) \quad\text{and}\quad \pi(\mathbf{x}) = (2\pi)^{-3/2} \int\limits_{-\infty}^{+\infty} d^3 k\, e^{i\mathbf{kx}} . \pi(\mathbf{k}). \qquad \text{A(18)}$$

Putting

$$\omega^2 = \mathbf{k}^2 + m_\theta^2 \qquad\qquad \text{A(19)}$$

we may introduce as creation- and annihilation-operators

$$a^*(\mathbf{k}) = \frac{1}{\sqrt{2\omega}}\,[\pi(\mathbf{k}) + i\omega\phi_\theta(\mathbf{k})], \qquad a(\mathbf{k}) = \frac{1}{\sqrt{2\omega}}\,[\pi(\mathbf{k}) - i\omega\phi_\theta(\mathbf{k})]. \quad \text{A(20)}$$

The commutation rules are:

$$[\pi(\mathbf{x}), \phi_\theta(\mathbf{x}')] = -i\delta^3(\mathbf{x}-\mathbf{x}'); \qquad [\pi(\mathbf{k}), \phi_\theta(\mathbf{k}')] = -i\delta^3(\mathbf{k}-\mathbf{k}'), \quad \text{A(21)}$$

$$[a(\mathbf{k}), a^*(\mathbf{k}')] = \delta^3(\mathbf{k}-\mathbf{k}'). \qquad\qquad \text{A(22)}$$

For the particle at the origin in the state N or V the operators ψ_N, ψ_N^* or ψ_V, ψ_V^* shall be defined as annihilation- and creation-operators, with the anticommutators:

$$\{\psi_V, \psi_V^*\} = 1, \qquad \{\psi_N \psi_N^*\} = 1, \qquad \{\psi_V, \psi_N^*\} = 0. \qquad \text{A(23)}$$

In these variables the Hamiltonian of the system is supposed to have the form:

$$H = m_V \psi_V^* \psi_V + m_N \psi_N^* \psi_N + \int \omega(\mathbf{k})\, a^*(\mathbf{k})\, a(\mathbf{k})\, d^3 k$$

$$-\frac{g_0}{\sqrt{4\pi}} \int^{\hat{\omega}} \frac{d^3 k}{\sqrt{2\omega}} [\psi_V^* \psi_N a(\mathbf{k}) + a^*(\mathbf{k})\, \psi_N^* \psi_V]. \qquad \text{A(24)}$$

m_V and m_N are the 'bare' energies of the states V and N. $\hat{\omega}$ is a cutoff-parameter, limiting the values of $\mathbf{k}$ in the integral by $\omega(\mathbf{k}) \leqslant \hat{\omega}$. Later the limiting case $\hat{\omega} \to \infty$ will be studied. g_0 is a real coupling constant.

A II-2 Solution of the Schrödinger equation in the sector $\begin{pmatrix} N+\theta \\ V \end{pmatrix}$

The Hamiltonian A(24) has two trivial solutions: If there is no particle at the origin, there is no interaction between the θ-particles; and the Hamiltonian is equivalent to that of a θ-boson field without interaction. If there is no θ-particle, but a particle at the origin in the state N, then again there is no interaction, and the energy of the system is m_N. (Without loss of generality we may define $m_N = 0$.)

The simplest non-trivial case is given by the situation $\begin{pmatrix} N+\theta \\ V \end{pmatrix}$, i.e. either the particle at the origin is in the V-state without θ-particles, or in the N-state with one θ-particle. The interaction term in A(24) connects these two possibilities. In this sector $\begin{pmatrix} N+\theta \\ V \end{pmatrix}$ a general state may be defined by

$$|\Phi\rangle = (c\psi_V^* + \psi_N^* \int \phi(\mathbf{k})\, a^*(\mathbf{k})\, d^3k)|0\rangle, \qquad \text{A(25)}$$

which shall be abbreviated by

$$|\Phi\rangle = \begin{cases} c \\ \phi(\mathbf{k}) \end{cases}. \qquad \text{A(26)}$$

Applying the Hamiltonian to Φ leads to

$$H\Phi = \Phi' = \begin{cases} c' \\ \phi'(\mathbf{k}) \end{cases}, \qquad \text{A(27)}$$

where

$$c' = m_V c - \frac{g_0}{\sqrt{4\pi}} \int^{\hat{\omega}} \frac{\phi(\mathbf{k})}{\sqrt{2\omega}} \, d^3k \qquad \text{A(28)}$$

$$\phi'(\mathbf{k}) = \omega\phi(\mathbf{k}) - \frac{g_0}{\sqrt{4\pi}} \frac{c}{\sqrt{2\omega}}. \qquad \text{A(29)}$$

As one sees from A(28) and (29), the interaction takes place only in the S-states. Therefore it is convenient to confine the calculations to these states and to put ($k = |\mathbf{k}|$)

$$\phi(k) = \frac{\psi(k)}{k\sqrt{4\pi}}. \qquad \text{A(30)}$$

The Schrödinger equation

$$H\Phi = E\Phi \qquad \text{A(31)}$$

then gives for the S-states

$$\left.\begin{aligned} (\omega - E)\,\psi(k) &= \frac{g_0 k}{\sqrt{2\omega}}\, c, \\[2em] (m_V - E)\,c &= g_0 \int_0^{\hat{\omega}} \frac{k}{\sqrt{2\omega}}\, \psi(k)\, dk. \end{aligned}\right\} \qquad \text{A(32)}$$

In the continuous spectrum $E = \sqrt{(m_\theta^2 + k_0^2)}$ we can put, according to A(32),

$$\psi(k) = \delta(k - k_0) - \frac{g_0 . kc}{(E - \omega + i\gamma)\sqrt{2\omega}}, \quad (\gamma \to 0) \qquad \text{A(33)}$$

where the term $i\gamma$ ($\gamma \to 0$) in the denominator leads to outgoing waves in the usual manner. Then from A(32)

$$c\left[m_V - E + g_0^2 \int^{\hat{\omega}} \frac{k^2\, dk}{2\omega(E - \omega + i\gamma)} \right] = \frac{g_0 k_0}{\sqrt{2\omega_0}}; \qquad \text{A(34)}$$

it is convenient to introduce the function $h(z)$ of the complex variable z:

$$h(z) = \frac{z - m_V}{g_0^2} + \int^{\hat{\omega}} \frac{k^2\,dk}{2\omega(\omega - z)}. \qquad\qquad \text{A(35)}$$

This is a regular analytical function in the z-plane cut along the real axis from $z = m_\theta$ to $+\infty$. Its values on the $\pm i$-side of the cut will be called $h^\pm$:

$$h^\pm(E) = \overline{h(E)} \pm \pi i \frac{k_0}{2}; \qquad\qquad \text{A(36)}$$

$\overline{h(E)}$ is real. From A(34) one gets:

$$h^+(E) \cdot g_0 c = -\frac{k_0}{\sqrt{2E}} \qquad\qquad \text{A(37)}$$

and

$$\psi(k) = \delta(k - k_0) + \frac{k k_0}{2\sqrt{(\omega E)}\,(E - \omega + i\gamma)\,h^+(E)}. \qquad\qquad \text{A(38)}$$

The phase of the scattered wave becomes

$$e^{2i\delta(E)} \equiv S(E) = 1 - \frac{\pi i k_0}{h^+(E)} = \frac{h^-}{h^+}, \qquad\qquad \text{A(39)}$$

and from A(36)

$$tg\,\delta(E) = -\pi \frac{k_0}{2\overline{h(E)}}. \qquad\qquad \text{A(40)}$$

In the discrete spectrum ($E < m_\theta$) Eqn. A(32) leads to

$$\psi(k) = \frac{g_0 k}{(\omega - E)\sqrt{2\omega}} \cdot c \qquad\qquad \text{A(41)}$$

and

$$h(E) = 0. \qquad\qquad \text{A(42)}$$

The discrete eigenstates are defined by the zeros of $h(z)$.

A II-3   Discussion of the function $h(z)$

The integral on the right-hand side of A(35) diverges linearly in the limit $\hat{\omega} \to \infty$. Therefore it is convenient to use the relation

$$\frac{1}{\omega - z} = \frac{1}{\omega} + \frac{z}{\omega^2} + \frac{z^2}{\omega^2(\omega - z)} \qquad\qquad \text{A(43)}$$

and to put

$$h(z) = a + bz + z^2 g(z) \qquad\qquad \text{A(44)}$$

Mathematical Appendix 143

where

$$a = -\frac{m_V}{g_0^2} + \int\limits^{\hat{\omega}} \frac{k^2\,dk}{2\omega^2}, \qquad b = \frac{1}{g_0^2} + \int\limits^{\hat{\omega}} \frac{k^2\,dk}{2\omega^3},$$ A(45)

$$G(z) = \int\limits_0^{\hat{\omega}} \frac{k^2\,dk}{2\omega^3(\omega - z)}.$$ A(46)

In $g(z)$ one can go to the limit $\hat{\omega} \to \infty$ and one gets in this case for the principal value $\overline{G(z)}$, putting $\zeta = z/m_\theta$:

$$2m_\theta\,\overline{G(z)} = \frac{\pi}{2\zeta^2} + \frac{1}{\zeta} + \begin{cases} \dfrac{\sqrt{(\zeta^2-1)}}{\zeta^2}\ln(-\zeta+\sqrt{(\zeta^2-1)}) & \text{for } \zeta \leqslant -1 \\[2ex] -\dfrac{\sqrt{(1-\zeta^2)}}{\zeta^2}\left[\dfrac{\pi}{2}+\arcsin\zeta\right] & \text{for } -1 \leqslant \zeta \leqslant 1 \\[2ex] -\dfrac{\sqrt{(\zeta^2-1)}}{\zeta^2}\ln[\zeta+\sqrt{(\zeta^2-1)}] & \text{for } 1 \leqslant \zeta. \end{cases}$$

A(47)

For finite values of $\hat{\omega}$ Eqn. A(44) defines a finite function $h(z)$ which may be used for determining the eigenvalues according to A(42). In the limit $\hat{\omega} \to \infty$, however, the constant b in A(45) and $h(z)$ could only stay finite for negative values of g_0^2, which would contradict the assumption A(24). Therefore the Lee model in its original form A(24) has no finite solution in the limit $\hat{\omega} \to \infty$.

Källén and Pauli[36] have however pointed out, that the Schrödinger equations A(27) to A(32) would retain their form, if the assumptions, §AII-1, of the Lee model would be changed with respect to the metric in Hilbert space. In this case one could introduce g_0 as an imaginary quantity and it would be possible to find finite solutions even in the limit $\hat{\omega} \to \infty$.

The necessary changes are: the norms of the V-state must be negative, therefore

$$\{\psi_V, \psi_V^*\} = -1.$$ A(48)

The new Hamiltonian is

$$H = -m_V\psi_V^*\psi_V + m_N\psi_N^*\psi_N + \int a^*(\mathbf{k})\,a(\mathbf{k})\,\omega\,d^3k$$

$$-\frac{g_0}{\sqrt{4\pi}}\int\frac{d^3k}{\sqrt{2\omega}}[-\psi_V^*\psi_N a(\mathbf{k})+a^*(\mathbf{k})\,\psi_N^*\psi_V],$$ A(49)

and A(25) should be replaced by

$$|\Phi\rangle = (-c\psi_V^* + \psi_N^* \int \phi(\mathbf{k})\, a^*(\mathbf{k})\, d^3k)|0\rangle. \qquad \text{A(50)}$$

The equations A(45) remain unchanged. Finite values of a and b can be obtained, if for $\hat\omega \to \infty$

$$g_0^2 \sim -\frac{2}{\ln\hat\omega}, \qquad m_V \sim -\frac{\hat\omega}{\ln\hat\omega}. \qquad \text{A(51)}$$

g_0 is now imaginary, the Hamiltonian A(49) is selfadjoint, but only 'pseudo-hermitian'. For eigenstates with a norm different from zero the eigenvalues of H are real according to 2(31).

For determining the properties of the function $h(z)$ in the limit $\hat\omega \to \infty$ we notice, that on the real axis $(d^2h/dz^2) > 0$ for $z < m_\theta$ and

$$\frac{dh}{dz}(-\infty) = -\infty, \qquad \frac{dh}{dz}(m_\theta) = +\infty.$$

The function $h(z)$ has just one minimum between $z = -\infty$ and $z = m_\theta$. Therefore it can have two zeros or no zeros on the real axis, and in the limiting case the two zeros combine to one zero, where the curve $h(z)$ touches the real axis: $h(z_0) = h'(z_0) = 0$. In this limiting case $1/h(z)$ has a double pole at the point $z = z_0$, and it is this double pole which can be used as analogon to the double pole in the approximative 2-point function 3(39). Finally if there are no zeros of $h(z)$ on the real axis, there will be two complex roots of the equation $h(z) = 0$, belonging to states with norms zero according to 2(31).

Here we will be interested only in the case of the double pole of $1/h(z)$, which will be assumed to be at the point $z = z_0 = E_0$; $\zeta_0 = z_0/m_\theta$. $h(z)$ then takes the form:

$$\overline{h(z)} = \frac{m_\theta}{2}\left\{\frac{1 - \zeta_0\zeta}{\zeta_0^2 - 1}s(\zeta_0) + \zeta - \zeta_0 + s(\zeta)\right\}, \qquad \text{A(52)}$$

where

$$s(\zeta) = \begin{cases} \sqrt{(\zeta^2 - 1)}\ln(-\zeta + \sqrt{(\zeta^2 - 1)}) & \text{for } \zeta \leqslant -1 \\[2mm] -\sqrt{(1 - \zeta^2)}\left(\dfrac{\pi}{2} + \arcsin\zeta\right) & \text{for } -1 \leqslant \zeta \leqslant 1 \\[2mm] -\sqrt{(\zeta^2 - 1)}\ln[\zeta + \sqrt{(\zeta^2 - 1)}] & \text{for } 1 \leqslant \zeta. \end{cases} \qquad \text{A(53)}$$

Figure 10 shows $h(z)$ for the special case $\zeta_0 = -1$.

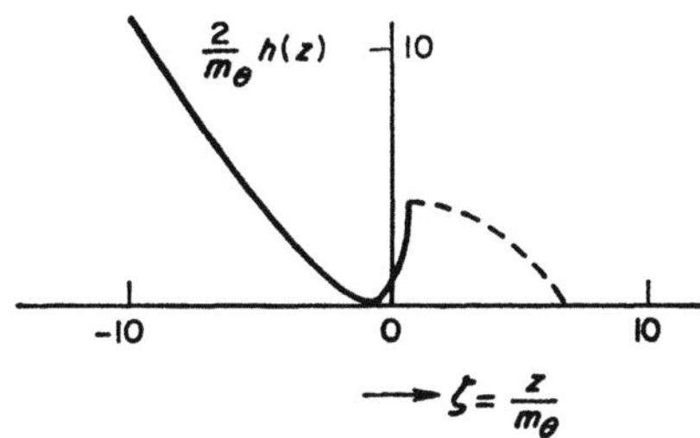

Fig. 10.

A II-4 The eigenstates connected with the double pole of $1/h(z)$

The double pole in $1/h(z)$ can be obtained as a limiting case from the case of the two zeros by varying the constant a in A(44). Before the limit is reached, the root of A(42) with the smaller value of ζ belongs to a state with negative norm, the other root with the higher value to a normal state. The question arises whether one gets one or two solutions of the Schrödinger equation in the limiting case and what the norm of these states will be. Before one reaches the limit, let the two solutions of the Schrödinger equations be Φ and $\Phi + \Delta\Phi$, the eigenvalues E and $E + \Delta E$. Then

$$\left.\begin{array}{r} H(\Phi + \Delta\Phi) = (E + \Delta E)(\Phi + \Delta\Phi) \\ H\Phi = E\Phi \end{array}\right\} \qquad \text{A(54)}$$

and by subtracting and neglecting terms of higher order

$$H\Delta\Phi = E\Delta\Phi + \Phi\Delta E$$

or

$$H\frac{\Delta\Phi}{\Delta E} = E\frac{\Delta\Phi}{\Delta E} + \Phi. \qquad \text{A(55)}$$

From A(54) and A(55) one learns that in the case of the double pole in $1/h(z)$ one has only *one* solution of the Schrödinger equation. But by combining this one solution with the eigenstates of the continuous spectrum one would *not* get the complete Hilbert space of the system (compare §2-6). One would for completeness have to add the dipole-ghost state

$$\Phi_{\text{Dip}} = \lim_{\Delta E \to 0} \frac{\Delta\Phi}{\Delta E}, \qquad \text{A(56)}$$

which is not an eigenstate of the Hamiltonian.

The norm of a discrete state fulfilling A(42) is, according to A(41),

$$\langle\Phi|\Phi\rangle = |c^2|\left[|g_0^2|\int\limits^{\hat{\omega}}\frac{k^2\,dk}{2\omega(\omega-E)^2}-1\right], \qquad \text{A(57)}$$

where the -1 in the bracket is due to the component with the V-state, the norm of which is negative. From A(35) and A(57) one gets

$$\langle\Phi|\Phi\rangle = |c^2 g_0^2|\frac{dh(E)}{dE} = |c^2 g_0^2|\,h'(E). \qquad \text{A(58)}$$

The product $|cg_0|$ remains finite in the limit $\hat{\omega}\to\infty$, as one can see, for example, from A(33) or A(37). For the state with the double root of A(42) we have $h'(E)=0$, therefore the norm of this state Φ_0 must be zero.

$$\langle\Phi_0|\Phi_0\rangle = 0. \qquad \text{A(59)}$$

Hence Φ_0 is a ghost state, like the state G_+ in 3(46). Both Φ_0 and Φ_{Dip} are orthogonal to all states of the continuous spectrum, but they are not orthogonal to each other. Φ_{Dip} is not uniquely determined by A(55) and A(56), since a multiplication of Φ by any function of E before going to the limit A(56) will generally change Φ_{Dip} by an additional term const. Φ_0. This term cannot be used to make Φ_0 and Φ_{Dip} orthogonal to each other, since the norm of Φ_0 vanishes, but it may be used to make the norm of Φ_{Dip} vanish as well. Therefore we will add to the definition of Φ_{Dip} the postulate

$$\langle\Phi_{\mathrm{Dip}}|\Phi_{\mathrm{Dip}}\rangle = 0. \qquad \text{A(60)}$$

Finally the metric tensor in the space of Φ_{Dip} and Φ_0 can be chosen as

$$
\begin{array}{c|cc}
 & \Phi_{\mathrm{Dip}} & \Phi_0 \\
\hline
\Phi_{\mathrm{Dip}} & 0 & 1 \\
\Phi_0 & 1 & 0
\end{array}
\qquad \text{A(61)}
$$

in close analogy to 3(48).

Since Φ_{Dip} fulfills the inhomogeneous equation

$$H\Phi_{\mathrm{Dip}} = E\Phi_{\mathrm{Dip}}+C\Phi_0 \qquad \text{A(62)}$$

the expectation value of the energy in the dipole ghost state is infinite:

$$\bar{H}_{\mathrm{Dip}} = \frac{\langle\Phi_{\mathrm{Dip}}|H|\Phi_{\mathrm{Dip}}\rangle}{\langle\Phi_{\mathrm{Dip}}|\Phi_{\mathrm{Dip}}\rangle} = \frac{C}{\langle\Phi_{\mathrm{Dip}}|\Phi_{\mathrm{Dip}}\rangle} = \infty. \qquad \text{A(63)}$$

Mathematical Appendix **147**

If one uses the Schrödinger picture in which the state vector varies as function of time according to

$$i\frac{\partial \Phi}{\partial t} = H\Phi, \qquad \text{A(64)}$$

A(54) and A(62) give

$$\Phi_0(t) = e^{-iE_0 t}\Phi_0(0),$$

$$\Phi_{\text{Dip}}(t) = e^{-iE_0 t}\Phi_{\text{Dip}}(0) - iCt\,e^{-iE_0 t}\Phi_0(0). \qquad \text{A(65)}$$

The last equation leads to a corresponding time-dependence of the matrix elements, which had been used in 3(45).

A II-5 Renormalization of ψ_V

The vacuum expectation value of the anticommutator

$$\langle 0|\{\psi_V\psi_V^*\}|0\rangle = \langle 0|\psi_V\psi_V^*|0\rangle = -1 \qquad \text{A(66)}$$

can be considered as a sum of the contributions from all intermediate states, the discrete states and the continuous spectrum. If the eigenfunctions of the states are properly normalized, the coefficient c in each eigenfunction is identical with the matrix element of ψ_V belonging to the transition from vacuum to this state. We consider first the case of the two separated roots of $h(z)=0$. Taking the coefficient c from A(58) for the discrete states and from A(57) for the continuous spectrum, Eqn. A(66) can be written as

$$\frac{1}{|g_0^2|}\left[\sum_i \frac{1}{h'(E_i)} + \int_0^\infty \frac{k^2\,dk}{2Eh^+(E)\,h^-(E)}\right] = -1. \qquad \text{A(67)}$$

In the limit $\hat{\omega} \to \infty$, $|g_0^2|$ vanishes. Therefore the sum over the discrete states must in this case compensate the contribution from the continuum. This can be seen by changing the path of integration in the second term in the bracket. We can call this term g^2:

$$g^2 = \int_0^\infty \frac{k^2\,dk}{2\omega h^+(\omega)\,h^-(\omega)} = \int_{m_\theta}^\infty \frac{k\,d\omega}{2h^+(\omega)\,h^-(\omega)} = \int_{m_\theta}^\infty \frac{d\omega}{2\pi i}\left(\frac{1}{h^-(\omega)} - \frac{1}{h^+(\omega)}\right)$$

$$= \int_C \frac{d\omega}{2\pi i h(\omega)} = -\sum_i \frac{1}{h'(E_i)}. \qquad \text{A(68)}$$

The path C of integration starts from $+\infty$ above the cut in the p_0-plane, goes to the branching point $\omega = m_\theta$ and from there below the cut to $+\infty$.

Deforming this path to a circle at infinity leaves residual integrals around the poles $1/h(\omega)$, which give rise to the terms $-1/h'(E)$.

Since all matrix elements c become infinite in the limit $\hat{\omega} \to \infty$, $g_0^2 \to 0$, it is convenient to renormalize the operator ψ_V by introducing

$$\psi_{Vr} = \frac{g_0}{g}\psi_V; \qquad \psi_{Vr}^* = \left(\frac{g_0}{g}\right)^* \psi_V = -\frac{g_0}{g}\psi_V. \qquad \text{A(69)}$$

Then, in the limit $\hat{\omega} \to \infty$ we have

$$\{\psi_{Vr}, \psi_{Vr}^*\} = 0. \qquad \text{A(70)}$$

If one uses a representation, in which the state vectors are constant and the operators are functions of the time t, then the anticommutator takes the form:

$$\langle 0|\{\psi_{Vr}(t), \psi_{Vr}^*(t')\}|0\rangle = \frac{1}{g^2}\left[\sum_l \frac{e^{iE_l(t'-t)}}{h'(E_l)} + \int_0^\infty \frac{k^2 e^{iE(t'-t)}\,dk}{2Eh^+(E)\,h^-(E)}\right]$$

$$= \frac{1}{2\pi i g^2}\int_{C'} \frac{d\omega}{h(\omega)}\, e^{i\omega(t'-t)}. \qquad \text{A(71)}$$

The path C' of integration is now given by Fig. 11 and corresponds to the conventional path for the Δ-function A(3). Formula A(71) applies also to the case of the double pole.

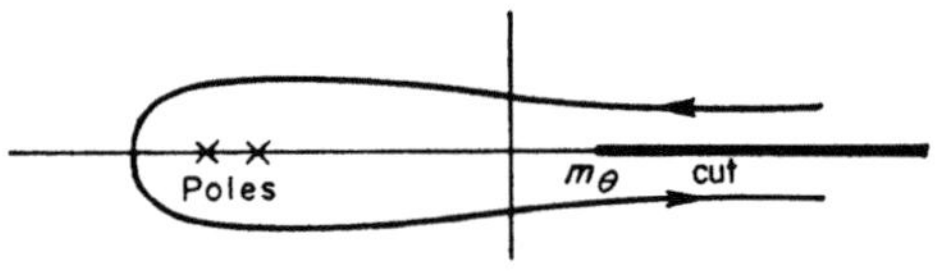

Fig. 11.

As one sees from A(68), A(69) and A(71), the renormalized operator ψ_{Vr} has finite matrix elements for the transition from the vacuum to any of the states A(26). The anticommutator $\{\psi_{Vr}, \psi_{Vr}^*\}$ vanishes at equal times. The contributions from the continuous spectrum are compensated by the contributions from the double pole, as in the approximative 2-point function 3(39). The function $h(\omega)$ plays the same rôle as $S^{-1}(p)$ in the relativistic field theory.

A II-6 The rôle of the ghost states in scattering problems

The description of physical phenomena requires a unitary S-matrix; therefore it is important to see in which way the Hilbert space with indefinite metric and its ghost states can be used for constructing such a unitary S-matrix.

We begin with the sector $\binom{N+2\theta}{V+\theta}$, in which two θ-particles may be scattered from the N-state and in which possibly a V-state of the ghost-state type could be left over after emission of one θ-particle.

The state vector in this sector can be represented by two wave functions

$$\Phi = \begin{cases} \phi(\mathbf{k}) \\ \phi(\mathbf{k}_1, \mathbf{k}_2). \end{cases} \qquad \text{A(72)}$$

From $H\Phi = E\Phi$ we get for the scattering states $E > 2m_\theta$

$$\phi(\mathbf{k}_1, \mathbf{k}_2) = \phi_0(\mathbf{k}_1\mathbf{k}_2) + \frac{g_0}{2\sqrt{(4\pi)}\,(\omega_1+\omega_2-E-i\gamma)}\left[\frac{\phi(\mathbf{k}_2)}{\sqrt{2\omega_1}} + \frac{\phi(\mathbf{k}_1)}{\sqrt{2\omega_2}}\right] \quad \text{A(73)}$$

(with $\gamma \to 0$), where $\phi_0(\mathbf{k}_1, \mathbf{k}_2)$ is an arbitrary function of $\mathbf{k}_1$ and $\mathbf{k}_2$ satisfying the condition

$$(\omega_1+\omega_2-E)\,\phi_0(\mathbf{k}_1, \mathbf{k}_2) = 0. \qquad \text{A(74)}$$

The second Schrödinger equation is

$$h^+(E-\omega)\,\phi(\mathbf{k}) = -\frac{1}{8\pi}\int \frac{d^3k'\,\phi(\mathbf{k}')}{\sqrt{(\omega\omega')}\,(\omega+\omega'-E-i\gamma)} - \phi_{0i}(\mathbf{k}) \quad \text{A(75)}$$

with

$$\phi_{0i}(\mathbf{k}) = \frac{2}{g_0\sqrt{4\pi}}\int \frac{d^3k'\,\phi_0(\mathbf{k}, \mathbf{k}')}{\sqrt{2\omega'}}. \qquad \text{A(76)}$$

We consider only the case of a total angular momentum zero. Then $\phi(\mathbf{k})$ is a function of the absolute value k only.

Eqn. A(75) indicates that $\phi(k)$ as a function of k has a singularity at the point $\omega = \omega_c = E - E_0$, where E_0 is the energy of the double pole. Since $h^+(E-\omega)$ has a double zero at this point, the behaviour of $\phi(k)$ depends on whether the right-hand side of A(75) vanishes or does not vanish at the critical point. If the right-hand side is different from zero, we get a behaviour of $\phi(k)$ of the type

$$\phi(k) = a\,\delta(E-E_0-\omega) + b\,\delta'(E-E_0-\omega) + \frac{\chi(k)}{(E-E_0-\omega)^2}. \qquad \text{A(77)}$$

Such solutions will certainly exist, but it will be difficult to find any physical interpretation for them. The Fourier transformation from momentum space to ordinary space shows that a solution of the type A(77) does not correspond to incoming and outgoing spherical waves of the form $(1/r)\exp(\pm ik_c r)$; it corresponds to expressions of the type $(\alpha + \beta r/r)\exp(\pm ik_c r)$ (compare similar expressions in §5-2), which cannot be used for the description of real phenomena.

A different type of solution is obtained if the right-hand side of A(75) vanishes at the critical point. Then we get

$$\phi(k) = a\,\delta(E - E_0 - \omega) + \frac{\chi(k)}{E - E_0 - \omega + i\gamma}. \qquad \text{A(78)}$$

It is very important to note that the arbitrary factor a of the δ-function in A(78) can be used to make the right-hand side of A(75) vanish at the critical point. Therefore one can actually find a group of solutions of A(75) of the behaviour A(78), but one cannot choose the constant a arbitrarily. Inserting A(78) into the condition that the right-hand side of A(75) shall vanish at $\omega = \omega_c = E - E_0$ gives

$$a\frac{k_c}{2(\omega_c - E_0)} = -\frac{1}{8\pi}\int \frac{d^3k'\,\chi(k')}{\sqrt{(\omega_c\omega')}\,(\omega_c + \omega' - E - i\gamma)(E - E_0 - \omega' + i\gamma)}. \qquad \text{A(79)}$$

(The last term $\phi_{0i}(k)$ in A(75) vanishes at $k = k_c$, since $\phi_{0i}(k)$ can only be different from zero if $m_\theta \leqslant \omega \leqslant E - m_\theta$.) The resulting integral equation for $\chi(k)$ is

$$\chi(k) = -\frac{(\omega_c - \omega)^2}{8\pi h^+(E - \omega)}$$

$$\times \int \frac{d^3k'\,\chi(k')}{\sqrt{(\omega\omega')}\,(\omega + \omega_c - E)(\omega + \omega' - E - i\gamma)(\omega_c + \omega' - E - i\gamma)}$$

$$-\frac{\omega_c - \omega}{h^+(E - \omega)}\phi_{0i}(k). \qquad \text{A(80)}$$

The right-hand side is automatically regular at the point $\omega = \omega_c$, therefore $\chi(k)$ is regular as well.

By means of a Fourier transformation one can go over to a representation of the state vector $|\Phi\rangle$ in functions in ordinary space

$$\Phi = \begin{cases} \phi(\mathbf{x}) \\ \phi(\mathbf{x}_1, \mathbf{x}_2). \end{cases} \qquad \text{A(81)}$$

On account of the pole on the right-hand side of A(78), the functions $\phi(\mathbf{x})$, $\phi(\mathbf{x}_1, \mathbf{x}_2)$ contain parts which behave like incoming or outgoing waves

Mathematical Appendix 151

of the type $(1/r)\exp(\pm ik_c r)$. A comparison of the solutions in the sector $\begin{pmatrix} N+2\theta \\ V+\theta \end{pmatrix}$ with those in the sector $\begin{pmatrix} N+\theta \\ V \end{pmatrix}$ shows, that these parts can be written as products of $(1/r)\exp(\pm ik_c r)$ with the wave function of the ghost state Φ_0 in $\begin{pmatrix} N+\theta \\ V \end{pmatrix}$. The norm of these contributions is zero, since the norm of Φ_0 is zero and the norm of a product state is equal to the product of the norms of the factors.

From this result one can see, that a unitary S-matrix can be constructed from the solutions of the type A(78). The asymptotic part of the wave function, which defines the S-matrix elements, contains waves of the type $N+2\theta$ and of the type $V+\theta$, but the latter only in the form $\Phi_0+\theta$. Asymptotic waves of the type $\Phi_{\mathrm{Dip}}+\theta$ do *not* occur. Since the N and θ states have positive norm, also the contributions $N+2\theta$ have positive norm. The contributions $\Phi_0+\theta$ have norm zero and are orthogonal to $N+2\theta$. The S-matrix connecting the incoming with the outgoing waves hence is a pseudo-unitary transformation matrix [on account of the pseudo-hermitian Hamiltonian A(49)] in a space with semi-definite metric. This is sufficient for stating that the S-matrix is unitary, and in the physical interpretation the contributions with norm zero can simply be omitted.

Translating this argument into the language of field theory (as applied in §3-4), one may say that the condition A(79) makes the vertex determining the interaction vanish at the point of the double pole. Thereby the double pole of $\phi(k)$ in A(75) is changed into a simple pole; consequently only the ghost state Φ_0, but not the dipole ghost state Φ_{Dip} can appear in the asymptotic waves, and the S-matrix refers to a space of semi-definite metric, which is sufficient for the proof of unitarity.

In the more general sectors $\begin{pmatrix} N+z\theta \\ V+(z-1)\theta \end{pmatrix}$ the situation is essentially the same, therefore the proof for the existence of solutions defining a unitary S-matrix shall not be repeated here.

For the completeness of the analysis of the model it may still be important to ask, whether in any of these sectors discrete stationary states exist which then in higher sectors could appear in the asymptotic states of a scattering process.

We will prove that such discrete states do not exist. The state vector in the sector $\begin{pmatrix} N+z\theta \\ V+(z-1)\theta \end{pmatrix}$ could be represented by

$$\Phi = \begin{cases} \phi(k_1 \ldots k_{z-1}) \\ \phi(k_1, k_2 \ldots k_z) \end{cases} \qquad \text{A(82)}$$

where the ϕ's are functions symmetrical under permutations of k_1, k_2....

For a discrete state the total energy should be smaller than zm_θ: $E < zm_\theta$. The Schrödinger equation, corresponding to A(75) for a discrete state ($\phi_{0i}=0$), has the form

$$\phi(k_1\ldots k_{z-1})h\left(E-\sum_1^{z-1}\omega_i\right) = -\frac{1}{4\pi}\int\frac{d^3k_z}{\sqrt{(2\omega_z)}\left(\sum_1^z\omega_l-E\right)}$$

$$\times\sum_{j=1}^{z-1}\frac{\phi(k_1\ldots k_{j-1}k_{j+1}\ldots k_z)}{\sqrt{2\omega_j}}\qquad\text{A(83)}$$

Multiplying this expression with $\phi^*(k_1\ldots k_{z-1})$ and integrating over $d^3k_1\ldots d^3k_{z-1}$, gives

$$\int d^3k_1\ldots d^3k_{z-1}|\phi(k_1\ldots k_{z-1})^2|h\left(E-\sum_1^{z-1}\omega_i\right)$$

$$= -\frac{1}{4\pi}\int_0^\infty d\alpha\int d^3k_1\ldots d^3k_z\frac{1}{\sqrt{2\omega_z}}e^{-\alpha\left(\sum_1^z\omega_l-E\right)}\qquad\text{A(84)}$$

$$\times\sum_{j=1}^{z-1}\frac{\phi(k_1\ldots k_{j-1}k_{j+1}\ldots k_z)}{\sqrt{2\omega_j}}\phi^*(k_1\ldots k_{z-1}).$$

Here the relation

$$\frac{1}{\sum_1^z\omega_i-E} = \int_0^\infty d\alpha\,e^{-\alpha\left(\sum_1^z\omega_l-E\right)}\qquad\text{A(85)}$$

has been used, which holds since $\sum_1^z\omega_i-E>0$.

Putting

$$A(k_1\ldots k_{z-2}) = \int d^3k_{z-1}\frac{\phi(k_1\ldots k_{z-1})}{\sqrt{2\omega_{z-1}}}e^{-\alpha\omega_{z-1}}\qquad\text{A(86)}$$

the right-hand side of A(84) can be written as

$$-\frac{z-1}{4\pi}\int d\alpha\int d^3k_1\ldots d^3k_{z-2}e^{-\alpha\left(\sum_1^{z-2}\omega_i-E\right)}|A(k_1\ldots k_{z-2})^2|,\qquad\text{A(87)}$$

when the symmetry of $\phi(k_1\ldots k_{z-1})$ under permutations of k_1, $k_2\ldots k_{z-1}$ is used.

Mathematical Appendix 153

The right-hand side of A(84) is therefore certainly negative. The left-hand side on the other hand is certainly positive, since $h(z)$ is never negative in the region $-\infty \leqslant z \leqslant m_\theta$. Hence Eqn. A(84) would lead to a contradiction if a discrete state existed. Therefore discrete states do not occur.

The model which has been analysed in the present chapter, had been introduced by Lee in order to study the consequences of the typical divergencies occurring in local relativistic field theories. The model may be called relativistic with respect to the behaviour of the θ-particles, but not really local; it shows the characteristic divergencies in the limit $\hat\omega \to \infty$ in A(45). Still it is very much simpler than a genuine relativistic field theory with interaction. In the model the divergencies enforce the introduction of the indefinite metric in Hilbert space. With this extension of the conventional quantum theory the mathematical scheme is consistent and does not give rise to any unexpected infinities. The scheme does allow the construction of a unitary S-matrix and permits a corresponding physical interpretation. Therefore it seems to be a suitable model for the mathematical structure of the non-linear spinor theory.

AIII. Calculation of Integrals in Momentum Space

The integral operators of the type ⌬ or ⌬ can conveniently be calculated by the methods developed by Feynman.[46] As an example we will explicitly evaluate the integral occurring in 5(11):

$$K_{\mu\nu} = -\int d^4p\, \frac{-g_{\mu\nu}(p, J-p) + p_\mu(J-p)_\nu + p_\nu(J-p)_\mu}{(p^2)^2\,(p^2 - \kappa^2)\,(J-p)^2}. \qquad \text{A(88)}$$

The main idea of Feynman's method consists in reducing the complicated integral to the standard form

$$\int \frac{d^4p}{(p^2 + a^2)^3}$$

by means of the relation

$$\int_0^1 \frac{dx}{[ax + b(1-x)]^2} = \frac{1}{ab} \qquad \text{A(89)}$$

or its derivatives. Since all integrals in the text are derived from time-ordered products of operators, the integration in p_0 can be carried out along the real axis, if p^2 is replaced by the limit $_{\delta \to 0}\,(p^2+i\delta)$. In this case one gets for the standard integral

$$\int \frac{d^4p}{(p^2+a^2)^3} = \frac{i}{2}\int d^4p \int_0^\infty d\alpha\,.\,\alpha^2\,e^{+i\alpha(p^2+a^2)}$$

$$= +\frac{\pi^2}{2}\int_0^\infty d\alpha\,e^{+i\alpha a^2} = \frac{i\pi^2}{2a^2}, \qquad \text{A(90)}$$

and by differentiation

$$\int \frac{d^4p}{(p^2+a^2)^4} = \frac{i\pi^2}{6a^4}; \qquad \int d^4p\,\frac{p^2}{(p^2+a^2)^4} = \frac{i\pi^2}{3a^2}. \qquad \text{A(91)}$$

The denominator in $K_{\mu\nu}$ in A(88) can be transformed by relations resulting from A(89) by differentiation

$$\frac{1}{(p^2)^2\,(p^2-\kappa^2)\,(J-p)^2}$$

$$= \int_0^1 \frac{2x\,dx}{[p^2-\kappa^2(1-x)]^3\,(J-p)^2}$$

$$= \int_0^1 dx \int_0^1 dy\,\frac{6xy^2}{[p^2-\kappa^2(1-x)\,y+J^2(1-y)-2Jp(1-y)]^4}$$

$$= \int_0^1 dx \int_0^1 dy\,\frac{6xy^2}{\{[p-J(1-y)]^2-\kappa^2(1-x)\,y+J^2(1-y)\,y\}^4}. \qquad \text{A(92)}$$

In the numerator of $K_{\mu\nu}$ in A(89) it is now convenient to replace p by $p'+J(1-y)$. Then in the integral the terms linear in p'_μ can be omitted and we get

$$K = -\int_0^1 dx \int_0^1 dy \int d^4p'\,\frac{6xy^2\{-g_{\mu\nu}[J^2(1-y)\,y-\frac{1}{2}p'^2]+2J_\mu J_\nu y(1-y)\}}{[p'^2-\kappa^2(1-x)\,y+J^2(1-y)\,y]^4}$$

$$= -i\pi^2 \int_0^1 dx \int_0^1 dy\,\frac{xy[-g_{\mu\nu}\kappa^2(1-x)+2J_\mu J_\nu(1-y)]}{[J^2(1-y)-\kappa^2(1-x)]^2} \qquad \text{A(93)}$$

or replacing x by $1-x$ and y by $1-y$:

$$K_{\mu\nu} = -i\pi^2 \int_0^1 dx \int_0^1 dy \frac{(1-x)(1-y)[-g_{\mu\nu}\kappa^2 x + 2J_\mu J_\nu y]}{(J^2 y - \kappa^2 x)^2}.$$

By partial integration:

$$K_{\mu\nu} = -\frac{i\pi^2}{J^2} \int_0^1 dx(1-x)\left[g_{\mu\nu} + \int_0^1 dy \frac{+g_{\mu\nu}\kappa^2 x + 2(1-2y)J_\mu J_\nu}{J^2 y - \kappa^2 x}\right]$$

$$= -\frac{i\pi^2}{J^2} \int_0^1 dx(1-x)\left\{g_{\mu\nu}\left(1 + \frac{\kappa^2 x}{J^2}\ln\frac{-J^2+\kappa^2 x}{\kappa^2 x}\right)\right.$$

$$\left. + \frac{4J_\mu J_\nu}{J^2}\left[-1+\left(\frac{1}{2}-\frac{\kappa^2 x}{J^2}\right)\ln\frac{-J^2+\kappa^2 x}{\kappa^2 x}\right]\right\}. \qquad \text{A(94)}$$

Putting $J^2/\kappa^2 = \lambda$ one finally arrives at:

$$K_{\mu\nu} = -\frac{i\pi^2}{2\kappa^2}\left\{g_{\mu\nu}\left[(1-\tfrac{2}{3}\lambda)\ln\lambda + \frac{1}{3\lambda} + \frac{2}{3} + \frac{1-3\lambda^2+2\lambda^3}{3\lambda^2}\ln|1-\lambda|\right]\right.$$

$$\left. + \frac{J_\mu J_\nu}{J^2}\left[\tfrac{2}{3}\lambda\ln\lambda - \frac{4}{3\lambda} - \frac{2}{3} - \frac{4-6\lambda+2\lambda^3}{3\lambda^2}\ln|1-\lambda|\right]\right\}, \qquad \text{A(95)}$$

which has been inserted in 5(11) and 5(12). In A(95) the absolute value under the logarithm shows that the principal value has been taken, as had been discussed in §5-1.

The other integrals like ⬡ are considerably more complicated, but can be evaluated along the same lines.

AIV. Functional Calculus for the Infinite Sets of Functions [33, 75, 76]

The infinite sets of functions introduced in §4-2 as representatives of the state vectors can be expressed by means of functionals which show the characteristic features of these sets in a very compact form.

In order to construct these functionals we introduce a 'field' $\xi(x)$, $\xi^*(x)$ which, like $\chi(x)$, is a Weylspinor in Lorentz space and a spinor in isospace. The transformation properties of $\xi(x)$ and $\xi^*(x)$ are however not supposed to be the same as those of $\chi(x)$; we assume, that in Lorentz space $\xi^*(x)$ transforms like $c\chi$ (with $\sigma_k = -c^{-1}\sigma_k^T c$), $\xi(x)$ like $\chi^* c^*$, in isospace $\xi(x)$

transforms like χ, $\xi^*(x)$ like χ^*; in this way $\chi(x)\xi^*(x)$ is invariant under Lorentz transformations and rotations in isospace.

The product of any two such ξ-variables is supposed to be a c-number, i.e. to commute with every other quantity. But we assume at the same time that $\xi(x)$ or $\xi^*(x)$ anticommute with the $\xi(y)$, $\xi^*(y)$, $\chi(y)$, $\chi^*(y)$ at any point y. This assumption is meant merely as a prescription for carrying out formal operations with the fields $\xi(x)$, without attempting to construct a Hilbert space in which these fields $\xi(x)$ could act as operators.

Then we define the functional

$$\tilde{T} = T e^{-i \int dx [\chi(x)\xi^*(x) + \xi(x)\chi^*(x)]}. \qquad \text{A(96)}$$

(T means, as usual, the time-ordered product.)

With the help of this functional the τ-functions of the text can be expressed by:

$$\tau(x_1 x_2 \ldots | y_1 y_2 \ldots) = \langle 0| T \chi(x_1) \chi(x_2) \ldots \chi^*(y_1) \chi^*(y_2) \ldots |\psi\rangle$$

$$= (-i)^m \cdot i^n \frac{\delta}{\delta \xi^*(x_1)} \cdot \frac{\delta}{\delta \xi^*(x_2)} \cdots \frac{\delta}{\delta \xi(y_1)}$$

$$\times \frac{\delta}{\delta \xi(y_2)} \cdots \langle 0|\tilde{T}|\psi\rangle \quad \text{for } \xi = \xi^* = 0. \qquad \text{A(97)}$$

m, n are the total numbers of variables χ resp. χ^* in the τ-function. In a similar way the functional $\tilde{\Phi}$ can be defined as:

$$\tilde{\Phi} = e^{\int \langle 0|T\chi(x)\chi^*(y)|0\rangle \xi^*(x)\xi(y) dx dy} \cdot \tilde{T}. \qquad \text{A(98)}$$

Then the ϕ-functions of 4(2) are given by

$$\phi(x_1 x_2 \ldots | y_1 y_2 \ldots) = (-i)^m i^n \frac{\delta}{\delta \xi^*(x_1)} \frac{\delta}{\delta \xi^*(x_2)} \cdots \frac{\delta}{\delta \xi(y_1)}$$

$$\times \frac{\delta}{\delta \xi(y_2)} \cdots \langle 0|\tilde{\Phi}|\psi\rangle \quad \text{for } \xi = \xi^* = 0. \qquad \text{A(99)}$$

This can be verified by inserting

$$\tilde{T} = e^{-\int \langle 0|T\chi(x)\chi^*(y)|0\rangle \xi^*(x)\xi(y) dx dy} \cdot \tilde{\Phi} \qquad \text{A(100)}$$

into A(97) and comparing the results with 4(2).

In a similar way the η-functions can be derived from

$$\tilde{H} = \ln \langle 0|\tilde{T}|0\rangle. \qquad \text{A(101)}$$

$$\eta(x_1 x_2 \ldots | y_1 y_2 \ldots) = (-i)^m i^n \frac{\delta}{\delta \xi^*(x_1)} \cdots \frac{\delta}{\delta \xi(y_1)} \ldots \tilde{H} \quad \text{for } \xi = \xi^* = 0.$$

$$A(102)$$

Again this relation can be checked by inserting

$$\langle 0 | \tilde{T} | 0 \rangle = e^{\tilde{H}}$$

into A(97) and comparing with 4(3).

Finally the ζ-functions are obtained from the functional

$$\tilde{Z} = \frac{\tilde{T}}{\langle 0 | \tilde{T} | 0 \rangle} \qquad A(103)$$

through

$$\zeta(x_1 x_2 \ldots | y_1 y_2 \ldots) = (-i)^m i^n \frac{\delta}{\delta \xi^*(x_1)} \cdots \frac{\delta}{\delta \xi(y_1)} \ldots \tilde{Z} \quad \text{for } \xi = \xi^* = 0.$$

$$A(104)$$

One can see at once from A(103) and A(104), that for transitions from vacuum to vacuum all ζ-functions vanish.

The fundamental field equation 3(1) can be expressed as a relation for the functional $\tilde{T}$:

$$\left(i\sigma^{\nu}_{\alpha\beta} \frac{\partial}{\partial x_\nu} \frac{\delta}{\delta \xi^*_{\beta\lambda}(x)} + l^2 \sigma^{\nu}_{\alpha\beta} \sigma_{\nu,\gamma\delta} : \frac{\delta}{\delta \xi^*_{\beta\lambda}(x)} \frac{\delta}{\delta \xi_{\gamma\mu}(x)} \frac{\delta}{\delta \xi^*_{\delta\mu}(x)} : \right) \tilde{T} = 0. \quad A(105)$$

The symbol $::$ is defined as in Eqn. 3(3):

$$: \frac{\delta}{\delta \xi^*(x)} \frac{\delta}{\delta \xi(x)} \frac{\delta}{\delta \xi^*(x)} : = \lim_{\delta \to 0, \, \epsilon \to 0} \left[\frac{\delta}{\delta \xi^*(x)} \frac{\delta}{\delta \xi(x+\delta)} \frac{\delta}{\delta \xi^*(x+\delta+\epsilon)} \right.$$

$$- F(x, x+\delta) \frac{\delta}{\delta \xi^*(x+\delta+\epsilon)}$$

$$\left. + F(x+\epsilon+\delta, x+\delta) \cdot \frac{\delta}{\delta \xi^*(x)} \right]$$

This equation may be used for instance, by inserting A(100) into A(105) for deriving the Tamm–Dancoff equation 4(14), which formulates the fundamental equation 3(1) in terms of ϕ-functions.

The functional calculus is convenient for showing the characteristic properties of the four different sets of functions which have been described in the text. It is however of not much help for the calculations necessary for determining eigenvalues or coupling constants. Methods which would use the functional calculus for the final numerical work have so far not been developed.

Unified Field Theory of Elementary Particles

AV. Analytical Properties and Numerical Values of some Important Functions

In connexion with the eigenvalue equations in Chapters 5, 7 and 8 some functions of the real positive variable λ have been defined, which are obtained as the principal values of certain Feynman integrals.

On the real axis they are the real parts of analytical functions of the complex variable λ. In the complex λ-plane there would be cuts along parts of the real axis. In the following lines we give the functions together with their expansions for $\lambda \ll 1$ and $\lambda \gg 1$, and figures and a numerical table which may be useful for numerical calculations.

In §5-1 the two functions $q_0(\lambda)$ and $q_1(\lambda)$ have been defined:

$$
\left.
\begin{aligned}
q_0(\lambda) &= \ln\lambda - \frac{1}{\lambda} - \frac{(1-\lambda)^2}{\lambda^2}\ln|1-\lambda| \\[2mm]
&\approx \ln\lambda - \frac{3}{2} + \frac{\lambda}{3} + \frac{\lambda^2}{12} + \dots \qquad \text{for } \lambda \ll 1, \\[2mm]
&\approx \left(\frac{2}{\lambda} - \frac{1}{\lambda^2}\right)\ln\lambda - \frac{3}{2\lambda^2} + \dots \quad \text{for } \lambda \gg 1.
\end{aligned}
\right\} \qquad \text{A(106)}
$$

$$
\left.
\begin{aligned}
q_1(\lambda) &= \left(1 - \frac{2}{3}\lambda\right)\ln\lambda + \frac{2}{3} + \frac{1}{3\lambda} + \frac{(1-\lambda)^2(1+2\lambda)}{3\lambda^2}\ln|1-\lambda|, \\[2mm]
&\approx \left(1 - \frac{2}{3}\lambda\right)\ln\lambda + \frac{1}{2} + \frac{8}{9}\lambda - \frac{\lambda^2}{4} + \dots \quad \text{for } \lambda \ll 1, \\[2mm]
&\approx \frac{1}{3\lambda^2}\ln\lambda + \frac{1}{\lambda} + \frac{5}{18\lambda^2} + \dots \qquad \text{for } \lambda \gg 1.
\end{aligned}
\right\} \qquad \text{A(107)}
$$

These two functions are closely related to the function $r(\lambda)$ introduced in §7-5 in connexion with the fermion eigenvalue equation. $r(\lambda)$ determines that part of the fermion mass which is due to the creation and annihilation of bosons.

$$
\left.
\begin{aligned}
r(\lambda) &= (2-\lambda)\ln\lambda + 1 + \frac{(1-\lambda)^2}{\lambda}\ln|1-\lambda|, \\[2mm]
&\approx (2-\lambda)\ln\lambda + \frac{3}{2}\lambda - \frac{\lambda^2}{3} + \dots \quad \text{for } \lambda \ll 1, \\[2mm]
&\approx \frac{1}{\lambda}\ln\lambda + \frac{3}{2\lambda} - \frac{1}{3\lambda^2} + \dots \qquad \text{for } \lambda \gg 1.
\end{aligned}
\right\} \qquad \text{A(108)}
$$

The three functions $q_0(\lambda)$, $q_1(\lambda)$ and $r(\lambda)$ are connected by the relations:

$$\left.\begin{array}{l} r(\lambda) = \dfrac{1}{2}q_0(\lambda) + \dfrac{3}{2}q_1(\lambda), \\[2ex] q_0(\lambda) = -\dfrac{1}{\lambda}r\left(\dfrac{1}{\lambda}\right), \\[2ex] q_1(\lambda) = \dfrac{2}{3}r(\lambda) + \dfrac{1}{3\lambda}r\left(\dfrac{1}{\lambda}\right), \end{array}\right\} \qquad \text{A(109)}$$

which can be derived immediately from A(106)–A(108). Fig. A1 gives the three functions in the range $0 < \lambda \leqslant 5$. Their numerical values are listed in the table at the end of this section.

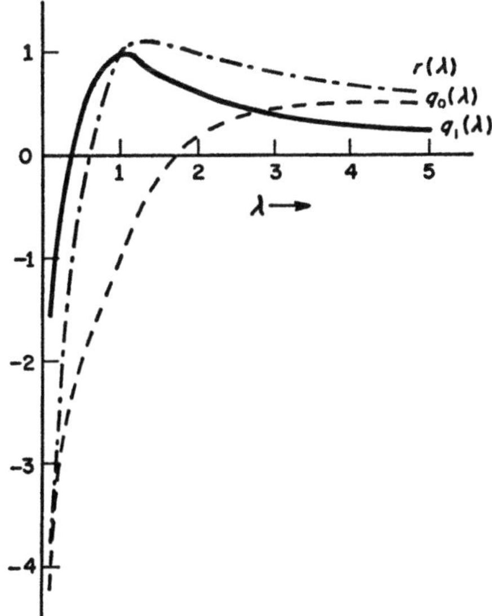

Fig. A1.

The functions $q_0(\lambda)$, $q_1(\lambda)$ and $r(\lambda)$ have been obtained from the 2-point function 3(39), in which a lepton double pole is assumed at the mass zero. This approximation is not sufficient for the eigenvalue equation of the photon or for the calculation of electromagnetic mass corrections. Therefore the functions $q(\lambda, \epsilon)$ have been defined which are derived from $q(\lambda)$ by transferring the lepton double pole from the mass value zero to the average mass value μ of the charged leptons $(\epsilon = \mu^2/\kappa^2)$. The expression

$$\frac{(\kappa^2)^2}{(p^2)^2(p^2 - \kappa^2)}$$

in the 2-point function should in this case be replaced by

$$\left(\frac{\kappa^2-\mu^2}{p^2-\mu^2}\right)^2 \frac{1}{p^2-\kappa^2}.$$

For calculating the resulting changes in the integral $K_{\mu\nu}$ of A(88), it is convenient to introduce the dimensionless variables $p'=p/\sqrt{J^2}$ and $j=J/\sqrt{J^2}$. Then A(88) can be written as

$$K\kappa^2 = \int d^4p' \; \frac{-g_{\mu\nu}(p',j-p')+p'_\mu(j-p')_\nu+p'_\nu(j-p')_\mu}{(p'^2)^2(p'^2\lambda-1)(j-p')^2}. \qquad \text{A(110)}$$

In this integral the expression

$$f(\lambda) = \frac{1}{(p'^2)^2(p'^2\lambda-1)} \qquad \text{A(111)}$$

should be replaced by

$$f(\lambda,\epsilon) = \left(\frac{1-\epsilon}{p'^2-(\epsilon/\lambda)}\right)^2 \frac{1}{p'^2\lambda-1} \qquad \text{A(112)}$$

It can easily be verified by differentiation that the functions $f(\lambda)$ and $f(\lambda,\epsilon)$ are connected by the relation

$$f(\lambda,\epsilon) = f(\lambda)-f\left(\frac{\lambda}{\epsilon}\right)+\frac{\lambda}{\epsilon}(1-\epsilon)\frac{df(\lambda/\epsilon)}{d(\lambda/\epsilon)}. \qquad \text{A(113)}$$

The same equation therefore holds for the functions q:

$$q(\lambda,\epsilon) = q(\lambda)-q\left(\frac{\lambda}{\epsilon}\right)+\frac{\lambda}{\epsilon}(1-\epsilon)\frac{dq(\lambda/\epsilon)}{d(\lambda/\epsilon)}. \qquad \text{A(114)}$$

This equation has been used in 8(23) for the calculation of $q_1(\lambda,\epsilon)$. More generally A(114) leads to

$$\left.\begin{aligned}
q_0(\lambda,\epsilon) &= \ln|\epsilon-\lambda|-\frac{(1-\lambda)^2}{\lambda^2}\ln|1-\lambda|-\frac{1-3\epsilon+2\epsilon^2}{\lambda} \\
&\quad +\frac{\epsilon}{\lambda^2}(3\epsilon-4\lambda-2\epsilon^2+2\lambda\epsilon)\ln\left|\frac{\epsilon-\lambda}{\epsilon}\right|, \\
&\approx \ln|\epsilon-\lambda|+1-\epsilon+\frac{\lambda}{\epsilon}+\ldots \quad \text{for } \lambda\ll\epsilon,\lambda\ll 1, \\
&\approx q_0(\lambda)+\frac{2\epsilon}{\lambda}-\frac{4\epsilon}{\lambda}\ln\left|\frac{\lambda}{\epsilon}\right|+\ldots \quad \text{for } \epsilon\ll\lambda,\epsilon\ll 1.
\end{aligned}\right\} \qquad \text{A(115)}$$

Mathematical Appendix 161

$$q_1(\lambda, \epsilon) = \left(1 - \frac{2}{3}\lambda\right)\ln|\epsilon - \lambda| + \frac{(1-\lambda)^2(1+2\lambda)}{3\lambda^2}\ln|1-\lambda|$$

$$+ \frac{1-\epsilon}{3\lambda}(1 - 2\epsilon + 2\lambda) - \frac{\epsilon^2}{\lambda^2}\left(1 - \frac{2\epsilon}{3}\right)\ln\left|\frac{\epsilon-\lambda}{\epsilon}\right|$$

$$\approx \left(1 - \frac{2}{3}\lambda\right)\ln|\epsilon - \lambda| + 1 - \epsilon + \frac{\lambda}{3\epsilon} + \frac{2\lambda}{3} + \dots \quad \text{for } \lambda \ll \epsilon, \lambda \ll 1,$$

$$\approx q_1(\lambda) - 2\frac{\epsilon}{\lambda} + \frac{2\epsilon}{3} + \dots \quad \text{for } \epsilon \ll \lambda, \epsilon \ll 1.$$

$$A(116)$$

The behaviour of the function $q_1(\lambda, \epsilon)$ for the values $\epsilon = 0, \frac{1}{4}, \frac{1}{2}, \frac{3}{4}$ is given in Fig. A2.

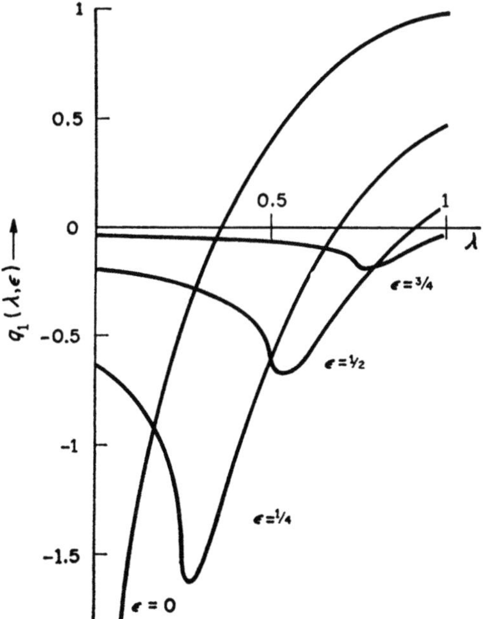

FIG. A2.

The curves have a vertical tangent at the point $\lambda = \epsilon$.

In connexion with the eigenvalue equations of the fermions 4(31), 5(27) and 5(33) the most important integral operator is —⟨⟩ which may be interpreted as representing that part of the fermion mass which is due to the creation and annihilation of fermion–antifermion pairs. This operator leads to the function $L(\lambda)$ of 5(29):

$$L(\lambda) = -i\left(\frac{\kappa}{\pi}\right)^4 \int d^4q\, d^4r\, \frac{(qr)(J, J-r-q)}{J^2(J-r-q)^2(q^2)^2(q^2-\kappa^2)(r^2)^2(r^2-\kappa^2)}.$$

$$\text{A(117)}$$

In terms of simpler analytical functions the integral $L(\lambda)$ can be represented in the following form:

$$L(\lambda) = -\frac{1}{24}\left(1+2\pi^2-\frac{1}{\lambda}\right) - \frac{1}{24}\left(12-8\lambda+\frac{\lambda^2}{2}\right)\ln\lambda$$

$$+\frac{1}{24\lambda^2}(7-9\lambda+15\lambda^2-\lambda^3)(1-\lambda)\ln|1-\lambda|$$

$$-\frac{\lambda^2-1}{2\lambda^2}f(\lambda) - \frac{1}{4}\left(6+\lambda+7\lambda^2-\frac{\lambda^3}{2}\right)g(\lambda) - \tfrac{1}{2}h(\lambda), \quad \text{A(118)}$$

where

$$f(\lambda) = -\tfrac{1}{8}\operatorname{arctg}^2\frac{(2-\lambda)\sqrt{[\lambda(4-\lambda)]}}{\lambda^2-4\lambda+2} \qquad \text{for } 0 \leqslant \lambda \leqslant 4,$$

$$= -\frac{\pi^2}{2}+\frac{1}{8}\ln^2\left|\frac{\lambda-2-\sqrt{[\lambda(\lambda-4)]}}{\lambda-2+\sqrt{[\lambda(\lambda-4)]}}\right| \quad \text{for } \lambda > 4;$$

$$\left.\vphantom{\begin{array}{c}1\\1\end{array}}\right\} \quad \text{A(119)}$$

$$g(\lambda) = -\frac{\sqrt{[\lambda(4-\lambda)]}}{12\lambda^2}\operatorname{arctg}\frac{(2-\lambda)\sqrt{[\lambda(\lambda-4)]}}{\lambda^2-4\lambda+2} \quad \text{for } 0 < \lambda \leqslant 4,$$

$$= \frac{\sqrt{[\lambda(\lambda-4)]}}{12\lambda^2}\ln\left|\frac{\lambda-2-\sqrt{[\lambda(\lambda-4)]}}{\lambda-2+\sqrt{[\lambda(\lambda-4)]}}\right| \qquad \text{for } \lambda > 4;$$

$$\left.\vphantom{\begin{array}{c}1\\1\end{array}}\right\} \quad \text{A(120)}$$

$$h(\lambda) = \sum_1^\infty \frac{\lambda^n}{n^2} \qquad\qquad \text{for } 0 \leqslant \lambda \leqslant 1$$

$$h(1) = \frac{\pi^2}{6}$$

$$h(\lambda) = -h\left(\frac{1}{\lambda}\right) - \tfrac{1}{2}\ln^2\lambda + \frac{\pi^2}{3} \quad \text{for } \lambda \geqslant 1.$$

$$\left.\vphantom{\begin{array}{c}1\\1\\1\\1\end{array}}\right\} \quad \text{A(121)}$$

A graph of the function $L(\lambda)$ has been given in §5-3. For small values of λ the first terms of the expansion are

$$L(\lambda) \approx \tfrac{1}{2} - \frac{\pi^2}{12} - \tfrac{1}{2}\ln\lambda + \ldots \quad \text{for } \lambda \ll 1. \qquad \text{A(122)}$$

At the point $\lambda = 1$ one gets

$$L(1) = -\frac{\pi^2}{6} + \frac{3\sqrt{3}}{16}\pi = -0\cdot6247\ldots. \qquad \text{A(123)}$$

Finally we give a numerical table of the functions $q_0(\lambda)$, $q_1(\lambda)$, $r(\lambda)$, $L(\lambda)$ in the range $0 < \lambda \leqslant 5$.

 Unified Field Theory of Elementary Particles

TABLE

λ	$q_0(\lambda)$	$q_1(\lambda)$	$r(\lambda)$	$L(\lambda)$
0·1	−3·7684	−1·5628	−4·2283	0·7093
0·2	−3·0391	−0·7277	−2·6110	0·2923
0·3	−2·5954	−0·2211	−1·6293	0·0392
0·4	−2·2669	0·1384	−0·9258	−0·1411
0·5	−2·0000	0·4091	−0·3863	−0·2782
0·6	−1·7703	0·6171	0·0405	−0·3854
0·7	−1·5641	0·7757	0·3815	−0·4702
0·8	−1·3726	0·8920	0·6518	−0·5367
0·9	−1·1880	0·9684	0·8585	−0·5878
1·0	−1·0000	1·0000	1·0000	−0·6247
1·1	−0·7948	0·9748	1·0648	−0·6471
1·2	−0·6063	0·9302	1·0922	−0·6560
1·3	−0·4427	0·8811	1·1003	−0·6542
1·4	−0·3030	0·8324	1·0972	−0·6444
1·5	−0·1842	0·7862	1·0872	−0·6288
1·6	−0·0832	0·7431	1·0731	−0·6090
1·7	0·0029	0·7033	1·0563	−0·5862
1·8	0·0763	0·6667	1·0382	−0·5616
1·9	0·1392	0·6331	1·0193	−0·5358
2·0	0·1931	0·6023	1·0000	−0·5094
2·1	0·2396	0·5740	0·9807	−0·4829
2·2	0·2797	0·5479	0·9616	−0·4566
2·3	0·3143	0·5238	0·9429	−0·4308
2·4	0·3443	0·5016	0·9246	−0·4056
2·5	0·3703	0·4811	0·9068	−0·3813
2·6	0·3929	0·4620	0·8895	−0·3579
2·7	0·4125	0·4443	0·8727	−0·3354
2·8	0·4296	0·4278	0·8565	−0·3141
2·9	0·4444	0·4124	0·8408	−0·2938
3·0	0·4572	0·3980	0·8256	−0·2746
3·1	0·4683	0·3845	0·8109	−0·2565
3·2	0·4780	0·3718	0·7968	−0·2394
3·3	0·4863	0·3600	0·7831	−0·2235
3·4	0·4934	0·3490	0·7699	−0·2087
3·5	0·4996	0·3382	0·7571	−0·1948
3·6	0·5048	0·3282	0·7447	−0·1820
3·7	0·5092	0·3188	0·7328	−0·1701
3·8	0·5128	0·3099	0·7213	−0·1591
3·9	0·5159	0·3014	0·7101	−0·1490
4·0	0·5183	0·2934	0·6993	−0·1397
4·1	0·5203	0·2858	0·6888	−0·1312
4·2	0·5218	0·2785	0·6787	−0·1233
4·3	0·5229	0·2716	0·6689	−0·1161
4·4	0·5236	0·2650	0·6593	−0·1094
4·5	0·5240	0·2587	0·6501	−0·1032
4·6	0·5241	0·2527	0·6411	−0·0974
4·7	0·5240	0·2470	0·6324	−0·0921
4·8	0·5236	0·2415	0·6240	−0·0871
4·9	0·5230	0·2362	0·6158	−0·0825
5·0	0·5222	0·2311	0·6078	−0·0782

Notations and Conventions

Units: $\hbar = c = 1$.

Metric in Lorentz space: $g_{00} = -g_{ii} = 1$.
(In earlier papers on the subject of the present book frequently the opposite metric $g_{00} = -g_{ii} = -1$ had been used.)

The Einstein summation convention is currently used in the text.

Indices: As a rule roman letter indices $i, j, k \ldots$ run from one to three, greek indices $\alpha, \beta, \ldots \lambda$ from one to two, greek indices $\nu, \mu, \ldots \rho$ from zero to three. Exceptions can easily be recognized in the text.

Co-ordinates: Space–time co-ordinates $x, y, z \ldots$.

$$x^\nu = (t, \mathbf{r}),$$

Energy-momentum vectors $p, q, r \ldots, k \ldots$

$$p^\nu = (E, \mathbf{p}).$$

Field operators: $\psi(x), \chi(x), \ldots$.

Representations of states:

$$\tau(x_1 x_2 \ldots | y_1 y_2 \ldots), \quad \phi(x_1 x_2 \ldots | y_1 y_2 \ldots)$$

$$\eta(x_1 x_2 \ldots | y_1 y_2 \ldots), \quad \zeta(x_1 x_2 \ldots | y_1 y_2 \ldots).$$

Functionals: $\tilde{T}, \tilde{\Phi}, \tilde{H}, \tilde{Z}$.

Parameters of continuous groups: $\alpha, \beta, \gamma \ldots$.

The asterisk * denotes the hermitian (or pseudo-hermitian) conjugate of a q-number, and the complex conjugate of a c-number.

Commutator:

$$[A, B] = AB - BA$$

$$\{A, B\} = AB + BA$$

165

tr A trace of the operator A.

$$\sigma^\nu = (1, \boldsymbol{\sigma})$$

$$\overline{\sigma^\nu} = (1, -\boldsymbol{\sigma})$$

$$\text{with } 1 = \begin{vmatrix} 1 & 0 \\ 0 & 1 \end{vmatrix} \text{ and } \boldsymbol{\sigma} \text{ the conventional Pauli matrices.}$$

$$\gamma^i = \begin{pmatrix} 0 & \sigma_i \\ \sigma_i & 0 \end{pmatrix}$$

$$\gamma^0 = \begin{pmatrix} 0 & 1 \\ -1 & 0 \end{pmatrix}$$

$$\gamma^5 = \begin{pmatrix} -1 & 0 \\ 0 & 1 \end{pmatrix} = i\gamma^0 \gamma^1 \gamma^2 \gamma^3.$$

2-point functions:

$$F(xy) = \langle 0 | T\chi(x)\chi^*(y) | 0 \rangle$$

$$F_0(xy) = \langle 0 | T\chi_0(x)\chi^*(y) | 0 \rangle, \quad \text{where } i\sigma_\nu \frac{\partial}{\partial x_\nu} \chi_0 = 0$$

$$G(xy) = \text{causal Green's function.}$$

(For a linear field theory the relation $F(xy) = -iG(xy)$ would hold.)
Symbols in Feynman diagrams:

$$F(xy) \qquad\qquad$$ 

$$G(xy) \qquad\qquad$$

Vacuum expectation values:

$$\tau(xyzu) \qquad\qquad$$ 

$$(\eta xyzu) \qquad\qquad$$ 

Matrix elements between vacuum and particle state:

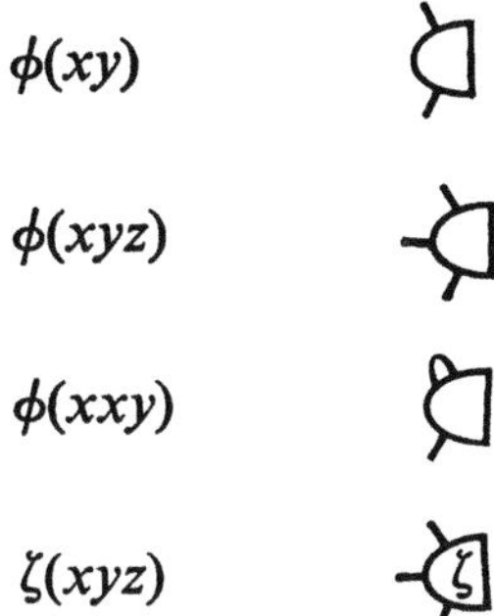

Each chapter of the book is divided into sections. Formulae are numbered through in each chapter separately; within the chapter they are referred to simply by their number between brackets.

TABLE OF ELEMENTARY PARTICLES [79]

BOSONS

Symbol	Mass in MeV	Isospin	Spin	P	G	C	$u = 1/2Y$	Mean life in sec.	Line width in MeV
γ	0		1	−		−	0	∞	0
π $\pi^\pm$	139·58	1	0	−	−		0	$2\cdot55 \times 10^{-8}$	$2\cdot58 \times 10^{-14}$
π^0	134·97	1	0	−	−	+	0	$1\cdot78 \times 10^{-16}$	$3\cdot7 \times 10^{-6}$
K $K^\pm$	493·8	1/2	0	−			$\pm 1/2$	$1\cdot23 \times 10^{-8}$	$5\cdot34 \times 10^{-14}$
K^0	497·7	1/2	0	−			$\pm 1/2$	$K_1\ 0\cdot88 \times 10^{-10}$ $K_2\ 5\cdot77 \times 10^{-8}$	$7\cdot47 \times 10^{-12}$ $1\cdot14 \times 10^{-14}$
η	548·8	0	0	−	+	+	0	$> 6\cdot6 \times 10^{-23}$	< 10
κ	725	1/2	0	+			$\pm 1/2$	$> 5\cdot5 \times 10^{-23}$	< 12
ρ	765	1	1	−	+	$-\,n$	0	$5\cdot3 \times 10^{-24}$	124
ω	782·8	0	1	−	−	−	0	$5\cdot5 \times 10^{-23}$	12
K^*	891·4	1/2	1	−			$\pm 1/2$	$1\cdot3 \times 10^{-23}$	49
X^0	958·6	0	0	−	+	+	0	$> 16 \times 10^{-23}$	< 4
ϕ	1019·5	0	1	−	−	−	0	2×10^{-22}	3·3
A_1	1072	1	1	+	−	$+\,n$	0	$5\cdot2 \times 10^{-24}$	125
C	1215	1/2 ?					$\pm 1/2$	$1\cdot1 \times 10^{-23}$	60
B	1220	1	1 ?		+	$-\,n$	0	$5\cdot2 \times 10^{-24}$	125
f	1253	0	2	+	+	+	0	$5\cdot6 \times 10^{-24}$	118
D	1286	0	1	+	+	+	0	$1\cdot6 \times 10^{-23}$	40
A_2	1324	1	2	+	−	$+\,n$	0	$7\cdot3 \times 10^{-24}$	90
κ^*	1405	1/2	2	+			$\pm 1/2$	$6\cdot9 \times 10^{-24}$	95
E	1420	0					0	$1\cdot1 \times 10^{-23}$	60
f'	1500	0	2	+	+	+	0	$8\cdot2 \times 10^{-24}$	80

The index n means that the C-property refers to the neutral particles.

 Unified Field Theory of Elementary Particles

TABLE OF ELEMENTARY PARTICLES[79]—*continued*

FERMIONS

Symbol		Mass in MeV	Isospin	Spin	P	$u = 1/2Y$	Mean life in sec.	Line width in MeV
	ν_e	0		1/2			∞	0
	ν_μ	0		1/2			∞	0
	$e^\pm$	0·5110		1/2			∞	0
	$\mu^\pm$	105·66		1/2			$2\cdot2 \times 10^{-6}$	3×10^{-16}
N	p	938·26	1/2	1/2	+	1/2	∞	0
	n	939·55	1/2	1/2	+	1/2	$1\cdot01 \times 10^{3}$	$6\cdot5 \times 10^{-25}$
	Λ	1115·44	0	1/2	+	0	$2\cdot61 \times 10^{-10}$	$2\cdot5 \times 10^{-12}$
Σ	Σ^+	1189·4	1	1/2	+	0	$0\cdot79 \times 10^{-10}$	$8\cdot3 \times 10^{-12}$
	Σ^0	1192·3	1	1/2	+	0	$<1\cdot0 \times 10^{-14}$	$>6\cdot6 \times 10^{-8}$
	Σ^-	1197·2	1	1/2	+	0	$1\cdot58 \times 10^{-10}$	$4\cdot2 \times 10^{-12}$
	N^*	1236	3/2	3/2	+	1/2	$5\cdot5 \times 10^{-24}$	120
Ξ	Ξ_0	1314·3	1/2	1/2	+	$-1/2$	$3\cdot05 \times 10^{-10}$	$2\cdot2 \times 10^{-12}$
	Ξ_-	1320·8	1/2	1/2	+	$-1/2$	$1\cdot75 \times 10^{-10}$	$3\cdot8 \times 10^{-12}$
	Y_2^*	1382·7	1	3/2	+	0	$1\cdot5 \times 10^{-23}$	44
	Y_0^*	1405	0	1/2	−	0	$1\cdot9 \times 10^{-23}$	35
	N^*	1518	1/2	3/2	−	1/2	$5\cdot5 \times 10^{-24}$	120
	Y_0^*	1520	0	3/2	−	0	$4\cdot1 \times 10^{-23}$	16
	Ξ^*	1530	1/2	3/2	+	$-1/2$	$0\cdot88 \times 10^{-22}$	7·5
	Y_1^*	1660	1	$\geqslant 3/2$		0	$1\cdot5 \times 10^{-23}$	44
	Ω^-	1675	0	3/2	+	-1	$1\cdot3 \times 10^{-10}$	$5\cdot1 \times 10^{-12}$
	N^*	1688	1/2	5/2	+	1/2	$6\cdot6 \times 10^{-24}$	100
	Y_1^*	1765	1	5/2	−	0	$0\cdot88 \times 10^{-23}$	75
	Y_0^*	1815	0	5/2	+	0	$1\cdot3 \times 10^{-23}$	50
	Ξ^*	1816	1/2	3/2	−	$-1/2$	$4\cdot1 \times 10^{-23}$	16
	N^*	1924	3/2	7/2	+	1/2	$3\cdot3 \times 10^{-24}$	200
	Ξ^*	1933	1/2	5/2	+	$-1/2$	$4\cdot7 \times 10^{-24}$	140
	Y_1^*	2065	1	7/2	+	0	$4\cdot1 \times 10^{-24}$	160
	N^*	2190	1/2	7/2	−	1/2	$3\cdot3 \times 10^{-24}$	200
	N^*	2360	3/2	9/2	−	1/2	$3\cdot3 \times 10^{-24}$	200
	N^*	2650	1/2	9/2	+	1/2	$3\cdot3 \times 10^{-24}$	200
	N^*	2825	3/2	11/2	+	1/2	$2\cdot5 \times 10^{-24}$	260

REFERENCES

1. W. Heisenberg, *Z. Physik*, **101**, 533, 1936; **113**, 61, 1939; *Nature (London)* **164**, 65, 1949.
2. E. Fermi, *Progr. Theoret. Phys.*, **5**, 570, 1950; *Phys. Rev.*, **81**, 683, 1951.
3. W. Heisenberg, *Two Lectures*, Cambridge Univ. Press, 1949; *Physica*, **19**, 897, 1953.
4. P. A. M. Dirac, *Proc. Roy. Soc. (London), Ser. A*, **117**, 610, 1928.
5. C. D. Anderson, *Science*, **76**, 238, 1932; P. M. S. Blackett and G. P. S. Occhialini, *Proc. Roy. Soc. (London), Ser. A*, **139**, 699, 1933.
6. H. A. Lorentz, *Proc. Acad. Sci. Amsterdam*, **6**, 809, 1904.
7. W. Heisenberg, *Z. Physik*, **77**, 1, 1932; **78**, 156, 1932.
8. M. Gell-Mann, *Phys. Rev.*, **92**, 833, 1953; T. Nakano and K. Nishijima, *Progr. Theoret. Phys.*, **10**, 581, 1953.
9. A. Pais, *Phys. Rev.*, **86**, 663, 1952; B. D'Espagnat and J. Prentki, *Nucl. Phys.*, **1**, 33, 1956; J. Schwinger, *Phys. Rev.*, **104**, 1164, 1956.
10. T. D. Lee and C. N. Yang, *Phys. Rev.*, **104**, 254, 1956; R. E. Marshak and E. C. G. Sudarshan, *Introduction to Elementary Particle Physics*, Interscience, New York, 1961.
11. K. Nishijima, *Progr. Theoret. Phys.*, **6**, 614, 1027, 1951; T. D. Lee and C. N. Yang, *Nuovo Cimento*, **3**, 749, 1956.
12. M. Gell-Mann, *Cal. Tech Rept.* CTSL-20, 1961; Y. Ne'eman, *Nucl. Phys.*, **26**, 222, 1961.
13. S. Okubo, *Progr. Theoret. Phys.*, **27**, 949, 1962.
14. M. Gell-Mann, *Proc. Intern. Conf. High Energy Phys. CERN* 1962, 805; V. E. Barnes *et al.*, *Phys. Rev. Letters* **12**, 204, 1964.
15. L. v. Lindern, *Z. Naturforsch.*, **11a**, 340, 1956; *Nuovo Cimento* **5**, 491, 1957.
16. N. N. Bogoliubov and D. V. Shirkov, *Introduction to the Theory of Quantized Fields*, Interscience, New York, 1959; S. S. Schweber, *An Introduction to Relativistic Quantum Field Theory*, Row, Peterson & Co., Evanston, 1961.
17. F. Bloch and A. Nordsieck, *Phys. Rev.*, **52**, 54, 1937.
18. J. A. Wheeler, *Phys. Rev.*, **52**, 1107, 1937; W. Heisenberg, *Z. Physik*, **120**, 513, 673, 1943; C. Møller, *Kong. Dansk Vid. Selsk. Mat. Fys. Medd.*, **23**, No. 1, 1945; **24**, No. 19, 1946.
19. J. Toll, *Thesis*, Princeton Univ., 1952; *Phys. Rev.*, **104**, 1760, 1956; M. Gell-Mann, M. L. Goldberger and W. Thirring, *Phys. Rev.*, **95**, 1612, 1954; M. L. Goldberger, *Phys. Rev.* **97**, 508, 1955; Y. Nambu, *Phys. Rev.*, **98**, 803, 1955; M. L. Goldberger, *Phys. Rev.*, **99**, 973, 1955; M. L. Goldberger, H. Miyazawa and R. Oehme, *Phys. Rev.*, **99**, 986, 1955; K. Symanzik, *Phys. Rev.*, **105**, 743, 1957; N. N. Bogoliubov, B. V. Medwedev and M. K. Polivanov, *Lecture notes*, (trans. Princeton), 1957, Inst. for Advanced Studies; *Fortschr. Physik*, **6**, 169, 1958; H. Lehmann, *Nuovo Cimento*, **10**, 579, 1958; **14**, *Suppl.*, 153, 1959.

20. G. Feldman and P. T. Matthews, *Phys. Rev.*, **102**, 1421, 1956; G. F. Chew, M. L. Goldberger, E. F. Low and Y. Nambu, *Phys. Rev.*, **106**, 1337, 1345, 1957.

21. P. Jordan and O. Klein, *Z. Physik*, **45**, 751, 1927; P. Jordan and E. Wigner, *Z. Physik*, **47**, 631, 1928.

22. R. Haag, Colloque internat. sur les problèmes de la théorie quantique des champs, Lille 1957; *Phys. Rev.*, **112**, 669, 1958; *Ann. Phys.*, **11**, 29, 1963.

23. W. Heisenberg, *Nachr. Akad. Wiss. Göttingen, IIa*, 111, 1953; *Z. Naturforsch.*, **9a**, 292, 1954.

24. W. Heisenberg, *Naturwiss.*, **50**, 3, 1963.

25. R. F. Streater and A. S. Wightman, PCT, *Spin and Statistics and All That*, Benjamin, New York, 1964; R. Jost, *The General Theory of Quantized Fields*, Am. Math. Soc. Publications, 1963; A. S. Wightman, *Phys. Rev.*, **101**, 860, 1956; A. S. Wightman, in: *Les Problèmes Mathematiques de la Théorie des Champs*, CNRS, Paris, 1959; A. S. Wightman, *Recent Achievements of Axiomatic Field Theory, Theoretical Physics*, IAEA, Vienna, 1963; R. Haag, *Dan. Mat. Fys. Medd.*, **29**, No. 12, 1955; R. Haag in: *Les Problèmes Mathematiques de la Théorie des Champs*, CNRS, Paris, 1959.

26. H. Lehmann, K. Symanzik and W. Zimmermann, *Nuovo Cimento*, **1**, 205, 1955; **2**, 425, 1955; **6**, 319, 1957.

27. K. Bleuler, *Helv. Phys. Acta*, **23**, 567, 1950; S. Gupta, *Proc. Phys. Soc. (London)* A, **63**, 681, 1950; A, **64**, 850, 1951.

28. W. Heisenberg, *Nucl. Phys.*, **4**, 532, 1957; K. L. Nagy, *Nuovo Cimento, Suppl.*, **17**, 92, 1960; L. K. Pandit, *Nuovo Cimento, Suppl.*, **10**, 157, 1959; S. Schlieder, *Z. Naturforsch.*, **15a**, 448, 460, 555, 1960; E. Scheibe, *Ann. Acad. Sci. Fennicae, Ser. A I*, 294, 1960.

29. H. P. Dürr, W. Heisenberg, H. Mitter, S. Schlieder and K. Yamazaki, *Z. Naturforsch.*, **14a**, 441, 1959; H. P. Dürr, *Z. Naturforsch.*, **16a**, 327, 1961.

30. G. Lüders, *Kong. Dansk Vid. Selsk. Mat. Fys. Medd.*, **28**, No. 5, 1954; *Ann. Phys.*, **2**, 1 (1957).

31. W. Pauli, *Niels Bohr and The Development of Physics*, McGraw-Hill, New York, 1955.

32. H. Umezawa and S. Kamefuchi, *Progr. Theoret. Phys.*, **6**, 543, 1951; G. Källén, *Helv. Phys. Acta*, **25**, 417, 1952; M. Gell-Mann and F. Low, *Phys. Rev.*, **95**, 1300, 1954; H. Lehmann, *Nuovo Cimento*, **11**, 342, 1954.

33. J. Schwinger, *The Theory of Coupled Fields*, Lecture Harvard, 1954 (unpubl.).

34. H. Mitter, *Nuovo Cimento*, **32**, 1789, 1964.

35. W. Heisenberg, *Z. Naturforsch.*, **10a**, 425, 1955; *Rev. Mod. Phys.*, **29**, 269, 1957.

36. T. D. Lee, *Phys. Rev.*, **95**, 1329, 1954; G. Källén and W. Pauli, *Kong. Dansk Vid. Selsk. Mat. Fys. Medd.*, **30**, No. 7, 1955.

37. W. Heisenberg, *Z. Phys.*, **144**, 1, 1956; *Nucl. Phys.*, **4**, 532, 1957; **5**, 195, 1958.

38. W. Heisenberg, *Proc. 1960 Intern. Conf. High Energy Phys. Rochester*, 851; *Proc. 1962 Intern. Conf. High Energy Phys. CERN*, 675.

39. Y. Nambu, *Proc. 1960 Intern. Conf. High Energy Phys. Rochester*, 857; *Phys. Rev.*, **117**, 648, 1960; Y. Nambu and G. Jona-Lasinio, *Phys. Rev.*, **122**, 345, 1961.

40. H. P. Dürr and W. Heisenberg, *Z. Naturforsch.*, **16a**, 726, 1961; *Nuovo Cimento*, **37**, 1446, 1965.

References   171

41. J. Goldstone, *Nuovo Cimento*, **19**, 154, 1961; J. Goldstone, A. Salam and S. Weinberg, *Phys. Rev.*, **127**, 965, 1962; S. Bludman and A. Klein, *Phys. Rev.*, **131**, 2364, 1963; A. Klein and B. W. Lee, *Phys. Rev. Letters*, **12**, 266, 1964; G. S. Guralnik, *Phys. Rev.*, **136B**, 1404, 1417, 1964; G. S. Guralnik, R. Hagen and T. W. Kibble, *Phys. Rev. Letters*, **13**, 585, 1964.
42. Y. Nambu, *Phys. Rev. Letters*, **9**, 214, 1964.
43. F. Bloch, *Z. Phys.*, **61**, 206, 1930.
44. Y. Nambu, *Progr. Theoret. Phys.*, **5**, 614, 1950; J. Schwinger, *Proc. Nat. Acad. Sci.*, **37**, 452, 455, 1951; M. Gell-Mann and F. Low, *Phys. Rev.*, **84**, 350, 1951; H. A. Bethe and E. E. Salpeter, *Phys. Rev.*, **84**, 1232, 1951.
45. E. I. Fredholm, *Acta Math.*, **21**, 365, 1903.
46. R. P. Feynman, *Phys. Rev.*, **76**, 749, 769, 1949.
47. E. Freese, *Z. Naturforsch.*, **8a**, 776, 1953.
48. I. Tamm, *J. Phys. (USSR)*, **9**, 449, 1945; S. M. Dancoff, *Phys. Rev.*, **78**, 382, 1950; F. J. Dyson, *Phys. Rev.*, **90**, 994, 1953; W. Zimmermann, *Nuovo Cimento Suppl.*, **11**, 43, 1954.
49. H. Stumpf, F. Wagner and F. Wahl, *Z. Naturforsch.*, **19a**, 1254, 1964.
50. J. Schwinger, *Phys. Rev.*, **82**, 664, 1951.
51. W. Zimmermann, *Nuovo Cimento*, **13**, 503, 1959.
52. E. Freese, *Nuovo Cimento*, **11**, 312, 1959.
53. J. Dhar and Y. Katayama, *Nuovo Cimento*, **36**, 533, 1965.
54. W. Heisenberg, H. Kortel and H. Mitter, *Z. Naturforsch.*, **10a**, 425, 1955.
55. H. Stumpf and H. Yamamoto, *Z. Naturforsch.*, **20a**, 1, 1965.
56. K. Yamazaki, *Nuovo Cimento*, **37**, 1143, 1965.
57. A. Scotti and D. Y. Wong, *Phys. Rev. Letters*, **10**, 142, 1963.
58. R. E. Cutkosky and M. Leon, *Phys. Rev.*, **135**, B 1445, 1964.
59. P. Cziffra and M. J. Moravcsik, *Phys. Rev.* **116**, 226, 1959.
60. T. D. Lee and C. N. Yang, *Phys. Rev.*, **119**, 1410, 1960.
61. H. N. Russell and F. A. Saunders, *Astrophys. J.*, **61**, 38, 1925.
62. W. Heisenberg, *Z. Phys.*, **49**, 619, 1928.
63. K. Johnson, *Conf. on High Energy Physics Dubna*, 1964.
64. H. P. Dürr, W. Heisenberg, H. Yamamoto and K. Yamazaki, *Nuovo Cimento*, **38**, 1220, 1965.
65. A. Sommerfeld, *Ann. Phys.*, **51**, 1, 1916.
66. J. C. Ward, *Phys. Rev.*, **78**, 182, 1950.
67. G. F. Chew and S. Mandelstam, *Nuovo Cimento*, **19**, 752, 1961.
68. T. Regge, *Nuovo Cimento*, **14**, 951, 1959; **18**, 947, 1960.
69. E. C. G. Sudarshan and R. E. Marshak, *Phys. Rev.*, **109**, 1860, 1958; R. P. Feynman and M. Gell-Mann, *Phys. Rev.*, **109**, 193, 1958.
70. W. Heisenberg, *Niels Bohr Memorial Meeting*, Copenhagen, 1963.
71. R. Hofstadter, *Ann. Rev. Nucl. Sci.*, **7**, 231, 1957.
72. P. Jordan and W. Pauli, *Z. Phys.*, **47**, 151, 1928.
73. J. Schwinger, *Phys. Rev.*, **75**, 651, 1949.
74. E. C. G. Stückelberg and D. Rivier, *Helv. Phys. Acta*, **22**, 215, 1949; D. Rivier, *Helv. Phys. Acta*, **22**, 265, 1949.
75. K. Symanzik, *Diss.* Univ. Göttingen, 1954; *Z. Naturforsch.*, **9a**, 809, 1954.
76. H. Rampacher, H. Stumpf and F. Wagner, *Fortschr. Physik*, **13**, 385, 1965.
77. A. Donnachie and J. Hamilton, *Phys. Rev.*, **133**, *B*, 1053, 1964; **138**, *B*, 678, 1965.

78. The connexion between gravitation and a unified quantum field theory has been discussed by W. Thirring, *Fortschr. Physik, VII*, **2**, 79–101, 1959; and from different points of view in a report by D. D. Jvanenko, New aspects of unified theory in 'Problems of gravitation', *2nd Soviet Conf. on Gravitation, Tiflis*, 20–28, April 1965.

79. A. H. Rosenfeld, A. Barbaro-Galtieri, W. H. Barkas, P. L. Bastien, J. Kirz and M. Roos, *UCRL*-8030-Part I, August 1965.

80. Compare e.g. L. de Broglie, D. Bohm, P. Hillion, F. Halbwachs, T. Takabayasi and J.-P. Vigier, *Phys. Rev.*, **129**, 438, 1963. L. de Broglie, F. Halbwachs, P. Hillion, T. Takabayasi and J.-P. Vigier, *Phys. Rev.*, **129**, 451, 1963. D. D. Ivanenko, *Izvestiya vysshikh uchebnykh zavedenii. Fizika*, No. **3**, 1965. D. F. Kurdgelaidze, *Cahiers de Physique*, **15**, 149, 1961.

Author Index

Subject Index

175

Physics Today *20*, No. 5, 27–33 (1967)

Nonlinear Problems In Physics

Interactions of matter and fields are generally nonlinear, so that nonlinear problems play a central role in physics. In fact because nonlinearity is so basic to nature, it is possible that even a theory as fundamentally linear as quantum theory may ultimately have to be replaced by a nonlinear one.

by Werner Heisenberg

NONLINEAR PROBLEMS in physics are difficult to discuss for several reasons. First, what are nonlinear problems? Practically every problem in theoretical physics is governed by nonlinear mathematical equations, except perhaps quantum theory, and even in quantum theory it is a rather controversial question whether it will finally be a linear or nonlinear theory. Therefore by far the largest part of theoretical physics is devoted to nonlinear problems.

Besides, it has been argued that every nonlinear problem is really individual; that is, it requires individual methods, usually very complicated and difficult methods, and it is rather improbable that one can learn from one nonlinear problem to solve another nonlinear problem.

Finally I have to emphasize that I am certainly not an expert in the field of nonlinear problems of mathematics or physics; but I have come across a few nonlinear problems on my way through physics, so I at least know some of the horrible difficulties and troubles which one meets in these problems.

In this situation I feel that the only thing I can do is not give a general survey of the field but rather tell about some of these problems that have passed my way in physics and see whether there are common features. In fact I do think that all these problems have some features in common—difficulties that are similar in different problems and also methods that one can use to solve the difficulties. So I hope you will allow me to go briefly through some of these problems and to see what are the common features.

I would like to start with a brief historical remark and to say a few words about point mechanics, and only then I will examine general problems of continuous media, and finally I will discuss to what extent quantum theory can be considered a linear or nonlinear theory.

The author is famous for his contributions to the development of quantum theory and for his recognition of the fundamental importance of the uncertainty principle that now bears his name. He received the Nobel Prize for this work in 1932. Since 1941 he has been director of the Max Planck Institute for Physics and Astrophysics, Munich. This article is based on a talk he gave last summer at The International School of Nonlinear Mathematics and Physics, Munich, directed by Norman J. Zabusky, Bell Telephone Laboratories, Whippany, N.J., and Martin D. Kruskal, Princeton University, Princeton, N.J.

Physics Today *20*, No. 5, 27–33 (1967)

Historical remarks

The course of history in physics usually proceeds from simple problems to more complicated problems. For the questions discussed here the simplest problem can be considered to be the solution of a homogeneous linear equation; next simplest is the solution of an inhomogeneous linear equation; and, finally, much more complicated is the solution of a nonlinear operation. Actually mathematical physics started 300 years ago with the law of inertia, which may be considered to be the solution of the homogeneous linear equation $d^2x/dt^2 = 0$ where x is the coördinate and t is the time. In the laws of free fall of Galileo, we find that he actually had solved an inhomogeneous linear equation, the force being the inhomogeneous term. Finally, Newton formulated the nonlinear problem because the equations of motion in Newton's mechanics definitely are nonlinear equations.

Point mechanics

Of course Newton was not in a position to find general solutions of nonlinear equations; so from the very be-

ginning he had to think about possible simplifications. Therefore, I might start by mentioning some of the most important simplifications that can be used, not only in the nonlinear problems of point mechanics, but actually in all nonlinear problems with which we have to deal.

Symmetry. There are mainly two different types of simplification that are important. The first comes from the symmetry of the problem. Physicists have learned from mathematicians that the symmetry of a problem as a rule produces a conservation law. All the conservation laws that we know in physics—conservation of energy, momentum, angular momentum etc.—rest upon fundamental symmetries in the underlying natural law.

For instance, Newton's equations of motion are invariant under the Galilean group, which is a continuous group of ten parameters. Therefore, one has ten conservation laws and these conservation laws can be used for the simplification of the problem. When Newton was able to solve the two-body problem in astronomy, it was by the use of symmetry. With two bodies there are 12 degrees of freedom because every body has 3 coördinates and 3 momenta; and of these 12 degrees of freedom, he could actually eliminate 10 by means of conservation laws, and the rest then was rather simple. Similar simplifications can be used in most problems.

Linearization. The next kind of simplification is linearization of the problem. This method has been very widely used and has actually been very important for the discussion of nonlinear problems. Let us assume that one has more or less by chance found simple solutions for the problem concerned, very special solutions—for instance, static solutions in which all the bodies are at rest, or as in hydrodynamics, stationary solutions in which the motion does not change with time; that is, the velocities at least are constant. In all these cases, one can study general solutions of the nonlinear problem in the neighborhood of the special solution. That is more or less the main object of perturbation theory. For small perturbations around a known motion one has linear equations to solve. Thereby one comes back to simpler mathematical problems for which one may be able to find the complete system of solutions. Such solutions can be studied for the whole time interval from minus infinity to plus infinity, and the solutions may be characterized by their asymptotic behavior in space and time.

If the special solution of the original system is a static or a stationary solution, then we have the special advantage that the coefficients in these linear equations of perturbation theory

NONLINEAR PHYSICS AND WERNER HEISENBERG

by Norman J. Zabusky

All physical processes are nonlinear if observed over a sufficiently long time or generated with a sufficiently intense source. For example, Einstein has remarked that Maxwell's equations in empty space represent formulations which are linear in form only because they are based on experience with very weak electromagnetic fields. "In general linear laws fulfill the superposition principle for their solutions but contain no assertions concerning the interaction of elementary bodies. The true laws cannot be linear, nor can they be derived from such."

It is too bad that this is a rather general feature of physical laws, because mathematical analysis of linear laws is relatively simple. Linearity implies solutions may be superposed; this principle permits reduction of a complex problem into simpler elements, a technique of enormous analytical power. On the other hand the elucidation of nonlinear phenomena is difficult. One often finds that methods of solution are highly restricted or "rigid" in that they are usually rendered useless if the problem posed is modified only slightly. The study of nonlinear problems has progressed slowly with few general results obtained compared to the progress of linear mathematics and physics.

During the past 50 years the major step forward in physics has been the development and application of quantum theory, a fundamentally linear theory. The name Werner Heisenberg is intimately connected with its basic concepts and further developments. It is probably not so well known that he also has had a recurrent interest in nonlinear physics problems.

His first published paper dealt with vortex motion, and two years later in 1924 he published a paper on the stability and turbulence of a flowing fluid and another on solutions for a nonlinear differential equation describing a viscous fluid. In 1936 he published a paper with H. Euler in which, using quantum theoretical methods, he derived the Lagrangian density for the electromagnetic field to the next, nonlinear, order. In 1948 three papers appeared on the statistical theory of turbulence and in 1953 a paper on "Meson Production as a Shockwave Problem." At that time his belief in the fundamental nature of nonlinear processes merged with his interest in applying quantum theory to elementary particles, and in 1953 he published a paper on quantization of nonlinear equations. This work has been followed by many papers by him and his associates on the quantum theory of nonlinear wave equations and nonlinear spinor theory.

* * *

Condensed from an introduction to the International School of Nonlinear Mathematics and Physics.

---

are constant in time, and it will be possible to look for solutions that behave like an exponential function of time, either $e^{i\omega t}$ or $e^{\alpha t}$.

If such a system of solutions has been found, one can at once answer questions of stability or instability. If only periodic solutions exist or only solutions that decrease in time, the system is stable—a perturbed solution starting at a given time will in the course of time not deviate strongly from the original simple solution. If, however, there are solutions of the linear equation that increase exponentially with time, the solution is not stable. This method of simplification can be applied even if the original solution is not a static solution. Many papers have been written on general perturbation theory and much insight has been gained by such methods.

Unpredictability. Beyond this point one comes very soon into the real difficulties of the nonlinear problems and I might first mention the difficulties in the case of point mechanics. For instance, consider the well known problem of three bodies in astronomy or the more complicated

Physics Today *20*, No. 5, 27–33 (1967)

problem of our planetary system. There one has of course applied perturbation theory, but one has very soon found that it is extremely diffi-

29 cult to know how far such a theory can work. It is true that the motion of the planets can be followed rather easily by perturbational methods for a finite amount of time, say for a few thousand years. But when time increases, perturbations may no longer be small.

Resonance effects have been observed owing to commensurabilities between the orbital frequencies of different planets, and in the course of time there may be a big effect, so that in certain cases it finally may be impossible to say whether the orbit will be periodical or whether the planet will move out of the system. Such questions can be extremely difficult to answer.

I might mention a most paradoxical result of this mathematical analysis—the theorem of Heinrich Bruns. He proved that even in an infinitesimaly close neighborhood of a point where the perturbation theory converges, there must always be other points where the perturbation theory diverges. So one can say that the points where the perturbation theory converges and those where it diverges form a dense manifold. This result suggests that after a very long time one can never know where the orbit finally will go. It is only for a finite time that we can really determine what will happen.

This is not only a mathematical result interesting for the astronomers: it may have very important practical applications. I might just mention a problem that troubled us about 15 years ago when the CERN proton-synchrotron was constructed. For the big CERN accelerator it was very important to know whether a proton running around the circle of the accelerator would be stable in the assumed orbit or not. One could select conditions such that there would be stable

oscillations around the orbit. But when there is some small perturbing defect for the orbit in the tube there may be resonances. The frequency of rotation of the proton around the machine and the frequency of oscillations around the orbit can become commensurable, and then there could arise difficulties similar to those in the astronomical problems.

One would expect the amplitude of the oscillation to increase if there is resonance, but then you come into the region where the problem is not linear and perturbation theory does not work. On account of the nonlinearity the frequency will be changed, and therefore the resonance will be stopped. In this way the nonlinear terms should exert a stabilizing influence until the motion has passed the critical region of resonance.

This was the hope. When numerical calculations were carried out, the result was that the particle may actually run around its orbit 10 000 times; that is, the stabilizing effect seems to work very well. First the amplitude increases, then the frequency gets out of resonance, then the amplitude decreases again, and so on. But after 10 000 revolutions the particle can run out of its orbit just as well. So you can get this kind of surprise, a final instability after a long time of apparent stability, in nonlinear problems, and that, I think, is a very characteristic feature of nonlinear problems. Therefore one might say, to use a very simple term, that nonlinear problems have a certain kind of unpredictability. One doesn't know how the solutions will behave after a very long time; I think that this may be a very general feature of nonlinear problems.

Statistical methods. What can one do if it is too difficult to study the single orbit, too difficult to get a kind of survey of all the possible solutions of such an equation? The way out is to investigate, not a single solution, but ensembles of solutions. One can

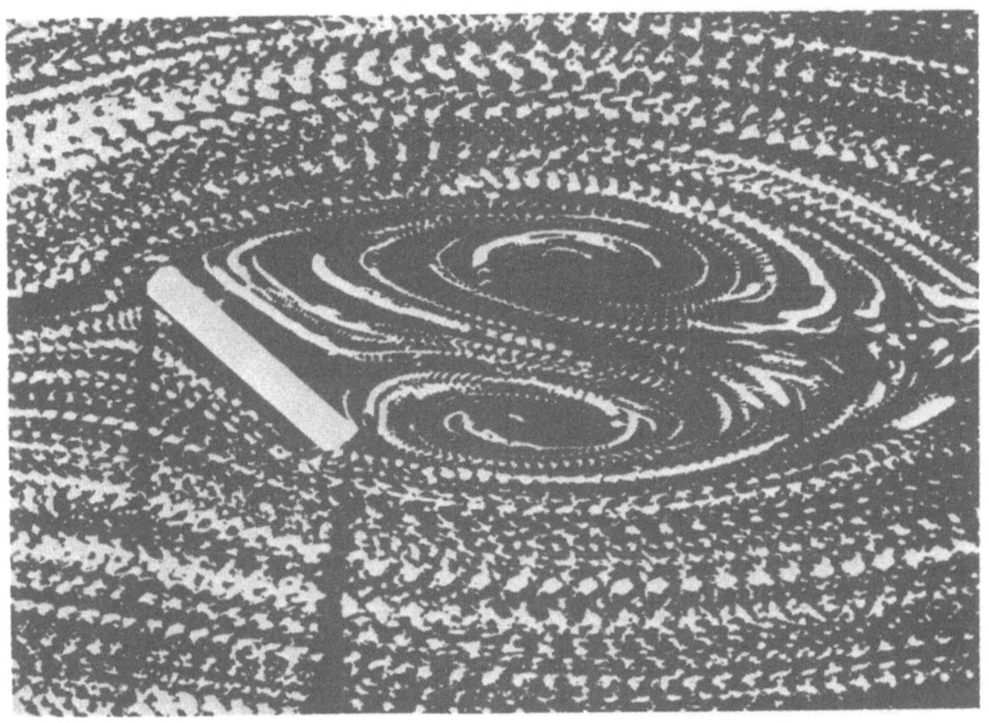

LAMINAR MOTION will be stable at small Reynolds numbers, but above certain values there are deviations for which the amplitude increases resulting in turbulent motion.

argue that it may not be necessary to know every detail of the solution, that one can be satisfied with an incomplete knowledge of the system. In this case one cannot ask: What will be the state of the system after a certain time? Rather one has to ask: What will be the probable state of the system after some time? This has been the line of research in statistical mechanics. If you follow it, you may get into the difficult problem whether an average over the time variable is equivalent to an average over the ensemble. I only mention the problem of the ergodic hypothesis; I cannot go into any details here.

From this brief review of point mechanics we can conclude that we have four characteristic features that we will probably find in many nonlinear problems. First, we can always use the *symmetries* to simplify our problems. The second, *linearization* of the problem, is possible around special solutions that are simple and may be found just by looking at nature. The third is that solutions have a kind of *unpredictability*—you never know what the solution will finally do, and it may not be possible to extend the solution beyond a certain time. And finally, one possible way out is *statistical methods;* we can be interested, not in special solutions, but in ensembles of solutions or in systems that we do not know completely.

Continuous media

But let me now come to those problems of present interest. They are the problems with infinitely many degrees

Physics Today *20*, No. 5, 27–33 (1967)

of freedom or problems concerning continuous media. Such problems, of course, exist in very great number—fluid dynamics, gas dynamics, elasticity, electromagnetism—electromagnetism is a nonlinear theory if the interactions with the bodies are taken into account—and finally gravitation.

And let me mention at once a complication which comes up again and again in this field of physics. Equations in which the field quantities depend on the continuous variables x, y, z and t are almost necessarily incorrect for the following reason: At very small distances one has to take atomic structure into account. Usually in hydrodynamics we need not do it. For example, in the Navier-Stokes equation we do not speak about molecular structure of a liquid, we just introduce a viscosity term. The same is true for elastic bodies, for problems of gases, and so on. Still, we have to remember that at very small distances something new will happen; something that is not described by the equation, and this has to be studied if the equation does not determine the course of events within the bigger dimensions. It is only in quantum field theory or in electromagnetism that the continuum may be a fundamental continuum. But even there our doubts arise, and there we come to the general question of whether quantum theory is a linear or a nonlinear theory.

Symmetry and linearization. Now we can try to see whether, in these many problems of fluid dynamics, elasticity, etc., one can use the same methods of simplification that I have mentioned in point mechanics. First of all, one will definitely always use the symmetry properties of the underlying system and the conservation laws. Then, one will use the method of linearization. How this linearization works I might just explain in one problem that I studied myself some time ago and in which I was strongly interested for many years—the stability of laminar hydrodynamical motion.

At that time one studied mostly incompressible fluids, say, the flow of water, and so started from the Navier-Stokes equation, which introduces, besides inertia, also the viscosity of the liquid. The question was whether a laminar flow is stable or unstable.

One knew from the experiments that the laminar motions were only very special solutions of these equations. The general solutions were much more complicated; there we have to do with the big field of turbulent motions that one knows from daily experience. The laminar motions are known, one can say, from inspection. We see how a liquid can flow through a tube or between two parallel plates, and we can easily find the solutions of the Navier-Stokes equations corresponding to these simple laminar motions. They had been derived long ago and described in the textbooks. But such a solution is sometimes not stable. It is well known that when a liquid flows through a tube and exceeds a certain speed, the type of motion changes completely. Instead of the smooth flow through the tube, we have all kinds of vortex motion appearing: Eddies are formed and dispersed again, and one can ask why that happens at a certain speed. This was a typical problem that could be solved by the method of linearization.

One could ask for motions in a very small neighborhood around the laminar motion, and since this latter motion was stationary, the coefficients in the linear equations for the perturbation were constant in time. So one could look for periodical solutions and then solve an ordinary eigenvalue equation as in quantum theory. One could use those mathematical methods that later proved useful in quantum theory, namely the investigation of asymptotic solutions, or one could introduce the modern techniques of applied mathematics including the use of electronic computers. The result was that for small Reynolds numbers all deviations from laminar motion would die down, but above a certain Reynolds number there are deviations for which the amplitude increases, and then one has an entirely different type of motion, namely turbulent motion, that cannot be treated by perturbation theory.

Unpredictability. But let me now come to really nonlinear problems in physics such as occur in gas dynamics and which are especially interesting. While in linear motion, for instance in linear wave motions like sound waves, the velocity is essentially independent of the amplitude and independent of the frequency of the waves, the situation is entirely different for a nonlinear wave motion. If, for instance, we have an initial discontinuity, then in the linear wave equation we would expect that this initial discontinuity, say across a surface, will be propagated with the velocity of sound like any other wave.

But in a nonlinear motion something different happens. This discontinuity may either disappear at once or may be propagated as a shock front—one or two shock fronts—not with the speed of sound but with supersonic speed. And not only that; it may be that out of a smooth motion in which there is no discontinuity at the beginning, a discontinuity will be formed later. I think that this fact corresponds in some way to the other fact mentioned before that nonlinear equations always have a tendency to diverge, that is, to reach points in time beyond which one cannot continue the solution.

One can say in a more mathematical language that nonlinear equations often do not admit solutions that can be continuously extended to any region where the differential equation remains regular. So the regularity of the equation does not guarantee the regularity of the solution, and you may after a finite time get to singularities in the solutions. If such a singularity occurs, the equation itself cannot determine what happens afterwards. At least the equation alone is not sufficient since we come to a point where the behavior of the substance in very small dimensions becomes important.

In hydrodynamics or gas dynamics the fundamental equation treats the gas or the fluid as a continuum, and this cannot be correct for very small distances; here we have to take the molecular structure into account. When our solutions become discontinuous, the molecular structure must play a role in the region of discontinuity.

Fortunately, in some cases in gas dynamics it is sufficient to know that in the region of discontinuity some irreversible processes must occur. The irreversibility will necessarily lead to an increase of entropy, and this result alone is sometimes sufficient to determine the course of the shock wave. This seems to be a rather lucky incident in gas dynamics. At least in principle I don't see any reason why it should be so. Fundamentally, one could imagine a need to go into the microscopic details and study the behavior of the atoms in order to say what happens. But, as I said before, sometimes one can be lucky and things go better than might be expected.

But even when we come to problems in which the atomic structure plays no role—problems that are continuous by their very nature—even then we may be confronted with the fact that solutions cannot be continued over a certain time, that solutions can disappear into nothing or arise from nothing. I would like to mention one example that has worried me for some time in quantum field theory. Let us compare a linear relativistic wave equation, namely the well known Dirac equation, with another equation that comes from the Dirac equation if one replaces the mass term of the Dirac equation by a nonlinear term.

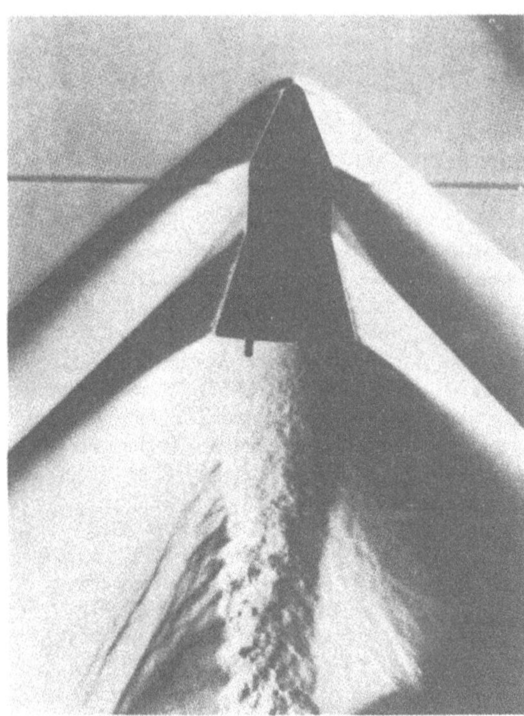

IN NONLINEAR MOTION a discontinuity may be propagated as a shock front, or one or two shock fronts, and not with the speed of sound but with supersonic speed.

These two equations are

$$\gamma_\nu \frac{\partial \psi}{\partial x_\nu} + \kappa\psi = 0 \qquad (1)$$

$$\gamma_\nu \frac{\partial \psi}{\partial x_\nu} + \psi(\bar\psi\psi) = 0 \qquad (2)$$

In a linear equation like equation 1 we know that when the solution has been zero everywhere for some time, it will never become different from zero.

It is of course true that solutions of the corresponding inhomogeneous equation can have a different behavior; we can construct the Green's function to this wave equation. Such a Green's function could then be either a retarded or an advanced Green's function and may be different from zero only in the future or only in the past. But for the homogeneous equation 1 the only solution that is zero for finite space-like distances and different from zero in the future or in the past is a solution given by the commutator in the quantum field theory; it

is identical with the difference between the retarded and the advanced Green's function. In a certain sense this function can be called a relativistically invariant solution of the homogeneous equation 1.

For equation 2, however, you have both types of solutions; that is, you can have solutions that are zero for spacelike distances and are different from zero for the past and the future, but you also can have solutions for the same equation that start, so to speak, from nothing. The wave function may be zero for the past and all of a sudden start to become nonzero at the singular point and develop like the retarded Green's function in spite of the fact that there is no inhomogeneous term involved here. This emphasizes very strongly the strange behavior of nonlinear equations. There may always be situations in which solutions of a nonlinear equation cannot be continued into the future or into the past. By the way, this problem plays an important role in the discussion of the later question whether quantum theory is linear or nonlinear.

Statistical methods. Finally, we can treat nonlinear problems in hydrodynamics and gas dynamics with statistical methods as we do in point mechanics. As I said before, it usually is not possible to follow the complete set of solutions of a nonlinear equation. Therefore, it may be convenient to study ensembles of solutions, that is, the statistical behavior of fluids. The general principle of such a statistical approach is that one is satisfied with incomplete knowledge and incomplete description of the system; starting from this point one studies the probable development. For instance in the theory of turbulence, one can start with only the knowledge of the general distribution of the eddies. One knows only the spectrum of the motion, the intensity distribution of the different frequencies, without knowing the phase relations between these different amplitudes. In this case one can only ask what is the probable development of this liquid in the course of time.

Again at this point one gets into difficulties at very small instances, and in a certain sense we meet what in quantum theory we call the "ultraviolet catastrophe," because physically what happens is this: We have a tur-

32 bulent motion, say in water, after we have produced by external motions some big eddies. These big eddies develop into smaller ones; so more and more smaller eddies are formed that get the energy. Finally the energy is dissipated into the infinitely many degrees of freedom that belong to the extremely small eddies. This process would go on to infinity if one did not introduce viscosity.

The concept of viscosity is, in this case, a nice way of getting around the fact that we get into the molecular region. Actually the energy is finally dissipated into thermal motion of the molecules, and this dissipation of energy into the motions of molecules can formally be replaced by including viscosity. But there may also be problems that can not be answered by the simple Navier-Stokes equation.

Generally I believe that in the problems of continuous media you will notice these four characteristic features already mentioned in connection with point mechanics. You will frequently use two essential simplifications, *symmetries* of the problem and *linearization*, which in many cases give a first insight into the problem. Then you will find the singularities, which result in the *unpredictability* of nonlinear solutions and frequently prohibit continuation of solutions beyond a certain point in time at which the singularity occurs. And finally, you will use *statistical methods* to get information not about a single system but about ensembles of systems, and thereby you will be able to predict either the probability for certain events or the probable course of events.

Quantum theory: linear or nonlinear?

Is quantum theory a linear or a nonlinear theory? There is no doubt that quantum mechanics in its conventional form is a linear theory. It is linear in the sense that although the operator equations are nonlinear, they can be fulfilled by solving Schrödinger equations, that is, by looking for certain transformation matrices, and these Schrödinger equations are definitely linear equations.

Linearity in quantum theory has a very deep, almost philosophical reason and is not just connected with some approximation. In quantum theory we do not deal with facts but with possibilities: The square of the wave function describes the probability, and the superposition of wave functions, the possibility of adding two solutions to get a new solution, is absolutely essential for the whole foundation of quantum theory. Therefore it defi-

Physics Today *20*, No. 5, 27–33 (1967)

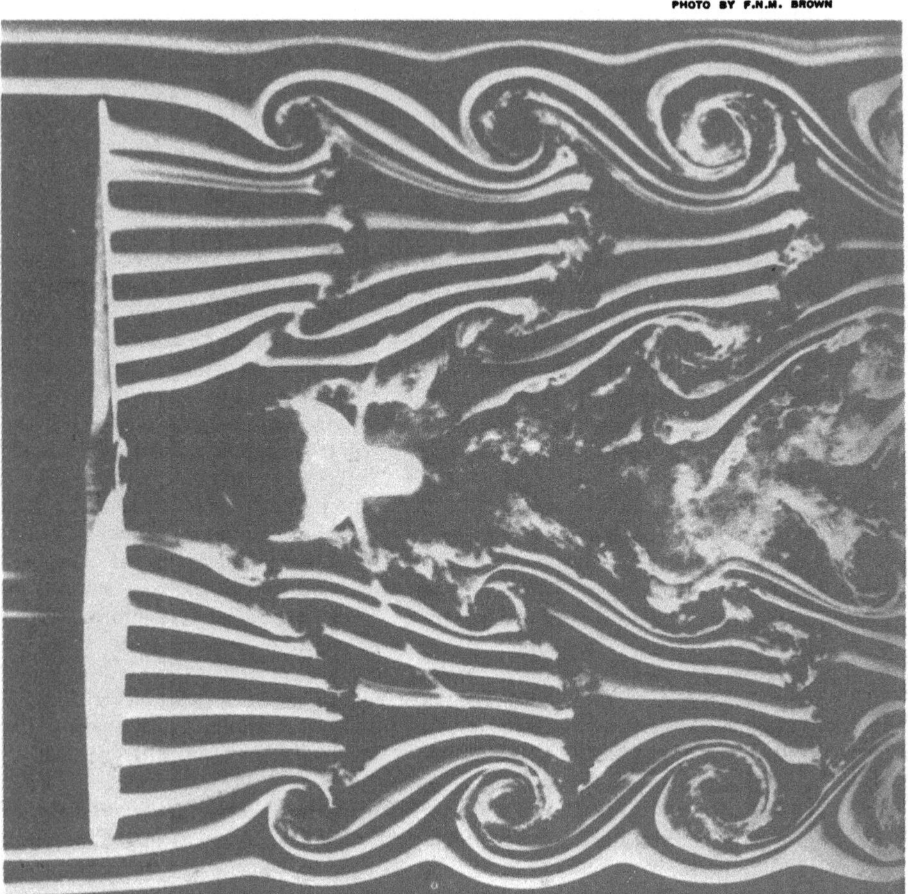

TURBULENCE in the wake of a propeller made visible by streams of smoke injected into the backwash of the propeller. Instead of smooth flow we have all kinds of vortex motion appearing.

nitely would be wrong to say that the linear character of quantum theory is approximate in the same sense as the linearity of Maxwell's equations is approximate. The linearity of the quantum-mechanical equations is essential for the understanding of quantum theory and for the interpretation of quantum theory as a statistical basis for calculating what happens to the atoms.

This fact raises interesting mathematical problems on which I shall just touch. Let us, for instance, calculate solutions of the three-body problem, say, the helium atom, by means of quantum mechanics. We know that we get the complete system of all solutions by solving the linear Schrödinger equation in the coördinate space of the three bodies in the helium atom. So one could say that in quantum mechanics we have really solved the three-body problem that never has been solved in classical mechanics completely; and that seems somewhat strange, because in the limit $h \to 0$ the quantum-mechanical solutions must, if prepared in a certain way, go over into the classical solutions. Therefore, it seems as if we could replace the solution of a nonlinear problem in classical theory by solving linear problems in quantum theory and then going to the limit $h \to 0$.

Let me discuss the connection between the two theories a little further. The limiting process would require that we start our solution with a wave packet in coördinate space. We would put the two electrons around the nucleus at some approximate positions, moving with an approximate ve-

locity so that the uncertainty relations are fulfilled, and of course these wave packets can be made smaller and smaller when we let the quantity h tend to zero, but for any finite value of h we would have finite wave packets. It is quite clear that the wave packet after a long time will disperse. It will become bigger and less dense and finally it will be spread out over the whole system. So we should always compare with this wave packet not a *single* solution of the classical problem, but instead an *ensemble* of solutions, in particular an ensemble belonging to the same set of uncertain initial conditions.

It is just the nonlinear character of the classical equations that makes the deviations between two neighboring solutions become bigger and bigger when time increases. In the classical theory, when we start with an ensemble of solutions belonging to a wave-packet-like distribution of initial conditions, after some time this wave packet will also spread out over large parts of coördinate space. So one can see that if we first let h and the size of the wave packet tend to zero, and then go to larger times, we do get a complete representation of the classical theory from quantum theory. If, however, we reverse the limiting processes, if we first go to infinite times and then let h tend to zero, the situation is quite different because then the wave packet has become infinitely big in every case.

In spite of this relation between quantum theory and the nonlinear

classical theory, quantum mechanics is definitely a linear theory.

But is this still true when we come to the theory of elementary particles, a theory that does not start from given elementary particles but which tries to understand and to derive them? In our institute we are interested in an attempt called the nonlinear spinor theory. But I wish to emphasize that this term "nonlinear" does not necessarily mean nonlinear in the sense I have been using the word, for the following reason: We start from an operator equation which looks more or less like equation 2, and therefore we can call it a nonlinear equation. But the operator equations in quantum theory always are nonlinear equations and the question is whether the solution of these operator equations corresponds to a series of linear equations or nonlinear equations.

Generally we have learned that in quantum mechanics we can replace the nonlinear operator equation by a differential equation (Schrödinger equation) or a system of equations that are linear. In the same sense one could presume also that a nonlinear equation such as equation 2 can be replaced by a Schrödinger equation that in this case would be not a linear differential or integral equation but actually a linear functional equation. Alternatively, one could replace it by a system of infinitely many differential equations with infinitely many unknowns, and all these differential equations would be linear. This would be true if such a theory could be quantized along the ordinary rules of quantum theory. In this case we would have between $\psi(x)$ and (x') commutation relations that are essentially given by a delta function in space at the time $t - t' = 0$.

But this is still a controversial problem. From our present knowledge we can be rather certain that the commutator of such an equation does not look like such a delta function. Such a conventional commutator has a strong

Physics Today *20*, No. 5, 27–33 (1967)

singularity, and this strong singularity just allows one to reduce a nonlinear problem of the operators to the linear problem of the Schrödinger equation. In field theory, however, we can take it for granted that in the center of the commutator we have no delta function because if we had a delta function, an equation like equation 2 would have no meaning. Therefore the singularity of the two-point function or the commutator at the origin is the real problem in any nonlinear field-operator equation like equation 2, and so we have to ask whether this problem can be treated with the help of delta functions in a similar way as in quantum mechanics.

If one formulates the equations for the commutator or the two-point function itself, these equations are extremely complicated nonlinear integral equations. If it should be necessary to solve these equations, the problem definitely would be a nonlinear problem, and at the very basis of quantum theory we would again come to nonlinear mathematics. The trouble is we do not know whether we really have to solve these equations.

It may be that the following procedure will be sufficient: We just guess approximate solutions for such a commutator and then put the chosen solutions into approximation schemes of the Tamm-Dancoff type; then of course in every finite approximation the result will depend upon whether we have taken a good or poor approximation of the two-point· function. There are, however, examples which tend to show that in extremely high approximations the errors that we make by not having chosen the correct 2-point function, become less and less

important, and it may be that even if we start with an incorrect commutator, at the end we get the right results in infinitely high approximation. Such at least is the situation for simpler problems in quantum mechanics that have been studied by Harald Stumpf and others.

Therefore it may be that again the actual treatment of nonlinear equations can be replaced by the study of infinite processes concerning systems of linear differential equations with an arbitrary number of variables, and the solution of the nonlinear equation can be obtained by a limiting process from the solutions of linear equations. This situation resembles the other one I mentioned where by an infinite process one can approach the nonlinear three-body problem in classical mechanics from the linear three-body problem of quantum mechanics.

Nonlinear progress

The present conclusion is that we do not know whether ultimately the fundamental problem in the quantum theory of elementary particles will be a nonlinear problem or a linear problem.

Finally I wish to emphasize again that the progress of physics certainly will depend to a large extent on the progress of nonlinear mathematics, of methods of solving nonlinear equations. It may still be that every such problem is individual and requires individual methods. Yet, as I have said, there are definitely some common features and therefore one can learn by comparing different nonlinear problems. □

In *Fundamental Problems in Elementary Particle Physics: Proceedings of the Fourteenth Conference on Physics at the University of Brussels, October 1967* (Interscience, London, New York, Sydney 1968) pp. 129–144

REPORT ON THE PRESENT SITUATION IN THE NON-LINEAR SPINOR THEORY OF ELEMENTARY PARTICLES

W. Heisenberg

Max-Planck-Institut für Physik und Astrophysik, München, Germany.

The non-linear spinor theory, as an attempt for a more fundamental theory of elementary particles, rests on the conviction that the complicated spectrum of elementary particles—as, for example, in old times the optical spectrum of the iron atom—must finally be deducible from an underlying natural law; and we hope it to be a simple law, whatever its mathematical form may be. From the experimental material available ten years ago, when Pauli and I worked on this problem, it looked as if the spinor equation

$$i\sigma_v \frac{\partial \chi}{\partial x_v} + \sigma_v : \chi\,(\chi^* \sigma^v \chi): \, = 0 \tag{1}$$

(this is the form given to the equation later by Dürr) could possibly be a sufficient frame, a suitable 'master equation' for such a theory. In the meantime, much new information has been collected by the experiments, and the theory has been developed in many details. Therefore a survey of the present situation may be useful.

The first part of my talk will be devoted to the mathematical structure of the theory, the second to the symmetry properties of the equation and their consequences with regard to the experiments; in the third part I will try to compare the methods and results of this theory with those of more conventional schemes.

1. The mathematical scheme

If one wants to give a mathematical meaning to a field equation like (1), the obvious example is quantum electro-dynamics. This latter theory is undoubtedly a working theory, even if its mathematical structure is not completely known; it gives very accurate results, e.g. for the Lamb shift. During the development of this theory in the early thirties we learnt that the postulates of quantum theory and those of special relativity cannot easily be reconciled. The requirement of local causality in special relativity together with the uncertainty relations of quantum theory may

129

cause divergencies, and the long range of the electromagnetic field (rest mass zero of the photon) introduces new problems. A price has to be paid, and it may be paid at different points. First the process of renormalization seems to be an essential part of the formalism. Furthermore one can either introduce an indefinite metric in Hilbert space, as Bleuler and Gupta have done or one can give up the manifest Lorentz covariance of the scheme and introduce the non local Coulomb forces, as in Dirac's theory of radiation; or one may introduce limiting processes as suggested by Källén. All these forms are equivalent. In any case, when the price has been paid, one can construct an approximation scheme, which in every step gives well defined, finite results. It is a characteristic feature of this perturbation theory that in every step the number of variables in the wave-functions is limited, but it increases indefinitely by going to higher and higher approximations. In principle a hydrogen atom may not only consist of proton and electron, it may also be composed of a proton, 2 electrons and 1 positron, or generally a proton, n electrons and $n - 1$ positrons. If all these infinite possibilities were to be included from the beginning, the equations of quantum electro-dynamics would probably not define a mathematical problem. But every step in the approximation scheme does define a mathematical problem, and the complete theory has a meaning, if this approximation scheme converges. A proof of convergence has not yet been given.

Taking this mathematical interpretation of quantum electro-dynamics as a model, one can try to give a mathematical meaning to Eqn. (1) in a similar fashion. If one studies the behaviour of the 2-point function

$$F(x - y) = \langle 0| \, \chi(x)\chi^*(y) \, |0\rangle \tag{2}$$

and the rôle played by it in every step of the approximation scheme, one sees that it cannot contain δ- or δ'-functions at the light cone—contrary to the conventional Schwinger functions—otherwise the occurring integrals could diverge. This cancelling of the δ- and δ'-functions can only be achieved by introducing an indefinite metric in Hilbert space. After doing this one may represent (2) by

$$\langle 0| \, \chi(x)\chi^*(y) \, |0\rangle = (2\pi)^{-4} \int \rho(\kappa^2) \, d(\kappa^2) \int d^4p \, e^{ip(x-y)} \frac{p_\nu \bar{\sigma}^\nu \cdot \kappa^4}{(p^2)^2(p^2 - \kappa^2)}. \tag{3}$$

For a general spectrum $\rho(\kappa^2)$ this representation is actually not less general than the usual representation given by Umezawa, Kamefuchi, Källén, or Lehmann. But for a mass spectrum consisting only of a few lines, or rapidly converging at high masses, it will generally state the existence of a regularizing dipole ghost at mass zero. Therefore in low approximations, when only a few masses can be considered, this representation (3) contains a significant statement concerning the behaviour of the mass spectrum at very small masses. This statement was, as a trial, suggested by the experimental situation and by an argument discussed at this conference by Chew. If one starts from a 'master equation', like (1), there is always the danger that some particles could appear as 'real elementary particles', while from the experiments we have good reason to believe, that 'every elementary particle consists of all other particles', i.e. that actually all particles are compound systems, are 'dressed up' by interacting with the others. In this respect I would agree completely with the philosophy outlined by Chew in the first part of his talk. In case of Eqn. (1) there is obviously the danger that the neutrino could appear as 'really elementary'. Therefore this has at once been excluded by putting a dipole ghost at mass zero, thereby giving the neutrino solution the norm zero. Hence there is in this approximation no real particle of mass zero which could interact with the others. This dipole ghost could therefore, in the lowest approximation, represent the lepton part of the spectrum, which does not take part in the strong interactions; and the hope would be, that in higher approximations the dipole ghost at mass zero would gradually develop into the more complicated real spectrum of leptons, which then interact electromagnetically or weakly with other particles. The consequences of a dipole ghost have been studied in detail with the help of the Lee model.

The actual approximation method can be constructed from different schemes, which have in common the fundamental assumption that, as in quantum electro-dynamics, in every finite approximation the number of variables in the wavefunctions is limited, that this number however increases indefinitely with going to higher and higher approximations. Every single step in the scheme gives well defined, finite results, and if the scheme converges it defines a solution of the problem. The convergence is expected to be much slower than in quantum electro-dynamics since in Eqn. (1) there is no weak perturbation term; again no proof of convergence has yet been given. In the early papers mostly the new Tamm–Dancoff method had been used. Recently considerable help has been obtained from the methods of many body physics, which have proved successful e.g. in the theory of solid bodies. Actually the problems of solid state physics are frequently similar to those of elementary particle physics, since 'excitons' and 'polarons' etc. are also particles 'dressed up' by interaction, and the spontaneous breakdown of symmetries by the ground state may happen equally well in both cases. It is especially the method of the Green's functions, developed by Schwinger and others, which can be used with success in both theories.

I might just mention two examples, one using the Tamm–Dancoff method the other one that of the Green's functions, in order to illustrate the practical applications. The following notation will be used:

$$\langle 0| \, \chi(x)\chi^*(y) \, |0\rangle = \ ; \qquad \langle 0| \, \chi(x)\chi(y)\chi^*(z)\chi^*(u) \, |0\rangle = $$

The irreducible part of will be called η .

Wave functions:

$$\langle 0| \, \chi(x) \, |\text{Fermion}\rangle = $$

$$\langle 0| \, \chi(x)\chi^*(y) \, |\text{Boson}\rangle = \tag{4}$$

Green's function of mass zero:

$$(2\pi)^{-4} \int d^4p \, e^{ip(x-y)} \frac{p_\nu \sigma^\nu}{p^2} = \ .$$

The lowest Tamm-Dancoff approximation for the boson eigenvalue equation is

$$= \qquad \text{or} \qquad =$$

$$\text{or} \qquad \left(1 - \right) = 0 \tag{5}$$

The lowest approximation for the 4-point function in the method of Green's function is

$$\eta \ = \ \eta \ + \tag{6}$$

with the solution

$$\eta \ = \ \frac{1}{1 - } \tag{7}$$

Equation (5) gives the masses of π- and η-meson in a fair approximation. The low mass of the pion, which had been mentioned as a major problem by Chew, comes out naturally from (5); and this result rests essentially on the dipole ghost at mass zero in (3). Equation (7) shows clearly the poles of the 4-point-function at the masses of the bosons and it gives, according to Dhar and Katayama, a value for the coupling of the bosons to the nucleons in reasonable agreement with the experiments.

The method of the Green's functions allows, at least in principle, the determination of the 2-point function (3) and the spectrum $\rho\,(\kappa^2)$ from Eqn. (1). Unfortunately already the lowest order approximation is too complicated for an explicit solution. Still one can see from the equation, that there cannot be δ- or δ'-functions at the light cone of (3), and that the asymptotic behaviour defined by the dipole ghost may go well together with the requirements of the equation.

The flexibility of quantum electro-dynamics with respect to the point where the price has to be paid for the reconciliation of quantum theory and relativity suggests the existence of other mathematical schemes interpreting Eqn. (1), which do not introduce an indefinite metric in Hilbert space, but replace it by the concept of non-local forces. A simple scheme of this type has been suggested by Dürr. Instead of the field operator $\chi(x)$ one may introduce another field operator $\psi(x)$ connected with $\chi(x)$ by the relation:

$$\psi(x) = \Box\chi(x) \quad \text{or} \quad \chi(x) = \Box^{-1}\psi(x) + \chi_0(x). \tag{8}$$

For the 2-point function $\langle 0|\ \psi(x)\psi^*(y)\ |0\rangle$ one gets from (3)

$$\langle 0|\ \psi(x)\psi^*(y)\ |0\rangle = (2\pi)^{-4}\int \rho(\kappa^2)\,d(\kappa^2)\int d^4p\,e^{ip(x-y)}\frac{p_\nu\bar{\sigma}^\nu\cdot\kappa^4}{p^2-\kappa^2}. \tag{9}$$

The dipole ghost at mass zero has disappeared and the spectrum $\rho(\kappa^2)$ could well be positive definite, since now the wave Eqn. (1), written in terms of $\psi(x)$, contains only a non local interaction ($\Box^{-1}$ is a non local operator), which allows the integrals in the approximation scheme to be finite, even if the 2-point function (9) contains δ- or δ'-functions at the light cone. It is true that the non local forces in the wave-equation would give rise to rather complicated problems concerning the boundary conditions, i.e. the in- and out-fields connected with (8), but there may well be a one-to-one correspondence between the solutions written in terms of an indefinite metric and those starting from the non local forces. Therefore the theory established by Eqn. (1) has probably the same kind of flexibility as quantum electro-dynamics.

2. The symmetry properties connected with Eqn. (1)

Equation (1) is invariant under the proper Lorentz group, including the two discrete operations PC or PCT, the isospin group SU_2, a gauge group which may represent the conservation of the baryonic number and finally the dilatation group. It is not invariant under SU_3 or higher groups of this type, and it does not immediately represent parity P, strangeness (or hypercharge) and lepton conservation. Furthermore the invariance of (1) under the dilatation group cannot lead to an invariance of the complete theory under this group, since an invariant 2-point function or commutator would imply a δ-function in (3), which is impossible. Therefore a mass scale must be introduced; then as a kind of compensation one can define parity, as has been pointed out by Dürr, since for a massive particle one may always define 'left-hand' and 'right-hand' wavefunctions. But in order to represent the empirical spectrum of elementary particles with regard to strangeness and SU_3 it will certainly be necessary to introduce new degrees of freedom.

The most natural way of doing this seems to be the assumption of a spontaneous breakdown of symmetry, the introduction of an asymmetrical groundstate, like in solid state physics; such a breakdown could on the one hand lead to an understanding of the electromagnetic violation of SU_2; on the other hand it would give an isospin property to the vacuum; it would therefore, without changing Eqn. (1), introduce new degrees of freedom from the vacuum, which could be just sufficient to account for hypercharge (or strangeness) and, in a very rough approximation, for SU_3 and the higher groups.

This general scheme has been followed up to some extent in recent years. The most important step was the application of the theorem of Goldstone in a somewhat generalized form (in a paper by Dürr, Yamamoto, Yamazaki and myself). In the special form given to the theorem in the first papers of Goldstone, Salam, Weinberg, Nambu and others, it states that the degeneracy of the vacuum will automatically lead to the existence of particles of mass zero, and thereby indirectly to long range forces. In more recent investigations of Higgs, Kibble and others, which have been reported by Dürr at this conference, the problem has been reversed by asking: if long range forces are assumed from the beginning, what are the consequences of a degeneracy of the vacuum? I need not discuss the answers again. In any case the investigations have revealed an intimate connexion between an asymmetry of the groundstate and long range forces. Therefore in the non linear spinor theory the problem could be formulated by asking: Is it possible that the equations for the Green's functions, which are themselves symmetrical under the isospin group SU_2, could have asymmetrical solutions, and that this asymmetry in isospace could be connected with the appearance of a boson-pole at mass zero, corresponding to the photon in its transformation properties?

W. HEISENBERG   135

This seems in fact to be the case. It turns out that a long range field of this type must be defined with regard to its transformation properties in isospace by a projection operator $\frac{1}{2}(Y + \tau_3)$, in accordance with the rule of Gell-Mann and Nishijima; this projection operator then distinguishes between neutral and charged particles. If the dipole of the charged leptons in (3) is moved from zero to finite masses—thereby indicating the asymmetry in isospace and an electromagnetic mass of the leptons—a boson-pole at mass zero and spin 1 can actually be established from (5) or (7). In that approximation, in which the baryon octet is represented by just one pole in (3), the average lepton mass is determined to be $\sim$40 MeV, and the coupling, i.e. the fine-structure constant has, according to (7), a value around $\frac{1}{120}$. Therefore it seems that the actual behaviour of nature in this field can be well imitated in the mathematical scheme of the non linear spinor theory.

At the same time the interaction between the particles and the ground-state—which may be studied e.g. by the model of a ferromagnet—can lead to the formation of strange particles i.e. particles, to which some isospin from the groundstate (a 'spurion') has been attached as has been pointed out by Biritz. In this way the hypercharge Y can be established by a gauge transformation in isospace of the groundstate. Finally the two lowest octets in the baryon and the boson spectrum seem to be a natural consequence of the symmetry in the interaction between particles and groundstate.

If this picture is correct, the three main interactions can be described in the following way. In the strong interaction the groundstate never takes up any property of the particles and vice versa; therefore in collision processes pairs of spurions and antispurions with isospin zero may be created or annihilated, but this makes no change in the groundstate. In electromagnetic processes spurion–antispurion pairs of the electromagnetic (photon) type may be transferred between the particles and the ground-state, and these pairs have with equal probability isospin 1 and 0; therefore in such processes isospin is transferred, but no hypercharge. Finally in weak interactions even single spurions (hence isospin and hypercharge) may be transferred. This would look like a process contradicting causality; but it may (according to Dürr) be compared with the Mössbauer effect where a momentum seems to be transferred at once on the whole crystal. In the latter case Weisskopf has demonstrated, that a deviation from causality cannot be observed.

The idea of using the interaction between particles and the groundstate for producing the strange particles (in a first approximation the two lowest octets) makes SU_3 appear as a secondary symmetry of dynamical origin.

In fact, if SU_3 were considered as a fundamental (i.e. exact) symmetry of the underlying natural law disturbed by an SU_3-asymmetrical vacuum, one would either expect Goldstone particles of mass zero and of undefined symmetry in SU_3, which interact strongly with the other particles, or one would have to assume a spectrum of the type studied by Higgs and Kibble, which both seem not to agree with the actual spectrum. Therefore it looks more natural to compare the SU_3-multiplets with the optical multiplets in the atomic spectra. These optical multiplets can be considered as the result of the group $O_3 \times O_3$ (independent rotation of the orbits and the spins of the electrons), which obviously is not a fundamental group; yet it may result approximately from dynamics in some parts of the spectrum.

On the other hand the classification of SU_3 as a secondary dynamical symmetry has consequences of considerable importance, which can be checked by the experiments. The most important one is the non-existence of quark-particles or of non integer electromagnetic charges. Only such (approximate) representations of SU_3, which can be obtained by multiplying the octet representation with another octet representation etc. could be expected as parts of the spectrum. Furthermore the algebra of the currents should represent very accurately $SU_2 \times SU_2$, but only with considerably minor accuracy $SU_3 \times SU_3$. All these results seem to be compatible with existing experimental evidence; but one may doubt whether or not future experiments could reveal the existence of quark-particles, which then would rule out Eqn. (1).

In any case the classification of SU_3 as secondary dynamical symmetry may be a controversial problem, and if there should be good arguments in favour of the opposite classification of SU_3 as fundamental symmetry, I hope that these arguments will be brought forward in the discussion.

Two more remarks should be added concerning the leptons and electrodynamics. In an approximation where the baryon octet is represented by one pole in (3), only the average lepton mass can be determined. If however the mass splitting in the octet is taken into account, there must be at least two different charged leptons corresponding in their symmetry to proton and Ξ. The particle connected with the proton gets a very small mass and should be identified with the electron, the other one gets a mass not very much smaller than the pion mass and corresponds to the muon. The muon mass has this rather high value in spite of being of purely electromagnetic origin. From this point of view one may call the leptons not 'real matter', but—as one might have done some hundred years ago— rather 'pure electricity' or 'electromagnetic singularities'.

Equation (1) has been interpreted mathematically along the lines of quantum electro-dynamics. Therefore it is not surprising that the mathematical scheme of quantum electro-dynamics seems to be contained completely in the more general scheme defined by (1). This can be seen in two ways. The operator $\chi^*(x)\sigma_\nu\chi(x)$ in (1) contains matrix elements connected with the creation or annihilation of photons (or more generally of electromagnetic field). If this part of the operator is called A_ν and treated separately from the rest of the operator, one recognizes in (1) the fundamental field equation of quantum electro-dynamics [the Dirac equation or more correctly Weyl equation for a (bare) electron without mass interacting with the electromagnetic field A_ν].

The other possibility is the translation of every special Feynman graph in quantum electro-dynamics into a corresponding graph in the formalism defined by (1). Any photon line in the electromagnetic formalism could in fact be replaced by the expression $\mathrel{<}\!\!\!\!\boxed{\eta}\!\!\!\!\mathrel{>}$ of the non linear spinor theory. Hence e.g. the graph for electron–electron or proton–proton scattering

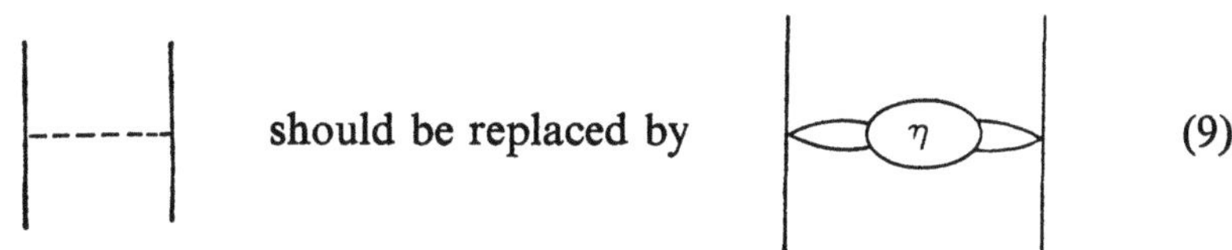

$$\text{should be replaced by} \qquad\qquad\qquad\qquad (9)$$

The latter graph contains, besides the electromagnetic forces, also short range forces which may be present besides the electromagnetic interactions, e.g. the proton–proton interaction by means of pions. Still the pole of $\mathrel{<}\!\!\!\!\boxed{\eta}\!\!\!\!\mathrel{>}$ at mass zero guarantees the equivalence of the two graphs with respect to electromagnetic forces in that approximation in which the coupling constants on both sides are equal. Hence e.g. the well known formula for the Lamb shift expressed in terms of the charge and mass of the electron should be a consequence of (1) as well as of quantum electro-dynamics.

Almost nothing has been done in the non linear spinor theory so far with respect to the weak interactions. The violation of parity and the weakness of the interaction in β-decay obviously suggest that this phenomenon should, like the electromagnetic forces, be considered under the viewpoint of an asymmetry of the groundstate. The same should be true for the still smaller PC-violating interaction in K_2-decay and finally for gravitation. But no serious attempt has yet been made in this direction. As one argument in favour of an incorporation of weak interactions into a theory given by the Eqns. (1) and (3), one should mention the fact that the interaction term in (1) has just the same form as the interaction term in β-decay (especially with respect to the $(1 + \gamma_5)$-factor corresponding to a 2-component theory), and the universality of weak interaction.

3. Comparison with methods and results of other theories

Equation (1) contains a contact interaction which in the 'Feynman graphs' of the practical applications (5)–(7) appears as a vertex point, where four fermion lines meet. This fundamental interaction will in higher approximations lead to more indirect interactions which are produced by the exchange of other particles. In this way the formal scheme will in higher approximations gradually approach that kind of scheme which one would use in order to describe the assumptions of the bootstrap mechanism or other more phenomenological methods: Any particle may be considered as a compound system of other particles, held together by forces, which again are particles taken in the 'crossed channel'.

This is clearly seen in a recent paper by Dürr and F. Wagner, which investigates the baryon eigenvalue equation in an approximation where all wavefunctions with more than three variables are omitted. This approximation enables us already to consider the baryon as composed of one fermion and one boson, and should therefore include higher resonance states of any angular momentum. Actually it leads to an eigenvalue equation of the Bethe–Salpeter type, where the two particles, boson and fermion, are held together by an exchange force produced by the exchange of a fermion. Calculations of this type have been carried out on a phenomenological basis many years ago in a low energy approximation by Chew and Low, for the Bethe–Salpeter equations for scalar particles in case of mass zero exchange particles by Wick, Cutkowsky and others. The recent investigations of Dürr and Wagner show how the Tamm–Dancoff approximations gradually become similar to the more conventional schemes. Actually the Bethe–Salpeter equation seems to lead to a series of resonance states, where in each series the angular momenta increase by two units from one member to the next, in other words to Regge trajectories. The only difference against the more conventional schemes can be seen in the existence of the contact interaction and in the fact that some baryons can be expressed by the fundamental operator $\chi(x)$ alone, others only by the combination of at least three such operators. Whether this fact will finally have some influence on the shape of the Regge-trajectories is still an open question. The existence of the contact interaction may be connected in the bootstrap method with the problem of the behaviour of the dispersion integrals at very high energies.

The general dynamical aspects and the problems of convergence in the non linear spinor theory have recently been investigated in papers by Stumpf, Wagner, Wahl, Rampacher and others; in these papers the methods described in the first part of my talk have been applied to old problems of quantum mechanics, especially the anharmonic oscillator, in order to compare the various methods.

The numerical evaluation of a single step in the approximation procedure of the field theory has been simplified in a paper by Dürr and Wagner by the elaboration of the 'Feynman diagrams' to precise mathematical formulae. In a further development of this technique one may hope that an equation written thereby in the form of a 'Feynman diagram' can be directly translated into a programme for a computer. For the single integrals the computers have already been used by Géhéniau, Mitter and Biritz with considerable success. Hence a further extension of these methods seems quite feasible. It should be emphasized that a development in this direction will be necessary, independently of what finally the theoretical formulation of elementary particle physics will be. The degree of complication in the eigenvalue equations of this part of physics cannot possibly be smaller than that in the theory of complicated atoms or molecules, since in both cases we have typical many body problems. Therefore one will finally have to rely on computers in order to cope with this degree of complication. The adaptation of any method that has proved successful in many body physics will certainly also be useful.

The general tendencies in the bootstrap method as reported by Chew and in the non linear spinor theory are very closely related. However, at some points a theory starting from a master equation like Eqn. (1) will contain information which is not too easily added to the fundamental assumptions of the bootstrap formalism; yet they seem to be necessary to form a complete theory. The first point is that Eqn. (1) states the fundamental groups, which are—except for the Lorentz group—not stated explicitly in the bootstrap method. I cannot believe that the constraints in this formalism should suffice to determine the groups.

Then the analytical behaviour of the S-matrix elements is defined precisely by the differential Eqn. (1), while only general concepts like 'maximum analyticity' are available in the bootstrap scheme. Finally the concept of a degenerate vacuum, which is natural for a theory defined by (1), cannot without additive complications be incorporated into a pure S-matrix theory. Still, the general tendency of considering all particles as compound systems is common for both theories and is immediately suggested by the experiments.

Coming back to the general situation in the non linear spinor theory at the present time, it is clear that we are still very far from a complete theory on this basis. However, the outcome of the experiments during the last ten years may allow the conclusion that, when looking for a master equation for elementary particle physics, one need not look for anything more complicated than Eqn. (1).

May I make one final remark concerning a criticism which has often been expressed: that it is a too ambitious programme to try the formulation of a theory for the complete spectrum of elementary particles. Looking back on the development of atomic physics 40 years ago, it is clear that it would have been much more difficult to formulate a theory for only some part, say the triplet part, of the iron spectrum, than a theory for the complete spectrum. Therefore I feel that it would be more ambitious in our time to construct a theory only for the hadrons, or only for the leptons etc. than a theory for the complete spectrum.

Discussion on the report of W. Heisenberg

R. E. Marshak. As someone who is sympathetic to your programme to develop a unified theory of hadrons and leptons, I feel that your master equation in terms of one four-component field ('urmaterie') does not do full justice to the suggestiveness of the lepton triplet (ν_e and $\bar{\nu}_\mu$ can be represented by one 4-component Dirac spinor and the other two are e and μ). Since the leptons are weakly interacting, they seem to be reflecting the attractiveness of starting with a master equation in terms of three fields (quark model). In your theory, you must work very hard to simulate broken SU_3 symmetry results (e.g. I am not sure that you can reproduce the Gell-Mann–Okubo formula for the baryon decuplet, the condition $2I + y = 0$ (mod. 2) etc.). Also, it would be difficult to understand the universality of the weak interactions between leptons and hadrons. Finally, if you do not believe one will observe 'urmaterie', I do not see why quarks should be found.

So much by way of comment. My question is whether in your theory, the muon mass has an electromagnetic origin. In the Baker–Johnson–Willey theory, the scale is not fixed for the electron mass but they hope to determine the (m_μ/m_e) ratio. So far they have not succeeded. I wonder how you can obtain such a large (m_μ/m_e) ratio by means of electromagnetism? My own inclination would be to connect m_μ with the hypercharge effect on mass for hadrons.

W. Heisenberg. First I would prefer to speak about 4 leptons (e, μ, ν_e, ν_μ) corresponding somehow to one-half of an octet; I cannot see any strong argument for speaking of a triplet and therefore in this sense referring to SU_3. But with respect to the question whether the muon mass is purely electromagnetic the situation in our theory is as follows: in that approximation, where the vacuum is isosymmetric (and when we therefore have no electromagnetic field), all lepton masses are zero. Hence they may afterwards all be called electromagnetic. When the vacuum is considered as asymmetric and when therefore the photons appear, those leptons that interact with the photon acquire their mass. But this mass is related to the baryon masses in the sense that each baryon needs supplementation (or you may call it compensation) by one (or possibly several) lepton dipoles, to make the photon of mass zero possible. The electron is the supplement of the photon, the muon that of the Ξ. In so far these masses are determined by the baryon masses, and the mass difference Ξ-proton has its counterpart in the different masses of e and μ.

With regard to the universality of weak interactions I would like to emphasize that it is at least a very natural basis for an understanding of this universality that baryons and leptons are created by the same field operator. The universality of electromagnetism (proton and positron have exactly the same charge) is due just to this feature of the theory. On account of the Ward identity the coupling constant (the charge) cannot depend on the amount of creation operator for a special kind of fermions, which is present in the universal field-operator. Therefore their charges are exactly equal. A similar situation could occur for the weak interactions, but this problem has not been worked out yet.

H. P. Dürr. I think we should not talk too much about leptons and weak interactions in connexion with the non-linear spinor theory at the moment. Too little has been done, up to now. Perhaps in ten years from now we will know a little more, and then we can come back to these points.

I want to comment on the possibility to consider $SU(3)$ as a fundamental symmetry broken only by the vacuum state. The first impression was that on the basis of the Goldstone theorem we would have to expect the occurrence of four scalar, strange massless particles if we only consider the violation of strong interactions. These scalar massless particles should interact strongly with hadrons and two of them should be charged. These particles we do not know experimentally, and we can be pretty certain that they do not exist. However, Higgs, Kibble, Brout, Englert and others have shown how the Goldstone theorem can be invalidated to some extent by coupling in gauge fields. In this case the strong breaking of $SU(3)$ would produce four massive strange vector mesons, so to say, half the octet. Generally speaking the spontaneous breakdown of a symmetry either produces mass zero particles or incomplete multiplets. Usually the four know strange vector mesons K^* are considered to be part of a vector-meson octet, including in addition the ρ and the ω. Hence one would have to find some other candidates for the incomplete multiplet to make such an interpretation acceptable.

I want to comment on another point. Heisenberg has compared the indefinite metric in the nonlinear spinor theory with the indefinite metric which occurs in the Gupta–Bleuler formalism of quantum electrodynamics. Now, it is rather clear that the indefinite metric in q.e.d. is of a much less dangerous type than the one used in the spinor theory, in particular, we can give a straightforward prescription how it can be avoided altogether. Nevertheless, there seems to be at least some formal similarity which I may express in the following way: in q.e.d. in the Bleuler–Gupta description we introduce two redundant fields, a longitudinal field connected with quanta of positive norm and a scalar field connected with quanta of negative norm. Taking the plus and minus combinations we obtain what I will shortly call a ghost couple, i.e. two ghost fields connected with quanta of zero norm, which, however, are not orthogonal to each other. One ghost, a 'good' ghost, obeys the Lorentz condition, the other, the 'bad' ghost, however, does not. We arrive at the physical theory by projecting out the 'bad ghost' by a subsidiary condition on the infields. Then the 'good ghosts' cannot do any harm because of their vanishing norm; they do not give contributions to any matrix elements and just reflect the gauge independence of the matrix elements. One can repeat the same construction for a massless spin two field couple to a conserved source function. Here one has to introduce 10 field operators where 8 fields are redundant. It can be shown that they constitute four ghost couples of the type described above. The four 'bad ghosts' are eliminated by the four Einstein conditions.

In a similar way the dipole ghost introduced in the nonlinear spinor theory constitute a ghost couple of this type, a 'good ghost' which is an energy eigenstate and a 'bad ghost' which is not. By the condition that physical states all must be eigenstates of the energy one can eliminate the bad ghosts, and hence arrive at a physically interpretable theory, provided that the bad ghosts do not produce bound states, which is hard to check.

E. C. G. Sudarshan. I find the equality of the electron charge and the proton charge to be extremely interesting and significant: it appears to be due to the use of the same description for both kinds of particles. Could you tell me how we can be assured that the electron or μ-meson would not get transmuted into a baryon? Needless to say the conservation of baryon number is also a very significant law which should not be violated without 'due processes of law'.

H. P. Dürr. The lepton conservation, I think, is a rather interesting point in our theory, but, I feel, also quite a hazardous one, which easily may prove to lack any basis. It actually states that the superselection rules we find in nature may not all arise from group theoretical conditions (gauge invariance) but also may arise from a peculiar analytical structure. In our case, baryons and leptons are not distinguished group theoretically, they transform under the same gauge transformation, and hence have the property to cancel each other's singularities. However, the leptons are dipoles and the baryons are poles in the Green's functions. Consequently the leptons, because of their vanishing norm (good ghost) have zero interaction with all particles, except with the photons, where the norm of the states is immaterial on the basis of Ward's indentity. Hence the baryon world is only linked to the lepton world by photons, and we therefore get a separate lepton and baryon conservation, if weak interactions later on do not mess up this separation. Of course, one has to be very careful about the statement that zero norm particles are coupled to photons. If we were to use Källén's approach to q.e.d. by employing a finite mass photon in which the zero limit is taken afterwards, then zero norm particles would not couple.

In *Proceedings of the Symposium on Basic Questions in Elementary Particle Physics, June 8–18, 1971,* ed. by H.-P. Dürr (Max-Planck-Institut für Physik und Astrophysik, Munich 1971) pp. 1–10

Review of Recent Progress [+])

by W. Heisenberg

The present conference is intended - like the conference at Feldafing six years ago - to give an opportunity for discussions on basic problems in elementary particle physics. May I start these discussions by going briefly through those basic problems in which, as I think, there has been much progress during the last six years.

You all know that at present still different physicists frequently have different hopes connected with elementary particle physics. Some of us may hope that one can construct directly the S-matrix from conditions of consistency, bootstrap mechanism and analyticity, others hope that we can work with field operators, either using the old axioms of Lehmann, Symanzik, Zimmermann and Wightman or an indefinite metric in state space, again others will hope that from the experiments on very high energies, say with the SLAC-accelerator or with the storage rings in CERN there will come entirely new discoveries and insights. Optimists may even hope that some quark particles or partons will be discovered in the future. We in our institute have connected our hopes mostly with a special non-linear spinor equation containing a number of groups which seem to be fundamental. But whatever these hopes may be - important are not the hopes but the questions to which these hopes lead. Therefore I would like to discuss these questions.

Perhaps I should first say, that obviously we all must finally look for a kind of natural law from which the spectrum of elementary particles can be derived, both the spectrum and the interactions of the particles. Just as we have in the course of years in old times understood the optical spectra of iron or helium, so it should be possible in our time to understand the spectrum of the elementary particles. The questions shall be divided essentially in two groups. There are the mathematical problems connected with the representation of the requested natural law: Which quantities shall be used for the definition of this law? Is the S-matrix sufficient or do we need field operators and in which way can we connect the field operators? The other part concerns just the group structure of this underlying natural law: What

---

[+]) Corrected version of the text taken from the tape record.

- 2 -

are the fundamental symmetries of the law? Is this natural law already uniquely determined by these symmetries or do we need more besides the symmetries?

So let me first discuss the problem of the mathematical representation.
There, I think, a lot of work has been done in these last six years concerning the possibility of using field operators in a relativistic theory with
interaction. Let me got briefly through the history. You know that already
thirty years ago Dirac and then later Pauli tried to say that we can use field
operators in such a theory, but then we must connect them with a state space
with indefinite metric. This line was followed in later papers of Bleuler and
Gupta on quantum electrodynamics. These authors started with Maxwell equations
and with a local interaction between the electromagnetic potentials and the
wavefunction of the electrons. Quantizing these equations they had to treat
all the four electromagnetic potentials on equal footing. That meant that
besides introducing the transversal photons they also had to introduce longitudinal photons and scalar photons. Then it followed from relativity that in
the state space the norm of the scalar photons would have to be negative which
meant an indefinite metric. The trouble with unitarity which Pauli had mentioned in his first paper could be avoided by a subsidiary condition corresponding to the Lorentz condition in classical electrodynamics. The scalar and
longitudinal photons could never come out as free particles, they played their
rôle only in the interaction, and thereby any trouble with the unitarity of
the S-matrix could be avoided. Actually Bleuler and Gupta could show that their
theory was equivalent to the old Dirac-Schwinger formalism in which one had
a non-local interaction - namely a non-retarded Coulomb field - and in which
therefore the equations did not look Lorentz-invariant. It could be demonstrated that the two theories are equivalent and are both invariant under
Lorentz transformations.

The next step in this direction was taken in connection with our attempts
on a unified field theory. We tried to assume that the two-point function of
the fermions should be regularized in such a way that the baryons are regularized by means of the leptons and the leptons by the baryons. Of course this
requires an indefinite metric in state space. In order to separate the leptons
from the baryons we had tentatively introduced a double pole at mass zero for
the representation of the leptons. The consequences of such a double pole
had been analyzed by means of the Lee model. In this model there are particles
called N and Θ , which can be bound together to form a particle V. Analytical-

- 3 -

ly this model can be constructed in such a way that the V particle becomes a
dipole just as in the 2-point function mentioned above. Then it could be shown
that with this dipole the difficulties for the unitary S-matrix can really be
avoided, because the dipole meant actually two 'ghost'-states of norm zero
which are not orthogonal to each other. One of the two states can be called
the good ghost and the other one the bad ghost, and one could arrange scattering states in such a way that the bad ghosts never are emitted or absorbed.
Then the good ghosts with norm zero would not appear in the physical interpretation. Actually the S-matrix is just a unitary scattering matrix between N-
and θ-particles.

At this point I think recently an important progress has been made in a
paper by Karowski in Hamburg. Karowski could show that in this Lee model -
like in the Gupta-Bleuler case - one can replace the mathematical scheme with
the dipole by another mathematical scheme in which one has a positive-defi-
nite metric, but at the same time a non-local interaction between N and θ
particle. This result is a further indication for the equivalence between a
local interaction plus indefinite metric and a non-local interaction and positive metric. This equivalence may be true in a much wider range of theories
than just in quantum electrodynamics in the Gupta-Bleuler case.

Especially interesting in this paper of Karowski is the following feature:
When one changes over to this model with only N and θ particles but no V
particles, and a non-local interaction, then - since the analytical structure
of this model is the same as that of the Lee model - there must be some analogue to the V particle, a bound state between N and θ . If one looks into
the mathematical details, one finds that actually these V particles have not
disappeared completely, on the contrary the corresponding eigenvalue equation has a root just at the energy of the V particle. This time it is not a
dipole but a pole, and that of course is necessary because in a theory with
positive metric there can only be a positive norm to every stationary state.
The V particle has become quite a normal particle represented by a pole with
a positive norm. At the same time, however, a selection rule has appeared
which makes it impossible that this V particle should ever be created or annihilated. This is of course exactly the behaviour one would hope for in connection with the two-point function of the fermions. The leptons may become
real particles with positive norm if there should be such an equivalent scheme

- 4 -

with non-local interaction and positive metric, but at the same time the analytical structure would provide a selection rule which makes it impossible that these leptons are created or annihilated. The connection between baryons and leptons could come in much later through the breaking of the isospin symmetry and the photons. So I think that the paper of Karowski has contributed a lot to make these things clearer.

There was a related paper by Dürr and Rudolph which stresses the similarity between the Gupta-Bleuler case on the one hand and the dipole case of the Lee model on the other. If one reformulates the Gupta-Bleuler theory by introducing the sum and the difference of the scalar and the longitudinal photon states then these new states would correspond exactly to the ghost states, the bad ghosts and the good ghosts of the Lee model, and the subsidiary condition of Gupta and Bleuler would mean that the bad ghosts are never emitted or absorbed. The good ghosts do not contribute to the expectation values. So the analogy between the Gupta-Bleuler case on the one hand and the Lee model and perhaps the non-linear spinor theory on the other is now rather close. I would like to mention still a few other papers on the indefinite metric. Sudarshan has tried to apply the indefinite metric in a more general way on quantum electrodynamics and its connexion with baryon physics. Normally such quantities as say the electromagnetic mass differences between the different components of an isospin multiplet would diverge, they would become infinite; but if - following the old ideas of Pauli and Villars - one introduces heavy masses to regularize the integrals, which means an indefinite metric, then they become finite and one can get sensible results. Similar attempts have then later been made by Lee and Wick, who also have taken up the indefinite metric recently. They tried to go a little bit further and to say: if these heavy masses are complex - which just means that they can decay into other fragments - then one can, simply by changing the old Feynman rules, get a universal prescription for avoiding all troubles with the unitarity of the S-matrix. This procedure again was recently critisized by Sudarshan and his collaborators, but I dare not go into these difficult mathematical problems. Whatever the answer may be, it looks as if in all these cases difficulties with unitarity can be avoided.

This however leads to a very interesting consequence. First of all it means that for the amplitudes or also for the kernel of eigenvalue equations

- 5 -

etc. we have functions which are only piecewise analytic, they are analytic in
a certain region, but then they break off and a new analytic function starts.
The analyticity changes at the threshold, where the ghost states can be created.
In consequence that means also deviations from the old concepts of causality.
You know that in relativistic quantum theory it is not so easy to define causality as it was in classical physics. In classical electrodynamics we can
construct waves with a sharp edge, and this sharp edge of the wave front shall
not be propagated with a higher velocity than the velocity of light. From this
condition you get all the properties of analyticity which are so well known.
But in quantum theory it is not possible to make a wave front with a sharp
edge because the frequencies are limited. We have only states of positive
energy and therefore we cannot use the whole spectrum of frequencies. Hence
such a wave front will have not a sharp edge but a kind of precursor which
usually would decrease exponentially. If, however, these functions of which
I have spoken are not analytic at the points where the ghosts can be created,
then, as Dürr and Seiler have demonstrated, there will be precursors which do
not decrease exponentially but will decrease by some power law. There you see
already some similarity between the idea of long range precursors and the
Coulomb force in quantum electrodynamics. Both introduce some kind of long
distance effects in the theory. Therefore I think, the most important progress
which has been made recently in this field is that one now has some idea in
which way causality might possibly be modified. It may be very difficult to
observe such effects directly; but indirectly, by observing the shape of the
amplitudes as function of energy near the threshold, it should be possible to
check experimentally the consequences of the indefinite metric and of deviations from conventional causality. In this respect I was very glad about the
papers of Sudarshan and Lee and Wick and others who have emphasized this experimental side of the problem. Perhaps I should just remind you that e.g.
those kernels of eigenvalue equations which we use in the non-linear spinor
theory offer examples for such functions which are only piecewise analytic.

Going now from the indefinite metric to more general mathematical questions
I would like to mention just briefly the problem of eigenvalue equations. I
had already emphasized in the beginning and I want to emphasize it again: a
theory of elementary particles means a theory of the spectrum of elementary
particles; so finally we must solve eigenvalue equations just like in the old

- 6 -

times when we had to calculate eigenvalues from Schrödinger equations. Some
progress has been made with regard to simple equations of the Bethe-Salpeter
type; may I quote just the papers of Schwartz and Zemach and of Mitter and
Ladanyi. But a theory of the spectrum, say of the type of the non-linear spinor
theory, will lead to much more general eigenvalue equations which can be written down in the form of Feynman graphs. My hope would be that after some years
numerical methods will be developed so far, that the Feynman graphs can simply
be fed into a big computer, which will do the rest and give the result.

In this connexion I would like to mention that in its complexity elementary particle physics can be compared to quantum chemistry. In this field we
cannot hope to get exact eigenvalues for say a benzene molecule or something
like that, but what we can hope is that we can find approximation methods which
can be put into the computers and thereby lead to approximate eigenvalues. The
degree of complication in elementary particle physics is roughly the same as
the degree of complication in quantum chemistry, and therefore we should also
make some effort to develop methods which can be used in connexion with the
big computers.

But let me now come to the second part of my talk namely to the group
theoretical structure of the natural law which underlies elementary particle
physics. I think one of the main advances during the last six years was the
recognition of the importance of the scale transformation which has been discussed by many authors. I may first mention papers of Kastrup and Mitter, then
of course Johnson's quantum electrodynamics about which he will, as I hope,
speak afterwards, papers of Brandt and Wilson and others. In all these papers
it has become clear that the scale transformation plays an important rôle in
the whole theory of elementary particles and that one can make use of this
transformation in many respects. Again I cannot go into any details and I hope
that during the discussion many of you will come back to the subject. But I
should perhaps mention that also in our non-linear theory this scale transformation has always played an important rôle since the fundamental field
equation is invariant under a scale transformation. Already some years ago
Mitter had suggested that the behaviour of the two-point function near the
light cone can possibly be determined by the scale transformation. It is clear
that the theory as a whole cannot be invariant under the scale transformation
because we have masses, and a mass means that we have set a measure for the
scale. Still it is possible that in the neighbourhood of the light cone the

- 7 -

masses are unimportant and we have nearly a behaviour which is scale-invariant. This may be modified by logarithmic terms like in quantum electrodynamics, but we need not go into these details. I only wanted to say that this approximate scale invariance in the neighbourhood of the light cone can help for deciding other problems which had not been solved so far. You know that in recent years there has, on the phenomenological side, been very much progress by means of the current algebra. One could construct currents, vector currents and axial vector currents, could assume exact conservation for the vector currents, and perhaps an approximate conservation for the axial vector currents etc. The currents and their algebra were connected with gauge invariance of the second kind. The problem was now whether by means of the scale group one can get some understanding of these phenomenological properties. You may remember that already in the very old times, about fourty years ago, Dirac had to calculate such quantities as currents or energy density from his theory. He found that all the conventional expressions were highly singular, but one could derive finite expressions by the following procedure: In the bilinear expression, say for a current, one took first the two operators at slightly different points. Then the expectation values of these quantities turned out to be highly singular when the distance goes to zero; from dimensional reasons one could see that the first term would be singular like one over the distance to the third power, then there was a term with one over the square of the distance and finally of course one came to the regular terms. So one had to subtract these highly singular parts in order to derive physical results. This kind of subtraction physics was not very satisfactory, but now it comes into a very different light under the new circumstances. First of all the wave function which we use in the non-linear spinor theory is of the dimension one over the square root of a length, so that these bilinear expressions have as highest singularity a reciprocal length. In the spinor equation itself the subtraction has already been made; it is indicated by the Wick product. Recently Dürr and Winter, applying the methods of Brandt and Wilson, have been able to use this scale invariance for deriving finite expressions of the correct dimension for the currents and for deciding questions about the conservation of vector currents or axial vector currents, about chirality symmetry and so on. Now again it is not my task here to go into these details, I hope Dürr will do it in his lecture.

Let me come to the next point, the rôle of the other groups which are contained in this non-linear spinor theory. We have, so far, when I spoke

- 8 -

about current algebra from our point of view, only the possibility to derive
such symmetries as SU_2 and $SU_2 \cdot SU_2$. SU_3 or $SU_3 \cdot SU_3$ are not contained in such
a simple spinor equation. Phenomenologically even SU_2 is not an exact group,
and it belonged to our program to explain these violations of the SU_2-symmetry
by an asymmetry of the groundstate. The idea was, that the theory is exactly
invariant under SU_2 but the groundstate called 'world' is not invariant. Thereby one direction in isospace is singled out and by this asymmetry of the groundstate we can explain the violations of the isospin group, the mass splitting
and the electromagnetic forces. When we discussed this problem six years ago
there were the following difficulties: The Goldstone theorem did require the
existence of particles of mass zero. We hoped that the photons could somehow
be the Goldstone particles. But this was certainly not correct in the first
instant; because the photons have the wrong symmetry with respect to the isospin group. If one wanted to use the analogy between the spin in ferromagnetism
and isospin in the world, one was of course inclined to say: the groundstate
of the world has a very big isospin in one direction, like a ferromagnet has
a very big spin in one direction and therefore there should be Goldstone particles corresponding to the magnons in the ferromagnet. But the magnons in the
ferromagnet have an angular momentum around the axis of symmetry and this
would mean that the photons should be charged which of course is not correct.
Now, here again I think, there has been much progress in recent years. First
by the papers of Higgs and Kibble who treated the problem in a slightly more
subtle manner; they started from a wave function and a gauge field, both vectors
in isospace, and a Hamiltonian symmetrical under the isogroup. If an interaction
was chosen, which led to a spontaneous breaking of symmetry so that one direction in isospin was singled out, then the restmass zero remained only for those
gauge field particles which had the isospin momentum zero around the axis, in
other words which were not charged. The other gauge field particles acquired
a finite mass, so that actually only the light quanta as a kind of more complicated substitute of the Goldstone particles would remain. The connexion
between the degeneracy of the groundstate and the existence of the electromagnetic field seems thereby established. In the meantime, together with
Wagner and Yamazaki I had tried to discuss the problem from a different point
of view. The ferromagnetic model lacks the symmetry between particle and antiparticle which is an essential feature of the real world. Therefore we wanted
to improve the model by introducing this symmetry and to look for the conse-

quences. We introduced two kinds of spins like in an antiferromagnet. Each atom gets two kinds of spins - the one we called σ-, the other τ-spin. All the σ-spins together form a ferromagnet and tend to be parallel. All the τ-spins have the same tendency, but there is an interaction between the σ and τ at each atom in such a way that these two spins try to put themselves antiparallel; hence in the lowest state we have very many σ-spins in one direction and very many τ-spins in the opposite direction. The Hamiltonian is symmetrical under exchange of σ and τ. In such a model the magnons have the angular momentum zero around the axis of symmetry and insofar they have just the properties which the photons have in isospace. But otherwise these magnons are more complicated than the photons, and they can obviously not imitate the Lorentz-properties of the photons. Still the ground-state of this model may be useful for understanding the problems of the SU_3-groups. If there is a finite interaction between σ and τ, this interaction will cause a zero point motion, so that the crystal can probably be described best by saying: there is a certain large fraction of all sigma spins, say the fraction $1 - \alpha$, which are parallel and also the fraction $1 - \alpha$ of the τ-spins, which are parallel in the opposite direction, but then there is a small fraction α where σ- and τ-spins are connected so that they form units of spin zero and they rotate according to their zero point motion. If other particles, say electrons with a σ-spin or positrons with a τ-spin, are brought into the crystal, the spin of these electrons will interact with the spins in the groundstate. This interaction can lead to a mass-splitting between the two spin directions. But a σ-spin of the electron may also be combined with a τ-spin in the groundstate to form a $(\sigma-\tau)$ pair of total spin zero. In this way 'strange' particles of spin zero may be produced. For the analogous case of the isospin in the real world this would probably mean, that besides the operator ψ_λ, creating fermions with isospin direction λ, one should also consider an operator ψ_g, which creates these fermions in such a way, that its isospin is combined with an isospin from the groundstate to form a saturated pair.

Of the two particles created by ψ_λ, the one would be charged, the other one neutral; because the one with its isospin in the 'wrong' direction would cause a tension in the groundstate, alias Coulomb field, the other one would, like the particle created by ψ_g, cause no tension. In this way the scheme

would be a replica of the Sakata model (ψ_N producing the nucleons and ψ_Λ
the Λ -particles) and could, by considering the dynamics of the system, represent the essential features of an approximate SU_3-symmetry.

If one tries to interpret SU_3 in this way, there is no room for quark-
particles, and therefore I find it satisfactory that so far no quarks have
been found. But obviously one should do further experiments to establish this
point more definitely.

Since in the case of electrodynamics the connexion between the degeneracy of the groundstate, the gauge invariance (of the second kind) and the
photons of mass zero seems to be well established, it is satisfactory that, —
following the earlier papers of Thirring and Weinberg, Dürr and Rudolph have
been able to get similar connexions in the case of gravitation. Here the degeneracy refers to the group of the translations in space and time. But again
I will leave this problem to further discussions in our meeting.

In conclusion I would like to review in a few words those points, where
I think that considerable progress has been made during the last six years.
The indefinite metric in Hilbert space is probably equivalent to the introduction of long range forces to which no corresponding particles ('quanta')
exist. It leads to slight deviations from the conventional form of causality,
which should be compatible with the experiments. Their consequences can possibly be observed in the behaviour of amplitudes near the thresholds for
'ghost'-production. The scale transformation belongs to a very important group
which can be used for discussing the behaviour of 2-point functions near the
light cone, for constructing currents and for deriving current algebra. The
degeneracy of the groundstate can be connected with the gauge invariance (of
the second kind). Its relevance for an understanding of electromagnetism and
gravitation has been clarified in many details.

Therefore generally I am quite optimistic for the further development
of the theory of elementary particle physics, and I hope to learn from your
discussions during our meeting how things are going on.

In *Problems of Theoretical Physics: A Memorial Volume to Igor Tamm* (Nauka, Moscow 1972) pp. 34–36

BEMERKUNGEN ZUR NEUEN TAMM — DANCOFF-METHODE

W. HEISENBERG

*Max-Plank-Institut für Physik und Astrophysik,
München, DBR*

Die von I. Tamm im Jahre 1945 entwickelte Approximationsmethode zur Lösung quantentheoretischer Probleme, die später zur sogenannten neuen Tamm — Dancoff-Methode weiterentwickelt worden ist [1], gehöhrt zu dem umstrittenen mathematischen Rüstzeug der heutigen Quantenfeldtheorie und damit der Theorie der Elementarteilchen. Trotz der Einwände, die im Lauf der Jahre gegen diese Methode — wie gegen jede andere der Feldphysik — geltend gemacht worden sind, hat sie in einigen Punkten ganz entscheidende Fortschritte gegenüber früheren Methoden bewirkt, und der vorliegende Festband kann als willkommener Anlaß betrachtet werden, die Gründe dafür noch einmal in einem kurzen Text zusammenzufassen und die grosse prinzipielle Bedeutung der Tamm — Dancoff-Methode zur approximativen Behandlung feldtheoretischer Probleme hervorzuheben.

In den ersten Jahrzehnten der Quantenfeldtheorie beschäftigte man sich vor allem mit der Quantisierung linearer Feldgleichungen, also mit Feldtheorien ohne innere Wechselwirkung. Die Wechselwirkung wurde, wenn sie überhaupt behandelt wurde, als kleine Störung aufgefaßt, die nach den Perturbationsmethoden der alten Quantenmechanik einbezogen werden konnte. Erst im Lauf der Jahre lernte man, besonders im Zusammenhang mit der Quantenelektrodynamik, daß die Wechselwirkung grundsätzlich keine kleine Störung ist. Die Selbstenergie des Elektrons erwies sich als unendlich groß, woraus man den unbequemen Schluß ziehen mußte, daß schon der Ausgangspunkt der gewählten Störungsrechnung unendlich falsch gewesen war. Auf die Geschichte dieser Physik der Divergenzen und der zu ihrer Abhilfe ersonnenen Renormierungsvorschriften kann hier natürlich nicht eingegangen werden. Als Endresultat kann man vielleicht vereinfachend festhalten, daß schon der Zustandsraum einer wechselwirkungsfreien Feldtheorie von dem Zustandsraum einer Theorie mit Wechselwirkung unendlich verschieden ist, daß es also gar keine echte unitäre Transformation geben kann, die von einem System von Basisvektoren im ersteren Zustandsraum zu einem im zweiten Zustandsraum führen kann. Wenn mann also mit der Existenz der Wechselwirkung ernst machen wollte, mußte man von dieser Physik der unendlichen Integrale und der fiktiven unitären Transformationen wegkommen. Dies ist zum ersten Mal durch die neue Tamm — Dancoff-Methode geleistet worden. In einer Theorie mit Wechselwirkung ist schon der Grundzustand unendlich verschieden von dem einer entsprechenden Theorie ohne Wechselwirkung. Die neue Tamm — Dancoff-Methode erlaubt, sich nur für die Matrixelemente der Feldoperatoren zu interessieren, die von dem wirklichen Grundzustand zu einem wirklichen angeregten Zustand führen. Damit ist die Möglichkeit geschaffen, Operatorgleichungen aufzuschreiben, bei denen jedes einzelne Matrixelement von vornherein endlich ist und wohl auch grundsätzlich physikalisch interpretiert werden kann. Ausserdem kann man bei diesem Verfahren gebundene Zustände einbeziehen und Eigenwertgleichungen für sie erhalten, was bei dem früheren Störungsverfahren unmöglich war; das sind die entscheidenden Fortschritte der neuen Tamm — Dancoff-Methode.

Die Einwände gegen diese Methode betreffen vor allem die Frage der Konvergenz. Bisher ist die Konvergenz nur in einem einzigen Falle, nämlich bei der Anwendung einer speziellen Form der Tamm — Dancoff-Methode auf den anharmonischen Oszillator, einigermaßen zwingend nachgewiesen worden [2]. Bei Anwendung anderer Formen dieser Methode auf das gleiche physikalische System ist die Konvergenz schon sehr zweifelhaft [3], und man muß daraus (und aus verschiedenen anderen Untersuchungen) allgemein schließen, daß die Konvergenz entscheidend von der Art abhängt, wie das System der Gleichungen jeweils in einem bestimmten Näherungßchritt abgebrochen wird. Es handelt sich also bestenfalls um eine bedingte Konvergenz, und die Vorschrift über die Art des Abbrechens ist für die Definition des mathematischen Problems entscheidend wichtig. Offenbar lassen sich verbesserte Vorschriften angeben, die zur Konvergenz führen [4]. Trotz solcher Schwierigkeiten können niedrige Näherungen schon ein qualitativ brauchbares Bild der zu erwartenden Lösungen geben. Die Frage, ob die neue Tamm — Dancoff-Methode zu einem exakten mathematischen Formalismus ausgebaut werden kann, wird einstweilen der zukünftigen Entwicklung überlassen werden müssen.

Es gibt aber noch ein weiteres wichtiges Argument, das der neuen Tamm — Dancoff-Methode einen gewissen Vorzug gegenüber anderen Approximationsmethoden einräumt. In einer Quantenfeldtheorie mit Wechselwirkung ist der Grundzustand kein einfacher, leicht in seinen Eigenschaften zu definierender Zustand wie in einer wechselwirkungsfreien Theorie; vielmehr sind die Eingenschaften des Grundzustands in ebenso komplizierter Weise durch die (in den Operatorgleichungen niedergelegte) Dynamik des Systems bestimmt wie die aller angeregten Zustände. In der neuen Tamm —

 W. Heisenberg

Dancoff-Methode kommt diese weitgehende Gleichberechtigung des Grundzustandes und der angeregten Zustände deutlich zum Ausdruck. Wenn z. B. der Grundzustand als Folge der Dynamik des Systems nicht die volle Symmetrie der Operatorgleichung besitzt, sondern entartet ist, so läßt sich auch dieser Sachverhalt leicht in der neuen Tamm — Dancoff-Methode formulieren. Als Beispiel sei die Behandlung des vereinfachten Modells der Supraleitung genannt [5].

Die Arbeit von Tamm aus dem Jahr 1945 hat also vielfältige Früchte getragen, und man kann kaum daran zweifeln, daß sie dies auch in Zukunft tun wird.

ЛИТЕРАТУРА

1. *I. Tamm.*— J. Phys. (USSR), 1945, **9**, 449; *S. M. Dancoff.*— Phys. Rev., 1950, **78**, 382; *F. J. Dyson.*— Phys. Rev., 1953, **90**, 994; *W. Zimmermann.*— Nuovo Cimento Suppl., 1954, **11**, 43.
2. *H. Stumpf, F. Wagner, F. Wahl.*— Z. Naturforsch., 1964, **19a**, 1254.
3. *Ch. Schwartz.*— Ann. Phys., 1965, **32**, 277.
4. *D. Maison, H. Stumpf.*— Z. Naturforsch., 1966, **21a**, 1829.
5. *K. Yamazaki.*— Nuclear Phys., 1961, **23**, 139.

In *Aspects of Quantum Theory,* ed. by A. Salam, E. Wigner (University Press, Cambridge 1972) pp. 129–136

Indefinite Metric in State Space

W. Heisenberg

In his Bakerian lecture in 1941 Dirac has suggested that in a relativistic quantum theory, use should be made of a state space with indefinite metric.[1] This was a rather revolutionary suggestion, since in unrelativistic quantum theory the state space could always be interpreted as a Hilbert space with positive metric, and the whole probabilistic interpretation of the formalism of quantum theory rested upon this assumption. Dirac's idea has been developed in the course of years in a great number of papers by many physicists, and it cannot be the intention of the present paper to give a more or less complete historical account of this development. Only its essential steps shall be briefly described and analysed, in order to arrive at conclusions about the significance of Dirac's suggestion in the present relativistic theory of elementary particles. A few years ago a rather comprehensive survey of this whole field was given in a book by Nagy,[2] which also quotes a large part of the related literature.

There were two main reasons which had led Dirac to his suggestion. One was the fact, that in a relativistic theory the indefinite metric looked more natural from a mathematical point of view; the other was the success of Dirac's theory of holes, in which it had turned out to be rather easy, by a reinterpretation of the formalism, to get rid of the negative energy values in the theory of the electron. In a similar way Dirac hoped to get rid of the negative probabilities, which formally seemed to enter through the indefinite metric.

The first point can be seen from the Klein–Gordon equation for particles of spin zero, where the charge density – contrary to Dirac's theory of spin $\frac{1}{2}$ particles – is indefinite; and from the commutator in quantum electrodynamics

$$[A_\mu(x)A_\nu(x')] = ig_{\mu\nu}D(x - x'), \tag{1}$$

where the opposite sign of g_{00} against g_{ii} indicates the indefinite metric in the corresponding state space. A more fundamental reason for this situation may be seen in the fact, that, disregarding the translations, the Lorentz-group – contrary to the Galilei-group – is a non-compact group. Any finite representation of a non-compact group requires a space with indefinite metric; though infinite representations with definite metric may exist.

The second point is more problematic. Briefly after Dirac's first paper Pauli had taken up Dirac's ideas with the intention of developing a general consistent interpretation of the new formalism.[3] This however did not seem possible, the negative probabilities could not generally be avoided; what remained were rather arbitrary rules which could not be fitted into a simple convincing scheme.

The next decisive progress was achieved in the well known reformulation of quantum electrodynamics by Gupta[4] and Bleuler[5]. These authors, starting from Maxwell's equations, treated the four components A_μ on equal footing and consequently introduced – besides the transversal photons – longitudinal and scalar photons; the latter according to (1) had a negative norm in the state space. A subsidiary condition, however, corresponding to the Lorentz condition $(\delta A_\mu/\delta x_\mu)=0$ in classical theory, prevented the longitudinal and scalar photons from appearing as free particles. They played their rôle only in the region of interaction, they were in fact responsible for the Coulomb-forces between charged particles, but they never appeared in the asymptotic region, and therefore not in the S-matrix. Thereby a new idea was introduced into quantum field theory: the total state space should be divided into two parts. The one contains all asymptotic states, which are physically allowed, and in this subspace the metric is positive. The other part, in which the metric can be indefinite, appears physically only in the interaction. The subsidiary condition, which projects the physical states out of the complete state space, guarantees the consistency of the partition independent of time, and the probabilistic interpretation of the S-matrix does not encounter any difficulties. Hence it is not surprising that the Gupta–Bleuler formalism can be transformed back into the Dirac–Schwinger formalism; it is mathematically equivalent to a theory with positive metric, where a long range interaction, the Coulomb force, has been introduced from the beginning. The indefinite metric appears here as a kind of luxury allowing a formulation of quantum electrodynamics which is manifestly covariant and contains only local interactions.

This idea of the two parts of state space and of the indefinite metric was taken up in the author's attempt for a unified field theory,[6] in order to formulate a consistent scheme which from the beginning does not lead to any infinities, divergent integrals etc. It was assumed that the two-point functions, on account of the interactions, do not contain dangerous singularities at the light cone. In the simplest approximation the two-point function for fermions was represented in momentum space by a function, that contained a pole at an average baryon mass and a dipole at mass zero. This latter assumption implied the introduction of an indefinite metric in state space, and the mass zero was chosen to include the leptons. The baryon function should be regularised by means of the leptons, the lepton function by the baryons. It was of course not clear at that time, in which way the dipole at mass zero could be connected with the leptons. The subsidiary condition, which should project the physical states out of the total state space, was formulated as the postulate, that only combinations of eigenstates of the total energy can appear as physical states.

A sufficient clarification of the dipole case and the corresponding subsidiary condition was obtained shortly afterwards by means of the Lee model.[7] Pauli and Källén had demonstrated, that for a sufficiently high (or infinite) cut-off energy the Lee model requires a state space with indefinite metric.[8] The parameters can be chosen such that in the eigenvalue problem for the 'V-particle' the two roots coincide and a dipole results. In this situation two states appear with norm zero, but not orthogonal to each other. The one (the 'good' ghost) is an eigenstate of the Hamiltonian, the other (the 'bad' ghost) is not. Solutions can be found in which only the good ghosts, not the bad ones, occur as incoming or outgoing particles, and the resulting S-matrix is unitary. Since even the good ghosts have the norm zero, they do not contribute to any expectation value, hence the V-particles are never created as physical states. The subsidiary condition can be formulated as indicated above; the physical states are those that can be constructed as combinations of eigensolutions of the Hamiltonian.

A similar situation arises, as could be shown by Pauli,[9] when the two roots in the eigenvalue equation of the V-particles are complex. At least in the low sectors of the Lee model (implying only one V-particle) the unitarity of the S-matrix is not endangered by such complex poles. On the other hand the case of the two real roots (one of which then belongs to a negative norm) seemed to lead to a non-unitary S-matrix.

These results gave rise to a number of papers concerning the general conditions under which a theory with a state space of indefinite metric can find a probabilistic interpretation. For the details we refer to Nagy's book.[2] A general argument of Sudarshan seemed to show that a fairly large class of theories of this kind should allow such an interpretation:[10] If incoming and outgoing waves can be defined, and an S-matrix that transforms from the one to the other, then it should be possible by a linear transformation to bring this S-matrix into its diagonal form. Some of the diagonal elements will then be of the type $e^{i\alpha}$, i.e. have the absolute value unity, some others won't. The states belonging to the first group can be defined as the physical states, the others as the rest. The unitarity of the S-matrix is then guaranteed by this definition. But of course the unitarity of the S-matrix is not the only condition which must be fulfilled by a reasonable quantum field theory.

So far the indefinite metric and the concept of the two state spaces had been used for two quite different purposes. In the Gupta–Bleuler method it had led to a formulation of quantum-electrodynamics which is manifestly covariant and avoids the non-local long-range interactions; in the Lee model and – possibly – in the unified field theory it had permitted a consistent theory of an interaction without (i.e. with an infinite) cut-off without infinities or divergencies.

A connexion between the two aspects is suggested by the conjecture, that the equivalence between the Gupta–Bleuler method and the Dirac–Schwinger formalism is not limited to quantum electrodynamics, that in modified form it plays an essential rôle in much more general cases and especially in the unified field theory of elementary particles. This conjecture has found the support of two recent papers, the one by Karowski,[11] the other one by Dürr and Rudolph.[12]

Karowski has been able to show that the Lee model, e.g. its dipole version, is mathematically equivalent to another model, in which only the N- and θ-particles, not the V-particles, are introduced from the beginning; they are connected by a non-local interaction of a special type, and the metric in state space is positive. Both models lead to exactly the same S-matrix for collisions between one N- and one or two θ-particles (the results can probably be extended to any number of θ-particles); the same analytical functions occur in both models. The V-particle appears in the second model only as bound state between N- and θ-particle, not as primary field, and this state is characterised by a pole, not a dipole. The norm connected with this pole is positive, as it should be in a state space with positive metric. Analytically this change is brought about by a factor in the eigenvalue equation for the bound state (as compared with that of the V-particle in the first model), which vanishes at the critical energy. Hence in this second model a genuine bound state of N and θ exists, with positive norm. But the analytical structure provides a selection rule which prevents the formation or annihilation of this bound state in any collision process between N- and θ-particles.

It is very tempting to use this mathematical structure as a model for the interpretation of the two-point function mentioned above, which as a first approximation had been tried in the unified field theory. The leptons had been represented by a dipole at mass zero. The system could possibly be mathematically equivalent to another one with a positive metric, not manifestly covariant and with a non local interaction. In this system the dipole at mass zero would automatically be replaced by a pole, the leptons would become genuine particles with positive norm, but the analytical structure would provide a selection rule, which in this approximation prevents the creation or annihilation of leptons, in agreement with the observations.

The close analogy between the Bleuler–Gupta method and the dipole case of the Lee model was further underlined in the paper by Dürr and Rudolph.[12] If in the Bleuler–Gupta formalism the wavefunctions of the longitudinal and the scalar photons are replaced by their sum and their difference, then the sum can be compared with the field of the good ghosts, the difference with that of the bad ghosts. The norm of both kinds of ghosts is zero, but their wavefunctions are not orthogonal to each other. The subsidiary condition, as in the Lee model, prevents the bad ghosts from appearing as free particles, while the good ghosts are permitted, but do not contribute to any expectation-value. Actually in electrodynamics the addition of good ghosts means just a change of gauge, which does not affect the expectation values of physical observables. There is one characteristic difference however: While in electrodynamics the contribution of the good ghosts is quite arbitrary, i.e. any change of gauge is permitted, this contribution is fixed in the Lee model by the condition, that the bad ghosts must not appear as incoming or outgoing waves. But aside from that the analogy is complete.

The Gupta–Bleuler method has been extended by Dürr and Rudolph to fields of higher spin,[13] especially to gravitation. Again a manifestly Lorentz invariant and local formulation of the theory requires a state space with indefinite metric. The partition of the complete state space into a subspace of physical states and the rest is again reflected in the appearance of 'good' and 'bad' ghosts. In the S-matrix the bad ghosts are eliminated by the subsidiary condition.

So far the analysis of the Gupta–Bleuler method and the Lee model has led to the conclusion that in principle a consistent relativistic quantum theory of interaction can be constructed by making use of a state space with indefinite metric, as Dirac had suggested. The state space must be divided into two parts, the subspace of the asymptotic physical states and the rest. The physical states are defined by a subsidiary condition. The 'non physical' states play their rôle only in the region of interaction; they represent – or produce indirectly – a non local interaction which is not 'quantised', in the sense that it is not connected with free particles which would be the quanta belonging to this field. In this sense neither the Coulomb field nor the Newtonian gravitation are quantised, they may be called 'classical fields' with sufficient caution.

These results have strengthened the confidence in the indefinite metric in state space, and recently in many papers this method has been applied to problems which apparently could not be solved by conventional methods. We mention the electromagnetic mass splitting in isospin multiplets and similar problems. Arons, Han and Sudarshan[14] and Lee and Wick[15] have attempted to develop quantum electrodynamics as a theory without infinities by using from the beginning the indefinite metric in state space. Both papers follow the general idea of Pauli and Villars, that heavy masses may be used to regularise the two-point functions of the electrons. Arons, Han and Sudarshan consider real, Lee and Wick, complex poles. The two papers differ by the formulation of the subsidiary condition. Lee and Wick assumed, that in the case of complex poles a simple change in Feynman's well known prescription for the path of integration should be sufficient to define the physical states. This assumption has been criticised by Sudarshan *et al.*[16] It seems that in higher approximations one cannot get around a careful analysis of the asymptotic behaviour of the various 'ghosts'.

This question is already closely connected with the problem of causality. A theory with indefinite metric in state space will certainly not obey those strict criteria of causality, which had been derived from the conventional axioms; but such a theory may still be compatible with the existing observations.

Recently the consequences of the indefinite metric for the problem of causality and for the analytic behaviour of the amplitudes have been discussed in several papers by Lee and Wick,[15] Sudarshan *et al.*[16] and Dürr and Seiler.[17]

The problem of causality is not so well defined in quantum theory as it had been in classical theory. In the latter, one could construct sharp wave fronts, and causality required that these wave fronts never travelled faster than with the velocity of light. In quantum theory it is not possible to construct such a wave front, since it would require the superposition of plane waves of all frequencies, negative as well as positive. But even for the photon field a coherent mixture of many photon states covers only the positive energy range, i.e. contains only positive frequencies. Hence the forward edge of any wavefront will not be sharp; it will, from its maximum value in the forward direction decrease exponentially or possibly by a power law or otherwise, and the form of this forward tail will depend on the analytical structure of the amplitudes as a function of energy and momentum. From the existing experimental evidence on the time order of events it will be difficult to tell which form of the tail is still permissible. But it may be possible to observe, directly or indirectly, the analytical structure of the amplitudes.

Taking the dipole case as an example it is easily seen that the amplitudes (or related functions) cannot be analytic at a possible threshold for the creation of ghost-states, because the boundary conditions must be different below and above the threshold. The amplitudes will be piece-wise analytic functions, which are continuous but not analytic at the threshold. Examples for such functions have been given in the non-linear spinor theory.[18] The consequences for the form of the 'acausal' precursors have been discussed by Dürr and Seiler.[17]

This then is the decisive new feature introduced by Dirac's indefinite metric in state space. The amplitudes will possibly not be global analytic functions – as one might have expected for a theory with purely local interaction and positive metric – but only piece-wise analytic functions, which are connected continuously at the threshold points for 'ghost'-production. Acausal precursors need not decrease exponentially; even a decreasing power law may be in agreement with the observations.

In the thirty years that have elapsed since Dirac's first paper on the indefinite metric in state space, the subject should have been clarified sufficiently to draw a conclusion about its applicability as a basis for the theory of elementary particles. The alternative, a theory with positive metric and purely local interaction, has so far not been developed in a convincing way; the author believes that the efforts in this direction are largely based on wishful thinking. Even if this could be demonstrated, it does not prove that the theory with indefinite metric can be built up to a consistent mathematical scheme. Since the mathematical problems involved are extremely difficult, it would probably be best to look for

                      W. HEISENBERG

experimental verifications of the indefinite metric. The analytical behaviour of the amplitudes in the S-matrix or of the kernels to eigenvalue equations, is at least in principle open to experimental investigation. Therefore the recent efforts of Lee and Wick to obtain experimental evidence on the problem point in the right direction. For the time being we have to be satisfied with the statement that all existing observations seem to be compatible with Dirac's hypothesis of the indefinite metric in state space.

REFERENCES

1. P. A. M. Dirac, *Proc. Roy. Soc. (London)* A180, 1 (1942).
2. K. L. Nagy, *State Vector Spaces with Indefinite Metric in Quantum Field Theory* (Akademiai Kiadó, Budapest: 1966).
3. W. Pauli, *Rev. Mod. Phys.* 15 (3), 175 (1943).
4. S. N. Gupta, *Proc. Phys. Soc. (London)* A LXIII, 681 (1950).
5. K. Bleuler, *Helv. Phys. Acta* XXIII, 567 (1950).
6. W. Heisenberg, *Z. Naturforsch* 9a, 292 (1954); *Z. Physik* 144, 1 (1956).
7. W. Heisenberg, *Nucl. Phys.* 4, 532 (1957).
8. W. Pauli and G. Källén, *Math. Phys. Medd.* 30, 7 (1955).
9. W. Pauli, *Proceedings of the Annual International Conference on High-Energy Physics*, CERN, p. 127 (Geneva: 1958).
10. E. C. G. Sudarshan, *Phys. Rev.* 123, 2183 (1961).
11. M. Karowski, *Z. Naturforsch* 24a, 510 (1969).
12. H. P. Dürr and E. Rudolph, *Nuovo Cimento X*, 62A, 411 (1969).
13. H. P. Dürr and E. Rudolph, *Nuovo Cimento X*, 65A, 423 (1969).
14. M. E. Arons, M. Y. Han and E. C. G. Sudarshan, *Phys. Rev.* 137, 4B, B1085 (1965).
15. T. D. Lee and G. C. Wick, *Phys. Rev. D*, 2, 6, 1033 (1970).
16. A. M. Gleeson, R. J. Moore, H. Rechenberg and E. C. G. Sudarshan, *Analyticity, Covariance and Unitarity in Indefinite Metric Quantum Field Theories*, CPT-77, AEC-26.
17. H. P. Dürr and E. Seiler, *Nuovo Cimento X*, 66A, 734 (1970).
18. W. Heisenberg, *Introduction to the Unified Field Theory of Elementary Particles* (J. Wiley and Sons, New York: 1966).

Die Naturwissenschaften *61*, 1–5 (1974)

The Unified Field Theory of Elementary Particles: Some Recent Advances

W. Heisenberg*

Max-Planck-Institut für Physik und Astrophysik, München

Progress in the unified field theory of elementary particles over the last seven years includes experimental verification of the scale group and its application in theoretical investigations; connecting the unified field theory with phenomenological current algebra; exploring the consequences of the indefinite metric for the problem of causality and the classification of leptons; elaboration of methods for the calculation of mass eigenvalues; and experimental verification of the predicted behavior of cross-sections at extremely high energies.

Before I report on recent progress it may be useful to describe briefly the intention and the main ideas behind what Pauli and the author in 1958 called the "unified field theory of elementary particles". With regard to the intention, I quote a statement made by Niels Bohr in one of his early papers [1] on atomic theory: "It should be emphasized that the theory is not intended to explain the relevant phenomena in the conventional way, but to establish a connection between facts not explainable in the present state of physics ...". We may compare the task of constructing such a theory with that of assembling a jig-saw puzzle from a jumble of different pieces: It is easier to find pieces that can be fitted together after one has looked at the picture on the outside of the box. The unpublished paper by Pauli and the author can be considered as an attempt to guess what the picture was before undertaking the hard work of putting the pieces together [2]. The different pieces are: strong, electromagnetic, weak, and superweak interactions, and gravitation; or hadrons, leptons, photons, gravitons etc. The main ideas on how to combine them, i.e. the "picture", are set out below.

1. Main Ideas of the Unified Field Theory

a) There is no fundamental difference between elementary particles and compound systems. The particles are defined by their symmetry, i.e. by their behavior during certain fundamental transformations, in the same way as atoms or molecules are, but not by their composition of smaller constituents. We may state this in the form of a paradox: Every particle consists of all other particles.

b) Fundamental transformations are those in which the underlying "natural law" defining the theory is invariant. The simplest way of formulating such a natural law is as a nonlinear equation between field operators. It should be nonlinear to represent interaction, and it should be a hyperbolical differential equation in order to represent causality. The following equation was chosen (in the form given to it by Dürr [3]) on the basis of the experimental material available 15 years ago:

$$i\sigma^\nu \frac{\partial \chi(x)}{\partial x^\nu} + \sigma^\nu: \chi(x)\left(\chi^*(x)\,\sigma_\nu\chi(x)\right): = 0. \qquad (1)$$

$\chi(x)$ is a 2-component spinor both in Lorentz space and in isospace.

The groups of transformations defined as fundamental by this equation are: Lorentz group, isospin group (SU_2), scale group, PC, and T. The equation does not include SU_3 and P. While the first two groups (Lorentz and SU_2) seem to be generally accepted as fundamental, the inclusion of the scale group and the exclusion of P and SU_3 have been to some extent controversial. The group SU_3 is considered in the present theory as a dynamical group that may be compared with the approximate dynamical group $O_3 \times O_3$ responsible for the multiplets in the optical spectra of atoms.

c) Eq. (1) defines a theory only when supplemented by an algebra of the $\chi(x)$ operators and by boundary conditions. The most important information in this respect is given by the 2-point function of the ground-state, which may be written in the form:

$$\left\langle 0\left|T\chi\left(x+\tfrac{\zeta}{2}\right)\chi^*\left(x-\tfrac{\zeta}{2}\right)\right|0\right\rangle = \frac{\bar{\sigma}_\nu\zeta^\nu}{\zeta^2 - i\varepsilon} \cdot f$$

$$= \int \frac{\bar{\sigma}_\nu p^\nu d^4 p}{(p^2 + i\varepsilon)^2}\, e^{ip\zeta} \cdot g \qquad (\varepsilon \to 0) \qquad (2)$$

* Der vorliegende Vortrag wurde im Frühjahr 1973 an verschiedenen Universitäten in den USA gehalten und wird daher hier in englischer Sprache wiedergegeben.

1

If the ground state were invariant under all fundamental transformations, f and g would simply be constants, as can be seen by a dimensional consideration. This assumption, however, could not lead to finite mass values and therefore cannot represent the experimental situation. Hence the ground state must be less symmetrical than Eq. (1), a situation well known from the quantum mechanics of solid bodies, e.g. ferromagnetism. f may depend on $\zeta^2 m^2$ and g on p^2/m^2, where m defines a mass scale (e.g. the mass of the proton), and on isospin or other variables. The ground state is thereby assumed to have physical properties that introduce new degrees of freedom. Arguments have been advanced, that the group PC can in this way be extended to $P \cdot C$, and SU_2 can be extended to $SU_2 \cdot U_1 = U_2$, which then allows SU_3 as an approximate dynamical group. The degeneracy of the ground state with respect to isospin is connected with the existence of the electromagnetic field, and its degeneracy with respect to the translational group with the gravitational field (in accordance with the general theorem that states that a degeneracy of the ground state implies the existence of long range forces or of particles of rest mass zero, or both).

d) A comparison of (2) with a Lehmann-Källén representation shows that the operator $\chi(x)$ must act in a state space with indefinite metric. This is an essential extension of quantum theory [4] which has been studied in connection with the Bleuler-Gupta version of quantum electrodynamics, with the Lee model, and other problems.

2. The Scale Group

Eq. (1) is invariant under the scale transformation:

$$\chi(x) \to \eta^{\frac{3}{2}} \chi(x\eta). \tag{3}$$

The ground state cannot be invariant under this transformation, since we know that finite mass eigenvalues exist (e.g. the mass of the proton). But the invariance of (1) under (3) suggests that for very large energies ($\gg m$) or very small distances ($\zeta^2 m^2 \ll 1$) the mass scale will become unimportant, i.e. the mass m will not enter. Therefore it had been assumed [5] that e.g. the cross-section for a shower of energy E (in the center-of-mass system), in which an energy $\geq \varepsilon$ is transmuted into secondaries, should under the conditions $E \gg m$ and $\varepsilon \gg m$ depend on E and ε in the form:

$$\sigma_\varepsilon \sim \frac{\text{const}}{\varepsilon^2} f\left(\frac{E}{\varepsilon}\right), \tag{4}$$

a prediction which seemed to fit the cosmic-ray data.

Recently deep inelastic electron–proton and electron–neutron scattering has been extensively studied with the help of the SLAC accelerator [6]; the results confirm the scaling property. If in a process

$$e + P \to e + N^* \quad \text{or}$$

both the mass M of the final hadronic system N^* and the momentum transfer $\sqrt{|t|}$ carried by the photon γ are very large compared with m yet in such a way that the ratio t/M^2 stays finite, then the hadronic part of the process turns out to depend on the ratio t/M^2 only and not on t and M^2 separately, thereby immediately suggesting the scale group. This result has been interpreted in the literature as indicating that the electric charge distribution inside the proton has a granular structure composed of pointlike elements. Such an interpretation, however, appears to overemphasize the particle aspect of the phenomena. The reduction to the scale group is more abstract but perhaps less misleading. The experimental establishment of the scale group is an important recent advance.

The relevance of this group has another interesting theoretical aspect: it brings the nonlinear spinor theory into the neighbourhood of quantum electrodynamics, in which the scale group, according to Johnson [8], plays a similar role. Both theories in this sense are of the renormalizable type and for the analysis of the Wick product $:\chi^*(x)\,\chi(x):$ or the "finite part" of similar products use can be made of recent investigations by Wilson [9], Brandt [10], and Zimmermann [11] on related problems in quantum electrodynamics. These new methods permit discussion of the question of whether or not Eq. (1) can be invariant under gauge transformations of the second kind (local gauge transformations) and whether they allow the construction of local current operators [12]. These are the basis of the so-called current algebra [13], which has been successfully applied in the phenomenological analysis of electromagnetic and weak interactions.

3. Current Algebra

The relations characteristic for current algebra are an extension of the relations familiar from quantum electrodynamics, well known for many years. The fundamental equations of quantum electrodynamics are invariant under the gauge transformation which multiplies the field operators creating charged particles by $e^{i\alpha}$. The generator for this transformation in Hilbert space is the charge Q, which is linked with the electromagnetic currents $j_\nu(x)$ through the equations:

$$Q = \int j_0(x)\, d^3x \quad \text{and} \quad \frac{\partial j_\nu(x)}{\partial x_\nu} = 0. \tag{5}$$

The "local" gauge transformation:

$$\psi(x) \to \psi(x)\, e^{i\alpha(x)} \quad \text{and} \quad A_\nu \to A_\nu + \frac{1}{e}\frac{\partial \alpha}{\partial x^\nu} \tag{6}$$

(A_ν is the electromagnetic potential, e the coupling constant) is generated by $\int \left(j_0(x) - \frac{\partial F_{\mu 0}}{\partial x_\mu}\right) \alpha(x)\, d^3x$. ($F_{\mu\nu}$ is the electromagnetic field.)

Current algebra [13] assumes that similar relations hold for any transformations of the gauge type which can be carried out in the state space by a generating operator corresponding to Q, even if the Hamiltonian is not invariant under the transformation; we mention SU_2, SU_3, Lorentz group. The currents and charges are then defined by their behavior under the possible transformations of the system and by their properties as generators. They are universal in the same sense as the electromagnetic currents, i.e. they do not refer to any specific particles by which they could be produced. Their properties define their algebra at a specified time.

These definitions already demonstrate a close relationship between the universal field operator $\chi(x)$ in (1) and the universal currents; the spinor $\chi(x)$ is related to a fermion field, the vector $j_\nu(x)$ to a boson field, but otherwise both are defined by their behavior under the transformations of the system without any reference to specific particles. Therefore it should be possible to derive from the universal field $\chi(x)$ the currents for the fundamental transformations of (1), Lorentz group and SU_2; (for SU_3 such a connection could be established only after a proper representation of the new degrees of freedom introduced by the ground state has been given).

The construction of the currents has recently been worked out in a paper by Dürr and Winter [14]. In the unified field theory as defined by (1) the currents take a slightly more complicated form than in a conventional canonical theory. If from the universal spinor field $\chi(x)$ in (1) a second spinor field $\psi(x)$ is derived by the equation:

$$\psi(x) = \sigma_\nu \frac{\partial}{\partial x_\nu} \chi(x), \tag{7}$$

which transforms under the scale group to

$$\psi(x) \to \eta^{\frac{3}{2}} \psi(x\eta), \tag{8}$$

the currents can be written essentially in the form

$$j_{\nu 0}(x) = \; : \psi^*(x)\, \sigma_\nu O \psi(x): , \tag{9}$$

where the operator O characterizes the behavior under the internal groups (e.g. SU_2).

If the parity operation P is introduced in connection with the existence of finite mass eigenvalues, i.e. with the degeneracy of the ground state under the scale transformation, two kinds of currents can be constructed: a vector current, and an axial vector current. Only the vector current obeys the relation $\frac{\partial j_\nu(x)}{\partial x_\nu} = 0$ if the Hamiltonian and the observables are invariant under the corresponding gauge transformation. By summation or subtraction of the two currents new kinds of currents with a specified "chirality" can be formed, and these are useful in the phenomenological description of the weak interaction. In this way a close connection has been established between the unified field theory and the phenomenological current algebra.

4. The Indefinite Metric in State Space

The consequences of the extension of the physical state space to a larger state space with an indefinite metric have recently been investigated in two directions. The possible changes in the causal behavior of the S-matrix elements have been analyzed in some detail, and the equivalence between a theory with indefinite metric and local interaction, and another corresponding theory with definite metric and nonlocal interaction has been studied in connection with the Lee model.

With respect to the first problem we mention the papers of Sudarshan *et al.* [15], Lee and Wick [16], Dürr and Seiler [17]. A conventional calculation of the electromagnetic mass splitting in isomultiplets leads to divergent integrals. In a theory with indefinite metric these integrals can be regularized by the assumption of ghost states, which should not appear as ingoing or outgoing particles. Under these circumstances, if unitarity is to be preserved, the S-matrix elements will as a rule not be analytic at the thresholds for the creation of the ghost states. Hence the S-matrix elements will not be global analytic functions but only piecewise analytic functions. This means that, besides the exponential precursors of a wave front, which are characteristic for any quantum field theory, there may be precursors that fall off according to a power law. The existence of such precursors, which would then be characteristic for the indefinite metric in state space, may well be compatible with the existing experimental evidence.

These results fit in well with the conjecture that there may be a general equivalence between local theories with indefinite metric and corresponding nonlocal theories with definite metric. The first example of such a situation had been given by the equivalence of the Gupta-Bleuler method in quantum electrodynamics and the Dirac-Schwinger formalism. A similar equivalence has recently been demonstrated for the Lee model by Karowski [18]. The Lee model with N, V, Θ particles, where the V particles appear as dipole ghosts, can be replaced by another model containing a nonlocal interaction between N and Θ particles, but based on a definite metric in state space. The S matrix in both models is identical. In this second case the V particle appears only as a bound state between N and Θ particle (with positive norm), but the analytical structure of the theory provides a selection rule forbidding any process by which this bound state could be created or annihilated. These results suggest that in the 2-point function (2) of the unified field theory the dipole at mass zero should be indentified with the leptons and that the analytical structure of the theory may provide a selection rule which prevents the production of leptons in that approximation in which only the strong interactions are considered. One may hope—but this part of the theory has not yet been worked out—that even in higher approximations two separate conservation laws would result, one for the baryon number (for the poles) and one for the lepton number (for the dipoles).

5. Consequences for the Classification of Leptons

If this result could be demonstrated, the leptons would be classified by only two good quantum numbers, electric charge Q and lepton number L: strangeness or hypercharge would not be good quantum numbers since they are not conserved in weak interactions. In the following table charge Q and lepton number L are given for the leptons (for the neutrinos, the conventional notation has been used following from $N \to P + e_- + \bar{\nu}_e$; $\mu_- \to e_- + \bar{\nu}_e + \nu_\mu$). The third line contains the empirical helicity, as it appears in the weak interaction.

Table 1

Leptons	μ_-	μ_+	e_+	e_-	$\bar{\nu}_e$	ν_e	ν_μ	$\bar{\nu}_\mu$
Q	-1	1	1	-1	0	0	0	0
L	1	-1	1	-1	1	-1	1	-1
hel	-1	1	1	-1	1	-1	-1	1

These quantum numbers provide selection rules. The transitions $\mu \to 3e$ or $\mu \to e + \gamma$ e.g. are forbidden, in agreement with the experimental evidence.

The breaking of isospin by the electromagnetic forces is sometimes expressed by saying that the photon, except for its mass, can be compared with a mixture of a ϱ_0 and an ω meson. In a similar way and again disregarding the mass, the positron can be compared with a mixture of proton and Σ^+ hyperon, the μ^- with a mixture of Ξ^- and Σ^-, the $\bar{\nu}_e$ with a mixture of neutron and $\frac{\Sigma^\circ + \Lambda}{\sqrt{2}}$, the ν_μ with a mixture of Ξ^0 and $\frac{\Sigma^\circ - \Lambda}{\sqrt{2}}$.

But it should be emphasized, that these assignments are speculative so long as the role of the dipoles in higher approximations has not been clarified.

6. Calculation of Mass Eigenvalues

The central problem of any theory of elementary particles, the determination of the mass eigenvalues, was treated in the earlier papers on the unified field theory by means of the so-called new Tamm-Dancoff method. This method is limited by the fact that practically only the lowest-order approximations can be calculated, higher orders being too complicated. The 2-point function which enters into this method must first be chosen by arguments of plausibility, and only later can it be checked by criteria of consistency. The convergence is uncertain. But other methods which might be fundamentally better were not then available.

In recent years several papers have been published with the intention of discussing or improving this situation. The intrinsic similarity between elementary-particle theory and many-body theory has been used by Bleuler *et al.* [19] for a general justification of the new Tamm-Dancoff method. Stumpf *et al.* [20] have developed a functional calculus which can replace equations between operators, like the fundamental Eq. (1), and have applied it to the calculation of eigenvalues and S-matrix elements, and to the discussion of conditions of symmetry. Saller [21] has tried to improve the 2-point function by the information obtained from the local gauge invariance and to study the consequences for the eigenvalues.

If one keeps in mind that particle physics can by its very nature ("each particle consists of all other particles") not be less complicated than quantum chemistry, one understands that progress in the direction of the calculation of eigenvalues can only be slow.

7. High-Energy Jets

The completion of the storage rings in the CERN laboratory in 1971 has provided the opportunity to study collision processes in which the energy in the center-of-mass system reaches values up to 50 GeV. Some of the experiments carried out with this new tool have given interesting informations about the behavior of cross-sections at very high energies. The total cross-section for collisions between two protons does not seem to stay constant at extremely high energies. Contrary to what had been widely assumed, it seems to increase asymptotically with the square of the logarithm of the energy. From the general picture of the particles described in the present talk, such an increase had been predicted [22] by the following argument.

An elementary particle is a roughly spherical object of diameter of the order 10^{-13} cm; its density in the outer parts falls off exponentially, and the exponent is related to the mass of the pion, since this is the lightest hadron (electromagnetic forces and leptons are neglected in this argument). In the center-of-mass system the two colliding protons look like discs, and the ratio of radius to thickness is given by their energy E/m in the center-of-mass system. We will assume that without interaction their centers would pass at a distance b (collision radius). The discs will become less transparent when their density increases on account of the Lorentz contraction with increasing energy. The production of secondaries in the collision of the two discs will require a minimum "blackness" at the points of overlapping. The radius inside which this minimum blackness is reached, will—on account of the exponential decrease of the density in the outer parts—increase with the logarithm of the energy. Hence also the critical collision radius b_c will increase in the same way

$$b_c \sim \mathrm{const} \cdot \ln \frac{E}{m}, \tag{10}$$

and for the total cross-section one gets

$$\sigma_\mathrm{tot} \sim \mathrm{const} \cdot \ln^2 \frac{E}{m}. \tag{11}$$

Later Froissart [23] was able to show that this asymptotic behavior at very high energies means the maximum rate of increase which is still compatible with the unitarity of the S matrix.

The recent experiments with the storage rings at CERN [24] seem to confirm the asymptotic behavior (11) of the total cross-section and thereby the general picture of elementary particles that underlies the unified field theory. They do not necessarily confirm Eq. (1) with its special group structure; other more complicated equations, including e.g. SU_3, could equally well lead to (11).

However, other recent experiments with the storage rings tend to show that SU_3 should not be included among the fundamental symmetries. Even at the highest energies no production of quark particles has been observed [25] so that their existence has become still less probable; the interpretation of SU_3 as an approximate dynamical group has thus gained in plausibility.

Quite generally from the experiments with the storage rings one gets the impression that here the asymptotic region has been reached, and that no fundamentally new phenomena are to be expected at these extreme energies. But this view is open to doubt so long as the behavior of weak interactions at very high energies has not been clarified.

In conclusion, coming back to our jig-saw puzzle, we seem justified in saying that in recent years a number of pieces have actually been put together and that they fit into a coherent picture, as we had hoped. Hence there is good reason to be optimistic with regard to the future development of elementary-particle physics.

4

1. Bohr, Niels: Drei Aufsätze über Spektren und Atombau; Vorwort zur 1. Aufl., Vieweg 1922
2. The first detailed paper on this subject was published by H. P. Dürr, W. Heisenberg, H. Mitter, S. Schlieder u. K. Yamazaki, Z. Naturforschung **14a**, 441 (1959). For later literature see W. Heisenberg, Introduction to the Unified Field Theory of Elementary Particles, Interscience publ. 1966
3. Dürr, H. P.: Z. Naturforschung **16a**, 327 (1961)
4. First discussed by P. A. M. Dirac, Proc. Roy. Soc. A **180**, 1 (1942)
5. Heisenberg, W.: Introduction to the Unified Field Theory of Elementary Particles, p. 131. Interscience publ. 1966
6. Kendall, H. W., Panofsky, W.: Scientific American **224**, 60 (1971)
7. Bjorken, J. D.: Phys. Rev. **179**, 1547 (1969)
8. Johnson, K.: Conf. on High-Energy Physics, Dubna 1964
9. Wilson, K. G.: Phys. Rev. **179**, 1499 (1969)
10. Brandt, R. A.: Ann. of Phys. **44**, 221 (1967); **52**, 122 (1969); Phys. Rev. **180**, 1490 (1969)
11. Zimmermann, W.: Comm. Math. Phys. **6**, 161 (1967); **8**, 66 (1968)
12. Dürr, H. P., Winter, N. J.: I, Nuovo Cim. X., LXXA, 467 (1970); II, Nuov. Cim. II, Vol. 7A, 461 (1972)
13. See e.g. Renner, B.: Current algebras and their applications. Pergamon press 1968
14. Dürr, H. P., Winter, N. J.: Nuovo Cim. Ser. 11, 7A, 461 (1972)
15. Arons, M. E., Han, M. Y., Sudarshan, E. C. G.: Phys. Rev. **137**. 4B, 13, 1085 (1965); Gleeson, A. M., Moore, R. J., Rechenberg, H., Sudarshan, E. C. G.: Analyticity, Covariance and Unitarity in Indefinite Metric of Quantum Field Theories CPT-77, AEC-26
16. Lee, T. D., Wick, G. C.: Phys. Rev. D., **2**, 6, 1033 (1970)
17. Dürr, H. P., Seiler, E.: Nuovo Cimento X, **66**A, 734 (1970)
18. Karowski, M.: Z. Naturforschung **24a**, 510 (1969)
19. Bleuler, K., Friederich, A., Petry, H. R., Schütte, D.: In: Quanten und Felder, p. 267. Braunschweig: Vieweg u. Sohn 1971
20. Stumpf, H.: Funktionale Quantentheorie, in Quanten und Felder, p. 189. Braunschweig: Vieweg u. Sohn 1971; Stumpf, H., Scheerer, K., Märtl, H. G.: Z. Naturforschung **25a**, 1556 (1970); Dammeier, K.: ibid. **27a**, 1041 (1972); Bauhoff, W., Scheerer, K.: ibid. **27a**, 1539 (1972)
21. Saller, H.: Nuovo Cim. **4**A, 404 (1971)
22. Heisenberg, W.: Vorträge über kosmische Strahlung, p. 155. Berlin: Springer 1953
23. Froissart, M.: Phys. Rev. **123**, 1053 (1961)
24. Amaldi, U., *et al.*: Physics Letters B **44**, 112 (1973)
25. Bott-Bodenhausen, M., *et al.*: ibid. B **40**, 693 (1972)

Received August 6, 1973

Physics Today *29*, No. 3, 32–39 (1976)

The nature of elementary particles

Because particle number is not conserved in high-energy interactions
it may be meaningless to ask about the constituent parts
of elementary particles; perhaps the central problem is dynamics.

Werner Heisenberg

The question, "What is an elementary particle?" must find its answer primarily in experiment, although it must also be confronted with philosphical considerations. I will therefore begin by giving a short survey of the important experimental results of the last fifty years. This survey will show that a critical unbiased study of these results already gives an answer to the question; theory, as we shall see, cannot add much to this answer.

Next I will deal with the philosophical problems that arise in connection with the concept of an elementary particle. It may be objected that in this question we should concentrate on physics rather than on philosophy. But this separation is not so simple. In fact, I believe that certain erroneous developments in particle theory—and I am afraid that such developments do exist—are caused by a misconception by some physicists that it is possible to avoid philosophical arguments altogether. Starting with poor philosophy, they pose the wrong questions. It is only a slight exaggeration to say that good physics has at times been spoiled by poor philosophy.

Finally I will discuss these problematic developments. Having witnessed similar mistakes in the development of quantum mechanics fifty years ago, I am in a position to make some suggestions to avoid such errors in the future. This will lead us to conclude on an optimistic note.

Particle number not conserved

Let us start with the experimental facts. Nearly fifty years have passed since P. A. M. Dirac predicted, on the basis of his theory of the electron, the existence of its antiparticle, the positron: A few years later the existence of these was demonstrated experimentally by Carl Anderson and P. M. S. Blackett. They produced them in pair creation, forming what we now call "antimatter" artificially.

This was a discovery of prime importance. Before this time it was assumed that there were two fundamental kinds of particles, electrons and protons, which, unlike most other particles, were

Physics Today *29*, No. 3, 32–39 (1976)

immutable. Therefore their number was fixed and they were referred to as "elementary" particles. Matter was seen as being ultimately constructed of electrons and protons. The experiments of Anderson and Blackett provided definite proof that this hypothesis was wrong. Electrons can be created and annihilated; their number is not constant; they are not "elementary" in the original meaning of the word.

The next important step was the discovery of artificial radioactivity by Frédéric Joliot and Irène Curie. From many experiments it became clear that an atomic nucleus can be transmuted into another nucleus by emitting particles, provided the laws of conservation of energy, angular momentum, electric charge and so on permit this transmutation. The transformation of energy into matter, predicted as a possibility very early in the theory of special relativity, has become recognized as a rather common phenomenon. Contrary to the earlier views, there was no conservation of particle number. However, there are physical properties that can be characterized by quantum numbers, for instance angular momentum and electric charge; these quantum numbers may assume positive or negative values and are subject to laws of conservation.

The 1930's brought a few other important experimental discoveries. In cosmic radiation very energetic particles are observed. A cosmic-ray particle, when colliding with another particle such as an atomic nucleus in a photographic emulsion, can produce a shower of many secondary particles. For some time many physicists believed that such showers can be produced only by cascades in the interior of a heavy nucleus. Later it became clear, however, that even the collision of two protons can lead to the production of many secondaries in one step. In the late 1940's Cecil Powell discovered the pions, which play the main role in this process of multiple production of particles. His results emphasized again that the transmutation of energy into matter is the

33 decisive process, and that it would be meaningless to speak about the "division" of the original particles. Experimentally, the concept of "dividing" had lost its meaning.

This new situation was confirmed again and again in the experiments of the 1950's and 1960's; many new particles of various lifetimes were discovered and there was no answer to the question, "What do these particles consist of?" A proton could be obtained from a neutron and a pion, or from a Λ hyperon and a kaon, or from two nucleons and one antinucleon, and so on. Could we therefore simply say a proton consists of continuous matter? Such a statement would be neither right nor wrong: There is no difference in principle between elementary particles and compound systems. This is probably the most important experimental result of the last fifty years.

This development convincingly suggests the following analogy: Let us compare the so-called "elementary" particles with the stationary states of an atom or a molecule. We may think of these as various states of one single molecule or as the many different molecules of chemistry. One may therefore speak simply of the "spectrum of matter." Experiments in the 1960's and 1970's with large accelerators have demonstrated that this picture fits elementary particles as well. Like the stationary states of atoms, the elementary particles can be characterized by quantum numbers—that is, by their behavior under certain transformations. The corresponding laws of conservation determine what transmutations are possible. For an excited hydrogen atom it is its behavior under rotation that determines whether it can fall into a lower

state with the emission of a light quantum. In the same way for a φ boson, it is its symmetry properties that determine whether it can disintegrate with the emission of a pion into a ρ boson.

The stationary states of atoms have very different lifetimes, and the same is true of particles. The ground state of an atom has an infinite lifetime, and there are many particles with this property, including the electron, the proton and the deuteron. But these stable particles are no more elementary than the unstable ones—the ground state of hydrogen is a solution of the same Schrödinger equation as any of the excited states. Similarly, the electron and the proton are no more elementary than the Λ hyperon.

During recent years experimental particle physics has applied itself to tasks similar to those of spectroscopy in the early 1920's. Just as at that time all stationary states of atoms were collected in large tables, the "Paschen-Götze" volumes, so nowadays the "reviews of particle properties" every year collect new or improved data on the masses and quantum numbers of particles. This kind of work corresponds to the astronomical surveys, and obviously every observer hopes occasionally to find an especially interesting object in his sector.

Two types of broken symmetry

There are nevertheless certain characteristic differences between the physics of atomic shells and particle physics. In the shells the relevant energies are small, so the typical features of the theory of relativity play no role, and their behavior can be described at least approximately by non-relativistic quantum mechanics. Hence in shell physics and in particle physics the underlying symmetry groups are different. The Galilean group in shell physics is replaced by the Lorentz group in particle physics. Other groups have to be added, such as the isospin group, which is isomorphic to SU_2; then SU_3, the

Physics Today *29*, No. 3, 32–39 (1976)

scale group and others. It is an important experimental proposition to determine all groups relevant to particle physics, and this problem has been solved to a large extent during the past twenty years.

From the physics of atomic shells we can learn that among those groups that describe only approximate symmetries in nature, two essentially different types can be distinguished. In the optical spectra of atoms, for example, the groups O_3 and $O_3 \times O_3$ play very different roles. The fundamental equations of quantum mechanics of atoms are strictly invariant under O_3. Therefore the stationary states with higher angular momenta are strictly degenerate, so that there are always several states with exactly the same energy. Only when external electromagnetic fields are applied do these states split and the well known fine structures of the Zeeman effect or the Stark effect appear.

A similar effect can be produced in systems in which the ground state is not invariant under rotations, such as a crystal or a ferromagnet: The two directions of electron spin in a ferromagnet do not belong to precisely the same energy. In this case, according to a well known theorem of Goldstone, bosons must also exist, with energy that tends to zero with increasing wavelength; in the case of ferromagnetism the spin waves of Bloch, "magnons", take the place of the Goldstone waves.

An entirely different situation is met with in the group $O_3 \times O_3$, which produces the well known multiplets in the optical spectra. The group $O_3 \times O_3$ is only an approximate symmetry, which comes about if the spin-orbit interactions become small in a certain part of the spectrum, so that the orbits and the spins of the particles can be rotated almost independently. The symmetry $O_3 \times O_3$ results from the dynamics of the system and is useful only in certain parts of the spectrum. Empirically the two types of broken symmetries can be distinguished by the existence or nonexistence of the Goldstone modes. If they are found, it is plausible to assume that the degeneracy of the ground state plays an important role.

If we apply the experience of shell physics to particle physics, the experi-

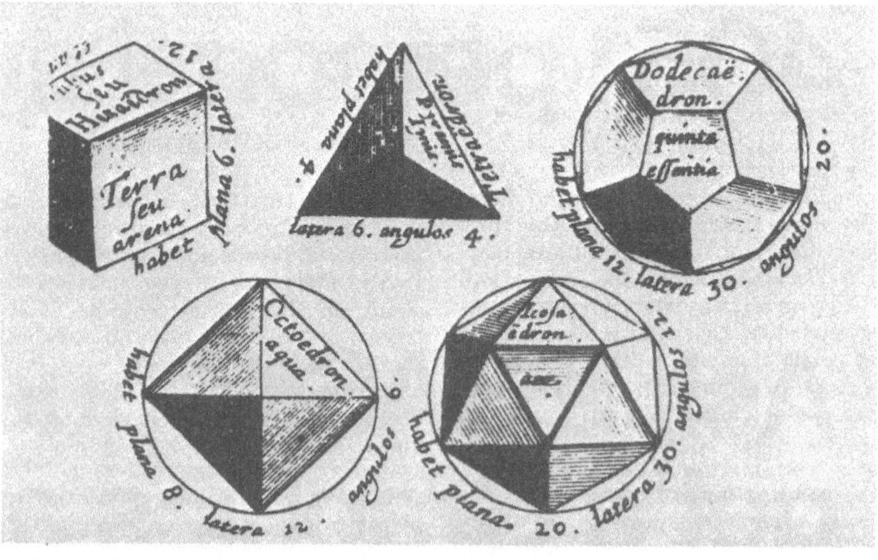

"[Plato's] forms are not matter themselves, but they make up matter. For the element earth, the characteristic body is the cube . . ."

ments suggest that the Lorentz group and the group SU_2 should be interpreted as fundamental symmetries of the underlying natural law. Electromagnetism and gravitation then appear as those long-range effects that, according to Goldstone, are connected with the broken symmetry of the ground state. The more complicated groups such as SU_3, SU_4, SU_6, $SU_2 \times SU_2$ or $SU_3 \times SU_3$, should be taken as dynamical symmetries, like $O_3 \times O_3$ in shell physics. We may doubt whether the dilatation group or scale group should be counted among the fundamental symmetries. They are disturbed by the existence of particles of finite mass and by the gravitational influence of the big masses in the universe. Because of their close mathematical relation to the Lorentz group they probably belong to the fundamental group. This particular coordination of the broken empirical symmetries with the two proposed types of broken symmetries is suggested by the existing experimental evidence, but it may not yet be finally settled. It is important to emphasize that, for any symmetry group arising out of the phenomenological analysis of the spectrum, the question must be asked—and if possible answered—to which of these two types it belongs.

Another special feature of shell physics should be mentioned. Among the optical spectra there are non-combining or weakly combining term systems such as the parahelium and the orthohelium spectrum. In particle physics one could perhaps compare the division of the fermion spectrum into baryons and leptons with this feature.

It is evident that the analogy between the stationary states of an atom or molecule on one hand and the particles of high-energy physics on the other is nearly complete, qualitatively answering the question about the nature of the "elementary particle." But only qualitatively. For the theoretician the further question arises whether this interpretation can be based on quantitative calculations. Here we must answer a preliminary question: What does it

mean to "understand a spectrum quantitatively"?

Dynamics and contingent conditions

A number of examples, both from classical physics and from quantum mechanics, teach the general procedure by which we acquire physical understanding. Let us think of the spectrum of elastic vibrations of a steel plate. If a qualitative theoretical interpretation is not enough, we have to start with the elastic properties of the steel plate, which should be represented mathematically. When this has been done, we must add boundary conditions that tell whether the plate is a circle or a square, and whether it is stretched in a frame or free. This knowledge then should be sufficient to calculate, at least in principle, the spectrum of the acoustic vibrations. It is true that, because of the high degree of complication, we are frequently unable to calculate all the vibrational frequencies precisely, but only the lowest ones—those with the smallest number of nodes.

Hence two elements are prerequisite to a quantitative understanding of the spectrum: a precise knowledge of the elastic behavior of the plate and the boundary conditions. The latter may be called "contingent" because they depend on the particular circumstances—the plate could have been cut differently. A similar case would be the electrodynamical vibrations of a cavity. Maxwell's equations determine the dynamical behavior, and the shape of the cavity defines the boundary conditions. Another comparable situation is met with in the optical spectrum of the iron atom. The Schrödinger equation for a system of one nucleus and 26 electrons determines the dynamical behavior, while the boundary condition determines that the wave function vanishes at infinity. If the atom were enclosed in a small box the spectrum would be different.

To relate these results to particle physics, our first problem must be to determine experimentally the dynamical properties of the system "matter" and to formulate them mathematically.

Physics Today *29*, No. 3, 32–39 (1976)

Then the contingent element, the boundary conditions, has to be added. These contain in this case statements about the so-called "empty space": the cosmos and its symmetry properties. In other words: The first step must be the attempt to formulate mathematically a natural law that defines the dynamics of matter. The second step is to determine the boundary conditions. Without such conditions the spectrum can not be defined. We might, for example, conjecture that inside a black hole the spectrum of elementary particles would be quite different from the spectrum in normal space—but unfortunately we can not check this by experiments!

Let me add a word about that decisive first step, the formulation of the governing dynamical law. There are pessimists among the particle physicists who believe that no such natural law defining the dynamics of matter exists. This view appears quite absurd to me. There must be some clearly defined dy-

namics of matter or there could be no spectrum; therefore a mathematical description should be possible. The pessimistic view implies that particle physics aims at no other goal than presenting an immense volume of particle data, a "super review of particle properties." In this super volume nothing could be understood since there is no dynamics of matter and so the volume would scarcely be read.

But I would like to emphasize strongly that I can not see any reason for such pessimism. A spectrum with sharp lines is observed, and this should imply a well defined dynamics of matter. The experimental results mentioned in the beginning give definite hints as to the fundamental invariances of the underlying natural law, and from the dispersion relations a lot is known about the degree of causality this law contains. Essential parts of the natural law therefore already belong to our definite knowledge. Because so many other spectra in physics have finally come to be understood quantitatively, this

should be also possible here—in spite of the high degree of complication. It is on account of this intricacy that I will not discuss here the special proposal that Wolfgang Pauli and I made many years ago for the mathematical formulation of the underlying law, one I still regard as having the best chance of being the correct one. It is more important to emphasize that the formulation of such a law is the unavoidable precondition for a quantitative understanding of the spectrum. Anything else can not be called understanding; it would be scarcely more than looking up the table of data, and theoreticians at least should not be content with that.

Philosophical problems

Taking this point of view, I am now going to discuss the philosophy that, whether consciously or unintentionally, has determined the direction of particle physics. For 2500 years philosophers and scientists have pondered the questions: "What happens if one tries to divide matter again and again? What are the smallest particles of matter?" Different philosophers have given very different answers, all of which have influenced the history of natural science. The best known answer is that of the philosopher Democritos: In the attempt to divide again and again one finally ends up with indivisible, unchangeable units, called atoms, of which all matter is composed. The positions and motions of the atoms determine the qualities of matter.

For Aristotle and his medieval successors, on the other hand, the concept of the smallest particle is not so well defined. It is true that for every kind of matter smallest particles are assumed—further division would change the characteristic qualities of the substance—but these smallest particles can be changed continuously like the substances themselves. Mathematically the substances can be divided *ad infinitum*. Matter is taken as continuous.

The clearest position against Democritos was taken by Plato. In his opinion the attempt to divide again and

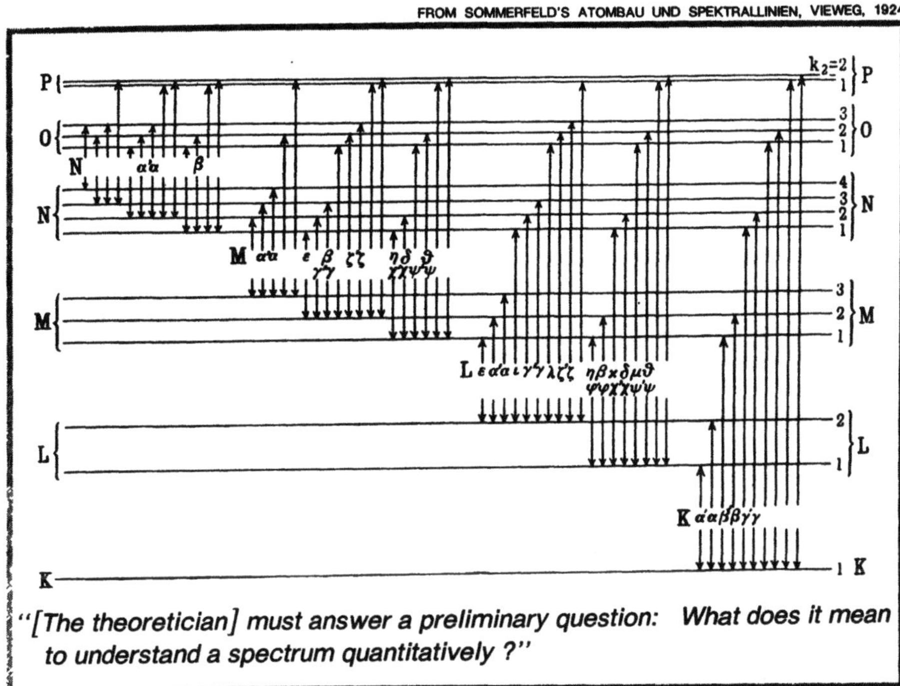

again results in mathematical forms: the regular bodies of stereometry, defined by their symmetry, and the triangles of which they are composed. These forms are not matter themselves, but they make up matter. For the element earth, for example, the characteristic body is the cube; for fire, the tetrahedron. What all these philosophies have in common is the attempt to deal with the antinomy of the infinitely small, which was discussed extensively by Immanuel Kant.

There have been even more naive attempts to rationalize this paradox. Some biologists have proposed the idea that within the seed of an apple an invisibly small apple tree is concealed, which again bears flowers and fruits; that the fruits again contain a still much smaller apple tree and so on *ad infinitum*. It was an amusing game, in the early days of the Bohr–Rutherford theory of the atom as small planetary system, to develop a similar idea: The electrons, the planets of the system, are inhabited by very small living bodies who build houses, cultivate their soil and study atomic physics—and find that their atoms are again small planetary systems, and so on in an unending progression.

In the background of every such fiction is Kant's antinomy: It is difficult to imagine that matter can be divided again and again, but it is equally difficult to imagine that this division must necessarily come to an end. As we now know, the paradox is caused by the erroneous assumption that our intuition can be applied to the smallest dimensions.

The strongest influence on the physics and chemistry of the last century undoubtedly came from the atomism of Democritos. This view allows an intuitive description of chemical processes on a small scale. Atoms can be compared with the mass points of Newtonian mechanics, and from this a satisfactory statistical theory of heat was developed. It is true that the atoms of the chemists turned out to be not mass points but small planetary systems, and the atomic nucleus likewise was a compound system formed of protons and

Physics Today *29*, No. 3, 32–39 (1976)

neutrons. Nevertheless the electron, the proton and possibly the neutron could, it seemed, be considered as the genuine atoms, the indivisible building blocks, of matter. In this way the atomism of Democritos became an essential part of the materialistic interpretation of the world during the last century: Easily understood and intuitively plausible, it determined the way of thinking of even those physicists who insisted on not dealing with philosophy. At this point let me substantiate my earlier statement that, in the physics of elementary particles of our time, good physics has sometimes been unconsciously spoiled by poor philosophy.

We can not avoid using a language bound up with the traditional philosophy. We ask: "What does a proton consist of? Can an electron be divided or is it indivisible? Is a photon simple or compound?" But all these questions are wrongly put, because words such as "divide" or "consist of" have to a large

extent lost their meaning. It must be our task to adapt our thinking and speaking—indeed our scientific philosophy—to the new situation created by the experimental evidence. Unfortunately this is very difficult. Wrong questions and wrong pictures creep automatically into particle physics and lead to developments that do not fit the real situation in nature. We will discuss these fallacies below.

But first a word should be added concerning the postulate that understanding requires a visual picture of the phenomena. Some philosophers have claimed that such pictures are the preconditions for understanding. For example, the philosopher Hugo Dingler of Munich held the view, in regard to the theory of relativity, that Euclidean geometry was the only possible correct geometry because we assume its correctness in building our measuring apparatus; on this latter point Dingler was right, of course. Therefore he argued that the experimental facts that are at the foundation of general relativity can not be described by a Riemannian, non-

Euclidean geometry, because this would lead to contradictions. Here the postulate seems to be overdrawn: It is enough to know that, within the dimensions of our apparatus, Euclidean geometry applies with sufficient accuracy.

We will have to accept the fact that experimental data on a very large or a very small scale do not necessarily produce pictures, and we must learn to do without them. We then come to recognize that the antinomy of the smallest dimensions is solved in particle physics in a very subtle manner, of which neither Kant nor the ancient philosophers could have thought: The word "dividing" loses its meaning.

If we wish to compare the results of present-day particle physics with any of the old philosophies, the philosophy of Plato appears to be the most adequate: The particles of modern physics are representations of symmetry groups and to that extent they resemble the symmetrical bodies of Plato's philosophy.

Wrong questions

My intention, however, is not to deal with philosophy but with physics. Therefore I will now discuss that development of theoretical particle physics that, I believe, begins with the wrong questions. First of all there is the thesis that the observed particles such as the proton, the pion, the hyperon consist of smaller particles: quarks, partons, gluons, charmed particles or whatever else, none of which have been observed. Apparently here the question was asked: "What does a proton consist of?" But the questioners appear to have forgotten that the phrase "consist of" has a tolerably clear meaning only if the particle can be divided into pieces with a small amount of energy, much smaller than the rest mass of the particle itself.

To demonstrate how a word that seems to be well defined can lose its meaning in special situations, I can not resist repeating a story that Niels Bohr loved to quote: A small boy enters a shop with two pennies in his hand and asks the grocer for two pennies' worth

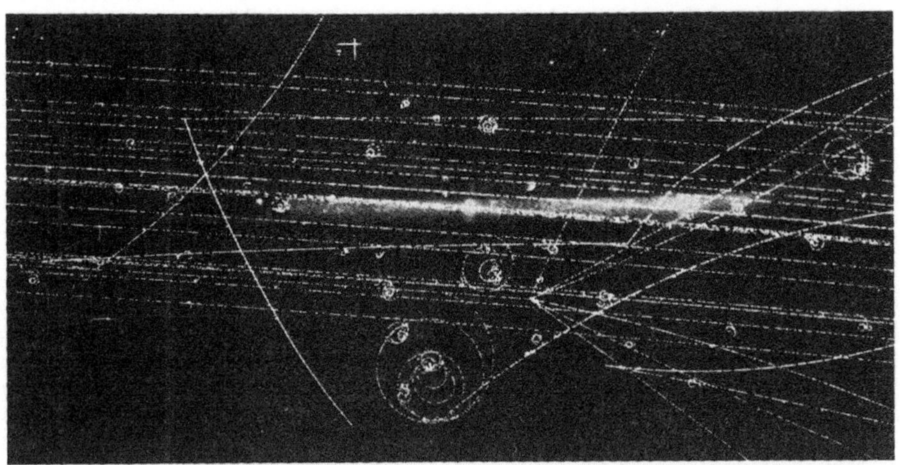

". . . there was no answer to the question, 'What do these particles consist of?'   A proton could be obtained from a neutron and a pion, or from . . ."

of mixed sweets. The grocer hands him two sweets and adds: "You can do the mixing yourself." The concept "consist of" for a proton has just as much meaning as the concept of "mixing" in the story of the boy.

At this point many readers may object that the quark hypothesis was derived from experimental material, namely from the empirical relevance of the SU_3 group, and that it has been successfully applied in the interpretation of many experiments even beyond the use of the SU_3 group. This can not be denied. But I would like to mention a counter-example from the history of quantum mechanics, of which I have been a witness; a counter-example that shows clearly the weakness of such arguments.

Before Bohr's theory of the atom many physicists believed that an atom must consist of harmonic oscillators. The optical spectrum contains sharp lines, and sharp lines can only be emitted by harmonic oscillators. The charges in these oscillators, however, would correspond to e/m values different from those of the electron, and it would be necessary to assume very many different oscillators, since there are very many lines in the spectrum.

Notwithstanding these difficulties Woldemar Voigt in Göttingen in 1912 developed a theory of the anomalous Zeeman effect of the D lines in the optical spectrum of sodium, on the following basis: He assumed two coupled oscillators which, without an external magnetic field, emitted the frequencies of the two D lines. The interaction between the two oscillators, and their coupling with an outer magnetic field could be arranged such that for weak fields the anomalous Zeeman effect, and for strong fields the Paschen–Back effect, could be reproduced correctly. For the intermediate range of medium fields the frequencies and intensities were described by long and complicated square roots, which, however, seemed to fit the experiments well.

Fifteen years later Pascual Jordan and I took the trouble to treat the same problem on the basis of the perturbational calculus of quantum mechanics. It came as a great surprise to us that we got exactly the old formulas of Voigt both for the frequencies and for the intensities, even for intermediate fields. The reason, as we understood later, was of purely formal mathematical nature. The perturbational calculus leads to a system of coupled linear equations and

Physics Today *29*, No. 3, 32–39 (1976)

the frequencies are determined by the eigenvalues of the system. A system of coupled harmonic oscillators in classical theory also leads to a system of coupled linear equations. Since in Voigt's theory the important parameters had been adjusted to the empirical situation, it was not strange that the results ended up the same in the two cases. But Voigt's theory has not contributed to our understanding of atomic structure.

Why was this attempt by Voigt so successful on the one hand, and so useless on the other hand? Because Voigt intended to discuss only the D lines, without taking notice of the complete spectrum. Voigt phenomenologically used a special aspect of the oscillator hypothesis but ignored the other problems and difficulties of the model; at least he consciously left them undefined. He did not really take the oscillator hypothesis seriously. In the same way I am afraid that the quark hypothesis is not really taken seriously today

by its proponents. Questions dealing with the statistics of quarks, the forces that keep them together, the reason why the quarks are never seen as free particles, the creation of pairs of quarks inside an elementary particle, are all left more or less undefined. If the quark hypothesis is really to be taken seriously it is necessary to formulate precise mathematical assumptions for the quarks and for the forces that keep them together and to show, at least qualitatively, that all these assumptions reproduce the known features of particle physics.

There should be no problem in particle physics to which these assumptions can not be applied. I do not know of any such attempt, and I am afraid that every attempt written down in precise mathematical language would easily lead to contradictions. My objections to the quark hypothesis can therefore be put in the form of questions:
▶ Does the quark hypothesis contribute more to the understanding of the particle spectrum than Voigt's hypothesis contributed in an earlier period to the

understanding of the structure of the atomic shells?
▶ Do we not find behind the quark hypothesis the old idea—refuted long ago by experiments—that simple and compound particles can be distinguished?

No more surprises?

Let us now discuss a few special questions. If the SU_3 group plays an important role in the structure of the spectrum—and this appears from the experimental evidence to be the case—then it is important to decide whether SU_3 is a fundamental symmetry of the underlying natural law or a dynamical symmetry, which by its very nature can be only approximately valid. If this decision is left open, the other assumptions concerning the dynamics of the system are left open as well, and no understanding can be gained. The higher symmetries such as SU_4, SU_6, SU_{12}, $SU_2 \times SU_2$ probably belong to the dynamical symmetries that may be useful for a phenomenological description; but their heuristic value could probably be compared with the heuristic value of cycles and epicycles in Ptolemy's astronomy. They give only very indirect information about the underlying natural law.

In the most important experimental results of recent years, bosons with relatively high mass (3–4 GeV) and long life have been discovered. Such states were to be expected, as has been especially emphasized by Hans–Peter Dürr. Whether their long life allows us to interpret them as being "composed of" other particles of long life is a difficult dynamical question, in which all the complications of many-body physics play a role. I would, however, consider it as useless speculation once again to introduce *ad hoc* new particles, of which these objects are assumed to "consist." This would again be the wrong question, which can not contribute to the understanding of the spectrum.

Recently, in the storage rings in Geneva and in the Batavia machine, total cross sections of collisions between protons at extremely high energies were measured. The result was that at very high energies the cross sections increase roughly as the square of the logarithm of the energy. This behavior had long ago been conjectured in the theory for the asymptotic region, independent of the nature of the 'particles. In the meantime the collisions between other particles have led to similar results, and this outcome strongly suggests that in the big accelerators the asymptotic region has already been reached—that even at highest energies no more surprises are to be expected.

New experiments generally can not be expected to yield a *deus ex machina* that suddenly leads to an understanding of the spectrum. The experiments of the last fifty years have already given qualitatively a quite satisfactory, consistent and complete answer to the question of the nature of elementary particles. The quantitative details can—as in quantum chemistry—be analyzed only in the course of years by much detailed work in physics and mathematics, far from being solved in a single step.

Therefore this article can be concluded with a more optimistic view of those developments in particle physics that promise success. New experimental results are always valuable, even if they only enlarge the data table; but they are especially interesting if they answer critical questions of the theory. In the theory one should try to make precise assumptions concerning the dynamics of matter, without any philosophical prejudices. The dynamics must be taken seriously, and we should not be content with vaguely defined hypotheses that leave essential points open. Everything outside of the dynamics is just a verbal description of the table of data, and even then the data table probably yields more information than the verbal description can. The particle spectrum can be understood only if the underlying dynamics of matter is known; dynamics is the central problem.

*   *   *

This article was adapted from a translation of the opening lecture of the German Physical Society's spring meeting, given 5 March 1975. The original version is published in the February 1976 issue of Naturwissenschaften. □

Die Naturwissenschaften 63, 63–67 (1976)

Cosmic Radiation and Fundamental Problems in Physics

W. Heisenberg † *

Max-Planck-Institut für Physik und Astrophysik, Institut für Physik, München

Cosmic-ray research has advanced our understanding of fundamental problems in physics, when concepts previously used are shown to have a limited range of applicability. Since cosmic rays contain information on the behaviour of matter in the smallest (elementary particles) and largest dimensions (the universe), they have been particularly valuable in testing the concepts of daily life in relation to their meaning in physics and in leading physicists to find new ones.

Cosmic radiation has—ever since its discovery some 60 years ago—played a very important rôle in the development of physics. It is a very interesting history which led from the first indication of rays coming from outer space to the earth to the discovery of very energetic particles in this radiation, of new particles with unexpected properties, of new fundamental symmetries in the laws of nature, and finally to a great wealth of information about the residual matter and magnetic fields in interstellar space, and about those processes which can possibly produce the cosmic radiation. But I will not follow this historical line.

I will try to confine my talk to those fundamental problems of physics which have been touched or essentially advanced by the progress of knowledge in cosmic radiation. It is the interaction between this very special field of physics and the fundamental problems, basic to all physics, in which I shall be interested in this talk. This interaction became visible for the first time in the early thirties, when cosmic radiation played an essential part in one of the most important discoveries in physics of this century, the discovery of the positron. It is true that this discovery has not been made primarily in cosmic-ray research. Dirac in his theory of the electron had predicted a positively charged counterpart to the electron, but the first convincing evidence for its existence was found in the cosmic radiation by Anderson, and by Blackett and Occhialini. The first cloud-chamber pictures of showers, where photons created pairs of electron and positron, and these particles again created photons when traversing matter, demonstrated beyond any

doubt the existence of the positrons and the validity of Dirac's theory. Shortly afterwards the positrons could also be seen in nuclear processes, that is in β-decay.

I should perhaps add a few words about the fundamental importance of this discovery. Up to that time, physicists had, perhaps more or less unconsciously, followed the philosophy of the old Greek philosopher Demokritos. When one tries to divide a piece of matter over and over again, one would—so was the conjecture—eventually end up with smallest parts of matter which could not be divided any further, and therefore were called atoms. These atoms were taken as indivisible, unchangeable units of matter, as the building blocks from which all matter is constructed, and the atoms—or as we would nowadays say: elementary particles—should by their relative position and motion determine the visible properties of the various kinds of matter. This whole picture, as plausible as it may seem, has been destroyed completely by the theory of Dirac and by its consequence, the discovery of the positron. The decisive point was not so much the existence of a new, hitherto unknown particle—many new particles have later been found without serious consequences for the foundations of physics—it was *the discovery of a new symmetry,* the particle-antiparticle conjugation, which was closely connected with the Lorentz group of special relativity, and with the transmutation from energy to matter and vice versa. In nonrelativistic physics the number of particles of any kind was a constant of motion like energy or momentum. In relativistic physics this number was not a good quantum number any more. A hydrogen atom, for example, did not necessarily consist of proton and electron, it may be taken as consisting of proton, two electrons and one positron,

<hr>

* Lecture given at the 14th International Cosmic Ray Conference, August 18, 1975, in Munich (due to the illness of the lecturer, the text was read by Prof. R. Lüst).

even if this latter configuration would only amount to a small relativistic correction of the complete wave function of hydrogen. One of the consequences of this situation was the conjecture that in a very energetic collision of two particles a larger number of new particles may be created, and these possibilities should be limited only by the laws of conservation of energy, momentum, isospin etc. It was again in cosmic radiation that this conjecture could be tested.

Actually, already in the late thirties Blau and Wambacher discovered in photographic plates, exposed at high altitudes the so-called stars, events in which from the same point in the plate a great number of tracks started. Apparently an atomic nucleus had been hit by an incoming, very energetic particle and it emitted, as a result of the collision, a number of different particles. The interpretation of these stars was not simple, since the beginning of the process could possibly be a kind of cascade in the nucleus, similar to the wellknown electron-positron cascades, followed by some evaporation of the nucleus. So these results did not immediately demonstrate the multiple production of particles in a collision of only two of them, which I mentioned as a conjecture. But in the course of time the cosmic-ray experiments could be refined, and after 15 years the occurrence of multiple production was definitely established.

These results meant that the concepts of "dividing" and "consisting of" have only a limited range of applicability. Just as in relativity the concept "simultaneous" or in quantum theory the concepts "position" and "velocity" can be applied only with characteristic restrictions, and loose their meaning when used uncritically at the wrong place, so also the concepts "dividing" or "consisting of" are well defined only in special situations. When a particle, by a small amount of energy, can be disintegrated into two or several parts, the restmass of which is very large compared with this small energy, then and only then may one say that the particle consists of these parts, can be divided into these parts. In all other cases the words "dividing" or "consisting of" have no well defined meaning. What actually happens in a very energetic collision of two particles is the creation of new particles out of the kinetic energy. Energy becomes matter by assuming the form of elementary particles. But again the distinction between "elementary" particle and "compound system" has no well defined meaning. Particles are stationary states of the physical system "matter". All these very important and fundamental results had their experimental basis in cosmic-ray research.

Another interesting result of cosmic-ray research was the discovery of the muon or μ-meson by Neddermayer and Anderson in 1937. This object was first mistaken as the particle which had been predicted by Yukawa as the material counterpart to the strong interaction between nucleons. But it soon turned out, that the interaction of muons with heavy particles like proton and neutron was much too small; the muon could not be responsible for the strong interaction in a nucleus. Rather, the muon appeared as a heavier brother of the electron, different from it only by its larger mass. The discovery of the muon did not cause as fundamental a change in the basis of physics as the discovery of the positron. But it revealed an interesting feature in the spectrum of particles. This spectrum is divided into *two only weakly combining termsystems,* the hadrons and the leptons. Such weakly combining termsystems are well known from the optical spectra of atoms. But whether the causes for such splitting are similar in both cases is still an open question. The muons constitute — besides the neutrinos — the most penetrating part of the cosmic radiation and therefore play an important rôle in determining the intensity of the cosmic radiation as a function of height in the atmosphere.

I should perhaps mention another rather odd case in which the muons helped to settle a very fundamental question. In our country just before the war the theory of relativity was not accepted by the political power, and it was especially the dilatation of time in moving bodies, which was criticized as absurd and pure theoretical speculation. There were even trials concerning the question whether the theory of relativity could be taught at Universities. In one of these discussions I could point out that the decay time of muons should depend on their velocity; muons which move almost with the velocity of light decay more slowly than those with smaller velocities; — this was the prediction of the theory of relativity. The experimental results confirmed this prediction; the dilatation of time could be observed directly and the way was open for courses on relativity. So I felt always grateful to the muons.

Shortly after the war Powell in Bristol discovered the pion, which plays a very important rôle in most cosmic-ray phenomena. This object fulfills all the conditions formulated by Yukawa for the material counterpart of strong interaction; it was, as was recognized later, not the only particle of this kind, but being the hadron of the smallest mass it was soon discovered in almost all events of very high energy. Besides that, the pion decays into muon and neutrino, so the origin of the muons had been clarified.

Like the muons the pion did not cause fundamental changes in the basis of physics. It just confirmed that the various particles are stationary states of the system matter, different through their different behaviour under the transformation of the fundamental group. The groups are more fundamental than the particles.

At that time, besides the Lorentz group of relativity, only the isospin group was known as fundamental. It had been found 1932 in connexion with nuclear physics; but it was first through the pions, that its fundamental character was fully understood. Cosmic-ray experiments on the pion demonstrated, that the isospin group is an exact symmetry for the strong interaction, and only the electromagnetic interaction and

64

weaker interactions break this symmetry. These results could be interpreted by assuming, that the natural law, underlying the spectrum of particles, is exactly invariant under the isospin transformation, and that the deviations from this symmetry are caused by an asymmetric, degenerate ground state. Similar situations are well known from quantum mechanics of solid bodies.

Almost simultaneously with the pion, other particles were discovered in the cosmic radiation — heavier than the pion and somewhat "strange" in their behaviour. They had a rather long lifetime of the order 10^{-10} sec and therefore their tracks could be observed in cloud chambers or in emulsions. But this long lifetime could not be understood, if only the known symmetries and corresponding quantum numbers (baryonic number, isospin, angular momentum) were considered; one would have expected a very much shorter lifetime, and insofar their behaviour was strange. The correct interpretation was given by Pais in 1952, when he introduced a new quantum number, called strangeness and the corresponding symmetry (or transformation property). So cosmic-ray research had led to a new symmetry group; and since, as I mentioned before, groups are more important than particles, this was again a very essential contribution to fundamental problems in physics.

There was general agreement among most physicists at that time that many more particles could be found, if objects with very short lifetime could be made observable. The particles are just stationary states of the system matter, therefore one had to expect many different particles, most of them of very short lifetime. Such objects could be observed only as so-called *resonance states*, and for that purpose better statistics was needed than cosmic-ray observations could offer. Fortunately for the particle physicists the first big accelerators had been built at that time and went into action, the cosmotron in Brookhaven, the bevatron in Berkeley and the proton synchrotron of CERN in Geneva. So for a long time to come, the important results in particle physics were obtained with the help of big accelerators, while cosmic-ray research turned its attention primarily to astrophysical problems. This development was unavoidable, but did not always meet the wishes of the particle physicists; it was a sad change.

The romantic time had gone, in which the study of cloud-chamber pictures in a mountain laboratory at high altitude could be combined with skiing and mountaineering, or in which balloon experiments could be started from a beautiful island in the Mediterranean with the help of an airplane and a military vessel from the Italian navy, as had been managed by our Italian friends. Certainly the warm sun of the Mediterranean has contributed to the scientific success of the experiments. But this gay time had now gone, and particle research had to be done in the "matter of fact" atmosphere of huge accelerator establishments.

In astrophysics cosmic radiation became a valuable new tool which promised information beyond that obtained from the visible or infrared light of the stars. The first problem was of course the origin of the radiation. Already Forbush recognized that some low-energy part of the cosmic radiation was occasionally emitted by the sun, by some turbulent phenomena at its surface. But it was understood very soon that a definite answer to the problem of the origin of cosmic rays required a thorough knowledge of the electromagnetic fields in the plasma between the stars, in our planetary system — here I may remind you of the solar wind, first discussed by Biermann —, in our galaxy and finally in extragalactic space. Investigations of these fields became a central part of astrophysics in recent years, and much information has been obtained with the help of the cosmic radiation. Concerning its origin the general opinion seems to be now that supernovae and their relics, the pulsars, are the main sources of high-energy cosmic rays. But I should not go into the details of astrophysics, but rather come back to my first question: where does cosmic radiation touch fundamental problems of physics?

I just mentioned the *pulsars*, which belong to the stars of highest density hitherto observed. Their matter density is comparable to the density of an atomic nucleus. They are held together by gravitational forces. Such stars raise two fundamental problems; one concerns the relation of gravitation to the other forces of interaction within matter, the other concerns the equation of state for matter of that — or still higher density. But before I come to these problems, I would like to mention a few valuable contributions of cosmic-ray research to very important problems in particle physics, even *during* the time of the big accelerators.

The particles of the cosmic radiation have energies up to 10^{19} eV, and it is obvious that such high energies cannot be reached by accelerators, at least not in the near future. Hence the collisions of particles at such extreme energies could be investigated only in the cosmic radiation; — and even if low intensity and poor statistics prevented accurate results, the question was put how the cross-section or other characteristics of showers should change with energy in the range of extremely high energies. Is there an asymptotic region, far above the energies of the common particles or resonance states, where no more spectacular new events or drastic changes are found or expected? The information obtained from the cosmic radiation on this problem was scarcely more than a vague indication; still, it stimulated theoretical investigations, which, more than 20 years ago, led to the conjecture, that the total cross-section for collisions between any hadrons should at high energies increase with the square of the logarithm of the energy. Hence there should be an asymptotic region; but the total cross-sections in this region should not be constant, they should increase logarithmically. This conjecture has been born out in recent experiments with the storage rings in CERN and with the Batavia accelerator.

The asymptotic region seems to start at an energy of the order of 10 GeV in the centre of mass system, and has been followed up to 50 GeV for proton-proton collisions in the storage rings in CERN. The essential contribution from the Batavia machine was the result, that the logarithmic increase could be observed also for collisions of pions or kaons with protons. This was a strong argument in favour of the assumption, that there is a general asymptotic region and that it had been reached in these experiments. For the purpose of understanding this asymptotic region it is sufficient to describe the particles as nearly spherical clouds of continuous matter without any reference to particles, of which these clouds could consist. This is satisfactory since the word "consisting of" has as a rule lost its meaning in particle physics.

Another problem has occupied the minds of the particle physicists during the last ten years. We know that the group SU_3 plays a rôle in the spectrum of particles as an approximate symmetry. The simplest representation of SU_3 is three-dimensional, therefore one could expect a triplet of particles corresponding to this representation; the electric charge of these particles would be $1/3$ or $2/3$ of the elementary charge, and they have been given the name "quarks". However such particles have never been observed in experiments with the big machines. Therefore it was suggested that the quarks might be rather heavy, held together by very big binding energies, hence the existing machines would not be sufficient to take them apart. At this point cosmic radiation turned out to be very helpful: because the energy of primaries in the cosmic radiation may be thousand or more times bigger than the maximum particle energy in a big machine. The fact that even in the cosmic radiation no quarks have been found, is a very strong argument in favour of their non-existence. If such results are final, it seems to me very difficult to give any well defined meaning to the statement: "a proton consists of three quarks", because neither the word "consists of" nor the word "quark" have a well defined meaning. How could such a sentence then be interpreted? The same scepticism is justified with respect to other particles, which have been predicted, but not found: W-mesons, partons, gluons, magnetic poles, charmed particles. If they cannot be observed, neither in big machines nor in the cosmic radiation, it is difficult to argue that they are good concepts in a phenomenological description. Here we meet a situation which is well known already from quantum mechanics. Our normal language induces us to ask questions which have no meaning; for example: "What is the orbit of an electron moving around an atomic nucleus?" Neither the word "orbit" nor the word "moving" are well defined on acccunt of the relations of uncertainty; hence the question has no meaning.

This leads me to a central problem, which is closely connected to experiences in cosmic radiation. But before I will discuss the empirical side. I would like to explain its fundamental importance in particle physics and physics in general.

From the experiments of the last decades we have learned, that the different particles are just different stationary states of the system matter. They are characterized by quantum numbers, or if you prefer, by their transformation properties under fundamental groups. Theoretical understanding of particle physics can only mean: an understanding of the spectrum of particles. A single line in the optical spectrum of iron cannot be understood; but the spectrum can be understood, it can be reduced to the Schrödinger equation of a system, containing 26 electrons and the iron nucleus.

The essential elements of the theoretical interpretation of a spectrum are well known and can be learned both from classical physics and from quantum mechanics. We can think of the elastic vibrations of a string, or of the electromagnetic vibrations in a cavity, or of the stationary states of an atom, for example the iron atom. In all cases we first need a precise statement about the dynamical properties of the system, and then we have to add the special boundary conditions. In the case of the string a precise mathematical formulation of the elastic and dynamic properties of the string is the first step; then, by stating where the string is fixed, we can calculate the spectrum of the vibrations. For the electromagnetic vibrations in a cavity Maxwell's equations define the dynamical properties of the system. The boundary conditions are given by the form of the cavity. On account of the complexity of the problem it will frequently not be possible to calculate the whole spectrum precisely; but for the lowest vibrations one should be able to get good approximations. For the iron atom the dynamical properties are defined by quantum mechanics, that is the Schrödinger equation. By the supplementary condition that the wave function must vanish at infinity, the stationary states are fixed. If the atom would be enclosed in a small box, the stationary states would be different.

Starting from these analogies it is clear that the first condition for understanding the particle spectrum is *a precise mathematical formulation of the dynamics of matter*. It is obvious that the word particle should not enter in this formulation. Because the particles are defined later on by combining the dynamics of the system matter with the boundary conditions; the particles are secondary structures. The spectrum of the particles may easily be different in our surrounding in the universe from that in the interior of a very dense neutron star, because the boundary conditions in both cases may be different. From this you see the fundamental importance of the dynamics of matter, and the question arises how we can get hold of its mathematical formulation.

Since the particle concept is not useful in this connexion, the group properties of the dynamical law must play a decisive rôle. The dynamical law of the vibrating string, for example is invariant under translations in time and translations along the string, and under rotations around the string. By the boundary conditions the second invariance is broken, the third as a rule is not broken. For the electromagnetic vibrations in a cavity the dynamical law is invariant under the full Lorentz group: the invariance is partially broken by the boundary conditions.

For the dynamics of matter some of the essential invariances are known: The Lorentz group and the isospin group SU_2. Also the scale group should possibly be counted under the fundamental invariances. But I should not go into the details of these symmetries of the dynamical law. I would rather like to come back to cosmic radiation. How can research in cosmic radiation or more generally in astrophysics contribute to our knowledge of the dynamics of matter?

First a word about causality. We know from the dispersion relations, that interaction in matter follows the law of causality. The exact mathematical formulation of this statement is perhaps not completely known, but we have good reason to believe that interaction can be formulated as local interaction, like, for example in quantum electrodynamics. The non-local Coulomb force is compatible with this statement. Starting from such a situation it is plausible that the study of matter of extremely high density should give the most direct information about this local interaction, and thereby about the dynamics of matter.

In a. neutron star the density is of the same order as in an atomic nucleus. At such densities it has still a meaning to say, that the nucleus consists of a number of nucleons. Because a small amount of energy—small compared with the restmass of a nucleus— is sufficient to take a proton or a neutron out of the nucleus. The nucleons are still far enough apart from each other in the nucleus—their energy of interaction therefore is small compared to their restmass. This is equally true in a neutron star, and therefore it has been possible to get estimates for the equation of state of such stellar matter. If, however, the density is considerably higher—e.g. in a star of larger mass contracted ' by gravitation—then the question, of which particles the star consists, has no well defined meaning. The space available for a particle would be smaller than its normal size, therefore it could not have its normal mass; the interaction is so strong, that particles would as a rule not be on their mass-shell. In other words, one could only speak of a mixture of all particles, and then it is more reasonable to speak of continuous matter. It is the dynamical behaviour of this continuous matter, which is the fundamental problem in particle physics.

If it should be possible to get more informations about the equation of state not only in neutron stars but especially in stars of still higher density, this would be extremely important for understanding the dynamical behaviour of matter. Whether observations in cosmic radiation or in wider fields of astrophysics will be more helpful I cannot judge. I only wanted to emphasize the importance of the problem.

There is one other special field in cosmic radiation where this problem of the dynamics of matter can be tackled from an entirely different side. If two particles of extremely high energy collide, then, in the first moment of the collision, one has a small disc of extremely dense matter, which then explodes and, by diminishing its density, finally disintegrates into many particles. This is the well known process of multiple production of particles which is, of course, the more interesting the higher the energy of the colliding particles has been. If the primary cosmic-ray particle has an energy of 10^6 GeV, then the density of the initial disc in the collision can be thousand times greater than in a neutron star.

The study of the behaviour of such cosmic-ray showers of extreme energy should therefore give valuable informations on the dynamics of matter. It is encouraging in this connection that already in the storage rings of CERN and in the Batavia machine one seems to have reached, or at least approached, the asymptotic region. For the initial phase of collisions in this region the primary particles can simply be pictured as clouds of continuous matter the density of which falls off exponentially at the surface. This model explains the logarithmic increase of the total cross-section as a function of increasing energy. I should still point to a characteristic difference between the two experiments, those on the stars of extreme density or those on the discs after the collision of very energetic particles. In the first case gravitation plays an important rôle, in the second it is unimportant. Therefore the two kinds of experiments can give two different kinds of relevant information.

Coming back in conclusion to the general questions mentioned in the beginning of my talk I should perhaps say that the special rôle of cosmic radiation in the whole field of physics rests on two facts. This cosmic radiation contains informations on the behaviour of matter in the smallest dimensions and it contributes to our knowledge about the structure of the Universe,—of the world in the largest dimensions. These two extreme ends are not accessible to direct observation, they can be investigated only by very indirect deductions, in which the concepts of daily life have to be replaced by other rather abstract new concepts; and only then we will learn, what such words as "extreme ends" or "infinity" can mean in relation to nature. In this sense cosmic radiation can—inspite of any changes in the style of the experiments—still be called a very romantic, a very inspiring science.

Received September 1, 1975

Bibliographical Citation List

1924   „Nichtlaminare Lösungen der Differentialgleichungen für reibende
        Flüssigkeiten",
        in *Vorträge aus dem Gebiete der Hydro- und Aerodynamik (Innsbruck 1922)*,
        ed. by Th. v. Kármán, T. Levi-Civita (Springer, Berlin) pp. 139–142 23
   –   Quantitatives über die Deformierbarkeit edelgasähnlicher Ionen.
        Verh. Dtsch. Phys. Ges. (3) *5*, 7–8 27
1926   Über quantentheoretische Kinematik und Mechanik.
        Math. Ann. *95*, 683–705 29
   –   Quantenmechanik.
        Die Naturwissenschaften *14*, 989–994 52
1928   „La mécanique des quanta" (with Max Born),
        in *Électrons et Photons: Rapports et Discussions du Cinquième Conseil de
        Physique Tenu a Bruxelles du 24 au 29 Octobre 1927 sous les Auspices de l'Institut
        International de Physique Solvay*, ed. by Institut International de Physique
        Solvay (Gauthier-Villars, Paris) pp. 143–184 58
   –   „Zur Quantentheorie des Ferromagnetismus",
        in *Probleme der modernen Physik: Arnold Sommerfeld zum 60. Geburtstage
        gewidmet von seinen Schülern*, ed. by P. Debye (S. Hirzel, Leipzig) pp. 114–122 ... 100
1929   Die Entwicklung der Quantentheorie 1918–1928.
        Die Naturwissenschaften *17*, 490–496 109
1930   *The Physical Principles of the Quantum Theory*
        (The University of Chicago Press, Chicago) 117
   –   Fortschritte in der Theorie des Ferromagnetismus.
        Metallwirtschaft *9*, 843–844 167
1932   "Contribution to Discussion on the Structure of Simple Molecules",
        in *Chemistry at the Centenary (1931) Meeting of the British Association for the
        Advancement of Science* (W. Heffer & Sons, Cambridge) pp. 247–248 171
1933   Über die Streuung von Röntgenstrahlen an Molekülen und Kristallen.
        Ergeb. Tech. Röntgenkd. *3*, 26–31 173
   –   „Considérations théoriques générales sur la structure du noyau",
        in *Structure et Propriétés des Noyaux Atomiques: Rapports et Discussions du
        Septième Conseil de Physique Tenu à Bruxelles du 22 au 29 Octobre 1933*, ed. by
        Institut International de Physique Solvay (Gauthier-Villars, Paris) pp. 289–335 . 179
1935   „Bemerkungen zur Theorie des Atomkerns",
        in *Pieter Zeeman, 1865 – 25 Mei – 1935, Verhandelingen op 25 Mei 1935.
        Aangeboden aan Prof. Dr. P. Zeeman* (Martinus Nijhoff, The Hague)
        pp. 108–116 238

934